Essentials of Fire Fighting and Fire Department Operations
5th Edition

P9-AFJ-136

Edited By
Carl Goodson, Senior Editor
and **Lynne Murnane**, Senior Editor

Brady/Prentice Hall Health
Upper Saddle River, New Jersey 07458

Validated by the International Fire
Service Training Association

Published by
Fire Protection Publications
Oklahoma State University

RECYCLABLE

The International Fire Service Training Association

The International Fire Service Training Association (IFSTA) was established in 1934 as a *nonprofit educational association of fire fighting personnel who are dedicated to upgrading fire fighting techniques and safety through training*. To carry out the mission of IFSTA, Fire Protection Publications was established as an entity of Oklahoma State University. Fire Protection Publications' primary function is to publish and disseminate training texts as proposed and validated by IFSTA. As a secondary function, Fire Protection Publications researches, acquires, produces, and markets high-quality learning and teaching aids as consistent with IFSTA's mission.

The IFSTA Validation Conference is held the second full week in July. Committees of technical experts meet and work at the conference addressing the current standards of the National Fire Protection Association and other standard-making groups as applicable. The Validation Conference brings together individuals from several related and allied fields, such as:

- Key fire department executives and training officers
- Educators from colleges and universities
- Representatives from governmental agencies
- Delegates of firefighter associations and industrial organizations

Committee members are not paid nor are they reimbursed for their expenses by IFSTA or Fire Protection Publications. They participate because of commitment to the fire service and its future through training. Being on a committee is prestigious in the fire service community, and committee members are acknowledged leaders in their fields. This unique feature provides a close relationship between the International Fire Service Training Association and fire protection agencies, which helps to correlate the efforts of all concerned.

IFSTA manuals are now the official teaching texts of most of the states and provinces of North America. Additionally, numerous U.S. and Canadian government agencies as well as other English-speaking countries have officially accepted the IFSTA manuals.

ISBN 978-0-13-515111-2   Library of Congress Control Number: 2007040446

First Edition, First Printing, January 2008   *Printed in the United States of America*

10   9   8   7   6   5   4   3

If you need additional information concerning the International Fire Service Training Association (IFSTA) or Fire Protection Publications, contact:

Customer Service, Fire Protection Publications, Oklahoma State University
930 North Willis, Stillwater, OK 74078-8045
800-654-4055   Fax: 405-744-8204

For assistance with training materials, to recommend material for inclusion in an IFSTA manual, or to ask questions or comment on manual content, contact:

Editorial Department, Fire Protection Publications, Oklahoma State University
930 North Willis, Stillwater, OK 74078-8045
405-744-4111   Fax: 405-744-4112   E-mail: editors@osufpp.org

Chapter Summary

Table of Contents

List of Tables

This fifth edition of the IFSTA **Essentials of Fire Fighting** is intended to serve as a primary text for the firefighter candidate or as a reference text for firefighters who are already on the job. This manual addresses the fire fighting objectives found in NFPA® 1001, *Standard for Fire Fighter Professional Qualifications*, 2008 Edition. The objectives in NFPA® 1001 for competencies in first aid and hazardous materials response are addressed in the IFSTA/Brady **Essentials of Fire Fighting and Fire Department Operations** manual.

Acknowledgment and special thanks are extended to the members of the IFSTA validating committee who contributed their time, wisdom, and knowledge to this manual.

IFSTA Essentials of Fire Fighting, 5th Edition Validation Committee

Chairman
Russell Strickland
Maryland Emergency Management Agency
Maryland Fire and Rescue Institute-University of Maryland
Reisterstown, MD

Vice-Chair
Wesley Kitchel
Santa Rosa Fire Department
Santa Rosa, CA

Secretary
Stephen Ashbrock
Madeira & Indian Hill Fire Department
Cincinnati, OH

Committee Members

Paul Boecker, III
Sugar Grove Fire District
Oswego, IL

Ronald Bowser
Maryland Fire and Rescue Institute
College Park, MD

John Brunacini
Fireground Command Training, Inc.
Phoenix, AZ

Kenneth Gilliam
Federal Aviation Administration
Orlando, FL

Russell Grossman
Iowa Fire Service Training Bureau
Ames, IA

Jerry Hallbauer
Kansas Fire & Rescue Training Institute
Lawrence, KS

Edward Hartin
Gresham Fire & Emergency Services
Gresham, OR

Michael Jepeal
Simsbury Fire Department
Tariffville, CT

Alan E. Joos
LSU Fire and Emergency Training Institute
Baton Rouge, Louisiana

Richard Karasaki
Honolulu Fire Department
Honolulu, HI

Committee Members (continued)

John Kenyon
Port Coquitlam Fire Rescue
Port Coquitlam, BC

John Leete
Fairfax County Fire & Rescue
Fairfax, VA

Dan Madrzykowski
National Institute of Standards and Technology
Gaithersburg, MD

Roy Paige
Los Angeles Fire Department
Los Angeles, CA

Harold Richardson
Yarmouth Fire Department
Yarmouth, Nova Scotia, Canada

Cary Roccaforte
Texas Engineering Extension Service/Emergency
Services Training Institute
College Station, TX

Anne Wieringa
Deadwood Volunteer Fire Department
Deadwood, SD

Ron Williams
Lake County Vo-Tech
Tavares, FL

Special thanks go to committee members Ed Hartin and Dan Madrzykowski for developing the Fire Behavior chapter.

Perhaps the most significant change from the 4th to 5th editions of **Essentials of Fire Fighting** is the partnership forged between IFSTA/FPP and Brady Publishing, a division of Pearson Education. Brady has a long and outstanding reputation in the fire publishing world and is the *premier* publisher of emergency medical training materials in North America and beyond. Teaming the two most dominant publishers of emergency responder training materials clearly will benefit the firefighters and agencies that use these materials. In addition to all of the other types of products that were available with the 4th edition of **Essentials**, the resources that Brady brings to this partnership allow us to develop a greatly expanded **Essentials** product line. New products made available by this partnership include the expanded **Essentials of Fire Fighting and Fire Department Operations** manual, the bonus CD in the back of this manual, additional on-line and electronic resources for instructors, Vango Notes® audio files , and a test generator on the curriculum CD. Special thanks go out to Brady Publisher Julie Alexander, Executive Editor Marlene Pratt, Senior Acquisitions Editor Stephen G. Smith, and all the Brady staff who worked to make this partnership happen.

It would not be possible to develop a manual of this type without the assistance and cooperation of numerous individuals, fire departments, and training agencies that rise above and beyond the call of duty to assist us with shooting the thousands of photographs needed to illustrate the concepts contained herein. Special thanks go to FPP staff members Jeff Fortney, Fred Stowell, Ed Kirtley, Mike Sturzenbecker, and Mike Wieder for shooting many of the new photographs in this edition. The following individuals and agencies provided extensive assistance on this manual:

The City of Los Angeles, California Fire Department Community Relations Unit and crews from Fire Stations 10 and 88, including members Battalion Chief Kwame Cooper, Captain/PIO Ernest Bobadilla, and Firefighters Michael T. Coffey, Armand Dabuet, Benjamin Fazeli, Mike Guzman, Robert Hinojosa, Jonith Johnson, Jr., Christian Pedini, Jordan Purrington, William B. Wenger, and Patrick Wilkinson.

The officers and firefighters of Allen Fire Department, Allen, Texas

The personnel of the Collin County Community College Live Fire Training Complex, McKinney, Texas

Training Officer Jon Neely, Assistant Training Officer Joe Elam, and the officers and firefighters of the Edmond Fire Department, Edmond, Oklahoma.

Scott Hebert, Randy Moore, and the instructors and staff of the Louis F. Garland Fire Training Academy, U.S. Department of Defense, Goodfellow AFB, San Angelo Texas

Fred Smith and the firefighters of the Goodfellow AFB Fire Department, Goodfellow AFB, San Angelo, Texas

Training Battalion Chief Ron Moore and the officers and firefighters of the McKinney Fire Department, McKinney, Texas

Bryan West, Dan Knott, and the staff of Oklahoma State Fire Service Training Professional Skills Center, Stillwater, OK

Chief Bradd Clark, Assistant Chief Chris Garrett, and the officers and firefighters of the Owasso Fire Department, Owasso, Oklahoma.

Chief Clark Purdy and the officers and firefighters of the Guymon Fire Department, Guymon, Oklahoma.

Wes Kitchel and the officers and firefighters of the Truck One, C Shift, Santa Rosa Fire Department, Santa Rosa, California.

Rex Mott and the officers and firefighters of Stillwater Fire Department, Stillwater, OK.

District Chief/Safety Officer Mike Mallory and the officers and firefighters of the Tulsa Fire Department, Tulsa, Oklahoma.

Rich Mahaney, Linn County Emergency Management Agency, Cedar Rapids, IA for providing most of the hazardous materials container photos used in the tables in Chapter 22.

In addition to the organizations listed above, the following individuals and organizations have also contributed information, photographs, or other assistance that made final completion of this manual possible:

Cory Ahrens, Washington Criminal Justice Training Commission, Burien, WA

Akron Brass Manufacturing Company, Wooster, OH

ALACO Ladder Company, Chino, CA

Alameda County (CA) Fire Department

Amerex Corporation, Trussville, AL

Ansul Inc., Marinette, WI

Sherry Arasim, Aloha, OR

Danny Atchley, Oklahoma City, OK

U.S. Agency for Toxic Substances and Disease Registry (ATSDR)

Badger Fire Protection, Charlottesville, VA

Bainbridge Island (WA) Fire Department

Steven Baker, New South Wales Fire Brigades, Australia

Kenneth Baum, Ron Bogardus, Albany, NY

Boring (OR) Fire and Rescue

George Braun, Gainesville (FL) Fire-Rescue

CDC Public Health Images Library

Cedar Rapids (IA) Fire Department

Cherry Hill (NJ) Fire Department

Clallam County Fire District 3, Sequim, WA

Tom Clawson, Technical Resources Group, Inc., Idaho Falls, ID

Conoco/Phillips, Ponca City, OK

Creston (IA) Fire Department

Crimson Fire, Brandon, SD

Detector Electronics Corporation, Minneapolis, MN

Harvey Eisner, Tenafly, NJ

Elkhart Brass Manufacturing Company, Elkhart, IN

Bob Esposito, Pennsburg, PA

John Evans, Exit Technologies, Boulder, CO

Federal Emergency Management Agency (FEMA), Washington, D.C.

Keith Flood, Santa Rosa (CA) Fire Department

Gary Friedel, Oklahoma State University, Fire Service Training

Richard (Dick) Giles, Stillwater, OK

Globe Manufacturing Company, LLC, Pittsfield, NH

Sam Goldwater, North Tree Fire International (NTFI)

Gresham (OR) Fire and Emergency Services

Judy Halmich, Joan Hepler, Ron Hiraki, Pierce County Fire District #5, Gig Harbor, WA

Honolulu (HI) Fire Department

Thomas A. Hughes, Ingalls (OK) Fire Department

IFE Hong Kong Branch, Kowloon, Hong Kong

Illinois Fire Service Institute

International Tanker Owners Pollution Federations Ltd., Houndsditch, London

Ron Jeffers, New Jersey Metro Fire Photographers Association

John Kenyon, Port Coquitlam (BC) Fire and Rescue

Kitsap County Fire District #18, Poulsbo, WA

Sheldon Levi, IFPA Centerville, VA

Darrel Levine,

Phil Linder, Richmond, British Columbia

Thomas Locke and South Union Volunteer Fire Company

Majestic Fire Apparel, Inc., Lehighton, PA

Dr. George McClary, Santa Rosa, CA

District Chief Chris E. Mickal, New Orleans (LA) Fire Department Photo Unit

Rick Montemorra, Mesa (AZ) Fire Department

Monterey County (CA) Fire Training Officers Association

Morning Pride/Total Fire Group, Inc., Dayton, OH

Ron Moore, McKinney (TX) Fire Department

National Fire Protection Association, Quincy, MA

National Institute of Standards and Technology (NIST), Washington, D.C.

Brett Noakes, Ripley (OK) Fire Department

Rebecca Noble, Olympia, WA

Robert H. Noll, Yukon (OK) Fire Department (R)

Oakland (CA) Fire Department

Pigeon Mountain Industries, LaFayette, GA

Brandon Poteet, Grayson County College, Denison, TX

Pyrotechnic Tool Company, St. Louis, MO

Paul Ramirez, Phoenix (AZ) Fire Department

Renton (WA) Fire Department

Dave Ricci, Santa Rosa (CA) Fire Department

RKI Instruments, Union City, CA

Rock-N-Rescue, Middlesex Twp (PA) Volunteer Fire Company

Tom Ruane, Peoria, AZ

The Sanborn Map Company, Inc., Colorado Springs, CO

San Francisco (CA) Fire Department

San Ramon Valley (CA) Fire District

U.S. Army Soldier and Biological Chemical Command (SBCCOM)

Peter Schecter, U.S. Virgin Islands

Jeff Seaton, San Jose (CA) Fire Department

Seattle (WA) Fire Department

Jerry Shacklett, Los Angeles County (CA) Fire Department

Shell Chemical Company

Glenn Speight, BlueWater Ropes, Inc., Carrollton, GA

Supersonic Air Knife, Inc., Allison Park, PA

Steve Taylor, Rescue Engineering Institute, Fort Wayne, IN

United States Department of Energy

Michael Watiker, Pickerington, OH

Williams Fire & Hazard Control, Inc.

Women in the Fire Service, Inc., Madison, WI

Wright Rescue Solutions, Panama City, FL

Ziamatic Corp., Yardley, PA

Last, but not least, gratitude also is extended to the following members of the Fire Protection Publications staff whose contributions made the final publication of this manual possible:

Mike Wieder, Assistant Director/Managing Editor

Ed Kirtley, IFSTA/Curriculum Projects Coordinator

Carl Goodson, Senior Technical Editor

Lynne Murnane, Senior Editor

Jeff Fortney, Senior Technical Editor

Cynthia Brakhage, Senior Editor

Fred Stowell, Senior Technical Editor

Barbara Adams, Senior Editor

Leslie Miller, Senior Editor

Beth Ann Fulgenzi, Curriculum Developer

Melissa Noakes, Curriculum Developer

Michelle Skidgel, Curriculum Developer

Ann Moffat, Production Coordinator

Clint Parker, Senior Graphics Designer

Lee Shortridge, Senior Graphics Designer

Ben Brock, Senior Graphics Designer

Missy Hannan, Senior Graphics Designer

Matt Miller, Graphics Technician

Overview of Essentials 5th Edition Components

The 5th edition of **Essentials of Fire Fighting** offers bold new learning and instructor packages for students. There are several new components that have been added to the learning and instructor packages, all designed to add to the success of the student. In addition, established components have been improved. A description of those components is provided in this section. For more specific details on any of the Essentials of Fire Fighting 5th edition family of products, call Fire Protection Publications at 800.654.4055 or Brady Books at 800.922.0579. You may also visit our websites at www.IFSA.org or www.bradybooks.com.

Companion Website

http://www.prenhall.com/ifsta

This online resource for Essentials of **Fire Fighting and Fire Department Operations, 5/e** offers a valuable supplement to the textbook, providing students and instructors with an interactive learning and teaching tool. Organized by chapter, you'll receive objectives for each chapter, distinguished by Firefighter I and Firefighter II classifications, review questions, key terms, chapter review, audio glossary, key industry weblinks and more!

Bonus CD

Student CD Rom. Bounded into every manual, this bonus CD contains multiple-choice questions, a tool inventory, glossary, case studies, vehicle tours, and a resume builder.

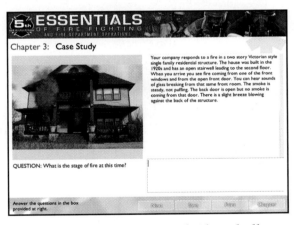

- Each chapter has a case study that challenges the student to apply the concepts from that chapter to a potential scenario faced by firefighters.

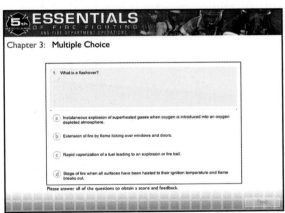

- Multiple-choice questions are provided for each chapter to test your knowledge and help prepare for tests and quizzes.

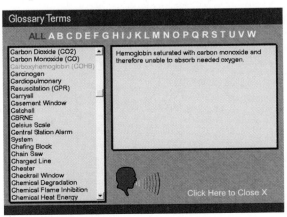

- The glossary provides a quick reference to all the key terms in the manual . . a helpful tool for reviewing for tests and quizzes.

Study Guide

- Heavy-duty binding allows the printed version of the study guide to lay flat on the table.

- The study guide is also available in the easy-to-use CD format.

- The study guide covers all the key concepts in each chapter and the corresponding requisite knowledge in the NFPA® 1001 standard.

- Firefighter I and Firefighter II questions are provided in separate sections within each chapter to help maximize student time.

- Multiple-choice questions are used so that review questions studied are similar to most promotional and certification tests.

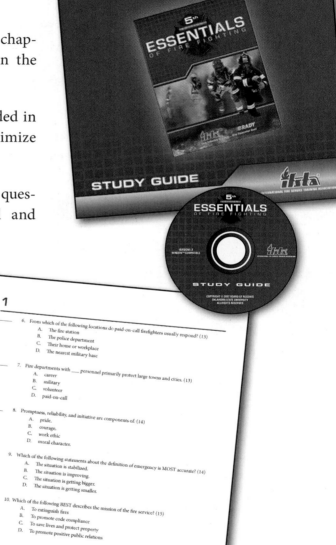

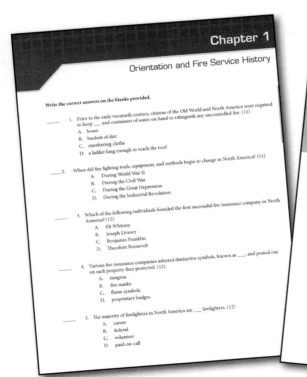

- Heavy-duty wire binding prevents pages from being lost and allows the handbook to lay flat.

- Stronger, more resilient paper enables handbook to be used by students and instructors during hands-on drills.

- All skills addressed in the manual are included in the skills handbook.

- Photos of skills being performed help students review and practice skills.

- Instructors are able to use the skills handbook as a reference when evaluating student performance.

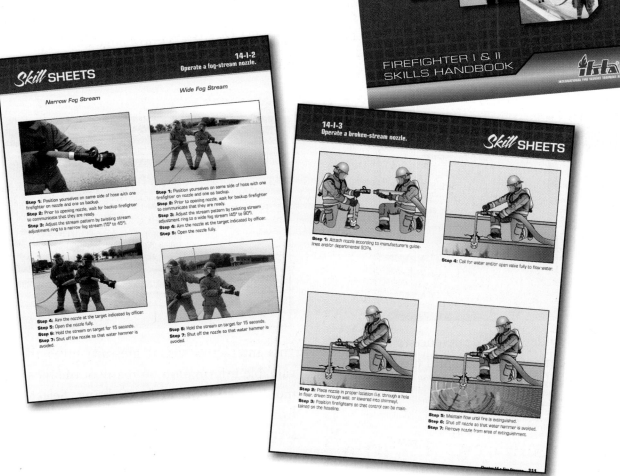

Bonus CD

Student CD Rom. Bounded into every manual, this bonus CD contains multiple-choice questions, a tool inventory, glossary, case studies, vehicle tours and a resume builder.

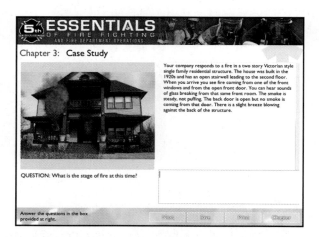

- Each chapter has a case study that challenges the student to apply the concepts from that chapter to a potential scenario faced by firefighters.

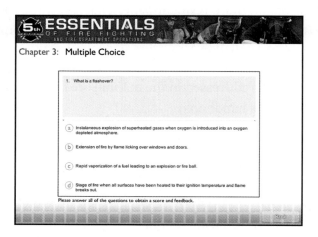

- Multiple-choice questions are provided for each chapter to test your knowledge and help prepare for tests and quizzes.

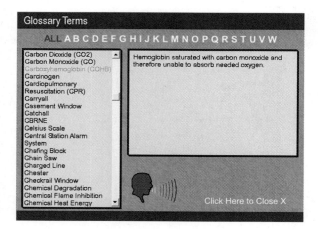

- The glossary provides a quick reference to all the key terms in the manual .. a helpful tool for reviewing for tests and quizzes.

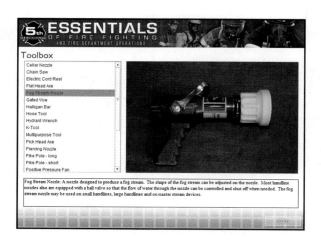

- This interactive visual glossary gives the user valuable information on many of the most commonly used tools of fire fighting. Select a tool from the menu, see a picture of it and read a description of it's intended uses.

Bonus CD

- Videos of actual fire fighting operations give the student a visual taste of the firefighter's world. The videos are used to reinforce the concepts and lessons addressed in the manual.

- Aerial, Brush, Pumper, Rescue and Tanker... take a tour of each of these vehicles and learn where the equipment can be found. Select front, rear, right side or left side, and roll over the hotspots for more information.

Curriculum CD

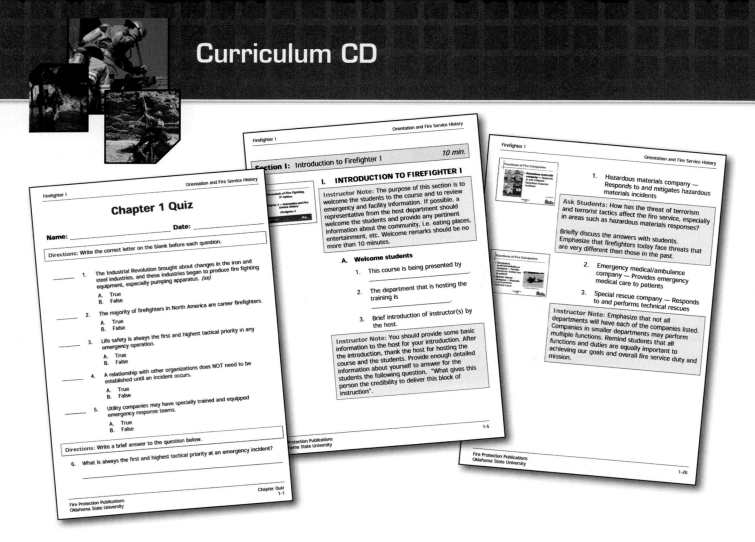

- Thumbnails of each Power-Point® slide are shown in the lesson plan for easy reference by the instructor.

- The learning objective for each section is listed in the lesson plan.

- Each lesson plan provides a detailed, easy-to-follow outline.

- For fast and easy instructor references, text pages are listed for the section being delivered.

- Instructor boxes provide tips for improving teaching effectiveness.

- Key concepts in each chapter are evaluated through quizzes and tests.

- All test questions are multiple choice.

- Lesson quizzes use several types of questions including multiple choice and true/false.

- To help the students during review, the page number in the manual from where the question was taken is listed.

- Answers are provided for all tests and quizzes.

- A test generator is provided on the curriculum CD so the instructor can select a specific number of questions for a chapter test, midterm test or final course test.

Introduction

Welcome to the **Essentials of Fire Fighting**, 5th edition! Fire Protection Publications (FPP), the International Fire Service Training Association (IFSTA), and Brady Publishing have partnered to bring you this new edition of our premier fire service training manual for entry-level firefighters. This manual has been prepared to assist you in meeting the Fire Fighter I and Fire Fighter II training and certification requirements of the National Fire Protection Association's (NFPA®) Standard 1001, *Standard for Fire Fighter Professional Qualifications*.

Since its original adoption in 1974, NFPA® 1001 has become widely accepted as the standard of measurement for all firefighters in North America and beyond. The acceptance and recognition of a national standard provides a baseline for professionalism in the fire service. It is the intent of FPP, IFSTA, and Brady Publishing to promote this professionalism by providing the most accurate and highest quality training materials to those who choose this challenging profession. That is why IFSTA developed the first edition of **Essentials of Fire Fighting** in 1978 and why we continue to develop new editions when the NFPA® standards are updated.

When considering this standard, keep in mind that NFPA® 1001 (and all NFPA® standards, for that matter) sets forth *minimum* requirements and any jurisdiction may exceed the specified requirements when applying the standard locally. This is perfectly acceptable. What is not considered acceptable is for any jurisdiction to reduce the requirements of the standard.

The early editions of NFPA® 1001 contained three levels of competence: Fire Fighter I, Fire Fighter II, and Fire Fighter III. These three levels were maintained through the first several revisions of the standard. However, while preparing the 1992 revision of the standard, the NFPA® 1001 committee determined that the information in the Fire Fighter III section was repetitive of many of the objectives in the lower levels of NFPA 1021, *Standard for Fire Officer Professional Qualifications*. Therefore, NFPA® 1001 was reduced to two levels: Fire Fighter I and Fire Fighter II. It should be noted that some state and local jurisdictions still use a Fire Fighter III designation in their training and certification programs. In most cases they have chosen to rename the NFPA® 1001 Fire Fighter II level to their own Fire Fighter III level to meet local needs.

To delineate between a Fire Fighter I and Fire Fighter II, both the NFPA® 1001 and the IFSTA **Essentials** committees decided that a *Fire Fighter I* is a person who is minimally trained to function safely and effectively as a member of a fire fighting team under direct supervision. A person meeting the requirements of Level I is by no means considered a "complete" firefighter. This is not accomplished until the objectives of both Levels I and II have been satisfied. A *Fire Fighter II* may operate under general supervision and may be expected to lead a group of equally or lesser trained personnel through the performance of a specified task.

The 2008 edition of NFPA 1001 continues the Job Performance Requirement (JPR) format introduced in the 1997 edition of the standard. *Job Performance Requirements* are based on a job task analysis that identifies either what a firefighter actually does on the job or should be capable of doing. This JPR format is used in all NFPA® professional qualifications standards.

The 2008 edition of NFPA® 1001 does not reflect a substantial overhaul of the 2002 version of the standard. The major changes that occurred in the 2008 edition can be summarized as follows:

- The 2008 edition requires Fire Fighter I personal to meet the requirements for First Responder at the Operational Level of NFPA® 472, *Standard for Competence of Responders to Hazardous Materials/Weapons of Mass Destruction Incidents*. In the 2002 edition of NFPA® 1001, the Fire Fighter I was only required to meet the First Responder at the Awareness Level and Fire Fighter II personnel needed to meet the Operational requirement. This change was instituted because a significant number of personnel stop their training and certification at the Fire Fighter I level, yet federal law requires all firefighters who respond to potential haz mat or terrorist incidents to be trained to the Operational Level.

- Requirements for public fire and life safety education activities were moved from the Fire Fighter I to Fire Fighter II level. This was not done to deemphasize the importance of these activities, but more so to provide a more equal balance of materials in the two levels and to reduce the amount of training time required for the Fire Fighter I level.

- Requirements for firefighters to be able to test fire hydrants were dropped from the standard. In most jurisdictions this is now a function of the water utility and it was felt that this was no longer a minimum requirement that was needed by *all* firefighters. In jurisdictions where firefighters still perform this task, they should be trained to do it correctly.

- Requirements for operating safely at roadway incidents were strengthened. This is one of the most hazardous locations in which firefighters operate and injuries and deaths at these locations are on the increase.

The information contained in the 5th edition of **Essentials** is intended to help firefighters meet these new requirements of NFPA® 1001. Because there is considerable overlap in the requirements of NFPA® 1001, it was not feasible to develop separate Fire Fighter I and Fire Fighter II sections in this manual. In many cases, the Fire Fighter II information would be difficult to interpret and use in a stand-alone format. For ease of reading, the material is presented in the most sound and logical manner possible.

The list of JPRs covered in each chapter is found at the beginning of each chapter, as are the chapter objectives based on the JPRs. More specific directions on where JPRs are covered within each chapter are contained in the curriculum materials for the instructor. NFPA® 1001 does not require that the objectives be mastered in the order in which they appear in the standard. Local agencies may choose the order in which they wish the material to be presented. **Appendix A** contains a guide showing the JPRs and the pages and chapter that relate to the requirements.

This edition of the manual also continues the use of skill sheets as a teaching vehicle for the students and instructor alike. The committee believes that separating the written text from the step-by-step procedures makes the manual easier to read. Therefore, skill sheets describing the step-by-step procedures for many of the tasks described in the text are found at the end of that chapter. Note that while the skill sheets contained within the manual do contain written information on all of the important steps in that skill, they do not have photos or illustrations for each step.

This was done in order to keep the manual to a reasonable size. The supplemental **Essentials of Fire Fighting Fire Fighter I and II Skills Handbook** contains all of the skill sheets covered in the **Essentials** manual, with additional photographs and illustrations to highlight all of the steps in the processes.

Scope and Purpose

The **Essentials of Fire Fighting** manual is intended to provide the firefighter candidate with the information needed to meet the fire-related performance objectives in NFPA® 1001, *Standard for Fire Fighter Professional Qualifications,* Fire Fighter I and II. In order to fully meet the requirements of NFPA® 1001, the firefighter must also meet some minimal first aid requirements and the requirements for First Responder at the Operational Level of NFPA® 472, *Standard for Competence of Responders to Hazardous Materials/Weapons of Mass Destruction Incidents.* The information needed to meet these requirements is covered in the IFSTA/Brady **Essentials of Fire Fighting and Fire Department Operations** manual. Fire agencies and instructors will need to determine which of these two manuals best suits the content of the course they are teaching. The first twenty chapters in the **Essentials of Fire Fighting and Fire Department Operations** manual are identical to the twenty chapters in the basic **Essentials of Fire Fighting** manual.

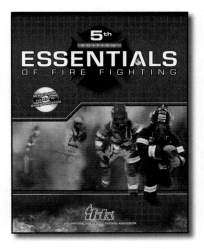

Essentials of Fire Fighting and Fire Department Operations contains three additional chapters that cover the first aid and hazardous material information. All study guide and instructor materials cover the full 23 chapters.

The methods shown throughout this manual have been validated by the International Fire Service Training Association (IFSTA) as accepted methods for accomplishing each task. However, they are *not* to be interpreted as the only acceptable methods of accomplishing a given task. Other methods of performing any task may be specified by a local agency. For guidance in seeking additional methods of performing a given task, the student or instructor may consult any of the IFSTA expanded-topic manuals (such as **Fire Hose Practices**) for more in-depth information on a particular topic.

Key Information

Various types of information in this manual are given in shaded boxes marked by symbols or icons. See the following definitions:

Information Plus Sidebar

Information Plus Sidebars give additional relevant information that is more detailed, descriptive, or explanatory than that given in the text.

Information

Information boxes give facts that are complete in themselves but belong with the text discussion. It is information that may need more emphasis or separation. They can be summaries of points, examples, calculations, scenarios, or lists of advantages/disadvantages.

Safety Alert

Provides additional emphasis on matters of safety.

Case History

A case history analyzes an event. It can describe its development, action taken, investigation results, and lessons learned. Illustrations can be included.

Key Information

Key information is a short piece of advice that accents the information in the accompanying text.

Personal Alert Safety System (PASS) — Electronic lack-of-motion sensor that sounds a loud tone when a firefighter becomes motionless. It can also be manually triggered to operate.

Key Terms. A key term is designed to emphasize key concepts, technical terms, or ideas that firefighters need to know. They are listed at the beginning of each chapter and the definition is placed in the margin for easy reference.

Three key signal words are found in the text: **WARNING, CAUTION,** and **NOTE.** Definitions and examples of each are as follows:

- **WARNING** indicates information that could result in death or serious injury to fire and emergency services personnel. See the following example:

> **WARNING!**
> Live-fire training must adhere to the requirements set forth in NFPA 1403, Standard on Live Fire Training Evolutions (Current edition).

- **CAUTION** indicates important information or data that fire and emergency service responders need to be aware of in order to perform their duties safely. See the following example:

> **CAUTION**
> Fire and emergency responders must be familiar with the physiological, emotional, and technological limitations caused by the use of respiratory protection equipment to prevent injury or death.

- **NOTE** indicates important operational information that helps explain why a particular recommendation is given or describes optional methods for certain procedures. See the following example:

NOTE: This information is based on research performed by the International City/County Management Association, Inc.

Referenced NFPA Standards and Codes

One of the basic purposes of IFSTA manuals is to allow fire and emergency services personnel and their departments to meet the requirements set forth by NFPA® codes and standards. These NFPA® documents are referred to throughout this manual. References to information from NFPA® codes are used with permission from National Fire Protection Association, Quincy, MA 02169. This referenced material is not the complete and official position of the National Fire Protection Association on the referenced subject, which is represented only by the standard in its entirety.

Chapter Contents

Key Terms

Orientation and Fire Service History

This chapter provides information that addresses the following job performance requirements of NFPA® 1001, *Standard for Fire Fighter Professional Qualifications (2008)*:

<u>NFPA® 1001 references for Chapter 1:</u>
5.1.1

Photo courtesy of District Chief Chris E. Mickal, NOFD Photo Unit.

Chapter Objectives

Fire Fighter I Objectives

1. Describe the history and culture of the fire service.

2. Describe the mission of the fire service.

3. Define fire department organizational principles.

4. Distinguish among functions of fire companies.

5. Summarize primary knowledge and skills the firefighter must have to function effectively.

6. Distinguish among the primary roles of fire service personnel.

7. Distinguish among policies, procedures, and standard operating procedures (SOPs).

8. Summarize components of the Incident Command System (ICS).

9. Distinguish among the functions of the major subdivisions within the ICS structure.

10. Define ICS terms.

11. Discuss fire service interaction with other organizations.

Orientation and Fire Service History

Whether as a volunteer or career firefighter, the path you have chosen is both challenging and rewarding. Working as a firefighter will require hard physical labor and may expose you to a high level of mental and emotional stress, and even mortal danger. Doing the job well requires a highly developed sense of personal dedication, a genuine desire to help people, and the ability to master a variety of skills and apply them when needed. While being a firefighter is very demanding, for those who are psychologically and physically up to the task, it can be the most rewarding job in the world!

Being a career firefighter may require you to work duty schedules unlike almost any other profession. Because fire departments must provide services around the clock every day, most career firefighters (and some volunteers) work a series of 24-hour shifts with one or more days off between work shifts. Other career firefighters work any of a variety of other schedules, some involving either a 10-hour day shift or a 14-hour night shift. An individual firefighter may work all days or all nights for weeks at a time before switching to the opposite shift. In the wildland fire community, it is not unusual for career firefighters to be on duty continuously for several consecutive 24-hour shifts during fire season. These long shifts can put a strain on firefighters and their families. From the start, both the firefighters and their families must accept the uniqueness of the profession and the challenges that come with it.

NFPA® 1001, *Standard for Fire Fighter Professional Qualifications*, requires those qualified at the Fire Fighter I level to know the following about the fire service:

- Organization of the fire department
- Role of the Fire Fighter I in the organization
- Mission of the fire service
- Department's standard operating procedures (SOPs)
- Department's rules and regulations
- Roles of other agencies as they relate to the fire department

This chapter addresses the history and culture of the fire service in North America; fire department organization; and the various positions, roles, and responsibilities found in a typical fire department. The chapter discusses the regulations governing the activities of fire departments and their firefighters. Also discussed are the possible roles that firefighters may assume in the Incident Command System (ICS) as defined by the U.S. National Incident Management System (NIMS). Finally, a discussion on interacting with other organizations will familiarize you with the types of agencies with which you may work at an emergency scene.

Fire Service History and Culture

It is important to be aware of the past in order to appreciate the fire service as it exists today and what it may become in the future. In addition, if you want to be a positive contributor to today's fire service and perhaps help to shape its future, you will need to appreciate the culture of the profession.

Whenever there is an emergency of any kind, the fire department is often the first emergency response organization called to the scene. The majority of fire departments in North America respond to a variety of emergencies, not just fires. They respond to medical emergencies, motor vehicle crashes, trench cave-ins, building collapses, aircraft crashes, tornadoes, earthquakes, hazardous materials incidents, civil disturbances, technical rescues, explosions, and terrorist attacks (**Figure 1.1**). The possibilities are limitless.

More than ever before, fire departments today reflect the populations of the communities they protect. Women and minorities are an integral and growing part of their staffs. Regardless of the composition of the department or an individual fire company, firefighters train together, work together, and live together in the fire stations. This interaction promotes closeness between members and they come to rely

Figure 1.1 Firefighters respond to many different types of emergencies. *Courtesy of Bob Esposito and District Chief Chris E. Mickal, NOFD Photo Unit.*

on each other. In short, they become a team — a team in which everyone looks out for everyone else. Because career firefighters are public employees and on-duty volunteers represent the community they have sworn to protect, they are expected to calmly evaluate emergency and nonemergency situations and work to bring them to successful conclusions. Firefighters are not extraordinary people — they are ordinary people who consciously put themselves in extraordinary situations. Firefighters have a high level of motivation as well as extensive training and sophisticated equipment. Despite their efforts, however, they may not be able to solve every problem to which they are called and cannot do everything at once. Therefore, they and the public must accept this reality.

Fire Service History

For thousands of years, people have been using fire to warm themselves, cook their food, and soften metals so they could be hammered or cast into new and useful shapes. However, they have also known that fire has the potential to injure and kill them and their livestock as well as to destroy their crops and homes. Wherever buildings were clustered together, a fire starting in one building could and often did spread to the building next to it and the ones beyond that until most or all of the adjacent buildings were destroyed. Because of this potential for destruction, villages, towns, and cities throughout the Near East, Europe, Scandinavia, and Britain began to require their citizens to keep a ladder long enough to reach the roof and containers (earthen jars or wooden buckets or barrels) of water on hand with which to extinguish any uncontrolled fire. These measures continued to be required of building occupants for centuries, not just in the Old World, but also in North America. In fact, these measures were required of homeowners in North America as late as the early twentieth century.

As the peoples of the Old World evolved from nomadic tribes of shepherds or hunter/gatherers into more stable agrarian societies, they built more substantial structures in which to live and to protect their livestock. Even though most of the buildings were constructed with walls of rock or other materials that wouldn't burn, most of the roofs were covered with thatched straw. Many of these buildings also had fireplaces for heating and cooking. Sparks from the fireplace chimneys would sometimes land on the dry thatch and ignite it. This happened so often that in Germany fires starting in thatched roofs came to be called "der rote Hahn," or "the Red Hen." Unfortunately, the means of extinguishing uncontrolled fires had not evolved beyond the bucket brigade in which people stood side by side in a line from a well or other water source to the burning building and passed buckets of water from one person to the next. The last person in line would throw the water onto the fire. This crude method would sometimes work; more often than not, however, the roof would burn off and the fire would consume most or all of what was inside the building.

When the Pilgrims and others immigrated to North America starting in the seventeenth century, they brought with them the construction and fire fighting methods of the countries they left behind. Unfortunately, the results were also similar. It wasn't until the Industrial Revolution beginning in the last quarter of the eighteenth century that fire fighting tools, equipment, and methods began to change in North America.

The Industrial Revolution affected many aspects of society, but especially the textile and steel industries. Textile mills sprang up in North America, particularly in the New England states. These mills produced cotton and linen cloth along with tremendous quantities of highly flammable lint and other debris that accumulated on and under the oiled wooden floors of these massive mill buildings. Because most of these buildings were lighted by oil lamps and eventually gas lights, sources of ignition were numerous. Even though mill buildings were usually constructed of brick and heavy timber, fires in these occupancies were frequent and sometimes catastrophic. However, the revolution in the iron and steel industries began to produce fire fighting tools and equipment — especially pumping apparatus — that would make fighting fires in the textile mills and similar occupancies feasible. These early pumpers and ladder trucks evolved into the modern fire apparatus that we see today.

It was during this period that cities and towns in North America began to organize fire companies and fire departments to protect them from the ravages of fire. These groups often evolved from organized social and fraternal clubs devoted to fighting fires. Among the first of these was a group organized by Benjamin Franklin in Philadelphia, PA. Franklin also founded the first successful fire insurance company in North America, which he called the Philadelphia Contributorship. This and other early insurance companies supported the local fire companies. The various companies adopted distinctive symbols — known as fire marks — and posted one on each property they protected **(Figures 1.2 a–d)**.

Fire Mark — Distinctive metal marker once produced by insurance companies for identifying their policyholders' buildings.

Figures 1.2 a-d Typical fire marks on old homes in Philadelphia.

Many of the early fraternal groups were made up of military veterans and the organizations adopted that rank structure, such as privates, corporals, sergeants, lieutenants, captains, and majors, with which their members were most familiar. Some fire departments today still use those ranks, but the majority of departments now use different designations except for the ranks of lieutenant and captain. Then, as today, the majority of firefighters in North America were volunteers.

The members of these organizations were very proud of their work and many became quite competitive — so much so that members of different groups arriving at the scene of a fire at the same time would sometimes fight each other for the right to fight the fire. As competitive and sometimes combative as these groups were, they eventually evolved into the volunteer and career fire departments that protect our cities and towns today.

Fire Service Culture

When you successfully complete your basic training course or recruit academy, if you are not already a member of a fire department you may soon become one. In some cases, individuals are accepted as members of volunteer departments before they receive their basic training. Many others enroll in basic fire training courses through local community colleges and other institutions to meet the eligibility requirements to apply for employment with career fire departments. Today, most firefighters serve in public departments that fall into one of four categories: volunteer, paid-on-call, career, or combination. Many firefighters serve in federal or military fire departments. Other firefighters serve in private fire brigades at industrial facilities.

Volunteer fire departments are found in communities with a population of a few hundred to those with populations in the tens of thousands. Volunteer fire departments and their firefighters greatly outnumber career departments and their firefighters. Some fire departments may use paid-on-call firefighters to staff their departments or to supplement their career firefighters. Like volunteers, paid-on-call firefighters usually respond from their homes or workplaces, but they receive reimbursement for each call they attend.

Fire departments with career personnel (salaried firefighters) primarily protect large towns and cities. Some cities and towns choose to operate combination departments — that is, those that combine full-time career firefighters with volunteers or paid-on-call members. Regardless of the type of department of which you are or may become a member, there are certain characteristics and behaviors that are fundamental to your success as a firefighter. Among these characteristics are integrity, moral character, work ethic, a sense of personal and professional pride, and courage.

Integrity

While the word *integrity* can be defined in different ways depending upon the context, in the fire service it means "obedience to the unenforceable." In other words, it means doing the right thing simply because it's right — not because someone has the power to force you to do it. Integrity can also mean "self–discipline," which means doing the right thing on your own without needing someone else to remind you. Integrity can be either personal or professional and is the basis for all of the other characteristics discussed in this section. An example of personal integrity would be if you were to find a wallet in a public place while off duty. The law does not require you to turn it in or attempt to contact the owner. But that's the right thing to do, and that's what a person with integrity would do. The same rules apply professionally. At fires and other emergencies, valuables belonging to the victims must be turned in to someone in authority or identified so that they may be protected in place. On calls you may find yourself inside someone's home or business when the person is absent or unconscious. There may be cash or other valuables lying about unprotected. As a firefighter, you are trusted by members of the public to protect them and their property to the fullest extent possible.

Moral Character

More than anything else, what *moral character* means in this context is truthfulness and honesty. As a firefighter, your coworkers, supervisors, and members of the public must be able to trust you and believe what you say — without question. You must be willing to speak the truth, even when it hurts. For example, if you made a mistake that might cause harm to someone or your department, you must have the strength of character to admit it and accept the consequences. In a 1997 study conducted by a respected research organization, the respondents said their trust in firefighters was second only to the trust they had in members of their own families. As a firefighter, you must strive to maintain that trust. Any violation of that trust will not only reflect on you but on all firefighters everywhere.

Work Ethic

Having a good work ethic is also critical to your success as a firefighter. It means doing what needs to be done without being told, doing what you are asked to do without complaint, doing it completely, and doing it to the best of your ability. Even tasks that are unpleasant or that seem to be relatively unimportant require your best effort. Cleaning the toilets at the fire station may seem like a menial task, but it is absolutely necessary to maintain decent living conditions within the station and to protect the health of all who are stationed there. A fire engine that is left dirty can hide a broken or missing part that might be critical to its operation; a thorough cleaning process would detect this. A tire that is left underinflated can overheat and blow out, disabling an otherwise operational vehicle on its way to an emergency, and endangering those riding in the vehicle or members of the public. A fire that is not extinguished completely can flare up and burn more than it did originally and perhaps result in injuries or fatalities to your fellow firefighters or members of the public. It is important to remember that throughout your fire service career, you will be judged by the quantity and quality of work that you do. Other components of your work ethic include promptness, reliability, and initiative.

Pride

Like the other characteristics just discussed, pride in yourself and your work are absolutely critical to your success as a firefighter. Pride in yourself starts with how you look and behave. If you look professional — that is, clean, well groomed, and with a clean and nicely pressed uniform, those with whom you come in contact will be favorably impressed and will take you more seriously. Likewise, if you conduct yourself in a friendly but businesslike manner, you will be more successful and bring honor to yourself and your department.

Courage

Fires and other emergencies can be very frightening situations — so much so that most average citizens may be frightened into inaction. Firefighters do not have that luxury. Firefighters are sworn to do whatever they can to save the lives of those who are in jeopardy and to protect their property if possible. In many cases, entering a situation from which others are fleeing takes courage. However, in this context, courage does not mean having a "macho" attitude and charging blindly into a fire or other dangerous situation. Rather, it means facing sometimes frightening situations with the training, experience, and self-discipline to assess the risks and take those that are appropriate — but in a controlled and rational way. See the IFSTA Principles of Risk Management in Chapter 2, Firefighter Safety and Health.

Fire Service Mission and Organization

By definition, an *emergency* is a situation that is *emerging* — in other words, it is not staying the same — it is getting bigger or getting worse. While many of the calls to which you and your fellow firefighters will respond are rather low-key situations, the *emergencies* to which you will be called are the equivalent of civil disturbances. That is, they start small but without some outside intervention, they can grow into full-scale riots involving everyone nearby, producing serious (perhaps fatal) injuries, and causing widespread property damage. Fire departments exist to intervene as quickly as possible in fires and other emerging situations.

Fire Service Mission

The mission of the fire service is *to save lives and protect property*, and this is also the mission of every fire department and firefighter. Obviously, this means saving people whose lives are threatened by a fire or other emergency, but it also includes protecting the lives of the firefighters involved in the incident.

One of the most efficient and cost-effective ways that fire departments can save lives and protect property is by preventing fires from starting in the first place. Most fire departments use the following two programs to prevent fires:

- Fire prevention and code enforcement
- Public education

Fire Prevention and Code Enforcement

Some fire departments have a separate division or bureau devoted almost entirely to preventing fires by enforcing local fire and building codes. These units often include personnel trained in fire cause determination and arson investigation. Many fire departments, whether or not they have a dedicated fire prevention bureau, involve their firefighters in fire prevention activities as well **(Figure 1.3)**. In that capacity, fire companies conduct fire safety and code enforcement inspections in businesses within their response districts. In addition to helping prevent fires, these inspections allow the firefighters to become familiar with the buildings and their contents. This knowledge can be very important when fighting fires in these buildings.

Public Education

Many fire departments have highly trained public education specialists who plan, organize, and in some cases deliver fire and life safety talks and demonstrations to local schools, businesses, and civic organizations. In other cases, firefighters are trained to deliver these talks and demonstrations **(Figure 1.4)**.

Figure 1.4 Firefighters often deliver public education messages to citizens.

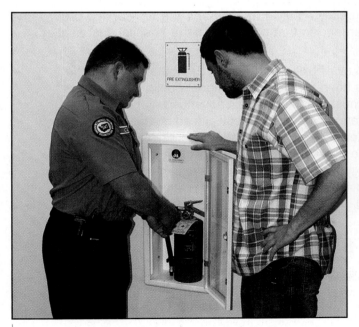

Figure 1.3 Many firefighters conduct code enforcement inspections.

Overview of Fire-Service-Based EMS

The role of the fire service in providing prehospital emergency medical care in the United States is rapidly increasing and has become a fundamental element of the mission of most fire departments. As a firefighter you will be an essential provider of emergency medical care in that system. Because of that you should understand the changing role of the fire service in (EMS). The following is taken from the white paper report **Prehospital 9-1-1 Emergency Medical Response: The Role of the United States Fire Service in Delivery and Coordination** *published by Fire-Service-Based EMS Advocates. The entire report is reprinted in* **Appendix B** *in the back of the manual.*

EMS is an essential component of the public services provided in the United States. In the United States the fire service has become a first-line medical responder for critical illness and injury in virtually every community in America. Regardless of whatever agency provides medical transportation services, the fire service is the agency that first delivers on-scene health care services under most true emergency conditions. Prehospital 9-1-1 emergency response, in support of community health, security, and prosperity, is not only a key function of each community, it has become, almost universally, a principal duty of the fire service as well. In addition, fire service-based EMS systems are strategically positioned to deliver time-critical response and effective patient care rapidly. Furthermore, fire-service-based EMS accomplishes this rapid first response while emphasizing responder safety, sending competent and compassionate workers and delivering cost-effective operations.

The protection of life and property has been the mission of the fire service for over 200 years, but the fire department of the 21st century is evolving into a multidisciplinary public safety department. It not only handles most aspects of public safety (beyond law enforcement security issues), but it also will continue to provide advances in emergency medical care and many developing public health needs such as preparations for pandemics, disasters, and weapons of mass effect. Firefighters must realize and understand the integral and evolving role that they play in their community's public safety services and public health system.

The fire service has formally been part of the 9-1-1 emergency care delivery system since EMS began in the late 1960's. Many of the original prehospital EMS providers were firefighters, who had "special" additional training in providing medical services during emergencies that occurred outside the hospital. Today, most firefighters receive emergency medical training and the fire service provides the majority of medical services during emergencies that occur out of the hospital, just as it has done for the past four decades.

The community-based fire station, with its ready availability of personnel 24 hours a day, coupled with the unique nature of medicine outside of the hospital, blends traditional public concepts and duties of the fire service with the potential for the most rapid delivery of advanced prehospital 9-1-1 emergency response and care. Traditionally, fire stations are strategically placed across geographic regions, typically commensurate with population densities and workload needs. This creates an all-hazard response infrastructure meeting the routine and catastrophic emergency needs of all communities regardless of the nature of the emergency. Accordingly, the fire service helps ensure the prosperity and security of all communities and providing prehospital 9-1-1 emergency medical care is consistent with its legacy.

Types of Fire-Service-Based EMS Systems

The fire service can be configured many ways to deliver prehospital 9-1-1 emergency medical care such as the following general configurations:

- Fire-service-based system using cross-trained/multirole firefighters. Firefighters are all-hazards responders, prepared to handle any situation that may arise at a scene including patient care and transport.

- Fire-service-based system using employees who are not cross-trained as fire suppression personnel. Single-role EMS-trained responders accompany firefighter first responders on 9-1-1 emergency medical calls.

- Combined system using the fire department for emergency response and a private or "third-service" (police, fire, EMS) provider for transportation support. Single-role emergency medical technicians and paramedics accompany firefighter first responders to emergency scenes to provide patient transport in a private or third-service ambulance.

While there are pros and cons to the various system approaches, the individual firefighter must always strive to ensure that his or her contribution provides the maximum benefit to the patient and minimum risk to the community.

Effectiveness of Fire-Service-Based EMS

The U.S. Fire-Service-based emergency response and medical care system is the most effective, coordinated system worldwide. The National Incident Management System (NIMS) and other nationally defined coordination plans ensure that fire-service-based 9-1-1 emergency response and medical care always provides skilled medical services to the patient regardless of the circumstances surrounding the location and condition of the patient. In addition, the fire service has the day-to-day experience and ability to work smoothly with other participants in the prehospital 9-1-1 emergency medical care arena: private ambulance companies, law enforcement agencies, health departments, public works departments, the American Red Cross, and other government and non-government agencies involved in medical care, disaster response, and patient services.

Prehospital 9-1-1 emergency patient medical care is a major part of the safety net for the American health care system. To its credit, the fire-service-based, prehospital 9-1-1 emergency patient medical care provides unconditional service to all members of our population. Therefore, the fire service must now become an integral part of the public health system and work closely with medical and public health experts to help alleviate unnecessary burdens on already overburdened hospital, medical, and public health systems. Already part of local government, the fire service may be best positioned to help provide important data to facilitate creating solutions to pressing health care public policy issues.

Above all, rapid response times are a pivotal advantage of fire-service-based, prehospital 9-1-1 emergency EMS systems. Now equipped with automated defibrillators to reverse sudden cardiac events, the fire apparatus, coupled with bystander CPR, has become one of the greatest life-saving tools in medical history. With stroke centers to treat stroke within the golden 3-hour period, cardiac catheterization centers to treat heart attack in the 90-minute door-to-balloon time, and trauma centers to treat hemorrhaging patients, time efficiency is a key component of the best designed EMS systems. A fire-service-based EMS system is capable of rapid multifaceted response, rapid identification and triage and transport to the appropriate facility.

In terms of the rapid delivery of emergency medical care in the out-of-hospital environment, fire departments have the advantage of having a free-standing army ready to respond anytime and anywhere. Prehospital, 9-1-1 emergency response in support of community prosperity and security is one of the essential public safety functions provided by the United States fire service. Fire-service-based EMS systems are strategically positioned to deliver time-critical response and effective patient care and scene safety. Fire-service-based EMS accomplishes this while emphasizing responder and patient safety, providing competent and compassionate workers, and delivering cost-effective operations.

Tactical Priorities

Despite the best efforts to prevent fires, they continue to occur. Many are caused by mechanical or electrical malfunctions, natural disasters, human carelessness, or criminal activity including acts of terrorism. To save lives and protect property after a fire starts or another type of emergency occurs, fire departments have developed a standard set of tactical priorities. In practice, most fire departments operate with the following priorities:

- Life safety
- Incident stabilization
- Property conservation

Life Safety

Life Safety — Refers to the joint consideration of the life and physical well-being of individuals, both civilians and firefighters.

Life safety is always the first and highest priority in any emergency operation. As mentioned earlier, life safety includes protecting the lives of the firefighters, the occupants of the burning building or vehicle, those in other life-threatening situations, and the lives of any nearby spectators. Life safety also includes saving the lives of pets and livestock.

Incident Stabilization

Emergency Operations — Activities involved in responding to the scene of an incident and performing assigned duties in order to mitigate the emergency.

Also mentioned earlier was the fact that most fires and other emergencies will continue to get worse until someone — usually the fire department — steps in to interrupt the growth of the situation. In other words, before we can extinguish a fire we must first control it; that is, stop it from spreading to uninvolved portions of the building or to other nearby structures. In some cases, it may mean allowing the original fire building to burn down in order to protect neighboring buildings. Fire department personnel must first prevent an emergency from getting any worse than it was when they arrived on scene. They do this by addressing the underlying problem: performing rescues, extinguishing the fire, and treating those who are injured.

Property Conservation

To the extent possible, and that means without putting firefighters in mortal danger, fire departments are committed to saving as much property as possible. For example, after assessing the risks involved, an Incident Commander (officer in overall charge of an emergency operation) may order firefighters into a burning building to spread salvage covers over furniture and other contents on the floor below the fire floor to protect it from being damaged by the water used to extinguish the fire. As a firefighter, you will be trained how to force entry into a locked building or to ventilate a burning building without doing unnecessary damage in the process. You will also be trained to overhaul a fire — that is, to search concealed spaces and the debris after the fire has been controlled. This is done to find and extinguish any hidden fire to prevent it from flaring up after firefighters leave the scene. You will also be trained to remove excess water from buildings after a fire, to control and contain spilled hazardous materials, and to neutralize certain chemicals.

Fire Department Organization

An organizational chart graphically illustrates the structure of the fire department and its chain of command. Small fire departments have a relatively simple chain of command while large departments may have a considerably more complex chart.

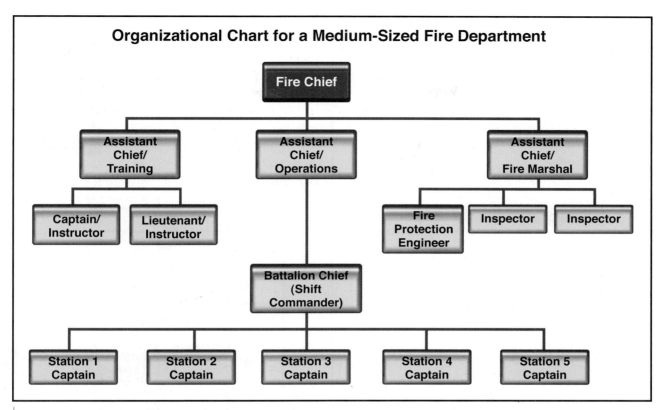

Organizational Chart for a Medium-Sized Fire Department

Figure 1.5 An organizational chart shows the structure of the fire department and its chain of command.

Figure 1.5 shows an organizational chart for a medium-sized fire department. It is meant to serve only as an example. Organizational charts vary from one fire department to another.

Organizational Principles

To function effectively as a member of a team, you will operate according to the following four basic organizational principles:

- Unity of command
- Span of control
- Division of labor
- Discipline

Unity of command. This means that you report to only one supervisor. Directly, each subordinate reports to one boss; indirectly, however, everyone reports to the fire chief through the chain of command (**Figure 1.6, p. 20**). The chain of command represents the lines of authority and responsibility from the highest level of the department to the lowest.

Span of control. This means that any officer can effectively supervise or manage only a certain number of individuals or groups. A rule of thumb in the fire service is that an officer can directly supervise three to seven firefighters effectively, with five being optimum, but the actual number can vary with the situation. Likewise, a chief officer can manage the same number of fire companies.

Chain of Command — (1) Order of rank and authority in the fire service. (2) The proper sequence of information and command flow as described in the Incident Command System.

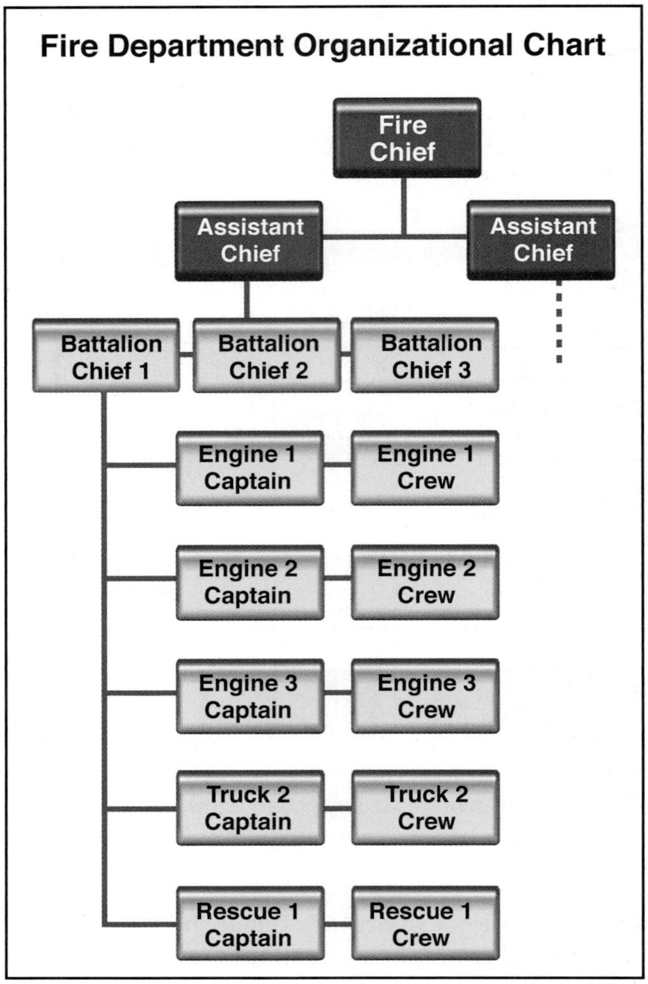

Fire Department Organizational Chart

Figure 1.6 Organizational charts identify reporting responsibilities.

Division of labor. This is the process of dividing large jobs into small jobs to make them more manageable. For example, if a multistory building needs to be searched, each floor may be assigned to a different fire company. In that way, each company has a reasonable workload and the entire building is searched quickly. Small jobs may then be broken down further and assigned to specific individuals. Division of labor is necessary in the fire service for the following reasons:

- To assign responsibility

- To prevent duplication of effort

- To make specific and clear-cut assignments

Discipline. In this context, *discipline* refers to both an organization's responsibility to provide the direction needed to satisfy its identified goals and objectives and the individual's responsibility to follow the direction given. In other words, discipline is setting the limits or boundaries for expected performance and enforcing them. This direction may come in the form of rules, regulations, or policies, but regardless of the term used, it must define acceptable performance and expected outcomes. The rules of the department must be clearly written and disseminated throughout the organization.

Discipline — Setting the limits or boundaries for expected performance and enforcing them.

For some the term *discipline* has a negative connotation — that is, it is perceived as *punishment*. This is a serious misperception. In the fire service, discipline can be negative or positive, preventive or corrective, and self-imposed (self-discipline) or imposed on a subordinate by a supervisor or manager. The highest and most positive form of discipline is self-discipline. As in the earlier discussion of integrity, self-discipline means doing the right thing simply because it is right, and because it's your job!

The word *discipline* comes from the root word *disciple* — a learner. One dictionary definition of discipline is "training that corrects." Therefore, the main purpose of discipline is to educate. In a fire department, discipline is intended to do the following:

- Educate and train.
- Correct inappropriate behavior.
- Provide positive motivation.
- Ensure compliance with established rules, regulations, standards, and procedures.
- Provide direction.

Fire Companies

The standard operating unit of a fire department is the *fire company* (commonly referred to as just a *company*), which is a group of firefighters assigned to a particular piece of fire apparatus or to a particular station. In most fire departments (but not all) a company consists of a company officer, a driver/operator, and one or more firefighters (**Figure 1.7**).

A fire company is organized, equipped, and trained for definite functions. The functions and duties of similar fire companies may vary in different localities because of the inherent hazards of the area, the size of the department, and the scope of the department's activities. A small fire department may have only one fire company to carry out all of the functions that normally would be performed by several companies in a larger fire department. The following lists the general descriptions of fire companies:

Figure 1.7 Company members work together on all emergency tasks. *Courtesy of Dick Giles.*

- *Engine company* — Deploys hoselines for fire attack and exposure protection (**Figure 1.8, p. 22**)
- *Truck (ladder) company* — Performs forcible entry, search and rescue, ventilation, salvage and overhaul, and utilities control and provides access to upper levels of a structure (**Figure 1.9, p. 22**)
- *Rescue squad/company* — Searches for and removes victims from areas of danger or entrapment and may perform technical rescues (**Figure 1.10, p. 22**)
- *Brush company* — Extinguishes wildland fires and protects structures in the wildland/urban interface (**Figure 1.11, p. 22**)
- *Hazardous materials company* — Responds to and mitigates hazardous materials incidents (**Figure 1.12, p. 23**)
- *Emergency medical/ambulance company* — Provides emergency medical care to patients (may provide transportation to a medical facility) (**Figure 1.13, p. 23**)
- *Special rescue company* — Responds to and performs technical rescues (**Figure 1.14, p. 23**)

Figure 1.8 Engine company personnel work as a team.

Figure 1.9 Truck company members are responsible for raising ladders and for performing forcible entry. *Courtesy of District Chief Chris E. Mickal, NOFD Photo Unit.*

Figure 1.10 The fire department is likely to be called to any emergency scene that results from accident or natural disaster. *Courtesy of District Chief Chris E. Mickal, NOFD Photo Unit.*

Figure 1.11 Wildland firefighters use specialized apparatus and techniques. *Courtesy of Monterey County Training Officers.*

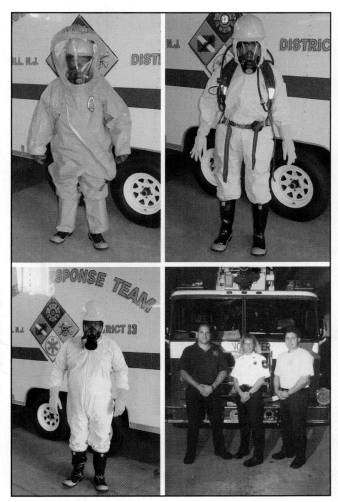

Figure 1.12 Firefighters wear different levels of protection when they respond to hazardous materials incidents.

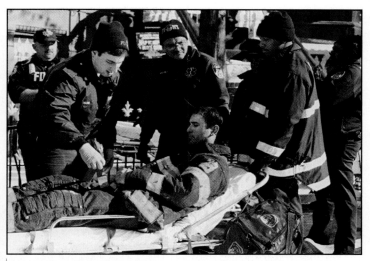

Figure 1.13 Most firefighters are also trained to provide emergency medical assistance. *Courtesy of District Chief Chris E. Mickal, NOFD Photo Unit.*

Figure 1.14 Some rescue companies, like this company practicing water rescue, are highly specialized. *Courtesy of Darrel Levine.*

Fire Department Personnel

Fire protection involves preventing, combating, and extinguishing fires; answering emergency calls; and operating and maintaining fire department equipment, apparatus, and quarters. These duties require extensive training in fire fighting, technical rescue, hazardous materials, and emergency medical care. Firefighters must operate apparatus and perform dangerous assignments under emergency conditions. Any of these situations may require strenuous physical activity in extremely hostile environments such as toxic smoke and very high temperatures (**Figure 1.15, p. 24**). Although fighting fires, performing rescue operations, and providing emergency medical care are the primary duties, in many departments a significant portion of a firefighter's time is spent conducting inspections, participating in training, and performing routine station duties.

Figure 1.15 Firefighters must perform in extremely hostile environments. *Courtesy of District Chief Chris E. Mickal, NOFD Photo Unit.*

Duties of Firefighter I and Firefighter II

In most fire departments, those trained to the Firefighter I level perform specifically assigned tasks under the direct supervision of a Firefighter II or a company officer. Those trained to the Firefighter II level may supervise a Firefighter I and perform a variety of tasks under the general direction of a company officer (**Figure 1.16**). To function effectively, a firefighter must have certain knowledge and skills including:

- Meeting the requirements of National Fire Protection Association (NFPA®) Standard 1001, *Standard for Fire Fighter Professional Qualifications*. Chapter 5 of the standard lists the requirements for Firefighter I, and Chapter 6 lists the additional requirements for Firefighter II.

- Knowing department organization, operation, and standard operating procedures (SOPs) (see Standard Operating Procedures section).

- Knowing the district or city street system and physical layout.

- Meeting minimum health and physical fitness standards.

- Meeting educational requirements established by the authority having jurisdiction (AHJ).

To make use of their basic skills and knowledge, firefighters must apply them in the performance of their duties. The following are some of the typical duties of a Firefighter I and a Firefighter II:

- Attending training courses; reading and studying assigned materials related to fire fighting, fire prevention, hazardous materials, and emergency medical care

- Responding to medical emergencies and other patient-care requests (**Figure 1.17**)

Figure 1.16 The company officer's job is to mentor new members and to train them to work as a team.

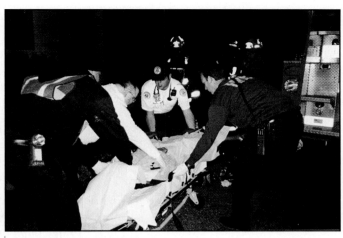

Figure 1.17 Knowing and applying skills learned during training are critical at the scene of an emergency. *Courtesy of Bob Esposito.*

- Responding to fire alarms as part of a company, operating fire fighting equipment, laying and connecting hose, maneuvering nozzles and directing fire streams, raising and climbing ladders, and using extinguishers and a variety of hand tools

- Ventilating burning buildings by opening windows and skylights, cutting holes in roofs or floors, and placing and operating ventilation fans

- Removing people from dangerous locations and administering first aid

- Performing loss control operations (reducing or eliminating loss and damage during and after a fire), which include spreading salvage covers, removing accumulated water, and removing burned debris

- Completing overhaul operations with the goal of ensuring total fire extinguishment

- Relaying instructions, orders, and information and identifying locations of alarms received from the communications dispatcher (telecommunicator)

- Following safety regulations to avoid injury while performing both routine and emergency duties

- Performing salvage operations that include removing valuable property from areas of danger or implementing means to protect it in place

- Ensuring the safekeeping and proper care of all fire department property

- Performing assigned fire inspections and checks of buildings and structures for compliance with fire prevention and life safety codes and ordinances

- Delivering fire and life safety talks and demonstrations to members of the public

Other Fire Department Personnel

Depending on local conditions and requirements, other specialized fire service personnel may be a part of the organization. Their duties and requirements vary depending on local needs and procedures. Following is a list of other positions among fire suppression personnel, their primary roles, and the NFPA® standard covering their professional qualifications:

- *Fire apparatus driver/operator* — Drives assigned fire apparatus to and from fires and other emergencies; operates pumps, aerial devices, or other mechanical equipment as required (NFPA® 1002, *Standard for Fire Apparatus Driver/Operator Professional Qualifications*) **(Figure 1.18)**.

- *Fire department officer* — Fulfills any of the following responsibilities, depending upon the size and structure of the fire department (NFPA® 1021, *Standard for Fire Officer Professional Qualifications*):

  — The fire chief is ultimately responsible for all operations within the fire department, including obtaining the funds needed to carry out its mission.

  — Fire department officers supervise a fire company in the station and at fires and other emergencies **(Figure 1.19)**. They may also supervise a group of fire companies in a specified geographical region of the city.

  — Other roles assigned include operations, training, personnel/administration, public information, fire prevention, resources, and planning **(Figure 1.20)**.

- *Fire department health and safety officer* — Oversees a fire department's occupational safety and health program (NFPA® 1500, *Standard on Fire Department Occupational Safety and Health Program,* and NFPA® 1521, *Standard for Fire Department Safety Officer*) **(Figure 1.21)**.

- *Fire department incident safety officer* — Monitors operational safety during emergency incidents (NFPA® 1521, *Standard for Fire Department Safety Officer*) **(Figure 1.22)**.

Figure 1.18 Apparatus driver/operators are responsible for the safety of those riding on the apparatus.

Figure 1.19 Fire officers must be able to perform a variety of supervisory/command functions at an emergency scene. *Courtesy of San Ramon Valley Fire District.*

Figure 1.20 Some fire officers are charged with providing information about the fire department and its operations during emergency and nonemergency operations.

Figure 1.21 One part of the department's health and safety officer job is to monitor any preemployment fitness tests. *Courtesy of Rick Montemorra.*

Figure 1.22 An incident safety officer monitors the safety of on-scene personnel.

To carry out the mission of the fire department, other personnel are also required. In many fire departments these positions are staffed by nonsworn civilian employees. The following list describes some of these personnel:

- *Communications personnel (telecommunicators)* — Receive emergency and nonemergency phone calls, process the information, dispatch units, establish and maintain a communications link to in-service companies, and complete incident reports (NFPA® 1061, *Standard for Professional Qualifications for Public Safety Telecommunicator*) **(Figure 1.23)**

- *Fire alarm maintenance personnel* — Maintain municipal fire alarm systems (NFPA® 72, *National Fire Alarm Code®*) **(Figure 1.24)**

Figure 1.23 Telecommunicators receive telephone calls and dispatch units. *Courtesy of Paul Ramirez, Phoenix FD, AZ.*

Figure 1.24 Fire alarm systems require regular maintenance. *Courtesy of Bob Noll.*

Figure 1.25 Some fire department members are qualified vehicle maintenance technicians.

Figure 1.26 Volunteers in some departments are trained in traffic control.

Figure 1.27 Some fire department personnel are trained to manage electronic data.

- *Apparatus and equipment maintenance personnel* — Maintain all fire department apparatus and portable equipment (NFPA® 1071, *Standard for Emergency Vehicle Technician Professional Qualifications*) **(Figure 1.25)**

- *Fire police personnel* — Assist law enforcement officers with traffic control, crowd control, and scene security at fires and other emergency operations **(Figure 1.26)**

- *Information systems personnel* — Manage the collection, entry, storage, retrieval, and dissemination of electronic databases such as fire reporting **(Figure 1.27)**

Special Operations Personnel

If fire departments only provided standard structural fire protection and emergency medical response to their communities, the positions discussed to this point in the chapter would be sufficient. However, many fire departments today provide a wide variety of special services to their communities. These services require special equipment and specially trained personnel. In many cases, these individuals serve as both regular firefighters and specialists in a particular discipline. The following list addresses some of the special operations found in many fire departments:

- *Airport firefighter* — Protects life and property, controls fire hazards, and performs general duties related to airport operations and aircraft safety (known as aircraft rescue and fire fighting [ARFF]) (NFPA® 1003, *Standard for Airport Fire Fighter Professional Qualifications*) **(Figure 1.28)**

- *Hazardous materials technician* — Handles hazardous materials and chemical, biological, radiological, nuclear, or explosive (CBRNE) emergencies (NFPA® 472, *Standard for Competence of Responders to Hazardous Materials/Weapons of Mass Destruction Incidents*) **(Figure 1.29)**

- *Technical rescuer* — Handles technical rescue situations such as high-angle (rope) rescue, trench and structural collapse, confined space entry, extrication operations, and cave or mine rescues (NFPA® 1006, *Standard for Rescue Technician Professional Qualifications*) **(Figure 1.30)**

- *Wildland firefighter* — Responds to and mitigates fires in outdoor vegetation including the wildland/urban interface (NFPA® 1051, *Standard for Wildland Fire Fighter Professional Qualifications*) **(Figure 1.31)**

Figure 1.28 Airport firefighters wear specialized gear.

Figure 1.29 Hazardous materials team members must wear clothing dictated by the situation. *Courtesy of Steven Baker.*

Figure 1.30 Technical rescuers must have highly specialized training.

Figure 1.31 Many firefighters are assigned to protect wildland areas. *Courtesy of Monterey County Training Officers.*

Fire Prevention Personnel

Because lives and property cannot be lost in a fire that never starts, one of the most effective and least costly ways of fulfilling the mission of the fire service is through fire prevention. Thus, it can be argued that the first duty of every member of the fire service is to prevent fires from starting. An effective fire prevention program decreases the need for suppression activities and thereby reduces the costs and risks of extinguishing fire. A chief officer typically heads the fire prevention division. Depending on the local organizational structure, this person may be called the chief in charge of fire prevention, the fire prevention officer, or the fire marshal. This individual may have subordinate officers to fill the various roles within the division. The fire prevention division generally includes four major positions:

- *Fire prevention officer/inspector* — Inspects a variety of occupancies to ensure code compliance; conducts technical and supervisory work in the fire prevention program (NFPA® 1031, *Standard for Professional Qualifications for Fire Inspector and Plan Examiner*) (**Figure 1.32**)

- *Fire and arson investigator* — Investigates fires and makes analytical judgments based on the physical evidence at the fire scene to determine the origin and cause of a fire (NFPA® 1033, *Standard for Professional Qualifications for Fire Investigator*) (**Figure 1.33**)

- *Public fire and life safety educator* — Makes presentations and conducts seminars to inform the public about fire hazards, fire causes, precautions, and actions to take during a fire (NFPA® 1035, *Standard for Professional Qualifications for Public Fire and Life Safety Educator*) (**Figure 1.34**)

- *Fire protection engineer/specialist* — Checks plans for proposed buildings to ensure compliance with local fire and life safety codes and ordinances; acts as a consultant to the fire department administration in the areas of fire department operations and fire prevention (**Figure 1.35**)

Emergency Medical Services (EMS) Personnel

Departments that provide EMS response may deploy trained medical responders on fire companies such as engines, trucks, or rescues (**Figure 1.36, p. 32**). These personnel may be trained to the first responder, emergency medical technician (EMT), or paramedic levels. The ambulance that responds to transport the victim may or may not be a part of the fire department.

The following list highlights the roles of personnel who are trained to the first responder, EMT, or paramedic levels. In many cases these duties are in addition to those of a firefighter:

- *First responder* — Stabilizes the victims of accidents or illnesses until more highly trained medical personnel arrive

- *Emergency medical technician (EMT)* — Provides basic life support (BLS) for the victims of accidents or illnesses

- *Paramedic* — Provides advanced life support (ALS) for the victims of accidents or illnesses

Training Personnel

The training that new firefighters receive is one of the most important aspects of their orientation and indoctrination. *A firefighter's training never ends.* New regulations, ideas, equipment, and tactics require new methods that must be learned. New materials and technology present challenges that never before existed. It is imperative that

Figure 1.32 Fire inspectors enforce local code requirements.

Figure 1.34 Public education specialists speak to citizens about fire and life safety.

Figure 1.33 Fire investigators determine the cause and origin of a fire.

Figure 1.35 Fire protection engineers check plans and act as consultants.

members of the fire service remain abreast of these changes. The following training personnel constantly improve and update the training program and deliver training (**Figure 1.37, p. 32**):

- ***Training officer/chief of training/drillmaster*** — Administers all fire department training activities (NFPA® 1041, *Standard for Fire Service Instructor Professional Qualifications*)

- ***Instructor*** — Delivers training courses to the other members of the department (NFPA® 1041, *Standard for Fire Service Instructor Professional Qualifications*) (**Figure 1.38, p. 32**)

Figure 1.36 Most firefighters respond to medical emergencies.

Figure 1.37 Training officers manage the department's training program.

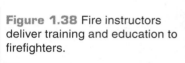

Figure 1.38 Fire instructors deliver training and education to firefighters.

Career Development

If you decide to make the fire service your career, you should be aware of the importance of *career development*. There are many career development opportunities that are above and beyond your daily, monthly, and annually required training. Staying active and current with national trends, industry standards, and professional qualifications are fundamentally important. Following are a few of the many educational opportunities available to you:

- College or university degree in fire science or technology, business administration, public administration — certificate of completion, associate, bachelor's or master's degree programs

- National Fire Academy courses

- Technical committee membership

- Curriculum design, development, and implementation
- Department accreditation and firefighter certification

Fire Department Regulations

The success of a fire department depends on the administration's commitment to ensuring that its members work for the benefit of the department and the public. The department must implement and enforce policies and procedures that specify how members are to cooperate and work effectively together. Anyone who becomes a member of a fire department must become familiar with the department's policies and procedures. Anyone who has questions about these policies and procedures should ask a supervisor to clarify any misunderstandings. This section introduces some of the policies and procedures that most departments generally follow.

Policies and Procedures

It is important to understand the difference between a policy and a procedure. A department *policy* is a guide to decision-making within an organization. Policy originates with or is approved by top management in the fire department. Policies define the boundaries within which the administration expects department personnel to act in specified situations.

A *procedure* is a written communication closely related to a policy. Whereas a policy is a guide to thinking or decision-making, a procedure is a detailed plan of action. A procedure describes in writing the steps to be followed in carrying out organizational policy for some specific, recurring problem or situation.

Orders and directives are used to implement departmental polices and procedures. Orders and directives may be either written or verbal. An *order* is based upon a policy or procedure, whereas a *directive* is not. Fire officers may issue both orders and directives **(Figure 1.39)**. Because orders are based on policies or procedures, compliance is mandatory. Directives are not based on policy; they are more in the nature of a request or suggestion. However, because of the seriousness of the situation on the fireground, both procedures and policies are generally considered as orders and must be followed.

Standard Operating Procedures

Some fire departments have a predetermined plan for nearly every situation they can conceive of occurring — both emergency and nonemergency. These plans are known as the department's *standard operating procedures (SOPs)*. Some fire departments use slightly different names such as Standard Operating Guidelines (SOGs) or General Operating Guidelines (GOGs), but regardless of the name used they are all the same thing. These procedures provide a standard set of actions that is the basis of every Incident Action Plan (IAP). The SOP format may vary considerably in different localities, but the principle is usually the same.

Even though there are obvious variations in how fires behave in different fuels and environments, most fires have more similarities than differences. These similarities are the basis for standard operating procedures. The Incident Commander (IC) is the individual in overall command of an incident. The IC knows departmental SOPs and can base a plan of action upon them. Some SOPs have built-in flexibility that allows, with reasonable justification, adjustments when unforeseen circumstances occur; others do not.

Policy — Guide to decision-making in an organization.

Procedure — A written communication closely related to a policy.

Figure 1.39 Fire officers may issue both orders and directives.

Standard Operating Procedures (SOPs) — Standard methods or rules in which an organization or a fire department operates to carry out a routine function. Usually these procedures are written in a policies and procedures handbook and all firefighters should be well versed in their content. A SOP may specify the functional limitations of fire brigade members in performing emergency operations.

In practice, the first fire companies that reach the scene usually implement the SOP. The SOP is primarily a means to start the emergency operation. It does not replace size-up, decisions based on professional judgment, evaluation, or command. In addition, there may be several SOPs from which to choose depending on the nature and scope of the emergency, its location, and the ability of first-in units to handle the incident.

An example of a typical SOP follows. This SOP is performed with crews wearing complete protective clothing and self-contained breathing apparatus (SCBA).

1. The first unit on the scene assumes command.

2. The first-arriving engine lays a supply line and attacks the fire.

3. The second-arriving engine pulls a backup line.

4. The first-arriving ladder truck performs necessary forcible entry, search, rescue, utilities control, and ventilation.

The SOP should follow the most commonly accepted order of fireground priorities:

• Life safety

• Incident stabilization

• Property conservation

The need to save lives in danger is always the first consideration. However, it may not be the first action taken. In many cases, the best way to protect the occupants of a burning building is by mounting a rapid and effective fire attack. In this approach, both life safety and incident stabilization are addressed simultaneously. Otherwise, once all building occupants have escaped or been rescued, attention is turned to stabilizing the incident. In this process, firefighters should make every reasonable effort to minimize damage to property. This can be accomplished through proper fire fighting tactics and good loss control techniques.

Following SOPs reduces chaos on the fire scene. All resources can be used in a coordinated effort to rescue victims, stabilize the incident, and conserve property. Operational procedures that are standardized, clearly written, and mandated to each department member establish accountability and increase command and control effectiveness. When all firefighters follow departmental SOPs, confusion is reduced. Firefighters will understand their duties and require a minimum of direction. SOPs also help prevent duplication of effort and uncoordinated operations because all positions are assigned and covered. The assumption and transfer of command, communications procedures, and tactical procedures are other areas that must be addressed by departmental SOPs.

Safety is the highest priority when writing SOPs. Requiring SCBA for all crews operating in potentially hazardous atmospheres is an example of a safety consideration. SOPs should be applied to all situations, including rescues and medical responses. They should be designed to limit personnel exposure to anticipated hazards such as communicable diseases. For example, SOPs may require personnel to use pocket masks when performing mouth-to-mouth resuscitation. They may also require personnel to wear medical exam gloves and safety glasses to prevent contact with bodily fluids during medical emergencies.

Standard operating procedures need not be limited to the emergency scene. Many departments prefer to carry out the administrative and personnel functions of the department through SOPs. These SOPs may include regulations on dress, conduct, vacation and sick leave, station assignments and duties, and other departmental policies.

National Incident Management System

As a firefighter you will be trained to function as a member of a team — your fire company. On every emergency call — large or small — your company will function along with other companies within an organizational system. In the event of a large-scale disaster such as a tornado, earthquake, hurricane, major fire or explosion, or an act of terrorism, your company may be joined with other companies to form larger tactical units. In 2003, Homeland Security Presidential Directive-5 was issued. HSPD-5 was entitled *Management of Domestic Incidents*, and it directed the U.S. Secretary of Homeland Security to develop and administer a National Response Plan. A major part of that plan is the National Incident Management System (NIMS). In the United States, NIMS is the mechanism by which large numbers of fire companies and larger units are organized and managed during emergency incidents.

NIMS provides a consistent nationwide template to enable all government, private-sector, and nongovernmental organizations to work together during domestic incidents.

In general, NIMS includes the following elements:

- Standardized organizational structures, processes, and procedures
- Standards for planning, training and exercising, and personnel qualification standards
- Equipment acquisitions and certification standards
- Interoperable communications processes, procedures, and systems
- Information management systems
- Supporting technologies — voice and data communications systems, information systems, data display systems, and specialized technologies

Historically, more than one incident management system has been used by fire departments in the U.S.; for example, one of the most widely used systems is the Fireground Command system (FGC), which was developed by the Phoenix (AZ) Fire Department. The other most commonly used system is the FIRESCOPE Incident Command System (ICS), which originated in Southern California.

The major difference between Fireground Command and FIRESCOPE ICS is that in FGC the term *Sector* can be used to designate either a *functional* or *geographical* assignment. In ICS, functional assignments are called *Groups* and geographical assignments are called *Divisions*. However, under the National Response Plan, all fire departments in the U.S. are required to use a NIMS-compliant Incident Command System, which does not recognize the use of the term *Sector*.

The NIMS-Incident Command System is designed to be applicable to incidents of all sizes and types. It applies to small, single-unit incidents that may last a few minutes as well as complex, large-scale incidents that can last for days or weeks involving several agencies and many mutual aid units.

Components of ICS

The ICS has a number of interactive components that provide the basis for clear communication and effective operations:

- Common terminology
- Modular organization
- Integrated communications
- Unified command structure

- Consolidated action plans
- Manageable span of control
- Predesignated incident facilities
- Comprehensive resource management

Overview of ICS

To understand the application of the ICS, firefighters should be familiar with the major subdivisions within the ICS structure: Command, Operations, Planning, Logistics, and Finance/Administration (**Figure 1.40**). There is a potential sixth function — Information and Intelligence — that may be needed on some incidents.

Command

Command includes the Incident Commander (IC) and the Command Staff (**Figure 1.41**). The IC is ultimately responsible for all incident activities, including the development and implementation of an Incident Action Plan (IAP). The IC has the authority to call resources to the incident and to release them from it. If the size and complexity of the incident require it, the IC may delegate authority to the Command Staff. Regardless of their actual rank, members of the Command Staff are referred to as Officers. The Command Staff consists of the following:

- **Safety Officer (SO)** — Monitors incident operations and advises the IC on all matters related to operational safety, including the health and safety of emergency responder personnel. The SO has emergency authority to stop and/or prevent unsafe acts during incident operations.

- **Liaison Officer (LNO)** — Point of contact for other governmental and nongovernmental agencies and private-sector organizations involved in the incident.

- **Public Information Officer (PIO)** — Responsible for interfacing with the public and media and/or other agencies with incident-related information requirements.

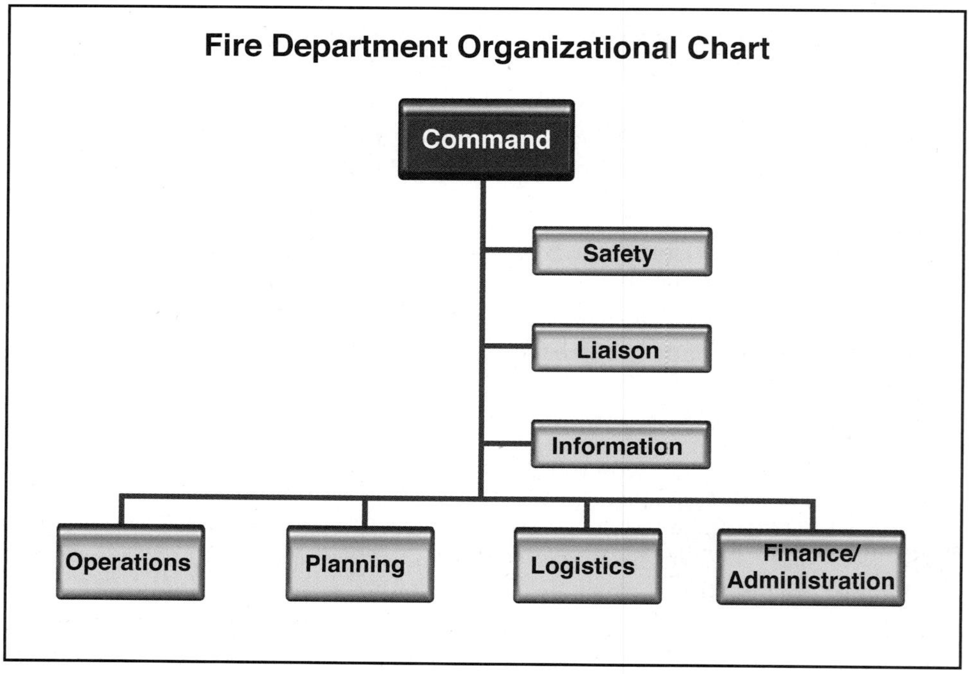

Incident Command System (ICS) — (1) System by which facilities, equipment, personnel, procedures, and communications are organized to operate within a common organizational structure designed to aid in the management of resources at emergency incidents. (2) Management system of procedures for controlling personnel, facilities, equipment, and communications so that different agencies can work together toward a common goal in an effective and efficient manner. (3) Recommended method of establishing and maintaining command and control of an incident. It is an organized approach to incident management, adaptable to any size or type of incident.

Figure 1.40 The five major areas within the NIMS-Incident Command System.

Fire Department Organizational Chart

- Command
  - Safety
  - Liaison
  - Information
  - Operations
  - Planning
  - Logistics
  - Finance/Administration

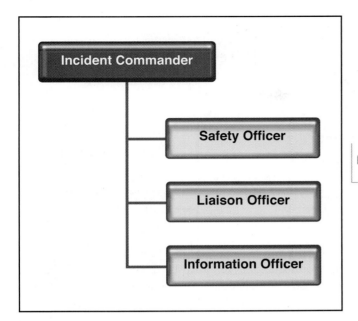

Figure 1.41 Components of the Command Staff.

General Staff

The General Staff is responsible for the functional aspects of the incident and is divided into Sections. Regardless of actual rank, the manager of each Section is referred to as a Chief. The General Staff consists of the following:

- *Operations Section Chief* (**Ops**) — Reports directly to the IC and is responsible for all activities focused on reducing the immediate hazard, saving lives and property, establishing situational control, and restoring of normal operations. Ops directs the tactical operations to meet the strategic goals and tactical objectives specified in the IAP. In large-scale operations, Ops is also responsible for Staging. Depending upon the size and complexity of the incident, the Operations Section may be subdivided into as many as ten Branches if necessary (**Figure 1.42**).

- *Planning Section Chief* (**Plans**) — Responsible for the collection, documentation, evaluation, and dissemination of incident situation information and intelligence to the IC. The Planning Section is responsible for tracking the status of all resources assigned to the incident and for developing the IAP for review and approval by the IC. Depending upon the size and complexity of the incident, the Planning section may expand to include the *Resource Unit, Situation Unit, Demobilization Unit*, and any technical specialists whose services are required (**Figure 1.43, p. 38**).

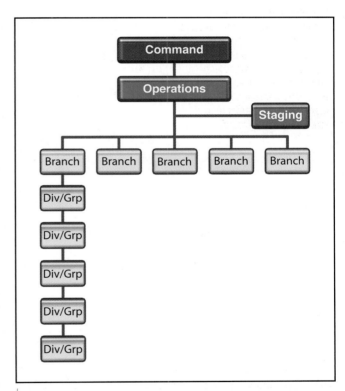

Figure 1.42 Components of the Operations Section.

- *Logistics Section Chief* (**Logistics**) — Responsible for all support requirements needed to facilitate effective and efficient incident management. This includes ordering resources from off-incident locations. There are two Branches within the Logistics section: the *Support Branch* and the *Service Branch*. The *Service Branch* includes medical, communications, and food services (**Figure 1.44, p. 38**). The *Support Branch* includes supplies, facilities, and ground support vehicle services.

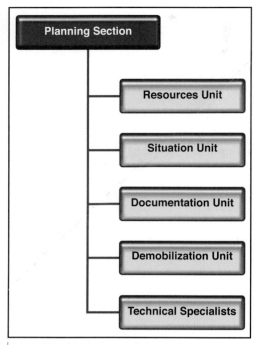

Figure 1.43 Components of the Planning Section.

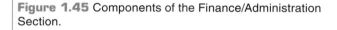

Figure 1.44 Components of the Logistics Section.

Figure 1.45 Components of the Finance/Administration Section.

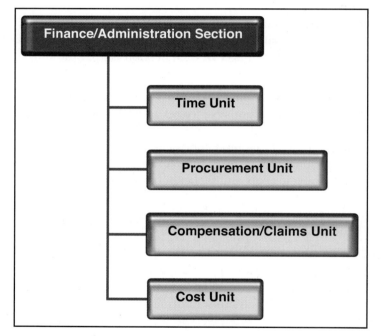

- *Finance/Administration Section Chief* (**Finance**) — Established when the agencies involved require finance and other administrative support services. Not all incidents require this section (**Figure 1.45**). Generally, the Finance/Administration Section is activated only on large-scale, long-term incidents. Day-to-day mutual aid responses are usually considered to be reciprocal and do not require interagency reimbursement.

- *Information/Intelligence Function* (**Intel**) — When required, this section is responsible for analyzing and sharing incident information and intelligence. This may include national security or other classified information as well as other operational information such as risk assessments, medical intelligence, weather information, geospatial data, structural designs, toxic contaminant levels, and other data from a variety of sources. This function may be assigned within the Command Staff, as a unit within the Planning Section, as a Branch within the Operations Section, or as a separate General Section.

ICS Terms

One of the major advantages to using ICS is standardized terminology. To prevent or reduce communications problems during emergency operations, firefighters should understand and use the following terms:

Assigned — Resources currently committed to an assignment.

Available — Resources that have checked in at the incident and are not currently assigned.

Branch — The organizational level between Divisions/Groups and the IC and Operations. It is managed by a Branch Director. At large, complex incidents, the number of Divisions and/or Groups may create span-of-control problems. Branches can be implemented to reduce these difficulties. Example: Fire Branch, EMS Branch, Law Enforcement Branch, Haz Mat Branch.

Command — The function of directing, ordering, and controlling resources by virtue of explicit legal, agency, or delegated authority. It is important that lines of authority be clear to all involved. Lawful commands by those in authority should be followed immediately and without question.

Command Post — Location from which all incident operations are directed. There is only one Command Post per incident. All agency representatives are located here. Planning function is performed here.

Division — A geographic designation assigning responsibility for all operations within a defined area. Divisions are assigned clockwise around a wildland fire or the exterior of a building and identified alphabetically. Division A is at the heel of a wildland fire or at the front (street address side) of an incident involving a building. Inside of buildings, Divisions are usually identified numerically by the floor or area to which they are assigned: First floor is Division 1, second floor is Division 2, etc. In a one-story building, the entire interior may be assigned as a Division (Interior Division) **(Figure 1.46)**. All groups operating within that specific geographic area report to that Division Supervisor. Organizationally, the Division level is between a Strike Team or other Operational Unit and a Branch.

Group — Functional designations (forcible entry, salvage, ventilation, etc.). When the Group's assigned function has been completed, it is available for reassignment.

Incident Action Plan — The written or unwritten plan for managing the emergency is the *Incident Action Plan* (IAP). A plan should be formulated for *every* incident. Small, routine incidents usually do not require a written plan; large,

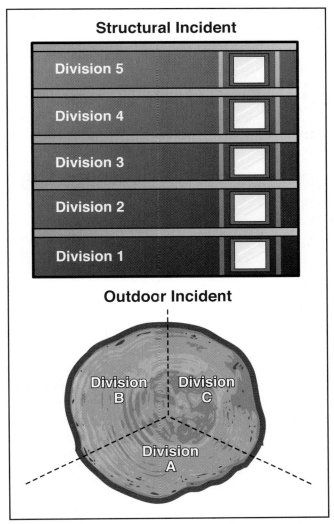

Figure 1.46 Examples of Divisions in the ICS.

complex incidents do. The plan identifies the strategic goals and tactical objectives that must be achieved to eliminate the problem.

Incident Commander (IC) — The officer at the top of the incident chain of command who is in overall charge of the incident. The IC is ultimately responsible for everything that takes place at the emergency scene. The IC is primarily responsible for determining the incident strategy, formulating or approving the Incident Action Plan, and coordinating and directing all incident resources to implement the plan and meet its stated objectives.

Out-of-Service — Resources not available for assignment.

Resources — All personnel and major pieces of apparatus on scene or en route and on which status is maintained. Resources may be individual companies, task forces, strike teams, or other specialized units. It is imperative that the status of these resources be tracked so that they may be assigned when and where they are needed without delay.

Resource Status — Resources assigned to a particular incident are in one of three status modes: available, assigned, or out-of-service.

Single Resource — Individual personnel and equipment items and the operators associated with them.

Strike Team — Set number or resources of the same kind and type with an established minimum number of personnel.

Strategic Mode — Determines the positions for companies operating on scene. The strategy is divided into two basic modes — offensive and defensive.

Supervisor — A supervisor is someone in command of a Division or a Group.

Task Force — Any combination of resources assembled in support of a specific mission or operational need.

Interacting with Other Organizations

As a firefighter, you may be required to interact with many different organizations that are a part of, or related to, the fire service. During major incidents, some interactions fall under the control of the Liaison Officer discussed earlier; others do not. The purpose of this section is to acquaint you with some of the organizations with which you may be required to interact.

Emergency Medical Services

If fire department personnel do not provide emergency medical services (EMS) or medical transportation (ambulance services), firefighters should establish a close working relationship with those who do. In some areas, fire departments provide medical transportation; in others the departments work very closely with private ambulance companies. Because one of the major functions of the fire department is the removal (and sometimes the initial treatment) of people trapped in wrecked vehicles and similar situations, it is important for firefighters to have an appropriate level of first-aid training even if their department does not do medical transport **(Figure 1.47)**. The level of training needed depends on the local EMS system and the department's SOPs. In many jurisdictions, fire department personnel are expected to primarily perform rescue and extrication functions. Beyond that, they provide only first responder medical treatment.

In most jurisdictions, once fire and EMS units are on the scene, EMS personnel are responsible for treating patients and fire/rescue personnel are responsible for scene safety and freeing trapped victims. Close coordination between the two groups is very important to avoid working against each other, wasting valuable time, and perhaps further endangering victims and rescuers.

Hospitals

In some areas, hospitals operate ambulances and provide EMS. In most areas hospitals do not provide such services, but in rare cases, hospital personnel may be called to the scene of an emergency **(Figure 1.48)**. This is most likely to occur during a mass-casualty incident. In such cases, hospital personnel are needed on-scene to assist in performing triage (sorting victims by the severity of their injuries) or conducting primary treatment of more seriously injured victims. An incident does not necessarily have to be large to require hospital personnel on the scene. Although quite rare, in some areas where EMS personnel are not trained to provide advanced life support, hospital personnel may be called to the scene to perform such functions as administering fluids and/or drugs intravenously while extrication operations are in progress.

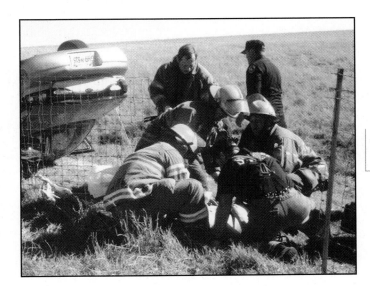

Figure 1.47 Firefighters must often deliver medical aid following vehicle crashes.

Figure 1.48 Hospital personnel may sometimes respond to the scene.

Another example where hospital personnel are needed is a serious entrapment where major medical procedures (such as amputation of a limb) may be the only way to free the victim.

Law Enforcement

It is important for law enforcement and firefighters to understand each other's roles and priorities and know what to expect from each other at the scene. Law enforcement personnel may be present at a fire scene, but their priorities are scene security and maintaining the flow of traffic during fire and rescue operations on streets and highways, investigating traffic accidents, and conducting arson investigations (**Figure 1.49**). When victims are either unconscious or otherwise unable to provide needed information, law enforcement personnel can often secure the information by using resources such as computer databases.

Firefighters may be called upon to assist law enforcement agencies in a variety of ways. These may include forcible entry to assist an investigation, emergency lighting to illuminate a crime scene, providing logistical support, or recovering a body.

Utility Companies

Many incidents involve utility providers (electricity, natural gas, or water) in some way, so it is important for firefighters to have a good working relationship with local utility company personnel. It is also important for responding fire units to coordinate with the utilities on mutual responses and to know what to do — and what not to do — until the utility crews arrive. In addition, utility companies may have specially trained and equipped emergency response teams that can greatly assist in rescue efforts (**Figure 1.50**).

Figure 1.50 Utility companies have emergency response teams.

Figure 1.49 Law enforcement personnel have specific duties and priorities at an emergency scene.

Media

As discussed earlier, NIMS-ICS includes a Public Information Officer (PIO) for dealing with the media. As a firefighter, you should not make comments or express opinions to members of the media but should direct them to the PIO. The media can play an important role in the delivery of news based on an incident. They also can be very helpful in diverting vehicular traffic away from the scene and alerting the public when large-scale evacuations are necessary.

Other Agencies

In addition to the agencies already mentioned, firefighters may interact with a number of other entities. These include public health departments, coroner/medical examiner's offices, and the Environmental Protection Agency (EPA), to name a few. Any potential contacts should be identified beforehand and a relationship established so that these agencies will be able to work more effectively with the fire department when an incident occurs.

Summary

The fire service in North America has a long and proud history dating back to the Pilgrims who landed in 1620. This early history provided the foundation upon which the fire service of today was built. Firefighters today are walking in the footsteps of those who came before them.

Most fire departments today protect their communities from far more than just the ravages of fire. They provide emergency medical services, perform technical rescues, handle hazardous materials leaks and spills, and provide a variety of other services to the public.

To uphold these proud traditions and to be a positive contributor to your department and your community, you must possess certain aptitudes and capabilities. You must be capable of and willing to learn the many and varied aspects of the firefighter's job in a modern society. You must be willing to follow orders and conduct yourself in a manner that will bring honor to yourself and your department. You must be capable of and willing to perform strenuous physical tasks in extremely hostile conditions. You must be willing to perform every assignment — even the most menial — to the best of your ability. As a volunteer or career firefighter, you will stand on the shoulders of all the firefighters who came before you. If you are willing and capable of doing all of these things and more, you can have the best job in the world — being a firefighter!

Fire Fighter I

1. What are the four categories of fire departments?

2. What are two necessary characteristics and behaviors of firefighters?

3. What is the mission of the fire service?

4. What is unity of command?

5. Name two types of fire companies and their functions.

6. Name three types of fire service personnel and describe their functions.

7. What are standard operating procedures (SOPs)?

8. What are the major subdivisions within the ICS structure?

9. Define the following ICS terms: *Command, Group, Strike Team, Supervisor.*

10. Name two organizations that fire service personnel may interact with.

Chapter Contents

Firefighter Safety and Health

This chapter provides information that addresses the following job performance requirements of NFPA® 1001, *Standard for Fire Fighter Professional Qualifications (2008)*:

NFPA® 1001 references for Chapter 2:

5.1.1

5.3.2

5.3.3

5.3.5

5.3.17

Photo courtesy of District Chief Chris E. Mickal, NOFD Photo Unit.

Chapter Objectives

Fire Fighter I Objectives

1. List ways to prevent firefighter injuries.

2. Discuss National Fire Protection Association standards related to firefighter health and safety.

3. Discuss Occupational Safety and Health Administration regulations.

4. Summarize the IFSTA Principles of Risk Management.

5. List the main goals of a safety program.

6. Discuss firefighter health considerations and employee assistance and wellness programs.

7. List guidelines for riding safely on the apparatus.

8. Discuss safety in the fire station.

9. Describe ways to maintain safety in training.

10. Explain how to maintain and service equipment used in training.

11. Discuss emergency scene preparedness.

12. Discuss emergency scene safety.

13. Summarize general guidelines for scene management including highway incidents, crowd control, and cordoning off emergency scenes.

14. Explain the importance of personnel accountability.

15. Summarize basic interior operations techniques.

16. Describe emergency escape and rapid intervention.

17. Respond to an incident, correctly mounting and dismounting an apparatus. (Skill Sheet 2-I-1)

18. Set up and operate in work areas at an incident using traffic and scene control devices. (Skill Sheet 2-I-2)

Firefighter Safety and Health

Case History

A fire department in New York state received a report of a propane leak in the basement of a three-story warehouse. A large propane tank had fallen from a forklift that was moving it. When the tank struck the concrete floor, the valve broke off, releasing liquid propane into the space. The forklift driver immediately called the fire department. Just as the first fire units arrived, the propane vapors reached an ignition source and a tremendous explosion followed. The explosion blew a ladder truck and two engines across the street. Five firefighters were killed and nine others seriously injured. Two civilians were killed and dozens more injured.

This incident clearly illustrates that there are no "routine" calls. Every emergency call is potentially lethal for firefighters. Fire fighting is one of North America's most dangerous civilian jobs, and job-related accidents can result in costly losses — the greatest loss being the death of a firefighter **(Figure 2.1)**. Other losses may include personnel unavailable for duty (due to injuries) and the cost of replacing them, damaged equipment (expensive to repair or replace), higher workers' compensation insurance rates, and legal expenses. In order to reduce these losses, it is necessary to prevent the accidents that cause them.

Preventing accidents saves lives and money. By knowing what injuries occur, where they occur, and why they occur, plans and procedures can be developed to recognize and mitigate factors that could injure firefighters performing their duties. However, along with your department, as a firefighter you also are responsible for your own safety on the job. An important part of fulfilling that responsibility is being aware of the causes of firefighter injuries and fatalities, the requirements of applicable safety standards, and the means available to minimize the risks.

Figure 2.1 Every emergency call is potentially lethal for firefighters. *Courtesy of Ron Jeffers.*

It is your responsibility to know and comply with your department's safety policies and procedures.

According to NFPA® 1001, those qualified at the Fire Fighter I level must know the following about firefighter safety and health:

- Provisions of the fire department's member assistance program
- Critical aspects of NFPA® 1500, *Standard on Fire Department Occupational Safety and Health Program*, as they apply to the Fire Fighter I

While task-specific safety procedures are included in each of the chapters that follow, this chapter presents an overview of firefighter safety and health. First, firefighter injury and fatality statistics are reviewed. Also discussed are the applicable standards and regulations that apply to firefighter safety and health. The IFSTA Principles of Risk Management are presented along with ways in which firefighters can protect themselves from avoidable risks during emergency and nonemergency activities. Finally, safety considerations while responding to, operating at, and returning from emergencies as well as safety in and around the fire station are discussed.

Firefighter Injuries and Fatalities

Figure 2.2 Proper nutrition and exercise help firefighters remain healthy.

Fighting fires, performing rescues, and delivering other emergency services is inherently dangerous work. According to statistics compiled by the U.S. Fire Administration (USFA), an average of about 100,000 firefighter injuries are reported in the U.S. each year. These injuries range from those that are relatively minor to those that require admission to a hospital. Many firefighter injuries can be prevented by effective training, maintaining company discipline and accountability (also called *team integrity*), following established safety-related SOPs, using personal protective clothing and equipment, and maintaining high levels of physical fitness. Physically fit firefighters are not only more productive but also less likely to suffer strains and sprains, which account for nearly 50 percent of all firefighter injuries. Physical exercise is also a very effective way of reducing stress and reducing the likelihood of heart attacks and strokes (**Figure 2.2**).

While the experience may vary slightly from year to year, USFA statistics for 2005 show that the majority of serious injuries to firefighters, as well as the greatest number of injuries, occurred on the fireground (52 percent). Another 15 percent occurred during nonfire emergency incidents. Other injuries occurred during nonemergency incidents, during training, while responding to or returning from calls, and in various other activities while on duty (**Figure 2.3**).

Excluding those firefighters who lost their lives in the World Trade Center on September 11, 2001, over the last two decades an average of more than 100 U.S. firefighters die each year in the line of duty. On a population-adjusted basis, the rate of firefighter fatalities in Canada is even higher.

In 2005, there were 115 on-duty firefighter fatalities in the United States. Of those, 34 were full-time career firefighters — 29.5 percent of the total. The others were volunteer, seasonal, and part-time firefighters (**Figure 2.4**). More than half of the fatalities (53.9 percent) were from stress or overexertion, including heart attacks, strokes, and the effects of extreme climatic heat exposure. Vehicle crashes claimed another 21.7 percent, and the rest were from a variety of other causes (**Figure 2.5**).

Firefighters have traditionally accepted job-related injuries and fatalities as part of the job. Knowing their vocation to be one of the most hazardous, many firefighters are resigned to occupational accidents, injuries, and fatalities. This attitude is compounded by the stereotypical image of the firefighter as heroic and fearless in the face of danger. However, most firefighter injuries are a direct result of preventable accidents. Firefighters should be too smart and too professional to take unnecessary risks.

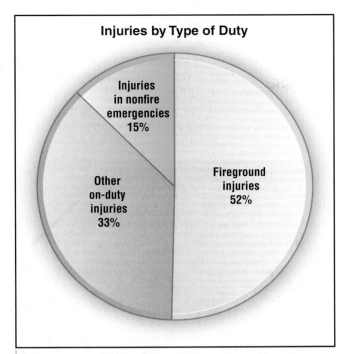

Injuries by Type of Duty

Injuries in nonfire emergencies 15%

Other on-duty injuries 33%

Fireground injuries 52%

Figure 2.3 Fireground injuries account for over half of all firefighter injuries annually.

Fatalities by Duty Status

Career Firefighter Fatalities 29.5%

Volunteer, Seasonal, Part-time Firefighter Fatalities 70.5%

Figure 2.4 Fatalities among volunteer firefighters is over twice the number suffered by career firefighters.

A large part of being smart enough to avoid unnecessary risks comes from knowing the safety and health standards and regulations that apply to the fire service.

The National Fallen Firefighters Foundation, a nonprofit organization, was created in 1992 by the United States Congress to lead a nationwide effort to honor America's fallen firefighters. One of the ways that the Foundation honors fallen firefighters is to strive to prevent line-of-duty deaths. The Foundation is committed to helping the U.S. Fire Administration meet its stated goal of reducing firefighter fatalities by 25 percent within 5 years and by 50 percent within 10 years. In 2004, the Foundation developed 16 Firefighter Life Safety Initiatives to provide the fire service with a blueprint for making changes. The Foundation sees fire service adoption of the life safety initiatives as a vital step in meeting these goals. The full text of the 16 Firefighter Life Safety Initiatives can be seen in **Appendix C**. For more information on the National Fallen Firefighters Foundation, visit its web site at www.firehero.org.

Firefighter Fatalities by Cause

Fatalities due to vehicle crashes 21.7%

Fatalities due to stress or overexertion 53.9%

Other causes 24.4%

Figure 2.5 The major cause of firefighter fatalities is cardiac arrest due to stress.

Safety Standards for the Fire Service

A number of occupational safety standards and regulations apply to U.S. fire departments and their members. Some of the more prominent ones are the standards and recommended practices published by the National Fire Protection Association (NFPA®), regulations of the Occupational Safety and Health Administration (OSHA), and the regulations found at Canada's National Occupational Health and Safety (CanOSH) web site.

Line-of-Duty Death (LODD) — Firefighter or emergency responder death resulting from the performance of fire department duties.

NFPA® 1500

In one way or another, many NFPA® standards relate to firefighter safety and health. However, NFPA® standards are consensus documents and are not law unless adopted by a state, provincial, or local governing body. Many governing bodies in both the U.S. and Canada have chosen to adopt one or more of the NFPA® standards into their laws and ordinances; many others have not. While there are a number of others, perhaps the single most important NFPA® standard dealing with firefighter safety and health is NFPA® 1500, *Standard on Fire Department Occupational Safety and Health Program.* This standard is the most comprehensive standard relating to firefighter safety and health. It specifies the minimum requirements for a fire department safety and health program **(Figure 2.6)**. This standard may be applied to any fire department or similar organization, public or private. It calls upon the fire department to recognize health and safety as official objectives of the department and to provide as safe and healthy a work environment as possible. The basic concept of NFPA® 1500 is to promote

Figure 2.6 NFPA® 1500 focuses on firefighter safety and health. One of the requirements is the establishment of a rapid intervention crew (RIC) at all structure incidents.

safety throughout the fire service regardless of individual status or type of organization. Because it is a minimum standard, any department or jurisdiction is free to exceed the requirements specified.

NFPA® 1500 reflects information that is contained in a long list of other NFPA® standards that relate to firefighter safety and health. Among the topics covered in NFPA® 1500 are the following:

- Safety and health-related policies and procedures
- Training and education
- Fire apparatus, equipment, and driver/operators
- Protective clothing and protective equipment
- Emergency operations
- Facility safety
- Medical and physical requirements
- Member assistance and wellness programs
- Critical incident stress management (CISM) program

Safety and Health-Related Policies and Procedures

To be in compliance with NFPA® 1500, fire departments must develop an organizational plan, a risk management plan, and a safety and health policy; define the roles and responsibilities of their members; establish a safety and health committee; keep records of all job-related accidents, illnesses, exposures, and fatalities; and appoint a designated department health and safety officer (HSO). The departmental safety and health program must address all anticipated hazards to which the members might be exposed. Besides the obvious hazards associated with fighting fires, the members may be exposed to hazardous materials releases, communicable diseases, energized electrical equipment, and the hazards of driving apparatus during emergency responses. The program must also include provisions for dealing with a variety of nonemergency issues such as alcoholism and drug abuse both on and off duty. To show the department's commitment to firefighter safety and health, the program must include appropriate SOPs. For example, one SOP would require that all personnel wear whatever personal protective clothing is appropriate for the nature and scope of the incident. The department administration must show its commitment to firefighter safety and health by following these SOPs themselves (leading by example) and enforcing these SOPs at all levels through the chain of command.

Training and Education

The goal of the department training and education program must be to prevent occupational deaths, injuries, and illnesses. In the fire service, *training* is usually understood to mean developing and maintaining job skills while *education* means learning new information, concepts, and procedures. To protect their firefighters, departments must train and educate their personnel to consider safety — their own and others' — in all emergency and nonemergency activities. Safety must be reinforced continually in training so that a "safety-first" orientation becomes second nature. In some departments, this concept is reinforced by posting reminders in every fire station saying, "Train as if your life depended on it — it does."

Fire Apparatus, Equipment, and Driver/Operators

NFPA® 1500 requires that safety and health be primary considerations in the design, specification, operation, maintenance, and repair of all fire department apparatus. Design considerations include providing restraint devices (seatbelts, etc.) for all personnel driving or riding in the apparatus. Apparatus must be designed to meet the requirements of the NFPA® standard applicable to the vehicle's intended role. NFPA® 1500 requires that all apparatus driver/operators be thoroughly trained before being allowed to operate the apparatus under emergency conditions.

With three exceptions (hose loading, tiller training, and some EMS operations), NFPA® 1500 requires that all personnel riding in the apparatus be seated and securely belted inside the cab or body of the vehicle whenever it is in motion. If the noise level on the apparatus exceeds 90 decibels (85 dB in Canada), hearing protection must also be worn. Those riding in the jump seats of older apparatus with safety bars or gates instead of fully enclosed cabs must also wear helmets and goggles.

Protective Clothing and Protective Equipment

To be in compliance, fire departments must provide all members with at least one set of protective clothing (preferably two) and protective equipment appropriate for the hazards to which they will be exposed. While covered in more detail in Chapter 5, basic protective clothing includes:

- Helmet
- Hood and/or shroud
- Coat
- Pants
- Safety shoes or boots
- Gloves
- Goggles or safety glasses
- Personal Alert Safety System (PASS) device

Safety Policy — Written policy that is designed to promote safety to departmental members.

Health and Safety Officer (HSO) — Member of the fire and emergency services organization who is assigned and authorized by the administration as the manager of the health and safety program and performs the duties, functions, and responsibilities in NFPA® 1521, *Standard for Fire Department Safety Officer.* This individual must meet the qualifications or approved equivalent of this standard.

Communicable Disease — Disease that is transmissible from one person to another.

Personal Alert Safety System (PASS) — Electronic lack-of-motion sensor that sounds a loud tone when a firefighter becomes motionless. It can also be manually triggered to operate.

Self-Contained Breathing Apparatus (SCBA) — Respirator worn by the user that supplies a breathable atmosphere that is either carried in or generated by the apparatus and is independent of the ambient atmosphere. Respiratory protection is worn in all atmospheres that are considered to be Immediately Dangerous to Life or Health (IDLH).

Protective equipment refers primarily to self-contained breathing apparatus (SCBA), supplied-air respirators (SARs), and other respiratory protection. It may also refer to body armor for those operating in areas that have a potential for violence or civil disturbances.

Emergency Operations

NFPA® 1500 requires that fire departments use an incident management system during all emergency operations. Most departments are now using the NIMS-ICS. Whatever system is used must include a risk management element or plan. Also required is a personnel accountability system. This section of the standard also limits emergency operations to those that can safely be conducted by the personnel available on scene (**Figure 2.7**). Rapid intervention for the rescue of firefighters in distress is required, along with rehabilitation (rehab) for firefighters during emergencies. This section also requires fire departments to develop procedures for limiting firefighter activities and exposure to violence during civil disturbances. Also included is a requirement for postincident analysis of all emergency operations.

Both NFPA® 1500 and 1561, *Standard on Emergency Services Incident Management System*, contain specific requirements regarding accountability of members, including the following:

- Written guidelines are required for the tracking and inventory of all members operating at an emergency incident.

- All members are responsible to actively participate in the accountability system.

- The IC is responsible for overall accountability and must initiate an accountability worksheet at the beginning of the incident and maintain it throughout the incident.

- The IC must maintain an awareness of the location and function of all assigned companies.

- Branch Directors and Division/Group Supervisors must supervise and account for all companies under their command.

- Company officers are responsible for all company members.

- Accountability appropriate to the size and complexity of the incident must be maintained through span-of-control requirements.

- Access to the scene must be controlled.

- The department must adopt a personnel accountability system and use it on every emergency incident.

- Procedures must be adopted for evacuating personnel from any area where imminent hazards are found.

- Appointment of an incident safety officer (ISO) is required.

Figure 2.7 When staffing is limited, the incident commander must balance rapid fire extinguishment or hazard control with the acceptable level of risk the firefighters are exposed to. *Courtesy of Dick Giles.*

Facility Safety

The standard sets minimum design requirements for fire department facilities (fire stations, training centers, administration buildings, and maintenance shops) that meet NFPA® 101, *Life Safety Code®*, and provide a means for cleaning, disinfecting, and storing infection control devices. It also requires that fire stations be designated as smoke-free environments. Inspection, maintenance, and prompt repair of facilities are also required.

Medical and Physical Requirements

NFPA® 1500 requires that firefighter candidates be medically evaluated to ensure that they are capable of performing the duties of a firefighter in that community. This section of the standard also prohibits any firefighter who is under the influence of alcohol or drugs from participating in any fire department operations. It requires fire departments to develop physical performance standards for those who participate in emergency operations. Also required is annual medical verification of continued fitness for duty. In addition, NFPA® 1500 requires the establishment of job-related fitness standards and a fitness program that allows members to develop and maintain the required level of fitness. A confidential health database must be established for each member. Fire departments are required to operate an infection control program. Finally, departments are required to designate a fire department physician to guide, direct, and advise members on their work-related health, fitness, wellness, and suitability to perform various duties. All on-the-job injuries and exposures are required to be reported and documented.

Member Assistance and Wellness Programs

Under this standard, fire departments are required to have a member assistance program to help firefighters and their immediate families deal with substance abuse, stress, and personal problems that have an adverse effect on the firefighter's job performance. Also required is the establishment of a wellness program to assist firefighters with health-related problems including quitting smoking and ceasing the use of other tobacco products. See Employee Assistance and Wellness Programs section.

Other NFPA® Standards

As mentioned earlier, many other NFPA® standards relate in some way to firefighter safety and health. In most cases, the title of each standard makes its content apparent. Other safety-related NFPA® standards are the following:

NFPA® 1250, *Recommended Practice in Emergency Service Organization Risk Management*

NFPA® 1401, *Recommended Practice for Fire Service Training Reports and Records*

NFPA® 1403, *Standard on Live Fire Training Evolutions*

NFPA® 1404, *Standard for Fire Service Respiratory Protection Training*

NFPA® 1410, *Standard on Training for Initial Emergency Scene Operations*

NFPA® 1451, *Standard for a Fire Service Vehicle Operations Training Program*

NFPA® 1521, *Standard for Fire Department Safety Officer*

NFPA® 1561, *Standard on Emergency Services Incident Management System*

NFPA® 1581, *Standard on Fire Department Infection Control Program*

NFPA® 1582, *Standard on Comprehensive Occupational Medical Program for Fire Departments*

NFPA® 1583, *Standard on Health-Related Fitness Programs for Fire Department Members*

NFPA® 1584, *Recommended Practice on the Rehabilitation of Members Operating at Incident Scene Operations and Training Exercises*

NFPA® 1851, *Standard on Selection, Care, and Maintenance of Protective Ensembles for Structural and Proximity Fire Fighting*

NFPA® 1951, *Standard on Protective Ensembles for Technical Rescue Incidents*

NFPA® 1971, *Standard on Protective Ensembles for Structural Fire Fighting and Proximity Fire Fighting*

NFPA® 1975, *Standard on Station/Work Uniforms for Fire and Emergency Services*

NFPA® 1981, *Standard on Open-Circuit Self-Contained Breathing Apparatus (SCBA) for Emergency Services*

NFPA® 1982, *Standard on Personal Alert Safety Systems (PASS)*

NFPA® 1983, *Standard on Life Safety Rope and Equipment for Emergency Services*

NFPA® 1999, *Standard on Protective Clothing for Emergency Medical Operations*

OSHA Regulations

In 1970, the Williams-Steiger Occupational Safety and Health Act became federal law in the United States. From this law came the Occupational Safety and Health Administration (OSHA). Operating under the U.S. Department of Labor, OSHA sets two overarching duties for employers:

- Furnish to each employee a place of employment that is free from recognized hazards that are likely to cause death or serious injury.

- Comply with the occupational safety and health standards contained within the OSHA regulations.

> Federal OSHA authority and regulations apply mainly to the private sector and do not cover employees of state and local governments, including career and some volunteer firefighters.

Industrial Fire Brigade — Team of employees organized within a private company, industrial facility, or plant who are assigned to respond to fires and emergencies on that property.

In terms of their applicability to firefighters, federal OSHA regulations apply only to federal employees who fight fires and to private-sector employees who fight fires — industrial fire brigades and incorporated volunteer fire companies. Although federal OSHA has no jurisdiction over local and state public-sector firefighters, the 24 states (and Puerto Rico and the U.S. Virgin Islands) operating OSHA-approved state plans do cover them. These state plans may differ from the federal standards but must provide equivalent protection. It is through these state plans that the "two-in/two-out rule" (discussed later in this chapter) applies to state and local government firefighters in those states.

Even in states where federal OSHA regulations do not apply, many fire departments and other local agencies choose to follow OSHA regulations because they are recognized national safety standards. As a firefighter, you must know and comply with the occupational health and safety regulations that apply in your particular jurisdiction. The following jurisdictions have adopted OSHA-approved plans:

Alaska	New Mexico
Arizona	New York
California	North Carolina
Connecticut	Oregon
Hawaii	Puerto Rico
Indiana	South Carolina
Iowa	Tennessee
Kentucky	Utah
Maryland	Vermont
Michigan	Virgin Islands
Minnesota	Virginia
Nevada	Washington
New Jersey	Wyoming

NOTE: The plans in Connecticut, New Jersey, New York, and the U.S. Virgin Islands cover public-sector (state and local government) employment only.

Federal OSHA regulations are contained in Title 29 of the *Code of Federal Regulations (CFR)*, which applies to labor (**Figure 2.8**). Chapter XVII of Title 29 contains the occupational safety and health requirements. Some of the more common OSHA regulations that fire departments and other emergency response agencies follow are:

- 29 *CFR* 1910.120, Hazardous waste operations and emergency response
- 29 *CFR* 1910.133, Eye and face protection
- 29 *CFR* 1910.134, Respiratory protection
- 29 *CFR* 1910.135, Head protection
- 29 *CFR* 1910.136, Foot protection
- 29 *CFR* 1910.137, Electrical protective equipment
- 29 *CFR* 1910.138, Hand protection
- 29 *CFR* 1910.146, Permit-required confined spaces
- 29 *CFR* 1910.156, Fire brigades
- 29 *CFR* 1910.1030, Bloodborne pathogens (requires employers to provide employees with immunization against Hepatitis B)
- 29 *CFR* 1910.1200, Hazard communication (material safety data sheets [MSDSs])
- 29 *CFR* 1926, Subpart P, "Excavations," covers operations in trenches

Figure 2.8 Most departments voluntarily comply with OSHA regulations protecting firefighters, including establishing a respiratory protection program that includes facepiece fit testing.

Knowing the content and focus of the foregoing standards and regulations will increase your awareness of what is required of you and your department during emergency operations. This critical awareness also comes from your department having a clearly stated risk management plan.

Canadian firefighters should refer to their provincial safety and health requirements. Although most Canadian requirements are based on NFPA® standards, additional Canadian Centre for Occupational Health and Safety (CCOHS) requirements may apply.

Risk Management

NFPA® 1500 requires that all incident management systems include a risk management plan. A risk management plan is an established set of criteria by which tactical decisions can be made based on an assessment of the benefits to be gained compared to the risks involved. Given these criteria, firefighters can perform a risk/benefit analysis of possible tactical options in a given emergency situation. The IFSTA Principles of Risk Management were developed after a review of a number of risk management plans being used in the fire service. The three most prominent plans considered were the Phoenix (AZ) Fire Department, NFPA® 1500, and the "Ten Rules of Engagement for Structural Fire Fighting" developed by the International Association of Fire Chiefs (IAFC). The IFSTA Principles of Risk Management are as follows:

- Activities that present a significant risk to the safety of members shall be limited to situations where there is a potential to save endangered lives.

- Activities that are routinely employed to protect property shall be recognized as inherent risks to the safety of members, and actions shall be taken to avoid these risks.

- No risk to the safety of members shall be acceptable when there is no possibility to save lives or property.

When applying these principles, there are three key points to keep in mind:

- Team integrity is vital to safety and must always be emphasized.

- No property is worth the life of a firefighter.

- Firefighters should not be committed to interior offensive fire fighting operations in abandoned or derelict buildings that are known or reasonably believed to be unoccupied.

You should also keep in mind that you did not start the fire, you did not put the victims in that situation, and you are not obligated to sacrifice yourself in a heroic attempt to extinguish a fire or save a victim — and especially not to recover a body.

In fact, protecting your life and those of your fellow firefighters is the highest fireground priority. The logic behind this position is simple: If you suffer a disabling injury while attempting to extinguish a fire or save a victim, you will be unable to help others who may be in mortal danger. In addition, other firefighters will have to respond to your injury and will therefore be unavailable to help the original victims.

One of the ways that the Phoenix (AZ) Fire Department applies its concept of risk management on a daily basis is through a list of key behaviors that are expected of all PFD personnel. Some of these behaviors apply to emergency operations; others do not. Even those that apply to nonemergency activities serve to help instill a *safety-first* mind-set in their personnel. These key behaviors are as follows:

- Think.
- Drive defensively.
- Drive slower rather than faster.
- Stop if you can't see at intersections.

- Don't run for a moving rig.
- Always wear your seatbelt/safety strap.
- Wear full protective clothing and SCBA.
- Don't *ever* breathe smoke.
- Attack with a sensible level of aggression.
- Always work under sector command — no freelancing.
- Keep the crew intact.
- Maintain a communications link to Command.
- Always have an escape route (hoseline/lifeline).
- Never go beyond your air supply.
- Use a big enough and long enough hoseline to accomplish the task.
- Evaluate the hazard — know the risk you are taking.
- Follow standard fireground procedures — know and be a part of the plan.
- Vent early and vent often.
- Provide lights for the work area.
- If it's heavy, get help.
- Always watch your fireground position.
- Look and listen for signs of collapse.
- Rehab fatigued companies — assist stressed companies.
- Pay attention at all times.
- Everybody takes care of everybody else.

Exhibiting these behaviors has the potential to prevent many firefighter injuries. In addition, they may help you survive to enjoy a service retirement instead of a disability retirement.

Fire Department Safety and Health Programs

To be in compliance with NFPA® 1500, every fire department must develop and implement a firefighter safety and health program. The success of these programs begins at the top of the fire department administrative chain **(Figure 2.9, p. 60)**. The administration's attitude toward safety is invariably reflected in the attitude of the supervising officers, which in turn affects the firefighters. The main goals of a safety program are as follows:

- Prevent human suffering, deaths, injuries, illnesses, and exposures to hazardous atmospheres and contagious diseases.
- Prevent damage to or loss of equipment.
- Reduce the incidence and severity of accidents and hazardous exposures.

To be effective, the safety program must be promoted and practiced at all levels throughout the organization. Safety requires compliance by everyone from the fire chief down to the newest recruit. Anyone who fails to comply with the rules of the safety program may influence others through their negative example. It is not enough to talk about and teach safety procedures; they must be practiced and enforced.

Safety Program — Program that sets standards, policies, procedures, and precautions to safely purchase, operate, and maintain the department's equipment and to educate employees on how to protect themselves from personal injury.

Figure 2.9 Firefighter safety must be a departmental priority.

Figure 2.10 Performing as a firefighter requires upper-body strength. Proper lifting, reaching, and carrying techniques help prevent back injuries.

Firefighter Health Considerations

Fire fighting is one of the most physically demanding and dangerous of all civilian occupations. You will need to be strong and in good shape to handle the physical demands of the job. You will need upper-body strength to perform such tasks as rescuing victims, raising ladders, handling hoselines, ventilating roofs, and forcing entry with heavy tools **(Figure 2.10)**. Muscular strength and aerobic endurance will be required to climb stairs, move rapidly down hallways, raise and climb ladders, and do all the other things required to control and extinguish fires — some of them many stories above street level. You will need to maintain your flexibility to be able to reach for equipment, manipulate ladders and heavy tools, work in awkward positions, or move victims onto ladders.

If you are required to perform wildland fire fighting, all of those same physical capabilities will be needed to climb hills, cut fire lines with hand tools, operate chain saws and other equipment, and work in very hostile environments. The following list contains information essential to maintaining your personal health as a firefighter:

- Stay informed about job-related health issues.

- Wear incident-appropriate personal protective equipment (PPE) and respiratory protection.

- Clean all PPE at least twice annually and remove heavy contamination after each use.

- Follow recommendations for vaccination against hepatitis B.

- Use precautions to avoid exposure to airborne and bloodborne pathogens.

- Use proper lifting techniques to avoid muscle strains and other related injuries.

- Use lifting tools or get help to assist with lifting heavy objects.

- Clean, disinfect, and store tools and equipment used in patient care.

- Maintain a regular exercise program to sustain physical fitness.

- Maintain a diet low in cholesterol, fat, and sodium.

- Reduce heart attack and stroke risk by maintaining blood pressure and cholesterol levels within acceptable limits.

- Reduce cancer risk by eliminating the use of all tobacco products.

- Have regular physicals and medical checkups.

Maintaining your physical fitness must be continuous throughout your fire service career.

The department is responsible for ensuring that measures are taken to limit the number of stress-related accidents and illnesses to which you are exposed. Department physical fitness and health programs are a good way of fulfilling this responsibility. For more information on firefighter fitness and health considerations, refer to NFPA® 1500 and the IFSTA **Fire and Emergency Occupational Safety, Health, and Wellness** manual.

Employee Assistance and Wellness Programs

An employee assistance program (EAP) is one way the fire department can help its members and their families. An EAP offers confidential assistance with problems that could adversely affect job performance. Some of the areas in which an EAP can assist are as follows:

- Alcohol abuse

- Drug abuse

- Personal and interpersonal problems

- Stress

- Depression

- Anxiety

- Divorce

- Financial problems

A wellness program can be critically important to firefighters who are struggling with other health-related problems. Among the areas with which a wellness program can be of assistance are:

- Nutrition

- Hypertension

- Cessation of tobacco use

- Weight control

- Physical conditioning program with peer fitness counselors

Employee Assistance Program (EAP) — Any one of many programs that may be provided by an employer to employees and their families to aid in solving work or personal problems.

These programs should be readily available to all members and their families. They should include referrals to appropriate professional services such as alcohol treatment (i.e., Alcoholics Anonymous), drug abuse intervention, supervised self-help groups, and other community services. The programs should provide counseling and education on health concerns **(Figure 2.11)**. They should allow members easy, yet confidential, access to counseling and professional help for any problems or concerns that may be interfering with their daily well-being. Pamphlets and flyers detailing services should be distributed throughout the department to make the program information widely accessible. Any service provided for departmental personnel should also be made available to family members.

Another significant area of employee assistance is critical incident stress management (CISM). Because the injuries suffered by some victims in fire and rescue incidents can be extremely gruesome and horrific, firefighters and any others who have to work directly with the victims should have a means of preventing or reducing the stress associated with these incidents. Because individuals react to and deal with extreme stress in different ways — some more successfully than others — and because the effects of unresolved stresses tend to accumulate, a critical incident stress debriefing should be made available as part of the department's CISM program **(Figure 2.12)**.

The process of managing this stress should actually start *before* firefighters enter the scene if it is known that conditions there are likely to produce psychological or emotional stress for the responders. This can be accomplished through a prebriefing process wherein the firefighters who are about to enter the scene are told what to expect so that they can prepare themselves.

If firefighters are required to work more than one shift in these conditions, they should go through a minor debriefing, sometimes called *defusing,* at the end of each work shift. To get the most benefit from the full debriefing session, they should attend within 72 hours of completing their work on the incident.

Defusing — Informal discussion with incident responders conducted after the incident has been terminated either at the scene or after the units have returned to quarters. Here commanders address possible chemical and medical exposure information, identify damaged equipment and apparatus that require immediate attention, identify unsafe operating procedures, assign information-gathering responsibilities to prepare for the postincident analysis, and reinforce the positive aspects of the incident.

Figure 2.12 As part of the department's CISM program, qualified stress managers should be available when needed.

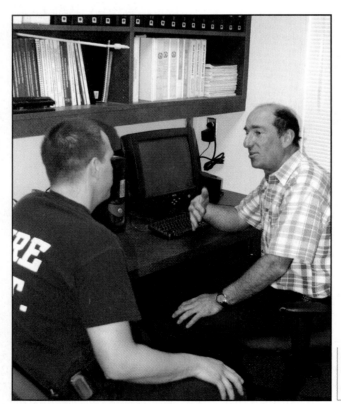

Figure 2.11 A wide range of services should be available to firefighters and their families through the employee assistance program.

Safety on the Apparatus

As a firefighter, one of the most common dangers to which you will be exposed is riding the apparatus to and from emergency calls. Statistically, this is one of the most hazardous activities of the job. In most fire departments, firefighters are required to have all of their protective clothing on when they enter the cab of the apparatus. Some departments allow their driver/operators to wait until after arrival at the scene to don their coats and helmets. As required by NFPA® 1500, all firefighters must ride in a seat within the cab and have their seatbelts fastened **(Figure 2.13)**. If sirens and engine noise levels exceed 90 decibels (85 dB in Canada), firefighters must also wear hearing protection **(Figure 2.14)**. Volunteers responding in their own vehicles should use defensive driving techniques to reduce the hazard.

WARNING!
Firefighters must NOT be allowed to ride anywhere on the outside of fire apparatus.

If it is necessary to ride in an unenclosed jumpseat, you should protect yourself by using the safety bars or safety gates provided and wear appropriate PPE **(Figure 2.15, p. 64)**. These bars or gates are not allowed on new apparatus but are acceptable on older apparatus still in service.

In most cases, you should use steps and handrails when mounting or dismounting the apparatus **(Figure 2.16, p. 64)**. Using the steps and handrails reduces the chances of accidentally slipping and falling from the apparatus. This is especially important if you already have SCBA strapped to your back when you dismount because the weight

Figure 2.13 All personnel must be seated with their seatbelts fastened before the apparatus is placed in motion.

Figure 2.14 Firefighters should protect their hearing through the use of earplugs, earmuffs, or radio communications head sets.

of the SCBA can make you more prone to falling backwards. One exception to the use of handrails is when dismounting an apparatus that is close to electrical wires. If the apparatus contacts the charged lines and you are in contact with the vehicle and the ground at the same time, you could be electrocuted. Therefore, if the apparatus is or may be in contact with energized electrical wires, always jump clear of the apparatus **(Figure 2.17)**. Steps for responding to an incident using correct procedures are described in **Skill Sheet 2-I-1**.

Figure 2.15 Older unenclosed apparatus must be equipped with safety gates or bars.

Figure 2.16 Steps and handrails make entry and exit safer.

Figure 2.17 It is important to keep your feet together as you jump clear.

Safety in Fire Stations and Facilities

In many fire departments, the firefighters' duties and activities center around the fire station and a significant portion of their on-duty time is spent there. Hazards in the fire station, not only endanger firefighters but can also endanger visitors who enter the station. Visitors are the responsibility of the fire department while they are in the building. Therefore, safe conditions must exist to limit the possibility of accidents and injuries (**Figure 2.18**).

Personnel Safety

Certain safety hazards are common to any fire station. Also, certain types of accidents are not limited to any specific location within a station. Improper lifting techniques and slip-and-fall accidents are two of the most common causes of injury.

Although back strains are the most common injuries related to improper lifting and carrying techniques, bruises, sprains, and fractures can also result. Improper lifting techniques not only cause personal injury, but they may also end in damaged equipment if it is dropped or improperly handled. Statistically, back injuries are the most expensive single type of accident in terms of workers' compensation, and they occur all too frequently.

All firefighters should be instructed in the correct method of lifting — that is, by keeping your back as straight as possible and lifting with your legs, not your back. You should not attempt to lift or carry an object that is too heavy or too bulky to safely handle by yourself; instead, you should get help to lift or carry it (**Figure 2.19**). Lifting and carrying heavy or bulky objects without help can result in serious strains and other injuries.

Other common accidents are slips, trips, and falls. Numerous factors contribute to these types of accidents. A slip, trip, or fall often results from poor footing. This can be caused by slippery surfaces, uneven surfaces, objects or substances on surfaces, inattention when climbing up or down stairs, and similar hazards. These accidents can easily result in minor or serious injuries as well as damaged equipment. One way to prevent such accidents is to stress good housekeeping. For example, floors should be kept clean and free from slipping hazards such as loose items and liquid spills (**Figure 2.20**). Aisles should be unobstructed and stairs should be well lighted. In addition to walking surfaces (floors, stair treads, and aisles), items such as handrails, slide poles, and slides must also be maintained in a safe condition.

Tool and Equipment Safety

Although tools and equipment are vital to your job as a firefighter, accidents can happen if you are not properly trained in the use and care of tools and equipment. Poorly maintained tools and equipment can be very dangerous and can result in costly accidents to firefighters in the station and at the emergency scene. NFPA® 1500 stresses the importance of safety in every aspect of tool and equipment design, construction, purchase, usage, maintenance, inspection, and repair.

Figure 2.18 Station visitors must be protected.

Figure 2.19 Proper lifting techniques, including the use of a second person or mechanical lifting device, prevent back injuries.

Figure 2.20 Floors should be kept clean and uncluttered.

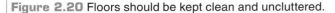

When working in a station shop or on the emergency scene, you must use appropriate personal protective equipment (PPE) such as gloves, vision protection, and hearing protection. Using PPE is fundamental for safe work practices. Although PPE does not take the place of good tool engineering, design, and use, it does provide some protection against hazards.

The most widely used tools in the station shop are hand tools and small power tools. Observe the following procedures when using hand and power tools:

- Wear appropriate personal protective equipment (PPE).
- Remove loose clothing and keep long hair clear of operating tool heads.
- Remove jewelry, including rings and watches.
- Select the appropriate tool for the job.
- Know the manufacturer's instructions and follow them.
- Inspect tools before use to determine their conditions.
- Do not use badly worn or broken tools.
- Provide adequate storage space for tools and always return them promptly to storage after use.
- Inspect, clean, and put all tools in a ready state before storing.
- Consult with and secure the approval of the manufacturer before modifying any tool.
- Use intrinsically safe tools when working in potentially flammable atmospheres such as around a vehicle's fuel system.
- Do not remove safety shields or modify the tool in any way that could compromise built-in safety devices.

Power Tools

Grinders, drills, saws, and welding equipment are commonly found in fire stations **(Figure 2.21)**. Improperly used, these tools can cause a serious or life-threatening injury. Whether the tool is driven by compressed air or electricity, it has a specific, safe method of operation that must be understood and followed. Only those firefighters who have read and who understand the tool manufacturer's instructions should be allowed to use power tools. These instructions must be readily accessible to firefighters.

Tool repairs should be made only by someone trained and authorized to repair the damaged tool. Depending on the department, this person may be someone within the fire department or an outside equipment dealer or repair agent. Keeping accurate records of repairs can help identify inherent deficiencies before the tool malfunctions and causes an injury.

Any electrical tool not marked "double insulated" should have a three-prong plug **(Figure 2.22)**. For firefighter safety, the third prong must connect to a ground while the tool is in use. Bypassing the ground plug in any way prevents it from protecting the user from injury or death caused by an electrical short.

Power Saws

Following a few simple safety rules when using power saws will prevent the most common accidents associated with these tools:

- Match the saw to the task and the material to be cut.
- Never force a saw beyond its design limitations.

Figure 2.21 Power tools must be equipped with safety shields and used in accordance with the manufacturer's instructions.

Figure 2.22 Three-prong plugs are designed to reduce shock hazards by providing a grounded circuit.

- Wear proper protective equipment, including gloves and vision and hearing protection.
- Remove loose clothing and contain long hair that could become entangled in the saw.
- Have hoselines in place when cutting materials that generate sparks.
- Avoid using power saws in potentially flammable atmospheres.
- Keep bystanders out of the work area.
- Follow manufacturer's procedures for proper saw operation.
- Allow gasoline-powered saws to cool before refueling.
- Keep blades and cutting chains well sharpened.
- Use extreme caution when operating any saw above eye level.

NOTE: In some departments, firefighters are required to maintain the landscaping at the fire stations. In these cases, firefighters must use the same safety procedures and protective equipment they use with other power tools. Vision and hearing protection must be worn along with gloves and safety boots or shoes.

Safety in Training

NFPA® 1500 requires that all personnel who may engage in structural fire fighting be adequately trained to do so. This training should reinforce safe practices until they become automatic. Other types of training are required on an "as-needed" basis. For example, training is required when new procedures or equipment are introduced.

Maintaining Personnel Safety

All personnel participating in training should be fully clothed in the appropriate protective gear. Raising ladders, deploying hose, or performing any other activity that simulates actual fireground conditions requires the use of PPE (structural fire fighting gear called *turnouts, bunker gear,* or *bunkers and SCBA*).

Figure 2.23 Horseplay is potentially dangerous and can result in injuries or equipment damage.

Safety in training is also influenced by your physical condition. Severe physical discomfort or illness can reduce trainees' abilities to perform demanding evolutions and may make them less alert and therefore more prone to injury. Trainees who have flu symptoms, colds, headaches, or other symptoms indicating severe physical discomfort or illness may be referred to a physician for evaluation prior to participation. In many departments, these trainees will not be allowed to resume training until cleared by a medical examination. Older trainees and those who appear to be in poor physical condition must be watched closely for signs of fatigue, chest pains, or unusually labored breathing during heavy exercise and physically demanding evolutions. Some trainees are reluctant to admit that they are physically unable to continue training. Nonetheless, if one member of a fire fighting team is impaired, it may start a chain reaction resulting in other team members being injured. Therefore, it is important that officers and instructors assess the physical condition of all participants before training begins.

It is also important that environmental conditions be taken into account during outside training. Trainees should be dressed appropriately for the weather as well as for the training. High winds may make ground ladder training unsafe. Rain, snow, or ice may make the training grounds too slippery for safety. In hot weather, trainees should be given appropriate breaks during which they may remove all or part of their protective clothing while they rest and rehydrate in the shade.

Horseplay and other unprofessional conduct cannot be allowed during training because it is distracting and can lead to accidents and injuries (**Figure 2.23**). Disruptive trainees should be removed from the class so they do not distract the other trainees.

Live Fire Training

A critically important aspect of training is that it be as realistic as possible because the way firefighters are trained is how they will perform in the field. However, adding realism sometimes increases the potential for accidents and injuries. Some fire departments include live fire exercises in specially designed burn buildings at training centers. Others conduct such training in abandoned buildings acquired for that purpose. To ensure that these exercises are safe, they should be conducted according to the requirements of NFPA® 1403, *Standard on Live Fire Training Evolutions*. Some of the key requirements of NFPA® 1403 are as follows:

- *Site preparation* — Before a structure can be used for live fire training, it must be cleared of all potential hazards to the participants. Such hazards include asbestos in any form, hazardous materials, structural deficiencies, pest infestations, etc. All utilities to the structure must be secured.

- *Safety* — A student-to-instructor ratio of 5:1 may not be exceeded. There must be a designated safety officer to oversee participant safety and an instructor-in-charge (acting as IC). Human beings cannot be used as simulated victims inside the structure, and fires may not be set in egress routes. Uniforms and PPE worn by participants must conform to NFPA® standards.

- *Prerequisite training* — All participants must have completed basic fire training including fire behavior, PPE, fire fighting tools, and tactical operations.

Live Fire or Burn Exercises — Training exercises that involve the use of an unconfined open flame or fire in a structure or other combustibles to provide a controlled burning environment.

Burn Building — Training structure specially designed to contain live fires for the purpose of fire-suppression training.

Evolution — Operation of fire service training or suppression covering one or several aspects of fire fighting. Also called *Practical Training Evolution*.

- *Water supply* — An adequate water supply must be provided based on the size of the structure and the type of training to be conducted. Separate water supplies are required for attack and backup lines.

- *Training plan* — A training plan must be prepared and a briefing held to review it with all involved **(Figure 2.24)**. The plan must state the purposes of the evolutions, personnel assignments, building features and egress routes, and a universally understood evacuation signal. Participants must be given a walk-through of the building before any fires are set.

- *Fuel* — Flammable liquids are prohibited in acquired structures or burn buildings not specifically designed for their use. Fuels used must have known burning characteristics and be as controllable as possible.

- *Ventilation* — Means must be provided to prevent uncontrolled flashover and backdraft. Ventilation may be provided by precut roof openings that are covered until needed.

This list highlights some of the essential requirements of NFPA® 1403. It is not intended to replace a thorough review of the standard. Anyone conducting live fire training in an acquired structure must be thoroughly familiar with the requirements of the standard and anyone participating must be familiar with and operate within the training plan.

Figure 2.24 Prior to any live fire training evolution, training officers must brief participants in the training plan and lead them through the burn structure indicating means of egress and building features.

Maintaining and Servicing Equipment

Equipment used for fire training evolutions must be in good condition. Items used frequently for training often wear out sooner than those used less frequently on emergency calls. Examples of frequently used training items are ropes, straps, buckles, and other harness parts that must be tied, fastened, and unfastened repeatedly. Tools with wooden handles can become worn and splintered when used repeatedly by trainees. All tools and equipment should be inspected before each drill to assess their condition and reliability. Records must also be maintained on all equipment used for training. Training equipment, like all fire fighting equipment, should be tested according to the manufacturer's instructions and applicable standards. On a weekly basis, many fire departments check and run all equipment that was not used during the previous week.

Emergency Operations

Regardless of how good and how thorough training exercises are, the participants — you and your fellow firefighters — must be able to perform safely at the emergency scenes. The following sections include some critical points that will help to ensure that type of performance.

Preparedness

Many of the injuries that occur during actual emergencies are caused by a *series of events*. In other words, accidents or injuries are seldom caused by one specific or single action, but by a series of little things that go wrong in succession, resulting in an injury or accident. Many accidents and injuries can be prevented if you and your fellow firefighters act in a professional manner by always being in a *ready state*. A ready state is defined as being and having all equipment prepared for immediate use.

At the beginning of every work shift, you should be in the proper uniform, physically rested, and mentally alert. You must ensure that all of the tools and equipment on your assigned unit are in place and in proper working order. Your PPE must be on the apparatus or in its proper location and in ready condition. Your SCBA must be fully functional and have a full air cylinder. All EMS equipment must be checked and restocked if necessary and all accountability equipment must be updated and in the proper place.

While in the fire station, never put yourself in a position that would delay an emergency response. This could cause the driver/operator to try to make up for the delay by driving too fast or taking other unnecessary risks while responding.

When an emergency alarm sounds, stay calm and listen to the dispatch information. Quickly but calmly mount the apparatus wearing the proper gear and belt yourself into your riding position before the apparatus starts to move.

If you are out of balance at any stage of an emergency, correct the situation before going on to the next stage of the incident. This will prevent the *snowball effect* that can create an almost irreversible chain of events leading to an accident.

> The public calls you and your fellow firefighters to deal with its emergencies. Firefighters must never operate in a manner that makes them part of the emergency or creates a new one.

Emergency Scene Safety

The officer of the first-arriving unit can begin to control the incident by assuming Command and coordinating all the actions of the subsequent arriving units by the use of an incident management system. After assuming Command, the initial Incident Commander (IC) must size up the *critical factors*. Critical factors are the basic items that the IC must consider when evaluating tactical situations. Critical factors include:

- Life safety hazard
- Nature and extent of the emergency
- Building type, arrangement, and access
- Resources
- Special hazards

The IC will then consider the most severe factors and the risk management plan to determine the incident's overall strategy. The strategy determines the location of the fire fighting forces and falls into either of two modes: offensive or defensive. Offensive operations are conducted inside the hazard zone; defensive operations are conducted outside the hazard zone.

Once the proper strategy has been determined, the IC will develop the Incident Action Plan (IAP) and base all operations around the completion of the tactical priorities within the chosen strategy. In a structure fire, the following are the Tactical Priorities in the proper order of their completion:

- Firefighter Safety (Life Safety)
- Rescue (Life Safety)

- Fire Control (Incident Stabilization)
- Loss Control (Property Conservation)

The objectives of each priority are reflected in the following benchmarks of completion:

- **_Personnel Accountability Report_** (PAR) — A report is made to the IC signifying that companies working in the hazard zone are all safe and accounted for.
- **_"All Clear"_** — The primary search has been completed and all savable occupants are out of the hazard zone.
- **_"Under Control"_** — The fire is controlled with the forward fire progress stopped, no additional units will be required, and there is no imminent danger to firefighters.
- **_"Loss Stopped"_** — Property conservation is complete.

Offensive fire operations are based around a controlled, aggressive interior search and fire attack. If available, ladder or truck companies will provide search or support work to assist interior units with completion of the tactical priorities. All companies working in the hazard zone must be assigned according to the fireground organizational structure and must work within the IAP with _absolutely no freelancing._ Company members inside the hazard zone must be within voice, vision, or physical contact with each other at all times. All companies working in the hazard zone must have at least one portable radio on the correct tactical channel. After a primary _All-Clear_ and the _Under Control_ benchmarks have been completed, all efforts must be focused on controlling the loss to the structure and its contents.

Defensive operations are based on determining the boundaries of the hazard zone along with the potential collapse zone and keeping _all_ companies out of these defined boundaries. The IC will need to identify cutoff points, set up master streams, get "all-clear" signals, and protect the exposed occupancies from any fire extension **(Figure 2.25)**. The highest priority on all defensive fires is firefighter safety.

Figure 2.25 Defensive operations may include the use of master stream appliances to control the fire and limit fire spread to exposures. _Courtesy of Danny Atchley, Oklahoma City Fire Department._

Firefighters should never be injured while working in a defensive mode.

In the vast majority of fires, quick action by firefighters to save lives and protect property is both reasonable and required. That is not to say that there are no risks involved; on the contrary, there is almost always some level of risk to firefighters. That's the nature of the job. But these risks can be minimized by following a few fundamental rules:

- Work within the incident action plan (IAP).
- Adequately assess the situation and maintain situational awareness.
- Wear appropriate PPE.
- Work together as a team.
- Follow all departmental SOPs.
- Maintain communications with team members and Command.
- Do a risk/benefit analysis for every action.
- Employ safe and effective tactics.
- Use a personnel accountability system.
- Have one or more rapid intervention crews (RICs) standing by.
- Set up Rehab on all extended operations.
- Use appropriate emergency escape techniques when needed.
- Maintain company discipline and team integrity while working in the hazard zone.

Rapid Intervention Crew (RIC) — Two or more fully equipped and immediately available firefighters designated to stand by outside the hazard zone to enter and effect rescue of firefighters inside if necessary. Also known as *Rapid Intervention Team.*

Perhaps the most important safety rule of all is *everyone looks out for everyone else.*

Scene Management

Proper scene management reduces congestion and confusion by reducing the number of people in or near the emergency scene. During incidents on streets and highways, scene management may also be critical to preventing firefighter injuries or fatalities. Crowd control is also essential to proper scene management. Procedures for scene management are illustrated in **Skill Sheet 2-I-2**.

Crowd Control — Limiting access to an emergency scene by curious spectators and other non-emergency personnel.

Highway Incidents

When vehicle fires or crashes occur on public streets or highways, it is critically important to protect the emergency responders and the victims from being struck by other vehicles using the roadway. The U.S. Department of Transportation (DOT) publishes guidelines for operating at highway emergencies entitled *Manual of Uniform Traffic Control Devices.* The following are consistent with its recommendations:

- Position fire apparatus to block oncoming traffic.
- Turn front wheels of blocking apparatus away from the emergency so that it will not be pushed into the emergency responders if it is struck from the rear.

- Set out traffic cones, signs, or other devices to detour traffic around the emergency scene **(Figure 2.26)**.

- Turn off lights that face opposing traffic to avoid blinding drivers or distracting them.

- Never walk with your back to the traffic.

- Wear reflective vests when PPE is not required.

At least one traffic lane next to the incident should be closed **(Figure 2.27)**. In some cases, it will be necessary to close more than one additional lane, perhaps the entire roadway. As a member of a crew on such an incident, exercise extreme caution when dismounting the apparatus to avoid being hit by a passing vehicle. Whenever possible, apparatus not directly involved in mitigating the problem should be moved to the shoulder or completely off the roadway.

Figure 2.26 Apparatus and warning devices can be combined to create a safe working area along streets and highways.

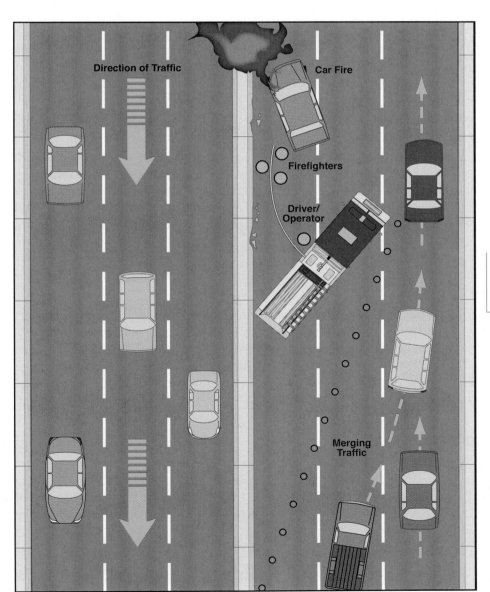

Figure 2.27 Closing at least one traffic lane will enable emergency personnel to operate more safely.

Crowd Control

If uncontrolled, spectators can wander about the scene and interfere with firefighters trying to do their jobs. In other cases, spectators can complicate the original emergency by injuring themselves and requiring attention by emergency responders. While crowd control is usually the responsibility of the law enforcement agency on the scene, it may sometimes have to be performed by firefighters or other emergency responders. It is the responsibility of the IC to ensure that the scene is secured and properly managed.

Even in the most remote locations, bystanders or spectators are often drawn to the scene by sirens, flashing lights, or smoke. They are often quite curious and try to get as close to the scene as possible. Some may be individuals who were involved in the incident but were not injured. All bystanders should be restrained from getting too close to the incident for their own safety and for that of victims and emergency personnel.

People at emergency scenes can become quite emotional, and they should be handled with care. This is particularly true when friends or relatives of the victims are at the scene. These particular bystanders are often difficult to deal with, and you must treat them with sensitivity and understanding. Relatives and friends of victims should be gently but firmly restrained from getting too close. They should be kept some distance from the actual incident but still within the cordoned area. While they may console each other, they should not be left entirely on their own. A firefighter or other responsible individual should stay with them until the victims have been removed from the scene.

Control Zones

In many cases, the best way to maintain scene security is by establishing control zones. This is often done by cordoning off the scene. Cordoning off the area will keep bystanders a safe distance from the scene and out of the way of emergency personnel. Many fire departments set up control zones commonly labeled "hot," "warm," and "cold" (**Figure 2.28**). There is no specific distance from the scene or area that should be cordoned off. The zone boundaries should be established by taking into account the area needed by emergency personnel to work, the degree of hazard presented by elements involved in the incident, wind and weather conditions, and the general topography of the area.

Cordoning off can be done with rope or fireline or caution tape tied to sign posts, utility poles, parking meters, or any other stationary objects available (**Figure 2.29**). Tape should not be tied to vehicles (especially emergency vehicles) because they may need to be moved during the incident. Once the area has been cordoned off, the boundary should be monitored to make sure unauthorized people do not cross the line.

Personnel Accountability

Every fire department must use some system of accountability that identifies and tracks all personnel working in the hazard zone at an incident. In structure fires, the hazard zone is anywhere within the collapse zone or anywhere that firefighters must have their SCBA operating in order to breathe. The accountability system should be SOP and all personnel trained in its use.

Accountability is vital in the event of a sudden and unexpected change in fire behavior or a structural collapse. If the IC does not know who is in the hazard zone, it is impossible to determine who and how many may be trapped inside. Sudden fire be-

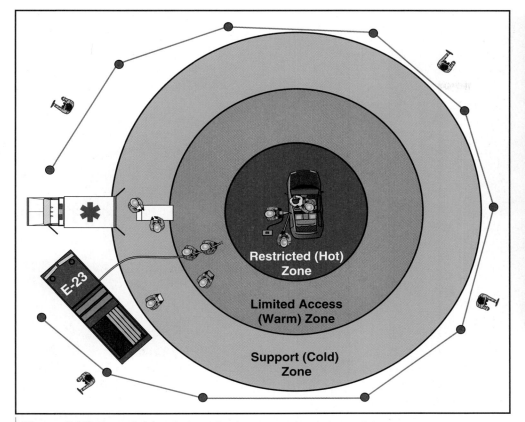

Figure 2.28 Control zones help maintain emergency scene safety.

Figure 2.29 A firefighter cordons off a scene with fireline tape.

havior changes such as a flashover (simultaneous ignition of room contents) can trap or injure firefighters (see Chapter 3, Fire Behavior). SCBA worn by firefighters in the hazard zone can malfunction. Firefighters can become lost or disoriented in mazes of rooms and corridors and run out of air. Firefighters have died because they were not known to be missing until it was too late. Company discipline (team integrity) is extremely important.

Passport System

A passport system, sometimes called a *tag* system, can aid in accounting for personnel within the hazard zone. Every company officer has a passport that lists the names of every member of that crew. Before entering the hazard zone, firefighters give their passports or tags to a designated Accountability Officer (AO) or supervisor (**Figure 2.30, p. 76**). The passports are then attached to a control board or personnel identification (ID) chart. Upon leaving the hazard zone, the officers collect their passports. This system enables Command to know exactly which companies and which personnel are operating in the hazard zone.

SCBA Tag System

An SCBA tag system provides even more accountability data on personnel inside a hazard zone. Before entering the hazard zone, officers give their tags to an AO who records time of entry and expected time of exit based on the pressure in the lowest-reading SCBA in the team (**Figure 2.31, p. 76**). The AO (or a designee) also does a brief check to ensure that all protective equipment is on and functional. This provides a greater degree of safety and accountability for those inside the hazard zone. Officers

Figure 2.30 A personnel accountability system tracks everyone in the hazard zone.

Figure 2.31 Officers give the accountability officer the SCBA readings.

leaving the hazard zone take back their tags so that the AO knows who is safely outside and who is still inside the hazard zone. On extended operations in the hazard zone, relief crews are sent in before the estimated time when interior crews will begin to run low on air.

Interior Operations

When firefighters are required to perform interior operations in burning buildings, they are in an inherently dangerous situation. Nonetheless, firefighters in this situation can be both safe and effective if they follow their department's SOPs, stay in contact with and follow the orders of their officers, and apply some basic interior operations techniques. Some of these techniques are as follows:

- Scan the outside of the building before entry to locate windows and doors that could be used as escape routes.

- Wear full PPE including SCBA and use an air management plan.

- Take the appropriate tools and equipment with you.

- Maintain team integrity by remaining in physical, voice, or visual contact with other members of your team.

- Remain in radio contact with Command or others outside the building.

- Take a hoseline or tag line (lifeline) with you into the hazard zone.

- Pay attention to your immediate surroundings.

One of the most important ways to pay attention to your surroundings is to maintain *situational awareness*. Situational awareness is your knowledge of the situation or environment around you. Like a soldier in combat, you need to develop the ability to recognize what is going on around you and know when there is a potential threat to your safety. To increase wildland firefighters' situational awareness, they attend a course called "Look Up, Look Down, Look Around." As a matter of personal survival in any emergency situation, you need to train yourself to be aware of your environment and the potential threats to your safety.

Another important way to increase your chances of surviving a life-threatening situation in a structure fire is to know your department's protocol for calling a Mayday. If you should find yourself separated from your company and trapped in a burning building, you need to call for help *immediately*! While you may find it embarrassing to call a Mayday, it is better to be embarrassed than dead!

Structure fires are uncontrolled and sometimes unpredictable situations. Because they are uncontrolled, they can sometimes deteriorate into life-threatening situations requiring you to take immediate action to escape.

Mayday — International distress signal broadcast by voice.

Emergency Escape

Sometimes called by the misnomer *self-rescue*, emergency escape techniques involve breaking through doors, windows, or walls if necessary to escape a life-threatening situation **(Figure 2.32)**. These situations can occur when there is a sudden and unexpected change in fire behavior (flashover, explosion, etc.) or a collapse of all or part of the structure. Emergency escape may also be necessary if you become lost or disoriented in a smoke-filled building and are in danger of running out of breathing air.

To help their personnel save themselves in life-threatening situations above grade, The New York City Fire Department (FDNY) equips each of its firefighters with a *personal safety system* (PSS). The PSS is intended to be used only once and then taken out of service. It consists of a belt-mounted nylon storage bag equipped with a heat sensor, an alloy steel hook attached to 50 feet (15 m) of life-safety rope with a stopper knot at the end, a descender, and a carabiner **(Figure 2.33, p. 78)**. This kit provides the means for FDNY firefighters to escape from and survive a situation six stories or more above the street. In some cases, however, firefighters may be unable to reach a window or balcony when injured or trapped. When they are unable to escape, a rapid intervention crew (RIC) will be needed.

Rapid Intervention

Both NFPA® 1500 and the OSHA regulations in 29 *CFR* 1910.134 state that whenever firefighters are in an atmosphere that is *immediately dangerous to life or health* (IDLH), including the inside of burning buildings when the fire is beyond the incipient or early growth stage, they must work in teams of two or more. In addition, at least two fully trained and equipped firefighters must be standing

Figure 2.32 Firefighters may have to create their own exits.

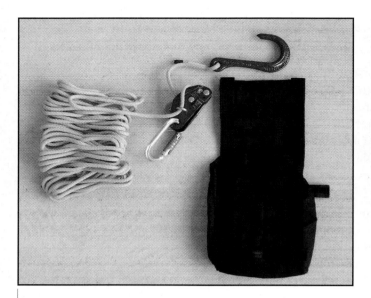

Figure 2.33 A PSS used to train firefighters. *Courtesy of Wright Rescue Solutions, Inc.*

Figure 2.34 The RIC must be prepared to enter the hot zone at any time.

by outside. The outside team must be ready to enter the burning structure at a moment's notice to rescue the entry team **(Figure 2.34)**. This has become known as the "two-in/two-out rule."

On the fireground, rapid intervention crew (RIC) members may be assigned other fireground support duties as long as they are able to abandon those other assignments (without endangering other firefighters) when they are needed to fulfill their primary function — to locate and rescue firefighters who have become lost, disoriented, injured, or trapped in a burning building. For more information on RIC, see Chapter 8, Rescue and Extrication.

Summary

As a firefighter, you will sometimes be put at some risk during training and emergency operations. It is your responsibility to maintain your physical and mental readiness to handle these situations. During fires and other emergencies, you may be ordered into inherently dangerous situations. In order to protect yourself, you must be aware of the hazards and risks involved. You must also be aware of how your department has chosen to manage those risks. Further, you must always maintain your situational awareness and be capable of applying the training you have received to successfully operate and survive in these hostile environments. You must always remember that along with your department and your officers, you are responsible for your safety. And, finally, you must remember the most fundamental of all firefighter safety rules: Everyone looks out for everyone else.

Fire Fighter I

1. What are three ways to prevent firefighter injuries?

2. What NFPA® standard specifies the minimum requirements for a fire department safety and health program?

3. What are the three IFSTA Principles of Risk Management?

4. What are three guidelines that can help firefighters maintain their personal health?

5. What can firefighters do to help prevent slips, trips, and falls in the fire station?

6. What are two safety rules for hand tools and small power tools?

7. What are two safety rules for power saws?

8. What NFPA® standard must live fire training exercises meet?

9. What is one type of personnel accountability system?

10. What are two basic interior operations techniques?

Step 1: Don appropriate personal protective clothing.

Step 2: Mount apparatus using handrails and steps per local procedures.

Step 3: Sit in a seat within the cab and fasten safety belt.

Step 4: Remain seated with safety belt fastened while vehicle is in motion.

Step 5: Unfasten safety belt and prepare to dismount when vehicle comes to a complete stop.

Step 6: Dismount apparatus using handrails and steps per local procedures.

NOTE: NFPA 1500 recommends that helmets not be worn in enclosed cabs on apparatus equipped with seats designed to protect firefighters' heads and necks during a collision.

Step 1: Don appropriate personal protective clothing.

Step 3: Set up established work areas.

Step 4: Perform tasks as directed to complete the assignment.

Step 5: Follow local procedures for highway incidents.

Step 2: Set up traffic cones and scene control devices appropriate to the assignment following local procedures.

Step 6: Remove traffic cones and scene control devices.

Chapter Contents

Fire Behavior

This chapter provides information that addresses the following job performance requirements of NFPA® 1001, *Standard for Fire Fighter Professional Qualifications (2008)*:

<u>NFPA® 1001 references for Chapter 3:</u>

5.3.8

5.3.10

5.3.11

5.3.12

5.3.16

Photo courtesy of District Chief Chris E. Mickal, NOFD Photo Unit.

Chapter Objectives

Fire Fighter I Objectives

1. Describe physical and chemical changes of matter related to fire.
2. Discuss modes of combustion, the fire triangle, and the fire tetrahedron.
3. Explain the difference between heat and temperature.
4. Describe sources of heat energy.
5. Discuss the transmission of heat.
6. Explain how the physical states of fuel affect the combustion process.
7. Explain how oxygen concentration affects the combustion process.
8. Discuss the self-sustained chemical reaction involved in the combustion process.
9. Describe common products of combustion.
10. Distinguish among classifications of fires.
11. Describe the stages of fire development within a compartment.
12. Summarize factors that affect fire development within a compartment.
13. Describe methods used to control and extinguish fire.

Case History

In 1999, firefighters in Washington, D.C. responded to a fire in a townhouse and observed thick, black smoke issuing from the open front door. When an engine company extended a hoseline into the first floor to locate and confine the fire, it found heavy smoke but little heat. The first-arriving truck company performed ventilation by opening windows at the front of the building and cutting holes in the roof. At the same time, engine and truck companies working at the rear of the building forced entry to the basement (ground level at the rear of the building). As they did so, air was sucked into the building and fire burning in the basement increased in size. Two firefighters died and one other was severely injured due to burns and exposure to intense heat.

Many people believe that fire is unpredictable. On the fireground, our ability to predict what will happen in the fire environment may be hampered by limited information, time pressure, and our level of understanding of fire behavior. In reality, there is no unpredictable fire behavior. As a firefighter, you need to understand the combustion process and how fire behaves in different materials and in different environments. Also, you need to know how fires are classified so that you can select and apply the most appropriate extinguishing agent. Most importantly, you need an understanding of fire behavior that permits you to recognize developing fire conditions and be able to respond safely and effectively to mitigate the hazards presented by the fire environment.

In 1994, firefighters in New York State responded to a fire in a three-story apartment building and observed smoke coming from the chimney of an apartment. A truck company opened a scuttle over the stairwell while engine companies extended hoselines into the building to investigate. The first-floor hose team that forced the apartment door observed a momentary rush of air into the apartment, followed by warm smoke pushing out the door. Within a short time, a large flame blew out from the upper part of the door and extended up the stairway. The crew on the first floor was able to duck under the flames and retreat back down the stairs, but the flames that now filled the stairway engulfed three firefighters working in the stairwell at the second-floor level. These three firefighters died from burns and exposure to intense heat.

In both this incident and in the one described at the beginning of the chapter, firefighters died because of extreme fire behavior. However, there are significant differences between the incidents. In the first incident, firefighters did not report any indicators of a significant fire before opening the door to the townhouse; in the second incident, it was quite obvious that there was a working fire in progress. These incidents also involved quite different fire behavior phenomena.

Firefighters arriving at the scene of any structure fire may be faced with a number of different problems — each demanding attention at the same time. The smoke and flames may be threatening the lives of trapped occupants. The fire may be threatening to spread to another structure or group of structures in an urban setting or in the wildland/urban interface. The room in which the fire started may be close to flashover. If the building is not properly ventilated, there may be potential for extreme fire behavior such as *flashover, backdraft,* or *smoke explosion.* To perform safely and effectively, you and your fellow firefighters must have at least a basic understanding of fire behavior. To understand the reaction we call *fire,* how it grows, and its products (products of combustion), we need to look at some basic concepts from physical science.

Physical science is the study of the physical world around us and includes the sciences of chemistry and physics. Such a study necessarily includes the laws related to matter and energy. This theoretical foundation must be translated into a practical knowledge of key fire behavior indicators. You need the ability to read a fire and recognize what is happening at present and predict potential fire behavior. This knowledge and skill is critical to safety during fire fighting operations.

According to NFPA® 1001, *Standard for Fire Fighter Professional Qualifications,* those qualified at the Fire Fighter I level must know the following about fire behavior:

- Fire behavior in a structure
- Products of combustion found in a structure fire
- Signs, causes, effects, and prevention of backdrafts
- Relationship of oxygen concentration to life safety and fire growth
- Methods of heat transfer
- Principles of thermal layering
- Classifications of fire
- Risks associated with each class of fire

This chapter integrates a discussion of physical science in the context of combustion and fire dynamics. This knowledge can help you interpret what you see on the fireground and recognize potential hazards, and it provides a basis for understanding fire control and ventilation tactical operations. The chapter also presents basic concepts related to combustion and fire development in structures. The same scientific principles and physical laws apply equally to other types of fire situations. Additional information must also be considered when dealing with situations such as vehicle fires, flammable liquid and gas fires, and wildland fires. These topics are addressed in additional detail in other chapters of this manual.

Science of Fire

You probably have a commonsense understanding of fire, heat, and temperature; however, it is important to consider these concepts from a somewhat more scientific perspective. Fire can take a variety of forms, but all involve a heat-producing chemical reaction between some type of fuel and oxygen or a similar substance. When anything burns, heat is generated faster than it can be dissipated, and this causes a significant increase in temperature.

Physical and Chemical Changes

As you look at the world around you, the physical materials you see are called *matter*. It is said that matter is the "stuff" that makes up our universe. *Matter* is anything that occupies space and has mass (weight). Matter can undergo many types of physical and chemical changes. In this chapter we will concentrate on those changes related to fire.

A physical change occurs when a substance remains chemically the same but changes in size, shape, or appearance. Examples of physical change are water freezing (liquid to solid) and boiling (liquid to gas). A chemical reaction occurs when a substance changes from one type of matter into another. This chemical change often involves the reaction of two or more substances to form other types of compounds. Oxidation is a chemical reaction involving the combination of oxygen (or similar types of substances) with other materials. Oxidation can be slow, such as the combination of oxygen with iron to form rust, or rapid, as in combustion of methane (natural gas). Oxygen is one of the more common elements on earth (our atmosphere is composed of 21 percent oxygen), and it reacts with many other elements found on the planet.

Chemical and physical changes almost always involve an exchange of energy. A fuel's potential energy is released during combustion and converted to kinetic energy. Reactions that give off energy as they occur are called *exothermic*. Fire is an exothermic chemical reaction called *combustion* that releases energy in the form of heat and sometimes light. Reactions that absorb energy as they occur are called *endothermic*. Converting water from a liquid to a gas (steam) requires the input of energy and is an endothermic physical reaction. As you will learn in subsequent chapters, converting water to steam is an important part of controlling and extinguishing fires.

Modes of Combustion

Combustion is a rapid and self-sustaining chemical process that yields heat and usually light. *Fire* is a form of combustion. Modes of combustion are differentiated based on where the reaction is occurring. In flaming combustion, oxidation involves fuel in the gas phase. This requires liquid or solid fuels to be converted to the gas phase or vaporized. When heated, both liquid and solid fuels will give off vapors that mix with oxygen and can burn, producing flames. Some solid fuels, particularly those that are porous and can char, can undergo oxidation at the surface of the fuel. This oxidation is nonflaming or smoldering combustion. Examples of nonflaming combustion include burning charcoal or smoldering fabric and upholstery (**Figure 3.1, p. 88**).

Fire Triangle

For many years, firefighters were taught that three components were needed for a fire to occur: oxygen, fuel, and heat. This relationship was represented by the *fire triangle* (**Figure 3.2, p. 88**). Remove any one of the three components and a fire cannot start — if burning, it will be extinguished. While this simple model is useful, it does not always provide a complete picture. The fire triangle provides a reasonable explanation of nonflaming or smoldering combustion.

Fire Tetrahedron

Flaming combustion is more accurately explained using the *fire tetrahedron* (**Figure 3.3, p. 88**), which is composed of the following four elements: oxygen, fuel, heat, and a self-sustained chemical chain reaction. Each component of the tetrahedron must be in place for flaming combustion to occur. If heat, fuel, or oxygen is removed from a

Matter — Anything that occupies space and has mass.

Exothermic Heat Reaction — Chemical reaction between two or more materials that changes the materials and produces heat, flames, and toxic smoke.

Endothermic Heat Reaction — Chemical reaction in which a substance absorbs heat energy.

Combustion — An exothermic chemical reaction that is a self-sustaining process of rapid oxidation of a fuel, that produces heat and light.

Figure 3.1 Glowing charcoal is an example of nonflaming combustion.

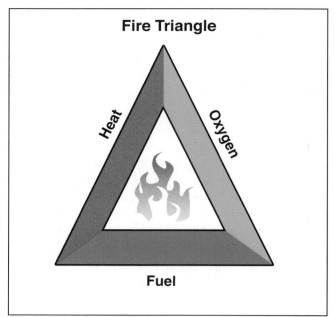

Figure 3.2 The fire triangle represents nonflaming combustion.

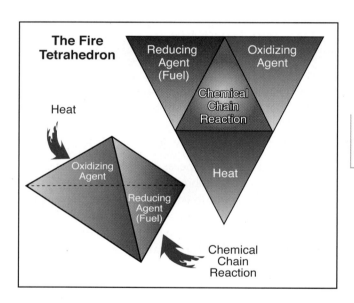

Figure 3.3 The fire tetrahedron represents flaming combustion.

Fire Tetrahedron — Model of the four elements/conditions required to have a fire. The four sides of the tetrahedron represent fuel, heat, oxygen, and chemical chain reaction.

fire, it will be extinguished (fire triangle). Flaming combustion will cease if the self-sustained chemical chain reaction of flaming combustion is inhibited or interrupted (fire tetrahedron); however, the fire may continue to smolder depending on the characteristics of the fuel. Each element of the fire tetrahedron is explained in greater detail in the following sections of this chapter.

Heat

Having a working knowledge of fire behavior requires an understanding of heat and temperature. Despite the fact that most people have a basic understanding of heat and temperature, these terms are often used interchangeably because the difference is not clearly understood.

Heat and Temperature

Heat is a form of energy, and energy exists in two states: potential and kinetic. *Potential energy* is the energy possessed by an object that may be released in the future. *Kinetic energy* is the energy possessed by a moving object. Heat is kinetic energy associated with the movement of the atoms and molecules that comprise matter. Before ignition, a fuel has potential chemical energy. When that fuel burns, the chemical energy is converted to kinetic energy in the form of heat and light. Temperature is a *measurement* of kinetic energy. Heat energy will move from objects of higher temperature to those of lower temperature. Understanding this movement is particularly important in both fire development and fire control tactics.

Energy is the capacity to perform work. Work occurs when a force is applied to an object over a distance or when a chemical, biological, or physical transformation is made in a substance **(Figure 3.4)**.

Although it is not possible to measure energy directly, it is necessary to measure the work that it does. In the case of heat, work means increasing the temperature of the substance.

The measure for heat energy is *joules* in the International System of Units (SI) or metric system. A *joule* is equal to 1 newton over a distance of 1 meter.

In the customary system, the unit of measure for heat is the British thermal unit (Btu). The *British thermal unit* is the amount of heat required to raise the temperature of 1 pound of water 1 degree Fahrenheit. While not used in scientific and engineering texts, the Btu is still frequently used in the fire service **(Figure 3.5, p. 90)**.

There are several different scales used to measure temperature; the most common are the *Celsius* and *Fahrenheit* scales. The Celsius temperature scale is used in the metric system while the Fahrenheit scale is used in the customary system. The freezing and boiling points of water provide a simple way to compare these two scales **(Figure 3.6, p. 90)**.

Potential Energy — Stored energy possessed by an object that can be released in the future to perform work.

Kinetic Energy — The energy possessed by a moving object.

Figure 3.4 The chemical transformation that produces fire is a form of energy. *Courtesy of District Chief Chris E. Mickal, NOFD Photo Unit.*

Increasing the temperature of a substance is work

This concept is readily applied to fire development as ignition requires increasing fuel temperature. It also applies to fire suppression where heat energy is transferred from hot gases and burning fuel to water from hose streams.

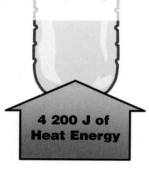

1 Kg (2.2 Lbs) of Water

Adding 4 200 Joules of heat energy to a kilogram of water increases its temperature by 1 degree Celsius

4 200 J of Heat Energy

This can be compared to the traditional unit of heat energy, the British Thermal Unit (Btu). Adding 1 Btu of heat energy to a pound of water increases its temperature by 1 degree Fahrenheit.

1 054.8 J = 1 Btu

Figure 3.5 An example of work is increasing the temperature of a substance.

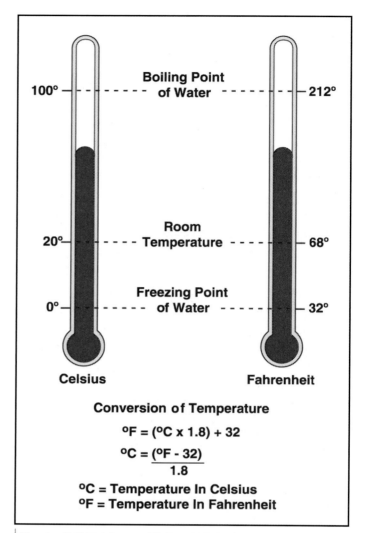

Boiling Point of Water — 100° / 212°

Room Temperature — 20° / 68°

Freezing Point of Water — 0° / 32°

Celsius

Fahrenheit

Conversion of Temperature

$$°F = (°C \times 1.8) + 32$$

$$°C = \frac{(°F - 32)}{1.8}$$

°C = Temperature In Celsius
°F = Temperature In Fahrenheit

Figure 3.6 Celsius and Fahrenheit scales.

Energy exists in many forms and can change from one form to another. In the study of fire behavior, the conversion of energy into heat is particularly important because *heat* is the energy component of the fire tetrahedron. When a fuel is heated, its temperature increases. Applying additional heat causes *pyrolysis* (the chemical decomposition of a substance through the action of heat) in solid fuels and *vaporization* of liquid fuels, releasing ignitable vapors or gases. A spark or other external source can provide the energy necessary for ignition, or the fuel can be heated until it ignites without a spark or other source. Once ignited, the process continues the production and ignition of fuel vapors or gases so that the combustion reaction is sustained (**Figure 3.7**).

There are two forms of ignition: *piloted ignition* and *autoignition*. Piloted ignition occurs when a mixture of fuel and oxygen encounter an external heat (ignition) source with sufficient heat energy to start the combustion reaction (**Figure 3.8**). Autoignition occurs without any external flame or spark to ignite the fuel gases or vapors. In this case, the fuel surface is chemically heated to the point at which the combustion reaction occurs. *Autoignition temperature* (AIT) is the temperature to which the surface of a substance must be heated for ignition and self-sustained

Autoignition Temperature — Same as ignition temperature except that no external ignition source is required for ignition because the material itself has been heated to ignition temperature.

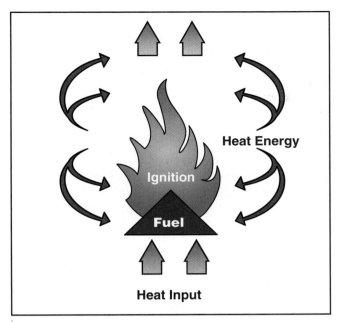

Figure 3.7 Heat sustains the combustion process.

Figure 3.8 Piloted ignition occurs in gasoline engines when fuel and oxygen mix in correct proportions in the carburetor and a spark from the spark plug ignites the mixture.

combustion to occur. The autoignition temperature of a substance is always higher than its piloted ignition temperature. While both piloted ignition and autoignition occur under fire conditions, piloted ignition is the most common.

Sources of Heat Energy

Chemical, mechanical, electrical, light, nuclear, and sound energy can all cause a substance to heat up by increasing the speed with which molecules are moving. Chemical, electrical, and mechanical energy are common sources of heat that result in the ignition of a fuel. Each of these sources is discussed in depth in this section.

Chemical heat energy is the most common source of heat in combustion reactions. When any combustible is in contact with oxygen, oxidation occurs. This process almost always results in the production of heat.

Self-heating, also known as spontaneous heating, is a form of chemical heat energy that occurs when a material increases in temperature without the addition of external heat. Normally, heat is produced slowly by oxidation and is lost to the surroundings almost as fast as it is generated. The process can be initiated or accelerated by an external heat source such as sunshine. In order for self-heating to progress to spontaneous ignition, the material must be heated to its autoignition temperature. For spontaneous ignition to occur, the following factors are required:

- The insulation properties of the material immediately surrounding the fuel must be such that the heat cannot dissipate as fast as it is being generated.

- The rate of heat production must be great enough to raise the temperature of the material to its ignition temperature.

- The available air supply (ventilation) in and around the material being heated must be adequate to support combustion.

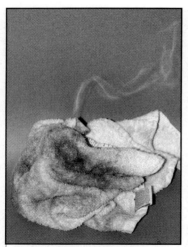

Figure 3.9 Under some conditions, oil-soaked rags will ignite spontaneously.

An example of a situation that could lead to spontaneous ignition would be one or more oil-soaked rags rolled into a ball and thrown into a corner. If the heat generated by the natural oxidation of the oil and cloth is not allowed to dissipate, either by movement of air around the rags or some other method of heat transfer, the temperature of the cloth could eventually increase enough to cause ignition **(Figure 3.9)**.

The rate of the oxidation reaction, and thus the heat production, increases as more heat is generated and held by the materials insulating the fuel. In fact, the rate at which most chemical reactions occur doubles with each 18°F (10°C) increase in the temperature of the reacting materials. The more heat generated and absorbed by the fuel, the faster the reaction causing the heat generation. When the heat generated by a self-heating reaction exceeds the heat being lost, the material may reach its ignition temperature and ignite spontaneously. **Table 3.1** lists some common materials that are subject to self-heating.

Electrical heat energy can generate temperatures high enough to ignite any combustible materials near the heated area. Electrical heating can occur in several ways, including the following:

- *Resistance heating* — When electric current flows though a conductor, heat is produced. Some electrical appliances, such as incandescent lamps, ranges, ovens, or portable heaters, are designed to make use of resistance heating **(Figure 3.10)**. Other electrical equipment is designed to limit resistance heating under normal operating conditions.

- *Overcurrent or overload* — When the current flowing through a conductor exceeds its design limits, it may overheat and present an ignition hazard. Overcurrent or overload is unintended resistance heating.

- *Arcing* — In general, an arc is a high-temperature luminous electric discharge across a gap or though a medium such as charred insulation. Arcs may be generated when a conductor is separated (such as in an electric motor or switch) or by high voltage, static electricity, or lightning.

- *Sparking* — When an electric arc occurs, luminous (glowing) particles can be formed and spatter away from the point of arcing. In electrical terms, *sparking* refers to this spatter, while an *arc* is the luminous electric discharge.

Table 3.1
Spontaneous Heating Materials and Locations

Type of Material	Possible Locations
Charcoal	Convenience stores Hardware stores Industrial plants Restaurants
Linseed oil-soaked rags	Woodworking shops Lumberyards Furniture repair shops Picture frame shops
Hay and manure	Farms Feed stores Arenas Feedlots

Figure 3.10 A common example of resistance heating occurs in the electric coils of a kitchen stove when the unit is turned on.

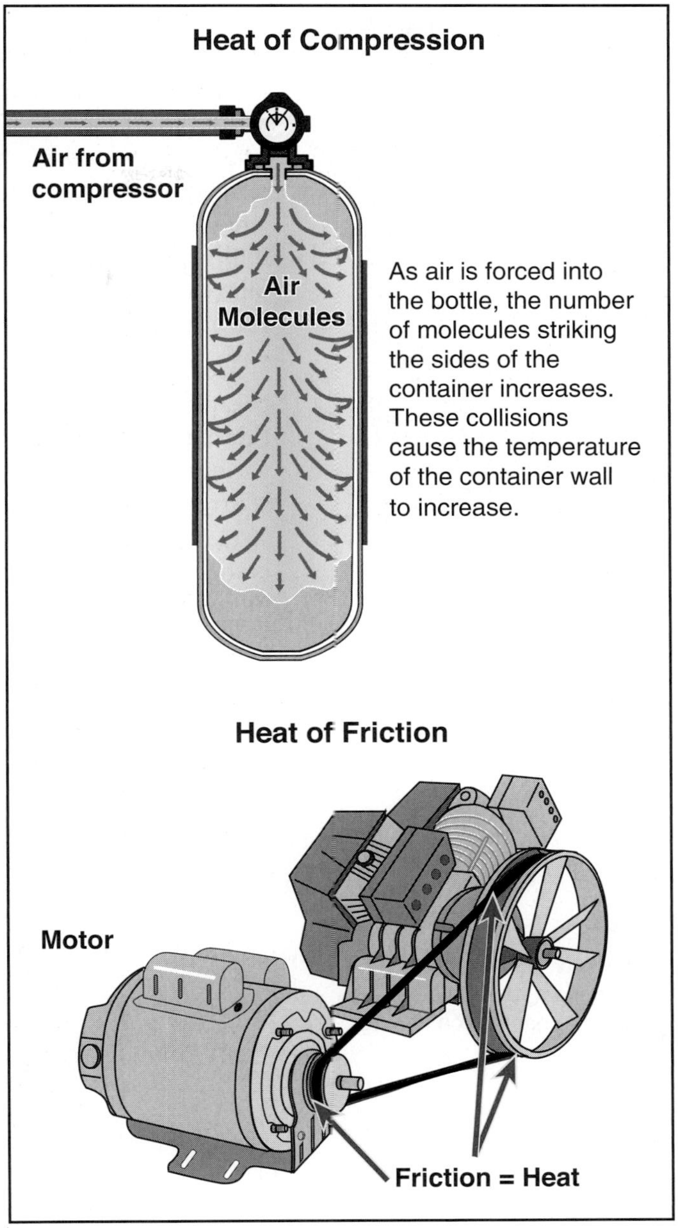

Heat of Compression

Air from compressor

Air Molecules

As air is forced into the bottle, the number of molecules striking the sides of the container increases. These collisions cause the temperature of the container wall to increase.

Heat of Friction

Motor

Friction = Heat

Figure 3.11 Examples of mechanical heat energy.

Mechanical heat energy is generated by friction or compression. The movement of two surfaces against each other creates *heat of friction*. This movement results in heat and/or sparks being generated. *Heat of compression* is generated when a gas is compressed. Diesel engines use this principle to ignite fuel vapor without a spark plug. The principle is also the reason that self-contained breathing apparatus (SCBA) cylinders feel warm to the touch after they have been filled **(Figure 3.11)**.

Transmission of Heat

The transmission or transfer of heat from one point or object to another is basic to the study of fire behavior. The transfer of heat from the initial fuel package (burning object) to other fuels in and beyond the area of fire origin affects the growth of any fire. Firefighters use their knowledge of heat transfer to estimate the size of a fire before attacking it and to evaluate the effectiveness of an attack. The definition of heat makes it clear that for heat to be transferred from one body to another, the two bodies must be at different temperatures.

Heat moves from warmer objects to those that are cooler. The rate at which heat is transferred is related to the temperature differential of the bodies and the thermal conductivity of the material involved. For any given substance, the greater the temperature differences between the bodies, the greater the transfer rate. The transfer of heat from body to body is measured as energy flow (heat) over time.

Heat can be transferred from one body to another by three mechanisms: *conduction, convection,* and *radiation.* Each of these is discussed in some detail in the following sections.

Conduction

Conduction is the transfer of heat within a body or to another body by direct contact. In other words, conduction is the heat flow through and between solids. Conduction results from increased molecular motion and collisions between the molecules of a substance to transfer energy through the substance. The more closely packed the molecules of a substance are, the more readily it will conduct heat. Conduction occurs when a material is heated as a result of direct contact with a heat source. For example, if a metal pipe is heated by a fire on one side of a wall, heat conducted through the pipe can ignite wooden framing components in the wall or nearby combustibles on the other side of the wall **(Figure 3.12)**.

Heat flow due to conduction is dependent on the area being heated, the temperature difference between the heat source and the material being heated, and the thermal conductivity of the material. **Table 3.2** shows the thermal conductivity of various common materials at the same ambient temperature (68°F/20°C). As you can see from the table, copper will conduct heat more than seven times more readily than steel. Likewise, steel is nearly forty times as thermally conductive as concrete. Wood is the least able to conduct heat of all these substances. Therefore, heat is transferred more readily through a steel-frame building than a wood-frame building.

Figure 3.12 Through conduction, materials remote from the heat source can be ignited.

Table 3.2 Thermal Conductivity of Common Substances		
Substance	**Temperature**	**Thermal Conductivity (W/mK)***
Copper	68°F (20°C)	386.00
Steel	68°F (20°C)	36.00 – 54.00
Concrete	68°F (20°C)	0.8 – 1.28
Wood (pine)	68°F (20°C)	0.13

*Watts per meter — Kelvin

Insulating materials retard the transfer of heat primarily by slowing conduction from one body to another. Good insulators are materials that do not conduct heat well. Because of their physical makeup, they disrupt the point-to-point transfer of heat energy. The best commercial insulators used in building construction are those made of fine particles or fibers with void spaces between them filled with a gas such as air. Gases do not conduct heat very well because their molecules are relatively far apart.

Convection

As a fire begins to grow, the air entrained into the fire is heated. The hot air and products of combustion become more buoyant and rise. *Convection* is the transfer of heat energy from a fluid (liquid or gas) to a solid surface. In the fire environment, this usually involves transfer of heat through the movement of hot smoke and fire gases. As with all heat transfer, the flow of heat is from the hot fire gases to the cooler structural surfaces, building contents, and air **(Figure 3.13)**.

> **Convection** — Transfer of heat by the movement of heated fluids or gases, usually in an upward direction.

Radiation

Radiation is the transmission of energy as an electromagnetic wave (such as light waves, radio waves, or X-rays) without an intervening medium **(Figure 3.14)**.

Thermal radiation results from temperature. All matter having a temperature above absolute zero radiate heat energy. Radiant heat becomes the dominant mode of heat transfer when the fire grows in size and can have a significant impact on the ignition of objects located some distance from the fire. Radiant heat transfer is also a significant factor in fire development and spread in compartments.

Wide ranges of factors influence radiant heat transfer. These factors include the nature of the surfaces, the distance between the surfaces, and the temperature difference between the heat source and materials being heated. Dark materials will emit

> **Radiation** —The transmission or transfer of heat energy from one body to another body at a lower temperature through intervening space by electromagnetic waves such as infrared thermal waves, radio waves, or X-rays.

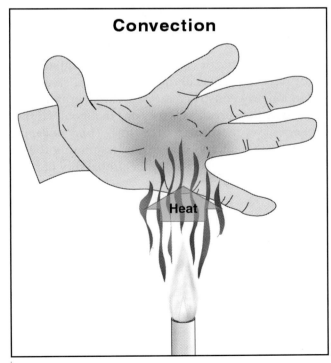

Figure 3.13 Heat rises through convection.

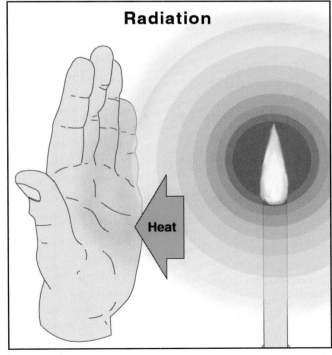

Figure 3.14 Radiant heat travels in all directions.

Figure 3.15 Fire spread is often the result of radiant heat. *Courtesy of NIST.*

and absorb heat more effectively than those of lighter color. Smooth or highly polished surfaces will reflect more radiant heat than those that are rough. Increasing distance reduces the effect of radiant heat. However, it is important to note that the temperature difference has a major impact on heat transfer through radiation. As the temperature of the heat source increases, the radiant energy increases by a factor to the fourth power. Doubling the temperature increases radiant heat by a factor of sixteen!

Because it is an electromagnetic wave, the energy travels in a straight line at the speed of light. The best example of heat transfer by radiation is the heat of the sun. The energy travels at the speed of light from the sun through space (a vacuum) until it collides with and warms the surface of the earth. Radiation is the cause of most exposure fires (fires ignited in fuel packages or buildings that are remote from the fuel package or building of origin) (**Figure 3.15**). As a fire grows, it radiates more and more energy in the form of heat. In large fires, it is possible for the radiated heat to ignite buildings or other fuel packages a considerable distance away. Radiated heat energy travels through vacuums and air spaces that would disrupt heat transfer by conduction and convection. Materials that reflect radiated energy will disrupt the transmission of heat.

Passive Agents

While the fire triangle consists of fuel, heat, and oxygen, other materials can have a significant impact on both ignition and how a fire develops. *Passive agents* are materials that absorb heat but do not participate actively in the combustion reaction. Fuel moisture (the water content of a combustible material) is a passive agent that slows the absorption of heat energy and retards the process of ignition and combustion. For example, a well-watered shrub will be slower to ignite than one that is dehydrated.

Relative humidity and fuel moisture are major considerations in wildland fire development, but the influence of passive agents can be important in structural fires as well. For example, a fire in a newly constructed wood-frame building (in which the wood is relatively green) may not spread as fast as one in an older building in which the framing members have dehydrated over time. This concept is important in understanding fire development and also in understanding the effectiveness of fire control tactics used to prevent or reduce the probability of rapid fire progress.

Fuel (Reducing Agent)

Reducing Agent — The fuel that is being oxidized or burned during combustion.

Fuel is the material or substance being oxidized or burned in the combustion process. In scientific terms, the fuel in a combustion reaction is known as the *reducing agent*. Fuels may be inorganic or organic. Inorganic fuels such as hydrogen or magnesium do not contain carbon. Organic fuels contain carbon. Most common fuels are organic, containing carbon along with other elements. These fuels can be further broken down into hydrocarbon-based fuels (such as gasoline, fuel oil, and plastics) and cellulose-based materials (such as wood and paper).

Two key factors influencing the combustion process are the physical state of the fuel and its distribution or orientation (horizontal or vertical). A fuel may be found in any of three states of matter: gas, solid, or liquid. For flaming combustion to occur, fuels must be in the gaseous state. Heat energy is required to change solids and liquids into gases.

Gaseous Fuel

Gaseous fuels such as methane (natural gas), hydrogen, acetylene, and others can be the most dangerous of all fuel types because they are already in the state required for ignition. Gases have mass but no definite shape or volume. A gas placed in a container will diffuse and completely fill the available space. When released from a container, gases will rise or sink, depending on their density relative to air. Gases that are lighter than air, such as methane, tend to rise. Those that are heavier than air, such as propane (liquefied petroleum gas), tend to sink.

Vapor density describes the density of gases in relation to air. Air has been assigned a vapor density of 1. Gases with a vapor density of less than 1 will rise while those having a vapor density of greater than 1 will sink. These densities presume that the gas and air are at the same temperature (generally specified as 68°F [20°C]). Heated gases expand and become less dense; when cooled they contract and become more dense. **Table 3.3** gives characteristics of some common flammable gases.

Table 3.3 Characteristics of Common Flammable Gases		
Material	**Vapor Density**	**Ignition Temperature**
Methane (Natural Gas)	0.55	(1,004°F) 540°C
Propane (Liquefied Petroleum Gas)	1.52	(842°F) 450°C
Carbon Monoxide	0.96	(1,128°F) 620°C

Source: *Computer Aided Management of Emergency Operations* (CAMEO)

Liquid Fuel

Liquids have mass and volume but no definite shape except for a flat surface. Liquids assume the shape of their container. When released, liquids will flow downhill and can pool in low areas. Just as gases are compared to air, the density of liquids is compared to that of water. *Specific gravity* is the ratio of the mass of a given volume of a liquid compared with the mass (weight) of an equal volume of water at the same temperature. Water has been assigned a specific gravity of 1. Liquids with a specific gravity less than 1, such as gasoline and most (but not all) flammable liquids, are lighter than water and will float on its surface. Liquids with a specific gravity greater than 1, such as epichlorohydrin (used in making plastics), are heavier than water. Therefore, water can be used to exclude oxygen from the burning liquid (**Figure 3.16, p. 98**).

In order to burn, liquids must be vaporized. *Vaporization* is the transformation of a liquid to vapor or gaseous state. At sea level, the atmosphere exerts a pressure of (14.7 psi (102.9 kPa). In order to vaporize, liquids must overcome the pressure exerted by the atmosphere. Vapor pressure is the pressure produced or exerted by vapors

Specific Gravity — Weight of a substance compared to the weight of an equal volume of water at a given temperature. A specific gravity less than 1 indicates a substance lighter than water; a specific gravity greater than 1 indicates a substance heavier than water.

Vaporization — Process of evolution that changes a liquid into a gaseous state. The rate of vaporization depends on the substance involved, heat, and pressure.

Flash Point — Minimum temperature at which a liquid gives off enough vapors to form an ignitable mixture with air near the liquid's surface.

Fire Point — Temperature at which a liquid fuel produces sufficient vapors to support combustion once the fuel is ignited. The fire point is usually a few degrees above the flash point.

released by a liquid. As a liquid is heated, vapor pressure increases along with the rate of vaporization. For example, a puddle of water eventually evaporates. When the same amount of water is heated on a stove, however, it vaporizes much more rapidly because there is more energy being applied. The rate of vaporization is determined by the vapor pressure of the substance and the amount of heat energy applied to it. The volatility or ease with which a liquid gives off vapor influences how easily it can be ignited.

Flash point is the temperature at which a liquid gives off sufficient vapors to ignite, but not sustain, combustion (piloted ignition). *Fire point* is the temperature at which sufficient vapors are being generated to sustain the combustion reaction. Flash point is commonly used to indicate the flammability hazard of liquid fuels. Liquid fuels that vaporize sufficiently to burn at temperatures under 100°F (38°C) present a significant flammability hazard (**Figure 3.17**).

The extent to which a liquid will give off vapor is also influenced by how much surface area is exposed to the atmosphere. In many open containers, the surface area of liquid exposed to the atmosphere is limited. If released, a liquid will flow onto the ground and pool in low areas. This action increases the surface area of the liquid exposed to the atmosphere and results in a proportional increase in the production of fuel vapors (**Figure 3.18**).

A number of other characteristics of liquids (in this case liquid fuels) are important to firefighters. Principal among these are density in comparison to water and ability to mix with water (solubility). Liquids such as hydrocarbon fuels (i.e., gasoline, diesel, and fuel oil) are lighter than water and do not mix with water. Other liquids (called *polar solvents*) such as alcohols (i.e., methanol and ethanol) will mix readily with water.

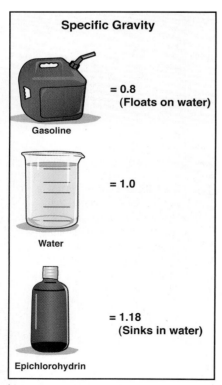

Figure 3.16 Some liquids will float on water and others will not.

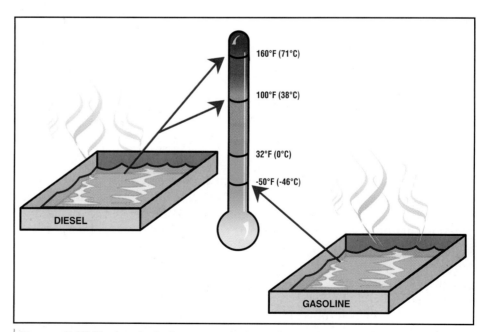

Figure 3.17 Flash point is the temperature at which a liquid gives off sufficient vapors to ignite. Liquids with low flash points are easily ignited. For instance, gasoline has a flash point of -50°F (-46°C) while diesel has a flash point between 100°F (38°C) and 160°F (71°C).

Figure 3.18 The greater the surface area, the larger the fire. *Courtesy of NIST.*

Solubility describes the extent to which a substance (in this case a liquid) will mix with water. This may be expressed in qualitative terms (i.e., slightly or completely) or as a percentage. Materials that are *miscible* in water will mix in any proportion.

Liquids lighter than water present a significant challenge when attempting to use water as an extinguishing agent. The volume of liquid (both water and oil) will increase as water is applied, potentially spreading the burning liquid. Water-soluble liquids present a different problem in that some water-based extinguishing agents, such as many types of fire fighting foam, will mix with the burning liquid, making them ineffective (See **Table 3.4** for more information). This necessitates the use of specialized extinguishing agents. This topic is discussed in greater detail in Chapter 14, Fire Streams.

Miscible — Materials that are capable of being mixed.

Solubility — Degree to which a solid, liquid, or gas dissolves in a solvent (usually water).

Table 3.4
Characteristics of Common Flammable and Combustible Liquids

Material	Water Soluble	Specific Gravity	Flash Point	Autoignition Temperature
Gasoline	No	0.72	(-36°F) -38°C	(853°F) 486°C
Diesel	No	>1.00	(125°F) 52°C	(410°F) 210°C
Ethanol	Yes	0.78	(55°F) 13°C	(689°F) 365°C
Methanol	Yes	0.79	(52°F) 11°C	(867°F) 464°C

Source: *Computer Aided Management of Emergency Operations (CAMEO)*

Solid Fuel

Solids have definite size and shape. Solids may also react differently when exposed to heat. Some solids (wax, thermoplastics, and metals) will readily change state and melt, while others (wood and thermosetting plastics) will not **(Figure 3.19, p. 100)**. Fuel gases and vapors are evolved from solid fuels by pyrolysis, which is the chemical decomposition of a substance through the action of heat. Simply stated, as solid fuels are heated, they begin to decompose and combustible vapors are given off. If there is sufficient fuel and heat, the process of pyrolysis generates sufficient quantities of burnable gases to ignite in the presence of sufficient oxygen (or another oxidizer).

Figure 3.19 Under a heat lamp, the wooden dowel (right) retains its shape while the candle (left) does not.

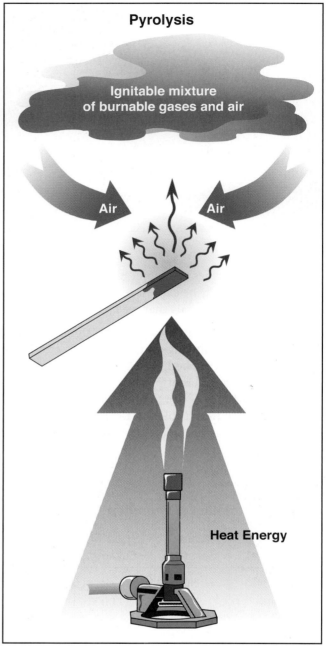

Pyrolysis

Ignitable mixture of burnable gases and air

Air   Air

Heat Energy

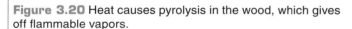

Figure 3.20 Heat causes pyrolysis in the wood, which gives off flammable vapors.

In a fire in a room or other enclosed area (compartment), the primary fuels commonly are solids such as wood, paper, or plastic. Pyrolysis must occur to generate the flammable vapors and gases required for combustion. When wood is first heated, water vapor is driven off as the wood dries. As heating continues, the wood begins to pyrolize and decompose into its volatile components and carbon (**Figure 3.20**).

Pyrolysis begins at temperatures below 400°F (204°C). This is lower than required for ignition of the vapors being given off, which ranges roughly from 1,000°F to 1,300°F (538°C to 704°C). **Table 3.5** outlines the pyrolysis effects within different temperature zones.

With synthetic fuels such as plastics, the process is similar. Unlike wood, though, plastics do not generally contain moisture that must be driven off by heat input before pyrolysis can occur.

Table 3.5
Pyrolysis Zones

	Temperature	Chemical Changes
Stage 1	Less than 392°F (200°C)	Moisture is released as the wood begins to dry; combustible and noncombustible materials are released into the atmosphere although there is insufficient heat to ignite them.
Stage 2	392° – 536°F (200°– 280°C)	The majority of the moisture has been released; charring has begun; the primary compound that is being released is carbon monoxide; ignition has yet to occur.
Stage 3	536° – 932°F (280°– 500°C)	Rapid pyrolysis takes place; combustible compounds are released and ignition can occur; charcoal is formed by the burning process.
Stage 4	Greater than 932°F (500°C)	Free burning exists as the wood material is converted to flammable gases.

Source: Adapted from NFPA® *Fire Protection Handbook*, 19th edition, Volume II, pages 8–35 and 36.

Unlike liquids or gases, solid fuels have a definite shape and size. This property significantly affects whether they are easy or difficult to ignite. The primary consideration is the surface area of the fuel in proportion to the mass, called the *surface-to-mass ratio*. One of the best examples is that of a large tree. To produce lumber, the tree must be felled and cut into a log. The surface area of this log is very low compared to its mass; therefore, the surface-to-mass ratio is low. The log is then sawn into planks. The result of this process is to reduce the mass of the individual planks compared to the log; the resulting surface area is increased, thus increasing the surface-to-mass ratio. The chips and sawdust produced as the planks are sawn into boards have an even higher surface-to-mass ratio. If the boards are milled or sanded, the resulting shavings or dust have the highest surface-to-mass ratio of any of the examples. As this ratio increases, the fuel particles become smaller (more finely divided), for example, shavings or sawdust as opposed to logs. Therefore, their ignitability increases tremendously (**Figure 3.21, p. 102**). As the surface area increases, more of the material is exposed to the heat and generates combustible pyrolysis products more quickly, making the fuel easier to ignite as surface-to-mass ratio increases.

The proximity and orientation of a solid fuel relative to the source of heat also affects the way it burns. For example, if you were to ignite one corner of a sheet of ⅛-inch (3 mm) plywood paneling that was laying horizontally (flat), the fire would consume the

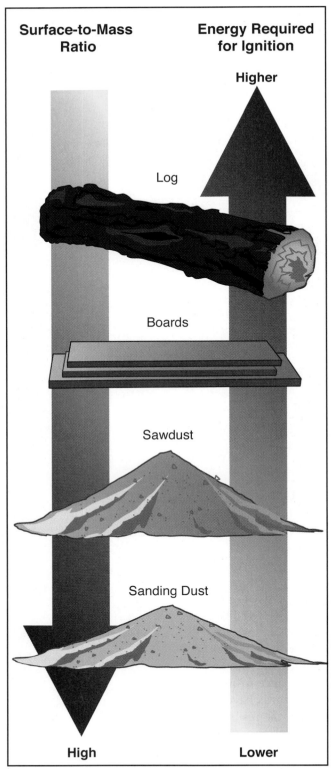

Surface-to-Mass Ratio

Energy Required for Ignition

Higher

Log

Boards

Sawdust

Sanding Dust

High

Lower

Figure 3.21 The more finely divided a material, the easier it is to ignite.

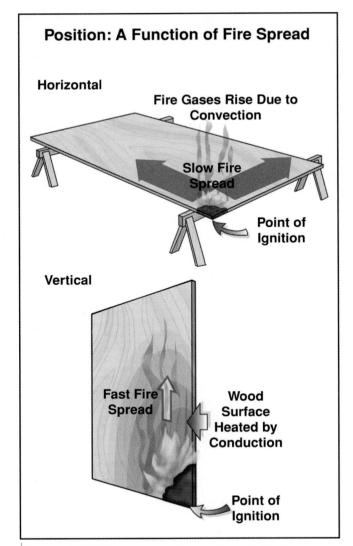

Position: A Function of Fire Spread

Horizontal

Fire Gases Rise Due to Convection

Slow Fire Spread

Point of Ignition

Vertical

Fast Fire Spread

Wood Surface Heated by Conduction

Point of Ignition

Figure 3.22 How a material is positioned affects the way it burns.

fuel at a relatively slow rate. The same type of paneling in a vertical position (standing on edge) burns much more rapidly. In this case, more heat is transferred to the solid fuel, speeding fire development (**Figure 3.22**).

Heat of Combustion and Heat Release Rate

The chemical content of any fuel influences both its *heat of combustion* and energy or *heat release rate* (HRR). The heat of combustion of a given fuel is the total amount of energy released when a specific amount of that fuel is oxidized (burned). In other words, different materials release more or less heat energy than others depending on their chemical makeup. Heat of combustion is usually expressed in kilojoules/gram (kJ/g). Many plastics, flammable liquids, and flammable gases contain more potential heat energy than wood. This is particularly significant to firefighters given the widespread use of synthetics as structural materials and finishes and in building contents.

Heat release rate (HRR) is the energy released per unit of time as a given fuel burns and is usually expressed in kilowatts (kW). HRR is dependent on the type, quantity, and orientation of the fuel (**Table 3.6**). The characteristics of the enclosure (if the fire is burning in a compartment) can also affect the heat release rate. In most fires, HRR varies over time — increasing as more fuel becomes involved and then falling as fuel is consumed.

Oxygen (Oxidizing Agent)

The primary oxidizing agent in most fires is the oxygen in the air around us. Normally, air consists of about 21 percent oxygen. In addition to oxygen, other materials can react with fuels in much the same way. These other materials are called *oxidizers*. Oxidizers are not combustible but like oxygen they will support combustion. **Table 3.7, p. 104,** lists some common oxidizers.

At normal ambient temperatures (70°F or 21°C), materials can ignite and burn at oxygen concentrations as low as 14 percent. When oxygen concentration is limited, the flaming combustion may diminish and combustion will continue in the surface

Table 3.6 Representative Peak Heat Release Rates (HRR) During Unconfined Burning		
Fuel Material	**Peak HRR in kilowatts**	**Common Locations for Material**
Small Wastebasket	4-18	Homes, businesses, shops
Cotton mattress	140-350	Homes, furniture stores, motels
Cotton easy chair	290-370	Homes, furniture stores, office buildings
Small pool of gasoline	400	Traffic crash, fuel stations
Dry Christmas tree	500-650	Homes, trash facilities, dumpsters, recycling sites
Polyurethane mattress	810-2630	Homes, furniture stores, motels, dormitories, jails
Polyurethane easy chair	1350-1990	Homes, furniture stores, motels
Polyurethane sofa	3120	Homes, furniture stores, motels, dormitories, office buildings

Adapted from NFPA® 921, 2004 edition

Table 3.7
Common Oxidizers

Substance	Common Use
Calcium Hypochlorite (granular solid)	Chlorination of water in swimming pools
Chlorine (gas)	Water purification
Ammonium Nitrate (granular solid)	Fertilizer
Hydrogen Peroxide (liquid)	Industrial bleaching (pulp and paper and chemical manufacturing)
Methyl Ethyl Ketone Peroxide	Catalyst in plastics manufacturing

Coutesy of Ed Hartin.

or smoldering mode. However, at high ambient temperatures, flaming combustion may continue at considerably lower oxygen concentrations. Surface combustion can continue at extremely low oxygen concentrations even when the surrounding environment is at a relatively low temperature.

Oxygen concentration in the atmosphere has a significant impact on both fire behavior and our ability to survive. Occupational Safety and Health Administration (OSHA) respiratory protection regulation, *29 CFR* (*Code of Federal Regulations*) 1910.134, defines an atmosphere having less than 19.5% oxygen in the air as being *oxygen deficient* and presenting a hazard to persons not wearing respiratory protection, such as SCBA, to provide a supply of fresh air. When the oxygen concentration in the atmosphere exceeds 23.5%, this regulation classifies the atmosphere as oxygen enriched and presenting an increased fire risk.

When the oxygen concentration is higher than normal, materials exhibit very different burning characteristics. Materials that burn at normal oxygen levels will burn more intensely and may ignite more readily in oxygen-enriched atmospheres. Some petroleum-based materials will autoignite (ignite spontaneously without an external heat source) in oxygen-enriched atmospheres. Many materials that do not burn at normal oxygen levels burn readily in oxygen-enriched atmospheres. One such material is Nomex® fire-resistant fabric, which is used to construct much of the protective clothing worn by firefighters. At normal oxygen levels, Nomex® does not burn. When placed in an oxygen-enriched atmosphere of approximately 31 percent oxygen, however, Nomex® ignites and burns vigorously.

Fires in oxygen-enriched atmospheres are difficult to extinguish and present a potential safety hazard to firefighters operating in those atmospheres. These conditions can be found in hospitals and other health care facilities, some industrial occupancies, and even private homes where occupants use breathing equipment containing pure oxygen.

For combustion to occur after a fuel has been converted into a gaseous state, it must be mixed with air (oxidizer) in the proper ratio. The range of concentrations of the fuel vapor and air (oxidizer) is called the *flammable (explosive) range*. The flammable range of a fuel is reported using the percent by volume of gas or vapor in air for the lower flammable limit (LFL) and for the upper flammable limit (UFL). The *lower flammable limit* is the minimum concentration of fuel vapor and air that supports combustion. Concentrations that are below the LFL are said to be *too lean* to burn. The *upper flammable limit* is the concentration above which combustion cannot take place. Concentrations that are above the UFL are said to be *too rich* to burn. Within the flammable range there is an ideal concentration at which there is exactly the amount of fuel and oxygen required for combustion (**Figure 3.23**).

Table 3.8 presents the flammable ranges for some common materials. The flammable limits for combustible gases are presented in chemical handbooks and documents such as the National Fire Protection Association (NFPA®) *Fire Protection Guide to Hazardous Materials*. The limits are normally reported at ambient temperatures and atmospheric pressures. Variations in temperature and pressure can cause the flammable range to vary considerably. Generally, increases in temperature or pressure broaden the range and decreases in temperature and pressure narrow it.

Flammable Range — The range between the upper flammable limit and lower flammable limit in which a substance can be ignited.

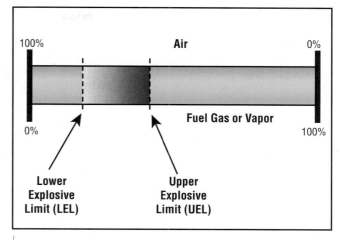

Figure 3.23 The air-to-fuel gas mixture must be between the lower and upper explosive limits to burn.

Table 3.8 Flammable Ranges of Common Flammable Gases and Liquids (Vapor)	
Substance	**Flammable Range**
Methane	5%–15%
Propane	2.1%–9.5%
Carbon Monoxide	12%–75%
Gasoline	1.4%–7.4%
Diesel	1.3%–6%
Ethanol	3.3%–19%
Methanol	6%–35.5%

Source: *Computer Aided Management of Emergency Operations (CAMEO)*

Self-Sustained Chemical Reaction

The self-sustained chemical reaction involved in flaming combustion is complex. Combustion of a simple fuel such as methane (natural gas) and oxygen provides a good example. Complete oxidation of methane results in production of carbon dioxide and water as well as release of energy in the form of heat and light. While this process seems to be quite simple, it is actually quite complex. As combustion occurs, the molecules of methane and oxygen break apart to form free radicals (electrically charged, highly reactive parts of molecules). Free radicals combine with oxygen or with the elements that form the fuel material (in the case of methane, carbon and hydrogen) producing intermediate combustion products (new substances), producing even more radicals and increasing the speed of the oxidation reaction (**Figure 3.24**). At various points in the combustion of methane, this process results in the production of carbon monoxide and formaldehyde, which are both flammable and toxic. When more chemically complex fuels burn, this process involves many different types of radicals and intermediate combustion products, many of which are also flammable and toxic.

Flaming combustion is one example of a chemical chain reaction. Sufficient heat will cause fuel and oxygen to form free radicals and initiate the self-sustained chemical reaction. The fire will continue to burn until the fuel or oxygen is exhausted or an extinguishing agent is applied in sufficient quantity to interfere with the ongoing reaction (**Figure 3.25**). In some cases, extinguishing agents deprive the combustion process of fuel, oxygen, or sufficient heat to sustain the reaction. *Chemical flame inhibition* is when a Halon-replacement extinguishing agent interferes with this chemical reaction, forms a stable product, and terminates the combustion reaction.

The self-sustained chemical reaction and the related rapid growth are the factors that separate flaming combustion from slower oxidation reactions. Slow oxidation reactions, such as the rusting of steel or the yellowing of paper, do not produce heat fast enough to reach ignition, and they never generate sufficient heat to become self-sustained.

Surface combustion also involves oxidation at the surface of a fuel material without initiation or continuation of the chemical chain reaction found in flaming combustion. Glowing charcoal briquettes is one example of this type of combustion. This distinction is important in that a surface combustion cannot be extinguished by

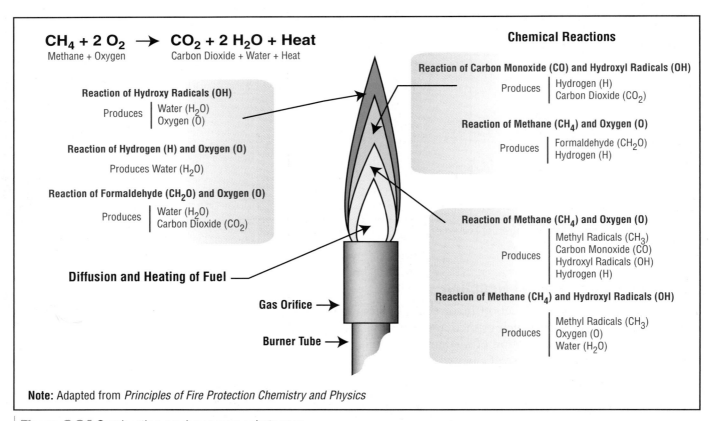

Figure 3.24 Combustion produces new substances.

Figure 3.25 A fire may continue to burn until all available fuel is consumed. *Courtesy of Gresham (OR) Fire and Emergency Services.*

chemical flame inhibition (because there are no flames and related chemical chain reaction). These fires must be extinguished by working on one of the sides of the fire triangle (heat, fuel, and oxygen).

Products of Combustion

As any fuel burns, its chemical composition changes. This change results in the production of new substances and the release of energy. In a fire, this energy is in the form of light and heat. At a very simple level, complete combustion of methane (natural gas) in air results in the production of heat, light, water vapor, and carbon dioxide. However, in a structure fire, multiple fuels are involved and limited air supply results in incomplete combustion. These factors result in extremely complex chemical reactions producing a wide range of combustion products including toxic and flammable gases, vapors, and particulates (**Figure 3.26**).

Figure 3.26 In fires, chemical reactions produce toxic smoke. *Courtesy of District Chief Chris E. Mickal, NOFD Photo Unit.*

Like the basic model of combustion provided by the fire triangle, describing products of combustion as heat, smoke, and sometimes light is deceptively simple. Of these three general types of products, heat and smoke have the most impact on firefighters.

The heat generated during a fire is one product of combustion that helps to spread the fire by preheating adjacent fuels and making them more susceptible to ignition. In addition, those lacking adequate protection from the heat may suffer burns, damage to their respiratory tract, dehydration, and heat exhaustion.

Carbon Monoxide (CO) — Colorless, odorless, dangerous gas (both toxic and flammable) formed by the incomplete combustion of carbon. It combines more than 200 times as quickly with hemoglobin as oxygen, thus decreases the blood's ability to carry oxygen.

Carbon Dioxide (CO₂) — Colorless, odorless, heavier than air gas that neither supports combustion nor burns. CO₂ is used in portable fire extinguishers as an extinguishing agent to extinguish Class B or C fires by smothering or displacing the oxygen.

While the heat energy from a fire is a danger to anyone directly exposed to it, toxic smoke causes most fire deaths. Smoke is an aerosol composed of gases, vapor, and solid particulates. Fire gases, such as carbon monoxide (CO), are generally colorless, while vapor and particulates give smoke its varied colors. Most components of smoke are toxic and present a significant threat to human life. The materials that make up smoke vary from fuel to fuel, but generally all smoke is toxic. **Table 3.9** lists some of the more common products of combustion and their toxic effects.

Carbon monoxide (CO) is a byproduct of the incomplete combustion of organic (carbon-containing) materials. This gas is probably the most common product of combustion encountered in structure fires. Exposure to CO is frequently identified as the cause of death for civilian fire fatalities and firefighters who have run out of air in their SCBA. Carbon monoxide acts as a chemical asphyxiant by binding with hemoglobin in the blood that transports oxygen throughout the body.

Hydrogen cyanide (HCN), produced in the combustion of materials containing nitrogen, is also commonly encountered in smoke, although at lower concentrations than CO. HCN also acts as a chemical asphyxiant but with a different mechanism of action. Hydrogen cyanide acts to prevent the body from using oxygen at the cellular level. HCN is a significant byproduct of the combustion of polyurethane foam, which is commonly used in furniture and bedding.

Carbon dioxide (CO₂) is a product of complete combustion of organic materials. It is not toxic in the same manner as carbon monoxide or hydrogen cyanide, but it acts as a simple asphyxiant by displacing oxygen. Carbon dioxide also acts as a respiratory stimulant, increasing respiratory rate.

These are only three of the more common products of combustion that can be hazardous to firefighters. It is important to remember that the toxic effects of smoke inhalation are not the result of any one gas; it is the interrelated effect of all the toxic products present.

Irritants in smoke are those substances that cause breathing discomfort and inflammation of the eyes, respiratory tract, and skin. Depending on the fuels involved, smoke will contain a wide range of irritating substances.

Because the substances in smoke from compartment fires can be deadly (either alone or in combination), firefighters must use SCBA for respiratory protection when operating in smoke. While the volume and density of smoke is considerably reduced during overhaul, this does not mean that the respiratory hazard has been eliminated. A research study conducted by the Phoenix (AZ) Fire Department determined that hazardous concentrations (above published short-term exposure limits) are likely to be present during overhaul.

Never breathe smoke! (See Chapter 2 for a list of other key safety behaviors.)

Given their personal protective clothing and SCBA, firefighters perceive smoke as less of a threat than heat and visible flames. This is not always a correct perception. Leaking fuel gases, such as methane and propane, are generally treated with a great deal of respect. Carbon monoxide, likely the most common fire gas, has both a lower ignition temperature and considerably wider flammable range than either of the two

Table 3.9
Common Products of Combustion and Their Toxic Effects

Acetaldehyde	Colorless liquid with a pungent choking odor, which is irritating to the mucous membranes and especially the eyes. Breathing vapors will cause nausea, vomiting, headache, and unconsciousness.
Acrolein	Colorless-to-yellow volatile liquid with a disagreeable choking odor, this material is irritating to the eyes and mucous membranes. This substance is extremely toxic; inhalation of concentrations as little as 10 ppm may be fatal within a few minutes.
Asbestos	A magnesium silicate mineral that occurs as slender, strong, flexible fibers. Breathing of asbestos dust causes asbestosis and lung cancer.
Benzene	Colorless liquid with a petroleum-like odor. Acute exposure to benzene can result in dizziness, excitation, headache, difficulty breathing, nausea, and vomiting. Benzene is also a carcinogen.
Benzaldehyde	Colorless-to-clear yellow liquid with a bitter almond odor. Inhalation of concentrated vapor is irritating to the eyes, nose, and throat.
Carbon Monoxide	Colorless, odorless gas. Inhalation of carbon monoxide causes headache, dizziness, weakness, confusion, nausea, unconsciousness, and death. Exposure to as little as 0.2% carbon monoxide can result in unconsciousness within 30 minutes. Inhalation of a high concentration can result in immediate collapse and unconsciousness.
Formaldehyde	Colorless gas with a pungent, irritating odor that is highly irritating to the nose; 50–100 ppm can cause severe irritation to the respiratory track and serious injury. Exposure to high concentrations can cause injury to the skin. Formaldehyde is a suspected carcinogen.
Glutaraldehyde	Light-yellow liquid that causes severe irritation of the eyes and irritation of the skin.
Hydrogen Chloride	Colorless gas with a sharp, pungent odor. Mixes with water to form hydrochloric acid. Hydrogen chloride is corrosive to human tissue. Exposure to hydrogen chloride can result in irritation of skin and respiratory distress.
Isovaleraldehyde	Colorless liquid with a weak, suffocating odor. Inhalation causes respiratory distress, nausea, vomiting and headache.
Nitrogen Dioxide	Reddish-brown gas or yellowish-brown liquid, which is highly toxic and corrosive.
Particulates	Small particles that can be inhaled and deposited in the mouth, trachea, or the lungs. Exposure to particulates can cause eye irritation and respiratory distress (in addition to health hazards specifically related to the particular substances involved).
Polycyclic Aromatic Hydrocarbons (PAHs)	PAHs are a group of over 100 different chemicals that generally occur as complex mixtures as part of the combustion process. These materials are generally colorless, white, or pale yellow-green solids with a pleasant odor. Some of these materials are human carcinogens.
Sulfur Dioxide	Colorless gas with a choking or suffocating odor. Sulfur dioxide is toxic and corrosive and can irritate the eyes and mucous membranes.

Source: *Computer Aided Management of Emergency Operations (CAMEO) and Toxicological Profile for Polycyclic Aromatic Hydrocarbons.*

Figures 3.27 a & b In addition to being toxic, carbon monoxide is flammable. *Courtesy of Gresham (OR) Fire and Emergency Services.*

most common fuel gases (methane and propane). This often-unrecognized hazard presents a significant threat to firefighters if not mitigated by effective fire control and ventilation tactics (**Figures 3.27 a and b**).

Flame is the visible, luminous body of a burning gas. When a burning gas is mixed with the proper amounts of oxygen, the flame becomes hotter and less luminous. The loss of luminosity is caused by a more complete combustion of the carbon. For these reasons, flame is considered to be a product of combustion. Of course, it is not present in those types of combustion that do not produce a flame such as smoldering fires or glowing charcoal.

Classification of Fires

The classification of a fire is important because each class of fire has its own requirements for extinguishment. A brief overview of the five classes of fire is discussed, along with how the classifications relate to the fire tetrahedron. More detailed information is provided in the chapters of this manual dealing with fire control and extinguishment.

Class A Fires

Class A fires involve ordinary combustible materials such as wood, cloth, paper, rubber, grass, and many plastics (**Figure 3.28**). The primary mechanism of extinguishment when dealing with Class A fires is cooling to reduce the temperature of the fuel to slow or stop the release of pyrolysis products.

Class B Fires

Class B fires involve flammable and combustible liquids and gases such as gasoline, oil, lacquer, paint, mineral spirits, and alcohol (**Figure 3.29**). Class B fires involving gases can be extinguished by shutting off the gas supply. Fires in Class B liquids can be extinguished with appropriately applied foam and/or dry chemical agents.

Class C Fires

Unlike the other classes, which are determined based on fuel type, Class C fires involve energized electrical equipment (**Figure 3.30**). Household appliances, computers, transformers, electric motors, and overhead transmission lines are typical

Figure 3.28 Class A fuels include a variety of materials. *Courtesy of Dave Ricci.*

Figure 3.29 Flammable liquids are Class B fuels.

Figure 3.30 Electrical energy defines Class C fires.

sources for Class C fires. However, electricity does not burn so the actual fuel in a Class C fire is usually insulation on wiring (Class A material) or lubricants (Class B materials). When possible, de-energize involved electrical equipment before beginning extinguishing efforts. Any extinguishing agent used before de-energizing the equipment must not conduct electricity.

Class D Fires

Class D fires involve combustible metals such as aluminum, magnesium, potassium, sodium, titanium, and zirconium **(Figure 3.31, p. 112)**. These materials are particularly hazardous in their powdered form. In the right concentrations, airborne metal dusts can cause powerful explosions given a suitable ignition source. The extremely high temperature of some burning metals makes water reactive and other common extinguishing agents ineffective. No single agent effectively controls fires in all combustible metals.

Class D materials may be found in a variety of industrial or storage facilities. It is essential to use caution when attempting to extinguish Class D fires because they can sometimes react violently to water and other substances (such as dry chemical extinguishing agents used for Class B fires) and may produce highly toxic smoke and vapors. Review information regarding a material and its characteristics before

Figure 3.31 Class D fires involve combustible metals.
Courtesy of NIST.

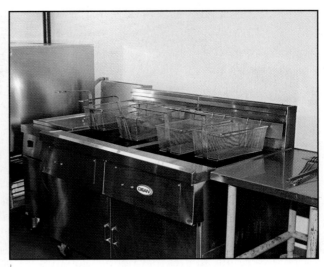

Figure 3.32 Class K fires involve cooking oils.

attempting to extinguish a Class D fire. The burning material should be isolated and treated as recommended in its material safety data sheet (MSDS) or in the *Emergency Response Guidebook (ERG)* from the U.S. Department of Transportation (DOT). All personnel operating near the material should be in full protective clothing including SCBA.

Class K Fires

Class K fires involve oils and greases normally found in commercial kitchens and food preparation facilities using deep fryers **(Figure 3.32)**. These fires require an extinguishing agent specifically formulated for the materials involved. Through a process known as *saponification*, these agents turn fats and oils into a soapy foam that extinguishes the fire.

Saponification — A phenomenon that occurs when mixtures of alkaline-based chemicals and certain cooking oils come into contact resulting in the formation of a soapy film.

Fire Development in a Compartment

For the purposes of this discussion, a *compartment* is an enclosed room or space within a building. The term *compartment fire* is defined as a fire that occurs within such a space. When a fire occurs in an unconfined area, much of the heat produced by the combustion reaction dissipates into the atmosphere through radiation and convection. However, when the fire is confined within a compartment, the walls, ceiling, floor, and other objects in the compartment absorb some of the radiant heat produced by the fire. Radiant heat energy that is not absorbed is reflected back, continuing to increase the temperature of the fuel and rate of combustion. Hot smoke and air heated by the fire become more buoyant and rise. Upon contact with cooler materials, such as the ceiling and walls of the compartment, heat is conducted to the cooler materials, raising their temperature. This heat transfer process raises the temperature of all materials in the compartment. As nearby fuel is heated, it begins to pyrolize (give off flammable vapor). Eventually the rate of pyrolysis can reach a point where flaming combustion can be supported and the fire extends.

When sufficient oxygen is available, fire development is controlled by the characteristics and configuration of the fuel. Under these conditions, the fire is said to be *fuel controlled*. As a fire develops within a compartment, development often reaches a point where it becomes limited by the available air supply. When fire development is limited by the air supply, the fire is said to be *ventilation controlled* **(Figure 3.33)**.

Fire development in a compartment may be described in terms of several stages; however, the boundaries between these stages are not always clearly defined (particularly outside the research laboratory). Despite this limitation, these stages provide a good framework for firefighters to understand fire development in a compartment. The stages include *incipient*, *growth*, *fully developed*, and *decay* **(Figure 3.34)**.

The stages illustrated are an attempt to describe the complex reaction that occurs as a fire develops in a space with no suppression action being taken. The ignition and development of a compartment fire is very complex and influenced by many variables. As a result, all fires may not develop through each of the stages described. The information is presented to depict fire as a dynamic event that is dependent on many factors for its growth and development.

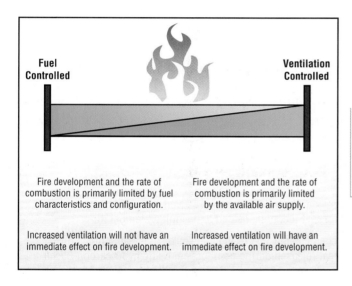

Figure 3.33 Fire development may be limited by the fuel characteristics or the availability of an air supply.

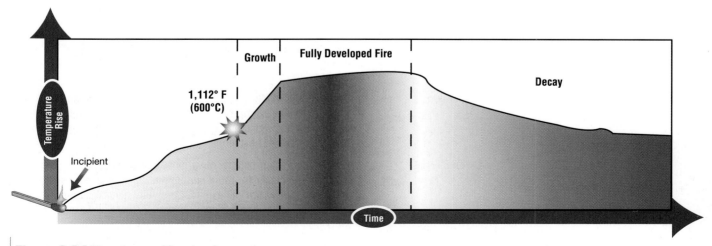

Figure 3.34 The stages of fire development.

Incipient Stage

The incipient stage starts with *ignition*. Ignition describes the point when the three elements of the fire triangle come together and combustion occurs. All fires occur as a result of some type of ignition. Ignition can be *piloted* (caused by a spark or flame) or *nonpiloted* (caused when a material reaches its autoignition temperature as the result of self-heating) such as spontaneous ignition. At this point, the fire is small and confined to the material (fuel) first ignited — and it may self-extinguish (go out on its own).

Once combustion begins, development of an incipient fire is largely dependent on the characteristics and configuration of the fuel involved (fuel-controlled fire). Air in the compartment provides adequate oxygen to continue fire development. During this initial phase of fire development, radiant heat warms adjacent fuel and continues the process of pyrolysis. A *plume* of hot gases and flame rises from the fire and mixes with the cooler air within the room (convection). As this plume reaches the ceiling, hot gases begin to spread horizontally across the ceiling in what fire-fighters have historically called *mushrooming*; however, in scientific or engineering terms it is referred to as forming a *ceiling jet*. Hot gases in contact with the surfaces of the compartment and its contents conduct heat to other materials (including additional fuel). This complex process of heat transfer begins to increase the overall temperature in the room. **Table 3.10** lists the factors that influence the development of fuel-controlled fires.

In this early stage of fire development, the fire has not yet influenced the environment within the compartment to a significant extent. The temperature, while increasing, is only slightly above ambient, and the concentration of products of

Table 3.10
Factors Influencing Development of a Fuel-Controlled Fire

Mass and Surface Area	The greater the surface area for a given mass of fuel, the easier it is for that fuel to be heated to its ignition temperature.
Chemical Content	The chemical makeup of the fuel has a significant impact on the heat released during combustion. Many hydrocarbon-based synthetic materials have a heat of combustion that is more than twice that of cellulose materials such as wood.
Fuel Load	The total amount of fuel available for combustion influences total potential heat release.
Fuel Moisture	While not a factor with all types of fuel, water acts as a thermal ballast, slowing the process of heating the fuel to its ignition temperature.
Orientation	Orientation in relation to the fire influences how heat is transferred. For example, a wood wall surface is heated by both convection and radiation, whereas the floor is more likely to be heated by radiant heat alone.
Continuity	Continuity is the proximity of various fuel elements to one another. The closer (or more continuous) the fuel is, the easier and more rapidly fire will extend. Continuity may be either horizontal (i.e., ceiling surface) or vertical (i.e., wall or rack storage).

Courtesy of Ed Hartin.

combustion is low. During the incipient stage, occupants can safely escape from the compartment, and the fire could be safely extinguished with a portable extinguisher or small hoseline.

It is essential to recognize that the transition from incipient to growth stage can occur quite quickly (in some cases in seconds) depending on the type and configuration of fuel involved as illustrated in **Figures 3.35 a–d.**

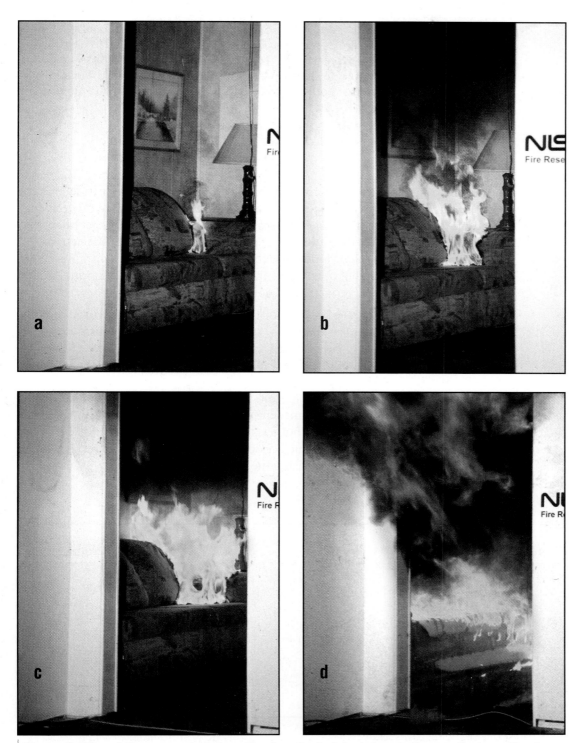

Figure 3.35 a–d Fires can quickly transition from the incipient stage to the growth stage. *Courtesy of NIST.*

Growth Stage

As the fire transitions from incipient to growth stage, it begins to influence the environment within the compartment. Likewise, the fire is influenced by the configuration of the compartment and the amount of ventilation. In a compartment fire, the ceiling and walls affect the plume of hot gases rising from the fire. The first impact is the amount of air that is entrained into the plume. Unconfined fires draw air from all sides and the entrainment of air cools the plume of hot gases rising from the fire, reducing flame length and vertical extension.

Growth Stage — The early stage of a fire during which fuel and oxygen are virtually unlimited. This phase is characterized by a rapidly increasing release of heat.

In a compartment fire, the location of the fuel package in relation to the compartment walls determines the amount of air that is entrained and thus the amount of cooling that takes place. Fires in fuel packages near walls can only entrain air from three sides. Fires in fuel packages in corners can only entrain air from two sides (**Figure 3.36**). Therefore, in both cases, the combustion zone expands vertically and higher plume temperatures result. This stretches the combustion zone even further, resulting in even higher plume temperatures. This factor significantly affects the temperatures in the developing hot-gas layer above the fire and the speed of fire development. In addition, as wall surfaces become hot, burning fuel receives more reflected radiant heat, further increasing the speed of fire development.

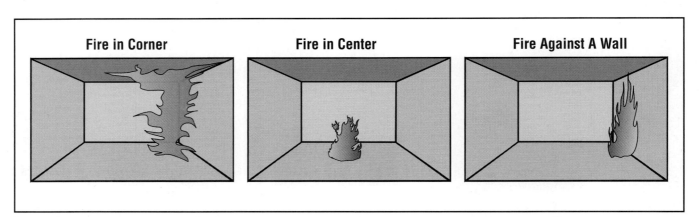

| Fire in Corner | Fire in Center | Fire Against A Wall |

Figure 3.36 Where a fire is located can affect its rate of growth.

Key Points for Firefighters:

- Gases expand and become less dense (more buoyant) than the surrounding air and will rise when heated.
- When gases are confined and heated, pressure increases. Based on this key point, increased pressure indicates higher temperatures.

Thermal Layering

When a fire develops in a compartment, heated products of combustion and entrained air become more buoyant than the surrounding air and rise to the ceiling in a plume. When these hot gases reach the ceiling, they mushroom (spread horizontally through the compartment as a ceiling jet). The gases continue to spread until they reach the walls of the compartment. As combustion continues, the depth of the gas layer then begins to increase. The difference in density between hot smoke and cooler air below causes them to separate into two distinct layers.

The *thermal layering* of gases is the tendency of gases to form into layers according to temperature. Other terms sometimes used to describe this tendency are *heat stratification* and *thermal balance*. The hottest gases tend to be in the top layer, while the cooler gases form the lower layer. In addition to the effects of heat transfer through radiation and convection described earlier, radiation from the hot gas layer also acts to heat the interior surfaces of the compartment and its contents (**Figure 3.37**).

As the volume and temperature of the hot gas layer increases, so does the pressure. Higher pressure in this layer causes it to push down within the compartment and out through any openings such as doors or windows.

The pressure of the cool gas layer is lower, resulting in inward movement of air from outside the compartment. At the point where these two layers meet as the hot gases exit through an opening, the pressure is neutral. The interface of the hot and cooler gas layers at the opening is commonly referred to as the *neutral plane* (**Figure 3.38**).

> **Thermal Layering (of Gases)** — Outcome of combustion in a confined space in which gases tend to form into layers, according to temperature, with the hottest gases found at the ceiling and the coolest gases at floor level.

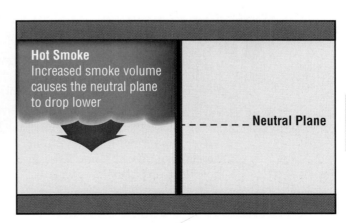

Figure 3.37 Thermal layering is also described as thermal balance because the fire gases form into layers according to temperature. *Courtesy of NIST.*

Figure 3.38 The neutral plane.

Hot Smoke
Increased smoke volume causes the neutral plane to drop lower

Neutral Plane

While the interface between the hot smoke and the cooler air below is sometimes referred to as the neutral plane, this neutral pressure only exists at openings where hot gases are exiting and cooler air is moving into the compartment. Whenever possible, it is desirable to maintain or raise the level of the hot gas layer above the floor to provide a more tenable environment for firefighters and trapped occupants. This requires effective application of fire control and ventilation tactics.

Isolated Flames

As the fire moves through the growth stage, pockets of flames may be observed moving through the hot gas layer above the neutral plane (**Figure 3.39**). Some refer to this phenomenon as *ghosting*. Combustion of these hot gases indicates that portions of the hot gas layer are within their flammable range and there is sufficient temperature to result in ignition. As these hot gases circulate to the outer edges of the plume, they find sufficient oxygen to ignite. This phenomenon is frequently observed prior to more substantial involvement of flammable products of combustion in the hot gas layer. Ghosting is classified as a *fire gas ignition* and may be an indicator of developing flashover conditions requiring immediate action by firefighters to prevent that from occurring.

Rollover/Flameover

The terms *rollover* (and less commonly *flameover*) describe a condition where the unburned fire gases accumulated at the top of a compartment ignite and flames propagate through the hot gas layer or across the ceiling. Like ghosting, rollover is a fire gas ignition; it is also a significant indicator of impending flashover. Rollover is distinguished from flashover by its involvement of only the fire gases at the upper levels of the compartment and not the other fuel packages within a compartment. Rollover may occur during the growth stage as the hot-gas layer forms at the ceiling of the compartment. Flames may be observed in the layer when the combustible gases reach their ignition temperatures. While the flames add to the total heat generated in the compartment, this condition is not flashover. Rollover will generally precede flashover, but it may not always result in flashover.

Flashover

Flashover is the rapid transition between the growth and the fully developed fire stages but is not a specific event such as ignition. Conditions for flashover are defined in a variety of different ways; however, during flashover, conditions in the compartment change very rapidly from partial to full involvement of the compartment. When flashover occurs, burning gases push out of openings in the compartment (such as a door leading to another room) at a substantial velocity (**Figure 3.40**).

Flashover can occur almost instantaneously. Any resulting rollover extending out of the compartment in which the flashover occurs can move faster than firefighters attempting to escape. The radiant heat generated by a flashover is not survivable for more than a few seconds, even when wearing full PPE and SCBA. The indicators of impending flashover are listed in **Table 3.11**.

Remember not to focus on any one indicator exclusively. Reading fire conditions requires that you view the big picture; that is, you evaluate as many indicators as possible.

Rollover — Condition in which the unburned combustible gases released in a confined space (such as a room or aircraft cabin) during the incipient or early steady-state stage accumulate at the ceiling level. These superheated gases are pushed, under pressure, away from the fire area and into uninvolved areas where they mix with oxygen. When their flammable range is reached and additional oxygen is supplied by opening doors and/or applying fog streams, they ignite and a fire front develops, expanding very rapidly in a rolling action across the ceiling.

Flashover — Stage of a fire at which all surfaces and objects within a space have been heated to their ignition temperature and flame breaks out almost at once over the surface of all objects in the space.

Table 3.11
Flashover Indicators

Building	Flashover can occur in all types of buildings. Building factors can influence how quickly a fire will reach flashover (e.g., fire load, ventilation profile, and thermal properties) and should be considered an integral part of ongoing risk assessment.
Smoke	Smoke indicators may or may not be visible from the exterior of the structure. However, smoke conditions indicating a developing fire are a warning sign of potential flashover conditions. After making entry, the presence of hot gases overhead and lowering of the hot gas layer are key indicators.
Air Flow	A strong bidirectional (air in and smoke out) air flow can be a significant indicator of flashover that will move in the direction of the opening. However, any air flow that shows air movement toward the fire can result in flashover.
Heat	Rapidly increasing temperature (although this is frequently a late indicator).
Flame	Isolated flames traveling in the hot-gas-layer (ghosting) or more substantially through the gas layer or across the ceiling (rollover). These flames may or may not be visible (without the use of a thermal imaging camera).

Courtesy of Ed Hartin.

Figure 3.39 Pockets of flame can be seen in the hot gas layer. *Courtesy of NIST.*

Figure 3.40 Flashover can happen almost instantaneously. *Courtesy of NIST.*

The intense heat produced by this extreme fire behavior phenomenon can result in disorientation as well as thermal burns. The most effective way for firefighters to manage this risk is by maintaining an awareness of developing fire conditions and controlling the fire environment through effective fire control and ventilation tactics.

WARNING!
Smoke is unburned fuel that is ready to ignite!

Maintain your *situational awareness* by continually observing your immediate surroundings. Look, listen, and feel what is going on around you.

- *Look* for any visible indicators — unusual fire behavior; smoke density, color, or movement; and look for the movement of fellow firefighters and others.

- *Listen* for any audible clues to what is happening in the immediate area — crackling fire sounds, hiss of escaping gas, sounds of building movement, faint calls for help, or the shriek of a PASS device.

- *Feel* what is happening around you — vibrations or structural movement; feel changes in fire intensity; feel floors as you progress for heat and locate any holes, open stairwells, or elevator shafts; feel walls for windows and doors.

Ceiling temperatures can be reduced through carefully considered fire control actions. Applying short bursts of water spray into the hot gas layer can be effective. Applying water directly onto whatever is burning can limit the release of unburned products of combustion as well as reduce ceiling temperature.

Ceiling temperature can also be reduced through calculated ventilation efforts. However, because of the potential to produce catastrophic changes in fire intensity and spread, ventilation must be carefully controlled and executed only when ordered.

While scientists and engineers define flashover in a variety of ways, the following is the most useful definition for firefighters: *When the temperature in a compartment results in the simultaneous ignition of all of the combustible contents in the space.* While no exact temperature is associated with this occurrence, a range from approximately 900°F to 1,200°F (483°C to 649°C) is widely accepted. This range correlates with the autoignition temperature of carbon monoxide (CO) of 1,128 °F (609°C), one of the most common gases produced by pyrolysis.

Just before flashover, several things are happening within the burning compartment: temperatures are rapidly increasing, additional fuel is becoming involved, and the fuel in the compartment is giving off combustible gases because of pyrolysis. As flashover occurs, the combustible materials in the compartment and the gases produced by pyrolysis ignite almost simultaneously. The result is full-room involvement.

An Alternative Path

Flashover does not occur in every compartment fire. Two interrelated factors determine whether a fire within a compartment will progress to flashover. First, the fuel must have sufficient heat energy to develop flashover conditions. For example, ignition of discarded paper in a small metal wastebasket may not have sufficient heat energy to develop flashover conditions in a large room lined with gypsum drywall. On the other hand, ignition of a couch with polyurethane foam cushions placed in the same room is quite likely to result in flashover. The second factor is ventilation.

A developing fire must have sufficient oxygen to reach flashover, and a sealed room may not provide enough. Heat release is limited by the available air supply. If there is insufficient natural ventilation, the fire may enter the growth stage but not reach the peak heat release of a fully developed fire (**Figure 3.41**).

While the heat release is reduced when the fire becomes ventilation controlled, the temperature in the compartment may continue to rise (although more slowly). When ventilation is increased (i.e., due to failure of window glazing or firefighters making entry), additional air will increase heat release (in some cases extremely rapidly).

It is important to recognize that most fires that grow beyond the incipient stage become ventilation controlled. Even when doors and/or windows are open, there is often insufficient air to allow the fire to continue to develop based on the available fuel. When windows are intact and doors are closed, the fire may move into a ventilation-controlled state even more quickly. While this reduces the heat release rate, fuel will continue to pyrolize, creating extremely fuel-rich smoke.

Fully Developed Stage

The fully developed fire stage occurs when all combustible materials in the compartment are burning. While this is occurring, the burning fuels in the compartment are releasing the maximum amount of heat possible for the available fuel and ventilation, producing large volumes of fire gases. In the fully developed stage, the fire is ventilation controlled in that heat release is dependent on the compartment openings, which are providing oxygen to support the ongoing combustion reaction and releasing products of combustion. Increases in the available air supply will result in higher heat release. During this stage, hot unburned fire gases are likely to flow from the compartment of origin into adjacent compartments or out through openings to the exterior of the building. These hot gases may ignite (fire gas ignition) as they enter a space where air is more abundant (**Figure 3.42**).

Decay Stage

A compartment fire will decay as the fuel is consumed or if the oxygen concentration falls to the point where flaming combustion can no longer be supported. Each of these situations can result in the combustion reaction coming to a stop, but decay

Fully Developed Stage — Stage of burning process where energy release is at maximum rate and is limited only by availability of fuel and oxygen.

Decay — Stage of fire development when fuel is consumed and energy release diminishes, and temperatures decrease. During this stage the fire goes from ventilation-controlled to fuel controlled.

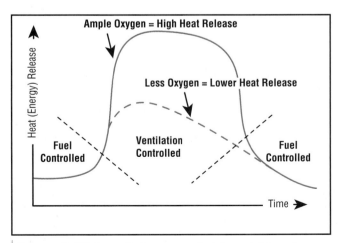

Figure 3.41 A ventilation-controlled fire exhibits a lower rate of heat release caused by an insufficient quantity of oxygen to feed the fire.

Figure 3.42 A fully developed fire. *Courtesy of Gresham (OR) Fire and Emergency Services.*

due to reduced oxygen concentration can follow a considerably different path if the ventilation profile of the compartment changes. Each of these scenarios will be examined in turn.

Consumption of Fuel

As the fire consumes the available fuel in the compartment and the rate of heat release begins to decline, it enters the *decay stage* or *hot-smoldering phase*. Assuming there is adequate ventilation, the fire again becomes fuel controlled. The heat release rate will drop, but the temperature in the compartment may remain high for some time. During this stage, the flammable products of combustion can accumulate within the compartment or adjacent spaces. If the products of combustion are within the flammable range, they can be ignited, resulting in a smoke explosion.

Limited Ventilation

When a compartment fire enters the decay stage due to a lack of oxygen, the rate of heat release will also decline. However, the continuing combustion reaction (based on available fuel and the limited oxygen available to the fire) may maintain an extremely high temperature within the compartment. Under these conditions, a large volume of flammable products of combustion can accumulate within the compartment. If these products are above their ignition temperatures, they can ignite explosively when mixed with additional air, resulting in a backdraft.

> It is important to recognize that this discussion of the stages of fire development examines fire behavior in a single compartment to illustrate fire progression. Actual conditions within a building composed of multiple compartments can vary widely.
>
> The compartment of origin may be in the fully developed stage while adjacent compartments may be in the growth stage. In addition, an attic or void space may be in a severely underventilated decay stage while adjacent compartments are in the growth or fully developed stage. This makes reading the fire and assessing the hazards presented by fire conditions a critical task for everyone working inside the burning building.

Backdraft

Backdraft — Instantaneous explosion or rapid burning of superheated gases that occurs when oxygen is introduced into an oxygen-depleted confined space. The stalled combustion resumes with explosive force. It may occur because of inadequate or improper ventilation procedures.

A ventilation-controlled compartment fire can produce a large volume of flammable smoke and other gases due to incomplete combustion. This mixture of flammable products can be well above its upper flammable limit. While the rate of heat release from a ventilation-controlled fire is limited, high temperatures are usually still present within the compartment. An increase in ventilation (such as opening a door or window) can result in a deflagration (explosively rapid combustion) called *backdraft*.

When potential backdraft conditions exist in a compartment, the space is filled with unburned fuel (smoke) that is at or above its ignition temperature and only lacks sufficient oxygen to burn. Making a horizontal opening provides the missing component (oxygen) and a backdraft results **(Figure 3.43)**.

A backdraft can occur without the creation of a horizontal opening. All that is required is mixing hot, fuel-rich smoke with air. Backdraft conditions can develop within a room, a void (such as an attic or a small room within a larger structure), or

Figure 3.43 Improper ventilation for the conditions may result in a backdraft.

within an entire building. Anytime a compartment or space contains extremely hot combustion products, potential for backdraft must be considered before creating any openings into the compartment.

To some degree, the violence of a backdraft is dependent on the extent to which the fuel/air mixture is confined. The more confined the deflagration, the more violent it will be.

According to a NIOSH report, two Illinois firefighters were killed and three others injured in 1998 during fire fighting operations in a tire shop. First-arriving firefighters found no smoke showing from the exterior of this large commercial building. The structure was of masonry construction with a bowstring truss roof. Entering the building, fire crews encountered some smoke in the showroom located at the front of the building and heavy black smoke in the upper area of the service bays. A truck company performing roof ventilation reported that shortly after they vented the roof, heavy smoke and then fire issued from the vent opening. When an overhead door was opened in the service area, the inflow of air mixed with the hot gas layer and resulted in a backdraft.

Despite these somewhat atypical conditions, fire involvement of the polystyrene insulation on the underside of the truss roof and combustion above a stable hot gas layer resulted in development of backdraft conditions.

While backdraft conditions are usually associated with enclosed spaces that are completely filled with hot products of combustion, this does not preclude the possibility of a backdraft in a compartment that is not filled with smoke (particularly in large-volume, high-ceiling compartments). A fire burning above the hot gas layer may develop backdraft conditions in the upper level of the compartment, even when conditions at floor level are only moderately affected.

As with flashover, it is critical to recognize the potential warning signs of backdraft conditions. Common indicators of the potential for a possible backdraft are shown in **Figure 3.44**.

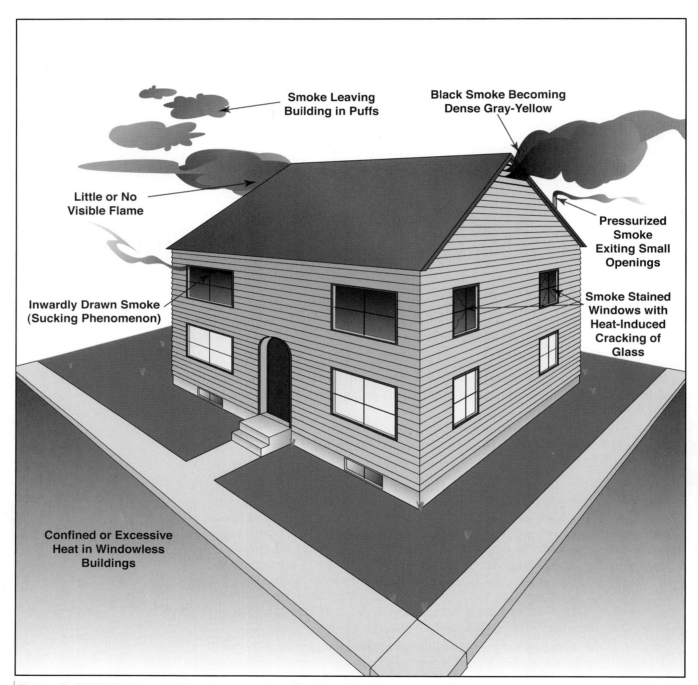

Figure 3.44 Firefighters must be able to recognize the visual indicators of a backdraft.

While pulsing smoke movement is commonly recognized as a backdraft indicator, raising and lowering of the neutral plane is not. However, when the compartment or structure is not full of smoke, changes in pressure and volume of smoke can raise and lower the neutral plane (rather than pulsing discharge of smoke from compartment or building openings).

It is often assumed (incorrectly) that a backdraft will always occur immediately or soon after making an opening into the building or involved compartment. Mixing of hot flammable products of combustion with air through the action of gravity currents, pressure differentials, and wind effects sometimes takes time. When potential backdraft conditions are encountered, firefighters should delay entry until actions are taken to change conditions inside the building or compartment (i.e., gas cooling with hose streams, vertical ventilation, etc.).

The effects of a backdraft can vary considerably depending on a number of factors including the volume of flammable products of combustion, degree of confinement, the speed with which fuel and air are mixed, and where ignition occurs.

Factors That Affect Fire Development

A number of factors influence fire development within a compartment. These factors include the following:

- Fuel type
- Availability and location of additional fuel in relation to the fire location
- Compartment geometry (volume and ceiling height)
- Ventilation and changes in ventilation
- Thermal properties of the enclosure
- Ambient conditions (wind, temperature, humidity, etc.)

Fuel Type

As discussed in the section of this chapter dealing with fuel, the type of fuel involved in combustion impacts both the amount of heat released and the time over which this occurs. In a compartment fire, the most fundamental fuel characteristics influencing fire development are mass and surface area. Combustible materials with high surface-to-mass ratios are much more easily ignited and will burn more quickly than the same substance with less surface area. In addition, many ordinary combustibles such as wood and paper are significantly influenced by fuel moisture. Water absorbs heat that would otherwise contribute to the process of pyrolysis.

Firefighters should be able to recognize potential fuels in a building or compartment and use this information to estimate the fire-growth potential for the building or space. Materials with high heat release rates, such as polyurethane foam-padded furniture or polyurethane foam mattresses, would be expected to burn rapidly once ignition occurs.

Availability and Location of Additional Fuel

A number of factors influence the availability and location of additional fuels. These include the configuration of the building, contents (nonstructural fire load), construction (structural fuel load), and location of the fire in relation to fuel that has not yet become involved.

Building configuration is the layout of the structure including the number of stories, avenues for fire spread, compartmentation, and barriers to fire spread. A building may have a high fire load but be highly compartmentalized with fire doors blocking the spread of hot smoke and fire gases. On the other hand, buildings with open floor plans or unprotected vertical shafts may provide the fire with access to fuel throughout the building (**Figures 3.45 a and b**).

The contents of a structure are often the most readily available source of fuel in a compartment fire. The quantity and nature of building contents significantly influence fire development. When contents have a high heat of combustion and heat release rate, both the intensity of the fire and speed of development will be greater. For example, synthetic furnishings, such as polyurethane foam, will begin to pyrolize rapidly under fire conditions (even when located some distance from the origin of the fire), speeding the process of fire development (**Figure 3.46**).

Type of construction influences fuel load as some types of building materials are combustible. For example, in wood frame buildings, the structure itself is a source of fuel. In addition to structural members, combustible interior finishes, such as wood paneling, can be a significant factor influencing fire spread (**Figure 3.47**).

> Combustible interior finishes have been a significant factor in a number of major fires. These include the Cocoanut Grove fire in Boston, Massachusetts, November 28, 1942. A total of 492 people lost their lives in this rapidly developing nightclub fire. Flammable decorations quickly spread the fire, and thick smoke and heat killed the occupants as they attempted to escape through the limited number of exits that were available.
>
> Just over 51 years later, 100 people lost their lives in another New England nightclub fire. Pyrotechnics ignited combustible interior finish materials (polyurethane foam sound insulation) at the Station Night Club in West Warwick, Rhode Island, resulting in extremely rapid fire development which trapped many of the buildings occupants.

Figure 3.45 a & b The interior configuration of a building can affect fire development. (a) The ability to close doors and contain a fire in a small compartment can limit spread and development. (b) Open floor plans lack any means of limiting the spread from one work area to another.

Figure 3.46 Synthetic materials, such as carpet and furniture coverings, can accelerate fire development.

Figure 3.47 Some interior finishes, including simulated wood paneling, can promote rapid fire spread.

The proximity (in relation to the fire) and continuity of contents and structural fuels also influence fire development. Fuels in the upper level of adjacent compartments will be more quickly pyrolized by the hot gas layer, and continuous fuels (such as those provided by combustible interior finishes) will rapidly spread the fire from compartment to compartment. Similarly, the location of the fire within the building will influence fire development. When the fire is located low in the building, such as in the basement or on the first floor, convected heat will cause vertical extension through unprotected stairways and vertical shafts. Fires originating on upper levels generally extend downward much more slowly (**Figure 3.48**).

Compartment Volume and Ceiling Height

All other things being equal, a fire in a large compartment will develop more slowly than one in a small compartment due to the greater volume of air and structural material that must be heated. Remember, though, that this large volume of air will support the development of a larger fire before ventilation becomes the limiting factor (**Figure 3.49, p. 128**).

A high ceiling may also mask the extent of fire development by allowing a large volume of hot smoke and other fire gases to accumulate at ceiling level, while conditions at floor level remain relatively unchanged. This situation is particularly hazardous because conditions can change rapidly if this hot gas layer ignites.

Ventilation

As discussed earlier in this chapter, for a fire to develop there must be enough air available to support burning beyond the incipient stage. Ventilation in a compartment significantly influences how the fire develops within the

Figure 3.48 Fire can spread rapidly upward in a multistory building. *Courtesy of District Chief Chris E. Mickal, NOFD Photo Unit.*

Figure 3.49 Fires may develop slowly in large, open structures, such as warehouses and aircraft hangars.

space. Preexisting ventilation is the actual and potential ventilation of a structure based on structural openings, construction type, and building ventilation systems. For the most part, all buildings exchange air inside the structure with that outside the structure. In some cases this is due to constructed openings such as windows and doors as well as leakage through cracks and other gaps in construction. In other cases, this air exchange is primarily through the heating, ventilating, and air-conditioning (HVAC) system.

When considering fire development, it is important to consider potential openings that could change the ventilation profile under fire conditions — windows are a good example. Under fire conditions windows can fail or doors are left open, increasing ventilation. When a fire develops to the point where it becomes ventilation controlled, the available air supply will determine the speed and extent of fire development and sometimes the direction of fire travel.

When a fire starts, existing ventilation conditions dictate the actual exchange of products of combustion inside the building or compartment with outside air. It is also critical to recognize the potential for changing ventilation conditions during fire fighting operations. Firefighters make additional ventilation openings when they open doors to enter a building or create other openings to vent the structure. Before taking any of these actions, they should consider the size, number, and arrangement of existing and potential ventilation openings and how these openings might affect the behavior of the fire.

Thermal Properties of the Enclosure

Thermal properties include insulation, heat reflectivity, retention, and conductivity. When a compartment is well-insulated, less heat is lost and more heat remains available to increase temperature and speed the combustion reaction. Similarly, surfaces that reflect heat return it to the combustion reaction and increase its speed. Materials such as masonry act as a heat sink and will retain heat energy, sustaining high temperatures for a long period of time. Other structural materials, such as steel, conduct heat readily. While not retaining heat to the same degree as masonry, these materials can transfer heat to other combustibles through conduction, spreading the fire (in some cases beyond the compartment or compartments already involved). Thermal windows (those with multiple panes) can also act to contain heat in a developing fire.

In 1999, six career firefighters in Massachusetts died after they became lost and ran out of breathing air in a vacant six-story, cold-storage building. The building contained six floors aboveground and a full basement, for a total of 94,176 square feet (8 749 sq m). The exterior walls were 18 inches thick (450 mm) and constructed of brick. Interior walls were covered with 6 to 18 inches (150 mm to 450 mm) of asphalt-impregnated cork or 4 inches of polystyrene and a flammable finish.

The same characteristics that made this structure an effective cold-storage warehouse contributed to fire development. Insulation that keeps heat out can also keep heat in. These materials contributed to the rapid fire development and severe smoke conditions in the building, disorienting the first two firefighters and the other four who attempted to locate and rescue them.

It is important to remember that dangerous fire behavior is often the result of multiple intersecting factors rather than one simple cause.

Ambient Conditions

While ambient temperature and humidity can have an impact on the ignitability of many types of fuel, these factors are less significant inside a structure (which is designed to minimize the impact of these factors on the occupants). High humidity and cold temperatures can impede the natural movement of smoke, however, and wind can be an extremely significant factor in fire development.

Strong winds can significantly influence fire behavior, particularly when ventilation changes. If a window fails or a door is opened on the windward side of the building, fire intensity and spread can increase significantly (**Figure 3.50, p. 130**).

Winds and Fire Growth

At approximately 4:54 a.m., smoky conditions were reported on the 10th floor (top floor) of an apartment building in New York State. The building was of fire-resistant construction, with concrete floors and ceilings and concrete block walls between the apartments forming the corridor. The door to the fire apartment was open to the corridor. The occupant of an apartment on the opposite side of the corridor opened his living room windows in an effort to clear out smoke from his apartment. Later the door from that apartment was opened to the corridor and remained open throughout the rest of the fire incident. Wind gusts were striking the side of the building closest to the room involved in fire.

At approximately 5:04 a.m., three firefighters entered the 10th floor elevator lobby and reported "light smoke" conditions. They then entered the fire corridor. Within 2 minutes, both of the stairwell doors on the roof had been propped open. At approximately 5:07 a.m., the roof team reported "high winds." About 3 minutes later (5:10 a.m.), they reported flames coming from the apartment window. Approximately 30 seconds later, a "Mayday" was transmitted.

Members of the fire attack team made several attempts to enter the corridor after the Mayday call. They were met with extreme heat conditions, and they were burned during their attempts. One firefighter described the flames coming out of the fire apartment as a "blowtorch" extending from floor to ceiling. Three firefighters trapped in the corridor between the upwind vent in the fire apartment and the downwind vent in the opposite side of the corridor died due to thermal injuries.

Figure 3.50 Wind can sometimes accelerate fire spread in structures. *Courtesy of Gresham (OR) Fire and Emergency Services.*

Impact of Changing Conditions

Structure fires can be dynamic, with ever-changing conditions. Factors influencing fire development can change as the fire extends from one compartment to another, providing additional fuel, or high temperatures cause windows to break and fall, increasing ventilation. Considering that most fires beyond the incipient stage are or will quickly become ventilation controlled, changes in ventilation are likely to be some of the most significant factors in changing fire behavior. In some cases these changes are caused by the fire acting on building materials, while in other cases fire behavior is influenced (positively or negatively) by tactical operations. Firefighters must recognize that opening a door to make entry for fire attack or search can have as much impact on ventilation as opening a window to purposefully vent smoke and products of combustion.

Fire Control Theory

Fire is controlled and extinguished by limiting or interrupting one or more of the essential elements in the combustion process (fire tetrahedron). Firefighters influence fire behavior by acting to reduce temperature, eliminating fuel or separating the fire from available fuel, changing the oxygen concentration, or interrupting the self-sustained chemical chain reaction (**Figure 3.51**).

Temperature Reduction

One of the most common methods of fire control and extinguishment is cooling with water. This process depends on reducing the temperature of a fuel to a point where it does not produce sufficient vapor to burn. Solid fuels and liquid fuels with high flash points can be extinguished by cooling. The use of water for cooling is also the most effective method available for the extinguishment of smoldering fires. To extinguish a fire by reducing its temperature, enough water must be applied to the burning fuel to absorb the heat being generated by combustion. Nonetheless, cooling with water cannot sufficiently reduce vapor production to extinguish fires involving low flash point flammable liquids and gases. Types of streams and extinguishing methods are discussed later in the manual.

Figure 3.51 The application of water can be used to reduce the temperature of a burning material. *Courtesy of District Chief Chris E. Mickal, NOFD Photo Unit.*

In addition to extinguishment by cooling, water can also be used to control burning gases and reduce the temperature of hot products of combustion above the neutral plane. This slows the pyrolysis process and reduces the potential for extreme fire behavior such as flashover.

Water absorbs significant heat as its temperature is raised, but it has its greatest effect when it is vaporized into steam. When water is converted to steam at 212°F (100°C), it expands approximately 1,700 times. This expansion rate emphasizes the importance of controlling steam production through good nozzle techniques, using an appropriate volume of water, and applying the water in the most effective form (fog, straight, or solid stream) based on existing conditions. Excess steam production can make it difficult to see and can increase the chances for steam burns.

Fuel Removal

Removing the fuel source effectively extinguishes any fire. The simplest method of fuel removal is to allow a fire to burn until all fuel is consumed. While this is not always the most desirable extinguishment method, it is sometimes appropriate. For example, fires involving pesticides or flammable liquid spills may create greater environmental harm if extinguished with water, creating substantial runoff. The best solution may be to allow the fire to burn, minimizing groundwater pollution. The fuel source may also be removed by stopping the flow of liquid or gaseous fuel by closing a valve or by removing solid fuels in the path of a fire. This is the preferred method of extinguishing pressurized gas fires.

Oxygen Exclusion

Reducing the oxygen available to the combustion process reduces a fire's growth and may totally extinguish it over time. In its simplest form, this method is used to extinguish rangetop fires when a cover is placed on a pan of burning grease. Flooding an area with an inert gas such as carbon dioxide displaces the oxygen and disrupts the combustion process. Oxygen can also be separated from some fuels by blanketing them with foam. Of course, neither of these methods works on those rare fuels that are self-oxidizing.

While not generally used for extinguishment in structure fires, limiting the fire's air supply (ventilation) can be a highly effective fire control action. The simplest example of this is when a building occupant closes the door to the fire room before leaving the building. This limits the air supply to the fire and can sometimes prevent flashover. By not opening doors and windows until fire control has been achieved, firefighters can control the air supply during fire fighting operations to limit fire growth.

Chemical Flame Inhibition

Extinguishing agents, such as some dry chemicals, halogenated agents (halons), and halon-replacement agents, interrupt the combustion reaction and stop flame production. This method of extinguishment is effective on gas and liquid fuels because they must flame to burn. These agents do not easily extinguish surface mode fires because they work on the chemical chain reaction of flaming combustion. The very high agent concentrations and extended periods necessary to extinguish smoldering fires make these agents impractical in these cases.

Summary

Many people believe that fire is unpredictable, but there is no unpredictable fire behavior. Our ability to predict what will happen in the fire environment is hampered by limited information, time pressure, and our level of fire behavior knowledge. As a firefighter, you need to understand the combustion process and how fire behaves in different materials and in different environments. You also need to know how fires are classified so that you can select and apply the most appropriate extinguishing agent. Most important, you and your fellow firefighters need to have an understanding of fire behavior that permits you to recognize developing fire conditions and be able to respond safely and effectively to mitigate the hazards presented by the fire environment.

Fire Fighter I

1. What are the four elements of the fire tetrahedron?

2. What are common sources of heat that result in the ignition of a fuel?

3. Define conduction, convection, and radiation.

4. What is flash point?

5. What are three hazardous products of combustion?

6. Describe the five classes of fire.

7. What are the stages of fire development in a compartment?

8. Define thermal layering, rollover, flashover, and backdraft.

9. What are the factors that influence fire development within a compartment?

10. How can fire be controlled and extinguished?

Chapter Contents

Building Construction

This chapter provides information that addresses the following job performance requirements of NFPA® 1001, *Standard for Fire Fighter Professional Qualifications (2008)*:

NFPA® 1001 references for Chapter 4:

5.3.4
5.3.10
5.3.12
6.3.2

Photo courtesy of Ron Jeffers.

Chapter Objectives

Fire Fighter I Objectives

1. Describe common building materials.

2. Describe construction types and the effect fire has on the structural integrity of the construction type.

3. Identify the primary strengths and weaknesses of construction types.

4. Describe dangerous building conditions created by a fire or by actions taken while trying to extinguish a fire.

5. Identify indicators of building collapse.

6. List actions to take when imminent building collapse is suspected.

7. Describe hazards associated with lightweight and truss construction.

Fire Fighter II Objectives

1. Describe the effects of fire and suppression activities on common building materials.

2. Describe items to be observed during size-up of a building.

3. Describe dangerous building conditions created by a fire or by actions taken while trying to extinguish a fire.

4. Identify indicators of building collapse.

5. Describe actions to take when imminent building collapse is suspected.

6. Describe building conditions that create additional risk in construction, renovation, and demolition.

Building Construction

Case History

According to an investigative report by the National Institute for Occupational Safety and Health (NIOSH), a fire in a Virginia auto parts store claimed the lives of two firefighters. The store was part of a single-story strip mall in which the individual occupancies were separated by masonry walls. The roof was constructed of 2 × 6-inch (50 mm by 150 mm) wooden trusses that spanned the 50-foot (15 m) width of the store. The roof also supported heating, ventilating, and air conditioning (HVAC) equipment weighing approximately 3,000 pounds (1 361 kg).

Store employees called 9-1-1 and reported hearing the sound of sparking and popping inside the fuse box. Upon arrival, the first engine company reported no smoke or fire visible. When smoke was seen coming from the roof and fire was discovered in the attic area, crews began an interior attack. Approximately 15 minutes after the arrival of the first engine, the roof collapsed without warning. The company officer and a firefighter from the first-arriving engine were trapped in the collapse debris and died of burns and smoke inhalation.

As the previous case history shows, failure to recognize the potential dangers of a particular type of construction and the effects that fire may have on it can be catastrophic for firefighters. For your safety and that of your fellow firefighters, you must have at least a basic knowledge of building construction. You need to know about construction materials, methods, and designs in general and those that are used in your area in particular. Knowledge of the various types of building construction and how fires react in each type gives you and your officers vital information to planning a safe and effective fire attack.

NFPA® 1001, *Standard for Fire Fighter Professional Qualifications*, requires those qualified at the Fire Fighter I level to know the following about building construction:

- Basic construction of doors, windows, and walls and the operation of doors, windows, and locks
- Indicators of potential collapse or roof failure
- Effects of construction type and elapsed time under fire conditions on structural integrity

The standard requires those qualified at the Fire Fighter II level to know the following about building construction:

- Dangerous building conditions created by fire and suppression activities
- Indicators of building collapse
- Effects of fire and suppression activities on wood, masonry, cast iron, steel, reinforced concrete, gypsum wallboard, glass, and plaster on lath

A basic understanding of construction is vital in being able to safely and effectively perform many fire fighting tasks such as forcible entry, ventilation, and overhaul. Understanding different construction types is also important in size-up and determining methods of attack.

This chapter introduces you to common building construction terms, materials, and methods. It also describes how various types of construction are classified, and how each type behaves in fires. The chapter also discusses some of the sights and sounds that indicate the possibility of structural collapse or other extraordinary events during interior fire fighting operations.

Construction Terminology

A variety of terms are commonly used in the construction trades. To help you better understand the discussions in the balance of this chapter, the most common construction terminology is presented in the following sidebar.

Common Construction Terminology

Assembly — Two or more interconnected structural components combined to meet a specific function or design requirement. Typical assemblies are roof trusses, wall frames, and doors including their frames.

Attic — An open space between the roof and ceiling of a building; most commonly found in single- and multifamily residential occupancies. Attics provide open spaces in which fires can burn undetected or spread throughout a structure.

Balloon Frame — A type of wood-frame construction in which the studs in exterior walls extend from the basement or foundation to the roof. This type of construction allows fires to spread – often undetected – from the basement to the attic through the hollow walls.

Bar Joist — A joist constructed of steel with bars in the vertical web space. A common structural component in office buildings and other commercial structures. Very high strength-to-weight ratio except when exposed to the heat of a fire – then early failure is likely.

Beam — A horizontal structural component subjected to vertical loads. Typical beams are steel or wooden I-beams or large-dimension wooden members.

Bowstring Truss — A roof assembly with a curved (arched) top chord and a horizontal bottom chord. These assemblies are very strong except when exposed to direct flame contact when catastrophic failure without warning may occur.

Butterfly Roof — A V-shaped roof in which the two sides slope toward a valley in the middle. An unusual type of roof that is rarely seen in cold climates where snow load is a factor.

Cantilever — A beam that is unsupported at one or both ends. Typically used to support balconies on apartments and some office buildings.

Chipboard — *See* Oriented Strand Board.

Chord — The main structural members of a truss as distinguished from diagonals. Chords span the open space between the upper and lower diagonal members in a truss assembly.

Cockloft — An open space between the roof and ceiling of a commercial or industrial building. Usually found under flat or nearly flat roofs. In a fire, these spaces act in much the same way as attics.

Column — A vertical supporting member. Columns may be wooden or steel posts. Steel posts often support lightweight roof assemblies, and if unprotected by surface insulation, steel posts may fail quickly in a fire.

Compression — Force that tends to push the mass of a material together. Bearing walls in a building are under compression from the weight of the roof and other materials above.

Course — Horizontal layer of masonry units. A row of bricks is an example of a course.

Curtain Board — Nonload-bearing interior wall extending down from a roof or ceiling to limit the horizontal spread of fire and heat. If curtain walls are penetrated by unprotected openings, fire can spread unchecked.

Curtain Wall — Nonload-bearing exterior wall used as a weather barrier but not for structural support. On many high-rise buildings, the outside walls (often sheet glass in frames) are curtain walls.

Decking — Planks or panels of plywood or oriented strand board (OSB) that form the substrate of a roof assembly. In vertical ventilation through a roof, the

Common Construction Terminology (continued)

decking must be removed from the ventilation opening to realize the full effect of the opening.

Drywall — Gypsum wall board. A fire-resistive wall covering also called *sheetrock™*.

Eave — The edge of a pitched roof that overhangs an outside wall. Attic vents in typical eaves provide an avenue for an exterior fire to enter the attic.

Engineered I-Beam — A wooden I-beam consisting of continuous wooden upper and lower chords separated by a web of OSB or similar sheet stock. Engineered I-beams are very strong and resist fire well.

Fire Door — A rated assembly consisting of a solid-core door, door frame, and hardware. Fire doors are used to confine a fire to one room or section of a building by closing a communicating opening when triggered by a fire. If fire doors are to function as designed, they must not be prevented from closing by being intentionally or inadvertently blocked open.

Fire Load — Total potential heat release if a building and its contents burned. The fire load of a fully stocked lumber yard is considerably higher than that of an empty building of the same dimensions.

Fire Wall — A rated assembly that extends from the foundation to and through the roof of a building to limit fire spread. Fire walls are intended to confine a fire to one room or section of a building. If they are penetrated by openings not protected with fire doors, fire can spread unchecked.

Flat Roof — A roof that is flat or nearly flat relative to the horizon. Many commercial buildings have flat roofs covered with tar and gravel or other weatherproof material. Flat roofs lend themselves to being opened for vertical ventilation.

Gable Roof — A pitched roof characterized by square-cut ends and sides that slope down from the ridge line to the eaves. These are the most common roof style on homes and other small buildings.

Gable Wall — A wall rising to meet a gable roof at the end of a building. These walls are found only at the ends of gable roofs and they often include an attic vent near the top of the wall.

Gambrel Roof — A roof characterized by a single ridge line from which roof sections on both sides of the ridge descend at two different pitches. These roofs are common on barns and other farm structures. Because of the differing angles of the slopes, gambrel roofs can make roof ladders difficult to use on them.

Girder — A horizontal structural member used to support beams or joists. Girders are almost always of larger dimension than the members they support.

Glue-Lam Beam — A wooden structural member composed of relatively short pieces of lumber glued and laminated together under pressure to form a long, extremely strong beam. Because of the mass of most glue-lam beams, they resist fire extremely well compared to other materials.

Gusset Plate — Wooden or metal plate used to connect structural members that are butted together; most often used in the construction of trusses. Many metal gusset plates are simply pressed into the wood and are subject to early failure if the plates warp from the heat of the fire. Gusset plates that are nailed or screwed to the members are much more reliable during a fire.

Gypsum Board — Interior finish material consisting of calcinated gypsum, starch, water, and other additives sandwiched between two sheets of specially treated paper; see drywall.

Header Course — Course of bricks laid with the ends facing outward. Because the ends of the bricks are smaller than the sides, a header course is easy to identify. Header courses are only used in unreinforced masonry, and this makes that type of construction easy to identify.

Hip Roof — A pitched roof in which the ends are all beveled so that there are no gable walls. A common roof style on many newer residences. Unlike gable roofs, in hip roofs the attic vents are only under the eaves or on the roof.

HVAC — Abbreviation for heating, ventilating, and air conditioning. Unless properly protected with automatic fire dampers, the ductwork associated with these systems can allow smoke and fire to spread throughout a building.

Interstitial Space — An accessible or inaccessible space between layers of building materials; an attic or cockloft sometimes used to house HVAC and other machinery. Like attics and cocklofts, unless properly protected, these spaces can allow fire to burn undetected or to spread throughout a building.

Joists — Horizontal structural members used to support a ceiling or floor. Drywall materials are nailed or screwed to the ceiling joists, and the subfloor is nailed or screwed to the floor joists.

Lamella Arch — An arch constructed of short wooden members connected in a specific geometric pattern. While rare in modern construction, these roof assemblies can still be found in many older buildings.

Mansard Roof — A roof characterized by steeply sloped facets surrounding a flat or nearly flat center section. Many remodeled buildings have false mansard roofs that consist of a fascia added to an existing flat roof. In some cases, the fascia forms a concealed space in which fire can burn undetected.

Common Construction Terminology (continued)

Mortar — A mixture of sand, cement, and water used to bond masonry units into a solid mass. The joints between bricks are filled with mortar. Mortar joints are sometimes the easiest to penetrate when a masonry wall must be breached.

Open Web Joist — A joist constructed with a web composed of materials such as bars or tubes that do not fill the entire web space. These are very common building assemblies because of their strength compared to their cost. When exposed to fire they lose their strength quickly.

Oriented Strand Board (OSB) — A wooden structural panel formed by gluing and compressing wood strands together under pressure. This material has replaced plywood and planking in the majority of construction applications. Roof decks, walls, and subfloors are all commonly made of OSB.

Parallel Chord Truss — A truss constructed with the top and bottom chords parallel. These trusses are used as floor joists in multistory buildings and as ceiling joists in buildings with flat roofs.

Parapet — A wall at the edge of some roofs. Most parapet walls range from a few inches (millimeters) to a few feet (meters) in height, but they can be high enough to require a ladder to reach the roof from the top of the wall. At night, parapet walls can be significant trip hazards for firefighters on the roof.

Party Wall — A wall shared by two adjoining buildings; usually a load-bearing wall that is also a fire wall. The failure of a roof assembly attached to a party wall can affect the structural integrity of the adjoining building.

Pitch — The ratio of rise-to-span of a roof assembly. The steeper the pitch, the greater the slip hazard unless roof ladders are used.

Pitched Roof — A roof that is sloped (pitched) to facilitate runoff. Pitched roofs range from those that are flat to those that are extremely steep, such as are common on some churches.

Plate — The top or bottom horizontal member of a frame wall. The sole plate is nailed or screwed to the subfloor, and the top plate is what the roof assembly rests on.

Platform Construction — Frame-type construction in which each floor interrupts the exterior studs forming an effective fire-stop at every floor. This is the most common type of construction used to frame modern residences and other small buildings.

Plywood — A wooden structural panel formed by gluing and laminating very thin sheets of wood together

under pressure. Plywood is still used in some applications but has been replaced in construction by OSB.

Rafters — Beams that span from a ridge board to an exterior wall plate to support roof decking. While it is important to cut away roof decking during vertical ventilation operations, cutting rafters can seriously weaken a roof and should be avoided whenever possible.

Rated Assembly — Two or more construction components combined to form an assembly that has a specific fire-resistance rating. A fire door is an example of a rated assembly as well as a wood-frame wall covered with a specified thickness of gypsum drywall.

Rebar — Short for reinforcing bar. These steel bars are placed in concrete forms before the cement is poured. When the concrete sets (hardens) the rebar within it adds considerable strength.

Reinforced Concrete — Concrete that has been poured into forms that contain an interconnected network of steel rebar.

Sawtooth Roof — A roof with a profile of vertical and sloping surfaces that resemble a saw blade. These roofs are common on older industrial buildings, but many are still in existence. The vertical walls in these roofs usually include many windows to allow light in. In some cases, the windows can be opened to provide natural ventilation. Removing these windows can sometimes provide adequate ventilation during a fire.

Sheathing — Plywood, OSB, or wooden planking (sometimes called *sheeting*) applied to a wall or roof over which a weather-resistant covering is applied. Most sheathing is relatively easy to penetrate for forcible entry or ventilation.

Shed Roof — A pitched roof that slopes in one direction only from the ridge.

Spalling — Degradation of concrete due to prolonged exposure to high heat. Water trapped within the concrete is vaporized by the heat and expands, causing the concrete to break apart. While spalling concrete can sound like gun fire, in most cases it is relatively harmless to firefighters.

Stud — A vertical structural member in a frame wall. Stud walls are the assemblies to which wall coverings are nailed or screwed. Studs can be made of either wood or light-gauge steel.

Tension — Force that tends to pull the mass of a material apart. Tension is what causes some roof assemblies to pull away from walls and fall inward.

Truss — A wooden or metal structural unit made up of one or more triangles in a flat plane. Because of the inherent strength of the triangles within its structure, when a truss is intact it is much stronger than the individual members of which it is made.

Common Building Materials

All materials react differently when exposed to the heat of a fire. Knowledge of how these materials react gives fire suppression personnel an idea of what to expect during fire fighting operations in buildings of a particular type of construction. This part of the chapter reviews the common materials found in building construction and explains how they react to fire involvement.

Wood

Wood is the most common building material used in North America and is the main component of a variety of structural assemblies. It may be used in *load-bearing walls* (those that support the weight of structural components above) or *nonload-bearing walls* (those that support only their own weight). Interior and exterior walls that support trusses and other roof systems are load-bearing walls. A *party wall* is a load-bearing wall shared by two adjacent structures **(Figure 4.1)**. A *partition wall* that simply divides two areas within a structure is an example of a nonload-bearing wall **(Figure 4.2)**. Some interior walls may also be load bearing, although this is often difficult to tell by just looking at them. This determination should be made during preincident planning surveys.

Load-Bearing Wall — Wall that is used for structural support.

Nonload-Bearing Wall — Wall, usually interior, that supports only its own weight.

Partition Wall — Interior non-load bearing wall that separates a space into rooms.

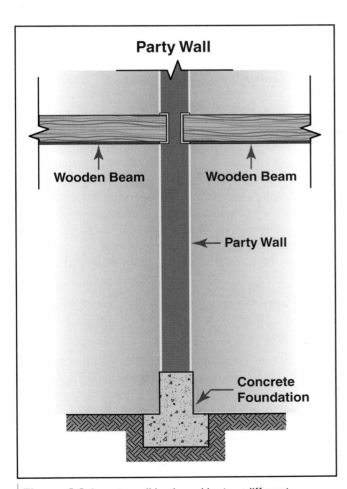

Figure 4.1 A party wall is shared by two different structures.

Figure 4.2 Interior partition walls may or may not be load-bearing.

The reaction of wood to fire conditions depends mainly on two factors: the size of the wood and its moisture content. The smaller the dimensions of the wood, the easier it is to ignite and the faster it will lose structural integrity. Large wooden beams, such as those used in heavy-timber construction, are difficult to ignite and retain their structural integrity even after prolonged exposure to direct flame impingement. Lumber of smaller dimensions needs to be protected by gypsum drywall or other insulation to increase its resistance to heat or fire.

The moisture content of the wood affects the rate at which it burns. Wood with a high moisture content (sometimes referred to as *green* wood) does not ignite as readily nor burn as fast as wood that has been kiln dried or dehydrated by exposure to air over a long period of time. In some cases, wood is pressure treated with fire retardants to reduce the speed at which it ignites or burns. However, fire retardants are not always totally effective in reducing fire spread.

Water used during extinguishing operations does not have a substantial negative effect on the structural strength of wood construction materials. Applying water to burning wood minimizes damage by stopping the charring process, which reduces wood's strength. Firefighters should check wood studs and structural members for charring to ascertain their structural integrity.

Newer construction often contains composite building components and materials that are made of wood fibers, plastics, and other substances joined by glue or resin binders. Such materials include plywood, particleboard, fiberboard, and paneling. Some of these products may be highly combustible, can produce significant toxic gases, or can rapidly deteriorate under fire conditions.

Masonry

Masonry includes bricks, stones, and concrete blocks **(Figure 4.3)**. Because masonry materials do not burn, a variety of masonry walls are used in the construction of *fire walls*. Fire walls are intended to provide separation that meets the requirements of a specified fire-resistance rating. Fire wall assemblies include the wall structure, doors, windows, and any other protected openings meeting the required protection-rating criteria. Fire walls may be used to separate two adjoining structures or two occupancies within the same structure to prevent the spread of fire from one to the other. Fire wall assemblies can also divide large structures into smaller portions and contain a fire to a particular portion of the structure. *Cantilever walls* extend beyond the structure that supports them **(Figure 4.4)**.

Concrete block walls may be load-bearing walls; however, many brick and stone walls are *veneer walls*, which are decorative and usually attached to the exterior surface of some type of load-bearing frame structure.

Masonry is minimally affected by fire and exposure to high temperatures. Bricks rarely show any signs of loss of integrity or serious deterioration. Stones may spall or lose small portions of their surface when heated. Concrete blocks may crack, but they usually retain most of their strength and basic structural stability. The mortar between the bricks, blocks, and stone may be degraded by heat and should be checked for signs of weakening **(Figure 4.5)**.

Rapid cooling, which can occur when water is used to extinguish a fire, may cause masonry materials to crack. This is a common problem when water is used to extinguish chimney flue fires. Masonry materials should be inspected for signs of this damage after extinguishment has been completed.

Green Wood – Wood with high moisture content.

Fire Wall — Fire-rated wall with a specified degree of fire resistance, built of fire-resistive materials and usually extending from the foundation up to and through the roof of a building, that is designed to limit the spread of a fire within a structure or between adjacent structures.

Masonry — Bricks, blocks, stones, and unreinforced and reinforced concrete products.

Cantilever Walls – Walls that extend beyond the structure that supports them.

Veneer Walls — Walls with a surface layer of attractive material laid over a base of common material

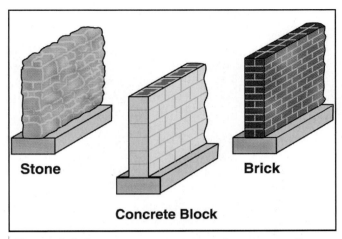
Figure 4.3 The most common types of masonry walls.

Figure 4.4 One type of cantilever wall.

Figure 4.5 A masonry wall with deteriorated mortar, indicated by the cracks.

Cast Iron

Cast iron is rarely used in modern construction; it typically is found only on older buildings **(Figure 4.6, p. 144)**. It was commonly used as an exterior surface covering (veneer wall). These large sections were fastened to the masonry on the front of the building. The cast iron stands up well to fire and intense heat, but it may crack or shatter when rapidly cooled with water. A primary concern from a fire fighting standpoint is that the bolts or other connections that hold the cast iron to the building can fail, causing these large heavy sections of metal to come crashing down.

Steel

Steel is the primary material used for structural support in the construction of large modern buildings **(Figure 4.7, p. 144)**. Steel structural members elongate when heated. A 50-foot (15 m) beam may elongate by as much as 4 inches (100 mm) when heated from room temperature to about 1,000°F (538°C). If the steel is restrained from movement at the ends, it buckles and fails somewhere in the middle. For all purposes, the failure of steel structural members can be anticipated at temperatures near or above 1,000°F (538°C). The temperature at which a specific steel member fails

Figure 4.6 Cast iron fascia typical of buildings constructed at the turn of the last century and found in many older cities.

Figure 4.7 A steel frame building under construction.

depends on many variables, including the size of the member, the load it is under, the composition of the steel, and the geometry of the member. For example, a lightweight, open-web truss will fail much quicker than a large, heavy I-beam.

From a fire fighting perspective, firefighters must be aware of the type of steel members used in a particular structure. Firefighters also need to determine how long the steel members have been exposed to heat; this gives an indication of when the members might fail. Another possibility for firefighters to consider is that elongating steel can actually push out load-bearing walls and cause a collapse (**Figure 4.8**). Water can cool steel structural members and stop elongation, which reduces the risk of a structural collapse.

Reinforced Concrete

Reinforced concrete is concrete that is internally fortified with steel reinforcement bars (rebar) or wire mesh (**Figure 4.9**). This gives the material the compressive strength of concrete along with the tensile strength of steel. While reinforced concrete does perform well under fire conditions, it can lose strength through spalling. Prolonged heating can cause a failure of the bond between the concrete and the steel reinforcement. Firefighters should look for cracks and spalling of reinforced concrete surfaces. This is an indication that damage has occurred and that strength may be reduced.

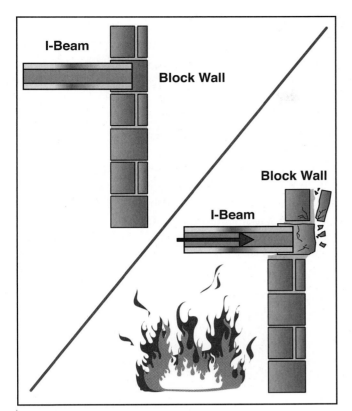

Figure 4.8 Expanding steel members can cause structural failure.

Figure 4.9 Exposed rebar in a concrete wall.

Figure 4.10 Wallboard is made of gypsum, a fire-resistant and fire-retardant material. The wallboard covers the wall studs, fiberglass insulation, and electrical, gas, and water services.

Gypsum

Gypsum is an inorganic product from which plaster and wallboards are constructed (**Figure 4.10**). It is unique because it has high water content, and the evaporation of this water absorbs a great deal of heat. The water content gives gypsum excellent heat-resistant and fire-retardant properties. Gypsum is commonly used to provide insulation to steel and wood structural members that are less adapted to high heat situations because it breaks down gradually under fire conditions. In areas where the gypsum has failed, the structural members behind it will be subjected to higher temperatures and could fail as a result.

Glass/Fiberglass

Glass is not typically used for structural support; it is used in sheet form for doors and windows (**Figure 4.11**). Wire-reinforced glass may provide some thermal protection as a separation, but for the most part conventional glass is not an effective barrier to fire extension. Heated glass may crack and shatter when it is struck by a cold fire stream.

Fiberglass is typically used for insulation purposes. The glass component of fiberglass is not a significant fuel, but the materials used to bind the fiberglass may be combustible and can be difficult to extinguish.

Figure 4.11 Many buildings, including modern high-rise structures, have glass exteriors.

Construction Classifications

Historically, a number of building codes were used throughout North America, but several of these codes have been combined. The most prominent codes used now are those published by the International Codes Council (ICC)®, NFPA® 5000, *Building Construction and Safety Code*® published by the National Fire Protection Association, and the National Building Code of Canada. In general, construction classifications are based on the types of materials used in the construction and on the fire-resistance ratings of major structural components.

It is important to remember that fire resistance ratings are a measure of how long structural assemblies will maintain their load-bearing ability under fire conditions, *not* of how easy or difficult it will be to fight a fire in that building. Most building codes have the same five construction classifications that are described in NFPA® 220, *Standard on Types of Building Construction*, but may use slightly different terms to name the classifications. The five types of building construction listed in NFPA® 220 include:

- Type I (Fire-resistive)
- Type II (Noncombustible)
- Type III (Ordinary)
- Type IV (Heavy timber)
- Type V (Wood frame)

Type I Construction

Also known as *fire-resistive construction*, Type-I construction maintains its structural integrity during a fire, and that is intended to allow occupants time to exit the building. Fire-resistive construction consists mainly of reinforced concrete with structural members, including walls, columns, beams, floors, and roofs, that are protected either by blown-on insulation or automatic sprinklers **(Figure 4.12)**. The fire-resistive compartmentation provided by partitions and floors tends to retard the spread of fire through the building. These features allow time for occupants to exit the building and firefighters to conduct interior fire fighting.

Fire-Resistive Construction — Another term for Type I construction; construction that maintains its structural integrity during a fire.

Because of the limited combustibility of the materials of construction, the primary fire hazards are the contents of the structure and the interior finishes. In a fire-resistive structure, firefighters are able to launch an interior attack with greater confidence than in a building that is not fire resistive. The ability of fire-resistive construction to confine the fire to a certain area can be compromised by openings made in partitions and by improperly designed and inadequately dampered heating and air-conditioning systems.

Strengths

- Resists direct flame impingement
- Confines fire well
- Little collapse potential from the effects of fire alone
- Impervious to water damage

Weaknesses

- Difficult to breach for access or escape
- Difficult to ventilate during a fire
- Massive debris following collapse
- Floors, ceilings, and walls retain heat

Type II Construction

Type II construction, also known as noncombustible construction, is made of the same materials as fire-resistive construction except that the structural components lack the insulation or other protection of Type I construction. Type II construction has a fire-resistance rating on all parts of the structure including exterior and interior load-bearing walls and building materials. All-metal buildings also fall into this classification **(Figure 4.13)**. Materials with no fire-resistance ratings, such as untreated wood, may be used only in limited quantities. Again, one of the primary fire protection concerns is the contents of the building. The heat buildup from a fire in the building can cause structural supports to fail. Another potential problem is the type of roof on the building.

Noncombustible or limited combustible construction buildings often have flat, built-up roofs. These roofs consist of a combustible or noncombustible roof deck covered by combustible felt, noncombustible insulation, and roofing tar **(Figure 4.14)**. Fire extension to the roof can eventually cause the entire roof to become involved and fail.

Noncombustible Construction — Another term for Type II construction; construction made of the same materials as fire-resistive construction except that the structural components lack the insulation or other protection of Type I construction

Strengths

- Almost as resistive to fire as Type I construction
- Confines fire well
- Almost as structurally stable as Type I construction
- Easier to vertically ventilate than Type I construction

Weaknesses

- Difficult to breach from access or escape
- Unprotected steel structural components can fail due to heat
- Roof systems less stable than Type I construction
- Steel components subject to weakening by fire
- Steel components subject to weakening by rust and corrosion
- Massive debris following collapse

Figure 4.12 This parking structure is a Type I building.

Figure 4.13 All-metal buildings are Type II construction.

Figure 4.14 Flat roofs are prone to failure if they are weakened by fire.

Type III Construction

Also known as *ordinary construction*, Type III construction requires that exterior walls and structural members be made of noncombustible or limited combustible materials such as concrete blocks or clay tile blocks. Interior structural members, including walls, columns, beams, floors, and roofs, are completely or partially made of wood **(Figure 4.15)**. The wood used in these members is of smaller dimensions than that required for heavy-timber construction. See the Type IV (Heavy Timber) Construction section that follows.

The primary fire concern specific to ordinary construction is the problem of fire and smoke spreading through concealed spaces. These spaces are between the walls, floors, and ceiling. Heat from a fire may be conducted to these concealed spaces through finish materials, such as gypsum drywall or plaster, or the heat can enter the concealed spaces through holes in the finish materials. From there, the heat, smoke, and gases may spread to other parts of the structure. If enough heat is present, the fire may actually burn within the concealed spaces and feed on the combustible construction materials in the space. These hazards can be reduced considerably by placing fire-stops inside these spaces to limit the spread of the combustion byproducts (heat, smoke, etc.).

Strengths

- Resists fire spread from the outside
- Relatively easy to vertically ventilate

Weaknesses

- Interior structural members vulnerable to fire involvement
- Fire spread potential through concealed spaces
- Susceptible to water damage

Type IV Construction

Also known as *heavy-timber construction*, Type IV construction requires that exterior and interior walls and their associated structural members be made of noncombustible or limited combustible materials. Other interior structural members, including beams, columns, arches, floors, and roofs, are made of solid or laminated wood with no concealed spaces **(Figure 4.16)**. This wood must have dimensions large enough to be considered heavy timber. These dimensions vary depending on the particular code being used.

Heavy-timber construction was used extensively in old factories, mills, and warehouses. Traditional heavy-timber construction is rarely used today in new construction except for decorative reasons. The use of heavy-timber construction with glue-lam beams is growing.

The primary fire hazard associated with heavy-timber construction is the massive amount of combustible contents presented by the structural timbers in addition to the contents of the building. Although the heavy timbers remain stable for a long period under fire conditions, they give off tremendous amounts of heat and pose serious exposure protection problems for firefighters.

Figure 4.16 Type IV construction uses heavy timbers.

Figure 4.15 A typical Type III building.

Strengths

- Resists collapse due to flame impingement of heavy beams
- Structurally stable
- Relatively easy to vertically or horizontally ventilate
- Relatively easy to breach for access or escape
- Manageable debris following collapse

Weaknesses

- Susceptible to fire spread from outside
- Potential for flame spread to other nearby structures
- Susceptible to rapid interior flame spread
- Susceptible to water damage

Type V Construction

Also known as *wood-frame construction*, Type V construction has exterior walls, bearing walls, floors, roofs, and supports made completely or partially of wood or other approved materials of smaller dimensions than those used for heavy-timber construction. Wood-frame construction is the type commonly used to construct the typical single-family residence or apartment house of up to seven stories. This type of construction presents almost unlimited potential for fire extension within the building of origin and to nearby structures, particularly if the nearby structures are also wood-frame construction **(Figure 4.17, p. 150)**. Firefighters must be alert for fire coming from doors or windows extending to the exterior of the structure.

Wood-Frame Construction — Another term for Type V construction; construction that has exterior walls, bearing walls, floors, roofs, and supports made completely or partially of wood or other approved materials of smaller dimensions than those used for heavy-timber construction.

Strengths

- Easily breached for access, ventilation, or escape
- Resistant to collapse from earthquakes due to light weight and flexibility
- Collapse debris relatively easy to manage

Weaknesses

- Susceptible to fire spread from outside
- Susceptible to rapid flame spread inside
- Susceptible to total collapse due to fire or explosion
- Susceptible to water damage

Figure 4.17 Wood-frame construction is Type V.

Figure 4.18 Walls constructed of hay bales have recessed windows and doors.

Nonstandard Construction

In many parts of North America, local building codes allow nonstandard buildings to be constructed under certain circumstances. These structures do not conform to any of the standard construction types listed in NFPA® 220. Some nonstandard structures are allowed to be built on large properties that are in very remote areas. Others are simply new construction concepts that are not yet recognized by national building codes. Even these innovative structures must conform to local zoning and land-use standards. One example of nonstandard construction that is allowed in some jurisdictions are homes in which the exterior walls are made of stacked hay bales (**Figure 4.18**). The weather side of the bales is sealed with stucco or similar materials. The interior walls are conventional wood-frame construction covered with gypsum drywall. The interior walls support the roof structure.

Other forms of nonstandard construction are allowed in certain jurisdictions. To protect yourself and your fellow firefighters, you must be aware of what types of structures are being built in your area of responsibility. In other words, you must pay attention to what is going on in your response district and make frequent pre-incident planning surveys of construction sites.

Firefighter Hazards Related to Building Construction

The reason that you and other firefighters need to have some knowledge of building construction and materials is to be able to relate that information to the fireground. You can use this knowledge to anticipate possible avenues of fire spread and to monitor building conditions for signs of structural instability. In a fire, immediately report any unusual building conditions that you observe to your supervisor or to command. Even though a safety officer may have been designated for the incident, it is the obligation of all personnel to constantly remain alert to their immediate surroundings. That is, they must maintain their *situational awareness* and be alert for unsafe conditions. Remember and practice what wildland firefighters call "look up, look down, look around." The following sections highlight some of the potential hazards related to building construction that affect firefighter safety.

Situational Awareness – Awareness of immediate surroundings.

Reading Existing Construction

The outward appearance of many modern structures can be deceiving because exterior finishes make these buildings appear to be more substantial than they really are. For example, some wood-frame structures have a veneer of polystyrene foam shaped to look like stone, masonry, or concrete over which a stucco finish is applied that makes them look like Type I buildings **(Figure 4.19)**. When reading (sizing up) a building, look for the following:

- Age of the building. Are there signs of weathering or deterioration?

- Construction materials. Is it a wood-frame, unreinforced masonry, all-metal, or concrete building?

- Roof type. Does it have an arched or lightweight roof?

- Renovations or modifications. Have rooms or entire sections been added? Have facades or false ceilings been added?

- Dead loads. Are there HVAC units, water tanks, or other heavy objects on the roof?

There are other items that are important to observe during size-up, such as the occupancy type, adjacent exposures, and fire conditions. These are not directly related to the hazards associated with building construction. For more information on these and other size-up considerations, see Chapter 15, Fire Control.

Figure 4.19 Exterior finishes can make a building appear to be constructed of more fire-resistive materials than it really is.

Dangerous Building Conditions

As a firefighter, you must be aware of the dangerous conditions created by a fire as well as dangerous conditions that may be created while trying to extinguish a fire. An already serious situation can be made much worse if firefighters fail to recognize the potential of the situation and take the wrong actions.

There are two primary types of dangerous conditions that may be posed by a particular building:

- Conditions that contribute to the spread and intensity of the fire
- Conditions that make the building susceptible to collapse

These two conditions are obviously related — conditions that contribute to the spread and intensity of the fire increase the likelihood of structural collapse. The following sections describe some of these conditions.

Fire Loading

Heavy Fire Loading — Presence of large amounts of combustible materials in an area or a building.

Fire load is the maximum heat that can be produced if all the combustible materials in a given area burn. Heavy *fire loading* is the presence of large amounts of combustible materials in an area or a building. The arrangement of materials in a building directly affects fire development and severity and must be considered when determining the possible duration and intensity of a fire.

Heavy content fire loading is perhaps one of the most critical hazards in commercial and storage facilities because the fire can overwhelm the capabilities of a fire sprinkler system (if present) and make it difficult for firefighters to gain access during fire suppression operations **(Figure 4.20)**. Proper inspection and code enforcement prior to an incident is the most effective defense against these hazards.

Combustible Furnishings and Finishes

Combustible furnishings and finishes contribute to fire spread and smoke production **(Figure 4.21)**. These two elements have been identified as major factors in the loss of many lives in fires.

Figure 4.20 Some buildings are likely to have a heavy fire load.

Roof Coverings

Roof coverings are the final outside layer that is placed on top of a roof deck. Common roof coverings include wooden and composition shingles, wooden shakes, rubber imitation tile, steel imitation shakes or tile, clay tile, slate, tin, and tar-and-gravel. The combustibility of a roof's surface is a basic concern to the fire safety of an entire community. Some of the earliest fire regulations ever imposed in North America related to combustible roof coverings because they were blamed for several conflagrations caused by flaming embers flying from roof to roof.

History has shown that wood shakes in particular, even when treated with fire retardant, can significantly contribute to fire spread. This is a particular problem in the wildland/urban interface fire where wood shake roofs have contributed to large fires **(Figure 4.22)**. Firefighters must use aggressive exposure protection tactics when faced with this type of fire.

Wooden Floors and Ceilings

Combustible structural components such as wood framing, floors, and ceilings also contribute to the fire loading in a building. Prolonged exposure to fire may weaken them and increase the chances of collapse.

Large, Open Spaces

Large, open spaces in buildings contribute to the spread of fire throughout. Such spaces may be found in warehouses, churches, large atriums, common attics or cocklofts, and theaters **(Figure 4.23)**. In these facilities, proper vertical ventilation (channeling smoke from a building at its highest point) is essential for slowing the spread of fire (see Chapter 11, Ventilation).

Building Collapse

Many firefighters have been seriously injured or killed by a structural collapse during interior fire fighting operations. The most infamous example was the collapse of the twin towers of the World Trade Center on September 11, 2001. This tragic incident was a classic example of collapse caused by fire weakening the structure.

Figure 4.21 Some furnishings and finishes can increase fire spread.

Figure 4.22 Combustible roof coverings allow fire to spread from one structure to another.

Figure 4.23 Some buildings, such as churches, have large open spaces that promote fire spread.

In virtually every case, structural collapse results from damage to the structural system of the building caused by the fire or by fire fighting operations. Knowledge of the types of construction and the ability to recognize them is important to your survival as a firefighter.

Some buildings, because of their construction and age, are more likely to collapse than others. For example, buildings of lightweight or truss construction will succumb to the effects of fire much quicker than a fire-resistive or heavy-timber building. Older buildings that have been exposed to weather and that have been poorly maintained are more likely to collapse than newer, well-maintained buildings **(Figure 4.24)**. Wooden structural components in older buildings may also dehydrate to the point that their ignition temperature decreases and their flame spread characteristics increase. Information on building age and construction type should be obtained when conducting inspections and documented in preincident plans.

Even the exterior of a burning building can collapse on firefighters. The walls of buildings, especially curtain walls, false fronts, marquees, parapet walls, and heavy signs can all come crashing down **(Figure 4.25)**. Other devices such as exterior fire escapes or trash chutes can also separate from the building and fall to the street. Although firefighters have no choice but to pass through the *collapse zone* (area extending horizontally from the base of the wall to one and one-half times the height of the wall) when entering or leaving the building, they should spend as little time there as possible **(Figure 4.26)**.

The longer a fire burns in a building, the more likely that the building will collapse. Fire weakens the structural support system until it becomes incapable of holding the weight of the structure above. The time it takes for this to happen varies with the fire severity, the type of construction, the presence or absence of protective insulation on structural members, the presence or absence of heavy industrial machinery on upper

Collapse Zone – The area extending horizontally from the base of the wall to one and one-half times the height of the wall.

Figure 4.25 Unsupported parapet walls have a high collapse potential.

Figure 4.24 Age, weather, and neglect can cause older buildings, even those advertised as fireproof, to collapse under fire conditions.

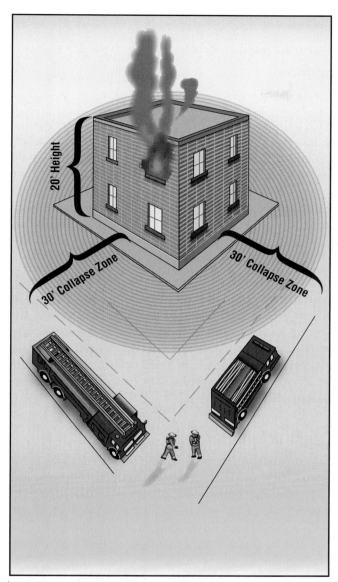

Figure 4.26 Firefighters must take extra caution when working within a potential collapse zone.

Figure 4.27 Obvious cracks in bearing walls are indicators of collapse potential.

Figure 4.28 Reinforcement stars may mean that the structure is of questionable stability. The stars are connected by tension rods to stars located on the opposite side of the buildings.

floors or on the roof, and the general condition of the building. You and your fellow firefighters should be aware of the following indicators of building collapse and be looking for them at every fire:

- Cracks or separations in walls, floors, ceilings, and roof structures (**Figure 4.27**)
- Evidence of existing structural instability such as the presence of tie rods and stars that hold walls together (**Figure 4.28**)
- Loose bricks, blocks, or stones falling from buildings
- Deteriorated mortar between the masonry
- Walls that appear to be leaning
- Structural members that appear to be distorted

- Fires beneath floors that support heavy machinery or other extreme weight loads
- Prolonged fire exposure to the structural members (especially trusses)
- Unusual creaks and cracking noises
- Structural members pulling away from walls
- Excessive weight of building contents

Fire fighting operations also increase the risk of building collapse. Improper vertical ventilation techniques can result in structural supports being cut that could weaken the structure. The water used to extinguish a fire adds extra weight to the structure and can weaken it. Water only a few inches (millimeters) deep over a large area can add many tons (tonnes) of weight to an already-weakened structure. For example, if a building that measures 50 × 100 feet (15 m by 30 m) has an accumulation of water 6 inches (150 mm) deep inside it, the structure must support an additional 78 tons (71 tonnes) of weight.

Immediate action must be taken if firefighters suspect that the collapse of a building is imminent or even likely. First, as you and the rest of your crew are exiting the building, Command and all others inside the building should be informed of the situation. Second, clear the collapse zone as soon as possible. No personnel or apparatus should be allowed to operate in the collapse zone except to cautiously place unmanned master stream devices and then immediately withdraw once they are in operation (see Chapter 15, Fire Control). Know and heed any evacuation or other emergency signals used by your department.

Lightweight and Truss Construction Hazards

One of the most serious building construction hazards facing firefighters today is the increased use of lightweight construction and trussed support systems (**Figure 4.29**). Lightweight construction is most commonly found in homes, apartments, small commercial buildings, and warehouses. Two of the most common types of lightweight construction involve the use of lightweight steel or wooden trusses. *Lightweight steel trusses* are made from long steel chords that are either straight or bent by as much as 90 degrees with either flat or tubular members in the web space (**Figure 4.30**). *Lightweight wooden trusses* are constructed of 2×3-, 2×4-, or 2×6- inch (50 mm by 75 mm, 50 mm by 100 mm, or 50 mm by 150 mm) lumber connected by gusset plates. *Gusset plates* may be either wooden or metal. Metal gusset plates, also called *gang nails*, are small metal plates (usually 18 to 22 gauge metal) with points or prongs that penetrate about ⅜-inch (10 mm) into the wood (**Figure 4.31**). Some jurisdictions require metal gusset plates to be corner-nailed to reduce the chances of the plates warping and pulling out during a fire.

Experience has shown that unprotected lightweight steel and wooden trusses can fail after 5 to 10 minutes of exposure to fire. These trusses can fail from exposure to heat alone without any flames. For steel trusses, 1,000°F (538°C) is the critical temperature. Unless they are corner-nailed, metal gusset plates in wooden trusses can warp and fail quickly when exposed to heat. Although both steel and wooden trusses may be protected with fire-retardant treatments to enhance their fire resistance, most lack this protection.

Wooden I-beams are also used in lightweight construction. They have fire characteristics similar to wooden trusses and similar precautions should be used when they are found in a structure (**Figure 4.32**).

Lightweight Steel Truss — Structural support made from a long steel bar that is bent at a 90-degree angle with flat or angular pieces welded to the top and bottom.

Lightweight Wood Truss — Structural supports constructed of 2- x 3-inch or 2- x 4-inch (50 mm by 75 mm or 50 mm by 100 mm) members that are connected by gusset plates.

Gusset Plates — Metal or wooden plates used to connect and strengthen the intersections of metal or wooden truss components roof or floor components into a load-bearing unit.

Gang Nail — Form of gusset plate. These thin steel plates are punched with acutely V-shaped holes that form sharp prongs on one side that penetrate wooden members to fasten them together.

Figure 4.29 Unprotected steel trusses fail quickly when exposed to high heat.

Figure 4.30 A typical metal bar joist used to support a roof deck.

Figure 4.31 Metal gusset plates connect wooden truss members where they intersect.

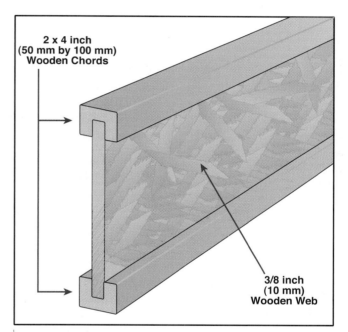

**2 x 4 inch
(50 mm by 100 mm)
Wooden Chords**

**3/8 inch
(10 mm)
Wooden Web**

Figure 4.32 Components of a typical wooden I-beam.

In 1988, a fire in a New Jersey automobile dealership cost the lives of five firefighters. A fire between the ceiling and the roof involved the 78-foot (26 m) long wooden bowstring trusses, which collapsed approximately 20 minutes after the fire crews started their interior attack. The roof collapsed on six firefighters, one of whom was able to escape. The others died of burn injuries.

Bowstring trusses are found in a great many older buildings. They are used in buildings that require large open spaces without supporting columns such as automobile dealerships, bowling alleys, factories, and supermarkets. Bowstring trusses should be suspected in any building with a rounded (arched) roof **(Figure 4.33, p. 158)**. Firefighters should be aware, however, that bowstring trusses are used in arched roofs but not all arched roofs contain bowstring trusses. Some bowstring trussed roofs cannot be seen from street level because of parapet walls or false fronts that have been added to the buildings. Therefore, the presence of bowstring trusses is best identified during preincident planning surveys.

All trusses are made up of one or more triangles (the strongest geometric shape known) and are designed to work as an integral unit. Some members are in *tension* (stresses that tend to pull things apart), and others are in *compression* (stresses that tend to press things together). One characteristic common to all types of trusses is that if one member fails, the entire truss is likely to fail. When a truss fails, the trusses next to it are likely to fail, and the domino effect can produce a total collapse almost instantaneously.

It is important that firefighters know which buildings in their district have truss roofs or floors. In many departments, firefighters are not allowed to enter or go onto the roofs of buildings that incorporate trusses in their construction if the buildings have been exposed to fire conditions for 5 to 10 minutes — which is often how long they have been exposed before the fire department arrives.

Construction, Renovation, and Demolition Hazards

For a variety of reasons, the risk of fire rises sharply when construction, renovation, or demolition is being performed in a structure. Contributing factors are the additional fire loads and ignition sources (such as open flames from cutting torches and sparks from grinding or welding operations) brought by building contractors and their associated equipment.

Buildings under construction are subject to rapid fire spread when they are partially completed because many of the protective features such as gypsum wallboard and automatic sprinklers are not yet in place **(Figure 4.34)**. Some firefighters think of buildings under construction with exposed wooden framing as the equivalent of a vertical lumberyard. The lack of doors or other barriers that would normally slow fire spread are also contributing factors to rapid fire growth.

Abandoned buildings or buildings that are being renovated or demolished are also subject to faster-than-normal fire growth. Breached walls, open stairwells, missing doors, and deactivated fire protection systems are all potential contributors. The potential for a sudden building collapse during fires in these buildings is also a serious consideration. Arson is also a factor at construction or demolition sites because of easy access into the building.

Due to the cost of new construction, renovating old buildings is becoming more common in many areas. Hazardous situations may arise during renovation because occupants and their belongings may remain in one part of the building while work continues in another. Fire detection or alarm systems may be taken out of service or damaged during renovation. If good housekeeping is not maintained, accumulations of debris and construction materials can block exits. This may impede occupants from escaping from the building in an emergency. The contractors performing the renovations do not always follow local building codes.

Even when codes are followed, problems for firefighters can be created. For example, in some cases a second roof (called a *rain roof*) is constructed over an older leaky roof that is beyond repair. This and other types of renovations can leave concealed spaces that are not apparent from the outside. Obviously, if firefighters attempt to open a rain roof to ventilate a building, they will find a second roof some distance below. As mentioned earlier in this chapter, when contractors add a mansard fascia

Figure 4.33 Many buildings have potentially dangerous arched roofs supported by bowstring trusses made of wood or metal.

Figure 4.34 Buildings under construction are extremely vulnerable to fire.

to a building, another type of concealed space can be created. As in all concealed spaces, fire can burn undetected or spread through the space to uninvolved areas of the building.

Because of the risks inherent in buildings under construction, being renovated, or being demolished, fire officers must do a quick risk/benefit analysis before ordering firefighters inside to fight a fire. If one or more construction workers are known to be trapped inside, the amount of risk that is justified goes up considerably. If, on the other hand, everyone has escaped the building and the structure itself is the only thing at risk, fire officers may choose to fight the fire defensively — from the outside.

Summary

A failure to recognize the potential dangers of a particular type of construction and the effects that fire may have on it can be catastrophic for firefighters. For your safety and that of your fellow firefighters, you must have at least a basic knowledge of building construction. You need to know about construction materials, methods, and designs in general, and those that are used in your area in particular. Knowledge of the various types of building construction and how fires react in each type gives you and your officers information that is vital to planning a safe and effective fire attack.

You need to know common building construction terms, materials, and methods. You also need to know how various types of construction are classified, and how each type behaves in fires. Finally, you need to know the sights and sounds that indicate the possibility of structural collapse or other extraordinary events during interior fire fighting operations.

Fire Fighter I

1. What are common materials found in building construction?

2. What are the five types of building construction listed in NFPA 220?

3. What are the strengths and weaknesses of the five building construction types?

4. What actions should be taken when imminent building collapse is suspected?

5. What hazards exist with lightweight and truss construction?

Fire Fighter II

1. What items should be observed during size-up of a building?

2. What is fire load?

3. List four indicators of building collapse.

4. What hazardous situations may arise from renovated buildings?

5. Why are buildings under construction a hazard?

Chapter Contents

Courtesy of Central Florida Fire Academy.

Firefighter Personal Protective Equipment

This chapter provides information that addresses the following job performance requirements of NFPA® 1001, *Standard for Fire Fighter Professional Qualifications* (2008):

NFPA® 1001 references for Chapter 5:

5.1.1

5.1.2

5.3.1

5.3.5

5.3.9

5.5.1

Chapter Objectives

Fire Fighter I Objectives

1. Describe the purpose of protective clothing and equipment.
2. Describe characteristics of protective clothing and equipment.
3. Summarize guidelines for the care of personal protective clothing.
4. List the four common respiratory hazards associated with fires and other emergencies.
5. Distinguish among characteristics of respiratory hazards.
6. Describe physical, medical, and mental factors that affect the firefighter's ability to use respiratory protection effectively.
7. Describe equipment and air-supply limitations of SCBA.
8. Discuss effective air management.
9. Distinguish among characteristics of air-purifying respirators, open-circuit SCBA, and closed-circuit SCBA.
10. Describe basic SCBA component assemblies.
11. Discuss storing protective beathing apparatus.
12. Summarize recommendations for the use of PASS devices.
13. Describe precautionary safety checks for SCBA.
14. Discuss general donning and doffing considerations for SCBA.
15. Summarize general items to check for in daily, weekly, monthly, and annual SCBA inspections.
16. Summarize safety precautions for refilling SCBA cylinders.
17. Discuss safety precautions for SCBA use.
18. Describe actions to take in emergency situations using SCBA.
19. Discuss operating in areas of limited visibility while wearing SCBA.
20. Discuss exiting areas with restricted openings under emergency conditions while wearing SCBA.
21. Don PPE and SCBA for use at an emergency. (Skill Sheet 5-I-1)
22. Doff PPE and SCBA and prepare for reuse. (Skill Sheet 5-I-2)
23. Inspect PPE and SCBA for use at an emergency incident. (Skill Sheet 5-I-3)
24. Clean and sanitize PPE and SCBA. (Skill Sheet 5-I-4)
25. Fill an SCBA cylinder from a cascade system. (Skill Sheet 5-I-5)
26. Fill an SCBA cylinder from a compressor/purifier. (Skill Sheet 5-I-6)
27. Perform emergency operations procedures for an SCBA. (Skill Sheet 5-I-7)
28. Exit a constricted opening while wearing standard SCBA. (Skill Sheet 5-I-8)
29. Change an SCBA cylinder — one-person method. (Skill Sheet 5-I-9)
30. Change an SCBA cylinder — two-person method. (Skill Sheet 5-I-10)

Firefighter Personal Protective Equipment

Case History

In 2002, Mark Noble, a nineteen-year veteran of the Olympia (WA) Fire Department, was diagnosed with brain cancer. After surgery to remove the tumor, Mark began a regimen of chemotherapy and radiation. During his treatment, Mark started researching the connection between firefighters and cancer. What he found was that firefighters are exposed to highly toxic substances in virtually every fire — especially during overhaul. These substances include asbestos, benzene, polycyclic aromatic hydrocarbons (PAH), and polychlorinated biphenyls (PCB), in addition to carbon monoxide (CO) and other well-known products of combustion. The toxic effects can accumulate with repeated exposures.

In his research, Mark found that when compared to the general adult population, firefighters are twice as likely to develop intestinal cancer, liver cancer, prostate cancer, and non-Hodgkin's lymphoma; 2.25 times as likely to develop malignant melanoma and 3 times as likely to develop other skin cancers; 2.5 times as likely to develop testicular cancer; 3 times as likely to develop bladder cancer and leukemia; 3.5 times as likely to develop brain cancer; and 4 times as likely to develop kidney cancer.

In 2005, at age 47, Mark Noble lost his battle with the brain cancer that he almost certainly developed because of the toxins to which he was exposed as a firefighter. Mark loved being a firefighter, but he said that if he had it to do over, he would wear his SCBA more and he would be more conscientious about hooking up apparatus exhaust collection hoses.

Permission granted by Mrs. Rebecca Noble and ERGOMETRICS & Applied Personnel Research, Inc., who produced a video interview with Mark during his final months. The video is available at www.ergometrics.org.

Firefighters require the best personal protective clothing and equipment available because of the hostile environments in which they perform their duties **(Figure 5.1, p. 166)**. Fighting fires, performing rescues, and delivering other emergency services are inherently dangerous activities. Even when fire departments provide the best protective clothing and equipment and their firefighters use it properly, firefighter safety is not guaranteed. Nonetheless, many firefighter injuries and illnesses can be prevented or their severity reduced if protective clothing and equipment are used conscientiously.

This case history illustrates the critical importance of using personal protective clothing and SCBA in every fire, and especially during the overhaul phase. As you read this chapter and those that follow, remember this case history and the lessons it has to offer you. Remember also the key safety behavior in Chapter 2 that said, "Don't *ever* breathe smoke." These points may help to keep you alive and healthy throughout your fire service career and for the rest of your life.

Personal Protective Clothing — Garments firefighters must wear to protect themselves while fighting fires, performing rescues, and delivering emergency medical services.

Personal Protective Equipment — Includes self-contained breathing apparatus (SCBA) or other respiratory protection and personal alert safety systems (PASS devices).

As described in Chapter 2, personal protective clothing refers to the garments firefighters must wear to protect themselves while fighting fires, performing rescues, and delivering emergency medical services. Personal protective equipment is usually taken to include self-contained breathing apparatus (SCBA) or other respiratory protection and personal alert safety systems (PASS). The combination of personal protective clothing and equipment forms the complete personal protective ensemble (PPE). All of the protective clothing and equipment discussed in this chapter is required by NFPA® 1500, *Standard on Fire Department Occupational Safety and Health Program*.

According to NFPA® 1001, *Standard for Fire Fighter Professional Qualifications*, those qualified at the Fire Fighter I level must know the following about SCBA:

- Conditions that require respiratory protection
- Uses and limitations of SCBA
- Components of SCBA
- Donning procedures
- Breathing techniques
- Emergency procedures used with SCBA
- Physical requirements for using SCBA

Those qualified at the Fire Fighter I level must also be capable of:

- Donning SCBA within one minute
- Controlling breathing
- Exiting restrictive passages while wearing SCBA
- Initiating emergency procedures if SCBA fails or runs out of air
- Replacing SCBA cylinders

There are no PPE-related requirements at the Fire Fighter II level.

The first part of this chapter discusses structural protective clothing. The items discussed in this section include:

- Helmets
- Protective hoods
- Protective coat and trousers
- Gloves
- Safety shoes or boots

Figure 5.1 The fireground is a very hostile environment. *Courtesy of District Chief Chris E. Mickal, NOFD Photo Unit.*

Personal protective clothing for use in wildland fire fighting as well as station wear is also discussed. The care and cleaning of personal protective clothing is discussed in the final part of this section.

The chapter continues with a discussion of protective clothing that is not part of NFPA® 1971, *Standard on Protective Ensembles for Structural Fire Fighting and Proximity Fire Fighting*, but is equally important to overall firefighter safety. These items include:

- Eye protection
- Hearing protection
- Self-contained breathing apparatus (SCBA)
- Personal alert safety systems (PASS)

The second part of this chapter provides an overview of respiratory protection. Included is information on the different types of protective breathing apparatus and personal alert safety systems (PASS). Reasons why protective breathing equipment must be worn and the general procedures for donning, doffing, inspecting, and maintaining respiratory protection are discussed. Changing and refilling air cylinders are also covered. The last portion of the chapter covers safety precautions and the use of SCBA during emergency situations.

Personal Protective Clothing

NFPA® 1971 specifies the minimum design, performance, and certification requirements for the structural fire fighting ensemble, as well as the test methods that must be used by manufacturers to demonstrate compliance. When operating at any emergency scene, firefighters must wear protective clothing and equipment suitable to that incident (**Figure 5.2**).

Figure 5.2 Firefighters must wear the most appropriate PPE for the hazard they face. *Courtesy of Dave Ricci, Santa Rosa (CA) Fire Department.*

Traditionally called *turnouts* or *bunker gear,* personal protective clothing is designed to inhibit the transfer of heat from the outside environment to your body. One limitation to this design is that turnouts also inhibit the transfer of heat *away* from your body. Body heat and moisture generated during work get trapped, creating a hot humid microclimate within the turnout clothing. Turnouts can significantly increase physiological stress during work by limiting the cooling effect of sweating, and thus increase heart rate, respiration rate, skin temperature, and core temperature. It is very important to remove your turnout coat during authorized breaks in Rehab to alleviate some of the heat stress and reduce your heart rate.

According to NFPA 1971, full PPE for structural fire fighting consists of the following:

- *Helmet* — Protects the head from impact as well as from scalding water and other products of combustion.

- *Protective hood* — Protects portions of the firefighter's face, ears, and neck not covered by the helmet or coat collar from heat.

- *Protective coat and trousers (garments)* — Protect trunk and limbs against cuts, abrasions, and burn injuries; protects from heat and cold, and provides limited protection from corrosive liquids.

- *Gloves* — Protect the hands from cuts, abrasions, and burn injuries.

- *Safety shoes or boots (footwear)* — Protect the feet from burn injuries and puncture wounds.

The following items are not part of NFPA 1971 requirements but are required by NFPA 1500:

- *Eye protection* — Protects the wearer's eyes from hazards encountered during structural fire operations such as flying particles or liquids.

- *Hearing protection* — Limits noise-induced hearing loss when firefighters engaged in structural fire fighting are exposed to extremely loud environments such as the use of power saws, pneumatic chisels, and gas-powered fans.

- *Self-contained breathing apparatus (SCBA)(respiratory protection)* — Protects the face and lungs from heat, smoke and other toxic products of combustion, and airborne contaminants; also provides some eye protection.

- *Personal alert safety system (PASS)* — Provides an audible means by which a lost, trapped, or incapacitated firefighter can be located.

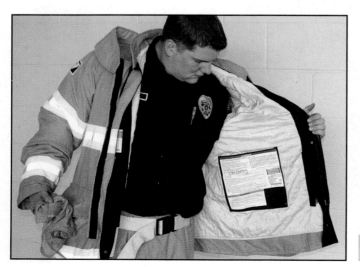

All components of the structural protective ensemble must have an appropriate product label for that component permanently and conspicuously attached (**Figure 5.3**). This label contains the following information:

"THIS STRUCTURAL FIRE FIGHTING PROTECTIVE . . . MEETS THE . . . REQUIREMENTS OF NFPA 1971, *STANDARD ON PROTECTIVE ENSEMBLES FOR STRUCTURAL FIRE FIGHTING AND PROXIMITY FIRE FIGHTING,* 2008 EDITION."

Figure 5.3 A typical product label in a turnout coat.

- Manufacturer's name, identification, or designation

- Manufacturer's address

- Country of manufacture

- Manufacturer's identification number or lot number or serial number

- Month and year of manufacture (not coded)

- Model name, number, or design

- Size or size range

- Principal materials of construction (coats, trousers, coveralls, hoods)

- Footwear size and width (boots)

- Cleaning precautions

All equipment worn by the firefighter should meet current applicable standards. Firefighters must understand the design and purpose of the various types of protective clothing and be especially aware of each garment's inherent limitations. The following sections highlight some of the important features of specific types of firefighter personal protective clothing.

Helmets

Historically, head protection was one of the first concerns for firefighters. The traditional function of the helmet was to shed water, not to protect from heat, cold, or impact. The wide brim, particularly where it extends over the back of the neck, was designed to prevent hot water and embers from reaching the ears and neck. Newer helmet designs perform this function as well as provide the following additional benefits:

- Protect the head from impact

- Protection from heat and cold

- Secondary protection of the face and eyes with a faceshield when SCBA or goggles are not required

- Colored helmets and removable shields can provide unit identification, rank identification, and help with accountability

Structural fire fighting helmets must have ear flaps or neck covers, which should always be used during fire fighting. Chin straps ensure that helmets stay in place upon impact (**Figure 5.4, p. 170**).

For secondary (supplemental) face and eye protection, faceshields are provided that attach to the helmet (**Figure 5.5, p. 170**). Faceshields alone do not provide adequate protection from flying particles or splashes and are intended to be used in combination with some other form of eye protection as described in the next section. NFPA 1500 requires that goggles or other appropriate primary eye protection be worn when participating in operations where protection from flying particles or chemical splashes is necessary. Most faceshields are designed to flip up and out of the field of vision when not in use. Some of these assemblies must be in the up position when protective breathing equipment is worn; others do not.

Helmet — Protective headgear worn by firefighters that provides protection from falling objects, side blows, the fire environment elements, and eye injuries.

Figure 5.4 The proper way to wear a helmet with chin strap engaged and ear flap down.

Figure 5.5 Some structural helmets have faceshields to protect the face and eyes.

Eye Protection

One of the most common injuries on the fireground is injury to the eyes. Eye injuries are not always reported because they are not always debilitating. Eye injuries can be serious, but they are fairly easy to prevent. It is important to protect your eyes on the fireground as well as while performing duties around the station. Eye protection for firefighters comes in many forms such as safety glasses, safety goggles, and SCBA masks **(Figure 5.6)**.

As a firefighter, you may encounter a variety of situations where eye protection greater than that provided by a helmet faceshield is needed, but respiratory protection is not. Some of these situations include medical calls where exposure to bodily fluids is possible, vehicle extrications, wildland fires, inspections in industrial occupancies, and station operations such as welding, grinding, or cutting.

Safety glasses and goggles protect against approximately 85 percent of all eye hazards. Several styles are available, including some that fit over prescription glasses. Firefighters who wear prescription safety eyeglasses should select frames and lenses that meet ANSI Standard Z87.1, *Practice for Occupational and Educational Eye and Face Protection,* for severe exposure to impact and heat.

In fire department facilities, including stations and maintenance areas, warning signs should be posted near operations requiring eye protection. Use of eye protection must be required through departmental standard operating procedures and enforced by supervisory personnel.

Hearing Protection

Firefighters are exposed to a variety of loud noises in the station, en route to the scene, and on the scene. Exposure to these sounds or a combination of sounds can produce hearing loss.

Figure 5.6 There are several forms of primary and secondary eye protection.

Figure 5.7 Intercom systems provide hearing protection and improve communication.

Figure 5.8 Earplugs and earmuffs provide hearing protection.

To comply with NFPA® 1500 and prevent hearing loss among their firefighters, fire departments are required to initiate hearing protection programs to identify, control, and reduce potentially harmful noise and/or provide protection from it. Eliminating or reducing noise levels is the best solution; however, this is often not entirely possible. Therefore, many departments provide hearing protection, standard operating procedures, training, and enforcement.

The most common use of hearing protection is for firefighters who ride apparatus that exceed maximum noise exposure levels of 90 dB in the U.S. and 85 dB in Canada. Intercom/ear protection systems provide a dual benefit because of their ability to reduce the amount of noise the ear is exposed to and at the same time allow the crew to communicate or monitor the radio **(Figure 5.7)**.

In some situations, earplugs or earmuffs may be used for hearing protection **(Figure 5.8)**. If single-use disposable earplugs are not available, each firefighter should be issued a personal set. Be aware that there are some potential hazards associated with using earplugs and earmuffs. For example, in a structural fire fighting

CAUTION
Noise-induced hearing loss is permanent hearing loss.

situation, ear plugs or earmuffs can interfere with your ability to communicate with your crew or to hear faint cries for help, the hiss of escaping gas, sounds that may indicate an imminent structural collapse, or critical commands or evacuation signals. For these reasons, wearing hearing protection during structural fire fighting is impractical and may even be dangerous. However, pump operators and those operating power tools should always wear hearing protection. As always, you should follow your departmental SOP on hearing protection.

Protective Hoods

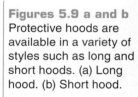

Protective Hood — Hood designed to protect the firefighter's ears, neck, and face from exposure to extreme heat. Hoods are typically made of Nomex®, Kevlar®, or PBI® and are available in long or short styles.

Protective hoods are designed to protect your ears, neck, and face from exposure to heat, hot embers, and debris. These hoods also cover areas not otherwise protected by the SCBA facepiece, helmet, ear covers, or coat collar. Hoods are typically made of fire-resistant material and are available in long or short styles (**Figures 5.9 a and b**). Protective hoods used in conjunction with the SCBA facepiece provide a higher level of protection. You must be careful to pull the hood up after your SCBA mask is secure so that it does not compromise the facepiece-to-face seal (**Figure 5.10**). Some departments advocate donning the hood before the turnout coat to help keep the bottom of the hood under the coat.

Figures 5.9 a and b Protective hoods are available in a variety of styles such as long and short hoods. (a) Long hood. (b) Short hood.

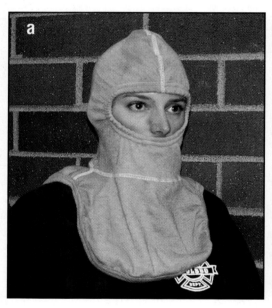

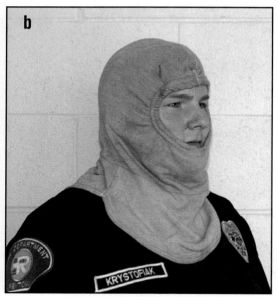

Figure 5.10 The SCBA facepiece must be secured before the hood is pulled up over it.

Turnout Coats

Turnout coats are used for protection in structural fire fighting and other emergency activities (**Figure 5.11**). NFPA 1971 requires that all protective coats used for structural fire fighting be made of three components: outer shell, moisture barrier, and thermal barrier (**Figure 5.12**). These barriers serve to trap insulating air that slows the transfer of heat from the outside to your body. They also provide limited protection from direct flame contact, hot water and steam, cold temperatures, and any number of other environmental hazards. Clearly, the construction and function of each component is important to your safety and should never be compromised. Removing the liner and wearing only the shell compromises the design of the coat, increases the likelihood of injuries, and makes you liable for using the coat differently than the manufacturer intended.

Some additional considerations regarding the use and limitations of PPE are as follows:

- Wearing PPE may increase your risk of heat stress.

- You may suffer burns with no warning.

- Structural PPE provides no chemical, biological, radiological, nuclear, explosive (CBRNE) protection.

- PPE will decrease your ability to feel ambient heat.

- Damaged PPE will put you at greater risk.

- Using all appropriate PPE (according to manufacturer's recommendations) is the only way to be properly protected.

Turnout Coat — Coat worn during fire fighting, rescue, and extrication operations.

WARNING!
All layers of the protective coat must be in place during any fire fighting operation. Failure to wear the entire coat and liner system during a fire may expose you to severe heat that could result in serious injury or death.

Figure 5.11 A typical structural turnout coat.

Figure 5.12 The turnout coat's layers must be intact for the garment to protect as intended.

Turnout coats have many features that provide additional protection and convenience to the wearer. Collars must be turned up to protect the wearer's neck and throat **(Figure 5.13)**. Wristlets prevent water, embers, and other foreign debris from entering the sleeves and provide protection for your wrists when your gloves are not sealed properly **(Figure 5.14)**. The closure system on the front of protective coats prevents water or fire products from entering through gaps between the snaps or clips **(Figure 5.15a)**.

Newer turnout coats are equipped with a Drag Rescue Device (DRD). The DRD is a built-in harness and hand loop at the back of the neck that permits a rescuer to grab and drag a downed firefighter **(Figure 5.15b)**. Turnout coats that meet NFPA stan-

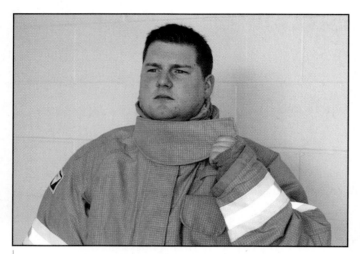

Figure 5.13 The collar must be turned up and fastened.

a

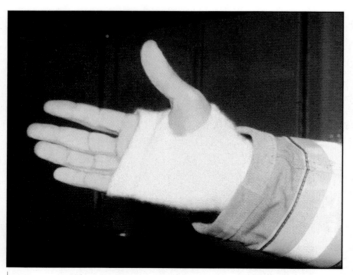

Figure 5.14 Wristlets enhance protection between the glove and coat sleeve.

b

Figures 5.15 a and b (a) The flap closure over the zipper keeps water and debris out of the coat. (b) coat with built in Drag Rescue Device (DRD).

dards should be cleaned according to manufacturer's specifications. Reflective trim should be maintained according to NFPA standards. Trim should not be obscured by pockets, patches, or storm flaps.

Turnout Pants

Protective trousers (turnout or bunker pants) are an integral part of your protective ensemble and are the only NFPA-compliant lower extremity covering. The fabric, moisture barrier, layering, and other considerations used in protective coats apply equally to protective trousers. Options such as leather reinforced knees and cuffs may increase the durability of protective trousers **(Figure 5.16)**. Some departments order gear without these reinforcements due to concerns about being able to adequately decontaminate leather that has come in contact with hydrocarbons and other toxic substances (hydraulic fluid, blood and other bodily fluids). Suspenders should be the heavy-duty type so that pants do not sag when they become wet **(Figure 5.17)**. It is very important that your turnout clothing fit properly. Turnout pants that are too tight may increase the risk of injury by restricting movement. Those that are too large will affect mobility, add extra weight, and may decrease the thermal protection of the clothing.

Turnout Pants — Pants worn during fire fighting operations. Also called bunker pants; night hitches.

Figure 5.17 Suspenders must support the weight of wet turnout pants.

Figure 5.16 Typical structural turnout pants with reinforced knees and cuffs.

Protective trousers that meet NFPA® standards should be cleaned according to manufacturer's specifications. Reflective trim should be maintained according to NFPA® standards.

Hand Protection

The most important characteristics of gloves are the protection they provide against heat, steam, or cold penetration and their resistance to cuts, punctures, and liquid absorption. Gloves must allow enough dexterity and tactile feel for the firefighter to perform the job effectively. If the gloves are too big or too awkward and bulky, you may not be able to grasp or manipulate small objects such as control knobs on portable radios (**Figure 5.18**). Gloves must fit properly and be designed to provide protection as well as to allow dexterity. Unfortunately, in order to provide protection, dexterity is often reduced.

Foot Protection

For structural fire fighting, safety boots and shoes are available in a variety of styles and materials. While some departments equip their firefighters with leather safety boots, the majority still rely on traditional rubber fire fighting boots for interior structural fire fighting. Whatever the style or material, fire fighting boots are designed to provide protection for firefighters' feet. The boots must fit well because ill-fitting boots can cause abrasions, blisters, and other painful and debilitating injuries.

There are numerous other potential hazards to the feet on the fire scene such as hot, contaminated water that sometimes accumulates on the floors of burning buildings. In addition, hot embers, nails, broken glass, and other sharp objects are common examples of fireground hazards that can produce foot injuries (**Figure 5.19**). Other potential hazards are imbedded glass shards contaminated with blood or other bodily fluids and heavy objects that can fall on a firefighter's foot, especially the toe area. Glass shards and metal fragments imbedded in boot soles can cause hand injuries when boots are being removed.

In addition to keeping your feet dry, footwear should provide protection against burns, punctures, cuts, and crushing injuries. Because of the nature of the work, you need to have the following two kinds of foot protection:

- Protective boots for fire fighting and similar emergencies (**Figure 5.20**)
- Safety shoes for station wear and other fire department tasks that include inspections, emergency medical responses, and similar tasks (**Figure 5.21**)

Many safety boots incorporate a stainless steel sole plate to protect against punctures; others have a steel insert. Most rubber fire fighting boots have insulation bonded to the shell of the boot. This makes the boots more comfortable, but also adds weight that can increase firefighter fatigue. Rubber boots have well-secured pull loops to make the boots easier to put on quickly.

In many fire departments, safety shoes or boots are part of their station uniform and they are required footwear while conducting inspections or doing work around the station. Safety shoes or boots usually have steel toes, puncture-resistant soles, or special inserts. Many provide good ankle support and are generally lighter in weight and less cumbersome than traditional rubber fire fighting boots.

Gloves — Part of the firefighter's protective clothing ensemble necessary to protect the hands.

Figure 5.18 Typical structural fire fighting gloves.

Safety Shoes — Protective footwear meeting OSHA requirements.

Figure 5.19 Boots are designed to protect from a variety of hazards including nails, broken glass, and sharp metal.

Figure 5.20 Typical rubber turnout boots used for structural fire fighting and other emergencies.

Figure 5.21 Typical leather work boots used for non-fire fighting tasks. Some leather boots are available and approved for fire fighting.

Figure 5.22 A fully outfitted wildland firefighter. The turnouts are lighter weight than those used for structural fire fighting.

Wildland Personal Protective Clothing

Personal protective clothing used for structural fire fighting is generally too bulky, too hot, and too heavy to be practical for use in wildland fire fighting. Specifications for wildland fire fighting personal protective clothing (sometimes called *brush gear*) and equipment are contained in NFPA® 1977, *Standard on Protective Clothing and Equipment for Wildland Fire Fighting*. Wildland personal protective clothing includes gloves, goggles, brush jackets/pants or one-piece jumpsuits, long-sleeve shirts, head and neck protection, and footwear **(Figure 5.22)**. Different forms of respiratory protection for wildland firefighters are also available.

Wildland fire fighting gloves are made of leather or other suitable materials and must provide wrist protection. They should be comfortable and sized correctly to allow for the greatest amount of dexterity.

The cuffs of the sleeves and the pants legs of protective clothing are closed snugly around the wrists and ankles. The fabric is treated cotton or some other inherently flame-resistant material. Underwear of 100 percent cotton should be worn under brush gear. If the sleeves of the jacket are a single layer of fabric, a long-sleeved T-shirt should be worn under the jacket. Socks should be made of natural fiber such as wool or cotton.

WARNING!
Firefighters should never wear clothing made of synthetic materials, such as nylon or polyester, when fighting a fire; these materials melt when heated and stick to the wearer's skin. This greatly increases the likelihood of major burn injuries.

Figure 5.23 A typical station uniform.

Hard hats or helmets with chin straps must be worn for head protection. Lightweight wildland helmets are preferred to structural helmets. They should be equipped with a protective shroud for face and neck protection. Goggles with clear lenses should also be worn.

What is deemed acceptable in footwear for wildland fire fighting varies in different geographical regions, but some standard guidelines apply in all areas. Lace-up safety boots with lug or grip-tread soles are most often used. Boots should be at least 8 to 10 inches (200 mm to 250 mm) high to protect the lower leg from burns, snakebites, and cuts and abrasions. Because the steel toes in ordinary safety boots tend to absorb and retain heat, they are not recommended for wildland fire fighting.

Station/Work Uniforms

Firefighter accident statistics show that certain types of clothing can contribute to on-the-job injuries. Certain synthetic fabrics, such as polyester, can be especially hazardous because they can melt during exposure to high temperatures. Some of the materials that have high temperature resistance are as follows:

- Organic fibers such as wool and cotton

- Synthetic fibers such as Kevlar® aramid fibers, Nomex® fire-resistant material, PBI® polybenzimidazole fiber, Kynol® phenolic resins, Gore-Tex® water repellent fabric, Orlon® acrylic fiber, neoprene, Teflon® fluorocarbon resins (nonstick coatings), silicone, and panotex

All firefighter station and work uniforms should meet the requirements set forth in NFPA® 1975, *Standard on Station/Work Uniforms for Fire and Emergency Services.* The purpose of the standard is to provide minimum requirements for work wear that is functional, will not contribute to firefighter injury, and will not reduce the effectiveness of outer protective clothing. Garments addressed in this standard include trousers, shirts, jackets, and coveralls, but not underwear **(Figure 5.23)**. Underwear made of 100 percent cotton is recommended.

The standard requires that no garment components will ignite, melt, drip, or separate when exposed to heat at 500°F (260°C) for 5 minutes. A garment meeting all requirements of the standard will have a notice to that effect permanently attached. It is important to note that while this clothing is designed to be fire resistant, it is not designed to be worn for fire fighting operations. Standard structural fire fighting clothing must always be worn over these garments when a firefighter is engaged in structural fire fighting activities. Wildland protective clothing, depending on design and local protocols, may be worn over station uniforms or directly over undergarments.

Note: It is important that the manufacturer's recommendations for the care and maintenance of personal protective clothing be followed whenever cleaning or repairing protective clothing. Failure to follow the recommendations may cause damage to the clothing, placing the firefighter at risk when working in hazardous environments, especially interior fire suppression operations.

Care of Personal Protective Clothing

In order for personal protective clothing to meet design requirements, it must be maintained according to the manufacturer's specifications. Any contamination of protective garments by hydrocarbons reduces their fire resistiveness. If any protective clothing becomes contaminated with flammable or combustible liquids, blood, or other bodily fluids, these garments should not be worn until properly laundered. Each piece of protective clothing has a particular manufacturer's recommended maintenance procedure that should be followed to ensure that the garment is ready for service.

Helmets should be properly cleaned and maintained to ensure their durability and maximum life expectancy. The following are guidelines for their proper care and maintenance:

- Remove dirt and soot from the shell. Dirt and stains absorb heat faster than the shell itself, thus exposing the wearer to more severe heat conditions.

- Remove chemicals, oils, and petroleum products from the shell as soon as possible **(Figure 5.24)**. Some chemicals may soften the shell material and reduce its impact resistance. These chemicals can make the helmet more electrically conductive. See the manufacturer's instructions for suggested cleansers.

- Replace helmets that do not fit properly **(Figure 5.25)**. A loose fit reduces the helmet's ability to resist the transmission of force.

- Replace helmets that are damaged. This includes leather helmets that have become cracked or brittle with age.

- Replace cracked, scratched, crazed, or cloudy faceshields.

- Inspect suspension systems frequently to detect deterioration. Replace if necessary.

- Consult the helmet manufacturer if a helmet needs repainting. Manufacturers can inform the department about the choice of paints available for a particular shell material.

- Remove from service any polycarbonate helmets that have come into contact with hydraulic oil. Some oils attack the polycarbonate material and weaken the helmet.

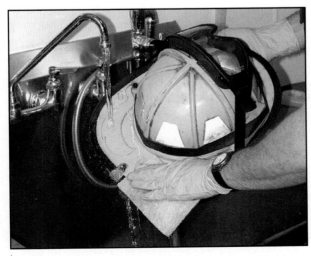

Figure 5.24 Helmets should be kept clean. The buildup of soot is both unhealthy and unsafe.

Figure 5.25 Proper fit is critical to the helmet performing as intended.

Respiratory Hazards — Any exposure to products of combustion, superheated atmospheres, toxic gases, vapors, or dust, or potentially explosive or oxygen-deficient atmospheres or any condition that creates a hazard to the respiratory system.

Oxygen-Deficient Atmosphere — An atmosphere containing less than 19.5 percent oxygen.

Hypoxia — Condition caused by a deficiency in the amount of oxygen reaching body tissues.

Cleanliness also affects the performance of protective coats, trousers, and hoods. The outer shells should be cleaned regularly and decontaminated after interior fire fighting and medical calls. Outer shells have better fire resistance when they are clean; dirty protective clothing absorbs more heat and may even become flammable after heavy contamination. Liners should be cleaned to remove contamination, grime, and perspiration.

After PPE has been washed it should be hung to dry in an area with adequate ventilation but not in direct sunlight. Ultraviolet light can degrade the fabric and reduce its protection. Follow the manufacturer's directions for cleaning. The directions are usually contained on a tag sewn to the garment. NFPA 1500 requires that protective clothing be cleaned through either a cleaning service or fire department facility that is equipped to handle contaminated clothing.

Gloves and boots should also be cleaned according to the manufacturer's instructions. NFPA® 1581, *Standard on Fire Department Infection Control Program*, further requires that personal protective clothing be cleaned and dried at least every six months in accordance with the manufacturer's recommendations.

Respiratory Protection

Considering the smoky and other toxic atmospheres in which firefighters must function, respiratory protection is critical. Failure to use the available respiratory protection equipment could lead to failed emergency operations, firefighter injuries, or firefighter fatalities. Well-trained firefighters must be aware of the respiratory hazards they may encounter, the situations in which wearing protective breathing apparatus is required, the procedures for donning and doffing the apparatus, and its proper care and maintenance.

Respiratory Hazards

While the toxic products of combustion can enter the body through inhalation, absorption, and ingestion, the means of most concern to firefighters is inhalation. When unprotected, the lungs and respiratory tract are highly susceptible to injury by the heat and gases encountered in fires. These toxic atmospheres are described as *immediately dangerous to life and health* (IDLH). By definition, an IDLH atmosphere is any that poses an immediate hazard to life or produces immediate, irreversible, debilitating effects on health. IDLH atmospheres represent toxic concentrations above which respiratory protection is required. OSHA considers the interior of a burning building to be an IDLH atmosphere. Never enter any potentially toxic atmosphere, such as the inside of a burning building, unless you are equipped with protective breathing apparatus (**Figure 5.26**).

There are four common respiratory hazards associated with fires and other emergencies. These hazards include the following:

- Oxygen deficiency
- Elevated temperatures
- Smoke
- Toxic atmospheres (with and without fire)

Oxygen Deficiency

The combustion process consumes oxygen while producing toxic gases that either physically displace oxygen or dilute its concentration. Oxygen deficiency can also occur in below-grade locations, chemical storage tanks, grain bins, silos, and other confined spaces. Other areas of potential hazard are rooms protected by total-flooding carbon dioxide extinguishing systems after discharge.

Figure 5.26 Respiratory protection is required in all situations involving IDLH. SCBA are designed and tested to provide protection during fire fighting operations. *District Chief Chris E. Mickal, NOFD Photo Unit.*

Table 5.1
Physiological Effects of Reduced Oxygen (Hypoxia)

Oxygen in Air (Percent)	Symptoms
21	None — normal conditions
17	Some impairment of muscular coordination; increase in respiratory rate to compensate for lower oxygen content
12	Dizziness, headache, rapid fatigue
9	Unconsciousness
6	Death within a few minutes from respiratory failure and concurrent heart failure

NOTE: These data cannot be considered absolute because they do not account for difference in breathing rate or length of time exposed.

These symptoms occur only from reduced oxygen. If the atmosphere is contaminated with toxic gases, other symptoms may develop.

Both OSHA and NFPA define an *oxygen-deficient atmosphere* as one containing less than 19.5 percent oxygen. When oxygen concentrations are below 18 percent, the human body responds by increasing its respiratory rate. The physiological effects of hypoxia (lack of oxygen) are shown in **Table 5.1**.

Some departments are equipped to monitor atmospheres and measure these hazards directly. When this capability exists, it should be used. Where monitoring is not possible or monitor readings are questionable, self-contained breathing apparatus should be worn.

Elevated Temperatures

Exposure to heated air can damage the respiratory tract, and if the air is moist, the damage can be much worse. Excessive heat taken quickly into the lungs can cause a serious decrease in blood pressure and failure of the circulatory system. Inhaling heated gases can cause pulmonary edema (accumulation of fluids in the lungs and associated swelling), which can cause death from asphyxiation. The tissue damage from inhaling hot air is not immediately reversible by introducing fresh, cool air. Prompt medical treatment is needed.

Pulmonary Edema — Accumulation of fluids in the lungs.

Asphyxiation — Condition that causes death because of a deficient amount of oxygen and an excessive amount of carbon monoxide and/or other gases in the blood.

WARNING!
Inhaling superheated air can cause serious injury or death.

Smoke

The smoke from a typical structure fire consists of a mixture of oxygen, nitrogen, carbon dioxide, carbon monoxide, finely divided carbon particles, and an assortment of products that have been released from the materials involved (**Figure 5.27**). The suspended particles provide a location for the condensation of some of the gaseous products of combustion, especially aldehydes and organic acids formed from carbon. Some of the suspended particles in smoke are merely irritating, but others may be lethal. The size of the particle determines how deeply into the unprotected lungs it will be inhaled.

Toxic Atmospheres Associated With Fire

Remember that a fire means oxygen deficiency and exposure to a combination of irritants and toxicants whose effects cannot be predicted accurately. In fact, the combination can have a synergistic effect in which inhaling two or more substances is more toxic or more irritating than would be expected if each were inhaled separately.

Inhaled toxic gases may have several harmful effects on the human body (**Figure 5.28**). Some of the gases cause impaired lung function, either immediately or delayed by hours or days. Other gases, such as carbon monoxide, have no directly harmful effect on the lungs but pass into the bloodstream and to other parts of the body and impair the oxygen-carrying capacity of the red blood cells.

The type and amount of toxic gases released at a fire vary according to four factors:

- Nature of the combustible
- Rate of heating
- Temperature of the evolved gases
- Oxygen concentration

Figure 5.27 Structure fires can produce tremendous volumes of smoke. *Courtesy of Dave Ricci.*

Figure 5.28 Many gases common to the fireground are toxic. Smoke inhalation is one of the leading causes of firefighter death and injuries and requires immediate medical attention and the administration of oxygen. *Courtesy of District Chief Chris E. Mickal, NOFD Photo Unit.*

Table 5.2 lists some of the most commonly found gases in fires. The IDLH concentrations are from the *National Institute for Occupational Safety and Health (NIOSH) Pocket Guide to Chemical Hazards.* These values were established to ensure that a worker could escape without injury or irreversible health effects from an IDLH exposure in the event of the failure of respiratory protection equipment.

Because more fire deaths result from carbon monoxide (CO) poisoning than from any other toxic product of combustion, a greater explanation of this toxic gas is necessary. This colorless, odorless gas is present with virtually every fire. When wood burns in an atmosphere of abundant oxygen, one of the normal combustion products is carbon dioxide (CO_2). When the oxygen supply is limited, however, there is insufficient oxygen available to combine with the carbon being produced to convert all of it to carbon dioxide. The result is an abundance of carbon monoxide.

Carbon monoxide is the result of incomplete combustion; therefore, the poorer the ventilation and the more inefficient the burning, the greater the quantity of carbon monoxide produced. A rule of thumb, although subject to much varia-

Carbon Monoxide Poisoning — Sometimes lethal condition in which carbon monoxide molecules attach to hemoglobin, decreasing the blood's ability to carry oxygen.

Table 5.2
Toxic Atmospheres Associated With Fire

Toxic Atmospheres	Sensibility	IDLH*	Caused By	Miscellaneous
Carbon Dioxide (CO_2)	Colorless; odorless	40,000 ppm**	Free-burning	End product of complete combustion of carboniferous materials
Carbon Monoxide (CO)	Colorless; odorless	1,200 ppm	Incomplete combustion	Cause of most fire-related deaths
Hydrogen Chloride (HCl)	Colorless to slightly yellow; pungent odor	50 ppm	Burning plastics (e.g., polyvinyl chloride [PVC])	Irritates eyes and respiratory tract
Hydrogen Cyanide (HCN)	Colorless; bitter almond odor	50 ppm	Burning of wool, nylon, polyurethane foam, rubber, and paper	Chemical asphyxiate; hampers respiration at the cellular and tissue level
Nitrogen Dioxide (NO_2)	Reddish-brown; pungent, acrid odor	20 ppm	Given off around silos or grain bins; also liberated when pyroxylin plastics decompose	Irritates nose and throat
Phosgene ($COCl_2$)	Colorless; odor of musty hay; tasteless	2 ppm	Produced when refrigerants such as Freon contact flame	Forms hydrochloric acid in lungs due to moisture

* Immediately dangerous to life and health — any atmosphere that poses an immediate hazard to life or produces immediate irreversible, debilitating effects on health
** Parts per million — ratio of the volume of contaminants (parts) compared to the volume of air (million parts)

Oxyhemoglobin — Combination of oxygen and hemoglobin.

Carboxyhemoglobin (COHb) — Hemoglobin saturated with carbon monoxide and therefore unable to absorb needed oxygen.

tion, is that the darker the smoke, the higher the carbon monoxide levels. Black smoke is high in particulate carbon and carbon monoxide because of incomplete combustion.

The blood's hemoglobin combines with and carries oxygen in a loose chemical combination called *oxyhemoglobin*. The most significant characteristic of carbon monoxide is that it combines with the blood's hemoglobin so readily that the available oxygen is excluded. The loose combination of oxyhemoglobin becomes a stronger combination called *carboxyhemoglobin* (COHb). In fact, carbon monoxide combines with hemoglobin about 200 times more readily than does oxygen. The carbon monoxide does not act on the body, but excludes oxygen from the blood and leads to eventual hypoxia (decreased oxygen levels) of the brain and tissues, followed by death if the process is not reversed.

Concentrations of carbon monoxide in air above five-hundredths of one percent (0.05 percent) (500 parts per million [ppm]) can be dangerous. When the level is more than 1 percent, unconsciousness and death can occur without physiological signs. Those exposed to CO can exhibit a variety of signs and symptoms including headaches, dizziness, nausea, vomiting, and cherry-red skin. These symptoms can occur at many concentrations, depending upon the dose and exposure, and should not be relied upon to provide a margin of safety. **Table 5.3** shows the toxic effects of different levels of carbon monoxide in air. These effects are not absolute because they do not take into account variations in breathing rate or length of exposure. Such factors could cause toxic effects to occur more quickly.

Table 5.3 Toxic Effects of Carbon Monoxide		
Carbon Monoxide (CO) (ppm*)	**Carbon Monoxide (CO) in Air (Percent)**	**Symptoms**
100	0.01	No symptoms — no damage
200	0.02	Mild headache; few other symptoms
400	0.04	Headache after 1 to 2 hours
800	0.08	Headache after 45 minutes; nausea, collapse, and unconsciousness after 2 hours
1,000	0.10	Dangerous — unconsciousness after 1 hour
1,600	0.16	Headache, dizziness, nausea after 20 minutes
3,200	0.32	Headache, dizziness, nausea after 5 to 10 minutes; unconsciousness after 30 minutes
6,400	0.64	Headache, dizziness, nausea after 1 to 2 minutes; unconsciousness after 10 to 15 minutes
12,800	1.28	Immediate unconsciousness; danger of death in 1 to 3 minutes

*ppm — parts per million

Measurements of carbon monoxide concentrations in air are not the best way to predict rapid physiological effects because the actual reaction is from the concentration of carboxyhemoglobin in the blood, causing oxygen starvation. Organs that require high oxygen levels, such as the heart and brain, are damaged early. The combination of carbon monoxide with the blood is greater when the concentration in air is greater. An individual's general physical condition, age, degree of physical activity, and length of exposure all affect the actual carboxyhemoglobin level in the blood. Studies have shown that it takes years for carboxyhemoglobin to dissipate from the bloodstream. People frequently exposed to carbon monoxide (such as firefighters) develop a tolerance to it, and they can function asymptomatically (without symptoms) with residual levels of serum carboxyhemoglobin that would produce significant discomfort in the average adult. The result is that firefighters may be suffering from the effects of CO exposure and be unaware of it.

Experiments have provided some comparisons relating air and blood concentrations to carbon monoxide. A 1-percent concentration of carbon monoxide in a room will cause a 50 percent level of carboxyhemoglobin in the bloodstream in 2½ to 7 minutes. A 5 percent concentration can elevate the carboxyhemoglobin level to 50 percent in only 30 to 90 seconds. A person previously exposed to a high level of carbon monoxide may have a delayed reaction because the newly formed carboxyhemoglobin may be traveling through the body. A person so exposed should not be allowed to use breathing apparatus or resume fire fighting activities until cleared by medical examination.

A hardworking firefighter may be incapacitated by a 1-percent concentration of carbon monoxide. The stable combination of carbon monoxide with the blood is only slowly eliminated by normal breathing. Administering pure oxygen is the most important element in immediate care. After an uneventful convalescence from a severe exposure, signs of nerve or brain injury may appear any time within three weeks. This is why an overcome firefighter who quickly revives should not be allowed to reenter a smoky atmosphere.

Toxic Atmospheres Not Associated With Fire

Hazardous atmospheres can be found in numerous situations in which fire is not involved. Many industrial processes use extremely dangerous chemicals to make ordinary items (**Figure 5.29**). For example, quantities of carbon dioxide would be stored at a facility where wood alcohol, ethylene, dry ice, or carbonated soft drinks are manufactured. Many other specific chemicals can be traced to numerous common products.

Many refrigerants are toxic and may be accidentally released, causing a rescue situation to which firefighters may respond. Ammonia and sulfur dioxide are two dangerous refrigerants that irritate the respiratory tract and eyes. Sulfur dioxide reacts with moisture in the lungs to form sulfuric acid. Other gases also form strong acids or alkalies on the delicate surfaces of the respiratory system.

Figure 5.29 Many industrial gases are highly toxic, flammable, or may act as asphyxiants.

An obvious location where a chlorine gas leak may be encountered is at a manufacturing plant or water treatment facility (**Figure 5.30, p. 186**); a not-so-obvious location is at a swimming pool or water park. Incapacitating concentrations can be found at either location. Chlorine is also used in manufacturing plastics, foam rubber, and synthetic textiles and is commonly found at water and sewage treatment plants.

Leaks are sometimes not at the manufacturing plant but occur during transportation of the chemical. Train derailments have resulted in container failures, exposing the public to toxic chemicals and gases. The large quantities of gases released can travel long distances.

Because of the likelihood of the presence of toxic gas or lack of oxygen, rescues in sewers, storm drains, caves, trenches, storage tanks, tank cars, bins, silos, manholes, pits, and other confined spaces require the use of self-contained breathing apparatus **(Figure 5.31)**. Workers have been overcome by harmful gases in large tanks during cleaning or repairs. Coworkers (without SCBA) have also been overcome while attempting a rescue. In addition, the atmosphere in many of these areas is oxygen deficient and will not support life even though there may be no toxic gas. For more information on confined spaces, see the IFSTA **Fire Service Search and Rescue** manual.

The manufacture, storage, and transport of hazardous materials have made virtually every area a potential site for a hazardous materials incident. Hazardous materials are routinely transported by vehicle, rail, water, air, and pipeline **(Figure 5.32)**. Firefighters need to be able to recognize when a chemical spill or incident is hazardous and know when to wear protective breathing apparatus. The United States Department of Transportation (DOT) defines a hazardous material as *"any substance which may pose an unreasonable risk to health and safety of operating or emergency personnel, the public, and/or the environment if it is not properly controlled during handling, storage, manufacture, processing, packaging, use, disposal, or transportation."*

Figure 5.30 Chlorine may be stored anywhere water is treated, including public and private swimming pools.

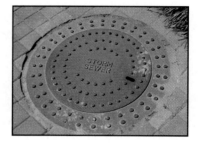

Figure 5.31 Confined spaces may contain toxic gases or have low oxygen content requiring the use of SCBA or SAR.

Figure 5.32 Hazardous materials incidents may occur anywhere, including industrial complexes, highways, marine facilities, or railroad lines.

Hazardous materials can range from chemicals in liquid or gaseous form to radioactive materials to etiologic (disease-causing) agents. Fire may complicate the hazards and pose an even greater danger. Many fire responses to industrial sites involve hazardous materials. Self-contained breathing apparatus should be a mandatory piece of protective equipment when dealing with hazardous materials situations.

Etiologic Agents — Living microorganisms, like germs, that can cause human disease; a biologically hazardous material.

When responding to a vehicle accident involving a commercial truck, the placard on the truck should serve as a warning that the atmosphere may be toxic and that self-contained breathing apparatus should be worn. In industrial facilities, placards and labels placed on containers provide warning of the dangerous materials inside. In many fire departments it is SOP to use binoculars to view these placards and labels from a safe distance until the involved material is identified.

Do not limit the use of self-contained breathing apparatus to transportation hazardous materials incidents. Common calls, such as natural gas leaks or carbon monoxide poisonings, may also require the use of self-contained breathing apparatus. *When in doubt, wear self-contained breathing apparatus!*

For more information on hazardous materials, see the IFSTA **Hazardous Materials for First Responders** and the *Hazardous Materials: Managing the Incident* manuals.

Protective Breathing Apparatus Limitations

To operate safely and effectively in hazardous atmospheres, firefighters must be aware of the limitations of their protective breathing apparatus. These limitations include those of the wearer, the equipment, and the air supply.

Limitations of Wearer

Several factors affect firefighters' ability to use respiratory protection effectively. These include physical, medical, and mental factors.

Physical factors

- *Physical condition* — Firefighters must be in sound physical condition in order to maximize the work that can be performed and to extend the air supply.

- *Agility* — Wearing some respiratory protection restricts firefighters' movements and can affect balance. Good agility will minimize these effects.

- *Facial features* — The shape and contour of the face can affect the wearer's ability to get a good facepiece-to-face seal.

NOTE: NFPA® 1500 prohibits wearing beards or facial hair that prevents a complete seal between the facepiece and the wearer's face. Spectacles (eyeglasses) are also prohibited if the side frames pass through the seal area.

Both NFPA® 1500 and OSHA CFR 1910.134 allow firefighters to wear soft contact lenses while using SCBA if the firefighter has demonstrated successful long-term (at least 6 months) use of contact lenses without any problems.

To avoid the problems associated with contact lenses while wearing SCBA and other supplied air respirators, some departments provide facepieces that will accommodate special prescription glasses inside the facepiece. Firefighters must know and comply with their department's policies regarding this issue.

Medical factors

- *Neurological functioning* — Good motor coordination is necessary for operating in protective breathing equipment.

- *Muscular/skeletal condition* — Firefighters must have the physical strength and size required to wear the protective equipment properly, and the strength and stamina to perform effectively while wearing the equipment.

- *Cardiovascular conditioning* — Good cardiovascular conditioning is needed to prevent heart attacks, strokes, or other related problems during strenuous activity.

- *Respiratory functioning* — Proper respiratory functioning will maximize the wearer's operation time in a self-contained breathing apparatus.

Mental factors

- *Adequate training in equipment use* — Firefighters must be knowledgeable in every aspect of respiratory protection use (**Figure 5.33**).

- *Self-confidence* — A belief in themselves and their abilities helps firefighters adapt to the changing conditions that sometimes occur when wearing protective breathing apparatus.

- *Emotional stability* — The ability to maintain control in claustrophobic and stressful environments will reduce the chances of serious mistakes being made.

Limitations of Equipment

In addition to being concerned about the limitations of the wearer, firefighters must also be aware of the limitations of the equipment.

- *Limited visibility* — The facepiece can reduce peripheral vision; facepiece fogging can reduce overall vision.

- *Decreased ability to communicate* — Unless it has built-in voice amplification, the facepiece can seriously hinder voice communication.

- *Increased weight* — Some types of protective breathing equipment can weigh 25 to 35 pounds (11 kg to 16 kg), depending on the model.

- *Decreased mobility* — The increase in weight and the restrictions caused by the harness straps reduce the wearer's mobility (**Figure 5.34**).

Limitations of Air Supply

Air supply is another factor to consider when assessing protective breathing apparatus limitations. Some limitations are based on the apparatus user whereas others are based on the actual supply of air in the cylinder.

- *Physical condition of user* — Those in poor physical condition expend their air supply faster than others performing the same tasks.

- *Degree of physical exertion* — The greater the physical exertion, the faster the air supply will be expended. This can result in a 30-minute air bottle only lasting 15 minutes under actual working conditions.

- *Emotional stability of user* — Firefighters who become excited increase their respiratory rate and use their air supply faster than a firefighters who stay calm.

- *Condition of apparatus* — Minor leaks and poor adjustment of regulators result in excess air loss.

Figure 5.33 Firefighters must be thoroughly trained on the proper use of their SCBA.

Figure 5.34 Wearing SCBA can limit range of motion.

- *Cylinder pressure before use* — If the cylinder is not filled to capacity, the amount of working time is reduced proportionately.

- *Training and experience of user* — Properly trained and highly experienced personnel are able to draw the maximum air supply from a cylinder.

Air Management

Historically, many fire departments trained their firefighters to continue working in a hazardous environment until the low-air alarm on their SCBA sounded. The alarm sounding was the signal to withdraw to a safe environment. However, the low-air alarm sounds when 25 percent of the cylinder air remains, which provides up to five minutes of air supply. That may be sufficient air supply for exiting a residence or other small building — unless you encounter something that impedes your egress. That amount of air is not sufficient for exiting from deep inside a large or complex structure. For your own safety and survival, you need to comply with the accountability system in use, maintain your situational awareness, and manage your air supply.

While your supervisor, the ISO, and the IC all are responsible for firefighter safety, the ultimate responsibility for your safety rests with you. You are responsible for managing your own air supply. There are three basic elements to effective air management:

- Know your *point of no return* (beyond 50 percent of the air supply of the team member with the lowest gauge reading)

- Know how much air you have (at all times)

- Make a conscious decision to stay or leave when your air is down to 50 percent

Except for a catastrophic event such as a structural collapse onto your team, or if you become separated from your team, the decision to stay or leave is always made by your supervisor and it is a team decision. Never leave your team in the hazard zone unless permitted to do so by your supervisor. If there are only two members in the hazard zone, both must leave at the same time.

Low-Pressure Alarm — Bell, whistle, or other audible alarm that warns the wearer when the SCBA air supply is low and needs replacement, usually 25 percent of full container pressure.

Point of No Return — That time at which the remaining operation time of the SCBA is equal to the time necessary to return safely to a nonhazardous atmosphere.

To maximize your air supply (and therefore your effectiveness) in fires and other hazardous situations, check your air supply status regularly. It is your responsibility to check your air supply at the following points:

- At the beginning of your shift (Make sure the cylinder is at least 90 percent full)
- When donning the SCBA and opening the cylinder valve
- While working (At 5-10 minute intervals and at key points – changing locations, finishing assignments)
- During egress from the hazard zone (At 2-3 minute intervals)
- When refilling or replacing a cylinder

Types of Respiratory Protection

Depending upon the nature of the atmosphere in which firefighters must function, a number of different types of respiratory protection are available. For atmospheres that contain normal levels of oxygen but are contaminated with airborne particulates such as a variety of dusts, filter masks (also called *air-purifying respirators*) are available. The most basic of these are simple surgical-type filter masks such as those used in emergency medical service. More sophisticated models have an air-purifying filter, canister, or cartridge **(Figure 5.35)**.

For atmospheres that are oxygen deficient, contaminated with smoke or other toxic materials, or both, firefighters must use some type of self-contained breathing apparatus (SCBA). There are two types of self-contained breathing apparatus used in the fire service: *open-circuit* and *closed-circuit.* Open-circuit SCBA uses compressed air; closed-circuit SCBA uses compressed oxygen. In open-circuit SCBA, exhaled air is vented to the outside atmosphere. In closed-circuit SCBA (also known as *rebreather* apparatus) the user's exhaled air stays within the system and is reused. Closed-circuit SCBA and open-circuit airline equipment are most often used in shipboard operations, extended hazardous materials incidents, and some rescue operations. Firefighters use open-circuit SCBA much more frequently than closed-circuit SCBA. Regardless of the type of SCBA used, thorough training in its use is essential.

Open-Circuit Self-Contained Breathing Apparatus

Several companies manufacture open-circuit SCBA, each with different design features or mechanical construction. Certain parts, such as cylinders and harness assemblies, are interchangeable; however, such substitution voids NIOSH and Mine Safety and Health Administration (MSHA) certification and is *not* a recommended practice. Substituting different parts may also void warranties and leave the department or firefighter liable for any injuries incurred.

Open-Circuit Self-Contained Breathing Apparatus — An SCBA that allows the wearer's exhaled air to be discharged or vented to the atmosphere.

There are four basic SCBA component assemblies:

- *Harness assembly* — A rigid frame with straps that hold the air cylinder on the firefighter's back
- *Air cylinder assembly* — Includes cylinder, valve, pressure gauge, and PASS device in some units
- *Regulator assembly* — Includes high-pressure hose with low-pressure alarm, by-pass valve, and a pressure reducing device
- *Facepiece assembly* — Includes facepiece lens, an exhalation valve, and a low-pressure hose (breathing tube) if the regulator is separate; in some cases, also includes voice amplification, head harness, or helmet mounting bracket

Harness assembly. The harness assembly is designed to hold the air cylinder on the firefighter's back as securely and comfortably as possible. Adjustable shoulder straps help to stabilize the unit and carry part of its weight. The waist straps are designed to help properly distribute the weight of the breathing apparatus to the hips **(Figure 5.36)**. A common problem is that some firefighters fail to buckle the waist straps. In other cases, waist straps are removed altogether. NIOSH and MSHA certify the entire SCBA unit, and removal of waist straps could void warranties and create a hazard for the user.

Air cylinder assembly. Because the cylinder must be strong enough to safely contain air pressure as high as 4,500 psi (31 000 kPa), it constitutes the main weight of the breathing apparatus **(Figure 5.37)**. The weight of air cylinders varies with each manufacturer and depends on the material used to fabricate the cylinder. Manufacturers offer cylinders of various sizes, capacities, and features to correspond to their varied uses in responses. Commonly found sizes of air cylinders used in the fire service include the following:

- 30-minute, 2,216 psi (15 280 kPa), 45 ft^3 (1 270 L) cylinders
- 30-minute, 4,500 psi (31 000 kPa), 45 ft^3 (1 270 L) cylinders
- 45-minute, 3,000 psi (21 000 kPa), 66 ft^3 (1 870 L) cylinders
- 45-minute, 4,500 psi (31 000 kPa), 66 ft^3 (1 870 L) cylinders
- 60-minute, 4,500 psi (31 000 kPa), 87 ft^3 (2 460 L) cylinders

Figure 5.35 A variety of air-purifying respirators are available. They must be certified for the specific type of hazard they are required to protect against.

Figure 5.37 A typical SCBA cylinder in a harness includes the shoulder and waist straps and the backpack frame.

Figure 5.36 It is important to buckle the waist strap. The loose end of the strap should be secured into the waist band to keep it from becoming caught when operating in restricted areas.

Depending upon the size of the cylinder and what it is made of, it can weigh from as little as 9.4 pounds (4.3 kg) to as much as 16.6 pounds (7.5 kg). When added to the weight of the rest of a firefighter's personal protective clothing, the weight of an SCBA cylinder increases physical stress during emergency operations.

Regulator assembly. Air from the cylinder travels through the high-pressure hose to the regulator. The regulator reduces the pressure of the cylinder air to slightly above atmospheric pressure and controls the flow of air to meet the respiratory requirements of the wearer (**Figure 5.38**). When the wearer inhales, a pressure differential is created in the regulator. The apparatus diaphragm moves inward, tilting the admission valve so that low-pressure air can flow into the facepiece. The regulator diaphragm is then held open, which creates the positive pressure. Exhalation moves the diaphragm back to the "closed" position. Newer SCBA units have regulators that fit into the facepiece (**Figure 5.39**). On older units, the regulator is on the firefighter's chest or waist strap.

Depending on the SCBA model, it will have control valves for normal and emergency operations. These are the mainline valve and the bypass valve (**Figure 5.40**). During normal operation, the mainline valve is fully open and locked if so equipped. The bypass valve is closed. On some SCBA, the bypass valve controls a direct airline from the cylinder in the event that the regulator fails. Once the valves are set in their normal operating position, they should not be changed unless the emergency bypass is needed (**Figure 5.41**).

Figure 5.38 Some older SCBA have harness-mounted regulators.

Figure 5.39 The majority of SCBA have facepiece-mounted regulators.

A remote pressure gauge that shows the air pressure remaining in the cylinder is mounted in a position visible to the wearer (**Figure 5.42**). This remote pressure gauge should read within 100 psi (700 kPa) of the cylinder gauge if increments are in psi (kPa). If increments are shown in other measurements, such as percents or fractions, both measurements should be the same (**Figure 5.43**). These pressure readings are most accurate at or near the upper range of the gauge's rated working pressures. Low pressures in the cylinder may cause inconsistent readings between the cylinder and

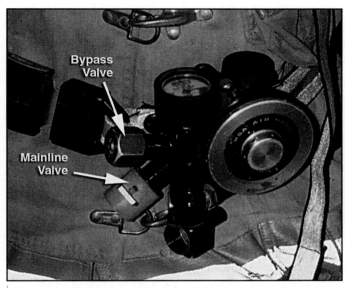

Figure 5.40 The mainline valve and bypass valve on this older belt-mounted regulator are identified by their shape, color and position.

Figure 5.41 The bypass valve on a facepiece-mounted regulator.

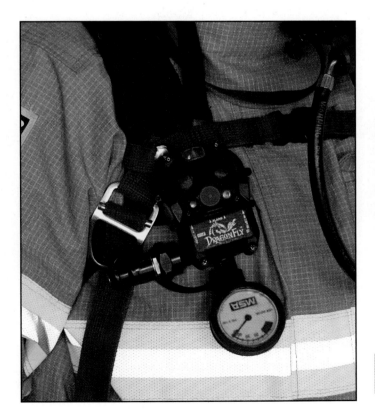

Figure 5.43 Compare the cylinder pressure gauge and remote gauge readings.

Figure 5.42 A typical remote pressure gauge mounted on the shoulder strap of the unit.

regulator gauges. If they are not consistent, rely on the lower reading and check the equipment for any needed repair before using it again. All units, depending on age have an audible and/or vibrating alarm that activates when the cylinder pressure decreases to approximately one-fourth of the maximum rated pressure of the cylinder, depending on the manufacturer **(Figure 5.44)**. Both of these devices will shut down with some air left in the cylinder and that gives the user about two minutes of air with which to exit. However, interior teams should have withdrawn to a clear atmosphere *before* the first low-pressure alarm sounds.

According to NFPA® 1981, *Standard on Open-Circuit Self-Contained Breathing Apparatus (SCBA) for Emergency Services,* all new SCBA must be equipped with a rapid intervention crew/universal air connection (RIC/UAC) to allow a cylinder that is low on air to be transfilled from another cylinder regardless of manufacturer. When the cylinders are connected, the air supply equalizes between them. Older SCBA can be retrofitted with the RIC/UAC, but it is not required. Thorough training in the use of this feature is required.

Facepiece assembly. A facepiece provides some protection from facial and respiratory burns and holds in the cool breathing air. The facepiece assembly consists of the facepiece lens, an exhalation valve, and a low-pressure hose to carry the air from the regulator to the facepiece if the regulator is separate **(Figure 5.45)**. The facepiece lens is made of clear safety plastic and is mounted in a flexible rubber facepiece. The mask is held snugly against the face by a head harness with adjustable straps, net, or some other arrangement **(Figures 5.46 a and b)**. Some helmets have a facepiece bracket that connects directly to the helmet instead of using a head harness **(Figure 5.47)**. The lens should be protected from scratches during use and storage. Self-adhesive clear plastic lens covers are available to protect the facepiece lens. Some facepieces have a speaking diaphragm to make voice communication clearer; others have built-in voice amplifi-

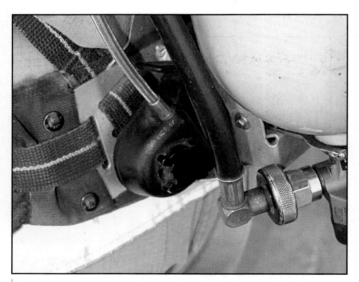

Figure 5.44 A low-pressure alarm is designed to activate when the cylinder pressure drops below one-fourth of the cylinder's rated capacity.

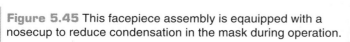

Figure 5.45 This facepiece assembly is eqauipped with a nosecup to reduce condensation in the mask during operation.

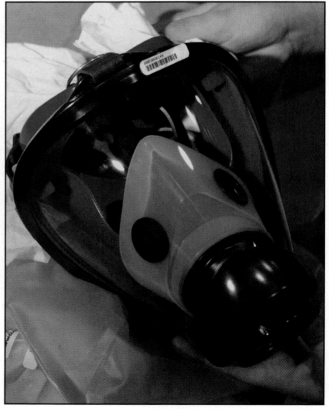

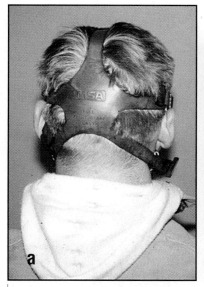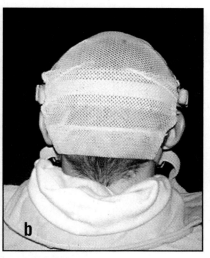

Figures 5.46 a and b Common types of head harness. (a) Facepiece with five-point harness. (b) Facepiece with mesh assembly.

Figure 5.47 Some helmets are designed to connect the facepiece directly.

cation. According to NFPA® 1981, all new SCBA must be equipped with a heads-up display (HUD). This feature displays the current cylinder pressure on the inside of the facepiece lens or mask-mounted regulator and allows the wearer to monitor the air supply without having to read an external gauge.

The facepiece for an SCBA with a harness-mounted regulator has a low-pressure hose, or breathing tube, attached to the facepiece with a clamp or threaded coupling nut. The low-pressure hose brings air from the regulator into the facepiece and is usually corrugated to prevent collapse or kinking when the wearer is working in close quarters, awkward positions, or in contact with a hard surface. Therefore, it must be kept free of kinks and away from contact with abrasive surfaces. This also applies to high-pressure airlines connecting to facepiece mounted regulators. (**Figure 5.48, p. 196**). Newer SCBA have no low-pressure hose because the regulator is attached directly to the facepiece (**Figure 5.49, p. 196**).

The exhalation valve is a simple, one-way valve that releases an exhaled breath without admitting any of the contaminated outside atmosphere. Dirt or other foreign materials can cause the exhalation valve to malfunction and allow air from the tank to escape and quickly deplete the air supply. Therefore, it is important that the exhalation valve be kept clean. It is also important that the exhalation valve be tested by the user during daily equipment checks and before entering a hazardous atmosphere (**Figure 5.50, p. 196**).

Just as with a malfunctioning exhalation valve, an improperly sealed facepiece can allow air to escape the mask and deplete the air supply. Once the air supply is depleted, the gases in the surrounding atmosphere can enter the mask. A fogged facepiece lens can obscure the vision of the wearer (**Figure 5.51, p. 196**). The differing temperatures inside and outside the facepiece can cause the moisture in exhaled breath to condense on the inside of the lens. When this occurs in a darkened roomor one filled with smoke or steam, the wearer is virtually blind. External fogging can also occur in moist air if the atmosphere condenses on the outside of a cooler facepiece lens.

Figure 5.48 Any hose supplying air to the facepiece regulator must be kept free of kinks and protected from potential damage.

Figure 5.49 Facepiece-mounted regulators are connected to the second-stage regulator by a high-pressure hose.

Figure 5.50 Testing the exhalation valve by sealing the regulator opening with the palm of the hand and inhaling.

Figure 5.51 A fogged facepiece obscures vision. Fogging may occur when the air in the facepiece is warmer than the ambient atmosphere.

Figure 5.52 Nosecups reduce facepiece fogging.

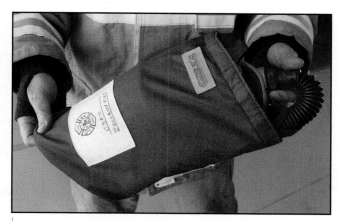

Figure 5.53 Facepieces should be protected in storage. Facepieces may be stored in bags provided with the unit or in specially designed pockets in the turnout coat.

External fogging can be removed by wiping the lens. One of the following methods can be used to prevent or control internal fogging of a lens.

- *Use a nosecup* — Facepieces can be equipped with a nosecup that deflects exhalations away from the lens **(Figure 5.52)**. However, if the nosecup does not fit well, it will permit exhaled air to leak into the facepiece and condense on the lens.

- *Apply an antifogging chemical* — Special antifogging chemicals recommended by the manufacturer can be applied to the lens of the facepiece. Some SCBA facepieces are permanently impregnated with an antifogging chemical.

When storing the facepiece, it may be packed in a case or stored in a bag or coat pouch **(Figure 5.53)**. Regardless of where the facepiece is stored, the straps should be left fully extended to facilitate donning.

Open-Circuit Airline Equipment

In some incidents, such as those involving hazardous materials releases or rescues that force personnel to remain in a contaminated atmosphere for extended periods, firefighters need a longer duration air supply than can be provided by standard open-circuit SCBA. In these situations, an air hose from a large air supply can be connected to a standard open-circuit SCBA regulator and facepiece. These systems, called *supplied-air respirators* (SAR), also include a 5- to 10-minute emergency egress cylinder called an *emergency breathing support system* (EBSS) **(Figure 5.54)**. Airline systems enable firefighters to travel up to 300 feet (100 m) from the air supply source, allowing them to work for hours if necessary without the encumbrance of a backpack. For more information on open-circuit airline equipment, see the IFSTA **Respiratory Protection for Fire and Emergency Services** manual.

Open-Circuit Airline Equipment — Airline breathing equipment that allows exhaled air to be discharged to the open atmosphere.

Figure 5.54 A typical supplied air respirator (SAR) used for confined space operations.

Closed-Circuit Breathing Apparatus

The use of closed-circuit breathing apparatus in the fire service is not as common as open-circuit breathing apparatus. However, closed-circuit breathing apparatus are sometimes used for shipboard operations and hazardous materials incidents because of their longer air supply duration **(Figure 5.55)**. Closed-circuit SCBA are available with durations of 30 minutes to 4 hours and usually weigh less than open-circuit units of similarly rated service time. They weigh less because a smaller cylinder containing pure oxygen is used. For more information on closed-circuit breathing apparatus, see the IFSTA **Respiratory Protection for Fire and Emergency Services** manual.

Storing Protective Breathing Apparatus

Methods of storing self-contained breathing apparatus vary from department to department. Each department should use the most appropriate method to facilitate quick and easy donning **(Figure 5.56)**. SCBA can be placed on the apparatus in seat mounts, side mounts, and compartment mounts, or stored in carrying cases. If placed in seat mounts, the SCBA should be arranged so that it may be donned without the firefighter having to unbuckle the seatbelt.

Personal Alert Safety Systems

The use of personal alert safety system (PASS) devices by all firefighters and rescuers wearing SCBA is mandatory under NFPA 1500. (The acronym PAD [*personal alert device*] is also used). A lost, incapacitated, or disoriented firefighter inside a burning building or other hazardous area may be in mortal jeopardy. PASS devices are designed to alert others that a firefighter has stopped moving and therefore may be in distress. These devices assist rescuers attempting to locate downed firefighters, even in total darkness or dense smoke. Some newer SCBA have built-in (integrated) PASS

Figure 5.55 A typical closed-circuit SCBA.

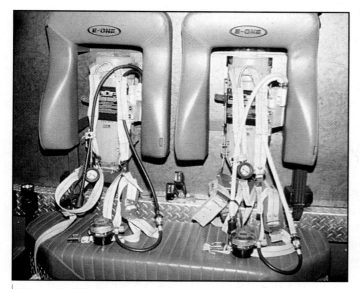

Figure 5.56 Some apparatus are designed with seat-mounted SCBA brackets.

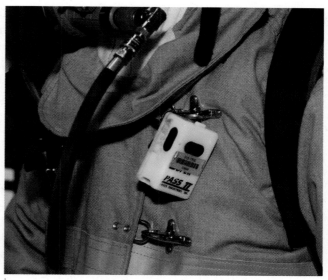

Figure 5.57 A detachable PASS device. The unit must be turned on prior to entering the hot zone.

devices that are activated when the main air supply valve is opened. This type of system can also be activated manually without opening the cylinder valve. Detachable PASS devices are about the size of a cell phone, are worn on the SCBA harness or turnout coat, and must be manually activated with a switch (**Figure 5.57**). If a firefighter wearing a PASS device should remain motionless for approximately 30 seconds, the device will emit a loud, pulsating shriek. The alert signal can also be started manually if the wearer is conscious and in trouble. Rescuers can follow the sound of the alert signal to locate the firefighter in distress.

PASS devices can save lives, but they must be maintained and used properly. The user must remember to turn on and test the device, whether built-in or detachable, before entering a hazard zone. Training classes should be conducted on techniques to be used when attempting rescue of a lost firefighter. Locating even the loud shriek of a PASS device in poor visibility conditions can be more difficult than expected because the sound reverberates off walls, ceilings, and floors. Some rescuers may be tempted to abandon established search procedures when they think they can tell the location from which the alarm sound is coming. Noise from the rescuer's SCBA operation and muffled hearing because of protective hoods also adds to the difficulty. Recommendations for use of PASS devices include the following:

- Use only PASS devices that meet the requirements of NFPA® 1982, *Standard on Personal Alert Safety Systems (PASS)*.

- Maintain the PASS device according to manufacturer's instructions and test it daily.

- Conduct realistic training with the PASS to teach firefighters how to initiate an alert signal and how to react appropriately to alert signals from others.

- Retrain semiannually with PASS devices.

- Check SCBA and PASS devices when coming on duty or before use to ensure proper operation.

- Train rescuers to listen for the distress sound by stopping in unison, controlling breathing, and lifting earflaps away from ears.

- Silence the PASS device to facilitate communications when a downed firefighter is located.

WARNING!
Scientifically controlled tests by NIST of both stand-alone and SCBA-integrated PASS devices have shown that the alarm signal produced may be significantly reduced at temperatures as low as 300°F (150°C).

Firefighters who become stranded in burning buildings may be protected from excessive heat for some time by their PPE and SCBA. Because the alarm signal from their PASS devices may be rendered barely audible by the heat, however, rescue teams may have difficulty locating them. This reinforces the need for firefighters to stay together in teams of two or more, maintain their situational awareness, and exit the hazard zone whenever conditions warrant.

Donning and Doffing Protective Breathing Apparatus

Several methods can be used to don self-contained breathing apparatus, depending on how the SCBA is stored. The most common methods used in the fire service include the over-the-head method, the coat method, donning from a seat, and donning from a rear mount or compartment mount. The steps needed to get the SCBA onto the firefighter vary somewhat with each method. There are also different steps for securing different brands and models of self-contained breathing apparatus.

Due to the varieties of SCBA, it is impossible to list step-by-step procedures for each manufacturer's model. Therefore, the information in this section is intended only as a general description of the different donning techniques. The wearer should follow manufacturer's instructions and local standard operating procedures for donning and doffing their particular SCBA.

General Donning Considerations

Regardless of the SCBA model or method of donning, several precautionary safety checks should be made prior to donning the SCBA. For departments that have daily shift changes, these checks may occur at shift change. The apparatus is then returned to the apparatus-mounted storage racks or the storage case. For departments that are unable to inspect their breathing apparatus daily, these checks should be made immediately prior to donning the SCBA no matter how it is stored:

- Check the air cylinder gauge to ensure that the cylinder is full. NFPA® 1404, *Standard for Fire Service Respiratory Protection Training,* recommends no less than 90 percent of cylinder capacity **(Figure 5.58)**.

- Check the remote gauge and cylinder gauge to ensure that they read within 100 psi (700 kPa) of the same pressure. Gauges not marked in increments of 100 psi (700 kPa) should be the same.

- Check the harness assembly and facepiece to ensure that all straps are fully extended **(Figure 5.59)**.

- Operate all valves to ensure that they function properly and are left in the correct position.

- Check your PASS device.

Once these checks are complete, the protective breathing apparatus may be donned using the most appropriate method.

Donning from a Storage Case or Apparatus Compartment

There are two donning methods that can be used to don SCBA that is stored in a case: the over-the-head method and the coat method. The coat method is most commonly used to don SCBA that is mounted vertically in an apparatus compartment. These donning methods require the SCBA to be positioned in front of the

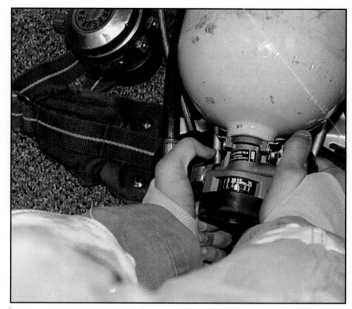

Figure 5.58 Check the cylinder gauge to be sure that the tank is full.

Figure 5.59 Check the condition of the facepiece harness and ensure the straps are fully extended.

firefighter (either in or out of the case) with all straps extended, ready to don. The steps for donning PPE and SCBA using these methods are described in **Skill Sheet 5-I-1**; the steps for doffing PPE and SCBA and preparing them for reuse are given in **Skill Sheet 5-I-2**.

Donning from a Seat Mount

Valuable time can be saved if the SCBA is mounted on the back of the firefighter's seat in the vehicle. By having a seat mount, firefighters can don SCBA while en route to an incident. Donning from a seat mount should only be done, however, if it can be safely accomplished without the firefighter having to unbuckle the seatbelt.

Seat-mounting hardware comes in three main types: lever clamp, spring clamp, or flat hook. Part of this hardware is a hold-down device for securing the SCBA to the bracket. A drawstring or other quick-opening bag should enclose the facepiece to keep it clean and to protect it from dust and scratches. (**NOTE:** Do not keep the facepiece connected to the regulator during storage. These parts must be separate to check for proper facepiece seal.)

The cylinder's position should match the proper wearing position for the firefighter. The visible seat-mounted SCBA reminds and even encourages personnel to check the equipment more frequently. Because it is exposed, checks can be made more conveniently. When exiting the fire apparatus, do so carefully because the extra weight of the SCBA on your back can make slips and falls more likely. Be sure to adjust all straps for a snug and comfortable fit.

CAUTION
Never breathe cylinder air when seated in the apparatus. This will deplete your air supply prior to entering the hazard zone.

WARNING!
Never stand to don SCBA while the vehicle is moving. Standing places both you and other firefighters in danger of serious injury in the event of a fall. NFPA 1500 requires firefighters to remain seated and belted at all times while the vehicle is in motion.

Donning from a Side or Rear Mount

Although it does not permit donning en route, some departments specify side- or rear-mounted SCBA on their apparatus. Although this type of mount requires a bit more time for donning than seat-mounted SCBA, it reduces the chances of slips and falls. Compared to SCBA stored in a carrying case, donning side- or rear-mounted SCBA saves time because the following steps are eliminated: removing the equipment case from the fire apparatus, placing it on the ground, opening the case, and picking up the unit. However, because these units are exposed to weather and physical damage, waterproof covers are desirable **(Figure 5.60)**.

If SCBA are mounted at the right height, firefighters can don them with little effort. Having the mount near the running boards or near the tailboard allows the firefighter to don the equipment while sitting. The donning steps are similar to those for seat-mounted SCBA.

Donning From a Compartment or Backup Mount

SCBA stored in a closed compartment can be ready for rapid donning **(Figure 5.61)**. Units mounted inside a compartment present the same advantages as side- or rear-mounted equipment but are protected from the weather. However, some compartment doors may interfere with a firefighter donning SCBA. Other compartments may be too high and make donning the SCBA difficult.

Some compartment mounts feature a telescoping frame that extends the equipment out of the compartment when needed **(Figure 5.62)**. One type of compartment mount telescopes outward, then upward or downward to proper height for quick donning.

The backup mount provides quick access to SCBA (some high-mounted SCBA must be removed from the vehicle and donned using the over-the-head or coat method). The procedure for donning SCBA using the backup method is similar to the method used for mounts from which the SCBA can be donned while seated.

Donning the Facepiece

The steps involved in donning most SCBA facepieces are very similar. One important difference in facepieces is that some use a rubber harness with adjusting straps while others use a mesh skullcap with adjusting straps. Different models from the

Figure 5.60 Externally-mounted SCBA units contained in protective covers can be donned quickly. *Courtesy of Ron Bogardus.*

Figure 5.61 Some apparatus have compartment-mounted SCBA. Spare air cylinders are also contained in the compartment.

same manufacturer may have a different number of straps. Another important difference is the location of the regulator. The regulator may be attached to the facepiece or mounted on the waist belt. The shape and size of facepiece lenses may also differ. Despite these variations, the uses and donning procedures for facepieces are essentially the same.

NOTE: Interchanging facepieces, or any other part of the SCBA, from one manufacturer's equipment to another makes any warranty and certification void.

When in operation, an SCBA facepiece cannot be worn loosely or it will not seal against the face properly. An improper seal will result in air escaping from the mask and thereby depleting the air supply more quickly.

Firefighters should not rely solely on tightening facepiece straps to ensure proper seal. NFPA® 1500 requires that each firefighter be fitted with a facepiece that conforms properly to the shape and size of his or her face **(Figure 5.63)**. For this reason, many SCBA are available with different-sized facepieces. Nosecups, if used, must also properly fit the firefighter.

The following are general considerations for donning all SCBA facepieces:

- No hair should come between the skin and the sealing surface of the facepiece.

- The chin should be centered in the chin cup and the harness centered at the rear of the head.

- Facepiece straps should be tightened by pulling them evenly and simultaneously to the rear. Pulling the straps outward, to the sides, may damage them and prevent proper engagement with the adjusting buckles. Tighten the lower straps first, then the temple straps, and finally the top strap if there is one.

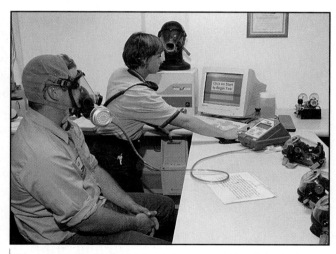

Figure 5.63 All SCBA facepieces must be fit-tested. A variety of facepiece sizes are available from SCBA manufacturers.

Figure 5.62 A telescoping frame mounted in an apparatus compartment makes donning easier.

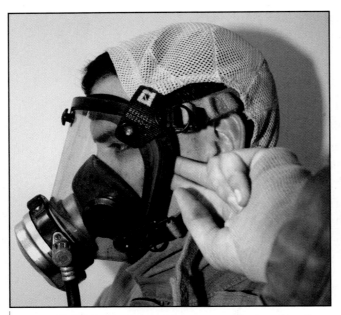

Figure 5.64 Check for positive pressure by breaking the seal between the facepiece and cheek.

Figure 5.65 The hood is worn over the head harness straps.

- The facepiece should be checked for proper seal and operation (exhalation valve functioning properly, all connections secure, and donning mode switch [if present] in proper position).

- Positive pressure should be checked by gently breaking the facepiece seal. This can be done by inserting two fingers under the edge of the facepiece (**Figure 5.64**). You should be able to feel air moving past your fingers. If you cannot feel air movement, remove the unit from service and tag it for repair.

- The hood must be worn over the facepiece harness or straps. All exposed skin must be covered and vision must not be obscured (**Figure 5.65**). No portion of the hood should be located between the facepiece and the face.

- The helmet should be worn with the chin strap secured. Helmets equipped with a ratchet adjustment should be adjusted so the helmet fits down properly on the head.

NOTE: Helmets are provided with adjustable chin straps so that the helmets remain on firefighters' heads during fire fighting operations. This is especially important while operating inside burning buildings.

Doffing SCBA

Doffing techniques differ for different types of SCBA. Generally, there are certain actions that apply to all SCBA when doffing:

- Make sure you are out of the contaminated area and that SCBA is no longer required.

- Discontinue the flow of air from the regulator to the facepiece.

- Disconnect the low-pressure hose from the regulator or remove the regulator from the facepiece, depending upon type of SCBA.

- Remove the facepiece.

- Remove the backpack assembly while protecting the regulator.
- Close the cylinder valve.
- Relieve pressure from the regulator in accordance with manufacturer's instructions.
- Extend all straps.
- Refill or replace the cylinder.
- Clean and disinfect the facepiece.

Inspection and Maintenance of Protective Breathing Apparatus

NFPA® 1404 and NFPA® 1500 require all SCBA to be inspected weekly, monthly, annually, and after each use (**Figure 5.66**). Inspecting SCBA for use at an emergency is described in **Skill Sheet 5-I-3**.

Daily/Weekly Inspections

Self-contained breathing apparatus requires proper care and inspection before and after each use to provide complete protection. Proper care includes making an inspection as soon as possible after reporting for duty. Some organizations may not be able to check the units every day. In this case, the SCBA should be checked at least weekly and before and after each use. While some of the following were listed under the Air Management section, they bear repeating here. The daily/weekly check should include:

Figure 5.66 SCBA must be inspected and cleaned regularly.

- Cylinder pressure — Cylinder should be filled to at least 90 percent of capacity.

- All gauges — The cylinder gauge and the remote gauge should read within 100 psi (700 kPa) of each other. Gauges not marked in increments of 100 psi (700 kPa) should read the same.

- Low-pressure alarm — The alarm should sound briefly when the cylinder valve is turned on and again as the pressure is relieved.

- All hose connections — Check to see that they are tight and free of leaks.

- Facepiece — Check to see that it is clean and in good condition.

- Harness system — Check to see that it is in good condition and straps are in the fully extended position.

- All valves — After checking the bypass valve, make sure it is fully closed.

- Built-in PASS devices — Check for proper operation.

Breathing apparatus should be cleaned and sanitized immediately after each use. Moving parts that are not clean may malfunction. A facepiece that has not been cleaned and sanitized may have an unpleasant odor and can spread germs to other department members who may wear the mask at a later time. An air cylinder with less air pressure than prescribed by the manufacturer renders the apparatus inefficient if not useless.

The facepiece should be thoroughly washed with warm water containing any mild commercial disinfectant and then rinsed with clear, warm water (**Figure 5.67**). Special care should be given to the exhalation valve to ensure proper operation. The air hose should be inspected for cracks or tears. The facepiece should be dried with a lint-free cloth or air dried.

Firefighters should avoid several actions that can damage SCBA:

- Do not use paper towels to dry the facepiece lens because the paper towel will scratch the plastic lens.

- Do not put water in or submerge low-pressure hoses on SCBA units with harness-mounted regulators.

Many departments now issue each firefighter a facepiece that is not shared with other firefighters. This eliminates the risk of spreading germs from one wearer to the next. Even though firefighters have their own facepieces, it is still important that they be cleaned after each use. The procedures to clean and sanitize PPE and SCBA are given in **Skill Sheet 5-I-4**.

Monthly Inspection and Maintenance

Monthly inspections should include removing the equipment from service and checking the following:

- All components for deterioration

- Leaks around valves and hose connections

- Operation of all gauges, valves, regulator, exhalation valve, and low-air alarm

Annual Inspection and Maintenance

Annual maintenance, testing, and repairs requiring the expertise of factory-certified technicians should be done in accordance with manufacturer's recommendations. This level of maintenance requires specialized training. The service provider must be able to disassemble the apparatus into its basic components and conduct tests using specialized tools and equipment generally not available to every fire department.

Air cylinders must be stamped or labeled with the date of manufacture and the date of the last hydrostatic test (**Figure 5.68**). According to both the U.S. Department of Transportation and Transport Canada, steel and aluminum cylinders must be tested every five years; composite cylinders every three years. Always empty cylinders before submitting them for servicing and testing.

Refilling Self-Contained Breathing Apparatus Cylinders

Air cylinders for self-contained breathing apparatus are filled from either a cascade system (a series of at least three, 300 cubic-foot [8.5 m³] cylinders) or directly from a compressor purification system (**Figures 5.69 a and b**). No matter how the cylinders are filled, the same safety precautions apply:

(1) Place the cylinder in a shielded fill station.

(2) Prevent the cylinder from overheating by filling slowly.

(3) Ensure that the cylinder is completely full but not overpressurized.

Skill Sheet 5-I-5 provides a sample procedure for filling an SCBA cylinder from a cascade system. **Skill Sheet 5-I-6** provides a sample procedure for filling an SCBA cylinder from a compressor/purifier.

Hydrostatic Test — A testing method that uses water under pressure to check the integrity of pressure vessels.

Figure 5.67 Facepieces must be washed and disinfected.

Figure 5.68 The air cylinder hydrostatic test date must be clearly visible.

NOTE: Some departments allow the air cylinders of SCBA units equipped with a RIC/UAV to be filled from a cascade system or compressor while still on the firefighter's back. To be in compliance with NFPA 1500, this may only be done if three conditions are met:

- NIOSH-approved fill options are used.

- The risk assessment process has identified procedures for limiting personnel exposure during the refill process and provides for adequate equipment inspection and member safety.

- There is an imminent life-threatening situation requiring immediate action to prevent the loss of life or serious injury.

Using Self-Contained Breathing Apparatus

Firefighters may have to wear SCBA in a variety of different environments — burning buildings, confined spaces, and areas contaminated with airborne hazardous materials. In addition to being familiar with the donning, operation, and doffing of protective breathing apparatus, firefighters must also be trained in safety considerations when using the equipment. The preceding sections of this chapter discussed why and how to operate self-contained breathing apparatus. This section covers safety precautions when using SCBA, emergency situations that may arise when using SCBA, and the use of SCBA in areas of obscured vision and restricted openings.

Safety Precautions for SCBA Use

Fire fighting can be a strenuous, demanding activity, so firefighters need to be in good physical condition. Although personal protective equipment is designed to

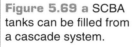

Figure 5.69 a SCBA tanks can be filled from a cascade system.

Figure 5.69 b SCBA tanks can be filled directly from a compressor.

protect firefighters, it can also work against them. The protective clothing required for structural fire fighting is as good at keeping heat in as it is at keeping heat out. It prevents body heat from escaping and restricts movement, both of which contribute to firefighter fatigue. These problems are intensified when self-contained breathing equipment is added. The difference between the weight of ordinary street clothes and fire fighting gear plus the SCBA unit has been measured at an extra 47 pounds (21 kg). The breathing unit alone can weigh from 25 to 35 pounds (11 kg to 16 kg), depending on size and type. Firefighters should be trained to recognize the signs and symptoms of heat-related illnesses that they can develop under these conditions. *Know your own capabilities and limitations!*

When using self-contained breathing apparatus, the following items should be remembered and observed for maximum safety:

- All firefighters who are certified to wear SCBA must be fit-tested annually or when new facepieces are issued to determine proper fit of the facepiece. Only firefighters with properly fitted facepieces will be allowed to operate in an IDLH atmosphere.

- Firefighters should closely monitor how they are feeling while wearing SCBA and rest when they become fatigued (**Figure 5.70**).

- Air supply duration will vary with the following:
  — Size of cylinder and beginning pressure
  — Firefighter conditioning
  — Task being performed
  — Level of training
  — Operational environment
  — Degree of stress
  — Other variables

- After entering a contaminated area, firefighters should not remove their breathing apparatus until they leave the contaminated area. Improved visibility does not ensure that the area is free of contamination. Before SCBA is removed, the atmosphere must be tested with properly calibrated instruments and found to be within safe limits.

- While in any IDLH atmosphere, firefighters must work in teams of two or more. Team members must remain in physical, voice, or visual contact with each other while in the hazardous area. Radio contact is *not* sufficient! If available, a thermal imaging camera (TIC) can help maintain contact.

While in the IDLH atmosphere, firefighters should check their air supply status frequently.

Emergency Situations

Emergencies created by malfunctioning protective breathing apparatus can be overcome in several ways. In all of these emergencies, conservation of air and immediate withdrawal from the hazardous atmosphere are of the utmost importance. **Skill Sheet 5-I-7** outlines emergency operations for SCBA.

Although SCBA regulators usually work as designed, they can malfunction. One method of using SCBA when the regulator becomes damaged or malfunctions is to intermittently open and close the bypass valve to allow air into the facepiece (**Figure 5.71**). Because the air is bypassing the regulator, it is under full cylinder pressure, so the bypass valve should be closed after each breath and then opened each time another breath is needed.

Figure 5.70 Firefighters should pace themselves when using SCBA. *Courtesy of Michael Watiker.*

The following are recommended actions in the event of an SCBA malfunction:

- Follow your departmental SOP for this type of situation.
- Do not panic! Panicking causes rapid breathing that uses more valuable air.
  — Control your breathing.
  — Alert other team members of your situation.
- Withdraw to a clear atmosphere:
  — With the other members of your team
  — Quickly but under control
  — Using any available exit opening, or creating one if necessary

While it is imperative that you stay in contact with the other members of your team, if you should become separated from them and you are lost or disoriented, the following actions are recommended:

- Use a portable radio (if you have one) to declare a Mayday!
- Follow your departmental SOP for this type of situation.
- Activate your PASS device.
- Stop and think:
  — How did you get to where you are?
  — Downstairs, upstairs?
  — Left turns, right turns?
- Hold your breath and listen:
  — For noise from other personnel
  — For hose and equipment operation
  — For sounds that indicate the location of fire
- Remember the different methods to find a way out:
  — Follow the hoseline out if possible (male coupling is closest to exit, female is closest to the fire) **(Figure 5.72)**.
  — Crawl in a straight line (hands on floor, move knee-to-hand).

Figure 5.71 The bypass valve can be used in an emergency to provide limited amounts of air when the low-pressure bell has activated.

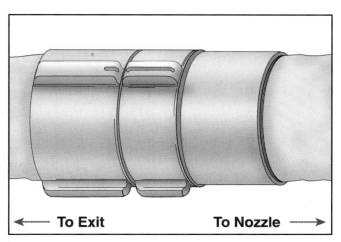

← **To Exit**   **To Nozzle** →

Figure 5.72 Fire hose couplings indicate direction. The male coupling indicates the direction to the exit.

— Crawl in one direction (all left-hand turns, all right-hand turns) once in contact with a wall.

— Call for directions, call out, or make noise for other firefighters to assist you.

— Break a window or breach a wall to escape if possible.

- Lie flat on the floor close to a wall so that you will be easier to find if you are exhausted or feel you may lose consciousness.

If you are separated from your team and are trapped in a room or area, take the following actions:

- Follow your departmental SOP for this type of situation.

- Use a portable radio (if you have one) to declare a Mayday and announce your last known location. Try to describe your surroundings – floor coverings, wall color, items in the room, etc.

- Activate your PASS device.

- Escape through any available opening or breach a wall to create a new one.

- Use your personal escape rope if a window is available.

- Control the door; close it or keep it closed (Unless it is an escape route).

- Place a flashlight on the floor with the light shining toward the ceiling.

- Slow your breathing as much as possible to conserve breathing air.

Emergency evacuation signals are used when the IC decides that all firefighters should immediately abandon a burning building or other hazard zone because conditions have deteriorated beyond the point of reasonable safety. All firefighters must be familiar with their department's method of sounding an evacuation signal. There are several ways this communication may be made. The two most common ways are to broadcast a radio message ordering interior crews to evacuate, or to sound the audible warning devices on the apparatus at the fire scene in a designated manner. The radio broadcast of an evacuation signal should be handled in a manner similar to that described for emergency traffic. The message should be broadcast several times to make sure that everyone hears it. The use of audible warning devices on apparatus, such as sirens and air horns, works when firefighters are in small structures but may not be heard by everyone when working in a large building. An evacuation signal is one of the actions that trigger a personnel accountability report (PAR) from all units on scene.

Personnel Accountability Report (PAR) — A roll call of all units (crews, teams, groups, companies, sectors) assigned to an incident. Usually by radio, the supervisor of each unit reports the status of the personnel within the unit at that time. A PAR may be required by SOP at specific intervals during an incident, or may be requested at any time by the IC or the ISO.

Special Uses of SCBA

In order to operate at maximum efficiency, firefighters must be able to operate effectively in areas of limited vision. They must also be able to negotiate tight passages without having to completely remove their breathing apparatus. The following sections address techniques for accomplishing these tasks.

Operating in Areas of Limited Visibility

In many instances where SCBA is required, firefighters will be operating in an area of limited visibility. Most interior fire attacks and many exterior attacks present firefighters with heavy smoke conditions that may reduce visibility to zero. Firefighters must learn techniques for safely moving about and performing critical tasks when vision is diminished.

One method of moving about in areas of limited visibility is crawling. Crawling is beneficial for several reasons. First, it allows firefighters to remain close to the floor and avoid the higher heat found closer to ceiling level. Crawling may also allow personnel to get below the thermal layer and increase visibility. Second, crawling allows firefighters to feel the floor in front of them with a tool as they move along. Probing with a tool prevents them from falling through holes burned or cut in the floor, falling down stairs or into elevator shafts, or running into furniture or other objects in their path. Crawling and probing also allows firefighters to feel for victims who may be lying on the floor, behind furniture, or in closets or shower stalls. If firefighters can see the floor, they may be able to move about using a crouched or "duck" walk **(Figure 5.73)**. This method is slightly faster than crawling but is more dangerous unless firefighters can clearly see the floor in front of them.

Any area of limited visibility is likely to also be an IDLH atmosphere. As mentioned earlier, firefighters in IDLH atmospheres must always operate in teams of two or more, and they should always have some sort of tag line that they can follow back to their point of entry. The tag line may be a hoseline, rope, or electrical cord. In the event that it becomes necessary to evacuate the structure in a hurry, the firefighters can turn around and follow the tag line to safety. If for some reason the team does not have a tag line or becomes separated from it, they should proceed to a wall and follow it until a door or window is found. If the team is trapped, tools will help them to break out or remove obstacles.

Figure 5.73 Stay low by crouching or crawling to increase visibility.

Tag Line — Nonload-bearing rope attached to an object to help steer it in a desired direction or act as a safety line.

Exiting Areas with Restricted Openings under Emergency Conditions

In an emergency, firefighters may need to exit through a restricted opening, that is, one that is too small to allow them to pass through while wearing SCBA in the normal manner. It may be necessary to slip out of the SCBA harness assembly while leaving the facepiece in place, exit the restricted area, and put the harness assembly back on. Remove only those parts of the SCBA necessary to exit the area. The specific procedures for removing the SCBA and maneuvering through an opening will be dictated by the type of equipment. Firefighters should be thoroughly familiar with their particular SCBA. Some things to keep in mind when using this technique include the following:

- Maintain contact with belt-mounted regulators at all times.

- Loosen straps as necessary to reduce your profile.

- Reduce your profile further by removing one or both backpack harness straps if absolutely necessary.

- Push the SCBA in front of you as necessary, maintaining control of the SCBA at all times.

Skill Sheet 5-I-8 gives the steps for exiting a constricted space while wearing SCBA.

Changing Cylinders

Firefighters are frequently required to change their SCBA cylinders during extended emergency operations. A salvage cover should be placed on the ground in the staging area to help protect spare cylinders. Cylinders that are empty should be segregated from the supply of cylinders that are serviced and ready for use. Cylinders that are damaged or otherwise out of service should be clearly marked as such and segregated from in-service cylinders. Changing cylinders can be either a one- or two-person job. **Skill Sheet 5-I-9** describes the one-person method for changing an air cylinder. When there are two people, the firefighter with an empty cylinder simply positions the cylinder so that it can be easily changed by the other firefighter (**Figures 5.74 a and b**). **Skill Sheet 5-I-10** describes the two-person method for changing an air cylinder.

Figure 5.74 a The second firefighter releases the empty cylinder when replacing the air cylinder without removing the SCBA completely.

Figure 5.74 b The second firefighter replaces the empty cylinder with a full one.

Summary

Because of the hostile environments in which they are often required to work, firefighters must have the best protective clothing and equipment available. But fighting fires, performing rescues, and delivering other emergency services are inherently dangerous activities. Even if their department furnishes firefighters with the latest personal protective equipment and they use it consistently and conscientiously, their safety is not guaranteed because all safety clothing and equipment has limitations. However, using protective clothing and equipment that is appropriate for the nature and scope of the incident gives firefighters the best chance of surviving without injury.

To use their protective clothing and equipment to the fullest, firefighters must be physically and psychologically fit. They must be aware of their equipment's capabilities and limitations — and their own. Firefighters must be thoroughly trained in the use of their protective gear and know how to cope with equipment failures during emergencies. Finally, they must be capable of maintaining their protective clothing and equipment so that it is ready when needed.

End-of-Chapter Review Questions

Fire Fighter I

1. What structural protective clothing is required by NFPA 1971?

2. List two guidelines for the proper care and maintenance of helmets.

3. What are the four common respiratory hazards associated with fires and other emergencies?

4. What factors affect firefighters' ability to use respiratory protection effectively?

5. When should firefighters check their air supply?

6. What are the four basic SCBA component assemblies?

7. What are PASS devices designed to do?

8. What checks should be made immediately prior to donning SCBA?

9. What should the daily/weekly check of protective breathing apparatus include?

10. What actions are recommended if a firefighter should become separated from team members?

Don Protective Clothing

Step 4. Don helmet.

Step 1: Don pants and boots according to manufacturer's guidelines, which includes suspenders in place.

Step 2: Don hood.

Step 3: Don coat, with closure secure and collar up.

Step 5. Don gloves.

Don SCBA: Over-the-Head Method

NOTE: The following are the general procedures for donning an SCBA. The specific SCBA manufacturer's recommendations for donning and use of the SCBA should always be followed.

Step 1: Position the SCBA with the valve end of the cylinder away from the body.

Step 2: Open cylinder valve fully.

Step 3: Check cylinder and regulator pressure gauges. Pressure readings within 100 psi OR needles on both pressure gauges indicate same pressure.

Step 5: Release the harness assembly and allow the SCBA to slide down the back.

Step 6: Fasten chest strap, buckle waist strap, and adjust shoulder straps.

Step 7: Don facepiece and check facepiece seal.

Step 4: Raise the SCBA overhead while guiding elbows into the loops formed by the shoulder straps.

Step 8: Connect air supply to facepiece.

Step 9: Activate PASS device.

Step 10: Don hood, helmet, and gloves.

Don SCBA: Coat Method

Step 1: Position SCBA with the valve end of the cylinder toward the body.

Step 2: Open cylinder valve fully.

Step 3: Check cylinder and regulator pressure gauges. Pressure readings within 100 psi OR needles on both pressure gauges indicate same pressure.

Step 4: Grasp the top of the left shoulder strap on the SCBA with the left hand and raise the SCBA overhead.

Step 5: Guide the left elbow through the loop formed by the left shoulder strap and swing SCBA around left shoulder.

Step 6: Guide the right arm through the loop formed by the right shoulder strap allowing the SCBA to come to rest in proper position.

Step 7: Fasten chest strap, buckle waist strap, and adjust shoulder straps.

Step 8: Don facepiece and check facepiece seal.

Step 9: Connect air supply to facepiece.

Step 10: Activate PASS device.

Step 11: Don hood, helmet and gloves.

Don SCBA: Seat-Mount Method

NOTE: Some department SOPs only allow seat-mounted SCBA or the facepiece to be donned upon arrival at the scene after the apparatus has stopped. Local procedures must be followed to ensure the safety of the firefighter. Specific steps for donning may vary by department according to local policy.

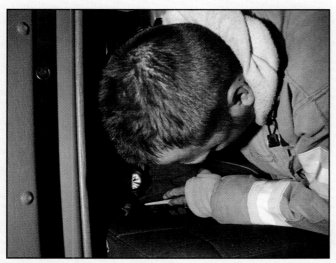

Step 1: Open cylinder valve fully.

Step 2: Check cylinder and regulator pressure gauges. Pressure readings within 100 psi OR needles on both pressure gauges indicate same pressure.

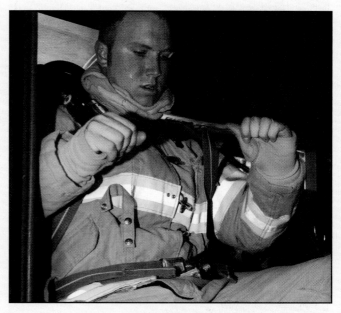

Step 4: Insert arms through shoulder straps.

Step 5: Fasten chest strap, buckle waist strap and adjust shoulder straps.

Step 3: Position body in seat with back firmly against the SCBA and release the SCBA hold-down device.

Step 6: Fasten seat belt before apparatus gets underway.

Step 7: Don facepiece and check facepiece seal.

Step 8: Connect air supply to facepiece.

Step 9: Activate PASS device.

Step 10: Don hood, helmet, and gloves.

Doff SCBA

Step 1: Remove SCBA.

Step 2: Close cylinder valve completely.

Step 3: Bleed air from high- and low-pressure hoses.

Step 4: Check air cylinder pressure and replace cylinder if less than 90% of rated capacity.

Step 5: Return all straps, valves and components back to ready state.

Step 6: Inspect SCBA and facepiece for damage and need for cleaning.

Step 7: Clean equipment as needed and remove damaged equipment from service and report to officer, if applicable.

Step 8: Place SCBA back in storage area so that it is ready for immediate use.

Doff PPE

Step 1: Remove protective clothing.

Step 2: Inspect PPE for damage and need for cleaning.

Step 3: Clean equipment as needed and remove damaged equipment from service and report to officer, if applicable.

Step 4: Place clothing in a ready state.

Inspect protective clothing

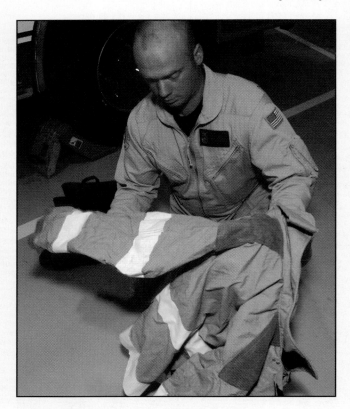

Step 1: Identify that all protective clothing is present: helmet, hood, coat and pants, suspenders (if applicable), boots, gloves.

Step 2: Inspect all articles of protective clothing for damage and cleanliness.

Step 3: Clean dirty components as necessary. If damage is found, remove from service and report damage to officer.

Step 4: Place protective clothing so it can be accessed quickly for donning in the event of a reported emergency.

5-I-3
Inspect PPE and SCBA for use at an emergency incident.

Skill SHEETS

Inspect SCBA

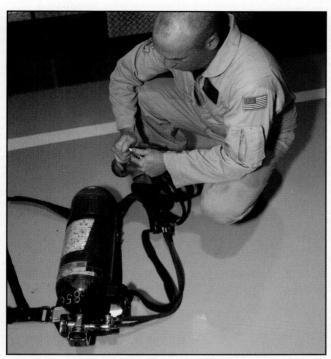

Step 1: Identify all components of SCBA are present: harness assembly, cylinder, facepiece, and PASS device.

Step 2: Inspect all components of SCBA for cleanliness and damage.

Step 3: If dirty components are found they are cleaned immediately. If damage is found, remove from service and report to officer.

Step 4: Check that cylinder is full (90%-100% of capacity).

Step 5: Open the cylinder valve slowly; verify operation of the low-air alarm and the absence of audible air leaks.

NOTE: On some SCBA, the audible alarm does not sound when the cylinder valve is opened.

Step 6: If air leaks are detected, determine if connections need to be tightened or if valves, donning switch, etc. need to be adjusted. Otherwise SCBA with audible leaks due to malfunctions shall be removed from service, tagged, and reported to the company officer.

Step 7: Check that gauges and/or indicators (i.e. heads-up display) are providing similar pressure readings. Manufacturers' guidelines determine the acceptable range.

Step 8: Check function (all modes) of PASS device.

Step 9: Don facepiece and check for proper seal.

NOTE: Not all facepieces are designed for a seal check without the regulator being attached and activated.

Step 10: Don regulator and check function by taking several normal breaths.

Step 11: Check bypass and/or purge valve, if applicable.

Step 12: Remove facepiece and prepare all components for immediate reuse.

Step 13: Place SCBA components so that they can be accessed quickly for donning in the event of a reported emergency.

Clean personal protective clothing (structural fire fighting)

Step 1: Clean all articles of protective clothing according to manufacturer's guidelines and departmental policies.

Step 2: Place all equipment in a manner and location so that it will dry.

Step 3: After equipment is dry, inspect for damage and place in a state of readiness.

Clean SCBA

Step 1: Prepare cleaning solution, buckets, etc. according to manufacturer's guidelines and departmental policies.

Step 2: Clean all components of SCBA unit according to manufacturer's guidelines and departmental policies.

Step 3: After equipment is clean, inspect for damage.

Step 4: Assemble components so they are in a state of readiness.

Step 5: Place all components in a manner and location so that they will dry.

NOTE: This skill sheet is only an example. The procedures outlined here may not be applicable to your cascade system. Always check the manufacturer's instructions before attempting to fill any cylinders.

Step 1: Check the hydrostatic test date of the cylinder.

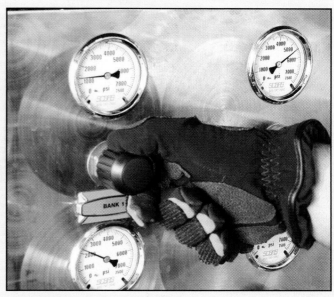

Step 4: Connect the fill hose to the cylinder and close bleed valve on fill hose.

Step 5: Open the SCBA cylinder valve.

Step 6: Open the valve at the fill hose, the valve at the cascade system manifold, or the valves at both locations if the system is so equipped.

NOTE: Some cascade systems may have a valve at the fill hose, at the manifold, or at both places.

Step 7: Open the valve of the cascade cylinder that has the least pressure but that has more pressure than the SCBA cylinder.

Step 2: Inspect the SCBA cylinder for damage such as deep nicks, cuts, gouges, or discoloration from heat. If the cylinder is damaged or is out of hydrostatic test date, remove it from service and tag it for further inspection and hydrostatic testing.

CAUTION: Never attempt to fill a cylinder that is damaged or that is out of hydrostatic test date.

Step 3: Place the SCBA cylinder in a fragment-proof fill station.

Step 8: Close the cascade cylinder valve when the pressures of the SCBA and the cascade cylinder equalize.

Step 9: Close the valve or valves at the cascade system manifold and/or fill line if the system is so equipped.

Step 10: Close the SCBA cylinder valve.

Step 11: Open the hose bleed valve to bleed off excess pressure between the cylinder valve and the valve on the file hose.

CAUTION: Failure to open the hose bleed valve could result in O-ring damage.

Step 12: Disconnect the fill hose from the SCBA cylinder.

Step 13: Remove the SCBA cylinder from the fill station.

Step 14: Return the cylinder to proper storage.

NOTE: This skill sheet is only an example. The procedures outlined here may not be applicable to your compressor/purifier system. Always check the compressor/purifier manufacturer's instructions before attempting to fill any cylinders.

Step 1: Check the hydrostatic test date of the cylinder.

Step 2: Inspect the SCBA cylinder for damage such as deep nicks, cuts, gouges, or discoloration from heat. If the cylinder is damaged or out of hydrostatic test date, remove it from service and tag it for further inspection and hydrostatic testing.

CAUTION: Never attempt to fill a cylinder that is damaged or that is out of hydrostatic test date.

Step 3: Place the SCBA cylinder in a fragment-proof fill station.

Step 4: Connect the fill hose to the cylinder and close bleed valve on fill hose.

Step 5: Open the SCBA cylinder valve.

Step 6: Turn on the compressor/purifier and open the outlet valve.

Step 7: Set the cylinder pressure adjustment on the compressor (if applicable) or manifold to the desired full-cylinder pressure. If there is no cylinder pressure adjustment, watch the pressure gauge on the cylinder during filling to determine when it is full.

Step 8: Open the manifold valve (if applicable), and again check the fill pressure.

Step 9: Open the fill station valve and begin filling the SCBA cylinder.

Step 10: Close the fill station valve when the cylinder is full.

Step 11: Close the SCBA cylinder valve.

Step 12: Open the hose bleed valve to bleed off excess pressure between the cylinder valve and valve on the fill station.

CAUTION: Failure to open the hose bleed valve could result in O-ring damage.

Step 13: Disconnect the fill hose from the SCBA cylinder.

Step 14: Remove the SCBA cylinder from the fill station and return the cylinder to proper storage.

Controlled Breathing Techniques

Step 1: Demonstrate a controlled breathing technique or skip breathing technique for two minutes.

Emergency Operation of SCBA

Step 1: Check regulator and open bypass valve. Close mainline if applicable.

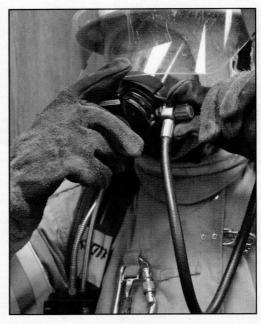

Step 2: Check main cylinder valve and verify it is fully opened.

Step 3: Check remote gauge or indicators, if applicable.

NOTE: When steps 1-3 do not correct problem, firefighter proceeds to Step 4.

Step 4: Use bypass valve to breathe.

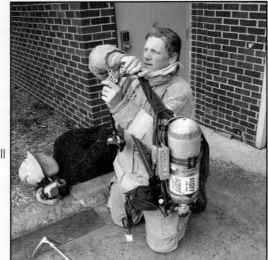

Step 5: Communicate with partner about situation and ask partner to call Mayday.

Step 6: Activate "alarm" mode on PASS device after Mayday is called.

Step 7: Exit hazardous atmosphere quickly.

Step 8: Notify Command after exiting building.

Step 9: Doff SCBA, tag unit, and remove from service.

Step 1: Don personal protective clothing and SCBA before entering hazardous atmosphere.

Step 2: Enter and negotiate obstacle course to the narrow passage or constricted exit. Maintain contact with wall or guideline/hoseline and team member (loud and clear communications). *(Lead team member):* Feel ahead with hand and tool.

Step 3: Reduce profile and attempt to pass through restriction.

Step 6: Pass through restricted opening while maintaining protection of full PPE.

Step 4: If unable to pass, notify Command of situation. Call a Mayday.

Step 5: If unable to pass, loosen parts of the SCBA harness or remove backpack completely as necessary.

NOTE: Some manufacturers do not recommend this procedure. Remember that this is a skill that simulates extreme conditions.

Step 7: Place SCBA back in correct position.

Step 8: Exit hazardous atmosphere and notify Command when safe.

Step 9: Doff SCBA and PPE when clear of hazardous atmosphere.

Step 1: Place the SCBA unit on a firm surface.

Step 2: Close the cylinder valve.

Step 3: Bleed air pressure from high- and low-pressure hoses.

Step 4: Disconnect the high-pressure coupling from the cylinder.

Step 5: Remove the empty cylinder from harness assembly.

Step 6: Verify that replacement cylinder is 90-100% of rated capacity.

Step 7: Check cylinder valve opening and the high-pressure hose fitting for debris.

Step 8: Place the new cylinder into the backpack.

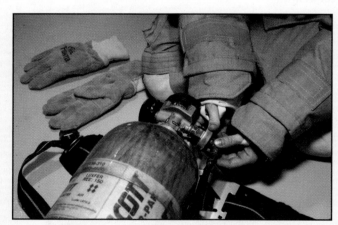

Step 9: Connect the high-pressure hose to the cylinder and hand-tighten.

Step 10: Slowly and fully open the cylinder valve and listen for an audible alarm and leaks as the system pressurizes.

NOTE: On some SCBA, the audible alarm does not sound when the cylinder valve is opened. You must know the operation of your own particular unit.

Step 11: If air leaks are detected, determine if connections need to be tightened or if valves, donning switch, etc. need to be adjusted. Otherwise SCBA with audible leaks due to malfunctions shall be removed from service, tagged, and reported to the officer.

Step 1: Disconnect the regulator from the facepiece or disconnect the low-pressure hose from the regulator.

Step 2: Position the cylinder for easy access by kneeling down or bending over.

Step 3: Fully close the cylinder valve.

Step 4: Release the air pressure from the high- and low-pressure hoses.

Step 5: Disconnect the high-pressure coupling from the cylinder.

Step 6: Remove the empty cylinder from harness assembly.

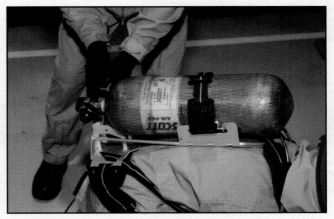

Step 7: Inspect replacement cylinder and ensure that cylinder is 90-100% of rated capacity.

Step 8: Place new cylinder into the harness assembly.

Step 9: Check the cylinder valve opening and the high-pressure hose fitting for debris, clearing any debris by quickly opening and closing cylinder valve.

Step 10: Connect the high-pressure hose to the cylinder and hand-tighten.

Step 11: Slowly open the cylinder valve fully and listen for an audible alarm and leaks as the system pressurizes.

Note: On some SCBA, the audible alarm does not sound when the cylinder valve is opened. You must know the operation of your own particular unit.

Step 12: Don regulator and take normal breaths.

Step 13: Check pressure reading on remote gauge and/or indicators and report reading.

Chapter Contents

Key Terms

Portable Fire Extinguishers

This chapter provides information that addresses the following job performance requirements of NFPA® 1001, *Standard for Fire Fighter Professional Qualifications* (2008):

<u>NFPA® 1001 references for Chapter 6:</u>

5.3.16

Chapter Objectives

Fire Fighter I Objectives

1. Describe methods by which agents extinguish fire.

2. List mechanisms by which portable extinguishers expel their contents.

3. Distinguish among classifications of fires and the most common agents used to extinguish them.

4. Describe types of extinguishers and their common uses.

5. Discuss extinguishers and agents for metal fires.

6. Explain the portable extinguisher rating system.

7. Describe factors to consider in selecting the proper fire extinguisher.

8. Describe items to check for immediately before using a portable fire extinguisher.

9. Describe the PASS method of application.

10. Summarize procedures that should be part of every fire extinguisher inspection.

11. Discuss damaged portable fire extinguishers and obsolete portable fire extinguishers.

12. Operate a stored pressure water extinguisher. (Skill Sheet 6-I-1)

13. Operate a dry chemical (ABC) extinguisher. (Skill Sheet 6-I-2)

14. Operate a carbon dioxide (CO_2) extinguisher. (Skill Sheet 6-I-3)

Portable Fire Extinguishers

Case History

Fire crews responded to a reported fire at an industrial structure at 3 a.m. The fire was located in the center of a 100,000 square foot facility. Firefighters entered the structure in full PPE including SCBA. Firefighters discovered the fire was contained to nine filter canisters in a large filtering system. Each filter canister contained magnesium and a hazardous material.

Because the fire involved flammable metal, the application of water was not feasible. Rather, firefighters successfully used two Class D and six A:B:C fire extinguishers to extinguish the fire. Firefighters continued to wear their PPE and SCBA during a difficult overhaul. Two firefighters did suffer mild heat exhaustion and one suffered a back injury. However, there were no serious injuries due to the use of PPE.

This case is an excellent example of firefighters using every tool at their disposal including portable fire extinguishers. This fire would have been difficult if not impossible to safely extinguish with the use of water. The other lesson from this incident is the importance of using all PPE including SCBA during fire situations, even when using portable fire extinguishers.

(Source: National Fire Fighter Near-Miss Reporting System)

Portable fire extinguishers are some of the most common fire protection appliances. They can be found in fixed facilities, such as homes and businesses, and on fire apparatus **(Figures 6.1 a and b, p. 234)**. Portable fire extinguishers are intended to be used on small fires in their incipient or early growth stage. In many cases, a portable extinguisher can control or extinguish a small fire in much less time than it would take to deploy a hoseline. In some metropolitan fire departments, one member on the initial attack team in high-rise fires carries either a water-type or a multipurpose portable extinguisher. If the fire proves to be still small, the extinguisher is used to put out the fire.

According to NFPA 1001, those qualified at the Fire Fighter I level must know the following about portable fire extinguishers:

- Classifications of types of fires
- Risks associated with each class of fire
- Operating methods of portable fire extinguishers
- Limitations of portable fire extinguishers

Fire Extinguisher — Portable fire fighting device designed to combat incipient fires.

 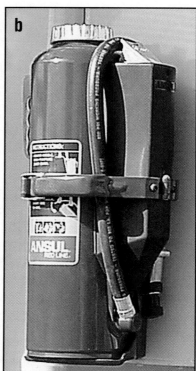

Figures 6.1 a and b Fire extinguishers may be located in fixed locations in facilities or on fire apparatus.

Those qualified at the Fire Fighter I level must be able to:

- Select appropriate extinguisher for size and type of fire
- Safely carry portable fire extinguishers
- Approach fire with portable fire extinguishers
- Operate portable fire extinguishers

There are no specific requirements related to portable fire extinguishers for those qualified at the Fire Fighter II level.

As a firefighter, it is important that you know the different types of portable fire extinguishers, their capabilities and limitations, and their safe and effective application. This chapter discusses the various types of portable fire extinguishers and how extinguishers are rated and inspected. Also covered are the steps involved in the selection and use of portable fire extinguishers. For additional information on the rating, placement (location), and use of portable extinguishers, see NFPA® 10, *Standard for Portable Fire Extinguishers*.

Types of Portable Fire Extinguishers

There are numerous types of portable fire extinguishers. This section highlights some of the common extinguishers used by fire service personnel. **Table 6.1** shows the operational characteristics of different types of portable fire extinguishers discussed in the following subsections. Every agent extinguishes fire by at least one of the following mechanisms:

- Smothering (oxygen exclusion)
- Cooling (reducing the burning material below its ignition temperature)
- Chain breaking (interrupting the chemical chain reaction)
- Saponification (forming an oxygen-excluding soapy foam)

Table 6.1
Operational Characteristics of Portable Fire Extinguishers

Extinguisher	Type	Agent	Fire Class	Size	Stream Reach	Discharge Time
Pump-Tank Water	Hand-carried; backpack	Water	A only	1 ½ - 5 gal (6 L to 20 L)	30-40 ft (9.1 m to 12.2 m)	45 sec to 30 min
Stored-Pressure Water	Hand-carried	Water	A only	1 ¼ - 2 ½ gal (5 L to 10 L)	30-40 ft (9.1 m to 12.2 m)	30-60 sec
Aqueous Film Forming Foam (AFFF)	Hand-carried	Water and AFFF	A & B	2 ½ gal (10 L)	20-25 ft (6.1 m to 7.6 m)	Approximately 50 sec
Halon 1211*	Hand-carried; wheeled	Halon	B & C	Hand-carried: 2 ½ - 20 lb (1 kg to 9 kg) Wheeled: to 150 lb (68 kg)	8-18 ft (2.4 m to 5.5 m) 20-35 ft (6.1 m to 10.7 m)	8-18 sec 30-44 sec
Halon 1301	Hand-carried	Halon	B & C	2 ½ lb (1 kg)	4-6 ft (1.2 m to 1.8 m)	8-10 sec
Clean Agent	Hand-carried	FE-36™	A, B, & C	1.4 to 15.5 lb. (.6 kg to 7 kg)	6-18 ft (1.8 m to 5.5 m)	9-14 sec
Carbon Dioxide	Hand-carried	Carbon Dioxide	B & C	2 ½ - 20 lb (1 kg to 9 kg)	3-8 ft (1 m to 2.4 m)	8-30 sec
Carbon Dioxide	Wheeled	Carbon Dioxide	B & C	50-100 lb (23 kg to 45 kg)	8-10 ft (2.4 m to 3 m)	26-65 sec
Dry Chemical	Hand-carried stored-pressure; cartridge-operated	Sodium bicarbonate, potassium bicarbonate, ammonium phosphate, potassium chloride	B & C	2 ½ - 30 lb (1 kg to 14 kg)	5-20 ft (1.5 m to 6.1 m)	8-25 sec
Multipurpose Dry Chemical	Hand-carried stored-pressure; cartridge-operated	Monoammonium phosphate	A, B, & C	2 ½ - 30 lb (1 kg to 14 kg)	5-20 ft (1.5 m to 6.1 m)	8-25 sec
Dry Chemical	Wheeled; ordinary or multipurpose		A, B, & C	75 - 350 lb (34 kg to 159 kg)	Up to 45 ft (13.7 m)	20 sec to 2 min
Dry Powder	Hand-carried; wheeled	Various, depending on metal fuel (this description for sodium chloride plus flow enhancers)	D only	Hand Carried: to 30 lb (14 kg) Wheeled: 150 lb & 350 lb (68 kg & 159 kg)	4-6 ft (1.2 m to 1.8 m)	28-30 sec
Wet Chemical	Hand-carried	Potassium Acetate	K only	2.5 gal (9.5 L)	8-12 ft (2.4 m to 3.6 m)	75-85 sec

Rating: Those larger than 9 lb (4 kg) capacity have small Class A Ratings (1-A to 4-A)

Table 6.2 lists the primary and secondary extinguishing mechanisms of various extinguishing agents. It is important to note that extinguishing agents that work by smothering are ineffective on materials that contain their own oxidizing agent. For example, a fire in magnesium or other combustible metals will flare up and intensify if water is applied.

Firefighters should not rely on privately owned extinguishers found in occupancies. These extinguishers may be inoperative as a result of improper maintenance, vandalism, or becoming obsolete. Responding firefighters should rely on the extinguishers carried on their apparatus. NFPA® 1901, *Standard for Automotive Fire Apparatus*, requires that pumping apparatus have two approved portable fire extinguishers with mounting brackets. These must be suitable for use on Class B and Class C fires. The

Extinguishing Agent — Any substance used for the purpose of controlling or extinguishing a fire.

Table 6.2
Extinguishing Agent Characteristics

Agent	Primary Mode	Secondary Mode
Water	Cooling	Oxygen depletion
Carbon Dioxide	Oxygen depletion	Cooling
Foam	Oxygen depletion	Vapor Suppression
Clean Agent	Chain inhibition	Cooling
Dry Chemical	Chain inhibition	Oxygen depletion
Wet Chemical	Oxygen depletion	Vapor suppression
Dry Powder	Oxygen depletion	Heat transfer cooling

stated minimum size requirement for a dry-chemical extinguisher is one with a rating of 80 B:C. The required rating for a carbon dioxide (CO_2) extinguisher is 10 B:C. Ratings represent the type of fire plus performance capability (see Extinguisher Rating System section). NFPA® 1901 also requires pumping apparatus to carry one 2½-gallon (10 L) or larger water extinguisher for use on Class A fires.

NOTE: Water-type extinguishers should be protected against freezing if they are going to be exposed to temperatures lower than 40°F (4°C). Freeze protection may be provided by adding antifreeze to the water or by storage in warm areas.

Expelling Agents

To be effective, every fire extinguisher must have some means of expelling the agent onto the fire. All portable fire extinguishers expel their contents by one of the following mechanisms:

- Manual pump
- Stored pressure
- Pressure cartridge

Fire Classifications

As discussed in Chapter 3, Fire Behavior, fires are classified according to the material that is burning. To choose the appropriate extinguisher for use on a given fire, you must first determine what is burning. Certain extinguishing agents are most effective on certain fuels (classes of fire). The five classes of fire are A, B, C, D, and K. The fire classification identifies the type of portable fire extinguisher that is most effective.

Class A

Class A fires involve ordinary combustibles such as textiles, paper, plastics, rubber, and wood. These fuels can be easily extinguished with water, water-based agents such as foam, or dry chemicals. Water is the most common extinguishing agent used by the fire service.

Class B

Class B fires involve flammable and combustible liquids, gases, and greases such as alcohol, cooking oils, gasoline, lubricating oils, and liquefied petroleum gas (LPG). Pressurized flammable liquids and gases are special fire hazards that should not

Dry Chemical — Extinguishing system that uses dry chemical powder as the primary extinguishing agent; often used to protect areas containing volatile flammable liquids.

be extinguished until the fuel gas is shut off. Special-hazard fires get larger as the fuel volume increases, such as when fuel storage tanks overflow or when pipes leak. Agents used to extinguish special hazard Class B fires are carbon dioxide (CO_2), dry chemical, and Class B foam.

Class C

Class C fires are Class A or Class B fires created by electrical energy. Because they are electrically conductive, water and water-based agents cannot safely be used on Class C fires until the electrical energy has been eliminated. Therefore, the recommended method of extinguishing Class C types of fires is to first turn off or disconnect the electrical power and then use the appropriate extinguisher, depending on what is burning.

Class D

Class D fires are those involving combustible metals and alloys such as lithium, magnesium, potassium, and sodium. Some common uses are in magnesium wheels and transmission components for automobiles and even some metal box springs. These types of fires can be identified by the bright white emissions from the combustion process.

Class D, dry powder extinguishers work best on these types of fires; however, do *not* confuse dry powder extinguishers with dry chemical units used primarily on Class B and Class C fires.

Class K

Class K fires involve combustible cooking oils. An example of these fuels is vegetable or animal fats and oils that burn at extremely high temperatures. While most of these fuels are found in commercial kitchens and industrial cooking facilities, they can also be found in private homes where high-temperature turkey fryers are used. Wet chemicals are used in the extinguishing systems and portable extinguishers for Class K fires.

Pump-Type Water Extinguishers

Pump-type water extinguishers are intended for use on small Class A fires only **(Figures 6.2 a - c, p. 238)**. There are several kinds of pump-type water extinguishers, but all operate in a similar manner. They are equipped with either a single- or double-acting pump.

Stored-Pressure Water Extinguishers

Stored-pressure water extinguishers, also called *air-pressurized water (APW) extinguishers* or *pressurized water extinguishers*, are useful for all types of small Class A fires. They are used often for extinguishing confined hot spots during overhaul operations **(Figure 6.3, p. 238)**.

Water is stored in a tank along with either compressed air or nitrogen. A gauge located on the side of the valve assembly shows when the extinguisher is properly pressurized **(Figure 6.4, p. 238)**. When the operating valve is activated, the stored pressure forces water up the siphon tube and out through the hose **(Figure 6.5, p. 238)**.

Class A foam concentrate is sometimes added to water extinguishers to increase their effectiveness **(Figure 6.6, p. 238)**. The addition of Class A foam serves as a wetting agent that aids in extinguishing deep-seated fires, such as those in upholstered furniture or vehicle seats, and wildland fires in densely matted vegetation.

Conductivity — The ability of a substance to conduct an electrical current.

CAUTION
The use of water or water-based agents on this type of fire will cause the fire to react violently, emit bits of molten metal, and possibly injure firefighters close by.

Alloy — Substance or mixture composed of two or more metals (or a metal and nonmetallic elements) fused together and dissolved into each other to enhance the properties or usefulness of the base metal.

Dry Powder — Extinguishing agent suitable for use on combustible metal fires.

Wet Chemical System — Extinguishing system that uses a wet-chemical solution as the primary extinguishing agent; usually installed in range hoods and associated ducting where grease may accumulate.

Figures 6.2 a and b Pump-type water extinguishers are available in a variety of designs. (a) The unit may be designed with the pump incorporated in the nozzle or (b) with the pump in the water tank. (c) A water vest extinguisher.

Figure 6.3 A typical stored-pressure water extinguisher. The unit must be inspected frequently to ensure that the proper air pressure is maintained in the tank.

Figure 6.4 The pressure gauge located on the top of the tank indicates the pressure available to expel the water.

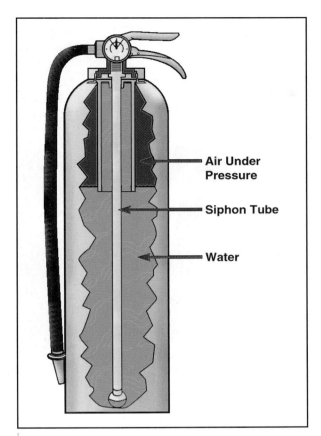

Figure 6.5 Cutaway of a stored-pressure water extinguisher.

Labels: Air Under Pressure, Siphon Tube, Water

Figure 6.6 Portable fire extinguishers containing Class A foam extinguishing agents are available. *Courtesy of Amerex.*

Water-Mist Stored-Pressure Extinguishers

Although very similar in appearance to standard stored-pressure water extinguishers, water-mist extinguishers use deionized water as the agent and nozzles that produce a fine spray instead of a solid stream. Because it is the impurities in water that make it electrically conductive, the deionized water also makes these Class A extinguishers safe to use on energized electrical equipment (Class C). The fine spray also enhances the cooling and soaking characteristics of the water and reduces scattering of the burning materials.

Wet Chemical Stored-Pressure Extinguishers

Also similar in appearance to standard stored-pressure water extinguishers, wet chemical (Class K) units are specifically designed to control and extinguish fires in deep fryers **(Figure 6.7, p. 240)**. These extinguishers contain a special potassium-based, alkaline agent formulated to cool and suppress fires in unsaturated cooking oils.

Aqueous Film Forming Foam (AFFF) Extinguishers

Aqueous film forming foam (AFFF) extinguishers are suitable for use on Class A and Class B fires. They are particularly useful in combating fires in or suppressing vapors from small liquid fuel spills.

Water Mist — In the fire service, water mist is associated with a fire extinguisher capable of atomizing water through a special applicator. Water-mist fire extinguishers use distilled water, while back-pump type water-mist extinguishers use ordinary water.

Aqueous Film Forming Foam (AFFF) — Synthetic foam concentrate that, when combined with water, can form a complete vapor barrier over fuel spills and fires and is a highly effective extinguishing and blanketing agent on hydrocarbon fuels. Also called light water.

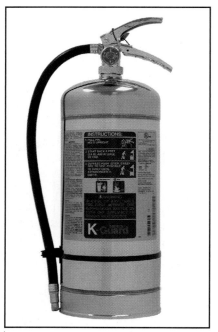

Figure 6.7 A typical Class K fire extinguisher. Class K extinguishers are found in commercial cooking facilities that operate deep fat fryers. *Courtesy of Ansul Corp.*

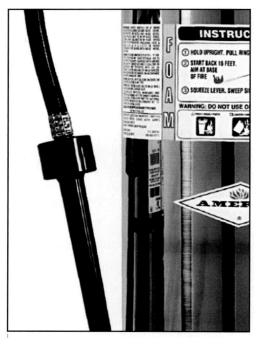

Figure 6.8 The air-aspirating nozzle of a foam extinguisher. *Courtesy of Amerex.*

Air-Aspirating Foam Nozzle — Foam nozzle especially designed to provide the aeration required to make the highest quality foam possible; most effective appliance for the generation of low-expansion foam.

AFFF extinguishers are different from stored-pressure water extinguishers in two ways. The AFFF extinguisher tank contains a specified amount of AFFF concentrate mixed with the water. It also has an air-aspirating nozzle that aerates the foam solution, producing a better-quality foam than a standard extinguisher nozzle provides (**Figure 6.8**).

The water/AFFF solution is expelled by compressed air or nitrogen stored in the tank with the solution. To prevent the disturbance of the foam blanket when applying the foam, it should not be applied directly onto the fuel; instead, it should be allowed to either gently rain down onto the fuel surface or deflect off an object (**Figures 6.9 a and b**).

Figures 6.9 a and b Two ways in which AFFF can be applied: raining down on the fuel or deflecting off an object.

When AFFF and water are mixed, the resulting finished foam floats on the surface of fuels that are lighter than water. The vapor seal created by the film of water extinguishes the flame and prevents reignition (**Figure 6.10**). The foam has good wetting and penetrating properties on Class A fuels but is ineffective on polar solvents (flammable liquids that are water-soluble) such as alcohol and acetone.

AFFF extinguishers are not suitable for fires in Class C or Class D fuels. They are not suitable for three-dimensional fires such as in fuel flowing down from an elevated point and fuel under pressure spraying from a leaking flange. They are most effective on static pools of flammable liquids.

NOTE: AFFF is corrosive and can remove paint from tools and apparatus.

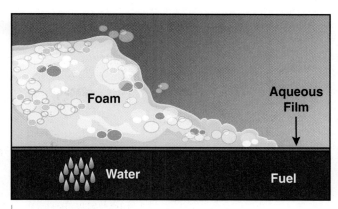

Figure 6.10 The thin film of AFFF floats ahead of the foam blanket providing a vapor seal over the fuel.

Clean Agent Extinguishers

Designed specifically as replacements for Halon 1211 (see Obsolete Fire Extinguishers section), these newer extinguishers use "clean agents" that are discharged as a rapidly evaporating liquid that leaves no residue. These clean agents include hydrochlorofluorocarbon (HCCF), hydrofluorocarbon (HFC), perfluorocarbon (PFC), or fluoroidiocarbon (FIC) (**Figure 6.11, p. 242**). These agents effectively cool and smother fires in Class A and Class B fuels, and the agents are nonconductive so they can be used on energized electrical equipment (Class C) fires. Pressurized with argon gas, clean agent extinguishers are approved by the U.S. Environmental Protection Agency (EPA).

Carbon Dioxide Extinguishers

Portable carbon dioxide (CO_2) fire extinguishers are found as both handheld units and wheeled units. CO_2 extinguishers are most effective in extinguishing Class B and Class C fires (**Figure 6.12, p. 242**). Because their discharge is in the form of a gas, they have a limited reach and the gas can be dispersed by wind. They do not require freeze protection.

Carbon dioxide is stored under its own pressure as a liquefied gas ready for release at anytime. The agent is discharged through a plastic or rubber horn on the end of either a short hose or tube. The gaseous discharge is usually accompanied by dry ice crystals or carbon dioxide "snow." This snow sublimes — changes into a gaseous form without becoming a liquid — shortly after discharge (**Figure 6.13, p. 242**). When released, the carbon dioxide gas displaces available oxygen and smothers the fire. Even though CO_2 discharges at subzero temperature, it has little if any cooling effect on fires. Carbon dioxide produces no vapor-suppressing film on the surface of the fuel; therefore, reignition is always a danger.

Carbon dioxide wheeled units are similar to the handheld units except that they are considerably larger (**Figure 6.14, p. 242**). Wheeled units are most commonly used in airports and industrial facilities. After being wheeled to the fire, the hose (usually less than 15 feet [5 m] long) must be deployed or unwound from the unit before use. The principle of operation is the same as in the smaller handheld units.

CAUTION

When carbon dioxide is discharged, a static electrical charge builds up on the discharge horn. Touching the horn before the charge has dissipated can result in a shock.

Figure 6.11 Clean agents are nonconductive and can be used on energized electrical equipment. *Courtesy of Ansul Corp.*

Figure 6.12 A typical handheld CO_2 fire extinguisher, distinguishable by the large discharge horn. *Courtesy of Ansul Corp.*

Figure 6.13 Carbon dioxide stream is visible in this photo.

Figure 6.14 A typical wheeled CO_2 fire extinguisher usually found at airports and refineries. *Courtesy of Badger Fire Protection.*

Dry Chemical Extinguishers

The terms *dry chemical* and *dry powder* are often incorrectly used interchangeably. Dry chemical agents are for use on Class A-B-C fires and/or Class B-C fires; dry powder agents are used on Class D fires only. Dry chemical extinguishers are among the most common portable fire extinguishers in use today. There are two basic types of dry chemical extinguishers: (1) regular B:C-rated and (2) multipurpose and A:B:C-rated (see Extinguisher Rating System section) **(Figures 6.15 a and b)**. Unless specifically noted in this section, the characteristics and operation of both types are exactly the same. The following are commonly used dry chemicals:

- Sodium bicarbonate
- Potassium bicarbonate
- Urea-potassium bicarbonate
- Potassium chloride
- Monoammonium phosphate

During manufacture, these agents are mixed with small amounts of additives that make the agents moisture-resistant and prevent them from caking. This process keeps the agents ready for use even after being stored for long periods, and it makes them free flowing.

Some dry chemical agents are not compatible with foam. Monoammonium phosphate and some sodium bicarbonate agents will break down the foam blanket when applied in conjunction with or after foam.

The dry-chemical agents themselves are nontoxic and generally considered quite safe to use. However, the cloud of chemicals may reduce visibility and create respiratory problems like any airborne particulate. Some dry chemicals are compatible with foam, but others will degrade the foam blanket. On Class A fires, the discharge should be directed at whatever is burning in order to cover it with chemical. When the flames have been knocked down, the agent should be applied intermittently as needed on any smoldering hot spots. Many dry chemical agents are corrosive to metals, so it may be better to use another agent such as carbon dioxide on them.

> **WARNING!**
> Never mix or contaminate dry chemicals with any other type of agent because they may chemically react and cause a dangerous rise in pressure inside the extinguisher.

Corrosive Materials — Gaseous, liquid, or solid materials that can burn, irritate, or destroy human skin tissue and can severely corrode steel. Also called Corrosives.

Figures 6.15 a and b There are two types of agents used in dry chemical extinguishers: (a) B:C rated extinguishers and (b) A:B:C-rated extinguishers. *(a) Courtesy of Ansul Corp.*

Handheld Units

There are two basic designs for handheld dry chemical extinguishers: *cartridge-operated* and *stored pressure* (**Figure 6.16**). The stored-pressure type is similar in design to the air-pressurized water extinguisher, and a constant pressure of about 200 psi (1 400 kPa) is maintained in the agent storage tank. Cartridge-operated extinguishers employ a pressure cartridge connected to the agent tank (**Figure 6.17**). The agent tank is not pressurized until a plunger is pushed to release the gas from the cartridge. Both types of extinguishers use either nitrogen or carbon dioxide as the pressurizing gas. Cartridge-operated extinguishers use a carbon dioxide cartridge unless the extinguisher is going to be subjected to freezing temperatures; in such cases, a dry nitrogen cartridge is used.

Figure 6.16 Dry chemical fire extinguishers may be of the cartridge-operated or stored-pressure types. *Courtesy of Ansul Corp.*

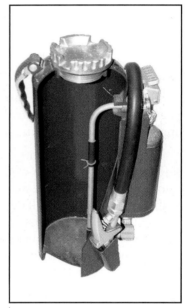

Figure 6.17 Cutaway of a cartridge-operated dry chemical fire extinguisher.

Figure 6.18 A typical wheeled dry chemical fire extinguisher. *Courtesy of Ansul Corp.*

Wheeled Units

Dry chemical wheeled units are similar to the handheld units but are on a larger scale (**Figure 6.18**). They are rated for Class A, Class B, and Class C fires based on the dry chemical in the unit (see Extinguisher Rating System section).

Operating the wheeled dry chemical extinguisher is similar to operating the handheld, cartridge-type dry chemical extinguisher. The extinguishing agent is kept in one tank and the pressurizing gas is stored in a separate cylinder. When the extinguisher is in position at a fire, the hose should first be stretched out completely (**Figure 6.19**). Once the agent storage tank and hose are charged, it can make removing the hose more difficult and the powder can sometimes pack in any sharp bends in the hose. The pressurizing gas should be introduced into the agent tank and allowed a few seconds to fully pressurize the tank before the nozzle is opened. The agent is applied in the same manner as described for the handheld, cartridge-type dry-chemical extinguishers.

Figure 6.19 The hose on the wheeled dry chemical extinguisher must be deployed before the extinguisher is pressurized.

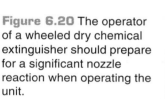

Figure 6.20 The operator of a wheeled dry chemical extinguisher should prepare for a significant nozzle reaction when operating the unit.

Figures 6.21 a and b Handheld and wheeled Class D fire extinguishers. *(b) Courtesy of Amerex Corp.*

Extinguishers and Agents for Metal Fires

Except for protecting nearby combustibles, the extinguishing agents discussed so far in this chapter are ineffective on Class D (combustible metal) fires.

Special extinguishing agents and application techniques have been developed to control and extinguish metal fires. No single agent will control or extinguish fires in all combustible metals. Some agents are effective against fires in several metals; others are effective on fires in only one type of metal. Some powdered agents can be applied with portable extinguishers, but others must be applied by either a shovel or a scoop. The appropriate application technique for any given dry powder is described in the manufacturer's technical sales literature. Firefighters should be thoroughly familiar with the information that applies to any agent carried on their emergency response vehicles.

Portable extinguishers for Class D fires come in both handheld and wheeled models **(Figures 6.21 a and b)**. Whether a particular dry powder is applied with an extinguisher or with a scoop, it must be applied in sufficient depth to completely cover the area that is burning to create a smothering blanket **(Figure 6.22, p. 246)**. The agent should be applied gently to avoid breaking any crust that may form over the burning metal. If the crust is broken, the fire may flare up and

Figure 6.22 To be effective, the dry powder must completely cover the area that is burning.

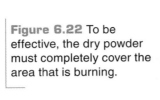

CAUTION

Water applied to a combustible metal fire results in a violent reaction that intensifies the combustion and causes bits of molten material to spatter in every direction, which can spread the fire.

Figure 6.23 Class D agent may be applied with a shovel.

Smothering — Act of excluding oxygen from a fuel.

expose more uninvolved material to combustion. Care should be taken to avoid scattering the burning metal. Additional applications may be necessary to cover any hot spots that develop.

If a small amount of burning metal is on a combustible surface, the fire should first be covered with powder. Then, a layer of powder 1 to 2 inches (25 mm to 50 mm) deep should be spread nearby and the burning metal shoveled onto this layer with more powder added as needed **(Figure 6.23)**. After extinguishment, the material should be left undisturbed until the mass has cooled completely. Only then should disposal be attempted.

Portable Fire Extinguisher Rating System

As discussed earlier in this chapter, portable fire extinguishers are classified according to the types of fire (A, B, C, D, or K) for which they are intended **(Figure 6.24)**. In addition to the classification represented by the letter, Class A and Class B extinguishers are also rated according to performance capability, which is represented by a number. The classification and numerical rating system is based on tests conducted by Underwriters Laboratories Inc. (UL) and Underwriters Laboratories of Canada (ULC). These tests are designed to determine the extinguishing capability for each size and type of extinguisher. The newest fire extinguisher classification, Class K, is used on cooking medium fires.

Class A Ratings

Class A portable fire extinguishers are rated from 1-A through 40-A. The Class A rating of water extinguishers is primarily based on the amount of extinguishing agent and the duration and range of the discharge used in extinguishing test fires. For a 1-A rating, 1¼ gallons (5 L) of water are required. A 2-A rating requires 2½ gallons (10 L) or twice the 1-A capacity. These ratings are based on test fires in various sizes of fuel cribs (**Figure 6.25**).

Class B Ratings

Extinguishers suitable for use on Class B fires are classified with numerical ratings ranging from 1-B through 640-B. The rating is based on the approximate square foot (square meter) area of a flammable liquid fire that a nonexpert operator can extinguish. The nonexpert operator is expected to extinguish 1 square foot (0.09 m²) for each numerical rating or value of the extinguisher rating.

Class C Ratings

There are no fire extinguishing capability tests specifically conducted for Class C ratings. Because electricity does not burn, extinguishers for use on Class C fires receive that letter rating because Class C fires are essentially Class A or Class B fires involving energized electrical equipment. The extinguishing agent is tested for electrical nonconductivity. The Class C rating confirms that the extinguishing agent will not conduct electricity. The Class C rating is assigned in addition to the rating for Class A and/or Class B fires.

Figure 6.24 Extinguishers are classified according to their intended use. Icons, such as the ones on this label, indicate that this is intended for Class A fires but prohibited for use on Class B and C fires.

Figure 6.25 Class A ratings are based on the ability of the extinguisher to control fires in wood cribs. *Courtesy of Ansul Corp.*

Figure 6.26 A typical faceplate on a Class D fire extinguisher.

Figure 6.27 A typical Class K fire extinguisher.

Class D Ratings

Test fires for establishing Class D ratings vary with the type of combustible metal being tested. The following factors are considered during each test:

- Reactions between the metal and the agent
- Toxicity of the agent
- Toxicity of the fumes produced and the products of combustion
- Time to allow metal to burn out without fire suppression efforts versus time to extinguish the fire

When an extinguishing agent is determined to be safe and effective for use on a combustible metal, the application instructions are included on the faceplate of the extinguisher, although no numerical rating is given (**Figure 6.26**). Class D agents cannot be given a multipurpose rating for use on other classes of fire.

Class K Ratings

Class K has been recognized by both UL and ULC since 1996 (**Figure 6.27**). Extinguishers suitable for this rating must be capable of saponifying (converting the fatty acids or fats to a soap or foam) vegetable oil, peanut oil, canola oil, and other oils with little or no fatty acids. Wet chemical agents containing an alkaline mixture, such as potassium acetate, potassium carbonate, or potassium citrate, work by suppressing the vapors and smothering the fire. Any of these agents capable of extinguishing a fire from a deep fryer using these light oils with a surface area of 2.25 square feet (0.2 m²) meet the minimum criteria for Class K rating.

Multiple Markings

Extinguishers suitable for more than one class of fire are identified by combinations of the letters A, B, and/or C or the symbols for each class. The three most common combinations are Class A-B-C, Class A-B, and Class B-C. All new portable fire extinguishers must be labeled with their appropriate markings. Any extinguisher not properly marked is not a listed unit and should not be used.

The ratings for each separate class of extinguisher are independent and do not affect each other. To better understand the rating system, a common-sized extinguisher, such as the multipurpose extinguisher rated 4-A 20-B:C, should extinguish a Class A fire that is 4 times larger than a 1-A fire, extinguish approximately 20 times as much Class B fire as a 1-B extinguisher, and extinguish a deep-layer flammable liquid fire of 20 square-feet (1.8 m²) in area. It must also be nonconductive so it is safe to use on fires involving energized electrical equipment.

Portable fire extinguishers are identified in two ways. One system uses geometric shapes of specific colors with the class letter shown within the shape. The second system, currently recommended in NFPA® 10, uses pictographs to make the selection of the most appropriate fire extinguishers easier. It also shows the types of fires on which extinguishers should *not* be used. Regardless of which system is used, it is important that the markings are clearly visible (**Table 6.3**).

Selecting and Using Portable Fire Extinguishers

When a small fire is discovered and the steps described earlier have been taken, the next steps involve selecting the proper extinguisher and using it correctly. There are a number of factors to consider when deciding which of the available fire ex-

Table 6.3
Classification of Fire

Class Name	Letter Symbol	Image Symbol	Description
Class A or Ordinary Combustibles	**A** Ordinary Combustibles		Includes fuels such as wood, paper, plastic, rubber, and cloth
Class B or Flammable and Combustible Liquids and Gases	**B** Flammable Liquids		Includes all hydrocarbon and alcohol based liquids and gases that will support combustion.
Class C or Electrical	**C** Electrical Equipment		This includes all fires involving energized electrical equipment.
Class D or Combustible Metals	**D**		Examples of combustible metals are: magnesium, potassium, titanium, and zirconium.
Class K or Kitchen	**K**		Includes unsaturated cooking oils in well-insulated cooking appliances located in commercial kitchens.

Reproduced with permission from Wayne State University, Detroit, MI.

tinguishers to choose for the situation as well as a number of things to remember when using the extinguisher you selected. The following section discusses all of these considerations.

Selecting the Proper Fire Extinguisher

Selection of the proper portable fire extinguisher depends on numerous factors:

- Classification of the burning fuel
- Rating of the extinguisher
- Hazards to be protected
- Size and intensity of the fire
- Atmospheric conditions
- Any life hazard or operational concerns
- Ease of handling extinguisher
- Availability of trained personnel

Select extinguishers that minimize the risk to life and property and are effective in extinguishing the fire. Because of their corrosive particulate residue, do not select dry chemical extinguishers for use in areas where highly sensitive computer equipment is located. The residue left afterwards could potentially do more damage to the sensitive electronic equipment than the fire. In these particular areas, clean agent or carbon dioxide extinguishers are better choices **(Figure 6.28)**.

Using Portable Fire Extinguishers

Portable extinguishers come in many sizes and types. While the operating procedures of each type of extinguisher are similar, firefighters should become familiar with the detailed instructions found on the label of the extinguisher. Fire extinguishers on emergency response vehicles must be inspected regularly to ensure that they are accessible and operable (see Inspecting Fire Extinguishers section). **Skill Sheets 6-I-1 through 6-I-3** detail general operating procedures for extinguishing Class A, Class B, and Class C fires.

Before attempting to use any fire extinguisher in an emergency, quickly check it first **(Figure 6.29)**. Such a check is necessary to ensure that the extinguisher is charged and operable. This check may protect you from injury caused by a defective or depleted extinguisher. If the extinguisher appears to be in working order, you can then use it to suppress a fire.

When inspecting an extinguisher immediately before use, check the following:

- External condition — no apparent damage
- Hose/nozzle — in place
- Weight — feels as though it contains agent
- Pressure gauge (if available) — in operable range

Escape Route — Pathway to safety. It can lead to an already burned area, a previously constructed safety area, a meadow that will not burn, or a natural rocky area that is large enough to take refuge without being burned. When escape routes deviate from a defined physical path, they must be clearly marked (flagged).

Figure 6.28 Clean agent extinguishers are used to protect computers, electronic equipment, and valuable archives, among other items.

Figure 6.29 Quickly check the extinguisher before attempting to use it.

After selecting the appropriate size and type of extinguisher for the situation, approach the fire from the windward side; that is, with the wind at your back (**Figure 6.30**).

PASS Method of Application

All modern fire extinguishers are operated in a similar manner. No modern fire extinguisher has to be inverted to operate; therefore, after performing the quick check described earlier, pick up the extinguisher by its handles and carry it to the point of application. Once in position to attack the fire, use the PASS method:

P – Pull the pin (breaking the thin wire or plastic seal)

A – Aim the nozzle (at whatever is burning)

S – Squeeze the handles together (to release the agent)

S – Sweep the nozzle back and forth (to cover the burning material)

Be sure the extinguishing agent reaches the fire — if it cannot, the agent is wasted (**Figure 6.31**). Smaller extinguishers require a closer approach to the fire than larger units, thus radiant heat may prevent you from getting close enough for the agent to reach the fire. Adverse winds also can limit the reach of an agent.

Operating an extinguisher close to the fire can sometimes scatter lightweight solid fuels or penetrate the surface of liquid fuels. Apply the agent from a point where it reaches but does not disturb the fuel. Releasing the handles will stop the flow of agent.

After the fire is knocked down, you may move closer to achieve final fire extinguishment. If extinguishment is not achieved after an entire extinguisher has been discharged onto the fire, withdraw and reassess the situation. If the fire is in a solid fuel that has been reduced to the smoldering phase, it may be overhauled using an appropriate tool to pull it apart. A charged hoseline can then be used to soak it well enough to achieve complete extinguishment. If the fire is in a liquid fuel, it may be necessary to either apply the appropriate type of foam through a hoseline or simultaneously attack the fire with more than one extinguisher.

Figure 6.30 Approach the fire from the windward side.

Figure 6.31 Make sure that the agent reaches the fire.

If more than one extinguisher is used simultaneously, work in unison with the other firefighters and maintain a constant awareness of each other's actions and positions. Lay empty fire extinguishers on their sides after use. This signals to others that they are empty and reduces the chance of someone taking one and approaching a fire with an empty extinguisher.

Inspecting Portable Fire Extinguishers

NFPA® 10 and most fire codes require that portable fire extinguishers be inspected at least once each year to ensure that they are accessible and operable. Verify that extinguishers are in their designated locations, that they have not been activated or tampered with, and that there is no obvious physical damage or other condition present that would prevent their operation. Servicing of portable fire extinguishers (and all other privately owned fire suppression or detection equipment) is the responsibility of the property owner or building occupant.

Although it is usually performed by the building owner or the owner's designee, firefighters should include extinguisher inspections in their building inspection program (**Figure 6.32**). When inspecting fire extinguishers, remember that there are three factors that determine the value of a fire extinguisher:

- Its serviceability
- Its accessibility
- Its simplicity of operation

NFPA® 10 describes procedures for the hydrostatic testing of extinguisher cylinders as required by both the U.S. Department of Transportation and Transport Canada. The test results must be affixed to the extinguisher shell. The hydrostatic test results on high- and low-pressure cylinders are recorded differently. Maintenance personnel should refer to NFPA® 10 for specific information on extinguisher testing and documentation.

The following procedures should be part of every fire extinguisher inspection:

- Check to ensure that the extinguisher is in a proper location and that it is accessible (**Figure 6.33**).
- Inspect the discharge nozzle or horn for obstructions. Check for cracks and dirt or grease accumulations.

Figure 6.32 Check the gauge to ensure that the unit is properly charged and the service tag to ensure that the inspections are being carried out.

Figure 6.33 Make sure that access to the extinguisher is not obstructed.

- Inspect the extinguisher shell for any physical damage.
- Check to see if the operating instructions on the extinguisher nameplate are legible.
- Check the locking pin and tamper seal to ensure that the extinguisher has not been discharged or tampered with.
- Determine if the extinguisher is full of agent and fully pressurized by checking the pressure gauge, weighing the extinguisher, or inspecting the agent level. If an extinguisher is found to be deficient in weight by 10 percent, it should be removed from service and replaced.
- Check the inspection tag for the date of the previous inspection, maintenance, or recharging.
- Examine the condition of the hose and its associated fittings.

If any of the items listed are deficient, remove the extinguisher from service and repair it in accordance with department policies. Replace the extinguisher with one that has an equal or greater rating.

NOTE: Regularly inspect extinguishers that are carried on fire apparatus.

Damaged Portable Fire Extinguishers

Damaged fire extinguishers can fail at any time and could result in serious injury to the user or anyone standing nearby. Leaking, corroded, or otherwise damaged extinguisher shells or cylinders should be discarded or returned to the manufacturer for repair **(Figure 6.34)**.

If an extinguisher shows only slight damage or corrosion but it is uncertain whether or not the unit is safe to use, it should be hydrostatically tested by either the manufacturer or a qualified testing agency. If allowed by departmental SOP, firefighters can replace leaking hoses, gaskets, nozzles, and loose labels.

Obsolete Portable Fire Extinguishers

American manufacturers stopped making inverting-type fire extinguishers in 1969. These included soda-acid, foam, internal cartridge-operated water and loaded stream, and internal cartridge dry-chemical extinguishers **(Figure 6.35)**. Manufacturing of extinguishers made of copper or brass with cylinders either soft soldered or riveted together was also discontinued at that time. Because of the toxicity of carbon tetrachloride and chlorobromomethane, extinguishers using these agents were prohibited in the workplace. OSHA regulations *29 CFR* 1910.157, Subpart "L" (c) (5), (dated September 12, 1980) required employers to permanently remove all of these obsolete extinguishers from service by January 1982. Occasionally, firefighters discover obsolete fire extinguishers in older buildings. If the occupant asks firefighters to dispose of an obsolete fire extinguisher, they should follow their departmental standard operating procedures.

Figure 6.34 Damaged extinguishers should be removed from service until repaired and tested.

CAUTION
Never attempt to repair the shell or cylinder of a defective fire extinguisher. Contact the manufacturer for instructions on where to have it repaired or replaced.

Figure 6.35 Obsolete fire extinguishers should be removed from service.

Halon Fire Extinguishers

Because of their ozone-depleting potential, halogenated extinguishing agents were included in the *Montreal Protocol on Substances that Deplete the Ozone Layer* adopted in 1987. This international treaty required a complete termination of the production and consumption of halogens by the year 2000. The only exceptions allowed under the agreement were for essential uses where no suitable alternatives are available. The United States unilaterally stopped producing halogens at the end of 1993, and research was begun on possible alternative extinguishing agents. The following information on halon extinguishers is included because these units may still be in service.

Halon is a generic term for halogenated hydrocarbons and is defined as *a chemical compound that contains carbon plus one or more elements from the halogen series (fluorine, chlorine, bromine, or iodine)*. While a large number of halogenated compounds exist, only a few are used to a significant extent as fire extinguishing agents. The two most common ones are Halon 1211 (bromochlorodifluoromethane) and Halon 1301 (bromotrifluoromethane).

Halogenated agents extinguish fire by interrupting the chain reaction of the combustion process. Halogenated vapor is nonconductive and effective on extinguishing surface fires in flammable and combustible liquids and energized electrical equipment. These agents are not effective on fires in self-oxidizing fuels such as combustible metals, organic peroxides, and metal hydrides. Although halons have long been used for the protection of internal combustion engines, their primary modern-day application is for the protection of sensitive electronic equipment such as computers. Halons are desirable in this application because unlike dry chemical agents, they do not leave any residue to contaminate the inner workings of the equipment. Although carbon dioxide leaves no residue, the extremely cold temperature of the agent can harm sensitive electronic components and may erase magnetic memory.

Halon 1211

Halon 1211 has been replaced by FE-36™ hexafluoropropane. Halon 1211 extinguishers were intended primarily for use on Class B and Class C fires. However, Halon 1211 extinguishers greater than 9 pounds (4 kg) in capacity also had a low Class A rating (1-A to 4-A, depending on size). Larger Halon 1211 extinguishers were found as wheeled units up to 150 pounds (68 kg).

Halon 1211 was stored in the extinguisher as a liquefied compressed gas, but nitrogen was added to the tank to increase discharge pressure and stream reach. Halon 1211 was discharged from an extinguisher in a clear liquid stream, giving it greater reach than a gaseous agent; nonetheless, the stream could be affected by wind when operated outside.

Halon 1301

Halon 1301 has now been replaced by FE-241™ chlorotetrafluoroethane and FM-200® heptafluoropropane. Halon 1301 was normally not used by itself in portable fire extinguishers because the agent was discharged as a nearly invisible gas highly susceptible to being affected by wind. In a confined space, such as a computer room, the agent's volatility allowed it to disperse faster than Halon 1211. For this reason and because it is effective at a lower concentration than Halon 1211, Halon 1301 was the agent of choice in most total-flooding systems using halogenated agents. For more information on flooding systems, see Chapter 16.

Summary

In many cases, fire extinguishers can control or extinguish small fires in less time than it takes to deploy a hoseline. However, even though portable fire extinguishers may be found in many of the homes, apartments, and businesses that you must enter to extinguish fires, you should only rely on those carried on your fire apparatus. To use fire extinguishers safely and effectively, you must know their capabilities and limitations — and your own — as well as the proper techniques for their application.

End-of-Chapter Review Questions

Fire Fighter I

1. What are the five classes of fire and what do they involve?

2. What fires are aqueous film forming foam (AFFF) extinguishers most effective on?

3. How do carbon dioxide (CO_2) portable fire extinguishers work?

4. What are the three most common combinations for extinguishers with multiple markings?

5. List three factors that affect the selection of the proper portable fire extinguisher.

6. What should be checked immediately before using a portable extinguisher?

7. What is the PASS method of application?

8. What procedures should be part of every fire extinguisher inspection?

Step 1: Size up fire, ensuring that it is safe to fight with an extinguisher.

Step 2: Pull pin at top of extinguisher to break the inspection band.

Step 3: Test to ensure proper operation.

Step 4: Carry extinguisher to within stream reach of fire.

Step 5: Aim nozzle toward base of fire.

Step 6: Discharge extinguishing agent and sweep slowly back and forth across entire width of fire.

Step 7: Cover entire area with water until fire is completely extinguished.

Step 8: Back away from the fire area.

Step 9: Tag extinguisher for recharge and inspection.

Step 1: Size up fire, ensuring that it is safe to fight with an extinguisher.

Step 2: Pull pin at top of extinguisher to break the inspection band.

Step 3: Test to ensure proper operation.

Step 4: Carry extinguisher to within stream reach of fire.

Step 5: Aim nozzle toward base of fire.

Step 6: Discharge extinguishing agent and sweep slowly back and forth across entire width of fire, avoiding splashing liquid fuels.

Step 7: Cover entire area with dry chemical until fire is completely extinguished.

Step 8: Back away from the fire area.

Step 9: Tag extinguisher for recharge and inspection.

Step 1: Size up fire, ensuring that it is safe to fight with an extinguisher.

Step 2: Pull pin at top of extinguisher to break the inspection band.

Step 3: Test to ensure proper operation.

Step 4: Carry extinguisher to within stream reach of fire.

Step 5: Aim nozzle toward base of fire.

Step 6: Discharge extinguishing agent and sweep slowly back and forth across entire width of fire.

Step 7: Cover entire area with gas cloud until fire is completely extinguished.

Step 8: Back away from the fire area.

Step 9: Tag extinguisher for recharge and inspection.

Chapter Contents

Chapter 7

Ropes and Knots

This chapter provides information that addresses the following job performance requirements of NFPA® 1001, *Standard for Fire Fighter Professional Qualifications (2008)*:

NFPA® 1001 references for Chapter 7:

5.1.1
5.1.2
5.5.1

Chapter Objectives

Fire Fighter I Objectives

1. Compare and contrast the characteristics of life-safety rope and utility rope.

2. Summarize criteria for reusing life-safety rope.

3. Describe rope materials.

4. Describe types of rope construction.

5. Summarize basic guidelines for rope maintenance.

6. Explain procedures for storing life-safety rope.

7. Describe webbing and webbing construction.

8. Describe parts of a rope and considerations in tying a knot.

9. Describe knot characteristics and knot elements.

10. Describe characteristics of knots commonly used in the fire service.

11. Select commonly used rope hardware for specific applications.

12. Summarize hoisting safety considerations.

13. Discuss rescue rope and harness.

14. Inspect, clean, and store rope. (Skill Sheet 7-I-1)

15. Coil and uncoil a rope. (Skill Sheet 7-I-2)

16. Tie the single overhand knot. (Skill Sheet 7-I-3)

17. Tie a bowline. (Skill Sheet 7-I-4)

18. Tie a clove hitch. (Skill Sheet 7-I-5)

19. Tie a clove hitch around an object. (Skill Sheet 7-I-6)

20. Tie a figure eight. (Skill Sheet 7-I-7)

21. Tie a figure-eight bend. (Skill Sheet 7-I-8)

22. Tie a figure eight on a bight. (Skill Sheet 7-I-9)

23. Tie a becket bend. (Skill Sheet 7-I-10)

24. Hoist an axe. (Skill Sheet 7-I-11)

25. Hoist a pike pole. (Skill Sheet 7-I-12)

26. Hoist a roof ladder. (Skill Sheet 7-I-13)

27. Hoist a dry hoseline. (Skill Sheet 7-I-14)

28. Hoist a charged hoseline. (Skill Sheet 7-I-15)

29. Hoist a power saw. (Skill Sheet 7-I-16)

Ropes and Knots

Case History

On November 15, 2006, a crew from a municipal fire department was drilling on rappelling methods from the top of a six-story training tower. A firefighter was positioned on the roof to assist other firefighters with their equipment during the drill. A bottom belay was used for safety during each rappel.

After several firefighters had rappelled, all of the eight plates (8s) were on the ground. Rather than wait for the proper equipment to be brought back to the roof, firefighters on the roof decided to use a brake bar rack in place of an eight plate. The rope was only placed through two bars on the rack. When the firefighter stepped over the edge of the roof, he immediately began falling without contro . The belay finally stopped him when he had fallen over 50 feet (15 m) and was within a few feet of striking the ground. An investigation identified that the cause of the incident was the inappropriate use of the brake bar rack during the rappel.

This case is an excellent example of the importance of every firefighter being familiar with the equipment used during rescue operations, especially ropes and rope equipment. The misuse of one piece of vital equipment almost resulted in the death of a firefighter. It is the responsibility of each firefighter to become familiar with and competent with the ropes, knots, and associated equipment used by the department.

(Source: National Fire Fighter Near-Miss Reporting System)

Rope is one of the oldest tools used by firefighters. Rope is useful for a variety of applications such as hoisting tools, gaining access to and rescuing people who are stranded above or below grade, stabilizing vehicles, and cordoning off areas. As a firefighter, you need to know the different types of rope so that you can choose the correct one for the job at hand. As used in this manual, the word *knot* includes all the knots, loops, bends, and hitches used in fire service applications. The ability to tie these knots correctly and quickly is crucial to the safety and effectiveness of rope-based operations.

NFPA® 1001, *Standard for Fire Fighter Professional Qualifications*, requires that those trained to the Fire Fighter I level know how to perform the following:

- Hoist tools and equipment with rope using the proper knot
- Know how to select the proper rope for a specific task
- Know how to properly maintain various types of ropes used by the fire department.

The standard further requires that those trained to the Fire Fighter II level be capable of the following:

- Rescue victims from motor vehicle accidents (which may involve stabilizing one or more vehicles with rope)

- Assist technical rescue teams (which may involve constructing or using rope rescue systems).

The knots discussed in this chapter are those most commonly used in fires and rescue operations. Local policies may require firefighters to know additional knots or to use different methods than are shown in this chapter. Firefighters must always know and follow their department's operating procedures and protocols. Any knots or applications that deviate from those shown in this chapter should be thoroughly tested under controlled conditions before being used in life-safety applications.

This chapter discusses the various types of rope and their applications along with rope construction and materials. The proper inspection, care, record keeping, and storage of fire service ropes are also discussed. Finally, the methods for hoisting tools and equipment are presented along with an introduction to rope rescue systems. For more information on ropes, knots, and rope rescue, consult the IFSTA **Fire Service Search and Rescue** manual.

Types of Rope and Their Uses

Fire service rope falls into two classifications: *life-safety rope* and *utility rope*. Life-safety rope is used to support rescuers and/or victims during actual incidents or training (see Rope Rescue section) **(Figure 7.1)**. Utility rope can be used in any other instance where rope is required (see Hoisting Tools and Equipment section).

All life-safety (rescue) rope must conform to NFPA® 1983, *Standard on Life Safety Rope and Equipment for Emergency Services*. This standard specifies that only rope of block creel construction (without knots or splices in fibers) using continuous filament virgin fiber for load-bearing elements is suitable for life-safety applications. Rope made of any other material or construction should not be used to support firefighters or victims.

According to NFPA® 1983, rope manufacturers must supply purchasers with information regarding use criteria, inspection procedures, maintenance procedures, and criteria for retiring life-safety rope from service. Manufacturers are further required to supply criteria to consider before a life-safety rope is reused in life-safety situations. Included in these criteria are the following conditions that must be met:

- Rope must not be visibly damaged.

- Rope must not show abrasions or have been exposed to high temperatures or direct flame contact.

- Rope has not been impact loaded (a force applied to a rope when it suddenly stops a falling load).

- Rope must not have been exposed to liquids, solids, gases, mists, or vapors from any chemical or material that can deteriorate rope.

- Rope must pass inspection made by a qualified person both before and after each use. Inspection procedures must follow the manufacturer's recommendations.

Life-Safety Rope — Rope that meets the requirements of NFPA® 1983, *Standard on Life Safety Rope and Equipment for Emergency Services*, and is dedicated solely for the purpose of constructing lines to be used for the raising, lowering, or supporting people during rescue, fire fighting or other emergency operations, or during training.

Utility Rope — Rope to be used in any situation that requires a rope — except life safety applications. Utility ropes can be used for hoisting equipment, securing unstable objects, and cordoning off an area.

The manufacturer must also provide the user with information about removing a life-safety rope from service when it does not meet all of the previously stated conditions or if there is any reason to doubt its safety or serviceability, such as when a rope has exceeded its shelf life.

Any life-safety rope that fails to pass inspection or that has been impact-loaded should be destroyed immediately. In this context, "destroy" means that it is altered in such a manner that it cannot be mistaken for life-safety rope and inadvertently used in that application again. Destroying a life-safety rope could include disposing of the rope entirely or removing the manufacturer's label, cutting the rope into shorter lengths, and clearly marking it as utility rope. Any rope that has been subjected to impact loading must have an entry made in its log because there is no way to determine by inspection if the rope has been impact loaded (see Maintaining a Rope Log section).

Utility rope can be used to hoist equipment, secure unstable objects, or cordon off an area. Although there are industry standards concerning the physical properties of utility rope, there are no standards set forth for utility rope applications; however, common sense should prevail in its use. Regularly inspect utility rope to see if it is damaged.

Rope Materials

Until the last half of the twentieth century, natural fiber rope was the primary type of rope used for rescue. Most natural fiber ropes are made of hemp (manila or sisal) or cotton **(Figure 7.2)**. After extensive experience, testing, and evaluation in recent years, natural fiber rope is no longer accepted for use in life-safety applications. It is acceptable to use natural fiber rope for utility purposes such as hoisting equipment or stabilizing objects.

Advances in synthetic rope construction have made its use preferable to natural fiber rope for life-safety applications. Synthetic fiber rope has excellent resistance to mildew and rotting, has excellent strength, and is easy to maintain. Unlike natural fiber rope, which is made of short overlapping strands of fiber, synthetic rope may feature continuous fibers running the entire length of the rope. **Table 7.1, p. 266,** provides the general characteristics of rope fibers. Firefighters should become familiar with the manufacturers' specifications and limitations for the ropes used in their department.

Natural Fiber Rope — Rope made of hemp or cotton used for utility purposes. Natural fiber rope is not accepted for use in life-safety applications.

Synthetic Fiber Rope — Rope featuring continuous fibers running the entire length of the rope; has excellent resistance to mildew and rotting, has excellent strength, and is easy to maintain.

Figure 7.1 Life-safety rope is used to support rescuers and victims.

Figure 7.2 Natural fiber ropes can be used for non-life-support applications, such as hoisting equipment and securing unstable objects. *Courtesy of Bluewater Ropes.*

Table 7.1
Rope Fiber Characteristics

Characteristics	Nylon	Polyester	Polypropylene	Polyethylene	Manila	Cotton	Kevlar® Aramid	H. Spectra® Polyethylene
Strength	3*	4*	5*	6*	7*	8*	2*	1*
Wet Strength vs. Dry Strength	85%	100%	100%	100%	115%	115%	90%	100%
Shock Load Ability	1*	3*	2*	4*	5*	6*	7*	7*
Floats or Sinks in Water (Specific Gravity)	Sinks (1.14)	Sinks (1.38)	Floats (0.92)	Floats (0.95)	Sinks (1.38)	Sinks (1.54)	Sinks (1.45)	Floats (0.97)
Elongation at Break (Approximately)	20–34%	15–20%	15–20%	10–15%	10–15%	5–10%	2–4%	<4%
Melting Point	480°F (249°C)	500°F (260°C)	330°F (166°C)	275°F (135°C)	Does not melt; chars at 350°F (177°C)	Does not melt; chars at 300°F (149°C)	Does not melt; chars at 800°F (427°C)	275°F (135°C)
Abrasion Resistance	3*	2*	7*	6*	4*	8*	5*	1*
Resistance: Sunlight Rot Acids Alkalis Oil & Gas	Good Excellent Poor Good Good	Excellent Excellent Good Poor Good	Poor Excellent Good Good Good	Fair Excellent Good Good Good	Good Poor Poor Poor Poor	Good Poor Poor Poor Poor	Good Excellent Poor Good Good	Good Excellent Excellent Excellent Excellent
Electrical Conductivity Resistance	Poor	Good	Good	Good	Poor	Poor	Poor	Good
Storage Requirements	Wet or Dry	Wet or Dry	Wet or Dry	Wet or Dry	Dry Only	Dry Only	Wet or Dry	Wet or Dry

* Scale: Best = 1; Poorest = 8
Source: Wellington Leisure Products, Inc.

Rope Construction

Two types of rope are used in life-safety situations: dynamic rope and static rope. Each type has its advantages and disadvantages because of different design and performance characteristics.

Dynamic (high-stretch) rope is used when long falls are a possibility, such as in recreational rock climbing. To reduce the shock of impact on both climbers and their anchor systems in falls, dynamic rope is designed to stretch without breaking. Because this elasticity is a disadvantage when trying to raise or lower heavy loads, dynamic rope is *not* considered practical for rescue or hauling applications.

Static (low-stretch) rope is used for most rope-rescue incidents. It is designed for low stretch without breaking. According to NFPA® 1983, low-stretch rope must not elongate more than 10 percent when tested under a load equal to 10 percent of its breaking strength. Static rope is used for rescue, rappelling, hauling, and where falls are not likely to occur or only very short falls are possible.

In NFPA® 1983, life-safety rope is further broken down into three categories: light-use, general-use, and throwline. According to the standard, light-use life-safety rope is ⅜-inch (9.5 mm) in diameter or greater, but less than ½-inch (12.5 mm). It is intended to support the weight of one person, must have a minimum breaking strength of almost 4,500 pounds (20 k/N), and a maximum safe working load limit of 300 pounds (136 kg).

General-use life-safety rope is ⁷⁄₁₆-inch (11 mm) in diameter or greater, but less than or equal to ⅝-inch (16 mm). It is intended to support the weight of two persons, must have a minimum breaking strength of nearly 9,000 pounds (40 k/N), and a maximum safe working load limit of 600 pounds (272 kg).

Throwline is ¹⁹⁄₆₄-inch (7 mm) in diameter or greater, but less than ⅜-inch (9.5 mm). It is used to tether rescuers during water rescues or to throw to a victim in the water, must have a minimum breaking strength of almost 3,000 pounds (13 k/N), and a maximum safe working load limit of 200 pounds (91 kg). The difference between these breaking strength and safe working load figures is based on providing a 15:1 safety factor in life-safety ropes.

NFPA® 1983 recognizes one additional category — *escape rope*. While not considered either life-safety or utility rope, escape rope is constructed in the same manner as life-safety rope. Escape rope must meet generally the same elongation, breaking strength, and safe working load requirements as throwline. However, escape rope is intended to be used one time only and then destroyed.

Although there are other types, the most common types of rope construction are kernmantle, laid, braided, and braid-on-braid, all of which must be of block creel construction **(Figure 7.3, p. 268)**.

Kernmantle Rope

Kernmantle, a jacketed rope, is composed of a braided covering or sheath (mantle) over a core (kern) of the main load-bearing strands **(Figure 7.4, p. 268)**. The core strands run parallel with the rope's length and work in conjunction with the covering, which increases the rope's stretch resistance and load characteristics. The load characteristics are also affected by the method of manufacture. The core is made of high-strength fibers, usually nylon; these account for 75 percent of the total strength of the rope. The sheath provides the rest of the rope's overall strength and protects the core from abrasion and contamination.

Dynamic Rope — Rope that stretches farther than a static rope stretches.

Static Rope — Rope that will stretch a relatively short distance under load.

Light-Use Rope — Life-safety rope that is ⅜-inch (9.5 mm) in diameter or greater, but less than ½-inch (12.5 mm) and is intended to support the weight of one person.

General-Use Rope — Life-safety rope that is ⁷⁄₁₆-inch (11 mm) in diameter or greater, but less than or equal to ⅝-inch (16 mm) and is intended to support the weight of two persons.

Throwline — Life-safety rope that is ¹⁹⁄₆₄-inch (7 mm) in diameter or greater, but less than ⅜-inch (9.5 mm) and is used to tether rescuers during water rescues or to throw a victim in the water.

Escape Rope — Rope (not considered life-safety or utility rope) that is ¹⁹⁄₆₄-inch (7.5 mm) in diameter or greater, but less than ⅜-inch (9.5 mm) and is constructed in the same manner as life-safety rope. It is intended to be used one time only and then destroyed.

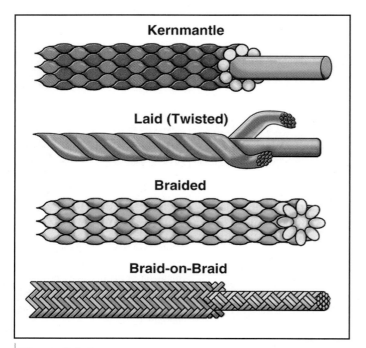

Figure 7.3 Common types of rope construction.

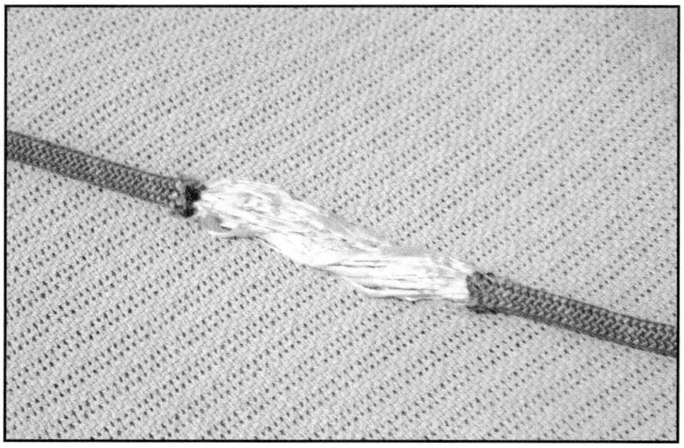

Figure 7.4 The sheath and core of kernmantle rope.

Kernmantle Rope — Rope that consists of a protective shield (mantle) over the load-bearing core strands (kern).

Kernmantle rope comes in both high-stretch and low-stretch types. High-stretch kernmantle is most commonly used as a sport rope for recreational rock or ice climbing. Low-stretch kernmantle is most commonly used as rescue rope where stretch is an undesirable characteristic.

Laid (Twisted) Natural or Synthetic Rope

Laid Rope — Rope constructed by twisting several groups of individual strands together.

Laid ropes, typical of most natural fiber ropes and some synthetic ropes, are constructed by twisting yarns together to form strands. Generally, three strands are twisted together to make the final rope **(Figure 7.5)**. How tightly these ropes are twisted and the type of fiber used determine the rope's properties. Twisted rope is susceptible to abrasion and other types of physical damage. Twisting a rope leaves all three load-bearing strands exposed along the entire length of the rope. Although this exposure allows for easy inspection, it also means that any damage immediately affects the rope's strength. Laid ropes are almost exclusively used as utility ropes.

Braided Rope

Braided Rope — Rope constructed by uniformly intertwining strands of rope together (similar to braiding a person's hair).

Although some braided ropes are made from natural fibers, most are of the synthetic variety. *Braided rope* is constructed by uniformly intertwining strands of rope together (similar to braiding a person's hair) **(Figure 7.6)**. Braided rope reduces or eliminates the twisting common to laid ropes. Because of its construction characteristics, the load-bearing fibers are subject to direct abrasion and damage. Like laid rope, braided rope is most commonly used as utility rope.

Braid-On-Braid Rope (Double Braid)

Braid-on-Braid Rope — Rope constructed with both a braided core and a braided sheath. The appearance of the sheath is that of a herringbone pattern.

Because braid-on-braid is a jacketed rope, it is often confused with kernmantle rope (see the previous section). *Braid-on-braid rope* is just what the name implies: It is constructed with both a braided core and a braided sheath. The sheath has a herringbone-pattern appearance **(Figure 7.7)**.

Figure 7.5 Typical laid rope. *Courtesy of BlueWater Ropes.*

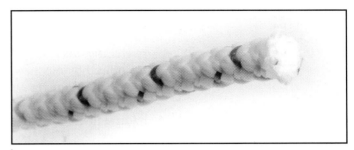
Figure 7.6 Typical braided rope. *Courtesy of BlueWater Ropes.*

Braid-on-braid rope is very strong. Half of its strength is in the sheath and the other half is in the core. A disadvantage of braid-on-braid rope is that it does not resist abrasion as well as kernmantle rope. Another disadvantage is that the sheath may slide along the inner core of the rope. Braid-on-braid rope is most often used in utility applications.

Figure 7.7 Typical braid-on-braid rope. *Courtesy of BlueWater Ropes.*

Rope Maintenance

In order for rescue rope to be ready and safe for use when needed, it must be properly maintained. This maintenance includes inspecting and cleaning the rope as well as maintaining a log of its use and maintenance history. **Skill Sheet 7-I-1** describes the procedures for inspecting, cleaning, and storing rope.

Inspecting Rope

Inspect all types of rope after each use. Unused ropes should be inspected at least annually. Inspect rope visually and by touch. When making inspections, use the following methods described for the various ropes and note any observations. Rope should be inspected for imbedded shards of glass, metal shavings, wood splinters, or other foreign objects that could cause damage. If any of these are found, the rope should be taken out of service. Document all inspections in the rope log.

Kernmantle Rope

Inspecting kernmantle rope for damage is somewhat difficult because the damage may not be obvious. The inspection can be performed by putting a slight tension on the rope while feeling for any lumps, depressions, or soft spots **(Figure 7.8)**. A temporary soft spot resulting from hard knots or sharp bends in the rope may be felt; however, the fibers within the core may realign themselves over time if the rope is undamaged. The only way to determine whether such a soft spot is damage or just temporarily misaligned core fibers is by carefully inspecting the outer sheath. Any damage to the outer sheath indicates probable damage to the core. The core of a kernmantle rope can be damaged without visible evidence on the outer sheath. If there is any doubt about the rope's integrity, it should be downgraded to utility status.

Figure 7.8 Rope should be inspected after each use.

In addition to inspecting rope for damage to the core and sheath, inspect the rope for irregularities in shape or weave, foul smells, discoloration from chemical contamination, roughness, abrasions, or fuzziness. A certain amount of fuzziness is normal and is not necessarily a cause for concern. If there is a great amount of fuzziness in one spot or if the overall amount is excessive based upon the inspector's judgment and experience, the rope should be downgraded.

Laid Rope

Inspect synthetic laid ropes for soft, crusty, stiff, or brittle spots; for areas of excessive stretching; for cuts, nicks, or abrasions; for chemical damage; for dirt or grease; and for other obvious flaws. Laid rope should be untwisted and checked internally for these flaws (**Figure 7.9**). In synthetic laid rope the presence of mildew does not necessarily indicate a problem; nonetheless, the rope should be cleaned and reinspected. In natural fiber rope, a foul smell might indicate rotting or mildew.

Braided Rope

Visually inspect braided rope for exterior damage such as heat sears (caused by friction or fire), nicks, and cuts. Also, visually inspect for excess or unusual fuzziness. Tactilely inspect for permanent mushy spots or other deformities.

Braid-On-Braid Rope

Inspect braid-on-braid rope for heat sears, nicks, and cuts. Also inspect for the sheath sliding on the core. If sliding is found, cut the end of the rope and pull off the excess material; then seal the end (**Figure 7.10**). Inspect for lumps that indicate core damage. A reduction in the rope's diameter may indicate a break in the core. Carefully examine any type of damage or questionable wear to the sheath.

Maintaining a Rope Log

Rope Log — A record that is kept by a department throughout a rope's working life. The date of each use and the inspection/maintenance records for the rope should be entered into the log, which should be kept in a waterproof envelope and placed in a pocket that is usually sewn on the side of the rope's storage bag.

When a piece of rescue rope is purchased, it must be permanently identified. Many departments identify new ropes by marking the ends with the unit number and the date it was placed in service. This can be done with a printed label sealed to the rope ends with a liquid compound made for this purpose. A record (rope log) must be started and kept throughout the rope's working life. Some departments mark each end of a new rope (A and B) so that firefighters can switch the ends in the bag to promote even wear. The date of each use and the inspection/maintenance records for the rope should be entered into the log. This information helps determine when the rope should be retired (**Figure 7.11**). The log should be kept in a waterproof envelope and placed in a pocket that is usually sewn on the side of the rope's storage bag (**Figure 7.12**) (see Storage of Life Safety Ropes section for more information).

Figure 7.9 Laid rope is inspected by untwisting it.

Figure 7.10 The sealed end of a braid-on-braid rope.

Oklahoma State University
Fire Service Training
Rope Log

Rope Type: _____ Rope Size: _____ Rope #: _____
Manufacturer: _____ Model: _____ Rope Color: _____
Purchased From: _____ Date: _____ Bag Color: _____

Date	Sign-Out	Use	Possible Damage/Comments	Sign-In

Figure 7.11 A typical rope log.

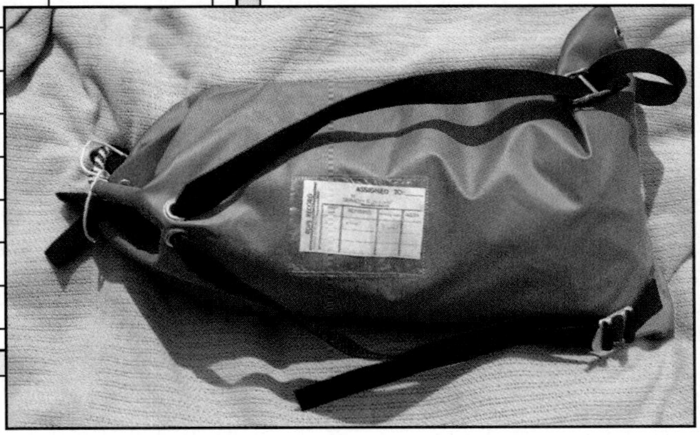

Figure 7.12 The rope log stays with the rope, in this case, attached to the exterior of a rope bag.

Cleaning Rope

Recommended methods of washing and drying rope vary with each manufacturer; therefore, it is always advisable to contact the manufacturer for specific cleaning and drying instructions for the type of rope or ropes in use. The following sections give some general guidelines for cleaning rope.

Natural Fibers

Natural fiber rope cannot be cleaned effectively because water cannot be used in the cleaning process. Water initially strengthens natural fiber rope; however, after continual exposure to wetting and drying, water weakens and damages the fiber. Wipe or gently brush the rope to remove as much of the dirt and grit as possible.

Synthetic Fibers

Cool water and mild soap are least likely to damage synthetic fiber ropes. Bleaches or strong cleaners should not be used. Some synthetic rope may feel stiff after washing, but this is not a cause for concern. There are three principal ways to clean synthetic rope: washing by hand, using a special rope-washing device, or placing it in a front-loading washing machine.

Washing by hand consists of wiping the rope with a cloth or scrubbing it with a brush and then thoroughly rinsing with clean water **(Figure 7.13, p. 272)**. Commercial rope-washing devices that can be connected to a standard faucet or garden hose are available **(Figure 7.14, p. 272)**. Rope is fed manually through the device and multidirectional streams of water clean all sides of the rope at the same time. These devices do an adequate job of cleaning mud and other surface debris from rope, but for a more thorough cleaning, rope should be washed in a washing machine.

Figure 7.13 Rope can be washed by hand.

Figure 7.14 A commercial rope washing device is effective for removing debris and dirt from the rope.

Front-loading washing machines without plastic windows are the best type to use for washing synthetic rope. Front-loaders that have a plastic window are not recommended because the plastic can cause enough friction with the rope during the spin cycle to damage the rope. Top-loaders may also damage the rope during agitation. The washer should be set on the coolest wash/rinse temperature available and only a small amount of mild soap, if any, added. The rope can be further protected by putting it in a mesh bag before placing it in the washer, or it can be coiled first (**Figure 7.15**).

Once the rope has been washed, it should be dried. It can be spread out on a hose rack *out of direct sunlight*, suspended in a hose tower, or loosely coiled in a hose dryer.

Storage of Life-Safety Rope

Rescue (life-safety) ropes can be stored in coils or in rope bags. Regardless of how rope is stored, where it is stored is of critical importance. Rescue rope should be stored in spaces or compartments that are clean and dry but have adequate ventilation. It should not be exposed to chemical contaminants, such as battery acid or hydrocarbon fuels, or the fumes or vapors of these substances. Rescue rope should not be stored in the same compartments where gasoline-powered rescue tools or the spare fuel for these tools are stored.

Bagging a Rope

The best method for storing kernmantle rope and other life-safety rope is to place it into a storage bag (**Figure 7.16**). The bag makes the rope easy to transport and protects the rope from contamination. An additional advantage of storing synthetic rope in a bag is that the rope can be deployed quickly by holding the end of the rope and throwing or dropping the bag. The weight of the rope inside the bag carries the bag toward the target and the rope pays out as the bag travels through the air. The bag may have a drawstring and shoulder straps for ease in carrying. Nylon or canvas bags are commonly used.

Figure 7.15 Front-loading washers without a plastic window are best for washing rescue rope.

Figure 7.16 Rescue ropes are often stored in storage bags to prevent damage in apparatus compartments and to increase the ease of carrying the rope.

Coiling/Uncoiling a Rope

Coiling rope so that it may be placed into service with a minimum of delay is very important in the fire service. An improperly coiled rope may become tangled and fail to uncoil, resulting in the failure of an evolution. Refer to **Skill Sheet 7-I-2** for the procedures for coiling and uncoiling a rope.

Webbing

Webbing is often used in conjunction with ropes. Most webbing is made from the same materials as synthetic rope, so the same precautions and maintenance procedures apply. The size of webbing needed varies with the intended use. Although 1-inch (25 mm) webbing is widely used in the fire service, most webbing used for lifting and pulling operations starts at about 2 inches (50 mm) in width.

There are two main types of webbing construction. One has a solid, flat design and the other (more common) has a tubular design. Both types look the same unless viewed at the ends **(Figures 7.17 a and b)**. Tubular webbing is also of two designs: a spiral weave and a chain weave. Overall, the spiral weave is stronger and more resistant to abrasion than the chain weave.

Webbing — Synthetic nylon, spiral weave, tubular material used for creating anchors, lashings, and for packaging patients and rescuers.

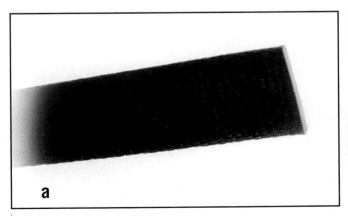

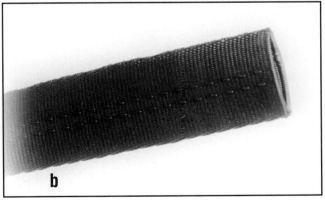

Figures 7.17 a and b Webbing can be distinguished by checking the ends: (a) flat design, and (b) tubular design. *Courtesy of Bluewater Ropes.*

Knots

Knots are used to join or connect ropes or webbing, form loops in ropes or webbing, or to attach ropes or webbing to objects. The ability to tie knots correctly and quickly is a critical part of fire and rescue operations. Improperly tied knots can be extremely hazardous to both rescuers and victims. The descriptions of how to tie knots include terms for parts of a rope **(Figure 7.18)**. The *running end* of a rope is the free end that is used for hoisting, pulling, or belaying. The *working end* of a rope is that which is tied to the object being raised, lowered, or stabilized. The *standing part* of a rope is that section between the working end and the running end.

All knots should be dressed after they are tied; that is, they should be tightened until snug with all slack removed. Even with these precautions, knots that are properly dressed and secure when tied to an object can sometimes loosen or fail because of repeated loading and unloading of the rope. One way to prevent such failures is to tie a safety knot in the tail (sometimes called the *bitter end*) of the working end of the rope. Safety knots include the single- and double-overhand knots. Other knots commonly used in the fire service are the bowline, half-hitch, clove hitch, figure-eight family, and becket bend (sheet bend).

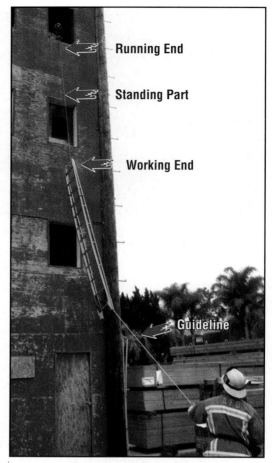

Figure 7.18 Standard rope terms used to describe knots.

Running End
Standing Part
Working End
Guideline

Running End — Part of the rope that is to be used for work such as hoisting, pulling, or belaying.

Working End — Part of the rope that is to be used in forming the knot. Also called Bitter End or Loose End.

Standing Part — That part of a rope between the working end and the running end.

 With all fire service knots, the key to competence and confidence is practice.

Elements of a Knot

To be suitable for use in rescue, a knot must be easy to tie and untie, be secure under load, and reduce the rope's strength as little as possible. A rope's strength is reduced to some degree whenever it is bent. The tighter the bend, the more strength is lost. Some knots create tighter bends than others and thereby reduce the rope's strength to a greater degree. Bight, loop, and round turn are names for the bends that a rope undergoes in the formation of a knot or hitch. Knots and hitches are formed by combining these elements in different ways so that the tight part of the rope bears on the working end to hold it in place. Each of these formations is shown in the following figures:

- The *bight* is formed by simply bending the rope back on itself while keeping the sides parallel (**Figure 7.19**).

- The *loop* is made by crossing the side of a bight over the standing part (**Figure 7.20**).

- The *round turn* consists of further bending one side of a loop (**Figure 7.21**).

Single/Double Overhand Safety Knots

As an added measure of safety, an overhand safety knot (often just called a *safety*) can be used when tying any type of knot (**Figure 7.22**). Although any properly tied knot should hold, it is best to provide the highest level of safety possible. Use of the overhand safety knot eliminates the danger of the end of the rope slipping back through the knot and causing the knot to fail. **Skill Sheet 7-I-3** describes the procedure for tying the single overhand knot.

Overhand Safety Knot — Knot used in conjunction with other knots to eliminate the danger of the running end of the rope slipping back through a knot, causing the knot to fail.

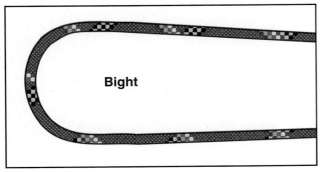

Figure 7.19 A bight in a rope.

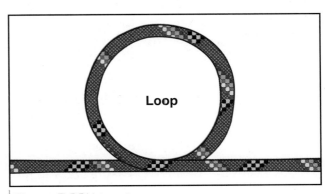

Figure 7.20 A loop in a rope.

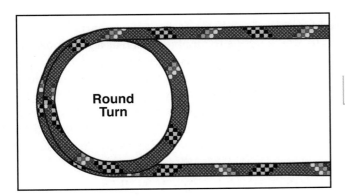

Figure 7.21 A round turn.

Figure 7.22 Two overhand safety knots are used to prevent the knot from slipping.

Bowline

Bowline Knot — Knot used to form a loop in natural fiber rope.

The bowline is one of the most important knots in the fire service. The bowline is easily tied and untied and is good for forming a single loop that will not constrict the object it is placed around. Firefighters should be able to tie the bowline in the open as well as around an object. The method shown in **Skill Sheet 7-I-4** is one way of tying the bowline, although other methods may be just as effective.

Half-Hitch

Half-Hitch — Knot that is always used in conjunction with another knot. The half-hitch is particularly useful in stabilizing tall objects that are being hoisted.

The half-hitch is particularly useful in stabilizing long objects that are being hoisted. The half-hitch is always used in conjunction with another knot or hitch. For example, one way of hoisting a pike pole is to tie a clove hitch around the pole and then tie a series of half-hitches along the length of the handle.

The half-hitch is formed by making a round turn around the object. The standing part of the rope is passed under the round turn on the side opposite the intended direction of pull (**Figure 7.23**). Several half-hitches can be applied in succession if required.

Figure 7.23 A half-hitch is used to stabilize an object during hoisting.

Clove Hitch

Clove Hitch — Knot that consists essentially of two half-hitches. Its principal use is to attach a rope to an object such as a pole, post, or hose.

The clove hitch may be formed by several methods. It consists essentially of two half-hitches. Its principal use is to attach a rope to an object such as a pole, post, or hoseline. Because it is highly susceptible to failure when repeatedly loaded and unloaded, the clove hitch is not regarded as suitable for use in anchoring a life-safety rope (or in a life-safety application). The clove hitch may be formed anywhere in the rope from either end to the middle. When properly applied, it withstands a steady pull in either direction without slipping. If the knot will be subjected to repeated loading and unloading, it should be backed up with an overhand safety knot.

The clove hitch, when formed by the method described in **Skill Sheet 7-I-5**, cannot be placed over an object that has no free end such as the center of a hoseline. Therefore, it is necessary to know how to tie the clove hitch around an object. **Skill Sheet 7-I-6** describes the procedure.

Figure-Eight Family of Knots

Family of Eight Knots — A series of rescue knots based on a figure-eight knot.

The figure-eight family of knots has gained increased acceptance and popularity for fire and rescue service applications. Several variations of the figure eight are commonly used such as the figure-eight bend, figure-eight follow through, figure eight on a bight, and double-loop figure eight.

Figure Eight

The figure eight is the foundation knot for the entire family of figure eights. It can also be used as a stopper knot so the rope will not pass through the grommet of a rope bag or through a rescue pulley **(Figure 7.24)**. Refer to **Skill Sheet 7-I-7** for the procedure for tying the figure-eight knot.

Figure-Eight Bend

Also known as the *Flemish Bend*, the figure-eight bend is used primarily on life-safety rope to tie ropes of equal diameters together. Refer to **Skill Sheet 7-I-8** for procedures on tying the figure-eight bend.

Figure Eight on a Bight

The figure eight on a bight is a good way to tie a closed loop in the end of a rope. It is tied by forming a bight in either end of the rope and then tying a simple figure eight with the bight in the doubled part of the rope. Refer to **Skill Sheet 7-I-9** for procedures on tying the figure eight on a bight.

Figure-Eight Follow-Through

The figure-eight follow through is used to secure a rope around an object. It starts off with a single figure eight. The end of the rope is then wrapped around an object and follows back through the single figure eight. The outcome is a figure eight on a bight that is around an object **(Figure 7.25)**.

Figure-Eight Knot — Knot used to form a loop in the end of a rope; should be used in place of the bowline knot when working with synthetic fiber rope.

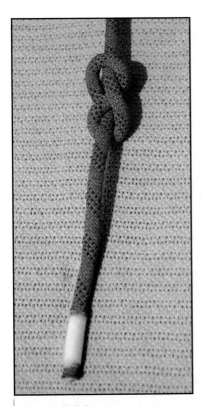

Figure 7.24 A figure-eight stopper is used to prevent the end of a rope from passing through an opening such as a pulley.

Figure 7.25 A figure-eight follow-through around an object.

Becket Bend (Sheet Bend)

The becket bend (or sheet bend) is used for joining two ropes of unequal diameters or joining a rope and a chain. It is unlikely to slip when the rope is wet. These advantages make it useful and dependable in fire service rope work; however, the becket bend is not suitable in life-safety applications. Refer to **Skill Sheet 7-I-10** for procedures on tying the becket bend.

Water Knot

The water knot is the preferred knot for joining two pieces of webbing or the ends of the same piece when a loop is needed. Similar to the figure-eight follow-through, the water knot is formed by tying a simple overhand knot in one piece or end and following it through in the reverse direction with another piece or end (**Figure 7.26**).

Rope Hardware

A variety of hardware items are used in conjunction with ropes and webbing, primarily in rope rescue. Some of the more commonly used hardware items are as follows:

- *Carabiner* — a metal snap link used to connect elements of a rescue system together (**Figure 7.27**).

- *Figure-eight plate (descender)* — used for rappelling or as a friction brake in lowering systems (**Figure 7.28**).

- *Brake bar rack (descender)* — used for rappelling or as a friction brake (**Figure 7.29**).

- *Ascender* — used to ascend a vertical rope (**Figure 7.30**).

- *Pulleys* — used in rescue systems to change the direction of pull or create mechanical advantage (**Figure 7.31**).

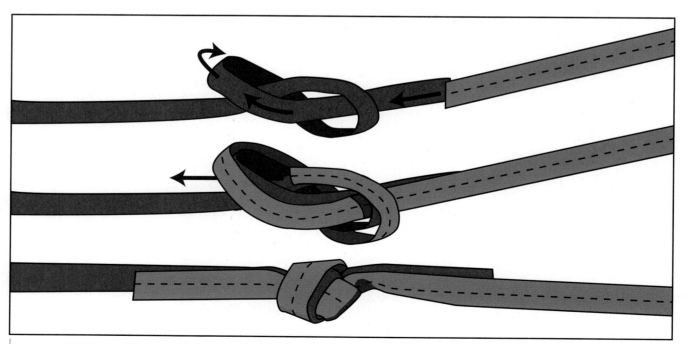

Figure 7.26 Process for tying a water knot in two sections of webbing. Also used to secure two ends of the same piece of webbing.

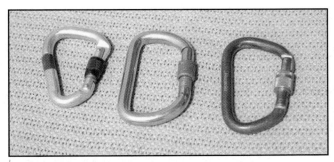

Figure 7.27 A variety of carabiners used with a rescue system.

Figure 7.28 A typical figure-eight plate used for rappelling.

Figure 7.29 A typical brake-bar rack or descender used for rappelling.

Figure 7.31 A double-sheave pulley is used to create mechanical advantage for hoisting.

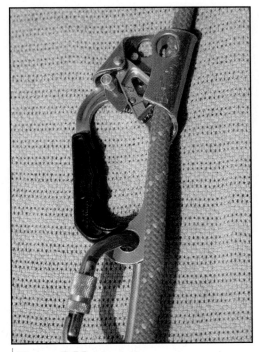

Figure 7.30 A typical ascender used for climbing a vertical surface with a rope rescue system.

Hoisting Tools and Equipment

Although ropes are also used for such tasks as stabilizing objects, one of their most common uses in the fire service is to raise or lower various tools and pieces of equipment from one elevation to another **(Figure 7.32, p. 280)**. A thorough knowledge of knots and hitches makes this a safe and efficient practice. Anything with a closed-type D-ring handle can be raised or lowered using a bowline or figure-eight bend. Hoisting pressurized cylinders, such as fire extinguishers or SCBA cylinders, is *not* recommended using this method.

Using the proper knots and securing procedures helps to prevent dropping the equipment. This avoids damage to the equipment and prevents possible injury to anyone standing below. Depending on local policy, a separate guide line may also be tied

Figure 7.32 A common rope application is hoisting tools and equipment.

Figure 7.33 Use a guideline to add safety and stability when hoisting tools and equipment with a rope.

to any of these pieces of equipment or the object may be tied in the center of the rope so that the hoisting rope also serves as the guide line. Firefighters on the ground use the guideline to prevent the equipment from striking the structure or other objects as it is raised (**Figure 7.33**). When one rope serves as both the guideline and the hoisting line, the knot or hitch-tying methods and methods of hoisting may vary. Keep safety in mind first, and then select the method of hoisting. The following sections discuss hoisting safety considerations and describe how to hoist various pieces of equipment.

Hoisting Safety Considerations

- Have solid footing and make necessary preparations before starting a hoisting operation.
- Use the hand-over-hand method to maintain control of the rope during a hoisting operation.
- Use an edge roller or padding to protect rope from physical damage when it must be pulled over sharp edges such as cornices or parapet walls (**Figure 7.34**).
- Work in teams to ensure firefighter safety when working from heights.

- Look to ensure that all personnel are clear of the hoisting area.
- Avoid hoisting operations near electrical hazards if possible. If this is not possible, use extreme caution.
- Secure the nozzles of any charged hoselines to prevent accidental discharge when hoisting.
- Use a guide line to help control the object being hoisted.

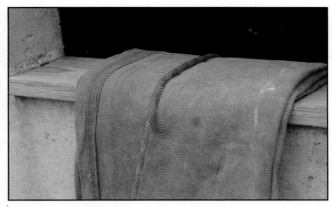

Figure 7.34 A salvage cover can be used as padding over windowsills and parapet walls to prevent damage to the rope.

Hoisting an Axe

The procedure for attaching and hoisting an axe is the same for either a pick-head axe or a flat-head axe. Refer to **Skill Sheet 7-I-11** for the procedure on hoisting an axe. The same rope can be used as both the hoist rope and as the guide line.

Hoisting a Pike Pole

To raise a pike pole (with the head up), tie a clove hitch near the butt end of the handle, followed by a half-hitch in the middle of the handle and another half-hitch around the head (**Figure 7.35**). **Skill Sheet 7-I-12** describes the steps for hoisting a pike pole.

Hoisting a Ladder

Tie a bowline or figure eight on a bight and slip it first through two rungs of the ladder about one-third of the way down from the top. After pulling that loop through, slip it over the top of the ladder (**Figure 7.36**). Refer to **Skill Sheet 7-I-13** for steps in hoisting a ladder.

Hoisting Hoselines

Hoisting hose is often the fastest and possibly the safest way of getting hoselines to upper levels. As with advancing hose up a ladder, it is easier and safer to hoist a dry hoseline (see **Skill Sheet 7-I-14**) than one that is charged; however, charged lines may also be hoisted. Whenever possible, bleed the pressure from a charged hoseline before hoisting it. If it is not possible to do so, use the procedures in **Skill Sheet 7-I-15** to hoist the charged hoseline. Care should be taken to reduce the possibility of damaging the coupling or the nozzle as the hoseline is being raised.

Figure 7.35 A clove hitch and two half hitches are used to hoist a pike pole.

Figure 7.36 One way to hoist a ground ladder to an upper level. A guideline is used to steady the ladder during the operation.

Hoisting Portable Fans

To hoist a smoke ejector, tie a bowline or figure-eight on a bight around two of the connecting rods between the front and back plates; on blowers, tie the knot through the carrying handle (**Figure 7.37**). This will be the hauling line. Attach a guideline to the bottom of the unit.

Hoisting a Power Saw

To hoist a rotary saw or chain saw, tie a bowline or figure-eight bend through the closed handle (**Figure 7.38**). Attach a guideline through the same handle. Refer to **Skill Sheet 7-I-16** for steps in hoisting a power saw.

Rescue Rope and Harness

When victims are located above or below grade and need to be rescued, the most efficient and sometimes the only means of reaching them and getting them to ground level may be by the use of ropes and rope systems. Rope rescue is a technical skill that requires specialized training. It may be necessary to lower rescuers into a confined space and to hoist a victim out with a mechanical advantage system made of rescue rope. Victims stranded on a rock ledge or on an upper floor of a partially collapsed building may have to be lowered to the ground with rescue ropes. The victims of vehicle accidents sometimes have to be transported up slopes in low-angle rescue operations.

Rescue rope is used for a variety of purposes. Rescue rope and harnesses are used to protect rescuers and victims as they move and/or work in elevated locations where a fall could cause injury or death. In combination with webbing, harness, and appropriate hardware, rope is the primary means for raising and lowering rescuers, equipment, and victims.

Rescue Harness

Ladder belts were once recognized as Class I harness, but no longer. Ladder belts are recognized only as positioning devices on ladders and for emergency escape. According to NFPA 1983, three classes of rescue harness are recognized in the fire service:

- Class I harness, also known as a *seat harness*, fastens around the waist and around the thighs or under the buttocks and is intended to be used for emergency escape with a load of up to 300 pounds (1.33 k/N) (**Figure 7.39**).

Figure 7.38 A power saw may be hoisted to upper levels using bowline and a guideline.

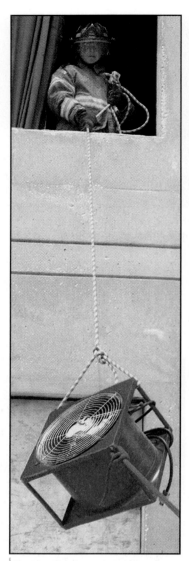

Figure 7.37 Heavy and bulky equipment, such as a smoke ejector, can be hoisted using utility ropes.

Rope Rescue — The use of rope and related equipment to perform rescue.

- Class II harness fastens in the same manner as Class I harness but is rated for up to a 600-pound (2.67 k/N) load. Class II harness looks exactly like Class I harness so the attached label must be used to verify its rating **(Figure 7.40)**.

- Class III harness, also known as *full body harness*, fastens around the waist, around the thighs or under the buttocks, and over the shoulders. Like Class II harness, Class III harness is rated for loads of up to 600 pounds (2.67 k/N) **(Figure 7.41)**.

Rope and appropriate hardware are also used to create a variety of mechanical advantage and safety systems. For more information on rope rescue, see the IFSTA **Fire Service Search and Rescue** manual.

Figure 7.39 A typical Class I or seat harness fastens around the waist and upper thighs.

Figure 7.41 A typical Class III harness is also referred to as a full body harness and is rated for loads up to 600 lbs (2.67 k/N).

Ladder Belt — Belt with a hook that secures the firefighter to the ladder.

Class I Harness — Harness that fastens around the waist and around the thighs or under the buttocks and is intended to be used for emergency escape with a load of up to 300 pounds (1.33 k/N). Also known as a seat harness.

Class II Harness — Harness that fastens around the waist and around the thighs or under the buttocks and is intended to be used for emergency escape with a load of up to 600 pounds (2.67 k/N). Class II harness looks exactly like Class I harness so the attached label must be used to verify its rating

Class III Harness — Harness that fastens around the waist, around the thighs or under the buttocks, and over the shoulders. Class III harness is rated for loads of up to 600 pounds (2.67 k/N). Also known as full body harness.

Figure 7.40 Example of the NFPA required label on a Class II harness. *Courtesy of Pigeon Mountain Industries.*

Summary

Rope is one of the oldest and most basic tools used by firefighters. It is used to stabilize vehicles and other objects, hoist tools and equipment, and allow firefighters to access and rescue victims who are stranded above or below grade or in bodies of water. Rope is also used to help firefighters escape from life-threatening situations.

To use rope safely and effectively during fires and rescue operations, you must know the various types of ropes and their applications. You must also be capable of tying a variety of knots and hitches quickly and correctly — and this takes practice. Finally, you must know how to inspect, clean, and store ropes so that they are ready for use when needed.

Fire Fighter I

1. What is the difference between life-safety rope and utility rope?

2. List three criteria that life-safety rope must meet before it is reused in life-safety situations.

3. Why is synthetic rope preferred for life-safety applications?

4. What two types of rope are used in life-safety situations?

5. Describe the most common types of rope construction.

6. How should the following types of rope be inspected: kernmantle rope, laid rope, braided rope, and braid-on-braid rope?

7. What are general guidelines for cleaning synthetic fiber rope?

8. What are the elements of a knot?

9. Describe commonly used rope hardware.

10. List four safety considerations for hoisting tools and equipment.

Clean Rope

Step 1: Clean the rope according to manufacturer's guidelines.

Step 2: Thoroughly rinse the rope.

Step 3: Dry the rope according to manufacturer's recommendations.

Inspect and Store Rope

Step 1: Using hands, inspect the entire length of the rope for soft, crusty, stiff, or brittle spots; areas of excessive stretching; cuts, nicks, and abrasions.

Step 2: Visually inspect the entire length of the rope for exterior nicks, cuts, dirt, embedded objects, and other obvious flaws, as well as for cleanliness.

Step 3: Determine if rope has been impact loaded, overloaded, chemically contaminated, or does not meet life-safety reuse requirements.

Step 4: Remove any flawed rope from service, disposing of it or labeling it as utility rope per local protocol.

Step 5: Record information in rope logbook.

Step 6: Store rope per local protocol.

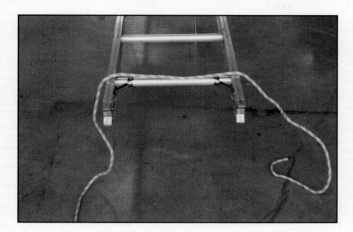

Step 1: Measure off and reserve a length of rope (about three times the distance between the standards — the structures used as a support) at the front of the rope to secure the coil when completed.

NOTE: A compact, finished coil can also be prepared using the beams of a ladder as standards.

Step 2: Drape this length of rope over the standards.

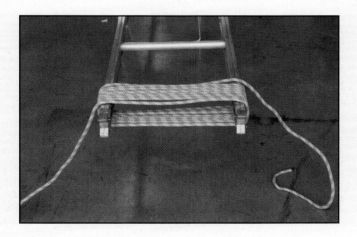

Step 3: Wrap the remaining rope around the standards until sufficient width is developed. It may be necessary to make two layers to coil all of the rope.

NOTE: Avoid making the coils too tight, which would make removal of the finished coil difficult.

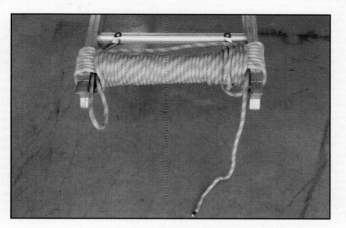

Step 4: Wrap the last portion of the rope around the loops.

Step 5: Fasten the end securely by tucking the end of the rope under the last wrap.

Step 6: Form a bight with the length measured in Step 1.

Step 7: Insert the bight through the end of the coil as shown.

Step 8: Place the end of the rope through the opposite end of the coil.

Step 9: Insert the end of the rope through the bight.

Step 10: Finish the coil by tucking the end of the rope next to the end in Step 5.

Step 11: Remove the coils from the standards.

Step 12: Untie the rope that is securing the coil.

Step 13: Begin uncoiling the rope.

Step 14: Continue to uncoil the rope while avoiding the creation of knots or tangles.

Step 1: Form a loop in the rope.

Step 2: Insert the end of the rope through the loop.

Step 3: Dress the knot by pulling on both ends of the rope at the same time.

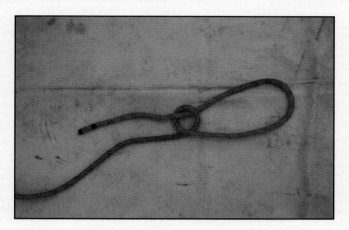

Step 1: Select enough rope to form the size of the loop desired.

Step 2: Form an overhand loop in the standing part.

Step 3: Pass the working end upward through the loop.

Step 5: Bring the working end completely around the standing part and down through the loop.

Step 4: Pass the working end over the top of the loop under the standing part.

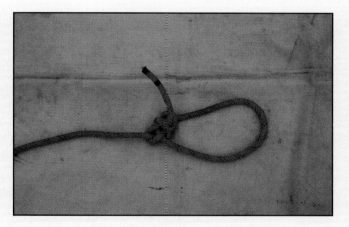

Step 6: Pull the knot snugly into place, forming an inside bowline with the working end on the inside of the loop.

Step 7: Secure the bowline with an overhand safety.

Step 1: Form a loop in your left hand with the working end to the right crossing under the standing part.

Step 2: Form another loop in your right hand with the working end crossing under the standing part.

Step 4: Hold the two loops together at the rope, forming the clove hitch.

Step 3: Slide the right-hand loop on top of the left-hand loop.

NOTE: This is the important step in forming the clove hitch.

Step 5: Slide the knot over the object.

Step 6: Pull the ends in opposite directions to tighten.

Step 1: Make one complete loop around the object, crossing the working end over the standing part.

Step 3: Pass the working end under the upper wrap, just above the cross.

Step 2: Complete the round turn about the object just above the first loop as shown.

Step 4: Set the hitch by pulling.

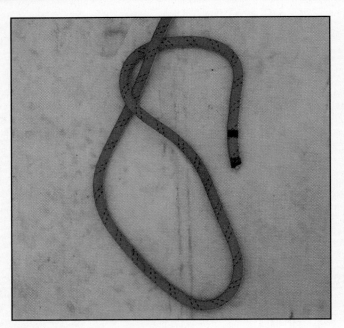

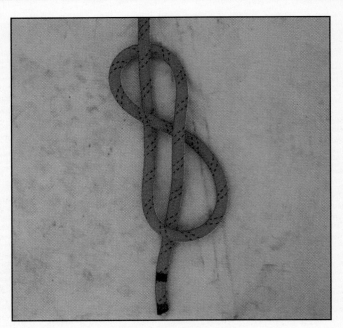

Step 1: Make a loop in the rope.

Step 2: Pass the working end completely around the standing part.

Step 3: Insert the end of the rope back through the loop.

Step 4: Dress the knot by pulling on both the working end and standing part of the rope at the same time.

7-1-8
Tie a figure-eight bend.

Step 1: Tie a figure-eight knot on one end of the rope.

Step 2: Feed the end of the other rope through the figure-eight knot in reverse. It should follow the exact path of the original knot.

Step 3: Use a safety knot, such as the overhand, with this knot.

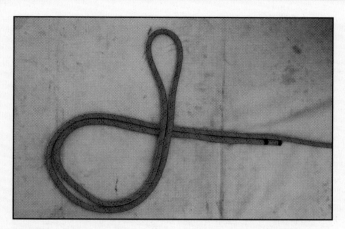

Step 1: Form a bight in the working end of the rope.

Step 2: Pass it over the standing part to form a loop.

Step 3: Pass the bight under the standing part and then over the loop and down through it; this forms the figure eight.

Step 4: Extend the bight through the knot to whatever size working loop is needed.

Step 5: Dress the knot.

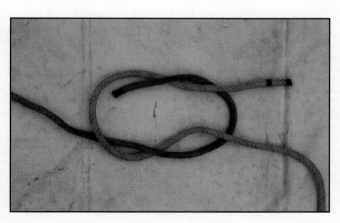

Step 1: Form a bight in one of the ends to be tied (if two ropes of unequal diameter are being tied, the bight always goes in the larger of the two).

Step 2: Pass the end of the second rope through the bight.

Step 3: Bring the loose end around both parts of the bight.

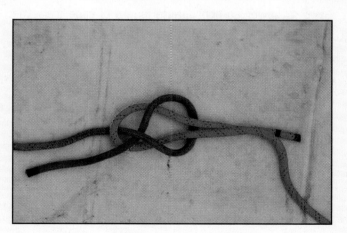

Step 4: Tuck this end under its own standing part and over the bight.

Step 5: Pull the knot snug.

Step 1: Lower an appropriate length of rope from the intended destination of the axe.

Step 2: Tie a clove hitch using the method shown in Skill Sheet 7-I-5.

NOTE: If the rope has a loop in the end, the loop may be used instead of a clove hitch.

Step 3: Slide the clove hitch down the axe handle to the axe head. The excess running end of the rope becomes the guideline.

Step 4: Loop the working end of the rope around the head of the axe and back up the handle.

Step 5: Tie a half-hitch on the handle a few inches (millimeters) above the clove hitch.

Step 6: Tie another half-hitch at the butt end of the handle.

Step 1: Lower an appropriate length of rope from the intended destination of the pike pole.

Step 2: Secure the rope to the pike pole pointing upward toward the end of the handle using a clove hitch.

Step 3: Leave enough excess running end so that it becomes the guideline.

Step 4: Tie a half-hitch or approved knot around the pike pole in the middle of the handle.

Step 5: Tie a second half-hitch or approved knot around the pike pole under the pike hook.

Step 6: Hoist the pike pole.

Step 1: Lower an appropriate length of rope from the intended destination of the ladder.

Step 2: Make a loop in the end of the rope using a bowline knot.

Step 3: Place the closed loop under the ladder and bring it up between the rung about one-third the distance from the hoisting end.

Step 4: Open the loop anc place it over the tip of the ladder.

Step 5: Arrange the standing part under the ladder rungs

Step 6: Tighten the loop around the beams, pulling the standing part of the rope up behind rungs toward ladder tip.

Step 7: Tie a guideline to the ladder.

Step 8: Hoist the ladder.

Step 1: Lower an appropriate length of rope from the intended destination of the hoseline.

Step 2: Fold the nozzle end of the hoseline back over the rest of the hose so that an overlap of 4 to 5 feet (1.2 m to 1.5 m) is formed.

Step 3: Tie a clove hitch, with an overhand safety knot, around the tip of the nozzle and the hose it is folded against so that they are lashed together.

Step 4: Place a half-hitch on the doubled hose about 12 inches (300 mm) from the loop end.

NOTE: With the ties properly placed, the hose will turn on the hose roller so that the coupling and nozzle will be on top as the hose passes over the roller.

Step 1: Lower an appropriate length of rope from the intended destination of the hoseline.

Step 2: Tie a clove hitch, with an overhand safety knot, around the hose about 1 foot (0.3 m) below the coupling and nozzle.

Step 3: Pass a bight through the nozzle handle and loop it over the nozzle so that the rope holds the nozzle shut while it is being hoisted.

Step 4: Tie a half-hitch around the nozzle to take the strain off the handle.

Step 1: Lower an appropriate length of rope from the intended destination of the power saw.

Step 2: Secure the rope to the handle of the power saw using a figure eight on a bight or a bowline knot. If a bowline is used, it must include an overhand safety knot.

Step 3: Leave enough excess running end so that it becomes the guideline.

Step 4: Hoist the power saw.

Chapter Contents

Courtesy of FEMA news photo.

Chapter 8

Rescue and Extrication

This chapter provides information that addresses the following job performance requirements of NFPA® 1001, *Standard for Fire Fighter Professional Qualifications* (2008):

NFPA® 1001 references for Chapter 8:

5.3.5

5.3.9

5.3.17

6.4.1

6.4.2

6.5.4

Chapter Objectives

Fire Fighter I Objectives

1. Distinguish between rescue and extrication operations.

2. Summarize safety guidelines for search and rescue personnel operating within a burning building.

3. Explain the objectives of a building search.

4. Describe primary search and secondary search.

5. Discuss conducting search operations.

6. Explain what actions a firefighter should take when in distress.

7. Describe actions that should be taken by a rapid intervention crew (RIC) when a firefighter is in distress.

8. Discuss victim removal methods.

9. Discuss emergency power and lighting equipment.

10. Conduct a primary and secondary search. (Skill Sheet 8-I-1)

11. Exit a hazardous area. (Skill Sheet 8-I-2)

12. Demonstrate the incline drag. (Skill Sheet 8-I-3)

13. Demonstrate the blanket drag. (Skill Sheet 8-I-4)

14. Demonstrate the webbing drag. (Skill Sheet 8-I-5)

15. Demonstrate the cradle-in-arms lift/carry — One-rescuer method. (Skill Sheet 8-I-6)

16. Demonstrate the seat lift/carry — Two-rescuer method. (Skill Sheet 8-I-7)

17. Demonstrate the extremities lift/carry — Two-rescuer method. (Skill Sheet 8-I-8)

18. Demonstrate the chair lift/carry method 1 — Two rescuers. (Skill Sheet 8-I-9)

19. Demonstrate the chair lift/carry method 2 — Two rescuers. (Skill Sheet 8-I-10)

20. Illuminate the emergency scene. (Skill Sheet 8-I-11)

Fire Fighter II Objectives

1. Discuss maintaining emergency power and lighting equipment.

2. Describe characteristics of hydraulic rescue tools.

3. Describe characteristics of nonhydraulic rescue tools.

4. Discuss cribbing for rescue operations.

5. Describe the characteristics of pneumatic tools.

6. Discuss lifting/pulling tools used in rescue operations.

7. Explain the size-up process for a vehicle incident.

8. Describe items to look for when assessing the need for extrication activities.

9. Discuss stabilizing vehicles involved in a vehicle incident.

10. List the three methods of gaining access to victims in vehicles.

11. List the most common hazards associated with wrecked passenger vehicles.

12. Explain the dangers associated with Supplemental Restraint Systems (SRS) and Side-Impact Protection Systems (SIPS).

13. Describe basic actions taken for patient management.

14. Describe patient removal.

15. Describe laminated safety glass and tempered glass.

16. Discuss removing glass from vehicles.

17. Explain considerations when removing vehicle roof and doors.

18. Describe common patterns of structural collapse.

19. Describe the most common means of locating hidden victims in a structural collapse.

20. Describe structural collapse hazards.

21. Describe shoring.

22. Discuss technical rescue incidents.

23. Service and maintain portable power plants and lighting equipment. (Skill Sheet 8-II-1)

24. Extricate a victim trapped in a motor vehicle. (Skill Sheet 8-II-2)

25. Assist rescue teams. (Skill Sheet 8-II-3)

Rescue and Extrication

Case History

On July 12, 2007, firefighters from the Guymon OK Fire Department responded to a one-vehicle rollover. The incident occurred on a gravel rural road. When firefighters arrived at the scene they found an SUV lying on its top in a field. From the damage observed, it was apparent the vehicle had rolled over several times before coming to rest. The driver was pinned under the vehicle with a roof support on top of her leg. She was conscious but critically injured.

Using ropes, air bags and other rescue tools firefighters quickly stabilized the vehicle. With the vehicle stabilized, firefighter/paramedics began advanced life support and called for a helicopter. Working closely with a local wrecker service trained in vehicle extrication methods, firefighters were able to use air bags and other tools to lift the vehicle off the patient. She was transported to a trauma center by helicopter and recovered from her injuries.

This incident is an example of how effectively firefighters, with proper training, can respond to an incident using common rescue tools and equipment. Even though the driver of the vehicle was critically injured, she survived due to the rescue skills of the firefighters.

Source: National Fire Fighter Near-Miss Reporting System

While the primary mission of the fire service is to save lives and property, rescue and extrication operations focus almost exclusively on saving lives. Because they are life-threatening situations, firefighters must be thoroughly prepared for any potential rescue and/or extrication situation to which they may be called. In this manual, a distinction is made between rescue and extrication.

Rescue incidents involve the removal of victims from entrapment by fires, terrain features, structural collapse, elevation differences, confined spaces, or any other situation not considered to be an extrication incident. *Extrication* incidents involve the disentanglement and removal of victims from vehicles or machinery.

Another factor of critical importance is the distinction between a *rescue* and a *body recovery* operation as described in Chapter 2, *Firefighter Safety and Health*. Rescues may involve putting firefighters at considerable risk in order to save a life; in body recoveries, the risks to firefighters should be minimized because there is no possibility of saving a life.

NOTE: Chapter 2 also describes the IFSTA Principles of Risk Management.

This chapter covers the basics of rescue and extrication equipment and techniques as required by NFPA® 1001. In the 2008 edition of NFPA® 1001, there are no specific rescue requirements for those qualified at the Firefighter I level, but they must be

Rescue — Saving a life from fire or accident; removing a victim from an untenable or unhealthy atmosphere.

Extrication — Incidents involving the removal and treatment of victims who are trapped by some type of man-made machinery or equipment.

capable of conducting search and rescue operations in a structure and setting up and operating within safe work areas such as at highway incidents. The standard requires those qualified at the Firefighter II level to also be capable of vehicle extrication operations and assisting technical rescue teams. For more information on extrication and rescue, see the IFSTA **Principles of Vehicle Extrication** and **Fire Service Search and Rescue** manuals.

Fireground Search and Rescue

The most fundamental duty of every fire department is to protect life and property from fire. Even though most fire departments in North America now provide a variety of other emergency services, the vast majority of search and rescue operations conducted by firefighters are on the fireground. Although too many people die in fires each year in the United States and Canada, many others are successfully rescued by firefighters.

Search Safety

Size-Up — Ongoing mental evaluation process performed by the operational officer in charge of an incident to evaluate all influencing factors and to develop objectives, strategy, and tactics for fire suppression before committing personnel and equipment to a course of action. Size-up results in a plan of action that may be adjusted as the situation changes. It includes such factors as time, location, nature of occupancy, life hazard, exposures, property involved, nature and extent of fire, weather, and fire fighting facilities.

While searching for victims in a fire, rescuers must always consider their own safety. Incident commanders (ICs) also must consider the hazards to which firefighters may be exposed while conducting search and rescue operations. Safety is the primary concern of rescuers because hurried, unsafe search and rescue operations may have serious consequences for rescuers as well as victims. Much of firefighter safety depends upon rescuers and their officers making a good initial size-up, continuing the size-up throughout the operation, and performing a risk/benefit analysis before each major step in the operation.

Personnel must be properly trained and equipped with the necessary tools to accomplish a rescue in the least possible time. For example, a rope is an essential search and rescue tool. It may be used as a tagline (sometimes called a *life line*) when conducting search and rescue operations in darkness or thick smoke. Because smoke is potentially flammable, some departments require search and rescue teams to take a charged hoseline with them into the building; others do not. When this is not an arbitrary requirement, a conscious decision must be made whether the search teams need to take protective hoselines with them in any particular situation. Other search and rescue tools include marking devices (to indicate which rooms have already been searched) and forcible entry tools (to aid in entry and egress and to enlarge the sweep area when searching) **(Figure 8.1)**.

Search Safety Guidelines

The following is a list of safety guidelines that should be used by search and rescue personnel operating within a burning building:

- Do not enter a building in which the fire has progressed to the point where viable victims are not likely to be found. Report to your supervisor any conditions that would appear to preclude viable victims.

- If backdraft conditions are apparent, attempt entry only after ventilation is accomplished.

- Work according to the Incident Action Plan (IAP). Do not freelance.

- Maintain radio contact with your supervisor (Command, Division or Group Supervisor, etc.), and monitor radio traffic for important information.

Figure 8.1 Search/rescue crews should always take forcible entry tools with them.

- Continuously monitor fire conditions that might affect the safety of your search team and other firefighters.

- Use the established personnel accountability system without exception.

- Be aware of the secondary means of egress established for personnel involved in the search.

- Wear full personal protective equipment, including self-contained breathing apparatus (SCBA) and personal alert safety system (PASS) device (and be sure the PASS device is turned ON).

- Work in teams of two or more and stay in constant contact with each other. Rescuers are responsible for themselves and each other.

- Search systematically to increase efficiency and to reduce the possibility of becoming disoriented.

- Stay low and move cautiously while searching where visibility is limited.

- Stay alert. Maintain situational awareness.

- Continuously monitor the structure's integrity and communicate any significant change.

- Check doors for excessive heat before opening them. (Use thermal imaging camera [TIC], apply water to door.)

- Mark entry doors into rooms and remember the direction turned when entering the room. To exit the building, turn in the opposite direction when leaving the room.

- Maintain contact with a wall, a hoseline, or a tagline when visibility is obscured. Working together, search team members can extend their reach by using ropes or straps.

- Have a charged hoseline at hand whenever possible when working on the fire floor (or the floor immediately above or below the fire) because it may be used as a guide for egress as well as for fire fighting.

Egress — Place or means of exiting a structure.

- Coordinate with ventilation teams before opening windows to relieve heat and smoke during search.

- Close the door, report the condition, and be guided by your supervisor's orders if fire is encountered during a search.

- Inform your supervisor immediately of any room or rooms that could not be searched, for whatever reason.

- Report promptly to the supervisor once the search is complete. Besides giving an "all clear" (primary or secondary search complete) report, also report the progress of the fire and the condition of the building.

Building Search

In structure fires — large and small — the primary way that fire departments meet their responsibility to protect the occupants is to conduct a thorough search of the building and rescue anyone who is in jeopardy but cannot escape without assistance. Regardless of how small a structure fire may look upon arrival, firefighters almost always search the building if it is reasonable and safe to do so. Even in relatively minor fires, there may be occupants in the building who are incapable of exiting on their own. Failing to locate a victim until after a "minor" fire is extinguished or, worse yet, missing a victim entirely is unprofessional and unacceptable. However, the decision to perform primary search is always based on a risk assessment and a determination that there may be savable victims.

While the initial size-up is the responsibility of the first-arriving officer, all firefighters should observe the exterior of the building and its surroundings as they approach. Careful observation may give them some indication as to the size and location of the fire, whether or not the building is likely to be occupied, the probable structural integrity of the building, and some idea of the amount of time it will take to effectively search the structure. Their initial exterior size-up will help them maintain their orientation within the building. They should identify their possible alternate escape routes (windows, doors, fire escapes) *before* they enter the building. Once inside, their specific location can sometimes be confirmed by looking out windows. The assessment of fire and building conditions must continue throughout the incident.

To obtain information about anyone who might still be inside and where they might be found, as well as to obtain information about the location and extent of the fire, firefighters should first question occupants who have escaped the fire **(Figure 8.2)**. Because neighbors may be familiar with occupants' habits and room locations, they may be able to suggest where occupants are likely to be found. They may also have seen an occupant near a window prior to the fire department's arrival. Information on the number and location of victims should be relayed to the IC and all incoming units. If possible, all information should be verified; in any case, firefighters should not assume that all occupants are out until the building has been searched.

If possible, the fire attack should be started simultaneously with any interior search operations. The safety of both the trapped occupants and the firefighters conducting the search is increased by attacking and controlling the fire. In some cases, fire control may need to precede search.

Figure 8.2 Occupants who have escaped the fire building can provide valuable information such as the location of the fire and any unaccounted for occupants.

Conducting a Search

There are two objectives of a building search: locating victims (searching for life) and obtaining information about the location and extent of the fire (assessing fire conditions). In most structure fires, the search for life requires two types of searches: primary and secondary. In some cases, a third type of search — rapid intervention — is needed to locate and rescue firefighters in distress. Rapid intervention is discussed later in this chapter.

A *primary search* is a rapid but thorough search that is performed either before or during fire suppression operations. Although the primary search is often carried out under extremely adverse conditions, it must be performed as soon as possible if there is a high probability that the structure is occupied. Because time is of the essence in these situations, some departments allow search teams to enter without taking a charged hoseline with them; others do not.

During the primary search, be sure to check the known or likely locations of victims as rapidly as conditions allow, moving quickly to search all affected areas of the structure as soon as possible. During this time the search team or teams can confirm that the fire conditions are as they appeared from the outside or report any surprises they may encounter.

A *secondary search* is conducted *after* the fire is under control and the greatest hazards have been controlled. The secondary search is a slower and more thorough search that attempts to ensure that no occupants were overlooked during the primary search. Ideally, the secondary search should be conducted by personnel other than those who conducted the primary search.

Primary Search

During the primary search, many fire departments follow a standard set of priorities that have been developed through experience with countless fires. These standard search priorities are as follows:

- Most severely threatened
- Largest numbers
- Remainder of hazard zone
- Exposures

During the primary search, always use the buddy system — working in teams of two or more. By working together, two rescuers can conduct a search quickly while maintaining their own safety. When searching in an immediately dangerous to life and health (IDLH) atmosphere, you must remain in physical, voice, or visual contact with other team members. The procedures for conducting a primary search are listed in **Skill Sheet 8-I-1**.

Primary search teams should always carry a radio, thermal imaging camera (TIC) if available, flashlight, and forcible entry tools with them whenever they enter a burning building and throughout the search **(Figure 8.3, p. 310)**. Valuable time will be lost if rescuers have to return to their apparatus to obtain basic tools and equipment. The tools used to force entry may be needed to force a way out of the building if rescuers become trapped. Some departments also require search teams to take a search rope with them when they enter the hazard zone. Some of these departments supply their teams with lighted search ropes that consist of a series of tiny white lights encased in a clear plastic sheath **(Figure 8.4, p. 310)**. Available in lengths of up to 300 feet (90 m) and carried in a canvas bag, the lighted rope provides at least some illumination to

> **WARNING!**
> Do not attempt to conduct interior search and rescue operations unless you are wearing appropriate personal protective clothing and equipment.

Figure 8.3 A search team equipped with a thermal imaging camera.

Figure 8.4 Lighted rope can increase safety. *Courtesy of Rock-N-Rescue and Middlesex VFC.*

Figure 8.5 Crawling on hands and knees requires more time but increases visibility and safety for the search team.

help rescuers see their surroundings and identify potential hazards. The rope operates on 120 volt AC power. The lighted rope's greatest value may be in that it makes the search rope easier to find when visibility is obscured.

Depending on conditions within the fire building, you may be able to search while walking upright or you may have to crawl on your hands and knees **(Figure 8.5)**. If there is only light smoke and little or no heat, walking is the fastest means of searching a building. Searching on your hands and knees (beneath the smoke) can increase visibility and reduce the chances of tripping or falling into open stairways or holes in floors. Crawling is much slower than walking, but it is usually noticeably cooler near the floor. Move up and down stairs on your hands and knees; when ascending, proceed head first, and when descending, proceed feet first.

If you can't see your feet because of the smoke, you shouldn't be walking upright.

When searching within a structure, move systematically from room to room, searching each room completely, while constantly listening for sounds from victims. On the fire floor, start your search as close to the fire as possible and then work back toward the entrance door. This allows your team to reach those in the most danger first — those who would be overtaken by any fire extension that might occur while the rest of the search was in progress. Because people who are a greater distance from the fire are in less immediate danger, they can wait to be reached as your team moves back toward safety. To reach a point nearest the fire, proceed as directly as possible from the entry point and pay out a tagline as you go. This will provide a way for you to remain oriented so you can find your way out quickly if fire conditions change rapidly during the search.

Because people instinctively seek any available shelter from a threatening fire, it is very important to search all areas such as bathrooms, bathtubs, shower stalls, closets, under beds, behind furniture, basements, attic rooms, and any other areas where occupants may attempt to seek shelter from the fire **(Figure 8.6)**. Search the perimeter of each room; because occupants may be overcome while trying to escape, always check behind doors and on the floor below the windows. As you move around the perimeter, extend your arms or legs or use the handle of a tool to reach completely under beds and other furniture **(Figure 8.7)**. After the perimeter has been searched, search the middle of the room.

During the primary search, visibility may be extremely limited, so use a TIC if you have one. If not, you may have to identify objects by touch — touch may provide the only clue to what type of room the team is in. Visibility being obscured by smoke should be reported through channels to the IC because it may indicate a need for additional ventilation. As you search, report essential information only (initial fire conditions found, significant changes in fire behavior or spread, etc.) to your supervisor or Command.

Figure 8.7 A tool handle extends a searcher's reach under beds and tables.

Figure 8.6 Every space large enough for someone to hide in must be searched. Small children may crawl under or behind furniture or into closets.

Rescue teams should maintain radio contact with their supervisor and periodically report their progress in accordance with departmental standard operating procedures (SOPs) **(Figure 8.8)**. During the primary search, negative information is just as important as positive information to ensure a complete search. For example, if the fire has spread farther than it appeared from the outside, if a number of trapped victims are found, or if the search has to be aborted for any reason, the officer in charge should be notified immediately. Informing the IC of any areas that have not been completely searched is especially important so that additional search teams can be assigned to these areas if necessary. Possible search and rescue organizational structures for small and large incidents are illustrated in **Figure 8.9**.

Search line system. For searching large or complex areas that are filled with smoke, some fire departments have adopted a primary search system that involves the use of a dedicated search line. The search line consists of 200 feet (60 m) of ⅜-inch (10 mm) rope with a Kevlar™ sheath for maximum abrasion protection and heat resistance. Every 20 feet (6 m) along its length, a 2-inch (50 mm) steel ring is tied into the search line **(Figure 8.10)**. Immediately after each ring, one or more knots are tied in the search line to indicate distance. After the first ring, one knot indicates 20 feet (6 m) from the beginning of the line. After the second ring, there are two knots indicating 40 feet (12 m) from the beginning of the line **(Figure 8.11)**. After the third ring, three knots are tied, and so on. The knots indicate the distance from the beginning of the line and they are always *after* the ring, so they provide a directional indication — knots toward the fire; rings toward the exit. The rings also provide an anchor point for lateral tethers. The tethers are 20 foot (6 m) lengths of ¼-inch (6 mm) rope with a Kevlar™ sheath. Each tether has a ¾-inch (19 mm) steel ring tied to one end, a knot at the midpoint, and either a nonlocking carabiner or a snap hook on the other end **(Figure 8.12)**. Each member of the search team carries one tether.

Carabiner — A steel or aluminum D-shaped snap link device for attaching components of rope rescue systems together. In rescue work, carabiners should be of a positive locking type, with a 5,000-pound (2 250 kg) minimum breaking strength. They are also called biners, crabs, or snap links.

Figure 8.8 The search team leader makes periodic reports to the IC on the search process and on interior conditions.

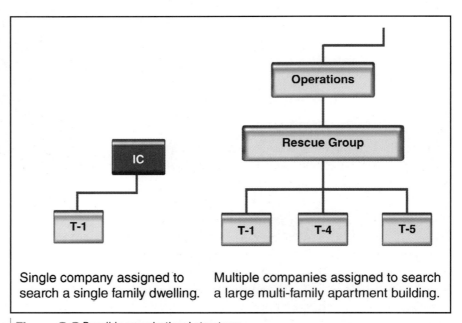

Single company assigned to search a single family dwelling.

Multiple companies assigned to search a large multi-family apartment building.

Figure 8.9 Possible organizational structures.

Figure 8.10 A large ring is tied in the search line every 20 feet (6 m). *Courtesy of Jeff Seaton.*

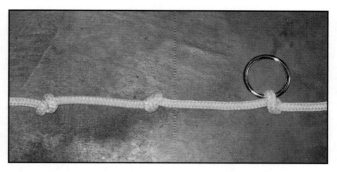

Figure 8.11 Each knot in the search line represents 20 feet (6 m) from the starting point. Note that this photo is representative of the ring and knots and is not to scale. *Courtesy of Jeff Seaton.*

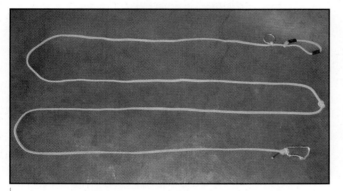

Figure 8.12 A typical lateral tether used in the search line system. *Courtesy of Jeff Seaton.*

Figure 8.13 The search line is attached to secure a fixed object about 10 feet (3 m) outside the entry point. A company identifier is left at the starting point. *Courtesy of Jeff Seaton.*

Implementing the search line system requires a team with a minimum of three members, although more is better. All members are in full protective ensemble. About 10 feet (3 m) outside the entry point to the area being searched, the end of the search line is tied to a fixed object about 3 feet (1 m) above the floor. A company identifier is left at that point **(Figure 8.13)**. Some departments using this system choose to station an attendant at the entry point to maintain communication with the search team and act as their air management timekeeper; others do not.

One member, usually called the "Lead," picks up the rope bag and enters the search area. The Lead is accompanied closely (shoulder to shoulder) by another member called the "Navigator." The Navigator is equipped with a hand light and TIC if available **(Figure 8.14, p. 314)**. The Navigator uses the light and/or the TIC to direct the Lead in the search area. One or more radio-equipped searchers follow them in. Each searcher carries a tether wrapped around one wrist and a forcible entry tool in the other hand **(Figure 8.15, p. 314)**. As they progress into the building, the search line pays out behind them and all team members maintain contact with the search line.

If it becomes necessary to search areas perpendicular to the search line, a searchers can snap a tether onto one of the 2-inch (50 mm) rings and begin searching laterally. Alternatively, the tether can be attached to the search line at any point between rings **(Figure 8.16, p. 314)**. Maintaining radio contact with the Navigator, the searcher pays out the tether as he or she progresses. Reaching the knot in the middle of the tether allows the searcher to make a 10 foot (3 m) diameter arc from the attachment point. Finding nothing, the searcher can progress an additional 10 feet (3 m) to the end of the tether. There, the searcher can sweep a 20 foot (6 m) arc.

If there is still more area to be searched beyond the first 20 foot (6 m) tether, a second searcher can attach his or her tether to the ring on the end of the first searcher's tether (**Figure 8.17**). This effectively doubles the area that can be searched from one point of attachment to the search line. As they return to the search line, the searchers wind their tethers around their wrists again. As the team progresses into the building, the Navigator keeps Command informed of their progress by reporting how many knots they are into the building, what the conditions are, and what they have found.

Figure 8.14 A fully equipped Navigator, left, accompanied by a Lead with search line rope bag, right. *Courtesy of Jeff Seaton.*

Figure 8.15 A typical search team member with a forcible entry tool and a personal tether. *Courtesy of Jeff Seaton.*

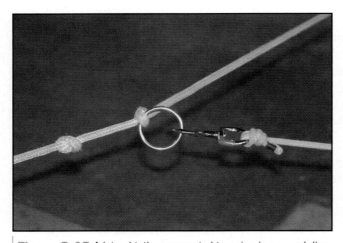

Figure 8.16 A lateral tether connected to a ring in a search line.

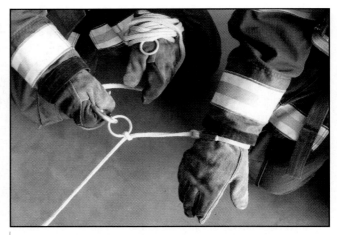

Figure 8.17 A second searcher attaches to the end of the first searcher's tether. *Courtesy of Jeff Seaton.*

Secondary Search

After the initial fire suppression and ventilation operations have been completed, personnel other than those who conducted the primary search are assigned to conduct a secondary search of the building. During the secondary search, speed is not as

Figure 8.18 The secondary search emphasizes thoroughness over speed.

important as thoroughness **(Figure 8.18)**. The secondary search is conducted just as systematically as the primary search to ensure that no rooms or spaces are missed. But the secondary search is conducted more slowly and carefully than the primary search. As in the primary search, any negative information, such as the fire beginning to rekindle in some area, is reported immediately. The procedures for conducting a secondary search are listed in **Skill Sheet 8-I-1**.

While each department must establish what constitutes "safe limits" in these situations, many departments use a meter reading of 50 ppm (parts per million) of carbon monoxide (CO). Any reading above 50 ppm CO means that firefighters inside the building must continue to wear their SCBA.

Searching Multistory Buildings

When searching in multistory buildings, the most critical areas are the fire floor, the floor directly above the fire, and the topmost floor **(Figure 8.19, p. 316)**. These floors should be searched first because this is where any remaining occupants will be in the greatest jeopardy due to rising smoke, heat, and fire. In addition, the majority of victims are likely to be found in these areas. Once these floors have been searched, the intervening floors should be checked.

During the primary search, unless they are part of the ventilation process, doors to rooms not involved in fire should be closed to prevent the spread of fire into these areas. Opening doors can disrupt ventilation efforts and can even spread the fire by drawing it toward the open doors. The exits, hallways, and stairs should be kept as clear as possible of unused hoselines and other equipment to facilitate the egress of occupants and to reduce tripping hazards **(Figure 8.20, p. 316)**.

Search Methods

When rooms, offices, or apartments extend from a center hallway, the search line system described earlier can be used. Otherwise, teams should be assigned to search both sides of the hallway. If two teams are available, each can take one side of the hallway. If there is only one search team, the team searches down one side of the hallway and back up the other side **(Figure 8.21, p. 316)**. During this or any search operation, it is critically important to control the access and egress openings and passageways to ensure that search teams can escape if fire conditions change rapidly.

CAUTION

Even though the interior of the building may appear to be free of smoke, firefighters conducting the secondary search must not remove their SCBA until the atmosphere has been sampled and found to be within safe limits.

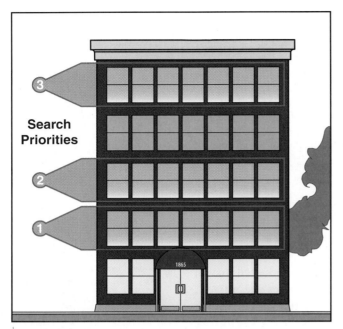

Figure 8.19 Typical search priorities beginning with the fire floor.

Figure 8.20 Hoselines in stairwells increase tripping hazards.

Entering the first room, the searchers turn right or left and follow the walls around the room until they return to the starting point. As rescuers leave the room, they turn in the same direction they used to enter the room and continue to the next room to be searched (**Figure 8.22**). For example, if they turned left when they entered the room, they turn left when they leave the room. When removing a victim to safety or to exit the building, rescuers must turn opposite the direction used to enter the room. It is important that rescuers exit each room through the same doorway they entered to ensure a complete search. This technique may be used to search most buildings, from a one-story, single-family residence to a large high-rise building.

In some departments, small rooms are searched by using a TIC to scan the room. If a TIC is not available, the procedure is for one member to stay at the door

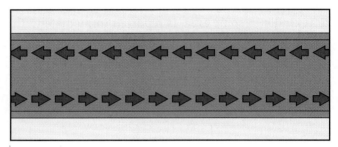

Figure 8.21 A single-team search pattern.

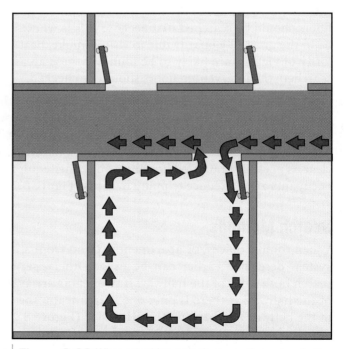

Figure 8.22 When searching a single room, always turn in the same direction. In this example, the team members keep the wall on their left side throughout the search.

while another member searches the room. The searcher remains oriented by maintaining a more-or-less constant dialogue with the member at the door. The searcher keeps the member at the door informed of the progress of the search. When the room search is completed, the two rejoin at the doorway, close and mark the door (see Marking Systems section), and proceed to the next room. When searching the next room, the partners exchange their roles of searching the room and waiting at the door.

This last method reduces the likelihood of rescuers becoming lost within the room, which reduces some of the stress of the situation. When searching relatively small rooms, this technique is often quicker than when both members search together because the searcher can move along more quickly without the fear of becoming disoriented.

Marking Systems

Several methods are used for marking searched rooms inside a burning building: chalk or crayon marks, masking or duct tape, specially designed door markers, and latch straps over doorknobs **(Figure 8.23)**. Latch straps also serve the secondary function of preventing other rescuers from being locked out of a room when the first search team is inside and needs help. Marking methods that might contribute to fire spread, such as blocking doors open with furniture, or methods that require subsequent searchers to enter the room to find the marker, are *not* recommended. Departmental SOPs usually dictate one accepted method of marking; however, all department personnel who may participate in a search must be trained to understand and use whatever method is selected. Regardless of the marking method used, the marks should be placed on the lower third of the wall or door so they can be seen below the smoke.

Some departments train their search teams to use a two-part marking system. They affix half of the mark when entering the room **(Figure 8.24)**. When the room search is finished, they complete the mark **(Figure 8.25)**. This avoids duplication of

Figure 8.23 A typical latch strap can be used to designate rooms that have been searched.

Figure 8.24 One slash indicates that the room is being searched. The slash may be made with tape, chalk, or any other substance that is visible in the dark.

Figure 8.25 A second slash is added when the room search is completed.

effort by alerting other rescue teams that the room is being or has been searched. If a search team fails to report and does not respond to calls for a personnel accountability report (PAR), this mark can serve as a starting point for others to begin looking for the missing team.

Building Search Safety

Every time firefighters respond to a fire, they know that the lives of building occupants may be in jeopardy. On arrival, a search is initiated as soon as possible to locate anyone still in the hazard zone and assess the degree to which they may be threatened. While rescuers must work quickly, they must also operate safely and with sound judgment if they are to fulfill their assignment and avoid becoming victims themselves.

As firefighters search a burning building, especially when visibility is limited because of smoke and/or darkness, they must always be alert for weakened or hazardous structural conditions, especially the floors. They should continually feel the floor in front of them with their hands or a tool to be sure that the floor is still intact (**Figure 8.26**). Otherwise, they may blindly crawl into an open elevator shaft, a stairway, an arsonist's trap, or a hole that may have burned through the floor.

Firefighters on or directly below the fire floor should also be alert for signs such as sagging floors or ceilings or the sound of structural members creaking or groaning that may indicate that the floor/ceiling assembly above them has weakened. Any of these signs should be immediately reported to Command!

When searching within a burning building, be very cautious when opening doors. Feel the top of the door and the doorknob to determine the heat level (**Figure 8.27**). If the door is hot to the touch, it should not be opened until a charged hoseline is in position. Do not remain in front of any door while it is being opened, especially outward-opening doors. Stay to one side, keep low, and open the door slowly. If there is fire behind the door, staying low allows the heat, flames, and combustion products to pass over your head. With outward-opening doors, stay on the hinge side (not the latch side) of the doorway so that the door can provide some protection if flames and heat erupt from the doorway opening. Some departments advocate applying a short burst of water fog above the door to cool the gases at the ceiling, then opening the

Figure 8.26 Searchers should check the floor ahead with a tool, such as an axe, to prevent stumbling into an opening or weakened floor structure.

Figure 8.27 Check closed doors for heat before opening them.

door and applying another short burst inside the room prior to entry. However, the risk to firefighters can be reduced by simply opening the door and allowing the fire to extend into the hallway at the ceiling level.

If an inward-opening door is difficult to open, do not kick the door to force it open because a victim may have collapsed behind it. Kicking the door may injure the victim further. Place a strap around the door knob to maintain control of the door and slowly push it open and check the area behind it for possible victims.

Firefighters in Distress

Even with the best incident command or accountability system in place, unusual circumstances can lead to a firefighter or a team of firefighters becoming trapped, lost, or disoriented within a burning structure. This can happen when the crew's escape route is cut off by unexpectedly rapid fire spread, sudden structural collapse, doors closing behind them, or firefighters straying from a hoseline or tagline and getting lost or running out of air.

According to a 2003 study of firefighter fatalities conducted by the San Antonio (TX) Fire Department, disorientation was the major factor. There were a number of other commonalities in the 17 incidents they studied. In every one of the fatal fires, the following were involved:

- Fire occurred in an enclosed structure.
- Firefighters used an aggressive interior attack.
- Prolonged zero visibility conditions developed.
- Firefighters were separated from their handlines or the lines were tangled.
- Company integrity was lost.
- Firefighters experienced disorientation.
- Firefighters experienced the disorientation sequence.

In this study, *prolonged zero visibility conditions* were defined as "heavy smoke conditions lasting longer than 15 minutes." In some cases, these conditions existed upon arrival; in others, they developed afterward. The *disorientation sequence* was described as, "A fire in an enclosed structure with smoke showing occurs. The arriving fire company initiates an aggressive interior attack to search for the seat of the fire. During the search, the seat of the fire cannot be located and conditions deteriorate with the production of heat, smoke, and prolonged zero visibility. As companies perform an emergency evacuation due to deteriorating conditions, handline separation occurs or tangled handlines are encountered. Disorientation then occurs as firefighters exceed their air supply, are caught in flashovers or backdrafts, or are trapped by the collapsing floor or roof. When the firefighters are not located quickly enough, the outcome is fatalities or serious injuries. The disorientation sequence usually plays out in a structure that does not have a sprinkler system or one that is operable."

If you become lost or disoriented in a burning building, try to remain calm. Giving in to excitement reduces your ability to think and react quickly. Excitement or disorientation may also cause you to expend your air supply faster than normal.

Mayday!

If you realize that you are in imminent life-threatening danger, immediately transmit Mayday! Mayday! Mayday! on your portable radio according to your local protocol (some departments may use a phrase different than Mayday). Activate your PASS device. Communicate your situation to your supervisor or Command. Give your last known location, unit number, name, assignment, and resources needed. If possible, communicate your air supply status and any actions you are taking. Remember, the sooner the IC and other crews know you need assistance the faster life-saving steps can be taken. When contact is made, describe your location as clearly as possible to help narrow the search pattern needed by those sent to rescue you. Stay in contact with Command and keep them informed of your situation and location.

If the situation is NOT life threatening, but you believe you require assistance, follow your local department protocols for that type of situation.

If you do not have a radio, try to retrace your steps to your original location. If you can locate a hoseline, you can crawl along it and feel the first set of couplings you come to. The female coupling is on the nozzle side of the set and the male is on the water source side of the set. The male coupling has lugs on its shank; the female does not — it has smaller lugs on the swivel **(Figure 8.28)**. Following the hoseline will lead you either to an exit or to the nozzle team. If there are loops in the hoseline or multiple hoselines at the same location, it can be confusing and may be more difficult to follow out of the hazard zone.

If retracing your steps is not possible, look for an exit from the building or at least from the area that is on fire. Search for an exit by locating a wall and crawling along it while sweeping the floor ahead of you with one hand and sweeping the wall with the other hand as high as you can reach without standing up **(Figure 8.29)**. Sweeping your hand back and forth on the floor in front of you will help you avoid any holes or other openings into which you might fall. Sweeping the walls as you crawl along will help you locate a window. If you find an outside window, you can signal for assistance by straddling the windowsill and turning on your PASS device, by signaling with your flashlight, by yelling and waving your arms, or by throwing objects out the window **(Figure 8.30)**. *Under no circumstances should you throw out your helmet or any other part of your protective ensemble.*

If conditions will not allow you to wait for help and you are not on an upper floor of a high-rise, you may be able to escape without assistance. If you have an escape rope, you can tie it to a desk, a heavy piece of furniture, or any solid object. You can also tie it to the middle of a Halligan or other tool placed across a lower corner of the window opening. Once the rope is adequately anchored, you can lower yourself to the roof of an adjacent building or the ground or sidewalk below. Some turnouts now have a harness integrated into them, which makes escape easier and safer **(Figures 8.31 a and b)**. If you do not have an escape rope, you may be able to safely drop from a second-story window by first removing your SCBA and hanging from the windowsill by your hands **(Figure 8.32)**. This will reduce your drop distance considerably.

Even if you do not locate a window, you will eventually find a door unless the door is blocked by collapse debris. If you find a door, secure it with a rope or strap attached to the door knob so it can be closed quickly. Check it for heat before opening it. If it is relatively cool to the touch, open it slowly but be prepared to close it again if there is fire or heavy smoke on the other side. If not, try to identify what is on the other side.

WARNING!

Dropping from any height while wearing SCBA may pull you over backwards and cause serious injury. If possible, remove your SCBA first.

Figure 8.28 Hose couplings can help firefighters find their way out. The male coupling indicates the direction to the exit.

Figure 8.29 Sweep the floor and the wall as you progress. Sweeping the wall will help to locate doors or windows that can be used for escape.

Figure 8.30 Straddle the windowsill and attract attention by waving or activating the PASS device.

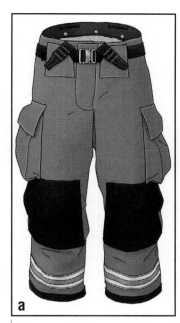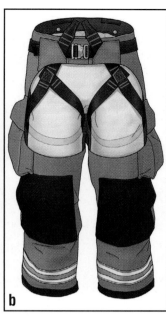

Figures 8.31a and b Some turnout pants have a built-in Class II harness. (a) The attachment points for the rescue rope are located at the waist of the pants. (b) Internally, the Class III harness wraps around the legs to form a seat that supports the wearer's legs. *Courtesy of Globe Manufacturing Co.*

Figure 8.32 The drop distance can be reduced by hanging from the windowsill.

If you feel shoes and clothing, you have probably found a closet, which is likely to be a dead end. Continue along the wall in the same direction until you find a window or another door.

If you cannot find your way out of a burning building through a door or window, you may be able to break through an interior or exterior wall. Breaching an interior wall may lead you to an exit path. Breaking through an exterior wall should allow you to escape from the building.

As with opening a door, before you attempt to breach an interior wall, feel it for heat. If the wall is relatively cool to the touch, you can create an opening large enough to pass through without having to remove any part of your protective ensemble. Most interior walls of either gypsum wallboard or lath and plaster over wood or metal studs may be breached. This can be done with almost any forcible entry tool, a piece of furniture, or if necessary, with your boot **(Figure 8.33)**.

All interior walls are not the same; the interior walls in public areas of some occupancies are covered with gypsum wallboard that is reinforced with a Lexan™ backing. This reinforced wallboard is highly resistive to being breached with forcible entry tools, but it can be done. Regardless of what type wall covering you are faced with, removing it from between two adjacent studs will only create an opening approximately 14 inches (350 mm) wide. To create a large enough opening through which to escape, you may have to remove sections on both sides of a stud and then remove the stud.

If you are unable to exit the structure using the above procedures and you do not have a functioning radio, activate your PASS device and move close to a wall because rescuers normally search along the walls before making sweeps of large interior areas. If you become exhausted or are close to losing consciousness, sit on the floor and lean against a wall to maximize the audible effects of your PASS device. This position also maximizes the chances for a quick discovery by rescue crews. If you cannot reach a wall, position your flashlight to shine toward the ceiling. This enhances the rescue crew's ability to locate you in the middle of a large room.

If you are trapped by a structural collapse or suffer an injury that prevents you from moving about, you will have fewer options than if you were merely lost or disoriented. Find a place of relative safety and activate your PASS device. If you are against a wall, tap or pound on the wall with a tool or any other hard object to alert anyone on the other side of the wall of your location **(Figure 8.34)**. Try to maintain your composure to maximize your air supply. The procedures for exiting a hazardous area are listed in **Skill Sheet 8-I-2**.

In a situation where you become disoriented or trapped, maintaining situational awareness is critical. Remain calm but alert. Monitor the environment for changing fire and smoke conditions or changes in other hazards. Monitor the level of air remaining in your SCBA cylinder. Follow your department's protocol for declaring a Mayday or emergency.

Rapid Intervention

Effectively, both NFPA 1500 and the OSHA regulations in 29 CFR 1910.134 require that a rapid intervention crew (RIC) be standing by whenever firefighters are in the hazard zone inside a burning building. The hazard zone is generally

Figure 8.33 Walls can be breached using a forcible entry tool or boot to create an escape path.

Figure 8.34 Try to attract attention to your location by hammering on the wall or activating your PASS device.

defined as any portion of a building that contains an IDLH atmosphere – one that requires firefighters to use their SCBA or take evasive action (crouching or crawling) to breathe smoke-free air. OSHA has determined that any interior space in which there is a fire beyond the incipient (early growth) stage is an IDLH environment.

A RIC is composed of at least two firefighters who are suitably equipped to enter the hazard zone to locate and rescue a firefighter in distress if necessary **(Figure 8.35, p. 324)**. In some cases, the IC will decide that the RIC needs to be composed of more than two firefighters. *Suitably equipped* means at least the same type and level of PPE as the interior fire fighting crews. RIC members may be assigned nonessential duties on the fireground such as setting up scene lighting or even operating a fire pump, as long as those duties can be abandoned immediately without putting other firefighters in jeopardy. In addition, these other allowable assignments must be in a location relatively close to the hazard zone. Further, they must not require such physical exertion that the RIC members' ability to perform search and rescue would be diminished. There may also be more than one RIC at a given fire, especially if interior crews entered the building at different points.

> **Rapid Intervention Crew (RIC)** — Two or more fully equipped and immediately available firefighters designated to stand by outside the hazard zone to enter and effect rescue of firefighters inside, if necessary. Also known as Rapid Intervention Team.

> A RIC composed of two members can search for and locate a firefighter in distress, but a rescue may require several more firefighters.

Figure 8.35 A RIC must be equipped and ready for entry.

A firefighter trapped in a burning building will almost certainly be using SCBA in order to breathe. Therefore, when entering the hazard zone, rescuers should take a spare SCBA cylinder for each downed firefighter along with a means of transferring that air to the firefighter in distress. If the downed firefighter's SCBA does not have a transfill connection, a complete SCBA unit should be taken in.

According to both NFPA® and NIOSH, so-called "buddy breathing" techniques are unreliable, more likely to produce two victims instead of one, and are not recommended. NIOSH recommends providing respiratory assistance and quickly removing the victim to a clear atmosphere.

Rescuers searching for a firefighter in distress should first try to establish radio contact. If conditions allow, the IC may order all pumps, generators, fans, and other noise-producing devices shut down briefly to increase the chances of hearing the distressed firefighter's PASS device. If these actions are unsuccessful, searchers should attempt to determine the downed firefighter's last known location and begin their search from that point.

When performing the search, a TIC should be used if available. The RIC should stop frequently, especially after moving from one room or area to another, and briefly remain silent. This will help them hear the downed firefighter's PASS device sounding. To be completely silent, crew members may even hold their breath for a few seconds when signaled to do so by the search team leader. This may allow the rescuers to hear faint calls for help or the sounds of the downed firefighter's SCBA exhalation valve operating.

Once the downed firefighter has been located, his or her air supply should be checked and a full cylinder connected if necessary. Then the firefighter should be medically evaluated before any move is attempted. The firefighter's level of consciousness and vital signs should be checked. If he or she is unable to walk, the rescuers should use any safe means possible to move the firefighter to safety. In most cases, the need to exit the hostile atmosphere overrides the need to stabilize injuries before moving the firefighter (**Figure 8.36**). If the firefighter has a functioning SCBA, carefully move the firefighter so as not to dislodge the mask. If the firefighter does not have a functioning SCBA, connect his or her mask to a functioning SCBA from the RIC kit or simply quickly remove the victim from the hazardous atmosphere (**Figure 8.37**).

Figure 8.36 Rapid extraction may override other considerations. Grasping the SCBA shoulder harness is a simple and efficient means for one or two firefighters to move an unconscious firefighter.

Figure 8.37 Provide additional air supply by connecting the RIC air cylinder to the downed firefighter's SCBA.

Tracking Devices

Some departments equip their firefighters with digital radio transceivers to help reduce the number of lost or disoriented firefighters and to aid in locating those who are lost. Originally developed as an aid in locating avalanche victims buried under several feet of snow, these devices have now been adapted for use in structure fires (**Figures 8.38 a and b**).

Radio transceivers are approximately the size of a PASS device and mount on the SCBA harness as do some PASS devices. Operating on 457 kHz, these units have a range of approximately 100 feet (30 m) and do not interfere with other on-scene radios.

Like PASS devices, these units are always turned on when entering a burning building. But, unlike PASS devices, the low-frequency radio transmission from these units is not blocked by walls, floors, or other solid objects. This signal is what allows fellow team members or a RIC to locate a downed firefighter from inside or outside the building. If a firefighter becomes separated from the team, the lost firefighter's transceiver can be switched from the standby mode to a search mode that helps locate an egress transmitter located near an exit. In the search mode, the unit displays distance and general direction to another transceiver or to the egress transmitter.

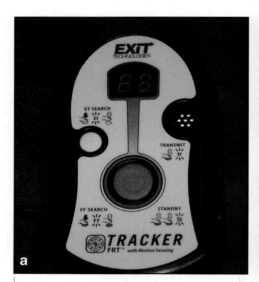 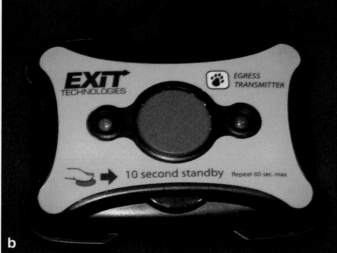

Figures 8.38 a and b Two models of firefighter tracking devices. *Courtesy of Exit Technologies.*

Removing Located Firefighters

Even though a two-firefighter RIC can locate a firefighter in distress, they are unlikely to be able to remove the firefighter from the hazard zone. In realistic scenarios conducted by major metropolitan fire departments, it was found that as many as 12 firefighters were needed to successfully move one downed firefighter from the hazard zone to a place of safety, and it took up to 20 minutes. Obviously, firefighters in distress who wait until the low-air alarm sounds before calling a Mayday will almost certainly be out of air by the time the RIC locates them. Given all of these factors, a RIC must do the following when searching for a missing firefighter:

* Proceed as quietly as possible and listen for the sound of a PASS device.

* Upon locating the firefighter, check the air supply status and provide a fresh supply as needed.

- Check vital signs if the firefighter is unconscious.

- Notify Command of the firefighter's location and status.

- Request assistance if the firefighter is trapped and/or injured.

- Attempt to mitigate any threats to the firefighter's life and safety while waiting for assistance. If necessary, move the individual for protection.

- Assist a firefighter who is uninjured and ambulatory to safety.

- Select and use an appropriate method of removal for an unconscious firefighter from those methods in this chapter.

Victim Removal

An uninjured victim or one with minor injuries may only require help to walk to safety — walking being probably the least demanding of all transportation methods. One or two rescuers may be needed, depending on how much help is available and the size and condition of the victim **(Figure 8.39)**.

Injured victims should not be moved before treatment is provided unless there is an immediate danger to the victim or to rescuers. Except when there is no other choice, never move a victim through the hazard zone. Emergency moves are necessary under the following conditions:

- There is fire or danger of fire in the immediate area.

- Explosives or other hazardous materials are involved.

- It is impossible to protect the accident scene.

- It is impossible to gain access to other victims who need immediate life-saving care.

- The victim is in cardiac arrest and must be moved to a different area (a firm surface for instance) so that rescuers can administer cardiopulmonary resuscitation (CPR).

Cardiopulmonary Resuscitation (CPR) — Application of rescue breathing and external cardiac compression used on patients in cardiac arrest to provide an adequate circulation and oxygen to support life.

Figure 8.39 Firefighters may have to assist a victim from an emergency scene.

The chief danger in moving a victim quickly is the possibility of aggravating a spinal injury. In an extreme emergency, however, the possible spinal injury becomes secondary to the goal of preserving life.

If it is necessary to perform an emergency move, the victim should be pulled in the direction of the long axis of the body — not sideways. Jackknifing the victim should also be avoided. If the victim is on the floor, pull on the victim's clothing in the neck or shoulder area (**Figure 8.40**). It may be easier to pull the victim onto a blanket and then drag the blanket.

It is always better to have two or more rescuers when attempting to lift or carry an adult. One rescuer can safely carry a small child, but two, three, or even four rescuers may be needed to safely lift and carry a large adult. An unconscious victim is always more difficult to lift; the person is unable to assist in any way and a relaxed body becomes "dead weight" (**Figure 8.41**). With an unconscious victim, it is critically important that his or her head and neck be supported and stabilized to prevent further injury.

It is not easy for inexperienced people to lift and carry a victim correctly. Their efforts may be uncoordinated and they usually need close supervision to avoid causing further injury to the victim. Rescuers helping to carry a victim should guard against losing their balance. They should lift as a team and with proper technique to avoid jostling the victim unnecessarily.

Lifting incorrectly is also one of the most common causes of injury to rescuers. Rescuers should always remember to keep their backs straight and lift with their legs, not their backs (**Figure 8.42**). If immobilization of a fracture is not feasible until the victim has been moved a short distance, one rescuer should support the weight of the injured part while others move the victim (**Figure 8.43**). There are a number of carries and drags that may be used to move a victim from an area quickly; these are described in the following sections.

Incline Drag

This drag can be used by one rescuer to move a victim up or down a stairway or incline and is very useful for moving an unconscious victim. **Skill Sheet 8-I-3** describes the method of performing the incline drag.

Blanket Drag

This drag can be implemented by one rescuer using a blanket, rug, or sheet. **Skill Sheet 8-I-4** describes the procedure for the blanket drag.

Webbing Drag

This drag can be implemented by one rescuer using a loop of 1-inch (25 mm) webbing. **Skill Sheet 8-I-5** describes the procedure for the webbing drag.

Cradle-in-Arms Lift/Carry

This lift and carry is effective for carrying children or very small adults if they are conscious. It is usually not practical for carrying an unconscious adult because of the weight, relaxed condition of the body, and the difficulty in supporting the head and neck. **Skill Sheet 8-I-6** describes the procedure for the cradle-in-arms lift/carry.

Seat Lift/Carry

This lift/carry can be used with a conscious or an unconscious victim and is performed by two rescuers. **Skill Sheet 8-I-7** describes the procedure for the seat lift/carry.

Three-Person Lift/Carry

Many victims are more comfortable when left in a supine position, and this lift/carry is an effective way to lift a victim who is lying down. The three-person lift/carry is often used for moving a victim from a bed to a gurney, especially when the victim is in cramped quarters.

Supine — Lying horizontal in a face upward position.

Figure 8.40 Grasping the clothing at the victim's shoulder, pull the victim toward the exit head first.

Figure 8.41 Before attempting to move an unconscious victim, support the victim's head and neck.

Figure 8.42 Rescuers should keep their backs straight and use their legs when lifting a victim. This photo illustrates the chair carry for a victim.

Figure 8.43 Using multiple rescuers, support any injured extremities.

Moving a Victim onto a Long Backboard or Litter

Occasionally, rescuers will have the advantage of being able to use some type of litter to remove a victim. There are many different types of litters such as the standard ambulance cot, army litter, scoop stretcher, basket litter, and long backboard. The long backboard is one of the most common types of litters used by fire service personnel. This section highlights the proper techniques for moving a victim onto a long backboard. Similar techniques should be used for moving people onto stretchers and basket litters.

Immobilizing a victim who is suspected of having a spinal injury on a long backboard requires four rescuers. One rescuer is needed to maintain in-line stabilization throughout the process and three rescuers are needed to actually move the victim to the board. It is critical that the victim with a suspected spinal injury be moved in such a way to avoid any unnecessary jolting or twisting of the spinal column. For this reason, the rescuer who applies and maintains in-line stabilization directs the other rescuers in their actions to ensure that the victim's head and body are moved as a unit. When dangers at the scene are life-threatening to the victim and rescuers, or the victim is not suspected of having a cervical spine injury and is just being relocated, this process may be performed with only two rescuers — one to maintain in-line stabilization and one to move the victim.

Extremities Lift/Carry

The extremities lift/carry can be used with either a conscious or an unconscious victim. This technique requires two rescuers. **Skill Sheet 8-I-8** describes the procedure for the extremities lift/carry.

Chair Lift/Carry

The chair lift/carry can be used with either a conscious or an unconscious victim. Be sure that the chair used is sturdy; do not attempt this carry using a folding chair. **Skill Sheets 8-I-9 and 8-I-10** describe two methods of performing the chair lift/ carry.

Rescue and Extrication Tools and Equipment

The skills and techniques required for rescue and extrication work can be mastered only when classroom and hands-on training delivered by competent instructors are combined with practice and experience. Although it is impossible to anticipate every extrication situation, rescue personnel will be best prepared if they are proficient with their tools and equipment. The following sections highlight some of the tools that are commonly used by firefighters who perform or assist in rescue and extrication operations.

Emergency Power and Lighting Equipment

Many rescue and extrication incidents occur at night or in buildings that are without electrical power. These conditions create the need to artificially light the scene, which allows for safer, more efficient operations. Firefighters must know how to operate the available emergency power and lighting equipment safely and efficiently. **Skill Sheet 8-I-11** describes procedures for illuminating emergency scenes.

Power Plants

An *inverter* is a step-up transformer that converts a vehicle's 12- or 24-volt DC current into 110- or 220-volt AC current. Inverters are used on emergency vehicles when small amounts of power are needed or when small electrically operated tools need to be used **(Figure 8.44)**. Advantages of inverters are fuel efficiency and low or nonexistent noise during operation. Disadvantages include limited power supply capability and limited mobility from the vehicle.

Generators are the most common power source used for emergency services; they can be portable or vehicle-mounted. Portable generators are powered by small gasoline or diesel engines and generally have 110- and/or 220-volt capacities **(Figure 8.45)**. Most portable generators are light enough to be carried by two people. They are extremely useful when electrical power is needed in an area that is not accessible to a vehicle-mounted system.

Vehicle-mounted generators usually have a larger power-generating capacity than portable units **(Figure 8.46)**. In addition to providing power for portable equipment, vehicle-mounted generators provide power for the floodlighting system on the vehicle. Vehicle-mounted generators can be powered by gasoline, diesel, or propane gas engines or by hydraulic or power take-off systems. Fixed floodlights are usually wired directly to the unit through a switch, and outlets are also provided for other equipment. These power plants generally have 110- and 220-volt output capabilities with capacities up to 50 kilowatts — sometimes greater. However, vehicle-mounted generators with a separate engine are noisy, making communication difficult near them. If not positioned downwind of an incident, the exhaust from these generators can contaminate the scene. Care must be used when refueling gasoline-powered units because spillage can affect a subsequent fire investigation.

Lighting Equipment

Lighting equipment can be divided into two categories: portable and fixed. Portable lights can be carried to and used in areas where vehicle-mounted lights are ineffective because of the distance between the vehicle and the scene, and/or because opaque obstructions are in the way. Portable lights generally range from 300 to 1,000 watts

Inverter — Auxiliary electrical power generating device. The inverter is a step-up transformer that converts the vehicle's 12- or 24-volt DC current into 110- or 220-volt AC current.

Generator — Auxiliary electrical power generating device. Portable generators are powered by small gasoline or diesel engines and generally have 110- and/or 220-volt capacities.

Figure 8.44 Apparatus-mounted inverters provide AC power for small electrical tools.

Figure 8.45 Example of a portable generator carried on a fire apparatus.

Figure 8.46 Vehicle-mounted generators provide greater electrical capacity for such things as emergency-scene lighting.

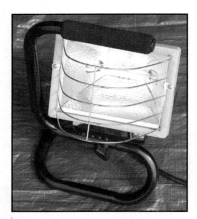

Figure 8.47 Portable floodlights provide interior and exterior scene lighting. They can be powered by apparatus-mounted or portable generators.

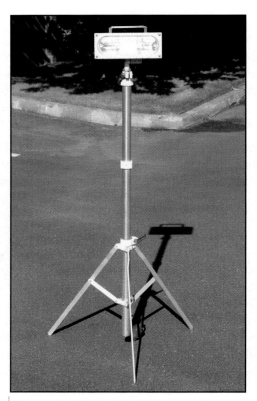

Figure 8.48 Portable lights mounted on telescoping stands can be used to illuminate many types of emergency scenes.

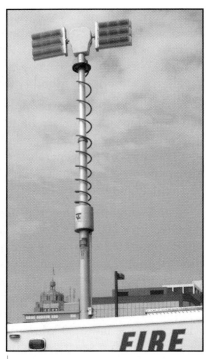

Figure 8.49 Many emergency vehicles are equipped with flood lights.

Figure 8.50 A hydraulically operated floodlight boom can provide adequate emergency scene lighting. *Courtesy of SVI Trucks.*

(**Figure 8.47**). They may be supplied with power by a cord from either a vehicle-mounted or portable power plant. The lights usually have handles for ease of carrying and large bases for stability. Some portable lights are mounted on telescoping stands, which allow them to be directed more effectively (**Figure 8.48**).

Fixed lights are mounted to a vehicle and their main function is to provide overall lighting of the emergency scene. Fixed lights are usually mounted so that they can be raised, lowered, or turned to provide the best possible angle. Often these lights are mounted on telescoping poles that allow both vertical and rotational movement (**Figure 8.49**). Some larger units include a hydraulically operated boom with a bank of lights (**Figure 8.50**). These banks of lights generally have a capacity of 500 to 1,500 watts per light. The amount of lighting should be carefully matched with the amount of power available from the power plant. Overtaxing the power plant gives poor lighting, may damage the power generating unit or the lights, and may restrict the operation of other electrical tools using the same power supply.

Auxiliary Electrical Equipment

A variety of other equipment may be used in conjunction with power plants and lighting equipment. Electrical cables or extension cords are necessary to conduct electric power to portable equipment. Cords may be stored in coils, on portable cord reels, or on fixed automatic rewind reels (**Figure 8.51**). Electrical cables should be waterproof, intrinsically safe, and have adequate insulation with no exposed wires. Twist-lock receptacles provide secure, safe connections as long as they are not immersed in water (**Figure 8.52**). Junction boxes may be used when multiple connections are needed (**Figure 8.53**). The junction box has several outlets and is supplied through one inlet from the power plant. All outlets should be equipped with ground-fault circuit interrupters and conform to NFPA 70E®, *Standard for Electrical Safety in the Workplace®*.

In situations where mutual aid departments frequently work together and have either different sizes or different types of receptacles (for example, one has two-prong; the other has three), adapters should be carried so that equipment can be interchanged (**Figure 8.54**). Adapters should also be carried to allow rescuers to plug their equipment into standard electrical outlets.

Figure 8.51 Light cord reels may be portable or permanently mounted in an apparatus compartment, as shown here.

Figure 8.52 The three-prong twist-lock adapter is designed to prevent the plug from slipping out of the receptacle.

Figure 8.53 Electrical junction boxes permit multiple lights, fans, or other electrical tools to operate off the same power source as long as the capacity of the source is not exceeded.

Figure 8.54 A variety of electrical adapters including twist-type and three-way male and female plugs.

Maintenance of Portable Power Plants and Lighting Equipment

Proper service and maintenance of portable power plants and lighting equipment is critical. Power plants and lighting equipment that have not been well maintained may fail at a crucial time during an operation. It is the responsibility of the firefighter to ensure that this equipment is well-maintained and operationally ready at all times. The general steps for servicing and maintaining portable power plants and lighting equipment are listed below. Refer to the manufacturer's instructions for each piece of equipment for specific directions and requirements.

- Review the manufacturer's service manual for specific directions. Gather the tools and service parts needed to perform the service and maintenance items.

- Carefully inspect the spark plug for damage, any visible corrosion, carbon accumulation or cracks in the porcelain. Check to make sure the spark plug wire is tight.

- If a spark plug is damaged, or if recommended in the service manual, replace the spark plug. Ensure proper gap prior to installing.

- Check equipment carburetor and identify any signs of leaks.

- Check fuel level and fill with fuel as necessary.

- If the fuel in the tank is more than three weeks old, empty the tank and replace with fresh fuel. Discard old fuel in approved container and manner. This step may not be necessary if fuel stabilizer is being used.

- Check oil level and replenish oil as needed.

- Start the generator and run any tests identified in the Operator Manual. If a problem is found with the generator, consult the manual to determine the proper action. Only certified service personnel or a licensed electrician should perform work on the generator portion of the power plant.

Arc — A luminous discharge of electricity across a gap. Arcs produce very high temperature.

- Inspect all electrical cords for frayed or damaged insulation or missing or bent prongs. Replace any plug with signs of arcing.

- Test the operation of the lighting equipment. This should include connecting each light to the generator one light at a time to prevent overloading. Avoid looking directly into the lights.

- Replace light bulbs as necessary. Shut off the power to the light before replacing the bulb. If possible, allow the bulb to cool before replacing. If the bulb must be replaced immediately, wear leather gloves to prevent being burned by the hot bulb. Discard faulty bulbs in an approved manner.

- Clean the work area and return all tools and equipment to the proper storage areas.

- Document the maintenance on the appropriate forms or records.

Skill Sheet 8-II-1 describes the steps for servicing and maintaining portable lighting equipment.

Hydraulic Tools

Hydraulics — Branch of fluid mechanics dealing with the mechanical properties of liquids and the application of these properties in engineering.

Rescue tools can be operated manually or by powered hydraulics. The development of powered hydraulic rescue tools has revolutionized the process of removing victims from various types of entrapments. The wide range of uses, speed, and superior power of these tools has made them the primary tools used in many rescue situations. Manual hydraulic tools operate on the same principles as powered hydraulic tools except that the hydraulic pump is manually operated by a rescuer pumping a lever.

Powered Hydraulic Tools

Powered hydraulic rescue tools are operated by hydraulic fluid pumped through special high-pressure hoses. Although there are a few pumps that are operated by compressed air, most are powered by electric motors or by two- or four-cycle gasoline engines. These units may be portable and carried with the tool, or they may be mounted on the vehicle and may supply power to the tool through a hose reel line **(Figure 8.55)**. Manually operated pumps are also available in case of a power unit failure **(Figure 8.56)**. Four basic types of powered hydraulic tools are used in rescue incidents: *spreaders, shears, combination spreader/shears,* and *extension rams.*

Spreaders. Powered hydraulic spreaders were the first powered hydraulic tools to become available to the fire/rescue service **(Figure 8.57)**. In combination with chains and adapters, they are capable of either pushing or pulling. Depending on the brand, this tool can produce tons of force at its tips. The tips of the tool may spread as much as 32 inches (800 mm) apart.

Shears. Hydraulic shear tools are capable of cutting almost any metal object that can fit between the blades, although some models cannot cut case-hardened steel **(Figure 8.58)**. The shears may also be used to cut other materials such as plastics or wood. Typical shears are capable of producing tons of cutting force and have an opening spread of approximately 7 inches (175 mm).

Figure 8.55 Hydraulic power units are available in a variety of styles and capacities.

Figure 8.56 Manual hydraulic pumps can provide a limited amount of power in the event the hydraulic unit fails.

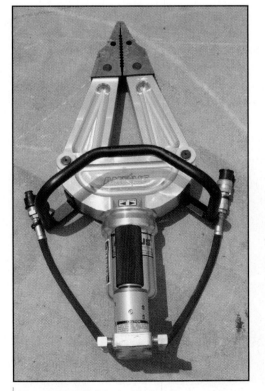

Figure 8.57 A typical hydraulic spreader can be used to extricate a victim in a vehicle accident, force entry into a structure, or lift a collapsed structural member.

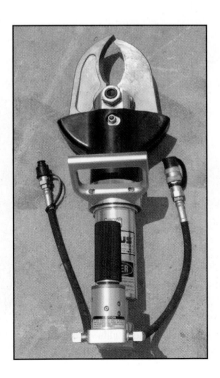

Figure 8.58 Typical hydraulic shears are designed to cut through most materials.

Combination spreader/shears. Most manufacturers of powered hydraulic rescue equipment offer a combination spreader/shears tool **(Figure 8.59)**. This tool consists of two arms equipped with spreader tips that can be used for pulling or pushing. The inside edges of the arms are equipped with cutting shears similar to those described in the previous paragraph. This combination tool is excellent for a small rapid-intervention vehicle or for departments where limited resources prevent the purchase of larger and more expensive individual spreader and cutting tools. However, the spreading and cutting capabilities of the combination tool are somewhat less than those of the individual units.

Extension rams. Extension rams are designed primarily for straight pushing operations, although they are effective at pulling as well. These tools are especially useful when it is necessary to push objects farther than the maximum opening distance of the hydraulic spreaders **(Figure 8.60)**. The largest of these extension rams can extend from a closed length of 3 feet (1 m) to an extended length of around 5 feet (1.5 m). They open with tons of pushing force. The closing force is about one-half that of the opening force.

Manual Hydraulic Tools

Two manual hydraulic tools are used frequently in extrication work: the *porta-power tool system* and the *hydraulic jack*. Some of the disadvantages of manual hydraulic tools are that they operate more slowly than powered hydraulic tools, have limited range of operation, and are labor-intensive. Some of the advantages are that they are relatively inexpensive, light weight, and can be used in areas inaccessible to powered units.

Porta-power tool system. The porta-power tool system is basically a commercial shop tool that has been adopted by the fire/rescue service **(Figure 8.61)**. It is operated by transmitting pressure from a manual hydraulic pump through a high-pressure hose to a tool assembly. A number of different tool accessories allow the porta-power tool system to be used in a variety of applications.

Figure 8.59 Combination spreader/shear units are available from most rescue tool equipment manufacturers.

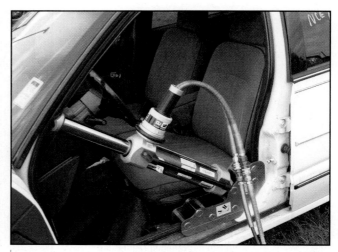

Figure 8.60 Hydraulic rams are effective for vehicle extrication activities.

The primary advantage of the porta-power tool over the hydraulic jack is that the porta-power tool has accessories that allow it to be operated in narrow places where the jack either will not fit or cannot be operated. The primary disadvantage of the porta-power tool is that assembling complex combinations of accessories and the actual operation of the tool is time-consuming.

Hydraulic jack. The hydraulic jack is designed for heavy lifting applications (**Figure 8.62**). It is also an excellent compression device for shoring or stabilizing operations (see Shoring section). Most hydraulic jacks have lifting capacities up to 20 tons (18 tonnes [t]), but units with a higher capacity are available.

Any kind of jack, hydraulic or otherwise, should have flat, level footing that may have to be constructed with cribbing. If a jack must be positioned on a soft surface, place a solid base of cribbing, a flat board, or a steel plate under the jack to prevent it from sinking into the surface (**Figure 8.63**).

Nonhydraulic Jacks

There are several kinds of jacks that can be considered hand tools because they do not operate with hydraulic power. Although these tools are effective for their designated purposes, they do not have the same lifting capabilities as hydraulic jacks. The following sections describe several of the nonhydraulic types of jacks. See the Hydraulic Jack section for safety guidelines when using any type of jack.

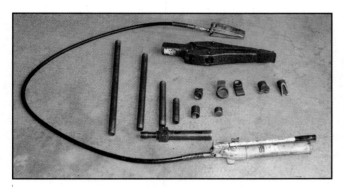

Figure 8.61 A typical porta-power set consists of a variety of accessories.

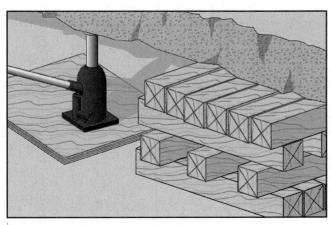

Figure 8.63 A jack must have a solid base plate provided by cribbing, flat piece of wood, or steel plate.

Figure 8.62 A heavy-duty hydraulic jack with handle used for lifting heavy objects.

Screw Jacks

Screw jacks can be extended or retracted by turning the threaded shaft. Jacks should be checked for wear after each use so that they are always in a state of readiness. They should also be kept clean and lightly lubricated, with particular attention paid to the screw thread. Footplates should also be checked for wear or damage.

Two commonly used types of screw jacks are the *bar screw jack* and the *trench screw jack*. Both jacks have a male-threaded stem (similar to a bolt) and a female-threaded component.

Bar screw jacks. Bar screw jacks are heavy-duty devices excellent for supporting collapsed structural members **(Figure 8.64)**. These jacks are normally not used for lifting; their primary use is to hold an object in place, not to move it. The jacks are extended or retracted as the shaft is rotated in the base. The shaft is turned with a long bar that is inserted through a hole in the top of the shaft.

Trench screw jacks. Because of their ease of application, durability, and relatively low cost, trench screw jacks are sometimes used to replace wooden cross braces in trench rescue applications. These devices consist of a swivel footplate with a stem that is inserted into one end of a section of 2-inch (50 mm) steel pipe (not to exceed 6 feet [2 m] in length) and a second swivel footplate with a threaded stem that is inserted into the other end of the pipe **(Figure 8.65)**. An adjusting nut (with handles) on the threaded stem is turned to vary the length of the jack and to tighten it between opposing members in a shoring (stabilizing) system.

Ratchet-Lever Jack

Also known as *high-lift* jacks, these medium-duty jacks consist of a rigid I-beam with perforations in the web and a jacking carriage with two ratchets on the geared side that fits around the I-beam **(Figure 8.66)**. One ratchet holds the carriage in position while the second ratchet works with a lever to move the carriage up or down.

Ratchet jacks can be dangerous because they are the least stable of all the various types of jacks. If the load being lifted shifts, ratchet-lever jacks may simply fall over, allowing the load to suddenly drop to its original position. Also, the ratchets can fail under a heavy load.

Shoring — General term used for lengths of timber, screw jacks, hydraulic and pneumatic jacks, and other devices that can be used as temporary support for formwork or structural components or used to hold sheeting against trench walls. Individual supports are called shores, cross braces, and struts.

WARNING!

Rescuers should never work under a load supported only by a jack. If the jack fails or the load shifts, severe injury or death could result. The load should also be supported by properly placed cribbing.

Figure 8.64 A bar screw jack is used to support an object once it has been lifted to the required height.

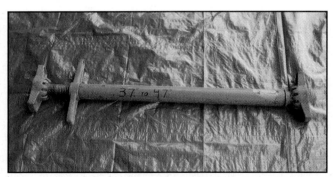

Figure 8.65 The trench screw jack is used to brace shoring on the sides of a trench to prevent collapse.

Figure 8.66 Ratchet-lever jacks, also known as high-lift jacks, have a variety of applications but are the least stable of all types of jacks.

Cribbing

Most rescue vehicles carry appropriately sized cribbing. Cribbing, essential in many rescue operations, is most commonly used to stabilize objects but has many other uses as well. Large wedges may be used to shim up loose cribbing (**Figure 8.67**). The wedges may be driven in with a mallet or a piece of cribbing.

When wood is selected for cribbing, it should be solid, straight, and free of such major flaws as large knots or splits. Various sizes of wood can be used, but the most popular are 2- × 4-inch (50 mm by 100 mm), 4- × 4-inch (100 mm by 100 mm), and 6- × 6-inch (150 mm by 150 mm) construction grade lumber. The length of the pieces may vary, but 16 to 24 inches (400 mm to 600 mm) are most common for the smaller dimension pieces. Larger pieces (6- × 6-inch [150 mm by 150 mm]) are usually 36 inches (900 mm) in length. The ends of the blocks may be painted different colors for easy identification by length. Other surfaces of the cribbing should be free of paint or any other finish because they can make the wood slippery, especially when it is wet. Individual pieces of cribbing may have a rope or webbing handle stapled to the end for easy carrying and safe removal from under objects (**Figure 8.68, p. 340**).

Cribbing made of plastic is also now in use. A full range of plastic cribbing components, including wedges, step chocks, struts, and pads for pneumatic lifting bags, are now available (**Figure 8.69, p. 340**). While considerably more expensive than wooden cribbing, plastic cribbing does not become contaminated by absorbing fuel, oil, and other substances. This characteristic and its innate durability make plastic cribbing last many times longer than wooden cribbing.

Cribbing can be stacked in a compartment with the grab handles facing out for easy access (**Figure 8.70, p. 340**). It can also be placed on end inside a storage crate (**Figure 8.71, p. 340**). Using storage crates makes it easier to carry several pieces of cribbing at the same time.

Cribbing — Varying lengths of hardwood, usually 4- × 4-inch (100 mm by 100 mm) or larger, used to stabilize vehicles and collapsed buildings during extrication incidents.

Figure 8.67 Rescue apparatus usually carry a variety of sizes of wedges and cribbing.

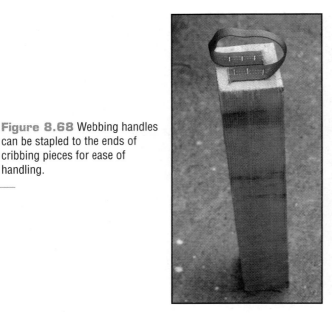

Figure 8.68 Webbing handles can be stapled to the ends of cribbing pieces for ease of handling.

Figure 8.69 Cribbing manufactured from recycled plastic provides longer life and is immune to contamination and splintering.

Figure 8.70 Cribbing should be stored in compartments with grab handles facing out for easy accessibility.

Figure 8.71 Cribbing can be stored in crates for easy transport and organization.

Pneumatic (Air-Powered) Tools

Pneumatic tools use compressed air for power. The air can be supplied by vehicle-mounted air compressors, apparatus brake system compressors, SCBA cylinders, or cascade system cylinders. While there are others, air chisels and pneumatic nailers are two of the most commonly used pneumatic tools.

Air Chisels

Pneumatic-powered chisels (also called *air chisels, air hammers,* or *impact hammers*) are useful in rescue and extrication work. Most air chisels operate at air pressures between 90 and 250 psi (630 kPa and 1 750 kPa). These tools come with a variety of interchangeable bits to fit the needs of almost any situation (**Figure 8.72**). In addition to cutting bits, special bits for such operations as breaking locks or driving in plugs are also available. Often used in vehicle extrication situations, these tools are good for cutting medium- to heavy-gauge sheet metal and for popping rivets and bolts. Cutting heavier-gauge metal requires more air at higher pressures.

> **WARNING!**
>
> Never use compressed oxygen to power pneumatic tools. Mixing pure oxygen with grease and oils found on the tools will result in fire or violent explosion.

Figure 8.72 Air chisels are used for cutting metal, removing rivets or bolts, and opening locks.

Figure 8.73 A pneumatic or air-operated nail gun, shown without its hose, can be used for constructing shoring and for applying materials to secure fire damaged roofs.

Pneumatic Nailers

Air-operated nail guns can be used to drive nails or heavy-duty staples into wood or masonry. They are especially useful for nailing wedges and other wooden components of shoring systems into place (**Figure 8.73**).

Other Pneumatic Tools

A number of other pneumatic tools are used in rescue operations. Among them are the impact tool, air knife, air vacuum, and the whizzer saw.

Impact Tool

Also called *impact wrenches*, these pneumatic or electric tools have a square drive onto which a socket can be snapped (**Figure 8.74**). The socket can then be applied to a nut or bolt head of the same size and the component can be tightened or loosened very quickly. These tools are ideal for disassembling machinery in which a victim is entangled.

Air Knife

These hand-held tools can blast away surface dirt with great efficiency (**Figure 8.75**). When operated at 90 to 100 psi (630 kPa to 700 kPa), the exit velocity of the air at the nozzle is close to Mach 2. Depending upon variables in air supply pressure, soil type and moisture content, air knives can almost instantly create a hole 1 foot (.3 m) in diameter and depth. Even though air knives can cut through dirt with ease, the air blast will not cut cables or damage underground pipes. The air to operate these tools is normally supplied by an air compressor. The only extra protection required by air knife operators is face and eye protection.

Figure 8.74 Pneumatic and electric impact wrenches can be used to disassemble machinery during industrial or agricultural rescue operations.

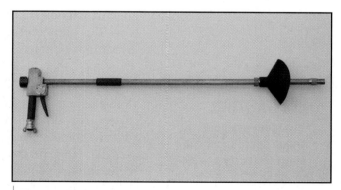

Figure 8.75 An air knife can be very useful for removing soil during a trench collapse operation. *Courtesy of Supersonic Air Knife, Inc.*

Air Vacuum

Closely related to the air knife, air vacuums are hand-held tools that can use the same air compressor as the air knife, but they operate on slightly less pressure (**Figure 8.76**). Consisting of a suction tube and a collector bag, air vacuums can pick up loose soil and rocks up to 2¾ inches (70 mm) in diameter. These tools can be used to vacuum loose soil from around victims buried in a trench collapse.

Figure 8.76 The air vacuum tool is used with the air knife to remove soil near a trench collapse victim. *Courtesy of Supersonic Air Knife, Inc.*

Whizzer Saw

The whizzer saw is an electric or air-driven cutting device with several advantages over other types of power saws (**Figure 8.77**). At about 2 pounds (0.9 kg), the whizzer weighs about one-tenth as much as a circular saw, so it is much more maneuverable. Operating at 20,000 rpm, its 3-inch (75 mm) Carborundum® blade cuts case-hardened locks and steel stock up to ¾-inch (19 mm) in thickness. The tool has a clear Lexan® blade guard to protect the operator and the victim. Driven by compressed air at 90 psi (630 kPa) from an SCBA cylinder with a regulator, the whizzer operates much more quietly than other power saws and will run approximately three minutes from a full cylinder. These saws are often used for delicate cutting operations such as removing rings from swollen fingers.

Lifting/Pulling Tools

Rescuers must sometimes lift or pull an object to free a victim. Several rescue tools have been developed to assist in this task. These include *tripods, winches, come-alongs, chains, pneumatic lifting bags,* and *block and tackle systems.*

Tripods

Rescue tripods are used to create an anchor point above a manhole or other opening. This allows rescuers to be safely lowered into confined spaces and rescuers and victims to be hoisted out of them (**Figure 8.78**).

Winches

Vehicle-mounted winches are excellent pulling tools. They can usually be deployed faster than other lifting/pulling devices, generally have a greater travel or pulling distance, and are much stronger. While some winches are located at the rear of the vehicle, most are mounted behind the front bumper (**Figure 8.79**). The three most common drives for winches are electric, hydraulic, and power take-off. Chains, cables, or combinations of the two are used for pulling.

Winches should be equipped with handheld, remote-control devices (**Figure 8.80**). These devices allow the operator to observe and control the pulling operation while standing outside the danger zone. The winch should be positioned as close to the object being pulled as possible so that if the cable breaks, there will be less cable to suddenly recoil.

Figure 8.77 Whizzer saws are used for fine cutting operations.

> **CAUTION**
> Whenever possible, the winch operator should stay farther away from the winch than the length of the cable from the winch to the load (Figure 8.81, p. 343). Winches should not be used with a live person load.

Figure 8.79
A vehicle-mounted winch can be used for moving heavy objects or for extricating the apparatus when it is stuck in soft soil.

Figure 8.78 Rescue tripods are used for lowering rescuers into a confined space and lifting out victims and rescuers.

Figure 8.80 Remote winch controls increase operator safety. *Courtesy of IFE Hong Kong Branch.*

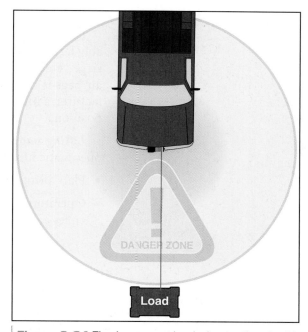

Figure 8.81 The danger zone in winch operation should be kept clear of all personnel. If the cable or chain breaks, it will snap into this area with disastrous results.

Come-Alongs

Another lifting/pulling tool often used in rescue is the come-along **(Figure 8.82)**. It is a portable cable winch operated by a manual ratchet lever. In use, the come-along is attached to a secure anchor point, and the cable is run out to the object to be moved. Once both ends are attached, the lever is operated to rewind the cable, which pulls the movable object toward the anchor point. The most common sizes or ratings of come-alongs are 1 to 10 tons (0.9 t to 9.1 t).

Figure 8.82 Manual come-alongs have limited lifting capacities.

Chains

Winches and come-alongs may use chains as part of a lifting/pulling system. Only alloy steel chains of the correct size should be used in rescue work (**Figure 8.83**). Alloy steel chains are highly resistant to abrasion, making them ideal for rescue and extrication work. Special alloys are available that are resistant to corrosive or hazardous atmospheres. Proof coil chain, also called *common* or *hardware chain*, is not suitable for rescue applications.

Pneumatic Lifting Bags

Lifting bags give rescuers the ability to lift or displace objects that cannot be lifted with other rescue equipment (**Figure 8.84**). There are three basic types of lifting bags: *high- pressure, medium-pressure,* and *low-pressure*. A fourth type of bag is used for sealing leaks but has little, if any, rescue application.

High-pressure bag. High-pressure bags consist of a tough, neoprene rubber exterior reinforced with steel wire or Kevlar™ aramid fiber. Deflated, the bags lie completely flat and are about 1 inch (25 mm) thick (**Figure 8.85**). They come in various sizes that range in surface area from 6 × 6 inches (150 mm by 150 mm) to 36 × 36 inches (900 mm by 900 mm). Depending on the size of the bags, they may inflate to a height of 20 inches (500 mm).

Low- and medium-pressure bags. Low- and medium-pressure bags are considerably larger than high-pressure bags and are most commonly used to lift or stabilize large vehicles or objects (**Figure 8.86**). Their primary advantage over high-pressure air bags is that they have a much greater lifting distance. Depending on the manufacturer, a lifting bag may be capable of lifting an object 6 feet (2 m) above its original position.

Lifting bag safety rules. Operators should follow these safety rules when using pneumatic lifting bags:

- Plan lifting operation before starting.

- Operators should be thoroughly familiar with the equipment — its operating principles and methods — and its capabilities and limitations.

Figure 8.83 Rescue chains must be constructed of alloy steel.

Figure 8.84 Pneumatic lifting bag systems have a variety of applications. The complete system consists of the bags, pressure regulators, hoses, and air cylinders.

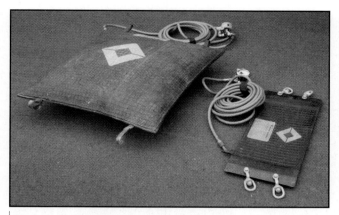

Figure 8.85 High-pressure lifting bags are constructed of neoprene rubber reinforced with steel wire or Kevlar™.

Figure 8.86 Typical low-pressure lifting bags are capable of lifting heavy objects, such as fuel tankers, as high as 6 feet (2 m).

- Operators should follow the manufacturer's recommendations for the specific system used.
- All components should be kept in good operating condition with all safety seals in place.
- Operators should have available an adequate air supply and sufficient cribbing before beginning operations.
- The bags should be positioned on or against a solid surface.
- The bags should be protected from sharp objects during inflation.
- The bags should be inflated slowly and monitored continually for any shifting.
- Rescuers should never work under a load supported only by air bags.
- The load should be continuously shored up with enough cribbing to adequately support the load in case of bag failure — **Remember, lift an inch, crib an inch.**
- When box cribbing is used to support an air bag, the top layer should be solid; leaving openings in the center may cause the bag to shift or rupture (**Figure 8.87**).
- Bags should not be allowed to contact materials hotter than 220°F (104°C).
- Bags should never be stacked more than two high. With the smaller bag on top, the bottom bag should be inflated first (**Figure 8.88, p. 346**). A single multicell bag is preferred (**Figure 8.89, p. 346**).

WARNING!
Plywood or other rigid material placed between a lifting bag and an object can be forcefully ejected if the bag distorts under pressure. Use only pliable material such as belting or a folded salvage cover between bags and objects being lifted.

CAUTION
Air bags should be inspected regularly and should be removed from service if any evidence of damage or deterioration is found.

Figure 8.87 Cribbing can be used to provide a solid base to support lifting bags.

Figure 8.88 Lifting bags can be stacked if necessary, but only two high with the larger bag on the bottom.

Figure 8.89 Multicell lifting bags, such as this three-cell unit, are safer than stacked bags.

Block and Tackle — Series of pulleys (sheaves) contained within a wood or metal frame. They are used with rope to provide a mechanical advantage for pulling operations.

Block and Tackle Systems

Because of their mechanical advantage in converting a given amount of pull to a working force greater than the pull, block and tackle systems are useful for lifting or pulling heavy loads **(Figure 8.90)**. A *block* is a wooden or metal frame containing one or more pulleys called *sheaves*. *Tackle* is the assembly of ropes used to multiply the pulling force. Nonrescue-rated block and tackle systems are suitable only for lifting or stabilizing objects and other utility applications. For more information on block and tackle systems see the IFSTA **Fire Service Search and Rescue** manual.

Vehicle Extrication

In most fire departments in the U.S. and Canada, the overwhelming majority of rescue incidents are vehicle extrications **(Figure 8.91)**. These incidents are the result of automobiles, motorcycles, trucks, trains, or aircraft colliding with each other or hitting bicycle riders, pedestrians, animals, or stationary objects. Because a victim who is trapped in or by a vehicle may be seriously injured, proper extrication procedures are essential to prevent further injury and to speed the victim's removal. It is also critical that firefighters coordinate with emergency medical personnel who are providing first aid to the victim. General procedures for extricating a victim trapped in a motor vehicle are found in **Skill Sheet 8-II-2**.

Figure 8.90 A rescue-rated block and tackle set used for lifting victims or rescuers.

Figure 8.91 Motor vehicle extrications are the most common type of rescue for fire departments. *Courtesy of Bob Esposito.*

Scene Size-Up

Scene size-up is essential to accomplishing a safe and efficient extrication operation. While incident size-up begins with the initial dispatch and continues throughout the incident, scene size-up begins as soon as the first emergency vehicle approaches the accident scene. Taking the time to carefully assess the scene can help prevent injury to rescuers, prevent further injuries to victims, clarify required tasks, and identify needed resources.

When arriving on the scene of a motor vehicle accident, the officer in charge should position the apparatus according to departmental SOP and the situation at hand. In general, it should be positioned close enough to the scene for the equipment and supplies to be readily available but not so close that it might interfere with other on-scene activities.

The first-arriving engine is usually positioned to provide a barrier to protect the scene from oncoming traffic, and cones, signs, flares, or other traffic control devices are set out to direct motorists around the scene (**Figure 8.92, p. 348**). Headlights and emergency lights should either be turned off or left on depending upon the situation and departmental SOP. The U.S. Department of Transportation (DOT) recommends that when headlights are not needed for scene illumination, they be turned off to avoid blinding or distracting drivers of oncoming vehicles.

If possible, at least one traffic lane in addition to the one or ones occupied by the vehicle or vehicles involved should be closed to nonemergency traffic. In some cases, it may be necessary to close the street or highway completely in one or both directions. Once law enforcement personnel are on scene, traffic control is their responsibility. Despite being given advance information by dispatchers during the response, firefighters face a number of unknowns when arriving at the scene. Rescuers should be observant as they approach the scene and consider the following:

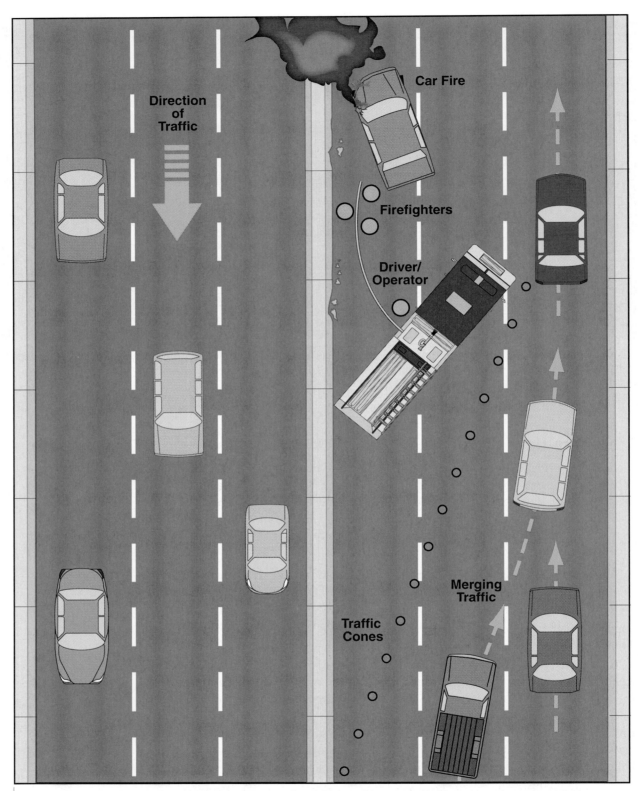

Figure 8.92 A fire apparatus positioned to protect a motor vehicle accident scene. Traffic cones and emergency lights are intended to alert motorists approaching the scene while the apparatus provides a safety barrier for victims and rescuers.

- What are the traffic hazards? What types of control devices are needed?

- How many and what types of vehicles are involved — Are any of them hybrids or do they use alternative fuels?

- Where and how are the vehicles positioned — Are they all on the roadway?

- How many victims are there and what is their status?

- Is there a fire or potential for a fire?

- Are there any hazardous materials involved?

- Are there any utilities, such as water, gas, or electricity, that may have been damaged? If so, are they posing a hazard to the victims or rescue personnel?

- Is there a need for additional resources?

Assessing the Need for Extrication Activities

At the scene, personnel should make a more thorough assessment of the situation before taking any action. Personnel should assess the immediate area around each vehicle and assess the entire scene in more detail. The rescuer who assesses each vehicle should be concerned with the number of victims in or around the vehicle and the severity of their injuries. The rescuer should also assess the condition of the vehicle, determine extrication tasks that may be required, and note any hazardous conditions that might exist. Ideally, there will be one rescuer to assess each vehicle involved in the incident but this may not be possible **(Figure 8.93)**. If there is only one rescuer available and more than one vehicle to survey, the rescuer must check each one separately and report the conditions in each vehicle to the IC.

While each vehicle is being checked, another rescuer should be assigned to survey the entire area around the scene **(Figure 8.94)**. This person should check to see if there are any other vehicles involved that may not be readily apparent (over an embankment, for example), any victims who have been thrown out of the vehicles, any damage to structures or utilities that present a hazard, or any other circumstances that warrant special attention.

Someone trained in first aid or more advanced emergency medical care should be assigned to determine the number of victims and the extent of their injuries, to tri-

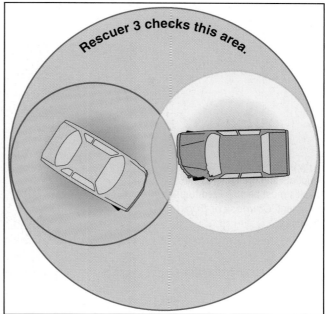

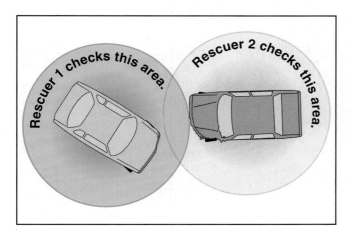

Figure 8.93 Rescuers thoroughly check around each involved vehicle.

Figure 8.94 A third rescuer checks the area around both vehicles.

age them, and to assess the nature and extent of their entrapment **(Figure 8.95)**. This information aids the IC in determining the resources needed to stabilize the incident and the order in which victims should be removed. Of course, more seriously injured victims must receive higher priority than those with minor injuries. Victims who are not trapped should be removed first to make more working

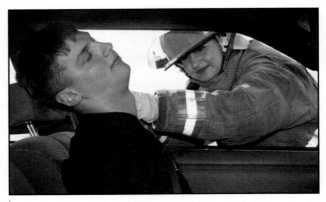

Figure 8.95 The condition and entrapment of victims should be assessed as soon as possible.

room for rescuers who are trying to remove those who are entrapped. As each assessment is completed, the rescuer should report findings to the IC.

Stabilizing the Vehicle

Following scene assessment, and before any other extrication activities are begun, rescuers must stabilize the vehicles involved **(Figure 8.96)**. This is vital to prevent further injury to the victims or possible injuries to rescuers. Proper stabilization refers to the process of using cribbing and shoring devices to provide support at key points between the vehicle and the road surface or the ground. The primary goal of stabilization is to prevent sudden or unexpected movement of the vehicle. Such movement can seriously aggravate the victim's injuries and could injure rescuers.

Vehicles can be found in a number of different positions following a collision. Inexperienced rescuers are sometimes tempted to test the stability of an overturned vehicle by pushing or pulling on it; however, they must be trained to resist this temptation because the slightest push in the wrong place may cause the vehicle

Figure 8.96 For the safety of victims and firefighters alike, vehicles must be stabilized to prevent movement. *Courtesy of Bob Esposito.*

to move. This is particularly true of vehicles that are on their sides or tops, resting partially over a cliff or embankment, or in danger of falling from a bridge or overpass (**Figure 8.97**).

Most vehicles involved in collisions remain upright. Rescuers must realize that even though the vehicle still has all its wheels on the ground, stabilization is still required. The vehicle should be stabilized with cribbing to prevent both vertical and horizontal movement.

Several methods can be used to prevent horizontal motion. The most common method is to chock the vehicle's wheels. It is most important to chock the wheels on the downhill side of a vehicle that is on a grade (**Figure 8.98**). If the vehicle is on level ground, chock the wheels in both directions (**Figure 8.99**). Chocking can be accomplished with standard wheel chocks, pieces of cribbing, or other similar-sized objects.

It may also be possible to use one or more of the vehicle's own mechanical systems to supplement the wheel chocks. This will depend on whether or not these systems are still operable. If possible, place automatic transmissions in the park position; place manual transmissions in gear. Set parking or emergency brakes.

There are numerous ways to prevent a vehicle from moving vertically. Jacks, pneumatic lifting bags, and cribbing are used most frequently for this purpose. Different types of jacks can be used to support the frame of the vehicle. One advantage of jacks is that they can be adjusted to the required height; one disadvantage is that they are

Chock — Wooden, plastic, or metal block constructed to fit the curvature of a tire; placed against the tire to prevent apparatus rolling.

CAUTION

Do not rely on mechanical systems, even if they are operable, as the sole means of stabilization. They should be used only with other stabilization measures.

Figure 8.97 This unstabilized vehicle can be easily tipped over, causing a hazard to both rescuers and victims.

Figure 8.98 Chock wheels on the downhill side to prevent vehicles from rolling down the incline.

Figure 8.99 On level ground, use cribbing to chock wheels in both directions.

time-consuming to place and may limit rescuer access. Pneumatic lifting bags can also be used for support. To be effective, at least two air-lifting bags are needed. They should be positioned either one on each side of the vehicle or one in the front and one in the rear (**Figure 8.100**). Regardless of which devices are used, they must be supplemented with cribbing and/or step chocks (**Figure 8.101**).

When cribbing is used for vehicle stabilization, a box formation is most often used to support and stabilize vehicles. It may be necessary to use wedges to ensure solid contact between the cribbing and the vehicle (**Figure 8.102**).

When using any of these stabilization methods, rescuers must take care to avoid placing any part of their bodies under the vehicle while placing the stabilizing device. There is always the possibility that the vehicle may drop unexpectedly, injuring or killing anyone beneath it. Cribbing pieces should be placed near the point of application and pushed into position with a tool or another piece of cribbing (**Figure 8.103**).

Figure 8.100 Air lifting bags are needed to fully stabilize the vehicle. One is placed on each side or at each end of the vehicle.

Figure 8.101 Cribbing is used to supplement the air bags.

Figure 8.102 Wedges are inserted between the cribbing and the vehicle to ensure complete contact. Wedges can also be used to support parts that are being cut or pried away, as shown here.

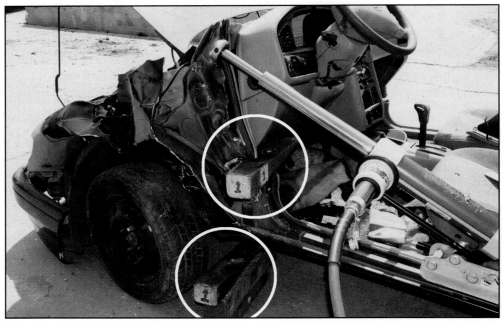

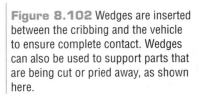

On occasion, vehicles will be found in positions other than upright, such as upside down, on their side, or on a slope. Under these circumstances, rescuers should use whatever means are available to stabilize the vehicle. Generally, a combination of cribbing, ropes, webbing, and chains are used to accomplish these types of stabilization tasks (**Figure 8.104**).

The final step in stabilizing a vehicle is to shut down the electrical power within the vehicle. Before shutting down the power, use it to lower power windows, unlock doors, or move seat backs.

Shutting down the power is critically important in all vehicles to eliminate one possible source of ignition. It is even more important in newer vehicles to eliminate power to the restraint systems (discussed in more detail later) such as front and side air bags, seat belt tensioners, and other devices intended to protect the occupants of the vehicle. In addition, hybrid vehicles also have high-voltage cables and components that represent an electrocution hazard for rescuers. The current in these systems ranges from 144 to more than 300 volts DC. Some hybrid vehicles can be identified by a distinctive nameplate (**Figure 8.105, p. 354**).

Turning off the ignition and removing the key eliminates power to both the conventional and the high-voltage systems. If the ignition is not accessible, power to both electrical systems and all restraint systems can be shut off by disconnecting or isolating the negative cables to the vehicle's 12-volt battery (**Figure 8.106, p. 354**). Cut the negative cable first, then the positive. Remove approximately 2 inches (50 mm) of each cable.

Gaining Access to Victims

In general, there are three methods of gaining access to victims in vehicles:

- Through a normally operating door
- Through a window
- By cutting away parts of the vehicle body

The simpler the required operation is, the better. When complex maneuvers are required to gain access into a vehicle, extrications become long, complicated, and ulti-

Figure 8.103 Push cribbing pieces under the load using a tool or separate piece of cribbing, never the hand.

Figure 8.104 In some cases, vehicles can be stabilized with utility rope.

Figure 8.105 The nameplate confirms this to be a hybrid vehicle. Firefighters must be aware of the potential shocking hazard that exists with these types of vehicles.

Figure 8.106 Isolating the battery cables does not eliminate all power to the restraint systems. The system may remain operational for up to 30 minutes after the loss of power.

mately more dangerous. For example, if a vehicle is not badly damaged, access may be obtained by simply opening an undamaged or operable door. When there is severe structural damage (for example, when the car roof is collapsed), gaining access can be a lengthy and complex process.

There are a number of additional potential hazards associated with wrecked passenger vehicles. The most common of these hazards are:

- Oil- and air-filled struts for hoods, trunk lids, and bumpers
- Fuel and other flammable liquids
- High-pressure tires
- Contents of trunk or vehicle interior

Supplemental Restraint System (SRS) and Side-Impact Protection System (SIPS)

Modern technology has added increased collision protection for vehicle occupants by means of Supplemental Restraint Systems (SRS) and Side-Impact Protection Systems (SIPS), commonly called *air bags* (**Figure 8.107**). These systems can be either electrically or mechanically activated. Although air bags have saved many lives, they have also added a potential rescuer safety hazard: accidental activation of the SRS or SIPS during extrication operations. These air bags can deploy with a speed of 200 mph (354 km/h) and exert a tremendous force.

An electrically operated restraint system receives its energy from the vehicle's battery and is designed to activate through a system of electronic sensors installed in the vehicle. These systems have a reserve energy supply that is capable of deploying an air bag even if the battery is disconnected or destroyed in the accident. When the bat-

WARNING!
The reserve energy supply can maintain sufficient voltage to deploy an air bag for up to 30 minutes after the battery has been disconnected. Stay out of the path of any undeployed air bag.

Figure 8.107 Firefighters must be aware that passenger vehicles and light trucks are equipped with air bags located in the steering wheel hub, passenger-side dash, and in some side moldings.

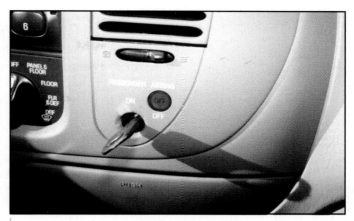

Figure 8.108 Some vehicles are equipped with a key-operated switch to disable the passenger side SRS.

tery is disconnected, the reserve energy supply will eventually drain, disarming the restraint system. Vehicle manufacturers have different time estimates on how long it takes for the reserve to deplete entirely.

Fire suppression or extrication activities are capable of accidentally activating electrically or mechanically operated restraint systems. For electrically operated systems, an electrical impulse during the extrication process may cause the air bag to deploy. There have been reports of rescuers being physically ejected from a vehicle with a connected battery when the "loaded" SRS was accidentally deployed during extrication operations. Personal protective equipment must be worn and extreme care taken when performing extrication operations on vehicles with SRS or SIPS.

On many vehicle models, the only method to prevent the accidental firing of electrical-type systems is to turn the ignition switch to the "off" position, disconnect both battery cables, and wait for the reserve power supply to drain down. Some vehicle models are equipped with a key-operated switch that disables and drains the reserve power to passenger-side air bags **(Figure 8.108)**.

Mechanically operated systems are sometimes used in SIPS design and do not require power from the vehicle's electrical system to activate. Therefore, these air bags may be deployed even if the battery has been disconnected. Simply applying pressure to the SIPS control unit, which may be located under the dash or in the center console, can deploy the bags. In these systems, disarming or preventing deployment of the air bags may require that the connection between the sensor and the air bag inflation unit be cut. How and where this is done is specific to each vehicle make and model.

Disentanglement and Patient Management

Rescuers should choose the easiest route available to gain access to a vehicle. They should try to open the doors normally, but if they are jammed, removing a window would be the next logical choice. Once access to the vehicle is gained, at least one rescuer with appropriate emergency medical training should enter the vehicle to begin stabilization of the patient and to protect the patient while disentanglement procedures are in progress. Rescuers working inside the vehicle must be wearing PPE that is appropriate to the hazards. Initial assessment and treatment should be done in accordance with local EMS protocols.

Once the patient's injuries have been assessed, treatment can begin simultaneously with preparation for removal from the vehicle. The most important point to remember is that the vehicle is removed from around the patient and not the reverse. Various parts of the vehicle, such as the steering wheel, seat, pedals, and dashboard, may trap the occupant. The situation should be assessed with the patient's safety foremost in the rescuer's mind.

Patient Removal

Packaging means that wounds have been dressed and bandaged, fractures have been splinted, and the patient's body has been immobilized to reduce the possibility of further injury **(Figure 8.109)**. Proper packaging protects the patient and facilitates the patient's removal. Once the path has been cleared and the patient has been properly packaged for removal, rescuers should cover sharp edges to prevent cutting themselves or the patient during the extraction process. Openings should be widened and edges padded with blankets, folded salvage covers, or duct tape. Openings should be wide enough so that the patient can be removed as smoothly as possible with as little jostling as possible **(Figure 8.110)**.

Packaging — Readying a patient for transport.

Figure 8.109 Victims must be properly packaged before they are removed from the vehicle.

Figure 8.110 A large enough opening must be created to facilitate patient removal.

Removing Glass

A common task required of rescuers during an extrication operation is removing glass from the vehicle. Glass may need to be removed to facilitate access to the passenger compartment or to reduce the hazard posed by remaining fragments of glass. To protect against any injuries from the loose or flying glass, rescuers should wear full protective equipment, including eye protection. The occupants inside the vehicle must be protected as well. Before discussing glass-removal techniques, it is important to understand the two primary types of glass used in vehicles: laminated *safety glass* and *tempered glass.*

Laminated safety glass. Safety or laminated glass is manufactured from two sheets of glass that are bonded to a sheet of plastic sandwiched between them (**Figure 8.111**). This type of glass is most commonly used for windshields and some rear windows. Impact produces many long, pointed shards with sharp edges. The plastic laminate sheet retains most of these shards and fragments in place. When broken, glass stays attached to the laminate and moves as a unit. This keeps the shards of glass from flying about. Some manufacturers have laminated an additional layer of plastic to the passenger-compartment side of the windshield. This provides greater protection from lacerations when impacted.

Tempered glass. Tempered glass is most commonly used in side windows and some rear windows. When struck, tempered glass is designed so that small lines of fracture are spread throughout the entire plate and the glass separates into many small pieces. This eliminates the hazard of long, pointed pieces of glass, but presents new problems, among them small nuisance lacerations to unprotected body parts and the entry of small pieces of glass into open wounds or the eyes.

Removing Laminated Glass

Removing windshields was once standard practice during many extrication operations but the design of many newer automobiles uses the windshield as a structural component. Removing the windshield from these vehicles seriously weakens the vehicle body and may cause it to collapse. Because it is difficult to tell which vehicles include this design, some departments have abandoned the practice of removing windshields. Whenever possible, leave the windshield intact.

If a laminated glass windshield or rear window must be removed, it will be somewhat more complicated and time-consuming than removing side or rear windows made of tempered glass. This is mainly because of the difference in glass types. Windshields and rear windows that are constructed of laminated glass will not disintegrate and fall out like tempered glass windows. Because more laminates are being added to windshields, it may not be as easy to chop through the windshields of newer vehicles. In this case, the best method for removing glass is with a saw (**Figure 8.112, p. 358**). The following common hand tools can be used to cut laminated glass:

- Air chisel
- Axe (standard or aircraft crash axe)
- Reciprocating saw
- Handsaw with a coarse blade
- Hay hook
- Windshield saw

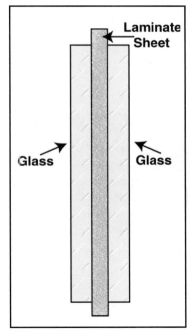

Figure 8.111 The basic construction of laminated glass consists of a layer of plastic between two layers of glass.

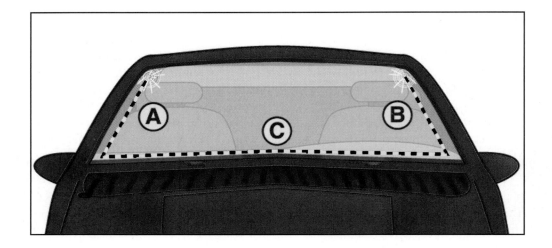

Figure 8.112 Cuts needed for total windshield removal. Using a window saw or other tool, make a cut on each side of the windshield and then across the bottom.

In older vehicles, total windshield removal is performed before the roof is laid back or removed. This method requires several rescuers — one on each side of the vehicle for cutting the windshield. As always, the firefighters must wear appropriate PPE, including eye protection. The passengers and rescuers inside the vehicle should be covered with a tarp or protective blanket. Two rescuers should hold the cover over the passenger(s) and rescuer inside the vehicle. A backboard can be added to protect people inside the vehicle from being struck by tools or loose glass.

Removing Tempered Glass

Removing side and rear windows constructed of tempered glass is a fairly simple task. These windows can easily be broken by either striking them with a sharp, pointed object in the lower corner of the window or by using a spring-loaded center punch pressed into the lower corner of the window. When using a center punch, the hand holding the punch should be supported by the opposite hand (**Figure 8.113**). This prevents rescuers from sticking their hands into the glass when it breaks and also prevents the center punch from coming in contact with a victim who may be close to the window. A standard center punch or Phillips screwdriver may also be used. It will need to be driven into the window with a hammer or mallet. A Halligan tool or the pick end of a pick-head axe will also work if nothing else is available.

When glass is broken using these methods, most of it will drop straight to the floor. If it is necessary to break a window to gain primary access to the victim, choose one as far away from the victim as possible.

One method commonly used to control broken glass is to apply a sheet of self-adhesive contact paper to the window before breaking the glass. This gives the window basically the same properties as laminated glass. Once the paper is applied, the window can be broken as previously described and most of the pieces of glass will stick to the paper, allowing the window to be removed as a unit.

Another method of controlling glass is to apply an aerosol spray adhesive that forms a laminated-type coating on the glass (**Figure 8.114**). This coating sets up in a matter of seconds and allows the glass to be broken and retained in a sheet. Then the glass can be removed in sheets instead of in little pieces (**Figure 8.115**).

When working with rear windows, rescuers must remember that some rear windows will be tempered and some will be laminated. If the window is not responding to removal techniques for tempered glass, it is probably laminated glass and will have to be removed in a manner similar to that for windshields.

Figure 8.113 Supporting the hand increases control of the center punch.

Figure 8.114 Adhesive sprayed onto the glass will help to contain it.

Figure 8.115 Adhesive-sprayed glass can be removed in one piece.

Removing the Roof and Doors

The disentanglement procedures used for any particular accident vary depending on the circumstances. A common evolution is the removal of the vehicle's roof to allow emergency medical personnel better access to victims. The designations A, B, and C are assigned to vehicle door posts from front to back (**Figure 8.116, p. 360**). The A-post is the front post area where the front door is connected to the body. The B-post is the post between the front and rear doors on a four-door vehicle or the post nearest to the door handle on a two-door vehicle. The C-post is the post nearest the handle on the rear door of a four-door vehicle. On a two-door vehicle, the rear roof post may be considered the C-post.   Removal can be done by either cutting all the roof posts and removing the roof entirely or by cutting only the front posts, cutting relief notches in the roof at the top of the rear door openings, and folding the roof back over the trunk (**Figure 8.117, p. 360**).

> **CAUTION**
> Locate air bag activating mechanisms before cutting door posts.

New materials such as plastics used in vehicle construction may prevent the roof from bending. In this case, the best method is to cut all roof posts and remove the entire roof. However, unibody vehicles are prone to collapse when their roofs are removed. Vehicles should be well supported before compromising the body of the vehicle. A third step block should be placed under the B-post of the vehicle (**Figure 8.118**).

Doors can be opened from the handle side or removed completely by inserting a spreader in the crack on the hinge side (**Figure 8.119**). If the outer door panel is made of plastic, rescuers may have to remove this outer skin to gain access to the metal frame. Rescuers may also need to remove interior plastic molding to check for curtain bag initiators or composite metal frame. Other techniques may be used to open doors by cutting hinges, breaking the latch mechanism (Nader pin), or compromising the door locks. For more information on these techniques, see the IFSTA **Principles of Vehicle Extrication** manual.

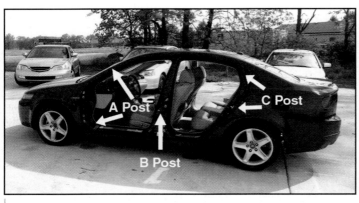

Figure 8.116 Standard doorpost identification that is common for all vehicles.

Figure 8.117 When removing a roof, the first cut is made in the A-post.

Figure 8.118 Because vehicles with unibody construction are prone to collapse when the roof is removed, a third step block supports the middle of the chassis.

Figure 8.119 One way of removing a door completely is to insert the tips of the spreader between the door and the B-post at the door latch.

Displacing the Dashboard

After a front-end collision, patients are often pinned under the steering wheel and/or wedged under the dashboard. To free the patient(s), rescuers should displace or roll the entire dashboard away from those in the front seat.

Displacing or rolling a dashboard can be accomplished in several ways. Each method used depends upon the vehicle's condition, the available tools, and the local policies and procedures. Once the door is removed, a relief cut should be made in the A-Post (**Figure 8.120**). An extension ram or jack can then be positioned in the doorway to displace or roll the dash forward (**Figure 8.121**). Cribbing or other suitable spacers should be inserted under the base of the A-post on unibody vehicles, or between the frame and the body on full-frame vehicles, to prevent the dashboard from returning to its original position. The extension ram or jack can be retracted and removed. This procedure can often be accomplished without removing the windshield, without removing or flapping a roof, and often by removing only one door.

Figure 8.120 Make a relief cut near the base of the A-post.

Figure 8.121 Push the dashboard assembly forward.

Technical Rescue Incidents

As a firefighter, you may participate in a variety of technical rescues as a member of a team. NFPA® 1670, *Standard on Operations and Training for Technical Search and Rescue Incidents*, defines technical rescue as, "The application of special knowledge, skills, and equipment to safely resolve unique and/or complex rescue situations." Technical rescues can include freeing victims from collapsed buildings, trench cave-ins, caves or tunnels, electrical contact, water and ice, industrial machinery, and elevators.

All firefighters should be educated about technical rescue so that they can recognize situations in which a technical rescue team is needed. They should also be familiar with their department's capabilities and limitations for handling technical rescue incidents. Firefighters may assist technical rescue personnel during these incidents by cordoning off the area and maintaining scene security, gathering the necessary tools and equipment in a staging area, and participating in the rescue operation under the supervision of a rescue technician.

The following sections provide information to help firefighters recognize situations that are beyond the rescue capabilities of most engine or ladder companies and where there is a need for technical rescue assistance. For additional information on technical rescue operations, see NFPA® 1670 and the IFSTA **Fire Service Search and Rescue** and **Technical Rescue for Structural Collapse** manuals. **Skill Sheet 8-II-3** outlines procedures for assisting rescue teams.

Rescue from Collapsed Buildings

Buildings may collapse because of damage from fire, extreme weather, earthquakes, explosions, or simply because an old or otherwise weakened structure yields to the force of gravity (**Figure 8.122**). The difficulty encountered in reaching a victim in a structural collapse depends upon conditions that are found. In some cases, uninjured or only slightly injured building occupants can make their way to the surface of the rubble and only need assistance in reaching a safe area. These individuals should be helped first. Next, those who are only lightly trapped by collapse debris should be extricated. Rescuing those who are heavily trapped and/or seriously injured deep within the rubble and debris requires the services of a technical rescue team. Structural collapse technicians know building construction and the technical rescue tools, equipment, and techniques needed to stabilize the rubble and perform rescues safely.

Types of Collapse

Structures collapse in predictable patterns. Being able to recognize these patterns can help rescuers make more informed decisions about the most likely location where viable victims may be trapped in the rubble and about the need for shoring and tunneling (see Shoring and Tunneling sections). The five recognized patterns of structural collapse are pancake, V-shaped, lean-to, A-frame, and cantilever.

Pancake collapse. This pattern of collapse is possible in any building where simultaneous failure of exterior walls results in the upper floors and the roof collapsing on top of each other such as in a stack of pancakes — thus, the name of this collapse pattern (**Figure 8.123**). The collapse of the World Trade Center towers was a graphic

Figure 8.122 Buildings collapse for a variety of reasons, including earthquakes, fires, and extreme weather.

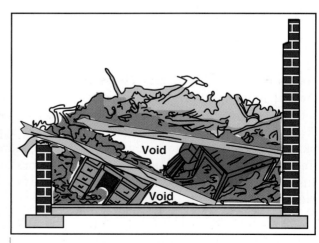
Figure 8.123 A pancake collapse may contain multiple voids that must be located and searched.

example of a pancake collapse. The pancake collapse is the pattern least likely to contain voids in which live victims may be found, but it must be assumed that there are live victims in the rubble until it is proven otherwise.

V-shaped collapse. This pattern of collapse occurs when the outer walls remain intact and the upper floors and/or roof structure fail in the middle **(Figure 8.124)**. This pattern offers a good chance of habitable void spaces being created along both outer walls.

Lean-to collapse. This pattern of collapse occurs when one outer wall fails while the opposite wall remains intact. The side of the floor or roof assembly that was supported by the failed wall drops to the floor, forming a triangular void underneath **(Figure 8.125)**.

A-frame collapse. This pattern of collapse occurs when the floor and/or roof assemblies on both sides of a center wall collapse into what might be seen as opposing lean-to collapses. This pattern offers a good chance of habitable void spaces on both sides of the center wall **(Figure 8.126)**.

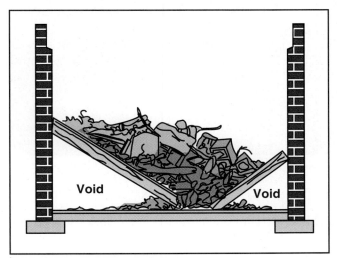

Figure 8.124 A V-shaped collapse creates voids along supporting exterior walls.

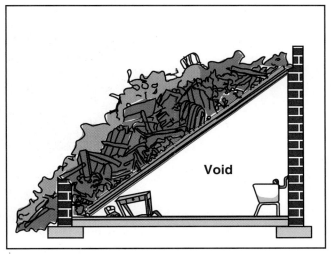

Figure 8.125 A lean-to collapse results in a large void that may contain trapped victims.

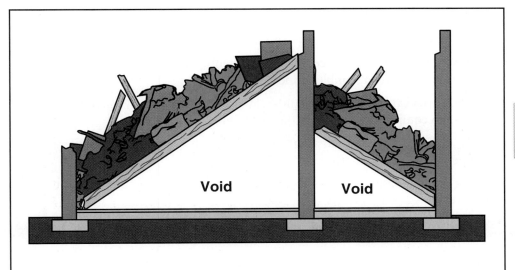

Figure 8.126 An A-frame collapse creates voids in two separate spaces along a common wall.

Cantilever collapse. This pattern of collapse occurs when one or more walls of a multistory building collapse, leaving the floors attached to and supported by the remaining walls (**Figure 8.127**). This pattern also offers a good chance of habitable voids being formed above and below the supported ends of the floors. This collapse pattern is perhaps the least stable of all the patterns and is the most vulnerable to secondary or subsequent collapse.

Figure 8.127 A cantilever collapse, represented here by the Murrah Federal Building in Oklahoma City, can result in habitable voids that must be searched.

Locating Hidden Victims

Concentrating on the areas most likely to contain victims, rescuers may use a variety of victim-locating devices and techniques. Some of these techniques require specialized tools and equipment; others do not. The most common means of locating hidden victims in the rubble of a structural collapse are as follows:

- Hailing — Calling out to elicit a response from hidden victims. This technique may be used in conjunction with physical and canine searches.

- Using seismic or short-distance radar devices that are capable of detecting the minute vibrations produced by victim movements.

- Using electronically enhanced acoustic listening devices that allow rescuers to hear otherwise imperceptible sounds.

- Using search cameras that are hand-held or robotic industrial-quality video cameras often equipped with lenses attached to fiber-optic cables.

- Using TICs that are similar to video cameras but are sensitive to differences in temperature (heat signature) rather than light.

- Using search dogs. Some dogs are trained to locate living victims; others to locate cadavers.

Structural Collapse Hazards

There are many actual and potential hazards involved in structural collapse rescue and they may take any of a wide variety of forms. Nonetheless, most of the hazards associated with this type of operation fall into one or both of two categories: environmental and physical.

Environmental hazards. Before rescuers can begin to search the rubble of a collapsed structure for victims, they may have to contend with a number of environmental problems in and around the collapse. Many of the secondary hazards — those created by the collapse or that developed after it — are environmental in nature. Most of the potential environmental hazards involve damaged utilities, atmospheric contamination, hazardous materials contamination, darkness, noise, fire, or temperature extremes and other adverse weather conditions.

Physical hazards. Physical hazards are those associated with working in and around piles of heavy, irregularly shaped pieces of rubble (some with sharp and jagged edges) that may suddenly shift or fall without warning. The primary physical hazards are those related to secondary or subsequent collapse, working in unstable debris, working in confined spaces (some of them below grade), working around exposed wiring and rebar, and dealing with heights.

> **⚠ CAUTION**
> Firefighters should not remove their SCBA – partially or fully – to enter structural collapse voids.

Shoring

Shoring is a general term used to describe any of a variety of means by which unstable structures or parts of structures can be stabilized **(Figure 8.128)**. It is the process of preventing the sudden or unexpected movement of objects that are too large to be moved in a timely manner and that may pose a threat to victims and/or rescuers. Shoring is not intended to move heavy objects but just to stabilize them. Stabilizing objects with shoring may involve using air bags and/or jacks, installing cribbing, constructing a system of wooden braces, or using a combination of these methods.

Figure 8.128 A shoring system is used to prevent the sudden movement of large objects.

Rescue from Trench Cave-Ins

Trench construction occurs in virtually every city and town in North America; in some large jurisdictions, it occurs almost daily somewhere within their boundaries. With all of this excavation going on, cave-ins are bound to occur, and they do. In some cases, the people killed in trench collapse incidents are would-be rescuers who fail to stabilize the trench before they enter it. They become additional victims when the trench caves in on them. Knowing how to make a trench safe to enter and taking the time to do it give both the victim and the rescuer the best chance for survival.

NOTE: The first major decision that must be made in a trench cave-in is whether the operation is a rescue or a body recovery. In some departments, the rule of thumb is if the victim is not visible, it is a body recovery operation.

Rescue operations depend on making the site as safe as possible by using shoring and cribbing to hold back other weakened trench walls and create a safe zone to protect rescuers and victims **(Figure 8.129, p. 366)**. Rescuers should not be sent into a trench unless they have been trained and equipped to safely conduct trench-rescue operations. During rescue operations, rescue apparatus, nonessential personnel, heavy equipment, and spectators should be kept well back from the trench lip to avoid causing subsequent cave-ins.

Figure 8.129 A typical pneumatic shore.

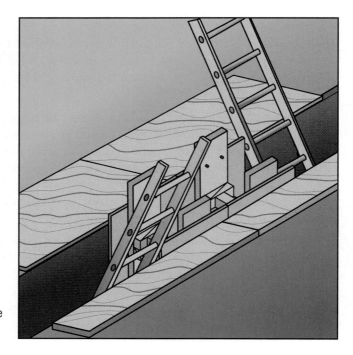

Figure 8.130 Exit ladders are used to create a safe work zone.

The following are safety precautions firefighters and officers must remember when they are involved in cave-ins and excavation rescues:

- Enter a trench only if you have specific trench-rescue skills.

- Do not enter a trench until it has been safely shored.

- Wear proper PPE when entering a trench for protection from physical, atmospheric, and environmental hazards associated with working in and around trenches.

- Wear self-contained breathing apparatus if a trench is found to be either oxygen-deficient or contaminated; otherwise, the trench will have to be mechanically ventilated before rescuers are allowed to enter.

- Place exit ladders in the trench at each end of the safe zone. Ladders should extend at least 3 feet (1 m) above the top of the trench **(Figure 8.130)**.

- Be careful with the tools being used in a trench to avoid injuring each other or the victim.

- Be aware of any other hazards that might exist at the scene such as underground electrical wiring, water lines, or toxic or flammable gases.

- Cordon off 100 feet (30 m) from the trench.

- Eliminate sources of vibration within 500 feet (150 m) of the trench (apparatus engines, etc.)

Confined Space Rescues

Fire fighting and rescue operations must sometimes be carried out in confined space locations that are below grade or otherwise without either natural or forced ventilation. OSHA defines a confined space as having the following characteristics:

- It is large enough and so configured that an employee can bodily enter and perform assigned work.

- It has limited or restricted means for entry and exit.

- It is not designed for continuous employee occupancy.

As you can see, trenches meet these characteristics, and most of the precautions listed for trench rescues also apply to rescues from confined spaces. Among the most common types of confined spaces to which firefighters may be called are the following:

- Tanks/vessels
- Silos/grain elevators
- Storage bins/hoppers
- Utility vaults/pits
- Aqueducts/sewers
- Cisterns/wells
- Coffer dams
- Storage tanks

As in trench rescues, firefighters without specific confined space rescue training are usually limited to performing non-entry rescues. They can also perform support functions outside of the space. Firefighters who do have confined space rescue training can enter and perform rescues that are within the scope of their training and experience.

The single most important factor in safely operating at these emergencies is recognition of the inherent hazards of confined spaces. The atmospheric conditions that may be expected include the following:

- Oxygen deficiencies
- Flammable gases and vapors
- Toxic gases
- Extreme temperatures
- Explosive dusts

In addition, physical hazards may also be present such as the following:

- Limited means of entry and egress
- Cave-ins or unstable support members
- Standing water or other liquids
- Utility hazards — electricity, gas, sewage

Plant or building supervisors or other knowledgeable people at the scene may provide valuable information on the number of victims and their probable location as well as on hazards that are present. Likewise, preincident plans of existing confined spaces in the fire department's jurisdiction reduce guesswork and should be referred to during operations in these locations. Firefighters should be ready to implement prearranged hazard mitigation plans, rescues, and extinguishment efforts without delay. These plans should include provisions for victim and rescuer protection, control of utilities and other physical hazards, communications, ventilation, and lighting. Electrical equipment such as flashlights, portable fans, portable lights, and radios should be intrinsically safe for use in flammable atmospheres.

Because of the hazards that may exist inside confined spaces, the command post and staging area must be established outside the hot zone. The staging area should be near, but not obstructing, the entrance. Firefighters must not be allowed to enter these spaces until an incident action plan (IAP) has been developed and communicated to on-scene personnel. An attendant must be stationed at the confined space entrance to track personnel and equipment entering and leaving the space.

The importance of wearing appropriate PPE and SCBA, using air monitoring, and using accountability systems cannot be overemphasized. Because the entrances to many confined spaces are very restrictive, it may be difficult for firefighters to en-

Figure 8.131 Supplied airline systems may be needed for rescues in confined spaces.

Figure 8.132 Rescuers entering a confined space must wear a lifeline.

ter while wearing their SCBA. To avoid this problem, especially in extended rescue operations, some departments equip their confined space rescuers with supplied air respirators in which breathing air is supplied through a hose up to 300 feet (100 m) long **(Figure 8.131)**.

Before firefighters enter a confined space, a lifeline will be attached to each harness **(Figure 8.132)**. This line must be constantly monitored at the entrance, and a RIC must be standing by whenever firefighters are working inside. A system of communication between inside and outside team members must be prearranged. While portable radios may work in some situations, they will not in others. Hard-wired telephones are usually the most unreliable.

If all else fails, a very basic method of signaling with the lifeline called the *O-A-T-H* method can be used. One tug on the lifeline represents the letter *O* which stands for *OK;* two tugs represent the letter *A* which stands for *Advance;* three tugs represent the letter *T* which stands for *Take-up* (eliminate slack); and four tugs represent the letter *H* which stands for *Help.* This system may be used for crude communication between the firefighters inside the space and the attendant outside. As with any tool or technique, this system must be practiced often to be effective when needed.

Air Monitoring

Atmospheric monitoring devices are available in a wide variety of models and configurations to match the particular environments in which they are intended to be used. The devices used must be capable of indicating the level of the gas or gases detected; most do so in percentages or parts per million (ppm) or both. The atmosphere is first checked for oxygen because most combustible-gas meters do not give an accurate reading in an oxygen-deficient atmosphere **(Figure 8.133)**. The results of these tests are used to determine whether and/or when it is safe for rescuers to enter, what type and level PPE is required, and the likelihood of finding viable victims inside. As long as a space continues to be occupied by rescuers or victims, the atmosphere within the space should be monitored. The monitoring device should be removed from time to time and recalibrated in clear air.

Accountability System

The purpose of the accountability system is to ensure that only those who are authorized and properly equipped to enter a hazardous area are allowed to do so and that both their location and their status are known as long as they remain inside the controlled zone. An accountability officer should check and record each entrant's name, mission, SCBA tank pressure, and estimated safe working time **(Figure 8.134)**. This procedure allows for the accounting of all team members and reduces the possibility of a member being unaccounted for after the safe working time limit has passed.

Rescues from Caves, Mines, and Tunnels

Although the local fire department may be called when someone is lost or injured in a cave, mine, or tunnel, most firefighters are not trained or equipped to perform these rescues. Rescue from these environments must be done by those who are familiar with the specific environment involved and who have the training and equipment needed. Unless they are specially trained to operate in these environments, fire/ rescue personnel usually confine their activities to aboveground support of other cave-rescue personnel.

Figure 8.133 Atmospheric monitoring is critical to firefighter safety in confined spaces.

Figure 8.134 Using an accountability system is vital during confined space operations.

Rescues Involving Electricity

Rescues involving energized power lines or equipment are some of the most common situations to which firefighters are called (**Figure 8.135, p. 370**). The frequency with which these situations occur should not be allowed to lull rescuers into a false sense of security — these situations are extremely dangerous. Improper actions by rescue personnel can result in their being injured or killed instantly. Whenever rescuers respond to any situation involving electricity, they should *always* do the following:

• Assume that electrical lines or equipment are energized. A power line in contact with a telephone line or a wire fence (out of sight of rescuers) can energize the entire length of the fence.

• Control the scene and deny unauthorized entry.

• Call for the power provider to respond, and stand by until they arrive.

• Allow only power company personnel to cut electrical wires.

Electrical wires on the ground can be dangerous without even being touched. Downed electrical lines can energize wire fences or other metal objects with which they come in contact. When an energized electrical wire comes in contact with the ground, current flows outward in all directions from the point of contact. As the current flows away from the point of contact, the voltage drops progressively. This is called *ground gradient*. Depending upon the voltage involved and other variables, such as ground moisture, this energized field can extend for several feet (meters) from the point of contact. A rescuer walking into this field can be electrocuted (**Figure 8.136, p. 370**). To avoid this hazard, rescuers should stay away from downed wires a distance equal to one span between poles until they are certain that the power has been shut off.

> **WARNING!**
> When you approach a downed power line to rescue a victim, if you feel a tingling in your feet, keep both your feet together and hop back away from the power line.

Figure 8.135 Downed wires should be located and marked.

Figure 8.136 Entering electrified areas can result in shock or electrocution.

Figure 8.137 Swift and turbulent water near low-head dams is particularly hazardous to both victims and rescuers.

Water and Ice Rescue

Virtually every jurisdiction in North America has the potential for water rescue and recovery operations. These situations can occur in swimming pools, ponds, lakes, rivers, streams, drainage canals, and other bodies of water such as low-head dams and water treatment facilities **(Figure 8.137)**. Areas subject to sustained freezing temperatures also have the potential for ice emergencies. Especially in drownings and other water-related incidents, it is important to repeat the distinction between rescues and recoveries. *Rescues* are situations where a victim is stranded or floundering, or has been submerged for a short period of time (usually less than half an hour). In these cases, the potential for saving the victim is real. *Recoveries* are situations where a victim has been submerged for such a long period of time that he or she is most probably deceased, and the goal of the operation is to recover the body.

All rescue personnel should wear appropriate PPE when operating at water and ice incidents. Standard firefighter turnout clothing is not acceptable. Proper PPE includes a water rescue helmet and an approved personal flotation device (PFD) **(Figure 8.138)**. For those who must work on the ice or in the cold water, thermal protective suits are also required **(Figure 8.139)**.

Water Rescue Methods

The following methods are used by those trained and equipped to rescue a victim during a water-related emergency.

- *REACH* — Extend a long-handled tool to the victim **(Figure 8.140)**.
- *THROW* — Throw a rope or flotation device with an attached rope to the victim **(Figure 8.141)**.
- *ROW* — Use a boat to retrieve the victim **(Figure 8.142)**.
- *GO* — Enter the water to reach the victim and pull him or her to safety **(Figure 8.143)**.

> **WARNING!**
> The ROW and GO rescue techniques should be attempted only by those who have been specifically trained in their application.

Figure 8.138 A full water rescue ensemble including wet suit, water rescue helmet, and personal flotation device.

Figure 8.139 Thermal protective suits are necessary for working in extremely cold water or performing ice rescues.

Figure 8.140 Any long-handled tool can be extended to a victim in the water.

Figure 8.142 When making a rescue from a boat use the reach or throw method when nearing the victim.

Figure 8.141 A rescuer throws a rope bag from shore.

Figure 8.143 Only personnel who are properly trained and equipped should enter the water to rescue a victim.

Hypothermia — Abnormally low or decreased body temperature.

Ice Rescue Methods

The steps in performing an ice rescue are designed to be as simple as possible because the rescuer has other factors to consider. One of those other factors is the unpredictability of the ice. Just because ice is thick does not mean that it is strong; *the victim in the water has demonstrated that the ice is weak.*

Rescue personnel must contend with the weather and its effect on those involved in the incident and on the scene. The victim will almost certainly be suffering the effects of hypothermia, so it is critical to have an advanced life support unit on scene to start immediate patient care. Another factor for ice rescuers to consider is that victims may not be able to help in their own rescue. With very cold or frozen hands, victims may not be able to grasp a rope or other aid, and with heavy, wet clothing, they may even have difficulty keeping their heads above water. With immersion in ice water, the body's core temperature can drop dramatically, and the victim's chances of survival may depend on how quickly he or she can get out of the water and into a warmer environment. The ice rescue protocol is as follows:

- Instruct the victim *not* to try to get out of the water until a rescuer says to do so.

- *REACH* — Implement only when the victim is close to solid ground and responsive and able to hold onto an aid.

- *THROW* — Allows the rescuer to span more distance while remaining on solid ground. The victim must be responsive and able to hold onto the aid.

- *GO* — Use when the victim is either too far from solid ground to use REACH or THROW or is incapable of grasping an aid **(Figure 8.144)**.

Figure 8.144 The GO method is used as a last resort in ice rescue operations. *Courtesy of Steve Taylor.*

Industrial Extrication

Industrial extrications can be among the most challenging rescue situations that firefighters face **(Figure 8.145)**. Because there are an endless number and variety of machines that have the potential to entrap victims, it is impossible to list specific techniques for victim removal. When surveying the situation, personnel should take into account the following:

- Medical condition and degree of entrapment of the victim

- Number of rescue personnel required

- Type and amount of extrication equipment needed

- Need for special personnel, equipment, or expert assistance

- Level of fire or hazardous material hazard that is present

- Scene safety

- Precautions necessary before securing power to the machine

These observations are critical to the rest of the incident. For example, if a victim is seriously entangled and in danger of bleeding to death, applying a tourniquet (which is likely to cost the victim his or her limb) is justified. In extremely rare cases, amputation by a doctor brought to the scene may be required to save the person's life.

Once the mechanism has been stabilized with cribbing, chains, or heavy-duty nylon webbing, the power should be shut off and secured to prevent inadvertent restoration of power. If it is obvious from the initial survey (size-up) that the problem is beyond the capability of the rescue team, outside expertise is required. In many cases, this will be plant personnel on site who are intimately familiar with the involved machinery. At the very least, plant maintenance personnel are usually good sources of information. In some cases, it may be necessary to bring in off-site sources, such as machinery manufacturers, for help. Ideally, these outside sources are identified during preincident planning.

Tourniquet — Any wide, flat material wrapped tightly around a limb to stop bleeding; used only for severe, life-threatening hemorrhage that cannot be controlled by other means.

CAUTION
If the machine in which the victim is entangled is still running, DO NOT turn it off until the mechanism has been stabilized. Turning the machine off may cause it to reverse itself or complete its cycle – either of which could harm the victim further.

Figure 8.145 Industrial rescues can be very challenging.

Elevator Rescue

Most elevator rescues are not true emergencies. They usually involve elevators that are stalled between floors because of a mechanical malfunction or power failure. In many cases, the situation can be resolved by having an elevator mechanic dispatched to the scene **(Figure 8.146), p. 374**. Unless there is a medical emergency in the elevator car or the car is threatened by fire, the best approach is to reassure the occupants that help is on the way and then wait for the elevator mechanic to arrive and return the elevator car to the nearest landing.

Figure 8.146 Unless there is a medical or fire emergency, elevator mechanics should be used to assist in the rescue of persons trapped on an elevator.

Figure 8.147 To reassure trapped occupants and reduce emotional stress, establish and maintain communication with trapped elevator occupants.

If the elevator mechanic can move the elevator car to a landing, passengers will be able to exit in the normal manner. Under *no* circumstances should firefighters alter the elevator's mechanical system in an attempt to move the elevator car. Only the elevator mechanic should perform adjustments to the mechanical system of the elevator installation.

If there is an emergency situation requiring immediate action or if the mechanical problem cannot be quickly fixed, it may be necessary to conduct an elevator rescue. These rescues require training in the use of proper rescue techniques. Only trained personnel should attempt elevator rescues.

Regardless of the type of situation, communication must be established with the passengers to assure them of their safety and that work is being done to free them. If a telephone or intercom is not available, shouting through the door near the stalled car may be sufficient for passing messages back and forth. Communication with the passengers is essential for their morale and mental state and should be established and maintained throughout the operation (**Figure 8.147**).

Escalator Rescue

Also called *moving stairways,* escalators are chain-driven mechanical stairways that move continuously in one direction (**Figure 8.148**). The steps are linked together and each step rides a track. The flexible rubber handrails move at the same rate as the stairs. The drive unit is usually located under the upper landing and is covered by a landing plate.

Many escalators have manual stop switches located on a nearby wall, at the base of the escalator, or at a point close to where the handrail goes into the newel base (**Figure 8.149**). Activating the switch stops the stairs and sets an emergency brake. The stairs should be stopped during rescues or when firefighters are advancing hoselines up or down the escalator stairway. As with the elevator, an escalator mechanic should be requested to assist in removing victims.

Figure 8.148 Escalators are found in many types of buildings.

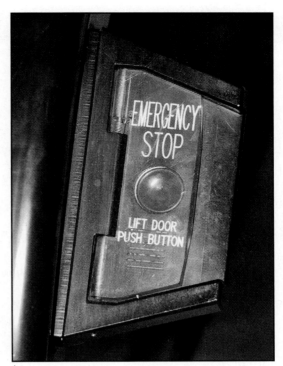

Figure 8.149 Most escalators are equipped with emergency stop buttons located at both ends of the unit.

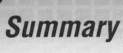

Summary

If you are to fulfill your obligations as a firefighter, you must be capable of performing basic rescue and extrication operations as a member of a team. You must be capable of conducting fireground search and rescue operations and vehicle extrication operations. You must also be capable of participating in a variety of technical rescue operations such as those involving collapsed structures, trench cave-ins, confined spaces, caves and tunnels, electrical contact, water and ice, industrial machinery entrapments, and stalled elevators. You should pursue specialized training in each of these areas. Finally, you must perform all of these duties while keeping the difference between a rescue and a body recovery operation in mind.

Fire Fighter I

1. List four guidelines that should be used by search and rescue personnel operating within a burning building.

2. What is a primary search?

3. What should primary search teams carry with them?

4. How can firefighters help operate safely while conducting building searches?

5. What should a firefighter in distress do?

Fire Fighter II

1. Describe powered hydraulic tools used in rescue incidents.

2. What are air chisels and pneumatic nailers commonly used for?

3. List four safety rules when using pneumatic lifting bags.

4. Why is stabilizing vehicles involved in incidents important?

5. What are the common means of locating hidden victims in the rubble of a structural collapse?

Conduct a Primary Search

Step 1: Confirm order with officer to conduct primary search.

Step 2: Size up structure to be searched.

Step 4: Identify rooms that have been searched.

Step 3: Search the structure using established search pattern.

Step 5: Remove any victims and inform IC of victim(s).

Step 6: Exit building when search is complete.

Step 7: Report to officer completion of primary search.

Conduct a Secondary Search

Step 1: Confirm order with officer to conduct secondary search and establish search pattern to be used.

Step 2: Size up structure to be searched.

Step 4: Identify rooms that have been searched.

Step 3: Search the structure using established search pattern.

Step 5: Remove any victims and inform IC of victim(s).

Step 6: Exit building when search is complete.

Step 7: Report to officer completion of secondary search.

Step 1: Size up environment and status of team.

Step 2: Determine that immediate exit by team is required.

Step 4: Move to safe area after exiting structure or hazardous area.

Step 5: Notify officer of situation using local procedures.

Step 3: Exit structure or hazardous area following guideline or hose.

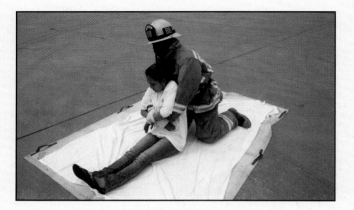

Step 1: Turn the victim (if necessary) so that the victim is supine.

Step 2: Kneel at victim's head.

Step 3: Support the victim's head and neck.

NOTE: If head or neck injury is suspected, provide appropriate support for head during movement.

Step 4: Lift the victim's upper body into a sitting position.

Step 5: Reach under the victim's arms.

Step 6: Grasp the victim's wrists.

Step 7: Stand. The victim can now be eased down a stairway or ramp to safety.

8-1-4
Demonstrate the blanket drag.

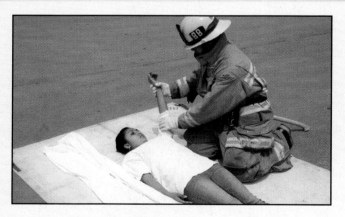

Step 1: Spread a blanket next to the victim, making sure that it extends above the victim's head.

Step 2: Kneel on both knees at the victim's side opposite the blanket.

Step 3: Extend the victim's nearside arm above his or her head.

Step 4: Roll victim against your knees.

Step 5: Pull the blanket against the victim, gathering it slightly against the victim's back.

Step 6: Roll victim gently onto the blanket.

Step 7: Straighten the blanket on both sides.

Step 8: Wrap the blanket around the victim.

Step 9: Tuck the lower ends around the victim's feet if enough blanket is available.

Step 10: Pull the end of the blanket at the victim's head.

Step 11: Drag the victim to safety.

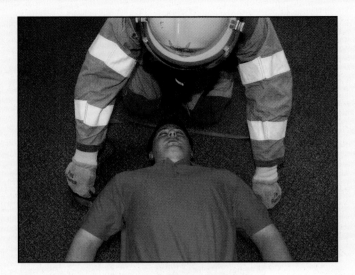

Step 1: Place the victim on his or her back.

Step 2: Slide the large webbing loop under victim's head and chest so the loop is even with their arm pits. Position the victim's arms so that they are outside the webbing.

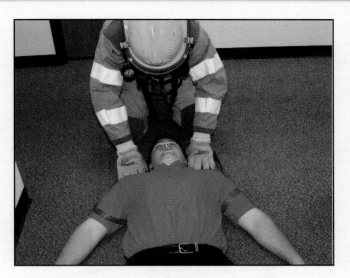

Step 4: Reach down through the large loop and under the victim's back and grab the webbing.

Step 5: Pull the webbing up and through the loop so that each webbing loop is drawn snugly around the victim's shoulders.

Step 3: Pull the top of the large loop over the victim's head so that it is just past their head.

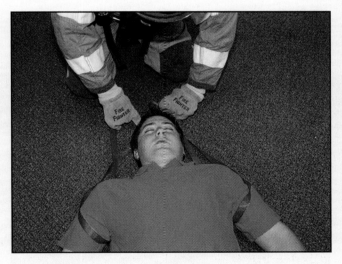

Step 6: Adjust hand placement on the webbing to support the victim's head.

Step 7: Drag the victim to safety by pulling on the webbing loop.

Skill SHEETS

8-1-6
Demonstrate the cradle-in-arms lift/carry — One-rescuer method.

Step 1: Place one arm under the victim's arms and across the back.

Step 2: Place the other arm under the victim's knees.

Step 3: Lift the victim to about waist height while keeping your back straight.

Step 4: Carry the victim to safety.

8-1-7
Demonstrate the seat lift/carry — Two-rescuer method.

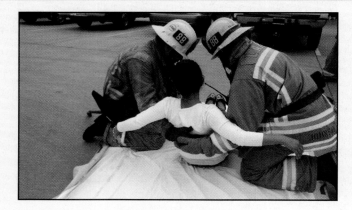

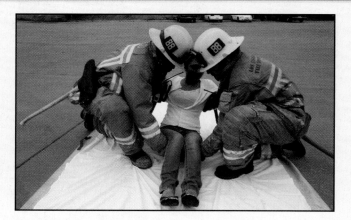

Step 1: Raise the victim to a sitting position.

Step 2: Link arms across the victim's back.

Step 3: Reach under the victim's knees to form a seat.

Step 4: Lift the victim using your legs. Keep your back straight while lifting.

Step 5: Move the victim to safety.

Step 1: *Both Rescuers:* Turn the victim (if necessary) so that the victim is supine.

NOTE: Keep head and neck stabilized during rolling to prevent spinal injury.

Step 2: *Rescuer #1:* Kneel at the head of the victim.

Step 3: *Rescuer #2:* Stand between the victim's knees.

Step 4: *Rescuer #1:* Support the victim's head and neck with one hand and place the other hand under the victim's shoulders.

Step 5: *Rescuer #2:* Grasp the victim's wrists.

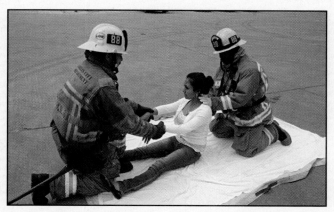

Step 6: *Rescuer #2:* Pull the victim to a sitting position.

Step 7: *Rescuer #1:* Push gently on the victim's back.

Step 8: *Rescuer #1:* Reach under the victim's arms and grasp the victim's wrists as Rescuer #2 releases them. Grasp the victim's left wrist with the right hand and right wrist with the left hand.

Step 9: *Rescuer #2:* Turn around, kneel down, and slip hands under the victim's knees.

Step 10: *Both Rescuers:* Stand and move the victim on command from Rescuer #1.

Skill SHEETS

8-1-9
Demonstrate the chair lift/carry method 1 — two rescuers.

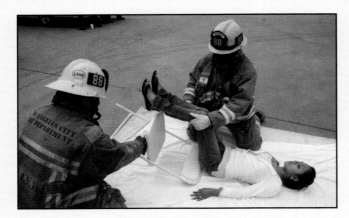

Step 1: *Both Rescuers:* Turn the victim (if necessary) so that the victim is supine.

NOTE: Keep head and neck stabilized during rolling to prevent spinal injury.

Step 2: *Rescuer #1:* Lift the victim's knees until the knees, buttocks, and lower back are high enough to slide a chair under the victim.

Step 3: *Rescuer #2:* Slip a chair under the victim.

Step 4: *Both Rescuers:* Raise the victim and chair to a 45-degree angle.

Step 5: *Both Rescuers:* Lift the seated victim with one rescuer carrying the legs of the chair and the other carrying the back of the chair.

8-1-10
Demonstrate the chair lift/carry method 2 — Two rescuers.

Step 1: *Rescuer #1:* Place the victim in a sitting position.

Step 2: *Rescuer #1:* Reach under the victim's arms and grasp the victim's wrists.

Step 3: *Rescuer #2:* Position the chair next to the victim.

Step 4: *Rescuer #2:* Grasp the victim's legs under the knees.

Step 5: *Both Rescuers:* Lift gently and place the victim onto the chair.

Step 6: *Both Rescuers:* Raise the victim and chair to a 45-degree angle.

Step 7: *Both Rescuers:* Lift the seated victim with one rescuer carrying the legs of the chair and the other carrying the back of the chair.

Step 1: Start generator per manufacturer's guidelines.

NOTE: If portable generator is used, two firefighters will lift and carry power source to assigned location before starting.

Step 2: Connect power cords to generator.

Step 3: Choose proper portable light for assigned task.

Step 4: Extend power cords to the area that needs illumination. Avoid pulling power cords over sharp objects or around tight bends that may cause damage to cord.

Step 5: Position portable light on stable surface and out of main traffic area so that work area is illuminated and firefighter's vision is not interrupted.

Step 6: Turn off generator per manufacturer's instructions.

Step 7: Dismantle lighting equipment and return to proper storage.

Step 1: Inspect equipment spark plug for damage, corrosion, carbon accumulation, or cracks in porcelain.

Step 2: Inspect spark plug wire and tighten connection, if needed.

Step 3: Replace equipment spark plug with spark plug recommended by manufacturer, and set to correct gap if inspection reveals damage or nonconformity.

Step 4: Check equipment carburetor, reporting any leaks found.

Step 5: Replace remaining fuel with fresh if fuel is three weeks old or older, and discard old fuel in approved manner and receptacle.

Step 8: Replenish oil as necessary.

Step 9: Inspect all electrical cords for frayed or damaged insulation or missing or bent prongs.

Step 6: Check fuel level and fill with fuel as necessary.

Step 7: Check oil level.

Step 10: Test operation of lighting equipment.

Step 11: Replace lightbulbs as necessary. Wear gloves to keep skin oil off bulbs.

Step 12: Discard faulty bulbs in the approved manner and in a receptacle.

Step 13: Clean work area and return equipment to proper storage, lifting properly to avoid back strain.

Step 14: Document service date and maintenance performed.

Rescue Preparation

Step 1: Confirm order with officer for rescue operation.

Step 2: Assess if the scene is safe.

Step 3: Stabilize vehicle (i.e. wheel chocks, cribbing, ropes, or other tools) prior to accessing patient.

Step 4: Assess the extrication methods that are required to access and extricate patient.

Step 5: Protect patients prior to beginning extrication procedures.

Windshield or Window Removal

Step 1: Confirm order with officer to remove windshield or window.

Step 2: Before starting work, plan the operation and determine the windows to be removed and the method of removing glass.

CAUTION: The window to be removed should be as far away from the passengers as possible to avoid injuring them.

Step 3: Check the area in which the work is to be done.

CAUTION: Operation of power saws can ignite flammable vapors. If flammable vapors are present, use an axe or other method that does not produce ignition sources.

Step 4: Remove glass to avoid causing further hazards or injuries.

Vehicle Door Removal

Step 1: Confirm order with officer to remove vehicle doors.

Step 2: Plan the operation before starting work.

Step 3: Isolate the door from other systems if necessary.

NOTE: The side-impact system can be de-energized but must always be done following manufacturer's instructions and only by experienced personnel.

Step 4: Prepare the area for operation of spreaders.

Step 5: Insert the spreader's tips between door and pillar aligned square with pressure points.

Step 6: Operate the spreaders until door is separated from hinges/locking mechanism.

Step 7: Move the door to area where it will not endanger others or interfere with operations.

8-II-2
Extricate a victim trapped in a motor vehicle.

Roof Removal

Step 1: Confirm order with officer to remove roof.

Step 2: Plan the operation (method of removing roof) before starting work.

Step 3: Check the area in which the work is to be done.

CAUTION: Contact between a saw blade and metal may produce sparks that will ignite flammable vapors.

Step 4: Remove the windshield *only if absolutely necessary to gain access to patient.*

CAUTION: On some models of modern cars, the windshield is part of the vehicle's structural integrity. If taken out, it can cause the vehicle to bend in the middle and shift on cribbing. Only remove the windshield if absolutely necessary.

Step 5: *Tool operator:* Cut all roof posts using tools as close to the vehicle roof as is practical.

CAUTION: Cutting roof posts can present hazards if not done properly. Do not cut near danger zones such as locations in which gas cartridges for side-impact airbags are located. For example, gas cartridges may be located low on C posts by the back window. The seat belt pretensioner inside a metal caisson cannot be cut and may cause damage to tools if attempted. Cut across the roof and across the car A posts and the B posts high near the roof.

CAUTION: Compromising the roof may cause the vehicle body to distort. Beware of shifts in the vehicle and ensure that stabilization is adequate f such a shift occurs.

Step 6: *Tool operator and assistance team:* Position four firefighters, one near each A post and one near each C post.

Step 7: *Tool operator and assistance team:* Lift the roof using legs, not back, and avoid twisting motions.

Step 8: *Tool operator and assistance team:* Move the roof to an area in which it will not endanger others or interfere with operations.

Prepare Vehicle for Steering Wheel and Column Removal

Step 1: Confirm order with officer to prepare vehicle.

Step 2: Plan the operation before starting work.

Step 3: Isolate the steering column and wheel from other systems if necessary.

Step 4: Remove the windshield or roof if necessary to gain access to steering column or steering wheel.

Step 5: Check the area in which the work is to be done.

CAUTION: Contact between the blade and metal may produce sparks that will ignite flammable vapors.

Step 6: Cut the component to be removed using hydraulic shears, reciprocating saw, or bolt cutters.

Step 7: Remove the component using care not to injure passengers and using proper lifting procedures.

Step 8: Move the component to an area in which it will not endanger others or interfere with operations.

Displace Dashboards

Step 1: Confirm the order with officer to displace the dashboard.

Step 2: Plan the operation before starting work by determining the method of removing the roof and the positioning of equipment.

Step 3: Remove windshield.

Step 4: *Equipment operators and assistance team:* Move or remove roof.

Step 5: Make a relief cut in A post.

Step 6: Position the extension rams or other tools to move dashboard.

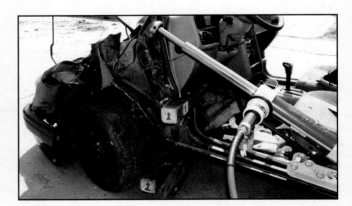

Step 7: Operate tools until dashboard is moved clear of passengers.

Step 8: Place cribbing or block in the relief cut to hold dashboard in displaced position, one on each side of vehicle.

Step 9: Remove the tools by relieving pressure.

Step 1: Confirm order with officer to assist rescue teams.

Step 2: Gather tools and equipment as directed.

Step 3: Provide assistance as requested or directed to rescue team members.

Step 4: Maintain situational awareness.

Step 5: Report to officer completion of assigned task.

Chapter Contents

Forcible Entry

This chapter provides information that addresses the following job performance requirements of NFPA® 1001, *Standard for Fire Fighter Professional Qualifications (2008):*

NFPA® 1001 references for Chapter 9:

5.3.4

5.5.1

Chapter Objectives

Fire Fighter I

1. Select appropriate cutting tools for specific applications.

2. Discuss manual and hydraulic prying tools.

3. Discuss pushing/pulling tools and striking tools.

4. Summarize forcible entry tool safety rules.

5. Describe correct methods for carrying forcible entry tools.

6. Summarize general care and maintenance practices for forcible entry tools.

7. Explain items to look for in sizing up a door.

8. Describe the characteristics of various types of wooden swinging doors.

9. Describe the characteristics of various types of metal swinging doors.

10. Describe the characteristics of various types of sliding doors, revolving doors, and overhead doors.

11. Explain how fire doors operate.

12. Describe the characteristics of basic types of locks.

13. Describe rapid-entry lockbox systems.

14. Describe methods of forcible entry through doors.

15. Describe methods of through-the-lock forcible entry for doors.

16. Explain action that can be taken to force entry involving padlocks.

17. Describe ways of gaining entry through gates and fences.

18. List hazards in forcing windows.

19. Describe types of windows and entry techniques.

20. Describe techniques for breaching walls.

21. Describe techniques for breaching floors.

22. Clean, inspect, and maintain hand tools and equipment. (Skill Sheet 9-I-1)

23. Clean, inspect, and maintain power tools and equipment. (Skill Sheet 9-I-2)

24. Force entry through an inward-swinging door — Two-firefighter method. (Skill Sheet 9-I-3)

25. Force entry through an outward-swinging door — Wedge-end method. (Skill Sheet 9-I-4)

26. Force entry using the through-the-lock method. (Skill Sheet 9-I-5)

27. Force entry using the through-the-lock method using the K-tool. (Skill Sheet 9-I-6)

28. Force entry using the through-the-lock method using the A-tool. (Skill Sheet 9-I-7)

29. Force entry through padlocks. (Skill Sheet 9-I-8)

30. Force entry through a double-hung window. (Skill Sheet 9-I-9)

31. Force entry through a window (glass pane). (Skill Sheet 9-I-10)

32. Force a Lexan® window. (Skill Sheet 9-I-11)

33. Force entry through a wood-framed wall (Type V Construction) with hand tools. (Skill Sheet 9-I-12)

34. Force entry through a masonry wall with hand tools. (Skill Sheet 9-I-13)

35. Force entry through a metal wall with power tools. (Skill Sheet 9-I-14)

36. Breach a hardwood floor. (Skill Sheet 9-I-15)

Forcible Entry

Case History

According to a NIOSH report, in March of 1998 units were dispatched to a reported fire in a one-story commercial building in California. The first-arriving firefighters saw light smoke coming from the building, and a ventilation crew proceeded to the roof and began opening up. Another crew began forcible entry into the front of the building through two metal-clad security doors. According to the report, it took from 7½ to 9 minutes to get these doors open. While three fire attack crews waited for the doors to be opened, fire conditions inside the building changed dramatically. Flames started issuing from the ventilation hole cut in the roof. Once the front doors were open, the crews advanced hoselines inside. Approximately 15 feet (5 m) inside the front door, they encountered heavy smoke and near-zero visibility. The crews advanced their lines 30 to 40 feet (10 m to 12 m) inside the building, but they could not locate the fire. Because conditions inside were deteriorating rapidly, company officers ordered their crews to exit the building. However, one of the officers became separated from his crew and remained inside. Approximately one minute later, a partial roof collapse blocked the front door, trapping the officer inside. A rapid intervention crew was sent in and they located the victim. CPR was started immediately and continued on the way to the hospital where the victim was pronounced dead. The medical examiner listed smoke inhalation and burns as the cause of death.

Modern society is understandably security conscious. Many businesses, industrial occupancies, and even private homes are more heavily secured than in the past **(Figure 9.1, p. 396)**. As shown in the foregoing case history, any delay in forcing entry can have tragic consequences. Despite the built-in security measures in these structures, firefighters must be able to gain access into these buildings quickly during fires, rescues, and sometimes even during odor investigations or alarm malfunctions. Forcible entry may be necessary to provide this access.

In the fire service, *forcible entry* refers to the variety of techniques used to gain access into a structure whose normal means of entry is locked or blocked. When properly applied, forcible entry efforts do minimal damage to the structure or structural components and provide quick access for firefighters. Forcible entry should not be used when normal means of access are readily available. However, firefighters may need to use forcible entry tools and techniques to breach a wall as a means of emergency egress (escape) from structures.

Forcible Entry — Techniques used by fire personnel to gain entry into buildings, vehicles, aircraft, or other areas of confinement when normal means of entry are locked or blocked.

To gain access as quickly as possible, minimize secondary damage, and reinforce the department's professional image, rule out other alternatives before forcing entry into a building.

Figure 9.1 Security bars on doors and windows can make forcible entry difficult. The bars can also prevent occupants and even firefighters from escaping a hazardous situation.

First, at every door or window, remember to **"Try before you pry"** (Figure 9.2). The door or window may be unlocked and can be opened in the normal manner. Second, especially on commercial and industrial occupancies, look for a lockbox near the main entrance **(Figure 9.3)**. Using a door key or numeric keypad combination found inside a lockbox avoids unnecessary damage to the property and may allow entry quicker than having to force a door or window. In some departments, lockbox information is maintained in the building pre-plan and/or computer-aided dispatch (CAD) data to speed access during incident operations.

When forcible entry is necessary, knowledge of the available tools and techniques increases a firefighter's effectiveness. The ability to use forcible entry tools and techniques quickly and effectively also demonstrates a firefighter's professionalism.

Remember that the purpose of built-in security barriers is to keep people out. To overcome these barriers, firefighters must have up-to-date knowledge of the types of barriers that will be encountered and their construction. Selection of the appropriate tool or set of tools is equally important. Forcible entry is not always easy, and the techniques must be practiced often. When forcible entry proves to be extraordinarily difficult at one point, it is sometimes necessary to force entry at two different points simultaneously. A complete and thorough understanding of the basic types of tools used in forcible entry, combined with practice in their application, ensures that the firefighters performing the task will be both safe and efficient.

NFPA® 1001 requires those qualified at the Fire Fighter I level be capable of forcing entry into a building using the proper PPE and forcible entry tools and techniques. To do this, they are required to know the following:

Figure 9.2 Remember to try before you pry. This is key to preventing unnecessary damage to the facility.

Figure 9.3 A lockbox containing keys to the facility provides a rapid, nondestructive means of entry.

- Basic construction of typical doors, windows, and walls
- Operation of doors, windows, and locks
- Dangers associated with forcing entry through doors, windows, and walls

  Those qualified at the Fire Fighter I level must also be capable of:

- Transporting and operating hand and power forcible entry tools
- Forcing entry through doors, windows, and walls

There are no additional structural forcible entry requirements at the Firefighter II level. The only additional forcible entry requirements relate to vehicle extrication incidents (see Chapter 8, Rescue and Extrication).

This chapter identifies the many tools that can be used for structural forcible entry operations. Their proper use, care, and maintenance are crucial to the success of any forcible entry operation. Also covered are the characteristics of various types of barriers that may have to be overcome to gain access to buildings during fires or other emergencies. Typical barriers are fences, gates, doors, windows, walls, and floors. Skill sheets are included to demonstrate actual forcible entry techniques. Tools and techniques for opening roofs are covered in Chapter 11, Ventilation.

Forcible Entry Tools

Before any type of forcible entry technique can be discussed, firefighters must know the characteristics, capabilities, and limitations of the tools available to perform the task. Selection of the proper tool may make the difference in whether any particular barrier is successfully overcome. This section describes the various categories of tools used for forcible entry operations. Also included in this section is information on the proper use, care, and maintenance of tools, all of which are crucial to the success of forcible entry operations. When using any of these forcible entry tools, always wear appropriate protective clothing.

Forcible entry tools can be divided into four basic categories:

- Cutting tools
- Prying tools
- Pushing/pulling tools
- Striking tools

Cutting Tools

There are many different types of cutting tools. Some of these tools are manually operated; others are powered. These tools are often specific to the types of materials they can cut and how fast they can cut them. No single cutting tool will safely and efficiently cut all materials. Using a cutting tool on materials for which it was not designed can damage the tool and endanger the operator. The following sections discuss the different types of cutting tools.

Axes

Axes are the most common types of cutting tools used by firefighters. There are two basic types of axes in common use, the pick-head axe and the flat-head axe (**Figure 9.4**). Smaller axes and hatchets are also available, but these tools are usually too lightweight for effective use in forcible entry operations. These smaller versions of pick-head and flat-head axes are sometimes used in overhaul and salvage operations, but they are inefficient for forcible entry.

Figure 9.4 Flat-head and pick-head axes are among the most common types of cutting tools.

Pick-head axe. Pick-head axes are available with either a 6-pound head or an 8-pound head (2.7 kg or 3.6 kg). Considered by many to be the most versatile forcible entry tool carried by firefighters, the pick-head axe can be used for cutting, prying, and digging. Handle sizes vary according to specifications, but they are made of either wood or fiberglass. This tool is very effective for chopping through wooden structural components, shingles and other roof coverings, aluminum siding, and other natural and lightweight materials. The pick end can be used to penetrate materials that resist being cut by the blade of the axe. The blade of the pick-head axe can also be used as a striking tool to break windows or as a prying tool to force some doors. Because of these attributes, pick-head axes are most often used in structural fire fighting operations.

Flat-head axe. Like the pick-head axe, the flat-head axe is available in either 6- or 8-pound (2.7 kg or 3.6 kg) head weights with either a wooden or fiberglass handle. It can also be used to chop through the same types of materials as the pick-head axe. The blade of the flat-head axe can be used for all the same purposes as that of a pick-head axe.

Unlike the pick-head axe, the flat-head axe can be used with other tools to force entry. The flat-head axe is used to strike the other tool, forcing the bit end into a door jam or window sill. The flat-head axe is commonly carried together with a Halligan bar as a set known as "irons." Irons are used in combination for a variety of forcible entry operations. Because of their versatility, flat-head axes are used in both structural and wildland fire fighting operations.

Metal Cutting Devices

A variety of tools and other devices are used by firefighters to cut through heavy-duty locks, metal-clad doors, window security bars, and similar items to gain access into buildings. The tools and devices include bolt cutters, cutting torches and flares, and manual or powered rebar cutters.

Bolt cutters. These metal cutting devices are used in forcible entry to cut bolts, iron bars, pins, cables, hasps, chains, and some padlock shackles (**Figure 9.5**). The continual advancement in security technology is limiting the use of the bolt cutter as a viable entry tool. Some high-security chains, hasps, and padlock shackles cannot be cut with bolt cutters. These materials shatter the cutting blades of the bolt cutter or cause the handles to fail due to the tremendous pressures that must be exerted by the operator. Other locks deny access to the shackle. Bolt cutters should not be used to cut case-hardened materials found in locks and other security devices. Face shields and eye protection must always be worn when using bolt cutters to prevent fragments of the cut material from striking the operator's face.

Rebar cutters. These hydraulic cutting tools are available in both powered and manual versions (**Figure 9.6**). The manual version is more labor-intensive, but it can be used in areas beyond the end of the hydraulic supply hose on powered units. These tools can be used to cut rebar (steel reinforcing bars used to increase the strength in concrete construction) when breaching a concrete wall, but they can also be used to cut security bars on windows or doors.

It may be necessary to use a cutting torch in situations where security devices resist being cut by bolt cutters, rebar cutters, or rotary saws (**Figure 9.7**). The torches commonly used by firefighters are the oxyacetylene cutting torch, oxygasoline cutting torch, burning bars, plasma cutters, and cutting flares. Training based on the manufacturer's recommendations specific to each cutting and burning device is necessary for safe and efficient operation.

Halligan Tool — Prying tool with a claw at one end and a spike or point at a right angle to a wedge at the other end.

CAUTION

Bolt cutters are not designed to cut electrical lines.

Oxyacetylene cutting torches. These devices are available in both hand-carried and wheeled units (**Figure 9.8**). They may be used to cut through heavy metal components that resist being cut by other forcible entry tools. In use, these torches preheat the metal to its ignition temperature and then burn a path in the metal with an extremely hot cone of flame created by introducing pure oxygen into it. While cutting, they generate a flame temperature of more than 5,700°F (3 149°C). These torches cut through iron and steel with relative ease.

The use of oxyacetylene cutting torches is diminishing in the fire service because of the safety concerns associated with them. Acetylene is an unstable gas with a wide flammability range (2.5 to 81 percent in air) that is both pressure and shock sensitive. To keep the gas stable and safe to use, the storage cylinders are filled with a porous substance called calcium silicate that prevents accumulations of free acetylene gas within the cylinder. The cylinders also contain liquid acetone in which the acetylene is dissolved. When the cylinder valve is opened, acetylene gas is liberated from the acetone and flows through the hose to the torch assembly.

Oxyacetylene Cutting Torch — A commonly used torch that burns oxygen and acetylene to produce a very hot flame. Used as a forcible entry cutting tool for penetrating metal enclosures that are resistant to more conventional forcible entry equipment.

Acetylene [C₂H₂] — Colorless gas that has an explosive range from 2.5 percent to 81 percent in air; used as a fuel gas for cutting and welding operations.

WARNING!
Keep acetylene cylinders in an upright position to prevent the loss of acetone. Losing acetone could cause the cylinder to explode. Never exceed 15 psi (105 kPa) operating pressure.

Figure 9.5 Bolt cutters can cut bolts, chains, hasps, and padlock shackles that are not case-hardened.

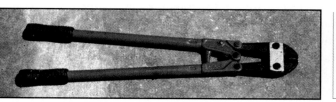

Figure 9.6 Hydraulic and manual rebar cutters can cut metal items that bolt cutters cannot.

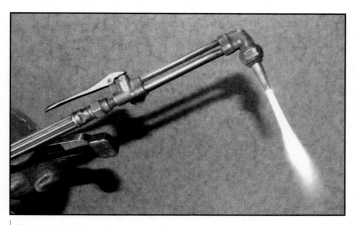

Figure 9.7 Cutting torches can be used to cut materials that other cutting tools cannot penetrate.

Figure 9.8 The use of portable oxyacetylene cutting torches is diminishing in the fire service. However, portable oxyacetylene kits like the one shown may still be found in service in some departments.

Figure 9.9 The use of oxygasoline cutting torches is increasing in the fire service.

Exothermic — Chemical reaction between two or more materials that changes the materials and produces heat, flames, and toxic smoke.

Some departments have replaced acetylene with methylacteylene-propadiene, stabilized (MPS), which is marketed as MAPP GAS® and APACHE GAS. MPS is less pressure sensitive than acetylene and less prone to explosive decomposition. The cutting equipment designed for use with acetylene is easily adapted for use with MPS.

Oxygasoline cutting torches. A relatively new cutting system, fueled by a mixture of gasoline and oxygen, is being used by a growing number of fire/rescue units in North America **(Figure 9.9)**. These systems use a conventional cutting torch and dual-hose configuration, but the fuel (gasoline) is delivered to the torch in liquid form. These torches produce a cutting flame in the range of 2,800°F (1 538°C) and are fully functional under water.

In addition to the ready availability of gasoline and its low cost compared to other fuels, the safety resulting from the fuel being delivered to the torch as a liquid is significant. Unlike torches that use gaseous fuels, in oxygasoline systems the flame cannot travel back through the hose.

Burning bars. Also called exothermic cutting rods, burning bars are ultra-high-temperature cutting devices capable of cutting through virtually any metallic, nonmetallic, or composite material. They cut through materials such as concrete or masonry that cannot be cut with an oxyacetylene or oxygasoline cutting torch, and they cut through metals much faster. The torch feeds oxygen through the rod to produce cutting temperatures in excess of 10,000°F (5 538°C). The cutting bars or rods range in size from ¼ to 1 inch (6 mm to 25 mm) in diameter and from 18 inches to 10 feet (450 mm to 3 m) in length **(Figure 9.10)**.

Plasma cutters. Plasma arc cutters are also ultra-high temperature metal-cutting devices that produce cutting temperatures as high as 25,000°F (13 871°C). These devices require a power supply as well as one of several compressed gases such as air, nitrogen, or a mixture of argon and hydrogen or argon and helium. As an electric arc melts the material being cut, a jet of gas is heated into plasma by the arc. The plasma is used to blow the molten material from the cutting area.

Cutting flares. Also available for cutting metal or concrete are exothermic cutting flares **(Figure 9.11)**. Approximately the size and shape of fuses or highway flares, cutting flares are also ignited in the same ways. Once these flares are ignited, they produce a 6,800°F (3 760°C) flame that will last from 15 seconds to two minutes depending upon the length and diameter of the flare used. The obvious advantages of cutting flares compared to other exothermic cutters are the absence of a hose or power cable and their light weight and portability.

Figure 9.10 A typical burning bar used to cut through concrete, nonmetallic, or composite materials.

Figure 9.11 A typical cutting flare that generates high temperature flames over a short period of time. *Courtesy of Pyrotechnic Tool Co.*

Handsaws

There are times when a power saw is unavailable or will not fit in the available work space. In these situations, a handsaw may be the best option. Handsaws commonly used in the fire service include the *carpenter's handsaw* (both rip cut and crosscut), *hacksaw, drywall saw,* and *keyhole saw* (**Figure 9.12**). Compared to power saws, handsaws are extremely slow. Knowing which saw is best suited to the cutting job at hand and using good handsaw technique will make firefighters more proficient in handsaw use.

Power Saws

Power saws are some of the most useful tools used in the fire service. Like any other tool, however, there are times when these saws should and should not be used. Several types of power saws are commonly used in the fire service including the *circular saw, rotary saw, reciprocating saw, chain saw,* and *ventilation saw.* Many power saws now available are able to run on both AC and DC power sources.

Circular saw. Originally designed for use in construction, these versatile electric saws have many applications in fire fighting and rescue operations (**Figure 9.13**). These saws are especially useful in situations where electrical power is readily available and heavier and bulkier power saws are too difficult to handle. Small battery-powered units are also available.

Rotary saw. The fire service version of this device is usually, but not always, gasoline powered with changeable blades (**Figure 9.14**). Different blades are available for cutting wood, metal, and masonry.

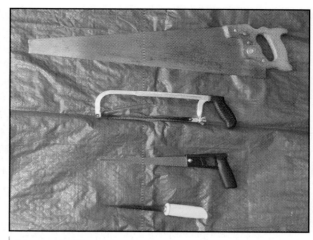

Figure 9.12 Typical hand saws used for cutting a variety of materials.

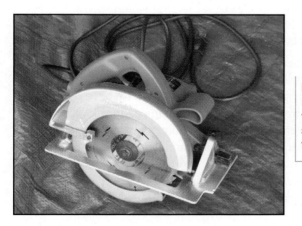

Figure 9.13 The circular saw requires an electric power source. A variety of blades are available for the saw depending on the type of material that must be cut.

Figure 9.14 A typical rotary saw is larger than the circular saw and does not require a separate power source.

CAUTION

Always use eye and hearing protection when operating any power saw. Do not force the saw (or any tool) beyond its design limits; the saw may be damaged and/or the operator injured. Never use a power saw in a flammable atmosphere. The saw's motor or sparks from the cutting operation can ignite a fire or cause an explosion.

CAUTION

When using a rotary saw to cut metal, have a charged hoseline or portable fire extinguisher close at hand because of the sparks produced in the cutting operation.

WARNING!

Never use a rotary saw to cut the shell of any tank that might contain flammable vapors.

Depending upon the saw used, the blades may spin at more than 6,000 rpm. Blades range from large-toothed blades for quick rough cuts to those with fine teeth for a more precise cut. Some blades are made specifically for cutting metal or concrete. Saw blades with carbide-tipped teeth are available and far superior to standard blades because they are less prone to dulling with heavy use. There are several manufacturers of rotary-type saws. The firefighter should be familiar with the type purchased by their department. Following both the manufacturer's recommendations and departmental operating procedures are imperative to maintaining a firefighter's personal safety when operating saws.

Reciprocating saw. The reciprocating saw is a very powerful, versatile, and highly controllable saw (**Figure 9.15**). This saw has a short, straight blade that moves in and out with an action similar to that of a handsaw. It can use a variety of blades for cutting different materials. When equipped with a metal-cutting blade, this saw is ideal for cutting sheet metal body panels and structural components on vehicles. Battery-powered reciprocating saws are also available.

Chain saw. The chain saw has been used for years by the logging industry (**Figure 9.16**). This handy, wood-cutting saw has also found a place in the fire service. These saws are also useful during natural disasters such as tornadoes and ice storms when trees and limbs must be cleared from streets and access routes, and structural collapse debris must be cut to free trapped victims.

One of the chain saw's most common uses is as a ventilation tool. When equipped with a carbide-tipped chain, depth gauge, and kickback protection, the saw makes fast cuts through many types of roofing materials.

Prying Tools

Prying tools are useful for opening doors, windows, locks, and moving heavy objects. Pry bars and other manually operated prying tools use the principle of the lever and fulcrum to provide mechanical advantage (**Figure 9.17**). Force applied to the tool's handle is multiplied by leverage to produce tremendous counter-force at the working end. The longer the handle, the greater the force produced.

Hydraulic prying tools can be either powered hydraulic or manual hydraulic. Both types receive their power from hydraulic fluid pumped through special high-pressure hoses. Although there are a few powered hydraulic pumps that are operated by compressed air, most are operated by electric motors or two- or four-cycle gasoline engines. Manual hydraulic tools operate more slowly than powered hydraulic tools and they are labor-intensive. However, they are usually smaller and lighter than powered units, so they are easier to carry.

Manual Prying Tools

A wide variety of manual prying tools are available to the fire service (**Figure 9.18**). Among the most common are the following:

- Crowbar
- Halligan bar
- Pry (pinch) bar
- Hux bar
- Claw tool
- Kelly tool
- Pry axe
- Flat bar (nail puller)
- Rambar

Lever — Device consisting of a bar turning about a fixed point (fulcrum), using power or force applied at a second point to lift or sustain an object at a third point.

Fulcrum — Support or point of support on which a lever turns in raising or moving something.

Figure 9.15 Reciprocating saws cut wood, metal, or glass. They are powerful, fast, and can be used in place of hand saws.

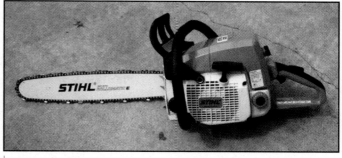

Figure 9.16 Chain saws are beneficial for cutting roofing materials when ventilating a structure.

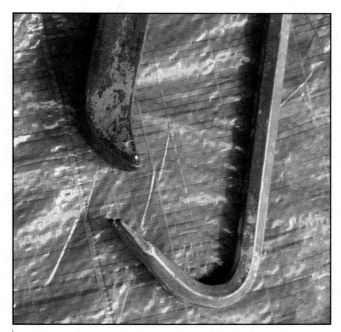

Figure 9.17 The curved end of a pry tool forms a fulcrum that increases the amount of leverage applied to the tool.

Figure 9.18 Various types of prying tools commonly found in fire service use.

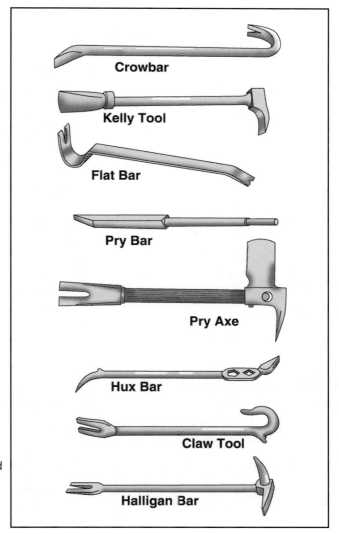

Crowbar

Kelly Tool

Flat Bar

Pry Bar

Pry Axe

Hux Bar

Claw Tool

Halligan Bar

Some departments call these tools by names other than those found in this manual. You should become familiar with the types and names of the tools carried on your apparatus. Some prying tools can also be used effectively as striking tools, although most cannot. You need to be familiar with the capabilities and limitations of each manual prying tool, such as which surfaces are used for prying and which surfaces may be used for striking. Efficiency in the use of a tool under emergency conditions is directly affected by the firefighter's familiarity with the tool. For safe and efficient operation of a tool, it should be used only for its intended purpose.

One of the most efficient of the manually operated prying tools is the rambar **(Figure 9.19)**. The tips of its forked and slightly curved head can be inserted between a door and the door frame, just above or just below the lock, and driven in by the sliding weight that surrounds the handle. Then, using the tool handle as a lever, the door will usually pop open **(Figure 9.20)**. A rambar can also be used to pry hasps and other surface-mounted locking mechanisms off doors and door frames.

One safety feature of the rambar compared to the Halligan bar is that the rambar does not need to be driven in with a flat-head axe or other striking tool. This eliminates the need for a second tool and the firefighter to operate it. It also eliminates the striking tool having to be swung in an arc near other firefighters, and the possibility of the striking tool missing its target and hitting the firefighter holding the Halligan.

Hydraulic Prying Tools

Hydraulic prying tools that can be operated by one person have proven to be very effective in extrication rescues. They are also useful in forcible entry situations. These tools are useful for a variety of different operations involving prying, pushing, or pulling. The rescue tools and hydraulic door opener are examples of hydraulic prying tools.

Rescue tools. The hydraulic spreader, most often associated with vehicle extrication, has some uses in forcible entry. Depending on the manufacturer, the tips on these tools can spread as much as 32 inches (800 mm). Their capability to exert force in either spreading or pulling makes them a valuable tool in some instances. The hydraulic ram is another hydraulic rescue tool that can be used for pushing and pulling. Although designed primarily for vehicle extrication, hydraulic rams have spreading capabilities ranging from 36 inches (900 mm) to an extended length of nearly 63 inches (1.6 m). In certain forcible entry situations, these tools may be invaluable. One use as a forcible entry tool is to place the ram inside a door frame to spread it apart far enough to allow the door to swing open **(Figure 9.21)**.

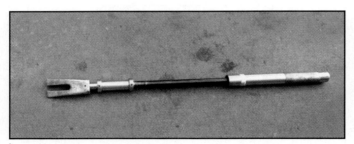

Figure 9.19 The rambar requires only one operator and is an efficient tool for opening locked doors.

Figure 9.20 When using the rambar to open a door, care must be taken to keep from pinching the hand between the slide and the stop on the bar.

Figure 9.21 Hydraulic rams can be used to open metal frame doors set in masonry walls.

Figure 9.22 Hydraulic door openers can be used on a variety of door types.

Hydraulic door opener. This manually-operated spreader is relatively lightweight. It consists of a hand pump and spreader device **(Figure 9.22)**. The spreader has intermeshed teeth that can be easily slipped into a narrow opening such as between a door and door frame. A few pumps of the handle cause the jaws of the spreader device to open, exerting pressure on the object to be moved. The pressure usually causes the locking mechanism or door to fail. These are extremely valuable tools when more than one door must be forced, such as in apartments or hotels.

Pushing/Pulling Tools

Another category of tools available for forcible entry use is the push/pull category. These tools have limited use in forcible entry, but in certain instances, such as breaking glass and opening walls or ceilings, they are the tools of choice. This category of tools includes the following:

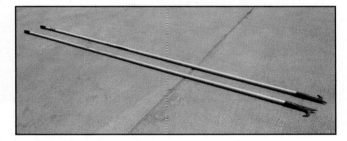

Figure 9.23 Pike poles are typically used for pulling down drop-in or Sheetrock™ ceilngs and opening interior walls.

- Standard pike pole **(Figure 9.23)**
- Clemens hook
- Plaster hook
- Drywall hook
- San Francisco hook
- Multipurpose hook
- Roofman's hook
- Rubbish hook **(Figure 9.24)**

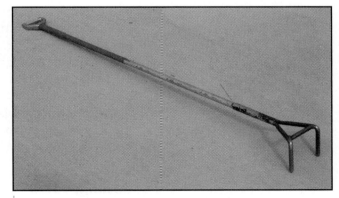

Figure 9.24 The rubbish hook is used for moving loose materials that contain hidden fires, such as trash, hay, or burned debris.

Figure 9.25 The proper method for pulling a ceiling with a pike pole.

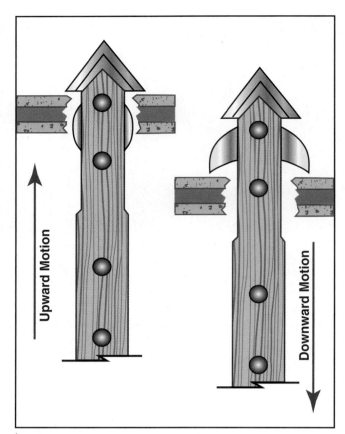

Figure 9.26 This illustrates the operation of a plaster hook. As the tip penetrates the plaster, the wings retract. When the wings have extended, a downward pull on the tool will pull the plaster or Sheetrock™ away from the studs.

Pike Pole — Sharp prong and hook of steel, on a wood, metal, fiberglass, or plastic handle of varying length, used for pulling, dragging, and probing.

Pike poles and hooks, which are available in various lengths, give firefighters a reach advantage when performing certain tasks **(Figure 9.25)**. When using a pike pole to break a window, the firefighter should stay upwind of the window and be positioned higher than the window so that falling glass will not slide down the handle toward the firefighter. The plaster hook has two knifelike wings that depress as the head is driven through a ceiling or other obstruction and reopen or spread outward under the pressure of self-contained springs **(Figure 9.26)**. With the exception of the roofman's hook, which is all metal, pike poles and hooks should not be used as prying tools. Their strength is in pushing or pulling, not prying. If a lever is needed, select the appropriate prying tool. Handles of pike poles and hooks are made of wood or fiberglass and may be broken by the application of inappropriate force.

Striking Tools

A striking tool is a very basic hand tool consisting of a weighted head attached to a handle **(Figure 9.27)**. Some examples of common striking tools are as follows:

- Sledgehammer (8, 10, and 16 pounds [3.6 kg, 4.5, and 7.3 kg])
- Maul

- Battering ram
- Pick
- Flat-head axe
- Mallet
- Hammer
- Punch
- Chisel

Figure 9.27 A variety of striking tools including sledgehammers, flat-head axe, and pick.

In some cases, a striking tool is the only tool required. In many other situations, the striking tool is used in conjunction with another tool to gain entry. As common as they are, striking tools are dangerous when improperly used, carried, or maintained. Striking tools can crush fingers, toes, and other body parts. Improperly maintained striking surfaces may cause chips or splinters of metal to fly into the air. Therefore, proper eye protection (safety glasses or goggles *in addition to* the helmet faceshield) must be worn when using striking tools.

Tool Combinations

There is no single forcible entry tool that provides the firefighter with the needed force or leverage to handle all forcible entry situations. In some cases, firefighters must combine two or more tools to accomplish a task. The types of tool combinations carried vary, depending on building construction, security concerns, tool availability, and other factors within a fire department and the area served.

The most important factor to consider is selecting the proper tools to do the job. Using tools in situations for which they are not designed can be extremely dangerous. Preincident surveys will help determine which tools will be required to force entry into a particular building or through a particular door in that building.

Tool Safety

Hand and power tools used in the fire service can be extremely dangerous if used incorrectly. Firefighters must become familiar with all the tools they will use, which includes reading and following all the manufacturer's guidelines as well as individual department standard operating procedures (SOPs) on tool safety. In atmospheres that could be explosive, extreme caution should be taken in the use of power and hand tools that may cause arcs or sparks. When tools are not in use, they should be kept in properly designated places on the apparatus **(Figure 9.28, p. 408)**. Check the location of tools carried on the apparatus and make sure they are secured in their holders. The following sections contain information concerning prying tool safety, safety information particular to rotary saws, and safety when using power saws in general.

Prying Tool Safety

As with other tools, using prying tools incorrectly creates a safety hazard. For example, it is not acceptable to use a "cheater" or to strike the handle of a pry bar with other tools. A *cheater* is a piece of pipe slipped over the handle of a prying tool to lengthen the handle, thus providing additional leverage **(Figure 9.29, p. 408)**. Use of a cheater can put forces on the tool that are greater than the tool was designed to withstand.

Figure 9.28 For ease of access, tools should be stored in their assigned locations. For safety and to reduce damage, tools should be secured in place.

Figure 9.29 So-called *cheaters* should never be used on tool handles. The use of a cheater is an unsafe practice that can result in personal injury and damage to the tool.

This can cause serious injury to the operator if the tool slips, breaks, or shatters. Furthermore, such action can damage the tool. If a job cannot be done with a particular tool, do not strike the handle of the tool; use a larger tool. Also, do not use a prying tool as a striking tool unless it has been designed for that purpose.

Rotary Saw Safety

Rotary saws must be used with extreme care to prevent injury from the rapidly rotating blade that continues to spin after the throttle has been released. Blades from different manufacturers may look alike, but they may not be interchangeable. Another hazard associated with the rotary saw is the twisting (*gyroscopic* or *torsion effect*) caused by the spinning blade. When using a rotary saw, it is important to start all cuts at full revolutions per minute (rpm) to prevent the blades from binding into the material. It is important to store blades in a clean, dry environment. Do not store composite blades in any compartment where gasoline fumes accumulate (such as where spare saw fuel is stored) because the hydrocarbons will attack the bonding material in the blades and make them subject to sudden and violent disintegration during use.

Safety with Other Power Saws

Following a few simple safety rules when using any type of power saw will prevent most typical accidents:

- Match the saw to the task and the material to be cut. Never force a saw beyond its design limitations.

- Always wear proper protective equipment, including gloves, hearing protection, and eye protection.

- Fully inspect the saw before and after use.

- Do not use any power saw when working in a flammable atmosphere or near flammable liquids.

- Maintain situational awareness.

- Keep unprotected and nonessential people out of the work area (**Figure 9.30**).
- Follow the manufacturer's guidelines for proper saw operation.
- Keep blades and chains well sharpened. A dull saw is more likely to cause an accident than a sharp one.
- Be aware of hidden hazards such as electrical wires, gas lines, and water lines.
- Start all cuts with the saw running at full rpm.

Carrying Tools

Firefighters must carry tools and tool combinations in the safest manner possible. Precautions should be taken to protect the carrier, other firefighters, and bystanders. When lifting heavy tools or other objects, always lift with your legs, not your back. Get help when transporting heavy tools. Some recommended safety practices for carrying specific tools are as follows:

Axes — If not in a scabbard, carry the axe with the blade away from the body. With pick-head axes, grasp the pick with a hand to cover it. Never carry an axe on the shoulder (**Figure 9.31**).

Prying tools — For safety, carry these tools with any pointed or sharp edges away from the body. With multiple surfaces, this will be somewhat difficult.

Figure 9.30 When operating any type of power saw, keep the work area clear of bystanders.

Figure 9.31 The safest way to carry a pick-head axe is to cradle the blade firmly under the arm and grasp the pick with a gloved-hand while holding the handle with the other hand.

Combinations of tools — Strap tool combinations together (**Figure 9.32**). Halligan bars and flat-head axes can be nested together and strapped. Short sections of old hose can be slipped over the handles of some tools and smaller prying tools inserted into the hose.

Pike poles and hooks — For safety, carry these tools with the tool head down, close to the ground, and ahead of the body when outside a structure (**Figure 9.33**). When entering a building, carefully reposition the tool and carry it with the head upright close to the body to facilitate prompt use (**Figure 9.34**). These tools are especially dangerous because they are somewhat unwieldy and can severely injure anyone accidentally jabbed with the working end of the tool.

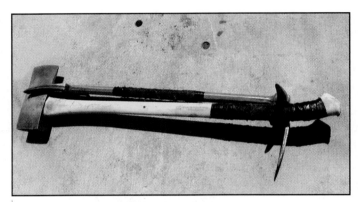

Figure 9.32 A typical set of irons made up of a flat-head axe and a Halligan bar.

Figure 9.33 Proper method of carrying a pike pole outside a structure.

Figure 9.34 Proper method of carrying a pike pole inside a structure.

Striking tools — Keep the heads of these tools close to the ground. Maintain a firm grip. Mauls and sledgehammers are heavy and may slip.

Power tools — Never carry a running power tool more than 10 feet (3 m). Transport the tool to the area where the work will be performed and start it there. Running power tools are potentially lethal weapons.

Care and Maintenance of Forcible Entry Tools

Well-maintained forcible entry tools are essential to the success of any forcible entry operation. Forcible entry tools will function as designed if they are properly maintained and kept in the best of condition. Tool failure on the fireground may have harsh consequences, including severe injury or death. Always read the manufacturer's recommended maintenance guidelines for all tools, especially power tools. The following sections describe some basic maintenance procedures for various forcible entry tools. Refer to **Skill Sheet 9-I-1** for procedures for cleaning, inspecting, and maintaining hand tools; refer to **Skill Sheet 9-I-2** for procedures for cleaning, inspecting, and maintaining power tools.

Wooden Handles

- Inspect the handle for cracks, blisters, or splinters (**Figure 9.35**).

- Sand the handle if necessary to eliminate splinters.

- Wash the handle with mild detergent and rinse; wipe dry. Do not soak the handle in water because it will cause the wood to swell.

- Apply a coat of boiled linseed oil to the handle to preserve it and prevent roughness and warping. Do not paint or varnish the handle.

- Check the tightness of the tool head.

- Limit the amount of surface area covered with paint for tool marking. Some departments use self-adhesive bar codes on the handle for identification.

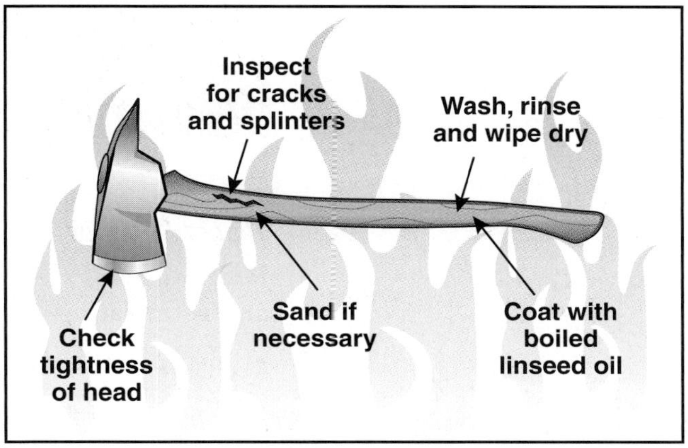

Figure 9.35 Care and maintenance of wood-handled tools includes these tasks.

Fiberglass Handles

- Wash the handle with mild detergent, rinse, and wipe dry.

- Check for damage or cracks.

- Check the tightness of the tool head.

Cutting Edges

- Inspect the cutting edge for chips, cracks, or spurs.

- Replace cutting heads when required.

- File the cutting edges by hand; grinding weakens the tool (**Figure 9.36, p. 412**).

- Sharpen blade as specified in departmental SOP. Some axe blades are intentionally left only semisharp to make them less prone to chipping.

Figure 9.36 File cutting edges by hand to achieve the proper angle and sharpness.

Plated Surfaces

Plated surfaces are those that are protected by chromium or other metal applied by an electroplating process.

- Inspect for damage.
- Wipe plated surfaces clean or wash with mild detergent and water.

Unprotected Metal Surfaces

Unprotected metal surfaces are the blades and other tool components that have not been electroplated to protect them from rust or corrosion.

- Keep free of rust.
- Oil the metal surface lightly. Light machine oil works best. Avoid using any metal protectant that contains methyl chloroform. This chemical may damage and weaken the handle.
- Do not paint metal surfaces – paint hides defects.
- Inspect the metal for chips, cracks, or sharp edges, and file them off when found.

Axe Heads

How well an axe head is maintained directly affects how well it will perform. If the blade is extremely sharp and ground too thin, pieces of the blade may break when cutting gravel roofs or striking nails and/or screws in roof decking or flooring. If the blade is too thick, regardless of its sharpness, it is difficult to drive the axe head through ordinary objects.

NOTE: DO NOT PAINT AXE HEADS! Painting hides faults in the metal. Paint also may cause the cutting surface to stick and bind.

Power Equipment

- Read and follow the manufacturer's instructions.
- Be sure that battery packs are fully charged and ready for immediate use.
- Inspect power tools periodically and ensure they will start manually.
- Check blades for damage or wear.
- Replace blades that are damaged or worn **(Figure 9.37)**.
- Check all electrical components (cords, etc.) for cuts or other damage.
- Ensure that all guards are functional and in place.
- Ensure that fuel is fresh. A fuel mixture may separate or degrade over time.

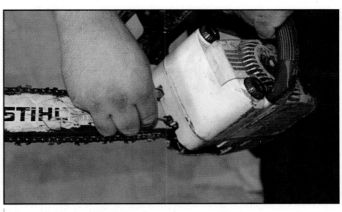

Figure 9.37 Replace chain saw blades as needed. Inspect blades for proper tightness.

Personnel should be familiar with the procedures for reporting tools and equipment that need to be repaired. Tools that are damaged or excessively worn should be removed from service, tagged, and sent to the proper authority for repair or replacement.

Door Size-Up and Construction Features

The primary obstacle firefighters face in gaining access into a building is a locked or blocked door. Some form of forcible entry is required in these situations. Size-up of the door is an essential part of the forcible entry task. Recognizing how the door functions, how it is constructed, and how it is locked are critical issues to successful forcible entry. From a forcible entry standpoint, doors function in one of the following ways:

- Swinging (inward, outward, or both)
- Sliding
- Revolving
- Overhead

Regardless of the type of door, firefighters should try the door to make sure that it is locked before using force. **Remember, "Try before you pry!"** If the door opens, there is no need for forcible entry. If it is locked, begin additional size-up. Look at the door and its immediate surroundings. Is there a glass panel in the door or a window beside it? If so, will breaking the glass allow a firefighter to reach in and unlock the door from the inside?

If there is no glass panel or side window, check to see if it is a swinging door or another type. If it is a swinging door, which way does it swing – in or out? An easy way to recognize which way a door swings is to look for the hinges. If you can see the hinges of the door, it swings toward you. If you cannot see the hinges, the door swings away from you. Both exterior and interior doors in residences and small office buildings usually swing inward. Exterior doors and interior exit doors in large commercial, public assembly, industrial, and other high-occupancy buildings usually swing outward, in the direction of exit travel. Some high-occupancy buildings have double-acting exterior doors that swing in both directions. If it is a sliding door, does it slide left or right? Does it roll up?

CAUTION

Only break or open windows when ordered to do so. Opening any window can adversely affect the ventilation operation and result in unwanted changes in fire behavior.

CAUTION

During forcible entry, maintaining control of the door is critical. Opening a door changes the ventilation profile and may have an adverse impact on fire behavior. If the door will be damaged to the point that it cannot be closed, steps should be taken to deal with subsequent changes in fire behavior. There may be times you want to block a door open to prevent it from closing against a hoseline.

There will be times when even the best size-up and forcible entry efforts are not successful. It is important not to get focused on one tool or technique. If the door does not open using the technique chosen, choose another. If the tool chosen does not open the door, choose another. Spending too much time forcing a door is counterproductive. If the door proves to be too well secured, look for another door.

After determining how a door functions, firefighters must understand which type of door it is. The type of door will determine the tools and techniques required to force it open. Exterior door types range from solid core wooden doors to high-security, metal-clad doors. In commercial occupancies, many interior doors are also solid core. In residential occupancies, most interior doors are hollow-core wooden doors.

The selection of forcible entry tools and techniques will also be affected by the type of lock installed in the door. Forcing locks is discussed later in this chapter.

Wooden Swinging Doors

There are three general types of wooden swinging doors: *panel, slab,* and *ledge.* Most exterior doors are the panel or slab types. The door is only one component of a door assembly. Doorjambs are the sides of the opening into which the door is fitted. Wooden swinging doors may have either rabbeted or stopped jambs (**Figure 9.38**). The rabbeted jamb is a doorstop milled into the casing that the door closes against. Complete door assemblies installed by contractors and do-it-yourselfers usually have rabbeted jambs. Door frames built on site usually include a piece of molding added to them to act as a doorstop. Unlike the rabbeted jamb, a nailed-on doorstop can be easily removed with prying tools, allowing firefighters easier access to the door latch.

Panel Doors

Wooden panel doors are made of solid wooden members inset with panels (**Figure 9.39**). The panels may be wood or other materials such as glass, Lexan® polycarbonate plastic, or Plexiglas™ acrylic to allow in light. These panels may be held in place by molding that can be removed for quick access.

Slab Doors

Slab doors are among the most common types in use. They are constructed in two configurations: *solid core* and *hollow core* (**Figure 9.40**). Most interior doors in newer residences are hollow core. The name may be misleading because it suggests that the entire core of the door is hollow, which is not accurate. The core or center portion of the door is made up of a grid of interlocking wooden strips that are glued in place. Panels of wood or imitation wood veneer are glued to both sides. Hollow-core doors are used because they are lightweight and relatively inexpensive. Most interior slab doors in newly constructed residences are hollow core, but exterior slab doors are usually solid core. Most slab doors do not have windows or other openings in them. Raised panels on slab doors are purely decorative.

Solid-core doors are much more substantial than hollow-core doors. The core of a solid-core door is constructed of some type of solid material. In very old homes, the doors may be made of thick planks that have been tongue-and-grooved together. Modern solid-core doors may be filled with a material used for insulation or soundproofing. Other doors may be filled with gypsum or a similar mineral material for fire resistance. In either case, the solid-core door is solid with a veneer covering. Solid-core doors are much heavier and more expensive than hollow-core doors. In high-crime areas, some panel doors have been replaced with heavier solid-core slab doors.

Rabbeted Jamb — Jamb into which a shoulder has been milled to permit the door to close against the provided shoulder.

Ledge Doors

Ledge doors, also known as *batten doors,* are found in warehouses, storerooms, barns, sheds, and similar structures **(Figure 9.41)**. Although many ledge doors are made commercially, firefighters will also find many homemade versions of this type of door. These doors are made of planks — often tongue and groove — fastened to horizontal and diagonal ledge boards. Ledge doors are generally locked with some type of surface locking mechanism, hasp, padlock, bolt, or bar. Hinges on this type of door are generally pin type, fastened with screws or bolts.

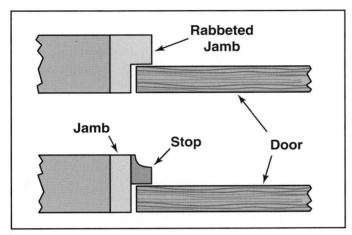

Figure 9.38 Typical rabbeted and stopped jambs.

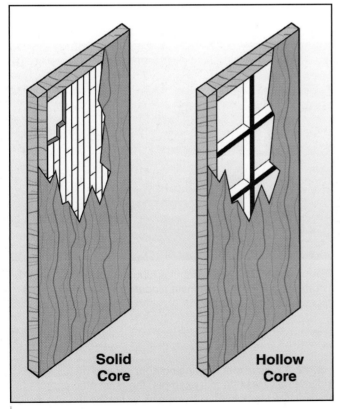

Figure 9.40 Slab doors are made of either solid or hollow core.

Figure 9.39 Wooden panel doors may have a variety of insets made of wood, glass, or plastic.

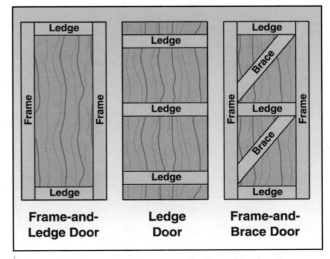

Figure 9.41 Ledge doors may be found in warehouses, barns, or sheds.

Metal Swinging Doors

Metal swinging doors are classified as *hollow metal, metal clad,* and *tubular.* Metal swinging doors are more difficult to force due to the materials of which they are constructed. Metal doors are most often set in a metal frame **(Figure 9.42)**. These doors are quite rigid and resist being penetrated by most forcible entry hand tools. When these doors are set in a metal frame in a concrete or masonry wall, power tools are almost always needed to open them.

The construction of metal doors varies depending upon their intended use. Metal-clad doors are solid wooden doors with a metal skin. Others are hollow shells filled with fire-resistive material. Tubular metal doors are constructed of seamless rectangular tube sections **(Figure 9.43)**. A slot is provided in the rectangular tube for glass or metal panels. The tube sections form a one-piece door with unbroken lines. These doors are found on exterior entry openings of modern buildings. Tubular doors are usually hung with conventional hardware, but some may use hardware based on the balance principle. This type of hardware consists of an upper and a lower arm, each connected by a concealed pivot. The arms and pivots are visible from the exterior side only. From the interior side, the balanced door resembles any other door.

Tubular aluminum doors with narrow stiles are also quite common. The panels of these doors are generally glass but some metal panels are used. Tubular aluminum doors are comparatively light in weight, strong, and quite rigid.

When ordered to force a metal door, consider the use of power tools, especially rotary saws or hydraulic tools. Do not spend a lot of time trying to force the door. If the door will not open after a few tries, move on to another site. In some cases, it may be easier to breach the wall next to a steel door rather than try to force the door itself.

Figure 9.43 Typical rectangular tubular metal doors found in commercial, office, and institutional structures as well as places of assembly.

Figure 9.42 A typical metal swinging door set in a metal frame in a masonry wall. In this example, the most efficient entry may involve removing the hinge pins or cutting the exposed hinges.

Sliding Doors

Most residential sliding doors travel either left or right; those in supermarkets and other retail businesses often travel in both directions in their frame and in the same plane as the wall in which they are set. Like hollow-core slab doors, *sliding doors* is a misleading name because these doors do not actually slide. Small roller or guide wheels in the base of these doors make them easy to move. Some interior sliding doors are *pocket doors* that move laterally into a "pocket" framed into the wall. Unlike other sliding doors, pocket doors are usually suspended from an overhead track. When this type of door is in the pocket, only the leading edge of the door is visible **(Figure 9.44)**.

The more common type of sliding door is the door assembly used in patio areas of residences or as doors to porches or balconies in houses, hotels, and apartments. The sliding section of glass doors usually moves past a stationary glass panel. In most cases, there is a lockable lightweight sliding screen door on the outside of the assembly **(Figure 9.45)**. The glass panels and sliding door are heavy glass window panels set in a metal or wood frame. These glass panels are often double-glazed glass, and doors installed in colder climates may even be triple-glazed. Some door assemblies may have tempered (safety) glass (glass heat treated to increase its strength and flexibility), which makes these doors very heavy and very expensive.

Patio sliding doors may sometimes be barred or blocked by a metal rod or similar device. These devices are commonly called *burglar blocks*. This feature can easily be seen from the outside, and it virtually eliminates any possibility of forcing the door

Figure 9.44 A partially closed pocket door found in a residential occupancy.

Figure 9.45 Exterior sliding doors can be difficult to force without breaking the glass.

without breaking the glass. However, in some installations it is possible to lift the sliding door out of its track, allowing the door to be removed without breaking the glass or damaging the frame. If it is necessary to break the glass, it should be broken using the techniques described in the Tempered Plate Glass Doors section later in this chapter.

Revolving Doors

A *revolving door* is made up of glass door panels that revolve around a center shaft **(Figure 9.46)**. The number of panels in the door varies with the manufacturer and the way the door is used. The revolving door turns within a metal or glass housing assembly that is open on each side to allow users entry and egress. The ends of the door panels are usually fitted with some type of large rubber weather stripping to help prevent the transfer of cold air into the building in winter or the loss of cool air in the summer.

Revolving doors may be locked in various ways; in general, they are considered difficult to force when locked. Usually there are swinging doors on either side of the revolving door. In some cases it is more effective to force entry through the swinging door than to try and force open a locked revolving door **(Figure 9.47)**.

All revolving doors are equipped with a mechanism that allows them to be locked open during an emergency. A problem for firefighters is that not all revolving doors can be locked open in the same way. Therefore, it is important that preincident surveys be conducted to locate revolving doors and to identify how their individual mechanisms work. Three basic types of mechanisms are used to lock revolving doors open: *panic- proof, drop-arm,* and *metal-braced.*

Figure 9.47 Forcing entry through an adjacent swinging door may be easier than attempting to force entry through the revolving door assembly.

Figure 9.46 A common style of revolving door found in many types of occupancies including hotels, shopping malls, and airport terminals.

Panic-Proof Type

This mechanism has a ¼-inch (6 mm) cable holding the door panels in place. The release mechanism is triggered by forces pushing in opposite directions on the panels (**Figure 9.48**).

Drop-Arm Type

The drop-arm mechanism has a solid arm passing through one of the panels. A pawl is located on the panel through which the arm passes. To collapse the system, press the pawl to disengage the arm and push the panel parallel with the next one. Repeat this procedure with the other panels.

Metal-Braced Type

This type of mechanism resembles a gate hook-and-eye assembly. To release the mechanism, lift the hook and fasten it back against the fixed panel. Hooks are located on both sides of the panel. Generally, the pivots are cast iron and easily broken by applying force to the panel at the pivot points.

Figure 9.48 Firefighters push on door panels to release them, creating a wide opening into the structure.

Overhead Doors

Overhead doors have a wide variety of uses. Many are used as residential and commercial garage doors; others are used as service doors at loading docks. Depending upon their intended function, overhead doors may be constructed of wood, metal, or fiberglass. Some overhead doors pose a formidable forcible entry problem. These doors are heavily secured, sometimes motor driven, and are usually spring-assisted. Forcible entry may be difficult but it is not impossible. Overhead doors are classified as follows: *sectional, tilt-slab, roll-up,* and *telescoping.*

Sectional

The sectional (folding) overhead door is usually not too difficult to force unless it is either motor driven or remotely controlled. If there is an external door lock handle, it is usually located in the center of the door. It controls two latches, one located on each side of the door. The lock and latch may also be located on only one side. These latches and locks are illustrated in **Figure 9.49**.

Sectional overhead doors may be forced by prying upward at the bottom of the door with a strong prying tool, but in some cases, removing a panel and turning the latching mechanism from the inside takes less time and results in less damage (**Figure 9.50, p. 420**). Some overhead doors may be locked with a padlock through a hole at either end of the bar, or the padlock may even be in the track. These locking systems may make it necessary to cut a hole in the door to gain access and remove the padlock.

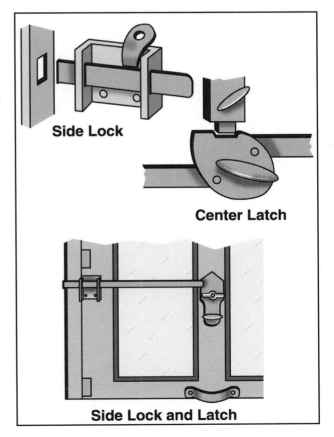

Side Lock

Center Latch

Figure 9.49 Types of overhead door latches found on commercial and residential overhead doors.

Side Lock and Latch

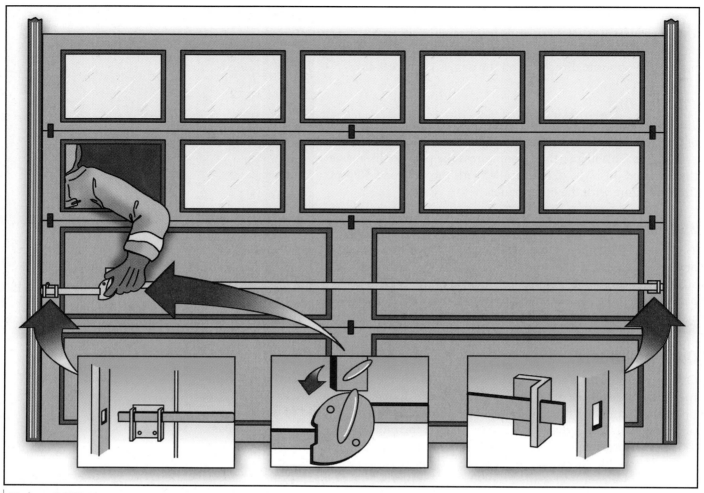

Figure 9.50 After removing a panel or pane of glass, reach through to open the latch.

Tilt-Slab

Pivoting or tilt-slab doors, sometimes called *awning doors,* can be sometimes difficult to force due to the design of the door. Tilt-slab doors made entirely of wood tend to be very heavy. Because the spring-operated lifting mechanism must pivot or tilt the door out and up, there must be a few feet (meters) of unobstructed space in front of the door to allow it to open **(Figure 9.51)**. The locking mechanisms on pivoting or tilt-slab doors are almost always located at one side or the other **(Figure 9.52)**. These mechanisms can sometimes be forcibly removed from the door entirely or the bolt in the lock can be cut with a rotary saw.

Roll-Up

Also called *sheet curtain* doors, steel roll-up doors are often used as high-security service doors. They are usually, but not always, locked on the inside. These doors may be manually operated without a lift mechanism, manually operated with a chain-hoist lift mechanism, or motor driven **(Figure 9.53)**. Regardless of the opening mechanism, many roll-up doors have a standard pedestrian door next to them **(Figure 9.54)**. It is often easier and faster to force entry through the pedestrian door and then use the lift mechanism to open the roll-up door. Heavy-gauge steel roll-up doors can be among the toughest forcible entry challenges faced by firefighters.

Figure 9.51 Tilt-slab overhead doors can be obstructed by parked vehicles on the outside or materials stored inside the garage area.

Figure 9.52 A typical locking mechanism mounted on the side of a tilt-slab gate.

Figure 9.53 A typical roll-up door seen from the outside.

Figure 9.54 Forcing entry through an adjacent swinging door and then using the roll-up door opening mechanism would be faster and less destructive.

Telescoping

A new type of industrial or institutional service door is becoming more common. This new type is called a telescoping door **(Figure 9.55, p. 422)**. These doors consist of a number of interlocking, inverted U-shaped metal sections. When the door is open, the sections are nested together at the top of the doorway opening **(Figure 9.56, p. 422)**. As the name implies, the door sections telescope into position as the door is closed.

Figure 9.55 A typical telescoping door. *Courtesy of Jerry Shacklett.*

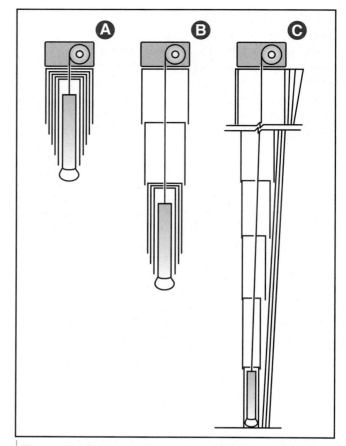

Figure 9.56 Stages of operation of the telescoping door: (a) door completely open, (b) door partially open, and (c) door completely closed.

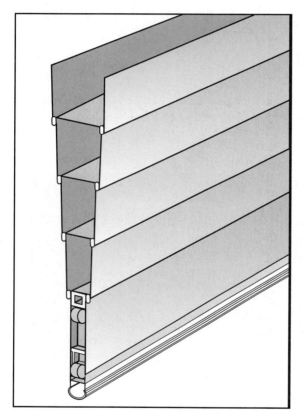

Figure 9.57 When telescoping doors are closed, each section interlocks with the one above and below it.

When a telescoping door is closed, it forms a barrier of hollow interlocking sections, each with a rectangular cross section (**Figure 9.57**). Like many roll-up doors, telescoping doors are operated by chain-hoist mechanisms at one side of the doorway opening. These mechanisms may be manually or electrically operated, but unlike roll-up doors, telescoping doors are raised and lowered by internal cables, and there are no springs or counterweights in the lifting mechanisms. However, just like roll-up doors, many telescoping doors have pedestrian doors beside them.

Forcing Entry Through Overhead Doors

Historically, one of the most common methods of cutting through a roll-up or sheet curtain door was to make a large triangular cut, sometimes called a *tepee* cut, in the center of the door. The purpose of the triangular cut was *not* to create an opening through which hoselines could be stretched, but only to allow firefighters access to the door-opening mechanism or to provide trapped firefighters an emergency exit. In making the triangular cut, however, the slats on both sides of the cut were often damaged or bent, which would cause the door to jam and not allow it to be opened fully. Therefore, this technique has fallen out of favor because making a triangular opening takes as long or longer to cut and creates a much smaller opening than the square or rectangular cut.

When overhead doors must be forced open, it is best to use a rotary saw to cut a square or rectangular opening about 6 feet (2 m) high and nearly the full width of the door **(Figure 9.58)**. Once firefighters have access to the interior, they should then use the lift mechanism to open the door fully.

Fire Doors

Fire doors are movable assemblies designed to cover doorway openings in rated separation walls in the event of a fire in one part of a building. A fire-door assembly includes the door, frame, and associated hardware **(Figure 9.59)**.

Types of standard fire doors include *horizontal and vertical sliding, single and double swinging,* and *overhead rolling* **(Figures 9.60 and 9.61)**. Fire doors may be manually, mechanically, or electrically operated. Fire doors may or may not be counterbalanced. Counterbalanced doors are a form of vertical-sliding doors that are generally used on openings to freight elevators, and they are mounted on the face of the wall inside the shaft **(Figure 9.62, p. 424)**. Counterbalanced doors are almost always manually operated.

WARNING!
All overhead doors should be blocked in the up or open position with a pike pole or other tool to prevent injury to firefighters should the built-in control device fail.

Figure 9.58 Access to an overhead door can be made by making a rectangular cut with a power saw. *Courtesy of Cedar Rapids (IA) Fire Department.*

Figure 9.59 Fire doors that close automatically are commonly found in schools, hospitals, and places of assembly.

Figure 9.60 An overhead rolling steel fire door separating two areas of a warehouse.

Figure 9.61 A horizontal sliding door used to protect an elevator hoistway.

There are two standard means by which fire doors operate: self-closing and automatic closing. When a *self-closing* type door is opened, it returns to the closed position immediately (**Figure 9.63**). *Automatic-closing* type doors normally remain open. They close only when the hold-open device releases the door because a fusible link has melted or due to activation of either a local smoke detector or a fire alarm system (**Figure 9.64**).

Self-closing swinging fire doors usually are installed in stairway enclosures and other areas where they must be opened and closed frequently (**Figure 9.65**). Automatic-closing swinging fire doors are usually installed in hallways, corridors, and other high-traffic areas and are normally held in the open position (**Figure 9.66**).

Figure 9.63 Self-closing door mechanism that returns the door to the closed position after use.

Figure 9.62 A counterbalanced fire door consisting of two sections that retracts above and below the opening.

Figure 9.65 A typical self-closing swinging fire door mechanism found on stairway doors.

Figure 9.64 The magnetic hold-open device releases the door when activated by a fire detection device or loss of power in the facility.

Vertical sliding fire doors are normally open but close automatically. They are employed where horizontal sliding or swinging fire doors cannot be used. Some vertical- sliding models utilize telescoping sections that slide into position vertically on side-mounted tracks; the sections are operated by counterweights.

Fire doors that slide horizontally are preferable to other types when floor space is limited. Horizontal sliding fire doors close automatically. Wheels attached to the tops of these doors travel on overhead tracks that are mounted on the fire wall at a slight angle. When a fusible link releases the door, gravity causes the door to roll into place across the opening **(Figure 9.67)**.

Overhead rolling fire doors may be installed where space limitations prevent the installation of other types. Like vertical sliding doors, overhead rolling doors are designed to close automatically. These doors have a barrel that is usually turned by a set of gears located near the top of the door on the inside of the building. This feature makes the door exceptionally difficult to force. Whenever possible, entrance to the building should be gained at some other point and the door operated from the inside.

Figure 9.66 Fire doors in busy corridors are normally open.

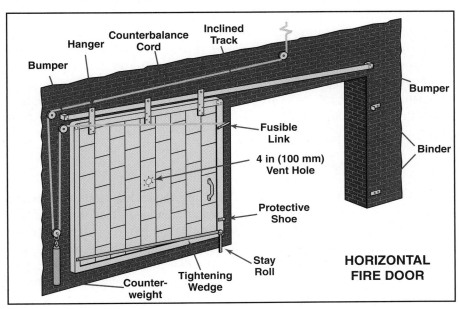

Figure 9.67 Horizontal sliding fire doors are closed by gravity when released by an activated fusible link.

Most interior fire doors do not lock when they close; they can be opened without using forcible entry techniques. Doors that are used on exterior openings may be locked; therefore, the lock must be forced.

A precautionary measure that firefighters should take when passing through an opening protected by a fire door is to block open the door to prevent its closing and trapping them. Fire doors have also been known to close behind fire attack crews and cut off the water supply in their hoselines.

Locks and Locking Devices

Locking devices vary from a simple horizontal bar resting in saddles on the inside of a set of double doors to very sophisticated mechanical and electronic locking devices. To perform an adequate forcible entry size-up, firefighters must have an understanding of the types of locks and locking devices that will be encountered during a fire or other emergency. Although locks are sold under a variety of brand names, they can be divided into four basic types: *mortise lock, bored (cylindrical) lock, rim lock,* and *padlock*.

Mortise Lock

This lock mechanism is designed to fit into a cavity in the door (**Figure 9.68**). It usually consists of a latch mechanism and an opening device (knob, lever, etc.). Older mortise locks have only the latch to hold the door closed, while other mortise locks have a bolt or bar (tang). When the mechanism is in the locked position, the bolt protrudes from the lock into a receiver that is mortised into the jamb. Newer mortise locks may also have larger and longer dead-bolt features for added security. Mortise locks can be found on private residences, commercial buildings, and industrial buildings.

Bored (Cylindrical) Lock

Bored locks are so named because their installation involves boring two holes at right angles to one another: one through the face of the door to accommodate the main locking mechanism and the other in the edge of the door to receive the latch or bolt mechanism. One type of bored lock is the key-in-knob lock.

The *key-in-knob lock* has a keyway in the outside knob; the inside knob may contain either a keyway or a button (**Figure 9.69**). The button may be a push button or a push and turn button. Key-in-knob locks are equipped with a latch mechanism that is locked and unlocked by both the key and, if present, by the knob button. In the unlocked position, a turn of either knob retracts the spring-loaded beveled latch bolt, which is usually no longer than ¾-inch (19 mm). Because of the relatively short length of the latch, key-in-knob locks are some of the easiest to pry open. If the door and frame are pried far enough apart, the latch clears the strike and allows the door to swing open.

Rim Lock

The rim lock is one of the most common locks in use today. It is best described as being surface-mounted and for this reason is used as an add-on lock for doors that already have other types of locks (**Figure 9.70**). This lock is found in all types of occupancies, including houses, apartments, and some commercial buildings. The rim lock can be identified from the outside by a cylinder that is recessed into the door in a bored latching mechanism fastened to the inside of the door and a strike mounted on the edge of the door frame.

Figure 9.68 Mortise locks, recessed in a door cavity, include dead bolts as well as latches.

Figure 9.69 Typical key-in-knob locks found in interior and exterior doors.

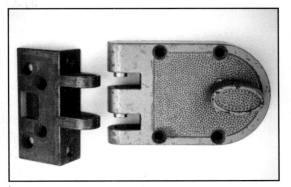

Figure 9.70 The interlocking dead bolt is one type of rim lock that is mounted on the inside of a door.

Figures 9.71 a-c Types of regular and heavy-duty padlocks used on gates, overhead doors of unoccupied spaces, and shed doors.

Padlock

Padlocks are portable or detachable locking devices **(Figure 9.71a-c)**. There are two basic types of padlocks: standard and heavy-duty. *Standard padlocks* have shackles of ¼ inch (6 mm) or less in diameter and are not case-hardened. *Heavy-duty padlocks* have shackles more than ¼ inch (6 mm) in diameter and are case-hardened. Many heavy-duty padlocks have what is called *toe and heel locking*. Both ends of the shackle are locked when depressed into the lock mechanism. These shackles will not pivot if one side of the shackle is cut. Both sides of the shackle must be cut in order to remove the lock.

Nondestructive Rapid Entry

The problem of gaining rapid entry through locked doors without doing damage has challenged fire departments for as long as locks have existed. Historically, many departments kept an inventory of keys to all the buildings in their response areas. While this can avoid damage from forcible entry, it also presents the problem of maintain-

Figure 9.72 Lockboxes allow rapid entry nondestructively. Facility keys are kept in the boxes, and the code or box key is provided to the fire department.

ing an inventory of keys and identifying the right key at the right time. Additionally, the possible liability placed on the department if a theft occurs in a building for which they have a key has led some departments to discontinue this practice.

Many of the problems presented by locked doors can be eliminated through the use of a rapid-entry lockbox system (**Figure 9.72**). All necessary keys or numeric key-pad combinations to unlock the building, storage areas, gates, and elevators are kept in a lockbox mounted at a high-visibility location on the building's exterior. Only the fire department carries a master key that opens all boxes in its jurisdiction.

Proper mounting is the responsibility of the property owner. The fire department should indicate the desired location for mounting, inspect the completed installation, place the building keys or combinations inside, and lock the box with the department's master key. Unauthorized duplication of the master key is prevented because the special key blanks are not available to locksmiths and cannot be duplicated with conventional equipment.

Conventional Forcible Entry Through Doors

Conventional forcible entry is the use of standard fire department tools to force open doors and windows to gain access. Once a firefighter has sized up a door, forcible entry can be performed if necessary (**Figure 9.73**). In this section, various methods of opening doors are discussed. Forcible entry through windows is discussed later in the chapter (see Forcing Windows section). If there are no glass panels in or next to the door and a door is definitely locked, the door will have to be forced open. In conventional forcible entry, a number of tools and tool combinations may be used. Many departments choose the rambar because it is versatile and requires only one person to use it effectively (**Figure 9.74**). Other departments choose the combination of the 8-pound (3.6 kg) flat-head axe and the Halligan bar, even though it is a two-person operation (**Figure 9.75**).

Figure 9.73 Conventional forcible entry may cause structural damage to locks, doors, and door frames.

Figure 9.74 One firefighter can effectively use a rambar to pry open a door.

Figure 9.75 Two firefighters can open a locked door with a Halligan bar and a flat-head axe.

Breaking Glass

One of the fastest and least destructive techniques for forcing locked doors is to break the glass in the door or the sidelight next to it. Once the glass is broken, the firefighter reaches inside and unlocks the door from the inside **(Figure 9.76, p. 430)**. In some situations, breaking the glass (or what appears to be glass) may be more difficult and costly. For example, tempered glass is very expensive, and Plexiglas™ and Lexan may resist being broken with conventional hand tools.

Because ordinary window glass will shatter into fragments with sharp cutting edges, the act of breaking glass must be done in a manner to ensure the safety of the firefighter **(Figure 9.77, p. 430)**. Firefighters should wear full protective equipment, especially hand and eye protection. If breaking the glass to gain access into a burning building, firefighters should wear SCBA and have a charged hoseline in place, ready to attack the fire. The techniques used for breaking both door glass and window glass are similar. If breaking the glass is the most appropriate method of entry, *do it!*

Forcing Swinging Doors

The most common type of door is one that swings at least 90 degrees to open and close. Most swinging doors have hinges mounted on one side that permit them to swing in one or both directions; others swing on pivot pins at the top and bottom of the door. Swinging doors can be either inward- or outward-swinging, or both. Double-acting swinging doors are capable of swinging 180 degrees. Forcing entry through all types of swinging doors involves basic skills, but requires practice to master.

Inward-Swinging Doors

Conventional forcible entry of single inward-swinging doors requires either one or two skilled firefighters, depending upon the tool or tools used **(Figure 9.78), p. 430**. For example, a single firefighter using a rambar can open most standard swinging doors; however, if a Halligan bar and flat-head axe are used, two firefighters are required. **Skill Sheet 9-I-3** describes one technique for forcing a single inward-swinging door.

Figure 9.76 Breaking small panes of glass is an inexpensive way to gain entry. Once the glass pane is broken and all the glass removed, reach in and unlock the door.

Figure 9.77 When breaking glass, always wear full personal protective clothing.

Figure 9.78 Some tools require two firefighters; others only one.

If the swinging door is metal or metal-clad in a metal frame set in a concrete or masonry wall, other forcible entry techniques may be needed. In some cases, a hydraulic door opener (often called a *rabbit tool*) can be used to force the door open **(Figure 9.79)**. In other cases, doors resist being pried open so it is necessary to cut around the lock in either of two ways. Using a rotary saw with a metal cutting blade, two intersecting cuts isolate the locking mechanism and allow the door to swing freely **(Figure 9.80)**. Three intersecting cuts can also produce the same result **(Figure 9.81)**.

Outward-Swinging Doors

Sometimes called *flush fitting doors,* outward-swinging doors present a different set of problems for firefighters. Because the hinges on outward-swinging doors are on the outside, it is often possible to use a nail set (or even a nail) to drive the hinge pins out of the hinges and simply remove the door **(Figure 9.82)**. If the bottoms of the hinges are solid so that the pins cannot be driven out, it may be possible to break the hinges off with a rambar or Halligan **(Figure 9.83)**. It may be possible to cut the hinges off with a rotary saw or an exothermic cutting device. Outward-swinging doors can also be forced by inserting the blade of a rambar or Halligan into the space between the door and the doorjamb and prying that space open wide enough to allow the lock bolt to slip from its keeper. **Skill Sheet 9-I-4** describes one procedure for making forcible entry through an outward-swinging door.

Figure 9.79 A hydraulic door opener or rabbit tool can be effective for opening inward-swinging doors.

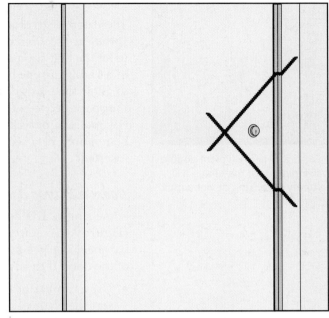

Figure 9.80 Using a rotary saw, make two cuts adjacent to the lock to allow the door to open.

Figure 9.81 Three cuts may be needed to open a metal or metal-clad door.

Figure 9.82 Some doors have exposed hinges with a hole in the bottom that allows the pin to be driven out. The door can then be pried out of the frame and removed.

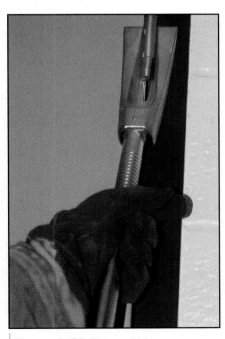

Figure 9.83 External hinges can sometimes be broken off with a Halligan or a rambar.

Special Circumstances

The basic techniques described earlier will work on most conventionally locked doors. There are circumstances where additional measures may need to be taken to force a door due to building construction features, door construction, or the presence of security hardware. A few of the doors needing additional forcing measures are *double-swinging doors, doors with drop bars,* and *tempered plate glass doors.*

Figure 9.84 A typical double-swinging door may have a variety of locking devices on it.

Double-Swinging Doors

These doors can present a problem depending on how they are secured (**Figure 9.84**). If they are secured only by a mortise lock, the doors can be pried apart far enough to let the bolt slip past the receiver. By inserting the blade of a rambar or wedge end of a Halligan-type bar between the doors, they can often be pried apart far enough to allow the bolt to clear the receiver. The blade of a rotary saw can be inserted into the space between the doors and the deadbolt cut. Some double doors have a security molding or weather strip over the space between the doors. This molding must be removed or a section cut away to allow the blade of the forcible entry tool to be inserted.

Doors with Drop Bars

Some single- and double-swinging entry doors are secured by a drop-bar assembly (**Figure 9.85**). A *drop-bar assembly* is a horizontal wooden or steel security bar (drop-bar) held in place across the door by wooden or metal stirrups attached to the inside of the door. If this type of door must be entered, try one of the following methods:

- Use a rambar or Halligan to spread the space between the double doors. Then insert the blade of a handsaw or other narrow tool into the opening and lift the bar up and out of the stirrups.

- Use a rotary saw to cut the exposed bolt heads that are holding the stirrups on the outside of the door. This will allow the drop-bar to fall away and the door to be opened.

- Insert the blade of a rotary saw into the space between the halves of double doors and cut the security bar (**Figure 9.86**).

Figure 9.85 Although drop bars are not permitted for use on doors that are equipped with panic hardware and designated as exits, you will find them in use. Their existence will not be apparent from the exterior of the door, however.

Figure 9.86 A drop bar can sometimes be cut with a rotary power saw after removing the weather strip over the gap between the doors.

Tempered Plate Glass Doors

In commercial, light industrial, and institutional occupancies, firefighters may be faced with metal-frame doors with tempered plate glass panels (**Figure 9.87**). These doors are heavy and very expensive. Tempered glass mounted in a metal door frame can be very difficult to break. Unlike regular plate glass, tempered glass resists heat; when broken, it shatters into thousands of tiny cube-like pieces.

If it becomes necessary to break through a tempered plate glass door, strike the glass at a bottom corner with the pick end of a pick-head axe. The firefighter should wear a suitable faceshield to protect against eye injury. Some departments place a shield made from a salvage cover as close to the glass as possible, and the blow is struck through the cover (**Figure 9.88**). Any remaining glass can then be scraped from the frame.

Tempered plate glass doors should be broken only as a last resort. Firefighters can also use the through-the-lock method (discussed in the next section) to open tempered plate glass doors as well as other doors.

Through-the-Lock Forcible Entry

The through-the-lock method is preferred for many commercial doors, residential security locks, padlocks, and high-security doors. This technique is very effective and does minimal damage to the door when performed correctly.

> **Tempered Glass** — Type of glass specially treated to become harder and more break-resistant than plate glass or a single sheet of laminated glass.

Figure 9.87 A typical metal frame door with tempered plate glass in a commercial occupancy.

Figure 9.88 If it is necessary to shatter the door glass, use a salvage cover to contain the broken glass.

Through-the-lock forcible entry requires a good size-up of both the door and the lock mechanism. Conventional forcible techniques should be used if the door and lock and suitable. If the door does not open with conventional forcible entry methods, the through-the-lock entry method can be used.

On some door locks, the lock cylinder can actually be unscrewed from the door. This is common on some storefront doors because it makes it easier for locksmiths to rekey the locks when occupancy changes. If the lock is not protected by a collar or shield, use the procedure described in **Skill Sheet 9-I-5**.

Removing the lock cylinder is only half the job. A key tool must then be inserted to open the lock as though firefighters had the key. The key tool is usually flat steel with a bent end on the cam end and a flat screwdriver shaped blade on the other end (**Figure 9.89**).

Like conventional forcible entry, the through-the-lock method requires patience and practice. Along with standard forcible entry striking and prying tools, special tools may also be needed for this forcible entry technique. Some examples of these special tools are the K-tool, A-tool, J-tool, and shove knife.

Figure 9.89
The key tool below is used to remove lock cylinders, with or without the K-tool.

K-Tool

The K-tool is useful in pulling all types of lock cylinders (rim, mortise, or tubular) (**Figure 9.90**). Used with a Halligan-type bar or other prying tool, the K-tool is forced behind the ring and face of the cylinder until the wedging blades take a bite into the cylinder (**Figure 9.91**). A metal loop on the front of the K-tool provides a slot in which to insert one end of the prying tool. The top of the prying tool must then be struck with a flat-head axe or other striking tool to set the K-tool. Once set, the prying tool is used to pull the K-tool and the lock cylinder from the door. Use the technique described in **Skill Sheet 9-I-6**.

When a lock cylinder is located close to the threshold or jamb, the narrow blade side of the K-tool will usually still fit behind the ring. Some sliding glass doors have little clearance, but only a ½-inch (13 mm) clearance is needed. Once the cylinder is removed, a key tool can be inserted into the hole to move the locking bolt to the open position.

Figure 9.90 The K-tool is designed to bite into the lock cylinder.

Figure 9.91 Attached to a Halligan, the K-tool is driven down onto the lock cylinder using a flat-head axe.

A-Tool

In many cases, the A-tool can accomplish the same job as the K-tool (**Figure 9.92**). The A-tool may cause slightly more damage to the door than a K-tool, but it will rapidly pull the cylinder. Many locks are manufactured with collars or protective cone-shaped covers over them to prevent the lock cylinder from being unscrewed. The A-tool was developed as a direct result of those lock design changes. The A-tool is a prying tool with a sharp notch with cutting edges machined into it. The notch resembles the letter *A*. This tool is designed to cut behind the protective collar of a lock cylinder and maintain a hold so that the lock cylinder can be pried out.

The curved head and long handle are then used to provide the leverage for pulling the cylinder. The chisel head on the other end of the tool is used when necessary to gouge out the wood around the cylinder for a better bite of the working head. When pulling protected dead bolt lock cylinders and collared or tubular locks, use the A-tool and the procedure described in **Skill Sheet 9-I-7**.

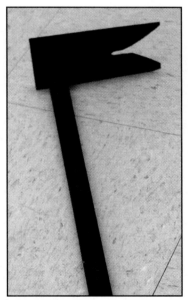

Figure 9.92 The A-tool is similar to the K-tool and is designed for removing lock cylinders.

J-Tool

The *J-tool* is a device made of rigid, heavy-gauge wire and designed to fit through the space between double-swinging doors equipped with panic hardware (**Figure 9.93, p. 436**). The J-tool is inserted between the doors far enough to allow the tool to be rotated 90 degrees in either direction. A firefighter can then pull the tool until it makes contact with the panic hardware. The firefighter then makes another sharp pull, and the tool should operate the panic hardware and allow the door to open.

Shove Knife

This flat-steel tool, resembling a wide-bladed putty knife with a notch cut in one edge of the blade, can provide firefighters rapid access to outward swinging latch-type doors (**Figure 9.94, p. 436**). When used properly, the blade of the tool depresses the latch, which allows the door to open. It is an ideal tool for opening doors that lead into smoke tower exit stairways.

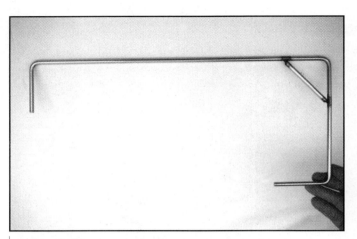

Figure 9.93 The J-tool is used to manipulate the bar on some types of panic hardware.

Figure 9.94 A shove knife works well on outward-swinging doors. The blade is inserted between the door and the frame and used to depress the latch.

Forcible Entry Involving Padlocks

Padlocks are portable locking devices used to secure a door, window, or other access. Padlocks range from those that are easily broken to the high-security types that are virtually impenetrable. Firefighters must be capable of forcing either the padlock itself or the device to which it is fastened. Conventional forcible entry tools can be used to break a padlock or detach the hasp to gain access. Additional tools are available to make forcible entry through padlocks easier. Some of these tools include the following (**Figure 9.95**):

- Duck-billed lock breaker
- Hammerheaded pick
- Locking pliers and chain
- Hockey puck lock breaker
- Bam-bam tool

Size-up of the lock is important. If the lock is small, with a shackle of ¼ inch (6 mm) or less and not case-hardened, forcible entry can be made using the techniques described in **Skill Sheet 9-I-8**. If these techniques fail to break the padlock, it is often easier to break the hasp or detach it from the door frame (**Figure 9.96**).

Special Tools and Techniques for Padlocks

If the shackle of the padlock exceeds ¼ inch (6 mm) and the lock, including body, is case-hardened, the firefighter faces a more difficult forcible entry task. Conventional methods of forcing padlocks may not work effectively. Firefighters may need to use either the duck-billed lock breaker or the bam-bam tool (**Figure 9.97**).

Duck-billed lock breaker. The *duck-billed lock breaker* is a wedge-shaped tool that will widen and break the shackles of padlocks, much like using the hook of a Halligan-type bar. This tool is inserted into the lock shackle and driven by a maul or flat-head axe until the padlock shackles break (**Figure 9.98**).

Bam-bam tool. This tool uses a case-hardened screw that is screwed into the keyway of the padlock. Once the screw is firmly set, a few sharp pulls on the sliding hammer will pull the lock tumbler out of the padlock body. The flat end of a key tool or a screwdriver can then be inserted to trip the lock mechanism (**Figure 9.99**).

NOTE: This method will *not* work on Master Locks, American Locks, and other high-security locks. These locks have a case-hardened retaining ring in the lock body that prevents the lock cylinder from being pulled out.

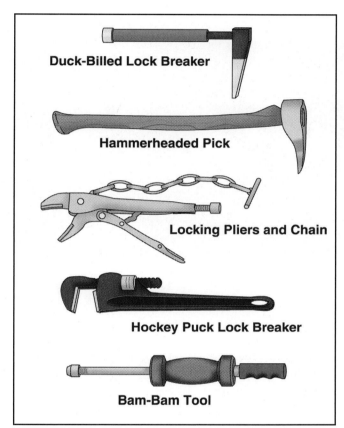

Figure 9.95 Various tools used for forcing padlocks open.

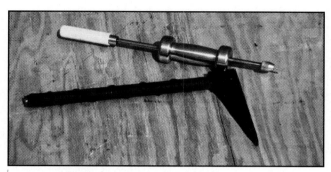

Figure 9.97 The bam-bam tool (above) and the duck-billed lock breaker (below) can be used to break padlocks.

Figure 9.98 The duck-billed lock breaker is driven into the lock shackle to break it.

Figure 9.96 Breaking or removing the hasp may be the most efficient way to open a door that is secured with a padlock.

Figure 9.99 The bam-bam tool is screwed into the keyway of the lock and then used to pull the cylinder out of the locks. This tool will not work on Master brand locks or high security locks.

Cutting Padlocks with Saws or Cutting Torches

Using either a rotary saw with a metal-cutting blade or a cutting torch may be the quickest method of removing some padlocks. High-security padlocks are designed with heel and toe shackles. Heel and toe shackles will not pivot if only one side of the shackle is cut. Cutting padlocks with a power saw or torch can be somewhat dangerous. One firefighter should stabilize the lock with a set of locking pliers and chain, and pull the lock straight out from the staple. A second firefighter cuts both sides of the padlock shackle with the saw or torch (**Figure 9.100**).

Figure 9.100 One firefighter steadies the lock by pulling on the chain attached by locking pliers to the padlock. The second firefighter cuts the lock with a rotary power saw.

Gates and Fences

Property owners, especially those with livestock and those concerned about break-ins, often take measures beyond protecting their homes and businesses with well-built and heavily secured doors and windows. One of these measures is to install fences, which presents special problems to firefighters.

Fences can be made of wood, plastic, masonry, barbed wire, chain-link wire fabric, or ornamental metal. Some are topped with barbed wire or razor ribbon. Fences may also be used to keep livestock or guard dogs on the premises; therefore, exercise extreme caution when entering a fenced area.

Barbed wire fences can be cut with bolt cutters to gain access. However, when cutting chain-link fences, it is much easier and faster to use a rotary saw (**Figure 9.101**). Wire fences should be cut near posts to facilitate repair after the incident, provide adequate space for fire apparatus access, and reduce the danger of injury from the recoil of wires when they are cut. An alternative method of opening a chain-link fence is to cut the wire bands holding the fence fabric to the posts. The fabric can then be laid on the ground. Fence gates are often secured with padlocks or chains; therefore, many of the techniques previously discussed will allow firefighters to gain access. When livestock are present, firefighters should be careful to close and latch any gates as soon as they pass through them.

A-frame ladders can be used to bridge masonry and ornamental metal fences (**Figure 9.102**). As always, adequate size-up and using the most efficient tools and techniques are important when forcing entry through fences and gates.

In some gated-communities, access through a secure gate may be the only way into the complex. The gates are usually controlled by electronic locks that are activated by a remote opener (similar to a garage door opener) or a keypad. There may also be a

Figure 9.101 A rotary saw can be used to cut through chain-link fencing quickly.

Figure 9.103 A lockbox or keypad speeds entry through a locked gate.

Figure 9.102 An A-frame ladder can be used to quickly bridge a metal fence without damaging the fence or cutting locks.

lockbox in the vicinity of the gate that contains an opener or keypad code (**Figure 9.103**). If necessary, and only if allowed by departmental SOP, entry may be forced by prying the gate open or by using the apparatus bumper to force the gate.

Forcing Windows

Forced entry can be made through windows even though they are not the best entry point into a burning building. Windows are sometimes easier to force than doors, and entry can be made through a window to open a locked door from the inside. As with doors, size-up of windows is critical to a successful forced entry. Breaking the glass is the most common technique, but this slows entry into the structure while the glass and frame are being cleared. Breaking the glass of the wrong window may also disrupt ventilation efforts, intensify fire growth, and draw fire to uninvolved sections of the building.

Breaking window glass on the fireground presents a multitude of hazards to both firefighters and civilians. Flying glass shards may travel great distances from windows on upper floors. Glass shards on floors make footing treacherous for movement of firefighters handling charged hoselines. Glass may shower victims inside the structure, causing additional harm. Wire glass is more difficult to break and remove than ordinary window glass because the wire prevents the glass from shattering and falling out of the frame. A sharp tool, such as the pick of an axe, can be used to chop wire glass out of its frame. Because double- or triple-glazed windows are expensive, firefighters must decide if the benefits of breaking the window outweigh the expense of replacing the windows. Multi-pane windows are also time-consuming to remove because the glass is held in place by a rubber cement that makes shard removal difficult.

Windows come in a variety of types and sizes. Some of the basic window styles include double-hung (checkrail) windows, hinged (casement) windows, projected (factory) windows, and awning or jalousie windows. Any of these window styles may also have Lexan® or other security panes as well as security bars or screens over them.

Double-Hung (Checkrail) Windows

The *checkrail,* or more commonly known as the *double-hung* window, has been an extremely popular window in building construction **(Figure 9.104)**. Structures hundreds of years old are fitted with double-hung windows. Manufactured in wood, metal, or vinyl, these windows are made up of two sashes. The top and bottom sashes are fitted into the window frame and operate by sliding up or down. Wood-frame double-hung windows are counterweighted for ease of movement. Newer double-hung windows not only move up and down, but can be tilted inward for cleaning. Double-hung windows may contain ordinary glass (single-, double-, or triple-pane), wire glass, or in certain circumstances, Plexiglas™ acrylic plastic or Lexan® plastic.

In most cases, double-hung windows are secured by one or two thumb-operated locking devices located where the horizontal frame members of the top and bottom sashes meet **(Figure 9.105)**. They may also be more securely fastened by window bolts. Replacement windows have two side-bolt-type mechanisms located on each side of the sash that, when operated, allow the window sash to tip inward.

Forcible entry techniques for double-hung windows depend on how the window is locked and the material of which the sash frames are made. The general technique for forcing a double-hung wood window is described in **Skill Sheet 9-I-9**.

NOTE: Metal windows are more difficult to pry. The lock mechanism will not pull out of the sash and may jam, creating additional problems. Use the same technique given for a wood-frame window, but if the lock does not yield with a minimal amount of pressure, it may be quicker to break the glass and open the lock manually.

Figure 9.105 A typical locking device on a double-hung window.

Figure 9.104 Double hung (checkrail) windows may be constructed of wood, metal, or vinyl.

In emergency situations where a window is the best means of access into a structure, valuable time can be saved and firefighter safety increased if the window glass is broken, the entire window area is cleared of all glass, and both top and bottom sashes are completely removed. This is especially true of metal double-hung windows. Glass removal can be accomplished safely with any long striking tool provided the firefighter is properly attired, the tool head is lower than the handle when the window is struck, and the firefighter is positioned upwind of the window. Removal of the sashes prevents any obstacles from snagging a firefighter's equipment or breathing apparatus when making entry. General techniques for forcing entry through windows with glass panes are described in **Skill Sheet 9-I-10**.

Hinged (Casement) Windows

Casement windows are hinged windows with wooden or metal frames. This type of window is sometimes called a *crank out window*, because the window is opened with a small hand crank (**Figure 9.106**). Casement windows should not be confused with *awning* or *jalousie* windows (see Awning and Jalousie Windows section), some of which are operated with a hand crank. Casement windows consist of one or two sashes mounted on side hinges that swing outward, away from the structure, when the window crank assembly is operated (**Figure 9.107**).

Locking devices vary for the casement window from simple thumb-operated devices to latch-type mechanisms. In addition to the locking devices, the casement window can only be opened by operating the crank mechanism. Double casement windows have at least four locking devices as well as two crank devices. Because this type of window is very difficult to force, firefighters should seek another means of entry if possible.

Projected (Factory) Windows

Projected windows are most often found in factories, warehouses, and other commercial and industrial buildings (**Figure 9.108**). These windows often have metal sashes with wire glass and function by pivoting at either the top or bottom. They are classified by the way that they swing when opened: projected-in, projected-out, or pivoted-projected. The most practical method of forcing factory-type windows is the same as that described for casement windows. The metal frames and wire glass make it difficult to effectively accomplish rapid forcible entry so firefighters

Figure 9.106 The hand crank is used to open and close casement windows. Some cranks are designed to fold back on themselves.

Figure 9.107 A casement window in the open position. Besides the crank, the window will have one or two locking latches for added security.

Figure 9.108 Projected (factory) windows may be found on industrial, commercial, and some types of residential occupancies.

should not enter through projected windows unless it cannot be avoided. In addition, these windows often have security bars or screens over the outside and inside to discourage entry.

Factory windows often cover a large area, but the movable window sections themselves are relatively small. Factory windows are usually located several feet (meters) off the floor so firefighters usually have to work from ladders when forcing these windows. If another entry point is not readily available and entry must be made, a rotary saw can be used to cut the window frame around the moveable section to enlarge the opening.

Awning and Jalousie Windows

Awning windows consist of large sections of glass about 1 foot (0.3 m) high and as long as the window width. They are constructed with a metal or wood frame around the glass panels, which are usually double-strength glass. Awning windows are hinged along the top rail, and the bottom rail swings out **(Figure 9.109)**.

Jalousie or *louvered windows* consist of small sections about 4 inches (100 mm) high and as long as the window width. The individual glass panes are held in the moveable frame only at the ends **(Figure 9.110)**. The operating crank and gear housing are located at the bottom of the window.

Figure 9.109 One type of awning window with the hinge along the top of each panel.

Entry through these windows requires the removal of several panes. Because most awning and jalousie windows are relatively small, they offer very restricted access even when all of the glass is removed. As an alternative, if entry must be made through a jalousie window, it may be faster and more efficient to cut through the wall around the entire window assembly and remove it.

Other Common Window Types

In addition to those already discussed, there are a number of other common window types in use in various regions. They are the hopper, tilt-turn, slider, and fixed types of windows.

A *hopper window* is similar to the awning except that it hinges at the bottom. A hopper is normally used for ventilation above a door or window, where protected by eaves.

The *tilt-turn window* is a fairly new type of window. It tilts out for ventilation but also can be opened fully for cleaning or as an emergency escape route.

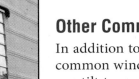

Figure 9.110 A typical jalousie window consisting of multiple panels 4 inches (100 mm) deep.

A *slider* or *gliding window* is made from two separate sashes – one is fixed, the other slides in a track. It is similar in design to the sliding glass patio door mentioned earlier.

A *fixed* or *picture window* is usually a relatively large solid glass unit that does not open. This type of window may be broken in an emergency, but this should be done only as a last resort. Because picture windows cover large expanses, breaking them will allow a great deal of air into or out of the building and may seriously affect ventilation efforts.

Hurricane Windows

Windows designed to resist hurricane-force winds use laminated glass with an advanced polymer. These windows are intended to help keep the building intact. An ionoplast layer is sandwiched between two layers of glass resulting in a laminated glass that is 100 times as rigid and five times as tear resistant as commonly used high-impact glass. Identifying these windows during preincident planning will help in the selection of the most effective tools and techniques that will be needed when forced entry is required.

High-Security Windows

Window manufacturers have responded to an increasing demand for the security of public and private property. A growing number of buildings have some form of high-security windows installed. To be most effective when forcing entry through high-security windows, firefighters must have identified these barriers during pre-incident planning so they know what types of tools will be needed to remove the security devices from any particular window. Some examples of high-security devices that have become common forcible entry problems are discussed in the following sections.

Lexan® Windows

In areas where broken windows and other vandalism is a recurring problem, fire-fighters may encounter window assemblies that have break-resistant plastic panes instead of glass. Lexan® is one of the plastics used in this application. Lexan® is 250 times stronger than safety glass, 30 times stronger than acrylic, and classified as self-extinguishing. Lexan® is virtually impossible to break with conventional forc-ible entry hand tools. Polycarbonate windows can be identified several ways. Tapping them gently with a tool produces a dull plastic sound different from glass. Lexan® windows will scratch much easier than glass. Large polycarbonate windows have a wavy surface and some obvious distortion along the sides.

Testing by an independent engineering firm proved that cooling Lexan® windows with carbon dioxide before striking them does not make them easier to break.

The following are two recommended techniques for forcing entry through Lexan® windows.

Using a rotary power saw with a carbide-tipped, medium-toothed blade (approximately 40 teeth), start all cuts at full rpm to avoid bounce and chatter of the saw. Large-toothed blades will skid off the surface, and smaller toothed blades will melt the Lexan® and cause the blade to bind. While cutting, wear goggles or other eye protection because the chips and shards produced are thrown out with considerable force. Cut as rapidly as possible without forcing the saw. Make the horizontal cuts first, then the vertical cuts. If a chain saw is used, it must be equipped with a carbide-tipped cutting chain.

Using striking or impact tools, such as sledge hammers, axes, rambars, or Halligans, can be effective if the entire pane can be punched through the frame. In many in-stallations, however, the Lexan® pane is bolted or riveted to the frame to prevent punch-through. General techniques for forcing entry through Lexan® windows are given in **Skill Sheet 9-I-11**.

Barred or Screened Windows and Openings

To avoid the expense of installing break-resistant plastic windowpanes, some building owners install metal bars or heavy-gauge metal screens over windows and sometimes door openings. Security bars and screen assemblies can present a significant barrier to entry for rescue and fire fighting. Security bars and screens may also prevent fire-fighters from escaping in an emergency.

Security screens may be permanently fixed, hinged at the top or side, or fitted into brackets and locked securely **(Figure 9.111)**. Those that are hinged can be opened easily if the lock is accessible and can be cut from the frame. If not, the screen fabric can be cut using a rotary saw. Once the screen is removed, the window must be forced as discussed earlier.

Other property owners choose to install heavy metal bars over their windows. These "burglar" bars vary in their types and construction, but their main feature is that they are difficult to force open **(Figure 9.112)**. Some of these bars are attached directly to the building; others are attached to the window frame. In some cases, forcing entry through burglar bars can be a difficult and time-consuming task. Forcible entry considerations for burglar bar installations are as follows:

- Shear off the bolt heads holding the screen frame or bar assembly if they are visible and accessible using either a rambar or the flat-head axe and Halligan bar combination **(Figure 9.113)**.

- Cut the bar assembly or its attachments to the structure using a rotary saw equipped with a metal cutting blade **(Figure 9.114)**.

- Cut the bars or screen frame using a hydraulic rebar cutter **(Figure 9.115)**.

Figure 9.111 Heavy-duty window screens are common security devices.

Figure 9.112 Security bars on windows can make entry difficult.

Figure 9.113 Bolt heads on security bars can be removed with a Halligan or rambar.

Figure 9.114 Cutting the attachment points of a security grill will allow the bars to be removed.

Figure 9.115 Window bars can be cut with a hydraulic rebar cutter.

Breaching Walls

During fire fighting operations, situations may arise where doors and windows are inaccessible or heavily secured, and it would be faster and more efficient to gain access (or exit) through the wall of a structure. Opening a hole in a wall is known as *breaching*. This action should be taken only after experienced firefighters with a thorough knowledge of building construction have sized up the situation and determined that breaching a particular wall is safe and will accomplish the purpose. For more information on wall breach as a means of escape, see Chapter 8.

Breaching load-bearing walls in a structure already weakened by fire can be very dangerous. Improperly locating the breach or removing too many structural components could cause a partial or total collapse. Walls also conceal electrical wiring, plumbing, gas lines, and other components of the building utilities (**Figure 9.116**). The area selected for the breach must be clear of all these obstructions. The techniques for breaching various types of walls are covered in the following sections and in **Skill Sheets 9-I-12 through 9-I-14.**

Plaster or Gypsum Partition Walls

As discussed in Chapter 4, interior walls may or may not be load-bearing. Some interior walls support the weight of floor or roof assemblies above; most do not. Both load-bearing and non-load-bearing interior walls are designed to limit fire spread. This fire resistance is provided by covering the wall with a variety of materials, including gypsum wallboard or lath-and-plaster over wooden or metal studs and framing (**Figure 9.117, p. 446**). Both lath-and-plaster and gypsum wallboard are often relatively easy to penetrate with forcible entry hand tools.

Figure 9.116 Be careful of plumbing and wiring when breaching walls.

Figure 9.117 Most interior walls are covered with gypsum wall board.

Reinforced Gypsum Walls

In some newer buildings, the interior walls in public access areas such as hallways, lobbies, and restrooms are covered with gypsum wallboard that is reinforced with Lexan®. Like other wallboard, reinforced wallboard is attached to the wall frame using drywall nails or screws. The Lexan® reinforcement is bonded to the back side of ordinary wallboard, so when reinforced wallboard is finished and painted, it looks exactly like any other wallboard. This wallboard is installed because of its resistance to being damaged by vandals; therefore, it is also resistant to being breached by firefighters using conventional forcible entry hand tools. If firefighters are to breach this material in a timely fashion, they must know that they are not dealing with ordinary wallboard and must bring the most appropriate tools — power saws — with them when they enter the building. The only way they will know that the wallboard is reinforced is by identifying it during preincident planning surveys.

Brick or Concrete Block Walls

Masonry walls can be difficult to breach during emergency operations. One tool that may be used is the battering ram. The battering ram is made of iron with handles and hand guards. One end is forked for breaking ordinary brick and concrete blocks, and the other end is rounded and smooth for battering doors and other types of walls (**Figure 9.118**). Two to four firefighters are required to use the battering ram. The firefighters must work together to swing the ram back and forth into the wall (**Figure 9.119**). Each time the ram strikes the wall, a little more masonry material is chipped away. Obviously, breaching a masonry wall with a battering ram can be a slow and labor-intensive operation; therefore, it is best suited for opening a hole in a wall through which water can be applied to a fire on the other side of the wall. While it is possible to use a battering ram to create an opening large enough for firefighters to pass through, it is impractical.

Power tools such as rotary saws with masonry blades or pneumatic or electric jackhammers are best for breaching masonry and concrete block walls (**Figure 9.120**). They are faster and usually require only one person to operate. If these tools are not available and a wall needs to be breached to allow water to be applied to a fire on the other side of the wall, a penetrating nozzle can be driven though the wall (**Figure 9.121**).

Concrete Walls

Breaching concrete walls is even slower and more labor-intensive than breaching masonry walls. In addition to the density of the material, it is often reinforced with steel rebar (**Figure 9.122, p. 448**). Therefore, breaching concrete walls should be done only

Battering Ram — Large metal pipe with handles and a blunt end used to break down doors or create holes in walls.

when absolutely necessary and no other alternative is available. A number of power tools can be used to breach concrete walls, but one of the fastest and most efficient is a chain saw equipped with a diamond-tipped chain (**Figure 9.123, p. 448**). Some of these chain saws are hydraulically powered (**Figure 9.124, p. 448**). If a chain saw is not available, a pneumatic jackhammer can be used.

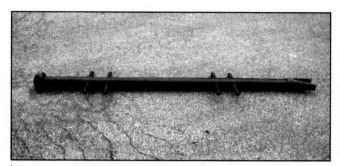

Figure 9.118 A typical battering ram used to breach many types of walls and doors.

Figure 9.119 At least two firefighters are required when using a battering ram to breach a masonry wall.

Figure 9.120 A rotary saw can be used to breach masonry walls.

Figure 9.121 A penetrating nozzle can be driven through a masonry wall. Once in place, the nozzle is charged and water is applied to the fire on the opposite side of the wall.

Figure 9.122 Most concrete walls contain rebar.

Figure 9.123 A chain saw with a masonry-cutting chain is very effective for breaching a concrete wall or cutting collapsed material during a rescue operation.

Figure 9.124 A typical hydraulically-powered chain saw.

Metal Walls

Metal walls are found in many buildings **(Figure 9.125)**. Prefabricated metal walls are common in both rural and urban settings, but given the right tools firefighters have little difficulty breaching these walls.

As with all other walls, a metal wall should be breached only after a careful size-up is made. These walls are usually constructed of overlapping light-gauge sheet metal panels fastened to metal or wooden studs. The panels may be attached by nails, rivets, bolts, screws, or other fasteners. Conventional forcible entry tools, such as an axe, rotary saw, or air chisel, cut these thin metal panels with relative ease **(Figure 9.126)**. Make sure no building utilities are located in the area selected for cutting. As mentioned earlier, have a charged hoseline or an appropriate fire

Figure 9.125 The use of all-metal buildings is becoming more common and may be found in use as warehouses, factories, retail outlets, and places of assembly.

Figure 9.126 An air-chisel can be used to breach a metal wall quickly.

extinguisher at hand when cutting metal with a rotary saw because of the sparks produced. As with overhead roll-up doors, it is best to cut a square or rectangular opening that is large enough for firefighters to pass through easily. Therefore, the opening should be at least 6 feet (2 m) tall and as wide as needed. If the wall must be breached to allow water to be applied to a fire on the other side of the wall, a penetrating nozzle can be driven through the metal siding.

Breaching Floors

There are almost as many kinds of floors and floor coverings as there are buildings, but subfloor construction is limited to either wood or concrete. Either of these two may be finished with a variety of covering materials. Concrete slab floors are quite common in residential, commercial, and industrial occupancies. Even upper floors of low- and high-rise buildings are often finished with lightweight concrete for its soundproofing qualities. The upper floors of multistory residences are usually wooden subfloors over wooden joists or I-beams.

It is not uncommon for a floor to be classified according to its covering instead of the material from which it is constructed. The feasibility of opening a floor during a fire fighting operation obviously depends upon how it was constructed and from what material. A wood floor does not in itself ensure that it can be penetrated easily. Many wood floors are laid over a concrete slab. The type of floor construction can and should be determined during preincident planning surveys. Some accepted and recommended techniques for opening wooden and concrete floors are offered in the following sections.

Wooden Floors

Wooden floor joists can be spaced from 12 to 24 inches (300 mm to 600 mm) apart depending upon the distance spanned and the dimensions of the lumber in the joists. Wooden I-beams are generally spaced 24 inches (600 mm) apart regardless of the span **(Figure 9.127, p. 450)**. The floor joists are covered by a subfloor consisting of either tongue-and-groove planks or sheets of plywood laid over the joists. Some wooden plank subfloors are laid diagonally to the joists, and the finished floor per-

Figure 9.127 The use of engineered wooden I-beams in building construction is growing in popularity.

pendicularly to the joists. Plywood subflooring is generally laid perpendicularly to the joists. The finish flooring, which may be sheet vinyl, ceramic tile, hardwood, or carpeting, is laid last. The procedure for opening a wooden floor is shown in **Skill Sheet 9-I-15.**

Before a floor is cut, carpets and rugs should be removed or rolled to one side. Some power saws make relatively neat cuts in wooden floors; others make rougher cuts. A circular saw makes the neatest cuts, but a chain saw may be faster. Because electrical service to a burning building is subject to being interrupted, it is better to supply power to electric saws from a portable generator carried on the fire apparatus.

Concrete Floors

All concrete floors are reinforced to some degree. The amount and type of reinforcement depends on where the floor is located and the loads it is designed to support. Unless there is a room or open space below the floor, there is rarely any reason to open a concrete floor. If a concrete floor must be opened, a number of tools can be used. While it is possible to open a concrete floor using sledgehammers and other hand tools, it is too slow and too labor-intensive to be practical. Concrete-cutting blades are available for most portable power saws, but the most efficient tool may be a pneumatic or hydraulic jackhammer (**Figure 9.128**). To help a jackhammer open a concrete floor even faster, the line to be cut is sometimes "stitch drilled" (**Figure 9.129**). A stitch drill can also be used to create an opening for a penetrating nozzle.

Figure 9.129 Stitch drilling concrete beforehand makes breaching with a jackhammer easier.

Figure 9.128 A typical hydraulic jack hammer can be used to breach concrete floors and cut into building collapse debris.

Summary

Forcible entry is the technique used by firefighters to gain access into a structure whose normal means of entry is locked or blocked. When properly applied, forcible entry efforts do minimal damage to the structure or structural components and provide quick access for firefighters. Forcible entry should not be used when a normal means of access is readily available. However, firefighters may need to use forcible entry tools and techniques to breach a wall as a means of escaping from a burning building.

When forcible entry or emergency escape are necessary, a thorough knowledge of forcible entry tools and techniques increases a firefighter's effectiveness and may save his or her life. In addition, performing forcible entry in a professional manner enhances the image of both the firefighter and the fire department.

End-of-Chapter Review Questions

Fire Fighter I

1. What are the four basic categories of forcible entry tools?

2. Why is the pick-head axe often used in structural fire fighting operations?

3. What tool is often used for ventilation purposes?

4. List three safety rules when using power saws.

5. List two basic maintenance procedures for the following: wooden handles, fiberglass handles, and power equipment.

6. What should firefighters do during door size-up?

7. What are the four basic types of locks?

8. What is conventional forcible entry?

9. What hazards are presented by breaking window glass?

10. When should a wall be breached?

Step 1: Wash tools with mild detergent or per manufacturer's guidelines, rinse, and wipe dry.

Step 2: Inspect tool handles for cracks, splinters, or other damage.

Step 3: Inspect tool head for tightness.

Step 4: Inspect working surface for dullness, damage, chips, cracks, or metal fatigue.

Step 5: Maintain wooden handles by repairing the loose tool heads, sanding the handle to eliminate splinters, and applying a coat of boiled linseed oil to the handle to preserve it and prevent roughness and warping.

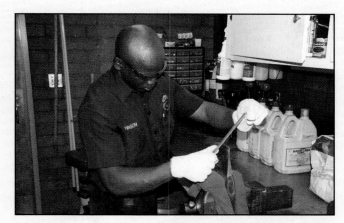

Step 6: Maintain cutting edges by sharpening blades by hand or as specified in departmental SOP.

Step 7: Maintain unprotected metal surfaces by keeping free of rust, filing chips, cracks, or sharp edges, and oiling the metal surface lightly.

Skill SHEETS

9-1-2
Clean, inspect, and maintain power tools and equipment.

NOTE: Always refer to the manufacturer's guidelines when cleaning, inspecting, and maintaining power tools and equipment.

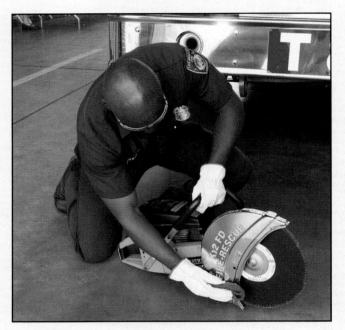

Step 1: Clean tools according to manufacturer's guidelines.

Step 5: Change a cutting blade on a power tool.

Step 6: Check fuel level in all power tools and fill as necessary.

Step 7: Check oil level in all tools and fill as necessary.

Step 8: Start all power tools and keep them running.

Step 9: Tag a tool that is out of service.

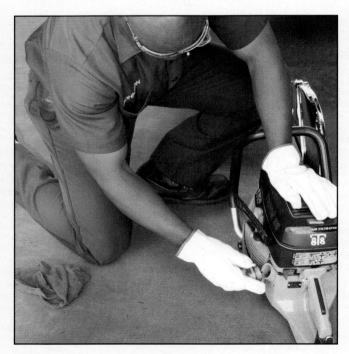

Step 2: Inspect tools for damage.

Step 3: Inspect parts for tightness and function.

Step 4: Inspect working surface for damage or wear.

9-1-3
Force entry through an inward-swinging door —
Two-firefighter method.

Skill SHEETS

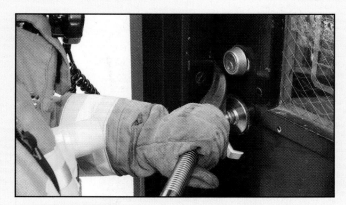

NOTE: Be sure to try before you pry.

Step 1: *Firefighter #1:* Place the fork of a Halligan bar just above or below the lock with the bevel side of the fork against the door.

Step 4: *Firefighter #2:* Drive the forked end of the tool past the interior doorjamb.

Step 5: *Firefighter #1:* Move the bar slowly perpendicular to the door being forced to prevent the fork from penetrating the interior doorjamb.

NOTE: If unusual resistance is met, remove the bar and turn it over. Begin again with the concave side of the fork now against the door.

Step 6: *Firefighter #1:* Make sure the fork has penetrated between the door and the doorjamb.

Step 2: *Firefighter #1:* Angle the tool slightly up or down.

Step 3: *Firefighter #2:* Strike the tool with the back side of a flat-head axe.

NOTE: Strike the tool only when Firefighter #1 calls for it.

Step 7: *Firefighter #1:* Exert pressure on the tool toward the door, forcing it open.

NOTE: If additional leverage is needed, Firefighter #2 can slide the head of the axe between the fork and the door.

CAUTION: The door may swing open uncontrollably when pressure is exerted on the Halligan bar. Maintain control of the door at all times. Placing locking pliers and a chain or a utility rope on the doorknob will allow the forcible entry team to maintain control of the door.

NOTE: Be sure to try before you pry.

Step 1: *Firefighter #1:* Place the wedge end of the Halligan bar just above or below the lock. If there are two locks, place the wedge between the locks.

Step 3: *Firefighter #1:* Pry down and out with the fork end of the tool.

Step 2: *Firefighter #2:* Strike the tool using a flat-head axe on the surface behind the wedge, driving the wedge into the space between the door and the jamb.

NOTE: Strike the tool only when directed by Firefighter #1.

NOTE: Be sure to try before you pry.

Step 1: Size up the door and lock.

Step 2: Place a set of locking pliers firmly on the lock cylinder.

NOTE: Make sure that the tool bites hard into the cylinder.

Step 4: Look inside the lock and identify the type of mechanism.

Step 5: Insert an appropriate key tool into the lock through the cylinder hole.

Step 6: Manipulate the tool to release the latching mechanism.

Step 7: Open the door.

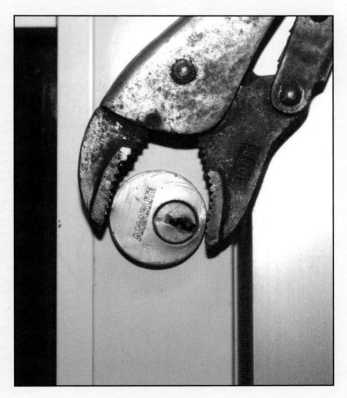

Step 3: Turn the lock cylinder counterclockwise to unscrew it from the door and remove it.

Skill SHEETS

9-1-6
Force entry using the through-the-lock method using the K-tool.

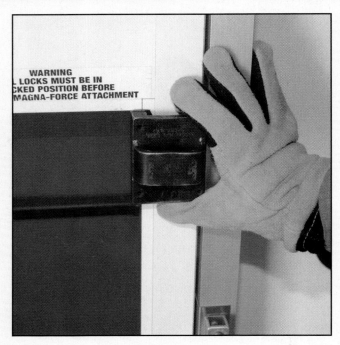

NOTE: Be sure to try before you pry.

Step 1: Size up the door and lock.

Step 2: Slide the K-tool down over the lock cylinder face.

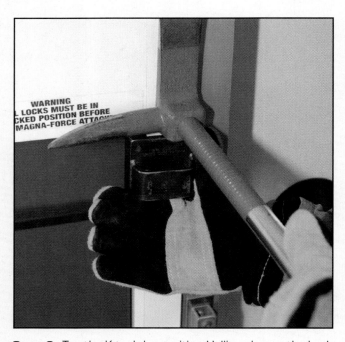

Step 3: Tap the K-tool down with a Halligan bar or the back of a flat-head axe.

Step 4: Insert the wedge end of the pry tool into the strap on the K-tool.

Step 5: Drive the K-tool further onto the cylinder.

Step 6: Pry UP on the tool handle to pull the cylinder.

Step 7: Insert a key tool through the cylinder hole to release the latching mechanism.

Step 8: Open the door.

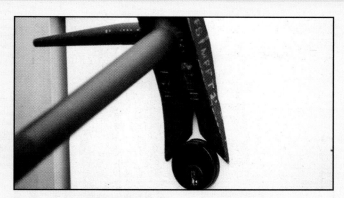

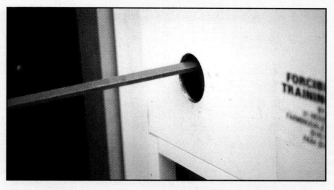

NOTE: Be sure to try before you pry.

Step 1: Size up the door and lock.

Step 2: Insert the V-notch of the A-tool between the lock cylinder and the door frame.

Step 4: Pry up on the tool and remove the lock cylinder.

Step 5: Insert the key tool into the lock through the cylinder hole.

Step 6: Manipulate the tool to release the latching mechanism.

Step 7: Open the door.

Step 3: Tap the A-tool firmly in place behind the lock cylinder.

Method One — Hook End

Step 1: *Firefighter #1:* Insert the hook of a Halligan bar into the shackle of the lock and pull the lock out away from the staple.

Step 2: *Firefighter #2:* Strike the Halligan bar sharply with a flat-head axe to drive the hook through the lock shackle and break it.

Method Two — Fork End

Step 1: Place the fork of the Halligan bar over the padlock shackle.

Step 2: Twist the lock until the shackle or the hasp breaks.

Method Three — Bolt Cutters

Step 1: Cut the shackle of the padlock, the chain, or the staple with bolt cutters.

NOTE: Do not attempt to cut case-hardened lock shackles with bolt cutters.

NOTE: Be sure to try before you pry.

Step 1: Insert the blade of an axe or other prying tool under the center of the bottom sash in line with the lock mechanism.

Step 2: Pry upward on the tool handle to force the lock.

Step 3: Push the lower sash upward to open the window.

Step 2: Break the window glass and use the tool to clean all the broken glass out of the frame once the glass has been broken.

NOTE: Be sure to try before you pry.

Step 1: Size up the situation.

NOTE: Be sure to try before you pry.

Step 1: Size up the situation.

Step 2: Check the saw blade.

Step 3: Ensure that the blade and hand guards are in place.

Step 4: Start and operate the rotary saw per manufacturer's instructions.

Step 5: Saw a hole in the window, making a triangular shape (three cuts) large enough to reach through.

NOTE: For a non-opening type window, saw around edges of entire pane so it can be removed.

Step 6: Turn off the saw.

Step 7: Open the window.

Skill SHEETS

9-1-12
Force entry through a wood-framed wall
(Type V construction) with hand tools.

Step 1: Confirm order with officer to force entry through wall.

Step 2: Size up the situation.

Step 3: Confirm with Command that utilities are off.

Step 4: Remove siding if necessary and locate stud.

Step 5: Cut an inspection hole (small triangle).

Step 6: Make cut utilizing inspection hole.

Step 7: Increase size of hole to allow the passage of fire-fighter (stud may be removed, if necessary).

Step 8: Utilizing the inspection hole, remove wall and insulation material with hand tool and place out of traffic area.

Step 9: Using hand tool, push inward and remove interior wall covering.

9-1-13
Force entry through a masonry wall with hand tools.

Skill SHEETS

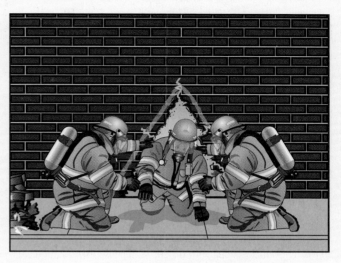

Step 1: Confirm order with officer to force entry through wall.

Step 2: Size up the situation.

Step 3: Confirm with Command that utilities are off.

Step 4: Determine one block to strike with tool and strike block until it is fractured.

Step 5: Systematically strike and fracture individual block in a triangle pattern until desired hole size is reached.

Step 7: Remove wall material and insulation and place out of traffic area.

Step 8: Using tool, push inward and remove interior wall covering, if needed.

Step 6: Using prying and/or striking tools, begin to remove the highest block first while moving downward and side to side.

Step 1: Confirm order with officer to force entry through wall.

Step 2: Size up the situation.

Step 3: Confirm with Command that utilities are off.

Step 4: Cut an inspection hole (small triangle).

Step 5: Locate studs (wall screws indicate) and cut hole near stud.

Step 6: Cutting a triangle, increase size of hole to allow the passage of firefighter.

Step 7: Remove wall material and insulation and place out of traffic area.

Step 8: Using tool, push inward and remove interior wall covering if needed.

Step 1: Determine the approximate location and size of hole.

Step 2: Sound the floor to determine the location and direction of the joists.

Step 3: Cut one side of the finished floor using angled cuts.

Step 4: Cut the other side of the finished floor in like manner.

Step 5: Remove the flooring between the cuts with the pick of the axe or other tool.

Step 6: Cut the subfloor using the same technique and angle cuts.

NOTE: Cut all four sides of the subfloor before removing it. If just a few boards or a small section of plywood are removed before the other cuts are made, the heat and smoke issuing from the first cuts may make completing the final cuts more difficult.

Chapter Contents

Ground Ladders

This chapter provides information that addresses the following job performance requirements of NFPA® 1001, *Standard for Fire Fighter Professional Qualifications* (2008):

NFPA® 1001 references for Chapter 10 :

5.3.6

5.3.9

5.3.11

5.3.12

5.5.1

Chapter Objectives

Fire Fighter I Objectives

1. Describe parts of a ladder.

2. Describe types of ground ladders used in the fire service.

3. Discuss materials used for ladder construction.

4. Discuss ladder maintenance and cleaning.

5. Summarize items to check for when inspecting and service testing ladders.

6. Summarize factors that contribute to safe ladder operation.

7. Discuss selecting the proper ladder for the job.

8. Summarize items to consider before removing and replacing ladders on apparatus.

9. Describe proper procedures to follow when lifting and lowering ground ladders.

10. Describe various types of ladder carries.

11. Explain proper procedures for positioning ground ladders.

12. Explain precautions to take before raising a ladder.

13. Describe various types of ladder raises.

14. Describe procedures for moving ground ladders.

15. Describe heeling and tying in ground ladders.

16. List guidelines for climbing ladders.

17. Describe methods for lowering conscious or unconscious victims down ground ladders.

18. Clean, inspect, and maintain a ladder. (Skill Sheet 10-I-1)

19. Carry a ladder — One-firefighter low-shoulder method. (Skill Sheet 10-I-2)

20. Carry a ladder — Two-firefighter low-shoulder method. (Skill Sheet 10-I-3)

21. Carry a ladder — Three-firefighter flat-shoulder method. (Skill Sheet 10-I-4)

22. Tie the halyard. (Skill Sheet 10-I-5)

23. Raise a ladder — One-firefighter method. (Skill Sheet 10-I-6)

24. Raise a ladder — Two-firefighter flat raise. (Skill Sheet 10-I-7)

25. Raise a ladder — Two-firefighter beam raise. (Skill Sheet 10-I-8)

26. Raise a ladder — Three- or four-firefighter flat raise. (Skill Sheet 10-I-9)

27. Deploy a roof ladder — One-firefighter method. (Skill Sheet 10-I-10)

28. Pivot a ladder — Two-firefighter method. (Skill Sheet 10-I-11)

29. Shift a ladder — One-firefighter method. (Skill Sheet 10-I-12)

30. Shift a ladder — Two-firefighter method. (Skill Sheet 10-I-13)

31. Leg lock on a ground ladder. (Skill Sheet 10-I-14)

32. Assist a conscious victim down a ground ladder. (Skill Sheet 10-I-15)

33. Remove an unconscious victim down a ground ladder. (Skill Sheet 10-I-16)

34. Select, carry and raise a ladder properly for various types of activities. (Skill Sheet 10-I-17)

Case History

In 1999, a California firefighter was seriously injured when the unsecured base of a ladder he was on slid away from the wall of a burning building. The 24-foot [7.3 m] straight ladder had originally been positioned against the top of an 18-foot [6 m]) wall but was later moved to a wall section that was only about 13 feet [4 m] high. In the second location, nearly half of the ladder's length was above the point of contact with the wall. When the firefighter climbed the ladder and leaned forward to place a chain saw onto the roof, the base of the ladder slipped and skidded away from the building. The spurs of the ladder caught in a crack in the paved surface and abruptly stopped the skid. The sudden stop threw the firefighter off the ladder onto the pavement below.

As the foregoing case history shows, a change in the situation may necessitate a change in the tools and equipment being used. It also shows how critical it is that all ground ladders be properly secured when anyone is climbing up or down on them. These and other critical factors in the safe and efficient use of fire service ground ladders are explained in this chapter.

Fire service ground ladders are ladders that are not permanently mounted on a piece of apparatus and are carried from the apparatus to the point of application by one or more firefighters. Ground ladders must then be raised into position by hand. Because they allow relatively quick and easy access to windows, balconies, rooftops, and other points above and below grade, these ladders are essential to the performance of many fireground and rescue operations. From both tactical and safety standpoints, it is crucial that firefighters know the characteristics and proper uses of ground ladders. While ladders are most often used to climb to elevated locations, they also provide a means by which firefighters can move victims down from elevated locations. In some cases, ground ladders provide the only means by which firefighters can escape from life-threatening situations.

According to NFPA® 1001, all those qualified at the Fire Fighter I level must know the following about ground ladders:

- Parts of a ladder
- Hazards associated with setting up ground ladders
- What constitutes a stable foundation for ladder placement
- Different ladder angles for various tasks
- Safety limits to the degree of angulation
- What constitutes a reliable structural component for top placement

This knowledge must also be translated into the development of certain basic ladder skills. NFPA® 1001 requires those qualified at the Fire Fighter I level (alone or as a member of a team) to be capable of the following:

- Carrying ground ladders
- Raising ground ladders
- Extending ground ladders and locking the fly
- Determining that a structural component (wall or roof) is capable of supporting a ladder
- Judging extension ladder height requirements
- Placing a ladder to avoid obvious hazards

There are no ladder-specific requirements at the Fire Fighter II level.

Fire service ladders are similar to any other ladder in shape and design; however, they tend to be built more substantially and are capable of withstanding heavier loads than those manufactured for use by private industry or the general public. Their use under adverse conditions requires that fire service ground ladders provide a margin of safety not usually expected of other ladders. NFPA® 1931, *Standard for Manufacturer's Design of Fire Department Ground Ladders,* lists the requirements for the design of ground ladders and the testing required of manufacturers. NFPA® 1932, *Standard on Use, Maintenance, and Service Testing of In-Service Fire Department Ground Ladders,* lists the use, maintenance, and testing requirements during a ladder's service life.

This chapter introduces the basic ladder parts and terminology common to most fire service ground ladders. The various types of ladders used in the fire service are reviewed. Also included are the proper care, carrying techniques, deployment, and use of fire service ground ladders. For more information on the subjects covered in this chapter, see the IFSTA **Fire Service Ground Ladders** manual or the IFSTA **Fireground Support Operations** manual.

Basic Parts of a Ladder

The discussion of fire service ground ladders begins with a description of the various parts of the ladder. Many of these terms apply to all types of ladders; others may be specific to a certain type of ladder (**Figures 10.1 and 10.2**).

- **Beam** — Main structural member of a ladder supporting the rungs or rung blocks
- **Bed section (base section)** — Lowest and widest section of an extension ladder; while the ladder is being raised or lowered, this section always maintains contact with the ground or other supporting surface
- **Butt (also called heel or base)** — Bottom end of the ladder; the end that is placed on the ground or other supporting surface when the ladder is positioned
- **Butt spurs** — Metal plates, spikes, or cleats attached to the butt end of ground ladder beams to prevent slippage
- **Dogs** — See *pawls.*
- **Fly Section** — Upper section(s) of extension or some combination ladders; the section that moves
- **Footpads** — Swivel plates attached to the butt of the ladder; usually have rubber or neoprene bottom surfaces

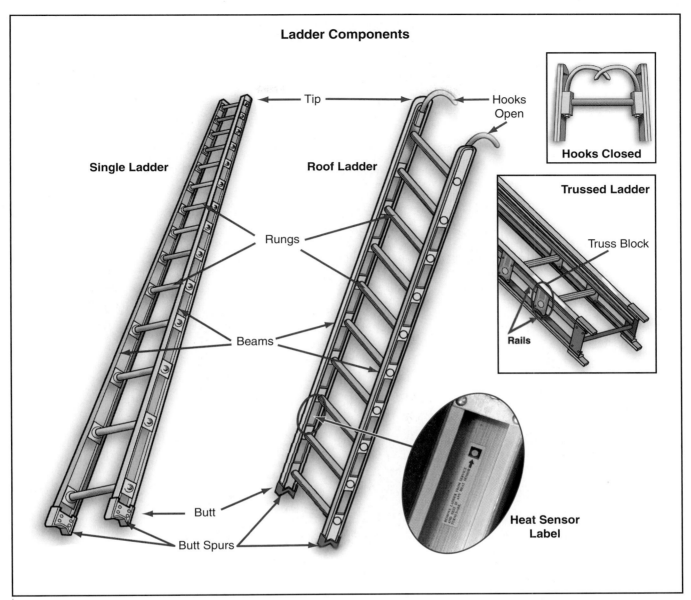

Figure 10.1 Basic parts of a straight ladder.

- **Guides** — Wood or metal strips, sometimes in the form of slots or channels, on an extension ladder that guide the fly section while being raised

- **Halyard** — Rope or cable used for hoisting and lowering the fly sections of an extension ladder; also called fly rope

- **Heat-sensor label** — Label affixed to the inside of each beam of each ladder section; a color change indicates that the ladder has been exposed to a sufficient degree of heat that it should be tested before further use

- **Heel** — See *butt*.

- **Hooks** — Curved metal devices installed near the top end of roof ladders to secure the ladder to the highest point on a peaked roof of a building

- **Locks** — See *pawls*.

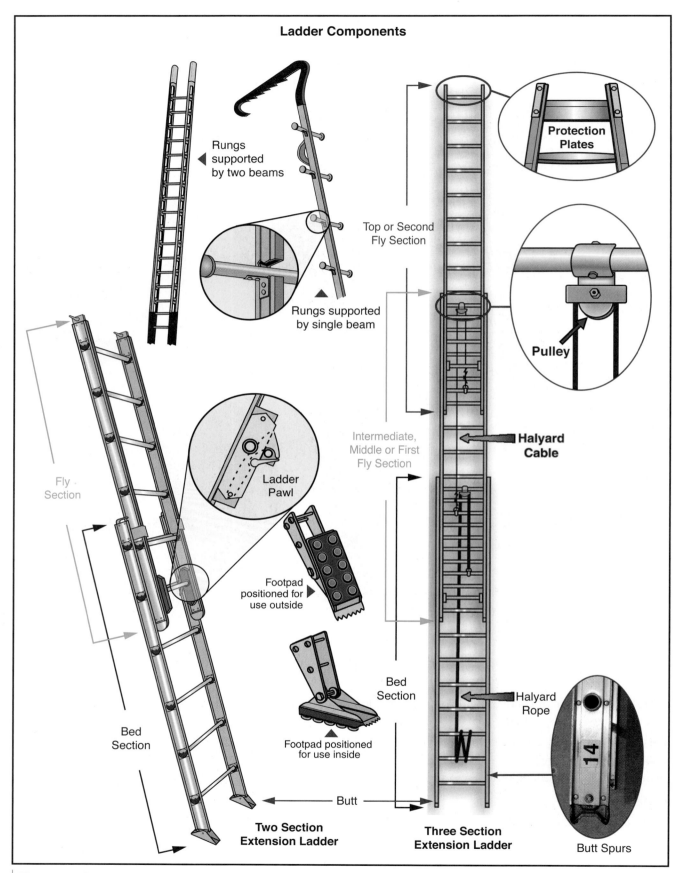

Ladder Components

Rungs supported by two beams

Rungs supported by single beam

Protection Plates

Top or Second Fly Section

Pulley

Fly Section

Ladder Pawl

Intermediate, Middle or First Fly Section

Halyard Cable

Footpad positioned for use outside

Bed Section

Bed Section

Footpad positioned for use inside

Halyard Rope

Butt

Two Section Extension Ladder

Three Section Extension Ladder

Butt Spurs

Figure 10.2 Extension ladders have additional components to enable them to be extended and secured.

- **Main section (also called bed or base section)** — Bottom section of an extension ladder
- **Pawls (also called dogs or ladder locks)** — Devices attached to the inside of the beams on fly sections used to hold the fly section in place after it has been extended
- **Protection plates** — Strips of metal attached to ladders at chafing points, such as the tip, or at areas where they come in contact with the apparatus mounting brackets
- **Pulley** — Small, grooved wheel through which the halyard is drawn on an extension ladder
- **Rails** — The two lengthwise members of a trussed ladder beam that are separated by truss or separation blocks
- **Rungs** — Cross members that provide the foothold for climbing; the rungs extend from one beam to the other except on a pompier ladder where the rungs pierce the single beam
- **Shoes** — See *footpads*.
- **Stops** — Wooden or metal pieces that prevent the fly section from being extended too far **(Figure 10.3)**
- **Tie rods** — Metal rods extending from one beam to the other **(Figure 10.4)**
- **Tip (top)** — Extreme top of a ladder
- **Truss block** — Spacers set between the rails of a trussed ladder; sometimes used to support rungs

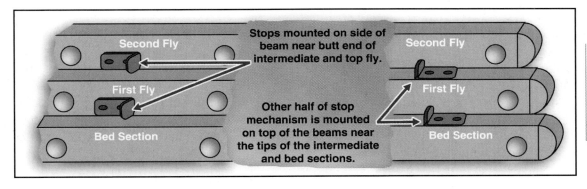

Second Fly

First Fly

Bed Section

Stops mounted on side of beam near butt end of intermediate and top fly.

Other half of stop mechanism is mounted on top of the beams near the tips of the intermediate and bed sections.

Second Fly

First Fly

Bed Section

Figure 10.3 Stops are used on extension and pole ladders to prevent overextension of the fly section(s).

Tie rod

Figure 10.4 Steel tie rods are used to help hold the beams of wood ladders together. They are located directly beneath the rungs at various levels.

Ladder Types

All of the various types of fire service ladders have names. However, some of the names describe their use and firefighters frequently refer to them that way — roof ladders, attic ladders, and so on (**Figure 10.5**). The weight of any ladder per unit of length varies with the materials of which it is constructed. Heavier ladders require more personnel to safely handle them. The descriptions in the following sections more clearly identify fire service ground ladders.

Single Ladders

Also called *wall ladders* or *straight ladders*, *single ladders* consist of only one section of a fixed length. Single ladders are most often identified by the overall length of the beams. For example, a firefighter may call for a 20-foot (6 m) straight ladder. They are often used for quick access to windows and roofs on one- and two-story buildings. Some single ladders are of the trussed type, a design intended to maximize their strength while reducing weight. Lengths of single ladders vary from 6 to 32 feet (2 m to 10 m) with the more common lengths ranging from 12 to 24 feet (4 m to 8 m).

Roof Ladders

Roof ladders are single ladders equipped with folding hooks that provide a means of anchoring the ladder over the ridge of a pitched roof or some other roof part. In position, roof ladders generally lie flat on the roof surface so that a firefighter may stand on the ladder for roof work. The ladder distributes the firefighter's weight and helps prevent slipping. Roof ladders may also be used as single wall ladders. Their lengths range from 12 to 24 feet (4 m to 8 m).

Folding Ladders (Attic Ladders)

Folding ladders are single ladders that are often used for interior attic access. They have hinged rungs allowing them to be folded so that one beam rests against the other. This capability allows them to be carried in narrow passageways and used in attic scuttle holes and small rooms or closets. Folding ladders are commonly found in lengths from 8 to 16 feet (2.5 m to 5 m) with the most common being 10 feet (3 m). NFPA® 1931 requires folding ladders to have footpads attached to the butt to prevent slipping on floor surfaces.

Extension Ladders

An *extension ladder* is adjustable in length. It consists of a base or bed section and one or more fly sections that travel in guides or brackets to permit length adjustment. Its size is designated by the full length to which it can be extended. Unlike single ladders, extension ladders can be adjusted to the specific length needed to access windows and roofs. Extension ladders generally range in length from 12 to 39 feet (4 m to 11.5 m).

Pole ladders (also called *Bangor ladders*) are extension ladders with poles that can be attached to the top of the bed sections for added leverage and stability when raising the ladders. NFPA® 1931 requires all extension ladders that are 40 feet (12 m) or longer to be equipped with these staypoles. Pole ladders are manufactured with two to four sections. Most modern pole ladders do not exceed 50 feet (15 m) in length.

Pole Ladder — Large extension ladder that requires tormentor poles to steady the ladder as it is raised and lowered.

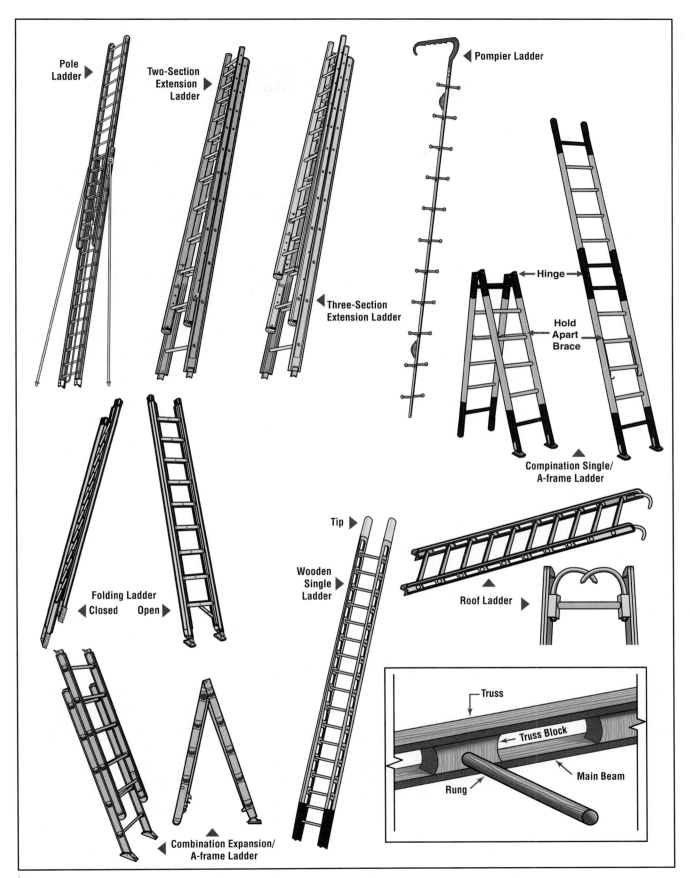

Figure 10.5 A number of ladder types are needed for specific jobs.

Combination Ladders

Combination ladders are designed so that they may be used as a self-supporting stepladder (A-frame) and as a single or extension ladder. Lengths range from 8 to 14 feet (2.5 m to 4.3 m) with the most popular being the 10-foot (3 m) model. The ladder must be equipped with positive locking devices to hold the ladder in the open position.

Pompier Ladders

Sometimes referred to as *scaling ladders, pompier ladders* are single-beam ladders with rungs projecting from both sides of the beam. These ladders have a large metal "gooseneck" projecting at the top for inserting into windows or other openings. Pompier ladders are used to climb from floor to floor, via exterior windows, on a multistory building. Lengths vary from 10 to 16 feet (3 m to 5 m).

Ladder Construction

Ground ladders are constructed of metal, wood, or fiberglass. Each material has certain advantages and disadvantages when made into ladders.

Metal:

- Good conductor of heat, cold, electricity
- Easy to repair
- Can suddenly fail when exposed to heat or flame
- Widest range of sizes

Wood:

- Highest cost of all ladders
- Heaviest per unit of length
- Retains strength when exposed to heat or flame
- Very durable

Fiberglass:

- Generally a poor conductor of electricity
- Can suddenly crack and fail when overloaded
- Can burn when exposed to flame

Ladder Inspection and Maintenance

Fire service ladders must be able to withstand considerable abuse – sudden overloading, exposure to temperature extremes, and being struck by falling objects. Regardless of what materials or designs are used for ladders, they must conform to NFPA® 1931. All ladders meeting NFPA® 1931 are required to have a certification label affixed to the ladder by the manufacturer indicating that the ladder meets the standard. All ground ladders should be tested before being placed in service, then annually and after any use that exposes them to high heat or rough treatment.

Maintenance

Before discussing ladder maintenance, it is important to understand the difference between maintenance and repair. *Maintenance* means keeping ladders in a state of usefulness or readiness. *Repair* means to either restore or replace that which is damaged or worn out. All firefighters should be capable of performing routine maintenance on ground ladders according to departmental SOP and the manufacturer's recommendations. Any ladders in need of repair require the service of a trained ladder repair technician. Ladders should be service tested before being placed in service, annually, and after any use that exposes them to high heat or rough treatment.

General maintenance requirements for ground ladders include the following guidelines:

- Keep ground ladders free of moisture.
- Do not store or rest ladders in a position where they are subjected to vehicle exhaust or engine heat **(Figure 10.6)**.
- Do not store ladders in an area where they are exposed to the elements.
- Do not paint ladders except for the top and bottom 18 inches (450 mm) of the beams for purposes of identification or visibility.

NFPA® 1932, *Standard on Use, Maintenance, and Service Testing of In-Service Fire Department Ground Ladders,* lists the general maintenance items that apply to all types of ground ladders.

Cleaning Ladders

Regular and proper cleaning of ladders is more than a matter of appearance. Accumulated dirt or debris from a fire may collect and harden to the point where ladder sections cannot function as designed. Therefore, it is recommended that ladders be inspected regularly and cleaned after every use.

A soft bristle brush and running water are the most effective tools for cleaning ladders **(Figure 10.7)**. Tar, oil, or greasy residues should be removed with mild soap and water or environmentally safe solvents according to departmental SOP and manufacturer's recommendations. Anytime a ladder is wet, whether after cleaning or use, wipe it dry. As they clean the ladder, firefighters should look for damage or wear. Any defects should be reported according to departmental standard operating procedures (SOPs). Where recommended by the manufacturer, occasional lubrication will maintain smooth operation of the ladder.

Maintenance — Keeping equipment or apparatus in a state of usefulness or readiness.

Repair — To restore or put together that which has become inoperable or out of place.

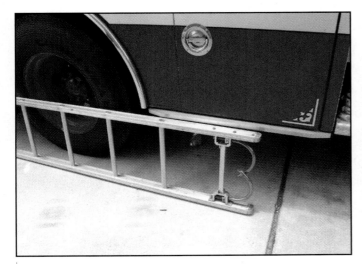

Figure 10.6 Do not place ladders near the apparatus exhaust pipe.

Figure 10.7 Regularly clean ladders to remove fire-scene debris and road grime.

Inspecting and Service Testing Ladders

NFPA® 1932 requires ladders to be inspected after each use and on a monthly basis. Because fire service ladders are subject to harsh conditions and physical abuse, it is important that they be service tested to ensure they are still fit for use. NFPA® 1932 serves as the guideline for ground ladder service testing. This standard recommends that only the tests specified be conducted either by the fire department or an approved testing organization. NFPA® 1932 further recommends that caution be used when performing service tests on ground ladders to prevent damage to the ladder or injury to personnel.

When inspecting ground ladders, some of the elements that should be checked on all types of ladders include the following:

- Heat sensor labels on metal and fiberglass ladders for a color change indicating heat exposure

NOTE: Heat sensors should be replaced when their expiration date is reached. Ladders without a heat sensor label may also show signs of heat exposure such as heavy carbon (soot) deposits or blistered paint on the ladder tips. Wooden ladders may show blistered or blackened varnish, and there may be discoloration on fiberglass ladders.

- Rungs for damage or wear **(Figure 10.8)**
- Rungs for tightness
- Bolts and rivets for tightness

NOTE: Bolts on wooden ladders should not be so tight that they crush the wood.

- Welds for any cracks or apparent defects
- Beams and rungs for cracks, splintering, breaks, gouges, checks, wavy conditions, or deformation

In addition to these general inspections, there are some other items that need to be checked, depending on the specific type of ladder being inspected. The following sections highlight some of these items.

Wooden Ladders/Ladders with Wooden Components

Look for the following when inspecting wooden ladders or ladders with wooden components:

- Areas where the finish has been chafed or scraped
- Darkening of the varnish (indicating exposure to heat)
- Dark streaks in the wood (indicating deterioration of the wood)
- Marred, worn, cracked, or splintered parts
- Rounded or smooth shoes
- Water damage

Roof Ladders

When inspecting roof ladders, make sure the roof hook assemblies operate with relative ease **(Figure 10.9)**. In addition, the assembly should not show signs of rust, the hooks should not be deformed, and parts should be firmly attached with no sign of looseness.

Heat Sensor Label — Label affixed to the ladder beam near the tip to provide a warning that the ladder has been subjected to excessive heat.

WARNING!
Remove from service and test any ladder that is subjected to direct flame contact or high heat or whose heat sensor label has changed color.

CAUTION
Any indication of deterioration of the wood is cause for the ladder to be removed from service until it can be service tested.

CAUTION
Any serious problems found should result in removal of the ladder from service pending service testing.

Figure 10.8 Inspect ladder rungs for damage, wear, and to ensure they are tight.

Figure 10.9 When inspecting roof ladders, ensure that the hooks fully open with ease.

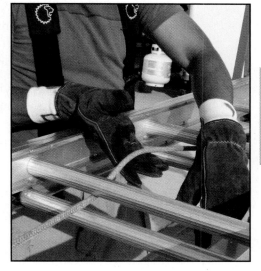

Figure 10.10 Inspect ladder halyards for damage caused by wear, such as frays or cuts.

Extension Ladders

The following must be checked when inspecting extension ladders:

- *Pawl assemblies.* The hook and finger should move in and out freely.

- *Halyard.* If damage or wear is found, the halyard should be replaced (**Figure 10.10**).

- *Halyard cable.* Check to see that it is taut when the ladder is in the bedded position. This check ensures proper synchronization of the upper sections during operation.

- *Pulleys.* Make sure they turn freely.

- *Ladder guides.* Check their condition and that the fly sections move freely.

- *Staypole toggles.* Check their condition and that they move freely. Detachable staypoles are provided with a latching mechanism at the toggle. This mechanism should be checked to be sure that it moves freely and latches properly.

If any discrepancies are found, the ladder should be removed from service until it can be repaired and tested. Ladders that cannot be safely repaired must be destroyed or scrapped for parts. **Skill Sheet 10-I-1** contains general procedures for cleaning, inspecting, and maintaining a ladder.

Handling Ladders

NFPA® 1901, *Standard for Automotive Fire Apparatus*, sets the minimum lengths and types of ladders to be carried on all pumper or engine companies. Each pumper must carry the following ladders:

Bedded Position — Extension ladder with the fly section(s) fully retracted.

WARNING!
Failure to remove a defective ladder from service can result in a catastrophic ladder failure that injures or kills firefighters.

- One straight ladder equipped with roof hooks
- One extension ladder
- One attic ladder

The standard does not specify the minimum lengths of these ladders. It does recommend that engines carry a 35-foot (11 m) extension ladder in areas where no ladder trucks are in service.

To be in compliance with NFPA® 1901, aerial apparatus must carry the following ladders:

- One 10-foot (3 m) or longer attic ladder
- Two roof ladders (equipped with hooks) at least 16 feet (4.9 m) long
- One combination ladder at least 14 feet (4.3 m) long
- One 24-foot (7.3 m) or longer extension ladder
- One 35-foot (10.7 m) extension ladder

Ladder Safety

Firefighter safety while carrying, raising, lowering, and working on ladders depends on a number of factors. Among the factors that contribute to safe ladder operation are the following:

- Developing and maintaining adequate upper body strength
- Wearing a full body harness with belay line when training on ladders
- Operating ladders according to departmental training and procedures
- Wearing protective gear, including gloves and helmet, when working with ladders
- Choosing the proper ladder for the job
- Using leg muscles, not back or arm muscles, when lifting ladders below the waist
- Using an adequate number of firefighters for each carry and raise
- Not raising any ladders to within 10 feet (3 m) of electrical wires
- Checking the ladder placement for the proper angle (**Figure 10.11**)
- Being sure that the hooks of the pawls are seated over the rungs (**Figure 10.12**)
- Being sure that the ladder is stable before climbing (both butts in contact, with the ground/roof ladder hooks firmly set)
- Being careful when moving ladders sideways
- Heeling the ladder or securing it at the top
- Climbing smoothly and rhythmically
- Not overloading the ladder (**Figure 10.13**) (one firefighter every 10 feet [3 m] or one per section)
- Tying in to ground ladders with a leg lock or ladder belt when working from the ladder (see Working on a Ladder section)
- Not relocating a positioned ladder unless ordered to do so
- Using ladders for their intended purposes only
- Inspecting ladders for damage and wear after each use

Belay — A climber's term for a safety line.

WARNING!
Use extreme caution with metal ladders near electrical power sources. Contact with power sources may result in electrocution of anyone in contact with the ladder.

WARNING!
Sliding down a ladder either feet first or head first – even in an emergency – is unsafe and may result in serious injury or death.

CAUTION

Set Up Ladder Properly To Reduce Slip And Overload Hazards. Follow These Instructions.

75° (approx.)

① Place Toes Against Bottom Of Ladder Siderails.

② Stand Erect.

③ Extend Arms Straight Out.

④ Palms Of Hands Should Touch Top Of Rung At Shoulder Level.

OUT ←

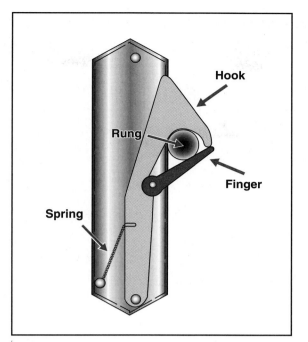

Figure 10.12 The hooks of the pawls must be securely fastened over the rung to prevent the fly section of an extension ladder from moving.

Figure 10.11 This instructional label affixed to most ladders shows in which direction the fly should be facing and helps firefighters ensure that the ladder is at a proper climbing angle.

Figure 10.13 Ladders that are overloaded are unsafe and may fail or be permanently damaged.

Selecting the Proper Ladder for the Job

Before raising ground ladders, firefighters must first select the proper ladder for the given job and then carry it to its location for use. Selecting the location may be affected by the needs of the situation, the ladders available, and wall heights and other building features. It is important that ladders be raised safely and smoothly to avoid injury to firefighters or damage to the ladder. Because speed is often required, movements should be smooth and controlled. Because more than one firefighter may be needed to raise ladders safely and efficiently, teamwork is also important. Individual and team proficiency in handling ladders is developed and maintained only with repetitive training.

Selecting a ladder to reach a specific point sometimes requires the ability to judge distance. The base of the ladder should be placed away from the building approximately one-quarter of the vertical distance from the ground to the point of contact with the wall. This will provide the optimum climbing angle of approximately 75 degrees. Depending upon the height of the foundation and other factors, a residential story averages about 10 feet (3 m), and the distance from the floor to the windowsill averages about 3 feet (1 m). A commercial story averages 12 feet (4 m) from floor to floor, with a 4-foot (1.2 m) distance from the floor to windowsill. **Table 10.1, p. 484** is a general guide that can be used in selecting ladders for specific locations.

Table 10.1
Ladder Selection Guide

Working Location of Ladder	Ladder Length
First-story roof	16 to 20 feet (4.9 m to 6.0 m)
Second-story window	20 to 28 feet (6.0 m to 8.5 m)
Second-story roof	28 to 35 feet (8.5 m to 10.7 m)
Third-story window or roof	40 to 50 feet (12.2 m to 15.2 m)
Fourth-story roof	over 50 feet (15.2 m)

Table 10.2
Maximum Working Heights for Ladders Set at Proper Climbing Angle

Designated Length of Ladder	Maximum Reach
10 foot (3.0 m)	9 feet (2.7 m)
14 foot (4.3 m)	13 feet (4.0 m)
16 foot (4.9 m)	15 feet (4.6 m)
20 foot (6.1 m)	19 feet (5.8 m)
24 foot (7.3 m)	23 feet (7.0 m)
28 foot (8.5 m)	27 feet (8.2 m)
35 foot (10.7 m)	34 feet (10.4 m)
40 foot (12.2 m)	38 feet (11.6 m)
45 foot (13.7 m)	43 feet (13.1 m)
50 foot (15.2 m)	48 feet (14.6 m)

Fascia — Flat horizontal or vertical board located at the outer face of a cornice or a broad flat surface over a storefront or below a cornice.

Parapet —Portion of the exterior walls of a building that extends above the roof. A low wall at the edge of a roof.

Rules of thumb for ladder length include the following:

- Extend the ladder (three to five rungs) beyond the roof edge to provide both a footing and a handhold for anyone stepping on or off the ladder (**Figure 10.14**).
- Place the tip of the ladder about even with the top of the window and to the windward (upwind) side of it to gain access to a narrow window or for ventilation (**Figure 10.15**).
- Place the tip of the ladder just below the windowsill when rescue from a window opening is to be performed (**Figure 10.16**).

NOTE: If a fascia or parapet extends more than 6 feet (2 m) above the roof, an additional ladder should be placed from the top of the fascia/parapet down to the roof to assist firefighters to and from the roof.

The next step is to determine how far various ladders will reach. Knowledge of the designated length of a ladder can help to answer this question. Remember that the designated length (normally shown on the ladder) is a measurement of the maximum extended length (**Figure 10.17**). This is *NOT THE LADDER'S REACH,* because ladders are set at angles of approximately 75 degrees for climbing. Therefore, the reach will be *LESS* than the designated length. One more thing needs to be considered: Single, roof, and folding ladders meeting NFPA® 1931 are required to have a measured length equal to the designated length. In the case of extension ladders, however, the maximum extended length may be as much as 6 inches (150 mm) LESS than the designated length.

Table 10.2, p. 486, provides information on the reach of various ground ladders when placed at the proper climbing angle. The following measurements should be noted when considering the information contained in Table 10.2.

- For lengths of 35 feet (11 m) or less, reach is approximately 1 foot (300 mm) less than the designated length.
- For lengths over 35 feet (11 m), reach is approximately 2 feet (600 mm) less than the designated length.

Ground ladders are mounted on fire apparatus in variety of ways depending on departmental requirements, type of apparatus and body design, type of ladder, type of mounting bracket or rack used, and manufacturer's preferences (**Figures 10.18 a–c,**

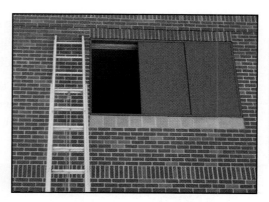

Figure 10.15 Place the ladder tip about even to the windward side when using the window to gain access to a narrow window or for ventilation.

Figure 10.16 When a window is used for rescue, place the ladder tip just below the windowsill.

Figure 10.14 A ladder raised to the roof should extend three to five rungs above the roof level.

p. 486). There are no established standards for the location and mounting of ground ladders on fire apparatus. Differences in how ladders are mounted make it necessary for each fire department to develop its own procedures for removing and replacing ground ladders on apparatus.

Removing and Replacing Ladders on Apparatus

Before removing ground ladders from apparatus, firefighters should be able to answer the following questions:

- What ladders (types and lengths) are carried and where are they carried on the apparatus?

- Are the ladders racked with the butt toward the front or toward the rear of the apparatus?

- Where ladders are nested together, can one ladder be removed leaving the other(s) securely in place? (In particular, can the roof ladder be removed from the side of the pumper and leave the extension ladder securely in place?)

- In what order do the ladders that nest together rack? (Extension ladder goes on first, roof ladder second, or vice versa?)

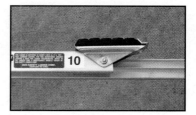

Figure 10.17 The designated ladder length is marked on the outside of both beams at the butt.

- Is the top fly of the extension ladder on the inside or on the outside when the ladder is racked on the side of the apparatus?
- How are the ladders secured?
- Which rungs go in or near the brackets when ladders are mounted on the side of apparatus? (Many departments find it a good practice to mark ladders to indicate when rungs go in or near the brackets as shown in **Figure 10.19**).

Figure 10.18a On most pumpers ground ladders are mounted on the right side of the apparatus.

Figure 10.18c Some ground ladders are mounted vertically in covered compartments in the apparatus and must be loaded/unloaded from the rear.

Figure 10.18b On aerial apparatus, ladders are loaded in a flat position from the rear of the apparatus.

Figure 10.19 Most ladders have marks on the beams to indicate where they should be placed on the apparatus mounting brackets.

Proper Lifting and Lowering Methods

Firefighters may be injured by using improper lifting and lowering techniques. Such injuries are preventable if the following procedures are used:

- Have adequate personnel for the task.

- Bend your knees, keeping your back as straight as possible, and lift with your legs, *NOT WITH YOUR BACK OR ARMS* (**Figure 10.20**).

- When two or more firefighters are lifting a ladder, they should lift on the command of a firefighter at the butt position who can see the entire operation (**Figure 10.21**). Any firefighter who is not ready should make it known immediately so that the lift will be delayed. Lifting must be done in unison.

- Reverse the procedure for lifting when it is necessary to place a ladder on the ground before raising it. Lower the ladder using your leg muscles. Also, be sure to keep your body perpendicular and feet parallel to the ladder so that when the ladder is placed it does not rest on your toes (**Figure 10.22**).

Ladder Carries

Once the ladder has been removed from its mounting, there are numerous ways it can be transported. The procedures for removing ladders when the ladders are mounted in a flat ladder bed (such as on a ladder truck) differ from those when the ladders are mounted on the side or top of an engine. Removal methods must reflect the indi-

Figure 10.21 When two or more firefighters lift a ladder, the firefighter at the butt gives the command to lift.

Figure 10.20 To prevent back injuries, always lift with your legs and not your back.

Figure 10.22 As the ladder is lowered to the ground, keep your body and toes parallel to the ladder.

vidual situation. Because there are many different types of apparatus and means of mounting ladders, all carries in this section are demonstrated from the ground. In most cases, the ladders are carried butt forward.

One-Firefighter Low-Shoulder Carry

Some single or roof ladders may be safely carried and raised by one firefighter. The low-shoulder carry involves resting the ladder's upper beam on the firefighter's shoulder, while the firefighter's arm goes between two rungs **(Figure 10.23)**. **Skill Sheet 10-I-2** shows the steps for performing the one-firefighter low-shoulder carry.

Two-Firefighter Low-Shoulder Carry

Although the two-firefighter low-shoulder carry may be used with single or roof ladders, it is most commonly used for 24-, 28- and 35-foot (8 m, 9 m, and 11 m) extension ladders. The two-firefighter low-shoulder carry gives firefighters excellent control of the ladder **(Figure 10.24)**. The forward firefighter places the free hand over the upper butt spur. This is done to prevent injury in case there is a collision with someone while the ladder is being carried. **Skill Sheet 10-I-3** describes the two-firefighter low-shoulder carry.

Three-Firefighter Flat-Shoulder Carry

The three-firefighter flat-shoulder carry is typically used on extension ladders up to 35 feet (11 m). This method has two firefighters, one at each end on one side of the ladder, and one firefighter on the other side in the middle **(Figure 10.25)**. **Skill Sheet 10-I-4** shows the procedure for carrying the ladder using the three-firefighter flat-shoulder carry.

Four-Firefighter Flat-Shoulder Carry

The same flat-shoulder method used by three firefighters for carrying ladders is used by four firefighters except that there is a change in the positioning of the firefighters to accommodate the fourth firefighter. When four firefighters use the flat-shoulder carry, two are positioned at each end of the ladder, opposite each other **(Figure 10.26)**.

Two-Firefighter Arm's Length On-Edge Carry

The two-firefighter arm's length on-edge carry is best performed with lightweight ladders **(Figure 10.27, p. 490)**. The two-firefighter arm's length on-edge carry is based on the fact that the firefighters are positioned on the bed section (widest) side of the ladder when it is in the vertical position.

Special Procedures for Carrying Roof Ladders

The procedures previously described are for carrying ladders butt forward. In some cases, a firefighter will carry a roof ladder with the intention of climbing another ground ladder and placing the roof ladder with hooks deployed on a sloped roof. In this situation, the firefighter should use the low-shoulder method and have the tip (hooks) forward **(Figure 10.28, p. 490)**.

Normally, the roof ladder is carried with the hooks closed to the foot of the first ladder. The firefighter sets the ladder down, moves to the tip, picks up the tip, opens the hooks, lays down the tip, returns to the midpoint, picks up the ladder, and re-

Figure 10.23 In a one-firefighter low-shoulder carry, the top beam rests on the firefighter's shoulder at the midpoint of the ladder.

Figure 10.24 In the two-firefighter low-shoulder carry, the butt end of the ladder is advanced toward the building.

Figure 10.25 The ladder is carried in a horizontal position in the three-firefighter flat-shoulder carry.

Figure 10.26 The four-firefighter flat-shoulder carry is used for extension ladders that are greater than 35 feet (11 m) in length.

Figure 10.27 Lightweight ladders can be easily carried with the two-firefighter arm's length on-edge carry.

Figure 10.28 The roof ladder may be carried with the tip forward and slightly depressed.

Figure 10.29 Open the roof hooks away from your body.

Figure 10.30 The roof ladder is carried with the hooks open facing outward.

sumes the carry (**Figure 10.29**). Or, the hooks may be opened at the apparatus before the carry is begun; they are turned outward in relation to the firefighter carrying the ladder (**Figure 10.30**).

Positioning Ground Ladders

Proper positioning, or placement, of ground ladders is important because it affects the safety and efficiency of the operations. The following sections contain some of the basic considerations and requirements for ground ladder placement.

Responsibility for Positioning

Normally, an officer designates the general location where the ladder is to be positioned and the task is to be performed, but personnel carrying the ladder frequently decide on the exact spot where the butt is to be placed. The firefighter nearest the butt is the logical person to make this decision because this end is placed on the ground to initiate raising the ladder. When there are two firefighters at the butt, the one on the

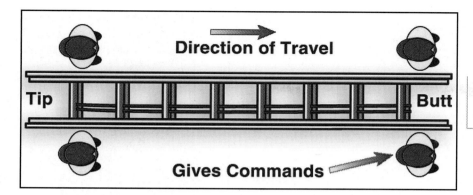

Figure 10.31 When multiple firefighters carry a ladder, the firefighter on the right side of the butt gives the commands.

right side is usually the one responsible for placement **(Figure 10.31)**. Because this guideline this may vary from one department to another, firefighters must always follow their department's protocols.

Factors Affecting Ground Ladder Placement

When placing ladders, there are two objectives to be met:

1. Position the ladder properly for its intended use.

2. Place the butt the proper distance from the building for safe and easy climbing.

There are numerous factors that dictate the exact place to position the ladder. If a ladder is to be used for positioning a firefighter to break a window for ventilation, place it alongside the window to the windward (upwind) side. The tip should be about even with the top of the window **(Figure 10.32)**. The same position can be used when firefighters need to climb in or out of narrow windows or direct hose streams into them.

If a ladder is to be used for entry or rescue from a window, usually the ladder tip is placed slightly below the sill **(Figure 10.33, p. 492)**. If the sill projects out from the wall, the tip of the ladder can be wedged under the sill for additional stability **(Figure 10.34, p. 492)**. If the window opening is wide enough to permit the ladder tip to project into it and still allow room beside it to facilitate entry and rescue, place the ladder so that two or three rungs extend above the sill **(Figure 10.35, p. 492)**.

Figure 10.32 Place the tip adjacent to the top of the window opening on the windward side.

Other ladder placement guidelines include the following:

- Ladder at least two points on different sides of the building **(Figure 10.36, p. 492)**.

- Avoid placing ladders over openings such as windows and doors where they might be exposed to heat or direct flame contact **(Figure 10.37, p. 493)**.

- Take advantage of strong points in building construction (such as the corners) when placing ladders.

- Raise the ladder directly in front of the window when it is to be used as a support for a smoke ejector removing cold smoke after a fire has been extinguished. Place the ladder tip on the wall above the window opening.

Smoke Ejector — Gasoline, electrically, or hydraulically driven blower (ducted fan) device used primarily to expel (eject) smoke from burning buildings although it is sometimes used to blow fresh air into a building to assist in purging smoke or other contaminants. May be used in conjunction with a flexible duct.

Figure 10.33 When the ladder is used for rescue from a window, place the tip just below the lower sill.

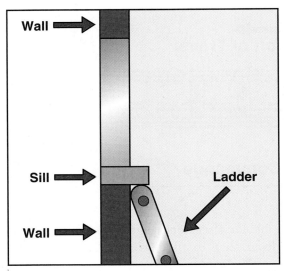

Figure 10.34 Additional stability is obtained by wedging the tip under the windowsill.

Figure 10.35 If the window is large enough, extend the ladder two to three rungs into the opening if possible.

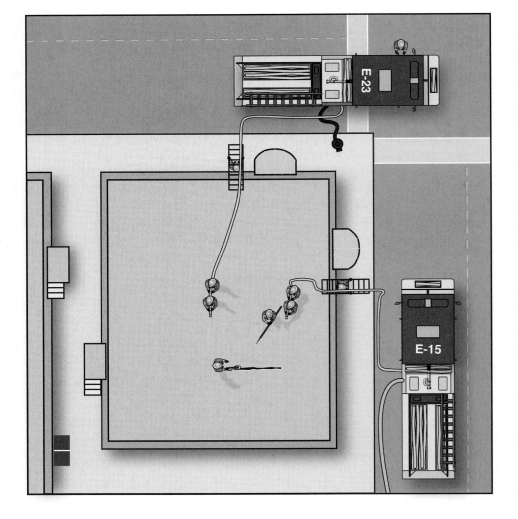

Figure 10.36 Ladder the building at two points on different sides of the building.

- Avoid placing ladders where they may come into contact with overhead obstructions such as wires, tree limbs, or signs (**Figure 10.38**).
- Avoid placing ladders on uneven terrain or on soft spots.
- Avoid placing ladders in front of doors or other paths of travel that firefighters or evacuees will need to use (**Figure 10.39**). Instead, place the ladder to the side of the opening.
- Avoid placing ladders where they may be exposed to flames or excessive heat.
- Avoid placing ladders on top of sidewalk elevator trapdoors or sidewalk deadlights. These areas may give way under the added weight of firefighters, their equipment, and the ladder (**Figure 10.40**).
- Do not place ladders against unstable walls or surfaces.

Figure 10.37 Do not place ladders in front of doors or windows where the ladder could be exposed to heat or direct flame contact.

Figure 10.38 Always be aware of overhead obstructions, such as power lines, wires, signs, or tree limbs.

Figure 10.40 Ladders should not be placed on top of trapdoors, grates, and manholes located on sidewalks or streets.

Figure 10.39 Do not place ladders in front of doors that may be used for entrance and egress during the operation.

Figure 10.41 The desired climbing angle for a ladder is 75 degrees.

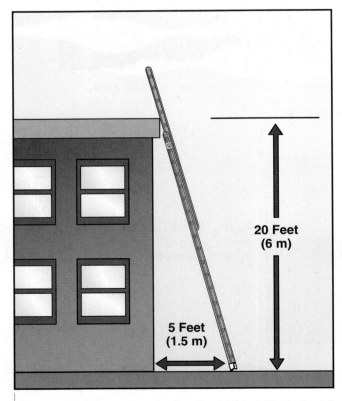

Figure 10.42 A ladder that is raised 20 feet (6 m) should have the base 5 feet (1.5 m) from the building.

With the exception of certain rescue situations, when the ladder has been raised and lowered into place, the desired angle of inclination is approximately 75 degrees (**Figure 10.41**). This angle provides good stability, places less stress on the ladder, provides the optimum climbing angle, and allows easy climbing because it permits the climber to stand perpendicular to the ground, at arm's length from the rungs. The distance of the butt end from the building establishes the angle formed by the ladder and the ground. If the butt is placed too close to the building, its stability is reduced because climbing tends to cause the tip to pull away from the building.

If the butt of the ladder is placed too far away from the building, the load-carrying capacity of the ladder is reduced, and it has more of a tendency to slip. If placement at such angles becomes necessary, either tie in or heel (steady) the bottom of the ladder at all times. (See Securing the Ladder section for tying in and heeling instructions.)

An easy way to determine the proper distance between the heel of the ladder and the building is to divide the working length (length actually used) of the ladder by 4. For example, if 20 feet (6 m) of a 28-foot (8.5 m) ladder is needed to reach a window, the butt end should be placed 5 feet (1.5 m) from the building (20 feet divided by 4 [6 m divided by 4]) (**Figure 10.42**). Note that only the length of ladder *used* to reach the window, not the ladder's overall length, is used in this calculation. Exact measurements are unnecessary on the fire scene.

Trained and experienced firefighters develop the ability to visually judge the proper positioning for the ladder. The proper angle can also be checked by standing on the bottom rung and reaching out for the rung in front. A firefighter should be able to grab the rung while standing straight up, with arms extended straight out (**Figure 10.43**). Newer ladders are equipped with an inclination marking on the outside of the beam whose lines become perfectly vertical and horizontal when the ladder is properly set (**Figure 10.44**).

General Procedures for Raising and Climbing Ladders

A properly positioned ladder becomes a means by which important rescue and fire fighting operations can be performed. To be most effective, teamwork, smoothness, and rhythm are necessary when raising and lowering fire department ladders. Before learning the techniques of raising ladders, firefighters should be aware of certain general procedures that affect these operations.

Figure 10.43 Check the climbing angle by standing on the bottom rung and reaching for the rung directly at shoulder level.

Figure 10.44 Labels affixed to the beams help firefighters achieve a proper climbing angle.

Transition from Carry to Raise

The methods and precautions for raising single and extension ladders are much the same. With the exception of pole ladders, it is not necessary to place the ladder flat on the ground prior to raising; only the butt needs to be placed on the ground (**Figure 10.45**). The transition from the carrying position to the raise can and should be done in one smooth and continuous motion.

Before raising a ladder, there are a number of things firefighters need to consider and precautions they must take. Some of the more important ones are contained in the sections that follow.

Electrical Hazards

A major concern when raising ladders is possible contact with live electrical wires or equipment, either by the ladder or by the persons who have to climb it. The danger with metal ladders has been stressed previously.

To avoid electrical contact hazards, care must be taken *BEFORE BEGINNING A RAISE* (**Figure 10.46, p. 496**). Look up to check for overhead electrical wires or equipment before making the final selection on where to place a ladder or what method to use for raising it. The Occupational

Figure 10.45 Place the butt of the ladder on the ground and raise the ladder in one smooth motion.

Safety and Health Administration (OSHA) requires that all ladders maintain a distance of at least 10 feet (3 m) from all energized electrical lines or equipment. This distance must be maintained at all times — raising the ladder, using the ladder, and lowering the ladder. In some cases, a ladder may come to rest a safe distance from electrical equipment but come too close to the equipment during the actual raise **(Figure 10.47 a)**. In these cases, an alternate method for raising the ladder, such as raising parallel to the structure as opposed to perpendicular, may be required **(Figure 10.47 b)**.

Position of the Fly Section on Extension Ladders

Each ladder manufacturer specifies whether the ladder should be placed with the fly in or out. This recommendation is based on the design of the ladder, the materials of which it is made, and the fly position at which the manufacturer's tests show it to be strongest. Failure to follow this recommendation could void the ladder's warranty if a failure or damage occurs.

In general, all modern metal and fiberglass ladders are designed to be used with the *FLY OUT* (away from the building) **(Figure 10.48)**. Wooden ladders that are designed with the rungs mounted in the top truss rail (the only type of wooden ladder

Figure 10.46 Before raising a ladder, always check for overhead electrical hazards such as wires and lights.

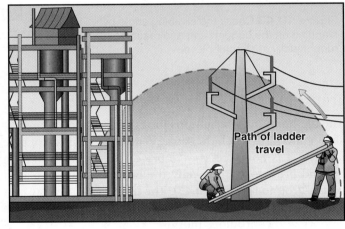

Figure 10.47 a Overhead electrical power lines may prevent the ladder from being raised perpendicular to the structure.

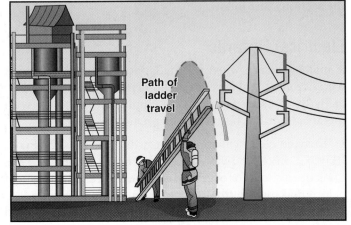

Figure 10.47 b Raising the ladder parallel to the wall of the building may avoid overhead electrical power lines.

still manufactured today) are intended to be used with the *FLY IN* (**Figure 10.49**). Again, consult departmental SOPs or the manufacturer of the ladder to determine the correct fly position.

Some departments have ladders that are intended to be used with the fly out but prefer that the firefighter extending the halyard be on the outside of the ladder. In this case, firefighters will need to pivot or roll the ladder 180 degrees (discussed later) after it has been extended.

Tying the Halyard

Once an extension ladder is resting against a building and before it is climbed, the excess halyard should be tied to the ladder with a clove hitch and an overhand safety (**Figure 10.50, p. 498**). This prevents the fly from slipping and prevents anyone from tripping over the rope. In rescue situations where speed is critical, it is not always necessary to wrap the excess halyard before tying off but it should be put out of the way. The same tie can be used for either a closed- or open-ended halyard. **Skill Sheet 10-I-5** describes the procedure for tying the halyard.

Figure 10.48 Metal extension ladders are deployed with the bed section toward the building and the fly section away from the building.

Figure 10.49 Wooden extension ladders are deployed with the fly toward the building.

Figure 10.50 The halyard should be secured to a rung with a clove hitch and safety.

Ladder Raises

There are numerous ways to safely raise ground ladders. These methods vary depending on the type and size of the ladder, number of personnel available to perform the raise, and weather and topography considerations. The raises discussed here represent only some of the more commonly used methods; there are many more.

NOTE: The following sections contain step-by-step information only for raising ladders. In every case, the procedure for lowering the ladder is to reverse the listed steps in the order given.

One-Firefighter Raise

One firefighter may safely raise single ladders and small extension ladders. The following procedures should be used to perform these raises.

One-Firefighter Single Ladder Raise

Single and roof ladders are generally light enough that one firefighter with sufficient upper body strength can usually place the butt end at the point where it will be located for climbing without heeling (steadying) it against the building or another object before raising (**Figure 10.51**). The steps described in **Skill Sheet 10-I-6** should be used to perform the one-firefighter raise.

One-Firefighter Extension Ladder Raise

One method of raising extension ladders with one firefighter is from the low-shoulder carry. When using the one-firefighter raise from the low-shoulder carry, the placement of the butt is important. In this instance, a building is used to heel the ladder to prevent the ladder butt from slipping while the ladder is brought to the vertical position (**Figure 10.52**).

Two-Firefighter Raises

Space permitting, it makes little difference if a ladder is raised parallel with or perpendicular to a building. If raised parallel with the building, the ladder must be pivoted after it is in the vertical position. Whenever two or more firefighters are involved in raising a ladder, the firefighter at the butt end, the *heeler*, is responsible for placing it at the desired distance from the building and determining whether the ladder will be raised parallel with or perpendicular to the building. He or she also gives commands during the operation. There are two basic ways for two firefighters to raise a

Figure 10.51 Place the butt on the ground at an appropriate distance from the building.

Figure 10.52 One firefighter can raise an extension ladder by placing the butt against the building while bringing it to the vertical position.

ladder: the flat raise and the beam raise. **Skill Sheet 10-I-7** describes the procedure for the two-firefighter flat raise. **Skill Sheet 10-I-8** shows the procedure for the two-firefighter beam raise.

Three-Firefighter Flat Raise

As the length of the ladder increases, the weight also increases. This means that more personnel are needed for raising the larger extension ladders (**Figure 10.53, p. 500**). Typically, ladders of 35 feet (11 m) or longer should be raised by at least three firefighters. Skill **Sheet 10-I-9** describes the procedure for flat-raising ladders with three or four firefighters.

To raise a ladder using the beam method with three firefighters, follow the same procedures for the two-firefighter flat raise. The only difference is that the third firefighter is positioned along the beam (**Figure 10.54, p. 500**. Once the ladder has been raised to a vertical position, follow the procedures described for the flat raise.

Four-Firefighter Flat Raise

When available, four firefighters can be used to better handle the larger and heavier extension ladders (**Figure 10.55, p. 500**). A flat raise is normally used, and the procedures for raising the ladder are similar to the three-firefighter raise except for the placement of personnel. A firefighter at the butt is responsible for placing the butt at the desired distance from the building and determining whether the ladder will be raised parallel with or perpendicular to the building. **Skill Sheet 10-I-9** describes the procedure for flat-raising ladders with three or four firefighters.

Placing a Roof Ladder

There are a number of ways to get a roof ladder in place on a pitched roof. Once a firefighter has carried the roof ladder to the location, it can be placed by either one or two firefighters. **Skill Sheet 10-I-10** shows the procedure for one firefighter to place a roof ladder in position.

Figure 10.53 More personnel are required to raise extension ladders that are greater than 35 feet (11 m).

Figure 10.54 In a three-firefighter beam raise, one person steadies the butt of the ladder while the other two raise it.

Figure 10.55 A four-firefighter flat raise is used for larger extension ladders due to increased weight and length.

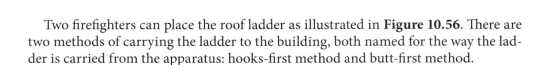

Two firefighters can place the roof ladder as illustrated in **Figure 10.56**. There are two methods of carrying the ladder to the building, both named for the way the ladder is carried from the apparatus: hooks-first method and butt-first method.

Procedures for Moving Ground Ladders

In some cases, the basic ladder-raising procedures previously described are not sufficient to get the ladder into its final position for use. In these cases it will be necessary to move the ladder slightly after it has been extended.

Figure 10.56 Two firefighters can easily deploy a roof ladder on a pitched roof.

Pivoting Ladders with Two Firefighters

Occasionally an extension ladder is raised with the fly in the incorrect position for deployment. When this happens, it is necessary to pivot the ladder. Any ladder flat-raised parallel to the building also requires pivoting to align it with the wall upon which it will rest. Use the beam closest to the building for the pivot. Whenever possible, pivot the ladder before it is extended.

The two-firefighter pivot may be used on any ground ladder that two firefighters can raise (**Figure 10.57**). The procedure described in **Skill Sheet 10-I-11** is for a ladder that must be turned 180 degrees to get the fly section in the proper position. The same procedure is used for positioning a ladder that was flat-raised parallel to the building. In this case, the beam nearest the building is used to pivot the ladder 90 degrees.

Shifting Raised Ground Ladders

Occasionally, circumstances require that ground ladders be moved while vertical. Because they are hard to control, shifting a ladder that is in a vertical position should be limited to short distances such as aligning ladders perpendicular to a building or to an adjacent window.

One firefighter can safely shift a single ladder that is 20 feet (6 m) long or less. The procedure for the one-firefighter shift is described in **Skill Sheet 10-I-12**. Because of their weight, extension ladders require two firefighters for the shifting maneuver described in **Skill Sheet 10-I-13**. Another way to shift a ladder a *short distance* from side to side is to lay the ladder into the building, slide the top of the ladder sideways, then pick up the butt and move it into position.

Securing the Ladder

For safety, ground ladders must be secured whenever firefighters are climbing or working from them. The process of securing a ground ladder may include any or all of the following:

Figure 10.57 Positioning a ladder may require that two firefighters pivot the ladder before placing it against the building.

- Make sure the ladder locks are locked (extension ladders only). This should have already been done before the ladder was placed against the structure.
- Tie the halyard with a clove hitch and an overhand safety (extension ladder only).
- Prevent movement of the ladder away from the building by heeling and/or tying in.

The two methods of securing a ladder discussed in this section are *heeling* and *tying in*.

Heeling

One way of preventing movement of a ladder is to properly heel, or foot, it. There are several methods for properly heeling a ladder.

One method is for a firefighter to stand beneath the ladder with feet about shoulder-width apart (or one foot slightly ahead of the other). The firefighter then grasps the ladder beams at about eye level and pulls backward to press the ladder against the building **(Figure 10.58)**. When using this method, the firefighter heeling the ladder must wear head and eye protection and not look up when there is someone climbing the ladder. The firefighter must be sure to grasp the beams and not the rungs, and stay alert for falling debris.

Another method of heeling a ladder is for a firefighter to stand on the outside of the ladder and chock the butt end with one foot **(Figures 10.59 a and b)**. With this method, either the firefighter's toes are placed against the butt spur or one foot is placed on the bottom rung. The firefighter grasps the beams and presses the ladder against the building. The firefighter heeling the ladder must stay alert for firefighters descending the ladder.

Figure 10.58 To ensure that the ladder does not move, one firefighter in full protective clothing should anchor the ladder by grasping the beams in both hands.

Figure 10.59 a The ladder may be heeled by standing in front of the ladder with one foot placed against the beam at the butt.

Figure 10.59 b The ladder may also be held with a foot on the bottom rung.

Tying In

Whenever possible, a ladder should be tied securely to a fixed object. Tying in a ladder is simple, can be done quickly, and is strongly recommended to prevent the ladder from slipping or pulling away from the building. Tying in also frees personnel who would otherwise be holding the ladder in place. A rope hose tool or safety strap can be used between the ladder and a fixed object (**Figures 10.60 a and b**).

Figure 10.60 a If personnel are not available to heel the ladder, it may be tied off to the structure near the bottom.

Figure 10.60 b The ladder should be secured to the building at the top to prevent it from being moved while personnel are working on the roof.

Climbing Ladders

Ladder climbing should be done smoothly and rhythmically. You should ascend the ladder so that there is the least possible amount of bounce and sway. This smoothness is accomplished if your knee is bent to ease the weight on each rung. Balance on the ladder will come naturally if your body is perpendicular to the ground. This occurs when the ladder is properly spaced from the building to create the optimum climbing angle.

The climb may be started after the climbing angle has been checked and the ladder is properly secured. Your eyes should be focused forward, with an occasional glance at the tip of the ladder. Keep your arms straight (horizontal) during the climb; this keeps your body away from the ladder and permits free knee movement during the climb (**Figure 10.61, p. 504**). When no equipment is being carried, place your hands on the rungs. Grasp the rungs with palms down and your thumbs beneath the rungs. Grasp alternate rungs while climbing (**Figure 10.62, p. 504**). Coordinate hand and foot movement so that the right hand and left foot are in contact with the ladder as you move the opposite hand and foot to the next rungs (**Figure 10.63, p. 504**).

If your feet should slip, your arms and hands are in a position to stop the fall. Progress upward using your leg muscles, not your arm muscles. Your arms and hands should not reach above your head while climbing because that will bring your body too close to the ladder.

Figure 10.61 Proper ladder-climbing techniques include keeping both the back and arms straight while climbing.

Figure 10.63 Move the opposite hand and foot at the same time.

Figure 10.62 Grasp alternate rungs while climbing.

Practice climbing slowly to develop form rather than speed. Speed develops with repetition after the proper technique is mastered. Too much speed results in lack of body control, and quick movements cause the ladder to bounce and sway.

Firefighters are often required to carry equipment up and down a ladder during fire fighting. This disrupts your natural climbing motion either because of the added weight on your shoulder or the need to use one hand to hold the tool. If a tool is carried in one hand, it may be desirable to slide your free hand under the beam while making the climb (**Figure 10.64**). This method permits constant hand contact with the ladder. Whenever possible, a utility rope should be used to hoist tools and equipment rather than carrying them up a ladder. One way to carry a pike pole up a ladder is to hook it over a rung as far above as you can reach. When you have climbed up to the level of the hook, repeat the process as necessary.

Working from a Ladder

Firefighters must sometimes work with both hands while standing on a ground ladder. Either a ladder belt or a leg lock can be used to safely secure the firefighter to a ladder while performing work (**Figure 10.65 a**). If a firefighter chooses to apply a leg lock on a ground ladder, the procedure in **Skill Sheet 10-I-14** should be used.

> **WARNING!**
> Do not exceed the rated capacity of a ladder. To avoid overloading the ladder, allow only one firefighter on each section of a ladder at the same time. Be careful about stressing ladders laterally.

Figure 10.64 When carrying a tool in one hand, slide the other hand up the underside of the beam.

Figure 10.65 a A leg lock can be used to secure a firefighter to the ladder while working.

Figure 10.65 b A ladder or safety belt may be used to secure a firefighter to the ladder while working from the ladder.

If a ladder belt is used, it must be strapped tightly around the waist. The hook may be moved to one side, out of the way, while a firefighter is climbing a ladder. After reaching the desired height, the firefighter returns the hook to the center and attaches it to a rung (**Figure 10.65 b**). According to NFPA® 1983, *Standard on Life Safety Rope and Equipment for Emergency Services*, a ladder belt is rated only as a positioning device for use on a ladder and does not meet the requirements of life safety harnesses.

Assisting a Victim Down a Ladder

When a ground ladder is intended to be used for rescue through a window, the ladder tip is raised to just below the sill. This makes it easier for a conscious victim to climb onto the ladder and for firefighters to lift an unconscious victim onto the ladder. The ladder is heeled, and all other loads and activity removed from it during rescue operations. Because even healthy, conscious occupants are probably unaccustomed to climbing down a ladder, they must be protected from slipping and perhaps falling. To bring victims down a ground ladder, at least four firefighters are needed: two inside the building, one or two on the ladder, and one heeling the ladder. The following evolutions can be performed with the ladder set at the normal climbing angle; however, these techniques work better if the ladder is set at a slightly steeper angle. The procedures for removing conscious and unconscious victims down ladders are given in **Skill Sheets 10-I-15 and 10-I-16**.

Several methods for lowering conscious or unconscious victims are as follows:

- Conscious victims can be lowered feet first (facing the building) onto a ladder (**Figure 10.66, p. 506**).

- An unconscious victim can be held on a ladder in the same way as a conscious victim except that the victim's body rests on the rescuer's supporting knee. (**Figure 10.67**). The victim's feet must be placed outside the rails to prevent entanglement. The rescuer grasps the rungs to provide a secure hold on the ladder and help to protect the victim's head from hitting the ladder.

WARNING!
Never use a leg lock on an aerial ladder. Extending or retracting the ladder could result in serious injury.

- Another way to lower an unconscious victim involves using the same hold by the rescuer described in the previous bullet, but the victim is turned around to face the rescuer **(Figure 10.68)**. This position reduces the chances of the victim's limbs catching between the rungs.

- An unconscious victim (facing the rescuer) is supported at the crotch by one of the rescuer's arms and at the chest by the other arm **(Figure 10.69)**. The rescuer may be aided by another firefighter.

- Another method of removing extraordinarily heavy victims requires two rescuers. Two ground ladders are placed side by side. One rescuer supports the victim's waist and legs. A second rescuer on the other ladder supports the victim's head and upper torso **(Figure 10.70)**.

- Small children who must be brought down a ladder can be cradled across the rescuer's arms **(Figure 10.71)**.

Figure 10.66 A conscious victim can be lowered feet first, facing the ladder, with the firefighter supporting the victim.

Figure 10.67 Unconscious victims are lowered down the ladder with their weight resting on the firefighter's knee.

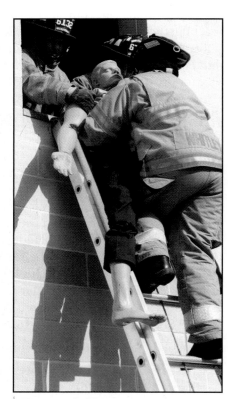

Figure 10.68 To prevent an unconscious victim's arms or legs from becoming caught in the ladder rungs, the victim is positioned facing the firefighter.

Figure 10.69 Another method to rescue an unconscious victim is for the firefighter to place one arm under the victim's arm and the other arm under the opposite leg.

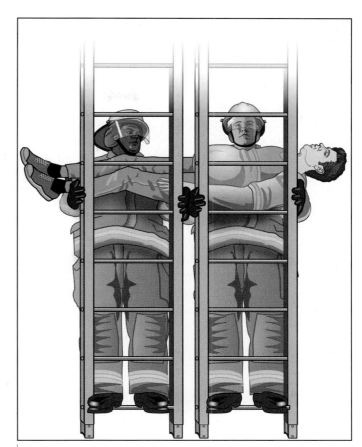

Figure 10.70 It may be necessary to use two ladders and two rescuers working side by side to remove an extraordinarily heavy victim.

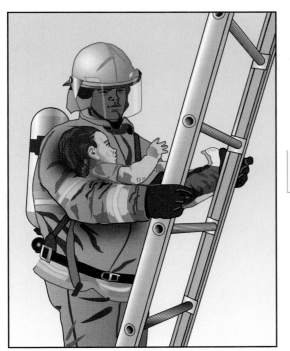

Figure 10.71 A small child is cradled across the firefighter's arms.

Summary

To be an effective and fully contributing member of your department, you must be able to safely carry, raise, extend, climb, and lower fire service ground ladders when needed. These ladders may be needed for fire fighting operations, rescues, or both. To use ladders safely and effectively, you must know the types of ladders available to you, along with their capabilities and limitations. You must know the parts of a ladder, the hazards associated with setting up ground ladders, what constitutes a stable foundation for ladder placement, proper angles for various ladder applications, safe limits related to degree of angulation, and what constitutes a reliable structural component against which a ladder can be placed. You must have all of this knowledge in order to safely use fire service ground ladders as well as know how to clean and inspect them after use. **Skill Sheet 10-I-17** describes procedures for selecting, carrying, and raising ladders properly.

Fire Fighter I

1. Describe the following types of ladders: roof ladders, folding ladders, extension ladders, combination ladders, and pompier ladders.

2. What are the advantages and disadvantages of metal, wood, and fiberglass construction for ladders?

3. List general maintenance guidelines that apply to all types of ground ladders.

4. What items should be checked when inspecting all types of ladders?

5. List four factors that contribute to safe ladder operation.

6. What questions should firefighters be able to answer before removing ground ladders from apparatus?

7. What procedures should be followed when lifting and lowering ladders?

8. List three ladder placement guidelines.

9. Describe methods of heeling a ladder.

10. What is the proper procedure for climbing a ladder?

Cleaning

Step 1: Place the ladder flat on the sawhorses, lifting and carrying appropriately.

Step 2: Clean all parts of the ladder with scrub brush and cleaning solution, removing greasy residues with approved cleaners.

Step 3: Rinse the ladder thoroughly with clean water.

Step 4: Dry the ladder thoroughly with clean, dry cloths.

Tool Maintenance

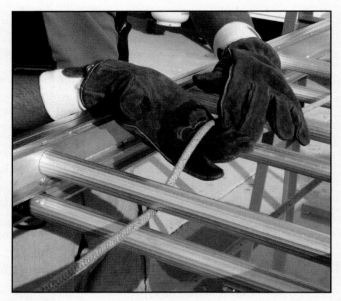

Step 5: Inspect each part of the ladder.

Step 6: Circle any defects found with chalk or grease pen.

Step 7: Inspect the ladder halyard (extension ladders).

Step 8: Inspect all movable parts (extension, roof, and pole ladders).

Maintenance

Step 9: Lubricate parts as needed and per manufacturer's guidelines.

Step 10: Replace halyard if necessary.

Step 11: Tag and remove from service for any conditions that cannot be corrected with cleaning, inspection, and simple maintenance. Notify officer.

Step 12: Record cleaning, inspection, and maintenance performed.

10-I-2
Carry a ladder — One-firefighter low-shoulder method.

Step 1: Position yourself at lifting point near the center of the ladder.

Step 2: Kneel beside the ladder.

Step 5: Stand up.

Step 6: Reposition yourself for carrying by pivoting toward ladder tip and inserting other arm through rungs.

Step 3: Grasp the ladder rung opposite your knee.

Step 4: Stand the ladder on edge.

Step 7: Position ladder for carrying with tip lowered slightly.

Step 8: Lower the ladder to the ground.

NOTE: Firefighter #1 is located near the butt end of the ladder. Firefighter #2 is located near the tip of the ladder.

Step 1: *Both Firefighters:* Kneel on the same side of the ladder facing the tip.

Step 2: *Both Firefighters:* Gasp a convenient rung with the near hand, palm forward.

Step 4: *Firefighter #1:* Give the command to "shoulder the ladder."

Step 5: *Both Firefighters:* Stand, using the leg muscles to lift the ladder.

Step 6: *Both Firefighters:* Tilt the far beam upward as the ladder and the firefighters rise.

Step 7: *Both Firefighters:* Pivot and place the free arm between two rungs with both firefighters facing the butt.

Step 3: *Both Firefighters:* Stand the ladder on edge.

Step 8: *Both Firefighters:* Place the upper beam on the shoulders.

10-I-4
Carry a ladder — Three-firefighter flat-shoulder method.

Skill SHEETS

Step 1: *Firefighters #1 & #2:* Kneel on one side of the ladder, one at either end, facing the tip.

Step 2: *Firefighter #3:* Kneel on the opposite side at midpoint, also facing the tip end.

Step 3: *All Firefighters:* Stand and lift the ladder.

Step 4: *All Firefighters:* Pivot toward the butt when the ladder is about chest high.

Step 5: *All Firefighters:* Place the beam onto the shoulders.

Step 1: Wrap the excess halyard around two convenient rungs.

Step 2: Pull the halyard taut.

Step 3: Hold the halyard between the thumb and forefinger with the palm down.

Step 4: Turn the hand palm up.

Step 5: Push the halyard underneath and back over the top of the rung.

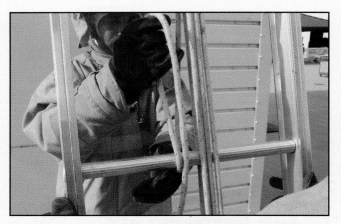

Step 6: Grasp the halyard with the thumb and fingers.

Step 7: Pull it through the loop, making a clove hitch.

Step 8: Finish the tie by making a half-hitch or overhand safety on top of the clove hitch.

CAUTION: Begin all ground ladder raises by first looking for overhead obstructions.

Single Ladder

Step 1: Visually inspect the work area.

Step 2: Lower the ladder butt to the ground with butt spurs against building wall.

Step 3: Position yourself to raise the ladder.

Step 4: Bring the ladder upright until it rests against the building using hand-over-hand method.

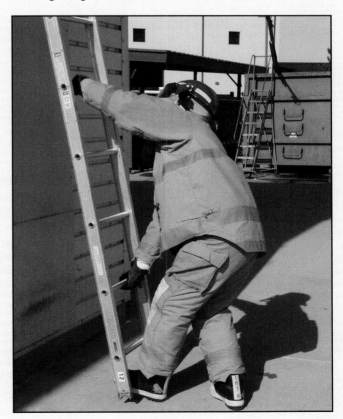

Step 5: Carefully move the ladder butt out from the building to the desired climbing angle.

Step 6: Lower the ladder, reversing the raising procedure.

Extension Ladder

Step 1: Visually inspect the work area.

Step 2: Lower the ladder butt to the ground with butt spurs against building wall and fly in.

Step 3: Position yourself to raise the ladder.

Step 4: Bring the ladder upright until it rests against the building using hand-over-hand method.

Step 5: Pull the ladder away from the building.

10-I-6 continued
Raise a ladder — One-firefighter method.

Skill SHEETS

Extension Ladder

Step 6: Balance ladder in a vertical position, one foot at butt of one beam, and ladder steadied with instep, knee, and leg.

Step 7: Extend the fly section. Pull halyard straight down to maintain ladder balance.

Skill SHEETS

10-I-6 continued
Raise a ladder — One-firefighter method.

Extension Ladder

Step 8: Engage the ladder locks at the desired elevation.

Step 9: Pivot the ladder if necessary until the fly faces out.

Step 10: Lower the ladder against the building.

Step 11: Tie off the halyard.

Step 12: Pull the ladder butt out from the building.

Step 13: Secure the ladder for climbing.

Step 14: Lower the ladder, reversing the raising procedure.

NOTE: Firefighter #1 is located near the butt end of the ladder. Firefighter #2 is located near the tip end of the ladder.

CAUTION: Visually check the area overhead for obstructions before bringing the ladder to a vertical position. Before stepping forward, visually check the terrain.

Step 1: *Firefighter #1:* Place the butt end on the ground.

Step 2: *Firefighter #2:* Rest the ladder beam on a shoulder.

Step 3: *Firefighter #1:* Heel the ladder by standing on the bottom rung.

Step 4: *Firefighter #1:* Crouch down to grasp a convenient rung or the beams with both hands.

Step 5: *Firefighter #1:* Lean back.

Step 6: *Firefighter #2:* Step beneath the ladder.

Step 7: *Firefighter #2:* Grasp a convenient rung with both hands.

Step 8: *Firefighter #2:* Advance hand-over-hand down the rungs toward the butt end until the ladder is in a vertical position.

Step 9: *Firefighter #1:* Grasp successively higher rungs or higher on the beams as the ladder comes to a vertical position until standing upright.

Step 10: *Both Firefighters:* Face each other.

Step 11: *Both Firefighters:* Heel the ladder by placing toes against the beams.

NOTE: When raising extension ladders, pivot the ladder to position the fly away from the building (fly in for wooden ladders) if it is not already in that position.

Step 12: *Firefighter #1:* Grasp the halyard.

Step 13: *Firefighter #1:* Extend the fly section with a hand-over-hand motion until the tip reaches the desired elevation.

Step 14: *Firefighter #2:* Grasp the beams.

Step 15: *Both Firefighters:* Lower the ladder gently onto the building.

NOTE: If the ladder has not yet been turned to position the fly in the out position, it can be done at this time.

Step 16: *Both Firefighters:* Tie the halyard.

NOTE: Firefighter #1 is located near the butt end of the ladder. Firefighter #2 is located near the tip end of the ladder.

Step 1: *Firefighter #1:* Place the ladder beam on the ground.

Step 2: *Firefighter #2:* Rest the beam on one shoulder.

Step 3: *Firefighter #1:* Place the foot closest to the lower beam on the lower beam at the butt spur.

Step 4: *Firefighter #1:* Grasp the upper beam with hands apart and the other foot extended back to act as a counterbalance.

Step 5: *Firefighter #2:* Advance hand-over-hand down the beam toward the butt until the ladder is in a vertical position.

CAUTION: Visually check the area overhead for obstructions before bringing the ladder to a vertical position. Before stepping forward, visually check the terrain.

Step 6: *Both Firefighters:* Pivot the ladder to position the fly away from the building (fly in for wooden ladders) if it is not already in that position.

Step 7: *Firefighter #2:* Grasp the halyard.

Step 8: *Firefighter #2:* Extend the fly section with a hand-over-hand motion until the tip reaches the desired elevation.

Step 9: *Both Firefighters:* Lower the ladder gently onto the building.

Step 10: *Both firefighters:* Tie the halyard.

NOTE: Firefighter #1 is located near the butt end of the ladder. Firefighters #2 and #3 are located near the tip end of the ladder. When four firefighters use the flat-shoulder carry, two are positioned at each end of the ladder, opposite each other.

Step 1: *Firefighter #1:* Place the ladder butt end on the ground. *Firefighters #2 and #3*: Rest the ladder flat on the shoulders.

Step 2: *Firefighter #1:* Heel the ladder by standing on the bottom rung or by placing the toes or insteps on the beam.

Step 3: *Firefighter #1:* Crouch down to grasp a convenient rung with both hands.

Step 4: *Firefighter #1:* Lean back.

CAUTION: Visually check the area overhead for obstructions before bringing the ladder to a vertical position. Before stepping forward, visually check the terrain.

Step 5: *Firefighters #2 and #3:* Advance in unison, with outside hands on the beams and inside hands on the rungs, until the ladder is in a vertical position.

NOTE: If necessary, the firefighters pivot the ladder to position the fly section away from the building. If using a wood ladder, the fly should be in toward the building.

Step 6: *Firefighters #2 and #3:* Place the inside of a foot against the butt spur.

Step 7: *Firefighters #2 and #3:* Steady the ladder with both hands on the beam.

Step 8: *Firefighter #1:* Grasp the halyard.

Step 9: *Firefighter #1:* Place the toe of one foot on the butt spur.

Step 10: *Firefighter #1:* Extend the fly section with a hand-over-hand motion until the tip reaches the desired elevation.

Step 11: *All Firefighters:* Lower the ladder gently onto the building.

Step 12: *Firefighter 1, 2, or 3:* Tie the halyard.

Step 1: Set the roof ladder down.

Step 2: Open the hooks.

Step 3: Face the hooks outward.

Step 4: Tilt the roof ladder up so that it rests against the other ladder.

Step 5: Climb the main ladder until your shoulder is about two rungs above the midpoint of the roof ladder.

Step 6: Reach through the rungs of the roof ladder.

Step 7: Hoist the ladder onto the shoulder.

Step 8: Climb to the top of the ladder.

Step 9: Lock into the ladder using a leg lock or life safety harness.

Step 10: Take the roof ladder off the shoulder.

Step 11: Use a hand-over-hand method to push the roof ladder onto the roof.

NOTE: Push ladder on the beam up the roof with hooks pointed away from extension ladder.

Step 12: Push the roof ladder up the roof until the hooks go over the edge of the peak and catch solidly.

NOTE: Remove the roof ladder by reversing the process.

Step 1: *Both Firefighters:* Face each other through the ladder.

Step 2: *Both Firefighters:* Grasp the ladder with both hands.

Step 3: *Appropriate Firefighter:* Place a foot against the side of the beam on which the ladder will pivot.

Step 4: *Both Firefighters:* Tilt the ladder onto the pivot beam.

Step 5: *Both Firefighters:* Pivot the ladder 90 degrees. Simultaneously adjust positions as necessary. Repeat the process until the ladder is turned a full 180 degrees and the fly is in the proper position.

NOTE: When firefighters become proficient in this maneuver, they may be able to pivot the ladder 180 degrees in one step.

Step 5: Shift grip on the ladder, one hand at a time, so that one hand grasps as low a rung as convenient, palm upward.

Step 6: Grasp a rung as high as convenient with the other hand, palm downward.

Step 1: Face the ladder.

Step 2: Heel the ladder.

Step 3: Grasp the beams.

Step 4: Bring the ladder outward to vertical.

Step 7: Turn slightly in the direction of travel.

Step 8: Visually check the terrain and the area overhead.

Step 9: Lift the ladder and proceed forward a short distance.

Step 10: Watch the tip as it is being moved.

WARNING! Do not attempt this procedure close to overhead wires.

Step 11: Set the ladder down at the new position.

Step 12: Switch grip back to the beams.

Step 13: Heel the ladder.

Step 14: Lower the ladder into position.

Step 1: *Both Firefighters:* Position on opposite sides of the ladder.

NOTE: Bring the ladder to vertical and make sure it is fully retracted.

Step 2: *Both Firefighters:* Position hands.

Step 3: *Both Firefighters:* Lift the ladder just clear of the ground.

Step 4: *Both Firefighters:* Watch the tip while shifting the ladder to the new position.

Step 5: *Both Firefighters:* Re-extend the ladder (if necessary).

Step 6: *Both Firefighters:* Lower the ladder gently into position.

Step 1: Climb to the desired height.

Step 2: Advance one rung higher.

Step 3: Slide the leg on the opposite side from the working side over and behind the rung to be locked in to.

Step 4: Hook foot either on the rung or on the beam.

Step 5: Rest on thigh.

Step 6: Step down with the opposite leg.

Skill SHEETS

10-I-15
Assist a conscious victim down a ground ladder.

Step 5: *Firefighters in building:* Lower the victim from the window to the rescuer on the ladder.

Step 1: Position the ladder.

Step 2: Secure the ladder.

Step 3: *Heeler:* Heel the ladder

Step 4: *Rescuer:* Climb the ladder.

Step 6: *Rescuer:* Position the victim for carrying.

Step 7: Descend the ladder.

Warning! An unconscious victim who regains consciousness while being rescued down a ladder may grab the ladder or the rescuer. This sudden reaction may cause the firefighter to lose his/her grip or footing and increase the risk of falling off the ladder. It is critical to observe the victim and be prepared to respond if the victim regains consciousness during the rescue.

Step 1: Position the ladder.

Step 2: Secure the ladder.

Step 3: *Heeler:* Heel the ladder.

Step 4: *Rescuer:* Climb the ladder.

Step 5: *Firefighters in building:* Lower the victim from the window to the rescuer on the ladder.

Step 6: *Rescuer:* Position the victim for carrying.

Step 7: Descend the ladder.

NOTE: Evaluator selects one of the tasks and explains to the firefighters which task has been selected.

- Ventilate a second-floor window from a ladder

- Perform a window rescue from a second floor window

- Access the roof of a building (1-story residential)

- Access the roof of a building (2-story commercial)

- Work from the ladder with a charged hoseline

Step 1: Select ladder that can perform task safely and effectively.

Step 2: Carry and place ladder according to appropriate skill sheet.

Step 3: Raise ladder according to appropriate skill sheet.

Step 4: Place ladder for ventilating a second-floor window.

Step 5: Place ladder for a rescue from a second-floor window (narrow-type window).

Skill SHEETS

10-I-17
Select, carry, and raise a ladder properly for various types of activities.

Step 6: Place ladder for access to the roof of a building (one-story residential).

Step 7: Place ladder for access to the roof of a building (two-story commercial).

Step 8: Place ladder so that a fire stream may be directed into a second-floor window.

NOTE: Handling hose and operating a nozzle may not have been taught and therefore this skill may be delayed until other skills are mastered.

Step 9: Lower and return to ladder staging area.

Step 10: Maintain communication throughout evolution.

Chapter Contents

Chapter 11

Ventilation

This chapter provides information that addresses the following job performance requirements of NFPA® 1001, *Standard for Fire Fighter Professional Qualifications (2008)*:

NFPA® 1001 references for Chapter 11:

    5.3.11

    5.3.12

Chapter Objectives

Fire Fighter I Objectives

1. Describe reasons for fireground ventilation.

2. List considerations that affect the decision to ventilate.

3. Discuss factors that are taken into account when deciding the need for ventilation.

4. Discuss vertical ventilation.

5. List safety precautions to observe when undertaking vertical ventilation.

6. List warning signs of an unsafe roof condition.

7. Discuss roof coverings and using existing roof openings for vertical ventilation purposes.

8. Discuss ventilation considerations for various types of roofs.

9. Describe trench or strip ventilation.

10. Explain procedures for ventilation of a conventional basement.

11. List factors that can reduce the effectiveness of vertical ventilation.

12. Discuss horizontal ventilation.

13. Discuss considerations for horizontal ventilation.

14. Distinguish between advantages and disadvantages of forced ventilation.

15. Describe negative-pressure ventilation.

16. Discuss positive-pressure ventilation.

17. Compare and contrast positive-pressure and negative-pressure ventilation.

18. Describe hydraulic ventilation.

19. List disadvantages to the use of hydraulic ventilation.

20. Explain the effects of building systems on fires or ventilation.

21. Ventilate a flat roof. (Skill Sheet 11-I-1)

22. Ventilate a pitched roof. (Skill Sheet 11-I-2)

23. Ventilate a structure using mechanical positive-pressure ventilation. (Skill Sheet 11-I-3)

24. Ventilate a structure using horizontal hydraulic ventilation. (Skill Sheet 11-I-4)

Ventilation

Case History

On August 9, 2005, a fire department responded to a fire in a two-story house. The call came in about 1 p.m. Upon arrival, firefighters found heavy smoke coming from the second floor but no visible flame. An attack team advanced a 1¾ inch (45 mm) line up the stairwell to the second floor to locate and extinguish the fire. Fire and smoke conditions suddenly increased and forced the team back to the landing on the stairwell. At that point ventilation had not been performed.

The company officer advised the Incident Commander (IC) of the situation and requested ventilation and a backup line. The officer then directed a firefighter to break the window on the landing. Immediately after breaking the window and allowing air to enter the heated atmosphere on the second floor, a fireball force knocked the team down the stairwell to the first floor. Fortunately, the firefighters were not injured. The IC ordered the house to be vertically ventilated. Once the ventilation opening was created the heat and smoke quickly left the second floor and the attack team was able to advance and extinguish the fire.

This incident provides two important lessons about ventilation. First, ventilation must be coordinated with fire attack. Any structure that requires ventilation must be ventilated before crews enter the structure to perform operations. This requires communication between the IC, ventilation team, and other crews operating on the fireground. Second, when done properly, ventilation can make a dramatic difference in the conditions inside the structure. What moments before was a deadly heated atmosphere is changed into an environment that allows firefighters in full personal protective equipment (PPE) to advance and extinguish the fire, rescue trapped citizens, and perform other vital operations. This chapter focuses on techniques for ventilating a structure safely and effectively.

Source: National Fire Fighter Near-Miss Reporting System

In the fire service, *ventilation* is the systematic removal of heated air, smoke, and fire gases from a burning building and replacing them with cooler air. The cooler air facilitates entry by firefighters for search and rescue and fire fighting operations. The importance of timely and effective ventilation and the resulting increase in firefighter safety cannot be overstated. Properly done, ventilation decreases the rate of fire spread and increases visibility so firefighters can locate the seat of the fire more quickly. Ventilation also decreases the danger to trapped occupants by channeling heat and toxic fire gases away from them and reduces the chances of flashover or backdraft. Timely and effective ventilation can also reduce the damage done to the structure and its contents.

Modern buildings have much more built-in mechanical ventilation than was typical of buildings erected before the second half of the twentieth century. Insulation, sealed windows, and other energy conservation measures increase the need for firefighters to ventilate burning buildings. In addition, energy-saving glass, insulated window and door frames, and full-building vapor barriers increase heat containment. This means that the heat from a fire is contained longer and flashover or backdraft conditions are more likely to develop than they are in a less well-insulated structure.

In addition, the use of plastics and other synthetic materials has dramatically increased the fuel load in all occupancies — new and old. Because of these synthetic materials, the products of combustion produced during fires are becoming more dangerous and are produced in larger quantities than ever before. For all these reasons, timely and effective ventilation becomes extremely important.

While there are no specific ventilation-related requirements in NFPA 1001 for those trained to the Fire Fighter II level, the standard requires those trained to the Fire Fighter I level to be capable of performing both horizontal and vertical ventilation of a structure as a member of a team. The knowledge required to perform horizontal ventilation includes the principles, advantages, limitations, and effects of horizontal ventilation accomplished by mechanical and hydraulic methods. In addition, firefighters must know the following information:

- Safety considerations when venting a structure
- Fire behavior in a structure
- Products of combustion found in a structure fire
- Signs, causes, effects, and prevention of backdrafts
- Relationship of oxygen concentration to life safety and fire growth

The skills required to perform horizontal ventilation are described as the ability to:

- Transport and operate ventilation tools and equipment
- Transport and set up ground ladders
- Safely break windows and door glass and remove obstructions

The knowledge required to perform vertical ventilation includes the following:

- Methods of heat transfer

Thermal Layering (of Gases) — Outcome of combustion in a confined space in which gases tend to form into layers, according to temperature, with the hottest gases are found at the ceiling and the coolest gases at floor level.

- Principles of thermal layering inside a burning building
- Techniques and safety precautions for venting flat and pitched roofs and basements
- Basic indicators of potential collapse or roof failure
- Effects on structural integrity of construction type and elapsed time under fire conditions
- Advantages and disadvantages of vertical and trench/strip ventilation

The skills required to perform vertical ventilation are described as the ability to:

- Transport and operate ventilation tools and equipment
- Select, carry, deploy, and secure ground ladders
- Deploy roof ladders on pitched roofs while secured to a ground ladder
- Carry ventilation tools and equipment up and down ground ladders
- Hoist ventilation tools to a roof
- Sound a roof for integrity
- Cut roofing and flooring materials
- Clear an opening with hand tools

This chapter covers the reasons for fireground ventilation and the considerations for deciding if, where, and how to ventilate. Also discussed are the basics of horizontal and vertical ventilation operations. Finally, the effects of building systems in fire situations are covered. These discussions encompass the four types of fire service ventilation: natural, mechanical, horizontal, and vertical. While there are four types of ventilation, in practice two types are always combined, such as natural/horizontal or natural/vertical; or mechanical/horizontal or mechanical/vertical.

While it is important for you to understand the theory and practice of tactical ventilation, as a firefighter you are not likely to be required to decide if ventilating a burning building is necessary. Nor will you have to decide the best way to ventilate the building or where to create the ventilation opening. These decisions are the responsibility of those in command of the operation. Under the supervision of your company officer, your responsibility will be to quickly and efficiently implement all or part of the ventilation element of the Incident Action Plan (IAP).

When the need for ventilation has been determined, the officer must consider the range of possible actions necessary to control the fire and weigh the potential benefits of each action against the potential risks to the firefighters who will have to carry out the action or actions chosen. To maximize safety, all firefighters assigned to ventilation must wear full protective clothing, including self-contained breathing apparatus (SCBA) **(Figure 11.1)**. In some cases, a charged hoseline will be needed for the protection of the ventilation team and to protect exposures. In other cases, the time and personnel needed to stretch a hoseline may render vertical ventilation impractical.

Figure 11.1 Full PPE is critical for ventilation team members. *Courtesy of District Chief Chris E. Mickal, NOFD Photo Unit.*

Reasons for Fireground Ventilation

Ventilating burning buildings helps firefighters meet the universal fireground objectives – life safety, incident stabilization, and property conservation – identified in Chapter 1. The ways in which tactical ventilation contributes to achieving these objectives are:

- Life safety
- Fire attack and extinguishment
- Fire spread control
- Reduction of flashover potential
- Reduction of backdraft potential
- Property conservation

Life Safety

Timely and effective ventilation increases firefighter safety by reducing the interior temperature and increasing visibility. By providing an escape path for the steam when water is vaporized, it reduces the chance of firefighters receiving steam burns when the water is applied to the fire. Ventilation also reduces the likelihood of sudden and hazardous changes in fire behavior such as flashover and backdraft.

Effective ventilation simplifies and expedites search and rescue operations by removing smoke and gases that endanger trapped or unconscious occupants. The replacement of heat, smoke, and gases with cooler, fresh air increases trapped victims' chances of survival and allows firefighters to locate unconscious victims faster **(Figure 11.2, p. 544)**.

WARNING!
Never direct a hose stream into a ventilation opening before the fire has been controlled. Doing so may compromise ventilation efforts and put interior crews in serious danger.

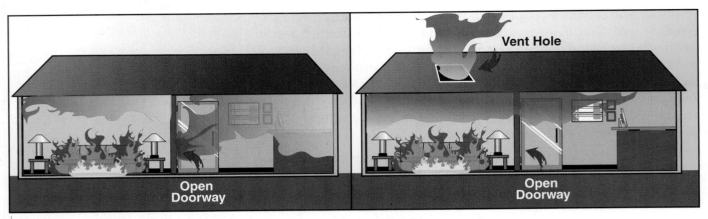

Figure 11.2 Ventilation improves conditions for both firefighters and building occupants.

Fire Attack and Extinguishment

When effective ventilation improves visibility inside a burning building, it permits firefighters to more rapidly locate the fire and proceed with extinguishment. When a ventilation opening is made in the upper portion of a building, a *chimney* effect (drawing air currents from throughout the building in the direction of the opening) occurs (**Figure 11.3**). For example, if this opening is made directly over a fire, it tends to localize the fire. If the opening is made elsewhere, it may contribute to fire spread by drawing the fire through uninvolved areas toward the ventilation opening (**Figure 11.4**). Therefore, ventilation *must* be closely coordinated with fire attack.

Fire Spread Control

Convection causes heat, smoke, and fire gases to travel upward to the highest point in a compartment until they are stopped by a roof or ceiling. As these products of combustion reach the ceiling, they spread laterally to involve other areas of the structure (**Figure 11.5**). While combustion gases flowing under a ceiling are technically called *ceiling jets*, in the fire service this process has historically been termed *mushrooming*. When these laterally-moving products of combustion reach a wall, they begin to bank down and fill the space with smoke, developing a hot gas layer. As the fire continues to burn, more smoke is produced, and the layer continues to increase in depth from the ceiling downward.

Effective ventilation of a building during a fire reduces the rate of smoke layer development and can reverse this process by exhausting more smoke from the structure than the fire is producing. This in turn reduces the rate at which fire will spread over an area by providing an escape path for the rising heated gases, at least for a short time. However, even with proper ventilation, if the fire is not extinguished soon after ventilation is accomplished, the increased supply of fresh air will feed the fire and allow it to grow and spread into attic and concealed spaces. Therefore, ventilation should occur only after hoseline crews are ready to move in and attack the fire (**Figure 11.6, p. 546**).

Reduction of Flashover Potential

Flashover is the transition between the *growth* and the *fully developed* fire stages. As an unventilated interior fire burns and flames, smoke, and hot gases extend across the ceiling, heat radiates back down until combustibles in the room are heated to their ignition temperatures. Once their ignition temperatures are reached, the contents

Chimney Effect — Created when a ventilation opening is made in the upper portion of a building and air currents throughout the building are drawn in the direction of the opening; also occurs in wildland fires when the fire advances up a V-shaped drainage swale.

Mushrooming — Tendency of heat, smoke, and other products of combustion to rise until they encounter a horizontal obstruction. At this point they will spread laterally until they encounter vertical obstructions and begin to bank downward.

of the entire room will ignite almost simultaneously. Anyone in the room when this occurs — even firefighters in full PPE — cannot survive for more than a few seconds. Ventilation helps to prevent flashover from occurring by removing or reducing the heat before it reaches the levels required for mass ignition.

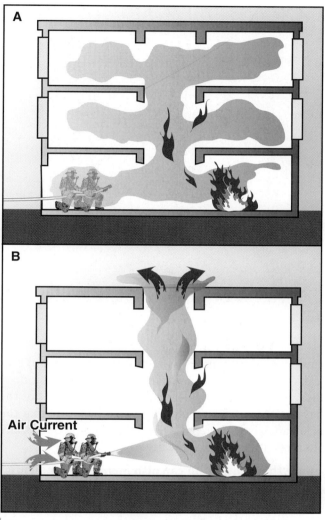

Figure 11.3 The chimney effect is common in high-rise building fires.

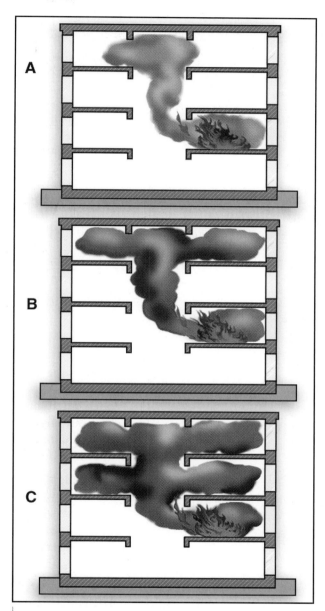

Figure 11.5 Lateral smoke spread will occur in unventilated structures.

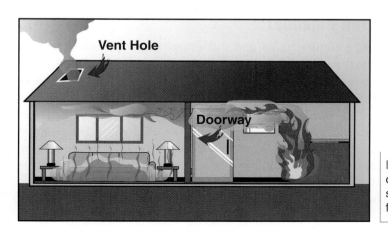

Figure 11.4 The incorrect use of vertical ventilation can spread the fire into unburned areas of the structure and endanger trapped occupants and firefighters.

Figure 11.6 Ventilation and fire attack efforts must be coordinated.

Opening a single door or window may not remove enough smoke and heat to prevent a flashover. If the fire breaks a window or an escaping occupant leaves a door open when exiting the structure, the fire may flash over prior to the arrival of the first-due engine. Where this has not occurred, firefighters need to be aware that the potential for a flashover or a backdraft exists. Therefore, ventilation should occur only after hoseline crews are ready to move in and attack the fire.

Firefighters must remain aware that ventilation can also *increase* **the potential for flashover.** If a fire is producing a significant amount of smoke (gaseous fuel) and heat, but is limited by the available oxygen (ventilation limited or fuel rich), ventilating the room may cause a rapid ignition of the hot fire gases in the upper layer, which could initiate a flashover. This process can occur quickly because the fire gases are hot and they have preheated the other fuels in the room.

Reduction of Backdraft Potential

When a fire is confined in a compartment and it does not break a window or burn through to the outside, the fire will reduce the oxygen level in the room until flaming combustion is no longer possible. Combustible materials in the room may continue to smolder and give off carbon monoxide and other unburned flammable gases. Without oxygen, these superheated gases will not ignite. In this very dangerous situation, if an air supply (which provides the necessary oxygen) is introduced by firefighters opening doors or windows, a violent deflagration (backdraft) occurs. Depending upon the degree of confinement, this deflagration may exhibit the characteristics of a true explosion. To prevent backdraft conditions from developing to direct the deflagration harmlessly, carefully controlled ventilation must be provided. In most cases, vertical ventilation is the safest way to vent a building with backdraft potential.

Firefighters must be aware of this potential and proceed cautiously in areas where excessive amounts of heat may have accumulated. As explained in Chapter 3, Fire Behavior, firefighters should be observant for the signs of possible backdraft conditions. If any signs of backdraft are present, firefighters should stay away from doors and windows until vertical ventilation has had the chance to reduce the severity of the situation. The signs of a potential backdraft include the following:

- Confinement and excessive heat
- Smoke-stained windows
- Smoke puffing at intervals from the building (appearance of breathing)
- Pressurized smoke coming from small cracks
- Little visible flame from the exterior of the building
- Black smoke becoming dense gray-yellow

Property Conservation

Rapid extinguishment of a fire reduces damage caused by water, heat, and smoke. Timely and effective ventilation helps firefighters extinguish interior fires faster and more efficiently. Smoke may be removed from burning buildings by natural or mechanical means. Natural ventilation involves opening doors and windows to allow natural air currents to move smoke and heat out of the building. Mechanical ventilation involves the use of fans **(Figure 11.7)**. Regardless of the method used, ventilation usually reduces smoke damage.

When smoke, gases, and heat are removed from a burning building, the fire can be confined to a specific area. Then, if there are sufficient personnel on scene, effective salvage operations can be started outside the immediate area of the fire even while fire control operations are being conducted.

Figure 11.7 Effective mechanical ventilation requires practicing as in this training evolution.

Natural Ventilation — Techniques that use the wind, convection currents, and other natural phenomena to ventilate a structure without the use of fans, blowers, or other mechanical devices.

Forced Ventilation — Any means other than natural ventilation. This type of ventilation may involve the use of fans, blowers, smoke ejectors, and fire streams. Also called Mechanical Ventilation.

Considerations Affecting the Decision to Ventilate

When weighing the need for ventilation, fire officers and their crews must consider the requirements of the IAP. As mentioned earlier, ventilation actions must be closely coordinated with fire attack efforts. Beyond that, officers must make a series of decisions that pertain to ventilation needs. These decisions, by the nature of fire situations, fall into the following order:

1. *Is there a need for ventilation at this time?* The need must be based upon the heat, smoke, and gas conditions within the structure as well as the structural conditions and the life hazard.

2. *Where is ventilation needed?* This answer requires a knowledge of construction features of the building, contents, exposures, wind direction, extent of the fire, location of the fire, top or vertical openings, and cross or horizontal openings. The location of victims probably will not be known until the building is searched, but if this information is available it can also affect the decision of where to ventilate.

3. *What type of ventilation should be used?* Horizontal (natural or mechanical)? Vertical (natural or mechanical)?

4. *Do fire and structural conditions allow for safe roof operations?* Has the roof been weakened by the fire? Are there heavy deadloads on the roof? Is the roof covered with snow and ice?

5. *Are the Vent Group personnel sufficiently trained and equipped to ventilate this building under these circumstances?* Do they have the necessary training and experience? Do they have the tools needed?

To answer these questions, fire officers have to evaluate several pieces of information and take into account numerous factors. These are detailed in the following sections.

Life Safety Hazards

Dealing with the danger to human life is of primary importance. The first consideration is the safety of firefighters and building occupants. Building occupants include people, pets, and livestock. Life hazards in a burning building are generally lower if the occupants are awake. On the other hand, if the occupants were asleep when the fire developed and are still in the building, a number of possibilities must be considered. First, the occupants may have been overcome by smoke and gases – some may still be alive; others may have perished. Second, they might have become lost in the building. Or, third, they may be alive and taking refuge in their rooms because their doors were closed. In any case, timely and effective ventilation will facilitate search and rescue operations. Depending on fire conditions, ventilation may be needed before search and rescue operations begin. Alternatively, if conditions warrant, the fire may need to be attacked first. Sometimes ventilation and search and rescue must be performed simultaneously.

In addition to the hazards that endanger occupants, there are potential hazards to firefighters and rescuers. The type of structure involved, its age, whether natural openings are adequate, and the need to cut through roofs, walls, or floors (combined with other factors) add more considerations to the decision-making process. The hazards that can be expected from the accumulation of smoke and gases in a building include the following:

- Visual impairment caused by dense smoke
- Presence of toxic gases
- Lack of oxygen
- Presence of flammable gases
- Possibility of backdraft
- Possibility of flashover

Visible Smoke Conditions

When first arriving at the scene, fire officers can make some ventilation decisions and other tactical decisions based on visible smoke conditions. Smoke accompanies most ordinary forms of combustion and it differs greatly with the substance of the materials being burned. The density of the smoke increases with the amount of suspended particles. Smoke conditions vary according to how burning has progressed. A developing fire must be treated differently than one in the decaying phase. Some fires that are in their early stage of development may give off smoke that is not very dense **(Figure 11.8)**. This smoke may become quite dense, however, if the fire involves home furnishings that contain polyurethane foam. When it is pyrolized, polyurethane foam produces significant quantities of carbon and toxic gases such as hydrogen cyanide.

As burning progresses and the fire exceeds the ventilation available, the combustion process becomes less efficient, resulting in increased quantities of unburned fuel in the form of gases and soot particles. This produces a darker and thicker smoke **(Figure 11.9)**. A large volume of very dense smoke may hide the fact that the structure itself has become involved with fire, reducing its structural integrity. ***It is important to remember that smoke is fuel that has not ignited.***

Pyrolysis (Pyrolysis Process or Sublimation) — Thermal or chemical decomposition of fuel (matter) because of heat that generally results in the lowered ignition temperature of the material. The pre-ignition combustion phase of burning during which heat energy is absorbed by the fuel, which in turn gives off flammable tars, pitches, and gases.

Figure 11.9 As a fire progresses, the smoke becomes denser. *Courtesy of Dick Giles.*

Figure 11.8 In the early stages of a fire, the smoke may not be very dense.

The Building Involved

Knowledge of the building involved is a great asset when decisions concerning ventilation are made. In addition to the location of the fire within the building, the building's age, type, and design must be considered in relation to the best ways to ventilate it. Other factors to be considered include the following:

- Number and size of wall openings

- Number of stories, staircases, elevator shafts, dumbwaiters, ducts, and roof openings

- Availability and involvement of exterior fire escapes and exposures (**Figure 11.10**)
- The extent to which a building is connected to adjoining structures
- The type of roof construction
- Built-in fire protection systems

If the results of preincident planning have been documented and stored in a way that makes them available to officers on the fireground, they may provide valuable information that affects ventilation decisions. For example, plans may show if and how buildings were altered or subdivided. These plans can also provide information concerning heating, ventilating, and air-conditioning (HVAC) systems as well as avenues of escape for smoke, heat, and fire gases.

High-Rise Buildings

The danger to occupants from heat and smoke is a major consideration in high-rise buildings. These buildings may be occupied by hospitals, hotels, apartments, or business offices. In any case, depending upon the day of the week and the time of day, a great number of people may be at risk.

Fire and smoke can spread rapidly through pipe shafts, stairways, elevator shafts, air-handling systems, and other vertical and horizontal openings. These openings contribute to a *stack effect* (natural movement of heat and smoke throughout a building), creating an upward draft and interfering with evacuation and ventilation (**Figure 11.11**).

Hot smoke and fire gases travel upward through a building until they reach the top of the building or they are cooled to the temperature of the surrounding air. When this equalization of temperature occurs, the smoke and fire gases stop rising

Stack Effect — Phenomenon of a strong air draft moving from ground level to the roof level of a building. Affected by building height, configuration, and temperature differences between inside and outside air.

Figure 11.10 When making ventilation and fire attack decisions, the availability and condition of fire escapes is an important factor to consider.

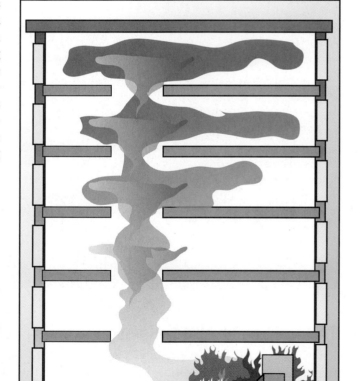

Figure 11.11 The stack effect is common in high-rise buildings.

and *stratify* (form layers) within the building. In some cases, these layers of smoke and fire gases will collect on floors below the top floor of the building. The smoke spread (mushrooming effect), which is usually expected at the top floor, does not occur in these buildings unless and until sufficient heat is added to cause the stratified smoke and fire gases to again move in an upward direction. This is sometimes called the *stack effect* or *stacking*. Strategies such as horizontal or vertical ventilation assisted by mechanical means must be developed to cope with the ventilation and life hazard problems inherent in stratified smoke.

Ventilation in a high-rise building must be carefully coordinated to ensure the safest and most effective use of personnel, equipment, and extinguishing agents. The personnel required for rescue and fire fighting operations in high-rise buildings is often four to six times as great as required for a fire in a typical low-rise building. In many instances, ventilation must be accomplished horizontally with the use of mechanical ventilation devices and the HVAC systems in the buildings. Protective breathing equipment will be in great demand, and the ability to provide large quantities of replacement SCBA cylinders must be addressed. The problems of communication and coordination among the various attack and ventilating teams become more involved as the number of participants increases.

Top ventilation in high-rise buildings must be considered during preincident planning. In many buildings, only one stairwell penetrates the roof. This vertical "chimney" can be used to ventilate smoke, heat, and fire gases from various floors while another stairway is used as the escape route for building occupants. However, during a fire, the doors on uninvolved floors must be controlled so occupants do not accidentally enter the involved stairway as they are evacuating. Before the doors on the fire floors are opened and the stair shaft is ventilated, the door leading to the roof must be blocked open or removed from its hinges (**Figure 11.12**). Preventing the door at the top of the shaft from closing ensures that it cannot compromise established ventilation operations.

Many elevator shafts penetrate the roof; under some conditions they may be used for ventilation. Using stairwells or elevator shafts for evacuation and ventilation simultaneously is ineffective and potentially life-threatening. It must be remembered that when ventilating the top of a stairwell, you will be drawing the smoke and heat to you or anyone else in the stairwell between the fire floor and the roof. This technique may make it difficult for the fire suppression team to make entry onto the fire floor. The safest and most effective technique may be to pressurize the stairways with positive pressure fans to confine the smoke on the floors. See the Positive-Pressure Ventilation section in this chapter.

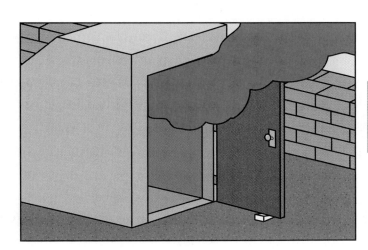

Figure 11.12 When used for ventilation, roof access doors should be blocked open or removed.

Basements and Windowless Buildings

Basement fires can be among the most challenging situations firefighters will face. Without effective ventilation, access into the basement is difficult because firefighters have to descend through the worst heat and smoke to get to the seat of the fire. Access to the basement may be via interior or exterior stairs, exterior windows, or hoistways. Many outside entrances to basements may be blocked or secured by iron gratings, steel shutters, wooden doors, or combinations of these for protection against weather and for security **(Figure 11.13)**. All of these features serve to impede attempts at natural ventilation. If a ventilation opening is made opposite the entry point, the fire and smoke can be pushed toward the opening by a positive-pressure ventilation (PPV) fan **(Figure 11.14)**.

NOTE: See the Positive-Pressure Ventilation section later in this chapter.

Many buildings, especially in business districts, have windowless wall areas. While windows may not be the most desirable means for entry into or emergency egress (escape) from burning buildings, they are an important consideration for ventilation. Windowless buildings complicate fire fighting and ventilation operations **(Figure 11.15)**. In some cases, creating the openings needed to ventilate a windowless building may delay the operation for a considerable time, allowing the fire to gain headway and flash over or create backdraft conditions.

Figure 11.13 Entrances to basements may be secured by wood or metal doors.

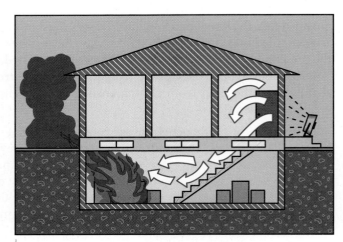

Figure 11.14 A PPV fan can be used to ventilate a basement or cellar.

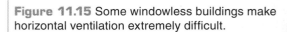

Figure 11.15 Some windowless buildings make horizontal ventilation extremely difficult.

Ventilating this type of building can be difficult, and the problems involved vary depending on the size, occupancy, configuration, and type of material from which the building is constructed. Windowless buildings usually require mechanical ventilation for the removal of smoke. Most buildings of this type are automatically cooled and heated through ducts (**Figure 11.16**). Mechanical ventilation equipment can sometimes effectively clear the area of smoke by itself; however, these systems are more likely to cause the spread of heat and fire.

Figure 11.16 Smoke and flames can travel through the building's HVAC ductwork.

Because it is difficult and time-consuming to ventilate these buildings, it is important to limit the steam generated by fire fighting efforts before ventilation is accomplished. To reduce the volume of steam produced by applying water onto fires in these buildings, some departments train their personnel to use solid bore nozzles to limit steam conversion.

Location and Extent of the Fire

Because of the time needed for a fire burning inside a building to be discovered, reported, and assistance to be dispatched, the fire may have traveled some distance throughout the structure by the time firefighters arrive. Therefore, first-arriving units must determine the size and extent of the fire as well as its location. Creating ventilation openings before the fire is located may spread the fire to uninvolved areas of the building and cut off escape routes for building occupants. The severity and extent of the fire depend upon a number of factors, including the type of fuel and the amount of time it has been burning, installed fire detection and suppression devices, and the degree of confinement. The phase to which the fire has progressed is a primary consideration in determining ventilation procedures.

Some of the ways vertical fire extension occurs are as follows:

- Through stairwells and elevators shafts by direct flame contact or by convected air currents (**Figure 11.17, p. 554**)
- Through partitions and walls and upward between the walls by flame contact and convected air currents, especially in balloon-frame construction (**Figure 11.18, p. 554**)
- Through windows or other outside openings where flame extends to other exterior openings and enters upper floors (commonly called *lapping* or *exterior flame spread*) (**Figure 11.19, p. 554**)
- Through ceilings and floors by conduction of heat through beams, pipes, or other objects that extend from floor to floor (**Figure 11.20, p. 554**)
- Through unprotected floor and ceiling openings where sparks and burning material fall through to lower floors or flames and hot gases spread to upper floors
- By the collapse of floor and roof assemblies
- Through voids, duct work, and concealed spaces

Selecting the Place to Ventilate

When selecting a place to ventilate, the ideal situation is one in which firefighters have as much information as possible about the fire, the building, and the occupancy. There is no rule of thumb for selecting the exact point at which to open a roof except "as directly over the fire as possible." Many factors have a bearing on where to ventilate, including the following:

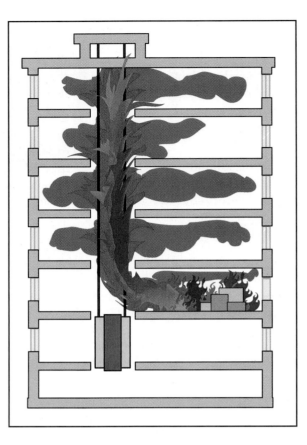

Figure 11.17 Elevator shafts can allow fire to spread from floor to floor.

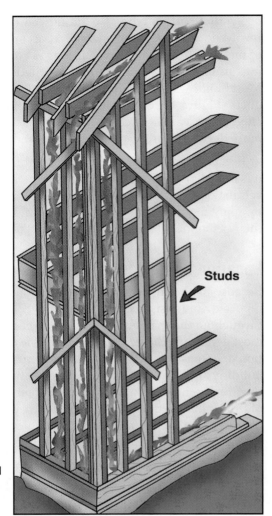

Studs

Figure 11.18 Fire can spread through wall spaces between exterior and interior walls, especially in balloon-type construction.

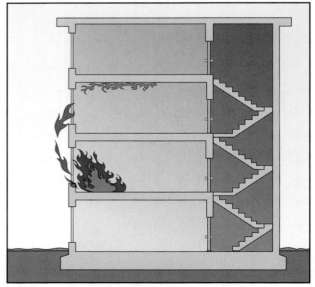

Figure 11.19 Fire can spread vertically by lapping up the exterior of the building from window to window.

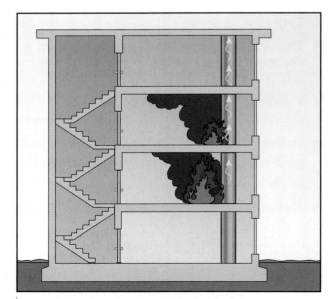

Figure 11.20 Fire can spread from compartment to compartment through conduction. Highly conductive materials such as metal pipes and steel structural members can spread the fire vertically and horizontally.

- Availability of existing openings such as skylights, ventilator shafts, monitors, and hatches that access the fire area (**Figure 11.21**)

- Location of the fire and the direction in which the IC wishes it to be drawn

- Type of building construction

- Wind direction

- Extent of progress of the fire and the condition of the building and its contents

- Indications of potential roof collapse

- Effect that ventilation will have on the fire

- Effect that ventilation will have on exposures

- Hose crew's state of readiness to enter and begin the fire attack

- Ability to protect exposures prior to actually opening the building (**Figure 11.22**)

Before ventilating a building, adequate resources (personnel and equipment) must be ready because the fire may immediately increase in intensity when the building is opened. These resources may be needed for both the building involved and other exposed buildings. As soon as the building has been opened to permit hot gases and smoke to escape, firefighters should make an effort to reach the seat of the fire for extinguishment if conditions permit this to be done safely. If wind direction permits, entrance should be made into the building as near the fire as possible. Charged hose-

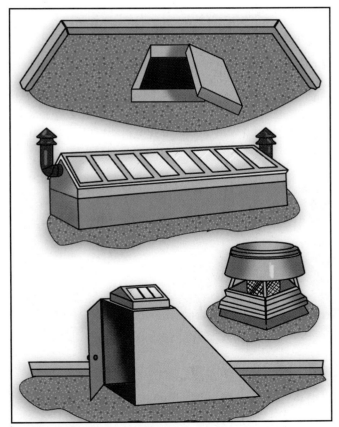

Figure 11.21 Existing roof openings may be used to assist in ventilation. Existing openings include (top to bottom) hatches, skylights, monitors, and roof access doors, among others.

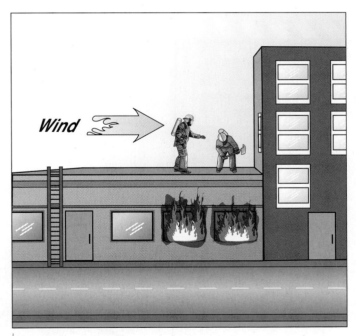

Figure 11.22 Venting one building may expose another to fire spread.

lines should be positioned at the entry point so that an interior attack can be started as soon as the building is ventilated. Charged hoselines should also be in place at critical points of exposure to prevent the fire from spreading.

Vertical Ventilation

Vertical ventilation generally means opening the roof or existing roof openings for the purpose of allowing heated gases and smoke to escape to the atmosphere. In order to properly ventilate a roof, firefighters must understand the basic types and designs of roofs. Many designs are used and their names sometimes vary on a regional basis.

In general, firefighters must deal with three prevalent types of roof shapes: *flat, pitched,* and *arched* (**Figure 11.23**). Some buildings have a combination of these roof designs. Some of the more common styles are the *gable, hip, gambrel, shed, mansard,* and *lantern* (**Figure 11.24**). Less common are *sawtooth* and *butterfly* roofs (**Figure 11.25**).

To maximize their safety and effectiveness during vertical ventilation operations, firefighters must know how the roofs in their response areas are constructed. During preincident inspections of new construction projects, they should observe and document the types of roof construction and roof coverings being used. In existing construction, they should look for roofs to which extra insulation has been added. They should also look for the use of lightweight building materials. Lightweight roofs consisting of wooden I-beam or lightweight truss construction are more susceptible to early roof collapse if involved in fire than roofs of more conventional construction. Information gathered during preincident planning, if properly documented, can alert firefighters to potential problems during vertical ventilation operations. For more information on roof construction, refer to Chapter 4, Building Construction.

The likelihood of roof collapse is affected by a variety of factors. Among these factors are the volume of fire, how long the fire has been burning, the type of construction, the level of protection (assembly rating), and the load on the roof. While preincident planning is beneficial, it is important for fire officers and their crews to carefully assess the need for and safety of a vertical ventilation operation or an interior fire attack.

Vertical Ventilation — Ventilating at the highest point of a building through existing or created openings and channeling the contaminated atmosphere vertically within the structure and out the top. Done with holes in the roof, skylights, roof vents, or roof doors.

WARNING!
Significant fire burning in the roof area of any structure with engineered trusses can cause roof collapse with little or no warning.

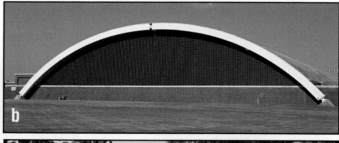

Figures 11.23 a-c The general shapes of most roofs: (a) flat, (b) arched, and (c) pitched.

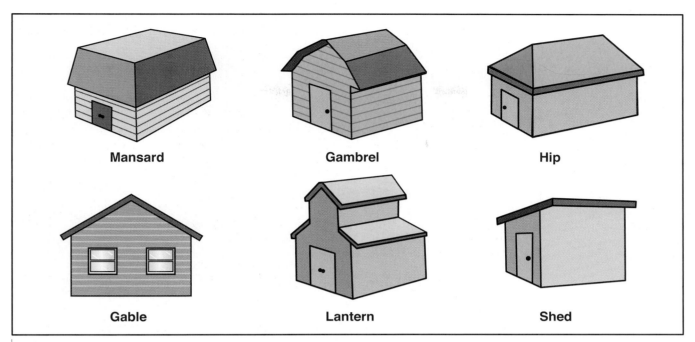

Figure 11.24 Common styles of roofs found in North America.

Mansard   Gambrel   Hip

Gable   Lantern   Shed

Figure 11.25 The sawtooth-type roof is generally found on industrial types of occupancies.

Vertical ventilation can be undertaken after the officer in charge has completed the following:

- Determined that vertical ventilation can be done safely and effectively.
- Considered the age and type of construction involved
- Considered the location, duration, and extent of the fire
- Observed safety precautions
- Identified escape routes
- Selected the place to ventilate
- Moved personnel and tools to the roof

The roof team should be in constant communication with their supervisor (**Figure 11.26, p. 558**). Portable radios are most adaptable to this type of communication. Responsibilities of the leader on the roof include the following:

- Ensuring that the roof is safe to operate on (sounding, visual observation)
- Ensuring that only the required openings are made
- Directing efforts to minimize secondary damage (damage caused by fire fighting operations)
- Coordinating the crew's efforts with those of firefighters inside the building
- Ensuring the safety of all personnel who are assisting in the opening of the building
- Ensuring that the team leaves the roof as soon as their assignment is completed

Safety Precautions

Some of the safety precautions that should be observed include the following:

- Check the wind direction with relation to exposures.
- Work with the wind at your back or side to provide protection while cutting the roof opening.
- Note the existence of obstructions or excessive weight on the roof.
- Provide a secondary means of escape for crews on the roof (**Figure 11.27**).
- Ensure that main structural supports are not cut while creating a ventilation opening.
- Guard the opening to prevent personnel from falling into it.
- Evacuate the roof promptly when ventilation work is complete.
- Use lifelines, roof ladders, or other means to prevent personnel from sliding and falling off the roof.

Roof Ladder — Straight ladder with folding hooks at the top end. The hooks anchor the ladder over the roof ridge.

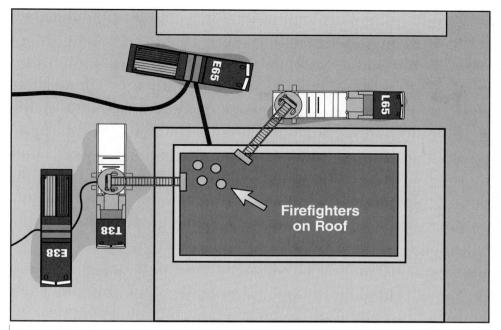

Figure 11.27 Two means of egress from the roof must be provided to increase vent team safety.

Figure 11.28 Check the structural integrity by sounding the roof with a pike pole or axe.

- Make sure that a roof ladder (if used) is firmly secured over the peak of the roof before operating from it.
- Exercise caution when working around electric wires and guy wires.
- Ensure that all personnel on the roof are wearing full PPE including SCBA, and that they are breathing SCBA air.
- Keep other firefighters out of the range of those who are swinging axes and operating power saws.
- Caution axe users to beware of overhead obstructions within the range of their swing.
- Start power tools on the ground to ensure operation but make sure they are shut off before hoisting or carrying them to the roof. Make sure that the angle of the cut is not toward the body.
- Extend ladders at least three to five rungs above the roof line and secure the ladder.
- When operating from aerial ladder platforms, the floor of the platform should be even with or slightly above roof level.
- Check the roof for structural integrity (by sounding) before stepping on it and continue sounding it throughout the operation; do not jump onto a roof (**Figure 11.28**).
- Always walk on bearing walls and strongest points of roof structure whenever possible – both before and after ventilating.
- Ensure that ceilings are punched through to enhance ventilation.

CAUTION

Roof ladders are not intended to be used on fire-weakened roofs. They are only meant to prevent slipping from a steep or slippery roof.

Sounding a Roof

Before stepping off a ladder, parapet wall, or other place of safety onto the roof of a burning building — especially if the roof surface is obscured by smoke or darkness — firefighters should sound the roof (if possible) by striking the roof surface with the blunt end of a pike pole, rubbish hook, or axe. When struck by a tool, some roofs will feel solid over structural supports and the tool will tend to bounce off the surface. Between the supports, the roof may feel softer and less rigid. The roof may also *sound* solid when struck over a rafter or joist and produce a hollow sound when struck between the supports. By practicing on structurally sound roofs, firefighters can learn to recognize the difference in the feel and the sound of supported and unsupported areas of a roof.

Remember, however, that roofs with several layers of composition shingles or other roof coverings may not respond to sounding as just described. They may sound quite solid when struck with a tool even though the roof supports may have been severely damaged by the fire. Also, roofs covered with tile or slate cannot be sounded; the tiles/slates must be removed to reveal the underlying structure.

Use preincident planning information to identify buildings that have roofs supported by lightweight or wooden trusses. These roofs may fail early in a fire and are extremely dangerous to be on or under.

Be aware of the following warning signs of an unsafe roof condition:

- Melting asphalt

- "Spongy" roof (a normally solid roof that springs back when walked upon)

NOTE: Some roofs are spongy with no fire involvement. Know the roofs in your district.

- Smoke coming from the roof

- Fire coming from the roof

Roof Coverings

In most cases, the first step in opening a roof for ventilation is to cut through or remove the roof covering. A *roof covering* is the part of the roof exposed to the weather. Roof coverings may be wooden shakes or shingles, imitation shakes or tiles made of molded metal or rubber, composition shingles or roll roofing, terracotta or concrete tile, slate, synthetic membrane, blown-on foam, or a built-up tar and gravel surface **(Figure 11.29)**. Some roof coverings are susceptible to ignition from sparks and burning embers falling on them; others are not. Some roof coverings have a coating of insulating material installed over them, making them thicker and harder to penetrate.

Existing Roof Openings

Existing openings can sometimes be used for vertical ventilation purposes because it is often quicker to enhance an opening than it is to cut a hole in the roof. These openings are rarely in the best location or large enough for adequate ventilation, however. In most cases they will simply supplement holes that have to be cut. In this context, existing roof openings refers to scuttle hatches, skylights, monitors, ventilating shafts,

CAUTION

Work in groups of at least two, but with no more personnel than absolutely necessary to get the job done.

Roof Covering — Final outside cover that is placed on top of a roof deck assembly. Common roof coverings include composition or wood shake shingles, tile, slate, tin, or asphaltic tar paper.

Figure 11.29 Various types of roof coverings.

and penthouse or bulkhead doors that can be used instead of or in addition to cutting a ventilation opening in a roof (**Figure 11.30**). Many existing roof openings will be locked or secured in some manner.

Scuttle hatches are normally square or rectangular and large enough to permit a person to climb through (**Figure 11.31, p. 562**). A scuttle hatch may be metal or metal-clad wood, and often does not provide an adequate opening for ventilation purposes. If skylights contain ordinary window glass, the glass panes can be removed or broken out with an axe or other tool. If they contain wired glass, Plexiglas acrylic plastic, or Lexan® plastic, the panes are very difficult to shatter but can be removed by dismantling part of the frame. The sides of a monitor may contain glass (which is easily removed) or louvers made of wood or metal (**Figure 11.32, p. 562**). If the side panels are hinged, the latch can be forced and the panel opened. If the top of the monitor is not removable, at least two sides should be opened to create the required draft. Penthouse or bulkhead stairway doors may be forced open in the same manner as other doors of the same type. The doors should then be blocked open or removed.

Figure 11.30 Skylights may be found on all types of occupancies and may be operable or fixed in place.

Figure 11.31 A typical scuttle hatch is not large enough to serve as a ventilation opening.

Figure 11.32 One type of roof monitor vent.

Roofs

When cutting a ventilation opening in a roof, there are a couple of critical points to bear in mind:

- A square or rectangular opening is easier to cut and easier to repair after the fire.

- One large opening, at least 4 × 4 feet (1.2 m by 1.2 m), is much better than several small ones (**Figure 11.33**).

Rotary saws, carbide-tipped chain saws, or ventilation saws are excellent for roof-cutting operations because they are faster and less damaging than axes or other manual cutting tools. The saw operator must have good footing and maintain control of the saw at all times. On a pitched roof, an axe, rubbish hook, or Halligan can be used to provide a secure foothold for the saw operator (**Figure 11.34**). In most cases, it is safest to turn off the saw when it is being transported to or from the point of operation – especially when moving up or down a ladder.

While a large square or rectangular opening is the type most frequently cut for vertical ventilation, other types of openings can be cut. Other types of openings used in vertical ventilation operations include kerf cuts, inspection cuts, and louver cuts (**Figure 11.35**). Always cut ventilation openings at or very near the highest point on the roof when possible. On peaked roofs, cut a few inches (mm) below the peak.

Flat Roofs

Flat roofs are commonly found on commercial, industrial, and apartment buildings, but they are common on many single-family residences as well. This type of roof may or may not have a slight slope to facilitate drainage. The flat roof is frequently penetrated by chimneys, vent pipes, shafts, scuttles, and skylights (**Figure 11.36**). The roof may be surrounded and/or divided by parapets; it may support water tanks, HVAC equipment, antennas, and other obstructions that may interfere with ventilation operations.

The structural part of a flat roof is generally similar to the construction of a floor assembly consisting of wooden, concrete, or metal joists covered with sheathing or decking. The decking may be sheets of plywood or oriented strand board (OSB), planks butted together, or corrugated metal. The decking is commonly covered with a layer of insulation, a waterproofing material (tar paper), and then mopped with hot asphalt roofing tar. Pea gravel or similar mineral material is then broadcast over the surface to increase weather resistance. Other finish materials may be roll roof-

Kerf Cut — A single cut the width of the saw blade made in a roof to check for fire extension.

Louver Cut or Vent — Rectangular exit opening cut in a roof, allowing a section of roof deck (still nailed to a center rafter) to be tilted, thus creating an opening similar to a louver. Also called Center Rafter Cut.

Figure 11.33 The ventilation opening must be large enough to be effective. Care must be taken to not cut the supporting rafters.

Figure 11.35 An inspection cut is used to locate roof supports. It is a small opening that will be enlarged once the rafters are located.

Figure 11.34 A pick-head axe can provide secure footing for one team member.

Figure 11.36 Flat roofs are usually penetrated by vent pipes, access doors or hatches, and other openings.

ing or a synthetic membrane. Over other flat roofs, a thin layer of reinforced or lightweight concrete is poured. Still others use precast gypsum or concrete slabs set within metal joists. The materials used in flat-roof construction determine what equipment will be necessary to create ventilation openings in the roof.

A procedure for opening a flat roof is suggested in the sequence given in **Skill Sheet 11-I-1**.

Pitched Roofs

There are a number of pitched roof styles. Among the most common are those elevated in the center along a ridge with a roof deck that slopes down to the eaves along the roof edges **(Figure 11.37, p. 564)**. Shed roofs are pitched along one edge and the deck slopes down to the eaves at the opposite edge. Most pitched-roof construction involves rafters or trusses that run from the ridge to the top of the outer wall at the eaves level. The rafters or trusses that carry the sloping roof can be made of wood or

Bowstring Truss — Lightweight truss design noted by the bow shape, or curve, of the top chord.

Figure 11.37 Many residential occupancies and some commercial occupancies have pitched roofs.

Figure 11.38 Solid sheathing will be covered with a weather-proof covering such as composition shingles.

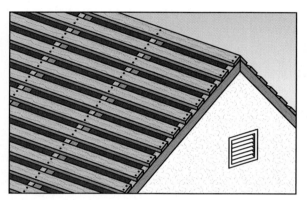

Figure 11.39 Typical skip sheathing consists of boards with open spaces at even intervals. A layer of roofing paper covered by shingles is attached to the skip sheathing.

Lamella Arch — A special type of arch constructed of short pieces of wood called lamellas.

metal. Over these rafters, the sheathing or decking material is applied at right angles. Sheathing is sometimes applied solidly over the entire roof (**Figure 11.38**). In other applications, sheathing consists of boards or planks set with a small space between them (**Figure 11.39**). This is commonly called *skip* sheathing. Pitched roofs usually have a covering of roofing paper applied before the finish is laid. The finish may consist of shingles, roll roofing, metal panels, slate, or tile.

Pitched roofs have a more pronounced downward slope than flat roofs, and in some cases may be quite steep. The procedures for opening pitched roofs are quite similar to those for flat roofs except that additional precautions must be taken to prevent slipping. Suggested steps for opening pitched roofs are given in **Skill Sheet 11-I-2**.

Other types of pitched roofs may require different opening techniques. For example, some slate and tile roofs may require no cutting. Slate and tile roofs can be opened by removing the individual pieces or using a large sledgehammer to smash the slate or tile pieces and the boards that support them. Metal roofs can be sliced open with an axe or rotary saw and peeled back.

Arched Roofs

Because they can span large open areas unsupported by pillars or posts, arched roofs are ideal for some types of occupancies. While there are several types of arched roofs, one is constructed using *bowstring* trusses. The lower chord of bowstring trusses may be covered with a ceiling to form an enclosed cockloft or attic space (**Figure 11.40**). These concealed spaces can hide a fire in progress long enough for the fire to seriously weaken the truss system. Bowstring trusses weakened by fire create a very dangerous situation for firefighters working either above or below the roof system.

Another type of arched roof is the Lamella or trussless arched roof. Lamella arched roofs are made up of relatively short boards of uniform length. These boards are beveled and bored at the ends, where they are bolted together at an angle to form an interdependent network of structural members (**Figure 11.41**). Because it is an arch rather than a truss, the roof exerts a horizontal reaction in addition to the vertical reaction on supporting structural components. To counteract these horizontal forces, Lamella roofs are supported by buttresses or tie rods or both. The same operational and safety considerations apply to Lamella roofs as bowstring truss roofs.

Another type of trussless arched roof uses massive arches of steel, concrete, or laminated wood buttressed into the ground at each end. These arches are connected to each other by horizontal members called *purlins*. Rafters between the purlins support the roof decking **(Figure 11.42)**. Because the roof support comes primarily from the arches, a hole of considerable size may be cut or burned through the roof without causing collapse of the roof structure.

Procedures for cutting ventilation openings in arched roofs are the same as for flat or pitched roofs except that there is no ridge over which to hook roof ladders, and the curvature of the roof prevents roof ladders from lying flat against the roof. Remember to sound the roof and walk only on the trusses and other strong points when possible. As soon as you are on the roof, cut an inspection hole to locate the arches and to observe the truss space and fire involvement before proceeding further.

Concrete Roofs

There are essentially two types of concrete roofs common in North America: precast and poured-in-place. Because precast concrete roof units can be fabricated off-site and hauled to the construction site ready for use, this type of roof construction is in widespread use. Precast roof slabs are available in many shapes, sizes, and designs **(Figure 11.43, p. 566)**.

Purlin — Horizontal member between trusses that supports the roof.

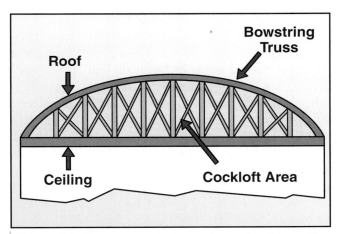

Figure 11.40 The space between the roof and the ceiling is known as the cockloft. This space can contribute to fire spread if ventilation is not properly performed.

Figure 11.42 Purlins connect the arches of a trussless roof.

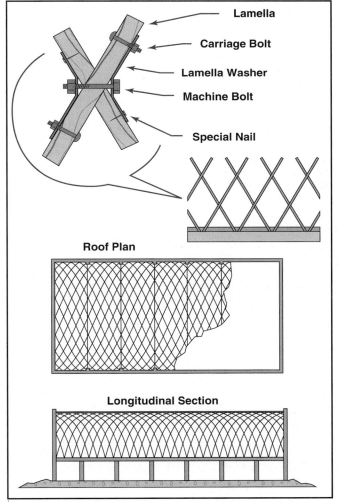

Figure 11.41 The typical Lamella arch pattern that permits the construction of arched roofs without trusses.

Figure 11.43 Precast concrete roof units may be found in many types of occupancies.

Figure 11.44 Some concrete roofs are poured in place on the construction site. *Courtesy of Ron Moore.*

Another type of precast concrete roof uses lightweight material made of gypsum plaster and portland cement mixed with aggregates, such as perlite, vermiculite, or sand. This material is sometimes referred to as lightweight concrete. Lightweight precast planks are manufactured from this material, and the slabs are reinforced with steel mesh or rods. Lightweight concrete roofs are usually finished with roofing felt and a mopping of hot tar to make them watertight.

Some lightweight concrete roof decks are also poured in place over permanent form boards, steel roof decking, paper-backed mesh, or metal rib lath. These lightweight concrete slabs are relatively easy to penetrate. Some types of lightweight concrete can be penetrated with a hammer-head pick or power saw with concrete blade. Heavier concrete roofs will require the jackhammer or diamond-tipped chain saw. Other builders choose to form and pour the concrete on the job **(Figure 11.44)**. Roofs of either precast or reinforced concrete are extremely difficult to break through, so opening them should be avoided whenever possible. Existing roof openings should be used to vertically ventilate buildings with heavy concrete roofs.

Figure 11.45 Corrugated metal roofs are found on many industrial and commercial occupancies and a few residential-type structures.

Metal Roofs

Metal roof coverings are made from several different kinds of metal and are constructed in many styles. Light-gauge steel roof decks can either be supported on steel frameworks or laid over an existing roof for security. Other types of corrugated roofing sheets are made from light-gauge cold-formed steel, galvanized sheet metal, and aluminum. The light-gauge cold-formed steel sheets are used primarily for the roofs of industrial buildings. Except when covered with lightweight concrete, corrugated galvanized sheet metal and aluminum are seldom covered with a roofing material, and the sheets can usually be pried from their supports **(Figure 11.45)**.

Metal cutting tools or power saws with metal cutting blades must be used to open metal roofs. Metal roofs on industrial buildings are often penetrated by roof openings such as skylights or hatches. Older buildings may have roofs that are made of large pieces of sheet metal laid over skip sheathing. These can be opened by cutting with a power saw, axe, or a large sheet-metal cutter similar to an old-fashioned can opener.

Trench or Strip Ventilation

Trench ventilation (also called *strip ventilation*) is used in a slightly different way than the conventional vertical ventilation techniques previously described. Conventional vertical ventilation is used primarily to remove heated smoke and gases from the structure and is best done directly above the fire. *Trench ventilation* is used to stop the spread of fire in a long, narrow structure. Trench ventilation is performed by cutting a large opening, or trench, that is at least 4 feet (1.2 m) wide and extends from one exterior wall to the opposite exterior wall **(Figure 11.46)**. This opening must be created well ahead of the advancing fire because the cut must be completed before the fire front reaches that point. Otherwise, the fire will simply burn past the unfinished trench cut and continue to spread throughout the building and endanger firefighters on the roof. In many cases, a large ventilation opening (sometimes called a *heat hole*) is cut between the trench cut and the fire. Its purpose is to vent as much of the heat and flames as possible before they reach the trench cut.

Conventional Basement Fire Ventilation

The importance of ventilation when attacking basement fires cannot be overstated. In the absence of built-in vents from the basement, heat and smoke from basement fires will quickly spread upward into the building **(Figure 11.47)**. This is especially true in buildings of balloon-frame construction where the wall studs are continuous from the foundation to the roof. There may be no fire-stops (wood or other solid material placed within a void to retard or prevent the spread of fire through the void) between the studs. In buildings of this type, the first extension of a basement fire will commonly be into the attic.

The likelihood of vertical extension of the fire may be reduced by direct ventilation of the basement during fire attack. In addition, unless the basement is vented in a location away from the point of entry, the fire attack team may be prevented from en-

Trench Ventilation — Defensive tactic that involves cutting an exit opening in the roof of a burning building, extending from one outside wall to the other, to create an opening at which a spreading fire may be cut off.

Figure 11.46 Trench or strip ventilation is used to stop the spread of fire in a long narrow structure.

Figure 11.47 Convection spreads fire vertically through a structure using open stairs, shafts, and the inside of walls that lack fire stops.

tering the basement by the heat, smoke, and fire venting through the entry opening. If a ventilation opening is made away from the point of entry, attack crews can enter with nozzles set on a wide fog pattern and push the heat and smoke out the ventilation opening. After the basement fire is confirmed to be extinguished, the attic may be vented to remove residual smoke.

Basement ventilation can be accomplished in several ways. If the basement has ground-level windows or even belowground-level windows in wells, horizontal ventilation can be employed effectively (**Figure 11.48**). If these windows are not available, interior vertical ventilation must be performed. Natural paths from the basement, such as stairwells and hoistway shafts, can be used to evacuate heat and smoke provided there is a means to expel the heat and smoke to the atmosphere without placing other portions of the building in danger (**Figure 11.49**). As a last resort, a hole may be cut in the floor near a ground-level door or window, and the heat and smoke can be forced from the hole through the exterior opening using fans (**Figure 11.50**).

Figure 11.48 Basement windows can be used for ventilation.

Figure 11.50 If necessary, a basement fire can be vented through a hole cut in the floor above.

Figure 11.49 Stairways can be used to vent a basement or cellar by using mechanical ventilation fans to draw the smoke up and force it out the first floor openings.

Precautions Against Upsetting Vertical Ventilation

When vertical ventilation is accomplished, the natural convection of the heated gases creates upward currents that draw the fire and heat in the direction of the upper opening. Fire attack teams take advantage of the improved visibility and less contaminated atmosphere to attack the fire at its lower point.

Elevated streams are sometimes used to extinguish sparks and flying embers from a burning building or to cool the thermal column of heat over a building (**Figure 11.51**). However, if handlines or master streams are directed downward into a ventilation opening or are improperly used to reduce the thermal column, the orderly movement of fire gases from the building is either reduced or reversed. This can force superheated air and gases back down on firefighters, perhaps causing serious injury or death. At the very least, it will contribute to the spread of fire throughout the structure.

Ventilation problems can be avoided by well-trained firefighters conducting a well-coordinated attack. Some common factors that can reduce the effectiveness of vertical ventilation are:

- Improper use of forced ventilation
- Indiscriminant window breaking
- Fire streams directed into ventilation openings
- Breaking skylights
- Explosions
- Burn-through of the roof, a floor, or a wall
- Additional openings between the attack team and the upper opening

Thermal Column — Updraft of heated air, fire gases, and smoke directly above the involved fire area.

WARNING!
Never direct any type of fire stream into a ventilation opening during offensive operations. This interrupts the ventilation process and places interior crews in serious danger.

Figure 11.51 Using elevated streams to cool the thermal column reduces the chances of secondary fires caused by flying embers.

Vertical ventilation is not the solution to all ventilation problems because there are instances where its application is impractical or impossible. In these cases, other strategies, such as the use of strictly horizontal ventilation, must be employed.

Horizontal Ventilation

Horizontal ventilation is the venting of heat, smoke, and gases through horizontal openings such as windows and doors. Structures that lend themselves to the application of horizontal ventilation include the following:

- Buildings in which the fire has not involved the attic area
- Involved floors of multistoried structures below the top floor, or the top floor if the attic is uninvolved
- Buildings so weakened by the fire that vertical ventilation is unsafe
- Buildings with daylight basements
- Buildings in which vertical ventilation is ineffective (cold smoke fires)

Although many aspects of vertical ventilation also apply to horizontal ventilation, a different procedure must be followed in horizontally ventilating a room, floor, cockloft, attic, or basement. The procedure to be followed will be influenced by the location and extent of the fire. Besides direct flame contact, some of the ways by which horizontal extension occurs inside a structure are as follows:

- Through wall openings by convected air currents
- Through corridors, halls, or passageways by convected air currents
- Through open space by radiated heat or by convected air currents
- In all directions by explosions or flash fires
- Through walls and interior partitions by direct flame contact
- Through walls by conduction of heat through beams, pipes, or other objects that extend through walls

Weather Conditions

Weather conditions must always be considered when setting up horizontal ventilation. Although wind is not the only weather phenomenon that can affect horizontal ventilation, it has the most potential influence. Wind can either aid or hinder horizontal ventilation. Wind direction is designated as either windward or leeward. The side of the building the wind is striking is *windward,* the opposite is *leeward* (**Figure 11.52**). When setting up natural horizontal ventilation, the entry opening must be made on the windward side of a burning building, and the exit opening on the leeward side. This makes the maximum use of the natural wind currents. When there is no wind, natural horizontal ventilation is less effective, but the buoyancy of the hot fire gases and the pressure caused by the fire will continue to force smoke out of the openings. In other instances, natural horizontal ventilation cannot be used because of the danger of wind blowing toward an exposure or increasing the intensity of the fire.

Exposures

When setting up horizontal ventilation, firefighters must consider both internal and external exposures. Internal exposures include the building occupants and any uninvolved rooms or portions of the building. Because horizontal ventilation does not release heat and smoke directly above the fire, some routing is necessary. The routes by which the smoke and heated gases would travel to exit the building may be the same corridors and passageways that occupants need for evacuation. Therefore, because horizontal ventilation may block the escape of occupants, the effects of horizontal ventilation on rescue and exit requirements must be considered.

Because horizontal ventilation causes heat, smoke, and sometimes fire to be discharged below the highest point of the building, there is also the danger that the rising gases will ignite portions of the building above the exit point. Heat and gases may be drawn into open windows or attic vents above the exit point (**Figure 11.53**). They may also ignite the eaves of the burning building or adjacent structures.

Unless it is done for the specific purpose of aiding in rescue, a building should *not* be ventilated until charged hoselines are in place at the entry point. Ideally, charged hoselines should also be ready at the intermediate point where fire might be expected to spread, and in positions to protect other exposures (**Figure 11.54**).

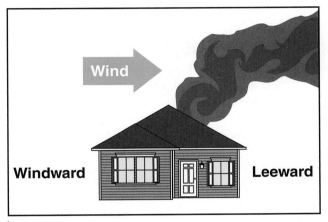

Figure 11.52 Knowledge of wind direction and how to describe it is extremely important when determining what type of ventilation to establish.

Figure 11.53 Fire may reenter above the fire floor.

Figure 11.54 Attack lines must be in position before ventilation begins. Communication between the hose teams and the ventilation teams is critical for effective fire suppression.

Daylight Basement Ventilation

Homes built on a slope may have what are called *daylight basements* (**Figure 11.55**). These are finished basements with large exterior windows, and usually one or more exterior doors. Daylight basements are generally much easier to ventilate than the conventional basements discussed earlier. Opening the exterior door or doors, breaking out the large windows, or both will provide adequate ventilation to allow an aggressive fire attack from the interior.

Precautions Against Upsetting Horizontal Ventilation

Opening a door or window on the windward side of a burning building before creating a ventilation exit opening on the leeward side may pressurize the building, intensify the fire, and cause the fire to spread to uninvolved areas. Firefighters should take advantage of air currents established by horizontal ventilation (**Figure 11.56**). If the established currents are blocked by a firefighter or other obstruction in the doorway, the positive effects of horizontal ventilation may be reduced or eliminated (**Figure 11.57**).

Forced Ventilation

To this point, only ventilation created by the natural flow of air currents and the currents created by the fire has been discussed. When natural ventilation is inadequate, firefighters must use *forced ventilation*. Forced ventilation is accomplished *mechanically* with fans or blowers or *hydraulically* with fog streams. Positive forced ventilation is used to blow fresh air into a building, and negative forced ventilation is used to expel smoke or other airborne contaminants from the building. Ideally, forced ventilation

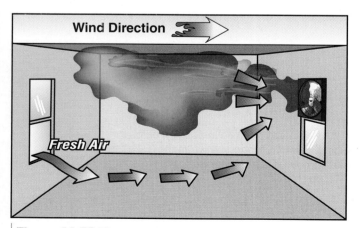

Figure 11.55 A residence with a typical daylight basement that is open on one side.

Figure 11.56 The prevailing wind can be used to assist in horizontal ventilation.

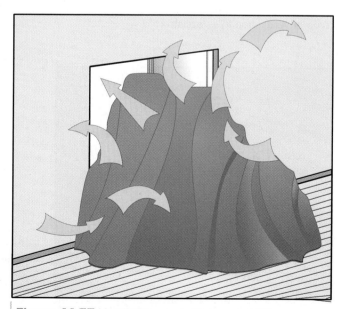

Figure 11.57 Ventilation can be hampered when equipment or materials block openings. Piles of furniture that are covered by a salvage cover can present such a problem.

Figures 11.58 a and b Common types of portable fans may be powered by (a) electricity or (b) gasoline engines.

should be used in conjunction with and take advantage of any natural ventilation caused by prevailing winds. While water from a hoseline can be used to expel smoke and other products of combustion from a building, the majority of forced ventilation operations are carried out using one or more portable fans or blowers. While there are portable blowers powered by hydraulic pressure, most portable fans used in forced ventilation are powered by electric motors or gasoline-driven engines (**Figures 11.58 a and b**).

Advantages of Forced Ventilation

The need for ventilation during interior fire fighting operations is clear. There are also situations that do not involve fire where contaminated atmospheres must be quickly and thoroughly cleared from a building or other confined space. For example, buildings filled with a flammable or toxic gas must be ventilated quickly but safely. In these and many other situations, forced ventilation is the best technique to use. Some of the advantages of using forced ventilation in fires and other situations include the following:

• It supplements and enhances natural ventilation

• It ensures more control of air flow

• It speeds the removal of contaminants

• It reduces smoke damage

• It promotes good public relations

Disadvantages of Forced Ventilation

As worthwhile as forced ventilation is, if it is misapplied or not properly controlled, it can cause a great deal of harm. Some of the disadvantages of improperly forced ventilation include the following:

• It may cause a fire to intensify and spread

• It depends upon a power source

• It requires special equipment

Negative-Pressure Ventilation — Technique using smoke ejectors to develop artificial circulation and to pull smoke out of a structure. Smoke ejectors are placed in windows, doors, or roof vent holes to pull the smoke, heat, and gases from inside the building and eject them to the exterior.

Churning — Movement of smoke being blown out of a ventilation opening only to be drawn back inside by the negative pressure created by the ejector because the open area around the ejector has not been sealed. Also called recirculation.

Negative-Pressure Ventilation (NPV)

The term *negative-pressure ventilation (NPV)* describes the oldest type of mechanical forced ventilation techniques: using fans (smoke ejectors) to develop artificial circulation or enhance natural ventilation to expel smoke from a structure. Fans are placed in windows, doors, or roof vent openings to exhaust the smoke, heat, and gases from inside the building to the exterior **(Figure 11.59)**.

In NPV, the fan should be positioned to exhaust in the same direction as the prevailing wind. This technique uses the wind to supply fresh air to replace that which is being expelled from the building. If the prevailing wind is too light to be effective, fans can be positioned on the windward side of the structure to blow air into the building. At the same time, the fans on the other side exhaust the smoke and other combustion products from the building.

If the open areas around a smoke ejector are not properly sealed, air can recirculate back into the building. Atmospheric pressure pushes air back through the open spaces in the doorway or window and pulls the smoke back into the room. This recirculation is called *churning*, and it reduces ventilation efficiency **(Figure 11.60)**. To prevent churning, cover the open area around the fan with a salvage cover or other material.

The flow of smoke and other gases to the exit opening should be kept as straight as possible. Every corner causes turbulence and decreases ventilation efficiency. Avoid opening windows or doors near the exhaust fan because this action can greatly reduce ventilation efficiency. Remove all obstacles to the airflow. Even a window screen will cut effective exhaust by half. Do not allow the intake side of the fan to become obstructed by debris, curtains, drapes, or anything else that can decrease the amount of intake air.

When ventilating potentially flammable atmospheres, only exhaust fans equipped with intrinsically safe motors and power cable connections should be used. Exhaust fans should be turned off when they are moved, and they should be carried by the

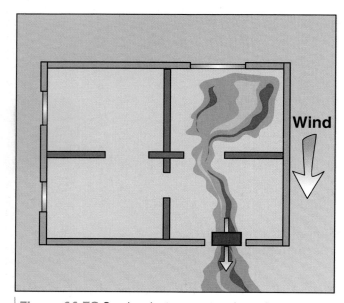

Figure 11.59 Smoke ejectors are used to exhaust smoke and heat from a building. The unit is placed in an opening in the exterior of the structure and used to pull the smoke out.

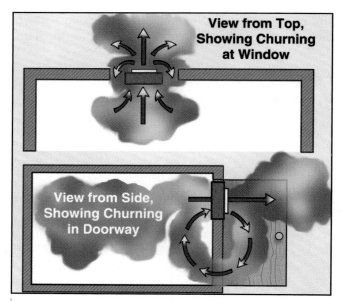

Figure 11.60 Churning is a condition that occurs when the area surrounding the fan is not sealed with a salvage cover. The smoke continues to be drawn back into the fire room rather than making a complete exit.

handles provided for that purpose. Before starting exhaust fans, be sure that no one is near the blades and that clothing, curtains, or draperies are not in a position to be drawn into the fan. The discharge stream of air should be avoided because of debris that may be picked up and blown by the venting equipment.

Positive-Pressure Ventilation (PPV)

Positive-pressure ventilation (PPV) is a forced ventilation technique that uses a high-volume fan to create a slightly higher pressure inside a building than that outside **(Figure 11.61)**. As long as the pressure is higher inside the building, the smoke within the building is forced through the ventilation exit opening to the lower-pressure zone outside.

The location where PPV is set up is called the *entry point*. Once that location is selected, an exit opening must be created opposite the entry point. The size of the exit opening varies with the size of the entry opening and the capacity of the blower used. The exit opening may be a window or doorway.

Once an exit opening has been created, a blower is placed about 4 to 10 feet (1.2 m to 3 m) outside the open entry point so that the cone of air from the blower completely covers the doorway opening **(Figure 11.62)**. The smoke is then expelled from the exit opening distant from the entry point that is the same size as or slightly smaller than the entry opening. To maintain the positive pressure inside, it is important that no other exterior doors or windows are opened during the positive-pressure ventilation operation.

Positive-Pressure Ventilation (PPV) — Method of ventilating a confined space by mechanically blowing fresh air into the space in sufficient volume to create a slight positive pressure within and thereby forcing the contaminated atmosphere out the exit opening.

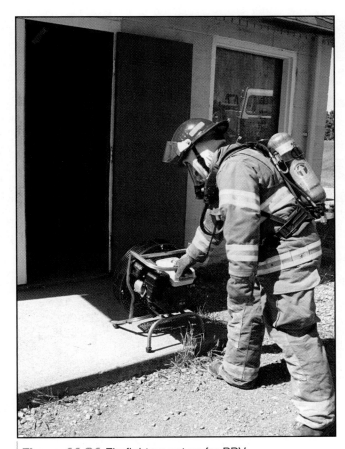

Figure 11.61 Firefighters set up for PPV.

Figure 11.62 The cone of air must cover the entire doorway opening.

By selectively opening and closing interior doors and exterior windows, it is possible to pressurize one room or area at a time (**Figure 11.63**). This process accelerates the removal of heat and smoke from the building. In some cases, the process of removing cold smoke after a fire has been extinguished can be accelerated even more by positioning a negative-pressure fan in the exit point (**Figure 11.64**).

When using PPV to ventilate a multistory building, it is best to apply positive pressure at the lowest point (**Figure 11.65**). Positive pressure is applied to the building at ground level through the use of one or more blowers. The positive pressure is then directed throughout the building by opening and closing doors until the building is totally evacuated of smoke.

If a single fan cannot provide enough flow, an additional PPV blower can be set up on an upper floor (**Figure 11.66**). Smoke can then be systematically removed one floor at a time (starting with the floor most heavily charged with smoke) by selectively opening exit points. This is accomplished by either cross-ventilating fire floors or directing smoke up a stairwell and out the stair shaft rooftop opening. Fans or blowers larger than those typically carried on an engine or ladder truck are also available for use in multistory and large-volume buildings. The procedure for ventilating using mechanical positive pressure is given in **Skill Sheet 11-I-3**.

Figure 11.63
Opening and closing interior doors at the proper time can help clear a building of smoke.

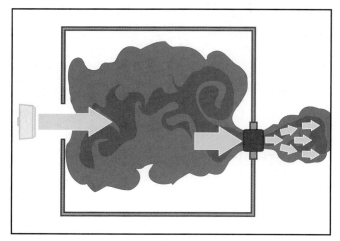

Figure 11.64
The simultaneous use of PPV and NPV fans can be effective in removing smoke.

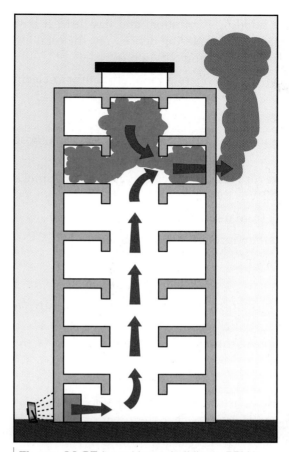

Figure 11.65 In multistory buildings, PPV is usually applied at the ground floor.

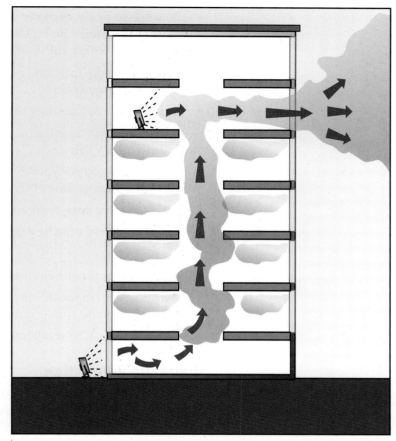

Figure 11.66 In some situations, PPV can be improved by adding a second blower to direct the smoke to the exit point.

Positive-pressure ventilation requires good fireground discipline, coordination, and tactics. The main problem in using positive-pressure ventilation in aboveground operations is coordinating the opening and closing of the doors in the stairwell being used to ventilate the building. Curious tenants will often stand in their doorways and thereby redirect the positive pressure away from the fire floor. To control openings or pressure leaks, put one person in charge of the pressurizing process. It is helpful to use portable radios and to have firefighters patrol the stairwell and hallways.

To ensure an effective PPV operation, take the following actions:

- Take advantage of existing wind conditions.
- Make certain that the cone of air from the blower covers the entire entry opening.
- Reduce the size of the area being pressurized to speed up the process by selectively opening and closing interior doors.
- Keep the size of the exit opening in proportion to the entry opening.
- Avoid creating horizontal openings by breaking glass or removing doors.

The advantages of PPV compared to NPV include the following:

- Firefighters can set up PPV without entering the smoke-filled environment.
- PPV is equally effective with either horizontal ventilation or vertical ventilation because it merely supplements natural ventilation currents.
- Removal of smoke and heat from a structure or vessel is more efficient.

- The velocity of air currents within a building is minimal and has little, if any, effects that disturb the building contents or smoldering debris. Yet, the total exchange of air within the building is faster than using NPV alone.
- Fans powered by internal combustion engines operate more efficiently in clean, oxygen-rich atmospheres.
- The placement of fans does not interfere with ingress or egress.
- The cleaning and maintenance of fans used for PPV is significantly less than those needed for NPV fans.
- PPV is effective in all types of structures or vessels, particularly at removing smoke from large, high-ceiling areas where NPV is ineffective.
- Heat and smoke may be directed away from unburned areas or paths of exit.
- Exposed buildings can be pressurized to reduce fire spread into them.

The disadvantages of PPV are as follows:

- An intact structure is required.
- Interior carbon monoxide levels may be increased if the exhaust from fans driven by internal combustion engines is allowed to enter.
- Hidden fires may be accelerated and spread throughout the building.

Hydraulic Ventilation

Hydraulic ventilation may be used in situations where other types of forced ventilation are unavailable. *Hydraulic ventilation* is used to clear the room or building of smoke, heat, steam, and gases after a fire has been controlled. This technique uses the air movement created by a fog stream to help draw the products of combustion out of the structure.

To perform hydraulic ventilation, a fog stream is set on a wide fog pattern that will cover 85 to 90 percent of the window or door opening from which the smoke will be pushed out. The nozzle tip should be at least 2 feet (0.6 m) back from the opening (**Figure 11.67**). The larger the opening, the faster the ventilation process will go. In fact, a master stream device can be set up in an open commercial or industrial doorway such as those on loading docks (**Figure 11.68**). The procedure for horizontal hydraulic ventilation is shown in **Skill Sheet 11-I-4**.

There are disadvantages to the use of hydraulic ventilation. These disadvantages include the following:

- There may be an increase in the amount of water damage within the structure.
- There will be a drain on the available water supply. This is particularly crucial in rural fire fighting operations where water shuttles are being used.
- In freezing temperatures, there will be an increase in the amount of ice on the ground surrounding the building.
- The firefighters operating the nozzle must remain in the heated, contaminated atmosphere throughout the operation.
- The operation may have to be interrupted when the nozzle team has to leave the area to replenish their air supply.

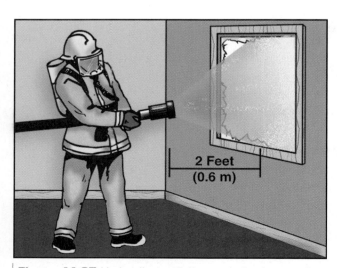

2 Feet (0.6 m)

Figure 11.67 Hydraulic ventilation uses the fog nozzle water stream to entrain or carry the smoke out of the building.

Figure 11.68 A master stream device can be used for large-scale hydraulic ventilation.

Effects of Building Systems on Fires

Many modern buildings have built-in HVAC systems that can significantly contribute to the spread of smoke and fire throughout a structure. These systems are usually controlled from a panel in a maintenance and operations center located somewhere in the building – often the basement. Wherever the controls are located, there is often a diagram of the duct system in the building along with information on smoke detection and fire suppression systems built into the HVAC ductwork. These systems are designed to shut down the HVAC system automatically when smoke or fire is detected in the ducts. Fire personnel should be familiar with the location and operation of controls that will allow them to shut down the HVAC system when necessary.

While firefighters may need to shut the HVAC system down during a fire, clearing the system of residual smoke and restoring it to operation are responsibilities of the building engineer or maintenance superintendent. Because an HVAC system may draw fire into the ducts along with the heat and smoke before it shuts down, firefighters should check combustibles adjacent to the ductwork for fire extension due to conduction.

Many other buildings, especially high-rises, shopping malls, and buildings with open atria are equipped with built-in smoke control systems. These systems are designed to confine a fire to as small an area as possible by compartmentalizing the building when smoke or fire is detected. This is achieved by the automatic closure of doors, partitions, windows, and fire dampers. Like HVAC systems, smoke control systems usually have a system diagram in the same location as the control panel. The panel should indicate where the alarm originated and which automatic closers were activated (**Figure 11.69**).

WARNING!
Only building engineers or maintenance superintendents should operate building systems to assist in ventilation. Incorrect use of these systems can cause severe damage to them and may create a more hazardous condition.

Figure 11.69 A smoke control system panel may be located in the fire control room, security office, or building mechanical office.

Summary

Ventilating a burning building allows heat, smoke, and other products of combustion to escape to the atmosphere. It also allows clear, cool air to be drawn into the building. This allows firefighters to see better, locate victims more easily, and find the seat of the fire sooner. Ventilation also limits fire spread and channels the heat and smoke away from any trapped victims.

To perform horizontal and vertical ventilation safely and effectively, firefighters must understand fire behavior and know the various ventilation methods. Firefighters must have a knowledge of roof construction and know how to create ventilation openings in flat and pitched roofs that have a variety of coverings.

Fire Fighter I

1. What are the reasons for fireground ventilation?

2. What are the signs of a potential backdraft?

3. What questions should be asked when deciding to ventilate?

4. List three factors that help determine where to ventilate.

5. What is the difference between horizontal ventilation and vertical ventilation?

6. List four safety precautions for vertical ventilation.

7. What is trench ventilation?

8. Why is conventional basement fire ventilation important?

9. What structures lend themselves to the application of horizontal ventilation?

10. Describe two ways to accomplish forced ventilation.

Step 1: Confirm order with officer to ventilate flat roof.

Step 2: Size up scene for any hazards.

Step 3: Select location for ventilation.

Step 4: Outline ventilation opening with pick on axe or other similar tool.

Step 5: Cut three-sided (triangular) inspection opening in roof to determine fire conditions.

Step 6: Cut roof deck parallel to a roof truss or support on side furthest away from ladder or escape route. This is cut #1.

Step 7: Cut roof deck on one side of opening perpendicular to the first cut — cut must intersect first cut in Step 6. This is cut #2.

Step 8: Cut roof deck on opposite side of cut made in Step 7 – cut must intersect cut made in Step 6. This is cut #3.

Step 9: Complete the ventilation hole by cutting between cut #2 and cut #3.

Step 10: Remove decking from the ventilation opening with axe, pike pole, or other sounding tool.

Step 11: Plunge through interior ceiling using pike pole working from upwind side of ventilation hole.

Step 12: Report to officer completion of assigned task.

Step 1: Confirm order with officer to ventilate pitched roof.

Step 2: Size up scene for any hazards.

Step 3: Select location for ventilation.

Step 4: Outline ventilation opening with pick on axe or other similar tool.

Step 5: Cut roof deck across the rafters on the high side of the roof parallel to the ridge.

Step 6: Cut roof deck on furthest side of ventilation opening perpendicular to the cut made in Step 5.

Step 7: Cut roof deck on opposite side of cut made in Step 6.

Step 8: Complete the ventilation opening by cutting between the bottom of the two parallel cuts made in Steps 6 and 7.

Step 9: Remove decking from the ventilation opening with axe or pike pole.

Step 10: Plunge down through the ceiling using pike pole working from upwind side of ventilation opening.

Step 11: Report to officer completion of assigned task.

Skill SHEETS

11-I-3
Ventilate a structure using mechanical positive-pressure ventilation.

Step 1: Confirm order with officer to ventilate structure.

Step 2: Place fan near entrance opening so that it will create a positive pressure within the structure.

Step 3: Start fan(s) and temporarily direct away from opening.

Step 4: Create exit opening approximately equal to or smaller than the "point of entry."

Step 5: Direct fan into point of entry so that cone of air covers opening.

Step 6: Determine if smoke is moving away from point of entry and toward exit. If not, discontinue use of fan and revaluate location of point of entry and exit and any obstructions of the flow of air.

Step 7: Clear smoke out of building.

NOTE: Instructor gives firefighters the following information:

- You are inside a room within a structure where you have extinguished a fire involving room and contents only.

- The ventilation crew has been unsuccessful.

- We need to ventilate this room with our hoseline.

- Find a window or door and open it.

Step 1: Extend nozzle outside of opening and open nozzle.

Step 2: Set the fog nozzle pattern wide enough to cover 85 to 90 percent of window or door opening.

11-I-4
Ventilate a structure using horizontal hydraulic ventilation.

Skill SHEETS

Step 3: Bring nozzle approximately 2 feet (0.6 m) inside building.

Step 4: Monitor progress of ventilation.

Chapter Contents

Water Supply

This chapter provides information that addresses the following job performance requirements of NFPA® 1001, *Standard for Fire Fighter Professional Qualifications (2008):*

NFPA® 1001 references for Chapter 12

5.3.15

6.5.3

Chapter Objectives

Fire Fighter I Objectives

1. Describe dry-barrel and wet-barrel hydrants.

2. Discuss fire hydrant marking and location.

3. Summarize potential problems to look for when inspecting fire hydrants.

4. Explain the process of fire hydrant testing.

5. Discuss alternative water supplies.

6. Discuss rural water supply operations.

7. Operate a hydrant.

8. Make soft-sleeve and hard-suction hydrant connections.

9. Connect and place a hard-suction hose for drafting from a static water source.

10. Deploy a portable water tank.

Fire Fighter II Objectives

1. List sources of water supply.

2. Describe the three methods of moving water in a system.

3. Discuss water treatment facilities.

4. Explain the operation of water storage and distribution systems.

5. Distinguish among the pressure measurements relevant to water supply.

6. Use a pitot tube.

Chapter 12

Water Supply

Case History

On May 27, 2005 a group of firefighters were on the 3rd floor of a 3-story garden apartment building performing overhaul. The crew had opened walls to extinguish hot spots when water pressure was lost. Other crews operating in the building also lost water pressure.

The fire rekindled and very quickly extended into the hallway. Firefighters were unable to contain the fire due to lack of water. Fire conditions soon forced the crew to retreat down the hallway to the stairwell. The fire that had also rekindled on the second floor had made the stairwell unusable as an escape route. Firefighters took refuge in an open apartment, closed the hallway and bedroom door and went onto the balcony. A crew raised a ladder to the balcony and firefighters were able to escape. Without a sustainable water supply, fire extended unchecked throughout the building and over 70 apartments were destroyed. Fortunately, all firefighters working inside the building at the time were able to escape without injury.

A sustainable water supply is essential to all fire fighting operations. Firefighters must be able to quickly establish a water supply that is dependable and will meet fire suppression requirements. Failure to establish such a water supply limits operational capability and places firefighters at risk.

(Source: National Fire Fighter Near-Miss Reporting System)

Despite the development of many new tools and techniques for fighting fires, water continues to be the primary extinguishing agent because of its availability, affordability, and effectiveness. Water has two other desirable characteristics: it is easily stored and can be conveyed long distances. These are also the fundamental characteristics of a water supply system – the storage and transfer of large quantities of water. Because water continues to be the primary extinguishing agent used by firefighters, it is important that they have a working knowledge of water supply systems.

At the Fire Fighter I level, NFPA® 1001 requires that firefighters be able to connect a fire department pumper to a water supply as a member of a team. This requires knowledge of the following:

- Loading and off-loading procedures for water tenders
- Fire hydrant operation
- Suitable static water supply sources

To carry out these operations, those qualified at the Fire Fighter I level must be able to:

- Deploy portable water tanks and associated equipment
- Fully open and close a hydrant

At the Fire Fighter II level, the firefighter must have knowledge of the sources of water supply for fire protection. The chapter also covers the skill of using a pitot tube.

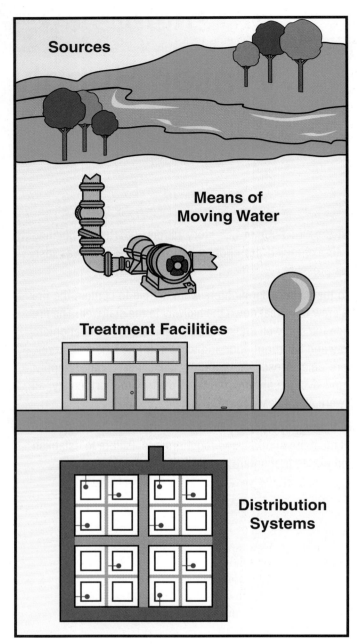

Sources

Means of
Moving Water

Treatment Facilities

Distribution
Systems

Figure 12.1 There are four components to any municipal water supply system.

This chapter discusses the principles of public and private water supply systems and the methods of moving water through these systems. Also included is a description of the components of a water distribution system and the different ways pressure is measured within a system. The chapter also explains the design of fire hydrants, how and where they are located, and how they are maintained. Finally, alternative water supplies such as lakes and ponds are discussed along with the methods of moving the water from the source to the fire by water shuttles and relay pumping.

Principles of Water Supply Systems

Many agricultural and industrial water supply systems serve only the properties on which they are located. In contrast, public and many private water supply systems serve large areas and share the water with many individual properties. When populations increase in rural areas, these communities are forced to develop or improve water sources and distribution systems.

In any given area, the water department may be a separate, city-operated utility or a regional or private water authority. Its principal function is to provide potable water (suitable for drinking) to the customers within the district's boundaries. In most cases, water department officials are considered the local experts in water supply. Fire departments work with their local water departments in planning fire protection coverage. In turn, water department officials recognize the vital interests that fire departments have in water supplies and the locations and types of fire hydrants.

The design of water supply systems may vary from region to region. However, all systems are composed of the following basic components, which are explored in the following subsections (**Figure 12.1**):

- Sources of water supply
- Means of moving water
- Water treatment facilities
- Water storage and distribution systems

Sources of Water Supply

The jurisdiction's primary water supply can be obtained from surface water, groundwater, or both. Although most water systems are supplied from only one source, there are instances where both types of sources are used. Examples of surface water supply are rivers, aqueducts, lakes, and reservoirs (**Figure 12.2**). In most cases, groundwater supplies come from water wells drilled into underground aquifers. A very few come from water-producing springs.

The amount of water that a community needs for both domestic use and fire protection can be calculated based on the history of consumption and estimates of anticipated needs. Cities and other water providers track their average and maximum daily water consumption over time. Some even track their peak hourly consumption. To this history, engineers add anticipated fire protection needs based on fire flow requirements within the jurisdiction's boundaries. To be considered adequate, a system must be capable of supplying the water needed for fire protection in addition to the domestic requirement. In most cities, the domestic/industrial requirements exceed that required for fire protection. However, in small towns the requirements for fire protection may exceed other requirements.

Means of Moving Water

There are three methods of moving water within a system:

- Direct pumping systems
- Gravity systems
- Combination systems

Direct Pumping Systems

While a few cities have direct pumping water systems that are dedicated for fire protection, most direct pumping systems are found in agricultural and industrial settings. Many rural fire departments are equipped to tap into agricultural irrigation systems when water is needed for fire fighting (**Figure 12.3**). In direct pumping systems, one or more pumps draw water from the primary source and transport it to the point of use. If the water is to be used for drinking and other domestic purposes, it is pumped to a filtration and treatment facility (**Figure 12.4, p. 596**). Some systems provide filtration and chlorination at the source (**Figure 12.5, p. 596**). From there, the chlorinated water is pumped directly into the distribution system. The main disadvantages of direct pumping systems are their total dependence on pumps (subject to mechanical failure) and on electricity (subject to power outages) to run the pumps. While emergency generators can prevent a total loss of power, the pumps and distribution piping are still vulnerable. Therefore, duplicate pumps and piping are necessary to ensure system reliability.

Direct Pumping System — Water supply system supplied directly by a system of pumps rather than elevated storage tanks.

Figure 12.2 If accessible, large rivers make good water supply sources.

Figure 12.3 Irrigation systems can be a source of water for fire fighting.

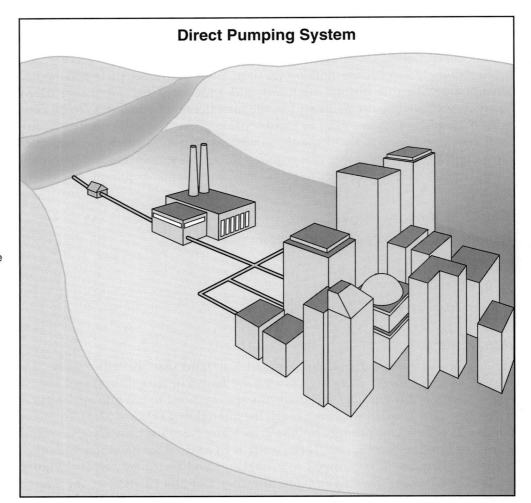

Direct Pumping System

Figure 12.4 A direct pumping system is used when the surface water source lacks the required elevation to create adequate pressure.

Figure 12.5 Water is chlorinated at the source.

Gravity Systems

Gravity System — Water supply system that relies entirely on the force of gravity to create pressure and cause water to flow through the system. The water supply, which is often an elevated tank, is at a higher level than the system.

A true gravity system uses a primary water source located at a higher elevation than the distribution system and delivers water to the system without the use of pumps. Gravity provides the pressure needed to transport the water to where it is needed **(Figure 12.6)**. However, gravity pressure is adequate only when the primary water source is located more than 100 feet (30 m) higher than the highest point in the water distribution system. The most common examples of true gravity systems are those supplied from an alpine lake or a mountain reservoir that supplies water to consumers below.

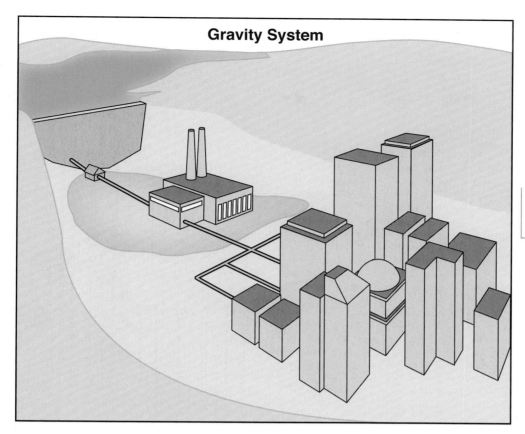

Gravity System

Combination Systems

The majority of communities in North America use a combination of the direct pumping and gravity systems. In most cases, water is pumped from the treatment facility to elevated storage tanks located near the point of use **(Figure 12.7, p. 598)**. Water is stored in these tanks and delivered to the consumer by gravity pressure. If the water storage capacity is adequate relative to consumption, an extended power outage or pump failure will not affect the reliability of the water supply or the availability of water to the consumer.

Many industrial facilities have their own combination water supply systems with elevated storage tanks. By prior agreement, these water supplies are often available to the local fire department in an emergency. Water for fire protection may be available to some communities from storage systems, such as underground cisterns (tanks used to store rainwater), that are considered a part of the distribution system. Fire department pumpers draft (draw water from a static source) from these sources and provide the pressure needed to transport the water to a fire.

Combination System — Water supply system that is a combination of both gravity and direct pumping systems. It is the most common type of municipal water supply system.

Water Treatment Facilities

Water intended for domestic use is treated to remove contaminants that may be detrimental to the health of those who use or drink it. Drinking water is most often treated by filtering out particulates and adding chlorine to kill bacteria and other organisms. In many communities, fluoride is also added to prevent tooth decay. From a fire protection standpoint, the main concern regarding water treatment facilities is that a mechanical breakdown, natural disaster, loss of power supply, or fire could disable the facility's pumps. Disabling all or part of a facility's pumping

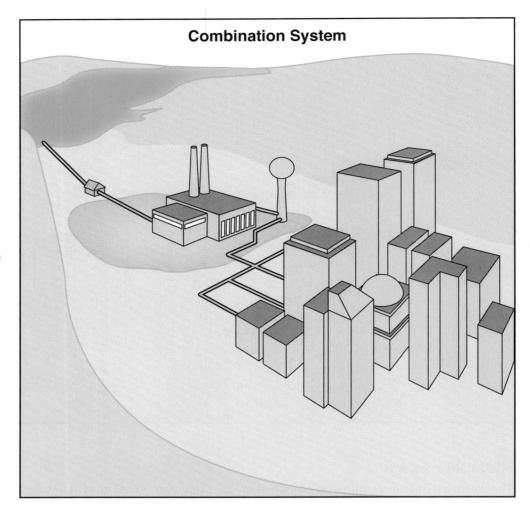

Combination System

Figure 12.7 A combination system.

capacity would seriously reduce the volume and pressure of water available for fire fighting operations. Fire departments must have contingency plans to deal with potential water supply shortfalls.

Another potential problem involving water treatment facilities has nothing directly to do with water supply. As part of the water purification process, these facilities store and use large quantities of liquid chlorine (**Figure 12.8**). Releases of this highly toxic and corrosive substance can have serious health consequences for anyone breathing the fumes or coming in contact with the liquid chlorine. An uncontrolled release can force an evacuation of the facility and areas downwind of it. For more information on dealing with chemical releases, see Chapter 23.

Water Storage and Distribution Systems

Water storage and distribution are critical parts of the overall water supply system. Water is received from the treatment facility and delivered to elevated storage tanks before being distributed throughout the area served (**Figure 12.9**). The ability of a water system to deliver a sufficient quantity of water at adequate pressure depends upon the capacity and elevation of the storage tanks and the condition and carrying capacity of the system's network of underground pipes, often called *mains*. When water flows through any pipe, its movement causes friction that reduces the water pressure.

Figure 12.8 Bulk storage of chlorine is a potential hazard.

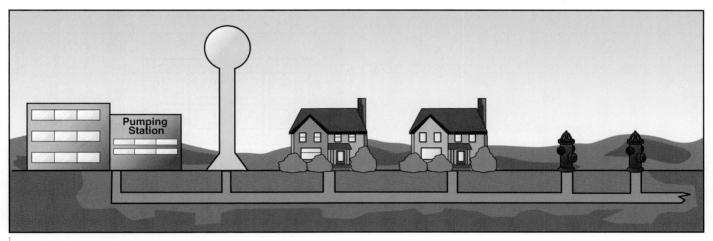

Pumping Station

Figure 12.9 A typical water distribution system.

The friction loss in water mains is increased by encrustations that can accumulate inside the mains over a period of years. Friction loss reduces the volume and pressure of water available from fire hydrants connected to the water distribution system. A fire hydrant that receives water from only one direction (also known as a *dead-end hydrant*) has a limited water supply **(Figure 12.10, p. 600)**. There is much less pressure loss in fire hydrants that are supplied from two or more directions. Fire hydrants that receive water from more than one direction are said to have *circulating feed* or a *looped line* **(Figure 12.11, p. 600)**. A distribution system that provides circulating feed from all directions is called a *grid system* **(Figure 12.12, p. 600)**. A grid system consists of the following components:

Primary feeders — Large mains, with relatively widespread spacing, that convey large quantities of water to various points in the system for distribution to secondary feeders and smaller mains

Secondary feeders — Network of intermediate-sized mains that subdivide the grid within the various loops of primary feeders and supply the distributors

Distributors — Grid arrangement of smaller mains serving individual fire hydrants and blocks of consumers

Circulating Feed — Fire hydrant that receives water from two or more directions.

Loop System — Water main arranged in a complete circuit so that water will be supplied to a given point from more than one direction. Also called circle system, circulating system, or belt system.

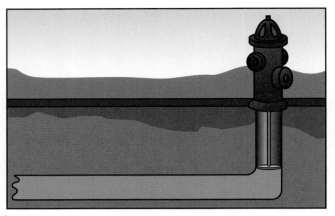

Figure 12.10 Dead-end hydrants receive water from only one direction.

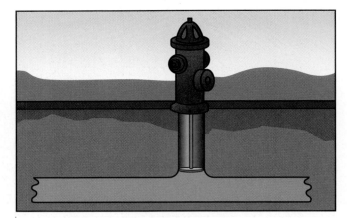

Figure 12.11 Circulating-feed hydrants receive water from two directions.

Figure 12.12 A typical grid system layout.

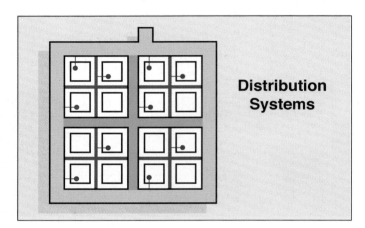

Distribution Systems

To ensure a sufficient water supply, two or more primary feeders should run from the source of supply to the high-risk and industrial districts of the community by separate routes. Similarly, secondary feeders connect the primary feeders and supply water from two directions to any point. This arrangement increases the capacity of the supply at any given point and ensures that a break in a feeder main will not completely cut off the supply.

According to the American Water Works Association, the recommended size for fire-hydrant supply mains is at least 6 inches (150 mm) in diameter in residential areas. These should be closely gridded by 8-inch (200 mm) cross-connecting mains at intervals of not more than 600 feet (180 m). In the business and industrial districts, the minimum recommended size is an 8-inch (200 mm) main with cross-connecting mains every 600 feet (180 m). Twelve-inch (300 mm) mains may be used on principal streets and in long mains not cross-connected at frequent intervals.

Water Main Valves

The function of a valve in a water distribution system is to provide a means for controlling the flow of water through the distribution piping. Valves should be located at frequent intervals in the grid system so that only small sections are cut off if it is necessary to isolate parts of the system for repairs (see **Figure 12.13**). Because the need to operate any particular valve in a water system may occur rarely, sometimes not for many years, valves should be operated at least once a year to keep them working.

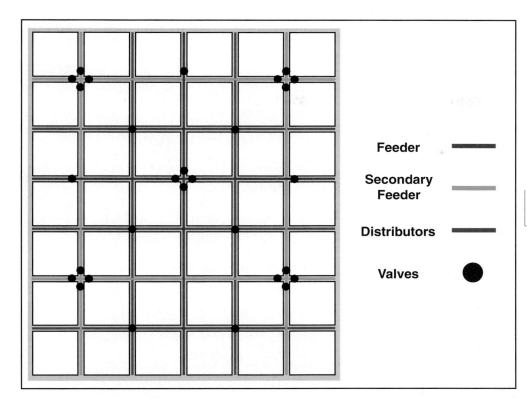

Figure 12.13 Typical grid system valve locations.

Feeder ▬▬▬▬

Secondary Feeder ▬▬▬▬

Distributors ▬▬▬▬

Valves ●

One of the most important factors in the operation of a water supply system is the water department's ability to promptly open valves when increased flow is needed during an emergency, or to close valves during a system breakdown. A well-run water utility has records of the locations of all valves, as well as the number of turns required to open or close each valve. Valves should be inspected and operated on a regular basis, although this is usually the responsibility of the water department and not the fire department.

Water supply system valves are broadly divided into two types — *indicating* and *nonindicating*. An indicating valve visually shows the position of the gate or valve seat — open, closed, or partially closed. Valves in private fire protection systems are usually of the indicating type. Two commonly used indicating valves are the *post indicator valve (PIV)* and the *outside stem and yoke (OS&Y) valve*. The post indicator valve is a hollow metal post that houses the valve stem. A plate attached to the valve stem inside this post has the words *OPEN* and *SHUT* printed on it so that the position of the valve is shown (**Figure 12.14, p. 602**). The OS&Y valve has a yoke on the outside with a threaded stem that opens or closes the gate inside the valve (**Figure 12.15, p. 602**). The threaded portion of the stem is visible when the valve is open and invisible when the valve is closed.

Nonindicating valves in a water distribution system are normally buried or installed in utility manholes. If a buried valve is properly installed, the valve can be operated aboveground through a valve box (**Figure 12.16, p. 602**). In other cases, a special socket wrench on the end of a reach rod is needed to operate the valve (**Figure 12.17, p. 602**).

Control valves in water distribution systems may be either gate valves or butterfly valves, and of the indicating or nonindicating types. Gate valves are usually the nonrising stem type; as the valve nut is turned by the valve key (wrench), the gate either rises or lowers to control the water flow (**Figure 12.18, p. 603**). Gate

OS&Y Valve — Outside stem and yoke valve; a type of control valve for a sprinkler system in which the position of the center screw indicates whether the valve is open or closed.

Post Indicator Valve (PIV) — A type of valve used to control underground water mains that provides a visual means for indicating "open" or "shut" position; found on the supply main of installed fire protection systems.

Gate Valve — Control valve with a solid plate operated by a handle and screw mechanism. Rotating the handle moves the plate into or out of the waterway.

Figure 12.14 The status of the PIV is visible through a small window.

Figure 12.15 A typical OS&Y valve.

Figure 12.16 Some water control valves are located below ground level.

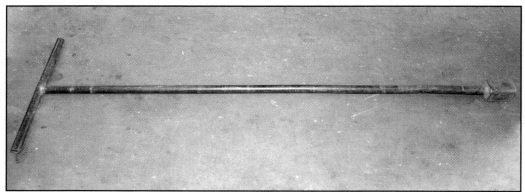

Figure 12.17 A typical water valve key.

Figure 12.18 A typical nonrising stem gate valve.

Figure 12.19 Butterfly valves are common in private water systems.

valves should be marked with a number indicating the number of turns necessary to completely open or close the valve. If a valve resists turning after fewer than the indicated number of turns, it usually means there is debris or another obstruction in the valve. Butterfly valves are tight closing, and they usually have a rubber or a rubber-composition seat that is bonded to the valve body. The valve disk rotates 90 degrees from the fully open to the tight-shut position **(Figure 12.19)**. The nonindicating butterfly type also requires a valve key. Its principle of operation provides satisfactory water control after long periods of inactivity.

The advantages of proper valve installation (spacing) in a distribution system are readily apparent. If valves are installed according to established standards, it normally will be necessary to close off only one or perhaps two fire hydrants from service while a single break in a main is being repaired. The advantage of proper valve installation is, however, reduced if all valves are not properly maintained and kept fully open. Friction loss is increased by valves that are only partially open. When valves are closed or partially closed, the condition may not be noticeable during periods of ordinary domestic water use. As a result, the impairment may not be known until the fire department experiences reduced water flow from a hydrant during a fire or until detailed inspections and fire flow tests are conducted.

Butterfly Valve — Type of control valve that uses a flat circular plate in the pipe which rotates ninety degrees across the cross section of the pipe to control flow.

Water Mains

Underground water mains are generally made of cast iron, ductile iron, asbestos cement, steel, polyvinyl chloride (PVC) plastic, or concrete. Whenever a main is installed, it must be the proper type for the soil conditions and pressures to which it will be subjected. When water mains are installed in unstable or corrosive soils, steel or reinforced concrete pipe may be used to give the strength needed. Some locations that may require extra protection include areas beneath railroad tracks and highways, areas close to heavy industrial operations, or areas prone to earthquakes.

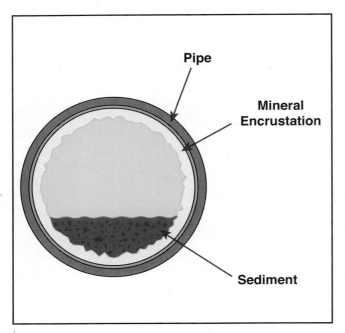

Figure 12.20 Water mains can become obstructed with encrustations and sediments.

Pipe

Mineral Encrustation

Sediment

The internal surface of the pipe, regardless of the material from which it is made, offers resistance to water flow. Some materials, however, have considerably less resistance to water flow than others. Personnel from the engineering division of the water department determine the type of pipe best suited for the particular conditions.

The amount of water able to flow through a pipe and the amount of friction loss created can also be affected by other factors. In addition to the encrustations mentioned earlier, sediments settle out of the water and collect in the bottom of the pipe where they solidify (**Figure 12.20**). Both of these conditions result in increased friction loss, a reduction in the size of the opening through which water can flow, and a proportionate reduction in the volume of water that can be supplied by the system.

Pressure Measurements

Technically, pressure is defined as *force per unit area*. In fire service hydraulics, pressure is the force that moves water through a conduit — either a pipe or a hose. In water supply systems, pressure is measured in different ways for different purposes. In this context, pressure is measured in pounds per square inch (psi) or kilopascals (kPa). Firefighters need to understand the following terms that identify the pressure measurements that are relevant to water supply:

- Static pressure
- Flow pressure
- Residual pressure

Static Pressure

The pressure in a public or private water system is that exerted by either pumps or gravity. When there is little or no water flow, the pressure that can be measured is called static pressure. *Static pressure* is stored potential energy that is available to force water through mains, valves, hydrants, and fire hose. Because there is always at least some water flow due to leaks in the system and domestic consumption, true static pressure is rarely found in a public water supply system. Therefore, static pressure is defined in this context as the normal pressure existing on a system before water is released from a hydrant (**Figure 12.21**).

Flow Pressure

When water is released from a hydrant or nozzle, the forward movement of the water stream exerts a pressure that can be measured with a pitot tube and gauge (**Figure 12.22**). *Flow pressure* is the forward velocity pressure at a discharge opening while water is flowing.

Residual Pressure

The term residual pressure represents the pressure left in the distribution system at a specific location when water is flowing (**Figure 12.23**). *Residual pressure* is that part of the total available pressure that is not used to overcome friction or gravity while

Static Pressure — (1) Potential energy that is available to force water through pipes and fittings, fire hose, and adapters. (2) Pressure at a given point in a water system when no water is flowing.

Flow Pressure — Pressure created by the rate of flow or velocity of water coming from a discharge opening.

Residual Pressure — Pressure at the test hydrant while water is flowing. It represents the pressure remaining in the water supply system while the test water is flowing and is that part of the total pressure that is not used to overcome friction or gravity while forcing water through fire hose, pipe, fittings, and adapters.

Figure 12.21 Static pressure on a hydrant – no water flowing.

Figure 12.22 A pitot tube measures flow pressure.

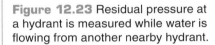

Figure 12.23 Residual pressure at a hydrant is measured while water is flowing from another nearby hydrant.

forcing water through pipes, valves, hydrants, or fire hose. Residual pressure also provides an indication of the availability of additional water.

Fire Hydrants

In general, all hydrant bonnets, barrels, and foot pieces are made of cast iron. The internal working parts are usually made of bronze, but valve facings may be made of rubber, leather, or composite materials. The two main types of fire hydrants used in North America are *dry-barrel* and *wet-barrel*. Even though they serve the same purpose, their designs and operating principles are quite different. Regardless of the type, all hydrants must be opened and closed slowly to prevent damage to hose, hydrants, and other equipment, or possible injury to firefighters. Opening a hydrant too fast may cause the hose connected to it to flail violently as the water pressure straightens out kinks and twists. If a hydrant is closed too fast, it is theoretically possible for it to cause a sudden increase in pressure (*water hammer*) within the water supply system. Refer to **Skill Sheet 12-I-1** for information on operating a hydrant and **Skill Sheet 12-I-2** for steps to make soft-sleeve and hard-suction hydrant connections.

Dry-Barrel Hydrants

Dry-barrel hydrants are installed in areas where prolonged periods of subfreezing weather are common **(Figures 12.24 a and b, p. 606)**. In a dry-barrel hydrant, the main valve is located below the frost line underground, and it prevents water from

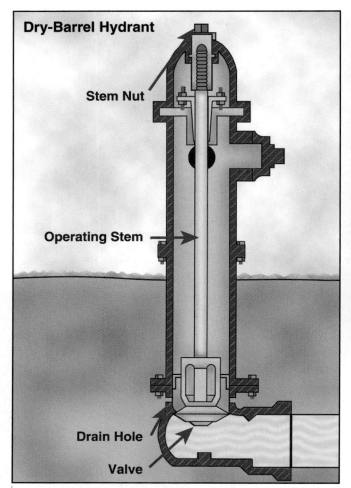

Figure 12.24 a Cutaway of a typical dry-barrel hydrant.

Figure 12.24 b One style of a dry-barrel hydrant.

entering the hydrant barrel. Normally, the hydrant barrel from the top of the stem down to the main valve is empty. This prevents the hydrant from being rendered inoperative by water in the barrel freezing during extended periods of subfreezing temperatures. When the stem nut is turned counterclockwise, the main valve moves downward allowing water to flow into the hydrant. As the main valve moves downward, a drain valve plate attached to the stem closes a drain hole located near the bottom of the hydrant, but allows water to flow past it into the hydrant barrel. When the hydrant is shut down by slowly turning the stem nut clockwise, the main valve rises and shuts off the flow of water into the hydrant barrel. At the same time, the drain valve plate rises, opening the drain hole. The water that remains in the hydrant barrel empties through the drain hole.

If a dry-barrel hydrant is not opened fully, the drain may be left partially open. The resulting flow through the drain hole can cause erosion of the soil around the base of the hydrant, sometimes called "undermining" the hydrant. Over time, this erosion can destroy the hydrant's support and cause it to leak badly. This can put the hydrant out of service and necessitate it being reinstalled. Therefore, it is important that dry-barrel hydrants be either completely open or completely closed.

When a dry-barrel hydrant is shut down, it is also important to verify that the water left in the hydrant barrel is draining out. This can be tested by taking the following steps:

- Close the main valve by turning the stem nut clockwise until resistance is felt; then turn it a quarter-turn counterclockwise.
- Cap all discharges except one.
- Place the palm of one hand over the open discharge (**Figure 12.25**)

If the hydrant is draining, a slight vacuum should be felt pulling the palm toward the discharge. If this vacuum is not felt, repeat the entire process and try again. If the hydrant still is not draining, notify the waterworks authority and have them inspect the hydrant because the drain hole is probably plugged. If this occurs in winter, the hydrant must be pumped out to prevent the water from freezing in the barrel before the hydrant is repaired or replaced. In some areas, the EPA has required drain holes to be closed to prevent contamination of the water supply. In these areas, dry-barrel hydrants must be pumped out after each use.

If water is seen bubbling up out of the ground at the base of a dry-barrel hydrant when the hydrant is fully open, a broken component in the hydrant barrel is allowing water to get past the drain opening. This hydrant should be reported to the water authority which will mark it out-of-service until it is repaired.

Figure 12.25 Check to see that the hydrant is draining.

Wet-Barrel Hydrants

Wet-barrel hydrants, also called *frost-free* or *California* hydrants, are usually installed in warmer climates where prolonged periods of subfreezing weather are uncommon. Wet-barrel hydrants have a horizontal compression-type valve at each outlet (**Figures 12.26 a and b, p. 608**). As the name implies, the hydrant barrel is always filled with water.

Fire Hydrant Marking

The rate of flow from individual fire hydrants varies for several reasons. First, the size of the main to which a hydrant is connected has a major impact on the rate of flow from that hydrant. Also, sedimentation and deposits within the water mains may decrease water flow. These problems develop over time so older water systems are more likely to experience a decline in the available flow than newer systems.

Fire officers can make better decisions regarding a fire attack if they at least know the flow rate available from individual hydrants near the fire. To aid them in that process, NFPA® has developed a system of marking fire hydrants to indicate a range of water flow available from any particular hydrant. In that system, fire hydrants are classified and color-coded as shown in **Table 12.1**. Because NFPA® standards are not laws, however, local color-coding may differ from that shown in the table. As always, firefighters must know and follow their departmental protocols.

Fire Hydrant Locations

Although the installation of fire hydrants is usually performed by contractors, water department personnel oversee and approve the installation. Decisions regarding the location, spacing, and distribution of fire hydrants are usually made by water department personnel based on recommendations from the fire department. In general, fire hydrants should not be spaced more than 300 feet (100 m) apart in high-value districts. A basic rule is to locate a hydrant at every other intersection so that every building on a given street is within one block of a hydrant. Additional intermediate

Wet-Barrel Hydrant —
Fire hydrant that has water all the way up to the discharge outlets. The hydrant may have separate valves for each discharge or one valve for all the discharges. This type of hydrant is only used in areas where there is no danger of freezing weather conditions.

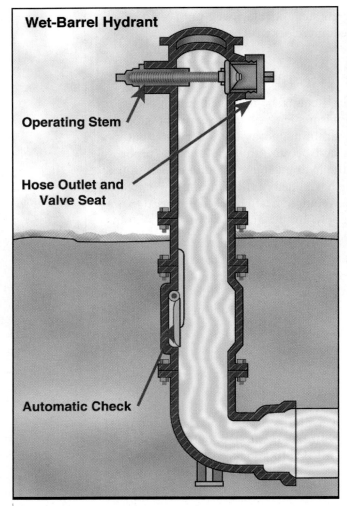

Wet-Barrel Hydrant

Operating Stem

Hose Outlet and
Valve Seat

Automatic Check

Figure 12.26 a Cutaway of a typical wet-barrel hydrant.

Figure 12.26 b One style of a wet-barrel hydrant.

hydrants may be required where distances between intersections exceed 350 to 400 feet (105 m to 120 m). Other factors that affect hydrant location and spacing include the types of construction, types of occupancies, building densities, the sizes of water mains, and required fire flows for occupancies within a given area.

Fire Hydrant Inspection and Maintenance

In most cities, the repair and maintenance of fire hydrants are responsibilities of the water department. However, even in these cities and many others, fire department personnel perform hydrant testing and inspections. When inspecting fire hydrants, firefighters should look for the following potential problems:

- Obstructions, such as sign posts, utility poles, weeds, bushes, or fences that might interfere with pumper-to-hydrant connections or with opening the hydrant valve
- Outlets that face the wrong direction for pumper-to-hydrant connections
- Insufficient clearance between outlets and the ground (**Figure 12.27**)
- Damage to the hydrant
- Rusting or corrosion
- Outlet caps missing or stuck in place with paint

Table 12.1 Hydrant Color Codes		
Hydrant Class	**Color**	**Flow**
Class AA	Light Blue	1,500 gpm (5 680 L/min) or greater
Class A	Green	1,000–1,499 gpm (3 785 L/min to 5 675 L/min)
Class B	Orange	500–999 gpm (1 900 L/min to 3 780 L/min)
Class C	Red	Less than 500 gpm (1 900 L/min)

Adapted from NFPA® 291, *Recommended Practice for Testing and Marking of Hydrants.*

Figure 12.27 Hydrant outlets that are too close to the ground may limit use of the hydrant.

- Stem nut that cannot be turned or turns feely with no visible result
- Obstructions (bottles, cans, rocks) inside the hydrant outlets
- Damp ground surrounding the hydrant or erosion indicating a drain valve leak
- Hydrants painted by property owners (caps adhered to threads by paint)

Maintenance, service, and testing of private hydrants in some gated communities can be an issue. The fire department responsible for these areas should have input into discussions about this issue.

Fire Hydrant Testing

Many fire departments are no longer responsible for the testing of hydrants within their jurisdictions. In fire departments that do routinely test the fire hydrants in their districts, firefighters may be required to assist in the testing process. The most basic test normally conducted on an annual basis is the flow test. While test procedures may vary from department to department, a common flow test method includes recording the static and residual pressure readings in addition to the flow test results. The process begins by selecting the hydrant to be flow tested, and then doing the following:

- Removing all outlet caps to verify that none is stuck
- Inspecting the outlet threads for damage or excessive wear
- Lubricating all outlet threads with graphite or food-grade grease
- Replacing all caps except one 2½-inch (65 mm) cap
- Connecting a cap-type pressure gauge to an outlet on a second hydrant nearby
- Turning the second hydrant on and recording the static pressure shown on the gauge

- Turning the test hydrant on fully and allowing the water to flow briefly to flush out any debris

- Using a pitot tube and gauge to measure the flow rate while the water is flowing at the first hydrant

- Recording the pitot gauge reading

- Taking and recording the residual pressure reading from the gauge connected to the second hydrant before shutting the test hydrant off

- Turning off the second hydrant, removing the gauge, and replacing the cap

- Turning off the test hydrant

- Testing for the vacuum created by an operating drain valve (dry barrel)

- Replacing the cap on the outlet

Repeat this procedure with each hydrant to be flow tested. To perform this test properly, it is necessary to know how to use a pitot tube.

Using a Pitot Tube

Firefighters who assist in flow testing hydrants must know how to use a pitot tube with an attached gauge. There are two methods of holding the pitot tube properly. The first is to grasp the pitot tube just behind the blade with the first two fingers and thumb of the left hand while the right hand holds the air chamber. The little finger of the left hand rests upon the edge of the hydrant outlet to steady the instrument (**Figure 12.28**). Another method is to hold the pitot tube with the fingers of the left hand and place the heel of that hand against the edge of the hydrant outlet to steady the pitot (**Figure 12.29**). The blade of the pitot can then be moved into the stream in a counterclockwise direction. The right hand once again steadies the air chamber. Flow test kits are also available for conducting hydrant tests. Using a "fixed-mount" pitot tube reduces the possibility of human error that may occur when using a handheld pitot tube (**Figure 12.30**). **Skill Sheet 12-II-1** shows the procedure for using the pitot tube.

NOTE: For more information on testing fire hydrants, see the IFSTA **Fire Inspection and Code Enforcement** manual.

Alternative Water Supplies

Even areas with good water systems should be surveyed for alternative supplies in case the water system fails or a fire occurs that requires more water than the system can supply. A public water system can be supplemented by an industry that has its own private water system. Fire department pumpers can draw water from many natural sources such as lakes, ponds, rivers, and even the ocean (**Figure 12.31**). Water is also found in swimming pools and farm stock tanks. And, as mentioned earlier in this chapter, some cities have underground cisterns at strategic locations. These cisterns collect rainwater and are an additional source of water for fire fighting.

The process of drawing water from a static source to supply a pumper is known as *drafting*. Almost any static source of water can be used if it is sufficient in quantity and not contaminated to the point of creating a health hazard or damaging the fire pump. The depth of the water source is an important consideration. Silt and debris can render a source useless by clogging strainers, by seizing (stopping) or damaging pumps, and by allowing sand and small rocks to enter attack lines and clog

Figure 12.28 A pitot may be supported with the little finger.

Figure 12.29 A pitot can be supported with the heel of one hand.

Figure 12.30 A fixed-mount pitot eliminates the need for manual support.

Figure 12.31 A large lake is a reliable water source.

fog nozzles. When drafting from any natural source, all hard intake hoses should have strainers on them. The intake hose should be positioned and supported so the strainer does not rest on or near the bottom of the source. A minimum of 24 inches (600 mm) of water above and below the hard intake strainer is usually needed for it to function properly (**Figure 12.32, p. 612**). However, floating strainers can draft water from sources as shallow as 24 inches (600 mm) deep (**Figure 12.33, p. 612**).

In some areas, dry hydrants are installed at static water sources to increase the water supply available for fire fighting. Dry hydrants are usually constructed of steel or PVC pipe with strainers at the water source and steamer ports to connect to the pumper. They are designed to supply at least 1,000 gpm [4 000 L/min]. More information can be found in NFPA® 1142, *Standard on Water Supplies for Suburban and Rural Fire Fighting.*

Fire departments should make every effort to identify, mark, and record alternative water supply sources during preincident planning. Consideration should be given to the effect that weather has on the amount of water available and the accesses to water sources. **Skill Sheet 12-I-3** details the steps for drafting from a static water source.

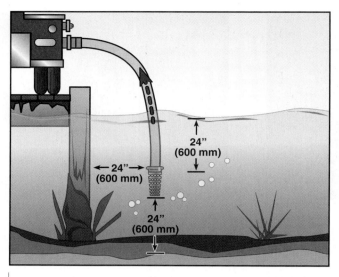

Figure 12.32 A minimum of 24 inches (600 mm) of water should surround the strainer.

Figure 12.33 Floating strainers make some ponds useable as water sources. *Courtesy of Ziamatic.*

Rural Water Supply Operations

Rural water supply operations often depend upon water shuttle operations with water tenders (mobile water supply apparatus) or relay pumping. For either type of operation to succeed, preincident planning and frequent practice are required. Adequate resources must be dispatched promptly, and an incident management system is necessary for control and coordination. The following subsections briefly highlight each of these operations. For more information on rural water supply operations, see the IFSTA **Pumping Apparatus Driver/Operator Handbook** and NFPA® 1142.

Water Shuttles

Water shuttles involve hauling water from a supply source (fill site) to portable tanks (dump sites) from which water may be drawn to fight a fire. In most cases, water shuttle operations are recommended for distances greater than ½ mile (0.8 km) or greater than the fire department's capability of laying supply hoselines. It is critical to have a sufficient number of water tenders to maintain the needed fire flow.

Critical to efficient water shuttle operations are fast-fill and fast-dump capabilities. Water supply officers should be positioned at both the fill and dump sites to coordinate these operations. Also important are traffic control, hydrant operations, hookups, and tank venting. If possible, water tender driver/operators should remain in their vehicles during filling/dumping operations.

There are three key components to water-shuttle operations:

- Attack apparatus at the fire (dump site)
- Fill apparatus at the fill site (unless self-filling vacuum tankers are used)
- Water tenders (mobile water supply apparatus) to haul water from the fill site to the dump site **(Figure 12.34)**

The dump site is generally located near the actual fire or incident. The dump site usually consists of one or more portable water tanks into which water tenders dump water before returning to the fill site (**Figure 12.35**). Apparatus attacking the fire may draft directly from the portable tanks, or other apparatus may draft from the tanks and supply the attack apparatus. Low-level intake devices permit use of most of the water in the portable reservoir (**Figure 12.36**).

Capacities of portable tanks range from 1,000 gallons (4 000 L) upward. When large flows must be maintained, multiple portable tanks are set up. One tank is needed for the attack pumper, and water tenders dump into the others. When two portable tanks are used, they can be interconnected through their drain fittings (**Figure 12.37**). If multiple portable tanks are needed, jet siphon devices can be used to transfer water from one tank to another (**Figure 12.38, p. 614**). A jet siphon uses a 1½-inch (38 mm) discharge line connected to the siphon. The siphon is then attached to a hard sleeve placed between two tanks (**Figure 12.39, p. 614**). Consult **Skill Sheet 12-I-4** for steps to take in deploying a portable water tank.

Some fire departments use jet-assist siphons to transfer water from one portable tank to another. Although 4-inch (100 mm) PVC or aluminum piping has been used for such devices, 6-inch (150 mm) units are more common. Using a ½-inch (13

Figure 12.34 A typical water tender.

Figure 12.35 Portable tanks allow water tenders to dump their loads quickly.

Figure 12.36 Low-level strainers will draft water down to depths of 2 inches (50 mm). *Courtesy of Ziamatic.*

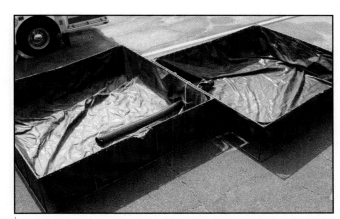

Figure 12.37 Portable tanks can be connected at their drain openings.

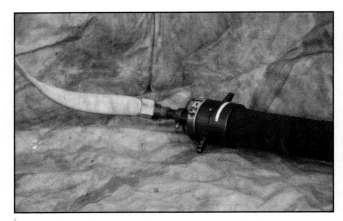

Figure 12.38 The jet siphon device requires water pressure from a small hose to operate.

Figure 12.39 When the siphon supply line is charged, water is transferred from one tank to another.

Portable Tank — Storage tank used during a relay or shuttle operation to hold water from water tanks or hydrants. This water can then be used to supply attack apparatus. Also called Catch Basin, Fold-a-Tank, Portable Basin, or Porta-Tank.

mm) jet nozzle supplied by a 1½-inch (38 mm) hose makes possible transfer flows of 500 gpm (2 000 L/min). Some departments merely add the jet to a length of hard intake hose.

The volume of water supplied in water shuttle operations can be calculated by timing the complete round trip including fill time, travel time, and dump time. Dividing the total gallons carried in each tender by this time indicates the gpm (L/min) being supplied.

In general, there are two types of portable water tanks. One is the collapsible or folding style that uses a square metal frame and a synthetic or canvas duck liner **(Figure 12.40)**. Another style is a round, self-supporting synthetic tank with a floating collar that rises as the tank is filled **(Figure 12.41)**. These frameless portable tanks are widely used in wildland fire fighting operations.

Figure 12.40 A typical square portable water tank.

Figure 12.41 A pumper drafts from a self-supporting portable tank.

Before opening a portable tank, a salvage cover or heavy tarp should be spread on the ground to help protect the liner once water is dumped into it. When the situation permits, portable tanks should be as level as possible to ensure maximum capacity, and positioned in a location that allows easy access from multiple directions but does not inhibit access of other apparatus to the fire scene. Ideally, portable tanks will be set up so that more than one water tender can offload at the same time.

There are four basic methods by which water tenders unload:

- Gravity dumping through large (10- or 12-inch [250 mm or 300 mm]) dump valves **(Figure 12.42)**

- Jet dumps that increase the flow rate **(Figure 12.43)**

- Apparatus-mounted pumps that off-load the water **(Figure 12.44)**

- Combination of these methods

NFPA® 1901, *Standard for Automotive Fire Apparatus,* requires that water tenders on level ground be capable of dumping or filling at rates of at least 1,000 gpm (4 000 L/min). This rate necessitates adequate tank venting and openings in tank baffles. Because the tank can be seriously damaged if done incorrectly, pumping the water from the tank of a water tender should only be done by a trained apparatus driver/op-

Figure 12.42 Some tenders are only capable of gravity dumping.

Figure 12.43 Some tenders are capable of jet-assisted dumping.

Figure 12.44 Some tenders are equipped with pumps.

erator. Gravity dumps may be activated by any firefighter, which relieves the driver/operator from having to exit the cab. This saves time in the overall process. Some gravity dumps can be activated from the cab, but others must be activated by a lever near the outlet.

To fill water tenders quickly, use the best fill site or hydrant available, large hoselines, multiple hoselines, and if necessary, a pumper for adequate flow. In some situations multiple portable pumps may be necessary. Both fill sites and dump sites should be arranged so that a minimum of backing (or maneuvering) of apparatus is required.

Relay Pumping

In some situations the water source is close enough to the fire scene that relay pumping can be used instead of water shuttles. Some departments use variations of a combination water shuttle and relay pumping to minimize congestion of apparatus at the fire scene. Two important factors must be considered regarding the establishment of a relay operation:

Relay Operation — Using two or more pumpers to move water over a long distance by operating them in series. Water discharged from one pumper flows through hoses to the inlet of the next pumper, and so on. Also called Relay Pumping.

- The water supply must be capable of maintaining the desired volume of water required for the duration of the incident.

- The relay must be established quickly enough to be worthwhile.

The number of pumpers needed and the distance between pumpers is determined by several factors such as volume of water needed, distance between the water source and the fire scene, size of hose available, amount of hose available, and pumper capacities. The apparatus with the greatest pumping capacity should be located at the water source. Large-diameter hose or multiple hoselines increase the distance and volume that a relay can supply because of reduced friction loss. A water supply officer must be appointed to determine the distance between pumpers and to coordinate water supply operations.

After considering these factors, a quick calculation must be made by the water supply officer in order to determine the distance between pumpers. It is important to know the friction loss at particular flows for the size hose being used. These figures can be made into a chart and placed on the pumper for quick reference. The best way to prepare for relay operations is to plan them in advance and to practice

Summary

Because water is still the primary fire extinguishing agent used by firefighters in North America, and because fires often occur considerable distances from major water sources, fire departments must develop ways to transport the available water from its source to where it is needed. Firefighters must know what water supply systems have been developed and what their responsibilities are when these systems are used. They must know about water sources, pumping systems, gravity systems, and the system of underground water mains used to distribute the water. They must know how to inspect, test, maintain, and use the fire hydrants in their communities. Finally, they must know how to transport large quantities of water when there is no water supply system or when the water supply system is inoperative.

Fire Fighter I

1. What is the difference between dry-barrel and wet-barrel hydrants?

2. How are fire hydrants marked?

3. What factors affect hydrant location and spacing?

4. List alternative water supplies.

5. What are the three key components of a water shuttle operation?

Fire Fighter II

6. List sources of water supply for a jurisdiction.

7. What are the three methods of moving water in a system?

8. Describe the components of a grid system.

9. What is static pressure, flow pressure, and residual pressure?

10. What are the two methods of holding a pitot tube properly?

Step 1: As a safety precaution, tighten hydrant outlet caps that will not be used.

Step 2: Turn outlet nut counterclockwise and remove the cap from one outlet.

Step 3: Open the hydrant fully.

Step 4: Close the hydrant fully.

Step 5: Replace cap on outlet.

Soft-sleeve connection (Hydrant Firefighter)

Step 1: Confirm order with officer to make hydrant connection.

Step 2: Remove necessary equipment from the pumper.

Step 3: Remove the hydrant cap by turning it counterclockwise and using a spanner wrench if the cap is tight.

Step 4: Inspect the hydrant for exterior damage and check for debris or damage in inside outlet.

12-I-2 continued
Make soft-sleeve and hard-suction hydrant connections.

Soft-sleeve Connection (Hydrant Firefighter)

Step 5: Place the hydrant wrench on hydrant nut with handle pointing away from outlet.

Step 6: (**NOTE:** If adapter is preconnected to the hose, proceed with Step 7). If adapter is not preconnected, place the reducer adapter on the hydrant, turning clockwise and making hand tight.

Step 7: Remove the intake hose from the pumper.

Step 8: Connect the intake hose to the pump intake, turning clockwise and making hand tight.

Skill **SHEETS**

12-I-2 continued
Make soft-sleeve and hard-suction hydrant connections.

Soft-sleeve Connection (Hydrant Firefighter)

Step 9: Stretch the intake hose to the hydrant, placing two full twists in the hose to prevent kinking.

Step 10: Make the hydrant connection to steamer outlet or outlet with adapter, turning clockwise and making hand tight.

Step 11: Open the hydrant slowly until hose is full.

Step 12: Tighten any leaking connections using rubber mallet or spanner wrench.

Hard-suction Connection (Hydrant Firefighter)

Step 1: Confirm order with officer to make hydrant connection.

Step 2: Remove the intake hose from the pumper.

Step 3: Connect the intake hose to the hydrant or apparatus (depending on local preference), turning connection clockwise and making hand tight.

Step 4: Connect opposite end to the hydrant or apparatus, turning connection clockwise and making hand tight.

Skill SHEETS

12-I-3
Connect and place a hard-suction hose for drafting from a static water source.

Step 1: Confirm order with officer to connect hose for drafting.

Step 2: Check the hard-suction couplings for dirt, debris, and worn gaskets.

Step 3: Connect the sections of hard-suction hose.

Step 4: Connect the strainer to one end of the hard-suction hose.

Step 5: Put the strainer into the water; if a barrel strainer, use the rope to maneuver the hose and to keep the strainer off the bottom.

Step 6: Prepare pump intake for coupling by removing pump intake cap and keystone intake valve from intake, if applicable.

Step 7: Connect the hard-suction hose to the pumper pump intake, aligning the sections and hand tightening in a clockwise direction.

Step 8: Tie up strainer rope (if used) to pumper or stationary object.

Step 9: Dismantle drafting equipment and return to proper storage on pumper per departmental SOPs.

Step 1: Remove the tarps from the apparatus.

Step 2: Carry the tarps to the planned location for the water reservoirs.

Step 3: Open the tarps and spread them flat on the ground.

Step 6: Connect the intake and discharge hoses to the jet siphon per manufacturer's instructions.

Step 7: Position the jet siphon properly to draw and discharge water, per manufacturer's instructions.

Step 4: Remove the portable tank, jet siphon, and manufacturer's setup instructions from the apparatus.

Step 5: Set up two portable tanks within departmental time limits, if specified.

Step 8: Dismantle the portable tanks.

Step 9: Shake and fold the tarps.

Step 10: Return equipment to the proper storage locations on the apparatus.

Step 1: Open the petcock or press the bleed button on the pitot tube. Close the petcock.

NOTE: The petcock is located on the bottom of the pitot tube. When properly drained and ready, the gauge needle should read zero.

Step 2: Edge the blade into the stream with the small opening or point centered in the stream and hold away from the orifice at a distance approximately half the diameter of the orifice.

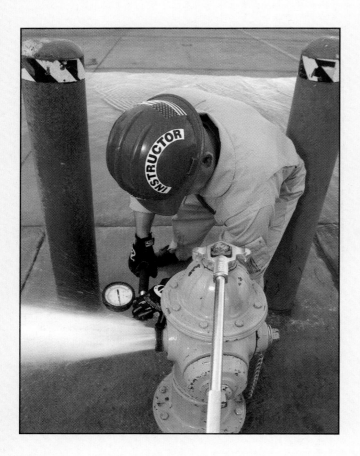

Step 3: Keep the air chamber above the horizontal plane passing through the center of the stream.

NOTE: This position increases the efficiency of the air chamber and helps avoid needle fluctuations.

Step 4: Record the velocity pressure reading from the gauge.

Chapter Contents

Courtesy of District Chief Chris E. Mickal, NOFD Photo Unit.

This chapter provides information that addresses the following job performance requirements of NFPA® 1001, *Standard for Fire Fighter Professional Qualifications* (2008):

NFPA® 1001 references for Chapter 13:

5.3.10

5.5.2

6.3.2

6.5.5

Chapter Objectives

Fire Fighter I Objectives

1. Discuss fire hose sizes.
2. Describe types of fire hose damage and practices to prevent such damage.
3. Discuss general care and maintenance of fire hose.
4. Distinguish between characteristics of threaded couplings and nonthreaded couplings.
5. Discuss care of fire hose couplings.
6. Describe the characteristics of hose appliances and tools.
7. Describe common hose rolls.
8. List general hose loading guidelines.
9. Describe common hose loads.
10. Describe hose load finishes.
11. Discuss preconnected hose loads for attack lines.
12. List guidelines when laying hose.
13. Describe the basic hose lays for supply hose.
14. Describe procedures for handling preconnected and other hose.
15. List general safety guidelines that should be followed when advancing a hoseline into a burning structure.
16. Discuss procedures for advancing hose.
17. Describe techniques for operating hoselines.
18. Inspect and maintain hose. (Skill Sheet 13-I-1)
19. Make a straight hose roll. (Skill Sheet 13-I-2)
20. Make a donut hose roll. (Skill Sheet 13-I-3)
21. Couple a hose. (Skill Sheet 13-I-4)
22. Uncouple a hose. (Skill Sheet 13-I-5)
23. Make the accordion hose load. (Skill Sheet 13-I-6)
24. Make the horseshoe hose load. (Skill Sheet 13-I-7)
25. Make the flat hose load. (Skill Sheet 13-I-8)
26. Make the preconnected flat hose load. (Skill Sheet 13-I-9)
27. Make the triple layer hose load. (Skill Sheet 13-I-10)
28. Make the minuteman hose load. (Skill Sheet 13-I-11)
29. Connect to a hydrant using a forward lay. (Skill Sheet 13-I-12)
30. Make the reverse hose lay. (Skill Sheet 13-I-13)
31. Advance the preconnected flat hose load. (Skill Sheet 13-I-14)
32. Advance the minuteman hose load. (Skill Sheet 13-I-15)
33. Advance the triple layer hose load. (Skill Sheet 13-I-16)
34. Advance hose — Shoulder load method. (Skill Sheet 13-I-17)
35. Advance hose — Working line drag method. (Skill Sheet 13-I-18)
36. Advance a line into a structure. (Skill Sheet 13-I-19)
37. Advance a line up and down an interior stairway. (Skill Sheet 13-I-20)
38. Advance an uncharged line up a ladder into a window. (Skill Sheet 13-I-21)
39. Advance a charged line up a ladder into a window. (Skill Sheet 13-I-22)
40. Extend a hoseline. (Skill Sheet 13-I-23)
41. Simulate the procedure for controlling a loose hoseline. (Skill Sheet 13-I-24)
42. Replace a burst line. (Skill Sheet 13-I-25)
43. Operate a charged attack line from a ladder. (Skill Sheet 13-I-26)

Fire Fighter II Objectives

1. Describe the characteristics of hose appliances and tools.
2. Explain service testing fire hose.
3. Discuss test site preparation for service testing fire hose.
4. List equipment necessary to service test fire hose.
5. Explain the service test procedure.
6. Service test fire hose. (Skill Sheet 13-II-1)

Case History

According to a U.S. Fire Administration report, a Georgia fire chief was killed in 2000 when he was struck by high-pressure water from a fire hose that was being service tested. The report stated that the hose had separated from a coupling and the force of the water propelled the chief into a fire apparatus parked nearby. He suffered a head injury, and despite CPR and other medical treatment begun at the station, the chief succumbed 45 minutes later at a local hospital.

As this case history shows, something as ordinary and familiar as fire hose can be lethal to firefighters. It underscores the fact that fire fighting is a dangerous occupation and firefighters must learn the proper (safe) way to use and handle the tools of the fire fighting trade. Remember this case history when you read about testing fire hose later in this chapter.

The term *fire hose* describes a type of flexible tube used by firefighters to carry water or other extinguishing agents under pressure from a source of supply to a point of application. Fire hose is one of the most used items in the fire service. To be reliable, fire hose should be constructed of the best materials available, used in an appropriate manner, and maintained according to the manufacturer's recommendations. Most fire hose is flexible, watertight, and has a smooth rubber or neoprene lining covered by a durable jacket. Depending on its intended use, fire hose is manufactured in a variety of configurations. While there are other types, the single-jacket, double-jacket, rubber single-jacket, and hard-rubber or plastic noncollapsible types are the most common (**Figure 13.1, p. 632**).

In general, fire hose is used as either supply hose or attack hose. Supply hose transports water or other agents from a hydrant or other source of supply to a pumper at the fire scene. Attack hose transports water or other agents, at increased pressure, from the pumper to a nozzle or nozzles or a fire department connection (FDC) for application onto the fire.

To fight fires safely and effectively, firefighters must know the capabilities and limitations of the various types of hose used in their departments. NFPA® 1001 requires those qualified at the Fire Fighter I level to know necessary precautions for the advancement and operation of hoselines on the fireground. They must also know the procedures for marking a defective hose and removing it from service. The skills required for Fire Fighter I include the following:

- Carrying hose

- Advancing charged and uncharged hoselines

- Extending hoselines

Fire Department Connection (FDC) — Point at which the fire department can connect into a sprinkler or standpipe system to boost the water flow in the system. This connection consists of a clappered siamese with two or more 2½-inch (65 mm) intakes or one large-diameter (4-inch [100 mm] or larger) intake.

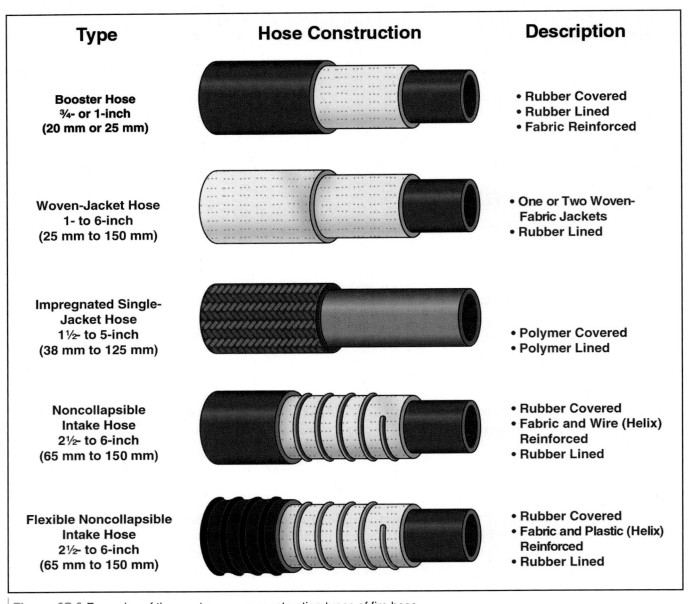

Type	Hose Construction	Description
Booster Hose **¾- or 1-inch** **(20 mm or 25 mm)**		• Rubber Covered • Rubber Lined • Fabric Reinforced
Woven-Jacket Hose **1- to 6-inch** **(25 mm to 150 mm)**		• One or Two Woven-Fabric Jackets • Rubber Lined
Impregnated Single-Jacket Hose **1½- to 5-inch** **(38 mm to 125 mm)**		• Polymer Covered • Polymer Lined
Noncollapsible Intake Hose **2½- to 6-inch** **(65 mm to 150 mm)**		• Rubber Covered • Fabric and Wire (Helix) Reinforced • Rubber Lined
Flexible Noncollapsible Intake Hose **2½- to 6-inch** **(65 mm to 150 mm)**		• Rubber Covered • Fabric and Plastic (Helix) Reinforced • Rubber Lined

Figure 13.1 Examples of the most common construction types of fire hose.

- Replacing burst hose sections
- Cleaning, inspecting, and returning fire hose to service
- Hose rolls and loads

Those qualified at the Fire Fighter II level must have knowledge of the appropriate selection of nozzles and hose for attack and of adapters and appliances used for specific fireground situations. Fire Fighter II requirements also include the procedures for safely conducting hose service testing. The Fire Fighter II level requires the ability to perform an annual service test on fire hose.

This chapter discusses fire hose sizes, causes of hose damage and its prevention, and general care and maintenance. Also discussed are the types of hose couplings and their care and use. The different types of appliances used with fire hose and the tools used in hose operations are discussed. Also included are step-by-step procedures for rolling hose, loading supply hose on apparatus, preparing finishes, and loading preconnected attack lines. The chapter discusses hose-lay procedures, hose-handling techniques, and advancing and operating hoselines. Finally, the chapter ends with a discussion of the procedures for service testing fire hose.

Fire Hose Sizes

Fire hose is produced in different diameters, each for a specific purpose. The size of a fire hose refers to its *inside diameter* (**Figure 13.2**). Fire hose is most commonly cut and coupled into pieces 50 or 100 feet (15 m or 30 m) long for ease of handling and replacement, but longer pieces are available. Pieces of fire hose are also referred to as *lengths* or *sections,* and they must be coupled together to produce a continuous hoseline.

Intake hose is used to connect a fire department pumper or a portable pump to a water source. There are two groups within this category: *soft intake hose* and *hard intake hose.* Soft intake hose is sometimes referred to as *soft sleeve* hose. Soft intake hose is used to transfer water from a pressurized water source, such as a fire hydrant, to the pump intake (**Figure 13.3**). Soft intake hose is available in sizes ranging from 2½ to 6 inches (65 mm to 150 mm) in diameter.

Hard intake hose (also called *hard suction*) is used primarily to draft water from a static source (**Figure 13.4, p. 634**). It is also used to siphon water from one portable tank to another, usually in a water shuttle operation. Some hard intake hose is constructed of a rubberized, reinforced material; others are made of heavy-duty corrugated plastic (**Figure 13.5, p. 634**). Some plastic intake hose is not rated for use on hydrants, and both types are designed to withstand the partial vacuum conditions created when drafting. Hard intake hose is also available in sizes ranging from 2½ to 6 inches (65 mm to 150 mm) in diameter.

NFPA® 1961, *Standard on Fire Hose,* lists specifications for fire hose; NFPA® 1963, *Standard for Fire Hose Connections,* lists specifications for fire hose couplings and screw threads. NFPA® 1901, *Standard for Automotive Fire Apparatus,* requires pumpers to carry 15 feet (4.5 m) of large soft intake hose or 20 feet (6 m) of hard intake hose,

Soft Intake Hose — Large diameter, collapsible piece of hose used to connect a fire pump to a pressurized water supply source; sometimes incorrectly referred to as soft sleeve hose.

Hard Intake Hose — A flexible rubber hose reinforced with a steel core to prevent collapse from atmospheric pressure when drafting; connected between the intake of a fire pump and a water supply and must be used when drafting. Also called hard suction hose.

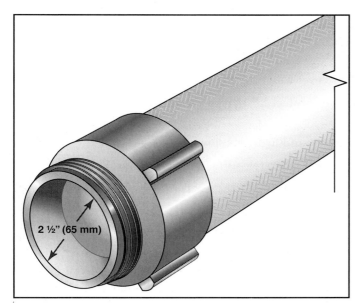

Figure 13.2 Fire hose sizes are based on the inside diameter.

2 ½" (65 mm)

Figure 13.3 A soft intake hose is used to connect a pump to a hydrant.

Figure 13.4 Hard intake hose is used for drafting water from a static water source, such as a portable tank, pond, or swimming pool.

Figure 13.5 Typical hard intake hose mounted on a fire apparatus.

800 feet (240 m) of 2½-inch (65 mm) or larger fire hose, and 400 feet (120 m) of 1½-, 1¾-, or 2-inch (38 mm, 45 mm, or 52 mm) attack hose. These lengths and sizes may be increased, depending on the needs of the department.

Causes and Prevention of Fire Hose Damage

There are many ways in which hose can be damaged during fire fighting operations. Because it is often difficult to protect hose from damage at fires, the most important factor relating to the longevity of fire hose is the care it gets after fires, in storage, and on the fire apparatus. Even if fire hose is constructed of quality materials, it may not withstand mechanical injury, prolonged direct flame contact, mildew and mold, and contact with caustic chemicals.

Mechanical Damage

Fire hose may be damaged in a variety of ways while being used at fires. Some common examples of mechanical damage are slices, rips, and abrasions on the coverings, crushed or damaged couplings, and cracked inner linings **(Figure 13.6)**. To prevent these damages, the following practices are recommended:

- Avoid laying or pulling hose over rough, sharp edges or objects.
- Use hose ramps or bridges to protect hose from vehicles running over it **(Figure 13.7)**.
- Open and close nozzles, valves, and hydrants slowly.
- Change position of folds in hose when reloading it on apparatus.
- Provide chafing blocks to prevent abrasion to hose when it vibrates near the pumper **(Figure 13.8)**.
- Avoid excessive pump pressure on hoselines.

Thermal Damage

Exposing fire hose to excessive heat or direct flame contact can char, melt, or weaken the outer jacket and dehydrate the rubber lining. Inner linings can also be dehydrated when hose is hung to dry in a drying tower for a longer period of time than is necessary or when it is dried in direct sunlight. Mechanical hose dryers eliminate this concern but extend the time required to dry large amounts of hose because of their limited capacity **(Figure 13.9)**. To prevent thermal damage, firefighters should conform to the following recommended practices:

- Protect hose from exposure to excessive heat or fire when possible.

- Do not allow hose to remain in any heated area after it is dry.

- Use moderate temperature for mechanical drying. A current of warm air is much better than hot air.

- Keep the outside of woven-jacket fire hose dry when not in use.

- Run water through hose that has not been used for some time to keep the liner soft.

- Avoid laying fire hose on hot pavement to dry.

- Roll dry hose in a straight roll for storage. This keeps the liner from drying out.

Figure 13.6 All types of fire hose are susceptible to mechanical damage including abrasions caused by rubbing against road surfaces, curbs, or apparatus hose compartments. Damaged hose should be inspected before being reused.

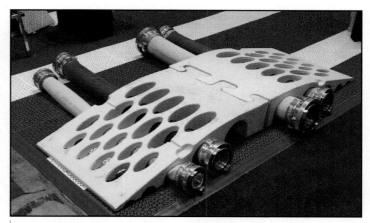

Figure 13.7 Hose ramps protect hose from being run over by vehicles.

Figure 13.8 Hose should be protected from abrasion due to vibration caused by pump pressure. A roll of fire hose can be used as an improvised chafing block.

Figure 13.9 Fire hose should not be dried in direct sunlight. Mechanical hose drying racks can be used to eliminate this problem.

- Prevent hose from coming in contact with, or being in close proximity to, vehicle exhaust systems.
- Use hose bed covers on apparatus to shield the hose from the sun (**Figure 13.10**).

NOTE: Hose can also be damaged by freezing temperatures. Wet or dry, hose should not be subjected to freezing conditions for prolonged periods of time.

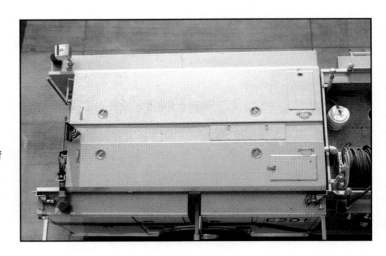

Figure 13.10 Hose bed covers protect fire hose from direct sunlight, rain, and snow. Covers may be constructed of metal or made of waterproof fabric

Figure 13.11 Mold or mildew can destroy woven-jacket fire hose.

Organic Damage

Rubber-jacket hose is not subject to damage caused by living organisms such as mildew and mold. Such damage is a potential problem on hose with a woven-jacket of cotton or other natural fiber if the hose is stored wet (**Figure 13.11**). Mildew and mold rot the fibers of the hose jacket, and this can result in the hose rupturing under pressure. The outer jacket of some woven-jacket fire hose is made of synthetic fibers such as Dacron™ polyester. These synthetic fibers resist organic damage. The outer jacket of some natural-fiber hose has been chemically treated to resist mildew and mold but such treatment is not always 100 percent effective. Some methods of preventing mildew and mold on natural-fiber woven-jacket hose are as follows:

- Remove all wet hose from the apparatus after a fire and replace with dry hose.
- Inspect, wash, and dry hose that has been contaminated in any way.
- Remove, inspect, sweep, and reload hose if it has not been unloaded from the apparatus during a period of six months. Make sure that the hose is folded at different points than when previously loaded.
- Inspect and test hose annually and after possible damage.

Chemical Damage

Certain chemicals and chemical vapors can damage the outer jacket on fire hose or cause rubber lining to separate from the inner jacket. When fire hose is exposed to petroleum products, paints, acids, or alkalis, it may be weakened to the point of bursting under pressure. Runoff water from a fire may also carry foreign materials that can damage fire hose. After being exposed to chemicals or chemical vapors, hose should be cleaned as soon as practical. Some recommended practices are as follows:

- Scrub hose thoroughly and brush all traces of acid contacts with a solution of baking soda and water. Baking soda neutralizes acids.

- Remove hose periodically from the apparatus, wash it with plain water, and dry it thoroughly.

- Test hose properly if there is any suspicion of damage (see Service Testing Fire Hose section).

- Avoid laying hose in the gutter or next to the curb where vehicles have been parked because of possible accumulations of oil from their mechanical components and acid from batteries (**Figure 13.12**).

- Dispose of hose according to departmental SOP if it has been exposed to hazardous materials and cannot be decontaminated.

Figure 13.12 When possible, protect fire hose from contamination due to fluids leaked or spilled from vehicles.

General Care and Maintenance of Fire Hose

If fire hose is properly cared for, its life span can be extended appreciably. The techniques of washing and drying and the provisions for storage are very important functions in the care of fire hose. The following sections highlight the proper care of fire hose. Refer to **Skill Sheet 13-I-1** for techniques in inspecting and maintaining hose.

Washing Hose

The method used to wash fire hose depends on the type of hose. Hard-rubber booster hose, hard intake hose, and rubber-jacket collapsible hose require little more than rinsing with clear water, although a mild soap may be used if necessary.

Most woven-jacket fire hose requires a little more care than those with rubber jackets. After woven-jacket hose is used, any dust and dirt should be thoroughly brushed or swept from it. If the dirt cannot be removed by brushing or sweeping, the hose should be washed and scrubbed with clear water and a stiff brush (**Figure 13.13, p. 638**).

When fire hose has been exposed to oil, it should be washed with a mild soap or detergent using common scrub brushes or straw brooms with a stream of water from a garden hose. Make sure that the oil is completely removed. The hose should then be rinsed thoroughly with clear water.

A hose-washing machine can make the care and maintenance of fire hose much easier (**Figure 13.14, p. 638**). The most common type washes almost any size of fire hose up to 3 inches (77 mm). The flow of water into this device can be adjusted as desired, and the movement of the water assists in propelling the hose through the device. The hoseline that supplies the washer with water can be connected to a pumper or used directly from a hydrant. Higher water pressure, obviously, gives better results.

A cabinet-type machine that washes, rinses, and drains fire hose is designed to be used in the station (**Figure 13.15, p. 638**). This type of machine can be operated by one person, is self-propelled, and can be used with or without detergents.

Drying Hose

Once fire hose has been washed, it should be dried before being stored. Woven-jacket hose must be thoroughly dried before being reloaded on an apparatus. The methods used to dry hose depend on the type of hose. Hose should be dried in accordance

Figure 13.13 Following use, hose can often be cleaned by rinsing alone. Heavy accumulations of dirt, fire debris, or oil will require the use of mild detergents and scrub brushes.

Figure 13.14 Jet-spray hose washers can make the job of cleaning the hose easier and faster.

Figure 13.15 Cabinet-type hose washers are designed to clean hose thoroughly. *Courtesy of Thomas Locke and South Union Volunteer Fire Company.*

with departmental SOP and manufacturer's recommendations. Hard-rubber booster hose, hard intake hose, and synthetic-jacket collapsible hose may be placed back on the apparatus while wet with no ill effects.

Figure 13.16 Clean, dry fire hose is stored in rolls on a rack.

Storing Hose

After fire hose has been adequately brushed, washed, and dried, it should be rolled and stored in suitable racks unless it is to be placed back on an apparatus (see Hose Rolls section). Hose racks should be located in a clean, well-ventilated room in or close to the apparatus room for easy access. Racks can be freestanding on the floor or mounted permanently on the wall (**Figure 13.16**). Mobile hose racks can be used to both store hose and move hose from storage rooms to the apparatus for loading.

Fire Hose Couplings

Fire hose couplings are made of durable materials and designed so that it is possible to couple and uncouple them quickly and with little effort. The materials used for fire hose couplings are generally alloys in various percentages of brass, aluminum, or magnesium. Couplings are made of these alloys because they are durable and will not rust. How well a fire hose functions depend to a large extent upon the condition and maintenance of its couplings. Firefighters should know the types of couplings with which they work.

Types of Fire Hose Couplings

Several types of couplings are used on fire hose. The most commonly used couplings are the *threaded* and *nonthreaded* types (**Figures 13.17 a and b**). Most attack hose is equipped with threaded couplings. Supply hose may be equipped with either nonthreaded or threaded couplings.

Threaded Couplings

Threaded couplings consist of two major components — one *male* and one *female*. The male coupling has external threads and the female has internal threads (**Figure 13.18**). One disadvantage with threaded couplings is that one male and one female coupling are necessary to make the connection unless a double-male or double-female adapter is used. These couplings can be either three-piece or five-piece types (**Figure 13.19**).

Figure 13.17a A typical threaded male hose coupling.

Figure 13.17b Nonthreaded hose couplings are usually found on supply hose and soft intake hose. Threaded adapters may be required for hydrants or pump intakes.

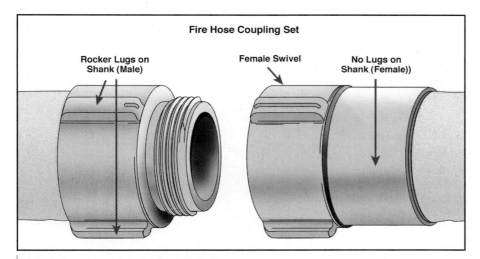

Fire Hose Coupling Set

Rocker Lugs on Shank (Male)

Female Swivel

No Lugs on Shank (Female))

Figure 13.18 The construction of threaded fire hose coupling set. The male coupling is fixed to the hose and does not swivel while the female coupling swivels on the fixed shank.

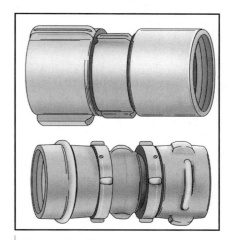

Figure 13.19 Typical three-piece and five-piece threaded hose coupling sets. The components of the couplings include the male, the female swivel, and the shank.

Some intake hose is equipped with two-piece female hose couplings on each end of the hose. Threaded hose couplings on large intake hose are equipped with extended lugs that provide convenient handles for attaching the intake hose to a hydrant outlet or pump intake (**Figure 13.20**).

The portion of the coupling that serves as the point of attachment to the hose is called the *shank* (also called the *tailpiece, bowl,* or *shell*). When the couplings are connected, the male half can be distinguished from the female by checking the shanks. As shown in Figure 13.18, the male coupling has lugs on its shank; the female coupling has lugs on the swivel.

The lugs aid in tightening and loosening the couplings. They also aid in grasping the coupling when making and breaking coupling connections. Connections may be made by hand or with *spanners* (special tools that fit against the lugs) (see Hose Tools section). There are three types of lugs: *pin, rocker,* and *recessed* (**Figures 13.21 a – c**). Although still available, pin-lug couplings are uncommon on new fire hose because of their tendency to snag when hose is dragged over objects. The majority of new threaded couplings have rounded *rocker* lugs. Couplings with rocker lugs are less prone to getting snagged than are those with pin lugs, and rocker lugs tend to slide over obstructions when the hose is moved on the ground or around objects. Hose couplings may be ordered with either two or three rocker lugs. Booster hose normally has couplings with recessed lugs, which are simply shallow holes drilled into the coupling. This lug design prevents abrasions that

Shank — Portion of a coupling that serves as a point of attachment to the hose.

Figure 13.20 A soft intake hose has female threaded couplings with extended lugs on each end of the hose section.

a

b

c

Figures 13.21 a-c There are three types of fire hose coupling lugs: (a) female coupling with pin lugs, (b) male and female couplings with rocker lugs, and (c) booster hose couplings with recessed lugs.

would occur if the hose had protruding lugs and was wound onto reels. These holes are designed to accept a special spanner wrench that can be used to couple or uncouple the hose **(Figure 13.22)**.

Another feature found on most threaded couplings is the Higbee cut and indicator. The *Higbee cut* is a special type of thread design in which the beginning of the thread is "cut" to provide a positive connection between the first threads of opposing couplings, which tends to eliminate cross-threading **(Figure 13.23)**. One of the rocker lugs on each half of the coupling has a small indentation, the *Higbee indicator*, to mark where the Higbee cut begins. This indicator aids in matching the male coupling thread to the female coupling thread, which is not readily visible.

Higbee Cut — Special cut at the beginning of the thread on a hose coupling that provides positive identification of the first thread to eliminate cross-threading.

Higbee Indicators — Notches or grooves cut into coupling lugs to identify by touch or sight the exact location of the Higbee Cut.

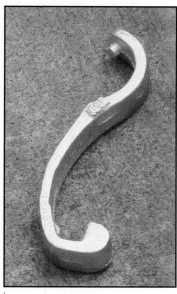

Figure 13.22 A combination spanner wrench used on recessed lug (top end) and rocker lug (bottom end) couplings.

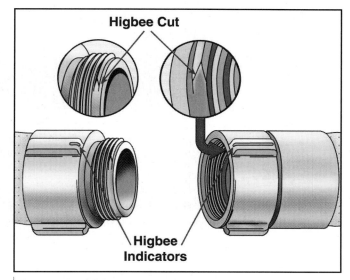

Figure 13.23 Higbee cuts and indicators found on threaded hose couplings.

Storz Couplings

Storz couplings are sometimes referred to as *sexless* couplings. Both terms mean that there are no distinct male or female components as threaded couplings have. All Storz couplings of the same size are identical and may be connected to each other. These couplings are designed to be connected and disconnected with a quarter turn. The locking components are lugs and slots built into the swivel rings of each coupling **(Figure 13.24, p. 642)**. When mated, the lugs of each coupling fit into the slots in the opposing coupling ring and then slide into locking position with a quarter turn. To be in compliance with NFPA 1963, Storz couplings on large-diameter hose (LDH) must have locking devices on them **(Figure 13.25, p. 642)**.

Storz Coupling — Nonthreaded (sexless) coupling commonly found on large-diameter hose.

Care of Fire Hose Couplings

Although fire hose couplings are designed to be durable, they can be damaged. On threaded couplings, the male threads are exposed when not connected and subject to being dented. The female threads are not exposed, but the swivel can be bent into an oval shape. When either threaded or unthreaded couplings are connected, there is

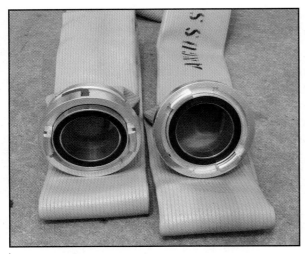

Figure 13.24 The locking components consisting of lugs and slots on Storz couplings.

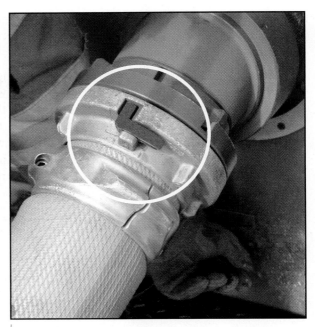

Figure 13.25 To disengage the Storz couplings, the locking device must be depressed before the couplings can be rotated.

less danger of damage to their parts during ordinary usage; however, they can be bent or crushed if they are run over by vehicles. This is reason enough to prohibit vehicles from running over fire hose. Some simple rules for the care of fire hose couplings are as follows:

- Avoid dropping and/or dragging couplings.
- Do not permit vehicles to run over fire hose.
- Inspect couplings when hose is washed and dried.
- Remove the gasket and twist the swivel in warm, soapy water.
- Clean threads to remove tar, dirt, gravel, and oil.
- Inspect gasket, and replace if cracked or creased.

Hose-washing machines will not clean hose couplings sufficiently when the coupling swivel becomes difficult to spin because of dirt or other foreign matter. The swivel part should be submerged in a container of warm, soapy water and worked forward and backward to thoroughly clean the swivel. The male threads should be cleaned with a stiff brush. It may be necessary to use a wire brush if threads are clogged by tar, asphalt, or other foreign matter.

The *swivel gasket* and the *expansion-ring gasket* are used with threaded fire hose couplings. The swivel gasket is used to make the connection watertight when female and male ends are connected (**Figure 13.26**). The expansion-ring gasket is used at the end of the hose where it is expanded into the shank of the coupling. These two gaskets are not interchangeable. The difference lies between their thickness and width. Swivel gaskets should occasionally be removed from the coupling and checked for cracks, creases, and general elastic deterioration. The gasket inspection can be made by simply pinching the gasket together between the thumb and index finger. This method usually discloses any defects and demonstrates the ability of the gasket to return to normal shape.

Figure 13.26 Always check for the presence of the swivel gasket prior to coupling hose sections together or attaching the hose to the pump.

Hose Appliances and Tools

A complete hose layout for fire fighting purposes includes one end of the hose attached to or submerged in a source of water and the other attached to a nozzle or sprinkler/standpipe connection. There are various devices used with fire hose, other than hose couplings and nozzles, to complete such a layout. These devices are usually grouped into two categories: *hose appliances* and *hose tools*. Appliances include valves and valve-controlled devices such as wyes, siameses, water thieves, large-diameter hose appliances, and hydrant valves, as well as fittings such as adapters and intake devices. Examples of hose tools include hose rollers, spanner wrenches, hose strap and hose rope tools, hose chain tools, hose ramps, hose jackets, blocks, and hose clamps. The following sections highlight some of the more common hose appliances and hose tools.

Figure 13.27 The ball valve is designed to provide an open waterway that minimizes friction loss when fully open.

Hose Appliances

A hose appliance is any piece of hardware used in conjunction with fire hose for the purpose of delivering water. A simple way to remember the difference between hose appliances and hose tools is that water flows through appliances but not through tools.

Valves

The flow of water is controlled by various valves in hoselines, at hydrants, and at pumpers. These valves include the following types:

Ball valves — Used in pumper discharges and gated wyes (**Figure 13.27**). Ball valves are open when the handle is in line with the hose and closed when it is at a right angle to the hose. Ball valves are also used in fire pump piping systems.

Gate valves — Used to control the flow from a hydrant. Gate valves have a baffle that is moved by a handle and screw arrangement (**Figure 13.28**).

Butterfly valves — Used on large pump intakes and incorporate a flat baffle that turns 90 degrees. Most are operated by a quarter-turn handle, but some can also be operated remotely by an electric motor. The baffle is in the center of the waterway and aligned with the flow when the valve is open (**Figure 13.29, p. 644**).

Clapper valves — Used in siamese appliances (as well as pump piping) to allow water to flow in one direction only. Clapper valves prevent water from flowing out of unused ports when one intake hose is connected and charged before the addition of more hose. The clapper is a flat disk hinged at the top or one side and that swings in a doorlike manner (**Figure 13.30, p. 644**).

Valve Devices

Valve devices allow the number of hoselines operating on the fire ground to be increased or decreased. These devices include wye appliances, siamese appliances, water thief appliances, large-diameter hose appliances, and hydrant valves.

Wye appliances. Certain situations make it desirable to divide a single hoseline into two or more lines. Various types of wye appliances are used for this purpose; however, all wyes have a single female inlet connection and two male outlets. Wyes that have valve-controlled outlets are called *gated* wyes. One of the most common wyes has a 2½-inch (65 mm) inlet that divides into two 1½-inch (38 mm) outlets, although many other combinations are available (**Figure 13.31, p. 644**). For high water volume operations, wyes with an LDH inlet and two 2½-inch (65 mm) outlets are used (**Figure 13.32, p. 644**).

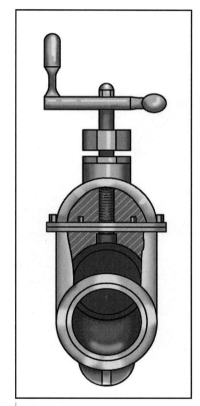

Figure 13.28 A gate valve, shown in a cutaway illustration, opens and closes the waterway with a gate that moves up and down as the valve handle is rotated.

Wye — Hose appliance with one female inlet and two or more male outlets, usually smaller than the inlet. Outlets are also usually gated.

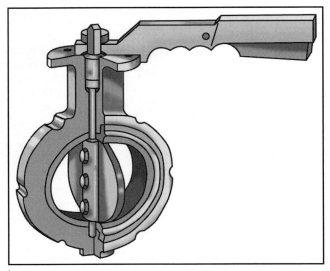

Figure 13.29 The butterfly valve has a baffle that pivots in the waterway to interrupt or permit the flow of water through the valve.

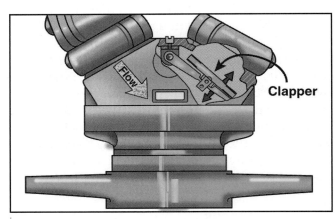

Figure 13.30 Cutaway of one type of clapper valve inside a Siamese appliance. Water pressure from the intake side of the valve closes the clapper on the unused open side of the valve.

Figure 13.31 A gated wye has one female inlet fed by a supply hose and two male outlets that can supply two smaller attack hoselines.

Figure 13.32 A large diameter hose line equipped with a Storz coupling used to supply a gated wye. *Courtesy of Sam Goldwater.*

Siamese — Hose appliance used to combine two or more hoselines into one. The siamese generally has female inlets and a male outlet and is commonly used to supply the hose leading to a ladder pipe.

Siamese appliances. Firefighters sometimes confuse siamese and wye appliances because of their similar appearance. Siamese appliances usually consist of two female inlets, with either a center clapper valve or two clapper valves (one on each side) and a single male outlet (**Figure 13.33**). Some siamese appliances are equipped with three clappered inlets; they are commonly called *triamese* appliances or *manifolds*. Siamese and triamese appliances are commonly used when LDH is not available to overcome friction loss in exceptionally long hose lays or those that carry a large flow. They are also used when supplying ladder pipes that are not equipped with a permanent waterway.

Water thief appliances. In operation, the water thief is similar to the wye appliance. Water thieves are most often used in wildland fire fighting operations. Because hoseline mobility is critical and water flow volume tends to be less than in structural fire fighting, the most common wildland water thief consists of a 1½-inch (38 mm) female inlet, a 1½-inch (38 mm) male outlet, and one valve-controlled 1-inch (25 mm) male outlet **(Figure 13.34)**. Larger volume water thief appliances are also available. They consist of an LDH inlet and outlet and two or more 2½-inch (65 mm) valve-controlled male outlets **(Figure 13.35).**

Large-diameter hose appliances. Some fire fighting operations require that water be distributed at various points along the main supply line. In these cases, an LDH water thief can be used. In other cases, when a large volume of water is needed near the end of the main supply line, an LDH manifold appliance can be used. A typical LDH manifold consists of one LDH inlet and three 2½-inch (65 mm) valve-controlled male outlets **(Figure 13.36)**. Depending on the locale and the configuration of the appliance, these devices are sometimes called *portable hydrants, phantom pumpers,* or *large-diameter distributors.*

Water Thief — Any of a variety of hose appliances with one female inlet for 2½-inch (65 mm) or larger hose and with three gated outlets, usually two 1½-inch (38 mm) outlets and one 2½-inch (65 mm) outlet.

Figure 13.33 A siamese has two female inlets and a single male outlet. The Siamese is used to provide larger quantities of water at lower friction loss.

Figure 13.34 One type of water thief used in wildland operations permits a smaller booster line to be attached to an existing attack line.

Figure 13.35 A large diameter Storz manifold or water thief permits multiple hoselines to be fed by a single large diameter supply line at various locations along the main LDH.

Figure 13.36 Older type large-diameter triamese designed for use with threaded hoselines.

Figure 13.37 A typical four-way hydrant valve used to provide increased pressure from low-pressure hydrants.

Figures 13.38 a-c Operation of the four-way hydrant valve.

Hydrant valves. A variety of hydrant valves are available for use in supply-line operations (**Figure 13.37**). Known by a variety of regional names, these valves are used when a forward lay is made from a low-pressure hydrant to the fire scene. The hydrant valve allows the original supply line to be connected to the hydrant and charged before the arrival of another pumper at the hydrant. By using the hydrant valve, additional hoselines may be laid to the hydrant, the supply pumper connected to the hydrant, and pressure boosted in the original supply line without having to interrupt the flow of water in the original supply line (**Figures 13.38 a-c**).

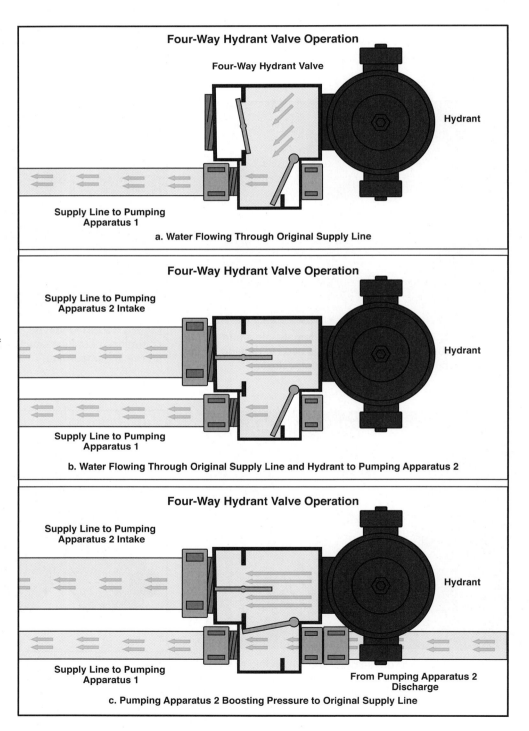

Fittings

Hardware accessories called *fittings* are available for connecting hose of different diameters and thread types. An *adapter* is a fitting for connecting hose couplings with dissimilar threads and the same inside diameter. The double-male and double-female adapters are among the most often used hose fittings (**Figures 13.39 a and b**). These adapters allow two male couplings or two female couplings of the same diameter and thread type to be connected. An increasingly common adapter is one used to connect a sexless coupling to a threaded outlet on a hydrant (**Figure 13.40**).

Other common hose fittings are *reducers* (**Figure 13.41**). They are used to connect a smaller hoseline to the end of a larger one. However, using a reducer limits the larger hose to supplying one smaller line only. Using a wye appliance allows the larger hose to supply two smaller ones.

Other common fittings include *elbows* that provide support for intake or discharge hose at the pumping apparatus (**Figure 13.42**). The threads on pump male discharge outlets are protected with *hose caps* (**Figure 13.43, p. 648**). Female inlets on some FDC are capped with *hose plugs* (**Figure 13.44, p. 648**).

Adapter — Fitting for connecting hose couplings with dissimilar threads but with the same inside diameter.

Fitting — Device that facilitates the connection of hoselines of different sizes to provide an uninterrupted flow of extinguishing agent.

Reducer — Adapter used to attach a smaller hose to a larger hose. The female end has the larger threads, while the male end has the smaller threads.

Figures 13.39 a and b Typical double-male (a) and double female (b) hose adapters used to connect similar types of hose couplings together.

Figure 13.40 In some areas, all hydrants are equipped with nonthreaded (sexless) adapters on the main discharge.

Figure 13.41 A reducer allows a smaller diameter hose to connect directly to a larger size hose.

Figure 13.42 Elbow adapters reduce stress on hose connected to the pump and prevent kinks in the hose. Elbows may be permanently attached to the pump intakes or discharges.

Figure 13.43 The threads on apparatus pump male outlets are protected by hose caps.

Figure 13.44 The female threads on inlets are protected by plugs. Besides protecting the female threads, the plugs prevent foreign objects from being placed in the inlet.

Figure 13.45 The attached rope helps keep the strainer off the bottom of the water source reducing the amount of debris and silt that can obstruct the strainer and reduce water flow.

Figure 13.46 Floating strainers keep the hard intake hose away from the bottom and reduce the amount of debris that may clog the strainer. *Courtesy of Ziamatic.*

Intake Devices

Intake strainers are devices attached to the drafting end of a hard intake (suction) to keep debris from entering the fire pump. Such debris can either damage the pump or pass through it to plug the nozzle. Do not rest strainers on the bottom of a static water source except when the bottom is clean and hard, such as the bottom of a swimming pool. To prevent a strainer from resting on the bottom of a lake or pond, a length of rope is tied to the eyelet on the end of the strainer and to the engine or another anchor point. Some departments keep this rope attached to the strainer **(Figure 13.45)**. Some departments also use floating strainers as a way to keep the intake strainer off the bottom of a static water source **(Figure 13.46)**.

Hose Tools

A variety of tools are used in conjunction with hoselines. The following sections highlight some of the more common ones: hose rollers, hose jackets, hose clamps, spanner wrenches, ramps and blocks, hose strap and hose rope tools, and hose chain. As mentioned earlier, water does not flow through hose tools.

Hose Roller (Hoist)

Hose can be damaged when dragged over sharp corners such as roof edges and windowsills. A device for preventing such damage is the *hose roller* (also know as *hose hoist*) **(Figure 13.47)**. These devices consist of a metal frame with two or more rollers. The notch of the frame is placed over the potentially damaging edge and the frame is secured with a rope or clamp. The hose can then be pulled over the rollers. This tool can also be used for protecting rope from similar edges.

Hose Jacket

When a section of hose ruptures, the entire hoseline is out of service until the section is replaced or the rupture temporarily closed. When conditions preclude shutting down the hoseline to replace the bad section, a *hose jacket* can sometimes be installed at the point of rupture. A hose jacket consists of a hinged two-piece metal cylinder **(Figure 13.48)**. The rubber lining of each half of the cylinder seals the rupture to prevent leakage. A locking device clamps the cylinder closed when in use. Hose jackets are made in two sizes: 2½ inches and 3 inches (65 mm and 77 mm). The hose jacket encloses the hose so effectively that it can continue to operate at full pressure. A hose jacket can also be used to connect hose with mismatched or damaged screw-thread couplings.

Hose Clamp

A hose clamp can be used to stop the flow of water in a hoseline for the following reasons:

- To prevent charging the hose bed during a forward lay from a hydrant
- To allow replacement of a burst section of hose without shutting down the water supply (see Replacing Burst Sections section)
- To allow extension of a hoseline without shutting down the water supply (see Extending a Section of Hose section)
- To allow advancement of a charged hoseline up stairs (see Advancing Hose Up a Stairway section)

Figure 13.47 Hose rollers prevent mechanical damage to hoselines that cross windowsills or wall parapets.

Figure 13.48 In an emergency, a hose jacket can be applied to a hose to temporarily repair a rupture or leak.

Based on the method by which they work, there are three types of hose clamps: *screw-down, press-down,* and *hydraulic press* (**Figures 13.49 a-c**). Unless applied correctly, a hose clamp can injure firefighters or damage the hose. Some general rules that apply to hose clamps are as follows:

- Apply the hose clamp at least 20 feet (6 m) behind the apparatus.

- Apply the hose clamp within 5 feet (1.5 m) from the coupling on the incoming water side.

- Center the hose evenly in the jaws to avoid pinching the hose.

- Close and open the hose clamp slowly to avoid water hammer.

- Stand to one side when applying or releasing any type of hose clamp (the operating handle or frame can snap open suddenly) (**Figure 13.50**).

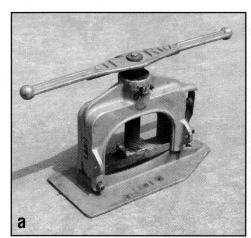

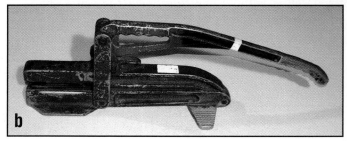

Figures 13.49 a-c Various types of hose clamps used to stop the flow of water in a hose, including (a) screw-down, (b) press-down, (c) and hydraulic press. *(c) Courtesy of Creston (IA) Fire Department.*

Figure 13.50 Place a foot on the base of the hose clamp, release the catch, and open the hose clamp carefully. Opening the device rapidly can cause water hammer in the hose.

Spanner, Hydrant Wrench, and Rubber Mallet

A variety of tools are used to tighten and loosen hose couplings. Some of the most common tools used for this purpose are the *spanner wrench,* or simply *spanner, hydrant wrench* (**Figure 13.51**), and the *rubber mallet.* Although the primary purpose of the spanner wrench is to tighten or loosen couplings, a number of other features have been built into some spanner wrenches:

- Wedge
- Opening that fits gas utility valves
- Slot for pulling nails
- Flat surface for hammering

Hydrant wrenches are primarily used to remove caps from fire hydrant outlets and to open fire hydrant valves. The hydrant wrench is usually equipped with a pentagonal opening in its head that fits most standard fire hydrant operating nuts. The lever handle may be threaded into the operating head to make it adjustable, or the head and handle may be of the ratchet type. The head may also be equipped with a spanner to help make or break coupling connections.

A *rubber mallet* is sometimes used to strike the lugs to tighten or loosen intake hose couplings. It is sometimes difficult to get a completely airtight connection with intake hose couplings even though these couplings may be equipped with long operating lugs. Thus, the rubber mallet is used to further tighten the connection.

Hose Bridge or Ramp

Hose bridges (also called *hose ramps*) help prevent damage to hose when vehicles must drive over it (**Figure 13.52**). They should be used wherever a hoseline is laid across a street or other area where it may be run over. Hose ramps can also be positioned over small spills to keep hoselines from being contaminated, and they can be used as chafing blocks (see following Chafing Block section).

Spanner Wrench — Small tool primarily used to tighten or loosen hose couplings.

Hydrant Wrench — Specially designed tool used to open or close a hydrant and to remove hydrant caps.

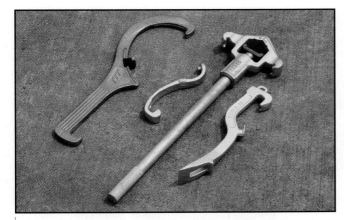

Figure 13.51 Various types of spanner and hydrant wrenches, including (left to right) large diameter pin-lug wrench, combination pin- and rocker-lug wrench, hydrant wrench, and spanner. Note that the spanner wrench has a pry end as well as an opening for use on gas shutoff valves.

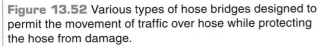

Figure 13.52 Various types of hose bridges designed to permit the movement of traffic over hose while protecting the hose from damage.

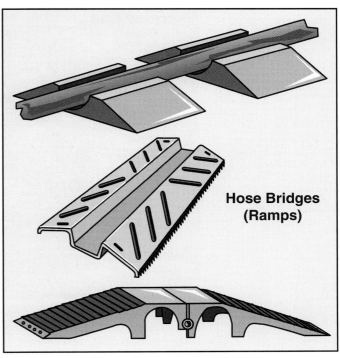

Hose Bridges (Ramps)

Chafing Block

Chafing blocks are devices that are used to protect fire hose where the hose is subjected to rubbing from vibrations **(Figure 13.53)**. Chafing blocks are particularly useful where intake hose comes in contact with pavement or curbs. At these points, wear on intake hose is most likely because vibrations from the pumper may keep the intake hose in constant motion. Chafing blocks may be made of wood, leather, or sections of old truck tires.

Hose Strap, Hose Rope, and Hose Chain

One of the most useful tools to aid in carrying or handling a charged hoseline is a *hose strap* **(Figure 13.54)**. Similar to the hose strap are the *hose rope* and *hose chain*. These devices can be used to carry and pull fire hose, but their primary value is to provide a more secure means to handle pressurized hose when applying water. Another important use of these tools is to secure hose to ladders and other fixed objects **(Figure 13.55)**.

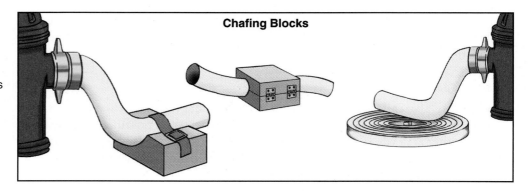

Figure 13.53 Chafing blocks prevent abrasions to hose jackets at points of contact with the ground where pump or water pressure vibrations are likely.

Chafing Blocks

Figure 13.54 Hose straps can be used to reduce stress on the nozzle operator and improve the mobility of a charged hoseline.

Figure 13.55 Hose straps have a variety of applications. They can be used to secure a hoseline to a ladder during interior operations.

Hose Rolls

There are a number of different methods for rolling fire hose, depending on whether it is intended to be used or stored. In all methods, care must be taken to protect the couplings. Some of the more common hose rolls are discussed in the following sections.

Straight Roll

The straight roll, which is the simplest of all hose rolls, starts at on end — usually at the male coupling. To complete the roll, roll the hose toward the other end **(Figure 13.56)**. When the roll is completed, the female end is exposed and the male end is protected in the center of the roll. The straight roll is commonly used for hose in the following situations:

- When loaded back on the apparatus at the fire scene
- When returned to quarters for washing
- When placed in storage (especially rack storage)

This method is also used for easy loading of the minuteman load (See Preconnected Hose Loads for Attack Lines section).

To indicate that a section of hose must be repaired and/or tested before being placed back in service, some departments use a variation of the straight roll. This roll is begun at the female coupling so that when the roll is completed, the male coupling is exposed. Another method is to roll the hose as just described and tie a knot in the exposed end. A tag is usually attached to the male coupling indicating the type and location of damage. This roll is also used when the hose is going to be reloaded on the apparatus for a forward (straight) lay. **Skill Sheet 13-I-2** describes the procedure for making the basic straight roll.

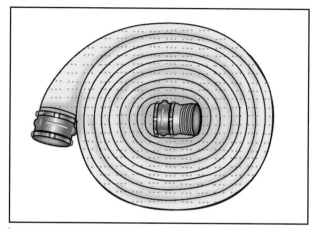

Figure 13.56 A straight roll consists of the male coupling at the center of the roll and the female end on the outside.

Donut Roll

The donut roll is commonly used in situations where hose is likely to be deployed for use directly from a roll **(Figure 13.57)**. The donut roll has certain advantages over the straight roll:

- The firefighter has control of both couplings, which protects them from damage.
- The hose rolls out easier with fewer twists or kinks.
- Holding both couplings facilitates connecting to other couplings (even in the dark).

When a section of fire hose needs to be rolled into a donut roll, the task can be performed by one or two firefighters. **Skill Sheet 13-I-3** describes a method used to make the donut roll.

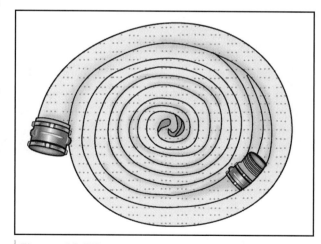

Figure 13.57 In the donut roll, the couplings are on the outside of the roll within 12 inches (300 mm) of each other and the male coupling protected by a portion of the hose.

Twin Donut Roll

The twin donut roll works well on 1½-inch (38 mm) and 1¾-inch (45 mm) hose, although 2-, 2½-, or 3-inch (50 mm, 65 mm, or 77 mm) hose can also be rolled in this manner **(Figure 13.58, p. 654)**. The purpose of this roll is to create a compact roll that can be easily transported and carried for special applications such as high-rise operations.

If the couplings are offset by about 1 foot (0.3 m) at the beginning, they can be coupled together after the roll is tied or strapped. This forms a convenient loop that can be slung over one shoulder for carrying while leaving the hands free. By offsetting the couplings at the beginning, they do not irritate the firefighter's shoulder when carried (**Figure 13.59**).

Self-Locking Twin Donut Roll

The self-locking twin donut roll is a twin donut roll with a built-in carrying loop formed from the hose itself (**Figure 13.60**). This loop locks over the couplings to keep the roll intact for carrying. The length of the carrying loop may be adjusted to accommodate the height of the person carrying the hose.

Coupling and Uncoupling Fire Hose

Coupling and uncoupling are simple procedures for connecting and disconnecting sections of hose with either threaded or sexless couplings. The need for speed and efficiency under emergency conditions requires that specific techniques for coupling and uncoupling hose be learned and practiced. Nozzles may be connected to or removed from the hose using the same methods as when coupling and uncoupling two sections of hose. **Skill Sheets 13-I-4 and 13-I-5** describe methods of coupling and uncoupling hose.

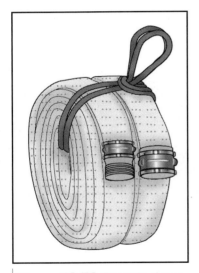

Figure 13.58 The twin-donut roll permits both the male and female couplings to be available for connection when deployed.

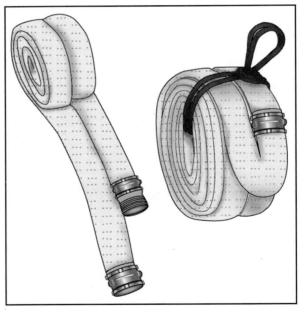

Figure 13.59 A modification of the twin-donut roll permits the couplings to be connected together to protect the threads. The hose strap or rope is used for carrying the roll.

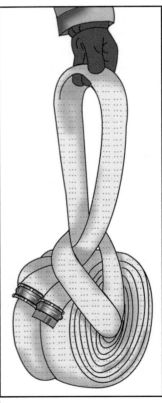

Figure 13.60 The self-locking twin-donut roll provides a built-in carrying strap.

Basic Hose Loads and Finishes

The term most commonly used to describe the hose compartment on a fire apparatus is *hose bed*. Hose beds vary in size and shape, and they are sometimes built for specific needs. In this manual, the front of the hose bed is that part of the compartment closest to the front of the apparatus, and the rear of the hose bed is that part of the compartment closest to the rear of the apparatus. Most hose beds have open slats in the bottom that allow air to circulate throughout the hose load. This feature helps to prevent mildew damage to woven-jacketed hose.

The hose beds on some fire apparatus are divided or separated by a vertical panel that runs from the front to the rear of the hose compartment (called a *split* bed) **(Figure 13.61)**. A split bed allows the apparatus to have hose loaded for both forward and reverse lays at the same time (See Supply Hose Lays section). Hose in a split bed should be stored so that both beds may be connected when a long lay is required **(Figure 13.62)**.

Another way to arrange hose is to "finish" a hose load with additional hose that can be quickly pulled at the beginning of a forward or reverse lay. *Finishes* are arrangements of hose, usually placed on top of a hose load, that are connected to the end of the load.

The following sections provide guidelines for loading hose and highlight the three most common loads for supply hoselines (*accordion, horseshoe,* and *flat*) along with hose load finishes.

Hose Bed — Main hose-carrying area of a pumper or other piece of apparatus designed for carrying hose.

Finish — Arrangement of hose usually placed on top of a hose load and connected to the end of the load. Also called Hose Load Finish.

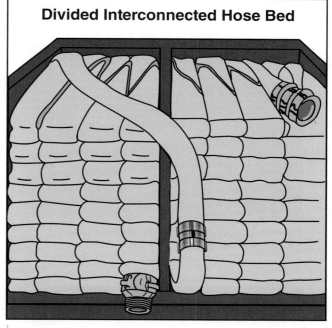

Divided Interconnected Hose Bed

Figure 13.62 Split hose bed loads permit the deployment of a single long supply line or two shorter supply lines.

Figure 13.61 This apparatus hose bed contains a split hose load divided by a hose bed baffle in the center of the apparatus and two separate beds for preconnected attack lines on either side.

Hose Loading Guidelines

Although the loading of hose on fire apparatus is not an emergency operation, it is a critical operation that must be done correctly. At a fire, properly loaded hose permits efficient and effective operations. The following general guidelines should be followed, regardless of the type of hose load used:

- Check gaskets and swivel before connecting any coupling.

- Keep the flat sides of the hose in the same plane when two sections of hose are connected **(Figure 13.63)**. Lugs on the couplings need not be aligned.

- Tighten the couplings hand-tight. Never use wrenches or excessive force.

- Remove kinks and twists from fire hose when it must be bent to form a loop in the hose bed.

- Make a short fold or reverse bend (called a *Dutchman*) in the hose during the loading process so that couplings are not too close to the front or rear of the hose bed and will not flip over when pulled out of the bed **(Figure 13.64)**. The Dutchman serves two purposes: (1) it changes the direction of a coupling and (2) it changes the location of a coupling.

- Load large-diameter hose (3½-inch [90 mm] or larger) with all couplings near the front of the bed. This saves space and allows the hose to lie flat.

- Do not pack hose too tightly. This puts excess pressure on the folds of the hose and may cause couplings to snag when the hose pays out of the bed. A general rule is that the hose should be loose enough to allow a gloved hand to be easily inserted between the folds **(Figure 13.65)**.

Dutchman — Extra fold placed along the length of a section of hose as it is loaded so that its coupling rests in proper position.

Figure 13.63 Keep the hose sections aligned with each other.

Figure 13.65 Do not pack hose too tightly in the bed. Hose should be loose enough to permit a gloved hand to be inserted between the folds.

Figure 13.64 The use of a reverse bend or Dutchman allows the coupling to be pulled smoothly from the hose bed without binding on other folds in the hose.

Accordion Load

The accordion load derives its name from the manner in which the hose appears after loading **(Figure 13.66)**. The hose is laid on edge in folds that lie adjacent to each other (accordionlike). The first coupling placed in the bed should be located to the rear of the bed. It can be placed on either side if the bed is not split. An advantage of this load is its ease of loading. Its simple design requires only two or three people to load the hose, although four people are best. Another advantage is that hose for shoulder carries can easily be taken from the load by simply picking up a number of folds and placing them on the shoulder. **Skill Sheet 13-I-6** shows the procedures for making an accordion load.

Horseshoe Load

The horseshoe load is also named for the way it appears after loading **(Figure 13.67)**. Like the accordion load, it is loaded on edge, but in this case the hose is laid around the perimeter of the hose bed, in a *U*-shaped configuration, working toward the middle. Each length is progressively laid from the outside of the bed toward the inside so that the last length is at the center of the horseshoe. The primary advantage of the horseshoe load is that it has fewer sharp bends in the hose than the accordion or flat loads.

A disadvantage of the horseshoe load occurs most often in wide hose beds — the hose sometimes comes out in a wavy, or snakelike, lay in the street or on the ground as the hose is pulled alternately from one side of a bed and then the other. This can result in an inefficient lay. Another disadvantage is that folds for a shoulder carry cannot be pulled as easily as with an accordion load. With the horseshoe load, two people are required to make the shoulder folds for the carry. As is the case with the accordion load, the hose is loaded on edge, which can promote wear on hose edges.

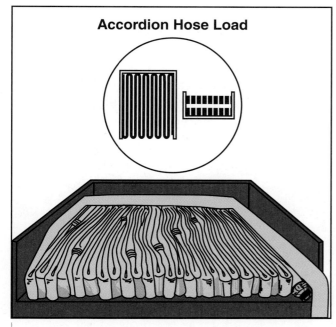

Figure 13.66 The accordion load is designed for ease of loading and of deployment using the shoulder carry.

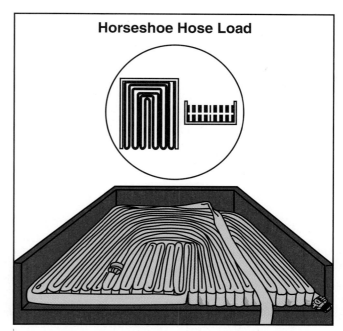

Figure 13.67 The horseshoe load results in fewer bends in the hose. As with any load where the hose is stored on its edge, mechanical wear along one edge can occur when the hose is deployed.

The horseshoe load does not work for large-diameter hose because the hose remaining in the bed tends to fall over as the hose pays off, which can cause the hose to become entangled.

In a single hose bed, the horseshoe load may be started on either side. In a split hose bed, lay the first length against the partition with the coupling hanging an appropriate distance below the hose bed. Determine this distance by estimating the anticipated height of the completed hose load so that the coupling can be connected to the last coupling of the load on the opposite side (crossover) and laid on top of the load. This placement allows easy disconnection of the couplings when the load must be split to lay dual lines. When one side is loaded for a reverse lay and the other is loaded for a forward lay (*combination load*), use an adapter to connect identical couplings. **Skill Sheet 13-1-7** describes the procedures for making the horseshoe hose load.

Flat Load

Of the three supply hose loads, the flat load is the easiest to load. It is suitable for any size of supply hose and is the best way to load large-diameter hose. As the name implies, the hose is laid so that its folds lie flat rather than on edge **(Figure 13.68)**. Hose loaded in this manner is less subject to wear from apparatus vibration during travel. A disadvantage of this load is that, like the accordion load, the hose folds contain sharp bends at both ends, which requires that the hose be reloaded periodically to change the location of the bends within each length to prevent damage to the lining.

In a single hose bed, the flat load may be started on either side. In a split hose bed, lay the first length against the partition with the coupling hanging an appropriate distance below the hose bed. Determine this distance by estimating the anticipated height of the hose bed so that the coupling can be connected to the last coupling of the load on the opposite side (crossover) and laid on top of the load. This placement allows easy disconnection of the couplings when the load must be split to lay dual lines. With a combination load, use an adapter to connect identical couplings. **Skill Sheet 13-I-8** demonstrates the proper method for making a flat load.

The flat load is better suited to large-diameter hose than either the accordion or horseshoe loads. Large-diameter hose can be loaded directly from the street or ground after an incident by straddling the hose with the pumper and driving slowly

Figure 13.68 A flat load in a split hose bed. Large-diameter supply hose is shown on the right and medium supply hose is stored on the left of the baffle. *Courtesy of Sam Goldwater.*

Figure 13.69 One way of loading large-diameter hose involves laying out the coupled hose sections along the street. The apparatus is then driven down the line of hose, straddling the hose between the wheels. Personnel at the rear of the apparatus feed the hose into the bed where others fold it into place.

Figure 13.70 One type of hose wringer used to expel air and water from large diameter hose.

forward as the hose is progressively loaded into the bed **(Figure 13.69)**. A hose wringer or roller can be used to expel the air and water from the hose as it is placed in the hose bed **(Figure 13.70)**. This creates a neat and space-efficient load of large-diameter hose.

While it is considered an unsafe practice to ride on a moving apparatus without safety restraints, NFPA 1500 does permit the loading of hose onto a moving apparatus as long as certain criteria are met and a local written policy is in place.

The hose load for large-diameter hose should be started 12 to 18 inches (300 mm to 450 mm) from the front of the hose bed. This extra space should be reserved for couplings, and all couplings should be laid in a manner that allows them to pay out without flipping them over **(Figure 13.71, p. 660)**. It may be necessary to make a short fold or reverse bend (Dutchman) in the hose to do this **(Figure 13.72, p. 660)**.

Hose Load Finishes

Hose load finishes are added to the basic hose load to increase the versatility of the load. Finishes are normally loaded to provide enough hose to make a hydrant connection and to provide a working line at the fire scene.

Finishes fall into two categories: those for forward lays (*straight finish*) and those for reverse lays (*reverse horseshoe, skid load finishes*). Finishes for forward lays are designed to facilitate making a hydrant connection and are not as elaborate as finishes for reverse lays. Finishes for reverse lays are designed to provide an adequate amount of hose at the scene for initial fire attack.

Figure 13.71 Some departments load large-diameter hose with all couplings near the front of the hose bed (toward the front of the apparatus). *Courtesy of Sam Goldwater.*

Figure 13.72 A Dutchman fold used in large-diameter hose to locate the coupling for proper deployment.

Straight Finish

A straight finish consists of the last length or two of hose flaked loosely back and forth across the top of the hose load. This finish is normally associated with forward-lay operation. A hydrant wrench, gate valve, and any necessary adapters are usually strapped to the hose at or near the female coupling (**Figure 13.73**).

Reverse Horseshoe Finish

This finish is similar to the horseshoe load except that the bottom of the *U* portion of the horseshoe is at the rear of the hose bed (**Figure 13.74**). It is made of one or two 100-foot (30 m) lengths of hose, each connected to one side of a gated wye. Any size attack hose can be used: 1½, 1¾, or 2½ inches (38 mm, 45 mm, or 65 mm). The smaller

Figure 13.73 A straight finish consists of the last section of hose folded across the hose bay. A hydrant wrench may be used to hold the hose in place by inserting it into the hose folds against the finish section.

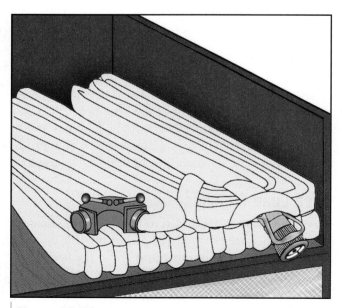

Figure 13.74 A reverse horseshoe finish with nozzle and gated wye attached to the hose. This arrangement creates a means for rapid deployment of an attack line.

sizes require a 2½- × 1½-inch (65 mm by 38 mm) gated reducing wye. The 2½-inch (65 mm) hose requires a 2½- × 2½-inch (65 mm by 65 mm) gated wye. A nozzle of the appropriate size is also needed for each attack line.

The reverse horseshoe finish can also be used for a preconnected line and can be loaded in two or three layers. With the nozzle extending to the rear, the finish can be placed over a shoulder and the opposite arm extended through the loops of the layers, pulling the hose from the bed for an arm carry. A second preconnected line can be bedded below when there is sufficient depth.

Skid Load Finish

A skid load finish consists of folding the last three lengths (150 feet/45 m) of 2½-inch (65 mm) hose into a compact bundle on top of the rest of the hose load. The load begins by forming three or more pull loops that extend beyond the end of the hose load. The rest of the hose, with nozzle attached, is accordion-folded across the hose used to form the pull loops in the hose bed (**Figure 13.75**).

Preconnected Hose Loads for Attack Lines

Preconnected hoselines are the primary lines used for fire attack by most fire departments. These hoselines are connected to a discharge valve and placed in an area other than the main hose bed. Preconnected hoselines generally range from 50 to 250 feet (15 m to 75 m) in length. Preconnected attack lines can be carried in the following places:

- Longitudinal beds (**Figure 13.76**)
- Raised trays
- Transverse beds (often called *cross lays*) (**Figure 13.77**)

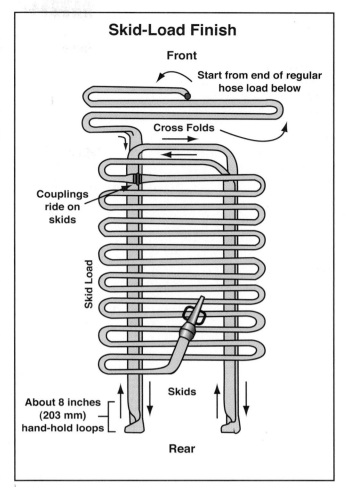

Figure 13.75 Example of the construction of a skid load finish.

Figure 13.76 A typical longitudinal hose bed. Large-diameter supply hose is located on the left and two attack lines are stored on the right of the hose bed. *Courtesy of Sam Goldwater.*

Figure 13.77 Transverse beds loaded with color-coded attack lines. Depending on local procedures, the hose may be loaded to deploy off one or both sides of the apparatus.

- Tailboard compartments
- Side compartments or bins
- Front bumper wells (sometimes called *jump lines*)
- Reels

Several different loads can be used for preconnected lines. The following sections detail some of the more common ones. Special loads to meet local requirements may be developed based on individual experiences and apparatus configurations.

Preconnected Flat Load

The *preconnected flat load* is adaptable for varying widths of hose beds and is often used in transverse beds **(Figure 13.78)**. This load is similar to the flat load for larger supply hose with two exceptions: (1) It is preconnected and (2) loops are provided to aid in pulling the load from the bed. Place the pull loops at regular intervals within the load so that equal portions of the load are pulled from the bed. The number of loops and the intervals at which they are placed depend on the size and total length of the hose. The procedures in **Skill Sheet 13-I-9** can be adapted for any type of hose bed.

Triple Layer Load

The *triple layer load* gets its name because the load begins with hose folded in three layers. The three folds are then laid into the bed in an S-shaped fashion **(Figure 13.79)**. The load is designed to be pulled by one person. A disadvantage with the triple layer load is that the three layers, which may be as long as 50 feet (15 m) each, must be completely removed from the bed before deploying the nozzle end of the hose. This could be a problem if another apparatus is parked directly behind the hose bed. While this hose load can be used for all sizes of attack lines, it is often

Figure 13.78 The preconnected flat load permits rapid deployment of an attack line. The loops at the end of the hose load are used for pulling either the entire load or a portion of it from the bed.

Figure 13.79 A triple layer load requires that the entire load be unloaded before the nozzle end can be deployed.

preferred for larger (2- and 2½-inch [50 mm and 65 mm]) lines that may be too cumbersome for shoulder carries. The procedures for making the triple layer load are given in **Skill Sheet 13-I-10**.

Minuteman Load

The *minuteman load* is designed to be pulled and advanced by one person (**Figure 13.80**). The primary advantage with this load is that it is carried on the shoulder, completely clear of the ground, so it is less likely to snag on obstacles. The load pays off the shoulder as the firefighter advances toward the fire. The load is also particularly well suited for a narrow hose bed. A disadvantage with the load is that it can be awkward to carry when wearing an SCBA. If the load is in a single stack, it may also collapse on the shoulder if not held tightly in place. The procedures for making the minuteman load are described in **Skill Sheet 13-I-11**.

Booster Hose Reels

Booster lines are rubber-covered hose that are usually carried preconnected and coiled on reels (**Figure 13.81**). These *booster hose reels* may be mounted in any of several places on the fire apparatus according to specified needs and the design of the apparatus. Some booster hose reels are mounted above the fire pump and behind the apparatus cab. This arrangement provides booster hose that can be unrolled from either side of the apparatus. Other booster hose reels are mounted on the front bumper of the apparatus or in rear compartments. Hand- and power-operated reels are available. Booster hose should be wound onto the reel one layer at a time in an even manner. This allows the maximum amount to be loaded and provides for the easiest removal from the reel.

Booster Hose or Booster Line — Noncollapsible rubber-covered, rubber-lined hose usually wound on a reel and mounted somewhere on an engine or water tender and used for the initial attack and extinguishment of incipient and smoldering fires. This hose is most commonly found in 1-inch (25 mm) diameters and is used for extinguishing low-intensity fires and mop-up.

WARNING!
Booster lines do not deliver a sufficient volume of water to protect firefighters if conditions within a burning structure suddenly deteriorate. Therefore, booster lines are not appropriate for interior fire fighting.

Figure 13.80 The minuteman load is designed to be pulled by and carried on the shoulder of one firefighter.

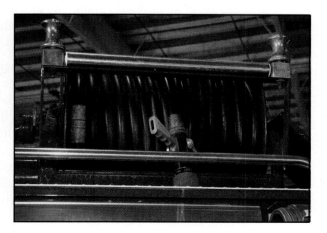

Figure 13.81 Booster hose reels are located on the top of the apparatus, as shown, or behind the front or rear bumper.

Supply Hose Lays

The three basic hose lays for supply hose are *forward lay* (also called *straight lay*), *reverse lay,* and *split lay* (sometimes called *combination lay*). Supply hose with threaded couplings is usually loaded for a forward lay. That means the hose is loaded into the bed so that when hose is laid, the end with the female coupling is toward the water source and the end with the male coupling is toward the fire. When hose is arranged in this manner, several hose-lay options are available. At the water source, hose can be connected to the male threads of a pumper discharge valve or to the male threads of a hydrant outlet. At the fire end, hose can be connected to the auxiliary intake valve of a pumper, into a fire department connection (FDC), or it can be connected directly to nozzles and appliances, all of which have female threads.

Hose-lay procedures vary from department to department, but the basic methods of laying hose remain the same. Hose is either laid forward from a water source to the incident scene, reverse from the incident scene to a water source, or split so the hose can be laid to and from the junction to the water source and the incident scene. These basic methods are presented to provide the foundation for developing hose lays that more specifically suit individual department needs.

Regardless of the method chosen, use the following guidelines when laying hose:

- Do not ride in a standing position when the apparatus is moving.

- Drive the apparatus at a speed no greater than one that allows the couplings to clear the tailboard as the hose leaves the bed — generally between 5 and 10 mph (8 km/h and 16 km/h).

- Lay the hose to one side of the roadway (but not in the gutter) so that other apparatus are not forced to drive over it.

Forward Lay

In a forward lay, hose is laid from the water source to the fire. This method is often used when the water source is a hydrant and the pumper must be positioned near the fire **(Figure 13.82)**. Hose beds set up for forward lays should be loaded so that the first coupling to come off the hose bed is female **(Figure 13.83)**. The operation consists of stopping the apparatus at the hydrant and permitting a firefighter to safely leave the apparatus and secure the hose. Then the apparatus proceeds to the fire laying either single or dual hoselines.

The primary advantage with this lay is that a pumper can remain at the incident scene so its hose, equipment, and tools are readily available if needed. The pump operator also has visual contact with the fire suppression operation and can better react to changes in the fire situation than if the pumper were at the hydrant.

One disadvantage with the forward lay is that if a long length of 2½- or 3-inch (65 mm or 77 mm) hose is laid, or if the hydrant has inadequate flow pressure, it may be necessary for a second pumper to boost the pressure in the line at the hydrant. The first pumper in this scenario must have used a four-way hydrant valve if the transition from hydrant pressure to pump pressure is to be made without interrupting the flow of water in the supply hose (see Using Four-Way Hydrant Valves section). Another disadvantage is that one member of the crew is temporarily unavailable for a fire fighting assignment because that person must stay at the hydrant long enough to make the connection and open the hydrant.

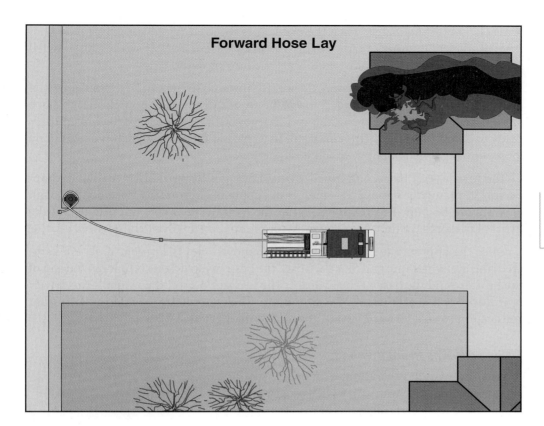

Forward Hose Lay

Figure 13.82 In the forward lay, the hose extends from the water source to the fire scene.

Figure 13.83 Hose loads that are intended for the forward lay have the female coupling loaded to come out of the hose bed first.

The firefighter who is going to make the hydrant connection (also known as *catching the plug* or *making the hydrant*) must know the following two primary skills: (1) the proper procedures for wrapping and connecting to the hydrant and (2) the operation of the hydrant valve if one is used.

Making the Hydrant Connection

The firefighter assigned to connect to the hydrant should have a combination spanner/hydrant wrench and a four-way hydrant valve (if part of local SOP) if these are not already preconnected to the supply line. Some departments choose to put all of these items in a pouch that is kept on the rear step (tailboard) of the apparatus. Other departments choose not to use hydrant valves. It is also desirable for the hydrant person to have a portable radio so that when the attack engine is ready to receive water, it may be sent immediately when the message is received. Many departments do not have a sufficient number of radios to do this so a visual or audible signal is used to tell the firefighter at the hydrant when to start the flow of water.

The use of such audible warning devices as horns or sirens can be a problem when other apparatus are responding to the scene. The hydrant person might hear a horn or siren from an arriving apparatus and charge the line before the driver/operator is ready to accept the water. This can result in a charged hose bed, which is useless, or a loose, flowing hose coupling, which is dangerous.

The first task to be accomplished when starting a forward lay is for the hydrant person to manually remove an amount of supply hose from the hose bed. Enough hose should be pulled to reach and wrap the hydrant. It is also important that the hydrant person have the necessary tools for making the hydrant connection.

Once the appropriate amount of hose is removed and the proper tools are gathered, the firefighter must anchor the hose. The best way to do this is to wrap the end of the hose around the base of the hydrant. The hydrant person then signals the driver/operator that it is safe to proceed to the fire. The procedures for making a hydrant connection from a forward lay are given in **Skill Sheet 13-I-12**.

Using Four-Way Hydrant Valves

A four-way hydrant valve allows a forward-laid supply line to be immediately charged and allows a later-arriving pumper to connect to the hydrant (**Figure 13.84**). The second pumper can then supply additional supply lines and/or boost the pressure to the original line. Typically, the four-way hydrant valve is preconnected to the end of the supply line. This allows the firefighter who is catching the hydrant to secure the valve and the hose to the hydrant in one action. There are several manufacturers providing four-way valves that have the same basic operating principles.

Figure 13.84 Connecting a four-way hydrant valve allows water supply pressure to be increased.

Reverse Lay

In a reverse lay, hose is laid from the fire to the water source. This method is used when a pumper must first go to the fire location so a size-up can be made before laying a supply line (**Figure 13.85**). It is also the most expedient way to lay hose if the apparatus that lays the hose must stay at the water source such as when drafting or boosting hydrant pressure to the supply line. Hose beds set up for reverse lays should be loaded so that the first coupling to come off the hose bed is male (**Figure 13.86**).

Laying hose from the incident scene back to the water source has become a standard method for setting up a relay pumping operation when using 2½- or 3-inch (65 mm or 77 mm) hose as a supply line. With long lays of this size hose, it is often necessary to place a pumper at the hydrant to increase the pressure in the supply hose. Of course, it is always necessary to place a pumper at a static water source. The reverse lay is the most direct way to supplement hydrant pressure and set up drafting operations.

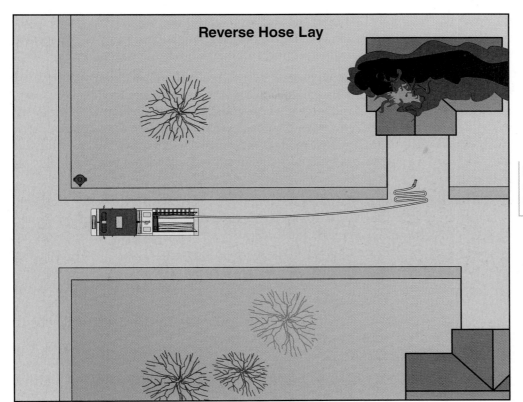

Reverse Hose Lay

Figure 13.85 The reverse lay is used to extend the fire hose from the fire to the water source.

Figure 13.86 The male coupling is usually located at the back of the hose bed in a reverse load. A nozzle or gated wye may be attached to the coupling.

A disadvantage with the reverse lay is that essential fire fighting equipment, including attack hose, must be removed and placed at the fire scene before the pumper can proceed to the water source. This causes some delay in the initial attack. The reverse lay also obligates one person, the pump operator, to stay with the pumper at the water source, thus preventing that person from performing other essential fireground activities.

A common operation involving two pumpers — an attack pumper and a water-supply pumper — calls for the first-arriving pumper to go directly to the scene to start an initial attack on the fire using water from its tank, while the second-arriving pumper lays the supply line from the attack pumper back to the water source. This is a relatively simple operation because the second pumper needs only to connect its just-laid hose to a discharge outlet, connect an intake hose, and begin pumping.

When reverse-laying a supply hose, it is not necessary to use a four-way hydrant valve. One can be used, however, if it is expected that the pumper will later disconnect from the supply hose and leave the hose connected to the hydrant. This may be desirable when the demand for water diminishes to the point that the second pumper can be made available for response to other incidents. As with a forward lay, using the four-way valve in a reverse lay provides the means to switch from pump pressure to hydrant pressure without interrupting the flow.

The reverse lay is also used when the first pumper arrives at a fire and must work alone for an extended period of time. In this case, the hose laid in reverse becomes an attack line. It is often connected to a reducing wye so that two smaller hoses can be used to make a two-directional attack on the fire (**Figure 13.87**). The reverse-lay procedures outlined in **Skill Sheet 13-I-13** describe how the second pumper lays a line from an attack pumper to a hydrant (**Figure 13.88**). They can be modified to accommodate most types of apparatus, hose, and equipment.

Making Hydrant Connections with Soft Intake Hose

Frequently, firefighters will assist pumper driver/operators in making hydrant connections following a reverse lay. Either soft or hard intake hose designed for hydrant operations may be used to connect to hydrants. Hard intake hose must be used when drafting from a static water supply source.

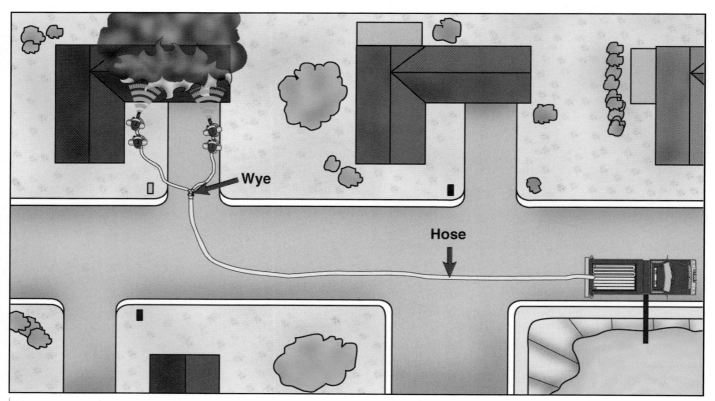

Figure 13.87 When a reverse hose lay is used, a gated wye and two attack lines can be supplied by the pumper at the water source.

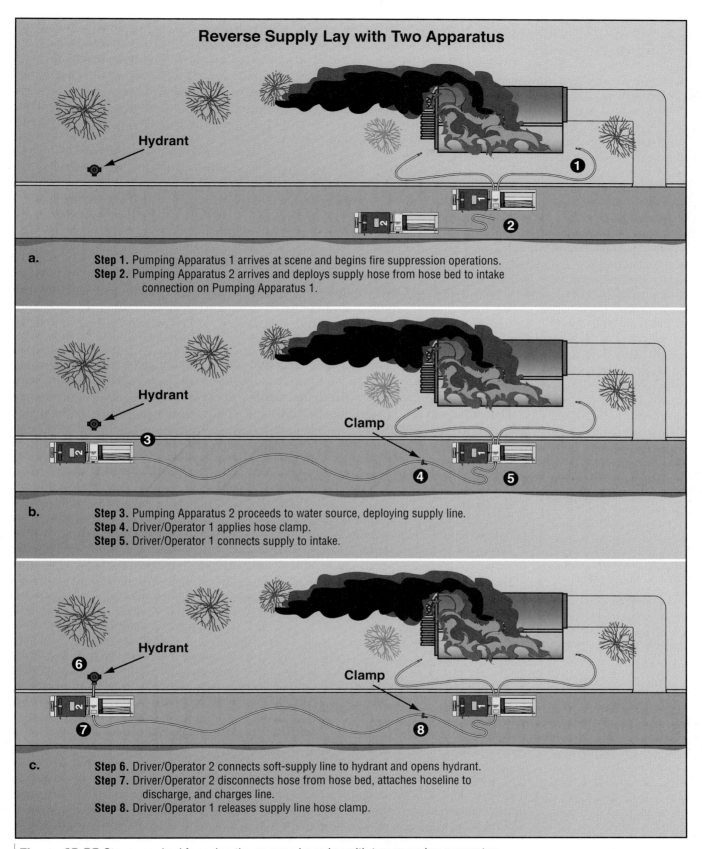

Reverse Supply Lay with Two Apparatus

Hydrant

a.

Step 1. Pumping Apparatus 1 arrives at scene and begins fire suppression operations.
Step 2. Pumping Apparatus 2 arrives and deploys supply hose from hose bed to intake connection on Pumping Apparatus 1.

Hydrant

Clamp

b.

Step 3. Pumping Apparatus 2 proceeds to water source, deploying supply line.
Step 4. Driver/Operator 1 applies hose clamp.
Step 5. Driver/Operator 1 connects supply to intake.

Hydrant

Clamp

c.

Step 6. Driver/Operator 2 connects soft-supply line to hydrant and opens hydrant.
Step 7. Driver/Operator 2 disconnects hose from hose bed, attaches hoseline to discharge, and charges line.
Step 8. Driver/Operator 1 releases supply line hose clamp.

Figure 13.88 Steps required for using the reverse hose lay with two pumping apparatus.

Figure 13.89 Some hydrants may not be equipped with a large steamer outlet. These hydrants can be used to supply two 2½-inch (65 mm) or 3-inch (70 mm) hose.

Not all hydrants have large steamer outlets capable of accepting direct connections from soft intake hose. Operations on hydrants equipped with two 2½-inch (65 mm) outlets require the use of two 2½- or 3-inch (65 mm or 77 mm) hoselines (**Figure 13.89**). These smaller intake hose can be connected to a siamese at the pump. It is more efficient to connect a 4½-inch (115 mm) or larger intake hose to a hydrant with only 2½-inch (65 mm) outlets. Such a connection is made by using a 4½-inch (115 mm) hose, or whatever size intake hose coupling is used, and connecting it to a 2½-inch (65 mm) reducer coupling.

Making Hydrant Connections with Hard Intake Hose

Connecting a pumper to a fire hydrant using hard intake hose may require coordination and teamwork because more people may be needed to connect hard intake hose than are needed to connect soft intake hose. Making hydrant connections with some hard intake hose is also considerably more difficult than making connections with a soft intake hose.

The first aspect that is important is the positioning of the pumper in relation to the hydrant. No definite rule can be given to determine this distance because not all hydrants are the same distance from the curb or road edge, and the hydrant outlet may not directly face the street or road.

Another determining factor is that while most apparatus have pump intakes on both sides, others may also have one at the front or rear. It is considered good policy to stop the apparatus with the intake of choice just short of the hydrant outlet. Depending on local protocols, the hard intake hose may be connected to either the apparatus or the hydrant first when making hydrant connections.

NOTE: If the hard intake is marked FOR VACUUM USE ONLY, do not use it for hydrant connections. This type of hard intake is for drafting operations only.

Split Lay

Split Lay — Hose lay deployed by two pumpers, one making a forward lay and one making a reverse lay from the same point.

The term *split lay* can refer to any one of a number of ways to lay multiple supply hose with a single engine. As described earlier in this chapter, dividing a hose bed into two or more separate compartments provides the most options for laying multiple lines. Depending upon whether the beds are set up for forward or reverse lays, lines can be laid in the following ways (assume that hose of the same diameter are in the two hose beds):

- Two lines laid forward
- Two lines laid reverse
- Forward lay followed by a reverse lay
- Reverse lay followed by a forward lay
- Two lines laid forward followed by one or two lines laid reverse
- Two lines laid reverse followed by one or two lines laid forward

One type of split lay is a hoseline laid in part as a forward lay and in part as a reverse lay. This is accomplished by one pumper making a forward lay from an intersection or driveway entrance toward the fire. A second pumper then makes a reverse lay from the point where the initial line was laid to the water supply source (**Figure 13.90**). Aavoid making the lay too long for the pump, hose size, and required flow.

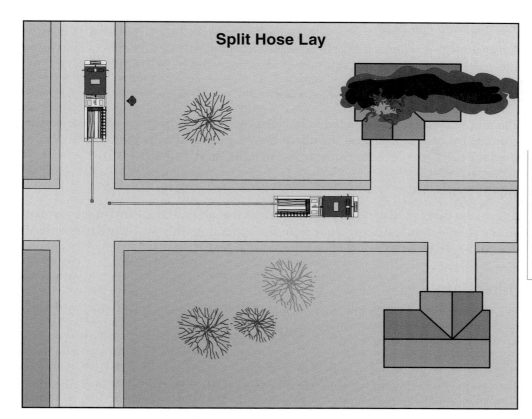

Split Hose Lay

Figure 13.90 An example of one form of split lay. Depending on the length of the hose lay and the available water pressure, the pumper at the hydrant may be required to connect to the hydrant and boost the hydrant pressure.

When using hose equipped with sexless couplings, the direction of the lay is unimportant. The hose may be laid in either direction with the same result. The only thing that firefighters and driver/operators must be concerned with is making sure that the proper adapters are present at each end of the lay to make the appropriate connections.

Clearly, there are many other split-lay options when the hose bed is divided. One of the most versatile arrangements is one in which one section of the hose bed contains large-diameter hose and the other section contains smaller diameter hose that can be used for either supply or attack. A pumper set up in this manner can lay LDH when the fire situation requires the pumper to lay its own supply line and work alone (laying it forward so the pumper stays at the incident scene). Firefighters can use small-diameter hose as a supply line at fires with less demanding water flow requirements as well as for attack lines on large fires. Therefore, a split hose bed gives fire officers the greatest number of choices when determining the best way to use limited resources.

Handling Hoselines

To effectively attack and extinguish a fire, hoselines must be removed from the apparatus and advanced to the location of the fire. The techniques used to advance hoselines depend on how the hose is loaded. Hoselines may be loaded preconnected to a discharge outlet or simply placed in the hose bed unconnected.

Preconnected Hoselines

The method used to pull preconnected hoselines varies with the type of hose load. The following sections describe the methods used to pull and carry preconnected hose from the loads described earlier in this chapter.

Preconnected Flat Load

Advancing the preconnected flat load involves pulling the hose from the compartment and walking toward the fire. This procedure is described in **Skill Sheet 13-I-14**.

Minuteman Load

The minuteman load is intended to be deployed without dragging any hose on the ground. The hose is flaked off the top of the shoulder as the firefighter advances toward the fire. This procedure is described in **Skill Sheet 13-I-15**.

Triple Layer Load

Advancing the triple layer load involves placing the nozzle and the fold of the first tier on the shoulder and walking away from the apparatus. This procedure is described in **Skill Sheet 13-I-16**.

Other Hoselines

The following procedures are used for handling hose that is not preconnected. This hose is usually 2½ inches (65 mm) or larger.

Wyed Lines

The reverse horseshoe finish and other wyed lines are normally used in connection with a reverse layout because the wye connection is fastened to the 2½- or 3-inch (65 mm or 77 mm) hose. The unloading process involves two operations that can be done consecutively by one person.

Shoulder Loads from Flat or Horseshoe Loads

Due to the way flat and horseshoe loads are arranged in the hose bed, it is necessary to load one section of hose at a time onto the shoulder.

Shoulder Loads from Accordion or Flat Loads

Because all of the folds in an accordion load and a flat load are nearly the same length, they can be loaded on the shoulder by taking several folds at a time directly from the hose bed. **Skill Sheet 13-I-17** describes the steps for shoulder loading and advancing hose that is loaded in either an accordion load or a flat load.

Working Line Drag

The working line drag is one of the quickest and easiest ways to move fire hose at ground level. Its use is limited by available personnel, but when adapted to certain situations, it is an acceptable method. **Skill Sheet 13-I-18** contains the procedure for advancing hose using the working line drag.

Advancing Hoselines

Once hoselines have been laid out from the attack pumper, they must be advanced into position for applying water onto the fire. The methods of deploying hose described so far work well if the firefighter is simply advancing hose over flat surfaces with no obstacles. Advancing hoselines is considerably more difficult when they must be deployed up or down stairways, from standpipes, up ladders, and deep into buildings. Hoselines can be deployed more easily before they are charged because water adds weight and pressure makes the hose stiffer. Because it is often unsafe to enter burning buildings with uncharged hoselines, however, firefighters must know how to handle both uncharged and charged lines. They must also know how to extend a hoseline by adding more hose. Finally, firefighters must know how to secure and replace a ruptured section of hose if necessary.

Advancing Hose Into a Structure

When advancing hose into a structure, firefighters must remain alert to potential dangers such as backdraft, flashover, and structural collapse. Firefighters should observe the following general safety guidelines when advancing a hoseline into a burning structure:

- Bleed air from charged hoselines before entering the building or fire area.
- Position the nozzle operator and all members of the hose team on the same side of the hoseline (**Figure 13.91**).
- Check doors for heat before opening.
- Stay low and avoid blocking ventilation openings such as doorways or windows.
- Chock self-closing doors open to keep the line from being pinched by the door.
- Always check for and remove kinks from the line.

Consult **Skill Sheet 13-I-19** for steps in advancing a line into a structure.

Figure 13.91 All members of the hose team should be on the same side of the hose.

Advancing Hose Up a Stairway

Because it is difficult to drag a charged hoseline up stairs, through doorways, and around corners, it should be advanced up stairways uncharged when conditions allow. The shoulder carry works well for stairway advancement because the hose is carried instead of dragged and is flaked out as needed. The minuteman load and carry is also excellent for use on stairways. When advancing hose up a stairway, lay the uncharged hose against the outside wall to keep the stairs as clear as possible and avoid sharp bends and kinks in the hose (**Figure 13.92**). It is easier to pull charged hose down stairs than up, so firefighters should flake excess hose up the stairs toward the floor above the fire floor. When the hoseline is charged, the weight of the water and gravity will make extending the excess hoseline easier. If possible, position a firefighter at every turn or point of resistance to aid in deployment of the hoseline.

Advancing Hose Down a Stairway

Advancing an uncharged (dry) hoseline down a flight of stairs is considerably easier than advancing a charged hose. However, advancing an uncharged line down stairs is recommended only when there is no fire present or it is very minor.

Advancing a charged hoseline down a stairway can be almost as difficult as advancing one up stairs. Because flaking excess hose down the stairway would obstruct the stairs, excess hose should be flaked outside the stairwell, such as in a hallway or room adjacent to the stairwell,

Figure 13.92 Hose should be laid close to the outside wall of the stairway. *Courtesy of Rick Montemorra.*

and firefighters positioned on the stairs to feed the hose down to the nozzle team. Firefighters must also be positioned at corners and pinch points. **Skill Sheet 13-I-20** gives steps for advancing a hoseline up and down an interior stairway.

Advancing Hose from a Standpipe

Figure 13.93 Firefighters assembling a typical hose bundle.

Getting hose to the upper floors of high-rise buildings can be a challenge for firefighters. While preconnected hoselines may be able to access fires on the lower floors, fires beyond the reach of these lines require different techniques. One approach is to have hose rolls or packs on the apparatus ready to carry aloft and connect to the building's standpipe system.

How these high-rise packs are constructed is a matter of local preference, but the most common are hose bundles that are easily carried on the shoulder or in specially designed hose packs complete with nozzles, fittings, and tools **(Figure 13.93)**.

Except in rare instances, firefighters are not allowed to use elevators in burning buildings, so hose must be brought to the fire floor over an aerial ladder or by an interior stairway. Regardless of how the hose is brought up, fire crews normally stop one floor below the fire floor and connect to the standpipe. If the standpipe connection is in an enclosed stairwell, it is acceptable to hook up on the fire floor. The standpipe connection is usually in or near the stairwell. By observing the floor below, firefighters can get a general idea of the layout of the fire floor.

Once at the standpipe connection, remove the outlet cap, check for foreign objects in the discharge, check the connection for the correct adapters (if needed), and connect the fire department hose to the standpipe **(Figures 13.94 a and b)**. Be alert for pressure-relief devices and follow department standard operating procedures for removal or connection. If 1½-, 1¾-, or 2-inch (38 mm, 45 mm, or 50 mm) hose is

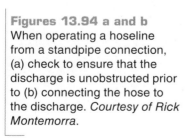

Figures 13.94 a and b When operating a hoseline from a standpipe connection, (a) check to ensure that the discharge is unobstructed prior to (b) connecting the hose to the discharge. *Courtesy of Rick Montemorra.*

used, it is a good practice to place a gated wye on the standpipe outlet. A 2½-inch (65 mm) attack line may also be used depending on the size and nature of the fire. While the standpipe connection is being completed, any extra hose should be flaked up the stairs toward the floor above the fire **(Figure 13.95)**.

During pickup operations after the fire, carefully drain the water contained in the hoselines to prevent unnecessary water damage. This can be accomplished by draining the hose down a floor drain, out a window, or down a stairway.

Advancing Hose Up a Ladder

In multistory buildings without standpipes, one of the safest ways to get hose to an elevated location is to carry it up the stairs in a bundle and drop the end over a balcony railing or out a window to connect to a water source. Another safe method is to hoist the hose and attached nozzle up to a window or landing using a rope (see Chapter 7, Ropes and Knots).

Figure 13.95 Flake the excess hose up the stairway toward the floor above the fire floor. *Courtesy of Rick Montemorra.*

Sometimes these methods cannot be used, however, and it is necessary to advance the hose up a ladder. Advancing fire hose up a ladder is easier and safer with an uncharged line. In most cases, the firefighter heeling the ladder can also help feed the hose to those on the ladder. If the hose is already charged with water, it may be advisable to drain the hose before advancing it up the ladder.

The best way to advance an *uncharged hoseline* up a ladder is to have the lead firefighter drape the nozzle or end coupling over the shoulder from the front on the side on which the hose is being carried. The lead firefighter advances up the ladder to the first fly section and waits until the next firefighter is ready to proceed. At this point, a second firefighter drapes a large loop of hose over the shoulder and starts up the ladder. If the ladder is a three-section ladder, a third firefighter may continue the process once the second firefighter reaches the first fly section **(Figure 13.96, p. 676)**. To avoid overloading the ladder, only one person is allowed on each section of the ladder. Rope hose tools or utility straps can also be used for this advancement **(Figure 13.97, p. 676)**. The hose can be charged once it has reached the point from which the fire attack will be made.

In those cases where it is absolutely necessary to advance a charged line up a ladder, firefighters should position themselves on the ladder within reach of each other. Each firefighter should be attached to the ladder with either a leg lock or ladder belt because both hands are required to move the charged line. The hose is then pushed upward from firefighter to firefighter **(Figure 13.98, p. 676)**. The firefighter on the nozzle takes the line into the window and the other firefighters support the hose by securing it to the ladder rungs with hose straps or utility straps. **Skill Sheets 13-I-21** and **13-I-22** describe procedures for advancing charged and uncharged lines up a ladder into a window.

It is sometimes necessary for firefighters to operate a hoseline from a ladder. The hoseline is first passed up the ladder as previously described. Secure the hose to the ladder with a hose strap at a point several rungs below the one on which the nozzle operator is standing **(Figure 13.99, p. 676)**. All firefighters on the ladder must use a leg lock or ladder belt to secure themselves to the ladder. The firefighter on the nozzle projects the nozzle through the ladder and holds it with a rope hose tool or similar aid. When the line and all firefighters on the ladder are properly secured, the nozzle can be opened **(Figure 13.100, p. 677)**.

> **WARNING!**
> Do not exceed the rated weight capacity of the ladder. If the hose cannot be passed up the ladder without exceeding the load limit, it should be hoisted up.

Figure 13.96 Firefighters advancing an uncharged fire hose up a ladder.

Figure 13.97 Rope hose tools can be used to carry the hose.

Figure 13.98 A charged hoseline may be pushed up a ladder when necessary.

Figure 13.99 Secure the hose to the ladder with a hose strap or rope hose tool.

Figure 13.100 When the hose is secured to the ladder, open the nozzle slowly.

Extending a Section of Hose

Occasionally, it becomes necessary for firefighters to extend the length of a hoseline with hose of the same size or smaller. **Skill Sheet 13-I-23** describes the procedures that may be used to extend hoselines.

Controlling a Loose Hoseline

A loose hoseline is one in which water under pressure is flowing through a nozzle, an open butt, or a rupture and is out of control. This is a very dangerous situation because the loose hoseline may flail about or whip back and forth. Firefighters and bystanders may be seriously injured or killed if they are hit by an uncontrolled and whipping hose or coupling. **Skill Sheet 13-I-24** gives steps for controlling loose hoselines.

Closing a valve at the pump or hydrant to turn off the flow of water is the safest way to control a loose line. Another method is to apply a hose clamp at a stationary point in the hoseline. It may also be possible to put a kink in the hose at a point away from the break until the appropriate valve is closed (**Figure 13.101**). To put a kink in a hose, form a loop in the line, press down on the top of the loop, and apply body weight to the bends in the hose. In most cases, this action will not completely stop the flow of water through the hose, but it will reduce the flow sufficiently for firefighters to safely gain control of the end of the hose.

NOTE: This procedure for placing a kink in hose does not apply to LDH due to its size and weight when charged.

Figure 13.101 If necessary, water flow can be stopped by kinking the hose.

Replacing Burst Sections

A hose clamp or a kink in the hose can also be used to stop the flow of water when replacing a burst section of hose. Two additional sections of hose should be used to replace any one bad section. This is necessary because hoselines stretch to longer lengths when under pressure; thus, the couplings in the line are invariably farther apart than the length of a single replacement section. Steps to take in replacing burst sections of hose are given in **Skill Sheet 13-I-25**.

Operating Hoselines

For their own safety, and to fight fires effectively, firefighters must know how to operate and control hoselines. While a number of different methods can be used, firefighters should use the method or methods taught by their departments. Some of the more common techniques are described in the following sections. **Skill Sheet 13-I-26** describes how to operate a charged attack line from a ladder.

Operating Small Handlines

The following methods can be used with booster lines and small handlines of 1½-, 1¾-, and 2-inch (38 mm, 45 mm, and 50 mm) hose.

One-Firefighter Method

Except during overhaul after a fire or for very small outdoor nuisance fires such as smoldering bark or a fire in a small trash can, firefighters are never assigned to operate an attack line alone. Whenever one firefighter is required to operate a small hose and nozzle, the hoseline should be straight for at least 10 feet (3 m) behind the nozzle. The firefighter should hold the nozzle with one hand and the hose with the other hand just behind the nozzle (**Figure 13.102**).

Two-Firefighter Method

With the exceptions just mentioned, two firefighters are the minimum number required for handling any attack line. Except for booster lines, two firefighters are usually needed when the nozzle must be advanced. The nozzle operator holds the nozzle with one hand and, unless the nozzle is equipped with a pistol grip, holds the hose just behind the nozzle with the other hand. The hoseline is then rested against the waist and across the hip. The backup firefighter takes a position on the same side of the hose about 3 feet (1 m) behind the nozzle operator. The second firefighter holds the hose with both hands and rests it against the waist and across the hip or braces it with the leg (**Figure 13.103**). The backup firefighter is responsible for keeping the hose straight behind the nozzle operator. During extended operations, either one or both firefighters may apply a hose strap or utility strap to reduce the effects of nozzle reaction.

Operating Large Handlines

The following methods can be used with large handlines of 2½- and 3-inch (65 mm and 77 mm) or larger hose.

One-Firefighter Method

Whenever a nozzle connected to a large handline is used, a minimum of two firefighters and preferably three should be used to operate the line. However, during exposure protection or overhaul operations, one firefighter may be assigned to operate a large handline alone if a master stream device is not available. First, a large loop is formed that crosses over the line about 2 feet (0.6 m) behind the nozzle. The firefighter then sits on the intersection where the hose crosses and directs the stream (**Figure 13.104**). To reduce fatigue during extended operations, the nozzle operator can either use a hose strap or utility strap looped over the shoulder, or reduce the nozzle flow if conditions allow. Except for limited lateral (side to side) motion, this method does not permit very much maneuvering of the nozzle.

Figure 13.102 A single firefighter can manage a small-diameter hoseline.

Figure 13.103 Two firefighters can safely control an operating hoseline.

Figure 13.104 One firefighter can control a large handline for an extended period.

Two-Firefighter Method

When only two firefighters are assigned to handle a large handline, they may need some means of anchoring the hose because of the nozzle reaction. The nozzle operator holds the nozzle with one hand and, unless the nozzle is equipped with a pistol grip, holds the hose just behind the nozzle with the other hand. The nozzle operator further secures the hoseline under his or her arm and allows the hoseline to rest against the waist or across the hip. The backup firefighter must serve as an anchor at a position about 3 feet (1 m) behind the nozzle operator. The backup firefighter may kneel and place the closest knee on the hoseline. This position prevents the hose from moving back or to either side. If the front part of the hose starts to move back or up, the backup firefighter is in a position to push it forward. In some situations, it is advantageous for the backup firefighter to face away from the nozzle and pull the hose toward the nozzle.

Another two-firefighter method uses hose rope tools or utility straps to assist in anchoring the hose. The nozzle operator loops a hose rope tool or utility strap around the hose a short distance from the nozzle and places the large loop across the back and over the outside shoulder. The nozzle is then held with one hand and the hose just behind the nozzle is held with the other hand. The hoseline is rested against the body. Leaning slightly forward helps control the nozzle reaction. The backup firefighter again serves as an anchor about 3 feet (1 m) back. The backup firefighter also has a hose rope tool or strap around the hose and the shoulder and leans forward to absorb some of the nozzle reaction.

Three-Firefighter Method

Several methods are used for three firefighters to control large handlines. In all cases, the positioning of the nozzle operator is the same as previously described for the two-firefighter method. The only differences are in the position of the second and third firefighters on the line. Some departments prefer the first backup firefighter to stand directly behind the nozzle operator, with the third firefighter kneeling on the hose behind the second firefighter. Another technique is for all firefighters to use hose straps and remain in a standing position; this methods allows for the most mobility (**Figure 13.105**).

Figure 13.105 Adding a third firefighter to the hoseline increases mobility and efficiency.

Regardless of the techniques used, firefighters operating fire hose nozzles and those backing them up must maintain their situational awareness and not limit their focus to what is directly in front of the nozzle. Changes in fire behavior, indicators of imminent structural collapse, and other warning signs can be missed if the nozzle team fails to monitor what is going on around them.

Service Testing Fire Hose

There are two types of tests for fire hose: *acceptance testing* and *service testing*. At the request of the purchasing agency, coupled hose is acceptance tested by the manufacturer before the hose is shipped. This type of testing is relatively rigorous and the hose is subjected to pressures much higher than those anticipated in the field. Fire department personnel should not attempt acceptance testing. Nonetheless, fire departments must service test hose periodically to ensure that it is still capable of performing as required. Guidelines for both types of tests are contained in NFPA® 1962, *Standard for the Inspection, Care, and Use of Fire Hose, Couplings, and Nozzles and the Service Testing of Fire Hose.*

Although a growing number of fire departments are choosing to have their fire hose tested by private contractors, many fire departments continue to test their own hose annually. The department firefighters often assist in this process. Fire department hose should also be tested after being repaired and after being run over by a vehicle.

Before being service tested, the hose should be examined for excessive wear or damage to the jacket, coupling damage, and defective or missing gaskets. If any defects are found, the hose should be tagged for repair. If damage is not repairable, the hose should be taken out of service.

Acceptance Testing (Proof Test) — Preservice tests on fire apparatus or equipment performed at the factory or after delivery to assure the purchaser that the apparatus or equipment meets bid specifications.

Service Test — Series of tests performed on apparatus and equipment in order to ensure operational readiness of the unit. These tests should be performed at least yearly or whenever a piece of apparatus or equipment has undergone extensive repair.

Test Site Preparation

Hose should be tested in a paved or grassy area or level ground with room enough to lay out the hose in straight runs, free of kinks or twists. The site should be protected from vehicular traffic. If testing is done at night, the area should be well lighted. The test area should be smooth and free from rocks and debris **(Figure 13.106)**. A slight grade to facilitate the draining of water is helpful. A water source sufficient for filling the hose is also necessary.

The following equipment is needed to service test hose:

- Hose testing machine, portable pump, or fire department pumper equipped with gauges certified as accurate within one year before testing

- Hose test gate valve

- Means of recording the hose numbers and test results

- Tags or other means to identify sections that fail

- Nozzles with shutoff valves

- Means of marking each length with the year of the test to easily identify which lengths have been tested and which have not without looking in the hose records

Service Test Procedure

Exercise care when working with hose, especially when it is under pressure. Pressurized hose is potentially dangerous because of its tendency to whip back and forth if a rupture occurs or a coupling pulls loose. To prevent this from happening, use a specially designed hose test gate valve **(Figure 13.107)**. This is a valve with a ¼-inch (6 mm) hole in the gate that permits pressurizing the hose but does not allow water to surge through the hose if it fails. Even when using the test gate valve, stand or walk near the pressurized hose only as necessary. See **Skill Sheet 13-II-1** for steps

> **CAUTION**
> All personnel operating in the area of the pressurized hose should wear at least a helmet as a safety precaution.

Figure 13.106 Hose should be tested in an area that is free from obstructions and traffic.

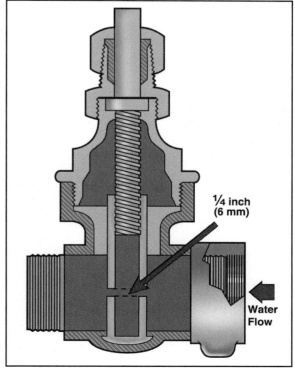

¼ inch (6 mm)

Water Flow

Figure 13.107 Cutaway of a hose test gate valve.

in service testing hose.

When using a fire department pumper, connect the hose to discharges on the side of the apparatus opposite the pump panel. Close all valves slowly to prevent water hammer in the hose and pump. Test lengths of hose should not exceed 300 feet (100 m) in length because longer lengths are more difficult to purge of air.

Laying large-diameter hose flat on the ground before charging it helps to prevent unnecessary wear at the edges. Stand away from the discharge valve connection when charging because the hose has a tendency to twist when it is filled with water and pressurized; this twisting could cause the connection to loosen.

Keep the hose testing area as dry as possible when filling and discharging air from the hose. During testing, this air aids in detecting minor leaks around couplings.

Summary

To fight fires safely and effectively, firefighters must know the capabilities and limitations of the various types of fire hose used in their departments. They must know the differences between supply hose and attack hose and how to use each one. Firefighters must know how to load hose onto apparatus. They must know how to make the various hose load finishes used in their department. Finally, firefighters must know how to maintain and test fire hose.

Fire Fighter I

1. List three methods to prevent each of the following types of fire hose damage: mechanical damage, thermal damage, organic damage, and chemical damage.

2. What is the difference between threaded couplings and nonthreaded couplings?

3. List common hose rolls.

4. List four general hose loading guidelines.

5. What are general safety guidelines when advancing hose into a burning structure?

Fire Fighter II

6. What are the following valves used for: ball valves, gate valves, butterfly valves, and clapper valves?

7. What are wye appliances and siamese appliances?

8. What is a hose clamp used for?

9. What are the two types of tests for fire hose?

10. What equipment is needed to service test hose?

Hand Cleaning

Step 1: Clean the coupling swivels of dirt and other foreign matter by submerging in warm, soapy water and working forward and backward.

Step 2: Clean the male threads if clogged with tar, asphalt, or other foreign material with stiff-bristled scrub brush or wire brush.

Step 3: Inspect hose couplings.

Step 4: Brush the length of the hose of accumulated dust and dirt one area at a time with a broom.

Step 5: Wash areas of hose that contain dirt not removed by brushing with booster hose and clear water.

Step 6: Scrub areas of hose that have been exposed to oil or grease with scrub brush and mild soap or detergent until all oil or grease is removed.

Step 7: Rinse the hose thoroughly with clear water.

Step 8: Inspect the hose for any remaining grease or oil stains or for frayed, snagged, or worn areas.

Step 9: Dry the hose out of the sun.

NOTE: If drying the hose in a hose-drying cabinet, set the thermostat to the drying temperature recommended by the cabinet manufacturer.

Step 10: Roll and store the hose after it has dried per departmental SOP.

Inspect Hose Couplings

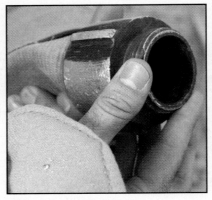

Step 1: Inspect the male couplings.

Step 2: Inspect the female couplings.

Step 3: Inspect the Storz couplings.

Replace a Hose Gasket

Step 1: Remove and discard old or damaged gasket in proper receptacle.

Step 2: Pick up new gasket with middle finger and thumb, and fold loop upward with index finger.

Step 3: Place gasket into swivel, large loop first, smoothing as necessary to seat.

Step 1: Lay out the hose straight and flat on a clean surface.

Step 2: Roll the male coupling over onto the hose to start the roll.

NOTE: Form a coil that is open enough to allow the fingers to be inserted.

Step 3: Continue rolling the coupling over onto the hose, forming an even roll.

NOTE: Keep the edges of the roll aligned on the remaining hose to make a uniform roll as the roll increases in size.

Step 4: Lay the completed roll on the ground.

Step 5: Tamp any protruding coils down into the roll with a foot.

Method One

Step 1: Lay the section of hose flat and in a straight line.

Step 2: Start the roll from a point 5 or 6 feet (1.5 m or 1.8 m) off center toward the male coupling.

Step 3: Roll the hose toward the female end. Leave sufficient space at the center loop to insert a hand for carrying.

Step 4: Extend the short length of hose at the female end over the male threads to protect them.

Method Two

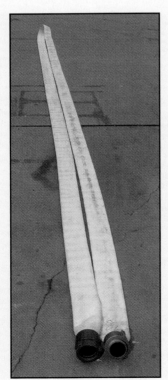

Step 1: Grasp either coupling end, and carry it to the opposite end.

NOTE: The looped section should lie flat, straight, and without twists.

Step 2: Face the coupling ends.

Step 3: Start the roll on the male coupling side about 2½ feet (0.8 m) from the bend (1½ feet [0.5 m] for 1½-inch [38 mm] hose).

Step 4: Roll the hose toward the male coupling.

Step 5: Pull the female side back a short distance to relieve the tension if the hose behind the roll becomes tight during the roll.

Step 6: Lay the roll flat on the ground as the roll approaches the male coupling.

Step 7: Draw the female coupling end around the male coupling to complete the roll.

Foot-Tilt Method

Step 1: Stand facing the two couplings so that one foot is near the male end.

Step 2: Place a foot on the hose directly behind the male coupling.

Step 3: Apply pressure to tilt it upward.

Step 4: Grasp the female end by placing one hand behind the coupling and the other hand on the coupling swivel.

Step 5: Bring the two couplings together, and turn the swivel clockwise with thumb to make the connection.

Two-Firefighter Method

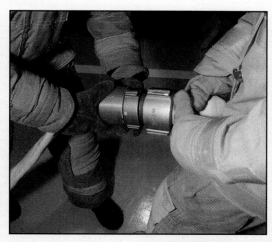

Step 1: *Firefighter #1:* Grasp the male coupling with both hands.

Step 2: *Firefighter #1:* Bend the hose directly behind the coupling.

Step 3: *Firefighter #1:* Hold the coupling and hose tightly against the upper thigh or midsection with the male threads pointed outward.

Step 4: *Firefighter #2:* Grasp the female coupling with both hands.

Step 5: *Firefighter #2:* Bring the two couplings together, and align their positions.

Step 6: *Firefighter #2:* Turn the female coupling counterclockwise until a click is heard. This indicates that the threads are aligned.

Step 7: *Firefighter #2:* Turn the female swivel clockwise to complete the connection.

Knee-Press Method

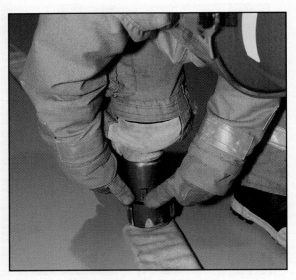

Step 5: Snap the swivel quickly in a counterclockwise direction as body weight is applied to loosen the connection.

Step 1: Grasp the hose behind the female coupling.

Step 2: Stand the male coupling on end.

Step 3: Set feet well apart for balance.

Step 4: Place one knee upon the hose and shank of the female coupling.

Two-Firefighter Method

Step 2: *Both Firefighters:* Keep arms stiff, and use the weight of both bodies to turn each hose coupling counterclockwise, thus loosening the connection.

Step 1: *Both Firefighters:* Take a firm two-handed grip on respective coupling and press the coupling toward the other firefighter, thereby compressing the gasket in the coupling.

Step 1: Lay the first length of hose in the bed on edge against the partition.

Step 2: Fold the hose at the front of the hose bed back on itself.

Step 3: Lay the hose back to the rear next to the first length.

Step 4: Fold the hose at the rear of the hose bed so that the bend is even with the rear edge of the bed.

Step 5: Lay the hose back to the front.

Step 6: Continue laying the hose in folds across the hose bed.

NOTE: Stagger the folds at the rear edge of the bed so that every other bend is approximately 2 inches (50 mm) shorter than the edge of the bed. This stagger may also be done at the front of the bed if desired.

Step 7: Angle the hose upward to start the next tier.

Step 8: Make the first fold of the second tier directly over the last fold of the first tier at the rear of the bed.

Step 9: Continue with the second tier in the same manner as the first, progressively laying the hose in folds across the hose bed.

Step 10: Make the third and succeeding tiers in the same manner as the first two tiers.

Step 11: Move to the opposite hose bed.

Step 12: Load the hose in the same manner as the first side.

Step 13: Connect the last coupling on top with the female coupling from the first side when the load is completed.

Step 14: Lay the connected couplings on top of the hose load.

Step 15: Pull out the slack so that the crossover loop lies tightly against the hose load.

Step 1: Place the coupling in a front corner of the hose bed.

Step 2: Lay the first length of hose on edge against the wall.

Step 3: Make the first fold at the rear even with the edge of the hose bed.

Step 4: Lay the hose to the front and then around the perimeter of the bed so that it comes back to the rear along the opposite side.

Step 5: Make a fold at the rear in the same manner as done before.

Step 6: Lay the hose back around the perimeter of the hose bed inside the first length of hose.

Step 7: Lay succeeding lengths progressively inward toward the center until the entire space is filled.

Step 8: Start the second tier by extending the hose from the last fold directly over to a front corner of the bed, laying it flat on the hose of the first tier.

Step 9: Make the second and succeeding tiers in the same manner as the first.

Step 1: Inspect the hose and hose couplings for damage.

Step 2: Place first coupling at a front corner of the hose bed.

Step 3: Lay the hose flat in the hose bed in a front-to-back fashion.

Step 4: Fold the hose back on itself (make a loop) and lay the hose in the opposite direction. Repeat until hose covers the bottom of the hose bed.

Step 5: Start second layer repeating Steps 3 and 4. Repeat until all hose is loaded.

Step 6: Finish hose load with donut roll or other finish as required by local protocol.

Step 1: Attach the female coupling to the discharge outlet.

Step 2: Lay the first length of hose flat in the bed against the side wall.

Step 3: Angle the hose to lay the next fold adjacent to the first fold and continue building the first tier.

Step 4: Make a fold that extends approximately 8 inches (200 mm) beyond the load at a point that is approximately one-third the total length of the load. This loop will later serve as a pull handle.

Step 5: Continue laying the hose in the same manner, building each tier with folds laid progressively across the bed.

Step 6: Make a fold that extends approximately 14 inches (350 mm) beyond the load at a point that is approximately two-thirds the total length of the load. This loop will also serve as a pull handle.

Step 7: Complete the load.

Step 8: Attach the nozzle and lay it on top of the load.

NOTE: Start the load with the sections of hose connected and the nozzle attached.

Step 1: Connect the female coupling to the discharge outlet.

Step 2: Extend the hose in a straight line to the rear.

Step 3: Pick up the hose at a point two-thirds the distance from the tailboard to the nozzle.

Step 4: Carry this hose to the tailboard.

Step 5: Using several firefighters, pick up the entire length of the three layers.

Step 6: Begin laying the hose into the bed by folding over the three layers into the hose bed.

Step 7: Fold the layers over at the front of the bed.

Step 8: Lay them back to the rear on top of the previously laid hose.

NOTE: If the hose compartment is wider than one hose width, alternate folds on each side of the bed. Make all folds at the rear even with the edge of the hose bed.

Step 9: Continue to lay the hose into the bed in an *S*-shaped configuration until the entire length is loaded.

Step 10: *Optional:* Secure the nozzle to the first set of loops using a rope or strap if desired.

NOTE: Some departments like to pull the loop at the end through the nozzle bale. This can be a problem if the line is charged before removing the loop from the bale. Once the line is charged, it may not be possible to pull the loop through the bale.

Step 1: Connect the first section of hose to the discharge outlet. Do not connect it to the other lengths of hose.

Step 2: Lay the hose flat in the bed to the front.

Step 3: Lay the remaining hose out the front of the bed to be loaded later.

NOTE: If the discharge outlet is at the front of the bed, lay the hose to the rear of the bed and then back to the front before it is set aside. This provides slack hose for pulling the load clear of the bed.

Step 4: Couple the remaining hose sections together.

Step 5: Attach a nozzle to the male end.

Step 6: Place the nozzle on top of the first length at the rear.

Step 7: Angle the hose to the opposite side of the bed and make a fold.

Step 8: Lay the hose back to the rear.

Step 9: Make a fold at the rear of the bed.

Step 10: Angle the hose back to the other side and make a fold at the front.

NOTE: The first fold or two may be longer than the others to facilitate the pulling of the hose from the bed.

Step 11: Connect the male coupling of the first section to the female coupling of the last section.

Step 12: Continue loading the hose to alternating sides of the bed in the same manner until the complete length is loaded.

Step 13: Lay the remainder of the first section in the bed in the same manner.

Step 1: Firefighter #1: Grasp a sufficient amount of hose to reach the hydrant.

Step 2: Firefighter #1: Step down from the tailboard and face the hydrant with all of the equipment necessary to make the hydrant connection.

Step 3: Firefighter #1: Approach the hydrant and loop the hydrant in accordance with SOPs.

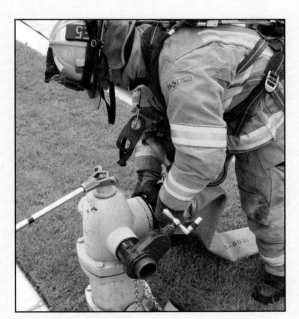

Step 4: Firefighter #1: Signal the driver/operator to proceed driving to the fire.

Step 5: Firefighter #1: Remove the cap from the hydrant.

Step 6: Firefighter #1: Place the hydrant wrench on the valve stem operating nut.

Step 7: Firefighter #1: Remove the hose loop from the hydrant.

Step 8: Firefighter #1: Connect the hose to the outlet nearest the fire.

Step 9: Firefighter #2: After completing the hose lay to the scene, apply the hose clamp on the supply line 20 feet (6 m) behind the apparatus.

Step 10: Firefighter #2: Give the signal to charge the line.

Step 11: Firefighter #2: Uncouple the hose from the bed (allowing enough hose to reach the pump inlet).

Step 12: Firefighter #2: Connect the hose to the pump.

Step 13: Firefighter #2: Release the hose clamp.

Step 14: Firefighter #1: Open the hydrant fully when the appropriate order or signal is given.

Step 15: Firefighter #1: Return to the apparatus, tighten leaking couplings, and push the hose toward the curb along the way.

Step 1: *Firefighter #1:* Pull sufficient hose to reach the intake valve on the attack pumper.

Step 2: *Firefighter #1:* Anchor the hose.

Step 3: *Firefighter #1:* Apply a hose clamp to the hose at the attack pumper.

Step 4: *Firefighter #2:* After the pumper stops at the water source, make an intake hose connection.

Step 5: *Firefighter #2:* Pull the remaining length of the last section of hose from the hose bed.

Step 6: *Firefighter #2:* Disconnect the couplings and return the male coupling to the hose bed.

Step 7: *Firefighter #2:* Connect the supply hose to a discharge valve.

Step 1: Put one arm through the longer loop.

Step 2: Grasp the shorter pull loop with the same hand.

Step 5: Walk toward the fire.

Step 3: Grasp the nozzle with the opposite hand.

Step 4: Pull the load from the bed using the pull loops.

Step 6: Proceed until the hose is fully extended.

Step 7: Conduct visual size-up of scene to identify hazards.

Step 1: Grasp the nozzle and bottom loops, if provided.

Step 2: Pull the load approximately one-third to one-half of the way out of the hose bed.

Step 3: Face away from the apparatus.

Step 4: Place the hose load on the shoulder with the nozzle against the stomach.

Step 5: Walk away from the apparatus, pulling the hose out of the bed by the bottom loop.

Step 6: Advance toward the fire, allowing the load to pay off from the top of the pile.

Step 7: Conduct visual size-up of scene to identify hazards.

Step 1: Place the nozzle and fold of the first tier over the shoulder.

Step 2: Face the direction of travel.

Step 4: Pull the hose *completely* out of the bed.

Step 5: Drop the folded end from the shoulder when the hose bed has been cleared.

Step 3: Walk away from the apparatus.

Step 6: Advance the nozzle.

Step 7: Conduct visual size-up of scene to identify hazards.

Step 1: *Firefighter #1:* Attach the nozzle to the end of the hose if desired.

NOTE: Assist other firefighters with loading hose on their shoulders.

Step 2: *Firefighter #2:* Position at the tailboard facing the direction of travel.

Step 3: *Firefighter #2:* Place the initial fold of hose over the shoulder so that the nozzle can be held at chest height.

Step 4: *Firefighter #2:* Bring the hose from behind back over the shoulder so that the rear fold ends at the back of the knee.

Step 5: *Firefighter #2:* Make a fold in front that ends at knee height and bring the hose back over the shoulder.

Step 6: *Firefighter #2:* Move forward approximately 15 feet (5 m).

Step 7: *Firefighter #3:* Position at the tailboard facing the direction of travel.

Step 8: *Firefighter #3:* Load hose onto the shoulder in the same manner as Firefighter #2, making knee-high folds until an **appropriate amount of hose is loaded.**

Step 9: *Firefighter #1:* Uncouple the hose from the hose bed, and hand the coupling to the last firefighter.

Step 1: Stand alongside a single hoseline at a coupling or nozzle.

Step 2: Face the direction of travel.

Step 3: Place the hose over the shoulder with a coupling in front, resting on the chest.

Step 4: Hold the coupling in place and pull with the shoulder.

Step 5: Position additional firefighters at each coupling to assist in advancing the hose.

Step 1: Confirm order with officer to advance a line into the structure.

Step 2: Unload the hose using accordion unload.

Step 3: Horseshoe shoulder the hose, all placing the hose on the same shoulder. Firefighters spaced about 12 feet (4 m) apart on same side of hose facing the nozzle with about 15 feet (5 m) to 20 feet (6 m) of hose between each firefighter.

Step 4: Start airflow in SCBA before approaching structure entrance or entering smoke environment.

Step 5: Advance the hose to building entrance but do not enter the building. Size up environment to identify hazards. Approach door from side opposite hinges.

Firefighter on nozzle

Step 6: Direct driver/operator to charge hoseline.

Step 7: Set the desired nozzle pattern and bleed air from hoseline.

Step 8: Confirm readiness to enter structure with officer.

Step 9: Enter the structure while staying low and maintaining spacing.

Step 10: Maintain situational awareness of the environment and fire conditions.

13-1-20
Advance a line up and down an interior stairway.

Skill SHEETS

Up Interior Stairs (Uncharged Line)

Step 1: Confirm order with officer to advance a line.

Step 2: Position for shouldering the hoseline by facing the nozzle with about 15 feet (5 m) to 20 feet (6 m) of hose between each firefighter.

Step 3: Place hose bundles on same shoulders per appropriate shoulder carry.

Step 4: Position stationary firefighters along the route and on the stairs at critical points (obstructions and corners) to help feed the hose and to keep the hose on the outside of the staircase.

NOTE: The last several firefighters can assume these stationary positions after their shoulder loads have payed out.

Step 5: Advance the hoseline up a flight of stairs against *outside* wall avoiding sharp bends and kinks and maintaining spacing between firefighters.

Step 6: Flake excess hose up the stairway leading to floor above fire to make fire floor advance easier and quicker.

Step 7: Lay the hose down the stairway along *outside* wall to fire floor.

Step 8: *Last firefighter:* After hose supply is depleted, advance and assist nozzle operator in removing kinks, pushing hose to outside wall of stairway as necessary.

 **SHEETS**

13-I-20 continued
Advance a line up and down an interior stairway.

Down Interior Stairs (Uncharged Hoseline)

Step 6: *Last firefighter:* After hose supply is depleted, advance and assist nozzle operator in removing kinks pushing push hose to outside wall of stairway as necessary.

Step 1: Confirm with officer order to advance a line.

Step 2: Position for shouldering the hoseline by facing the nozzle with about 25 feet (7.5 m) to 30 feet (9 m) of hose between each firefighter.

Step 3: Place hose bundles on same shoulders per appropriate shoulder carry.

Step 4: Position stationary firefighters along the route and at top of the stairs at critical points (obstructions and corners) to help feed the hose and to keep the hose on outside of the staircase.

NOTE: The last several firefighters can assume these stationary positions after their shoulder loads have payed out.

Step 5: Advance the hoseline down a flight of stairs against *outside* wall, avoiding sharp bends and kinks and maintaining spacing between firefighters.

13-I-20 continued
Advance a line up and down an interior stairway.

Skill SHEETS

Up Interior Stairs (Charged Hoseline)

Step 1: Confirm with officer order to advance line.

Step 2: Advance the line using the working line drag.

Step 3: Position stationary firefighters along the route and at top of the stairs at critical points (obstructions and corners) to help feed the hose and to keep the hose on outside of the staircase.

Step 4: Advance up the stairs against *outside* wall, avoiding sharp bends and kinks, maintaining spacing between firefighters, and using working drag to one floor above fire floor.

Step 5: Make a large loop on floor above fire floor to provide excess line for fire floor advancement.

Step 6: Advance the hose down the stairway to the fire floor, using working drag.

Step 7: *Last firefighter:* After hose supply is depleted, advance and assist nozzle operator in removing kinks pushing push hose to outside wall of stairway as necessary.

13-I-20 continued
Advance a line up and down an interior stairway.

Down Interior Stairs (Charged Hoseline)

Step 1: Confirm with officer order to advance line.

Step 2: Use the working drag to advance the line.

Step 3: Position stationary firefighters along the route and at top of the stairs at critical points (obstructions and corners) to help feed the hose and to keep the hose on outside of the staircase.

Step 4: Advance down the stairs against *outside* wall, avoiding sharp bends and kinks, maintaining spacing between firefighters, using working drag to one floor above fire floor.

Step 5: *Second firefighter:* After all hose is advanced, advance and assist nozzle operator to push hose to outside wall of stairway.

Step 1: Confirm order with officer to advance line.

Step 2: Position firefighters all on same side of hose, all facing the nozzle, with about 10 feet (3 m) between each firefighter.

Step 3: Place the line over your shoulders.

Step 4: Climb the ladder.

Step 5: *Nozzle firefighter:* Enter the window, laying down nozzle in window before entering.

Other firefighters

Step 6: Lock in.

Step 7: Feed the hose to nozzle firefighter until nozzle firefighter has advanced to desired location and signals you to stop.

Step 8: Secure the hose to the top rung of the ladder with a hose strap tool or utility strap, tying a clove hitch if using a utility strap.

Step 9: *Firefighter nearest top:* Advance up the ladder to back up the nozzle firefighter.

Step 1: Confirm order with officer to advance line.

Step 2: Position with one firefighter heeling ladder and remaining firefighters on same side of hose facing nozzle spaced about 6 to 8 feet (2 m to 2.4 m) apart.

Step 3: Climb the ladder, firefighter who will operate the nozzle first and others as turn comes.

Step 4: Lock in with leg lock or Class I safety harness, leaving hands free to control and advance the hose.

Step 5: *Firefighters below:* Feed the hose to the nozzle firefighter.

Step 6: *Nozzle firefighter:* Enter the window.

Step 7: *Firefighter on ladder:* Advance up the ladder maintaining appropriate distance from each other.

Step 8: *Firefighter on ladder:* Lock in when backup firefighter is in position opposite the window, using leg lock or Class I harness.

Step 9: *Backup firefighter:* Enter the window.

Step 10: *Firefighters below:* Feed the hose to nozzle and backup firefighters until signaled to stop.

Step 11: *Firefighters on ladder:* Secure the hose to the ladder.

Step 1: Bring additional sections of hose as needed to the nozzle end of the hoseline.

Step 2: Open the nozzle slightly.

Step 3: Apply a hose clamp approximately 5 feet (1.5 m) behind the nozzle OR call for hoseline to be shut down at the pump panel.

Step 6: Reattach the nozzle.

Step 7: Recharge the hoseline by slowly releasing the hose clamp or calling for the line to be charged.

Step 8: Check nozzle pattern and bleed air from hoseline.

Step 4: Remove the nozzle.

Step 5: Add the new section(s) of hose.

Skill SHEETS

13-I-24
Simulate the procedure for controlling a loose hoseline.

WARNING! Do not perform this procedure with an actual loose line. The unpredictability of the line makes the risk of injury to the trainee too great.

Step 4: Pin the end of the hose to the ground to control the line until second firefighter signals that water flow is halted.

Step 1: Lower faceshield.

Step 2: Position by hoseline, 30 to 40 feet (10 m to 12 m) from butt or nozzle and facing nozzle or butt.

Step 3: Move forward cautiously toward uncontrolled end of hose using a crawling motion.

13-I-25
Replace a burst hoseline.

Step 1: Call for hoseline to be shut down or use hose clamp to stop flow.

Step 2: Retrieve two sections of replacement hose.

Step 3: Remove burst section of hose.

Step 4: Couple replacement sections of hose into hoseline using two sections of hoseline to ensure the line will reach objective.

Step 5: Charge hoseline or remove hose clamp.

Step 6: Confirm hoseline is again in operation with driver/operator or officer.

Step 1: Advance the hoseline up the ladder using the proper procedure for either an uncharged line or a charged line.

Step 4: Secure the hose to the top or closest ladder rung with a rope hose tool or utility strap.

Step 2: When at desired elevation, lock in using leg lock or Class I harness, leaving both hands free.

Step 3: Position the nozzle through the rungs extending it at least 1 foot (0.3 m) beyond rungs.

Step 5: Open the nozzle slowly to reduce the effects of nozzle reaction and water hammer.

Step 1: Connect a number of hose sections (check the gaskets before connecting) into test lengths of no more than 300 feet (100 m) each.

Step 2: Use a spanner to tighten the connections between the sections.

Step 3: Connect an open test gate valve to each discharge valve.

Step 4: Use a spanner to tighten each connection.

Step 5: Connect a test length to each test gate valve.

Step 6: Use a spanner to tighten each connection.

Step 7: Tie a rope, hose rope tool, or hose strap to each test length of hose 10 to 15 inches (250 mm to 375 mm) from the test gate valve connections.

Step 8: Secure the other end to the discharge pipe or other nearby anchor.

Step 9: Attach a shutoff nozzle (or any device that permits water and air to drain from the hose) to the open end of each test length.

Step 10: Fill each hoseline with water with a pump pressure of 50 psi (350 kPa) or to hydrant pressure.

Step 11: Open the nozzles as the hoselines are filling.

Step 12: Hold nozzles above the level of the pump discharge to permit all the air in the hose to discharge.

Step 13: Discharge the water away from the test area.

Step 14: Close the nozzles after all air has been purged from each test length.

Step 15: Make a chalk or pencil mark on the hose jackets against each coupling.

Step 16: Check that all hose is free of kinks and twists and that no couplings are leaking.

NOTE: Any length found to be leaking from *BEHIND* the coupling should be taken out of service and repaired before being tested.

Step 17: Retighten any couplings that are leaking at the connections.

NOTE: If the leak cannot be stopped by tightening the couplings, depressurize, disconnect the couplings, replace the gasket, and start over at Step 10.

Step 18: Close each hose test gate valve.

Step 19: Increase the pump pressure to the required test pressure given in NFPA 1962.

Step 20: Closely monitor the connections for leakage as the pressure increases.

Step 21: Maintain the test pressure for the time specified in your departmental SOP.

Step 22: Inspect all couplings to check for leakage (weeping) at the point of attachment.

Step 23: Slowly reduce the pump pressure after 5 minutes.

Step 24: Close each discharge valve.

Step 25: Disengage the pump.

Step 26: Open each nozzle slowly to bleed off pressure in the test lengths.

Step 27: Break all hose connections and drain water from the test area.

Step 28: Observe marks placed on the hose at the couplings.

NOTE: If a coupling has moved during the test, tag the hose section for recoupling. Tag all hose that has leaked or failed in any other way.

Step 29: Record the test results for each section of hose.

Chapter Contents

Fire Streams

This chapter provides information that addresses the following job performance requirements of NFPA® 1001, *Standard for Fire Fighter Professional Qualifications (2008):*

NFPA® 1001 references for Chapter 14:

5.3.10

6.3.1

Courtesy of District Chief Chris E. Mickal, NOFD Photo Unit.

Chapter Objectives

Fire Fighter Objectives I

1. List methods that are used with fire streams to reduce the heat from a fire and provide protection to firefighters and exposures.

2. Discuss the extinguishing properties of water.

3. Describe friction loss.

4. Define water hammer.

5. Distinguish among characteristics of fire stream sizes.

6. Discuss types of streams and nozzles.

7. Discuss handling handline nozzles.

8. Describe types of nozzle control valves.

9. List checks that should be included in nozzle inspections.

10. Operate a solid-stream nozzle. (Skill Sheet 14-I-1)

11. Operate a fog-stream nozzle. (Skill Sheet 14-I-2)

12. Operate a broken-stream nozzle. (Skill Sheet 14-I-3)

Fire Fighter Objectives II

1. Describe the suppression characteristics of fire fighting foam.

2. Define terms associated with types of foam and the foam-making process.

3. Discuss how foam is generated.

4. Discuss foam concentrates.

5. Describe methods by which foam may be proportioned.

6. Discuss foam proportioners.

7. Discuss foam delivery devices.

8. List reasons for failure to generate foam or for generating poor-quality foam.

9. Describe foam application techniques.

10. Discuss hazards associated with foam concentrates.

11. Place a foam line in service — in-line eductor. (Skill Sheet 14-II-1)

Fire Streams

Case History

On March 31, 2006, a fire department responded to a structure fire shortly after noon. On arrival, firefighters found moderate smoke coming from the structure but no flames were visible. After coordinating with the ventilation team, an attack crew entered Side A of the structure with a handline and attempted to attack the fire. The crew withdrew when their fire stream was inadequate to extinguish the fire. Ventilation was increased on Side B and another attack was attempted. This time the crew was able to reach the room of origin but simply did not have enough flow to control the fire. Fire and smoke conditions forced firefighters to withdraw a second time. Finally, a master stream was used through the front door of the structure in a defensive mode and the fire was brought under control after a few minutes.

Firefighters at this incident learned firsthand the importance of having an adequate fire stream for the size and location of the fire. Regardless of the type of fire stream being used to attack the fire, the firefighter must be able to get enough volume of water at the seat of the fire to cool the fuel and extinguish the fire. Firefighters should not hesitate to use all of the tools in their toolbox when selecting the proper handline and fire stream, including the use of 2½- or 3-inch (65 mm or 77 mm) handlines at fully involved residential fires. The old adage about selecting the proper fire stream is still true today: "Big fire, big water."

Source: National Fire Fighter Near-Miss Reporting System

A *fire stream* can be defined as a stream of water or other extinguishing agent after it leaves a fire hose nozzle until it reaches the desired target. Because fire streams are created in different forms for different applications, it is not possible to sharply define what constitutes the "perfect fire stream." From the time a stream of water or extinguishing agent leaves the nozzle, it passes through the air and is influenced by its velocity, by gravity, by wind, and by friction with the air. The condition of the stream when it leaves the nozzle is influenced by operating pressures, nozzle design, nozzle adjustment, and the condition of the nozzle orifice.

Fire Stream — Stream of water or other water-based extinguishing agent after it leaves the fire hose and nozzle until it reaches the desired point.

Fire streams are used to reduce the heat from a fire and provide protection to firefighters and exposures through the following methods:

- Applying water or foam directly onto burning material to reduce its temperature
- Applying water or foam over an open fire to reduce the temperature so firefighters can advance handlines
- Reducing high atmospheric temperature
- Dispersing hot smoke and fire gases from a heated area
- Creating a water curtain to protect firefighters and property from heat
- Creating a barrier between a fuel and a fire by covering the fuel with a foam blanket

According to NFPA® 1001, *Standard for Fire Fighter Professional Qualifications*, those qualified at the Firefighter I level must know the following:

- Principles of fire streams
- Water application methods
- Types, design, operation, effects and flow capabilities of nozzles

In addition, they must be capable of:

- Opening, closing, and adjusting nozzle flow and discharge patterns
- Preventing water hammer when shutting down nozzles

At the Firefighter II level, they must also know the following:

- How foam prevents and controls a hazard
- How fire fighting foam is generated
- Causes of poor foam generation and corrective measures
- Characteristics, uses and limitations of fire fighting foams
- Use of fog nozzles and other types to apply foam
- Hazards associated with foam application

At the Firefighter II level, they must also be capable of:

- Preparing a foam concentrate supply for use
- Assembling foam stream components
- Applying foam using various techniques

This chapter focuses on several aspects of water and foam fire streams. The first portion of the chapter includes the elements required for the production of water fire streams, the different types of streams, and the different types of nozzles used to produce them. The second part of the chapter focuses on the basic principles related to fire fighting foams, such as how and why foam works, types of foam concentrates, the general characteristics of foam, how foam is mixed (proportioned) with water, application equipment, and foam application techniques.

Extinguishing Properties of Water

Water can extinguish fire in several ways. The primary way is by cooling, which absorbs the heat from a fire. Another way is smothering by diluting or excluding oxygen. When heated to its boiling point, water absorbs heat by converting into the gas phase called *water vapor* or *steam* (vaporization). The visible form of steam is called *condensed steam* (**Figure 14.1**).

Heat absorption involves both the heat required to raise the temperature of a substance and the additional heat required to change state (**Figure 14.2**). Specific Heat is the amount of heat energy required to raise the temperature of a specified mass of a substance by one degree. In SI this would be the temperature required to raise the temperature of 1 gram (g) of a substance 1°C; in the customary system this would involve raising the temperature of 1 pound (lb) of a substance 1°F. Energy is also required to change a substance from a liquid to a gas phase. This is called Latent Heat of Vaporization.

Complete vaporization does not happen the instant water reaches its boiling point because that amount of heat must be maintained until the entire volume of water is vaporized. When the water in a fire stream is broken into small particles or droplets, such as from a fog nozzle, it absorbs heat and converts into steam more rapidly than it

Vaporization — Process of evolution that changes a liquid into a gaseous state. The rate of vaporization depends on the substance involved, heat, and pressure.

Latent Heat of Vaporization — Quantity of heat absorbed by a substance at the point at which it changes from a liquid to a vapor.

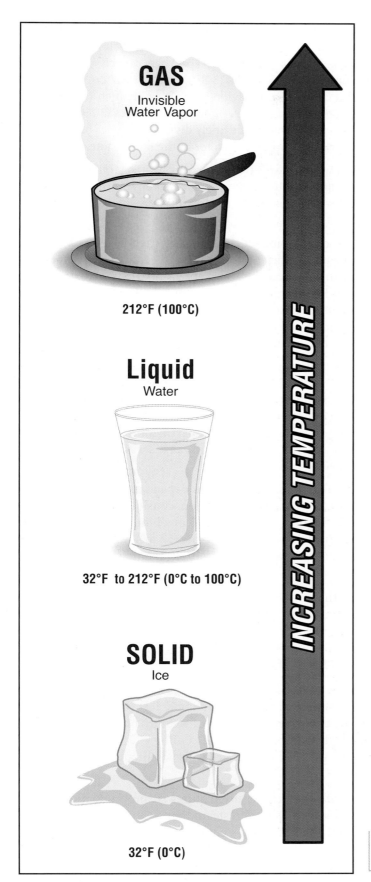

GAS
Invisible
Water Vapor

212°F (100°C)

Liquid
Water

32°F to 212°F (0°C to 100°C)

SOLID
Ice

32°F (0°C)

INCREASING TEMPERATURE

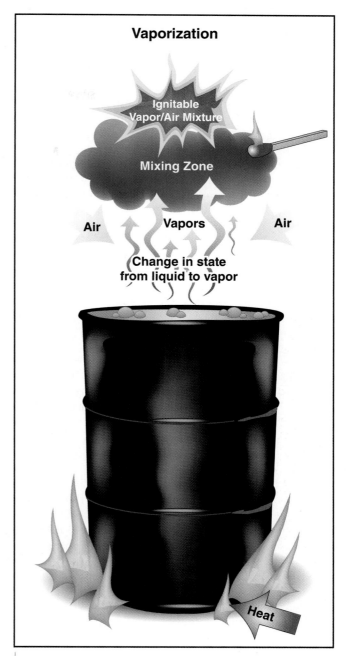

Vaporization

Ignitable
Vapor/Air Mixture

Mixing Zone

Air   Vapors   Air

Change in state
from liquid to vapor

Heat

Figure 14.2 Vaporization occurs when a sufficient quantity of heat is applied to an ignitable liquid. As the vapors mix with air they reach a point at which they can ignite in the presence of an ignition source.

Figure 14.1 Water exists in one of three physical states: solid, liquid, or gas.

would in a compact form, such as from a solid-bore nozzle. This occurs because more of the water's surface is exposed to the heat. For example, 1 cubic inch (1 638.7 mm³) of ice dropped into a glass of water takes some time to absorb its capacity of heat. This is because a surface area of only 6 square inches (3 870 mm² or 38.7 cm²) of the ice is exposed to the water. If that cube of ice is divided into ⅛-cubic inch (204.8 mm³) cubes and dropped into the water, a surface area of 48 square inches (30 967 mm² or 309.7 cm²) of ice is now exposed to the water. The finely divided particles of ice absorb heat more rapidly. This same principle applies to water in the liquid state.

NOTE: The efficiency with which a fire stream absorbs heat is largely dependent upon the surface area of the water introduced into the heated environment. This can be demonstrated by the effects of water discharged through a solid stream nozzle compared to a fog nozzle.

Because the solid stream is more compact than the fog stream, it has a smaller surface area and will absorb heat less efficiently. Of course, there are other important differences that make it inaccurate to say that one type of nozzle is superior to the other under all circumstances. For example, a fire stream is ineffective unless it can reach the fire with enough water to absorb the heat being produced. In some situations, a fog stream might be evaporated before reaching the burning fuel and a solid stream is required just to reach the seat of the fire.

Another characteristic of water that is sometimes useful in extinguishing a fire is its expansion capability. At 212°F (100°C), water expands approximately 1,700 times its original volume when converted to steam **(Figure 14.3)**. The amount of expansion varies with the temperature, so in hotter atmospheres water expands to even greater volumes of steam. This expansion helps cool the compartment by driving heat and smoke away. The steam is also dangerous because it can cause serious burn injuries to firefighters and building occupants.

To a large extent, the volume of steam produced is dependent on the amount of water applied. To illustrate the effects of steam expansion, consider the following two examples. Both cases involve a fire in a 10- × 20-foot (3 m x 6 m) compartment with a 10-foot (3 m) ceiling for a 2,000 cubic feet total volume (57 m³) which has a hot gas layer at approximately 500°F (260°C) and use a fog nozzle at a flow rate of 150 gpm (600 L/min).

Example 1: In this example, the nozzle operator flows water continuously for one minute. Over this time, 28 cubic feet (0.79 m³) of water is discharged and vaporized. This amount of water expands to approximately 48,000 cubic feet (1 359 m³) of steam. This is enough steam to fill a room approximately 10 feet (3 m) high, 50 feet (15 m) wide, and 96 feet (29 m) long **(Figure 14.4)**.

Example 2: In this example, the nozzle operator discharges two short pulses of water fog into the hot gas layer. Over this short time, 0.33 cubic feet (0.009 m³) of water expands to approximately 792 cubic feet (22.43 m³) of steam. This volume of steam is less than half the volume of the compartment. The cooling effect of this water application will also cause the hot gas layer to contract, thus potentially raising the level of the hot gas layer **(Figure 14.5)**.

Effective extinguishment with water generally requires steam production. Water absorbs much more heat when it is converted to steam than when it is simply heated to its boiling point. When fighting a fire inside a compartment, it is critical to recognize that water must be applied in the correct form and amount to achieve effective fire control and maintain a tenable working environment. Skillful nozzle operation and coordination of fire attack and ventilation are critical elements in this process **(Figure 14.6, p. 722)**.

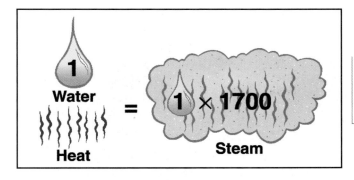

Figure 14.3 Water expands 1,700 times its original volume when converted to steam at 212°F (100°C).

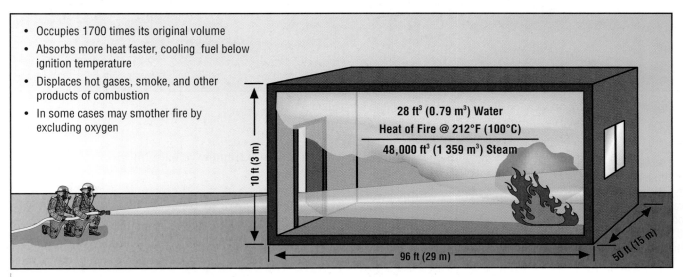

- Occupies 1700 times its original volume
- Absorbs more heat faster, cooling fuel below ignition temperature
- Displaces hot gases, smoke, and other products of combustion
- In some cases may smother fire by excluding oxygen

28 ft³ (0.79 m³) Water
Heat of Fire @ 212°F (100°C)
48,000 ft³ (1 359 m³) Steam

10 ft (3 m)

96 ft (29 m)

50 ft (15 m)

Figure 14.4 Converting water to steam is the basis of the indirect attack method.

Hot Gas Layer

Figure 14.5 Heat absorbed by the water spray causes the hot gas layer to contract.

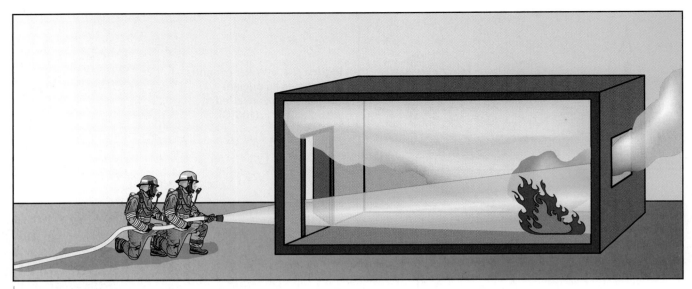

Figure 14.6 Water application can be used to assist the ventilation process.

Several characteristics of water that are extremely valuable for fire extinguishment are as follows:

- Water is readily available and relatively inexpensive.
- Water has a greater heat-absorbing capacity (high specific heat) than most other common extinguishing agents.
- Water changing into steam requires a relatively large amount of heat (high latent heat of vaporization).
- Water can be applied in a variety of ways — solid stream, fog stream (including straight, narrow, and wide pattern), or broken stream.

Pressure Loss or Gain

To produce effective fire streams, it is necessary to know how a number of factors affect pressure loss and gain. Two of these factors are friction loss and elevation pressure. Some pressure losses are due to friction caused by the water flowing through a conduit such as water mains, apparatus piping, hose, and appliances. A pressure loss or gain results when water flows uphill or downhill.

Friction Loss

As it relates to fire streams, friction loss can be defined as *that part of total pressure that is lost while forcing water through pipes, fittings, fire hose, and adapters.* When water flows through hose, couplings, and appliances, the water molecules rub against the insides of these items and friction is produced. The friction slows the water flow and reduces its pressure. The loss of pressure in a hoseline between a pumper and the nozzle (excluding pressure changes due to elevation) is the most common example of friction loss. Friction loss in a hoseline can be measured by inserting in-line gauges at different points in a hose layout **(Figure 14.7)**. The difference in the pressures between gauges when water is flowing through the hose is the friction loss for the length of hose between those gauges for that rate of flow.

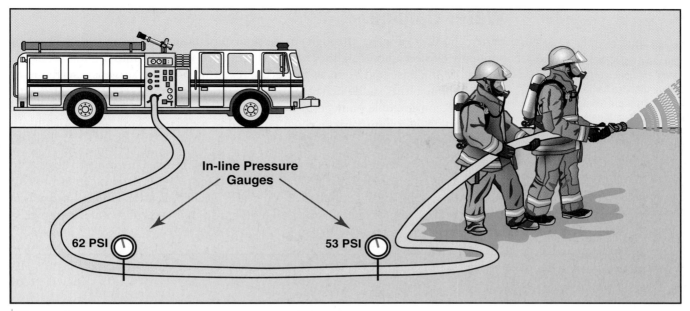

Figure 14.7 To measure friction loss between the pumper and the nozzle, insert in-line pressure gauges at intervals in the hoseline.

There is a practical limit to the velocity or speed at which water can travel through a hoseline. If the velocity is increased beyond this limit, the friction becomes so great that the water in the hoseline is agitated by resistance. Certain characteristics of hose layouts, such as hose size and length of the hose lay, also affect friction loss. In general, the smaller the hose diameter and the longer the hose lay, the higher the friction loss at a given pressure and flow volume.

Friction loss in fire hose is increased by any and all of the following:

- Rough linings in fire hose
- Damaged hose couplings
- Kinks or sharp bends in hose
- More adapters than necessary
- Hoselines longer than necessary
- Hose diameter too small for the volume needed

Elevation Loss/Gain

In this context, elevation refers to the position of a nozzle above or below the pumping apparatus. Elevation pressure refers to a gain or loss in hoseline pressure caused by gravity when there is a difference in elevation. When a nozzle is *above* the fire pump, there is a *pressure loss* (**Figure 14.8**). When the nozzle is *below* the pump, there is a *pressure gain*.

Figure 14.8 Elevation pressure loss occurs when the nozzle is above the level of the pump. *Courtesy of District Chief Chris E. Mickal, NOFD Photo Unit.*

Water Hammer

When the flow of water through fire hose or pipe is suddenly stopped, such as by suddenly closing a nozzle, a shock wave is produced when the moving water reaches the end of the hose and bounces back. The resulting pressure surge is referred to as *water hammer*. This sudden change in the direction creates excessive pressures that can cause considerable damage to water mains, plumbing, fire hose, hydrants, and fire pumps. Water hammer can often be heard as a distinct clank, very much like a hammer striking a pipe.

To prevent water hammer when water is flowing, nozzles, hydrants, valves, and hose clamps should generally be closed slowly **(Figure 14.9)**. One exception to this rule is when applying short pulses of water fog. Brief application (about 1 second) at flow rates up to approximately 150 gpm (600 L/min) will not cause water hammer.

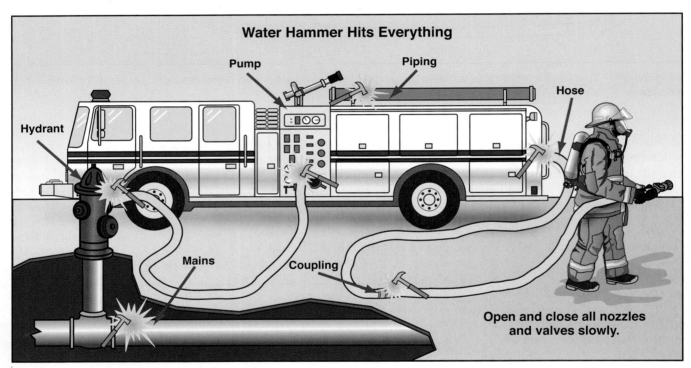

Water Hammer Hits Everything

Pump · Piping · Hose · Hydrant · Mains · Coupling

Open and close all nozzles and valves slowly.

Figure 14.9 Water hammer can be very damaging.

Fire Stream Patterns and Nozzles

A fire stream is identified by its size and type. The size refers to the volume of water flowing per minute; the type indicates the specific pattern or shape of the water after it leaves the nozzle. Fire streams are classified into one of three sizes: *low-volume streams, handline streams,* and *master streams.* The rate of discharge of a fire stream is measured in gallons per minute (gpm) or liters per minute (L/min).

Low-volume stream — Discharges less than 40 gpm (160 L/min) including those fed by booster lines. Typically supplied by ¾-inch (20 mm), 1-inch (25 mm), or 1½-inch (38 mm) hoselines.

Handline stream — Supplied by 1½- to 3-inch (38 mm to 77 mm) hose, with flows from 40 to 350 gpm (160 L/min to 1 400 L/min). Nozzles with flows in excess of 350 gpm (1 400 L/min) are not recommended for handlines. Typically supplied by 1½-inch (38 mm) to 3-inch (77 mm) hoselines.

Master stream — Discharges more than 350 gpm (1 400 L/min) and is fed by 2½- or 3-inch (65 mm or 77 mm) hoselines or large-diameter hoselines connected to a master stream nozzle. Master streams are large-volume fire streams.

The volume of water discharged is determined by the design of the nozzle and the water pressure at the nozzle. To be effective, a fire stream must deliver a volume of water sufficient to absorb heat faster than it is being generated. Fire stream patterns must deliver a sufficient volume of water onto whatever is burning. If the heat-absorbing capability of a fire stream does not exceed the heat output from the fire, extinguishment by cooling is impossible.

The type of fire stream indicates a specific pattern or shape of water stream. In general, the pattern must be compact enough for the majority of the water to reach the burning material. Effective fire streams must meet or exceed the critical rate of flow. They must also have sufficient reach to put water where it is needed to cool hot fire gases or burning fuel. There are three major types of fire stream patterns: *solid, fog,* and *broken* (**Figure 14.10**). The stream pattern may be any one of these in any size classification.

To produce an effective fire stream, regardless of type and size, several things are needed. All fire streams must have a pressuring device (pump), hose, an agent (water), and a nozzle (**Figure 14.11**). The following sections more closely examine the different types of streams and nozzles.

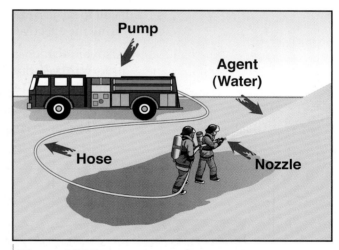

Figure 14.11 The components of a fire stream.

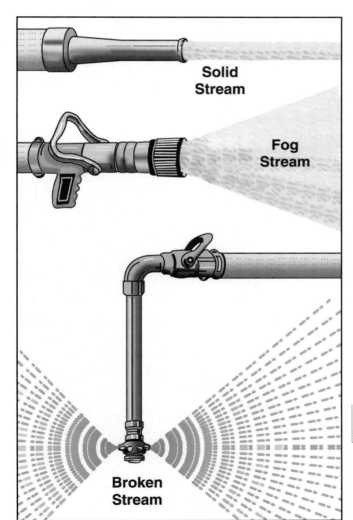

Figure 14.10 The most common fire stream patterns are solid, fog, and broken.

Figure 14.12 A typical solid-bore (solid stream) nozzle.

Solid Stream

A *solid stream* is a fire stream produced from a fixed orifice, solid-bore nozzle (**Figure 14.12**). Solid-stream nozzles are designed to produce a stream as compact as possible with little shower or spray. A solid stream has the ability to reach areas that other streams might not reach. The reach of a solid stream can be affected by gravity, friction of the air, and wind. Remember that solid streams are different from "straight streams" produced by fog nozzles.

Solid-stream nozzles are designed so that the shape of the water in the nozzle is gradually reduced until it reaches a point a short distance from the outlet (**Figure 14.13**). At this point, the nozzle becomes a smooth cylinder whose length is from 1 to 1½ times its inside diameter. The purpose of this short, cylindrical section is to give the water its round shape before discharge. A smooth-finish waterway contributes to both the shape and reach of the stream. Alteration or damage to the nozzle can significantly alter stream shape and performance.

The velocity of the stream is a result of the nozzle pressure. This pressure and the size of the discharge opening determine the flow from a solid-stream nozzle. When solid-stream nozzles are used on handlines, they are usually operated at 50 psi (350 kPa) nozzle pressure. Most solid-stream master stream devices are operated at 80 psi (560 kPa).

The extreme limit at which a solid stream of water can be classified as a good stream cannot be sharply defined and is, to a considerable extent, a matter of judgment. It is difficult to say just exactly where the stream ceases to be good. Observations and tests covering the effective range of fire streams classify effective streams as follows:

- A stream that does not lose its continuity until it reaches the point where it loses its forward velocity (*breakover*) and falls into showers of spray that are easily blown away (**Figure 14.14**)

- A stream that is cohesive enough to maintain its original shape and attain the required height even in a light, gentle wind (breeze).

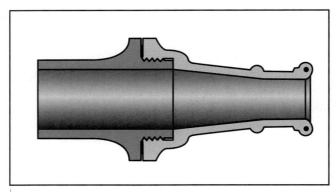

Figure 14.13 Solid-stream nozzles incorporate a slight narrowing near the tip.

Figure 14.14 The breakover point of a stream can be observed when it begins to lose its forward velocity. *Courtesy of Major Danny Atchley, Oklahoma City Fire Department.*

Point of Breakover

Flow Rate

The flow from solid-stream nozzles depends on the velocity of the stream resulting from the pump pressure and the size of the orifice. Some solid-stream nozzles are equipped with a single-size tip for a single flow rate and others have stacked tips to provide varied flows. When using nozzles equipped with a stacked tip, do not simply default to the smallest tip; remove low-flow tips before placing the nozzle in operation if higher flows are required. Changing the flow rate requires that the nozzle be shut off and the tip changed (**Figure 14.15**). **Table 14.1** shows the flows available through various size solid-bore tips at a constant pressure.

Some combination nozzles are equipped with a solid tip built into the nozzle shutoff. Using the solid-stream tip requires removal of the combination tip. Steps for using a solid stream are given in **Skill Sheet 14-I-1**.

Advantages of Solid Streams

- May maintain better interior visibility than other types of streams
- May have greater reach than other types of streams
- Operate at reduced nozzle pressures per gallon (liter) than other types of streams, thus reducing the nozzle reaction
- May be easier to maneuver due to lower operating pressures (**Figure 14.16, p. 728**)
- Have greater penetration power than other types of streams
- Are less likely to disturb normal thermal layering of heat and gases during interior structural attacks than other types of streams
- Are less prone to clogging with debris
- Produce less steam conversion than fog nozzles
- Can be used to apply compressed-air foam

Figure 14.15 To change the flow rate, the tip size must be changed.

Table 14.1 (Customary)
Flow in gpm from Various Sized Solid Stream Nozzles

Nozzle Pressure in psi	Nozzle Diameter in Inches									
	1	1⅛	1¼	1⅜	1½	1⅝	1¾	1⅞	2	2¼
50	209	265	326	396	472	554	643	740	841	1065
55	219	277	342	415	495	581	674	765	881	1118
60	229	290	357	434	517	607	704	810	920	1168
65	239	301	372	451	537	631	732	843	958	1215
70	246	313	386	469	558	655	761	875	994	1260
75	256	324	399	485	578	678	787	905	1030	1305
80	264	335	413	500	596	700	813	935	1063	1347

Figure 14.16 A solid stream operates at lower pressures than other types of streams.

Disadvantages of Solid Streams

- Do not allow for different stream pattern selections
- Provide less heat absorption per gallon (liter) delivered than other types of streams
- Hoselines more easily kinked at corners and obstructions

Fog Stream

A *fog stream* is a fine spray composed of tiny water droplets. The design of most fog nozzles permits adjustment of the tip to produce different stream patterns from the nozzle (**Figure 14.17**). Water droplets, in either a shower or spray, are formed to expose the maximum water surface for heat absorption. The desired performance of fog stream nozzles is judged by the amount of heat that a fog stream absorbs and the rate by which the water is converted into steam or vapor. Fog nozzles permit settings of *straight stream, narrow-angle fog,* and *wide-angle fog* (**Figure 14.18**). A straight stream is a pattern of the adjustable fog nozzle, whereas a solid stream is discharged from a solid-bore nozzle.

A wide-angle fog pattern has less forward velocity and a shorter reach than the other fog settings (**Figure 14.19**). A narrow-angle fog pattern has considerable forward velocity and its reach varies in proportion to the pressure applied. Fog nozzles should be operated at *their designed* nozzle pressure. Of course there is a maximum reach to any fog pattern, which is true with any stream. Once the nozzle pressure has produced a stream with maximum reach, further increases in nozzle pressure have little effect on the stream except to increase the volume.

There are five factors that affect the reach of a fog stream:

- Gravity
- Water velocity
- Fire stream pattern selection
- Water droplet friction with air
- Wind (**Figure 14.20**)

The interaction of these factors on a fog stream results in a fire stream with less reach than that of a straight or solid stream. The more negative factors there are, the less reach the stream is likely to have. This shorter reach is the reason fog streams are sometimes less useful for outside, defensive fire fighting operations. On the other hand, fog streams are well-suited for fighting interior fires. See **Skill Sheet 14-I-2** for procedures in using fog streams.

Figure 14.17 A typical fog-stream nozzle.

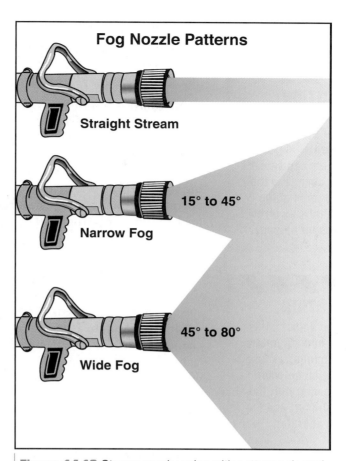

Fog Nozzle Patterns

Straight Stream

Narrow Fog 15° to 45°

Wide Fog 45° to 80°

Figure 14.18 Stream reach varies with pattern selected. An automatic fog nozzle set on straight stream will provide the greatest reach while the wide pattern provides the greatest dispersion of water droplets to absorb heat.

Figure 14.19 A wide-angle fog pattern provides protection to the nozzle operator.

Figure 14.20 Fog streams are more susceptible to wind influence than solid streams. *Courtesy of District Chief Chris E. Mickal, NOFD Photo Unit.*

Wind

Waterflow Adjustment

It is often desirable to control the rate of water flow through a fog nozzle, such as when the water supply is limited. Two types of nozzles provide this capability: *manually adjustable* nozzles and *automatic (constant-pressure)* nozzles.

Manually adjustable nozzles. The rate of discharge from a manually adjustable fog nozzle can be changed by rotating the selector ring — usually located directly behind the nozzle tip — to a specific gpm (L/min) setting **(Figure 14.21)**. Each setting provides a constant rate of flow as long as there is adequate nozzle pressure. The nozzle operator has the choice of making flow-rate adjustments either before opening the nozzle or while water is flowing. Depending upon the size of the nozzle, the operator may adjust flow rates from 10 gpm to 250 gpm (40 L/min to 1 000 L/min) for handlines and from 300 gpm to 2,500 gpm (1 200 L/min to 10 000 L/min) for master streams. Most of these nozzles also have a "flush" setting to rinse debris from the nozzle.

Automatic (constant-pressure) nozzles. Constant-pressure nozzles automatically vary the rate of flow to maintain a reasonably constant nozzle pressure through a specific flow range **(Figure 14.22)**. Obviously, a certain minimum nozzle pressure is needed to maintain a good spray pattern. With this type of nozzle, the operator can change the rate of flow by opening or closing the shutoff valve (see Nozzle Control Valves section). Automatic nozzles allow the nozzle operator to vary the flow rate while maintaining a consistent nozzle pressure.

Automatic nozzles for handlines are designed for low flows such as 10 gpm (40 L/min) to 125 gpm (500 L/min), mid-range flows such as 70 gpm (280 L/min) to 200 gpm (800 L/min), or high flows such as 70 gpm (280 L/min) to 350 gpm (1 400 L/min). Automatic master stream nozzles are typically designed to flow between 250 gpm (1 000 L/min) and 1,000 gpm (4 000 L/min), although larger nozzles are available.

Figure 14.21 A fog nozzle with a manually adjustable flow.

Figure 14.22 A typical constant-flow fog nozzle.

Nozzle Pressure

Combination nozzles are designed to operate at different pressures. The designed operating pressure for most combination nozzles is 100 psi (700 kPa); nozzles with a designed operating pressure of 75, 50, or even 45 psi (525, 350, or 315 kPa) are also available. Although these nozzles have less nozzle reaction compared to nozzles designed to operate at 100 psi (700 kPa), droplet size is much greater, fog pattern density is lower, and the stream has less velocity (**Figure 14.23**).

Advantages of Fog Streams

- The discharge pattern of fog streams can be adjusted to suit the situation (see Chapter 15, Fire Control).

- Fog streams can aid ventilation (see Chapter 11, Ventilation).

- Fog streams reduce heat by exposing the maximum water surface for heat absorption.

- Fog streams can provide protection to firefighters with a wide fog pattern.

Disadvantages of Fog Streams

- Fog streams do not have as much reach or penetrating power as solid streams.

- Fog streams are more affected by wind than are solid streams.

- Fog streams may disturb thermal layering in a room or compartment if applied incorrectly.

- Fog streams may push air into the fire area, thus intensifying the fire.

Broken Stream

A *broken stream* is one that has been broken into coarsely divided drops. While a solid stream may become a broken stream past the point of breakover, a true broken stream takes on that form as it leaves the nozzle. A cellar nozzle is an example of a broken stream nozzle (**Figure 14.24**). In some cases, the effects of a broken stream can be produced by impinging straight or solid streams so they break up over the fire. Steps for using a broken stream are given in **Skill Sheet 14-I-3**.

Advantages of Broken Streams

- Coarse drops absorb more heat per gallon (liter) than a solid stream.

- Have greater reach and penetration than a fog stream.

- Can be effective on fires in confined spaces such as attics, wall spaces, and underground vaults.

Disadvantages of Broken Streams

- May have sufficient continuity to conduct electricity so are not recommended for use on Class C fires.

- Stream may not reach some fires.

Nozzle Pressure — Velocity pressure at which water is discharged from the nozzle.

Figure 14.23 A typical constant-flow, low-pressure combination nozzle. *Courtesy of Elkhart Brass Manufacturing Company.*

Figure 14.24 A cellar nozzle is useful for providing a broken stream.

Handling Handline Nozzles

Because of the differing designs of handline nozzles, each one handles somewhat differently when operated at the recommended pressure. Those with variable patterns may handle differently in different settings. The water pattern produced by the nozzle setting may affect the ease with which a particular nozzle is operated. At or above standard operating pressures, fire stream nozzles are not always easy to control.

Handling Solid-Stream Nozzles

When water flows from a nozzle, the reaction is equally strong in the opposite direction, thus a force pushes back on the person handling the hoseline (nozzle reaction). This reaction is caused by the velocity, flow rate, and discharge pattern of the stream, which acts against the nozzle and the curves in the hose, sometimes making the nozzle difficult to handle. Increasing the nozzle discharge pressure and flow rate increases nozzle reaction.

Handling Fog Stream Nozzles

When water is discharged at angles from the center line of the nozzle, the reaction forces may counterbalance each other and reduce the nozzle reaction. This balancing of forces is the reason why a nozzle set on a wide-angle fog pattern can be handled more easily than a straight-stream pattern.

Nozzle Control Valves

Nozzle control (shutoff) valves enable the operator to start, stop, or reduce the flow of water while maintaining effective control of the nozzle. These valves allow nozzles to open slowly so that the operator can adjust as the nozzle reaction increases; they also allow nozzles to be closed slowly to prevent water hammer. There are three main types of nozzle control valves: *ball, slide,* and *rotary control.*

Ball Valve

Ball valves are the most common nozzle control valves and provide effective control during nozzle operation with a minimum of effort. The ball, perforated by a smooth waterway, is suspended from both sides of the nozzle body and seals against a seat **(Figure 14.25)**. The ball can be rotated up to 90 degrees by moving the valve handle (also called a *bale*) backward to open it and forward to close it. With the valve in the closed position, the waterway is perpendicular to the nozzle body, effectively blocking the flow of water through the nozzle. With the valve in the open position, the waterway is in line with the axis of the nozzle, allowing water to flow through it.

Although the nozzle will operate in any position between fully closed and fully open, operating it with the valve in the fully open position gives maximum flow and performance. When a ball valve is used with a solid-bore nozzle, the turbulence caused by a partially open valve may affect the desired stream or pattern.

Ball Valve — Valve having a ball-shaped internal component with a hole through its center that permits water to flow through when aligned with the waterway.

Slide Valve

The cylindrical slide valve control seats a movable cylinder against a shaped cone to turn off the flow of water **(Figure 14.26)**. Flow increases or decreases as the shutoff handle is moved to change the position of the sliding cylinder relative to the cone. This stainless steel slide valve controls the flow of water through the nozzle without creating turbulence. The pressure control then compensates for the increase or decrease in flow by moving the baffle to develop the proper tip size and pressure.

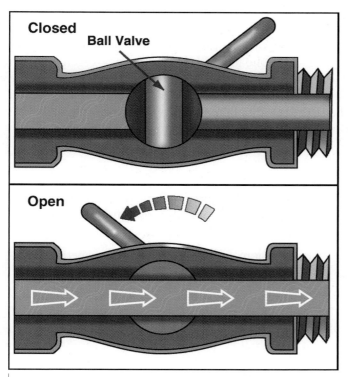

Figure 14.25 How a typical ball valve nozzle operates.

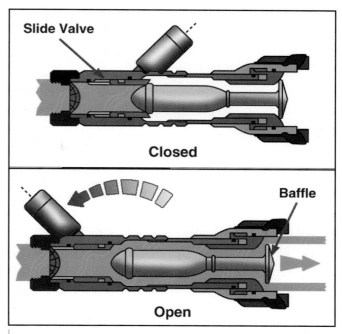

Figure 14.26 How a typical slide valve nozzle operates.

Rotary Control Valve

Rotary control valves are found only on rotary control fog nozzles (**Figure 14.27**). They consist of an exterior barrel guided by a screw that moves it forward or backward, rotating around an interior barrel. A major difference between rotary control valves and other control valves is that rotary control valves also control the discharge pattern of the stream.

Maintenance of Nozzles

Nozzles should be inspected periodically and after each use to make sure they are in proper working condition. This inspection includes the following checks:

- Swivel gasket for damage or wear. Replace worn or missing gaskets.

- External damage to the nozzle.

- Internal damage and debris. When necessary, thoroughly clean nozzles with soap and water using a soft bristle brush (**Figure 14.28**).

- Ease of operation of the nozzle parts. Clean and lubricate any moving parts that are sticking according to manufacturer's recommendations.

- Pistol grip (if applicable) is secured to the nozzle.

Figure 14.27 A typical rotary control fog nozzle. *Courtesy of Elkhart Brass Manufacturing Company.*

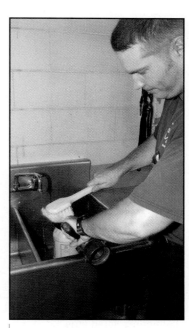

Figure 14.28 To function properly, nozzles must be kept clean according to the manufacturer's instructions.

Fire Fighting Foam

In general, fire fighting foam works by forming a blanket of foam on the surface of burning fuels — both liquid and solid. The foam blanket excludes oxygen and that stops the burning process. The water in the foam is slowly released as the foam breaks down. This action provides a cooling effect on the fuel. With liquid fuels, the foam blanket also prevents or reduces the release of flammable vapors from the surface of the fuel. Fire fighting foam extinguishes and/or prevents fire in several ways:

- *Separating* — Creates a barrier between the fuel and the fire
- *Cooling* — Lowers the temperature of the fuel and adjacent surfaces
- *Smothering* (sometimes referred to as *suppressing*) — Suppresses the release of flammable vapors and therefore reduces the possibility of ignition or reignition **(Figure 14.29)**
- *Penetrating* — Lowers the surface tension of water and allows it to penetrate deep-seated fires.

As discussed in Chapter 6, Portable Fire Extinguishers, the majority of fire fighting foams are either Class A foams intended for use on flammable solids, or Class B foams intended for use on flammable liquids. Both types of foams are reviewed in this section.

Water alone is not always effective as an extinguishing agent; therefore, under certain conditions foam is needed. Even fires that can be fought successfully using plain water may be more effectively fought if a fire fighting foam concentrate is added. Class B foam is especially effective on the two basic categories of flammable liquids: *hydrocarbon fuels* and *polar solvents*.

Hydrocarbon fuels, such as crude oil, fuel oil, gasoline, benzene, naphtha, jet fuel, and kerosene, are petroleum-based combustible or flammable liquids that float on water. Fire fighting foam is effective as an extinguishing agent and vapor suppressant on these Class B liquids because it can float on the surface of these fuels.

Separating — Act of creating a barrier between the fuel and the fire.

Cooling — Reduction of heat by the quenching action or heat absorption of the extinguishing agent.

Smothering — Act of excluding oxygen from a fuel.

Polar Solvents — Flammable liquids that have an attraction for water, much like a positive magnetic pole attracts a negative pole; examples include alcohols, ketones, and lacquers.

Figure 14.29 Foam fire streams can be very effective on flammable liquid fires.

Polar solvents, such as alcohols, acetone, lacquer thinner, ketones, esters, and acids, are flammable liquids that mix readily with water. Special formulations of fire fighting foam are effective on these fuels.

Specialized foams are also used for acid spills, pesticide fires, confined- or enclosed-space fires, and deep-seated Class A fires. In addition to regular fire fighting foams, there are special foams designed solely for use on unignited spills of hazardous liquids. These special foams are necessary because unignited chemicals have a tendency to either change the pH of water or remove the water from fire fighting foams, thereby rendering them ineffective.

Before discussing types of foams and the foam-making process, it is important to understand the following terms:

Foam concentrate — Raw foam liquid before the introduction of water and air

Foam proportioner — Device that introduces foam concentrate into the water stream to make the foam solution

Foam solution — Mixture of foam concentrate and water before the introduction of air

Foam (finished foam) — Completed product after air is introduced into the foam solution

> **Foam** — Extinguishing agent formed by mixing a foam concentrate with water and aerating the solution for expansion; for use on Class A and Class B fires. Foam may be protein, synthetic, aqueous film forming, high expansion, or alcohol type. Also known as Finished Foam.

How Foam Is Generated

Foams in use today are of the mechanical type and must be *proportioned* (mixed with water) and *aerated* (mixed with air) before they can be used. Foam concentrate, water, air, and mechanical agitation are needed to produce quality fire fighting foam (**Figure 14.30**). These elements must be present and blended in the correct ratios. Removing any element results in either no foam production or poor-quality foam.

Aeration is needed to produce an adequate amount of foam bubbles to form an effective foam blanket. Proper aeration produces uniform-sized bubbles to provide a longer-lasting blanket. A good foam blanket is required to maintain an effective cover over either Class A or Class B fuels for the period of time required. Even though the foam bubbles dissipate, a residual foam layer is still there.

Figure 14.30 Foam concentrate, water, air, and mechanical agitation combine to form an effective fire stream.

Foam Expansion

Foam expansion refers to the increase in volume of a foam solution when it is aerated. This is a key characteristic that must be considered when a foam concentrate for a specific application is chosen. The method of aerating a foam solution results in varying degrees of expansion that depend on the following factors:

- Type of foam concentrate used

- Accurate proportioning (mixing) of the foam concentrate in the solution

- Quality of the foam concentrate

- Method of aspiration

Depending on its purpose, foam can be described by three types: *low-expansion, medium-expansion,* and *high-expansion.* NFPA® 11, *Standard for Low-, Medium-, and High-Expansion Foam* states that low-expansion foam has an air/solution ratio up to 20 parts finished foam for every part of foam solution (20:1 ratio). Medium-expansion foam is most commonly used at the rate of from 20:1 to 200:1 through hydraulically operated nozzle-style delivery devices. In the high-expansion foams, the rate is from 200:1 to 1,000:1.

Foam Concentrates

To be effective, foam concentrates must match the fuel to which they are applied. Class A foams are not designed to extinguish Class B fires. Class B foams designed solely for hydrocarbon fires will not extinguish polar solvent fires regardless of the concentration at which they are used. Many foams that are intended for polar solvents may be used on hydrocarbon fires, but this should not be attempted unless the manufacturer of the particular concentrate specifically says this can be done. This incompatibility factor is why it is extremely important to identify the type of fuel involved before applying foam. **Appendix D** highlights each of the common types of foam concentrates and application techniques.

Class A Foam

Foams specifically designed for use on Class A fuels (ordinary combustibles) are increasingly used in both wildland and structural fire fighting. Class A foam is a special formulation of hydrocarbon surfactants. These surfactants reduce the surface tension of water in the foam solution. Reducing surface tension allows for better water penetration, thereby increasing its effectiveness. Aerated Class A foam coats and insulates fuels, preventing pyrolysis and ignition.

Class A foam may be used with fog nozzles, aspirating foam nozzles, medium- and high-expansion devices, and compressed air foam systems (CAFS) **(Figure 14.31)**. Class A foam concentrate has solvent characteristics and is mildly corrosive. It is important to thoroughly flush equipment after use. For more information on Class A foam, refer to the IFSTA **Principles of Foam Fire Fighting** manual.

Class B Foam

Class B foam is used to prevent the ignition of or to extinguish fires involving flammable and combustible liquids **(Figure 14.32)**. It is also used to suppress vapors from unignited spills of these liquids. There are several types of Class B foam concentrates; each type has its advantages and disadvantages. Class B foam concentrates are manufactured from either a synthetic or protein base. Protein-based foams are derived from animal protein. Synthetic foam is made from a mixture of fluorosurfactants. Some foam is made from a combination of synthetic and protein bases.

Class A Foam — Foam specially designed for use on Class A combustibles. These foams are becoming increasingly popular for use in wildland and structural fire fighting. Class A foams, hydrocarbon-based surfactants, are essentially wetting agents that reduce the surface tension of water and allow it to soak into combustible materials easier than plain water.

Class B Foam — Foam fire-suppression agent designed for use on un-ignited or ignited Class B flammable or combustible liquids.

Figure 14.31 Class A foam can be delivered through a variety of nozzle types.

Figure 14.32 Class B foam is designed to control flammable liquids fires. *Courtesy of Williams Fire & Hazard Control, Inc.*

Like Class A foam, Class B foam may be proportioned into the fire stream through a fixed system, an apparatus-mounted system, or by portable foam proportioning equipment. Proportioning equipment is discussed later in this chapter (see Foam Proportioners section). Foams such as aqueous film forming foam (AFFF) and film forming fluoroprotein foam (FFFP) may be applied either with standard fog nozzles or with air-aspirating foam nozzles (see Foam Delivery Devices section). The rate of application — namely, the minimum amount of foam solution that must be applied — for Class B foam varies depending on any one of several variables:

- Type of foam concentrate used

- Whether or not the fuel is on fire

- Type of fuel (hydrocarbon/polar solvent) involved

- Whether the fuel is spilled or contained in a tank (**Figure 14.33, p. 738**)

   NOTE: If the fuel is in a tank, the type of tank will have a bearing on the application rate.

- Whether the foam is applied via either a fixed system or portable equipment

Unignited spills do not require the same application rates as ignited spills because radiant heat, open flame, and thermal drafts do not attack the finished foam as they would under fire conditions. To be most effective, a blanket of foam 4 inches (100 mm) thick should be applied to the fuel surface. In case the spill does ignite, however, be prepared to flow at least the minimum application rate for a specified amount of time based on fire conditions.

Foam concentrate supplies should be on the fireground at the point of proportioning before application is started (**Figure 14.34, p. 738**). Once application has started, it should continue uninterrupted until extinguishment is complete. Stopping and re-starting may allow the fire to consume whatever foam blanket has been established.

Film Forming Fluoroprotein Foam (FFFP) — Foam concentrate that combines the qualities of fluoroprotein foam with those of aqueous film forming foam.

Figure 14.33 Foam may be applied to flammable liquids that are ignited or unignited. The rate of application depends on whether the fuel is burning or not.

Figure 14.34 To keep an uninterrupted supply on the scene, some apparatus carry additional 5-gallon (20 L) containers of foam concentrate. *Courtesy of Harvey Eisner.*

Because polar solvent fuels have differing affinities for water, it is important to know application rates for each type of solvent. These rates also vary with the type and manufacturer of the foam concentrate selected. Foam concentrate manufacturers provide information on the proper application rates as listed by UL. For more complete information on application rates, consult NFPA® 11, the foam manufacturer's recommendations, and the IFSTA **Principles of Foam Fire Fighting** manual.

Specific Application Foams

Numerous types of foams are available for specific applications according to their properties and performance. Some are thick and viscous and form a tough, heat-resistant blanket over burning liquid surfaces; others are thinner and spread more rapidly. Some foams produce a vapor-sealing film of surface-active water solution on a liquid surface. Others, such as medium- and high-expansion foams, are used in large volumes to flood surfaces and fill cavities.

Foam Proportioning

Proportioning — Mixing of water with an appropriate amount of foam concentrate to form a foam solution.

The term *proportioning* is used to describe the mixing of water with foam concentrate to form a foam solution (**Figure 14.35**). Most foam concentrates can be mixed with either fresh or salt water. For maximum effectiveness, foam concentrates must be proportioned at the specific percentage for which they are designed. This percentage rate varies with the intended fuel and is clearly marked on the outside of every foam container (**Figure 14.36**). Failure to follow this procedure, such as trying to use 6% foam at a 3% concentration, will result in poor-quality foam that may not perform as desired.

Most fire fighting foam concentrates are intended to be mixed with 94 to 99.9 percent water. For example, when using 3% foam concentrate, 97 parts water mixed with 3 parts foam concentrate equals 100 parts foam solution (**Figure 14.37**). For 6% foam concentrate, 94 parts water mixed with 6 parts foam concentrate equals 100 percent foam solution.

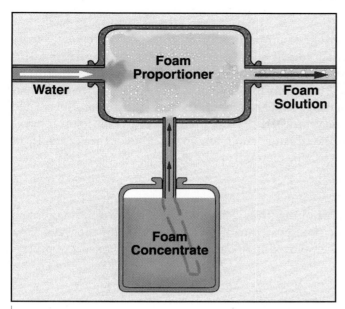

Figure 14.35 Foam concentrate is combined with water to form foam solution.

Figure 14.36 Labels on foam containers indicate the proper mixing percentage with water.

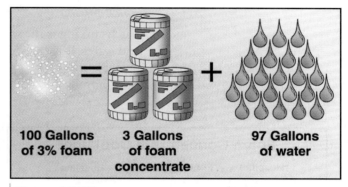

100 Gallons of 3% foam = **3 Gallons of foam concentrate** + **97 Gallons of water**

Figure 14.37 A 3% foam solution consists of 3 gallons of foam concentrate and 97 gallons of water.

Class A foams are an exception to this percentage rule. The proportioning percentage for Class A foams can be adjusted (within limits recommended by the manufacturer) to achieve specific objectives. To produce a dry (thick) foam suitable for exposure protection and fire breaks, the foam concentrate can be adjusted to a higher percentage. To produce wet (thin) foam that rapidly sinks into a fuel's surface, the foam concentrate can be adjusted to a lower percentage.

Class B foams are mixed in proportions from 1% to 6%. Some multipurpose Class B foams designed for use on both hydrocarbon and polar solvent fuels can be used at different concentrations, depending on which of the two fuels is burning. These concentrates are normally used at a 3% rate for hydrocarbons and 6% for polar solvents. Newer multipurpose foams may be used at 3% concentrations regardless of the type of fuel. Medium-expansion Class B foams are typically used at either 1½%, 2%, or 3% concentrations. Follow the manufacturer's recommendations for proportioning.

A variety of equipment is used to proportion foam. Some types are designed for mobile apparatus and others are designed for fixed fire protection systems. The selection of a proportioner depends on the foam solution flow requirements, available water pressure, cost, intended use (truck, fixed, or portable), and the agent to be used. Proportioners and delivery devices (foam nozzle, foam maker, etc.) are engineered to work together. Using a foam proportioner that is not compatible with the delivery device (even if the two are made by the same manufacturer) can result in unsatisfactory foam or no foam at all (see Foam Proportioners and Foam Delivery Devices/Generating Systems section).

There are four basic methods by which foam may be proportioned:

- Induction
- Injection
- Batch-mixing
- Premixing

Induction

The induction (eduction) method of proportioning foam uses the pressure energy in the stream of water to induct (draft) foam concentrate into the fire stream. This is achieved by passing the stream of water through an *eductor,* a device that has a built-in venturi **(Figure 14.38)**. Within the venturi is a separate orifice that is attached via a hose to the foam concentrate container. The pressure differential created by the water going through the venturi creates a suction that draws the foam concentrate into the fire stream. In-line eductors and foam nozzle eductors are examples of foam proportioners that work by this method.

Injection

The injection method of proportioning foam uses an external pump or head pressure to force foam concentrate into the fire stream at the correct ratio for the water flow. These systems are commonly employed in apparatus-mounted or fixed fire protection system applications.

Batch-Mixing

Batch-mixing is the simplest and most accurate method of mixing foam concentrate and water. It is commonly used to mix foam within a fire apparatus water tank or a portable water tank **(Figure 14.39)**. Batch-mixing is common with Class A foams but less common with Class B foams.

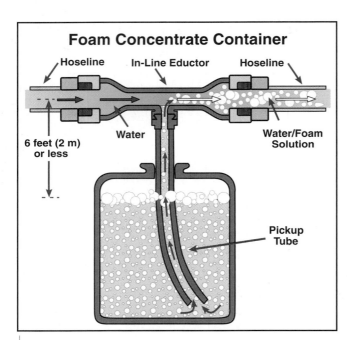

Figure 14.38 The principle of in-line eduction. As water flows past the pickup tube orifice, foam concentrate is drawn into the water stream to create the foam solution.

Figure 14.39 When batch-mixed, foam concentrate is poured into the water tank of the apparatus. Care must be taken when pouring the foam solution because it can be very corrosive to metal parts.

Batch-mixing may not be effective on large incidents because when the tank becomes empty, the foam attack lines must be shut down until the tank is completely filled with water and more foam concentrate is added. Another drawback of batch-mixing is that Class B concentrates and tank water must be circulated for a period of time to ensure thorough mixing before the solution is discharged. The time required for mixing depends on the viscosity and solubility of the foam concentrate. After the incident, all water tanks in which foam was batch-mixed must be thoroughly flushed with plain water because foam left in the tanks can corrode them.

Premixing

Premixing is one of the more commonly used methods of proportioning. With this method, premeasured portions of water and foam concentrate are mixed in a container. Typically, the premix method is used with portable extinguishers, wheeled extinguishers, skid-mounted twin-agent units, and vehicle-mounted tank systems (**Figure 14.40**).

Figure 14.40 Foam solution is premixed in twin agent units. *Courtesy of Ansul.*

In most cases, premixed solutions are discharged from a pressure-rated tank using either a compressed inert gas or air. An alternative method of discharge uses a pump and a nonpressure-rated atmospheric storage tank. The pump discharges the foam solution through piping or hose to the delivery devices. Premix systems are limited to a one-time application. When used, the tank must be completely emptied and then refilled before it can be used again.

> **Premixing** — Mixing premeasured portions of water and foam concentrate in a container. Typically, the premix method is used with portable extinguishers, wheeled extinguishers, skid-mounted twin-agent units, and vehicle-mounted tank systems.

Proportioners, Delivery Devices, and Generating Systems

In addition to a pump to supply water and fire hose to transport it, there are two other pieces of equipment needed to produce a foam fire stream: a *foam proportioner* and a *foam delivery device (nozzle or generating system)*. The proportioner and delivery device/system must be compatible to produce usable foam (**Figure 14.41**). Foam proportioning simply introduces the appropriate amount of foam concentrate into the water to form a foam solution. A foam-generating system/nozzle adds the air into foam solutions to produce finished fire fighting foam. The following sections detail the various types of foam proportioning devices commonly found in portable and apparatus-mounted applications and various foam delivery devices (nozzles/generating systems).

Figure 14.41 A self-educting master stream device in operation. *Courtesy of Conoco/Phillips.*

Foam Proportioners

Foam proportioners may be portable or apparatus-mounted. In general, foam proportioning devices operate by one of two basic principles:

* The pressure of the water stream flowing through a restriction creates a venturi action that inducts (drafts) foam concentrate into the water stream (see In-line Foam Eductors section).

* Pressurized proportioning devices inject foam concentrate into the water stream at a specified ratio and at a higher pressure than that of the water.

Portable Foam Proportioners

Portable foam proportioners are the simplest and most common foam proportioning devices in use today. Two types of portable foam proportioners are *in-line foam eductors* and *foam nozzle eductors*.

In-line foam eductors. The in-line eductor is the most common type of foam proportioner used in the fire service (**Figure 14.42**). This eductor is designed to be directly attached to the pump panel discharge outlet or connected at some point in the hose lay. When using an in-line eductor, it is very important to follow the manufacturer's instructions about inlet pressure and the maximum hose lay between the eductor and the appropriate nozzle.

In-line eductors use the Venturi Principle to draft foam concentrate into the water stream. As water at high pressure passes through a restricted opening, it creates a low-pressure area on the outlet side of the eductor. This low-pressure area creates suction because of the venturi effect. The eductor pickup tube is connected to the eductor at this low-pressure point. A pickup tube submerged in the foam concentrate draws concentrate into the water stream, creating a foam water solution (**Figure 14.43**). The foam concentrate inlet to the eductor should not be more than 6 feet (2 m) above the liquid surface of the foam concentrate. If the inlet is too high, the foam concentration will be very lean or foam may not be inducted at all (**Figure 14.44**).

Foam nozzle eductors. The foam nozzle eductor operates on the same basic principle as the in-line eductor; however, this eductor is built into the nozzle rather than into the hoseline (**Figure 14.45**). As a result, its use requires the foam concentrate

Venturi Principle — Physical law stating that when a fluid, such as water or air, is forced under pressure through a restricted orifice, there is an increase in the velocity of the fluid passing through the orifice and a corresponding decrease in the pressure exerted against the sides of the constriction. Because the surrounding fluid is under greater pressure (atmospheric), it is forced into the area of lower pressure. Also called Venturi Effect.

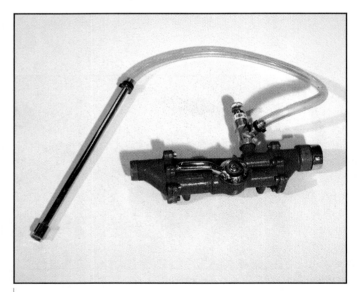

Figure 14.42 A typical in-line foam eductor. This unit can be connected to the pump discharge on the apparatus pump panel.

Figure 14.43 The pickup tube of the in-line foam eductor is inserted into the foam container.

to be available where the nozzle is operated. If the foam nozzle is moved, the foam concentrate must also be moved. The logistical problems of relocation are magnified by the gallons (liters) of concentrate required. Use of a foam nozzle eductor can also compromise firefighter safety. Firefighters cannot always move quickly, and they may have to leave their concentrate supplies behind if they are required to retreat for any reason.

Apparatus-Mounted Proportioners

Foam proportioning systems are commonly mounted on structural, industrial, wildland, and aircraft rescue and fire fighting apparatus (ARFF), as well as on fire boats **(Figure 14.46)**. Three types of the various apparatus-mounted foam proportioning systems are *installed in-line eductors, around-the-pump proportioners,* and *balanced pressure proportioners.* For more information on apparatus-mounted proportioners, refer to the IFSTA **Principles of Foam Fire Fighting** manual.

Compressed-Air Foam Systems (CAFS)

Many newer structural engines are equipped with CAFS. In these systems, a standard centrifugal pump supplies the water, a direct-injection foam-proportioning system mixes foam solution with the water on the discharge side of the pump, and

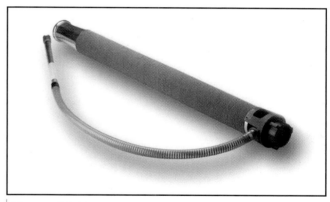

Figure 14.45 A typical foam nozzle eductor. *Courtesy of Akron Brass Company.*

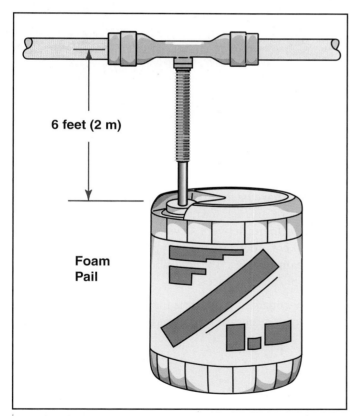

6 feet (2 m)

Foam Pail

Figure 14.44 The eductor must not be more than 6 feet (2 m) above the surface of the foam concentrate.

Figure 14.46 Most ARFF vehicles are equipped with large foam systems.

an onboard air compressor adds air to the mix before it is discharged from the engine. Unlike other foam systems, with CAFS the hoseline contains the finished foam. Among the advantages of using compressed-air foam are the following:

- Stream reach is considerably longer than with other foam systems
- Hoselines are lighter than those full of water or foam solution **(Figure 14.47)**
- Foam produced is very durable
- Foam produced adheres well to vertical surfaces

Some of the disadvantages are as follows:

- CAFS adds expense to the purchase and maintenance of the apparatus
- Hose reaction can be significant
- Additional training is required for firefighters and driver/operators

Figure 14.47 Hoselines filled with compressed air foam are lighter and easier to handle than those filled with water or foam solution.

Foam Delivery Devices (Nozzles/Generating Systems)

Once the foam concentrate and water have been mixed to form a foam solution, the foam solution must then be mixed with air (aerated) and delivered to the surface of the fuel. Nozzles/generating systems (foam delivery devices) designed to discharge foam are sometimes called foam makers. There are many types of devices that can be used, including standard water stream nozzles. The following paragraphs highlight some of the more common foam application devices.

NOTE: Foam nozzle eductors are considered portable foam nozzles but they have been omitted from this section because they were covered earlier in the Portable Foam Proportioners section.

Handline Nozzles

IFSTA defines a handline nozzle as "*any nozzle that one to three firefighters can safely handle and that flows less than 350 gpm (1 400 L/min)*." Most handline foam nozzles flow considerably less than that amount. The following sections detail the handline nozzles commonly used for foam application.

Solid-bore nozzles. The use of solid-bore nozzles is limited to certain types of Class A applications. In these applications, the solid-bore nozzle provides an effective fire stream with maximum reach capabilities **(Figure 14.48)**. Solid-bore nozzles are most often used with compressed air foam systems (CAFS).

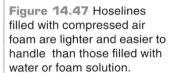

Handline Nozzle — Any nozzle that can be safely handled by one to three firefighters and flows less than 350 gpm (1 400 L/min).

Figure 14.48 Solid-stream nozzles can be used to apply compressed air foam.

Figure 14.49 An air-aspiration foam nozzle in operation.

Fog nozzles. Either fixed-flow or automatic fog nozzles can be used with foam solutions to produce a low-expansion, short-lasting foam. These nozzles break the foam solution into tiny droplets and use the agitation of water droplets moving through air to achieve foaming action. Their best application is when used with regular AFFF and Class A foams. These nozzles cannot be used with protein and fluoroprotein foams. Fog nozzles may be used with alcohol-resistant AFFF foams on hydrocarbon fires but should not be used on polar solvent fires. This is because insufficient aspiration occurs to handle the polar solvent fires. Some nozzle manufacturers have foam aeration attachments that can be added to the end of the nozzle to increase aspiration of the foam solution. See the IFSTA **Principles of Foam Fire Fighting** manual for more information.

Air-aspirating foam nozzles. The most effective appliance for the generation of low-expansion foam is the air-aspirating foam nozzle. The air-aspirating foam nozzle inducts air into the foam solution by a venturi action **(Figure 14.49)**. This nozzle is especially designed to provide the aeration required to make the highest quality foam possible. These nozzles must be used with protein and fluoroprotein concentrates. They may also be used with Class A foams in wildland applications. These nozzles provide maximum expansion of the agent. The reach of the stream is less than that of a standard fog nozzle.

Medium- and High-Expansion Foam Generating Devices

Medium- and high-expansion foam generators produce a foam that is semistable with high air content. For medium-expansion foam, the air content ranges from 20 parts air to 1 part foam solution (20:1) to 200 parts air to 1 part foam solution (200:1). For high-expansion foam, the ratio is 200:1 to 1,000:1. There are two basic types of medium- and high-expansion foam generators: the *water-aspirating type nozzle* and the *mechanical blower.*

Water-aspirating type nozzle. The water-aspirating type nozzle is very similar to the other foam-producing nozzles except it is much larger and longer **(Figure 14.50, p. 746)**. The back of the nozzle is open to allow air flow. The foam solution is pumped through the nozzle in a fine spray that mixes with air to form a medium-expansion foam. The end of the nozzle has a screen or series of screens that further breaks up the foam and mixes it with air. These nozzles typically produce a foam with lower air volume than do mechanical blower generators.

Fog Nozzle — Nozzle that can provide either a fixed or variable spray pattern. The nozzle breaks the foam solution into small droplets that mix with air to form finished foam.

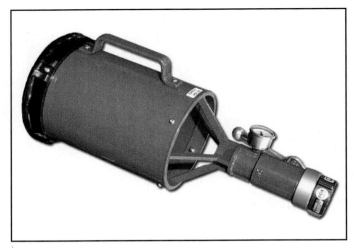

Figure 14.50 Water-aspirating nozzles are used to deliver medium-expansion foam.

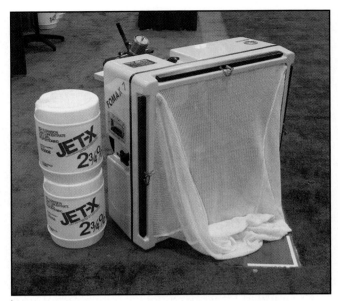

Figure 14.51 A mechanical blower is used to produce high-expansion foam.

Figure 14.52 Foam nozzles, eductors, and hose must be cleaned after each use. Inadequate flushing and cleaning of foam equipment can result in its failure.

In-Line Eductor — Eductor that is placed along the length of a hoseline.

Mechanical blower generator. A mechanical blower generator is similar in appearance to a smoke ejector **(Figure 14.51)**. It operates on the same principle as the water-aspirating nozzle except the air is forced through the foam spray by a powered fan instead of being pulled through by water movement. This device produces a foam with a high air content and is typically associated with total-flooding applications. Its use is limited to high-expansion foam.

Assembling a Foam Fire Stream System

To provide a foam fire stream, a firefighter or apparatus driver/operator must be able to correctly assemble the components of the system in addition to locating problem areas and making adjustments. There are a number of reasons for failure to generate foam or for generating poor-quality foam. The most common reasons for failure are as follows:

- Eductor and nozzle flow ratings do not match so foam concentrate cannot induct into the fire stream.
- Air leaks at fittings cause a loss of suction.
- Improper cleaning of proportioning equipment causes clogged foam passages **(Figure 14.52)**.
- Nozzle is not fully open, restricting water flow.
- Hose lay on the discharge side of the eductor is too long, creating excess back pressure and causing reduced foam pickup at the eductor.
- Hose is kinked and stops flow.
- Nozzle is too far above the eductor, which causes excessive elevation pressure.
- Mixing different types of foam concentrate in the same tank results in a mixture too viscous to pass through the eductor.

Skill Sheet 14-II-1 describes the steps for placing a foam line in service using an in-line eductor.

Foam Application Techniques

It is important to use the correct techniques when applying foam from handline or master stream nozzles. If incorrect techniques are used, such as plunging the foam into a liquid fuel, the effectiveness of the foam is reduced. The techniques for applying foam to a liquid fuel fire or spill include the *roll-on method, bank-down method,* and *rain-down method.*

Roll-On Method

The roll-on method directs the foam stream on the ground near the front edge of a burning liquid spill (**Figure 14.53**). The foam then rolls across the surface of the fuel. Firefighters continue to apply foam until it spreads across the entire surface of the fuel and the fire is extinguished. It may be necessary to move the stream to different positions along the edge of a liquid spill to cover the entire pool. This method is used only on a pool of liquid fuel (either ignited or unignited) on the open ground.

Bank-Down Method

The bank-down method may be employed when an elevated object is near or within the area of a burning pool of liquid or an unignited liquid spill. The object may be a wall, tank shell, or similar vertical structure. The foam stream is directed onto the object, allowing the foam to run down and onto the surface of the fuel (**Figure 14.54, p. 748**). As with the roll-on method, it may be necessary to direct the stream onto various points around the fuel area to achieve total coverage and extinguishment of the fuel. This method is used primarily in dike fires and fires involving spills around damaged or overturned transport vehicles.

Rain-Down Method

The rain-down method is used when the other two methods are not feasible because of the size of the spill area (either ignited or unignited) or the lack of an object from which to bank the foam. The rain-down method is also the primary manual appli-

Figure 14.53 Roll-on foam application for use with ignited or unignited flammable liquids spills.

cation technique used on aboveground storage tank fires. This method directs the stream into the air above the fire or spill and allows the foam to float gently down onto the surface of the fuel **(Figure 14.55)**. On small fires, the nozzle operator sweeps the stream back and forth over the entire surface of the fuel until the fuel is completely covered and the fire is extinguished. On large fires, it may be more effective for the firefighter to direct the stream at one location to allow the foam to collect there and then float out from that point.

Figure 14.54 In the bank-down method, foam is applied to a vertical surface and allowed to run down and onto the surface of the fuel.

Figure 14.55 The rain-down method of application. This method is generally used with large flammable liquids fires or spills.

Foam Hazards

Foam concentrates, either at full strengths or in diluted forms, pose minimal health risks to firefighters. In both forms, foam concentrates may be mildly irritating to the skin and eyes. Affected areas should be flushed with water. Some concentrates and their vapors may be harmful if ingested or inhaled. Consult the manufacturer's material safety data sheets (MSDS) for information on any specific foam concentrate.

Most Class A and Class B foam concentrates are mildly corrosive. Although foam concentrate is used in small percentages and in diluted solutions, follow proper flushing procedures to prevent damage to equipment.

The primary environmental impact is the effect of the finished foam after it has been applied to a fire or liquid fuel spill. The biodegradability of a foam is determined by the rate at which environmental bacteria cause it to decompose. This decomposition process results in the consumption of oxygen by the bacterial action. In a waterway, the subsequent reduction in oxygen can cause damage by killing fish and other aquatic creatures. The less oxygen required to degrade a particular foam, the better or the more environmentally friendly the foam is when it enters a body of water.

The environmental impact of foam concentrates varies. Each foam concentrate manufacturer can provide information on its specific products. In the United States, Class A foams should be approved by the USDA Forest Service for environmental suitability. The chemical properties of Class B foams and their environmental impact vary depending on the type of concentrate and the manufacturer. Generally, protein-based foams are safer for the environment. Consult the various manufacturers' data sheets for environmental impact information.

Material Safety Data Sheet (MSDS) — Form provided by the manufacturer and blender of chemicals that contains information about chemical composition, physical and chemical properties, health and safety hazards, emergency response procedures, and waste disposal procedures of the specified material.

Summary

To fight fires safely and effectively, firefighters must know the capabilities and limitations of all the various nozzles and extinguishing agents available in their departments. They must understand the effects that wind, gravity, velocity, and friction with the air have on a fire stream once it leaves the nozzle. Firefighters must know what operating pressure their nozzles require and how the nozzles can be adjusted during operation. Finally, firefighters must know the differences between the classes of foam, how to generate foam, and how to apply foam most effectively.

Fire Fighter I

1. What are the ways that water can extinguish fire?

2. Define friction loss, elevation loss/gain, and water hammer.

3. What factors can increase friction loss in fire hose?

4. What are the three size classifications of fire streams?

5. What is the difference between a solid stream and a fog stream?

Fire Fighter II

6. What are the ways that fire fighting foam extinguishes and/or prevents fire?

7. Describe types of foam concentrates.

8. What are the methods by which foam may be proportioned?

9. What are the types of portable foam proportioners and how do they work?

10. Describe the techniques used to apply foam.

Step 1: Position yourselves on same side of hose with one firefighter on the nozzle and one as backup.

Step 2: Prior to opening nozzle, wait for backup firefighter to communicate that they are ready.

Step 3: Aim the nozzle at the target indicated by officer.

Step 4: Open the nozzle fully.

Step 5: Hold the stream on target for 15 seconds.

Step 6: Shut off the nozzle so that water hammer is avoided.

Straight Stream

Step 1: Position yourselves on same side of hose with one firefighter on nozzle and one as backup.

Step 2: Prior to opening nozzle, wait for backup firefighter to communicate that they are ready.

Step 3: Twist the stream adjustment ring to adjust the stream pattern to a straight stream.

Step 4: Aim the nozzle at the target indicated by officer.

Step 5: Open the nozzle fully.

Step 6: Hold the stream on target for 15 seconds.

Step 7: Shut off the nozzle so that water hammer is avoided.

Narrow Fog Stream

Step 1: Position yourselves on same side of hose with one firefighter on nozzle and one as backup.

Step 2: Prior to opening nozzle, wait for backup firefighter to communicate that they are ready.

Step 3: Adjust the stream pattern by twisting stream adjustment ring to a wide fog stream (15° to 45°).

Step 4: Aim the nozzle at the target indicated by officer.

Step 5: Open the nozzle fully.

Step 6: Hold the stream on target for 15 seconds.

Step 7: Shut off the nozzle so that water hammer is avoided.

Wide Fog Stream

Step 1: Position yourselves on same side of hose with one firefighter on nozzle and one as backup.

Step 2: Prior to opening nozzle, wait for backup firefighter to communicate that they are ready.

Step 3: Adjust the stream pattern by twisting stream adjustment ring to a narrow fog stream (45° to 80°).

Step 4: Aim the nozzle at the target indicated by officer.

Step 5: Open the nozzle fully.

Step 6: Hold the stream on target for 15 seconds.

Step 7: Shut off the nozzle so that water hammer is avoided.

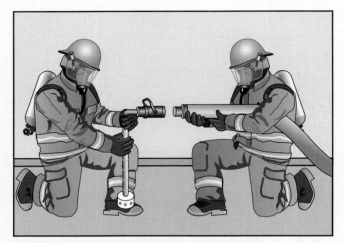

Step 1: Attach nozzle according to manufacturer's guidelines and/or departmental SOPs.

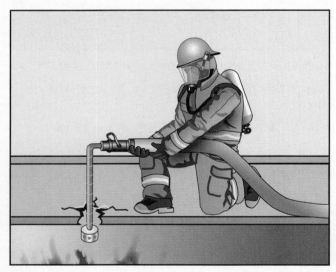

Step 2: Place nozzle in proper location (i.e. through a hole in floor, driven through wall, or lowered into chimney).

Step 3: Position firefighters so that control can be maintained on the hoseline.

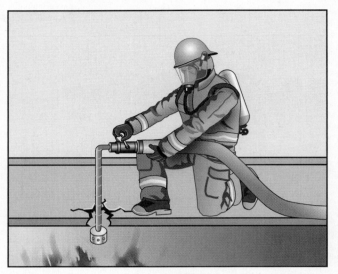

Step 4: Call for water and/or open valve fully to flow water.

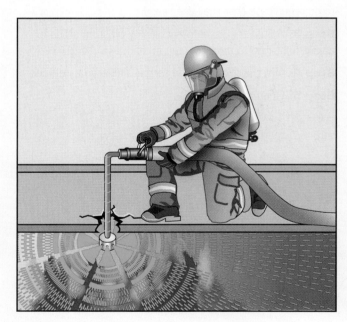

Step 5: Maintain flow until fire is extinguished.

Step 6: Shut off nozzle so that water hammer is avoided.

Step 7: Remove nozzle is removed from area of extinguishment.

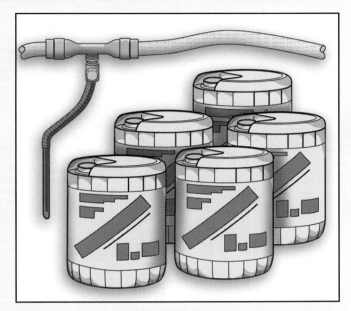

Step 1: Confirm order with officer to place line in service.

Step 2: Select the proper foam concentrate for the burning fuel involved.

Step 3: Place the foam concentrate at the eductor.

Step 4: Open enough buckets of foam concentrate to handle the task.

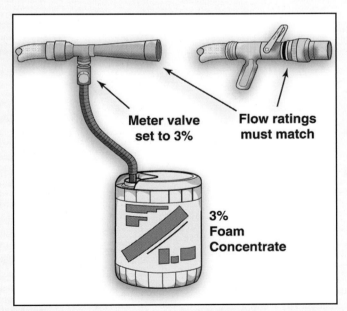

Meter valve set to 3%

Flow ratings must match

3% Foam Concentrate

Step 5: Check the eductor and nozzle for hydraulic compatibility (rated for the same flow).

Step 6: Adjust the eductor metering valve to the same percentage rating as that listed on the foam concentrate container.

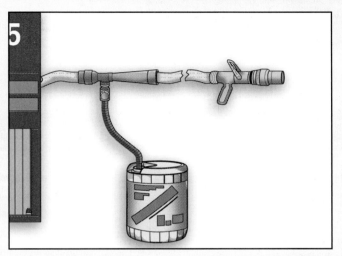

Step 7: Attach the eductor to a hose capable of efficiently flowing the rated capacity of the eductor and the nozzle.

NOTE: If the eductor is attached directly to a pump discharge outlet, make sure that the ball valve gates are completely open. In addition, avoid connections to discharge elbows. Any condition that causes water turbulence will adversely affect the operation of the eductor.

Step 8: Attach the attack hoseline and desired nozzle to the discharge end of the eductor. Avoid kinks in the hose.

Step 9: Place the eductor suction hose into the foam concentrate.

Step 10: Open nozzle fully.

Step 11: Increase the water-supply pressure to that required for the eductor. Be sure to consult the manufacturer's recommendations for the specific eductor.

Step 12: Report to officer completion of assigned task.

Chapter Contents

Fire Control

This chapter provides information that addresses the following job performance requirements of NFPA® 1001, *Standard for Fire Fighter Professional Qualifications (2008):*

NFPA® 1001 references for Chapter 15:

5.3.7
5.3.8
5.3.10
5.3.16
5.3.18
5.3.19
6.1.1
6.1.2
6.2.2
6.3.1
6.3.2
6.3.3

Courtesy of FEMA News Photo.

Chapter Objectives

Fire Fighter I Objectives

1. Describe initial factors to consider when suppressing structure fires.

2. Summarize considerations prior to entering a burning building.

3. Explain the gas cooling technique.

4. Describe direct attack, indirect attack, and combination attack.

5. Discuss deploying master stream devices.

6. Describe aerial devices used to deliver elevated master streams.

7. Describe actions and hazards associated with suppressing Class C fires.

8. List electrical hazards and guidelines for electrical emergencies.

9. Discuss responsibilities of companies in structural fires.

10. Explain actions taken in attacking fires in upper levels of structures.

11. Explain actions taken in attacking fires below-ground in structures.

12. Discuss structure fires in properties protected by fixed systems.

13. Explain actions taken when attacking a vehicle fire.

14. Explain actions taken when attacking trash container fires.

15. Explain actions taken when performing fire fighting and rescue operations in confined spaces.

16. Summarize influences on wildland fire behavior: fuel, weather, and topography.

17. Describe parts of a wildland fire.

18. List wildland protective clothing and equipment.

19. Describe methods used to attack wildland fires.

20. List ten standard fire fighting orders when fighting wildland fires.

21. Attack a structure fire — Exterior attack. (Skill Sheet 15-I-1)

22. Deploy and operate a master stream device. (Skill Sheet 15-I-2)

23. Turn off building utilities. (Skill Sheet 15-I-3)

24. Attack a structure fire (above, below, and grade Level) — Interior attack. (Skill Sheet 15-I-4)

25. Attack a passenger vehicle fire. (Skill Sheet 15-I-5)

26. Extinguish a fire in a trash container. (Skill Sheet 15-I-6)

27. Attack a fire in stacked/piled materials. (Skill Sheet 15-I-7)

28. Attack a ground cover fire. (Skill Sheet 15-I-8)

Fire Fighter II Objectives

1. Summarize considerations for hoseline selection.

2. Discuss stream selection.

3. Discuss suppressing Class B fires.

4. Explain why bulk transport vehicle fires are difficult incidents.

5. Discuss control of gas utilities.

6. Discuss command at structural fires.

7. Extinguish an ignitable liquid fire. (Skill Sheet 15-II-1)

8. Control a pressurized flammable gas container fire. (Skill Sheet 15-II-2)

9. Establish Incident Command and coordinate interior attack of a structure fire. (Skill Sheet 15-II-3)

Fire Control

Case History

In 1996 a metropolitan fire department responded to a report of an electrical problem in an auto parts store located in a strip mall. The store was 6,000 ft² (557 m²) with masonry walls, wood truss roof, and roof-mounted HVAC units — a type of building scenario common in most communities.

The initial report was of a sparking fuse box. The first crew at the scene did not find smoke in the interior. Firefighters soon discovered black smoke coming from the roof at the rear of the store. It was apparent there was heavy fire in the attic area of the store. The officer deployed a 1¾-inch (45 mm) handline into the rear door of the structure while the firefighter donned SCBA. Moving into the store, the crew attempted to make an attack but was unable to get the 1¾-inch (45 mm) fire stream into the attic and on the seat of the fire. Before another crew could provide help getting into the ceiling, the roof collapsed on the crew, killing both firefighters.

Effective fire control requires firefighters to select a hoseline and nozzle large enough to provide the amount of water needed to reach and penetrate the fire and then cool the fuel. Close coordination with other operations is necessary in order for that to occur, but results can be disastrous when the hoseline selected is too small, the improper fire stream is selected, or fire control is not executed in conjunction with other fireground operations such as ventilation and overhaul.

This chapter addresses proper hoseline selection, selection of the appropriate fire stream, and the various types of attacks that can be used by the fire control team.

Source: National Fire Fighter Near-Miss Reporting System

The success or failure of a fire fighting operation often depends upon the skill and knowledge of the personnel in the initial attack team. If a fire is discovered and reported promptly and the response is not delayed, a well-trained team of firefighters with an attack plan and properly applied water or other extinguishing agent can contain most fires in their early stages. However, failure to make a well-coordinated attack on a fire while it is still relatively small can result in it rapidly developing to the point that it overwhelms the on-scene resources and burns out of control. Loss of control of the fire can result in increased damage and increased risk to firefighters and civilians alike **(Figure 15.1, p. 760)**.

It is important that all firefighters be thoroughly trained in the tactics used by their departments and in the use of all the tools and equipment with which they are provided. The safe and efficient use of tools and techniques is further enhanced when the companies that frequently work together train together **(Figure 15.2, p. 760)**.

The need to follow established safety procedures and to wear personal protective equipment (PPE) during fire control operations cannot be overemphasized. In addition to protecting firefighters from heat and other hazards, helmets, gloves, turnout gear, boots, protective hoods, and breathing apparatus permit firefighters to work in close to the fire (**Figure 15.3**). Firefighters must work in teams of two or more whenever they are operating inside burning buildings. In addition to watching out for each other, all team members must remain alert for a number of potentially hazardous conditions, such as the following:

- Flashover or backdraft conditions
- Imminent building collapse

Figure 15.1 Losing control of a fire can be costly and dangerous. *Courtesy of District Chief Chris E. Mickal, NOFD Photo Unit.*

Figure 15.2 Joint training exercises for units that respond together increase emergency scene effectiveness.

Figure 15.3 The proper use of personal protective equipment increases firefighter safety.

- Fire behind, below, or above the attack team

- Kinks or obstructions to the hoseline

- Holes, weak stairs, or other fall hazards

- Suspended loads on fire-weakened supports

- Hazardous or highly flammable commodities likely to spill

- Electrical shock hazards

According to NFPA® 1001, those trained to the Fire Fighter I level must know the following about fire control:

- Principles of fire streams as applied to interior and exterior structure fires, vehicle fires, and wildland fires

- Hazardous building conditions created by fire **(Figure 15.4)**

- Principles of exposure protection

- Application of each size and type of attack line

- Attack and control techniques for fires above, below, and at grade level

- How to expose hidden fires

In addition, those trained to the Fire Fighter I level must be capable of performing the following operations:

- Applying water using direct, indirect, and combination attacks

- Advancing and operating charged attack lines

- Attacking fires above, below, and at grade level

- Attacking a passenger vehicle fire

- Attacking a fire in stacked and piled materials

- Extinguishing a fire in a trash container

- Attacking a ground cover fire

- Turning off building utilities

- Locating and suppressing interior wall and subfloor fires

Figure 15.4 Through preincident surveys, firefighters must be familiar with the types of heavy roof loads such as heating, ventilation, and air conditioning units that can cause roof collapse.

NFPA® 1001 requires that those trained to the Fire Fighter II level know the following about fire control:

- Most appropriate hose and nozzles for attacking various types of fires

- Hazardous building conditions created by fire and fire suppression activities

- Indicators of structural instability and building collapse

- Effects of fire and fire suppression activities on various common building materials

- Suppression techniques for various types of structural fires

In addition, those trained to the Fire Fighter II level must be capable of performing the following:

- Choosing attack techniques for fires above, below, and at grade level

- Evaluating and forecasting the growth and development of a fire

- Extinguishing an ignitable liquid fire with foam

- Controlling a flammable gas cylinder fire

- Coordinating an interior attack of a structure fire

This chapter discusses some of the common techniques for fighting different types of fires. Hazards peculiar to certain situations are also discussed. Finally, basic tactics for commonly encountered fire scenarios are reviewed.

Suppressing Structure Fires

To be successful, a fire attack on a burning structure must be coordinated with rescue operations, forcible entry, ventilation, and utilities control. As the operation progresses, fire attack must also be coordinated with loss control, cause determination, and recovery efforts. Firefighters must be trained and equipped to carry out any assignment they are given in a timely manner. When fighting any fire, large or small, exterior or interior, firefighters should always work as a team under the direction of a supervisor. Depending on the conditions at the fire scene, the Incident Commander (IC) will decide if the fire attack is to be conducted as an offensive or defensive operation; this strategy may be changed as conditions change. For example, the IC may decide that immediate rescue or exposure protection may be more important than fire attack.

Coordination between crews performing different functions is crucial. For example, ventilating a fire before attack lines are in place and forcible entry personnel are ready may result in the unwanted spread of fire due to the increase in air movement through the structure. When properly coordinated, the ventilation effort substantially aids the entry of attack teams and/or search and rescue teams.

Situational Awareness

One of the most critical aspects of coordination between crews — and of your personal safety and survival — is maintaining your *situational awareness*. As discussed in Chapter 2, Firefighter Safety and Health, your safety and that of your fellow firefighters can depend on being aware of everything going on around you. The opposite of situational awareness is *tunnel vision* where firefighters become so focused on fire fighting or other operational assignments that they fail to sense changes in their environment.

Maintain your situational awareness by looking up, looking down, and looking all around. Listen for new or unusual sounds and feel vibrations or other movements. Communicate to your supervisor, the other members of your crew, or Command any changes in the situation of which you become aware. Entire crews have died because one member failed to communicate critical information about the fire environment.

Teams advancing hoselines should carry equipment that may be needed to force interior doors, check concealed spaces for fire extension, or for emergency exit (**Figure 15.5**). This equipment should include at least a portable radio, hand light, pike pole, and forcible entry tools of some type. Before entering the building or the fire area, the person at the nozzle should bleed the air from the line by opening the nozzle slightly. Opening the handle slightly while waiting for water to arrive hastens this process. The operation of the nozzle should also be checked by testing the range of stream patterns or by setting a proper pattern for the attack based on the conditions found.

Figure 15.5 Carrying forcible entry tools while advancing hoselines increases firefighter safety and effectiveness.

Figure 15.6 Extinguish fire at or near the entrance before advancing into the structure. *Courtesy of Dick Giles.*

Sometimes a fire is confined to an upholstered chair or sofa or a mattress. These fires may be controlled with a small amount of water or a fire extinguisher. The burning item can then be carried outside for overhaul and complete extinguishment.

When the structure or major contents are involved in fire, firefighters should wait at the building entrance, staying low and out of the doorway, until the fire officer gives the order to advance. Before entry, extinguish fires showing in any fascia or soffit, boxed cornices, other exterior overhangs, or around entry or egress points (**Figure 15.6**). Whenever possible, approach and attack the fire from the unburned side to keep it from spreading throughout the structure. Therefore, if a fire is burning in rooms on the street side of the structure, initial entry should be made from the side or rear of the building when possible.

Once the fire has been contained, determine the area of origin and protect any evidence before overhaul and extinguishment operations are begun. Breathing apparatus must still be worn during overhaul and extinguishment due to the presence of toxic fire gases (**Figure 15.7**). Any valuables found during overhaul should be turned in to a supervisor for safekeeping.

Depending on the nature and size of the fire, firefighters may use a direct, indirect, or combination method of attacking the fire. Hoseline selection and stream selection are also made when fire attack is conducted.

Hoseline Selection

While there are other fire control technologies available that involve the application of dry chemicals, foam, and other extinguishing agents, water is still the agent most often used to

Figure 15.7 Due to the continuing presence of toxic atmosphere, firefighters must wear SCBA during overhaul operations.

extinguish structure fires. Water application will be successful only if the amount of water applied is sufficient to cool the burning fuels. For example, using a booster line may allow water to be applied sooner than would be possible by deploying a larger attack line, but an insufficient volume of water for the size of the fire may not only delay extinguishment but also expose firefighters to danger from advancing flame fronts. Use booster lines during overhaul for small nuisance fires, such as those smoldering in leaf litter or decorative bark or fires burning in trash cans. In some heavily involved structure fires, even 1½-inch (38 mm) or 1¾-inch (45 mm) hoselines may not be sufficient to safely and effectively attack the fire.

Hoseline selection should be dependent upon fire conditions and other factors such as the following considerations:

- Fire load and material involved
- Volume of water needed for extinguishment
- Stream reach needed
- Number of firefighters available to advance hoselines
- Need for speed and mobility
- Tactical requirements
- Ease of hoseline deployment
- Potential fire spread

Table 15.1 gives a simple analysis of hose stream characteristics and is not meant to replace the judgment of fire personnel in selecting hoselines.

Stream Selection

Stream selection is dictated by the fire situation and the capabilities of the nozzle being used. As its name implies, a solid-stream nozzle projects water in a more-or-less solid mass. Combination (fog) nozzles can project water in a range of patterns from a narrow straight stream to a wide-angle fog.

NOTE: For additional information on nozzle capabilities, see Chapter 14, Fire Streams.

Straight and solid streams provide the greatest reach, allowing water to be applied directly onto burning fuel from a greater distance (**Figure 15.8, p. 766**). However, they have limited capabilities for cooling hot gases or providing protection from radiant heat.

Combination fog nozzles provide streams that are effective at cooling hot gases but have less reach and penetration than straight or solid streams (**Figure 15.9, p. 766**). Wide fog patterns can also protect firefighters from radiant heat, but they can be affected by the wind.

Although converting water to steam is critical to heat absorption, excess steam production can obscure vision and inflict steam burns, even through turnout clothing. For this reason it is vital that firefighters use appropriate water application methods. These methods are discussed later in this chapter.

Making Entry

Before entering a burning building, every member of the crew should conduct a quick size-up and maintain a high level of situational awareness. Entry size-up begins well before reaching the entry point and may include visual observations

Table 15.1
Hose Stream Characteristics

Size in (mm)	GPM (L/min)	Reach (Maximum) ft (m)	No. of Persons on Nozzle	Mobility	Control or Damage	Control of Direction	When used	Estimated Effective Area
1½ inches (38 mm)	40–125 gpm (160 L/min to 500 L/min)	25–50 feet (8 m to 15 m)	1 or 2	Good	Good	Excellent	• Developing fire—still small enough or sufficiently confined to be stopped with relatively limited quantity of water.	One to Three Rooms
1¾ inches (45 mm)	40–175 gpm (160 L/min to 700 L/min)	25–50 feet (8 m to 15 m)	2	Good to Fair	Good	Good	• For quick attack. • For rapid relocation of streams. When personnel are limited. • When ratio of fuel load to area is relatively light. • For exposure protection.	
2 inches (50 mm)	100–250 gpm (400 L/min to 1 000 L/min)	40–70 feet (12 m to 21 m)	2 or 3	Fair	Fair	Good	• When size and intensity of fire are beyond reach, flow, or penetration of 1½-inch (38 mm) line. • When both water and personnel are ample.	One Floor or More— Fully Involved
2½ inches (65 mm)	125–350 gpm (500 L/min to 1 400 L/min)	50–100 feet (15 m to 30 m)	2 to 4	Fair to Poor	Fair	Good	• When safety of crew dictates. • When larger volumes or greater reach are required for exposure protection.	
Master Stream	350–2,000 gpm (1 400 L/min to 8 000 L/min)	100–200 feet (30 m to 60 m)	1	Poor to None (Aerial master streams can be good)	Poor	Good	• When size and intensity of fire are beyond reach, flow, or penetration of handlines. • When water is ample, but personnel are limited. • When safety of personnel dictates. • When larger volumes or greater reach are required for exposure protection. • When sufficient pumping capability is available. • When massive runoff water can be tolerated. • When interior attack can no longer be maintained.	Large Structures— Fully Involved

Adapted from Joe Batchler, Maryland Fire & Rescue Institute.

Figure 15.8 Straight or solid streams allow water to be applied to the fire from a safe distance.

Figure 15.9 Fog streams protect firefighters from radiant heat and are effective at cooling hot fire gases.

Figure 15.10 Control of the door into a burning structure is critical. Use a utility rope or hose strap looped around the doorknob to keep control of the door.

and the use of a thermal imaging camera or infrared heat-sensing device. The following are other pre-entry considerations that are critical to firefighter safety and effectiveness:

- Reading fire behavior indicators
- Understanding the crew's tactical assignment
- Identifying potential emergency escape routes (other doors, windows, etc.)
- Assessing forcible entry requirements
- Identifying hazards (overhead wires, structural instability, etc.)
- Verifying that radios are working, on the right channel, and being received

If a door to the fire area must be opened, all members of the hose team should stay low and to one side of the doorway. Check the door for heat before opening it by using a thermal imager or applying a small amount of water spray to the surface of the door. If the door is very hot, the water will evaporate and convert to steam. Excessive heat may be obvious from the smoke, air flow, and other fire behavior indicators. Firefighters must maintain control of the door as it is opened so it can be closed again if necessary. To control the door, firefighters can loop a utility strap over the doorknob **(Figure 15.10)**. Remember that smoke escaping at the top of the doorway is unburned fuel. Be ready to cool this hot smoke to prevent its ignition.

Gas Cooling

Gas cooling is not a fire extinguishment method but simply a way of reducing the hazard presented by the hot gas layer. This technique is effective when faced with a *shielded fire;* that is, one you cannot see from the doorway because objects are shielding it. In these situations, you cannot apply water directly onto the burning material without entering the room and working under the hot gas layer.

The hot gas layer accumulating in the upper levels of the compartment can present a number of problems for you and the other members of the hose team. Remember that **smoke is fuel**, and it may transition to rollover, flashover, or a smoke explosion at any time. In addition, hot smoke radiates heat to furniture and other combustibles in the compartment. This increases pyrolysis which adds more flammable fuel to the gas layer. Cooling the hot gas layer mitigates these hazards by slowing the transfer of heat to other combustibles and reducing the chances of the overhead gases igniting.

You can cool the hot gas layer by applying short pulses of water fog into it. With the nozzle set on a 40- to 60- degree fog pattern, direct it upward toward the gas layer and quickly open and close it in one- to two-second pulses **(Figure 15.11)**. Remember that your intent is to cool the gases, *not* to produce a large volume of steam. When water droplets begin to fall out of the overhead smoke layer, it means that the gases have been cooled and you can stop spraying water into the smoke. If the fire continues to burn unchecked, the gas layer will regain its heat and the gas-cooling technique may have to be repeated.

The gas-cooling technique should be repeated as necessary while the hose team advances under the gas layer toward the fire. In narrow hallways, the fog pattern may need to be restricted. In large-volume compartments, the duration of the pulses may need to be increased slightly.

Smoke — Visible products of combustion resulting from the incomplete combustion of carbonaceous materials and composed of small particles of carbon, tarry particles, and condensed water vapor suspended in the atmosphere, which vary in color and density depending on the types of material burning and the amount of oxygen.

Figure 15.11 The hot gas layer can be cooled with short pulses of water fog as the attack team enters the structure. *Courtesy of Dick Giles.*

Direct Attack

The most efficient use of water on free-burning fires is made by a direct attack on the fire — usually from a solid or straight stream. The water is applied in short bursts directly onto the burning fuels (often called "penciling") until the fire "darkens down" **(Figure 15.12, p. 768)**. Another effective technique (often called "painting") is to cool hot surfaces to slow or stop the pyrolysis process by gently applying water and al-

Direct Attack (Structural) — Attack method that involves the discharge of water or a foam stream directly onto the burning fuel.

Figure 15.12 Attacking the fire with a straight stream reduces the chances of upsetting the thermal layering.

lowing it to run over the hot material. Water should not be applied long enough to upset the thermal layering (sometimes called *thermal balance*); the steam produced will begin to condense, causing the smoke to drop rapidly to the floor and move sluggishly thereafter.

Indirect Attack

Indirect Attack (Structural) — Directing fire streams toward the ceiling of a room or building in order to generate a large amount of steam. Converting the water to steam absorbs the heat of the fire and cools the area sufficiently for firefighters to safely enter and make a direct attack on the fire.

When firefighters are unable to enter a burning building or compartment because of the intense heat inside, an indirect attack can be made from outside the compartment through a window or other small opening **(Figure 15.13)**. This method of attack is not ideal where building occupants may still be inside or where the spread of fire to uninvolved areas cannot be contained. However, this may be the only method of attack possible until temperatures are reduced.

To make an indirect attack on the fire, a fog stream is introduced through a small opening and directed at the ceiling where the heat is most intense. The heat converts the water spray to steam, which fills the compartment and absorbs the bulk of the heat. Once the fire has been darkened down and the space has been ventilated, hoselines can be advanced and water applied directly onto whatever is burning.

Direct and Indirect Attack

The difference between a direct attack and an indirect attack is the location from which the attack is started. An indirect attack is started from *outside* the compartment, and a direct attack is started from *inside* it. **Skill Sheet 15-I-1** describes how to attack a structure fire with an exterior attack.

Combination Attack — Battling a fire by using both a direct and an indirect attack. This method combines the steam-generating technique of a ceiling level attack with an attack on the burning materials near floor level.

Combination Attack

A combination attack uses the heat-absorbing technique of cooling the hot gas layer followed by a heat-reducing direct attack on the materials burning near the floor level. The attack starts with short bursts, known as *penciling,* from a penetrating fog stream directed into the hot gas layer at the ceiling level (gas cooling) **(Figure 15.14)**.

Figure 15.13 An indirect attack through a window or small opening in the structure can be effective. *Courtesy of Dick Giles.*

Figure 15.14 A penetrating fog stream can be used to cool and reduce the hot gas layer. *Courtesy of Central Florida Fire Academy.*

Then the attack switches to a straight stream, known as *painting*, to attack the combustibles burning near the floor level. Remember that applying water to cold smoke does not extinguish the fire and may cause unnecessary water damage and disturbance of the thermal layering.

Deploying Master Stream Devices

Master streams are usually deployed in situations where the fire is beyond the effectiveness of handlines or there is a need for fire streams in areas that are unsafe for firefighters. The three main uses for a master stream are as follows:

- Direct fire attack
- Backup handlines that are already attacking the fire from the exterior
- Exposure protection

Master stream devices must be properly positioned to apply an effective stream on a fire. Even though the master stream can be adjusted up and down and right and left, once the line is in operation it must be shut down if the device is to be moved. Moving a master stream device can be a time-consuming process, and the device is out of operation during that time. When a master stream is directed into a building, the device must be positioned close enough to the building that its stream can reach the fire. This is particularly important when using a fog nozzle because fog streams do not have the stream reach and penetration of solid streams.

The second aspect of master stream placement is the angle at which the stream enters the structure. Firefighters should aim the stream so it enters the structure at an upward angle and allow it to deflect off the ceiling or other overhead objects **(Figure 15.15, p. 770)**. This angle makes the stream break up into smaller droplets that rain down on the fire, providing maximum extinguishing effectiveness. Streams that enter the opening at a perfectly horizontal angle are not as effective. Streams that enter the opening at too low an angle could result in a loss of control of the master stream device and hoseline.

Master Stream — Large-caliber water stream usually supplied by siamesing two or more hoselines into a manifold device or by fixed piping that delivers 350 gpm (1 400 L/min) or more.

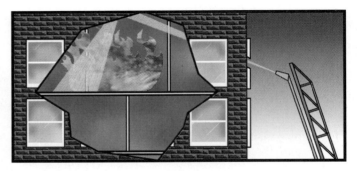

Figure 15.15 The master stream is deflected off the compartment ceiling and rains down onto the fire in small droplets.

It is also desirable to place the master stream device in a location that allows the stream to cover the most surface area of the building. This placement allows firefighters to change the direction of the stream and to direct it into more than one opening if necessary. This is particularly important in situations where there is a large volume of fire and a limited number of master stream devices.

Supplying Master Streams

Master stream devices flow a minimum of 350 gpm (1 400 L/min) which can mean high friction loss in supply hose. Therefore, except for small quick-attack devices that are designed to operate from a single 2½-inch (65 mm) line, it is not practical to supply master stream appliances with anything less than two 2½-inch (65 mm) hoselines **(Figure 15.16)**. Conventional master stream devices may be temporarily supplied by one 2½-inch (65 mm) line while setting up additional ones. Larger flows will require a third 2½-inch (65 mm) or large-diameter supply line. Some master stream devices are equipped to handle one large-diameter (4-inch [100 mm] or larger) supply line. When possible, it is best to supply these devices with a maximum of 100 feet (30 m) of hose in order to reduce the amount of friction loss.

Because master stream devices are used primarily in defensive fire fighting operations, it often desirable to shut down handlines to keep them from reducing the water supply available for master streams. As in all other situations, firefighters must be guided by their departmental SOPs regarding the simultaneous operation of master streams and handlines.

Staffing Master Stream Devices

Except for apparatus-mounted deck guns, it usually takes a minimum of two firefighters to deploy a master stream device and supply water to it, but this job can be accomplished more quickly if more firefighters are available. Once a portable master stream device is in place, it can be operated by one firefighter **(Figure 15.17)**. When water is flowing, at least one firefighter should be stationed at the master stream device at all times. An exception would be when the device is being used in hazardous positions such as close to a fire-weakened structure or near liquefied petroleum gas (LPG) storage tanks. The firefighter tending the device can change the direction of the stream when required and prevent the device from moving. The movement of master stream devices is caused by the pressure in the hoselines, but it is easily controlled by one firefighter.

Some situations may be too dangerous to have firefighters stationed at the master stream device; therefore, the device should be securely anchored in position, the desired stream developed, and personnel withdrawn a safe distance. If the device starts to move, firefighters should decrease pressure at the supply source to stop any movement.

Figure 15.16 Most master-stream devices require at least two 2½ inch (65 mm) supply lines.

Figure 15.17 Once set up, one firefighter can operate a portable master stream device.

Elevated Master Streams

Under a variety of circumstances, you may be assigned to operate an elevated master stream device or to support such an operation. Elevated master-stream devices are used to apply water to the upper stories of multistory buildings, either in a direct fire attack or to supply handlines. Elevated master streams are delivered by a number of different types of apparatus called *aerial devices* — those equipped with hydraulically operated ladders or booms. While there are others, the most common types of aerial devices are *quints*, *aerial ladders*, *aerial platforms*, and *water towers*.

Quints

These versatile apparatus can be described as either engines (pumpers) equipped with a hydraulically operated extension ladder or as aerial apparatus equipped with a pump **(Figure 15.18)**. The main ladders on most quints range in length from 50 feet (15 m) to 75 feet (25 m). All quints have waterways preplumbed to their pumps; therefore, the only external support they need for elevated stream operations is a water supply. When needed, the main ladder on a quint can be used for rescuing people (including firefighters) from exterior windows, ledges, and rooftops that are within the reach of the main ladder.

Figure 15.18 A quint is an aerial apparatus with a pump and limited capacity water tank as well as fire hose for use as supply and attack lines.

Aerial Ladders

Aerial ladders are apparatus equipped with hydraulically operated extension ladders **(Figure 15.19, p. 772)**. Most aerial ladders manufactured in North America range in length from 50 feet (15 m) to 135 feet (41 m), but some manufactured in Europe can reach 300 feet (100 m). Newer aerial ladders are equipped with built-in waterways that supply a master stream nozzle **(Figure 15.20, p. 772)**. These units need only a supply of water at the required operating pressures for elevated master stream operations. Some older aerial ladders are not equipped with built-in waterways and when an elevated master stream is needed, a ladder pipe must be attached to the end of the ladder and a supply hose laid up the ladder. The master stream nozzles of both types of apparatus can be operated by firefighters posi-

Aerial Ladder — A rotating, power-operated (usually hydraulically) ladder mounted on a self-propelled automotive fire apparatus.

Figure 15.19 Aerial apparatus in operation at a multiple-alarm fire. *Courtesy of District Chief Chris E. Mickal, NOFD Photo Unit.*

Figure 15.20 The built-in waterway telescopes as the aerial ladder is extended.

Aerial Ladder Platform — A power-operated (usually hydraulically) aerial device that combines an aerial ladder with a personnel-carrying platform supported at the end of the ladder.

Articulating Aerial Platform — Aerial device in which the structural member support (two or more booms) is hinged and operates in a folding manner. A passenger-carrying platform is attached to the working end of the device.

tioned at the ladder tip or remotely from the ground. When needed, aerial ladders can also be used for rescuing people (including firefighters) from exterior windows, ledges, and rooftops that are within the reach of the ladders.

Aerial Platforms

Aerial platforms are available in two configurations: *aerial ladder platforms* and *articulating aerial platforms*. Aerial ladder platforms are essentially aerial ladders with attached platforms (**Figure 15.21**). Articulating aerial platforms consist of a platform attached to the end of a hinged articulating boom. Articulating aerial platforms manufactured in North America range in length from 55 to 85 feet (17 m to 26 m). All articulating booms are equipped with built-in waterways and some have narrow escape ladders attached to the booms (**Figure 15.22**). Like aerial ladders, articulating aerial platforms can be used for rescuing people from upper-story windows and other locations.

Water Towers

Water towers are engines (pumpers) equipped with hydraulically operated booms that are dedicated to applying water (**Figure 15.23**). Most range from 50 to 130 feet (15 m to 40 m) in length, and some have narrow escape ladders attached to the booms. Unlike the other two types of aerial devices, water towers are not designed for rescue operations. See **Skill Sheet 15-I-2** for information on deploying and operating a master stream device.

Figure 15.21 A typical aerial ladder platform is used to apply an elevated fire stream when handlines will not work. *Courtesy of District Chief Chris E. Mickal, NOFD Photo Unit.*

Figure 15.22 An articulated aerial platform can be used for elevated fire streams as well as access to upper floors.

Figure 15.23 Water tower apparatus are primarily pumpers with booms that are used to provide elevated fire streams. *Courtesy of District Chief Chris E. Mickal, NOFD Photo Unit.*

Suppressing Class B Fires

As described in Chapter 6, Class B fires are those that involve flammable and combustible liquids and gases (**Figure 15.24, p. 774**). *Flammable liquids* are those that have flash points of less than 100°F (38°C); examples are gasoline and acetone. Flammable liquids can be ignited without being preheated. *Combustible liquids* are those with flash points higher than 100°F (38°C); examples are kerosene and vegetable oil. Combustible liquids must be heated above their flashpoint before they can be ignited. Flammable and combustible liquids can be further divided into hydrocarbons (that do not mix with water) and polar solvents (that do mix with water).

Firefighters must exercise caution when attacking large fires involving flammable and combustible liquids (**Figure 15.25, p. 774**). The first precaution is to avoid standing in pools of fuel or runoff water because there may be fuel floating on top of the water. Protective clothing can absorb fuel in a "wicking" action, which can lead to skin irritation and even to the clothing catching on fire if an

Figure 15.24 Class B fires, such as this controlled training fire, emit large amounts of heat and dense black smoke.

Figure 15.25 Coordination between attack teams is required to safely extinguish large Class B fires.

ignition source is present. Even if the wicking action does not occur, extreme danger exists if the pool of liquid ignites. In addition, benzene in petroleum product fumes is a known carcinogen. Therefore, PPE soaked with flammable or combustible liquids must be removed from service until it is thoroughly cleaned according to the manufacturer's recommendations.

Unless the leak can be stopped, do not extinguish fires burning around relief valves or piping (**Figure 15.26**). Simply try to contain the pooling liquid, if any, until the flow can be stopped. Unburned vapors are usually heavier than air and form pools or pockets of gas in low areas where they may ignite. Firefighters must attempt to control all ignition sources in a leak area. Vehicles, smoking materials, electrical fixtures, and sparks from steel tools can provide an ignition source sufficient to ignite flammable liquid vapors. An increase in the intensity of sound or fire issuing from a relief valve may indicate that the vessel is overheating and rupture is imminent. Firefighters should not assume that relief valves are sufficient to safely relieve excess pressures under severe fire conditions. Firefighters have been killed by the rupture of both large and small flammable liquid vessels that have been subjected to flame impingement.

Boiling Liquid Expanding Vapor Explosion (BLEVE) — Rapid vaporization of a liquid stored under pressure upon release to the atmosphere following major failure of its containing vessel. The failure of the containing vessel is the result of over-pressurization caused by an external heat source causing the vessel to explode into two or more pieces.

If pressure vessels containing flammable liquids, such as propane, are heated by direct flame contact, they can rupture and suddenly release their contents. The release and subsequent vaporization of these flammable liquids can result in a *boiling liquid expanding vapor explosion* (BLEVE). A BLEVE produces a violent explosion that sends large pieces of the tank flying in all directions and a huge fireball with radiant heat sufficient to incinerate anything near the site. The most common cause of a BLEVE is when flames contact the tank shell above the liquid level or when insufficient water is applied to keep a tank shell cool. When attacking these fires, apply water to the upper portions of the tank, preferably from unattended master-stream devices (**Figure 15.27**).

Applying foam is the method most often used to control flammable liquid fires (see Chapter 14, Fire Streams). Class B fire fighting techniques are also needed for fires in gas utility facilities and highway accidents involving fuel tankers. Water can be applied in several forms (cooling agent, mechanical tool, and crew protection) to control Class B fires. Extinguishing an ignitable liquid fire is described in **Skill Sheet 15-II-1**.

Figure 15.26 Flammable liquid fires should be allowed to burn until leaks can be controlled.

Figure 15.27 Because of the potential for BLEVEs, unattended master stream devices should be used to cool the tank.

Using Water to Control Class B Fires

Even though water alone is an ineffective extinguishing agent for Class B fires, it can be used in various ways for safely controlling them. These techniques require a basic understanding of the properties of Class B fuels and the effects water has on them. The important thing for firefighters to remember is that hydrocarbons (gasoline, kerosene, and other petroleum products) do not mix with water, and polar solvents (alcohols, lacquers, etc.) do mix with water. These differences affect how water can be used to control and even extinguish Class B fires.

Cooling Agent

Water can be used as a cooling agent to control Class B fires and to protect exposures. Water without foam additives is not particularly effective on lighter petroleum distillates (such as gasoline or kerosene) or alcohols. However, by applying water in droplet form in sufficient quantities to absorb the heat produced, fires in the heavier oils (such as raw crude) can be extinguished.

Water is most useful as a cooling agent for protecting exposures. To be effective, water streams need to be applied so they form a protective water film on the exposed surfaces. This applies to materials that might weaken or collapse such as metal tanks or support beams. Water applied to burning storage tanks should be directed above the level of the contained liquid to achieve the most efficient use of the water.

Mechanical Tool

Water from hoselines can be used to move Class B fuels (burning or not) to areas where they can safely burn or where ignition sources are more easily controlled. Class B fuels must *never* be flushed down storm drains or into sewers. Firefighters should use appropriate fog patterns for protection from radiant heat and to prevent "plunging" the stream into the liquid. Plunging a stream into burning flammable liquids causes increased production of flammable vapors and greatly increases fire intensity. Slowly play the stream from side to side and "sweep" the fuel or fire to the desired location. Take care to keep the leading edge of the fog pattern in contact with the fuel surface or the fire may flash under the stream and flash back around the attack crew.

Through the use of fog streams, water may also be used to dissipate flammable vapors. Because fog streams aid in dilution and dispersion, these streams can be used to influence the movement of the vapors to a desired location.

Crew Protection

Hoselines can be used as crew protection for teams advancing to shut off liquid or gas control valves (**Figure 15.28**). Coordination and slow, deliberate movements provide relative safety from flames and heat. Although one hoseline can be used for crew protection, two lines with a backup line are preferred for fire control and safety.

When pressure vessels containing flammable liquids or gases are exposed to flame impingement, apply solid streams from their maximum effective reach until relief valves close. This can best be achieved by lobbing a stream along the top of the vessel so that water runs down both sides (**Figure 15.29**). This film of water cools the vapor space of the tank, the tank shell, and the steel supports under the tanks. Tank supports must be cooled to prevent their collapse.

Hose teams can then be advanced under wide protective fog patterns to make temporary repairs or shut off the fuel source. Provide a backup line that is supplied by a separate pump and water source to protect firefighters in case other lines fail or additional tank cooling is needed. Approaches to storage vessels exposed to fire should be made at 45-degree angles to the tanks — never from the ends of the vessels or at right angles to them.

Figure 15.28 Water fog patterns are used to protect hoseline crews as they advance to shut off or control the leak. *Courtesy of District Chief Chris E. Mickal, NOFD Photo Unit.*

Figure 15.29 Direct streams so water runs down the sides to cool vessels.

Bulk Transport Vehicle Fires

Follow preincident plans for transportation emergencies to reduce life loss, property damage, and environmental pollution. The techniques of extinguishment for fires in vehicles transporting flammable fuels are similar in many ways to fires in flammable fuel storage facilities. The difficulties posed by the amount of fuel available to burn, the possibility of vessel failure, and danger to exposures are similar with both types of fires. The major differences in fires in vehicles transporting flammable fuels and fires in flammable fuel storage facilities include the following:

- Increased life-safety risks to firefighters from traffic

- Increased life-safety risks to passing motorists

- Reduced water supply

- Difficulty in identifying the products involved

- Difficulty in containing spills and runoff

- Weakened or damaged tanks and piping caused by the force of collisions

- Instability of vehicles

- Additional concerns posed by the location of the incident (civilians and structures exposed)

Although a serious accident may bring traffic to a halt, many incidents are handled with traffic passing the scene at near-normal speeds. At least one lane of traffic in addition to the incident lane should be closed during initial emergency operations. Avoid using road flares because of the possibility of igniting leaking fuels. When law enforcement personnel are unavailable, one or more firefighters trained in traffic control should be assigned to direct traffic and control access to the scene.

Whenever possible, position fire apparatus uphill and upwind to take advantage of topography (land surface configuration) and weather conditions and protect firefighters from traffic and other hazards **(Figure 15.30)**. Firefighters should exit the apparatus and work as much as possible from the side away from traffic. In addition, apparatus driver/operators should turn the wheels of vehicles parked to protect firefighters so that the apparatus cannot be pushed into them if it is struck by another vehicle.

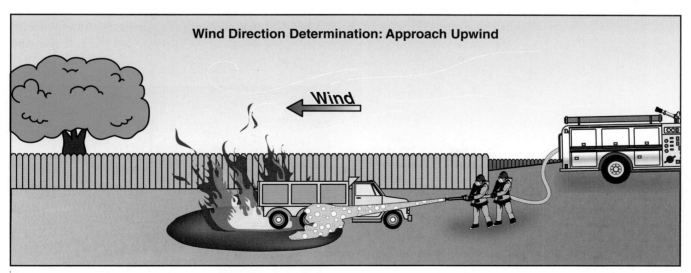

Wind Direction Determination: Approach Upwind

Figure 15.30 Position apparatus uphill and upwind whenever possible.

The techniques of approaching and controlling leaks or fires involving vehicles are the same as for storage vessels. Additionally, firefighters should be aware of the possibility of vehicle tires failing, which could cause a flammable load to shift suddenly. Crews need to know the status and limitations of their water supply. It may also be necessary to protect trapped victims with hose streams until they can be rescued.

Firefighters must determine the exact nature of cargos as soon as possible from bills of lading, manifests, placards, or the drivers of the transport vehicles. Unfortunately, cases exist where these items could not be found, placards were either wrong or obscured, and drivers were unable to identify their cargos. In these instances, contact should be made with the shippers or manufacturers responsible for the vehicles.

Control of Gas Utilities

Every firefighter should have a working knowledge of the hazards and correct procedures for handling incidents involving natural gas and liquefied petroleum gas (LPG). Many houses, manufactured homes, and businesses use natural gas or LPG for cooking, heating, or industrial processes. Natural gas is also used as a fuel for buses and other motor vehicles.

Natural Gas

Natural gas in its pure form is *methane*, which is flammable but nontoxic. When delivered to customers, natural gas may contain trace amounts of ethane, propane, butane, and pentane. Natural gas is lighter than air so it tends to rise and diffuse in the open. While natural gas is nontoxic, it is classified as an asphyxiant because it may displace normal breathing air in a confined space and lead to suffocation. Natural gas has no odor of its own, but a very distinctive odor (mercaptan) is added by the utility. It is distributed from gas wells to its point of use by a vast network of surface and subsurface pipes **(Figure 15.31)**. Gas pressure in these pipes ranges from 1,000 psi (7 000 kPa) in the distribution network to 0.25 psi (2 kPa) at the point of use. However, the pressure is usually below 50 psi (350 kPa) in local distribution piping.

Natural gas is explosive in concentrations between 5 and 15 percent in air. Natural gas may also be compressed, stored, and shipped in cylinders marked as compressed natural gas (CNG). Natural gas is also shipped and stored as a liquid (LNG) and is subject to BLEVE in this form.

Contact the local utility when any emergency involving natural gas occurs in its service area. The local utility will provide an emergency response crew equipped with special (nonsparking) tools, maps of the distribution system, and the training and

////////

CAUTION

Natural gas that leaks underground in wet soil can lose its odorant and become difficult to detect without instruments.

Figure 15.31 Inline control valves can be used to shut off the flow of gas to the leak. Control valves are visible in this photo of a natural gas substation.

experience needed to help control the flow of gas. The response of these crews is usually prompt, but the time may be extended in rural areas or in times of great demand such as following an earthquake, hurricanes, or other large-scale event.

Liquefied Petroleum Gas (LPG)

Also known as *bottled gas*, LPG refers to fuel gases stored in a liquid state under pressure. While there are two main gases in this category — *butane* and *propane* —propane is the most widely used. Propane is used primarily as a fuel gas in campers, manufactured homes, agricultural applications, rural homes, and businesses. It is also used as a fuel for motor vehicles. Propane gas contains small quantities of butane, ethane, ethylene, propylene, and isobutane or butylene. Propane gas has no natural odor of its own, but a very distinctive odor is added. The gas is nontoxic, but it is classified as an asphyxiant because it may displace normal breathing air in a confined space and lead to suffocation.

Liquefied Petroleum Gas (LPG) — Any of several petroleum products, such as propane or butane, stored under pressure as a liquid.

LPG is about one and one-half times as heavy as air so it will sink to the lowest point possible. The gas is explosive in concentrations between 1.5 and 10 percent. LPG is shipped from its distribution point to its point of usage in cylinders and in tanks on cargo trucks. It is stored in cylinders and tanks near its point of use, and then the tank or cylinder is connected by steel piping and copper tubing to the appliances the gas serves. The supply of gas into a structure may be stopped by shutting a valve at the tank **(Figure 15.32)**. An LPG leak will produce a visible cloud of vapor that hugs the ground **(Figure 15.33)**. This cloud of unburned gas may be dissipated by a fog stream of at least 100 gpm (400 L/min).

All LPG cylinders and tanks can BLEVE when exposed to intense heat or open flame. As described earlier, a BLEVE can be prevented by cooling the tank with large volumes of water applied to the top of the tank. Information on controlling a pressurized flammable gas container fire can be located in **Skill Sheet 15-II-2**.

Flammable Gas Incidents

Incidents involving both CNG and LPG distribution systems are most often caused by excavation equipment striking underground pipes and causing a break. When these breaks occur, contact the utility company immediately. Even if the gas has not yet ignited, apparatus should approach from and stage on the upwind side — that is, on the side from which the wind is blowing. Firefighters should be prepared in the event of an explosion and accompanying fire. The first concerns are the evacu-

Figure 15.32 On most LPG tanks, the shutoff valve is located on top.

Figure 15.33 Because it is heavier than air, liquefied petroleum gas hugs the ground.

ation of the area immediately around the break, evacuation of the area downwind, and elimination of ignition sources. Service connections near the break may have been damaged; therefore, check surrounding buildings for the odor of gas inside. Firefighters should follow their departmental SOPs regarding crimping off a gas line to stop a leak. If gas is burning, the flame *should not be extinguished*. If necessary, use hose streams to protect exposures.

In many structure fires, but not all, one important task is locating the gas meter and turning off the gas supply to the involved building. In some industrial and institutional occupancies, critical equipment and processes depend upon an uninterrupted supply of natural gas. For example, the emergency generators in some hospitals are fueled by natural gas. In the majority of homes and businesses served by natural gas, the meter is located outside the building and is often visible from the street (**Figure 15.34**). However, in some occupancies, the gas meter is located inside the building.

If the gas meter is involved in fire, the firefighters assigned to turn off the gas should be protected by a hoseline set on a wide fog pattern. The flow of gas into the building can be stopped by turning the cutoff valve to the closed position, which is at a right angle to the pipe (**Figure 15.35**).

Petcock

Figure 15.34 Natural gas meters may be located near both commercial and residential occupancies. The shutoff valve (petcock) is usually located on the gas pipe on the supply side of the meter.

Figure 15.35 The flow of gas can be stopped by closing the valve with a crescent or spanner wrench.

Suppressing Class C Fires

Fires involving energized electrical equipment (Class C fires) are quite common. One major safety hazard in Class C fires is the failure of firefighters to recognize the danger and take appropriate steps to protect themselves. However, when the electrical equipment is de-energized, these fires can be handled with relative ease. Once the electrical power is turned off, these fires may self-extinguish or they will fall into either Class A or Class B fires if they do continue to burn. Extraordinary electrical hazards (potential Class C fires) can be found in railroad locomotives, telephone relay switching stations, and electrical substations.

In many commercial and high-rise buildings, electrical power is necessary to operate elevators, air-handling equipment, and other essential systems. Electrical power to the entire building should not be shut off until ordered. If power is shut off to the entire building or any device in it, the main power switch should be locked- and tagged-out to prevent it from being turned back on before it is safe to do so (**Figure 15.36**). If lockout/tagout devices are not available, assign a firefighter equipped with a portable radio to tend the switch until power can be safely restored.

When handling fires in delicate electronic or computer equipment, clean extinguishing agents, such as Halotron®, should be used to prevent further damage to the equipment. Multipurpose dry-chemical agents are very effective at extinguishing Class C fires, but some are chemically reactive with electrical components. All of these agents create a considerable cleanup problem. Using water on energized equipment is inappropriate because of the inherent shock hazard. However, if water *must* be used, apply it from a distance in the form of a fog or spray stream.

Class C fire suppression techniques are also needed for fires involving transmission lines and equipment, underground lines, and commercial high-voltage installations. In addition, departmental operating procedures should clearly state the responsibilities for controlling electrical power, the dangers of electric shock, and the guidelines for electrical emergencies.

Transmission Lines and Equipment

A relatively small number of electrical emergencies involve fires in electrical substations, transmission lines, and associated equipment (**Figure 15.37**). Electrical power lines sometimes break and start fires in grass and other vegetation. Whether or not a fire is started, an area equal to one span between poles should be cordoned off around the break. If a fire does start, firefighters should wait for the fire to burn away from the break a distance equal to one span before extinguishing the fire. To reduce the risk to life and property in these incidents, it is SOP in many fire departments to notify the responsible electrical utility, control the scene, and deny entry until utility personnel arrive. For maximum safety, only utility personnel should cut electrical wires.

WARNING!
Assume that all power lines are energized until confirmed otherwise by the power company.

Figure 15.36 When it is necessary to shut off electrical service to a structure, the main power switch should be locked and tagged.

Figure 15.37 Electrical substations are rarely involved in fire.

Fires in electrical transformers are relatively common. Some older transformers can present a serious health and environmental hazard because of coolant liquids that contain PCBs (polychlorinated biphenyls). These liquids are flammable because of their oil base, and they are carcinogenic (cancer-causing). Even though transformers containing PCBs have been outlawed for many years and most have been replaced by less hazardous types, firefighters should assume that any transformer contains PCBs until proven otherwise.

Use a dry chemical or carbon dioxide extinguisher to extinguish fires in transformers at ground level. Allow pole-top transformer fires to burn until utility personnel can extinguish the fire with a dry-chemical extinguisher from an aerial device. If the fire is burning through a wooden cross-arm that would cause the power line to fall, it is SOP in some fire departments to extinguish the fire with a fog stream. Firefighters must always follow their departmental SOPs in these situations.

Underground Transmission Lines

Underground transmission systems consist of conduits and vaults below grade. The most serious hazards these systems present are explosions caused by fuses blowing or short-circuit arcing that ignites accumulated gases. Such explosions may blow utility access covers a considerable distance. Obviously, this is a danger to the public as well as firefighters. In these situations, firefighters should keep the public at least one block (300 feet [100 m]) away from the site and make sure that apparatus is not positioned over a utility access cover (**Figure 15.38**).

Figure 15.38 Never position apparatus over electrical utility access covers.

Because fire suppression actions can be taken from outside an electrical utility vault, firefighters should not enter one except to attempt a rescue. Because of the configuration of underground vaults, there is a higher-than-normal risk of backdraft conditions developing in these spaces. Firefighters should use pike poles or other long-handled tools to remove utility access covers. Once the cover has been removed, firefighters can simply discharge carbon dioxide or dry chemical into the utility vault and replace the cover. Water is not recommended for extinguishing underground vault fires because of its electrical conductivity. The water would also create puddles on the surface that could increase the electrical shock hazard.

Commercial High-Voltage Installations

Many commercial and industrial complexes have electrical equipment that requires current in excess of 600 volts. One obvious clue to this situation is *High-Voltage* signs on the doors of vaults or fire-resistive rooms housing equipment such as transformers or large electric motors (**Figure 15.39**). Some transformers use flammable coolants that are themselves hazardous. Water (even in the form of fog) should not be used in this situation because of the damage it may cause to electrical equipment not involved in the fire.

Figure 15.39 Transformers may be located outside the structure, as shown, or in electrical and mechanical spaces within the structure. In either case, warning signs will be posted on the unit or the door into the space.

Because of toxic chemicals used in plastic insulation and coolants, smoke from these fires is an additional hazard. Firefighters should enter these installations only when rescue operations require it. Entry personnel must wear full PPE including self-contained breathing apparatus, and wear a tag line monitored by an attendant outside the enclosure. A rapid intervention team (RIC) is also required in these situations. Entrants should search with a clenched fist or the back of the hand to prevent the reflex action of grabbing energized equipment if it is touched accidentally (**Figure 15.40**). Even the smoke in these situations may be highly toxic; therefore, follow appropriate decontamination procedures afterward.

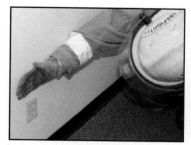

Figure 15.40 If vision is obscured, search with the back of your hand to prevent grasping an electrified wire or component.

Controlling Electrical Power

In many structure fires it is advantageous for electrical power to remain on to provide lighting, power ventilation equipment or fire pumps, or operate other essential systems. This is a decision to be made by the Incident Commander in consultation with the Incident Safety Officer. Firefighters must be able to control the flow of electricity into structures where emergency operations are being performed (**Figure 15.41, p. 784**). When a fire involves only one area of a structure, it would be counterproductive to shut off power to the entire building. When the building becomes damaged to the point that service is interrupted or an electrical hazard is created, however, power should be turned off at the main panel — preferably by a power utility employee. As always, firefighters must follow their departmental SOP.

In some residential and commercial installations, removing the electric meter does not completely stop the flow of electricity. Firefighters should be alert for installations with emergency power capabilities such as auxiliary generators (**Figure 15.42, p. 784**). In such cases, removing the meter or turning off the master switch does not turn off the power entirely. For more information on turning off building utilities, see **Skill Sheet 15-I-3**.

Clandestine drug labs and indoor marijuana-growing operations (Grow Ops) are often wired illegally to the main power supplies of adjacent buildings (**Figure 15.43, p. 784**). This is done both to steal electricity and to prevent the power company from noticing a huge increase in the use of electrical power at a given address. Firefighters should be cautious when called to one of these operations because of the following possible hazards:

- Electrical hazards due to makeshift wiring

- Volatile chemicals that may be toxic and/or flammable

- Booby traps set by the occupants

WARNING!
It is SOP in some fire departments to pull the electrical meter to turn off the electrical power in residential fires. This is an ineffective and potentially dangerous practice. IFSTA does not recommend it.

CAUTION
Before cutting into walls and ceilings that may contain electrical wiring or gas piping, verify with the Incident Commander that electrical and gas utilities have been shut off.

Figure 15.41 Downed electrical wires pose an immediate threat to fire fighting operations. *Courtesy of District Chief Chris E. Mickal, NOFD Photo Unit.*

Figure 15.42 Turning off normal electrical power may cause an emergency generator to start and reenergize the building.

Figure 15.43 Firefighters should be aware that some electrical panels may contain illegal wiring that is used to steal electricity or is the result of a clandestine operation or process. Such wiring is often unsafe. *Courtesy of John Kenyon.*

Electrical Hazards

To avoid injuries to themselves and to protect electrical equipment, firefighters should be familiar with electrical transmission systems and their hazards. While electrocution is usually associated with high-voltage equipment, conventional residential current is sufficient to deliver a fatal shock. In addition to reducing the risk of injury or electrocution, controlling electrical power reduces the danger of igniting adjacent combustibles or accidentally energizing electrical equipment.

The consequences of electrical shock can include the following:

- Cardiac arrest
- Ventricular fibrillation
- Respiratory arrest

- Involuntary muscle contractions
- Paralysis
- Surface or internal burns
- Damage to joints
- Ultraviolet arc burns to the eyes

Factors most affecting the seriousness of electrical shock include the following:

- Path of electricity through the body
- Degree of skin resistance — wet (low) or dry (high)
- Length of exposure
- Available current — amperage flow
- Available voltage — electromotive force
- Frequency — alternating current (AC) or direct current (DC)

Guidelines for Electrical Emergencies

The first rule in dealing with electrical hazards is to assume that all electrical wires and devices are energized until proven otherwise. The following list contains some other tips to help deal with electrical emergencies. The list is not all-inclusive but gives principles that should be considered to maintain a safe working environment for personnel:

- Establish an exclusion zone equal to one span in all directions from downed power lines (**Figure 15.44**).

- Be aware that other wires may have been weakened by a short circuit and may fall at any time.

- Wear full protective clothing and use only tested and approved tools with insulated handles.

- Guard against electrical shocks, burns, and eye injuries from electrical arcs.

- Wait for utility workers to cut any power lines.

- Use lockout/tagout devices when working on electrical equipment (**Figure 15.45**).

- Be very careful when raising or lowering ladders near power lines.

- Do not touch any vehicle or apparatus that is in contact with electrical wires (**Figure 15.46, p. 786**).

- Jump clear of any apparatus (keeping both feet together) that may be energized by contact with power lines.

- Do not use solid and straight streams on fires in energized electrical equipment.

- Use fog streams with at least 100 psi (700 kPa) nozzle pressure on energized electrical equipment.

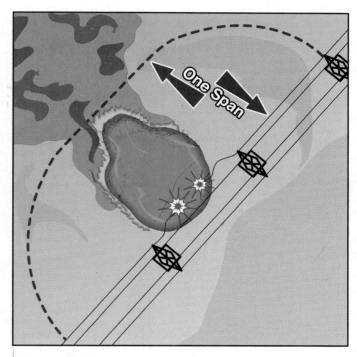

Figure 15.44 A one-span exclusion zone should be established around a downed power line. A span is the distance between power poles.

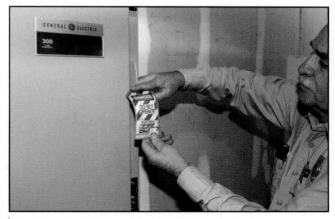

Figure 15.45 A typical lockout/tagout device includes a padlock and warning sign.

- Be aware that wire mesh or steel rail fences can be energized by wires outside your field of view.

- Where wires are down, heed any tingling sensation felt in the feet and back away.

- Avoid *ground gradient* hazards by maintaining a large safety zone around downed electrical wires.

Ground gradient is produced by an electrical current passing from a downed power line through the ground along the path of least resistance (from highest to lowest potential). In this situation, current flows outward in concentric circles from the point of contact with the ground (**Figure 15.47**). It is common for downed power lines to

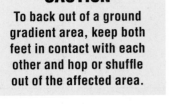

CAUTION

To back out of a ground gradient area, keep both feet in contact with each other and hop or shuffle out of the affected area.

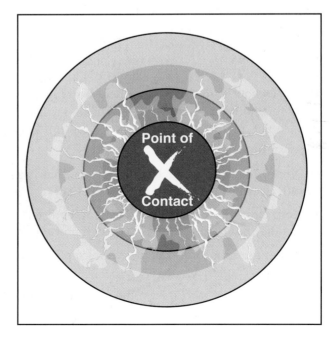

Figure 15.46 Natural disasters can result in many energized power lines being on the ground or near apparatus and personnel. *Courtesy of District Chief Chris E. Mickal, NOFD Photo Unit.*

Figure 15.47 Voltage drops as it spreads away from the source.

discharge their electrical current through surface objects several feet (meters) from their point of contact with the ground. The higher the voltage, the greater the area affected. Just walking in a normal stride through a ground gradient area is dangerous for firefighters because of the difference in potential between the firefighter's feet and the ground. Likewise, dragging a hoseline, ladder, pike pole, or other object in the area of a downed wire also risks injury from a ground gradient condition.

Suppressing Class D Fires

Class D fires (combustible metals) present the dual problem of burning at extremely high temperatures and being reactive to water. Directing hose streams at burning metal can result in the violent decomposition of the water and subsequent release of flammable hydrogen gas. Small metal chips or metal dust are more reactive to water than are large castings and other finished products. When a combustible metal is burning, water is only effective for protecting nearby exposures by keeping them below their ignition temperatures.

In some situations, if a Class D extinguishing agent is not available, firefighters can simply protect the exposures and allow the metal to burn out. Class D extinguishing agents can be manually shoveled or scooped onto the burning metal or sprayed on by Class D fire extinguishers in sufficient quantity to completely cover the burning metal.

Combustible metal fires can be recognized by a characteristic brilliant white light that is given off until an ash layer covers the burning material. Once this layer has formed, it may appear that the fire is out. Firefighters should not assume that these fires are extinguished just because flames are not visible. It may be an extended period of time before the area or substance cools to safe levels. Combustible metal fires are very hot — greater than 2,000°F (1 093°C) — even after they appear to be out.

Company-Level Fire Tactics

As discussed in Chapter 1, the standard tactical priorities for any emergency are *life safety, incident stabilization*, and *property conservation*, and these apply to the fireground. However, firefighters should not confuse the order of the priorities with a sequence of operations. Although the order of these priorities remains the same, the actions taken on the fireground may or may not be performed in that order. For example, in some situations the best way to protect trapped occupants may be to control the fire first and then perform rescues, or to protect the occupants in place while fighting the fire. Controlling the fire also contributes to both incident stabilization and property conservation. Likewise, when well-trained firefighters perform forcible entry, ventilation, and other fireground support activities in ways that avoid unnecessary property damage, they contribute to the property conservation effort.

Fires in Structures

Fires inside structures can be some of the most challenging that firefighters will face. Rescue, exposure protection, forcible entry, ventilation, confinement, and extinguishment functions must be performed in a coordinated way for the operation to be both safe and successful (**Figure 15.48, p. 788**). The following information describes a typical response to a fire in a residential structure and details the typical responsibilities of each unit involved.

Figure 15.48 Fireground operation requires careful coordination between crews. *Courtesy of Bob Esposito.*

First-Due Engine Company

If smoke or fire is visible as the first-due engine approaches the scene, it may be departmental SOP for that unit to stop and lay a supply line from a hydrant or from the end of the driveway into the scene. The hydrant can be opened and the supply line charged as soon as a hose clamp is applied at the scene. In other departments it is SOP to give the officer in charge of the first-due engine the option of laying a supply line or proceeding directly to the scene and initiating a quick attack if it appears that this strategy could result in saving lives and property.

Even though the size-up of the fire actually began much earlier, the first-arriving company officer will conduct a rapid initial assessment of the situation. In that quick mental evaluation, the officer must answer certain questions:

- Are there occupants in need of immediate rescue?
- Are only the contents involved or is the structure burning?
- Are there exposures threatened by the fire?
- Are there sufficient resources on scene or en route to handle this situation?

The answers to these questions will dictate the actions taken by the first-due engine company. If there are too few resources assigned to this fire, the officer will immediately call for more resources to be dispatched. If by taking immediate action the company can save one or more lives, it will do so even if there are not enough firefighters on scene to form a rapid intervention crew (RIC). If there are no obvious and immediate life-safety concerns and the fire is threatening to extend to another nearby structure, the officer may order lines pulled to apply water to the exposure. A deck gun may be put into operation to keep the exposure cool while lines are pulled for a direct attack on the fire.

Given a relatively small interior fire, commonly called a *room-and-contents fire,* the company officer on the first-arriving engine usually assumes command of the incident and transmits a report on conditions to the communications center. The report confirms the exact location, describes the conditions at the scene, and if necessary, includes a call for additional needed resources. The officer then orders whatever immediate actions are dictated by the answers to the questions just discussed. Depending on the present and expected behavior of the fire, the first-arriving engine company may need to perform search and rescue, exposure protection, or fire attack.

Once the location of the fire is known, the first engine company will position the initial attack hoseline to cover the following priorities:

- Intervene between trapped occupants and the fire.
- Protect rescuers.
- Protect primary means of egress.
- Protect interior exposures (other rooms).
- Protect exterior exposures (other buildings) **(Figure 15.49)**.
- Initiate extinguishment from the unburned side.
- Operate master streams.

Second-Due Engine Company

Unless otherwise assigned, the second engine company must first make sure that an adequate water supply is established to the fireground. Depending on the situation, it may be necessary to finish a hose lay started by the first engine company, lay an additional line, or connect to a hydrant to support the original or additional lines that have been laid **(Figure 15.50)**. The need to pump the lines from the hydrant depends on local factors; these include the size of the hose being used, the distance from the hydrant to the scene, and the water pressure available in the system.

Once the water supply has been established, the second company proceeds according to the following priorities:

- Back up the initial attack line.
- Protect secondary means of egress.
- Prevent fire extension (confinement).
- Protect the most threatened exposure.
- Assist in extinguishment.
- Assist with fireground support company operations.

Figure 15.49 Firefighters use a master stream to form a water curtain to protect passengers from radiant heat.

Figure 15.50 The second-due unit is responsible for ensuring an adequate water supply at the emergency scene.

Fireground Support Company

If a support company (ladder truck, quint, or rescue unit) has been dispatched to the fire, it may arrive with or after the first engine company. Once at the scene, this company is responsible for performing the following tasks (among other things) in the order dictated by the situation:

- Forcible entry
- Search and rescue
- Property conservation (salvage)
- Ladder placement
- Ventilation
- Scene lighting
- Utilities
- Checking for fire extension
- Operating elevated fire streams (**Figure 15.51**).
- Overhaul

These functions may be performed by engine personnel when support companies are not available. Initially, the support company may be assigned to check the outside of the building for victims needing immediate rescue and raise the ladders needed for rescues or roof access for ventilation. They may be assigned to force entry into the building to facilitate simultaneous fire attack and search-and-rescue operations inside the building. Depending upon the situation, teams may first search areas closest to the fire or they may search areas that are most likely to be inhabited. In most cases, the search priorities are the following:

- Most severely threatened
- Largest number threatened
- Remainder of fire area
- Exposures

Regardless of the pattern used, searches must be conducted systematically and in accordance with departmental SOPs.

In addition to search and ventilation operations, it is SOP in some departments for support company personnel to assist engine companies in making the fire attack. In some cases, support company personnel equipped with forcible entry tools accompany the hose team as they advance toward the seat of the fire. In other cases, support company assistance can be limited to the placement of ground ladders for fire attack, setting up scene lighting, and similar exterior functions (**Figure 15.52**). In other situations, a support company can use a master stream to attack the fire in what is called a *blitz attack*. Blitz attacks must be coordinated with other operations to both prevent injury to firefighters inside and to avoid spreading the fire to uninvolved parts of the building. Interior attack teams can be injured by steam or forced to retreat because of poorly directed outside master streams. Elevating devices such as aerial ladders, platforms, and ladder towers can also be used to conduct blitz attacks on upper floors and as substitute standpipes from which engine company personnel can advance hoselines.

Support company personnel are also commonly assigned to locate and secure the building's utilities. Shutting down the electrical power to the building (if not needed to run fire pumps or operate elevators) can reduce the risk of electrical

Blitz Attack — To aggressively attack a fire from the exterior with a large diameter (2½-inch [65 mm] or larger) fire stream.

shock (**Figure 15.53**). Turning off the gas supply to the building may eliminate a major source of fuel to the fire. Shutting off the water supply (if not needed for operating sprinklers) can dramatically reduce water damage within the building (**Figure 15.54**).

Rapid Intervention Crew (RIC)

The Incident Commander (IC) — often with the help of an Incident Safety Officer (ISO) — continually evaluates the incident scene for any possible safety concerns that may develop. However, it is

Figure 15.52 Fireground support companies may be responsible for laddering the building. *Courtesy of District Chief Chris E. Mickal, NOFD Photo Unit.*

Figure 15.51 Fireground support companies may be responsible for providing elevated master streams.

Figure 15.53 A firefighter shuts off main power switch to a building to prevent injuries.

Figure 15.54 Water shutoff valves for commercial structures may be located in the street or sidewalk in front of the structure.

not always possible to predict when an emergency situation (trapped or injured fire-fighters) or equipment failure (SCBA malfunction) will occur. When these situations arise, the Incident Commander must be prepared to immediately deploy rescue personnel or rapid intervention crews to assist other firefighters. With these situations in mind, NFPA® 1500, *Standard on Fire Department Occupational Safety and Health Program*, requires fire departments to ". . . provide personnel for the rescue of members operating at emergency incidents . . . ."

The exact number of rapid intervention crews is determined by the IC during the initial phases of the incident. Crews are added as necessary if the incident escalates or the number of operations increases. This allows flexibility in RIC composition based on the type of incident and numbers of personnel on scene. Each rapid intervention crew consists of two or more members wearing appropriate personal protective clothing, equipment, a radio, and equipped with any special rescue tools and equipment necessary to effect rescue of other emergency personnel **(Figure 15.55)**. Although individual RIC members may be assigned other emergency scene duties, they must be prepared to drop whatever they are doing and deploy immediately if needed.

Chief Officer/Incident Commander

Upon arriving at the scene, a chief officer may choose to assume command from the original IC and take responsibility for all on-scene operations **(Figure 15.56)**. Alternatively, if the original IC has the incident well organized and reasonable progress toward incident stabilization is being made, the chief officer may choose to assume another role in the Incident Command structure. For example, he or she may choose to act as liaison with other involved entities such as technical experts, utility crews, and members of the media. **Skill Sheet 15-II-3** discusses how to establish Incident Command and coordinate interior attack of a structure fire.

Figure 15.55 The RIC is fully equipped and ready to respond in the event an interim attack crew requires assistance or rescue. *Courtesy of District Chief Chris E. Mickal, NOFD Photo Unit.*

Figure 15.56 Depending on the situation, the Chief Officer may either assume Command on arrival or handle other duties.

Fires in Upper Levels of Structures

Fires in upper levels of structures, such as high-rise buildings, can be very challenging. The typical residential response consisting of two or three engines and one truck is usually inadequate for these incidents because many firefighters are required to conduct large-scale evacuations, carry fire fighting equipment to upper levels, and mount a sustained fire attack. In many cases, tools and equipment must be hand carried up many flights of stairs. In most fire departments, elevators are not used to transport fire crews to the fire floor because the fire may damage the elevator or its controls and strand the firefighters in the elevator car or deposit them directly onto the fire floor. If the building has low-rise elevators that do not serve the fire floor, departmental SOPs may allow firefighters to use them. Some departments allow freight elevators to be used to transport fire fighting tools and equipment to Staging, which is normally located two floors below the fire floor.

Typically, the fire attack is initiated from the floor below the fire floor **(Figure 15.57)**. Firefighters may wish to look at the floor below to get a general idea of the layout of the fire floor. If the standpipe is in a location that is protected from the fire, hoselines may be connected to the standpipe on the fire floor. Extra hose should be flaked up the stairs above the fire floor so that it will feed more easily into the fire floor as the line is advanced.

In addition to attacking the fire directly, crews should be checking floors above the main fire floor for fire extension and any victims who may have been unable to escape. The staging of extra equipment and personnel is usually established two floors below the fire floor.

Personnel must exercise caution in the streets around the outside perimeter of a high-rise building on fire. Glass and other debris falling from many stories above the street can severely damage equipment, cut hoselines, and injure or kill firefighters. To minimize the danger from falling objects or debris, the area should be cordoned off and safe paths of entry to the building identified. Conditions dictate how large an area needs to be cordoned off.

Figure 15.57 In multistory structures, the fire attack is often launched from the floor below the fire.

Fires Belowground in Structures

Structure fires belowground can expose firefighters to extremely hostile conditions. To avoid these conditions, it may be possible to control the fire without entering the basement. A thermal imaging camera or an infrared heat detector can be used on the floor above the fire to locate the seat of the fire. Once the fire is located, a hole is cut in the floor and a cellar nozzle is inserted through the hole into the basement **(Figure 15.58, p. 794)**. After a few minutes of operation, a reduction in heat should be detectible. The basement can then be vented and a nozzle crew sent in to complete extinguishment.

If a cellar nozzle is not available, firefighters may have to enter the burning basement. To do this, firefighters may have to descend an interior stairway to the basement level **(Figure 15.59, p. 794)**. These stairways can act as chimneys for the smoke, flames, and superheated gases being given off by the fire. To reduce this exposure as much as possible, a ventilation opening is created at the end of the basement opposite the entry stairway. A fire attack team descends the stairs behind the protection of a wide-angle fog pattern. Once in the basement, the team slowly advances toward the ventilation opening, pushing the smoke and flames before them. For this tactic to work effectively, the ventilation opening must be large enough to allow the smoke and heat to escape the basement.

Good ventilation techniques are extremely important when fighting fires that are belowground. If there are no basement windows, a ventilation opening can be cut in the ground-level floor near a window. A smoke ejector can be placed in the window to draw the smoke and heat from the basement and vent it to the outside. However, this action can draw smoke and flames through the smoke ejector and may damage it.

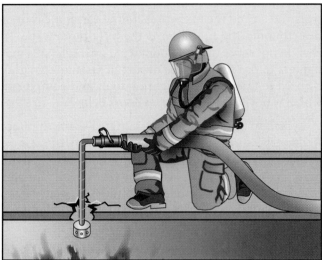

Figure 15.58 A cellar nozzle can be used effectively on basement fires.

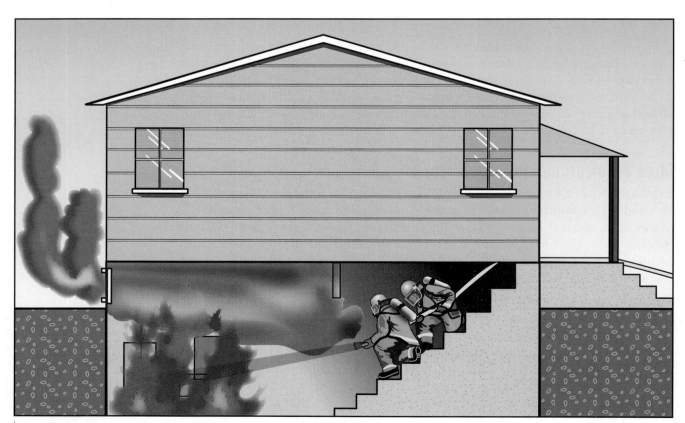

Figure 15.59 Fire attack into a basement may have to be made from the basement stairs.

In basement fires, heavy objects on the floor above the fire can increase the chances of floor collapse due to loss of structural strength in supporting members. Unprotected steel girders and other supports elongate when exposed to temperatures of 1,000°F (538°C) or more. These have been known to push over walls during a fire. The longer steel supports are subjected to fire, the more likely they are to fail, regardless of their configuration. **Skill Sheet 15-I-4** describes how to attack a structure fire above, below, and at grade level with an interior attack.

Fires in Properties Protected by Fixed Systems

Firefighters should familiarize themselves with the fixed fire extinguishing systems in buildings protected by their department **(Figure 15.60)**. Supporting these systems can be critical during a fire and such support should be a high priority. These systems include the following:

- Sprinkler systems
- Carbon dioxide systems
- Standpipe systems
- Clean-agent systems
- Dry-chemical hood systems
- Wet-chemical systems
- Foam systems

Some of the dangers involved when dealing with fires in occupancies with fixed fire extinguishing systems include the following:

- Oxygen depletion following activation of carbon dioxide flooding systems
- Poor visibility
- Energized electrical equipment
- Toxic environments

WARNING!
In any basement fire, firefighters must beware of the hazards of a weakened main floor and the danger of structural collapse.

Figure 15.60 Fire fighting crews should be familiar with the fixed fire suppression systems in their response areas in order to support them for maximum fire control.

Standard operating procedures used at these occupancies are often contained in a preincident plan. This plan includes a detailed description of the construction features, contents, protection systems, and surrounding properties. Some preincident plans specify the procedures for each engine/truck company to follow according to the conditions it finds. A building map showing water supplies, protection system connections, and engine/truck company placement is an integral part of the plan and must be updated regularly to reflect changes affecting fire department operations.

When a fire is burning in a building equipped with an automatic sprinkler system, support company personnel are often used to manage the system's operation. While these personnel must always follow their departmental SOPs regarding what they can and cannot do to the sprinkler system, some of the possible actions they may take are as follows:

- Assign a radio-equipped firefighter to the OS&Y valve to close or reopen it as ordered and to prevent it from being closed prematurely **(Figure 15.61, p. 796)**.
- Install wooden wedges or sprinkler stops to halt the flow of water from open sprinklers **(Figure 15.62, p. 796)**.

Figure 15.61 A firefighter equipped with a radio should be assigned to the sprinkler control valve.

Figure 15.62 Wooden wedges can be installed to temporarily stop the flow of water from an open sprinkler.

- Replace open sprinklers to allow the system to be restored to normal (if allowed by SOPs).
- Restore the sprinkler system to normal (if allowed by SOPs).
- Monitor the building after the fire has been extinguished and while waiting for the owner or designee to restore the sprinkler system.

Figure 15.63 Due to the types of materials used in the construction of modern vehicles, firefighters should wear full PPE including SCBA. *Courtesy of Bob Esposito.*

Vehicle Fires

Fires in small passenger vehicles such as automobiles, minivans, and sport utility vehicles are among the most common types of fires to which firefighters are called. These fires should be handled with the same degree of care as are structure fires. Firefighters should be wearing full PPE and breathing air from their SCBA (**Figure 15.63**).

The attack line should be at least a 1½-inch (38 mm) hoseline. Booster lines do not provide the protection or rapid cooling needed to effectively and safely fight a vehicle fire. Attack the fire from the side and from upwind and uphill when possible. If needed, deploy a backup line as soon as possible. Portable extinguishers can suppress some fires in the vehicle's engine compartment or electrical system.

Basic Procedures

For their own safety, one of the first actions firefighters should take is to establish a safe working zone by following U.S. DOT guidelines for protecting the scene from vehicular traffic **(Figure 15.64)**. These guidelines were discussed in Chapter 2, Firefighter Safety and Health. Once scene safety has been established, firefighters can focus on saving the vehicle occupants and fighting the fire. In doing so, they must remember to stay out of the potential travel path of the front and rear bumpers. Many bumpers incorporate hydraulic or pneumatic struts intended to absorb the shock of minor collisions **(Figure 15.65)**. If these struts are heated by the fire, they can explode and catapult the bumper to which they are attached considerable distances. Likewise, the struts used to support the engine hood (bonnet) and trunk lid (boot) can also be launched from the vehicle with tremendous force **(Figure 15.66)**. A firefighter standing in the path of travel could be injured or killed. The basic procedures for attacking a fire in an occupied vehicle are as follows:

- Position a hoseline between the burning vehicle and any exposures.

- Attack the fire from a 45-degree angle to the long axis of the vehicle.

- Extinguish any fire near the vehicle occupants first.

- Issue an "all clear" when all occupants are out of the vehicle.

- Extinguish any ground fire around or under the vehicle.

- Extinguish any fire remaining in or around the vehicle.

Apply water to cool combustible metal components that are not burning but are exposed to fire nearby. If combustible metal components do become involved, apply large amounts of water to protect adjacent combustibles while applying Class D extinguishing agent to the burning metal.

Figure 15.64 Guidelines for highway scene safety are defined by the Department of Transportation (DOT).

Figure 15.65 Firefighters should be aware of the potential hazard that exists with bumper struts when they are exposed to fire. *Courtesy of Ron Moore.*

Figure 15.66 Trunk lid and hood support struts may become projectiles when heated.

Figure 15.67 A piercing or drive-in nozzle can be used to extinguish engine fires if the hood cannot be opened.

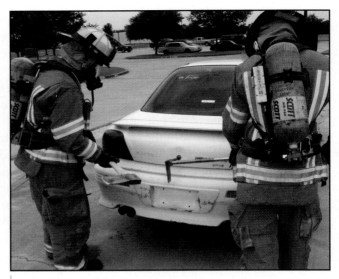

Figure 15.68 The trunk lid lock can be broken with the tip of a Halligan tool or pick-head axe.

In many engine compartment fires, the fire must be controlled before the hood can be opened. One control method is to use a piercing nozzle through areas such as the hood, fenders, or wheel wells (**Figure 15.67**). Another method is to make an opening large enough for a hose stream to be introduced. In some cases, a Halligan tool can be used to make an opening between the hood and the fender. A straight stream or narrow fog can then be directed into the opening.

When attacking a fire in the passenger compartment, use the most appropriate nozzle and pattern for the situation. Attempt to open the door, but if it is locked, the driver may have the key. If normal entry is not possible, break a window and attack the fire with a medium fog pattern. If normal access to the trunk is not possible, use the spike from a Halligan tool or a pick-head axe to knock the lock barrel out of place (**Figure 15.68**). Insert a screwdriver or similar object into the remaining locking mechanism and turn to release the lock.

The following three methods can be used for fires in the undercarriage:

1. If there is a hazard in getting close to the vehicle, use a straight stream from a distance to reach under the vehicle.

2. If the vehicle is on a hard surface such a concrete or asphalt, direct the stream downward and allow the water to deflect up toward the underside of the vehicle.

3. Open the hood and direct the stream through the engine compartment.

Once the fire has been controlled, overhaul should be conducted as soon as possible to check for extension and hidden fires. Other overhaul considerations include disconnecting the battery, securing air bags (Supplemental Restraint System [SRS] or Side-Impact Protection System [SIPS]), and cooling fuel tanks and any intact sealed components.

Hazards

In addition to the hazards associated with any other type of fire, there are hazards specific to vehicle fires. Catalytic converters, used to reduce harmful vehicle exhaust emissions, can act as an ignition source to dry grass or other fuels under the vehicle. The external temperature of a catalytic converter on a properly operating engine is about 1,300° F (704° C); on a poorly tuned engine, operating temperatures can reach as high as 2,500° F (1 371°C). A vehicle's interior components are mainly plastic, which burns rapidly at high temperatures and emits toxic gases. Another hazard associated with modern vehicles is air bags that can deploy from the steering wheel, dashboard, or door of the vehicle. Hybrid vehicles also incorporate high-voltage cables and components (**Figure 15.69**).

Firefighters should not assume that any vehicle is without extraordinary hazards such as saddle fuel tanks, LPG or CNG tanks, alternative fuel tanks, explosives, or hazardous materials. Delivery vans and even small passenger vehicles are used to transport tiny medical isotopes and other radioactive materials. Military vehicles could be carrying munitions or other hazardous cargos. For more information on how to attack a passenger vehicle fire, see **Skill Sheet 15-I-5**.

WARNING!
Do not cut or contact any orange-colored electrical cables or components in hybrid vehicles. These are high-voltage systems and electrocution is possible.

Figure 15.69 In hybrid vehicles, orange-colored cables and components contain high-voltage electrical power.

Trash Container Fires

The possibility of exposure to toxic products of combustion is ever-present when dealing with fires in trash containers. The refuse may include hazardous materials or plastics that give off highly toxic smoke and gases. Aerosol cans and batteries, which may explode, may also be present. For this reason, full PPE and SCBA should be worn when attacking any trash container fire (**Figure 15.70**).

The size of the attack line used depends on the size of the fire and its proximity to exposures. Fires in small piles of trash, garbage cans, and small containers can often be extinguished with a booster line. Larger piles, larger containers, and fires close to exposures should be attacked with at least a 1½-inch (38 mm) line. In some cases, even master streams may be needed to keep trash container fires from spreading to adjacent structures. Once the fire has been controlled, it may be possible to use standard overhaul techniques to complete the extinguishment. However, it may be advantageous to attack the fire using Class A foam. In some departments, it is SOP to use a master stream to flood the container with water to drown any fire that might be hidden in the contents. This technique can present containment problems if the water used to fill the container becomes contaminated with a hazardous substance. Extinguishment of fire in a trash container is discussed in more detail in **Skill Sheet 15-I-6**. For more information on attacking a fire in stacked or piled materials, such as are found in dumpsters, see **Skill Sheet 15-I-7**.

Figure 15.70 When attacking trash bin fires, firefighters should wear full PPE and use their SCBA.

Fires in Confined Spaces

Fire fighting operations must sometimes be carried out in locations that are below grade or otherwise without either natural or forced ventilation. Underground vaults and other structures, caves, sewers, storage tanks, and trenches are just a few examples of these spaces (**Figure 15.71**). The single most important factor in safely operating at these emergencies is recognition of the inherent hazards of confined spaces. The atmospheric conditions that may be expected include the following:

- Oxygen deficiencies
- Flammable gases and vapors
- Toxic gases
- Extreme temperatures
- Explosive dusts

  In addition, the following physical hazards may also be present:

- Limited means of entry and egress
- Cave-ins or unstable support members
- Standing water or other liquids
- Utility hazards — electricity, gas, sewage

Plant or building supervisors or other knowledgeable people at the scene may provide valuable information on the fire, its probable location, and hazards that are present. Likewise, preincident plans of existing confined spaces in the fire department's jurisdiction reduce guesswork and should be referred to during operations in these locations. Firefighters should be ready to implement prearranged hazard mitigation plans, rescues, and extinguishment efforts without delay. These plans should include provisions for victim and rescuer protection, control of utilities and other physical hazards, communications, fire extinguishment methods, ventilation, and lighting. Electrical equipment such as flashlights, portable fans, portable lights, and radios should be intrinsically safe for use in flammable atmospheres.

Because of the hazards that may exist inside these spaces, the command post and staging area must be established outside the hot zone. The staging area should be near, but not obstructing, the entrance. Firefighters must not be allowed to enter these spaces until an Incident Action Plan (IAP) has been developed and communicated to on-scene personnel. An attendant must be stationed at the confined space entrance to track personnel and equipment entering and leaving the space.

Fires in confined spaces may also be attacked indirectly with penetrating nozzles, cellar nozzles, or distributor nozzles (**Figure 15.72**). Due to the difficulty of venting heat from some confined spaces, firefighters may tire more quickly and consume their air supply faster. In these conditions, firefighters should be relieved before they are out of air. An effective air-management system should be part of the IAP to prevent firefighters from advancing into confined spaces farther than their air supplies will safely allow.

Wildland Fire Techniques

Wildland fires include those in weeds, grass, field crops, brush, forests, and similar vegetation (**Figure 15.73**). Because they are unconfined, wildland fires have characteristics that are not comparable to fires in burning buildings. The three main influences on wildland fire behavior are: *fuel*, *weather*, and *topography*. Of these, weather is the

Figure 15.71 Fires in confined spaces, such as grain storage bins, can be very difficult to extinguish.

Figure 15.72 Supplied airline systems may be needed for rescues in confined space incidents.

Figure 15.73 Wildland fires can occur in remote areas or near populated areas. *Courtesy of National Interagency Fire Center (NIFC).*

most significant. Because local topography, fuel types, water availability, and predominant weather patterns vary from one region to another, the tools and techniques used to control wildland fires also vary. Despite local differences, however, there are many similarities and these are discussed here.

Once a wildland fire starts, burning can be rapid and continuous. For their own survival, firefighters must learn how wildland fires behave in a variety of conditions and how temperature and humidity, winds, fuel types, and terrain features influence flame lengths and rate of fire spread.

Fuel

Fuels are generally classified by grouping those with similar burning characteristics together. This method classifies wildland fuels as subsurface, surface, and aerial fuels **(Figures 15.74 a–c, p. 802).**

- *Subsurface fuels* — Roots, peat, and other partially decomposed organic matter that lie under the surface of the ground

- *Surface fuels* — Needles, duff, twigs, grass, field crops, brush up to 6 feet (2 m) in height, downed limbs, logging slash, and small trees on or immediately adjacent to the surface of the ground

WARNING!
Fighting wildland fires can be very dangerous. Many firefighters have been seriously injured or killed while working in very light fuels and during the mop-up phase of an operation.

Figures 15.74 a-c The types of fuels found in wildland areas include subsurface fuels such as roots, surface fuels including grass and shrubs, and aerial fuels including limbs and foliage.

- *Aerial fuels* — Suspended and upright fuels (brush over 6 feet [2 m], leaves and needles on tree limbs, branches, hanging moss, etc.) physically separated from the ground's surface (and sometimes from each other) to the extent that air can circulate freely between them and the ground

The following factors affect the burning characteristics of fuels:

- *Fuel size* — Small or light fuels burn faster than heavier ones.

- *Compactness* — Tightly compacted fuels, such as hay bales, burn slower than those that are loosely piled.

- *Continuity* — When fuels are close together, the fire spreads faster because of heat transfer. In patchy fuels (those growing in clumps), the rate of spread is less predictable than in continuous fuels.

- *Volume* — The amount of fuel present in a given area (its volume) influences the fire's intensity and the amount of water needed to achieve extinguishment.

- *Fuel moisture content* — As fuels dry out, they ignite easier and burn with greater intensity (amount of heat produced) than those with a higher moisture content.

Weather

All aspects of the weather have some effect upon the behavior of a wildland fire. Some weather factors that influence wildland fire behavior are the following:

- *Wind* — Fans the flames into greater intensity and supplies fresh air that speeds combustion; very large-sized fires create their own winds.

- *Temperature* — Has effects on wind and is closely related to relative humidity; primarily affects the fuels as a result of long-term drying.

- ***Relative humidity*** — Impacts greatly on dead fuels that no longer draw moisture from their root system but only from the surrounding air.

- ***Precipitation*** — Largely determines the moisture content of live fuels. Dead flashy fuels (those easily ignited) may dry quickly; large dead fuels retain this moisture longer and burn slower.

Topography

Topography refers to the features of the earth's surface, and it has a decided effect upon fire behavior. The steepness of a slope affects both the rate and direction of a wildland fire's spread. Fires will usually spread faster uphill than downhill, and the steeper the slope, the faster the fire spreads (**Figure 15.75**). Other topographical factors influencing wildland fire behavior are the following:

Topography — Physical configuration of the land or terrain.

- ***Aspect*** — The compass direction a slope faces (aspect) determines the effects of solar heating. In North America, full southern exposures receive more of the sun's direct rays and therefore more heat. Wildland fires typically burn faster on southern exposures.

- ***Local terrain features*** — Features such as canyons, ridges, ravines, and even large rock outcroppings may alter airflow and cause turbulence or eddies, resulting in erratic fire behavior.

- ***Drainages (or other areas with wind-flow restrictions)*** — These steep ravines are terrain features that create turbulent updrafts causing a chimney effect. Wind movement can be critical in *chutes* (narrow V-shaped ravines) and *saddles* (depression between two adjacent hilltops). Fires in these areas can spread at an extremely fast rate, even in the absence of winds, and are potentially very dangerous.

Parts of a Wildland Fire

The names used to identify the parts of a typical wildland fire are shown in **Figure 15.76, p. 804**. Every wildland fire will contain at least two or more of these parts.

- ***Origin.*** The *origin* is the area where the fire started, and the point from which it spreads. The origin is often next to a trail, road, railroad, or highway, but one caused by lightning strikes or campfires may be located anywhere.

- ***Head.*** The *head* is the part of a wildland fire that spreads most rapidly. The head is usually found on the opposite side of the fire from the area of origin and in the direction toward which the wind is blowing. The head burns intensely and usually does the most damage. Usually, the key to controlling the fire is to control the head and prevent the formation of a new head.

- ***Finger.*** *Fingers* are long narrow strips of fire extending from the main fire. They usually occur when the fire burns into an area that has both light fuel and patches of heavy fuel. Light fuel burns faster than the heavy fuel, which gives the finger effect. When not controlled, these fingers can form new heads.

Figure 15.75 Wildland fires spread faster upslope than down. *Courtesy of Tony Bacon.*

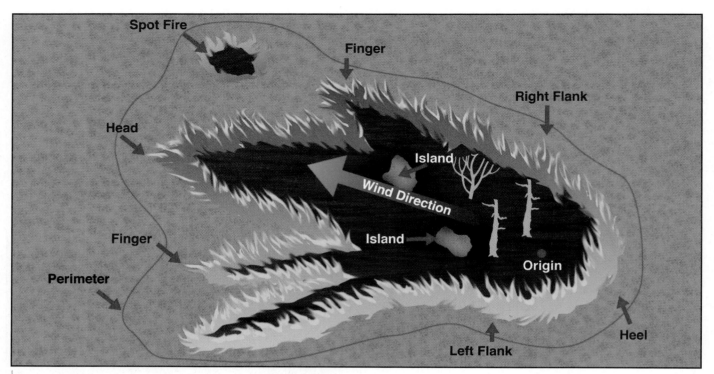

Figure 15.76 Terms used to describe the parts of a wildland fire.

- *Perimeter.* The *perimeter* is the outer boundary, or the distance around the outside edge of the burning or burned area. Also commonly called the *fire edge.* Obviously, the perimeter continues to grow until the fire is suppressed.

- *Heel.* The *heel,* or *rear,* of a wildland fire is the side opposite the head. Because the heel usually burns downhill or against the wind, it burns slowly and quietly and is easier to control than the head.

- *Flanks.* The *flanks* are the sides of a wildland fire, roughly parallel to the main direction of fire spread. The right and left flanks separate the head from the heel. It is from these flanks that fingers can form. A shift in wind direction can change a flank into a head.

- *Islands.* Patches of unburned fuel inside the fire perimeter are called *islands.* Because they are unburned potential fuels for more fire, they must be patrolled frequently and checked for spot fires (see following paragraph).

- *Spot fire.* *Spot fires* are caused by flying sparks or embers landing outside the main fire. Spot fires present a hazard to personnel (and equipment) working on the main fire because they could become trapped between the two fires. Spot fires must be extinguished quickly or they will form a new head and continue to grow in size.

- *Green.* The area of unburned fuels next to the involved area is called the *green.* While the term refers to the color of some of the fuels in the area, the "green" may not be green at all. The green does not necessarily indicate a safe area. It is simply the opposite of the burned area (the black) (see following paragraph) **(Figure 15.77).**

- *Black.* The opposite of the green — the *black* — is the area in which the fire has consumed or "blackened" the fuels. The black can sometimes be a relatively safe area during a fire but can be a very hot and smoky environment.

Figure 15.77 Areas of unburned vegetation are called *the green*. Areas that have been recently burned are called *the black*.

Wildland Protective Clothing and Equipment

Before attacking a wildland fire, firefighters need to be wearing wildland fire protective clothing because standard structural turn-out clothing is inappropriate and can even be dangerous. When fighting a wildland fire, firefighters should wear PPE that meets the requirements of NFPA 1977, *Standard on Protective Clothing and Equipment for Wildland Fire Fighting*. NFPA 1500, *Standard on Fire Department Occupational Safety and Health Program*, specifies the minimum PPE for firefighters to participate in wildland fire fighting **(Figure 15.78)**. This standard requires firefighters to be equipped with the following:

- Helmet with eye protection and neck shroud

- Flame retardant shirt and pants (or one-piece jumpsuit)

- Protective footwear (sturdy boots without steel toes)

- Gloves

- Fire shelter (in crush-resistive case)

In addition, most wildland fire agencies also provide firefighters with a canteen or bottled water and a backpack or web belt for carrying extra gear. Firefighters carry fusees, extra food, water, clean socks, and other items in these packs.

Figure 15.78 A wildland firefighter in full PPE.

Attacking the Fire

The methods used to attack wildland fires revolve around perimeter control. The control line may be established at the burning edge of the fire, next to it, or at a considerable distance away **(Figure 15.79, p. 806)**. The objective is to establish a control line that completely encircles the fire with all the fuel inside rendered harmless.

The *direct* and *indirect* approaches are the two most basic methods for attacking wildland fires. A *direct attack* is action taken directly against the flames at its edge or closely parallel to it. The *indirect attack* is used at varying distances from the advancing fire. Starting from an anchor point (road, highway, body of water, previous burn), a line is constructed some distance from the fire's edge and the unburned intervening fuel is burned out. This method is generally used against fires that are either *too hot, too fast*, or *too big* for a direct attack **(Figure 15.80, p. 806)**.

Figure 15.79 Wildland firefighters establish a control line in order to contain a slow moving wildland fire.

Figure 15.80 Burning out is one form of indirect attack. The one shown is anchored on a roadway.

Direct Attack (Wildland) — Operation where action is taken directly on burning fuels by applying an extinguishing agent to the edge of the fire or close to it.

Because a wildland fire is constantly changing, it is quite possible to begin with one attack method and switch to another. As with any other type fire, size-up must be continued throughout a wildland fire so that these adjustments can be made when required. More detailed information on attacking a ground cover fire can be located in **Skill Sheet 15-I-8**.

Ten Standard Fire Fighting Orders

The fire behavior characteristics listed earlier are not the only fire conditions that are potentially dangerous to fire personnel, just the most common ones. Other situations can put firefighters at risk. Studies of firefighter deaths led to the development of the Ten Standard Fire Fighting Orders. In every case where a firefighter was killed while fighting a wildland fire, it was shown that one or more of the Ten Standard Orders had been ignored. Violating one or more of these orders may result in firefighter deaths.

The orders are guidelines that help firefighters identify and avoid high-risk situations. Every firefighter should know and follow them. Being able to recite them is commendable, but putting them into practice is more important. Each order should be considered separately so firefighters will recognize when it applies during a fire and respond appropriately.

Fire Orders

1. Keep informed on fire weather conditions and forecasts.
2. Know what the fire is doing at all times.
3. Base all actions on current and expected behavior of the fire.
4. Identify escape routes and safety zones, and make them known.
5. Post lookouts when there is possible danger.
6. Be alert, keep calm, think clearly and act decisively.
7. Maintain prompt communications with your forces, your supervisor, and adjoining forces.
8. Give clear instructions and ensure that they are understood.
9. Maintain control of your forces at all times.
10. Fight fire aggressively, providing for safety first.

Summary

Attacking fires early in their development is an important aspect of a successful fire fighting operation. Likewise, selecting and applying the most effective fire attack strategy and tactics are also important. Failing to do any of these things can result in a fire growing out of control, an increase in fire damage and loss, and possibly in firefighter injuries. Firefighters need to know how to use the fire fighting tools and techniques adopted by their departments. They need to know how to safely and effectively attack and extinguish structure fires, vehicle fires, refuse fires, and wildland fires.

Fire Fighter I

1. What initial actions should firefighters take when suppressing a structural fire?

2. What are the differences among a direct attack, an indirect attack, and a combination attack?

3. When are master streams usually deployed?

4. What are three guidelines for electrical emergencies?

5. What are the parts of a wildland fire?

Fire Fighter II

6. What are three factors to consider in hoseline selection?

7. When would combination fog nozzles be used?

8. What is a boiling liquid expanding vapor explosion (BLEVE)?

9. What are the major differences in fires in vehicles transporting flammable fuels and fires in storage facilities?

10. What questions does the company officer of the first-due engine company ask when conducting a rapid initial assessment of the situation?

Step 1: Confirm order with officer to attack fire.

Step 2: Don all PPE and SCBA prior to entering the hot zone.

Step 3: Check nozzle pattern prior to approaching structure.

Step 4: Advance the hose near the structure with all firefighters on same side of hose.

Step 5: Extinguish the fire with an indirect pattern through a window or door by directing fire stream at ceiling and moving stream back and forth until fire is extinguished.

Step 6: Maintain situational awareness.

Step 7: Inform officer that fire is extinguished. Wear full PPE and SCBA protection until clear of hot zone.

Step 1: Confirm order with officer to deploy master stream device.

Step 2: Remove needed tools and appliances from the apparatus.

Step 3: With assistance, remove the monitor from the apparatus, using proper lifting techniques.

Step 4: Carry the monitor unit to the set-up area.

Step 5: Position the monitor on a solid, level surface.

Step 6: Secure monitor according to manufacturer's guidelines.

Step 7: Adjust the nozzle to the proper elevation.

Step 8: Secure the anchor lock, if applicable.

Step 9: Extend hoseline to the monitor.

Step 10: Connect the hoselines to the monitor unit.

Step 11: Tighten the swivel couplings using spanner wrenches.

Step 12: Check the tip size, ensuring proper tip for situation, or select desired fog pattern stream.

Step 13: Signal the pumper driver/operator to charge the line.

Step 14: Steady the monitor.

Step 15: Adjust the direction of water flow as necessary.

Step 16: Operate master stream device by aiming stream in correct direction and hitting designated target.

Step 1: Confirm order with officer to turn off utilities.

Step 2: Locate and shut off electricity at main service panel.

Step 3: Locate natural gas meter and/or LPG/CNG storage tank/cylinder and shut off.

Step 4: Locate water meter box and shut off water meter.

Step 5: Report to officer completion of assigned task to officer.

Skill SHEETS

15-1-4

Attack a structure fire (above, below, and grade level) — Interior attack.

Grade Level Fire Attack

Step 1: Confirm order with officer to attack fire.

Step 2: Prior to entry, check nozzle pattern and bleed air from hoseline.

Step 3: Size up environment for hazards.

Step 4: Extinguish burning fascia, boxed cornices, or other doorway overhangs as necessary before entering.

Step 5: Advance hoseline into the structure.

Step 6: Maintain situational awareness.

Step 7: Extinguish fire with a direct, indirect, or combination attack as directed by officer.

Step 8: Report to officer completion of assigned task.

15-I-4 continued
Attack a structure fire (above, below, and grade level) — Interior attack.

Skill SHEETS

Above Grade Fire Attack

Step 1: Confirm order with officer to attack fire.

Step 2: Prior to entry, check nozzle pattern and bleed air from hoseline.

Step 3: Size up environment for hazards.

Step 4: Extinguish burning fascia, boxed cornices, or other doorway overhangs as necessary before entering.

Step 5: Advance hoseline into the structure.

Step 7: Maintain situational awareness.

Step 8: Extinguish fire with a direct, indirect, or combination attack as directed by officer.

Step 9: Report to officercompletion of assigned task.

Step 6: Advance hoseline up stairwell to fire floor. If possible, lay extra hoseline in stairwell above fire floor.

NOTE: If fire is in high-rise or multistory commercial structure, fire attack may be from a standpipe connection.

**SHEETS**

15-I-4 continued
Attack a structure fire (above, below, and grade level) — Interior attack.

Below Grade Fire Attack

Step 5: Advance hoseline into the structure.

Step 6: Advance hoseline down stairwell into the basement.

Step 1: Confirm order with officer to attack fire.

Step 2: Prior to entry, check nozzle pattern and bleed air from hoseline.

Step 3: Size up environment for hazards.

Step 4: Extinguish burning fascia, boxed cornices, or other doorway overhangs as necessary before entering. Ventilate the basement before entry.

Step 7: Maintain situational awareness.

Step 8: Extinguish fire with a direct, indirect, or combination attack as directed by officer.

Step 9: Report to officer completion of assigned task.

Step 1: Confirm order with officer to attack passenger vehicle fire.

Step 2: Lay out attack line for fire attack, while using appropriate personal protective clothing and SCBA.

Step 3: Charge attack line.

Step 4: Advance attack line to vehicle from upwind and uphill (if possible). Use a fog pattern for protection.

Step 5: Extinguish any fire under vehicle or in line of approach using a straight stream.

NOTE: Fire attack should begin with passenger compartment if on fire. If not, attack fire in engine compartment first.

Step 6: Extinguish fire in passenger compartment.

Step 7: Extinguish fire in engine compartment.

Step 8: Extinguish fire in trunk.

Step 9: Overhaul hidden and smoldering fires.

Step 10: Report to officer completion of task to officer or supervisor.

Step 1: Confirm order with officer to extinguish fire.

Step 2: Set the nozzle flow to a straight stream.

Step 3: Open the nozzle fully, briefly, and aim stream to side to test fog pattern and expel air.

Step 4: Size up environment for hazards.

Step 5: Advance to the trash container from uphill and upwind.

Step 6: Cool outside of container and any exposures.

Step 8: Perform overhaul.

Firefighter #2: Break up material and probe with pike pole for hot spots.

Firefighter #1: Extinguish hot spots.

Step 7: Attack the fire with a medium fog pattern until it is knocked down.

Step 1: Confirm order with officer to attack fire.

Step 2: Size up environment for hazards.

Step 3: Check nozzle pattern and bleed air from hoseline.

Step 4: Check for threat to exposures and cool as necessary.

Step 5: Advance to position to make fire attack.

Step 6: Extinguish fire with straight stream.

Step 7: Overhaul debris using pike pole or trash hook.

Step 8: Report to officer completion of assigned task.

Step 1: Confirm order with officer to attack fire.

Step 2: Size up environment for hazards.

Step 3: Position at perimeter of hot zone and approach from the burned area (black).

Step 4: Approach flame edge and apply water with handline or extinguisher or use hand tools.

Step 5: Maintain situational awareness.

Step 6: Extinguish fire, while monitoring weather, fire, and smoke conditions.

Step 7: Mop up hot spots.

Step 8: Exit hazard area to safe zone.

Step 9: Report to officer completion of assigned task.

Step 1: Confirm order with officer to extinguish fire.

Step 2: Size up incident scene for hazards.

Step 3: Verify foam type and concentration are appropriate for fuel and fire conditions.

Step 4: Verify attack line is functioning and ready for attack.

Step 5: Extend hoseline to point of fire attack, upwind and uphill.

Step 6: Extinguish fire by applying foam solution as directed.

Step 7: Maintain situational awareness.

Step 8: Report to officer completion of assigned task.

15-II-2
Control a pressurized flammable gas container fire.

Step 1: Confirm order with officer to extinguish fire.

Step 2: Size up incident scene for hazards.

Step 3: Prepare hoselines and nozzles for fire attack.

Step 4: Cool cylinder or storage tank with a straight stream.

Step 5: Extend hoselines to isolate control valve, upwind and uphill.

NOTE: If firefighters are unable to push flame away from the valve, the attack team should withdraw immediately to a safe location and continue to cool the container.

Step 6: Maintain situational awareness.

Step 7: Close control valve completely.

Step 8: Cool container from safe distance.

Step 9: Report to officer completion of assigned task.

15-II-3
Establish Incident Command and coordinate interior attack of a structure fire.

Skill SHEETS

Step 1: Confirm order with officer.

Step 2: Size up incident scene on arrival.

Step 3: Transmit initial report over radio.

Step 4: Establish Incident Command.

Step 5: Identify incident objectives and strategies.

Step 6: Assign available resources to tasks.

Step 7: Request additional resources as required.

Step 8: Monitor progress of assignments.

Step 9: Maintain situational awareness of incident by evaluating fire and structural conditions.

Step 10: Provide briefing to senior officer who is assuming Command.

Chapter Contents

Fire Detection, Alarm, and Suppression Systems

This chapter provides information that addresses the following job performance requirements of NFPA® 1001, *Standard for Fire Fighter Professional Qualifications (2008)*:

NFPA® 1001 references for Chapter 16:

5.3.14

6.5.3

Chapter Objectives

Fire Fighter I Objectives

1. List functions of fire detection, alarm, and suppression systems.

2. Discuss general automatic sprinkler protection and types of coverage.

3. Describe control valves and operating valves used in sprinkler systems.

4. Describe major applications of sprinkler systems.

5. Discuss operations at fires in protected properties.

6. Operate a sprinkler system control valve.

7. Manually stop the flow of water from a sprinkler.

8. Connect hoseline to a sprinkler system FDC. *

Fire Fighter II Objectives

1. Describe types of heat detectors.

2. Describe types of smoke detectors/alarms.

3. Explain how flame detectors and fire-gas detectors operate.

4. Discuss combination detectors and indicating devices.

5. Describe types of automatic alarm systems.

6. Discuss supervising fire alarm systems and auxiliary services.

7. Describe the operation of an automatic fire sprinkler system.

8. Discuss water supply for sprinkler systems.

9. Describe major applications of sprinkler systems.

Fire Detection, Alarm, and Suppression Systems

Case History

One of the most effective fire safety tools is the automatic sprinkler. According to a 2005 report by the National Fire Protection Association, in 97% of the incidents where wet-pipe sprinklers activated, the fire was contained by ten sprinklers or less. When sprinklers were effective in controlling the fire – which is most of the time – 83% of the fires were controlled by two sprinklers or less. The impact a working sprinkler has on fire loss is significant. In residential structures such as single-family homes and apartments, the property damage from a fire was half that in a structure with sprinklers compared a structure without sprinklers.

The most important statistic deals with life safety. The number of fire deaths in structures protected by automatic sprinklers is 86% lower than those without sprinklers. This same level of protection also impacts the number of firefighter deaths. In 2006, not a single firefighter in the United States died at a fire in a structure protected by working automatic sprinklers.

In summary, automatic fire sprinklers are a proven technology. The use of sprinklers reduces property damage when a fire occurs and dramatically decreases the chances a resident or firefighter will die in a fire.

Source: National Fire Protection Association U.S. Experience with Sprinklers and Other Fire Extinguishing Equipment

There are a number of reasons for installing fire detection, alarm, and suppression systems in buildings and other properties. Each of these systems is designed to fulfill specific needs. The following are recognized functions:

- To notify occupants of a facility to take necessary evasive action to escape the dangers of a hostile fire

- To summon organized assistance to initiate or to assist in fire control activities

- To initiate automatic fire control and suppression systems and to sound an alarm

- To supervise fire control and suppression systems to ensure that operational status is maintained

- To initiate a wide variety of auxiliary functions involving environmental, utility, and process controls (including control of elevators)

Individual fire detection, alarm, and suppression systems may incorporate one or all of these features. Such systems may include components that operate mechanically, hydraulically, pneumatically, or electrically, but most state-of-the-art systems operate electronically.

Despite advances in other forms of fixed fire suppression systems, automatic sprinklers remain the most reliable form for commercial, industrial, institutional, residential, and other occupancies. It is clear that many fires controlled by sprinklers

Automatic Sprinkler System — System of water pipes, discharge nozzles, and control valves designed to activate during fires by automatically discharging enough water to control or extinguish a fire. Also called sprinkler system.

Automatic Suppression Systems — Sprinkler, standpipe, carbon dioxide, and halogenated systems, as well as fire pumps, dry chemical agents and their systems, foam extinguishers, and combustible metal agents that sense heat, smoke, or gas and activate automatically.

result in less business interruption and less water damage than those that have to be extinguished by traditional fire-attack methods. In fact, data compiled by FM Global Research (formerly Factory Mutual) indicates that about 70 percent of all fires are controlled by the activation of five or fewer sprinklers.

According to NFPA 1001, those qualified at the Firefighter I level must know the following about these systems:

- Operations at properties protected by automatic sprinkler systems
- How to stop the flow of water from an individual sprinkler
- How to identify the main control valve on an automatic sprinkler system

The standard also requires them to be able to perform the following:

- Stop the flow of water from an individual sprinkler using wedges or stoppers
- Operate the main control valve on an automatic sprinkler system

The standard requires those qualified at the Firefighter II level to also know:

- Fundamentals of fire suppression and detection systems

At the Firefighter II level they must also be able to:

- Identify the components of fire suppression and detection systems

The first part of this chapter discusses the most common types of fire detection and alarm systems and devices in use in North America. The second part of the chapter describes automatic sprinkler systems. Factors to consider during fires at protected properties are also given.

Types of Alarm Systems

The most basic alarm system is designed to only be initiated manually – that is, by pulling a handle **(Figure 16.1)**. While these systems are properly termed *protected premises fire alarm systems*, they are more commonly called *local warning systems*. These systems are installed in some small school buildings and other public properties. In these systems, the signal only alerts building occupants of the need to evacuate the premises; it does *not* notify the fire department. Therefore, when a local alarm system is activated, it is still necessary for someone to dial 9-1-1 to alert the fire department.

A wide variety of optional features can expand the capabilities of a local warning system. For example, automatic fire detection devices may be added to enable the system to detect the presence of a fire and to initiate a signal. These features are discussed in the following sections.

There are four basic types of automatic alarm-initiating devices. They are designed to detect heat, smoke, fire gases, or flame. The following sections describe the most common types of devices in use.

Heat Detectors

While there are several different designs of heat detection devices, there are just two basic types: fixed-temperature devices and rate-of-rise detectors.

Fixed-Temperature Heat Detectors

Systems using fixed-temperature type heat detectors are relatively inexpensive compared to other types of systems. They are also least prone to false activations.

Protected Premises Alarm System — (1) Alarm system that alert and notifies only occupants on the premises of the existence of a fire so that they can safely exit the building and call the fire department. If a response by a public safety agency (police or fire department) is required, an occupant hearing the alarm must notify the agency. (2) Combination of alarm components designed to detect a fire and transmit an alarm on the immediate premises.

Figure 16.1 Operating a local alarm system alerts occupants of the building but does not notify the fire department.

Depending on where they are installed, however, fixed-temperature heat detectors can also be the slowest to activate of all the various types of alarm-initiating devices.

Fixed-temperature heat detectors activate when they are heated to the temperature for which they are rated. If the ambient temperature in the room where they are installed is relatively low — below freezing, for example — a fire in the room would burn undetected until it raised the temperature of a heat detector to its rated activation temperature. Depending upon the size of the room, a fire could burn for quite some time before activating a fixed-temperature heat detector.

Because heat rises, heat detectors are installed in the highest portions of a room, usually on the ceiling **(Figure 16.2)**. The heat detectors installed should have an activation temperature rating slightly above the highest ceiling temperatures normally expected in that space. Heat detectors rated at 135°F–174°F (58°–79°C) are common for living spaces. Attics and other areas subject to elevated temperatures may have detectors rated at 200°F (94°C) or more.

The various types of fixed-temperature devices discussed in this section activate by one or more of three mechanisms:

- Expansion of heated material (fusible link; bimetallic)
- Melting of heated material (frangible bulb)
- Changes in resistance of heated material (continuous line)

Fusible devices/frangible bulbs. While these two devices are more commonly associated with automatic sprinklers, they are also used in fire detection and signaling systems. The operating principles of these devices are identical to the fusible links and frangible bulbs used with automatic sprinklers; only their applications differ (see Sprinklers section under Automatic Sprinkler Systems).

A *fusible device* is normally held in place by a solder with a known melting (fusing) temperature. Under normal conditions, the device holds a spring-operated contact inside the detector in the open position **(Figure 16.3)**. When a fire raises the ambient temperature to the fusing temperature of the device, the solder melts, allowing

Ambient Temperature — Temperature of the surrounding environment.

Fixed-Temperature Heat Detector — Temperature-sensitive device that senses temperature changes and sounds an alarm at a specific point, usually 135°F (57°C) or higher.

Frangible Bulb — Small glass vial fitted into the discharge orifice of a fire sprinkler. The glass vial is partly filled with a liquid that expands as heat builds up. At a predetermined temperature, vapor pressure causes the glass bulb to break, causing water to flow.

Fusible Device — (1) Connecting link device that fuses or melts when exposed to heat. Used in sprinklers, fire doors, dampers, and ventilators. (2) Two-piece link held together with a metal that melts or fuses at a specific temperature. Also known as Fusible Link.

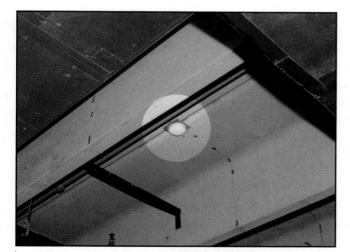

Figure 16.2 A fixed-temperature heat detector installed at the highest point of the ceiling.

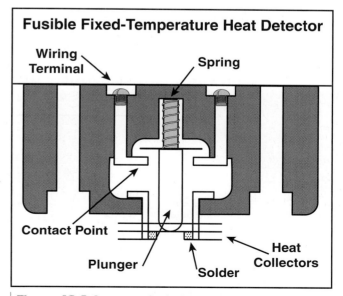

Figure 16.3 Cutaway of a fusible heat detector.

the spring to close the contact points. This action completes the alarm circuit, which initiates an alarm signal. Some of these detectors may be restored by replacing the fusible device; others require the entire heat detector to be replaced.

A *frangible bulb* in a detection device holds electrical contacts apart, much in the way that a fusible device does. The little glass vial (frangible bulb) contains a liquid with a small air bubble. The bulb is designed to break when the liquid is heated to a predetermined temperature. As the liquid in the bulb is heated, it expands and absorbs the air bubble. When the rated temperature is reached, the bulb fractures and falls out, and the contacts complete the circuit to initiate an alarm. In order to restore the system, the entire detector must be replaced. While detectors of this type are still in service, their manufacture has been discontinued.

Continuous line detector. Most of the detectors described in this chapter are of the *spot* type; that is, they are designed to detect heat only in a relatively small area surrounding the specific spot where they are installed. However, *continuous line detection devices* can detect heat over a linear area parallel to the detector.

One such device consists of a cable with a conductive metal inner core sheathed with stainless steel tubing **(Figure 16.4)**. The inner core and the sheath are separated by an electrically insulating semiconductor material that keeps them from touching but allows a small amount of current to flow between the two. This insulation loses some of its electrical resistance capabilities at a predetermined temperature anywhere along the line. When this happens, the current flow between the two components increases, thus initiating an alarm signal through the system control unit. This type of detection device restores itself when the ambient temperature is reduced.

Another type uses two insulated wires with an outer covering. When the rated temperature is reached, the insulation melts and allows the two wires to touch. This completes the circuit and initiates an alarm signal through the system control unit **(Figure 16.5)**. To restore this type of line detector, the fused portion of the wires must be cut out and replaced with new wire.

Bimetallic detector. One type of bimetallic detector uses two metals that have different thermal expansion characteristics. Thin strips of the metals are bonded together, and one or both ends of the strips are attached to the alarm circuit. When heated, one metal expands faster than the other, causing the strip to arch or bend. The deflection of the strip either makes or breaks contact in the alarm circuit, initiating an alarm signal through the system control unit. Another type of bimetallic detector utilizes a snap disk and microswitch **(Figure 16.6)**. Most bimetallic detectors will reset automatically when cooled. After a fire, however, they need to be checked to ensure that they were not damaged.

Bimetallic — Strip or disk composed of two different metals that are bonded together; used in heat detection equipment.

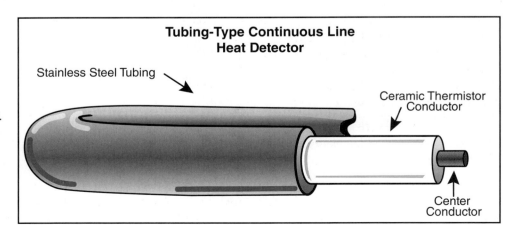

Figure 16.4 Tubing-type continuous line heat detector.

Tubing-Type Continuous Line Heat Detector

Stainless Steel Tubing

Ceramic Thermistor Conductor

Center Conductor

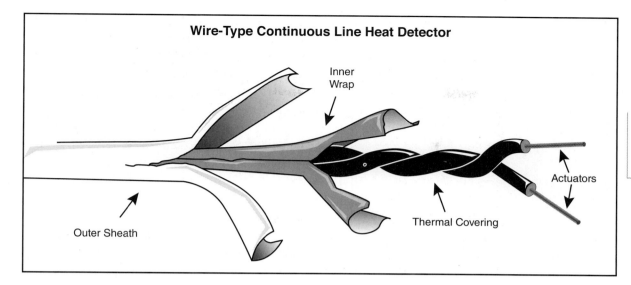

Wire-Type Continuous Line Heat Detector

Inner Wrap

Outer Sheath

Thermal Covering

Actuators

Figure 16.5 Wire-type continuous line heat detector.

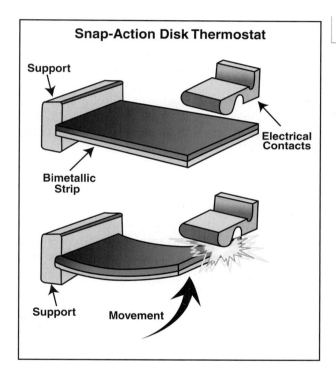

Snap-Action Disk Thermostat

Support

Electrical Contacts

Bimetallic Strip

Support

Movement

Figure 16.6 Bimetallic heat detector.

Rate-of-Rise Heat Detectors

A *rate-of-rise heat detector* operates on the assumption that the temperature in a room will increase faster from a fire than from normal atmospheric heating. Typically, rate-of-rise heat detectors are designed to initiate a signal when the rise in temperature exceeds 12° to 15°F (-11°C to -9°C) in one minute. Because the alarm is initiated by a sudden rise in temperature regardless of the initial temperature, an alarm can be initiated at a room temperature far below that required for initiating a fixed-temperature device.

Rate-of-Rise Heat Detector — Temperature-sensitive device that sounds an alarm when the temperature changes at a preset value, such as -11°C to -9°C per minute.

Most rate-of-rise heat detectors are reliable and not subject to false activations. If they are not properly installed, however, they can be activated under nonfire conditions. For example, if a rate-of-rise heat detector is installed just inside an exterior door in an air-conditioned building, opening the door on a hot day can initiate an alarm because of the influx of heated air. Relocating the detector farther from the doorway should alleviate the problem.

There are several different types of rate-of-rise heat detectors in use; all automatically reset if they are undamaged. The different types are discussed in more detail in the following paragraphs.

Pneumatic rate-of-rise spot detector. A pneumatic spot detector is the most common type of rate-of-rise detector in use (**Figures 16.7 a** and **b, p. 830**). It consists of a small dome-shaped air chamber with a flexible metal diaphragm in the base. A small

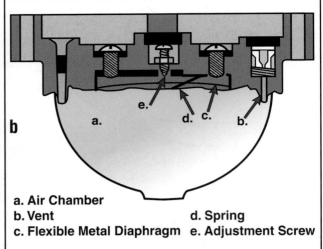

a. Air Chamber
b. Vent   d. Spring
c. Flexible Metal Diaphragm   e. Adjustment Screw

Figures 16.7 a and b Pneumatic rate-of-rise spot detector, (a) exterior and (b) interior components.

metering hole allows air to enter and exit the chamber during the normal rise and fall of atmospheric temperature and barometric pressure. In the heat of a fire, the air within the chamber expands faster than it can escape through the metering hole. This expansion causes the pressure within the chamber to increase, forcing the metal diaphragm against contact points in the alarm circuit and initiating an alarm signal. This type of detector is often combined in one unit that also has fixed-temperature capability.

Pneumatic rate-of-rise line detector. The spot detector monitors a small area surrounding its location; a line detector can monitor large areas. A *line detector* consists of a system of tubing arranged over a wide area of coverage (**Figure 16.8**). The space inside the tubing acts as the air chamber described in the preceding paragraph on spot detectors. The line detector also contains a diaphragm and is vented. When any portion of the tubing is subjected to a rapid increase in temperature, the detector functions in the same manner as the spot detector. The tubing in this system must be limited to about 1,000 feet (300 m) in length. The tubing is normally arranged in rows that are not more than 30 feet (10 m) apart and 15 feet (5 m) from walls.

Rate-compensated detector. This detector is designed for use in areas normally subject to regular temperature changes that are slower than those under fire conditions. The detector consists of an outer metallic sleeve that encases two bowed struts; the struts have a slower expansion rate than the sleeve (**Figure 16.9**). The bowed struts have electrical contacts on them. In the normal position, these contacts do not come together. When the detector is heated rapidly, the outer sleeve expands in length. This expansion reduces the tension on the inner strips and allows the contacts to come together, thus initiating an alarm signal through the system control unit.

If the rate of temperature rise is fairly slow, such as 5°F (-15°C) per minute, the sleeve expands slowly enough to maintain tension on the inner strips. This tension prevents unnecessary system activations. However, regardless of the rate of temperature increase, when the surrounding temperature reaches a predetermined point, an alarm signal will be initiated.

Thermoelectric detector. This rate-of-rise detector operates on the principle that when two wires of dissimilar metals are twisted together and heated at one end, an electrical current is generated at the other end. The rate at which the wires are heated

Rate-Compensated Heat Detector — Temperature-sensitive device that sounds an alarm at a preset temperature, regardless of how fast temperatures change.

Figure 16.8 Pneumatic rate-of-rise line detector.

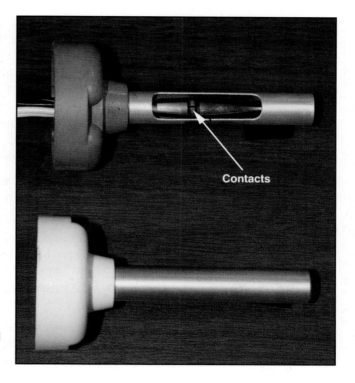

Contacts

Figure 16.9 Stages of operation of a rate compensated detector.

determines the amount of current that is generated. These detectors are designed to "bleed off" or dissipate small amounts of current, which reduces the chance of a small temperature change activating an alarm unnecessarily. Rapid changes in temperature result in larger amounts of current flowing and activation of the alarm system.

Smoke Detectors/Alarms

Before discussing the various types of smoke detectors, a distinction must be made between devices that merely detect the presence of smoke or other products of combustion and those that both detect and sound an alarm. In most cases, the devices installed in nonresidential and large multi-family residential occupancies are only capable of detecting the presence of smoke and must transmit a signal to another device that sounds the alarm. These units are called *smoke detectors* (**Figure 16.10**). The devices that are installed in single-family residences and smaller multi-family residential occupancies are self-contained units capable of both detecting the presence of smoke and sounding an alarm. These units are called *smoke alarms* (**Figure 16.11**).

Smoke Detector — Alarm-initiating device designed to actuate when visible or invisible products of combustion (other than fire gases) are present in the room or space where the unit is installed.

Smoke Alarm — a device designed to sound an alarm when the products of combustion are present in the room where the device is installed. The alarm is built into the device rather than being a separate system.

Figure 16.10 Smoke detectors that are typically found in residential occupancies may be single station battery operated or part of a monitored alarm system.

Figure 16.11 Typical smoke alarm commonly found in residences.

Because a smoke detector can respond to smoke or other products of combustion generated very early in the growth stage – well before sufficient heat is produced to initiate an alarm – it can initiate an alarm much more quickly than a heat detector can. For this reason, the smoke detector is the preferred type of detector in many types of occupancies. The two basic types of smoke detectors -- *photoelectric* and *ionization* -- are described in the following sections, along with a discussion of power sources for smoke alarms.

Photoelectric Smoke Detectors

A *photoelectric smoke detector,* sometimes called a *visible products-of-combustion detector*, uses a photoelectric cell coupled with a tiny light source. The photoelectric cell functions in two ways to detect smoke: *beam application* and *refractory application*.

The *beam application* type of smoke detector uses a beam of light that focuses across the area being monitored and onto a photoelectric cell. The cell constantly converts the beam into electrical current, which keeps a switch open. When smoke obscures the path of the light beam, the required amount of current is no longer produced, the switch closes, and an alarm signal is initiated (**Figure 16.12**).

The *refractory photocell* uses a light beam that passes through a small chamber at a point away from the light source. Normally, the light does not strike the photocell and no current is produced; this allows the switch to remain open. When smoke enters the chamber, it causes the light beam to be refracted (scattered) in all directions. A portion of the scattered light strikes the photocell, causing current to flow. This current causes the switch to close and transmits the alarm-initiation signal (**Figure 16.13**).

A photoelectric smoke detector works satisfactorily on all types of fires and automatically resets when the atmosphere is clear. Photoelectric smoke alarms are generally more sensitive to smoldering fires than are ionization detectors.

Photoelectric Smoke Detector — Type of smoke detector that uses a small light source, either an incandescent bulb or a light-emitting diode (LED), to detect smoke by shining light through the detector's chamber. Smoke particles reflect the light into a light-sensitive device called a photocell.

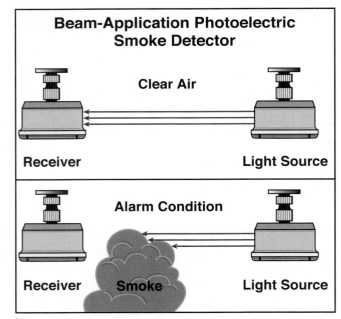

Figure 16.12 Operating principle of the beam-application smoke detector.

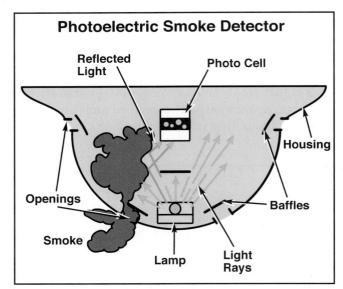

Figure 16.13 Operating principle of the refractory photoelectric smoke detector.

Ionization Smoke Detectors

During combustion, minute particles and aerosols too small to be seen by the naked eye are produced. These invisible products of combustion can be detected by devices that use a tiny amount of radioactive material (usually americium) to ionize air molecules as they enter a chamber within the detector. These ionized particles allow an electrical current to flow between negative and positive plates within the chamber. When the particulate products of combustion (smoke) enter the chamber, they attach themselves to electrically charged molecules of air (ions), making the air within the chamber less conductive. The decrease in current flowing between the plates transmits an alarm-initiating signal (**Figure 16.14**).

Ionization detectors respond satisfactorily to most fires but these types of detectors generally respond faster to flaming fires than to smoldering ones. They automatically reset when the atmosphere has cleared.

Ionization Detector — Type of smoke detector that uses a small amount of radioactive material to make the air within a sensing chamber conduct electricity.

Power Sources

Either batteries or household current can power residential smoke alarms. Battery-operated alarms offer the advantage of easy installation — a screwdriver and a few minutes are all that are needed. Battery-operated models are also independent of house power circuits and can therefore operate during power failures. This feature is especially important when the cause of the fire was a malfunction in the house wiring. Newer smoke alarms may come equipped with lithium batteries rated for a 10-year service life (**Figures 16.15 a and b**).

Firefighters should be aware of any state/province or local laws that deal with smoke alarms. Such legislation, in addition to specifying minimum installation requirements for given occupancies (including homes), may designate the power source to be used. Laws requiring hard-wired units were precipitated by statistics showing a growing lack of maintenance in battery-operated alarms (depleted batteries not being replaced). Consequently, many codes requiring alarms in newly constructed

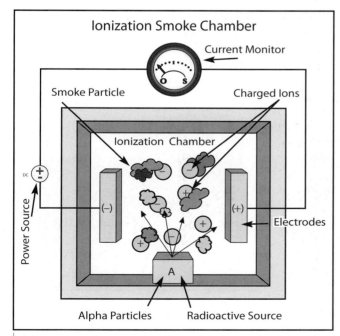

Figure 16.14 Operating principle of the ionization smoke detector.

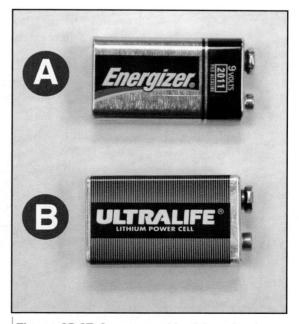

Figure 16.15 Common residential smoke alarms (top) one-year battery (bottom) lithium 10-year battery.

homes specify 110-volt, hard-wired units. Alarms powered by household current are usually more reliable. In some rural areas and areas with high thunderstorm occurrence, power failures may be more frequent and battery-operated units may be more reliable.

It is critical that the specific battery type recommended by the alarm's manufacturer be used for replacement. The batteries should be changed at least twice a year or more often if necessary. One way firefighters can get citizens to remember when to change smoke alarm batteries is by suggesting the change be made in the spring and fall at the same time clocks are reset for daylight savings time or returned to standard time.

Flame Detectors

There are three basic types of flame detectors (sometimes called *light detectors*):

- Those that detect light in the ultraviolet wave spectrum (UV detectors) **(Figure 16.16)**

- Those that detect light in the infrared wave spectrum (IR detectors) **(Figure 16.17)**

- Those that detect both types of light

While these types of detectors are among the most sensitive used to detect fires, they are also prone to being activated by nonfire conditions such as welding, sunlight, and other sources of bright light. To combat this problem, flame detectors are usually positioned in areas where these other light sources are unlikely. They are positioned so they have an unobstructed view of the protected area. If their line of sight is blocked by an opaque object, they will not activate.

Because some single-band IR detectors are sensitive to sunlight, they are usually installed in fully enclosed areas. To reduce the likelihood of false alarms, most IR detectors are designed to require the flickering motion of a flame to initiate an alarm.

Ultraviolet detectors are virtually insensitive to sunlight, so they can be used in areas not suitable for IR detectors. They are not suitable for areas where arc welding is done or where intense mercury-vapor lamps are used.

Flame Detectors — Detection and alarm devices used in some fire detection systems (generally in high-hazard areas) that detect light/flames in the ultraviolet wave spectrum (UV detectors) or detect light in the infrared wave spectrum (IR detectors). Also called Light Detectors.

Figure 16.16 UV flame detectors are designed to detect ultraviolet light waves.

Figure 16.17 Infrared flame detectors are extremely sensitive. *Courtesy of Detector Electronics Corp.*

Fire-Gas Detectors

When a fire burns in a confined space, it changes the composition of the atmosphere within the space. Depending on the fuel, some of the gases released by a fire may include the following:

- Water vapor (H_2O)
- Carbon dioxide (CO_2)
- Carbon monoxide (CO)
- Hydrogen chloride (HCl)
- Hydrogen cyanide (HCN)
- Hydrogen fluoride (HF)
- Hydrogen sulfide (H_2S)

Only water vapor, carbon dioxide, and carbon monoxide are released from all fires. Other gases released vary with the specific chemical makeup of the fuel. Therefore, it is only practical to monitor the levels of carbon dioxide and carbon monoxide for general fire detection purposes **(Figure 16.18)**. This type of detector will initiate an alarm signal somewhat faster than a heat detector but not as quickly as a smoke detector.

Of more importance than the speed of response is the fact that a fire-gas detector can be more discriminating than other types of detectors. A fire-gas detector can be designed to be sensitive only to the gases produced by specific types of hostile fires and to ignore those produced by friendly fires. This detector uses either semiconductors or catalytic elements to sense the gas and transmit the signal to initiate the alarm. Compared to the number of other types of detectors, few fire-gas detectors are in use.

Fire-Gas Detector — Device used to detect gases produced by a fire within a confined space.

Combination Detectors

Depending on the design of the system, various combinations of the previously described means of detection may be used in a single device. These combinations include fixed temperature/rate-of-rise heat detectors, combination heat/smoke detectors, and combination smoke/fire-gas detectors **(Figure 16.19)**. The different combinations make these detectors more versatile and more responsive to fire conditions.

Combination Detector — Alarm-initiating device capable of detecting an abnormal condition by more than one means. The most common combination detector is the fixed-temperature/rate-of-rise heat detector.

Figure 16.18 Fire-gas detectors are quicker than heat detectors for initiating an alarm. *Courtesy of RKI Instruments, Inc.*

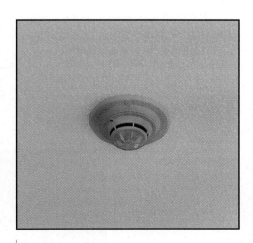

Figure 16.19 A combination heat/smoke detector is both versatile and responsive to fire conditions.

Indicating Devices

A variety of audible and visible alarm-indicating devices are also in use. Some produce a loud signal to attract attention in high-noise areas; some generate an electronic tone that is audible in almost any type of environment. Some systems employ bells, horns, or chimes; others use speakers that broadcast prerecorded evacuation instructions (**Figures 16.20 a-c**).

To accommodate special circumstances or populations, such as those who must wear hearing protection because of very high noise levels in their work areas, visual alarm indicators that employ a high-intensity clear strobe may be used. These indicators may be used singularly or in combination with other alarm devices (**Figure 16.21**). Appropriate strobe devices may also be used to meet the requirements of the Americans with Disabilities Act (ADA) in areas where there may be people with hearing impairments.

Americans with Disabilities Act (ADA) of 1990 - Public Law 101-336 — A federal statute (US) intended to remove barriers — physical and otherwise — that limit access by individuals with disabilities.

Fire Department Responses to Heat/Smoke Alarm Calls

When any of the systems just discussed initiates an alarm, a full structure fire response will usually be dispatched to that address. If there is no sign of fire or smoke when the first fire department unit arrives, a radio-equipped member of that company should be sent to the alarm system control panel to silence the local alarm bell or other alerting device while the investigation into the source of the alarm is being conducted. Silencing the alerting device reduces the level of noise at the scene and helps to calm any occupants who may be frightened by all of the commotion or agitated by the incessant ringing or other alarm signal. In most cases, the alerting device can be silenced with the push of a button or the flip of a switch at the control panel (**Figure 16.22**).

The system should *not* be reset or turned off until firefighters determine the cause of the alarm activation. The member assigned to silence the alerting device should remain at the control panel to monitor it during the investigation in case a subsequent alarm is initiated (**Figure 16.23**). If a subsequent alarm is initiated – especially one from a different location than the first – the IC should be notified immediately. Heat/smoke detector actuation may also cause fire doors in some residential buildings to close automatically. In that case, the doors should remain closed until the source of the alarm has been located.

Once the source of the alarm has been located and either neutralized or determined to be innocuous, such as a transient system malfunction, the system can be reset. If the system was damaged during the incident, it may have to be turned off and placed out of service. In that case, all occupants of the building, the building owner or manager, and the fire department communications/dispatch center must be notified that the system is inoperative.

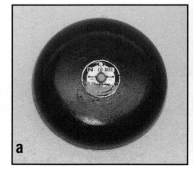
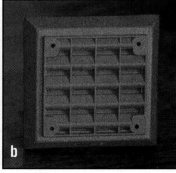

Figures 16.20 a-c
Typical alarm-indicating devices include (a) bells, (b) sirens, and (c) horns.

Figure 16.21 Strobe alarm-indicating devices are intended to alert occupants who have hearing deficiencies.

Figure 16.22 Most fire alarm systems can be silenced at the fire control panel.

Figure 16.23 A firefighter with a radio should monitor the alarm panel when determining the cause of the alarm activation.

Automatic Alarm Systems

Under some circumstances, insurance carriers may require occupancies to have a system that will transmit a signal to an off-site location for the purpose of summoning organized assistance in fighting a fire. This signal produces an automatic response upon activation of the local alarm at the protected premises. Various brands of alarm systems do this signaling with dedicated wire pairs, leased telephone lines, fiberoptic cable, or wireless communication links.

Fire alarm and supervisory systems may be installed to complement either wet-pipe or dry-pipe sprinkler systems (see Applications of Sprinkler Systems section). Devices capable of sensing a sudden increase or decrease in pressure can detect movement or flow of water; others are actuated by the actual movement of water within a pipe (**Figure 16.24, p. 838**). This movement would indicate that either a sprinkler had opened in response to a fire or that water was leaking from a broken pipe. To minimize false alarms but still maintain appropriate sensitivity, the detection device is set to respond to flow equal to that of a single sprinkler.

Auxiliary Systems

There are three basic types of auxiliary systems: the local energy system, the shunt system, and the parallel telephone system.

Local energy systems. Local energy systems are used only in those communities that are served by a municipal fire-alarm-box system. A *local energy system* is an auxiliary alarm system within an occupancy that is attached directly to a hardwired or radio-type municipal fire alarm master box (**Figure 16.25, p. 838**). When an alarm activates in the protected occupancy, it trips the alarm box to which it is attached and transmits an alarm to the fire alarm center. An alarm can be initi-

Automatic Alarm — (1) Alarm actuated by heat, gas, smoke, flame-sensing devices, or waterflow in a sprinkler system conveyed to local alarm bells or the fire station. (2) Alarm boxes that automatically transmit a coded signal to the fire station to give the location of the alarm box.

Figure 16.24 Sprinkler systems are equipped with waterflow alarm devices that activate when the system operates.

Figure 16.25 Local energy systems connect to municipal fire alarm systems.

Auxiliary Alarm System — System that connects the protected property with the fire department alarm communications center by a municipal master fire alarm box or over a dedicated telephone line.

ated by manual pull stations, automatic fire detection devices, or waterflow devices. Each community has its own requirements for local energy systems; some do not allow them at all.

Shunt systems. S*hunt systems* are those in which the municipal alarm circuit extends (is "shunted") into the protected property. When an alarm is initiated on the premises, whether manually or automatically, the alarm is instantly transmitted to the alarm center over the municipal system.

Parallel telephone systems. *Parallel telephone systems* do not interconnect with a municipal alarm circuit. Instead, they transmit an alarm from the protected property directly to the alarm center over a municipally controlled telephone circuit that serves no other purpose.

Remote Station Systems

Remote station systems are similar to auxiliary systems but are connected to the fire department telecommunication center directly or through an answering service by some means other than the municipal fire alarm box system. This connection can be done by leased telephone line or, where permitted, by a radio signal on a dedicated frequency. A remote station is common in localities that are not served by central station systems described later.

Remote Station Alarm System — System in which alarm signals from the protected premises are transmitted over a leased telephone line to a remote receiving station with a 24-hour staff; usually the municipal fire department's alarm communications center.

A remote station system may transmit either a coded or a noncoded signal. A noncoded system is only allowable where a single occupancy is protected by the system. Up to five facilities may be protected by one coded system. These facilities usually have a common connection to the remote station **(Figure 16.26)**. A remote station system must have the ability to transmit a trouble signal to the fire alarm center when the system becomes impaired. This type of system may not have local alarm capabilities if evacuation is not the desired action in the event of a fire.

Depending on local policy, the fire department may allow another entity such as the police department to monitor the remote station (**Figure 16.27**). This situation is particularly common in communities that have volunteer fire departments whose stations are not constantly staffed. In these cases, it is important that police dispatchers be appropriately trained in the actions to take upon receiving a fire-alarm signal.

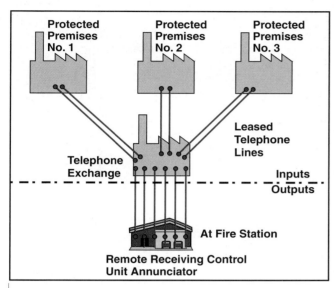

Figure 16.26 A remote receiver connects protected properties directly to the fire communications center.

Figure 16.27 Police communications personnel may monitor remote station systems.

Proprietary Systems

Proprietary systems are used to protect large commercial and industrial buildings, high-rise buildings, and groups of commonly owned buildings in a single location such as a college campus or industrial complex (**Figure 16.28, p. 840**). Each building or area has its own system that is wired into a common receiving point somewhere on the facility. The receiving point must be in a separate structure or in a part of the structure that is remote from any hazardous operations. The receiving station is constantly staffed by representatives of the occupant who are trained in system operation and the actions to take when an alarm is received (**Figure 16.29, p. 840**). The operator should be able to summon the fire department either through the system or by telephone.

> **Proprietary Alarm System —** Fire protection system owned and operated by the property owner.

Modern proprietary systems can be very complex with a wide range of capabilities. Some of these capabilities include the following:

- Transmitting coded-alarm and trouble signals
- Monitoring building-utility controls
- Monitoring elevator status
- Monitoring fire and smoke dampers
- Performing security functions

Central Station Systems

Central station systems are very similar to proprietary systems. The primary difference is that instead of having the alarm-receiving point monitored by the occupant's representative on the protected premises, the receiving point is at an off-site, contracted

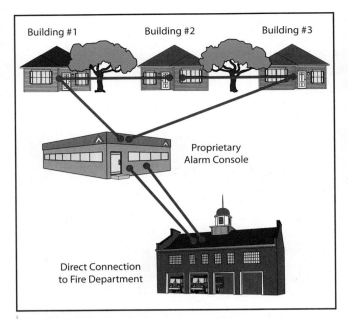

Figure 16.28 Components of a proprietary system.

Figure 16.29 A proprietary system is monitored by trained personnel. *Courtesy of Paul Ramirez.*

service point called a *central station* (**Figure 16.30**). Typically, the central station is an alarm company that contracts with individual customers. When an alarm is initiated at a contracting occupancy, central station employees take that information and initiate an appropriate emergency response. This response usually includes calling the fire department and representatives of the protected occupancy. The alarm systems at the protected property and the central station are most commonly connected by supervised telephone lines. All central station systems should comply with the requirements of NFPA® 72, *National Fire Alarm Code®.*

Supervising Fire Alarm Systems

Unlike ordinary electrical distribution systems that provide power to the public, fire alarm systems are designed to be self-supervising. This means that anytime the system is not operating normally, a distinct trouble signal is generated to attract attention to the system problem (**Figure 16.31**). This may happen when the system switches to battery power because of a utility power outage or when there is a break in a detector or notification circuit.

Many older systems operated with closed, supervised circuits in which a tiny current was constantly flowing. A detector would initiate a signal by closing contact points to create a short circuit. In the same way, if a break occurred in one of the detector circuits, a trouble signal would be initiated because the supervisory current would be interrupted.

Many newer systems incorporate some of the latest microelectronic components. Some of these systems have built-in microprocessors programmed to initiate an internal diagnostic test at specified intervals. The results are recorded on a printer or displayed on a computer screen (**Figure 16.32**). The same printer or screen is also used to present alarm information.

The sounds of the alarm and trouble signals may differ with each brand of system in use. Firefighters should familiarize themselves with both signals from each brand of system installed in their area.

Many fixed fire suppression systems depend on a signal from a manual pull station or from an automatic fire detection device to trigger the suppression system **(Figure 16.33)**. The control panel must be specifically listed by a testing laboratory for this purpose and all control circuits must be supervised. Depending on the application, a number of specific features may be incorporated into the alarm system. NFPA 12, *Standard on Carbon Dioxide Extinguishing Systems*, requires a predischarge alarm.

Predischarge Alarm — Alarm that sounds before a total flooding fire extinguishing system is about to discharge. This gives occupants the opportunity to leave the area.

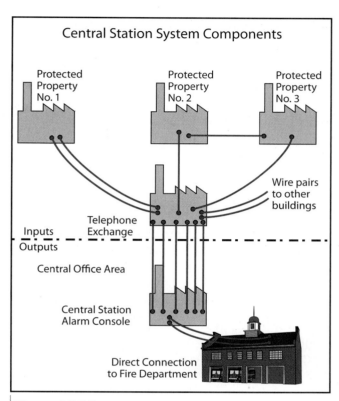

Figure 16.30 Components of a central station system.

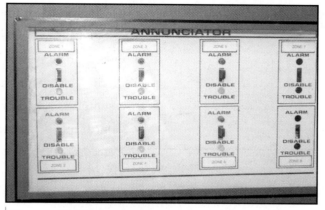

Figure 16.31 Supervised alarm systems generate a distinct trouble signal when a malfunction occurs.

Figure 16.32 Some self-testing systems periodically record test results on a printout.

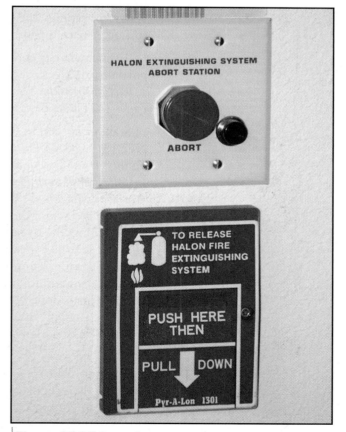

Figure 16.33 Some fire suppression systems are designed to operate when the alarm system is activated.

Auxiliary Services

Although the primary objective of a fire detection and alarm system is to save lives and reduce property loss in the event of a fire, technological improvements have made contemporary emergency signaling systems capable of much more. Systems that integrate process and environmental controls, security, and personnel-access controls are now common. The following are some of the other auxiliary services available:

- Shutting down or altering airflow in heating, ventilating, and air-conditioning (HVAC) systems for smoke control

- Closing smoke or fire-rated doors and dampers

- Facilitating evacuation by increasing air pressure in stairwells to exclude smoke

- Overriding elevator controls

- Monitoring operation of commercial incinerator management systems

- Monitoring refrigeration systems and cold-storage areas

- Controlling personnel access to hazardous process or storage areas

- Detecting combustible or toxic gases

Automatic Sprinkler Systems

Automatic sprinkler protection consists of a series of sprinklers (sometimes called *sprinkler heads*) arranged so that the system will automatically distribute sufficient quantities of water directly onto a fire to either extinguish it or hold it in check until firefighters arrive **(Figure 16.34)**. Water is supplied to the sprinklers through a system of piping. The sprinklers can either extend from exposed pipes or protrude through the ceiling or walls from hidden pipes.

There are two general types of sprinkler coverage: complete sprinkler coverage and partial sprinkler coverage. A *complete sprinkler system* protects the entire building. A *partial sprinkler system* protects only certain areas such as high-hazard areas, exit routes, or places designated by code or by the authority having jurisdiction.

Standards such as NFPA® 13, *Standard for the Installation of Sprinkler Systems,* NFPA® 13D, *Standard for the Installation of Sprinkler Systems in One- and Two-Family Dwellings and Manufactured Homes,* and NFPA® 13R, *Standard for the Installation of Sprinkler Systems in Residential Occupancies up to and Including Four Stories in Height,* are the primary guides used for installing sprinkler protection in most occupancies. These standards have requirements on the spacing of sprinklers in a building, the size of pipe to be used, the proper method of hanging the pipe, and all other details concerning the installation of a sprinkler system. These standards specify the minimum design area that should be used to calculate the system. This area is the maximum number of sprinklers that might be expected to activate **(Figure 16.35)**. This is done because installing the piping and other components required to adequately supply 500 or 1,000 operating sprinklers would be prohibitively expensive. Thus, the design of the system is based on the assumption that only a portion of the sprinklers will operate during a fire.

The automatic sprinkler and all component parts of the system should be listed by a nationally recognized testing laboratory such as Underwriters Laboratories Inc. or FM Global. Automatic sprinkler systems are recognized as the most reliable of all fire protection devices and it is essential for firefighters to understand the basic system and the operation of pipes and valves. Firefighters should also know the various applications of sprinkler systems along with their effects on life safety.

Figure 16.34 Automatic sprinkler systems are designed to contain or control the fire until firefighters arrive at the scene.

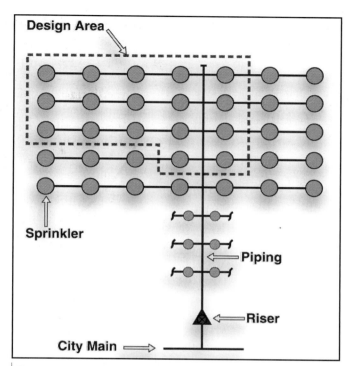

Figure 16.35 The automatic sprinkler system is designed to provide an adequate water supply to a specific number of sprinklers operating at the same time.

In general, reports reveal that only in rare instances do automatic sprinkler systems fail to operate. When failures are reported, the reason is rarely because of failure of the actual sprinklers. A sprinkler system may not perform properly because of the following:

- Partially or completely closed main water control valve
- Interruption to the municipal water supply
- Damaged or painted-over sprinklers
- Frozen or broken pipes
- Excess debris or sediment in the pipes
- Failure of a secondary water supply
- Tampering and vandalism
- Sprinklers obstructed by objects stacked too close **(Figure 16.36)**

Effects of Sprinkler Systems on Life Safety

The safety of building occupants is enhanced by the presence of a sprinkler system because it discharges water directly onto a fire while it is still relatively small. Because the fire is extinguished or controlled in the early growth stage, combustion products are limited. Sprinklers are also effective in the following situations:

- Preventing fire spread upwards in multistory buildings
- Protecting the lives of occupants in other parts of the building

There are also times when sprinklers alone are not as effective, such as in the following situations:

- Fires are too small to activate the sprinkler system.
- Smoke generation reaches occupants before the sprinkler system activates.
- Sleeping, intoxicated, or handicapped persons occupy the fire building.

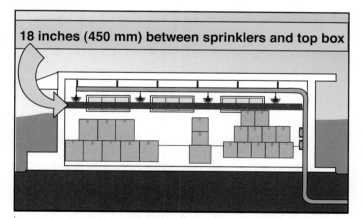

Figure 16.36 There should be at least 18 inches (450 mm) of clearance between sprinklers and stored materials.

Sprinkler System Fundamentals

The principal parts of an automatic sprinkler system are illustrated in **Figure 16.37**. The system starts with a water main and continues into the control valve. The *riser* is the vertical piping to which the sprinkler valve, one-way check valve, fire department connection (FDC), alarm valve, main drain, and other components are attached. The *feed main* is the pipe connecting the riser to the cross mains. The *cross mains* directly service a number of branch lines on which the sprinklers are installed. Cross mains extend past the last branch lines and are capped to facilitate flushing. System piping decreases in size from the riser outward. The entire system is supported by hangers and clamps.

Along with discussions of the various types of sprinklers, control valves, operating valves, and waterflow alarms, an explanation of how sprinkler systems are supplied with water is also given in the following sections. The various applications of sprinkler systems (dry-pipe, wet-pipe, preaction, deluge, and residential) are also described.

Riser — Vertical water pipe used to carry water for fire protection systems above ground such as a standpipe riser or sprinkler riser.

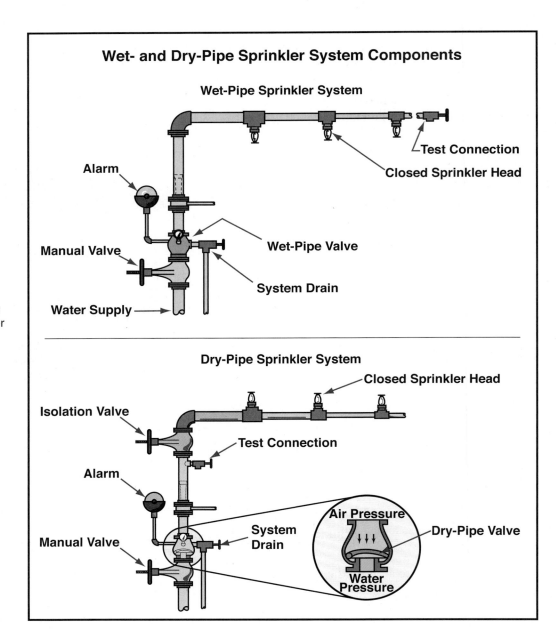

Figure 16.37 Components of wet- and dry-pipe sprinkler systems.

Sprinklers

Sprinklers discharge water after the release of a cap or plug that is activated by some heat-responsive element such as a fusible link **(Figure 16.38)**. This sprinkler may be thought of as a fixed-spray nozzle that is opened individually in response to heat. There are numerous types and designs of sprinklers.

Sprinklers are commonly identified by the temperature at which they are designed to operate. Usually this temperature is identified either by color-coding the sprinkler frame arms, by using different colored liquid in bulb-type sprinklers, or by stamping the temperature into the sprinkler itself (see **Table 16.1**).

Three of the most commonly used release mechanisms to activate sprinklers are fusible links, frangible bulbs, and chemical pellets. Each of these sprinkler mechanisms fuse or open in response to heat **(Figures 16.39 a–c, p. 846)**.

Fusible link. The design of a sprinkler using a fusible link involves a frame that is screwed into the sprinkler piping. Two levers press against the frame and a cap over the orifice from which the water flows. The fusible link holds the levers together until the link is melted by the heat of a fire, after which the water pressure pushes the levers and cap out of the way. Water then flows from the orifice and strikes the deflector attached to the frame. The deflector converts the standard ½-inch (13 mm) stream into water spray for more efficient extinguishment.

A quick-response mechanism was developed to enhance life safety. This specially designed fusible link offers increased surface area to collect the heat generated by a fire faster than a standard fusible-link sprinkler. This results in a faster opening of the sprinkler and quicker extinguishment of the fire.

Sprinkler — Waterflow device in a sprinkler system. The sprinkler consists of a threaded nipple that connects to the water pipe, a discharge orifice, a heat-actuated plug that drops out when a certain temperature is reached, and a deflector that creates a stream pattern suitable for fire control.

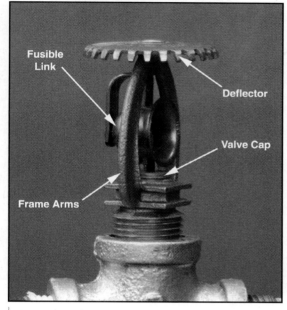

Figure 16.38 Components of a fusible link sprinkler.

Table 16.1
Sprinkler Temperature Ratings, Classifications and Color Coding

Color Coding	Temperature Classification	Temperature Rating °F	Temperature Rating °C
Uncolored or Black	Ordinary	135-170	57-77
White	Intermediate	175-225	79-107
Blue	High	250-300	121-149
Red	Extra High	325-375	163-191
Green	Very Extra High	400-475	204-246
Orange	Ultra High	500-575	260-302

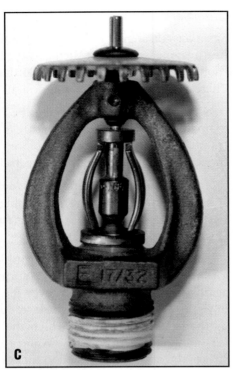

Figures 16.39 a-c Three types of sprinklers: (a) fusible-link, (b) frangible-bulb, and (c) chemical-pellet sprinkler.

Frangible bulb. Some sprinklers incorporate a small bulb filled with liquid and an air bubble to hold the orifice shut. The air bubble prevents false activations due to normal rise and fall of atmospheric temperature. In a fire, heat expands the liquid until the bubble is absorbed into the liquid. This increases the internal pressure until the bulb shatters at the proper temperature. The breaking temperature is regulated by the amount of liquid and the size of the bubble in the bulb. The liquid is color-coded to designate the designed breaking temperature (refer to **Table 16.1**). When the bulb shatters, the valve cap is released.

Chemical pellet. A pellet of solder, under compression, within a small cylinder melts at a predetermined temperature, allowing a plunger to move down and release the valve cap parts.

Sprinkler Position

Sprinklers are installed in three basic positions: pendant, upright, and sidewall. Sprinklers designed for installation in a particular position cannot be interchanged with those designed for use in other positions because they will not provide the proper spray pattern and coverage in a different position. There are also special-purpose sprinklers used in other applications.

Pendant. The *pendant* sprinkler, the most common type in use, extends down from the underside of the piping. This sprinkler sprays a stream of water downward into a deflector that breaks the stream into a hemispherical (umbrella-shaped) pattern (**Figure 16.40**).

Upright. The *upright* sprinkler is screwed into the top of the piping and discharges water into a solid deflector that breaks it into a hemispherical spray pattern that is redirected toward the floor. The standard upright sprinkler cannot be inverted for use in the hanging or pendant position because the sprinkler would produce an ineffective spray pattern if used in this manner (**Figure 16.41**).

Pendant Sprinkler — Automatic sprinkler designed for placement and operation with the head pointing downward from the piping.

Upright Sprinkler — Sprinkler that sits on top of the piping and sprays water against a solid deflector that breaks up the spray into a hemispherical pattern that is redirected toward the floor.

Figure 16.40 A pendant sprinkler in a protective cage may be found in storage areas or industrial occupancies.

Figure 16.41 An upright sprinkler with frangible-bulb activating mechanism.

Figure 16.42 Sidewall sprinklers are used in small rooms where the piping is mounted in the wall rather than the ceiling.

Sidewall. The *sidewall* sprinkler extends from the side of a pipe and is used in small rooms where the branch line runs along a wall. It has a special deflector that creates a fan-shaped pattern of water **(Figure 16.42)**.

Special-purpose. *Special-purpose* sprinklers are used in specific applications because of their unique characteristics. For example, special-purpose sprinklers with corrosive-resistant coatings are designed to be installed in areas with corrosive atmospheres. Other special-purpose sprinklers are designed for certain specific applications, such as being recessed in the ceiling to blend in with the room decor.

Sprinkler Storage

A storage cabinet to house spare sprinklers and a sprinkler wrench is usually installed near the sprinkler riser and main shut-off valve. Normally, these cabinets hold a minimum of six sprinklers and a sprinkler wrench in accordance with NFPA 13 and 13D **(Figure 16.43, p. 848)**. In many jurisdictions, the job of changing sprinklers must be performed by representatives of the building's occupants who are qualified to perform work on sprinkler systems. In other jurisdictions, firefighters are allowed to replace fused or damaged sprinklers to restore the system to service sooner. Firefighters must follow their departmental SOPs in this regard.

Control Valves

Every sprinkler system is equipped with a main water control valve. *Control valves* are used to turn off the water supply to the system in order to replace sprinklers, perform maintenance, or interrupt operations. These valves are located between the

> **Sidewall Sprinkler —** A sprinkler designed to be positioned at the wall of a room rather than in the center of a room. It has a special deflector that creates a fan-shaped pattern of water that is projected into the room away from the wall. Also called Wall Sprinkler.

Figure 16.43 The sprinkler storage cabinet is located near the main control valve and contains a quantity of the various types of sprinklers installed in the facility as well as a sprinkler wrench.

Figure 16.44 Local ordinances may require that sprinkler system control valves be locked in the open position.

source of water supply and the sprinkler system. The control valve is usually located immediately under the sprinkler alarm valve, the dry-pipe or deluge valve (see Dry-Pipe System and Deluge System sections), or outside the building near the sprinkler system it controls. Sprinkler control valves are either secured in the open position with a chain and padlock or electronically supervised to make sure they are not inadvertently closed (**Figure 16.44**).

Most main water control valves are of the indicating type and are manually operated. An *indicating control valve* is one that shows at a glance whether it is open or closed. There are four common types of indicator control valves used in sprinkler systems: outside stem and yoke (OS&Y), post indicator, wall post indicator, and post indicator valve assembly (PIVA).

Outside stem and yoke (OS&Y) valve. This valve has a yoke on the outside with a threaded stem that opens and closes the gate inside the valve housing. The threaded portion of the stem is visible beyond the yoke when the valve is open and not visible when the valve is closed (**Figure 16.45**).

Post indicator valve (PIV). The PIV is a hollow metal post that houses the valve stem. Attached to the valve stem is a movable plate with the words *OPEN* or *SHUT* visible through a small glass window on the side of the housing. When not in use, the operating handle is locked to the valve housing (**Figure 16.46**).

Wall post indicator valve (WPIV). A WPIV is similar to a PIV except that it extends horizontally through the wall with the target and valve operating nut on the outside of the building (**Figure 16.47**).

Post indicator valve assembly (PIVA). The PIVA does not use a plate with words *OPEN* and *SHUT* as does a PIV. Instead, a PIVA uses a circular disk inside a flat plate on top of the valve housing (**Figure 16.48**). When the valve is open, the disk is perpendicular to the surrounding plate. When the valve is closed, the disk is in line with the plate that surrounds it. Unlike the PIV or WPIV, the PIVA is operated with a built-in crank.

Operating Valves

Sprinkler systems employ various valves such as the alarm test valve, inspector's test valve, and main drain valve. The alarm test valve is located on a pipe that connects the supply side of the alarm check valve to the retard chamber (void that contains

OS&Y Valve — Outside stem and yoke valve; a type of control valve for a sprinkler system in which the position of the center screw indicates whether the valve is open or closed. Also known as outside screw and yoke valve.

Post Indicator Valve (PIV) — A type of valve used to control underground water mains that provides a visual means for indicating "open" or "shut" position; found on the supply main of installed fire protection systems.

Figure 16.45 An OS&Y is the most common type of control valve. The valve shown here is equipped with an alarm device that will activate when the valve is closed.

Figure 16.46 Three post indicator valves (PIV) in the open position. The word *Open* or *Shut* will appear in the glass window on the post.

Figure 16.47 Wall post indicator valves (WPIV) chained in the open position, electronically monitored, and adjacent to the fire department connection.

Figure 16.48 A typical post indicator valve assembly (PIVA).

excess water from momentary water pressure surges) **(Figure 16.49, p. 850)**. This valve is provided to simulate the actuation of the system by allowing water to flow into the retard chamber and operate the waterflow alarm devices.

An inspector's test valve is located in a remote part of the sprinkler system **(Figure 16.50, p. 850)**. The inspector's test valve is equipped with the same size orifice as one sprinkler and is used to simulate the activation of one sprinkler. The water from the inspector's test valve normally discharges outside the building.

Figure 16.49 A typical quarter-turn alarm test valve in the closed position.

Figure 16.50 Inspectors test valves are typically labeled and located at the most remote point on the system.

Figure 16.51 Waterflow alarms are typically located on the exterior of the building and include a system drain valve.

Every sprinkler system riser has a main drain valve. The primary purpose of the main drain is to allow sprinkler service personnel to drain water from the system for maintenance purposes. Because a large volume of water will flow when the main drain valve is opened, it can also be used to check the system water supply. For a description of how to operate a sprinkler system control valve, see **Skill Sheet 16-I-1**.

Waterflow Alarms

Actuation of fire alarms by sprinkler systems is triggered when water flows through the system. Sprinkler water flow alarms are normally operated either hydraulically or electrically. The hydraulic alarm is a local alarm used to alert the personnel in a sprinklered building or a passerby that water is flowing in the system **(Figure 16.51)**. This type of alarm is operated by a water motor that drives a local alarm gong. An electric waterflow alarm may also be employed to alert building occupants and can be configured to notify the fire department.

Water Supply

The majority of sprinkler systems have a water supply of adequate volume, pressure, and reliability. A minimum water supply has to deliver the required volume of water to the highest sprinkler in a building at a residual pressure of 15 psi (105 kPa). The minimum flow depends on the hazard to be protected, the occupancy, and the building contents. A connection to a public water system that has adequate volume, pressure, and reliability is a good source of water for automatic sprinklers. This type of connection is often the only water supply available.

In most cases, the water supply for sprinkler systems is designed to supply only a fraction of the sprinklers actually installed on the system. If a large fire occurs or a pipe breaks, the sprinkler system will need an outside source of water and pressure to do its job effectively. This additional water and pressure can be provided by a pumper that is connected to the sprinkler fire department connection **(Figure 16.52)**. Fire department connections for sprinklers usually consist of a siamese with at least two 2½-inch (65 mm) female connections with a clapper valve in each connection **(Figure 16.53)**, or one large-diameter connection that is attached to a clappered inlet.

Figure 16.52 In addition to the primary water supply, a sprinkler system can be supplied through a fire department connection (FDC).

Figure 16.53 Each inlet of the fire department connection has a clapper valve inside to permit the use of one or two hoselines.

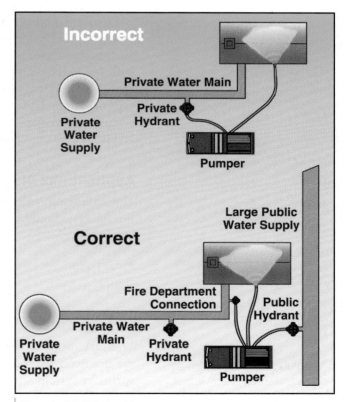

Figure 16.54 The incorrect and correct ways to support a sprinkler system with a pumper.

Sprinkler fire department connections should be supplied with water from pumpers that have a capacity of at least 1,000 gpm (4 000 L/min) or greater. A minimum of two 2½-inch (65 mm) or larger hoses should be attached to the FDC. Whenever possible, fire department pumpers supplying attack lines should operate from hydrants connected to mains other than the main supplying the sprinkler system (**Figure 16.54**).

After water flows through the fire department connection into the system, it passes through a check valve. This valve prevents water flowing from the sprinkler system from flowing back into the fire department connection; however, it does allow water from the fire department connection to flow into the sprinkler system (**Figure 16.55**). The proper direction of water flow through a check valve is usually indicated by an arrow on the valve or by the appearance of the valve casing. A ball-drip valve may be installed at the check valve and FDC; it is designed to keep the valve and connection dry and operating properly during freezing conditions.

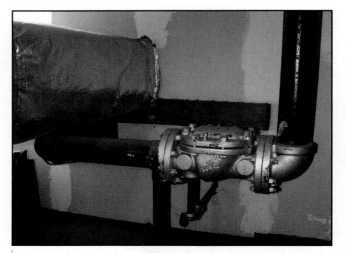

Figure 16.55 The check valve is located below the level of the FDC.

Figure 16.56 Firefighters should familiarize themselves with automatic sprinkler systems in their response areas.

Departmental preincident plans may identify the pressure at which a sprinkler system should be supported as well as the pressure needed for special circumstances. Such a plan cannot be established until fire department personnel become familiar with sprinklered properties in their jurisdiction. A standard plan of operation should cover the buildings in the department's jurisdiction, including type of occupancy, type of system, and extent of the system. Therefore, a preincident survey is a prerequisite for a plan of operation (**Figure 16.56**). A thorough knowledge of the public water system is also important, including knowing the volume and pressure available.

Applications of Sprinkler Systems

The following sections highlight the major applications of sprinkler systems. Firefighters should have a basic understanding of the operation of each.

- Wet-pipe
- Dry-pipe
- Preaction
- Deluge
- Residential

Wet-Pipe Systems

Wet-pipe sprinkler systems are used in locations where temperatures below 40°F (4°C) are not expected. A wet-pipe sprinkler system is the simplest type of automatic fire sprinkler system and generally requires little maintenance. This system contains water under pressure at all times. It is connected to a public or private water supply so that a fused sprinkler will immediately discharge a water spray in the area and actuate an alarm. This type of system is usually equipped with an alarm check valve that is installed in the main riser adjacent to where the feed main enters the building (**Figure 16.57a**). Newer wet-pipe sprinkler systems may not have an alarm check valve. Instead they may have a backflow prevention check valve and an electronic flow alarm. These are sometimes referred to as *straight stick* systems (**Figure 16.57b**).

A wet-pipe sprinkler system may be equipped with a retarding device, commonly called a *retard chamber,* as part of the alarm check valve. This chamber catches excess water that may be sent through the alarm valve during momentary water pressure surges. This reduces the chances of false alarm activations.

Wet-Pipe Sprinkler System — Fire-suppression system is built into a structure or site; piping contains either water or foam solution continuously; activation of a sprinkler causes the extinguishing agent to flow from the open sprinkler.

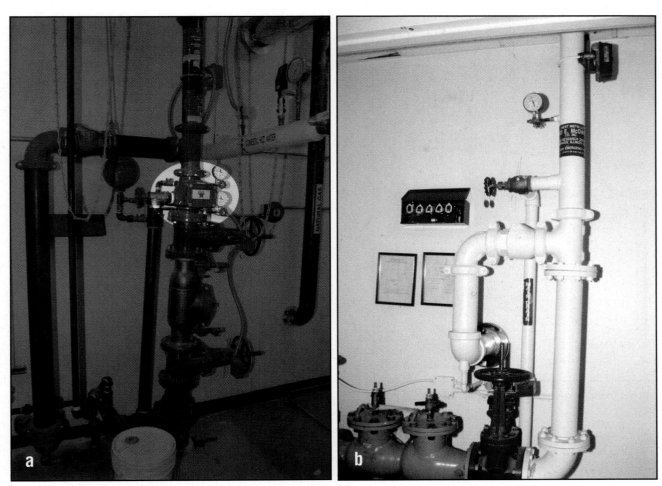

Figures 16.57 a and b Two types of wet-pipe sprinkler systems: (a) a system equipped with an alarm check valve and (b) a straight stick wet-pipe system.

Dry-Pipe Systems

Dry-pipe sprinkler systems are used in locations where the piping may be subjected to temperatures below 40°F (4°C). All pipes in dry-pipe systems are pitched (sloped) to help drain the water in the system back toward the main drain. In these systems, air under pressure replaces water in the sprinkler piping above the *dry-pipe valve* (device that keeps water out of the sprinkler piping until a fire actuates a sprinkler). When a sprinkler fuses, the pressurized air escapes first. Then the dry-pipe valve automatically opens to permit water into the piping system (**Figure 16.58, p. 854**).

A dry-pipe valve is designed so that a small amount of air pressure above the dry-pipe valve will hold back a much greater water pressure on the water supply side of the dry-pipe valve. This is accomplished by having a larger surface area on the air side of the clapper valve than on the water side of the valve. The valve is equipped with an air-pressure gauge above the clapper and a water-pressure gauge below the clapper. The required air pressure for dry-pipe systems is usually about 20 psi (140 kPa) above the trip pressure. Under normal circumstances, the air pressure gauge will read a pressure that is substantially lower than the water-pressure gauge. If the gauges read the same, the system has been tripped and water has been allowed to enter the pipes. **Figure 16.59, p. 854,** illustrates the dry-pipe valve in both the standby and fire positions. Dry-pipe systems are equipped with either electric or hydraulic alarm-signaling equipment.

Dry-Pipe Sprinkler System — Fire-suppression system that consists of closed sprinklers attached to a piping system that contains air under pressure. When a sprinkler activates, air is released that activates the water or foam control valve and fills the piping with extinguishing agent. Dry-pipe systems are often installed in areas subject to freezing.

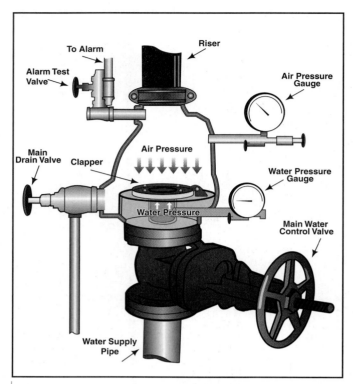

Figure 16.58 Components of a dry-pipe sprinkler system control valve.

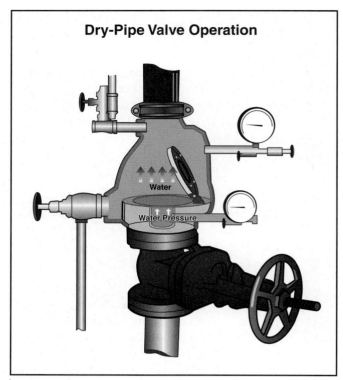

Figure 16.59 The dry-pipe valve shown in the operating position.

In a large dry-pipe system, several minutes could be lost while the air is being expelled from the system. Standards require that a quick-opening device be installed in systems that have a water capacity of over 500 gallons (2 000 L). An accelerator is one type of quick-opening device. The basic purpose of this device is to accelerate the opening of the dry-pipe valve. This allows water to be admitted into the sprinkler system quicker, resulting in quicker sprinkler discharge.

Preaction Systems

Preaction sprinkler systems are dry systems that employ a deluge-type valve (see Deluge System section), fire detection devices, and closed sprinklers. This type of system is used when it is especially important to prevent water damage, even if pipes are broken. The system will not discharge water into the sprinkler piping except in response to either smoke- or heat-detection system actuation.

Fire detection devices operate a release located in the system actuation unit. This release opens the deluge valve and permits water to enter the distribution system so that water is ready when the sprinklers fuse. When water enters the system, an alarm sounds to give a warning before the opening of the sprinklers.

Deluge Systems

Deluge sprinkler systems are similar to dry-pipe systems in that there is no water in the distribution piping before system activation. They differ from dry-pipe systems in that the sprinklers have no fusible links and do not function as fire detection devices. In a deluge system, all sprinklers are open all the time. This means that when the system is activated and water enters the piping, the water will discharge from all of the sprinklers simultaneously. The flow of water into the system is con-

Preaction Sprinkler System — Fire-suppression system that consists of closed sprinklers attached to a piping system that contains air under pressure and a secondary detection system; both must operate before the extinguishing agent is released into the system; similar to the dry-pipe sprinkler system.

Deluge Sprinkler System — Fire-suppression system consisting of piping and open sprinklers. A fire detection system is used to activate the water or foam control valve. When the system activates, the extinguishing agent expels from all sprinkler heads in the designated area.

trolled by a deluge valve. The operation of the deluge valve is controlled by fire detection devices (heat, smoke, or flame detectors) installed in the area protected by the system.

Normally installed in high-hazard occupancies such as aircraft hangars, deluge systems are designed to quickly supply a large volume of water to the protected area. These systems are sometimes used to discharge foam or other extinguishing agents in occupancies containing flammable liquids and other volatile fuels. Other partial deluge systems, in which some sprinklers are open and some are not, are also installed in certain occupancies.

Residential Systems

Residential sprinkler systems are installed in one- and two-family dwellings. These systems are designed to give occupants of the dwelling a chance to escape and to prevent total fire involvement in the room of origin. These systems are covered by NFPA 13D and may be either wet- or dry-pipe systems.

Residential sprinkler systems employ quick-response sprinklers (**Figure 16.60**). This type of sprinkler is available in both conventional and decorative models. There are several types of piping systems that can be used in this type of system (steel, copper, plastic). Residential sprinkler systems must have a pressure gauge (to check air pressure on dry-pipe systems and water pressure on wet-pipe systems), a flow detector, and a means for draining and testing the system (**Figure 16.61**). These systems can either be connected directly to the public water supply or to the dwelling's domestic water system. A control valve is required to turn off the water to the sprinkler system and to the domestic water system if they are connected. If the sprinkler system is supplied separately from the domestic water system, the sprinkler control valve must be supervised in the open position (**Figure 16.62**).

Figure 16.60 A typical quick-response sprinkler used in residential sprinkler systems.

Figure 16.61 A residential sprinkler pressure gauge.

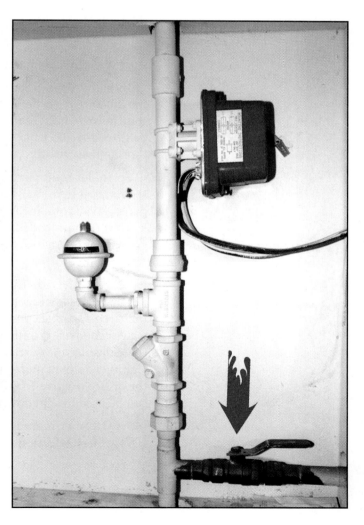

Figure 16.62 A typical quarter-turn control valve on a residential sprinkler system.

Residential sprinkler systems operate in the same manner as other wet-pipe or dry-pipe systems. Some residential systems may be equipped with a fire department connection (usually a 1½-inch [38 mm] connection), while others have no FDC. **Skill Sheet 16-I-2** describes how to manually stop the flow of water from a sprinkler.

Operations at Fires in Protected Properties

Several important factors must be considered when fighting fires in occupancies that have activated sprinkler systems. In addition to normal fire fighting operations, one of the early-arriving pumpers should connect to the Fire Department Connection (FDC) in accordance with the preincident plan **(Figure 16.63)**. Firefighters should make every effort to supply adequate water to an operating sprinkler system. The water supply may have to be conserved for this purpose by limiting the use of hoselines from the water supply system serving the sprinkler system. A second water supply for hoselines should be established if necessary. To a point, the effectiveness of sprinklers can be improved by increasing the pressure on the system. To learn how to connect hoseline to a sprinkler system FDC, refer to **Skill Sheet 16-I-3**.

Sprinkler system control valves must be open for proper operation; check to see that they are. If possible, observe the discharge from sprinklers in the area of the fire and maintain pressure at the pumper to adequately supply the sprinkler system.

Sprinkler control valves should not be closed until fire officers are convinced that further operations will simply waste water, produce heavy water damage, or hamper the progress of final extinguishment by fire fighting personnel. Premature closure of the control valve could lead to a dramatic increase in the intensity of a fire. When a sprinkler control valve is closed, a firefighter with a portable radio should be stationed at the valve in case it needs to be reopened should the fire flare up **(Figure 16.64)**.

In some departments, it is SOP to plug open sprinklers rather than shutting the entire system down by closing the main control valve; in others, the system is shut down. Firefighters must follow their departmental SOP. Pumpers should not be disconnected from the FDC until after extinguishment has been confirmed by a thorough overhaul.

Sprinkler equipment should be restored to service before leaving the premises. All sprinkler system maintenance should be performed by representatives of the occupant who are qualified to perform work on sprinkler systems. For liability purposes, it is not recommended that fire department personnel install or remove system components or operate any valves.

Firefighters may be required to stop the flow of water from a single sprinkler that has been activated. This may even be necessary after the main water control valve has been closed because residual water in the system will continue to drain through the open sprinkler until the system is drained below that level. To stop the flow of water, wooden sprinkler wedges may be inserted between the discharge orifice and the deflector and forced together by hand until the flow is stopped **(Figure 16.65)**. Commercially manufactured stoppers are also available that can be inserted to plug the orifice **(Figure 16.66)**. If available, control valves in stairwells and other locations can be used to shut down only part of the system.

NOTE: For more information on the systems discussed in this chapter, see the IFSTA **Fire Detection and Suppression Systems** manual.

Figure 16.63 According to the department's SOP, the first-arriving pumper should connect to the FDC of the automatic sprinkler system.

Figure 16.64 A firefighter with a hand-held radio should be assigned to stand by to reopen the sprinkler control valve if necessary.

Figure 16.65 Wooden wedges can be used to stop the flow of water from a sprinkler.

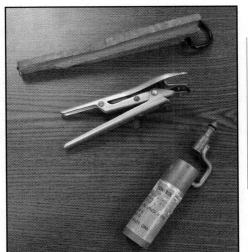

Figure 16.66 Commercially manufactured sprinkler tongs and stops should be provided in the sprinkler control kit carried on all apparatus.

Summary

Many of the buildings to which firefighters are called for emergency operations are protected partially or fully by automatic fire detection and/or suppression systems. Some of these systems are strictly local alarms — that is, they only ring a bell or activate another type of audible or visible signaling device to alert building occupants of the need to immediately evacuate the premises. Other systems detect fires, alert the occupants, and notify the fire department all at the same time. Still other systems, such as automatic sprinklers or flooding systems, alert the occupants, alert the fire department, and apply water or other extinguishing agents to a fire to control or extinguish it. Firefighters need to familiarize themselves with the types of systems installed in their area of responsibility. They should also learn the locations of these systems as well as the capabilities and limitations of each type of system.

Fire Fighter I

1. What are the functions of fire detection, alarm, and suppression systems?

2. What are the two general types of sprinkler coverage?

3. What is the function of control valves?

4. What is the difference between a wet-pipe system and a dry-pipe system?

5. When should sprinkler control valves be closed?

Fire Fighter II

6. How does a photoelectric smoke detector work?

7. How do flame detectors work?

8. Describe the three basic types of auxiliary automatic alarm systems.

9. What is a proprietary automatic alarm system?

10. Describe commonly used release mechanisms to activate sprinklers.

OS&Y

Step 1: Confirm order with officer to operate valve.

Step 2: Close the OS&Y valve by turning it clockwise until the valve is fully closed and the stem is flush with the wheel.

Step 3: Open the OS&Y valve by turning it counterclockwise until fully opened.

Step 4: Back off the OS&Y valve one-quarter turn clockwise.

PIV

Step 1: Confirm order with officer to operate valve.

Step 2: Unlock the PIV wrench from the PIV body.

Step 3: Position the PIV wrench on stem nut.

Step 4: Close the PIV valve, turning it clockwise slowly until the target window indicates CLOSED or SHUT.

Step 5: Open the PIV valve, turning it counterclockwise until fully open and target window indicates OPEN.

Step 6: Back off the PIV valve, turning it clockwise one-quarter turn ensuring that the target window remains OPEN.

Step 7: Replace and lock the wrench onto the PIV body.

Wedge

Step 1: Place and climb ladder safely to within reach of the sprinkler.

Step 2: Insert the wedges between the sprinkler arms, flat sides against sprinkler.

Step 3: Drive the wedges into the sprinkler with the heel of the hand until water flow stops.

Clamp-Type Sprinkler Tongs

Step 1: Place and climb ladder safely, to within reach of sprinkler.

Step 2: Insert the tongs into the sprinkler between the arms.

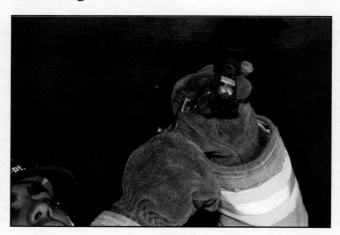

Step 3: Open the tongs (by clamping the handles together) until water flow stops.

Step 4: Lock the tongs in open position, with the keeper pulled as far as it will go toward end of handles.

Swivel-Type Sprinkler Tongs

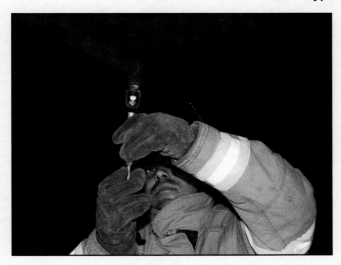

Step 1: Place and climb the ladder safely to within reach of the sprinkler.

Step 2: Insert the tongs into the sprinkler between the arms.

Step 3: Open the tongs with handle swiveled upward.

Step 4: Turn the locking knob clockwise to lock the tongs in the open position.

Step 1: Confirm order with officer to connect line.

Step 2: Extend hoselines to sprinkler connection, with the male thread toward sprinkler connection.

Step 5: Inspect the fire department connection for debris, check threads and check and replace gasket, if necessary.

Step 3: Lay down hose fittings at sprinkler connection, gently folding back hose so that male threads do not hit pavement.

Step 4: Remove caps from fire department connection.

Step 6: Connect hoselines to outlet with lowest fitting first.

NOTE: There may be more than one row of siamese fire service connections. Connect first to the center outlet on the bottom row.

Step 7: Tighten connections with spanner wrench.

Step 8: Report to officer completion of assigned task.

Chapter Contents

Loss Control

This chapter provides information that addresses the following job performance requirements of NFPA® 1001, *Standard for Fire Fighter Professional Qualifications (2008)*:

<u>NFPA® 1001 references for Chapter 17:</u>

- 5.3.14
- 5.3.13
- 5.5.1
- 5.3.10

Chapter Objectives

Fire Fighter I Objectives

1. Explain the philosophy of loss control.

2. Discuss planning and procedures for salvage operations.

3. Describe salvage covers, salvage cover maintenance, and equipment used in salvage operations.

4. Summarize basic principles of salvage cover deployment.

5. Summarize methods used to catch and route water from fire fighting operations and cover openings using salvage covers.

6. Discuss overhaul operations.

7. Describe tools and equipment uw3e in of overhaul.

8. Discuss fire safety during overhaul.

9. Summarize the overhaul process.

10. Discuss locating hidden fires.

11. Clean, inspect, and repair a salvage cover. (Skill Sheet 17-I-1)

12. Roll a salvage cover for a one-firefighter spread. (Skill Sheet 17-I-2)

13. Spread a rolled salvage cover — One-firefighter method. (Skill Sheet 17-I-3)

14. Fold a salvage cover for a one-firefighter spread. (Skill Sheet 17-I-4)

15. Spread a folded salvage cover — One-firefighter method. (Skill Sheet 17-I-5)

16. Fold a salvage cover for a two-firefighter spread. (Skill Sheet 17-I-6)

17. Spread a folded salvage cover — Two-firefighter balloon throw. (Skill Sheet 17-I-7)

18. Construct a water chute without pike poles. (Skill Sheet 17-I-8)

19. Construct a water chute with pike poles. (Skill Sheet 17-I-9)

20. Construct a catchall. (Skill Sheet 17-I-10)

21. Locate and extinguish hidden fires. (Skill Sheet 17-I-11)

Loss Control

Case History

If overhaul operations are not properly conducted, greater damage can occur following the initial incident. This was the case in a fire in a 2-story detached residential structure. The fire was attacked from the exterior and confirmed under control by the first-in truck a short time after the application of both water and foam. Additional crews established positive-pressure ventilation, made entry into the home from the main floor front entrance, checked the garage area for fire extension. Thermal imaging cameras were used along with firefighter 'know-how' and experience to access fire extension. The firefighter responsible for checking the attic for extension from within the home did so through the attic access, which provided a view from the middle of the home through to the front fire area. Direct access was not made into the attic to confirm if there were any hot spots or if additional overhaul was required; no visible flame or smoke was noted upon distant visual inspection. Once the fire was determined to be under control, the home was turned over to the homeowner.

Later that afternoon, a worker saw smoke coming from the roof area at the front of the home. A second alarm in the same day was dispatched to the home. The first-arriving truck reported heavy smoke when approaching from a distance. It was confirmed that the roof was fully involved and an interior attack was begun. Defensive operations, with a focus on protecting neighboring homes, were all that could be done at this stage of the fire. With the roof nearly burnt away, the home was subjected to extensive water damage. As a result, success was determined based on protecting all exposures at this fire scene.

Source: National Fire Fighter Near-Miss Reporting System

Loss control is an important component of fire department service delivery. Nonetheless, it is but one aspect of *customer service* — which is at the core of everything fire departments do. Most fire incidents abound with customer-service opportunities, including loss control. Many members of the public have a mental image of firefighters as being destructive — kicking in doors, chopping holes in roofs, or breaking windows — and they are pleasantly surprised to see firefighters spreading salvage covers over furniture, covering vent holes to keep rain out, and securing their property to prevent trespassing and possible looting. These positive loss-control actions greatly enhance the image of firefighters and the fire service.

> **Customer Service** — Quality of an organization's relationship with individuals — both internal and external — who have contact with the organization.

According to NFPA 1001, those trained to the Fire Fighter I level must know the following about loss control:

- Purpose of property conservation and its value to the public
- Methods used to protect property
- Types of and uses for salvage covers
- Types of water application devices most effective for overhaul

- Water application methods that limit water damage
- Tools and techniques used to expose hidden fires
- Dangers associated with overhaul
- Obvious signs of area of origin or signs of arson
- Reasons for protection of fire scene

The standard requires those trained to the Fire Fighter I level to be capable of the following:

- Conserving property as a member of a team
- Clustering furniture
- Deploying covering materials
- Rolling and folding salvage covers for reuse
- Constructing water chutes and catchalls
- Removing water
- Covering doors, windows, floor openings, and roof openings
- Separating, removing, and relocating charred material to a safe location
- Deploying and operating an attack line
- Removing floor, ceiling, and wall components without compromising structural integrity
- Applying water for maximum effectiveness
- Recognizing and preserving obvious signs of area of origin and arson.
- Locating, exposing, and controlling hidden fires
- Evaluating for complete extinguishment

There are no additional loss-control requirements for those trained to the Fire Fighter II level.

This chapter explains the philosophy of loss control and focuses on minimizing damage to structures and their contents during and after fire control operations. The chapter also details two of the most effective means of loss control: salvage operations and overhaul operations. Planning, procedures, and equipment are also discussed.

Philosophy of Loss Control

The philosophy of *loss control* is to minimize damage and provide customer service through effective mitigation and recovery efforts before, during, and after an incident. Because of its value to the affected property owners, loss control builds goodwill within the community. A fire department often receives words of appreciation and praise in the news media and letters of thanks from citizens for saving their property and cherished possessions. This praise gives firefighters a feeling of accomplishment, particularly when the appreciation comes from those whose belongings were saved.

Performing proper salvage and overhaul — two aspects of loss control — is of significant importance to both firefighters and property owners/occupants because they are the most effective means of loss control. Properly applied suppression techniques plus prompt and effective use of good salvage and overhaul procedures will minimize the total losses. Effective salvage procedures coordinated with a thorough and systematic overhaul will also facilitate prompt restoration of the property to full use.

Loss Control — The practice of minimizing damage and providing customer service through effective mitigation and recovery efforts before, during, and after an incident.

Loss Control Risk Analysis — The process in which specific potential risks (potential because the incident has not occurred yet) are identified and evaluated. The goal of this process is to develop strategies to minimize the impact of these risks.

Salvage consists of those operations associated with fire fighting that aid in reducing primary and secondary damage during fire fighting operations. Primary damage is that caused by the fire; secondary damage is that caused by the suppression activities. Both can be minimized through effective salvage efforts. Some of these damages cannot be avoided because of the need to do forcible entry, apply water to vent the building, and search for fires throughout a structure. Salvage starts as soon as adequate personnel are available and may be done simultaneously with fire attack if resources permit.

Overhaul consists of those operations involved in searching for and extinguishing hidden or remaining fires. Protecting the scene after the fire and preserving evidence of the fire's origin and cause are components of overhaul; however; the tools and techniques involved are discussed in Chapter 18, Protecting Fire Scene Evidence. To the extent possible based on fire conditions, do not start overhaul operations until the fire is under control, the fire cause has been determined, and any evidence has been identified and protected. In other words, delay overhaul operations until the fire scene investigator authorizes it.

Salvage

In this context, *salvage* is defined as methods and operating procedures associated with fire fighting by which firefighters attempt to save property and reduce further damage from water, smoke, heat, and exposure during or immediately after a fire by removing property from a fire area, by covering it, or other means. Proper salvage operations involve early planning, knowing the procedures necessary to do the job, and being familiar with the various tools and equipment used. Some improvising can be done when equipment is limited. Final parts of salvage are protecting damaged property from the weather and from trespassers.

Planning

Efficient salvage operations require planning and training for fire officers and firefighters **(Figure 17.1)**. Standard operating procedures (SOPs) should be developed to address early and well-coordinated salvage operations. Special preincident plans should be developed for buildings with high-value contents that are especially susceptible to water and smoke damage.

In residential occupancies, preincident plans should include covering upholstered furniture, bedding, and other water-absorbent objects. Plans should also be made to protect items such as photographs, important documents, computer equipment, and artwork that could be of particular importance to the residents. Protecting these items may involve covering them, moving them to a dry unaffected area in the house, or removing them altogether.

In commercial occupancies, preincident plans should reflect an awareness of the value of contents vital to business survival. If their accounts receivable (money owed to the business) records are lost, most businesses — even those fully insured — will never recover. Therefore, protecting computers and filing cabinets in the business offices can be more important to the survival of the business than protecting merchandise and other contents.

Figure 17.1 Firefighters must train in the use of salvage covers and floor runners to perform salvage operations efficiently.

Salvage — Methods and operating procedures associated with fire fighting by which firefighters attempt to save property and reduce further damage from water, smoke, heat, and exposure during or immediately after a fire by removing property from a fire area, by covering it, or other means.

Overhaul — Those operations conducted once the main body of fire has been extinguished that consist of searching for and extinguishing hidden or remaining fire, placing the building and its contents in a safe condition, determining the cause of the fire, and recognizing and preserving evidence of arson.

Fire departments can facilitate salvage efforts before a fire incident by working with the loss-control representatives of various local businesses. Identifying critical records and components needed for business continuation and suggesting implementation of continuous loss-control measures, such as stock protection, benefits both the business firm and the fire department.

Salvage Procedures

Salvage operations can often be started at the same time as fire attack if on-scene resources permit. For instance, the contents of the room(s) immediately below the fire floor can be covered while fire suppression operations are being conducted on the floor above. Contents of a room can be gathered and covered quickly before a ceiling is pulled. Catching debris with a cover also saves time and effort in cleanup as well as leaving a more professional appearance.

When possible, group building contents into compact piles that can be covered with a minimum of salvage covers. Grouping contents into piles allows more items to be protected than if they were covered in their original position. If possible, group household furnishings in the center of the room when arranging for salvage (**Figure 17.2**). In many cases, one standard salvage cover will protect the contents of one residential room. If the floor covering is a removable rug, slip the rug out from under furniture as each piece is moved, and roll the rug to make it easier to move.

A dresser, chest, or high object may be placed at the end of a bed. If there is a rolled rug, place it on top to serve as a ridge pole, allowing water to run off both sides of the covered furniture. Other furniture can be grouped close by; pictures, curtains, lamps, and clothing can be placed on the bed. It may sometimes be necessary to place a salvage cover into position before some articles are placed on the bed. In this event, the bed and furniture will be protected until other items are placed under the cover.

Furniture sitting on wet carpet will absorb water and can be damaged even though it may be well covered. To prevent this damage, raise the furniture off the wet floor with water-resistant materials. Precut plastic foam blocks are ideal but canned goods from the kitchen can also be used (**Figure 17.3**).

Figure 17.2 Move furniture together so that it is easier to cover, preferably in the center of the room.

Figure 17.3 Use small blocks of plastic to elevate furniture and prevent water damage.

Commercial occupancies present challenges for firefighters who are trying to perform salvage functions **(Figure 17.4)**. The actual arranging of contents to be covered may be limited when large stocks and display features are involved. Display shelves are frequently built to the ceiling and directly against the wall. This construction feature makes contents difficult to cover. When water flows down a wall, it naturally comes into contact with shelving and wets the contents. Contents stacked too close to the ceiling also present a salvage problem. Ideally, there should be enough space between the stock and ceiling to allow firefighters to easily apply salvage covers.

Stock should be stored on pallets, and a common obstacle to efficient salvage operations is a lack of skids or pallets under stock that is susceptible to water damage **(Figure 17.5)**. Some examples of contents that are perishable include food, materials in cardboard boxes, feed, paper, and other dry goods. When the number of salvage covers is limited, it is good practice to use available covers for water chutes and catch-alls even though the water must be routed to the floor and cleaned up afterward.

Firefighters must be extremely cautious of high-piled stock such as boxed materials or rolled paper that has become wet at the bottom. The wetness often causes the material to expand and push out interior or exterior walls. Wetness also reduces the strength of the material and may cause the piles to collapse **(Figure 17.6)**. Some rolls of paper can weigh a ton (900 kg) or more. A roll of paper that fell on firefighters could seriously injure or kill them.

Chute — Salvage cover arrangement that channels excess water from a building. A modified version can be made with larger sizes of fire hose.

Figure 17.4 Protecting stock in commercial occupancies can be difficult.

Figure 17.5 Stock on pallets is less vulnerable to water damage.

Figure 17.6 Water soaking into rolls of newsprint could weaken them enough to cause collapse of the stack.

Large quantities of water may be removed by locating and cleaning clogged drains, removing toilet fixtures, creating scuppers, making use of existing sanitary piping systems, or employing chutes made of salvage covers, plastic, or other available materials. Water left on cabinets and other horizontal surfaces may ruin finishes over a period of hours. Cabinets and tabletops can be wiped off quickly and easily with disposable paper towels. This simple service can save the building owner/occupant a great deal of potential loss.

Salvage Covers and Equipment

Salvage covers are made of waterproof canvas materials or vinyl and are manufactured in various sizes. These covers have reinforced corners and edge hems into which grommets are placed for hanging or draping. Synthetic covers are lightweight, easy to handle, economical, and practical for both indoor and outdoor use (**Figure 17.7**). Many departments use disposable plastic covers. Heavy-duty plastic is available on rolls and can be rolled out to cover large areas, or covers can be cut from the rolls in different sizes to serve various cover needs. Firefighters must be familiar with the salvage covers used in their departments.

Sizes of Commercial Salvage Covers and Floor Runners

Salvage Covers

9 x 12 feet (3 m x 4 m)

12 x 14 feet (4 m x 4.6 m)

12 x 18 feet (4 m x 6 m)

14 x 18 feet (4.6 m x 6 m)

Floor Runners

Widths: 24 inches (600 mm), 27 inches (675 mm), 30 inches (750 mm), 36 inches (900 mm)

Lengths: 10 feet (3 m), 20 feet (6 m), 30 feet (10 m), 40 feet (13 m), 50 feet (15 m)

A variety of equipment is used along with salvage covers. Every firefighter needs to know how to use these pieces of equipment and how to properly maintain salvage covers.

Figure 17.7 Synthetic salvage covers are lightweight, easy to handle, and can be used in or out of doors.

Salvage Cover Maintenance

Proper cleaning, drying, and repairing of salvage covers increases their service life. Ordinarily, the only cleaning required for canvas salvage covers is wetting or rinsing with a hose stream and scrubbing with a broom. Extremely dirty or stained covers may be scrubbed with a detergent solution and then thoroughly rinsed **(Figure 17.8)**. Permitting canvas salvage covers to dry when they are dirty is not a good practice. After carbon and ash stains have dried, a chemical reaction takes place that rots the canvas. Foreign materials are difficult to remove when dry, even with a detergent. Canvas salvage covers should be completely dry before they are folded and placed in service. This practice is essential to prevent mildew and rot. There is no particular objection to outdoor drying of salvage covers except that the wind tends to blow and whip them about. (**NOTE:** Long-term exposure of canvas to sunlight will result in damage from ultraviolet rays. Drying in direct sunlight may degrade the material over time.) **Skill Sheet 17-I-1** describes how to clean, inspect, and repair a salvage cover.

Synthetic salvage covers do not require as much maintenance as canvas ones. Synthetic covers may be folded wet, but it is usually better to let them dry first so they will not mildew **(Figure 17.9)**.

After salvage covers are dry, examine them for damage. Holes can be located by placing three or four firefighters side-by-side along one end. Have the firefighters pick up the end and pass it back over their heads while walking toward the other end, looking up at the underside of the cover **(Figure 17.10, p. 874)**. Light will show through even the smallest holes. Mark any holes with chalk or a marking pen **(Figure 17.11, p. 874)**. Use chalk for canvas salvage covers and a marking pen for vinyl covers. The holes can be repaired by placing duct tape or mastic tape over them or by patching with iron-on or sew-on patches, depending upon the material.

Figure 17.8 Salvage covers should be cleaned with soap and water and a stiff broom after each use.

Figure 17.9 To prevent mildew, salvage covers should be completely dry before folding.

Salvage Equipment

For conducting salvage operations at a fire, salvage equipment should be located in a readily accessible area on the apparatus. Individual fire department SOPs dictate on which apparatus the salvage equipment is carried and who performs the primary salvage operations on the fire scene. This avoids delay in beginning salvage operations, although salvage and loss control are everyone's responsibilities.

Keep smaller tools and equipment in a specially designated salvage toolbox or other container to make them easier to carry. Loss-control materials and supplies may be kept in a plastic tub and brought into the structure early in the fire event

Figure 17.10 Inspect salvage covers for damage, such as holes and tears.

Figure 17.11 Chalk or a marking pen is used to mark the location of holes or tears that require patching.

(Figure 17.12). The materials and supplies are then readily available for loss-control activities. The tub itself provides a useful water-resistant container to protect items such as computers, pictures, and other water-sensitive materials.

The following is a list of typical salvage equipment that should be carried on the apparatus. The use of this equipment, however, is not limited to salvage operations. Some of these items are discussed in the paragraphs that follow.

- Electrician's pliers
- Sidecutters
- Various chisels
- Tin snips
- Tin roof cutter (can opener)
- Adjustable wrenches
- Pipe wrenches
- Hammer(s)
- Sledgehammer
- Hacksaw
- Crosscut handsaw
- Heavy-duty stapler and staples
- Linoleum knife
- Wrecking bar
- Padlock and hasp
- Hinges
- Screwdriver(s)
- Battery-operated power tools
- Hydraulic jack
- Assortment of nails
- Assortment of screws

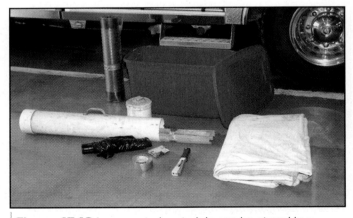

Figure 17.12 Loss control materials can be stored in a plastic tub for easy transport.

- Plastic sheeting
- Wooden laths
- Wooden wedges (for sprinkler stops)
- Soft wood plugs
- Sawdust (for diking)
- Mops
- Squeegees
- Scoop shovels
- Brooms
- Mop buckets with wringers
- Automatic sprinkler kit
- Water vacuum
- Submersible pump and discharge hose
- Sponges
- Chamois
- Paper towels
- Assortment of rags
- 100-foot (30 m) length of electrical cable with locking-type connectors, 14-3 gauge or heavier
- Pigtail ground adapters, 2 wire to 3 wire, 14-3 gauge or heavier with 12-inch (300 mm) minimum length
- Approved ground fault interruption device
- Salvage covers
- J-hooks
- S-hooks
- Floor runners
- Duct tape
- Plastic bags
- Cardboard boxes with tape dispenser
- Styrofoam™ blocks
- Rope
- Assortment of bungee cords

Automatic sprinkler kit. The tools found in a sprinkler kit are needed when fighting fires in buildings protected by automatic sprinkler systems. These tools are used to stop the flow of water from an open sprinkler before it has had time to completely drain after system shutdown. A flow of water from an open sprinkler can do considerable damage to merchandise on lower floors after a fire has been controlled in a commercial building. Sprinkler tongs or stoppers and wooden sprinkler wedges are suggested tools for a sprinkler kit. The SOPs of some departments may require additional equipment necessary to restore a system to service. Use caution when replacing sprinklers; some types require use of a special wrench to prevent damage to the sprinkler. Some fire departments allow firefighters to replace fused or damaged sprinklers; others do not. Firefighters must follow their departmental SOPs.

Carryalls (debris bags). *Carryalls,* sometimes referred to as *debris bags* or *buckets,* are used to carry debris, catch falling debris, and provide a water basin for immersing small burning objects **(Figure 17.13)**.

Floor runners. Costly floor coverings can be ruined by mud and grime tracked in by firefighters. These floor coverings may be protected by using floor runners. Floor runners can be unrolled from an entrance to almost any part of a building. Commercially prepared vinyl-laminated nylon floor runners are lightweight, flexible, tough, heat and water resistant, and easy to maintain **(Figure 17.14)**.

Dewatering devices. Dewatering devices are used to remove water from basements, elevator shafts, and sumps **(Figure 17.15)**. Fire department pumpers should not be used for this purpose because they are intricate and expensive machines and are not intended to pump the dirty, gritty water found in such places. Trash-type pumps are best suited for salvage operations. Use a jet-siphon device or a submersible pump for the removal of excess water. These devices can be moved to any point where a line of hose can be placed and an outlet for water can be provided.

Figure 17.14 Floor runners are used to protect carpet and hardwood floors.

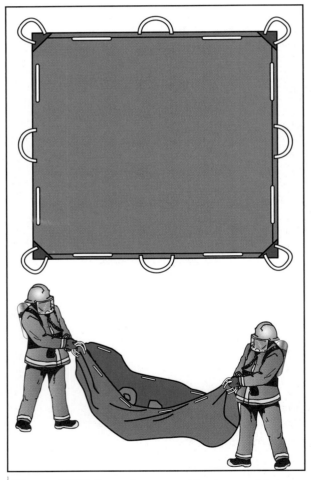

Figure 17.13 Carryalls are used to remove debris from a structure. Loops located on the sides are used as handles when carrying the debris.

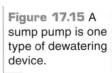

Figure 17.15 A sump pump is one type of dewatering device.

Water vacuum. One of the easiest and fastest ways to remove water is the use of a water-vacuum device. It can be used to dewater floors, carpets, and other areas where the water is not deep enough to be picked up by a submersible pump or siphon ejector. The water vacuum appliance consists of a tank (worn on the back or placed on wheels) and a nozzle. Backpack-type tanks normally have a capacity of 4 to 5 gallons (15 L to 20 L) and can be emptied by simply pulling a lanyard that empties the water through the nozzle or through a separate drain hose **(Figure 17.16)**. Floor models on rollers may have capacities up to 20 gallons (80 L) **(Figure 17.17)**.

J-hooks. These simple devices are designed to be driven into walls or wooden framing to provide a strong point from which to hang things. In the fire service, they are most often used to hang salvage covers on walls to protect wall-mounted book cases and other shelving units **(Figure 17.18)**.

S-hooks. S-hooks are most often used for the same purpose as J-hooks, but in a slightly different way. S-hooks cannot be driven into walls or framing but must have a horizontal ledge (such as the top of a secure shelving unit or structural member) from which to hang **(Figure 17.19)**.

Figure 17.16 A backpack-type water vacuum is an effective means for removing water from carpets and other types of flooring.

Figure 17.17 A floor model water vacuum.

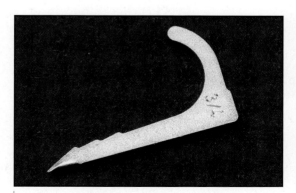

Figure 17.18 A typical J-hook used to hang salvage covers from walls or over vertical openings.

Figure 17.19 S-hooks are used to hang salvage covers from rails or furniture.

Folding/Rolling and Spreading Salvage Covers

One of the key factors in successful salvage operations is the proper handling and deployment of salvage covers. The following sections highlight basic principles of salvage cover deployment by one or two firefighters.

One-Firefighter Spread with a Rolled Salvage Cover

The principal advantage of the one-firefighter salvage cover roll is that one person can quickly unroll a cover across the top of an object and unfold it (**Figure 17.20**). **Skill Sheet 17-I-2** describes the procedure for two firefighters to roll a salvage cover so that one firefighter can spread it. A salvage cover rolled for a one-firefighter spread may be carried on the shoulder or under the arm. Use the steps described in **Skill Sheet 17-I-3** for a one-firefighter spread with a rolled salvage cover.

One-Firefighter Spread with a Folded Salvage Cover

Some departments prefer to carry salvage covers folded as opposed to rolled (**Figure 17.21**). The procedures in **Skill Sheet 17-I-4** highlight steps for folding a salvage cover for one-firefighter deployment. Two firefighters are needed to make this fold and they will be performing the same functions simultaneously. This folded salvage cover may be carried in any manner; however, carrying it on the shoulder is convenient. The steps described in **Skill Sheet 17-I-5** can be used when one firefighter spreads a folded salvage cover.

Two-Firefighter Spread with a Folded Salvage Cover

Some of the largest salvage covers cannot be easily handled by a single firefighter. These covers can be folded for two-firefighter deployment. The procedure in **Skill Sheet 17-I-6** can be used to make the two-firefighter fold. The most convenient way to carry this fold is on the shoulder with the open edges next to the neck. It makes little difference which end of the folded cover is placed in front of the carrier because two open-end folds will be exposed. Position the cover so that the carrier can grab the lower pair of corners and the second firefighter can grab the uppermost pair (**Figure 17.22**).

Figure 17.20 The one-firefighter spread uses a rolled cover, which requires that the cover be unrolled before it is unfolded.

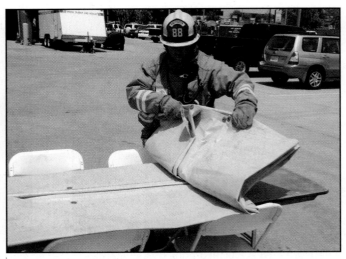

Figure 17.21 An alternative method to the rolled method is the folded method.

Figure 17.22 When two firefighters deploy a folded salvage cover, the person carrying it grasps the lower corners while the second person grasps the upper corner.

Figure 17.23 The balloon throw is one of the most common methods of deploying the salvage cover.

The balloon throw is the most common method for two firefighters to deploy a large salvage cover. The balloon throw gives better results when sufficient air is pocketed under the cover. This pocketed air gives the cover a parachute effect that floats it in place over the article to be covered (**Figure 17.23**). The steps described in **Skill Sheet 17-I-7** are for making the balloon throw.

Improvising With Salvage Covers

In addition to simply covering building contents, salvage covers may also be used to catch and route water from fire fighting operations or other structural flooding situations. The following section details some of these special situations.

Removing Water with Chutes

Using a chute is one of the most practical methods of removing water that comes through the ceiling from upper floors. Water chutes may be constructed on the floor below fire fighting operations to drain runoff through windows or doors (**Figure 17.24**). Some fire departments carry prepared chutes, approximately 10 feet (3 m) long, as regular equipment, but others find it more practical to construct chutes when and where needed using floor runners or one or more covers. Effective water diversion is limited only by the imagination of the firefighters. Plastic sheeting, a heavy-duty stapler, and duct tape can be used to construct water diversion chutes. **Skill Sheets 17-I-8** and **17-I-9** describe the procedures for constructing water chutes.

Constructing a Catchall

A catchall is constructed from a salvage cover that has been placed on the floor to hold small amounts of water (**Figure 17.25, p. 880**). The catchall may also be used as a temporary means to control large amounts of water until chutes can be constructed to route the water to the outside. Properly con-

Figure 17.24 A water chute may be constructed with a salvage cover or floor runner wrapped around two pike poles.

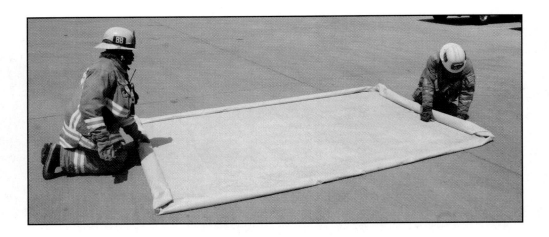

Catchall — Retaining basin, usually made from salvage covers, to impound water dripping from above.

structed catchalls will hold several hundred gallons (liters) of water and often save considerable time during salvage operations. Place the cover into position as soon as possible, even before the sides of the cover are rolled. Two people are usually required to prepare a catchall to make more uniform rolls on all sides. The steps required to make a catchall are shown in **Skill Sheet 17-I-10**.

Splicing Covers

When objects or groupings are too large to be covered by a single cover, or when long chutes or catchalls need to be made, it will be necessary to splice covers with watertight joints.

Splicing a Chute to a Catchall

As stated earlier, when a catchall is prepared and positioned to catch water draining from above, some method should be prepared well in advance to remove accumulated water before the catchall overfills. If available, submersible pumps may be used. However, if the water accumulation is slow, this operation requires constant attention and turning off the pump when not needed. Another method of water removal is a water chute spliced to the catchall. An advantage to this system is that as soon as water accumulates it is drained to the outside.

Covering Openings

One of the final parts of salvage operations is the covering of openings to prevent further damage to the property by weather. Cover any doors or windows that have been broken or removed with plywood, heavy plastic, or some similar materials to keep out rain **(Figure 17.26)**. Plywood, hinges, a hasp, and a padlock can be used to fashion a temporary door. Cover openings in roofs with plywood, roofing paper, heavy plastic sheeting, or tar paper. Use appropriate roofing nails if roofing, tar paper, or plastic is used. Tack down the edges with laths between the nails and the material.

Figure 17.26 Exterior openings must be secured before leaving the premises.

Overhaul

Overhaul involves those operations conducted once the main body of fire has been extinguished. These procedures consist of searching for and extinguishing hidden or remaining fire, placing the building and its contents in a safe condition, determining the cause of the fire, and recognizing and preserving evidence of arson (**Figure 17.27**). As discussed earlier in this chapter, overhaul should not start until authorized by the IC or fire cause investigator. Once the order is given, firefighters attempt to put the building, its contents, and the fire area in as safe and habitable a condition as possible and protected from the elements.

One of the first considerations during overhaul is safety. After a fire has been brought under control, there is time to plan and organize overhaul activities. This should be done in such a way as to provide the highest possible degree of safety to firefighters and others who might be allowed on the scene. The steps required to establish safe conditions include the following:

- Inspecting the premises

- Developing an operational plan

- Providing needed tools and equipment

- Eliminating or mitigating hazards (including securing the building's utilities)

One of the most common and most dangerous threats to firefighters during overhaul operations are the toxic gases that continue to be produced until all fire has been extinguished and the fire area thoroughly ventilated and allowed to cool. Wearing appropriate PPE, including respiratory protection, must be a routine part of the overhaul operation and should be enforced by the Incident Safety Officer (ISO).

////////

CAUTION

While performing overhaul and extinguishing hidden fires, firefighters must wear proper protective clothing including self-contained breathing apparatus (SCBA) until the atmosphere has been proven safe for a lower level of protection by reliable testing methods (Figure 17.28).

Figure 17.28 Before SCBAs can be removed, the atmosphere in the fire building must be monitored to ensure that it is safe.

Figure 17.27 Overhaul is a critical phase in every structure fire. It involves the extinguishment of any hidden or remaining fires including embers that are mixed in with fire debris.

Figure 17.29 Monitor firefighters for signs of fatigue, exhaustion, and heat stress.

Other hazards to firefighters during overhaul are present in many forms. Firefighters may be injured in falls through fire-weakened floors. When any potentially hazardous areas are identified, they should be marked or barricaded immediately. Firefighters can be injured by stepping on broken glass, nails, or other sharp objects. Because fire debris must often be manipulated by hand during overhaul, firefighters are susceptible to cuts, punctures, and thermal burns if gloves are not worn. Likewise, eye protection is critical if injuries to the eyes are to be avoided. Injuries, such as strains and sprains, during overhaul operations can be prevented through physical conditioning and by practicing safe lifting techniques and good body mechanics.

Another preventable cause of injury is fatigue (**Figure 17.29**). Exhausted firefighters are more susceptible to injury than those who are rested. Firefighters who were not directly involved in the rescue and fire control efforts should conduct overhaul operations if on-scene resources allow.

Salvage operations performed during fire fighting will directly affect any overhaul work that may be needed later. Many of the tools and equipment used for overhaul are the same as those used for other fire fighting operations. Some of the tools and equipment used specifically for overhaul, along with their uses, may include the following:

- *Pike poles and plaster hooks* — Open ceilings to check on fire extension.
- *Axes* — Open walls and floors.
- *Prying tools* — Remove door frames, window frames, and baseboards.
- *Power saws, drills, and screwdrivers* — Install temporary doors and window coverings.
- *Carryall, buckets, and tubs* — Carry debris or provide a basin for immersing smoldering material.
- *Shovels, bale hooks, and pitchforks* — Move baled or loose materials.
- *Thermal imaging camera* — Check void spaces and look for hot spots.

Carryall — Waterproof carrier or bag used to carry and catch debris or used as a water sump basin for immersing small burning objects.

Overhaul operations should be visually directed by a supervisor/officer not directly engaged in overhaul tasks. If a fire investigator is on the scene, this individual should be involved in planning and supervising the overhaul activities.

Fire Safety During Overhaul

Although charged hoselines should always be available for the extinguishment of hidden fires, the same size lines used to bring the fire under control are not always necessary during overhaul. Fire department pumpers can often be disconnected from hydrants at this time; however, departmental SOPs may dictate that at least one supply line be left in place as a precaution. Typically, 1½-inch (38 mm) or 1¾-inch (45 mm) attack lines are used for overhaul. However, some departments use 1-inch (25 mm) cotton forestry lines in 100-foot (30 m) lengths for overhaul operations, keeping larger attack lines available if needed.

During minor overhaul operations, water fire extinguishers or booster hoses may be used to extinguish small fires. At least one attack line should still be available in the event of a flare up. Regardless of the type of hose being used, place the nozzle so that it will not cause additional water damage. Do not allow water damage from leaking hoselines. Tighten or replace leaking couplings. Using a 100-foot (30 m) hoseline as the first section on attack lines greatly reduces the chances that any couplings other than those at the nozzle would even be inside a building. Using a booster line accomplishes the same goal.

To protect themselves during overhaul operations, firefighters must continue to maintain their situational awareness and their focus on safety. The following are some of the additional safety considerations during overhaul operations:

- Continue to work in teams of two or more.
- Maintain awareness of available exit routes.
- Maintain a RIC throughout the operation.
- Monitor personnel for the need for rehab.
- Beware of hidden gas or electrical utilities.
- Continue using the accountability system until the incident is terminated.

Locating Hidden Fires

Before starting a search for hidden fires, evaluate the condition of the area to be searched. The intensity of the fire and the amount of water used for its control are two important factors that affect the condition of the building. The intensity of the fire determines the extent to which structural members have been weakened. The amount of water used determines the additional weight placed on floors and walls due to the absorbent properties of the building contents. These factors should be considered along with appropriate measures to ensure firefighter safety during overhaul.

When evaluating the stability of a fire-damaged structure, firefighters should look for such other indicators of possible loss of structural integrity as the following:

- Weakened floors due to floor joists being burned away
- Concrete that has spalled due to heat
- Weakened steel roof members (tensile strength is affected at about 500°F (260°C)
- Walls offset because of elongation of steel roof supports
- Weakened roof trusses due to burn-through of key members
- Mortar in wall joints opened due to excessive heat
- Wall ties holding veneer/curtain walls melted from heat
- Heavy storage on mezzanines or upper floors
- Water pooled on upper floors
- Large quantities of wet insulation

Firefighters can often detect hidden fires by sight, touch, sound, or electronic sensors **(Figure 17.30, p. 884)**. The following are some of the indicators for each:

- Sight
  - Discoloration of materials
  - Peeling paint
  - Smoke emissions from cracks
  - Cracked plaster
  - Rippled wallpaper
  - Burned areas

Figure 17.30 Heat detectors can aid in locating hidden fire in concealed spaces.

Figure 17.31 Feeling walls and other surfaces with the back of the hand can sometimes reveal hidden fires.

- Touch — Feel walls and floors for heat with the back of the hand (**Figure 17.31**).
- Sound
  - Popping or cracking of fire burning
  - Hissing of steam
- Electronic sensors
  - Thermal (heat) signature detection with thermal imaging camera
  - Infrared heat detection

Overhaul Procedures

Typically, overhaul begins in the area of most severe fire involvement. The process of looking for fire extension should begin as soon as possible after the fire has been declared under control. The overhaul plan can then be systematically carried out. If it is found that the fire extended to other areas, firefighters must determine through what path it traveled (concealed wall spaces, unsealed pipe races, etc.). When floor beams have burned at their ends where they enter a party wall, overhaul the ends by flushing the voids in the wall with water. Also check the far side of the wall to see whether the fire or water has come through. Thoroughly check insulation materials because they can harbor hidden fires for a prolonged period. Usually it is necessary to remove the material in order to properly check it or extinguish fire in it.

An understanding of basic building construction will assist the firefighter in searching for hidden fires. When the fire has burned around windows or doors, there is a possibility that fire remains within the frames or casings. Open these areas to ensure complete extinguishment by simply pulling off the molding to expose the inner parts of the frame or casing (**Figure 17.32**). When fire has burned around a

Figure 17.32 Removing molding from around doors and windows can reveal hidden fire.

combustible roof or cornice, it is advisable to open the cornice and inspect for hidden fires. When dealing with balloon construction, check the attic and basement for fire extension.

NOTE: Refer to Chapter 4 to review information about building construction.

When concealed spaces below floors, above ceilings, or within walls and partitions must be opened during the search for hidden fires, move the furnishings of the room to locations where they will not be damaged. If it is not possible to move the contents, protect them with salvage covers. Remove only enough wall, ceiling, or floor covering to ensure complete extinguishment. Weight-bearing members should not be disturbed.

As discussed in Chapter 9, Forcible Entry, when opening concealed spaces, you need to consider whether the space contains electrical wiring, gas piping, or plumbing. Look for electrical outlets, gas connections, or water faucets before cutting into walls. Consideration should also be given to the restoration of the area. During overhaul operations, openings must often be made in construction to check for extension and allow extinguishment. When conditions allow, you should make neat, planned openings to ensure extinguishment but demonstrate workmanlike professionalism for future restoration.

Ceilings may be opened from below using a pike pole or other appropriate overhaul tool. To open some plaster ceilings, you must first break the plaster and then pull off the lath. Some plaster ceilings have chicken wire in the plaster. When these ceilings start to come down, they may fall in one very large piece. Some newer plaster ceilings are backed with gypsum wallboard instead of wooden lath. Metal or composition ceilings may be pulled from the joists in a like manner.

When pulling any ceiling, do not stand directly under the area to be opened. Always position yourself between the area being pulled and the doorway to keep the exit route from being blocked with falling debris. When pulling ceilings, always wear full protective clothing including eye and respiratory protection.

Frequently, small burning objects are uncovered during overhaul. Because of their size and condition, it is more effective to submerge entire objects in containers of water than to drench them with hose streams. Bathtubs, sinks, lavatories, and wash tubs are all useful for this purpose. Remove larger furnishings, such as mattresses, stuffed furniture, and bed linens, to the outside where they can be thoroughly extinguished. Remember that all scorched or partially burned articles may prove helpful to an investigator in preparing an inventory or determining the cause of the fire so these articles should be protected if possible.

The use of wetting agents such as Class A foam is of considerable value when extinguishing hidden fires. The penetrating qualities of wetting agents facilitate complete extinguishment in cotton, upholstery, and baled goods. The only way to ensure that fires in bales of rags, cotton, hay, etc., are out is to break them apart. Remove burning mattresses and upholstered furniture to the outside of the structure. Doing so reduces the mess inside and makes extinguishment easier. See **Skill Sheet 17-I-11** for more information on how to locate and extinguish hidden fires.

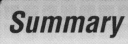

Summary

Customer service is the core of everything that fire departments do. Loss control is an important component of fire department service delivery and is but one aspect of customer-service opportunities. The philosophy of loss control is to minimize damage to structures and their contents during and after fire control operations. Salvage and overhaul operations are two of the most effective means of loss control. Salvage consists of those operations associated with fire fighting that aid in reducing primary and secondary damage during fire fighting operations. Overhaul involves those operations conducted once the main body of fire has been extinguished that consist of searching for and extinguishing hidden or remaining fire, placing the building and its contents in a safe condition, determining the cause of the fire, and recognizing and preserving evidence of arson. Planning, procedures, and equipment are essential for effective loss-control operations.

Fire Fighter I

1. What is the difference between salvage and overhaul?

2. List five items used in salvage operations.

3. How can water be removed from fire fighting operations using salvage covers?

4. When should overhaul start?

5. List three indicators of possible loss of structural integrity.

6. What are visual indicators of hidden fires?

7. What sounds may indicate a hidden fire?

8. Where does overhaul typically begin?

Step 1: Wash salvage cover with clean water and detergent by using a scrub brush.

Step 2: Rinse thoroughly with clean water.

Step 4: Inspect salvage cover.

Step 3: Hang to dry.

Step 5: Mark holes with chalk or marker.

Step 6: Patch according to manufacturer or departmental guidelines.

17-I-2
Roll a salvage cover for a one-firefighter spread.

Skill SHEETS

NOTE: Two firefighters must make initial folds to reduce the width of the cover to form this roll. Steps 1 through 8 are performed simultaneously by both firefighters on opposite sides of the cover. Steps 9 through 12 may be performed by both firefighters who are stationed at the same end of the roll.

Step 1: Grasp the cover with the outside hand midway between the center and the edge to be folded.

Step 2: Place the other hand on the cover as a pivot midway between the outside hand and the center.

Step 4: Grasp the cover corner with the outside hand.

Step 5: Place the other hand as a pivot on the cover over the outside fold.

Step 6: Bring this outside edge over to the center, and place it on top of and in line with the previously placed first fold.

Step 7: Fold the other half of the cover in the same manner by using Steps 1 through 6.

Step 3: Bring the fold over to the center of the cover. This creates an inside fold (center) and an outside fold.

Step 8: Straighten the folds if they are not straight.

Step 9: Fold over about 12 inches (300 mm) at each end of the cover to make clean, even ends for the completed roll.

Step 10: Start the roll by rolling and compressing one end into a tight compact roll; roll toward the opposite end.

Step 11: Tuck in any wrinkles that form ahead of the roll as the roll progresses.

Step 12: Secure the completed roll with inner tube bands or Velcro® straps or tie with cords.

Step 1: Start at one end of the object to be covered.

Step 2: Unroll a sufficient amount to cover the end.

Step 3: Unroll toward the opposite end and let the rest of the roll fall into place at the other end.

Step 4: Stand at one end.

Step 5: Grasp the open edges where convenient, one edge in each hand.

Step 6: Open the sides of the cover over the object by snapping both hands up and out.

Step 7: Open the other end of the cover over the object in the same manner.

Step 8: Tuck in all loose edges at the bottom.

NOTE: In addition to covering stacks of objects, the rolled cover may be used as a floor runner. Just unroll the cover and spread it out as wide as necessary.

NOTE: Two firefighters must make initial folds to reduce the width of the cover. Steps 1 through 7 are performed simultaneously by both firefighters on opposite sides of the cover. Steps 8 through 13 may be performed by both firefighters who are stationed at the same end of the fold.

Step 1: Grasp the cover with the outside hand midway between the center and the edge to be folded.

Step 2: Place the other hand on the cover as a pivot midway between the outside hand and the center.

Step 4: Grasp the cover corner with the outside hand.

Step 5: Place the other hand as a pivot on the cover over the outside fold.

Step 6: Bring this outside edge over to the center, and place it on top of and in line with the previously placed first fold.

Step 3: Bring the fold over to the center of the cover. This will create an inside fold (center) and an outside fold.

Step 7: Fold the other half of the cover in the same manner by using Steps 1 through 6.

Step 8: Straighten the folds if they are not straight.

17-I-4 continued
Fold a salvage cover for a one-firefighter spread.

Step 9: Grasp the same end of the cover, with the cover folded to reduce width.

Step 10: Bring this end to a point just short of the center.

Step 11: Use one hand as a pivot and bring the folded end over and place on top of the first fold.

Step 12: Fold the other end of the cover toward the center, leaving about 4 inches (100 mm) between the two folds.

Step 13: Place one fold on top of the other for the completed fold; the space between the folds now serves as a hinge.

Skill SHEETS

17-I-5
Spread a folded salvage cover —
One-firefighter method.

Step 1: Lay the folded cover on top of and near the center of the object to be covered.

Step 2: Separate the cover at the first fold.

Step 3: Select either end and continue to unfold the salvage cover by separating the next fold.

Step 7: Unfold the other end of the cover in the same manner over the object.

Step 8: Stand at one end.

Step 9: Grasp the open edges where convenient, one edge in each hand.

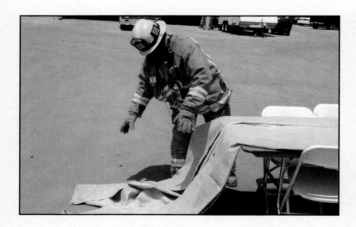

Step 4: Unfold this same end toward the end of the object to be covered.

Step 5: Grasp the end of the cover near the center with both hands to prevent the corners from falling outward.

Step 6: Bring the end of the cover into position over the end of the object being covered.

Step 10: Open the sides of the cover over the object by snapping both hands up and out.

Step 11: Open the other end of the cover over the object in the same manner.

Step 12: Tuck in all loose edges at the bottom.

17-I-6
Fold a salvage cover
for a two-firefighter spread.

Skill SHEETS

NOTE: Two firefighters must make initial folds to reduce the width of the cover. Steps 1 through 11 are performed simultaneously by both firefighters. Steps 12 through 19 are performed by the respective firefighters. Steps 20 through 23 are performed simultaneously by both firefighters.

Step 1: Grasp opposite ends of the cover at the center grommet with the cover stretched lengthwise.

Step 2: Pull the cover tightly between each firefighter.

Step 3: Raise this center fold high above the ground.

Step 4: Shake out the wrinkles to form the first half-fold.

Step 10: Stretch that part of the cover being folded tightly between each firefighter.

Step 11: Make the quarter-fold by folding the open edges over the folded edge.

Step 5: Spread the half-fold on the ground.

Step 6: Smooth the half-fold flat to remove the wrinkles.

Step 7: Stand at each end of the half-fold and face the cover.

Step 8: Grasp the open-edge corners with the hand nearest to these corners.

Step 9: Place the corresponding foot at the center of the half-fold, making a pivot for the next fold.

Step 12: *Firefighter #1:* Stand on one end of the quarter-fold.

Step 13: *Firefighter #2:* Grasp the opposite end and shake out all the wrinkles.

Step 14: *Firefighter #2:* Carry this end to the opposite end, maintaining alignment of outside edges.

Skill SHEETS

17-I-6 continued
Fold a salvage cover
for a two-firefighter spread.

Step 15: **_Both Firefighters:_** Place the carried end on the opposite end, aligning all edges.

Step 16: **_Both Firefighters:_** Position at opposite ends.

Step 17: **_Firefighter #2:_** Stand on the folded end of the cover.

Step 18: **_Firefighter #1:_** Shake out all wrinkles.

Step 19: **_Firefighter #1:_** Align all edges.

Step 20: Grasp the open ends and use the inside foot as a pivot for the next fold.

Step 21: Bring these open ends over and place them just short of the folded center fold.

Step 22: Continue this folding process by bringing the open ends over and just short of the folded end.

NOTE: During this fold, the free hand may be used as a pivot to hold the cover straight.

Step 23: Complete the operation by making one more fold in the same manner.

Step 24: Bring the open ends over and to the folded end using the free hand as a pivot during the fold.

17-I-7
Spread a folded salvage cover —
Two-firefighter balloon throw.

Skill SHEETS

Step 1: Stretch the cover along one side of the object to be covered.

Step 2: Separate the last half-fold by grasping each side of the cover near the ends.

Step 3: Lay the side of the cover closest to the furniture on the ground.

Step 4: Make several accordion folds in the inside hand.

Step 5: Place the outside hand about midway down the end hem.

Step 6: Place the inside foot on the corner of the cover to hold it in place.

Step 7: Pull the cover tightly between each firefighter.

Step 8: Swing the folded part down, up, and out in one sweeping movement in order to pocket as much air as possible.

Step 9: Pitch or carry the accordion folds across the object when the cover is as high as each firefighter can reach. This action causes the cover to float over the object.

Step 10: Guide the cover into position as it floats over the object.

Step 11: Straighten the sides for better water runoff.

Step 1: Open the salvage cover.

Step 2: Lay the cover flat at the desired location.

Step 4: Turn the cover over.

Step 5: Adjust the chute to collect and channel water by elevating one end.

Step 6: Extend the other end out a door or window.

Step 3: Roll the opposite edges of the salvage cover toward the middle until there is a 3-foot (1 m) width between the rolls.

Step 1: Open the salvage cover.

Step 2: Lay the cover flat at the desired location.

Step 3: Place pike poles at opposite edges of the salvage cover with the pike extending off the end of the cover.

Step 4: Roll the edges over the pike poles toward the middle until there is a 3-foot (1 m) width between the rolls.

Step 5: Turn the cover over, keeping the folds in place.

Step 6: Place the chute to collect and channel water.

NOTE: This can be done by hooking the pike poles over a ladder rung or similar object.

Step 7: Extend the other end out a door or window.

Step 1: Open the salvage cover.

Step 2: Lay the cover flat at the desired location.

Step 3: Roll the sides inward approximately 3 feet (1 m).

Step 4: Lay the ends of the side rolls over at a 90-degree angle to form the corners of the basin.

Step 5: Roll one end into a tight roll on top of the side roll and form a projected flap.

Step 6: Lift the edge roll.

Step 7: Tuck the end roll to lock the corners.

Step 8: Roll the other end in a like manner.

Step 9: Lock the corners.

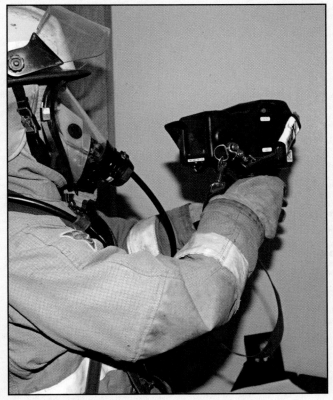

Step 1: Confirm order with officer to overhaul.

Step 2: Locate area(s) with potential hidden or smoldering fire.

Step 3: Remove ceiling and wall covering and insulation beginning with area closest to hidden or smoldering fire.

Step 4: Extinguish hidden and smoldering fires with small handline.

Step 5: Report to officer completion of assigned task.

Chapter Contents

Protecting Fire Scene Evidence

This chapter provides information that addresses the following job performance requirements of NFPA® 1001, *Standard for Fire Fighter Professional Qualifications* (2008):

NFPA® 1001 references for Chapter 18:

5.1.1
5.3.8
5.3.13
6.3.4

Chapter Objectives

Fire Fighter I Objectives

1. Describe signs and indications of an incendiary fire.
2. Summarize important observations to be made en route, after arriving at the scene, and during fire fighting operations.
3. Discuss firefighter conduct and statements at the scene.
4. Explain firefighter responsibilities after the fire.
5. Discuss protecting and preserving evidence.

Fire Fighter II Objectives

1. Discuss the roles of firefighters and investigators at investigations.
2. Summarize important observations to be made en route, after arriving at the scene, and during fire fighting operations.
3. Discuss firefighter conduct and statements at the scene.
4. Explain firefighter responsibilities after the fire.
5. Explain how legal considerations affect firefighters during operations that may involve incendiary evidence.
6. Discuss protecting and preserving evidence.
7. Protect evidence of fire cause and origin.

Protecting Fire Scene Evidence

Case History

Arson continues to be a significant portion of the fire problem in the United States. In 2005, over 31,000 structure fires were intentionally set. These fires caused over $664 million dollars in damage. Worst of all, 315 people were killed in these fires. As you can see, arson fires continue to be a serious element of the fire problem.

Prosecuting the crime of arson is challenging. Generally, much of the evidence is destroyed by the fire itself. Successful prosecution of the arsonist begins with you, the firefighter. While you may not be involved in the actual investigation, you do have the ability to make a critical contribution to it by ensuring that any evidence that is present is preserved. Without that evidence being protected, arson investigators may have a difficult time linking the cause of the fire to the suspected arsonist. This chapter discusses the methods used by firefighters to identify and then preserve potential evidence.

Source: National Fire Protection Association *Fire Loss in the United States During 2005 – Full Report.*

The majority of fires that occur each year in North America are of accidental or natural origin. *Accidental fires* are those resulting from unintentional acts or omissions (failures to act). *Natural causes* are such things as lightning, spontaneous heating, and friction between dehydrating logs that are stacked on top of each other. However, arson fires (also called *incendiary* fires) are the leading cause of fire loss, and they are the cause of one out of eight fire fatalities. Because the prosecution must prove that the perpetrator acted both *willfully* (intentionally) and *maliciously* (with criminal intent), only 2 percent of suspected incendiary fires lead to convictions.

Most fire departments investigate all fires to which they respond in order to determine the cause and origin of the fire **(Figure 18.1, p. 908)**. The cause of a fire is a combination of three factors:

- Fuel that ignited
- Form and source of the heat of ignition
- Act or omission that helped to bring these two factors together

In order for a fire cause determination to be made, evidence at the scene must be protected. A fire officer, fire investigator, or firefighter trained in collecting and preserving evidence collects and analyzes the evidence to determine the exact cause.

Knowing the causes of fires helps prevent fires in the future. When a new or different accidental fire cause is identified, requirements can be added to existing fire codes, or new codes adopted, that will help to reduce or eliminate future fires from

Incendiary — (1) A fire deliberately set under circumstances in which the responsible party knows it should not be ignited. (2) Relating to or involving a deliberate burning of property.

that specific cause. When an arson fire is investigated and the source of ignition identified, the case can be pursued and the perpetrator tried and convicted. Heavily publicizing the results — a lengthy prison term — helps to deter other prospective arsonists. As the number and severity of fires decrease, so do losses of life and adverse economic impacts.

As a firefighter on the scene, you may have an opportunity to observe evidence of fire cause and to assist in the fire cause determination effort. The observations of firefighters on scene help investigators determine how a fire started and why it behaved as it did. On-scene firefighters can also be the first link in the chain of evidence that may become a critical point in subsequent litigation.

The fire department is responsible for responding to and extinguishing fires as quickly as possible; however, the results of the fire fighting operation may impair an investigator in conducting a fire origin and cause determination investigation. The firefighters' actions may move evidence from its original location or completely sweep it away **(Figure 18.2)**. Therefore it is extremely important that firefighters take whatever precautions they can to protect evidence and keep it intact while fighting a fire.

Information gathered at the scene can be of critical importance to the fire investigator. This is especially important because investigators are seldom present during the early stages of a fire suppression operation. Before the investigator arrives on scene, firefighters may have moved building contents and otherwise disturbed or destroyed fire scene evidence in the process of controlling the fire. Because evidence can be disturbed or destroyed, firefighters must make every effort to recognize and protect anything that could point to the origin and cause

Figure 18.1 Firefighters may assist investigators in determining the cause and origin of fires. *Courtesy of Sheldon Levi, IFPA.*

Figure 18.2 Valuable evidence can be compromised by firefighters' actions on the scene.

of a fire. While fighting the fire, firefighters may notice indicators of the cause or origin of the fire that are later destroyed by the fire or fire suppression operations. Firefighters must recognize the importance of these observations and be prepared to report them as accurately as possible to the fire investigator.

NFPA® 1001 requires those qualified at the Fire Fighter I level to know the following about fire scene evidence preservation:

- Signs of area of origin
- Signs of arson
- Reasons for protecting fire scene

The standard also requires those qualified at the Fire Fighter I level to be able to:

- Recognize and preserve obvious signs of area of origin.
- Recognize and preserve signs of arson.

NFPA® 1001 requires those qualified at the Fire Fighter II level to also know:

- Methods to assess origin and cause
- Types of evidence
- Means of protecting evidence
- Roles of firefighters, criminal investigators, and insurance investigators in fire investigations
- Effects and problems associated with removing property or evidence from the scene

The standard also requires those qualified at the Fire Fighter II level to be capable of:

- Locating the fire's origin area
- Recognizing possible fire causes
- Protecting evidence

This chapter contains information on the responsibilities of the firefighter and the fire investigator. Observations that the firefighter can make en route, upon arrival, and during and after the fire that could assist in a subsequent fire investigation are also included. The chapter also discusses steps for securing the fire scene and protecting evidence. Finally, the chapter covers the firefighter's conduct at the scene and legal considerations. For more information, refer to the IFSTA **Introduction to Fire Origin and Cause** manual.

Roles and Responsibilities

Determining the cause of a fire may be a relatively simple procedure that requires only the expertise of the emergency responders on the scene. More complex fire incidents or possible crime scenes often require the additional assistance of law enforcement personnel and qualified fire investigators at the local, state, or provincial and federal level. The guidelines for initiating each of these steps vary among jurisdictions and according to the nature of the incident. It is critical that firefighters be aware of the guidelines and/or regulations in their areas for contacting law enforcement personnel, calling in fire investigators, and following guidelines in all matters relating to legal issues.

The Firefighter

In most jurisdictions, the fire chief has the legal responsibility for determining the cause and origin of a fire. The fire chief delegates this authority to the fire officers and firefighters at the scene to determine the true and specific cause of the fire. Proper training enables firefighters to recognize and collect important information by observing the fire and its behavior during the response, upon arrival, when entering the structure, and while locating and extinguishing the fire. Better than anyone else, the first-arriving firefighters are in a position to observe unusual conditions that may indicate an incendiary fire.

First-arriving firefighters should make mental notes of the following:

- Vehicles and people present in the area
- Status of doors and windows (locked or open)
- Evidence of forced entry by anyone other than firefighters
- Contents of the rooms — are they in usual order, ransacked, or unusually bare
- Indications of unusual fire behavior or more than one point of origin

During fireground operations, firefighters must be aware that what they do and how they do it can affect the determination of the origin and cause of the fire. Having an alert and open mind combined with performing judicious and careful overhaul might also uncover or preserve important evidence that would otherwise be lost.

The Investigator

Fire marshals, fire inspectors, or other members of a fire prevention bureau may be responsible for conducting investigations beyond the determination of fire origin and cause **(Figure 18.3)**. In many departments, fire investigators are also sworn peace officers who are authorized to carry weapons and make arrests. Firefighters may be questioned by an investigator or asked to assist in some aspect of an investigation.

Some fire departments have special fire investigation or arson squads. In other departments, fire department and law enforcement personnel work together. In some localities the police department has sole responsibility for handling an arson investigation. In other areas, the responsibility for cause determination and investigation lies with the state fire marshal or some other state agency rather than with local agencies. Private companies may conduct separate investigations when a fire involves their property, or an insurance company investigator may conduct the investigation.

On the Scene: Observations and Conduct

In the early stages of a fire fighting operation, firefighters may have to make mental notes of some important observations until they can report them. For example, signs of forced entry by someone other than firefighters may not be reportable until after the fire is controlled. At the earliest opportunity, however, remember to report to the proper authority any observations of unusual conditions.

Observations En Route

As a firefighter, your responsibility for gathering information begins when the alarm is received. Some of the things you should notice are the following:

- *Time of day* — Are people and circumstances at the scene as they normally would be at this time of day? For example, if a fire is in a dwelling at 3 a.m., the building occupants would probably be wearing night clothes, not street clothes. If a fire is in

Point of Origin — Exact physical location where the heat source and fuel come in contact with each other and a fire begins.

Figure 18.3 Trained fire investigators may be required to conduct thorough investigations if the company officer cannot determine the cause and origin of the fire.

an office building well after working hours, the owner or employees may need to explain why they are present at that hour.

- *Weather and natural hazards* — Is it hot, cold, or stormy? Is there heavy snow, ice, high water, or fog? If the outside temperature is high, the furnace in the structure would not be operating. If the outside temperature is low, the windows normally would not be wide open. Arsonists sometimes set fires during inclement weather because the fire department's response may be delayed.

- *Barriers* — Are there any barriers such as barricades, fallen trees, cables, trash bins, or vehicles blocking access to hydrants, sprinkler and standpipe connections, streets, or driveways? These situations could indicate an attempt to impede fire-fighters' access and delay fire suppression efforts **(Figure 18.4)**.

- *People leaving the scene* — Are people leaving the scene in haste? Most people are intrigued by a fire and will remain in the area to watch **(Figure 18.5)**. Noticing people leaving the scene by vehicle or on foot may be an important observation to an investigation. If a vehicle is seen leaving the scene (especially at high speed), note the color of the vehicle, as many details about it as possible, and especially the license plate number. If possible, note how many occupants are in the vehicle. If someone is seen leaving the scene on foot, try to remember that person's attire, general physical appearance, and any peculiarities such as trying to leave unde-tected, walking briskly, or looking over his or her shoulder.

Figure 18.4 Arsonists may try to impede fire department access to the fire scene by obstructing roadways.

Figure 18.5 Most people are curious about fires in progress and will remain to watch firefighters at work.

Observations Upon Arrival

Additional information that firefighters may notice while carrying out their duties include the following:

- *Time of arrival and extent of fire* — Note the extent of fire involvement at the time of arrival. Observe the color and movement of smoke and flames.

- *Wind direction and velocity* — Note wind direction and velocity. These factors may have a great effect on the natural path of fire spread.

- *Doors or windows locked or unlocked* — Note the position and condition of doors and windows upon arrival. Before opening doors and windows, determine whether they are locked, unlocked, or show any signs of forced entry such as broken glass or damaged frames **(Figure 18.6, p. 912)**. In some cases the insides of windows may be covered with blankets, paint, or paper to delay discovery of the fire.

Trailer — Combustible material, such as rolled rags, blankets, newspapers, or flammable liquid, often used in intentionally set fires in order to spread fire from one point or area to other points or areas.

- *Location of the fire* — Observe the location of the fire. This information may help to identify the area of origin. Also note whether there were separate, seemingly unconnected fires. If so, the fire might have been set in several locations or spread by trailers (combustible material used to spread fire from one area to another).

- *Containers or cans* — Note metal or plastic containers found inside or outside the structure. They may have been used to transport accelerants.

- *Burglary tools* — Note any tools such as pry bars or screwdrivers found in areas away from workshops. They may have been used to break into the facility (**Figure 18.7**).

- *Familiar faces* — Notice familiar faces in the crowd of bystanders — people who are seen at numerous fires in the area. They may be fire buffs, or they may be habitual firesetters.

Figure 18.6 Windows broken prior to the fire may indicate forced entry into the structure.

Figure 18.7 Hand tools found outside a door or window may indicate illegal entry.

Observations During Fire Fighting Operations

As the operation continues, firefighters should continue to observe conditions that may lead to the determination of the fire cause:

- *Unusual odors* — Note any unusual odors. Even though firefighters wear SCBA during interior fire fighting and overhaul operations, they may smell unusual odors outside the burning building. In addition, odors may cling to turnout gear and be detectible after firefighters come out of the smoke.

- *Abnormal behavior of fire when water is applied* — Observe how the fire behaves when water is applied to it. Flashbacks, reignition, several rekindles in the same area, and an increase in the intensity of the fire may indicate possible accelerant use. Water applied to a burning liquid accelerant may cause it to splatter, allowing flame intensity to increase and the fire to spread in several directions. Water applied to fires involving ordinary combustibles usually reduces flame spread.

- *Obstacles hindering fire fighting* — Note whether any doors are nailed shut or furniture is placed in doorways and hallways to hinder fire fighting efforts (**Figure 18.8**). Holes may be cut in the floors that not only hinder fire suppression activities but also spread the fire.

- *Incendiary devices* — Note any pieces of glass, fragments of bottles or containers, and metal parts of electrical or mechanical devices. Most incendiary devices (any device designed and used to start a fire) leave evidence of their existence (**Figure 18.9**). More than one device may be found, and sometimes a malfunctioning device can be found during a thorough search.

- *Trailers* — Note combustible materials such as rolled rags, blankets, newspapers, or ignitable liquid (trailer) that could be used to spread fire from one point to another. Trailers usually leave char or burn patterns and may be used with incendiary ignition devices (**Figure 18.10**).

- *Structural alterations* — Observe if there appear to be alterations to the structure: removal of plaster or drywall to expose wood; holes made in ceilings, walls, and floors; and fire doors secured in an open position. All of these methods can be used to allow a fire to spread quickly through the structure.

Incendiary Device — Material or chemicals designed and used to start a fire.

Figure 18.8 Arsonists may also try to hinder firefighters inside of buildings by piling furnishing in hallways or against doors.

Figure 18.9 Incendiary devices, such as Molotov cocktails, or parts of them, may be found at the scene.

Figure 18.10 Trailers are used to spread fire through a structure.

- *Fire patterns* — Note the fire's movement and intensity patterns. These can trace how the fire spread, identify the original ignition source, and determine the fuel(s) involved. Carefully note any areas of irregular burning or locally heavy charring in areas of little fuel **(Figure 18.11)**.

- *Heat intensity* — Look for evidence of high heat intensity, especially in relation to other areas of the same room. This may indicate the use of accelerants or intentionally disconnected gas lines. However, other factors may contribute to variations in heat intensity. One of these factors is synthetic materials, such as polyurethane, that may produce abnormally high heat intensity and may be confused with the use of accelerants.

- *Availability of documents* — Note anyone who conveniently produces insurance policies, inventory lists, deeds, or other legal documents that would normally be locked away. This may indicate that the fire was planned.

- *Fire detection and protection systems* — Check for evidence of tampering or intentional damage if fire detection and suppression systems and devices are inoperable **(Figure 18.12)**.

- *Intrusion alarms* — Check intrusion alarms to see whether they have been tampered with or intentionally disabled.

- *Location of fire* — Note any possible ignition sources in the area of the fire. Fires burning in areas remote from normal ignition sources demand an explanation. Some examples are fires in closets, bathtubs, file drawers, or in the center of the floor.

- *Personal possessions* — Note anything that might suggest that preparations were made for a fire: absence or shortage of clothing, furnishings, appliances, food, and dishes; absence of personal possessions such as diplomas, financial papers, and toys; absence of items of sentimental value such as photo albums, special collections, wedding pictures, and heirlooms; absence of pets that would ordinarily be in the structure **(Figure 18.13)**. (**NOTE:** Do not read too much into a shortage of material possessions. A person's economic status may dictate his or her lifestyle, and some people just do not have as much as others.)

- *Household items* — Note whether any major household items appear to be removed or replaced with those of lesser value or of inferior quality. Check to see whether major appliances were disconnected or unplugged and determine why they were in this condition.

- *Equipment or inventory* — Look for obsolete equipment or inventory, fixtures, display cases, equipment, and raw materials.

- *Business records* — Determine whether important business records are out of their normal places and left where they would be endangered by fire. Check safes, fire-resistant files, etc., to determine whether they are open and exposing the contents.

Conduct and Statements at the Scene

Firefighters and their officers should obtain as much information as possible pertaining to a fire. The owners or occupants of the property should be allowed to talk freely if they are inclined to do so. Valuable information is sometimes gathered this way. Firefighters should *not* attempt to interrogate a potential arson suspect unless they are trained and authorized to do so. In most cases, that is the job of a trained fire investigator.

Figure 18.11 Areas that are more heavily burned may be an indicator of the origin of the fire.

Figure 18.12 Smoke detectors may have been disabled to delay detection of the fire.

Figure 18.13 Look for evidence that clothing or other valuables were removed before the fire.

Firefighters should refrain from expressing personal opinions to anyone (even other firefighters) about the probable cause. These opinions could be overheard by the property owner, news media, or other bystanders who could consider such statements as fact. Unauthorized remarks that are published or broadcast can be very embarrassing to the fire department. Such remarks can also impede the efforts of an investigator to prove malicious intent as the fire cause. A sufficient reply to any question concerning cause is *The fire is under investigation.*

After the investigator arrives, firefighters should make their statements only to this individual. Any public statement regarding the fire cause should be made only after the investigator and ranking fire officer have agreed to its accuracy and validity and have given permission for it to be released.

Responsibilities After the Fire

Firefighters should report all of their observations concerning the fire to their supervisor as soon as possible. If asked, each firefighter should write a chronological account of important circumstances personally observed. This documentation should take place as soon as reasonably possible after the incident while observations are still fresh.

Firefighters should not discuss their observations with other crew members until after documentation is complete. This helps to avoid the *group statement* tendency where members of the crew discuss the events and come to a consensus about what happened. A written account will be valuable if the firefighter must testify in court later. Cases often come to trial years after an incident and memory alone is not always reliable. Hearsay should be reported to the investigator for validation. For example: *The neighbor told me he saw the lights flickering for a few days before the fire.* This is hearsay, but it may be very helpful to the investigator.

Improperly done, overhaul operations can be detrimental to the fire cause investigation. Some departments take great pride in their overhaul work and boast that they leave a building neater, cleaner, and more orderly than it was before the fire. This thoroughness in overhaul is admirable, but in some cases it destroys evidence of how a fire started. To avoid disturbing or destroying evidence, delay overhaul operations until the origin and cause of the fire have been determined. Once critical evidence has been identified and protected, overhaul operations can begin (**Figure 18.14**). Limit or postpone nonessential overhaul operations until the IC or whoever is in charge of the investigation authorizes it.

Figure 18.14 Once the investigator completes his or her work, firefighters can continue to preserve evidence and control the scene.

Securing the Fire Scene

The best efforts to avoid disturbing evidence of the cause of a fire are wasted unless the building and premises are properly secured until an investigator has finished evaluating the evidence exactly as it appears at the scene. Firefighters should take care not to contaminate the scene while operating power tools, hoselines, or other equipment. For example, firefighters can inadvertently contaminate the scene if they spill gasoline while refilling power tools and then track that back into the scene. This can result in false positive indications by arson dogs.

The fire department has the authority to deny access to any building (even to the owner) during fire fighting operations and for a reasonable length of time after fire suppression is terminated. If an investigator is not immediately available, the fire department or other lawful authority should keep control of the premises until all evidence has been collected (**Figure 18.15**). Mark, tag, and photograph evidence at this time. Once all fire personnel leave the scene, a search warrant or written consent to search will be required to reenter the premises. The job of tagging and photographing evidence may be given to law enforcement personnel, depending on local policies and personnel availability. Regardless of who is assigned this task, it should be carried out only by personnel trained in evidence collection and preservation.

Fire personnel should not allow anyone to enter a fire scene without the investigator's permission, and if allowed in, an authorized individual should escort the subject. Keep a written log of anyone other than fire personnel who enters the scene during fireground operations or the investigation. The log should show the person's name, the time of entry, the time of departure, and a description of any items the person took from the scene.

The premises can be secured and protected in several ways with the use of few personnel. Areas that are fenced can be monitored by one person at a locked gate. At large fire scenes, law enforcement personnel or private security guards may be used. In some cases, all doors, windows,

Figure 18.15 A law enforcement agency may be needed to secure the fire scene.

or other openings can be covered with plywood or similar material. Cordoning off the area also can help secure the scene during fire fighting operations. With the area cordoned, bystanders are kept at a safe distance from the incident and out of the way of emergency personnel. Once the emergency phase of the operation is completed, a more substantial barrier such as a chain-link fence may be needed for long-term scene security.

Cordoning can be accomplished with rope or specially designed fire and police line tape. It may be attached to signs, utility poles, parking meters, or any other stationary objects readily available. Do not attach tape to any vehicle that may need to be moved during the incident. Once in place, law enforcement personnel should monitor the line to make sure people do not cross it. Be aware of seemingly innocent persons (including curious people and the press) attempting to cross a line. Escort out anyone in the cordoned area who is not a part of the operation. Record the time, location, and his or her identity for future reference.

Legal Considerations

As previously discussed, firefighters may remain on the location as long as necessary, but once they leave they may be required to get a search warrant to reenter the scene. Requiring firefighters to obtain a search warrant to reenter the scene is based on the case of Michigan vs. Tyler (436 U.S. 499, 56 L.Ed.2d 486 [1978]). The U.S. Supreme Court held in that case that *once in a building [to extinguish a fire], firefighters may seize [without a warrant] evidence of arson that is in plain view . . . [and] officials need no warrant to remain in a building for a reasonable time to investigate the cause of a blaze after it has been extinguished.*

The Court agreed, with modification, with the Michigan State Supreme Court's statement that *[if] there has been a fire, the blaze extinguished and the firefighters have left the premises, a warrant is required to re-enter and search the premises, unless there is consent . . . .*

The impact of these decisions seems to be that if there is incendiary evidence, the fire department should leave at least one person on the premises until an investigator arrives **(Figure 18.16)**. To leave the premises, return later without a search warrant, and make a search might be enough to make prosecution impossible or for an appellate court to overturn a conviction.

Each department must comply with the legal opinions that affect its jurisdiction in this regard. These opinions or interpretations can be obtained from such persons as the district attorney or state attorney general. The fire department should write an SOP concerning these opinions.

Figure 18.16 Legally, the fire scene must remain in the possession of the fire department and under their control until the investigation is complete.

Protecting and Preserving Evidence

Firefighters should protect any evidence they find by keeping it untouched and undisturbed until an investigator arrives. They should not gather or handle evidence unless it is absolutely necessary in order to preserve it. If firefighters handle or procure evidence, they then become a link in the chain of custody for that evidence. Firefighters must accurately document all actions associated with that evidence as soon as possible and remember that it may be necessary for them to appear in court at some later date. Firefighters must know and follow their departmental SOP on evidence gathering and preservation.

Evidence must remain undisturbed except when absolutely necessary for the extinguishment of the fire. Firefighters must avoid trampling over possible evidence and obliterating it. The same precaution applied to the excessive use of water may help avoid similar unsatisfactory results. Human footprints and tire marks must be protected. Cardboard boxes placed over prints can prevent otherwise clear prints from being degraded before they are photographed or plaster casts made (**Figure 18.17**). Protect completely or partially burned papers found in a furnace, stove, or fireplace by immediately closing dampers and other openings. Leave charred documents found in containers such as wastebaskets, small file cabinets, and binders that can be moved easily. Keep these items away from drafts.

After evidence has been properly collected by an investigator, debris may be removed. Remove charred materials to prevent the possibility of rekindle and to help reduce smoke damage. Unburned materials can be separated from the debris and cleaned. Debris may be shoveled into large containers such as buckets, tubs, or bags to reduce the number of trips between the fire area and the debris collection location. Dumping debris onto streets, sidewalks, or shrubbery can cause poor public relations. If a backyard or alley not visible from the street is not available, it may be necessary to dump the debris in a driveway until it can be removed permanently. Dumping debris on inexpensive plastic tarps makes it easier to clean up, protects the drive or yard, and is good for public relations. **Skill Sheet 18-II-1** discusses protection of evidence of fire cause and origin.

Figure 18.17 Footprints and tire marks can be preserved by covering them with a cardboard box.

Summary

Before an investigation into the origin and cause of a fire can be conducted, there must be evidence to evaluate. As a firefighter, one of your most important responsibilities is to avoid disturbing or destroying evidence while fighting the fire. In the area of origin — as determined by how the fire behaved, burn patterns on walls, depth of char, and other clues — you must use appropriate caution when spraying water, moving debris, and even walking around. Once the area of origin is known, a more thorough investigation can be conducted to determine the exact cause of the fire. As a firefighter, you might be assigned to determine the cause of the fire; more likely, you may be assigned to assist your supervisor or a fire investigator in making that determination. If the fire origin and cause investigation reveals evidence of arson, the property becomes a crime scene and must be treated as one. You and your fellow firefighters must cooperate fully with whoever is assigned to investigate the crime.

Fire Fighter I

1. What observations should be made en route?

2. What observations should be made upon arrival?

3. What observations should be made during fire fighting operations?

4. What actions should firefighters take after a fire?

5. Why is protecting evidence important?

Fire Fighter II

6. What individuals may have responsibilities in a fire investigation?

7. What are some observations that may indicate a fire was incendiary?

8. What is one way that a fire scene can be secured?

9. Why should the fire department leave at least one person on the premises of a scene until the investigator arrives?

10. What are ways that evidence can be protected?

Step 1: Protect potential evidence.

Step 2: Preserve evidence as necessary.

Step 3: Move evidence as necessary.

Step 4: Record information about evidence.

Step 5: Provide evidence and records to investigator before leaving incident site.

Chapter Contents

Fire Department Communications

This chapter provides information that addresses the following job performance requirements of NFPA® 1001, *Standard for Fire Fighter Professional Qualifications (2008)*:

NFPA® 1001 references for Chapter 19:

5.2.1
5.2.2
5.2.3
6.2.1
6.2.2

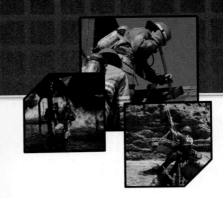

Chapter Objectives

Fire Fighter I Objectives

1. Describe communication responsibilities of the firefighter.
2. Summarize necessary skills for fire department communications.
3. Describe basic communications equipment used in telecommunications centers.
4. Describe basic business telephone courtesies.
5. Explain how a firefighter should proceed when receiving emergency calls from the public.
6. Describe types of public alerting systems.
7. Describe procedures that the public should use to report a fire or other emergency.)
8. Discuss ways of alerting fire department personnel to emergencies.
9. Summarize guidelines for radio communications.
10. Describe information given in arrival and progress reports.
11. Explain the purpose of tactical channels.
12. Discuss calls for additional resources and emergency radio traffic.
13. Discuss evacuation signals and personnel accountability reports.
14. Handle business calls and reports of emergencies. (Skill Sheet 19-I-1)
15. Use a portable radio for routine and emergency traffic. (Skill Sheet 19-I-2)

Fire Fighter II Objectives

1. Summarize guidelines for radio communications.
2. Describe information given in arrival and progress reports.
3. Explain the purpose of tactical channels.
4. Discuss calls for additional resources and emergency radio traffic.
5. Discuss evacuation signals and personnel accountability reports.
6. Summarize the information in incident reports.
7. Create an incident report. (Skill Sheet 19-II-1)

Fire Department Communications

Case History

One of the keys to both the success of fireground operations and to the safety of firefighters performing those operations is effective communication. Over the last 20 years there have been numerous fires where communication problems contributed to the deaths of firefighters. Several of these incidents involved inadequate radio communications equipment. In a few cases firefighters had difficulty operating communications equipment properly after the situation had become critical.

A metropolitan fire department's internal investigation of a fire in 1993 that resulted in the death of two firefighters acknowledged the importance of firefighters knowing how to use portable communications equipment (radios). The report recommended better training on incident command for firefighters and officers with an emphasis on fireground communications. This recommendation is not unique for this fire; similar recommendations were made following investigations into other fires involving communication problems. Every individual firefighter must focus on mastering communication procedures used by their organization during emergency operations. Those skills may very well save his or her life someday.

It is critically important that fire alarms or calls for help be handled expediently and accurately. History has proven that a failure to quickly communicate the need for help, either from the public to the communications center or from the communications center to the responding companies, can result in fires and other emergencies becoming larger or more severe. Fire department communications are a critical factor in the successful outcome of any incident.

Fire department communications include the methods by which the public can notify the communications center of an emergency, the methods by which telecommunicators can notify the proper fire fighting forces, and the methods by which information is exchanged between individuals and units at the scene of an emergency. Firefighters must also know how to handle routine communications, including nonemergency calls for business purposes or public inquiries made directly to the station.

NFPA® 1001, *Standard for Fire Fighter Professional Qualifications*, requires those qualified at the Fire Fighter I level to know the following about fire department communications:

- Procedures for reporting an emergency
- Departmental standard operating procedures for taking and receiving calls
- Radio codes or procedures

- Information needed by dispatch center
- Procedures for answering nonemergency telephone calls

Those qualified at the Fire Fighter I level must also be able to do the following:

- Operate fire department communications equipment including radios, station telephone, and intercom.
- Relay information.
- Record information.

Those qualified at the Fire Fighter II level must also know the following:

- Fire department radio communication procedures
- Standard operating procedures for alarm assignments
- Content requirements for basic incident reports
- Purpose and usefulness of accurate reports
- Consequences of inaccurate reports
- How to obtain necessary information
- Coding procedures

Those qualified at the Fire Fighter II level must also be able to:

- Operate fire department computers and other communications equipment.
- Complete a basic incident report.
- Determine necessary codes.
- Proofread reports.

This chapter provides information on the basics of fire department communications. It describes the role of the telecommunicator (also called a *dispatcher* or *fire alarm dispatcher*) and provides an overview of the communications center and the basic equipment found in such a center (**Figure 19.1**). The chapter also describes the procedures for receiving nonemergency telephone calls, receiving reports of emergencies, and alerting fire department personnel. The last section of the chapter describes the use of the incident report.

Communications Center Personnel

Most of the members of the public who contact a telecommunicator are not having a good day! The person calling is probably experiencing some kind of difficulty or problem that is upsetting enough that they want assistance. Because of this, telecommunicators need to be skilled in customer service and in communicating with those under stress.

Role of the Telecommunicator

The telecommunicator's role is different from but just as important as other emergency personnel. In calls for emergency service, time is of the essence. Given the generally accepted time period of 1 minute to initiate dispatch, telecommunicators must determine required actions very quickly. Time lost in the dispatch function cannot be "made up" by the responders.

In most jurisdictions, telecommunicators are full-time professional communications specialists, not firefighters. In some smaller rural communities, though, firefighters also serve as telecommunicators. In some cases, there is a drop line directly from the Public Safety Answering Point (PSAP) to the station. The PSAP simply

Public Safety Answering Point (PSAP) — Any location or facility at which 9-1-1 calls are answered either by direct calling, rerouting, or diversion.

forwards emergency calls to the station and the on-duty officer performs the role of the call taker/dispatcher. In other cases, calls made to the local emergency number go directly to a fire phone in the station and the first firefighter to answer the phone takes the information.

Telecommunicators must process calls from unknown and unseen individuals who are usually calling under stressful conditions **(Figure 19.2)**. Telecommunicators must be able to obtain complete, reliable information from the caller and prioritize requests for assistance. These decisions and the ability to quickly and accurately carry out the total dispatch function are often a matter of life and death to citizens.

Once the necessary information is gathered from the caller, telecommunicators must dispatch the emergency responders needed to stabilize the incident and mitigate the problem. In order to provide timely response, telecommunicators must know where emergency resources are in relation to the reported incident as well as their availability status **(Figure 19.3)**. Whenever possible, the appropriate unit closest to the incident should be dispatched on emergency responses. Telecommunicators need to know not only which units to assign but also how to alert them. During an

Figure 19.1 The fire department telecommunications center is the initial link between the citizen reporting the emergency and the responding units and personnel. *Courtesy of Paul Ramirez.*

Figure 19.3 The telecommunicator must know the status of all fire and emergency services resources in order to assign units efficiently.

Figure 19.2 The emergency situation may cause the individual reporting the incident to be agitated. It may be a challenge to gather information from callers who are experiencing an emergency.

incident, telecommunicators must stay in contact with the Incident Commander (IC) to receive requests for information and/or additional resources. A telecommunicator's job does not stop when the incident is terminated; records must be kept of each request for assistance and detail how each one was handled.

Customer Service

The consumer of emergency services is the general public. These customers expect and are entitled to professional service. In the majority of cases, a telecommunicator is the first member of the emergency response organization with whom the public has contact during an emergency. Therefore, it is critically important that this contact be a positive and successful experience.

On a daily basis, telecommunicators receive calls from any number of people in the community seeking assistance or information. Some of these contacts include requests for information regarding various social services such as homeless shelters, emergency financial assistance, and counseling services. These calls may come from the victims of crimes, fires, or personal difficulties. Calls may come from people who do not know any other way to get assistance, such as in the case of a power outage. As the customer's first contact with emergency services, telecommunicators must project a sense of both competence and genuine concern to the caller. The telecommunicator is the voice of the agency; this initial contact is often how people form their impression of the entire organization.

With many requests for assistance, telecommunicators refer callers to other appropriate persons or agencies. If a nonemergency call comes in over the 9-1-1 system or another locally used emergency line, the customer must be transferred or referred to another number to have the service request processed. Once they are using a nonemergency line, telecommunicators can provide necessary information to the customer about the appropriate agencies in the area **(Figure 19.4)**. Telecommunicators are not responsible for evaluating the appropriateness of requests for assistance. All requests are referred to the appropriate person or agency.

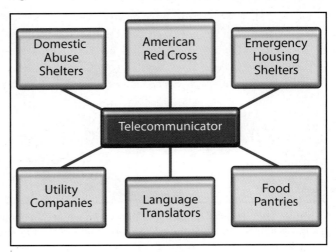

Figure 19.4 Telecommunicators may put callers in touch with other agencies depending on the nature of the emergency. They may also request that some of these agencies respond to assist the fire department.

Telecommunicator Skills

NFPA® 1061, *Standard for Professional Qualifications for Public Safety Telecommunicator*, contains the minimum job performance requirements for fire alarm dispatchers. An annex to NFPA® 1061 states that firefighters used as telecommunicators should meet the requirements of that standard. The following section provides basic information to assist firefighters in receiving and processing information at the station.

An important telecommunicator skill is to be able to maintain a positive attitude throughout the communication process. It is also vitally important for telecommunicators to be able to work effectively with and interact with other members of their team. The annex material in NFPA® 1061 suggests several other traits or personal characteristics that a telecommunicator should possess. They include the following abilities:

- Adjust to various levels of activity.
- Handle multitasking.
- Make decisions and judgments based on common sense and values.
- Maintain composure.
- Form conclusions from disassociated facts.
- Handle criticism.
- Remember and recall information.
- Deal with verbal abuse.
- Function under stress.
- Maintain confidentiality.

These traits and characteristics are necessary if telecommunicators are to be effective. They must routinely try to gather critical information from callers who may be hysterical, impaired by alcohol or drugs, or unable to provide a definite address because they are not calling from a familiar location. In addition, telecommunicators should also be skilled in communications and map-reading.

Communication Skills

The communication skills required by telecommunicators include the following:

- *Basic reading skills* — Ability to read and understand written material so that basic policies, instructions, and directions can be given in writing and be understood.

- *Basic writing skills* — Ability to create clear and understandable reports, memos, and letters. Keyboarding and computer literacy are also needed when a written description of an event or a problem is required. The reports generated may be used by the media, the courts, or the general public.

- *Ability to speak clearly* — Ability to enunciate with proper grammar and sentence structure. Telecommunicators should know how to control voice tone and speed.

- *Ability to follow written and verbal instructions* — Ability to read or listen to instructions and then execute those directions without further assistance. Telecommunicators are responsible for taking directions from a number of different sources through their job functions. In addition, they must know local procedures as well as local, state, and federal radio regulations.

In many jurisdictions, telecommunicators must also be bilingual. Immigrant populations may speak little or no English. Even those who have acquired a limited English vocabulary may revert to their native language under the pressure of an emergency.

Map Reading

In this age of computers and all things electronic, it is still critical for telecommunicators to be able to look at a map and locate specific points **(Figure 19.5, p. 930)**. Telecommunicators in agencies that have wildland responsibilities must be able to

Figure 19.5
To be efficient, telecommunicators must be able to read maps whether they are wall-sized, reproduced in map books, or stored on computers.

read maps laid out in townships, ranges, and sections. Many of the newest Computer-Aided Dispatch (CAD) systems contain sophisticated mapping displays. These maps can be displayed on a screen to help telecommunicators advise responding units of the best route or to locate the closest unit to a call.

Use of an Automatic Vehicle Locating (AVL) system increases the need for a telecommunicator to be able to read and use maps. In this technology, the location of a fire department unit is displayed on a map as the vehicle moves along the streets. Instead of reducing the need for using maps, the opposite is the case with this technology.

The ability to read maps is also necessitated by the enormous growth in such wireless communications devices as cellular phones. Future standards will require that a 9-1-1 call placed by a wireless phone provide X and Y coordinates for the location of the caller. This information will be translated by computers and displayed either as an address or on a map for the telecommunicator to see. This requirement may also include a Z coordinate (the altitude of the caller) to help identify if they are in a high-rise building or on a mountainside. All of this information will be displayed to the telecommunicator in some form of graphical representation that is similar to a map.

Communications Center

A communications center is the nerve center of emergency response. It is the point through which nearly all information flows, is processed, and then acted upon. The communications center houses the personnel and equipment to receive alarms and dispatch resources. Depending on the size and communications requirements of the department, the communications center may be located in a fire station or in a separate building. In some jurisdictions, the fire communications center will be part of a larger, joint communications center for all emergency services (**Figure 19.6**).

A communications center may be equipped with a variety of equipment depending on local capabilities. Some of the more common pieces of communications equipment include the following:

- Two-way base radio for communicating with mobile and portable radios at the emergency scene
- Tone-generating equipment for dispatching resources (**Figure 19.7**)
- Telephones for handling both routine and emergency phone calls
- Direct-line phones for communications with hospitals, utilities, and other response agencies

Figure 19.6 Fire department dispatching may be done from a joint communications center.

Figure 19.7 Tone-generating equipment is a common component of communications equipment. It is used to create the audio sound that alerts firefighters that a dispatch is being sent to their unit or station.

- Computers for dispatch information and communications
- Recording systems or devices to record phone calls and radio traffic
- Alarm-receiving equipment for municipal alarm box systems and private fire alarm systems

Telecommunicators must be able to operate fire department communications equipment. The following sections describe some of the basic communications equipment used in telecommunications centers.

Alarm-Receiving Equipment

Fire alarms may be received from the public in different ways: public alerting systems (See Public Alerting Systems section) and private alarm systems. Fire detection and alarm systems were covered in Chapter 16, Fire Detection, Alarm, and Suppression Systems. Receiving alarms from the public is covered later in this chapter.

Telephones

It is difficult to imagine life today without the telephone and all of its related services. The telephone is used to transmit voice messages, computer information, and documents **(Figure 19.8, p. 932)**. Telephones have grown to resemble computers, and computers today provide many of the same functions historically provided by the telephone.

The public telephone system is the most widely used method for transmitting fire alarms. In many areas such as outlying suburbs or rural settings, it is the only method of rapid communication. A major advantage of telephones is that the telecommunicator can ask the caller about the nature of the emergency and obtain the address or the callback number.

Commercial phone systems. Commercial phone systems access the public switch network. This means that when the phone is taken off the cradle or a button is depressed, the caller hears a dial tone. While many people visualize phone service as basically a residential single-line service, there are in fact vast numbers of commercial phone systems that offer access to multiple phone lines and provide features such as hold, caller identification, call-waiting notification, conference calling, speaker phones, and other features.

Direct lines. Direct lines differ from the normal phone lines in that they do not have access to the public switch network and do not have a dial tone. The line is directly connected between point A and point B. When one party picks up the phone, it immediately starts ringing at the other end. No numbers are dialed and there are no choices for the caller to make. Common applications for these lines would be between the telecommunications center and a fire station or a hospital. These may be just a button on the phone that a telecommunicator depresses to activate the circuit. In some cases, these types of circuits connect emergency communications centers with power plants, airport control towers, or weather services. Another common application for direct lines is to support signals from alarm systems and radio systems.

Some fire departments have drop lines from high-risk occupancies directly into the station. This is especially prevalent in so-called *one-industry* or *company* towns.

TDD/TTY/Text phones. A special communications device has been designed to allow the hearing- or speech-impaired community to communicate over the telephone system **(Figure 19.9)**. *Telecommunications device for the deaf (TDD), Teletypewriter (TTY)*, and *text phone* are phones that can visually display text. These names are interchangeable and denote a device that permits communications with the fire department by the hearing- or speech-impaired in the local jurisdiction. The current term used most often is *text phone*, because it is the most descriptive of the actual functioning of the device. Every firefighter who answers an emergency phone must have a basic understanding of the requirements for "equal access" to 9-1-1. These requirements can be downloaded from the following link: http://www.usdoj.gov/crt/ada/911ta.htm

Figure 19.8 Telephones are the most common means of receiving emergency calls in the telecommunications center.

Figure 19.9 Telephones designed for communicating with the hearing impaired are standard equipment in telecommunications centers.

Wireless (Cellular). Telephone consumers are not dependent only on the phones in their homes; they can now take a phone with them. Cellular phones rely on wireless technology to receive and transmit voice or digital information. The level of wireless service and information provided by the phone itself may be different from jurisdiction to jurisdiction. Even though a call to 9-1-1 made anywhere within range of a cell tower will be routed to the nearest PSAP, the nearest PSAP may not be in the same jurisdiction as the emergency being reported. This can be a problem when the PSAP and the emergency are separated by a large natural barrier such as a river, canyon, or a deep gorge. In addition, callers who do not know their location or are unable to describe it can make it difficult to identify where emergency resources are needed.

A related technology is domestic telephone service over the Internet provided by cable companies. When properly designed and functioning normally, there should be no difference between calls made or received over these systems and the hard-wired systems they are intended to replace.

Fax Machines

Fundamentally, a facsimile (fax) machine converts an image, text, or a diagram into digital signals. These digital signals are sent over a communications medium, usually a telephone line, although radio is another possible medium. At the receiving end, the other fax machine converts the digital signals into a facsimile of the original image or text. While many of these machines are stand-alone machines, which often double as telephone devices, others are built into computers. A computer-generated document does not need to be converted to a hard copy (paper) — the file is digitized and sent over the transmission medium. At the other end is either a stand-alone fax machine or another computer with a fax modem connected to it to convert the document.

Radios

The purpose of fire department radio communications is to provide a means by which all elements of the organization can communicate with each other. Telecommunications centers are equipped with powerful base radios, emergency vehicles are equipped with mobile radios, and individuals are equipped with portable radios **(Figure 19.10, p. 934)**. Radio equipment provides a method of transmitting information to and receiving it from other field units, the communications center, or the incident commander. This information can be task-related (*Command, Engine 7; we need an additional supply line to support Truck 37's ladder pipe.*) or the information can be a direct order based upon the decision of the incident commander (*Alarm, Main Street Command; transmit a third alarm. All incoming units report to staging at 5ᵗʰ and Main.*).

Individuals who operate radio equipment should realize that all radio transmissions can be monitored by the news media and the public **(Figure 19.11, p. 934)**. Any messages transmitted via radio could appear on the front page of tomorrow's newspaper. Therefore, radio operators must always use self-discipline and good judgment to avoid embarrassing themselves or the department. When transmitting over a radio, plan exactly what you intend to say before you key the microphone, and do not use slang or jargon. In most cases, it is also inappropriate to use anyone's name in a radio message.

Figure 19.10 Fire and emergency services units in the field communicate with each other and the telecommunications center via portable radios.

Figure 19.11 You should be aware that fire department and other emergency radio frequencies are monitored by the public. Self-discipline and good judgment should be used when making radio transmissions.

Computer-Aided Dispatch (CAD)

In some jurisdictions, computers are programmed to perform many dispatch functions. The term *computer-aided dispatch (CAD)* is also known as *computer-assisted Dispatch*. Both imply that the telecommunicator is assisted or aided in the performance of his or her duties by a computer system. Some departments have found that using a CAD system can shorten response times or enable dispatchers to handle a greater volume of calls. CAD can also reduce the amount of radio traffic between telecommunicators and responding units.

CAD systems are available in various designs to meet the needs of the telecommunications center and the departments it supports. A CAD system can be as simple as one that retrieves run card information or as complex as one that is programmed to select and dispatch units, determine the fastest route to the scene of an emergency, monitor the status of units, and transmit additional information via mobile data terminals. All of these functions can appropriately be handled by the computer to assist the telecommunicator.

Small organizations with only a couple of pieces of apparatus may not need a CAD system, or they may not require a complex system capable of tracking multiple stations and dozens of units. On the other hand, it is difficult to imagine a major operation involving dozens of units and many calls a day functioning without some type of computer assistance.

Recording Information

Recording information communicated during emergency operations provides a more-or-less permanent record of those transmissions. Two methods of recording information for future review by telecommunicators or other authorized personnel are voice recorders and radio logs.

Voice Recorders

Voice recorders document emergency telephone calls, radio traffic, and dispatching information as well as providing an accurate account of operations (**Figure 19.12**). They protect the department and its members in case of litigation when questions are raised about communications and operations. They also document such evidence as dispatch time and company arrival on the scene.

Figure 19.12 The telecommunications center records all telephone and radio communications and retains the tapes for a specific period of time.

Telephone lines connected to a recording device in the alarm center offer several benefits to the telecommunicator. If the caller hangs up or is disconnected, the information received can be played back. The recording device also is important when callers are so excited that they cannot be understood or when they speak a foreign language.

The recording devices run either continuously or intermittently. The continuous type operates even when no transmissions are taking place; the intermittent units run only when traffic is on the air. Because they run all the time, continuous units use more tape and are more expensive to operate than intermittent types. Intermittent units can miss the beginning of a transmission because they are actuated when a message is transmitted, and it takes a moment for the recording to begin. If an operator speaks before recording begins, the recorder misses the first part of the message. Operators can overcome the problem by using proper procedures: Pause after keying the microphone and before speaking. Recorders should be capable of instant playback. Equipment should also automatically record the time of the call.

Radio Logs

Radio logs are used to record the incident and location of each activity being performed by a public safety unit. This is basically a manual system written on paper (**Figure 19.13**). It is usually a chronological recording of each and every activity that has been reported or dispatched over the radio. In addition to the time of the incident, there is generally an entry as to the location and the nature of the incident, along with a notation of which unit(s) responded to this call. By reviewing the current entries, a telecommunicator can determine which units are currently on assignments and which ones are not. A typical series of entries might read as follows:

Figure 19.13 The radio log is the traditional means of recording emergency radio transmissions at the telecommunications center.

- 1827 hours: Alarm box 263, Engine 12, Engine 9, Ladder 6, Battalion 2 assigned to 3723 E. Main, Sue's Flower Shop

- 1829 hours: Engine 12 on-scene, light smoke visible

- 1830 hours: Dispatch call letters

- 1831 hours: Battalion 2, Engine 9, Ladder 6 on-scene

- 1844 hours: Battalion 2 transmitted control of fire, placed Engine 9 in-service
- 1857 hours: Battalion 2, Engine 12, ladder 6 in-service, returning to quarters
- 1901 hours: Battalion 2, Engine 12, Ladder 6 in-quarters
- 1902 hours: Engine 9 in-quarters

Receiving Nonemergency Calls from the Public

It is vital that any call to a communications center be treated as a possible emergency until it is determined otherwise. Telecommunicators must be able to differentiate between those calls that are emergencies and those that are not. Nonemergency service calls can be handled directly by the telecommunicator, referred to someone else in the fire department, or referred to other agencies.

Most business calls are received by telephone. Each department will have its own procedures and greeting for answering business calls. For this reason, it is important to know and follow your departmental SOPs on telephone procedures. The following list includes some basic business telephone courtesies:

- Answer calls promptly.
- Be pleasant and identify the department or company and yourself. For example, *Good morning, Station 16, Firefighter Jones speaking.* It is SOP in some departments to add, *How may I help you?*
- Be prepared to record messages accurately by including date, time, name of caller, caller's number, message, and your name.
- Never leave the line open or a caller on hold for an extended period of time.
- Post the message or deliver the message promptly to the person to whom it is directed.
- Terminate calls courteously. **Always allow the caller to hang up first.**

Handling business calls and reports of emergencies is discussed in **Skill Sheet 19-I-1**.

Receiving Emergency Calls from the Public

One of the most critical periods for telecommunicators is when an alarm is received. Telecommunicators should be trained to get the necessary information quickly to start responding units on their way **(Figure 19.14)**. When a citizen calls, the telecommunicator should proceed in the following manner:

- Identify the agency (for example, *Metro 9-1-1, what is your emergency?*).
- Control the conversation and ask questions to get the information needed. Ask questions in an assertive voice. Follow the department's SOPs.
- Gather information that describes the type of emergency and its location:
  — Incident location (2110 Maple Street, Highway 101 at College Avenue, etc.)
  — Type of incident/situation (Fire, car crash, person with severe chest pains, etc.)
  — Number of people injured or trapped, if any (How many, how trapped, etc.)
- Make sure to get the exact location of the victims (In a building, vehicle, creek, etc.).

Figure 19.14 Telecommunicators must be able to acquire necessary information and process it rapidly and efficiently.

Units can be dispatched at this point. Delays in starting the dispatch can increase response time.

- If it is safe to do so, keep the caller on the line and get the following information:
  — Name
  — Location if different from the incident location
  — Callback phone number
  — Address

- Ask the caller if it is safe to remain on the phone with you so that you can obtain all the necessary information.

Record the answers to these questions on the emergency alarm report used by your department. Maintain communications with all units until the call has been terminated.

Public Alerting Systems

Public alerting systems are those systems that may be used by anyone to report an emergency. These systems include telephones, two-way radios circuit boxes, telephone fire alarm boxes, and radio alarm boxes. Walk-in reports are also accepted.

Telephone

Depending on the locality, the fire department emergency number may be 9-1-1, a 7-digit number, or "0" for the operator. Emergency telephone number stickers placed directly on the telephone keep citizens from having to look up the number in an emergency **(Figure 19.15)**. There are generally two types of 9-1-1 service: basic and enhanced.

All 9-1-1 calls go to the nearest PSAP. Basic 9-1-1 service can be as fundamental as dialing 9-1-1 and the phone rings at the communications center if that is the local PSAP. Basic 9-1-1 can also have additional features, the most common of which are *called party hold*, *forced disconnect*, and *ringback*.

- *Called party hold* is a feature that allows a telecommunicator to maintain access to a caller's phone line. As long as the telecommunicator does not hang up or disconnect, the system will maintain control of the caller's phone line and keep it open. Callers who hang up and then try to place another call will discover that they are still connected to the telecommunicator.

Figure 19.15 Citizens should be encouraged to place emergency telephone numbers on their telephones to eliminate the time required to look them up.

- *Forced disconnect* is, in a way, the reverse of called party hold for telecommunicators. When the called party (telecommunicator) hangs up after receiving a 9-1-1 call, the caller can keep the line open for a short period of time. To prevent the 9-1-1 line from being tied up indefinitely, such as when a caller loses consciousness before hanging up, the forced disconnect feature drops the call out of the system and frees the line for the next caller.

- *Ringback* is a feature that allows the telecommunicator to call back a caller's phone after he or she has hung up.

Some basic 9-1-1 systems offer one of the features of an enhanced system: automatic number identification (ANI). This feature displays the calling party's phone number on a screen at the telecommunicator's position.

Some jurisdictions are equipped with enhanced 9-1-1 (E-9-1-1) systems. These systems combine telephone and computer equipment (such as CAD) to provide the telecommunicator with instant information such as the caller's location and phone number, directions to the location, and other information about the address. As soon as the telecommunicator picks up the phone, the computer displays the location from which the call is being made through automatic location identification (ALI), which uses Global Positioning System (GPS) data. Business extensions (PBX systems) may not allow the caller's exact location to be displayed on the computer. This system allows help to be sent even if the caller is incapable of identifying his or her location. Wireless telephones will not activate the E-9-1-1 system ALI.

Radio

On occasion, an emergency may be reported by radio. This type of report is most likely to come from fire department personnel or other government workers who happen upon an emergency. The firefighter or telecommunicator monitoring the radio should gather the same kind of information that would be taken from a telephone caller. Once all the necessary information is received, additional resources can be dispatched if required.

Some fire departments also monitor citizens band (CB) radio frequencies for reports of emergencies. The universal frequency for reporting emergencies, and the one most commonly monitored by emergency providers, is CB Channel 9. Reports taken over CB radio should be handled in the same way as those received by telephone; however, in place of the callback number, the caller's radio "handle" (call sign) or designation should be recorded.

Walk-Ins

From time to time, a citizen will walk into a fire station and report an emergency that has just occurred in the vicinity of the station (**Figure 19.16**). Whoever greets the citizen should ascertain the location and type of incident. It is very important to get the reporting party's name, address, and telephone number — especially their cell phone number. Being able to contact the reporting party may be the only way that critical information about the incident can be obtained during any subsequent investigation.

Once the information is obtained, local policy dictates the next step. Some departments require the company taking the report to first notify the communications center by phone before responding. Other jurisdictions allow personnel in the company to immediately start their response and radio the communications center with information on the incident while en route. As always, you must follow your departmental SOP in this regard.

Automatic Location Identification (ALI) — Enhanced 9-1-1 feature that displays the address of the party calling 9-1-1 on a screen for use by the public safety telecommunicator. This feature is also used to route calls to the appropriate public safety answering point (PSAP) and can even store information in its database regarding the appropriate emergency services (police, fire, and medical) that respond to that address.

Citizens Band (CB) Radio — Low-power radio transceiver that operates on frequencies authorized by the Federal Communications Commission (FCC) for public use with no license requirement.

Figure 19.16 Periodically fires and other emergencies are reported directly to the fire department by walk-ins to the station.

Wired Telegraph Circuit Box

Historically, many cities installed street corner alarm boxes to allow citizens to report a fire when businesses were closed and there were no public telephones in the neighborhood (**Figure 19.17**). These boxes were connected to a wired telegraph circuit that was connected to all fire stations in the jurisdiction and terminated in the alarm (communications) center. Before fire apparatus were equipped with mobile radios, units were dispatched by and communicated with the alarm center with a series of coded signals sent over this circuit. The signals were transmitted using a telegraph key inside the box. Some cities still maintain these systems.

The boxes are operated by pressing a lever on the door of the box that starts a spring-wound mechanism inside. The rotating mechanism transmits a code by opening and closing the telegraph circuit. Each box transmits a unique number code to identify its location. These systems are extremely reliable but also extremely limited. The only information transmitted is the location of the box and nothing about the nature of the emergency. These boxes are also notorious for malicious false alarms. Because of the availability of public telephones and cellular phones, the need for these systems has greatly diminished. Given the expense of testing telegraph systems and the reduced need for them, many cities have eliminated them.

Figure 19.17 Some cities still have hard-wired fire alarm boxes located on street corners and connected directly to the telecommunications center.

Telephone Fire Alarm Box

In this type of system, each fire alarm box is equipped with a telephone for direct voice contact with a telecommunicator. Some municipalities use a combination of both telegraph and telephone-type circuits. This provides the best of both systems. The pull-down hook is used to send the coded signal, and a telephone is included for additional use (**Figure 19.18**).

Radio Fire Alarm Box

A radio alarm box contains an independent radio transmitter with a battery power supply (**Figure 19.19**). Some systems include a small solar panel at each box for recharging the unit's battery. Others feature a spring-wound alternator to provide power when the operating handle is pulled.

There are different types of radio boxes. Activating the alarm in radio boxes alerts the telecommunicator by an audible signal, visual light indicator, and a printed record indicating the location. In addition to a red alarm-light indicator, some systems have a different-colored light that indicates a test or tamper signal. A time clock within the box permits the alarm to test itself every 24 hours. If the box is damaged or tampered with, the tamper light comes on and gives the box location. Some boxes are numbered, and this number also appears on the communications center display panel, informing the telecommunicator of the box involved and its location.

Figure 19.19 Radio fire alarm boxes are self-contained units that may be found along highways or streets.

Figure 19.18 Some fire alarm systems incorporate emergency telephones or intercoms.

When activated by the incoming radio signal, the printing devices in some systems print the date, the time of day in 24-hour time, the message sent by the box, the box number, and a coded signal that indicates the strength of the battery within the box. Some radio alarm boxes are designed to allow a person to select fire, police, or ambulance service. Some radio alarm boxes that are located along roads, highways, and in rural areas have two-way communications capabilities. Telecommunicators answering these radio alarm box reports should gather the same information as they would by telephone.

24-Hour Clock

The 24-hour clock is a convention of time keeping in which the day runs from midnight to midnight and is divided into 24 hours, numbered from 0 to 23. This method of keeping time is also known as *military time* or *Army time*. In spoken language, the hour is followed by the word *hundred.* For example, 9 am would be written 09:00 and spoken as "oh nine hundred," 2 pm would be written 14:00 and spoken as "14 hundred," 10 pm would be written 20:00 and spoken as "20 hundred." 4:30 pm would be spoken as "1630," etc.

Procedures for Reporting a Fire or other Emergency

To educate the public, the department's fire and life safety education program should include information on how to report an emergency correctly. The public should be trained to report emergencies using the methods given in the following sections.

By Telephone

- Dial the appropriate number:

  — 9-1-1

  (or)

  — Fire department 7-digit number

  (or)

  — "0" for the operator

- State the address where the emergency is located.

  (or)

- If no address, give the nearest cross streets or describe nearby landmarks.

- Give the telephone number from which you are calling.

- State the nature of the emergency **(Figure 19.20)**.

- State your name and location.

- Stay on the line if requested to do so by the telecommunicator.

From a Fire Alarm Telegraph Box

- Send signal as directed on the box.

- If safe to do so, stay at the box until firefighters arrive so you can direct them to the emergency.

Figure 19.20 Firefighters train children for the proper method of calling 9-1-1.

Figure 19.21 Follow directions on the pull station – lift lid and pull handle.

From a Local Alarm Box

- Send signal as directed on the box (**Figure 19.21**).

- Notify the fire department by telephone using the guidelines given earlier.

Alerting Fire Department Personnel

Various ways are used to alert firefighters of an emergency and to indicate its location. Some departments use a system of bells or other sounding devices, others use radio transmissions, and still others have an automatic, computer-operated system. One of the factors that determine the way fire stations and personnel are alerted is whether or not the station is staffed.

Some fire departments try to give information from preincident plans about the location of the alarm to fire companies as they respond. Some departments transmit the information by radio; others have the information available in individual fire vehicles on transparencies or microfiche. Some departments "pre-alert" their stations by transmitting the address of a call while researching the rest of the dispatch information; units respond if the address is in their district. The balance of the information is transmitted while the emergency vehicles are en route. This system reduces the amount of time it takes for units to leave the station, thus reducing the overall response time.

Figure 19.22 Fire departments with computer-aided dispatch systems have printers in each fire station.

Staffed Stations

Technological advances have added modern alerting systems to accompany the more traditional types. These types of alerting systems include the following:

- Computerized line printer or terminal screen with alarm (**Figure 19.22**)

- Voice alarm

- Teletype

- House bell or gong (**Figure 19.23**)

- House light

- Telephone from telecommunicator on secure phone line

Figure 19.23 Most fire stations are equipped with a house bell or gong that is part of the alarm alerting system.

- Telegraph register
- Radio with tone alert
- Radio/pagers

Unstaffed Stations

In order to facilitate as quick a response as possible from unstaffed stations, methods of simultaneously notifying all personnel are used. These systems include the following:

- Pagers **(Figure 19.24)**
- Cellular telephones and other devices with text-messaging capabilities
- Home electronic monitors
- Telephones
- Sirens **(Figure 19.25)**
- Whistles or air horns

Pagers and home electronic monitors are activated by tone signals that are sent over radio waves. The advantage of pagers is that firefighters can carry them wherever they go. Home monitors and telephones require the firefighter to be at home to receive notification. With the widespread availability of pagers, home monitors have been eliminated from most fire departments.

Pager — Compact radio receiver used for providing one-way communications.

A paging system is a transmitter on a given frequency that will activate a specific pager or specific group of pagers, which are really just miniature receivers. A pager, or receiver, is set on a specific radio frequency and given an address of some specific tones, codes, or frequency. Individual tones, such as for a chaplain or a fire investigator, can also be programmed. When the pager receives its codes, it turns on and alerts

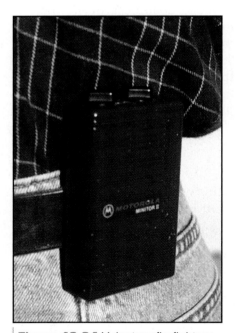

Figure 19.24 Volunteer firefighters may wear pagers in order to be alerted.

Figure 19.25 Small jurisdictions may use sirens to alert their volunteers.

the wearer by tone, light, or vibration. The pager will then either open the speaker for a voice message or display the alpha numeric message sent to it. Most pagers have an alert setting that only activates if that pager's tone is transmitted. Usually when a number of different departments or public safety agencies share the same dispatch frequency, it is desirable to set pagers to the alert setting to avoid hearing unwanted radio traffic.

Sirens, whistles, and air horns are most commonly employed in small communities. These devices produce a signal that everyone in the community can hear. This is both an advantage and a disadvantage. Civilians will be aware that emergency traffic may be on the streets; however, some may also be inclined to follow the apparatus and congest the emergency scene.

Radio Communications

In the United States, all radio communication is under the authority of the Federal Communications Commission (FCC). Fire departments that operate radio equipment must hold radio licenses from the FCC. Depending on the radio system in a particular locality, one license may cover several departments that operate a joint system. Local department rules should specify who is authorized to transmit on the radio. It is a federal offense to send personal or other unauthorized messages over a designated fire department radio channel. More information on use of a portable radio for routine and emergency traffic can be located in **Skill Sheet 19-I-2**.

Radio Procedures

To be in compliance with the National Incident Management System (NIMS), most emergency response agencies in the U.S., including fire departments, have eliminated the ten-codes and other local terminology that were the norm for decades. Many of these codes were necessary when radio equipment was not as technologically advanced as that currently available. Most fire departments now use plain English without codes of any kind. Others, especially those in the wildland fire community, use *clear text*. Clear text is a standardized set of fire-specific words and phrases. For more information on clear text, see the IFSTA **Wildland Fire Fighting for Structural Firefighters** manual. Departmental SOPs specify such things as test procedures, local protocols, and time limits on radios.

Telecommunicators can transmit more than their words by vocal inflection; that is, if they allow their excitement or other emotions to be reflected in their voice, those same emotions can affect the firefighter receiving the message. Excitement in a telecommunicator's voice can make a routine call sound like an emergency. Most agencies prefer that telecommunicators maintain a professional demeanor with a steady voice level — never too excited. There are a number of keys to this approach and some guidelines are as follows:

- Use a moderate rate of speaking — not too slow or too fast — focused on clear understanding. This includes not using pauses or verbal fillers such as *ah* or *um* during the dispatch.

- Use a moderate amount of expression in speech — not a monotone and not over-emphasized — with carefully placed emphasis. Avoid excitement or shouting over the radio and be careful to articulate properly. Strive for the correct pronunciation of words.

Clear Text — Use of plain English, including certain standard words and phrases, in radio communications transmissions.

- Use a vocal quality that is not too strong or weak. Finish every comment, and avoid trailing off near the end of the transmission. Keep the pitch in a midrange — not too high or too low. Avoid using slang or regional expressions, and strive for a good voice quality.

- Keep things such as gum and candy out of your mouth. Be confident in what you say, and position the microphone appropriately to make the best use of the system.

- Be concise and to the point; do not talk around the issue. Give the information required in a logical and complete manner that best addresses the service requested.

- Think about what you want to say before keying the microphone.

One of the ways that telecommunicators are evaluated is by their effectiveness on the radio. In the real world, gathering information from sometimes emotional or confused reporting parties can be difficult and stressful. While time is critical in this process, so is accuracy. It is necessary to provide responders with the most accurate location where help is needed. The guiding principle is to transmit accurate information as quickly as possible.

Another difficult area for telecommunicators is to use enough words but not too many in order to give a clear picture of the event or service request. Responders find it frustrating when a telecommunicator uses more words than are necessary to describe a request for service. The reverse can also be a problem. Being sent to a "smell of smoke" and finding a building fully involved in fire is an example of insufficient communication.

An additional but often forgotten factor when transmitting information and orders via radio is that *only essential information* should be transmitted. For example, consider the difference in the following two radio communications:

> **Radio Communication 1:** *This is Lieutenant Thompson on Engine 57 portable. I need another truck company at this location, Box 1333, for additional personnel.*

> **Radio Communication 2:** *Communications, Engine 57 portable — dispatch 1 truck company to Box 1333.*

Everyone on the fireground should follow two basic rules to control communications. First, units must identify themselves in every transmission as outlined in the standard operating procedures. Second, the receiver must acknowledge every message by repeating the essence of the message to the sender.

Example:

> **Engine 4:** *Communications, Engine 4. We are on scene and have a Dumpster® fire. Engine 4 can handle. Return all other units.*

> **Communications:** *Engine 4, I copy you have a Dumpster® fire, and you will handle. Other units can be canceled.*

Requiring the receiver to acknowledge every message ensures that the message was received and understood. This feedback can also tell the sender that the message was not understood and further clarification is necessary.

Other important considerations that should be remembered include the following:

- Do not transmit until the frequency is clear.

- Think about what you're going to say before keying the microphone.

- Remember that any unit working at an emergency scene has priority over any routine transmission.

- Do not use profane or obscene language on the air.

- All radio frequencies are monitored and every transmission will become part of the official record on the incident. They may also be used in local news coverage.

- Hold the radio/microphone 1 to 2 inches (25 mm to 50 mm) from your mouth when transmitting **(Figure 19.26)**.

On the emergency scene, consider the following additional items:

- Avoid laying the microphone on the seat of the vehicle because the button may be pressed inadvertently.

- Do not touch the antenna when transmitting. Radio frequency burns might result.

Figure 19.26 To achieve clear radio communication, firefighters must practice proper communications techniques.

Arrival and Progress Reports

The first company to arrive at the scene of an emergency should use the radio to provide a description of the conditions found **(Figure 19.27)**. This process is often referred to as a *report on conditions* or a *situation report*. Each department may have its own format for these reports. A good report establishes a time of arrival and informs other responding units of what actions might be needed upon their arrival. The following is a typical report on conditions:

> *Communications, Engine 611 is on scene at Third Street and Railroad Avenue. We have a two-story, wood-frame residence; there is an all-clear on the structure, but smoke is showing from the second-story windows. Engine 611 is attacking with two handlines. Engine 613, establish Railroad Avenue Command on arrival.*

Or like this:

> *Communications, Engine 611 is on scene at Broadway and 58th Street; nothing showing. Engine 611 is investigating.*

Figure 19.27 The first-arriving unit should provide a clear and complete description of the situation.

Communications, Battalion 7, we have a two-story, wood-frame residence with heavy fire in the first-floor rear. We have an all-clear on the structure. Battalion 7 is establishing Farmers Lane Command at 4th and Farmers.

Other departments use an initial report format that includes the following:

- Situation found

- Action(s) taken/actions to be taken

- Command status

The *situation found* includes the type of situation, construction, etc. *Action(s) taken/to be taken* includes the actions being taken by the on-scene company or to be taken by units en route. The *Command status* includes specifying who is in command, naming the incident, and the location of the command post when applicable.

Some situations require more detail in a report on conditions than others. Depending upon the nature and scope of the emergency, the following information may need to be included when giving a report on conditions:

- Address, if other than the one initially reported

- Building and occupancy description

- Nature and extent of fire or other emergency

- Attack mode selected

- Rescue and exposure problems

- Instructions to other responding units

- Location of Incident Command position

- Establishing Command

- Water supply situation (if applicable)

Once fire fighting operations have begun, it is important that the communications center be continually advised of the actions taken at the emergency scene. Such progress reports should indicate the following as applicable:

- Transfer of Command **(Figure 19.28)**

- Change in command post location

- Progress (or lack of) toward incident stabilization

- Direction of fire spread

- Exposures by direction, height, occupancy, and distance

- Any problems or needs

- Anticipated actions

Figure 19.28 Although transfer of Command should take place in person, the transfer should be reported over the radio to all units at the scene and to the telecommunications center.

Tactical Channels

Routine, day-to-day incidents can usually be handled on a single channel, but larger incidents may require using several channels to allow for clear and timely exchanges of information. Separate channels may be needed for command, tactical, and support functions.

When one radio channel is primarily used for dispatching, it is often necessary to assign a different channel for an incident. This allows the IC to have a relatively clear communications channel with the communications center and another clear channel to the fireground officers without having units "talking over each other" (interruptions caused by other transmissions). Tactical channels are most often used for large incidents such as structure fires. Small routine incidents such as fire alarm investigations and vehicle fires usually do not require the use of a tactical channel.

In many departments, units are initially dispatched on the primary dispatch channel. Upon arrival on the scene, they switch to an assigned tactical channel to communicate with the IC. Some of the roles the telecommunicator plays, depending on the nature and scope of the operation are as follows:

- Assign a tactical frequency for the management of the operation or the incident.

- Ensure that additional responding units are aware of the assigned tactical channel.

- Notify other agencies and services of the incident and the need for them to respond or take other appropriate action.

- Provide updated information that affects the incident.

Calls for Additional Resources

At some fires, it may be necessary to call for additional resources. Normally, only the Incident Commander may strike multiple alarms or order additional resources. Depending on who arrives first, the Incident Commander may be a company officer, chief officer, or a firefighter.

All firefighters need to know the local procedure for requesting additional resources. They must also be familiar with alarm signals (multiple or special alarms) and know what resources each additional alarm will include; that is, the number and types of units that respond to each additional alarm.

When multiple alarms are struck for a single fire, maintaining communications with each unit becomes more difficult as radio traffic increases. To reduce the load on the communications center, a radio-equipped, mobile communications vehicle can be used at large fires (**Figure 19.29**).

Alarm Assignment —
Predetermined number of fire units assigned to respond to an emergency.

Figure 19.29 At major emergency incidents, specialized command vehicles provide efficient on-scene communications.

When firefighters function as part of a team, they must be able to communicate the need for team assistance through the fire department's communications system. The designated supervisor must be in constant communications with the team and follow an incident management system with local SOPs. Some of these communications might be requests for additional personnel or special resources or to notify others on the fireground of any hidden hazards.

NOTE: In many rural areas, fire departments have only one frequency available to them. In these areas, it is important for all local departments to train together and to agree on the radio procedures to be used during emergency operations.

Emergency Radio Traffic

For a variety of reasons, it may be necessary to broadcast emergency traffic (urgent message) over the radio. Telecommunicators at communications centers are better equipped than on-scene personnel to hear weak signals from portable radios. When firefighters radio that they are in distress, telecommunicators can make a significant difference in firefighter survival.

When it is necessary to transmit emergency traffic, the person transmitting the message should make the urgency clear. For example: *Dispatch, Battalion 1, Emergency Traffic!* At that point, the telecommunicator should give an attention tone (if used in that system), advise all other units to stand by, and then advise the caller to proceed with the emergency message. After the emergency communication is complete, the telecommunicator notifies all units to resume normal or routine radio traffic.

Evacuation Signals

Evacuation signals are used when the IC decides that all firefighters should immediately withdraw from a burning building or other hazardous area because conditions have deteriorated beyond the point of reasonable safety. All firefighters should be familiar with their department's method of sounding an evacuation signal. There are several ways this communication may be made. The two most common are to broadcast a radio message on the incident frequency ordering them to evacuate and to sound audible warning devices on the apparatus at the fire scene in a prescribed pattern.

The radio broadcast of an evacuation signal is usually handled in a manner similar to that described for emergency traffic. The message is broadcast several times to make sure everyone hears it. The use of audible warning devices on apparatus, such as sirens and air horns, will work outside small structures, but they may not be heard by everyone working in a large building. Audible warnings using sirens or air horns can also be confused with those being used by units arriving at the scene.

Personnel Accountability Report (PAR)

A PAR is a systematic way of confirming the status of any unit (company, group, or division, etc.) operating at an incident. When a PAR is requested, every supervisor must verify the status of those under his or her command and report it. In conditions where visual confirmation is impaired by smoke or darkness, company officers and team leaders may have to rely on touch (physical contact) or hearing (voice contact — NOT radio) to verify each member's status. Others in the chain of command (Branch Directors, Section Chiefs, etc.) must rely on radio reports from their subordinates.

Command can request a PAR at any time, but certain benchmarks are also used. A PAR is usually requested when:

- The incident is declared under control.

- There is a change in strategy (such as from offensive to defensive).

- There is a sudden catastrophic event (flashover, backdraft, etc.).

- There is an emergency evacuation.

- A firefighter is reported missing or in distress.

Incident Reports

Every time a fire unit responds to an incident, proper reports must be completed. The National Fire Incident Reporting System (NFIRS), developed by the United States Fire Administration (USFA), outlines the necessary information needed to complete incident reports. NFIRS is a personal computer (PC) based system that uses the Internet to transfer data from each state to the federal database. This is a flexible system that allows various types of data to be entered. Each piece of data has a specific code corresponding to the information entered. All 50 states participate in the NFIRS.

From legal, statistical, and record-keeping standpoints, reports are a vital part of the emergency. Reports must be filled out completely and in terminology that non-fire service personnel can understand because reports are available to the public. Insurance companies frequently request copies for their records. To learn how to create an incident report, refer to **Skill Sheet 19-II-1**.

The following information should be included in incident reports:

- Fire department name, incident number, district name/number, shift number, and number of alarms

- Names and addresses of the occupant(s) and/or owner(s)

- Type of structure, primary use, construction type, and number of stories

- How the emergency was reported (9-1-1, walk-in, radio, etc.)

- Type of call (fire, rescue, medical, etc.)

- Action that was taken (investigation, extinguishment, rescue, etc.)

- Property use information (single-family dwelling, paved public street, etc.)

- Number of injuries and/or fatalities

- Number of personnel who responded and type of apparatus that responded

- How and where the fire started

- Method used to extinguish the fire

- Estimated cost of damage

- Remarks/comments (usually a narrative of the incident is written by the officer in charge)

Most fire departments enter this information into databases at the state and national level. This information is used to evaluate the needs of the department and the community it protects, which can improve the level of service delivered by the department. Reports also justify budget requests, code enforcement, and resource allocations.

Incident reports can also be entered into a computer by the officer in charge. This eliminates the need for the handwritten report that requires someone else to tabulate the data collected.

National Fire Incident Reporting System (NFIRS) — One of the main sources of information (data, statistics) about fires in the United States; under NFIRS, local fire departments collect fire incident data and send these to a state coordinator; the state coordinator develops statewide fire incident data and also forwards information to the USFA.

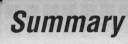

Summary

Fire alarms or calls for help must be handled expediently and accurately. If they are not, incidents can increase in size and severity. Fire department communications are a critical factor in the successful outcome of any incident. There is a direct connection between fireground communications and fireground safety. The better the communications, the safer the incident. When communications break down, safety breaks down as well.

Fire department communications include the methods by which the public can notify the communications center of an emergency, the methods by which telecommunicators can notify the proper responding units, and the methods by which information is exchanged between individuals and units at the scene of an emergency. As a firefighter, you must know how to handle both emergency and routine communications, including nonemergency calls for business purposes or public inquiries made directly to the station.

Fire Fighter I

1. What communication skills are necessary for fire department communications?

2. What is Computer-Aided Dispatch (CAD)?

3. List three basic business telephone courtesies.

4. What actions should be taken when receiving an emergency call from a citizen?

5. How should the public report a fire or other emergency using a telephone?

Fire Fighter II

6. List three guidelines for speaking over the radio.

7. What information should be given in an arrival report?

8. Why are tactical channels used?

9. When are evacuation signals given?

10. What information should be given in an incident report?

Skill SHEETS

19-I-1
Handle business calls and reports of emergencies.

Receive a Business Call

Step 1: Answer telephone promptly.

Step 2: Determine reason for call.

Step 3: Respond to caller's request or need.

Step 4: End call.

Step 5: Post message as required.

Receive a Report of an Emergency

Step 1: Answer telephone promptly.

Step 2: Gather information on nature of emergency.

Step 3: Provide life safety directions if caller is at immediate risk.

Step 4: Gather information on caller.

Step 5: Transfer information according to local procedures.

Step 6: End call according to local procedures.

Routine Traffic

Step 1: Rotate the selector knob to assigned frequency.

Step 2: Monitor for radio traffic until air is clear.

Step 3: Hold the microphone in transmit position 1 to 2 inches (25 mm to 50 mm) from your mouth at a 45-degree angle.

Step 4: Depress the transmit button, holding down until through with transmission.

Step 5: Transmit a routine traffic message using department codes and SOPs.

Skill SHEETS

19-I-2 continued
Use a portable radio for routine and emergency traffic.

Emergency Traffic/Call a Mayday

Step 1: Rotate the selector knob to assigned frequency.

Step 7: After transmitting Mayday, activate PASS alarm and follow departmental guidelines on positioning or actions.

Step 2: Hold the microphone in transmit position 1 to 2 inches (25 mm to 50 mm) from your mouth at a 45-degree angle.

Step 3: Depress the transmit button, holding down until through with transmission.

Step 4: Announce "emergency traffic" (or department's standard emergency traffic break-in message), interrupting air traffic as necessary.

Step 5: Transmit emergency traffic message following department's SOPs, using department codes.

Step 6: Repeat message until Command verifies information given.

er notes and other information on the incident.

S cord information on incident report form (writ-
or ronic version) used by department.

Step 3: Review incident report and make corrections or revisions as needed.

Step 4: Finalize and process report according to department policy.

Chapter Contents

Fire Prevention and Public Education

This chapter provides information that addresses the following job performance requirements of NFPA® 1001, *Standard for Fire Fighter Professional Qualifications (2008)*:

NFPA® 1001 references for Chapter 20:

6.5.1
6.5.2
6.5.3

Chapter Objectives

Fire Fighter II Objectives

1. Describe a survey and an inspection.

2. Discuss the fire prevention activities of reviewing community data and code enforcement.

3. Summarize common fuel and heat-source hazards.

4. Discuss common fire hazards and why they increase the likelihood of a fire.

5. Summarize special fire hazards in commercial, manufacturing, and public-assembly occupancies.

6. Summarize target hazard properties.

7. Discuss personal requirements and equipment requirements for conducting inspections.

8. Discuss scheduling and conducting fire inspections.

9. Discuss the benefits of preincident planning surveys.

10. Explain how a preincident planning survey is conducted.

11. Explain the purpose of a residential fire safety survey.

12. Summarize guidelines for conducting residential fire safety surveys.

13. Summarize common causes of residential fires.

14. Summarize items to address when conducting residential fire safety surveys.

15. Describe the basic steps in presenting fire and life-safety information.

16. Discuss fire and life-safety presentation topics.

17. Discuss fire station tours.

18. Prepare a preincident survey. (Skill Sheet 20-II-1)

19. Conduct a residential fire safety survey. (Skill Sheet 20-II-2)

20. Make a fire and life-safety presentation. (Skill Sheet 20-II-3)

21. Conduct a fire station tour. (Skill Sheet 20-II-4)

22. Make a fire and life safety presentation. (Skill Sheet 20-II-3)

23. Conduct a fire station tour. (Skill Sheet 20-II-4)

Fire Prevention and Public Education

Case History

The goal of most fire safety education programs is to equip citizens with the knowledge and skills to react properly when a fire or other emergency occurs. On April 26, 2005, the importance of fire safety education was again demonstrated when Maria Diaz of Plano, Texas, reacted to a fire in her neighbor's apartment. Maria went next door to investigate when she heard the smoke alarm sounding and discovered her friend's kitchen was on fire. She quickly went back to her apartment, got her fire extinguisher, and safely put out the fire.

Maria's quick reaction to the fire situation was a result of a fire safety education program she had attended that was sponsored by the Plano Fire Department. Firefighters had conducted a hands-on fire extinguisher class in cooperation with the local Even Start classroom and Lowe's Home Improvement Center. The participants had received no-cost fire extinguishers as part of the Lowe's Heroes program.

Source: Home Safety Council® Expert Network

Firefighters frequently respond to fires and other emergencies that would not have happened had the individual causing the problem complied with local fire prevention laws, codes, and ordinances. The requirements contained in these regulations are a collection of solutions to problems that people have experienced in the past. In other words, once the origin and cause of an accidental fire were determined, legislation was then adopted that would prevent citizens from creating that same problem in the future **(Figure 20.1)**. In addition, before firefighters or citizens can initiate any kind of corrective action, they must be able to recognize and properly interpret the potential risk or condition involved. Gaining this knowledge often requires education.

Most fire departments focus their public education efforts on helping their citizens to reduce known hazardous conditions and prevent dangerous acts before they develop into emergencies. This may be accomplished in a number of ways such as conducting public presentations, distributing safety information, writing newspaper articles, providing public service announcements (PSAs), or posting informational displays in high-traffic areas **(Figure 20.2, p. 960)**. Proper utilization of the news media during or just after a preventable accident has occurred can sometimes pay public awareness dividends. On such occasions, normally the department's public information officer or other fire officer works with media contacts **(Figure 20.3, p. 960)**.

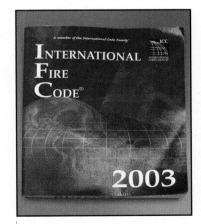

Figure 20.1 The *International Fire Code®* is an example of model codes that are adopted by the local jurisdiction and used by fire prevention and code enforcement inspectors.

PUBLIC SERVICE ANNOUNCEMENTS

SUBJECT: Fire Prevention
FROM: Your Fire Department
PROGRAM: Christmas Holiday Season

(20-Second Announcement)

Decided what to give the children this Christmas? Whatever toys you select, make sure they're safe. Look for the UL label of fire and shock safety on all electrical and heat producing toys. Choose chemistry sets carefully. Supervise play with toys involving fuels and chemicals. Protect your youngsters from fire!
****

(30-Second Announcement)

Is anyone in your family taking part in a Christmas pageant or choir recital during this holiday season? Costumes and choir robes are often made of loosely woven material which is highly combustible. No one wearing such a garment should ever get close to a flame or source of heat. If candles are to be carried, they should be electric candles. Don't let fire turn Christmas joy into tragedy!
****

Figure 20.2 Public service announcements are a cost-effective method for getting vital fire and life safety information to the public.

Figure 20.3 Fire department personnel should develop positive relations with media representatives to promote community awareness of fire prevention activities.

According to NFPA® 1001, *Standard for Fire Fighter Professional Qualifications*, there are no specific requirements for those qualified at the Fire Fighter I level regarding fire prevention and public education. However, those qualified at the Fire Fighter II level must know the following:

- Organizational policies and procedures
- Common causes of fires and how to prevent them
- Importance of fire safety surveys and public fire education programs
- Referral procedures
- Informational materials and how they are used
- Departmental SOPs for giving fire station tours

  Those qualified at the Fire Fighter II level must also be able to:

- Perform a fire safety survey in a private dwelling.
- Identify fire and life safety hazards and recommend corrective actions.
- Refer unresolved issues to the proper authority.
- Complete required forms.
- Present fire safety information to station visitors or small groups.
- Use prepared materials.
- Document presentations.
- Prepare a preincident survey.

This chapter provides information to help you perform fire prevention and public fire education activities. It begins with an overview of fire prevention and the various types of fire hazards. The next section discusses the two types of fire safety surveys:

the preincident survey and the residential fire safety survey. The last part of the chapter discusses public fire and life safety education. It provides information on fire safety topics that you may be asked to provide to the public during presentations or as part of a fire station tour.

Fire and Life Safety Contacts

Firefighters and fire inspectors contact property owners and occupants in two general ways for nonemergency visits: *surveys* and *inspections*. Each of these contacts serves different but equally important purposes.

Surveys

Firefighters survey properties within their response districts to either gather information or impart information. Firefighters survey commercial, institutional, industrial, and multifamily residential properties to gather information for preincident planning (see Preincident Planning Surveys section). If these surveys are conducted while the buildings are under construction, it helps firefighters to identify building construction techniques and their strengths and weaknesses, possible avenues of fire spread, points at which fire spread can be halted, and other features that may be important should a fire occur. These surveys also help to familiarize firefighters with the contents, manufacturing processes, and layouts of the buildings. Also noted during preincident surveys are potential occupant loads, means of egress, built-in fire protection devices and systems, and fire department access. All of these items should be discussed among the crew members regarding how fire department operations could be affected during emergencies. These items can be documented in a binder or computer database that can be a critical reference during a fire or other emergency.

During voluntary residential fire safety surveys (see Residential Fire Safety Surveys section), firefighters provide information to the occupants. Firefighters use their expertise and experience to advise the occupants on how they can keep their residences free of common fire and life safety hazards.

Inspections

Inspections are made by fire service personnel to ensure compliance with applicable fire and life safety code requirements (see Fire Prevention section). In some jurisdictions, code enforcement inspections are made only by fire inspectors; in others, company officers and their crews are assigned to carry out these duties. Fire company personnel usually enforce the most common code requirements such as housekeeping, fire extinguisher accessibility and serviceability, overloaded or illegal extension cords, and egress obstructions. When they notice a more serious code violation such as makeshift wiring or hazardous operations (welding and cutting, for example) in a building not zoned for such activities, they pass that information on to the fire prevention bureau for further action.

Fire Prevention

Code enforcement inspections of commercial, institutional, and industrial occupancies are conducted to ensure that citizens have a safe physical environment in which to work, study, worship, or play. To be effective in the inspection process, firefighters must be thoroughly familiar with the requirements of the applicable fire and life safety codes and ordinances in their jurisdiction.

Figure 20.4 Annual fire prevention inspections include checking the inspection and maintenance dates on portable fire extinguishers and seeing that fire extinguishers are acceptable.

Fire incident records contain critical information about the fire history of a community and can be helpful with fire prevention efforts. A wealth of information can be obtained by studying previous incidents, reviewing data obtained from various fire reports, and comparing statistical data. Such a review helps identify major fire causes and may suggest possible solutions.

An important fire prevention activity is the code enforcement inspection. In some jurisdictions, these inspections are conducted by fire inspectors with special training in code requirements for various types of occupancies **(Figure 20.4)**. In these jurisdictions, fire inspectors are usually trained to meet the objectives found in NFPA® 1031, *Standard for Professional Qualifications for Fire Inspector and Plan Examiner*. In other jurisdictions, fire company personnel are assigned to inspect ordinary mercantile and light industrial occupancies. Regardless of who makes the inspection, any unsafe conditions that are found can be documented and communicated to building owners or occupants. A plan of correction can then be worked out with the person responsible for making the corrections.

NOTE: Additional guidance regarding inspection practices can be found in the IFSTA **Fire Inspection and Code Enforcement** manual.

Fire Hazards

A *fire hazard* is a condition that increases the likelihood of a fire starting or would increase the extent or severity of a fire if one did start **(Figure 20.5)**. Basic fire chemistry suggests that fire cannot occur without a sufficient heat source. Once the fire starts, it will not continue to burn without a sufficient fuel supply, oxygen supply, and a self-sustained chemical reaction (fire tetrahedron). Therefore, hazardous fire conditions can be prevented by eliminating one or all of these elements.

Control of the oxygen supply is only practical under special circumstances because the air we breathe contains roughly 21 percent oxygen. The self-sustained chemical chain reaction can be interrupted only after a fire starts. Control of the fuel supply can be managed to some extent by manipulating the arrangement of the fuel within the building. However, of all the combustion components, heat sources are the most manageable. Any heat source may be dangerous. If heat sources are kept separated from fuel supplies, the condition usually remains safe. Not all fuel supplies can be ignited easily, but misuse of any fuel under extreme heat conditions can lead to a fire. Common fuel and heat source hazards are included in the following test.

Fuel Hazards

- Ordinary combustibles such as wood, cloth, or paper
- Flammable and combustible gases such as natural gas, liquefied petroleum gas (LPG), and compressed natural gas (CNG)
- Flammable and combustible liquids such as gasoline, oils, lacquers, or alcohol **(Figure 20.6)**
- Chemicals such as nitrates, oxides, or chlorates
- Dusts such as grain, wood, metal, or coal
- Metals such as magnesium, sodium, or potassium
- Plastics, resins, and cellulose

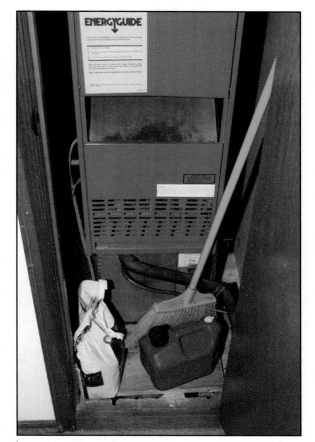

Figure 20.6 Fire prevention codes prohibit the storage of flammable liquids in a furnace room or enclosure.

Figure 20.5 Paint spray booths are potential fire hazards. Building and fire codes may require that such items be protected by sprinkler or other fire suppression systems.

Heat Source Hazards

- *Chemical heat energy* — Materials being improperly stored can result in chemical heat energy. Materials may come in contact with each other and react (oxidizer and reducing agent), or they may decompose and generate heat.
- *Electrical heat energy* — Poorly maintained electrical appliances, exposed wiring, and lighting are sources of electrical heat sources.
- *Mechanical heat energy* — Moving parts on machines, such as belts and bearings, are a source of mechanical heating **(Figure 20.7)**.
- *Nuclear heat energy* — Heat that is created by fission; however, this is not commonly encountered by most firefighters.

Figure 20.7 Mechanical malfunctions may result in fires caused by mechanical heat energy. Proper inspection and maintenance can help to prevent these types of fires.

NOTE: See Chapter 3, Fire Behavior, for a more in-depth discussion of heat source hazards.

Common Fire Hazards

The term *common* might be misleading, but in this context it refers to the probable frequency of a hazard being found, not to the severity of the hazard. A *common fire hazard* is a condition that is prevalent in almost all occupancies and increases the likelihood of a fire starting. Firefighters need to be alert to the following common hazards:

- Obstructed electrical panels

- Poor housekeeping and improper storage of combustible materials

- Defective or improperly used heating, lighting, or power equipment

- Improper disposal of floor-cleaning compounds

- Misuse of fumigation substances and flammable or combustible liquids

Poor housekeeping can make maneuvering through an area difficult and impede the path of egress. Poor housekeeping can also increase the fire load in an area and increase the chances that a flammable or combustible material may come in contact with an ignition source. It can also hide fire hazards in the clutter.

Improperly functioning heating, lighting, or other electrical equipment can provide an ignition source for nearby combustibles. A number of fires have been started in recent years by sheer fabrics being draped over lamps as a decoration. Floor cleaning compounds, fumigating substances, and other flammable and combustible liquids that are improperly used and stored can provide a volatile fuel source if an ignition source is present.

Common fire hazards also have a personal component. The term *personal* takes into consideration the individual traits, habits, and personalities of the people who work, live, or visit the occupancy, structure, or property in question. *Personal fire hazards* refer to those common hazards caused by the unsafe acts of individuals. Personal hazards, often considered intangible, are always present. A comprehensive program geared toward public awareness, fire and life safety education, and good safety practices can reduce the hazards caused by unsafe personal acts. One example of a personal fire hazard would be the habit of smoking in bed.

Common Fire Hazard — Condition likely to be found in almost all occupancies and generally not associated with a specific occupancy or activity.

Special Fire Hazards

A *special fire hazard* is one that arises as a result of the processes or operations that are characteristics of the individual occupancy. Commercial, manufacturing, and public-assembly occupancies each have particular special fire hazards. Typical special hazards are listed in the following sections.

Commercial Occupancies

- Lack of automatic sprinklers or other relevant fixed fire protection systems

- Change of occupancy that exceeds the use for which permits were issued

- Display or storage of large quantities of combustible products

- Mixed varieties of contents

- Difficulties in entering occupancies during closed periods

- Illegal building additions

- Illegal storage

- Storage aisles incorrect distance apart

- Fire department connection obstructed

- Storage obstructing sprinklers

- Existence of party walls, common attics, cocklofts, and other open voids in multiple occupancies **(Figure 20.8)**

Manufacturing

- High-hazard processes using volatile substances, oxidizers, or extreme temperatures

- Flammable liquids in dip tanks, ovens, and dryers in addition to those used in mixing, coating, spraying, and degreasing processes

- High-piled storage of combustible materials **(Figure 20.9)**

- Operation of vehicles, fork trucks, and other trucks inside buildings (use, storage, and servicing hazards)

- Large, open areas

- Large-scale use of flammable and combustible gases

- Lack of automatic sprinklers or other fixed fire protection systems

Public Assembly

- Lack of automatic sprinklers, detection systems, or fire notification systems

- Large numbers of people present, perhaps exceeding posted occupant limits

- Insufficient, obstructed, or locked exits

- Materials stored in paths of egress

- Highly combustible interior finishes

- Inadequate or inoperative fire extinguishers

- Inadequate or inoperative exit lighting

Figure 20.8 Commercial occupancies, such as this strip mall, may have continuous attic spaces that will permit a fire to extend throughout the structure.

Figure 20.9 Warehouses may contain large quantities of combustibles in high-piled storage arrangements.

Target Hazard Properties

A *target hazard* is any structure in which there is a greater-than-normal potential for the loss of life or property from a fire (**Figure 20.10**). These occupancies should receive special attention during surveys. Some examples of target hazards include the following:

- Lumberyards
- Bulk oil storage facilities
- Shopping malls
- Hospitals
- Theaters
- Nursing homes
- Rows of frame tenements
- Schools
- High-rise hotels/ condominiums
- Large public assemblies — concert halls, stadiums, etc.

Figure 20.10 Because of the high life-loss potential and the immobility of some patients, hospitals are considered target hazards.

Fire Inspections

Every firefighter engaged in fire prevention efforts such as conducting code enforcement inspections must be capable of meeting with property owners or occupants and clearly communicating code requirements and discussing possible remedies to observed code violations. Firefighters who are technically competent can provide a valuable service to the public in addition to conveying a favorable public impression of the fire department.

It is especially important that firefighters convey only technically accurate information during inspections. If they are unsure about a specific code requirement or the answer to a technical question, they should tell the property owner or occupant that they will research it and convey the answer to them.

A firefighter's ability to conduct inspections competently will improve with study, experience, and on-the-job training. The skills required for translating visual information into written reports and drawings will also improve with time and practice. The company officer, fire prevention officer, and others can assist when situations require highly technical explanations. Firefighters must be willing to ask for help from the company officer or other available experts.

Personal Requirements

Many members of the public assume that all firefighters are experts on everything relating to fire. They believe that every firefighter is professionally qualified to discuss all aspects of fire prevention and give reliable advice on how fire safety hazards can be corrected. Although firefighters cannot be expected to know the answers to all questions, they should be well-informed about fire and life safety issues and be prepared to refer citizens to additional sources of information.

When performing any public fire prevention activity, firefighters should present a well-groomed, neat appearance. Uniforms should be clean and in good condition. A neat appearance is a professional appearance, which helps to gain the respect of the citizens and enhance the fire department's public image **(Figure 20.11)**.

Equipment Requirements

To adequately conduct a fire inspection in some occupancies, firefighters will need certain special equipment. First of all, they may have to do considerable research before conducting the inspection. To inspect these occupancies, firefighters will need reference material (local fire codes, technical information about the specific occupancies, etc.) in addition to appropriate clothing and equipment. To inspect some occupancies, firefighters may be required to wear safety glasses, hearing protection, hard hats, gloves, and other specialized clothing **(Figure 20.12)**.

Before a survey can be conducted, the survey team must be provided with the proper tools and equipment needed to gather and document the information necessary to develop a complete preincident plan. The equipment needed may vary depending upon what is specified in the organization's guidelines and the nature of the occupancy to be surveyed. However, most preincident survey kits include the following supplies:

- *Writing equipment* — Tablet, pens, pencils, pencil sharpener, erasers, clipboard, and facility survey forms

- *Drawing equipment* — Engineering or graph paper, straightedge, and a copy of the NFPA standard symbols or those used by the organization

- *Other equipment* — Flashlight, water-pressure gauge, camera, and measuring tape or rangefinder

Members of the survey team will need at least one portable radio if they remain in service during the survey.

Figure 20.11 Personnel making fire prevention and code enforcement inspections must maintain a professional appearance.

Figure 20.12 Firefighters may need special clothing and equipment while making fire prevention inspections.

Scheduling Fire Inspections

Every fire department officer is faced with balancing multiple demands competing for the firefighter's time. Fire departments cannot choose the time when fires or other emergencies require their attention, but within certain limits, they can choose the time and place to conduct fire prevention activities. One of these limits is that, in most cases, fire inspections must be conducted during normal business hours. Therefore, many fire departments schedule specific times and days of the week for fire prevention activities.

In some fire departments, company officers are required to contact the building owners or occupants ahead of time to schedule an inspection **(Figure 20.13)**. The company officer informs the owner or occupant of the purpose of the fire inspection and asks which day and time would be most convenient. This helps to ensure that the building owner or occupant will be present and available at that time. Scheduling also allows fire inspections to be conducted at a time that will be least disruptive of the business operation, even though it may have to be after normal business hours.

In other departments, inspections are conducted in a systematic, block-by-block manner so scheduling or prearrangement is usually not necessary. This system works well with ordinary mercantile occupancies (retail shops, offices, etc.) but is less effective with large complexes such as industrial properties.

Conducting Fire Inspections

Departmental standard operating procedures (SOPs) usually specify how fire inspections are to be conducted in various occupancies. Firefighters must be familiar with the department's policy and clearly understand the inspection process. In many cases, this will require a review of the fire code requirements for the specific occupancy scheduled to be inspected.

During the inspection, firefighters must remember that their conduct — good or bad — will reflect directly on the department. Members of the inspection team must make an earnest effort to create a favorable impression upon the property owner because this helps to establish a positive and cooperative relationship **(Figure 20.14)**.

Figure 20.13 Scheduling inspections with the building owner or occupant ensures that the inspection will be convenient to all parties.

Figure 20.14 Being professional and businesslike produces positive results during fire prevention inspections.

The team should enter the premises at the main entrance and contact the individual with whom the inspection was scheduled.

The company officer should introduce the team, briefly review the inspection process, and answer any questions the occupant may have. The occupant or a representative should accompany the inspection team and guide them during the entire process. The guide can help obtain access to all areas of the building and answer the team's questions.

The inspection team should ask that all locked rooms or closets be opened for inspection. If the guide is reluctant to open any room, the team leader should tactfully explain why it is necessary to see these areas. For example, if the guide says, *There is nothing in this locked room,* the team leader might say, *Yes, we understand, but knowledge of the size, shape, and construction features of the room is important.* If admission to an area or room is refused because of a confidential or secure process, the team leader may suggest that the process be covered or screened to permit the survey to continue. If the inspection team is denied access to any rooms or areas, that fact should be reported to the fire marshal or the fire prevention officer so that an inspection warrant may be sought or other appropriate action can be taken.

In most occupancies, the inspection team starts the inspection from the outside of the structure, looking for any possible hazards and making an overall observation of the property and the activities being conducted there. The survey team then moves to the interior of the structure. Some of the specific items they should look for include the following:

- Access for fire apparatus and personnel to the structure and its fire protection equipment.
- Building name and type of business
- Emergency contact numbers for building owner/manager
- Address numbers clearly visible from the street
- Locations of fire department connections
- Cleanliness, maintenance, and good housekeeping
- Portable fire extinguishers in place, properly mounted, operable, and unobstructed
- Exits and exit passageways clear and unobstructed
- Exit signs and emergency lighting operable
- Flammable liquids and other hazardous materials properly stored
- Sources of ignition properly isolated from combustible materials

Depending upon the individual occupancy and the type of business being conducted there, the specific code requirements may differ. For example, if large quantities of hazardous materials are stored or used on the premises, the code may require that a marking system such as that outlined in NFPA® 704, *Standard System for the Identification of the Hazards of Materials for Emergency Response,* be affixed to the outside of the building.

Regardless of the type of occupancy and the specific code requirements that apply, each item inspected — and especially each code violation noted — should be explained to the person accompanying the inspection team. If one or more code violations are found, a plan of correction must be agreed upon between the inspection team and the occupant. In most cases, the plan will specify a reasonable period of time in which the violation or violations must be corrected and when a follow-up in-

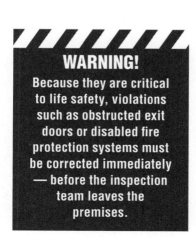

WARNING!
Because they are critical to life safety, violations such as obstructed exit doors or disabled fire protection systems must be corrected immediately — before the inspection team leaves the premises.

spection will be made to verify that the corrections have been made satisfactorily. The form and content of the plan of correction should be clearly defined in departmental SOPs covering the inspection process.

In most departments, the occupant or representative is required to sign an inspection form acknowledging that the fire inspectors have explained any violations and corrective measures. A copy of the signed form is left with the occupant. Finally, the inspection team leader should thank the occupant for cooperating and leave a business card with the offer to answer any questions that the occupant may have regarding the inspection results.

Fire Safety Surveys

Fire safety surveys include *preincident planning surveys* of public and commercial occupancies and *residential fire safety surveys*. As mentioned earlier, preincident planning surveys are for gathering information, and residential fire safety surveys are for imparting information.

A preincident planning survey in public or commercial occupancy allows firefighters to gather information about conditions that might affect future emergency operations in that building — both positively and negatively. Having this information allows firefighters to develop plans that minimize a building's deficiencies and maximize its strengths.

In most jurisdictions, single-family residences (houses and apartments) are not subject to the requirements of local fire codes to the same degree as are businesses, institutions, and industrial properties. In these jurisdictions, residential fire safety surveys are offered as a public service. Homeowner or occupant participation is voluntary. When requested by a homeowner or occupant, firefighters survey the residence and point out potential fire and life safety hazards and suggest corrective actions. In some jurisdictions, residential fire safety surveys may be conducted as part of a house-to-house fire prevention program.

Firefighters need a number of personal and technical skills to conduct fire safety surveys properly. Interpersonal skills may include those needed for communication, mitigation, facilitation, negotiation, or mediation. Technical knowledge and skills require firefighters to understand building construction, fire and life safety requirements, fire code requirements, common and special hazards, building utilities, energy systems, and various fire protection appliances and systems.

Preincident Planning Surveys

Preincident planning surveys allow firefighters to gather information about the structure under ideal conditions: when it is not on fire. This information can provide firefighters with critical information that might not be available during an active fire. Through these surveys firefighters become aware of a building's construction and layout, contents, hazardous materials storage, special processes, fire detection and suppression systems, fire load (also known as fuel load), and other pertinent details. This knowledge can greatly improve emergency operations and substantially improve both firefighter and citizen safety. Firefighters create maps, drawings, photographs, and written notes during preincident surveys. This documented information helps firefighters achieve the following:

- Become familiar with structures in the district, their uses, and their associated hazards (**Figures 20.15 a and b**).
- Recognize existing hazards.

- Visualize how standard tactics may or may not apply in various occupancies.

- Develop new tactics if necessary.

Fuel Loads

The term *fuel load* represents the bulk of fuel available to burn and generally refers to the contents of a building. Because the materials used in the construction of most modern commercial and mercantile buildings contribute relatively little fuel to a fire, the major fuel sources are furnishings and other building contents. Because different fire control procedures may be used depending upon what is burning, knowing the combustibles in a building can have a profound effect on the tactics and strategies employed during a fire and on firefighter safety.

It may be impossible to identify a building's contents after a fire starts, so it is imperative that this information be gathered during preincident surveys. This practice is especially important in buildings where toxic, highly flammable, or explosive materials are stored or used. During preincident surveys, company officers should be sure to document the existence of large quantities of the following materials:

- Plastics

- Aerosols

- Compressed gases

- Explosives

- Flammable and combustible liquids

- Combustible dusts

- Corrosive/water-reactive materials

More information on preparing a preincident survey can be found in **Skill Sheet 20-II-1.**

Figures 20.15 a and b Preincident planning surveys help firefighters become familiar with the types of occupancies in their response areas, including (a) lumberyards and (b) apartment complexes.

Conducting Preincident Planning Surveys

After the initial meeting with the owner (or designee), the survey team should return to the outside of the building to make general observations, complete preliminary notes, and take photographs. In some occupancies permission to take photographs may have to be obtained beforehand. These exterior observations and photographs make the survey of the interior easier and provide the necessary information for drawing the exterior walls on a floor plan (layout of each floor of a building). Detailed drawings and/or satellite photos of the exterior may also be available from the owner.

Firefighters should note the location of fire hydrants, fire department connections, post indicator valves (PIVs), water supply sources, and fire alarm control panels **(Figure 20.16)**. The type of building construction, height, occupancy, and proximity of adjacent exposures should be noted. Also noteworthy is the area surrounding the occupancy, the accessibility to all sides of the property, and the condition of the streets. Such factors become extremely important when considering fire apparatus response and positioning. Firefighters should check and include the following in the preliminary notes:

- Are address numbers sufficiently visible?

- Are all sides of the building accessible? **(Figure 20.17)**

- Is there a building setback or other barriers to aerial device operations?

- Are there trees and shrubs that could make access difficult or hide the FDC?

- Are there barred windows or security doors?

- What is the location of utility controls? (shutoffs)

- Are there power lines or other overhead obstructions that would restrict the use of ladders?

When the survey of the exterior is completed, the survey team should go directly to the roof or basement and proceed with a systematic survey. It does not matter if the team starts on the roof and works downward or starts in the basement and works up. From a practical standpoint, however, many firefighters find it helpful and less confusing to start on the roof. Regardless of which procedure is used, progression should be planned so that the firefighters can systematically look at each floor in succession.

If floor plan drawings are not available from the building owner, firefighters will have to create them. To conduct a thorough survey, firefighters must take enough time to make notes and take photographs of observed hazards and unsafe conditions. Drawings of the interior layout, high-hazard areas, egress routes, and important features should be made (or upgraded on existing drawings) or photographed. The drawings are particularly important when survey information is used to develop written preincident plans or is entered into computer-aided dispatch (CAD) systems. A complete set of notes, photographs, and well-prepared drawings of the building provides dependable information from which a complete report can be written (see Making Maps and Drawings and Photographs sections) **(Figure 20.18)**.

In large or complex buildings, it may be necessary to make more than one visit to complete the survey. If the property includes several buildings, each should be surveyed separately. It is a good idea to start on the roof of the highest building so you can get a good overall view. As you work your way down in the building, complete a drawing of each floor before proceeding to the next floor.

If a floor plan used on a previous survey is available, the survey can proceed more rapidly. Make sure to record any changes that have been made and update the floor plan drawings accordingly. Taking the time to discuss the survey results as well as any fire and life safety concerns with the owner or occupant is also beneficial.

Figure 20.16 Surveys include the location of hydrants and fire protection systems components such as PIVs.

Figure 20.17 It is important to document anything that might interfere with fire fighting operations. Access to some structures may be limited due to overhangs and proximity to other buildings.

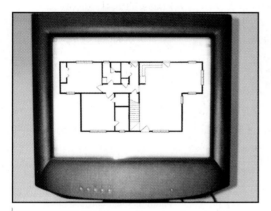
Figure 20.18 Computers are invaluable for creating floor and site plans and storing survey information.

Making Maps and Drawings

Maps that contain information regarding building construction, fire protection systems, occupancy, fire loading (fuel loading), special hazards, and other details are very helpful to firefighters. Large occupancies or complexes may already have maps that were prepared by their insurance carriers. These maps normally use some form of common map symbols **(Figure 20.19, p. 974)**.

For buildings where existing maps are unavailable or outdated, firefighters should include a simple plot plan drawing that shows the general arrangement of the property with respect to streets and other buildings. They should also note any other important features and information that might affect fire fighting tactics in that occupancy. In many cases, this drawing is the most important product of the survey so it should be made with neatness and accuracy.

A clipboard and a ruler can be helpful aids when making drawings. Data should be recorded by using common plan symbols as much as possible. Engineering or graph paper can make it easier to produce a drawing that is to scale **(Figure 20.20, p. 975)**. The use of computerized Geographic Information System (GIS) or other electronic mapping programs can save hours and should be used where available.

By using common symbols on a floor plan, firefighters can show the type of construction, thickness of walls, partitions, openings, roof types, parapets, and other important features. In addition, fire protection devices, water mains, automatic sprinkler control valves, and other potentially important features of fire protection should be included **(Figure 20.21)**.

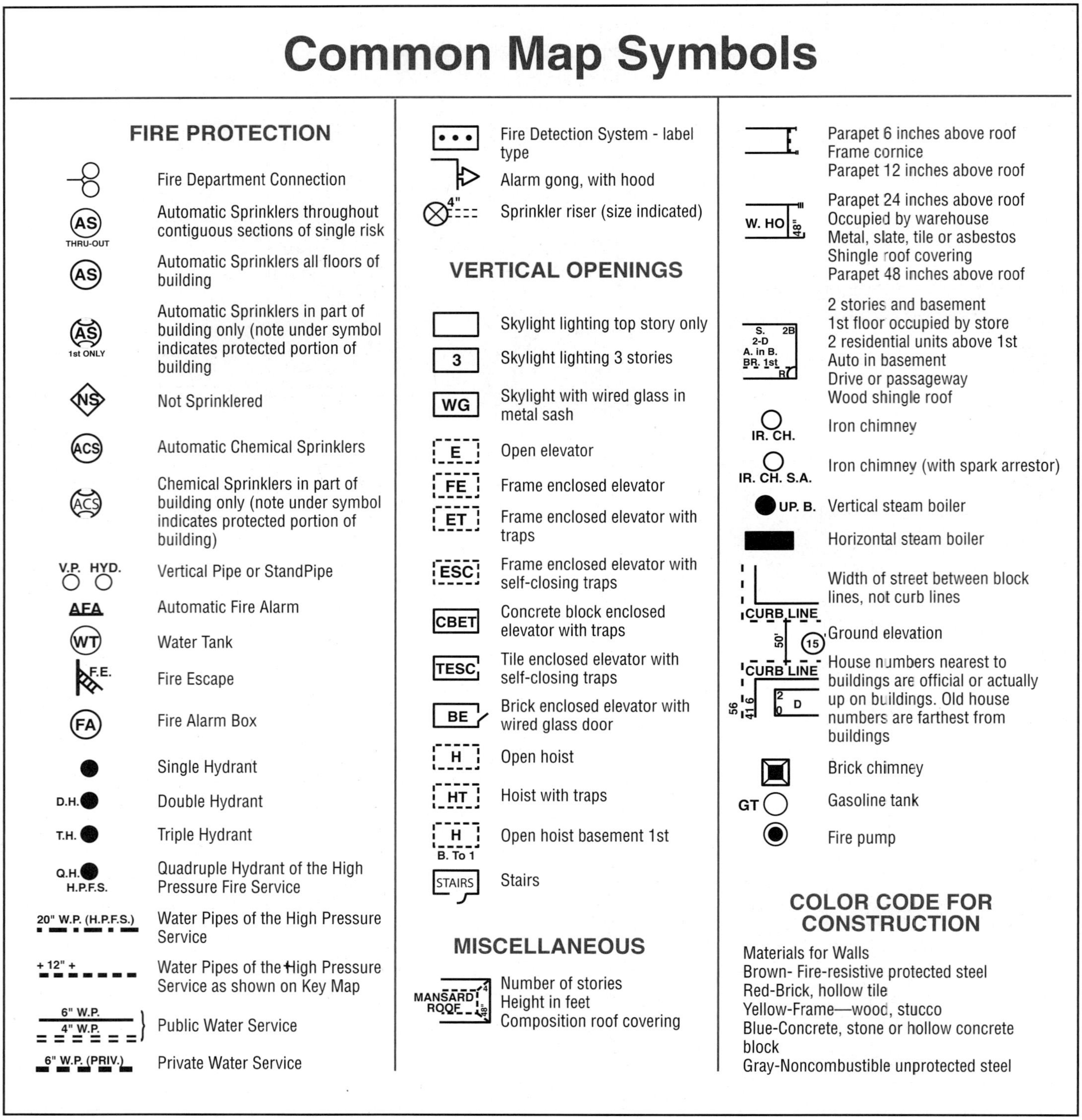

Common Map Symbols

FIRE PROTECTION

	Fire Department Connection
AS THRU-OUT	Automatic Sprinklers throughout contiguous sections of single risk
AS	Automatic Sprinklers all floors of building
AS 1st ONLY	Automatic Sprinklers in part of building only (note under symbol indicates protected portion of building
NS	Not Sprinklered
ACS	Automatic Chemical Sprinklers
ACS	Chemical Sprinklers in part of building only (note under symbol indicates protected portion of building)
V.P. HYD.	Vertical Pipe or StandPipe
AFA	Automatic Fire Alarm
WT	Water Tank
F.E.	Fire Escape
FA	Fire Alarm Box
●	Single Hydrant
D.H. ●	Double Hydrant
T.H. ●	Triple Hydrant
Q.H. ● H.P.F.S.	Quadruple Hydrant of the High Pressure Fire Service
20" W.P. (H.P.F.S.)	Water Pipes of the High Pressure Service
+ 12" +	Water Pipes of the High Pressure Service as shown on Key Map
6" W.P. 4" W.P.	Public Water Service
6" W.P. (PRIV.)	Private Water Service

● ● ●	Fire Detection System - label type
	Alarm gong, with hood
⊗ 4"	Sprinkler riser (size indicated)

VERTICAL OPENINGS

	Skylight lighting top story only
3	Skylight lighting 3 stories
WG	Skylight with wired glass in metal sash
E	Open elevator
FE	Frame enclosed elevator
ET	Frame enclosed elevator with traps
ESC	Frame enclosed elevator with self-closing traps
CBET	Concrete block enclosed elevator with traps
TESC	Tile enclosed elevator with self-closing traps
BE	Brick enclosed elevator with wired glass door
H	Open hoist
HT	Hoist with traps
H B. To 1	Open hoist basement 1st
STAIRS	Stairs

MISCELLANEOUS

MANSARD ROOF	Number of stories Height in feet Composition roof covering

	Parapet 6 inches above roof Frame cornice Parapet 12 inches above roof
W. HO	Parapet 24 inches above roof Occupied by warehouse Metal, slate, tile or asbestos Shingle roof covering Parapet 48 inches above roof
S. 2B 2-D A. in B. BR. 1st R	2 stories and basement 1st floor occupied by store 2 residential units above 1st Auto in basement Drive or passageway Wood shingle roof
IR. CH.	Iron chimney
IR. CH. S.A.	Iron chimney (with spark arrestor)
● UP. B.	Vertical steam boiler
▬	Horizontal steam boiler
CURB LINE	Width of street between block lines, not curb lines
15	Ground elevation
CURB LINE	House numbers nearest to buildings are official or actually up on buildings. Old house numbers are farthest from buildings
▣	Brick chimney
GT ○	Gasoline tank
◉	Fire pump

COLOR CODE FOR CONSTRUCTION

Materials for Walls
Brown- Fire-resistive protected steel
Red-Brick, hollow tile
Yellow-Frame—wood, stucco
Blue-Concrete, stone or hollow concrete block
Gray-Noncombustible unprotected steel

Figure 20.19 The use of common map symbols adds clarity to the map and saves time. *Courtesy of The Sanborn Map Company.*

A sectional elevation drawing of a structure, consisting of a cross-sectional or cutaway view of a particular portion of a building along a selected imaginary line, may be needed to show elevation changes, mezzanines, balconies, or other structural features **(Figure 20.22)**. The easiest sectional view to portray is to establish the imaginary line along an exterior wall. This view theoretically removes the exterior wall and exposes such features as roof construction, floor construction, fire walls, parapets, basements, attics, and other items that may be difficult to show on a floor plan.

Establishing the imaginary line along an exterior wall may not always show the section of the building that is most important. In this case, it may be better to divide the building near the center or along a line where a separate wing is attached to the main structure. From the preliminary drawing and notes, a permanent drawing can be made for inclusion in a preincident plan, future reference, and classroom study. The permanent drawing should be drawn to scale.

Photographs

If permitted by the building owner or occupant, photographs can show important details that even accurate drawings cannot. Such detail may be needed for developing preincident plans. Photographs can quickly and easily record a tremendous amount

Figure 20.20 Using graph paper makes accurate drawings easier to create.

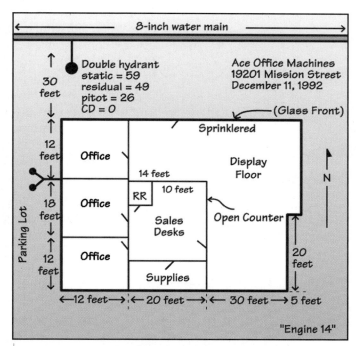

Figure 20.21 A basic floor plan drawing providing information on the building layout and additional site information.

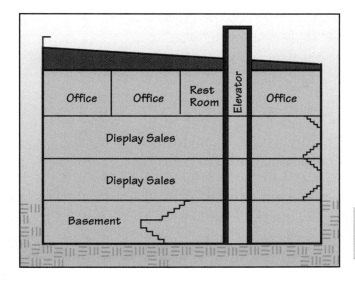

Figure 20.22 Sectional elevation views show the relationship of multiple floors and penetrations such as elevator, pipe shafts, and atriums.

Figure 20.23 If possible, a photograph taken from an elevated position can help to illustrate equipment and penetrations of the roof deck.

of information, especially if taken from more than one angle. One view that is often especially good from a fire fighting standpoint is from an elevated position (**Figure 20.23**). An adjoining building, elevated tower, or an aerial apparatus can be used for this purpose. Interior and close-up photographs can also be very effective aids in making a complete preincident plan. When possible and permitted, a video of the facility can be made for training purposes.

Residential Fire Safety Surveys

On average, annual U.S. fire statistics show that more than 70 percent of all fires and the vast majority of civilian fire casualties occur in residences. Most fire departments invest considerable resources in programs intended to improve safety in the home (**Figure 20.24**). As mentioned earlier, fire safety surveys in existing residential occupancies (particularly one- and two-family dwellings) can only be conducted on a voluntary basis. Most codes require inspections of structures that house three or more families (hotels, motels, inns, and residential healthcare facilities), but surveys of other-than-common areas in these structures may still be at the discretion of the occupants. Apartment complexes can be inspected, although inspection of the individual units is voluntary. Conducting a residential fire safety survey is discussed in **Skill Sheet 20-II-2**.

Figure 20.24 Firefighters can provide residential occupants with information on fire and life safety and on the location and operation of smoke detectors.

When residential fire safety surveys are conducted as part of an organized public awareness and education program, a great deal of advanced planning and publicity is necessary to gain full acceptance by the community. It must be made clear that the program is a *fire prevention activity* and not a *code enforcement activity*. In other words, the firefighters are coming to make family members aware of safety hazards, not to cite them for code violations. When firefighters enter the home to conduct a residential fire safety survey, their main objectives include the following:

- Preventing accidental fires
- Improving life safety conditions
- Helping the owner or occupant to understand and improve existing conditions

By conducting residential safety surveys, fire departments reap numerous other benefits in addition to the primary one of reducing loss of life and property. Citizens get to know and trust their firefighters. When residents get to know their firefighters, they gain an appreciation for the many services their firefighters provide. Safety surveys allow the residents to feel that their fire department really cares about their welfare. An increase in goodwill translates into support at budget time or during community fundraising events.

In addition to generating positive public relations and community support, safety surveys in the home increase fire awareness and the interest in public education efforts. During a safety survey may be the best time for firefighters to do the following:

- Distribute fire prevention literature.
- Promote programs such as *Exit Drills In The Home (EDITH)*.
- Check for or distribute emergency telephone number stickers.
- Discuss smoke alarms and residential sprinkler options.
- Provide information about CO detectors.
- Provide information on the prevention of burns (**Figure 20.25**).

Firefighters can discuss and expand on the information in the literature. In some cases, they can add emphasis and make the information more relevant by relating it to a recent or locally famous fire.

Some fire departments also give special cards or slips to compliment homeowners whose dwellings are found in a fire-safe condition. Other cards saying *We're sorry we missed you,* are used to notify absent households that firefighters were in the neighborhood conducting safety surveys. The cards list the appropriate telephone number that homeowners can call if they would like to schedule a safety survey. In addition, some departments use the "time change/battery change" as an opportunity to promote surveys for the elderly and handicapped by also offering to replace batteries in their smoke alarms.

Firefighters also gain valuable information when performing residential safety surveys. They become better acquainted with home construction, occupancy conditions, local development trends, streets, hydrants, and water supply locations. Notes on these items and other useful information should be made and discussed during training sessions. Finally, firefighters can become familiar with special needs of the residents. The elderly, infirm, and incapacitated have specific needs including text telephones and telecommunication devices for the deaf, visual alerting systems, and sign language interpreters. Firefighters can also provide information for other community services that these citizens may not be familiar with.

While these fringe benefits are helpful, the primary reason for conducting surveys is to reduce hazards associated with loss of life and property. In some departments, it is SOP for their on-duty firefighters to check every residence they enter for whatever reason, for smoke alarms. If none are found, they volunteer to install one.

Figure 20.25 Target groups, such as the elderly, can be provided with home safety information during voluntary residential inspections.

Firefighter Responsibilities

Because residential fire safety surveys can reduce the number of fire incidents and deaths occurring in homes, it is the responsibility of the survey team to do their jobs conscientiously. The public expects firefighters to be subject matter experts on home fire and life safety issues. Therefore, they must be prepared to answer a wide variety of technical questions. When conducting residential safety surveys, firefighters should use the following guidelines:

- Conduct surveys in teams of two.

- Dress and act professionally.

- Introduce yourself, your partner, and provide proper identification (**Figure 20.26**).

- Explain the survey procedure.

- Maintain a courteous and businesslike attitude at all times.

- Focus on preventing fires and eliminating threats to life safety.

- Compliment the occupants when favorable conditions are found.

- Offer constructive suggestions for correcting or eliminating hazardous conditions.

- Survey all rooms including the garage.

- Discuss the survey results with the owner or occupant and answer any questions.

- Thank the owners or occupants for the invitation into their homes.

- Keep the survey confidential; do not share the results with any outside entity.

If no one is at home, leave appropriate materials between the storm door and the front door or partially beneath the doormat. Remember that the U.S. Postal Service prohibits the use of mail boxes for unstamped materials.

Residential Fire Causes

Firefighters conducting residential surveys should look for the most common causes of fires. It is a good idea to complete a survey form for each residence and provide a copy to the occupant. If the form includes a checklist of common hazards, it can serve as a guide to summarize the survey results. Some common fire causes involve the following:

- Malfunctioning heating appliances and water heaters

- Combustibles too close to heating appliances or lamps

- Unsafe cooking procedures

- Smoking materials

- Overloaded extension cords and multiple-outlet devices

- Exposed electrical wiring

- Defective electrical appliances (**Figure 20.27**)

- Improper use of combustible or flammable liquids

- Poor housekeeping

- Untended candles

Figure 20.26 Residential surveys should be well publicized before they are made. If the resident does not want a home survey, fire and life safety information can be left with them.

Figure 20.27 Overheated electrical equipment may result in fires.

Firefighters must know the common causes of home fires in order to conduct effective residential surveys and to educate citizens about these dangerous conditions. For the homeowner or tenant, the residential fire safety survey provides a valuable life safety service. Some of the questions that firefighters should ask themselves as they conduct residential fire and life safety surveys are in the sections that follow:

Interior Surveys

- **Combustible materials** — Are clothing, upholstered furniture, cardboard boxes, papers, and other combustible materials stored close to heating registers or heat-producing appliances? Do they block exits?

- **Appliances** — Are the units operating properly? Are electrical cords in good condition? Are surge protectors being used?

- **Electrical wiring and equipment** — Is there old, frayed, or exposed wiring or improperly installed electrical conductors? Are extension cords overloaded? Are multiple-outlet devices being used? Are receptacles located near sinks the ground fault indicator (GFI) types? Check for unprotected lightbulbs or improperly maintained equipment such as exhaust fans covered with dust and lint.

- **Portable heating units** — Is the equipment listed with Underwriters Laboratories (UL), FM Global (FM), or another recognized testing laboratory? Is it adequately separated from combustible materials?

- **Woodstoves or fireplaces** — Is the unit properly installed and clear of combustibles? Is the vent pipe in good condition? Recommend regular chimney cleaning and maintenance.

- **Heating fuel** — Is wood or other fuel properly stored? How do residents dispose of the ashes?

- **General housekeeping practices** — Does the occupant use ash trays for smoking materials? Are matches and lighters kept out of the reach of children? Are candles and other items with open flames used safely? Are exhaust vents and dryer vents cleaned of lint regularly?

Figure 20.28 Work areas not kept clean are a fire hazard. Shop areas that contain metal shavings, oily rags, and dust particles can cause fire hazards.

- *Smoke alarms* — Are there smoke alarms? If so, recommend testing on a regular basis. For battery-operated units, recommend changing the battery whenever the clocks are reset because of Daylight Savings Time. If not, offer to install one (if that is department policy) or suggest where one may be obtained.

- *Electrical distribution panels* — Is there proper circuit protection and clearance?

- *Gas appliances* — Is there proper clearance from combustible materials? Is there an automatic gas control safety device and a manual supply line shutoff? Are the vent pipes in good condition?

- *Oil-burning units* — Are annual service records available and are oil burners, flue pipes, supply tanks, and piping in good condition?

- *Furnaces* — Is the unit properly installed with proper clearance from combustibles? Are dust and lint accumulations present? Is the vent pipe in good condition?

- *Water heaters* — Is the unit properly installed with earthquake strapping? Check for carbon stains (soot) around the fire box cover because they may indicate a malfunctioning unit. Check the water temperature settings to protect against scalds and burns. The recommended water temperature setting is usually 130°F (55°C). Where small children and invalids reside, the setting is recommended to be no higher than 120°F (49°C).

- *Shop or work rooms* — Are work areas neat and clean? Are materials stored in a safe and orderly manner? **(Figure 20.28)**

- *Accumulated waste* — Are there stacks of paper, discarded furniture, old rags, and improperly stored combustibles?

- *Flammable liquids* — Are flammable sprays, chemicals, and other dangerous solutions properly stored and out of the reach of children?

Exterior Surveys

- *Roof* — Does the roof appear to be in good condition? Does it have combustible wooden shingles or shakes? Is there an accumulation of dry leaves or needles?

- *Chimneys and spark arrestors* — Are chimneys and spark arrestors in good condition? **(Figures 20.29 a and b)**

- *Yard and porch areas* — Is there dry vegetation within 30 feet (10 m) of the house? Is firewood stacked against the house? Are combustible items stored in carports or under porches or overhangs? **(Figure 20.30)**

- *Barbecues and fuel* — Are barbecues used under balconies, porches, or other overhangs? Are they gas-fired or charcoal-burning? Is fuel stored properly? Are ashes from BBQs or wood stoves disposed of properly?

- *Outside waste burners* — Are outside waste burners (incinerators) in use? Are they allowed or is their use restricted by local ordinances?

- *Garages, sheds, barns, and outbuildings* — Are dangerous chemicals or other substances (swimming pool chlorine, propane cylinders, charcoal, lighter fluid, gasoline, ammonium nitrate fertilizer, and pesticides) properly stored? Are paints, turpentine, and similar liquids stored properly?

- *Flammable liquids and gases* — Are flammable liquids being stored and in what quantity? If so, recommend that they be kept in a safety cabinet or stored in a detached outside storage area **(Figure 20.31)**. Gasoline, propane, and other similar flammable liquids and gases should never be brought into a dwelling.

Remind the occupants that flammable liquids should never be used, inside or outside, for cleaning or other purposes that would expose the user and\or structure to explosive vapors.

- **Lightning protection** — Are lightning rods in place and in good condition? Recommend that the system components of fixed lightning protection systems on structures be tested periodically.

- **Security devices** — Are there security devices or animals that might hinder firefighters trying to enter? Are there window bars, security fences, or guard dogs?

- **Power lines** — Are tree limbs close to electric power or service lines?

Figures 20.29 a and b Chimneys should be inspected for (a) cracks or damage to the exterior and (b) the presence of spark arresters on the chimney cap.

Figure 20.30 Dry combustibles, such as firewood, should not be stored too close to a structure.

Figure 20.31 Flammable liquids should be stored properly.

Home Safety Issues

In addition to performing a residential fire safety survey of the premises, firefighters should also provide occupants with fire and life safety awareness information. Firefighters should discuss with occupants the value of making safe practices a way of life. One of the most important of these practices is maintaining clear and unobstructed exit pathways so they can escape if a fire should start. This can be especially important at night when occupants may have to find their way out in the dark. In addition, they should have two exits available to them. A related safe practice is keeping a working flashlight by the bed.

Another important area of safe practices concerns water. For example, drownings can be prevented by never leaving infants or toddlers alone in or near a bathtub, wading pool, or swimming pool — even for a few seconds. Remembering to turn pans full of hot liquids so their handles are over the range and inaccessible by small children can prevent scald injuries.

Public Fire and Life Safety Education

Educating citizens of all ages to recognize potential hazards and take appropriate corrective action is a critically important fire department function. Teaching fire survival techniques such as *stop, drop, and roll* or *crawl low under smoke* can change behavior and thereby prevent loss of life and property due to fire and other causes

Figure 20.32 During fire education programs, children should be taught the basics of fire safety.

(**Figure 20.32**). After a fire starts, homeowners do not have time to develop an escape plan. If they have been educated about fire and life safety, they can apply that knowledge to save themselves and their families. The firefighter's role as a public fire and life safety educator is one of the most rewarding and one of the most valuable to your community.

Effective injury prevention initiatives have four parts that work together to reduce death and injuries. These are education, engineering, enforcement, and economic incentive, commonly called the *Four E's*. Public fire and life safety education is a key component to reducing fire deaths and injuries in any community, and all firefighters are part of this educational delivery system. An additional E is sometimes included in the prevention equation. This is emergency response. Effective emergency response reduces the extent of an injury and prevents further harm.

General Considerations

Although the circumstances surrounding each presentation may have some differences, there are general considerations that apply to the preparation and delivery of information regarding life safety. No matter who is the audience or how brief the presentation, messages must be accurate, positive, and targeted to the specific audience in order to be effective.

Accurate Messages

The information a firefighter conveys to children or adults concerning the elimination of fire hazards, escaping a fire, testing a smoke alarm, installing a child safety seat correctly, wearing a bicycle helmet, or any other life safety message must be accurate. A firefighter must know the right information as well as the best way to communicate it. Conveying wrong information can be deadly.

NOTE: For more information about fire and life safety education topics, consult the IFSTA **Fire and Life Safety Educator** manual.

There are several sources available to assist each fire department in verifying the accuracy of the content of fire and life safety programs and presentations. A number of these sources also have prepared information that the firefighter can use for a variety of targeted audiences. For example, the Home Safety Literacy Project provides information designed to help convey messages to readers of all levels of reading ability. The following is a partial list of organizations that have developed fact sheets and information that firefighters can use for fire and life safety presentations:

- U.S. Fire Administration http://www.usfa.dhs.gov
- NIST, Building and Fire Research Laboratory http://www.fire.nist.gov
- Home Safety Council® http://www.homesafetycouncil.org
- Safe Kids http://www.usa.safekids.org
- American Red Cross http://www.redcross.org
- U.S. Consumer Product Safety Commission http://www.cpsc.gov
- Centers for Disease Control and Prevention (CDC) http://www.cdc.gov
- International Fire Service Training Association/ Fire Protection Publications (IFSTA/FPP) http://www.ifsta.org
- Underwriters Laboratories (UL) http://www.ul.com
- National Fire Protection Association (NFPA) http://www.nfpa.org
- Ready America http://www.ready.gov

> A number of organizations provide free information to help firefighters in presenting safety messages. For example, the Home Safety Literacy Project (part of the Home Safety Council) is designed to provide fire and life safety messages to people of all reading levels.

Positive Messages

Presenting fire and life safety information requires basic knowledge of effective educational methods. One of the most important of these methods is to present positive information. People are much more likely to take action to prepare for an emergency such as stocking a home disaster kit or testing their smoke alarm if they are given the facts without being exposed to the reality of the emergency. Adults want to know what to do without being scared.

The importance of delivering positive messages has a second contributing factor that applies to fire and life safety information. In an emergency such as a home fire, people must act quickly. Therefore, it is vital that they remember the correct behaviors. These behaviors need to be taught clearly.

Positive Messages		Negative Messages
"Crawl low under smoke."	instead of	"Do not stand up in smoke."
"Get out! Get out!"	instead of	"Do not hide in the closet."
"Call 9-1-1."	instead of	"Do not call the operator."

Targeted Messages

Realistically, everyone's attention span is limited to only a few minutes, so it is important to limit the information you present to messages that have been determined to be priorities for the people you are talking to. This is called *targeting the message*. Fire and life safety educational messages can be selected based on the age of the audience, the time of year, the fire hazard particular to a season, or a recent fire incident that is of interest in your community. There also may be an emerging issue such as security bars on windows or an increase in fires ignited by candles or Dumpster® fires after school.

A single presentation will not work for all age or ethnic groups. The audience may consist of the following people:

- Preschoolers
- School children
- Homeowners
- Apartment tenants
- People with physical impairments
- Public- and private-sector employees
- Medical and nursing facility personnel

Knowing your audience helps you to prepare a presentation that is specific to their needs.

Presenting Fire and Life Safety Education for Adults

There are certain times when adults seem to be more receptive to making changes in their homes and in their behavior. These are "teachable moments" when fire and injury prevention concepts may have a greater impact. These include a move into a new home or apartment, the arrival of a new family member, or when a tragedy has struck close to home. To take advantage of these moments, fire departments may seek to partner with others in the community to find and deliver injury prevention and life safety information to adult audiences. Effective partners might be a prenatal class at a hospital, a meeting of potential homebuyers sponsored by a realtor, or a support group of adoptive parents. Some fire departments have community education in a neighborhood where there was a recent fire. This provides an opportunity for the neighbors to ask questions and for the fire department to discuss important fire and life safety concepts such as the importance of installing and testing smoke alarms.

Although the following information is not sufficient to make you an accomplished speaker or instructor, it presents some basic concepts that will help you present fire and life safety information effectively to individuals and small groups. When making a fire and life safety presentation, you can take certain steps to make sure that all the information is presented and that the audience can perform basic fire and life safety skills such as calling the fire department or testing a smoke alarm. All presentations should follow the basic four-step method of instruction.

Basic Four-Step Method of Instruction

Preparation. The first step in making an effective presentation is preparation. Preparation includes learning the message that you are planning to deliver, practicing the presentation, and knowing your audience. Planning involves gathering the necessary audiovisual aids, handout materials, and information about the topic. Practicing the presentation requires that you actually make the presentation to a small audience. Other crew members or family members can help you with this.

Preparation also involves motivating the audience to learn. Preparation involves getting the participants' attention by letting them know why the material is important to them. Arousing curiosity, creating interest, and developing a sense of personal involvement on the part of the participants are all parts of preparation. One way to do this with adults, for example, is to relate a success story of a family who escaped a home fire because their smoke alarm woke them up and they used the family's escape plan. The use of child-restraint systems in vehicles is extremely effective in saving lives as well.

Presentation. The second step is to actually present facts and ideas to the audience. Presentation involves explaining information, using visual aids (smoke alarm, telephone for dialing 9-1-1, fire alarm pull station, etc.), and demonstrating techniques (stop, drop, and roll; crawl low in smoke; alert others of an emergency, etc.).

> Remember that relating a frightening event or showing pictures of a fire or an emergency incident is not an acceptable method to use as an attention-getter or in any part of your presentation.

Presentations are also an opportunity to promote fire and life safety and build community relations by rewarding the participants with inexpensive pens, pencils, refrigerator magnets, etc., with safety messages on them. These rewards are especially effective with small children. To learn more about making a fire and life safety presentation, see **Skill Sheet 20-II-3.**

Answering Questions

Sometimes people have challenging questions that may require an answer that is more in-depth or more technical than you are prepared to give. Referring such questions to your fire department inspector, fire marshal, or fire chief is appropriate. It is also acceptable to say, "That is a good question. I do not know the answer, but I will get back to you with the information you need." Then you can find additional information, verify it for accuracy, and communicate the answer.

In most circumstances it is not appropriate for firefighters to make assessments or to offer personal opinions concerning local fire and building codes, the cause of a fire or another emergency incident that is currently under investigation, or fire department rules and regulations when interacting with the public.

Application. In the third step — perhaps the most important one — the participants are given the opportunity to use or apply the information that they have been presented (**Figure 20.33, p. 986**). In this step, the audience members practice using the new ideas, information, techniques, and skills while you observe them and tactfully correct any mistakes that are made. For example, the participants might demonstrate how to report a fire; perform the stop, drop, and roll technique; or test a smoke alarm. In other situations, they may apply their new knowledge by answering oral questions.

Evaluation. Evaluation means assessing or determining the effectiveness of the fire and life safety presentation or program and identifying parts of the program that need to be modified or improved. An evaluation that measures reductions in

Motivation — Internal process in which energy is produced by needs or expended in the direction of goals. Motivation usually occurs in someone who is interested in achieving some goal.

Figure 20.33 Firefighters can use the demonstration approach for presenting facts of fire and life safety education. Participants should be allowed to apply the new information.

fire incidents or fire deaths and injuries requires collecting and analyzing data for multiple years. This assessment is almost certainly part of your fire department's participation in a fire incident reporting system through your state or province.

Evaluation can also mean measuring educational gain — what did the students learn? It can also mean measuring behavior change — how many participants tested their smoke alarm after the presentation? It is common for fire and life safety educators to hear or see the evidence of their program's successes. For example, a parent may call and report that the children knew the home escape plan and quickly went to the designated meeting place after a smoke alarm sounded.

Firefighters are not usually responsible for designing the fire and life safety evaluation step, but they will be part of the team that collects, records, and reports information. Having evidence that your fire and life safety programs have saved lives is the greatest reward for all fire and life safety educators and will ensure that these lifesaving activities are supported.

Presenting Fire and Life Safety Information for Young Children

Young children are especially at risk for fire deaths and burn injuries. The rate of deaths from home fires for preschool children is more than double relative to their numbers in the population. A common definition of young children is from birth to 8 years old. These children are in child care centers and Pre-K thru 2nd grade classrooms. Because of the high risk of this population, fire and life safety presentations for this group are emphasized in this chapter.

It is important to remember some of the important characteristics and needs of children:

- Remember that children often interpret information literally. For example, a child who knows that the letter carrier brings the mail might think that a firefighter brings fire. Similarly, remember that sarcasm and teasing are not appropriate for young children.

- Children have limited attention spans, so a presentation that takes 15 minutes or less is ideal.

- Remain flexible when presenting information to young children because they may fidget or bring up additional topics.

- When you are in a classroom, decide with the teacher ahead of time how you will handle questions. A good plan is to have the children and the teacher prepare a list of questions before you arrive.

Figure 20.34 Firefighters should demonstrate the proper techniques for concepts such as *Stop, Drop, and Roll.*

Figure 20.35 During home inspections, homeowners should be encouraged to develop and practice an escape plan.

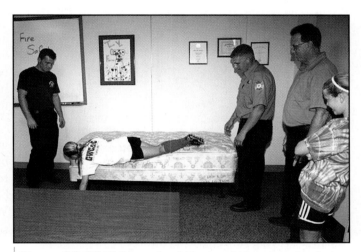

Figure 20.36 Children should be taught how to roll out of the bed onto the floor.

Figure 20.37 Occupants should be taught to feel doors for heat before opening them. They should place the back of their hand against the door.

- Do not place candles near curtains or flammable fabrics.
- Keep candles out of reach of children and away from pets.
- Separate candles by at least 4 inches (100 mm).
- Do not move lighted candles.
- Never leave lighted candles unattended.
- During a power outage, use flashlights or battery-operated lights, not candles.
- Use dripless candles and new self-extinguishing candles.

Smoke Alarms

As discussed earlier in this chapter, an important part of conducting home safety surveys is to communicate the importance of smoke alarms. Therefore, it is essential that firefighters have a good understanding of various residential smoke alarms.

NOTE: Refer to Chapter 16 for a discussion of smoke alarms and how they work.

Smoke alarms may be battery-operated units for local alerting or part of a security alarm system that is hardwired and monitored by a central alarm and reporting station. The resident should be advised of any inspection or maintenance needs for the alarm system and directed to contact the monitoring company for service.

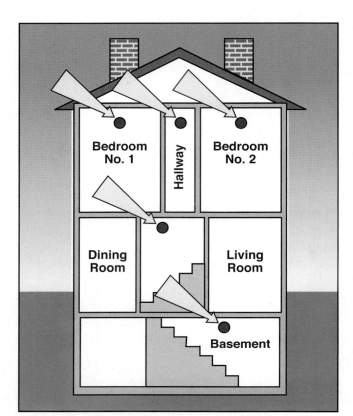

Figure 20.38 Smoke detectors should be located on every level of the structure.

Smoke alarm location. A smoke alarm in every room would provide the fastest detection times, but this may not be economically feasible for many residents. What is usually recommended is the placement of a smoke alarm in every bedroom and at every level of the living unit **(Figure 20.38)**. When following this "every-level" concept, the installer should consider locations such as hallways, stairways, and normal exit routes.

Smoke alarms should always be installed according to the manufacturer's recommendations. As a minimum, one should be mounted in the hallway outside each sleeping area and between the sleeping area and other rooms in the house. The alarm should be located close enough to the bedrooms that it can be heard when the bedroom door is closed. They are usually mounted on the ceiling at the highest point in the occupied area of the home. If a ceiling mount is not possible, the alarm should be positioned as high on the walls as possible, no closer than 4 inches (102 mm) and no farther than 12 inches (305 mm) from the ceiling, but not within a dead-air space **(Figure 20.39)**.

Maintenance and testing. Smoke alarms should be maintained and tested in accordance with the manufacturer's instructions. Maintenance is usually just a matter of keeping the alarm clean and free of dust by occasional vacuuming. Smoke alarms should never be disabled because of nuisance alarms.

The test buttons on some alarms may only check the device's horn circuit, so being aware of the manufacturer's smoke-test procedure is vital to maintaining a functional unit. Only when the unit incorporates a test button that simulates smoke or

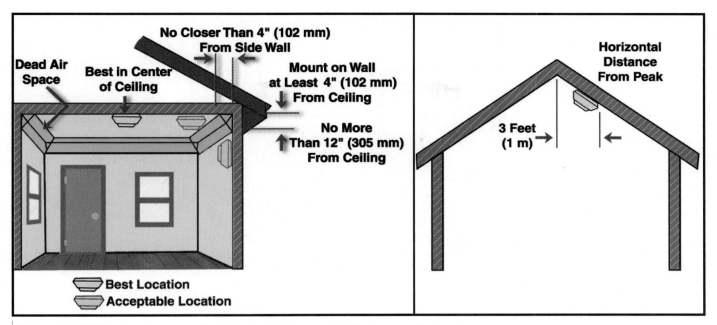

Figure 20.39 Smoke detectors should be installed according to the manufacturer's instructions.

checks the unit's sensitivity can the "smoke test" be eliminated **(Figure 20.40)**. Smoke alarms with test buttons that simulate smoke or check sensitivity are recommended for residences where the physically impaired or elderly live and smoke testing cannot realistically be conducted.

Aerosol spray (canned smoke) is also available for smoke-alarm testing. Some aerosol sprays can contaminate the ionization chamber, resulting in nuisance alarms. These types of sprays should only be used if they are listed as being acceptable to the smoke alarm manufacturer.

Carbon Monoxide Detectors

When discussing smoke alarms with homeowners, it is also a good idea to discuss the need for and installation of carbon monoxide detectors. As discussed in Chapter 3, Fire Behavior, carbon monoxide (CO) is an invisible, odorless, colorless, and tasteless gas that is also toxic. Even though CO detectors are similar in many ways to smoke alarms, they cannot be used interchangeably. While CO detectors should always be installed according to the manufacturer's recommendations, they are usually installed in the same locations as smoke alarms **(Figure 20.41)**. The maintenance requirements for most CO detectors are similar to those for smoke alarms.

Fire Station Tours

In most fire departments, firefighters are required to give tours of the fire station to interested citizens. These may be spur-of-the-moment visits from people who walk in off the street or scheduled visits from organized groups. Station tours for groups of children during Fire Prevention Week (the week including October 9th) are quite common **(Figure 20.42, p. 992)**. Fire station tours are discussed in **Skill Sheet 20-II-4**.

These tours are more than just an opportunity to enhance the department's public image. They are important opportunities to send a strong safety message and

Figure 20.40 Testing smoke alarm operation is critically important to ensure proper operation.

Figure 20.41 Carbon monoxide detectors are an additional safety precaution for residential occupancies.

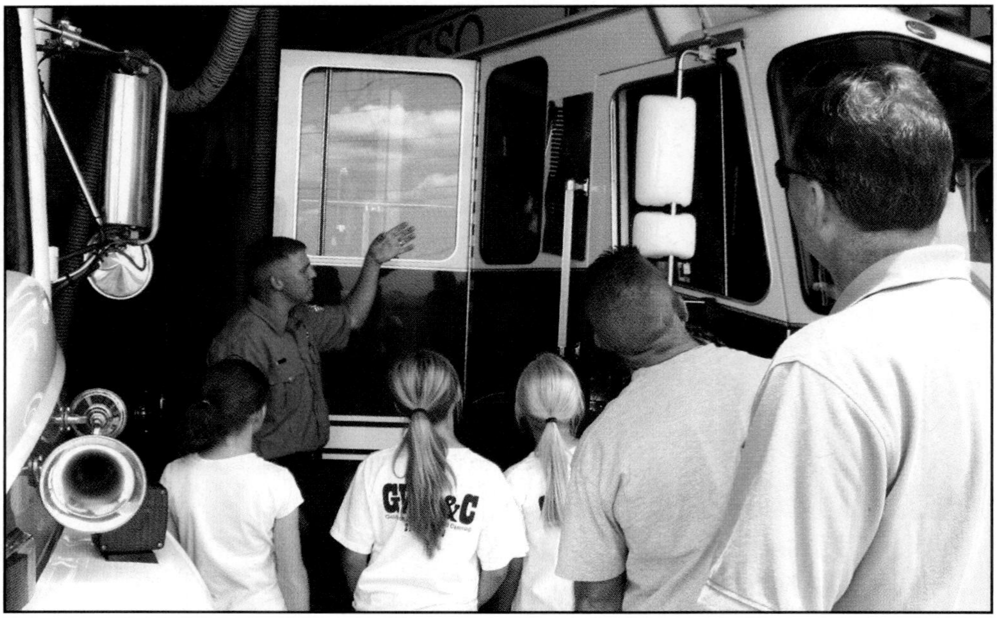

Figure 20.42 Fire station tours serve a dual purpose, allowing the community to meet the personnel who protect them and as an opportunity for firefighters to distribute fire and life safety information.

CAUTION

One of the first things to do when visitors are in the station is to provide safety instructions at the beginning of the tour about what to do and where to go if an alarm sounds during the tour.

to distribute fire and life safety awareness literature. Whenever citizens are in the station, firefighters should be dressed appropriately and should conduct themselves with courtesy and professionalism. The impressions firefighters will make on any visitors will be strong ones, so all television sets should be turned off and other activities should be as positive as possible. The goal is to project a professional workplace atmosphere to the visitor.

Firefighters should answer all questions courteously and to the best of their ability. Fire safety and prevention information can also be passed along to visitors at this time. While some departments allow visitors to climb on apparatus or don equipment items, many others do not. Many parents love to photograph their children wearing fire helmets and other gear; however, this can also allow head lice and other organisms to be passed from one person to another. This activity also exposes children to smoke and fire residue and other contaminates that are unhealthy. Also, some fire helmets are too heavy for some children's neck muscles to adequately support and this can lead to injuries. An alternative approach is to maintain a set of clean, used protective clothing specifically for children to wear.

Visitors, especially children, should never be allowed to roam around the fire station unescorted. Visiting groups should be met by an assigned firefighter or officer. Special care should be taken to protect curious children or other individuals around shop areas or slide poles. All groups should be kept together and, if necessary, larger groups rearranged into smaller groups with a firefighter assigned to each.

Equipment and apparatus should be demonstrated with appropriate caution to ensure that no one is endangered. Positioning firefighters at strategic locations will help to prevent young visitors — and even older ones — from straying into hazardous areas during demonstrations. Many fire departments prohibit taking visitors up on elevating platforms or aerial ladders. Appropriate caution should be exercised when blowing sirens in the presence of children because the decibels produced can be harmful to their hearing.

Remember that station mascots (dogs, cats, etc.) can be potential safety and liability hazards. Excited animals have been known to strike out and bite or scratch visitors; therefore, many fire departments restrict the presence of animals during sta-

Summary

These inspections are sometimes conducted by fire inspectors but are often conducted by company officers and their crews. When noncompliance is found, a plan of correction is agreed to by the property owner or occupant and the fire department.

Public education is the process of teaching members of the public how they can protect themselves from fires and other contingencies. One approach is to educate the public about fire and life safety hazards and how these hazards can be mitigated. Educating the public can involve firefighters making fire safety presentations to small groups, distributing fire safety literature, preparing fire safety messages for broadcast by the media, and by conducting station tours for individuals and groups.

Recognizing that the most conscientious code enforcement and public education efforts will not eliminate all uncontrolled fires, fire departments must prepare to suppress those fires that do start. One way to help firefighters control these fires most effectively and most safely is to increase their familiarity with the buildings in the districts. The most common way of accomplishing this allow fire crews to conduct frequent preincident planning surveys of target hazards and other buildings.

Fire Fighter II

1. What is the difference between a survey and an inspection?

2. List four common fire hazards.

3. List five items that should be identified in an inspection.

4. What items should firefighters check in a preincident planning survey?

5. What are the main objectives of a residential fire safety survey?

6. What are five questions that firefighters should ask themselves when conducting interior and exterior residential fire safety surveys?

7. What are the steps in presenting fire and life safety information?

8. What are some important characteristics and needs of children when presenting fire and life safety information?

9. List common topics for fire and life safety presentations.

10. What precautions should be taken when giving fire station tours?

Step 1: Contact the business owner or manager to gain permission to conduct the survey.

Step 2: Record initial observations of the outside of the building.

Step 3: Prepare a sketch of the building, streets, hydrants, etc.

Step 4: Calculate and record hydrant fire flow.

Step 5: Survey the interior of the building beginning on the lowest floor or roof.

Step 6: Record any features or conditions related to life safety and fire suppression.

Step 7: Draw floor plan of building to include all pertinent information from Step 6.

Step 8: Discuss results of survey with owner/manager.

Step 9: Disseminate completed preincident plan to other companies and stations according to local protocols.

Step 1: Gather equipment and informational materials required to conduct the survey.

Step 2: Contact the resident.

Step 5: Identify fire hazards and recommend appropriate solutions to the resident.

Step 3: Explain the purpose and benefits of the survey to the resident.

Step 4: Conduct survey of the residence.

Step 6: Discuss general fire safety information with the resident.

Step 7: Conclude survey.

Step 8: Record information on the survey in an appropriate department database.

Step 1: Determine the audience and fire or life safety topic to be taught.

Step 2: Select location, date, and time for the presentation.

Step 3: Review lesson outline and obtain necessary equipment and materials.

Step 4: Notify the group or audience of the presentation details.

Step 5: Conduct the presentation according to the lesson outline.

Step 6: Return equipment and materials according to department policy.

Step 7: Record information about the presentation in an appropriate department database.

Step 1: Determine characteristics of the group touring the station.

Step 2: Select appropriate fire safety message(s) to be presented during the tour.

Step 3: Select written materials, handouts, etc. to be distributed during the tour.

Step 4: Reconfirm the date and time of the tour with the group point of contact.

Step 5: Inspect station in preparation for the tour.

Step 6: Welcome the group to the station.

Step 7: Conduct the tour of the station.

Step 8: Conclude the tour.

Step 9: Record information about the station tour in an appropriate department database.

Chapter Contents

Basic Prehospital Emergency Medical Care for Firefighters

This chapter provides information that addresses the following job performance requirements of NFPA® 1001, Standard for Fire Fighter Professional Qualifications (2008):

NFPA® 1001 references for Chapter 21:

4.3

Chapter Objectives

1. Discuss the importance of body substance isolation (BSI).

2. Describe the components of personal protective equipment.

3. Discuss diseases of concern.

4. Describe laws that relate to infection control.

5. Explain the importance of immunizations.

6. Describe the physiological aspects of stress.

7. Describe types of stress reactions.

8. Summarize causes of stress.

9. List signs and symptoms of stress.

10. Explain various ways to deal with stress.

11. Describe scene safety considerations at hazardous materials incidents and rescue operations.

12. Describe actions required when responding to scenes involving violent or dangerous situations.

13. Discuss the circulatory system.

14. List the links in the chain of survival.

15. Explain actions to be taken before resuscitation.

16. Discuss rescue breathing.

17. Describe the steps of cardiopulmonary resuscitation (CPR).

18. Describe the CPR techniques for an infant patient.

19. Describe the CPR techniques for a child patient.

20. Describe the CPR techniques for an adult patient.

21. Discuss indications of effective CPR and when CPR may be interrupted.

22. Summarize when not to begin or to terminate CPR.

23. Summarize actions taken when clearing an airway obstruction.

24. Describe the main components of the circulatory system.

25. Differentiate between arterial, venous, and capillary bleeding.

26. Describe the steps for controlling external bleeding.

27. Discuss internal bleeding.

28. Describe types of shock.

29. Describe the signs of shock.

30. Describe the steps for managing shock.

Basic Prehospital Emergency Medical Care for Firefighters

Case History

According to a February 2007 report by the National Fire Protection Association (NFPA®), firefighters in the United States suffered over 38,500 injuries during the period of 2001–2004. Of those, 28,790 were considered minor while almost 10,000 were considered moderate or severe. Most of these injuries (82%) occurred during extinguishment and support activities. Seven out of ten injuries occurred at residential structure fires. The injuries varied from thermal burns to sprains; lacerations to cardiac symptoms. Some of the injuries required basic care at the scene while others required the firefighter to be transported to the hospital for advanced medical care. These statistics clearly highlight the risks associated with fire fighting and the importance of having prehospital emergency medical care available immediately available to care for firefighters when injuries do occur.

Source: Patterns of Firefighter Fireground Injuries, National Fire Protection Association, February 2007

It is apparent from the NFPA® report cited above that fire fighting is a dangerous profession. Even during the best fireground operations events can happen that result in firefighters being injured. While the majority of the injuries are not serious, many are life-threatening and require immediate emergency medical care. This is especially true in cases of cardiac events. For those injured firefighters, the difference between life and death may very well be the medical care provided by another firefighter at the scene. Every firefighter must be able and prepared to provide emergency medical care to their peers and citizen victims at emergency operations. The skills and information presented in this chapter will prepare you to provide emergency medical care at an incident when the need arises.

While this may be your first introduction to emergency medical care, it will most likely not be your last. As you learned in Chapter 1, the majority of firefighters in the United States and Canada are involved in some level of prehospital emergency medical care. If your department is one of those, your emergency medical training is just beginning. You may go on to be an Emergency Medical Technician-Basic (EMT-B) or Paramedic and provide exceptional patient care for citizens in your community. The information in this chapter will introduce you to some of the important skills used by EMTs.

According to NFPA® 1001, all those qualified at the Fire Fighter I level must meet minimum emergency medical care requirements, including:

- Infection control
- Cardiopulmonary resuscitation
- Bleeding control
- Shock management

Further emergency medical care training may be required by your department in order to meet certification requirements as a First Responder or EMT-B.

Infection Control
Body Substance Isolation (BSI) Precautions

Pathogens — The organisms that cause infection such as viruses and bacteria.

Diseases are caused by pathogens—organisms that cause infection—such as viruses and bacteria. Pathogens may be spread through the air or by contact with blood and other body fluids. *Bloodborne pathogens* can be contracted by exposure to the patient's blood and sometimes other body fluids, especially when they come in contact with an open wound or sore on the firefighter's hands, face, or other exposed parts including mucous membranes, such as those in the nose, mouth, or eyes. Even minor breaks in the skin, such as those found around fingernails, can be enough for a pathogen to enter your body. *Airborne pathogens* are spread by tiny droplets sprayed during breathing, coughing, or sneezing. These particles can be absorbed through your eyes or when you inhale.

Because it is impossible for a firefighter or other prehospital emergency medical care provider to identify patients who carry infectious diseases just by looking at them, all body fluids must be considered infectious and appropriate precautions taken for all patients at all times.

BSI — A form of infection control based on the presumption that all body fluids are infectious.

Equipment and procedures that protect you from the blood and body fluids of the patient—and the patient from your blood and body fluids—are referred to as body substance isolation (BSI) precautions. BSI precautions also are referred to as infection control. For each situation you encounter, it is important to apply the appropriate precautions. Taking too few will clearly increase your risk of exposure to disease. Too many precautions can potentially alienate the patient and reduce your effectiveness.

The Occupational Safety and Health Administration (OSHA) has issued strict guidelines about precautions against exposure to bloodborne pathogens. Under the OSHA guidelines, employers and employees share responsibility for these precautions. Employers must develop a written exposure control plan and must provide emergency care providers training, immunizations, and proper personal protective equipment (PPE) to prevent transmission of disease. The employee's responsibility is to participate in the training and to follow the exposure control plan.

Personal Protective Equipment (PPE) — Equipment that protects the firefighter from infection and/ or exposures to the dangers of emergency operations.

There is also a requirement for all agencies to have a written policy in place in the event of an exposure to infectious substances. Any contact such as a needle-stick or contact with a potentially infectious fluid must be documented. Refer to your local policy for reporting an exposure incident. Most plans call for baseline testing of the exposed person immediately following the exposure and periodic follow-up testing. Additionally, federal legislation has made it possible for emergency care providers to be notified if a patient with whom they have had potentially infectious contact turns out to be infected by a disease or virus such as tuberculosis (TB), hepatitis, or HIV (the virus associated with AIDS).

While BSI issues may seem intimidating—especially if you are just beginning your training—remember that it is possible, by following the proper precautions, to have a long and safe career in the fire service free from infection and disease.

Personal Protective Equipment

Protect yourself from all possible routes of contamination, or introduction of disease or infectious materials. Follow BSI guidelines and wear the appropriate personal protective equipment on every call **(Figure 21.1)**. Different types of protective equipment are discussed in the sections that follow.

Contamination — The introduction of dangerous chemicals, diseases, or infectious materials.

Protective Gloves

Latex, vinyl, or other synthetic gloves should be used whenever there is the potential for contact with blood and other body fluids. This includes actions such as controlling bleeding, suctioning, artificial ventilation, and CPR. *Make sure that you have the gloves on before you come in contact with a patient*. Do not wait. You might get distracted and forget the gloves or you may accidentally become contaminated. Be sure to change gloves between patients.

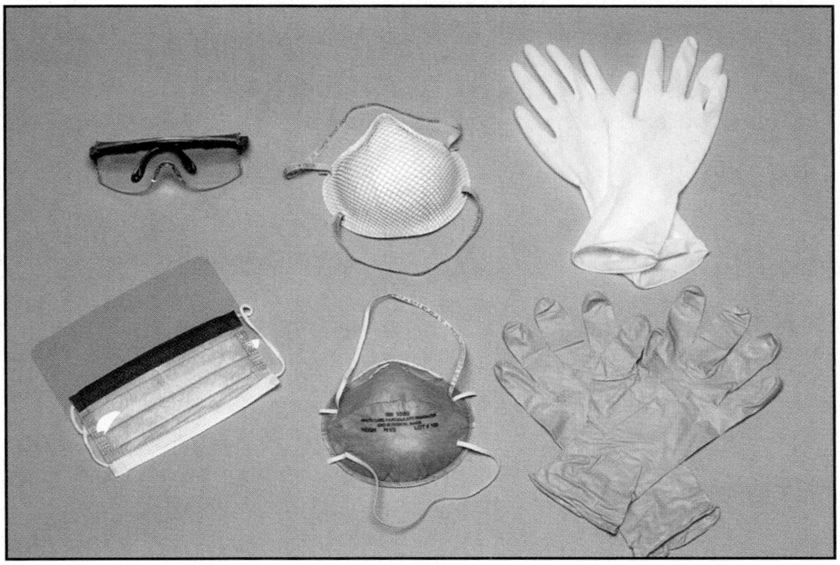

Some firefighters and patients are allergic to latex. The allergic reactions can be serious. Vinyl and other non-latex gloves are available. If your patient is known to be allergic to latex, use non-latex gloves. (You also will find that many other medical supplies such as oxygen masks, tubing, and some tapes contain latex. Be aware of the potential problems these may cause and use substitute materials when appropriate and possible.)

Figure 21.1 Always wear BSI equipment to prevent exposure to contagious disease.

A different type of glove must be worn when you clean the ambulance and soiled equipment. This glove should be heavyweight and tear-resistant. The force and type of movements involved in cleaning can cause lightweight gloves to rip, exposing your hands to contamination.

Handwashing

Even though you wear gloves when assessing and caring for patients, you must still wash your hands after patient contacts when gloves are removed. There are two methods of hand cleaning. These are:

- *Handwashing.* When soap and water is available, vigorous handwashing is recommended **(Figure 21.2)**. Wash your hands after each patient contact (even if you were wearing gloves) and whenever they become visibly soiled.

- *Alcohol-based hand cleaners.* These cleaners are considered effective by the CDC—except when hands are visibly

Figure 21.2 Careful, methodical handwashing is effective in reducing exposure to contagious disease.

Figure 21.3 Alcohol-based hand cleaners are effective and often available when soap and water are not.

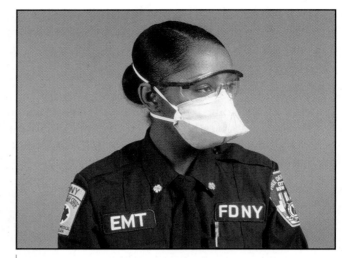

Figure 21.4 Wear a NIOSH-approved respirator when you suspect a patient may have tuberculosis.

CDC (Center for Disease Control) — U.S. government agency for the collection and analysis of data regarding disease and health trends.

soiled or when anthrax is present—and are often available when soap and water are not **(Figure 21.3)**. The alcohol helps kill microorganisms. Place the amount of hand cleaner recommended by the manufacturer in one palm and rub it so it covers your hands. Rub until dry.

Eye Protection

The mucous membranes surrounding the eyes are capable of absorbing fluids. Wear eye protection to prevent splashing, spattering, or spraying fluids from entering the body through these membranes. Protective eyewear should provide protection from the front and the sides. There are various types of eyewear on the market. Goggles provide acceptable protection but can fog and become uncomfortable. Goggles are not necessary because acceptable substitutes are available. If you wear prescription eyeglasses, clip-on side protectors are available. Some companies offer protective eyewear that resembles eyeglasses. These are much easier than goggles to use and carry, and they do not fog up or cause excessive perspiration.

Masks

In cases where there will be blood or fluid spatter, wear a surgical-type mask. In cases where tuberculosis (a disease that is carried by fine particles in the air) is suspected, an N-95 or a high efficiency particulate air (HEPA) respirator approved by the National Institute for Occupational Safety and Health (NIOSH) is the standard **(Figure 21.4)**. In some jurisdictions, when a patient is suspected of having an infection spread by droplets (such as measles), a surgical-type mask may be placed on the patient if he is alert and cooperative.

Gowns

A gown is worn to protect clothing and bare skin from spilled or splashed fluids. Arterial (spurting) bleeding is an indication for a gown. Childbirth and patients with multiple injuries also often produce considerable amounts of blood. Any situation that would call for the use of a gown would also require gloves, eye protection, and a mask.

It is a good idea not only to use personal protective equipment yourself but also to be an advocate for its use; that is, to encourage members of your crew and others to

WARNING!

When you cover a patient's mouth and nose with a mask of any kind, use caution. The mask reduces your ability to visualize and protect the airway. Monitor respirations and be prepared to remove the mask if necessary.

use appropriate protective equipment. In addition to the moral and ethical obligation to do so, you will be helping to keep your crew in good health.

Always properly remove and discard protective garments after use, carry out disinfection and cleaning operations, and complete all reporting documentation regarding infection control.

In the 1980s an increased emphasis was placed on BSI. Diseases such as AIDS and hepatitis brought a startling reality to the health-care professions. As a result, gloves, protective eyewear, and masks were used with increasing frequency. These items were worn as a "precaution," even when not technically necessary.

Today, providers have a more realistic attitude about BSI. Think back to your last physical examination, for example. Your physician most likely examined your eyes and ears, palpated your abdomen, and performed other examinations without gloves. Why is this? Because it is not necessary to wear gloves if your skin and the skin of your patient is intact. Additionally, wearing gloves while you exit the ambulance, carry equipment, and enter a scene may cause the gloves to rip before you reach the patient, rendering them ineffective.

Common sense should be the rule. To protect yourself from disease:

- Follow BSI guidelines as outlined in this chapter and in the rules and policies of your organization.

- Always have BSI equipment immediately available on your person and in your kits.

- If in doubt, wear gloves. This applies to incidents such as motor vehicle collisions where there is broken glass. Even if you do not see blood at the scene, remember it can still be present but hidden under clothes. In contrast, if your skin is intact and your patient has no open wounds, you will not need gloves unless the patient's condition changes.

Diseases of Concern

The communicable diseases we are most concerned with are caused by bloodborne and airborne pathogens, such as viruses, bacteria, and other harmful organisms. Bloodborne pathogens are contracted by exposure to an infected patient's blood, especially exposure through breaks in the non-infected person's skin. Airborne pathogens are spread by tiny droplets sprayed when a patient breathes, coughs, or sneezes. These droplets are inhaled or are absorbed through the non-infected person's eyes, mouth, or nose.

Although there are many communicable diseases, three are of particular concern: hepatitis, tuberculosis, and HIV/AIDS.

- *Hepatitis* is an infection that causes an inflammation of the liver. It comes in at least four forms: hepatitis A, B, C, and Delta. Hepatitis A is acquired primarily through contact with stool. The other forms are acquired through contact with blood and other body fluids. The virus that causes hepatitis is especially hardy. Hepatitis B has been found to live for many days in dried blood spills, posing a risk of transmission long after many other viruses would have died. **For this reason, it is critical for you to assume that any body fluid in any form, dried or otherwise, is infectious until proven otherwise.** Hepatitis B can be deadly. Before hepatitis B vaccine was available, the virus (HBV) killed hundreds of health-care workers every year in the United States, more than any other occupationally acquired infectious disease. There is no cure, but an effective vaccine is available and is generally required of health-care workers. As a result of the vaccination, the greatest hepatitis threat from blood exposure is now Hepatitis C—a form for which there is no vaccination.

- *Tuberculosis (TB)* is an infection that sometimes settles in the lungs and can be fatal. It was thought to be largely eradicated, but in the late 1980s it made a comeback. TB is highly contagious. Unlike many other infectious diseases, it can spread through the air. Health-care workers and others can become infected even without any direct contact with a carrier. Because it is impossible for the firefighter to determine why a patient has a productive cough, it is safest to assume that it could be the result of TB and that you should take the necessary respiratory precautions. This is especially true in institutions such as nursing homes or homeless shelters where there is an increased risk of TB.

- *AIDS (acquired immune deficiency syndrome)* is the name for a set of conditions that result when the immune system has been attacked by HIV (human immunodeficiency virus) and rendered unable to combat certain infections adequately. Although advances are being made in the treatment of HIV/AIDS, no cure has been discovered at the time of publication of this text. However, HIV/AIDS presents far less risk to health-care workers than hepatitis and TB, because the virus does not survive well outside the human body. This limits the routes of exposure to direct contact with blood by way of open wounds, intravenous drug use, unprotected sexual contact, or blood transfusions. Puncture wounds into which HIV is introduced, such as with an accidental needle stick, are also potential routes of infection. However, less than half of one percent of such incidents result in infection, according to the U.S. Occupational Safety and Health Administration (OSHA), compared to 30% for the hepatitis B virus (HBV). The difference is due to the quantity and strength of HBV compared to HIV.

Table 21.1
Communicable Diseases

Disease	Mode of Transmission	Incubation
AIDS (acquired immune deficiency syndrome)	HIV-infected blood via intravenous drug use, unprotected sexual contact, blood transfusions, or (rarely) accidental needle sticks. Mothers also may pass HIV to their unborn children.	Several months or years
Chickenpox (varicella)	Airborne droplets. Can also be spread by contact with open sores.	11 to 21 days
German measles (rubella)	Airborne droplets. Mothers may pass the disease to unborn children.	10 to 12 days
Hepatitis B and C	Blood, stool, or other body fluids, or contaminated objects.	Weeks to months, depending on type
Meningitis, bacterial	Oral and nasal secretions.	2 to 10 days
Mumps	Droplets of saliva or objects contaminated by saliva.	14 to 24 days
Pneumonia, bacterial and viral	Oral and nasal droplets and secretions.	Several days
Staphylococcal skin infections	Direct contact with infected wounds or sores or with contaminated objects.	Several days
Tuberculosis (TB)	Respiratory secretions, airborne or on contaminated objects.	2 to 6 weeks
Whooping cough (pertussis)	Respiratory secretions or airborne droplets.	6 to 20 days

Hepatitis B, TB, and HIV/AIDS are the communicable diseases of greatest concern because they are life-threatening. However, there are many communicable diseases to which emergency response and other health-care personnel may be exposed. **Table 21.1** lists common communicable diseases, their modes of transmission, and their incubation periods (the time between contact and first appearance of symptoms).

Emerging Diseases and Conditions

In recent years there have been both worsening of some diseases and discovery of others. West Nile virus, which originally was limited to Europe and Africa, has spread throughout the United States. It is spread by mosquitoes, causing flu-like symptoms in mild cases and infections of the brain and meninges in severe cases. Severe Acute Respiratory Syndrome (SARS) has caused concern worldwide and appears to be a new virus. SARS appears to be spread through respiratory droplets, by coughing, sneezing, or touching something contaminated and then touching your nose or eyes. It is a respiratory virus with symptoms that include fever, dry cough, and difficulty breathing. Protection against SARS in a patient-care setting includes frequent handwashing and the use of gloves, gowns, eye protection, and an N-95 respirator.

Interestingly, some diseases are also on the decline. For example, vaccines have significantly reduced the numbers of cases of chicken pox and epiglottitis in children.

Diseases of concern will change during the time you are an EMT-B. Your county or state health agencies will issue warnings and advisories on these diseases. You can also look for information from the Centers for Disease Control at www.cdc.gov.

Infection Control and the Law

Scientists have identified the main culprits in the transmission of many deadly infectious diseases—blood and body fluids. Firefighters and other health-care workers have been recognized as having a higher-than-usual exposure and, therefore, a higher risk of contracting these unwanted infections.

Congress and federal agencies have responded by taking several steps to ensure the safety of people who are in high-risk positions. In particular, the Occupational Safety and Health Administration (OSHA) of the U.S. Department of Labor, and the Centers for Disease Control and Prevention (CDC) of the U.S. Department of Health and Human Services, have issued standards and guidelines for the protection of workers whose jobs may expose them to infectious diseases. The Ryan White CARE Act establishes procedures by which emergency response workers may find out if they have been exposed to life-threatening infectious diseases. These legal protections are described in more detail in the next sections.

Occupational Exposure to Bloodborne Pathogens

In 1992, the OSHA standard on bloodborne pathogens took effect. It mandates measures employers of emergency responders must take to protect employees who are likely to be exposed to blood and other body fluids. One of the basic principles behind the standard is that infection control is a joint responsibility between employer and employee. The employer must provide training, protective equipment, and vaccinations to employees who are subject to exposure in their jobs. In return, employees must participate in an infection exposure control plan that includes training and proper workplace practices.

Without the active participation of both the employer and the employee, any workplace infection control program is destined to fail. Be sure your system has an active and up-to-date infection exposure control plan and that you and your fellow firefighters follow it carefully at all times. Consult with your state OSHA representative to make sure that specific hazards are identified and corrected.

Contact the U.S. Department of Labor to request the booklet "Occupational Exposure to Bloodborne Pathogens: Precautions for Emergency Responders—OSHA 3106 1998," an overview of the standard. For details regarding how to develop an infection control plan, request Title 29 Code of Federal Regulation 1910.1030 for the complete text of the standard and all requirements regarding occupational exposure to bloodborne pathogens. Critical elements of the standard are summarized below:

- *Infection exposure control plan.* Each emergency response employer must develop a plan that identifies and documents job classifications and tasks in which there is the possibility of exposure to potentially infectious body fluids. The plan must outline a schedule of how and when the bloodborne pathogen standards will be implemented. It also must include identification of the methods used for communicating hazards to employees, post-exposure evaluation, and follow-up.

- *Adequate education and training.* Firefighters must be provided with training that includes general explanations of how diseases are transmitted, uses and limitations of practices that reduce or prevent exposure, and procedures to follow if exposure occurs.

- *Hepatitis B vaccination.* Employers must make the hepatitis B vaccination series available free of charge and at a reasonable time and place.

- *Personal protective equipment* must be of a quality that will not permit blood or other infectious materials to pass through or reach a firefighter's work clothes, street clothes, undergarments, skin, eyes, mouth, or other mucous membranes. This equipment must be provided by the employer to the firefighter at no cost and includes but is not limited to protective gloves, face shields, masks, protective eyewear, gowns and aprons, plus bag-valve masks, pocket masks, and other ventilation devices.

- *Methods of control.* Engineering controls remove potential infectious disease hazards or separate the firefighter from exposure. Examples include pocket masks, disposable airway equipment, and puncture-resistant needle containers. Work practice controls improve the manner in which a task is performed to reduce risk of exposure. Examples include the proper and safe use of personal protective equipment, proper handling, labeling, and disposal of contaminated materials, and proper washing and decontamination practices.

- *Housekeeping.* Clean and sanitary conditions of the emergency response vehicles and work sites are the responsibility of both the firefighter and the employer. Procedures include the proper handling and proper decontamination of work surfaces, equipment, laundry, and other materials.

- *Labeling.* The standard requires labeling of containers used to store, transport, or ship blood and other potentially infectious materials, including the use of the biohazard symbol (**Figure 21.5**).

- *Post-exposure evaluation and follow-up.* Firefighters must immediately report suspected exposure incidents—including mucous membrane or broken-skin contact with blood or other potentially infectious materials—that result from the performance of an employee's duties. (See **Figure 21.6** for a model plan based on the Ryan White CARE Act, which is described below.)

Figure 21.5 The biohazard symbol must be included with warning labels for containers used to ship blood or other potentially infectious materials.

Ryan White CARE Act

In 1994, the CDC issued the final notice for the Ryan White Comprehensive AIDS Resources Emergency (CARE) Act Regarding Emergency Response Employees. This

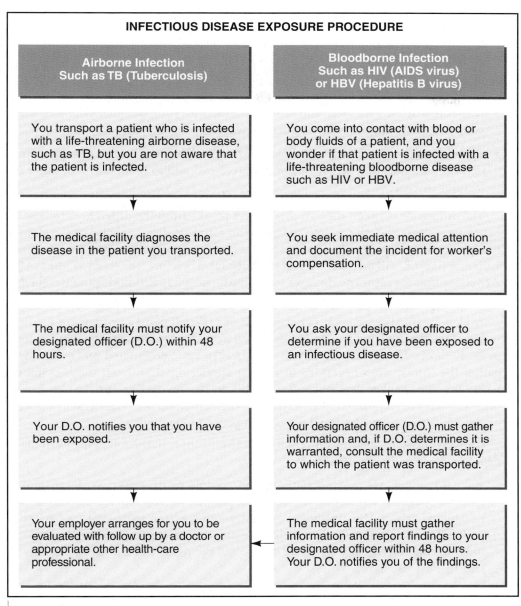

INFECTIOUS DISEASE EXPOSURE PROCEDURE

Airborne Infection Such as TB (Tuberculosis)

You transport a patient who is infected with a life-threatening airborne disease, such as TB, but you are not aware that the patient is infected.

↓

The medical facility diagnoses the disease in the patient you transported.

↓

The medical facility must notify your designated officer (D.O.) within 48 hours.

↓

Your D.O. notifies you that you have been exposed.

↓

Your employer arranges for you to be evaluated with follow up by a doctor or appropriate other health-care professional.

Bloodborne Infection Such as HIV (AIDS virus) or HBV (Hepatitis B virus)

You come into contact with blood or body fluids of a patient, and you wonder if that patient is infected with a life-threatening bloodborne disease such as HIV or HBV.

↓

You seek immediate medical attention and document the incident for worker's compensation.

↓

You ask your designated officer to determine if you have been exposed to an infectious disease.

↓

Your designated officer (D.O.) must gather information and, if D.O. determines it is warranted, consult the medical facility to which the patient was transported.

↓

The medical facility must gather information and report findings to your designated officer within 48 hours. Your D.O. notifies you of the findings.

Figure 21.6 Under the Ryan White CARE Act, there is a procedure for finding out and following up if you have been exposed to a life-threatening disease.

federal act, which applies to all 50 states, mandates a procedure by which emergency response personnel can seek to find out if they have been exposed to potentially life-threatening diseases while providing patient care. Emergency response personnel referred to in this act include firefighters, law enforcement officers, EMTs, and other individuals such as First Responders who provide emergency aid on behalf of a legally recognized volunteer organization.

The CDC has published a list of potentially life-threatening infectious and communicable diseases to which emergency response personnel can be exposed. The list includes airborne diseases such as TB, bloodborne diseases such as hepatitis B and HIV/AIDS, and uncommon or rare diseases such as diphtheria and rabies.

The Ryan White CARE Act requires every state's public health officer to designate an official within every emergency response organization to act as a "designated officer." The designated officer is responsible for gathering facts surrounding possible emergency responder airborne or bloodborne infectious disease exposures. Take time to learn who is the designated officer within your organization.

Two different notification systems for infectious disease exposure are defined in the act:

- *Airborne disease exposure.* You will be notified by your designated officer when you have been exposed to an airborne disease.

- *Bloodborne or other infectious disease exposure.* You may submit a request for a determination as to whether or not you were exposed to bloodborne or other infectious disease.

The difference between the two procedures results from the differences in how an exposure is most likely to be detected. With an airborne disease such as TB, you may not realize that the patient you have cared for and transported was infected. However, a disease like TB will be diagnosed at the hospital. Therefore, the Ryan White CARE Act states that for airborne diseases like TB, the hospital will notify the designated officer, who will notify you.

A bloodborne disease such as hepatitis B or HIV/AIDS may or may not be diagnosed at the hospital, but you will know if you have had contact with a patient's blood or body fluids. If so, you can submit a request to your designated officer who will gather the information necessary to request a determination from the hospital on whether you have been exposed and then will notify you of the result. In either case, once you have been notified of an exposure, your employer will refer you to a doctor or other health-care professional for evaluation and follow-up.

Consider how the Ryan White CARE Act would apply in the following example of exposure to an airborne pathogen:

As a firefighter, you treat a patient who complains of weakness, fever, and chronic cough. The next day you receive a phone call from your organization's designated officer, who informs you that you have been exposed to a patient with TB. The designated officer helps you arrange an appointment with a doctor who can determine if you have contracted the disease and arrange for early treatment if you have.

According to CDC guidelines, exposure to airborne pathogens may occur when you share "air space" with a tuberculosis patient. So, if a medical facility diagnoses that patient as having the airborne infectious disease, it must notify the designated officer within 48 hours. The designated officer must then notify the emergency care workers of disease exposure. Finally, the employer must schedule a post-exposure evaluation and follow-up.

Consider a possible exposure to a bloodborne pathogen:

You are called to treat an unconscious woman. Pink, frothy sputum trickles from her mouth. Breathing is labored. A hypodermic needle lies beside her. While you are suctioning the patient, your eye shield slips and fluids from her mouth splash into your eyes. You report the incident immediately after the call is completed. Your designated officer follows up, and you learn that you have not been exposed to a life-threatening bloodborne disease.

Under CDC guidelines, after contact with the blood or body fluids of a patient you have transported, you may submit a request for a determination to your designated officer. The designated officer must then gather information about the possible exposure. If the information indicates a possible exposure, the officer forwards the information to the medical facility where the patient is being treated. If the patient can be identified, medical records are reviewed to determine if the patient has a life-threatening disease. The medical facility then must notify your designated officer of their findings in writing within 48 hours after receiving the officer's request. The designated officer must notify you, and you will be directed by your employer to a health-care professional for a post-exposure evaluation and follow-up as appropriate.

It is important to note that the Ryan White CARE Act does not empower hospitals to test patients for bloodborne diseases at the request of the emergency worker or designated officer. Rather, they can only review the patient's medical records to see if evidence of a bloodborne or other disease exists. Thus a patient could be infected, but if the records reveal no testing or known indication of the presence of such an infection, the hospital can only report that "no evidence of bloodborne infection could be detected."

Tuberculosis Compliance Mandate

Thousands of new cases of TB are reported in the U.S. each year. Hundreds of health-care workers have been infected or exposed. Of particular concern is multi-drug resistant TB (MDR-TB), which does not respond to the usual medications. In 1994 CDC issued guidelines for treating a suspected or confirmed TB patient. OSHA has announced it will enforce those guidelines as if they were OSHA rules and will also require employers of health care workers to follow OSHA's respiratory standard (12.10.139) until a final rule on tuberculosis is in place. This standard describes the selection and proper use of different kinds of respirators including those classified as N-95 or HEPA.

Study the guidelines as summarized below. Learn to recognize situations in which the potential of exposure to TB exists. Those at greatest risk of contracting and transmitting TB are people who have suppressed immune systems, including people with HIV/AIDS and elderly patients such as those living in nursing homes. Patients who have TB may have the following signs and symptoms: productive cough (coughing up mucus or other fluid) and/or coughing up blood, weight loss and loss of appetite, lethargy and weakness, night sweats, and fever. It is safest to assume that any person with a productive cough may be infected with TB.

When the potential exists for exposure to exhaled air of a person with suspected or confirmed TB, OSHA requires that you wear a NIOSH-approved N-95 or high efficiency particulate air (HEPA) respirator. You are required to wear an N-95 or HEPA respirator when you are:

- Caring for patients suspected of having TB. High-risk areas include correctional institutions, homeless shelters, long-term care facilities for the elderly, and drug treatment centers.

- Transporting an individual from such a setting in a closed vehicle. If possible, keep the windows of the ambulance open and set the heating and air conditioning system on the non-recirculating cycle.

- Performing high-risk procedures such as endotracheal suctioning and intubation.

Remember to take all recommended infection control precautions, including hand-washing and using personal protective equipment and barrier devices such as pocket masks or bag-valve masks for rescue breathing. Properly dispose of contaminated equipment and materials, and decontaminate all surfaces, clothing, and equipment.

Immunizations

Immunizations against many diseases are available. Most people receive tetanus immunizations either routinely or after certain injuries. There is currently an immunization available to prevent hepatitis B. It may be available through your fire department, through a local physician, or from the local health department.

While there is no current immunization against tuberculosis, a test called a *purified protein derivative (PPD) test* can detect exposure. Firefighters are often given this test during routine or employment screening physicals. If the test determines that you have been exposed to tuberculosis, seek treatment and follow-up from a doctor or other health-care professional. Firefighters involved in prehospital emergency medical care should be checked for exposure to TB on a regular basis (usually yearly).

Some fire departments and medical facilities may require immunizations for measles and other common communicable diseases. Consult your instructor, your medical director, or your personal physician for more information on your current status and local protocols for immunizations.

Emotional Stress

Take a minute to think about the last time you told someone that you felt "stressed out." How did you feel? Did you feel tense, as if every muscle was tight, every nerve on edge? Were your palms sweaty and your stomach in knots? Was your heart pounding and did you have a lump in your throat? Did you have trouble sleeping or always feel exhausted no matter how much sleep you had the previous night? What was going on in your life at that time? Were you preparing for a big exam? Was a family member's illness causing you to worry? Did you feel torn between the demands of family, work, and school? Were you worried about your financial state, wondering how you would cover some large unexpected expense? Were you about to change jobs or were you going through a divorce? How you manage these and other stressors is critical to your well-being.

Physiologic Aspects of Stress

During the Middle Ages and for much of the next two or three hundred years, people viewed the mind and body as separate entities. In the last third of the twentieth century, however, medical science began to give increasing scrutiny to how the mind and body work together and influence each other. In fact, today it would be hard to find anyone who denies that there is a connection between mind and body or that stress plays a role in illness.

Stress — A state of physical and/or psychological arousal to a stimulus.

Stress is a widely used term in today's society. It is derived from a word used in the 1600s (*stresse*, a variation of *distresse*), which meant acute anxiety, pain, or sorrow. Today, doctors and psychologists generally define stress as a state of physical and/or psychological arousal to a stimulus. Any stimulus is capable of being a stressor for someone, and stressors vary from individual to individual and from time to time.

Many agree that stress poses a potential hazard for fire service personnel. However, it is important to recognize that stress is a normal part of life and, when managed appropriately, does not have to pose a threat to your well-being. As a firefighter you will be routinely exposed to stress-producing agents or situations. These stressors may be environmental factors (e.g., noise, inclement weather, unstable wreckage), your dealings with other people (e.g., unpleasant family or work relationships, abusive patients or bystanders), or your own self-image or performance expectations (e.g., worry over your expertise at specific skills or guilt over poor patient outcomes).

Ironically these stress-causing factors may be some of the same things that first attracted you to the fire service, such as an atypical work environment, an unpredictable but varied workload, dealing with people in crisis, or the opportunity to work somewhat independently. How you manage these stressors is critical to your survival as an EMS provider as well as in life.

Dr. Hans Selye (a Canadian physician and educator who was born in Austria) did a great deal of research in this area and found that the body's response to stress (*general adaptation syndrome*) has three stages.

During the first stage (*alarm reaction*), your sympathetic nervous system increases its activity in what is known as the "fight or flight" syndrome. Your pupils dilate, your heart rate increases, your bronchial passages dilate. In addition, your blood sugar increases, your digestive system slows, your blood pressure rises, and blood flow to your skeletal muscles increases. At the same time, the endocrine system produces more cortisol, a hormone that influences your metabolism and your immune response. Cortisol is critical to your body's ability to adapt to and cope with stress.

In the second stage (*stage of resistance*), your body systems return to normal functioning. The physiologic effects of sympathetic nervous system stimulation and the excess cortisol are gone. You have adapted to the stimulus and it no longer produces stress for you. You are coping. Many factors contribute to your ability to cope; these include your physical and mental health, education, experiences, and support systems, such as family, friends, and coworkers.

Exhaustion, the third stage of the general adaptation syndrome, occurs when exposure to a stressor is prolonged or the stressor is particularly severe. During this stage, the physiologic effects described by Selye include what he called the stress triad: enlargement (hypertrophy) of the adrenal glands, which produce adrenaline, wasting (atrophy) of lymph nodes, and bleeding gastric ulcers. At this point the individual has lost the ability to resist or adapt to the stressor and may become seriously ill as a consequence. Fortunately, most individuals do not reach this stage.

Types of Stress Reactions

Three types of stress reactions are commonly encountered: acute stress reactions, delayed stress reactions, and cumulative stress reactions. Any of these may occur as a result of a *critical incident*, which is any situation that triggers a strong emotional response. An *acute stress reaction* occurs simultaneously with or shortly after the critical incident. A *delayed stress reaction* (also known as post-traumatic stress disorder) may occur at any time, days to years, following a critical incident. A *cumulative stress reaction* (also known as *burnout*) occurs as a result of prolonged recurring stressors in our work or private lives.

Post-Incident Stress — Psychological stress that affects emergency responders after returning from a stressful emergency incident.

Acute Stress Reaction

Acute stress reactions are often linked to catastrophes, such as a large-scale natural disaster, a plane crash, or a coworker's line-of-duty death or injury. Signs and symptoms of an acute stress reaction will develop simultaneously or within a very short time following the incident. They may involve any one or a combination of the following areas of function: physical, cognitive (the ability to think), emotional, or behavioral. These are signs that this particular situation is overwhelming your usual ability to cope and to perform effectively. It is important to keep in mind that they are ordinary reactions to extraordinary situations. They reflect the process of adapting to challenge. They are normal and are not a sign of weakness or mental illness.

Some of these signs and symptoms require immediate intervention from a physician or mental health professional, while others do not. As a rule, any sign or symptom that indicates an acute medical problem (such as chest pain, difficulty breathing, or abnormal heart rhythms) or an acute psychological problem (such as uncontrollable crying, inappropriate behavior, or a disruption in normal, rational thinking) are the kinds of problems that demand immediate corrective action. These are the same kinds of problems that alert us to a potentially dangerous situation when we see them in a patient, and they should trigger the same response when exhibited by us or our coworkers. Helping people is not just about taking care of your patient; it is also always about taking care of each other and yourself.

As previously mentioned, some signs and symptoms associated with an acute stress reaction may not require intervention. For instance, you may feel nauseated, tremulous, or numb after working a cardiopulmonary arrest, particularly if your patient is close to your age. You may feel confused or have trouble concentrating or difficulty sleeping after working at a particularly bloody crash scene or a prolonged extrication. You may find that you have no appetite for food or cannot get enough to eat. If not too severe or long-lasting, these responses are uncomfortable but probably not dangerous, since they pose no immediate threat to your health, safety, or well-being.

Remember that you are not losing your mind if you exhibit signs and symptoms of stress after a critical incident. You are merely reacting to an extraordinary situation. Remember, too, that there is nothing wrong with you if you do *not* experience any symptoms after such an incident. This, too, is common. In other words, a wide range of responses is normal and to be expected.

Delayed Stress Reaction

Like an acute stress reaction, a delayed stress reaction, also known as *post-traumatic stress disorder (PTSD)*, can be triggered by a specific incident; however the signs and symptoms may not become evident until days, months, or even years later. This delay in presentation may make it harder to deal with the stress reaction since the individual has seemingly put the incident behind him and moved on with his life. Signs and symptoms may include flashbacks, nightmares, feelings of detachment, irritability, sleep difficulties, or problems with concentration or interpersonal relationships.

It is not uncommon for persons suffering from PTSD to seek solace through drug and alcohol abuse. Because of the delay and the apparent disconnect between the triggering event and the response, the victim of PTSD may not understand what is causing the problems. PTSD requires intervention by a mental health professional.

Cumulative Stress Reaction

Cumulative stress reaction, or burnout, is not triggered by a single critical incident, but instead results from sustained, recurring low-level stressors—possibly in more than one aspect of one's life—and develops over a period of years.

The earliest signs are subtle. They may present as a vague anxiety, progressing to boredom and apathy, and a feeling of emotional exhaustion. If problems are not identified and managed at this point, the progression will continue. Now the individual will develop physical complaints (such as headaches or stomach ailments), significant sleep disturbances, loss of emotional control, irritability, withdrawal from others, and increasing depression. Without appropriate intervention, the person's physical, emotional, and behavioral condition will continue to deteriorate, with manifestations such as migraines, increased smoking or alcohol intake, loss of sexual drive, poor interpersonal relationships, deterioration in work performance, limited self-control, and significant depression.

At its worst, cumulative stress may present as physical illness, uncontrollable emotions, overwhelming physical and emotional fatigue, severe withdrawal, paranoia, or suicidal thoughts. Long-term psychological intervention is critical at this stage if the individual is to recover.

The ultimate key to preventing or managing cumulative stress lies in seeking balance in our lives.

Causes of Stress

Emergencies are stressful. That is their nature. While most EMS calls are considered "routine," some calls seem to have a higher potential for causing excess stress on EMS providers (**Figure 21.7**). They include the following:

- *Multiple-casualty incidents.* A multiple-casualty incident (MCI) is a single incident in which there are multiple patients. Examples range from a motor-vehicle crash in which two drivers and a passenger are injured to a hurricane that causes the injury of hundreds of people.

- *Calls involving infants and children.* Involving anything from a serious injury to sudden infant death syndrome (SIDS), these calls are known to be particularly stressful to all health-care providers.

- *Severe injuries.* Expect a stress reaction when your call involves injuries that cause major trauma or distortion to the human body. Examples include amputations, deformed bones, deep wounds, and violent death.

Figure 21.7 Emergencies are often stressful for EMS providers. *Charles H. Porter IV/SYGMA.*

Multiple-Casualty Incident (MCI) — An emergency involving multiple patients.

- *Abuse and neglect.* Cases of abuse and neglect occur in all social and economic levels of society. You may be called to treat infant, child, adult, or elder abuse victims.

- *Death of a coworker.* A bond is formed among members of the public services. The death of another public-safety worker—even if you do not know that person—can cause a stress response.

Stress may be caused by a single event or it may be the cumulative result of several incidents. Remember that any incident may affect you and coworkers differently. Two firefighters on the same call may have opposite responses. Try never to make negative judgments about another person's reaction.

Stress may also stem from a combination of factors, including problems in your personal life. One common cause of stress is people who "just don't understand" the job. For example, your fire service organization may require you to work on weekends and holidays. Time spent on call may be frustrating to friends and family members. They may not understand why you cannot participate in certain social activities or why you cannot leave a certain area. You might get frustrated too, because you cannot plan around the unpredictable nature of emergencies. Then, after a very trying or exciting call, for instance, you may wish to share your feelings with a friend or someone you love. You find instead that the person does not understand your emotions. This can lead to feelings of separation and rejection, which are highly stressful.

Signs and Symptoms of Stress

There are two types of stress: eustress and distress. Eustress is a positive form of stress that helps people work under pressure and respond effectively. Distress is negative. It can happen when the stress of a scene becomes overwhelming. As a result, your response to the emergency will not be effective. Distress also can cause immediate and long-term problems with your health and well-being.

The signs and symptoms of stress include irritability with family, friends, and co-workers; inability to concentrate; changes in daily activities, such as difficulty sleeping or nightmares, loss of appetite, and loss of interest in sexual activity; anxiety; indecisiveness; guilt; isolation; and loss of interest in work.

Dealing with Stress

Lifestyle Changes

There are several ways to deal with stress. They are called "lifestyle changes":

- *Develop more healthful and positive dietary habits.* Avoid fatty foods and increase your carbohydrate intake. Also reduce your consumption of alcohol and caffeine, which can have negative effects, including an increase in stress and anxiety and disturbance of sleep patterns.

- *Exercise.* When performed safely and properly, exercise helps to "burn off" stress. It also helps you deal with the physical aspects of your responsibilities, such as carrying equipment and performing physically demanding emergency procedures.

- *Devote time to relaxing.* Try relaxation techniques, too. These techniques, which include deep-breathing exercises and meditation, are valuable stress reducers.

In addition to the changes you can make in your personal life to help reduce and prevent stress, there also are changes you can make in your professional life. If you are in an organization with varied shifts and locations, consider requesting a change to a different location that offers a lighter call volume or different types of calls. You may also want to change your shift to one that allows more time with family and friends.

There are many types of help available for firefighters and others experiencing stress. Seek them out. It is not a sign of weakness. There are many professionals who can help you deal with stress, and much of the care may be covered by health insurance policies or employee assistance programs.

Scene Safety
Hazardous Materials Incidents

Many chemicals are capable of causing death or life-long complications even if they are only briefly inhaled or in contact with a person's body. Many of these materials are commercially transported, often by truck or rail. Such materials are also often stored in warehouses and used in industry. When there is an accident or when containers begin to leak, a hazardous-material incident may occur, which will pose serious dangers for you as a firefighter as well as for others who are in the vicinity. When you face an emergency involving such materials, remember—you will not be able to help anyone if you are injured. Exercise caution.

The primary rule is to maintain a safe distance from the source of the hazardous material. Make sure your ambulance or other emergency vehicle is equipped with binoculars. They will help you identify placards, which are placed on vehicles, structures, and storage containers when they hold hazardous materials **(Figure 21.8)**. These placards use coded colors and identification numbers that are listed in the *Emergency Response Guidebook (ERG)* developed by the U.S. Department of Transportation, Transport Canada, and the Secretariat of Communications and Transportation of Mexico. This reference book should be placed in every vehicle that responds to or may

Emergency Response Guidebook (ERG) — Manual provided by the U.S. Department of Transportation that aids emergency response personnel in identifying hazardous materials placards. It also gives guidelines for initial actions to be taken at hazardous materials incidents.

respond to a hazardous-material incident. It provides important information about the properties of the dangerous substance as well as information on safe distances, emergency care, and suggested procedures in the event of spills or fire. (The *ERG* is also available on the internet at http://hazmat.dot.gov/pubs/erg/gydebook.htm.)

NOTE: In Canada, the equivalent manual is *Dangerous Goods Guide to Initial Emergency Response* (IERG).

Your most important roles at the scene of a hazardous material incident include recognizing potential problems, taking initial actions for your personal safety and the safety of others, and notifying an appropriately trained hazardous-materials response team. Do not take any actions other than those aimed at protecting yourself, patients, and bystanders at the scene. An incorrect action can cause a bigger problem than the one that already exists.

Figure 21.8 Placards with coded colors and identification numbers must be on vehicles and containers to identify hazardous materials.

The hazardous-materials response team is made up of specially trained technicians who will coordinate the safe approach and resolution of the incident. Each wears a special suit that protects the skin. A self-contained breathing apparatus (SCBA) is also required because of the strong potential for poisonous gases, dust, and fumes at a hazardous-material incident. You will generally not be required to wear personal protective equipment of this sort unless you have been specially trained to be part of a hazardous-materials response team. Instead, you will remain at a distance until the team has made the scene safe.

Rescue Operations

Rescue operations include rescuing or disentangling victims from fires, auto collisions, explosions, electrocutions, and more. As with hazardous materials, it is important to evaluate each situation and ensure that appropriate assistance is requested early in the call. Depending on the emergency, you may need the police, fire department, power company, or other specialized personnel. Never perform acts that you are not properly trained to do. Do your best to secure the scene. Then stand by for the specialists.

As you work in rescue operations or on patients during a rescue operation, you will need personal protective equipment that includes turnout gear (coat, pants, and boots), protective eyewear, helmet, and puncture-proof gloves.

Violence

As a firefighter, you will be called to scenes involving violence. Your first priority—even before patient care—is to be certain that the scene is safe. Dangerous persons or pets, people with weapons, intoxicated people, and others may present problems you are not prepared to handle. Learn to recognize those occasions. If the dispatcher knows that violence is or potentially may be present, he or she will advise you not to approach the scene until it is safe. The dispatcher may name a certain location where you should wait, or stage, a location that is far enough from the scene to be safe but near enough that you can respond as soon as the scene has been secured. Remember, it is the responsibility of the police to secure a scene and make it safe for you to perform your fire service duties.

Three words sum up the actions required to respond to danger: plan, observe, and react.

Plan

Many firefighters work together to prevent dangerous accidents and know what to do as a team when danger strikes. Scene safety begins long before the actual emergency. Plan to be as safe as possible under all circumstances. Factors to be addressed include the following:

- *Wear safe clothing.* Nonslip shoes and practical clothing will not only help you provide emergency care more efficiently, but they also help you to respond to danger without unnecessary restrictions. For personal protection, have reflective clothing available if you will be near traffic or in areas where it is important for you to be visible.

- *Prepare your equipment so it is not cumbersome.* You will be carrying your first-response kit into emergencies. If it is too heavy or bulky, it will take your attention away from the careful observation of the scene as you approach and slow you down if retreat becomes necessary. Many practical containers of reasonable size and weight are available.

- *Carry a portable radio whenever possible.* A radio allows you to call for help if you are separated from your vehicle.

- *Decide on safety roles.* If there will be more than one firefighter on any call, tasks should be split up. For example, one firefighter can obtain vital signs while another applies oxygen. One role that is frequently underused is that of observer. The firefighter directly involved in patient care should always be aware of, but will have trouble constantly monitoring, the surroundings. The firefighters who are not directly involved with patient care will be better able to actively observe for such things as weapons, mechanisms of injury, medications, and other important information.

Observe

Remember that it is always better to prevent a dangerous situation than to deal with one. If you observe or suspect danger, call the police. Do not enter the scene until they have secured it.

Observation begins early in the call. Observe the neighborhood as you look for house or building numbers. As you near the scene, turn off your lights and sirens to avoid broadcasting your arrival and attracting a crowd.

As you approach an emergency scene, notice what is going on. Emergencies are very active events. In situations where you notice an unusual silence, a certain amount of caution is advisable. In addition, observe for the following:

- *Violence.* Any indication that violence has occurred or may take place is significant. Signs include broken glass or overturned furniture, arguing, threats, or other violent behavior.

- *Crime scenes.* Try not to disturb a crime scene except as necessary for patient care. Make every effort to preserve evidence.

- *Alcohol or drug use.* When people are under the influence of alcohol and other drugs, their behavior may be unpredictable. You also may be mistaken for the police because you drove up in a vehicle with lights and sirens.

- *Weapons.* If anyone at the scene (other than the police) is in possession of a weapon, your safety is in danger. Even weapons that are only in view of a hostile person are a potential problem. Remember that almost any item may be used as a weapon. Weapons are not limited to knives and guns. If you observe or suspect the presence of any kind of weapon, notify the police immediately.

- *Family members.* Emotional or overwrought family members are often capable of violence or unpredictable behavior. Even though you are there to take care of a loved one, the violence may be directed at you.

- *Bystanders.* Many people gather at the scene of an accident (or anywhere an emergency vehicle parks). Sometimes you will find a bystander or group of bystanders beginning to show aggressive behavior. If this happens, call for the police. In some settings, it may be necessary to place the patient in the ambulance and leave the scene rather than wait for the police.

- *Perpetrators.* A perpetrator of a crime may still be on the scene—in sight or in hiding. Do not enter a crime scene or a scene of violence until police have secured it and told you it is safe to do so.

- *Pets.* While most domestic animals are not dangerous in ordinary circumstances, the presence of an animal at the emergency scene poses problems. Even friendly animals may become defensive when you begin to treat the owner. Animals also can be very distracting, interfere with patient care, and cause falls while you are lifting and moving the patient. No matter what the pet owner says ("He won't hurt you. He's very friendly."), it is usually best to have pets placed securely in another room.

Keep in mind that the vast majority of emergency calls will go by uneventfully. Nothing in this text is intended to create fear or paranoia. However, when a call does pose some kind of threat, you must be prepared to recognize the subtle and not-so-subtle signs that can warn you before danger strikes.

Reacting to Danger

Observation has provided the critical information needed about the danger. The next step is knowing how to react. The three Rs of reacting to danger are retreat, radio, and reevaluate.

It is not part of your responsibilities as a firefighter to subdue a violent person or wrestle a weapon away from anyone. To *retreat* from such dangers is a clear and justified course of action. Note that some ways of retreating are safer than others:

- *Flee.* Get far enough away so you will have time to react should the danger begin to move toward your new position. Place two major obstacles between you and the danger. If the dangerous person gets through one of the obstacles, you have a built-in buffer with the second.

- *Get rid of any cumbersome equipment.* In the event you must flee from the scene, do not get bogged down by your equipment. Discard all of it if this will enhance your ability to get away. Use equipment to your benefit.

- *Take cover and conceal yourself* (**Figure 21.9**). Taking cover is finding a position that protects your body from projectiles, such as behind a brick wall. Concealing yourself is hiding your body behind an object that cannot protect you, such as a shrub. Find a position that will both conceal **and** protect you.

a

b

Figure 21.9 (a) Concealing yourself is placing your body behind an object that can hide you from view. (b) Taking cover is finding a position that both hides you and protects your body from projectiles.

When fleeing danger, your best option is to use distance, cover, and concealment to protect yourself. Do not return to the scene until the police have secured it.

The second R of reacting to danger is *radio*. The portable radio is an important piece of safety equipment. Use it to call for police assistance and to warn other responding units of the danger. Speak into it clearly and slowly. Advise the dispatcher of the exact nature and location of the problem. Specify how many people are involved and whether or not weapons were observed. Remember, the information you have about the scene must be shared as soon as possible to prevent others from encountering the same danger.

Finally, the third R of reacting to danger is *reevaluate*. Do not reenter a scene until it has been secured by the police (**Figure 21.10**). Even then, be aware that where violence has been, it may begin again. Emergencies are situations packed with stress for families, patients, responders, and bystanders. Maintain a level of alert observation throughout the call. Occasionally you may find weapons or drugs while you are assessing the patient. If that happens, stop what you are doing and radio the police immediately. After the call, document the situation on your run report. Occasionally, the danger may cause delays in reaching the patient. Courts have held this delay acceptable, provided there has been a real and documented danger.

The next sections address cardiopulmonary resuscitation (CPR). Cardiac events are a common killer of firefighters during operations and other events. By knowing how to promptly recognize a cardiac event and then take effective action, you will increase the patient's chances of survival.

Introduction to Basic Cardiac Life Support
The Circulatory System

The circulatory system is responsible for delivering oxygen and nutrients to the body's tissues. It also is responsible for removing waste from the tissues. Its basic components are the heart, arteries, veins, capillaries, and blood.

Figure 21.10 Never enter a scene that is potentially violent until the police have secured it and told you it is safe. *Courtesy of Craig Jackson/In the Dark Photography.*

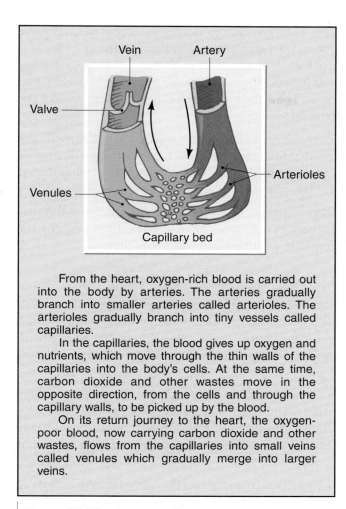

From the heart, oxygen-rich blood is carried out into the body by arteries. The arteries gradually branch into smaller arteries called arterioles. The arterioles gradually branch into tiny vessels called capillaries.

In the capillaries, the blood gives up oxygen and nutrients, which move through the thin walls of the capillaries into the body's cells. At the same time, carbon dioxide and other wastes move in the opposite direction, from the cells and through the capillary walls, to be picked up by the blood.

On its return journey to the heart, the oxygen-poor blood, now carrying carbon dioxide and other wastes, flows from the capillaries into small veins called venules which gradually merge into larger veins.

Figure 21.11 Arteries, capillaries, and veins.

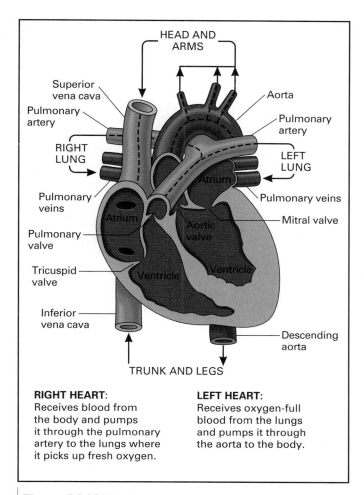

RIGHT HEART:
Receives blood from the body and pumps it through the pulmonary artery to the lungs where it picks up fresh oxygen.

LEFT HEART:
Receives oxygen-full blood from the lungs and pumps it through the aorta to the body.

Figure 21.12 The heart.

The heart is a hollow, muscular organ about the size of a fist. It lies in the lower left central region of the chest between the lungs. It is protected in the front by the ribs and sternum (breastbone). In the back it is protected by the spinal column. The heart contains four chambers. The two upper chambers are the left and right atria. The two lower chambers are the left and right ventricles. The septum (a wall) divides the right side of the heart from the left side. The heart also contains several one-way valves that keep blood flowing in the correct direction.

The circulatory system contains blood vessels. The vessels transport blood throughout the body. Arteries transport blood away from the heart. Veins carry blood back to the heart. The tiny capillaries allow for the exchange of gases and nutrients between the blood and the cells of the body (**Figure 21.11**).

How the Heart Works

The heart is like a two-sided pump (**Figure 21.12**). The left side receives oxygenated blood from the lungs and then pumps it to all parts of the body. The right side receives deoxygenated blood from the body and then pumps it into the lungs. In the lungs the blood will pick up incoming oxygen to be reoxygenated and release carbon dioxide for exhalation.

The blood is kept under pressure and in constant circulation by the heart's pumping action. In a healthy adult at rest the heart contracts between 60 and 80 times per

Pulse — The rhythmic beats caused as waves of blood move through and expand the arteries.

minute. The pulse is a sign of the pressure exerted during each contraction of the heart. Every time the heart pumps a wave of blood is sent through the arteries. This wave is felt as a pulse. It can be palpated, or felt, most easily where a large artery lies over a bone close to the skin. Sites include: the carotid pulse in the neck, the brachial pulse in the underside of the upper arm, the radial pulse in the thumb side of the wrist, and the femoral pulse in the upper thigh.

The pulse is felt most easily over the carotid artery on either side of the neck. The carotid artery should be palpated (felt) first when a patient is unresponsive. The radial artery on the thumb side of the inner surface of the wrist is also easy to feel. It should be palpated first when the patient is responsive.

The heart, lungs, and brain work together closely to sustain life. The smooth functioning of each is critical to the others. When one organ cannot perform properly the other two are handicapped. If one fails, the other two will soon follow.

When the Heart Stops

Respiratory Arrest — When breathing stops completely.

Clinical death occurs when a patient is in respiratory arrest and cardiac arrest. Immediate CPR slows the death process and is critical to patient survival. However, if a patient is clinically dead for 4 to 6 minutes, brain cells begin to die. After 8 to 10 minutes without a pulse, irreversible damage occurs to the brain, regardless of how well CPR is performed.

Cardiac Arrest — The sudden, abrupt loss of heart function.

There are many reasons why a heart will stop. They include heart disease, stroke, allergic reaction, prolonged seizures, and other medical conditions. The heart may also stop because of a serious injury. In infants and children, respiratory problems are the most common cause of cardiac arrest. This is why airway care is so important in young patients.

Chain of Survival — The four critical elements that impact the survival of cardiac arrest patients: early access, early CPR, early defibrillation, and early advanced care.

The patient in respiratory and cardiac arrest has the best chance of surviving if all of the links in the chain of survival come together **(Figure 21.13)**. This "chain" as identified by the American Heart Association (AHA) contains four links:

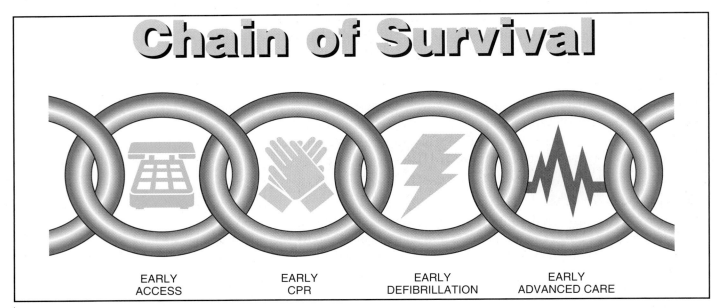

Figure 21.13 The chain of survival. *Courtesy of American Heart Association.*

- *Early access.* Lay people must activate the Emergency Medical System (EMS) immediately. The use of a universal access telephone number helps to speed system access. 9-1-1 is the most common universal access telephone number.
- *Early CPR.* Family members, citizens, and firefighters must be trained in CPR and begin as soon as possible. CPR will help to sustain life until the next step.
- *Early defibrillation.* Defibrillation is the process by which an electrical current is sent to the heart to correct fatal heart rhythms. The earlier defibrillation can be performed the more likely the survival of the patient.
- *Early advanced care.* Advanced care, or the administration of medications and other advanced therapies, must start as soon as possible. This can be done by paramedics at the scene or by prompt transportation to the emergency department.

As a firefighter you have an important role in the chain of survival. You can provide early CPR and, if permitted by your organization, defibrillation.

The principle of CPR is to oxygenate and circulate the blood of the patient until defibrillation and advanced care can be given. Any delay in starting CPR increases the chances of nervous system damage and death. The faster the response, the better the patient's chances of survival. Survival rates improve when the time between cardiac arrest and the delivery of defibrillation and other advanced measures is short.

Before Beginning Resuscitation

When a patient's breathing and heartbeat stop, *clinical death* occurs. This condition may be reversible through CPR and other treatments. However, when the brain cells die, *biological death* occurs. This usually happens within 10 minutes of clinical death and it is not reversible. In fact, brain cells will begin to die after 4 to 6 minutes without fresh oxygen supplied from air breathed in and carried to the brain by circulating blood. *Cardiopulmonary resuscitation (CPR)* consists of the actions you take to revive a person—or at least temporarily prevent biological death—by keeping the person's heart and lungs working.

Before you begin resuscitation (rescue breathing or CPR), you must take certain steps, including assessing the patient, activating EMS, positioning the patient, and assuring an open airway. These first steps are listed in **Table 21.2**.

Assessing the Patient

Patient assessment is crucial. Never initiate resuscitation without first establishing that the patient needs it. The required assessments are often described as determining unresponsiveness, breathlessness, and pulselessness—or the ABCs (airway, breathing, and circulation). As **Table 21.2** shows, these categories overlap.

Determining Unresponsiveness

When you encounter a patient who has collapsed, your first action is to determine unresponsiveness. Tap or gently shake the patient (being careful not to move a patient with possible spinal injury) and shout, "Are you okay?" The patient who is able to respond does not require resuscitation.

If the patient is unresponsive, immediately activate EMS ("phone first") unless the patient's condition is likely caused by a problem other than heart disease, e.g., submersion, injury, or drug overdose. If the patient is a child or an infant, activate EMS after 2 minutes of resuscitation ("phone fast") unless you have reason to think the

Table 21.2
Basic Life Support Sequence

ABCs	Assessment	Actions	Special Considerations
	Determine unresponsiveness ("Are you okay?")	*If unresponsive:* • Activate EMS. • Position patient.	*If alone with child/infant:* Activate EMS after 5 cycles (2 minutes) of resuscitation.
Airway	**Airway open?**	*If unresponsive, assume airway is or may become compromised.* • Open the airway.	*If no trauma, use head-tilt, chin-lift.* *If trauma is suspected, use jaw thrust.*
Breathing	**Determine breathlessness** (Look, listen, feel)	*If no breathing:* • Provide 2 ventilations.	*If breathing is present:* Continue care as necessary. *If first ventilation is unsuccessful:* Reposition head and try again. *If ventilations are still unsuccessful:* Follow airway obstruction clearance procedures.
Circulation	**Determine pulselessness** (Carotid pulse) Check for pulse	*If no pulse:* • Begin chest compressions (CPR).	*In an infant:* Feel for brachial pulse. *If pulse is present but breathing is absent:* Continue rescue breathing. *In infants and children with pulse less than 60/min: Begin chest compressions.*

patient's condition is caused by heart disease. Then position the patient and open the airway with the head-tilt, chin-lift or the jaw-thrust maneuver.

Determining Breathlessness

Determine breathlessness by the look-listen-feel method. Place your ear beside the patient's nose and mouth with your face turned toward the patient's chest. Look for chest rise and fall. Listen and feel for escape of air from the mouth or nose. The patient who is breathing adequately does not require resuscitation.

If the patient is not breathing, provide 2 ventilations (as explained later).

Determining Pulselessness

Determine pulselessness by feeling for the carotid artery in an adult or a child, by feeling for the brachial artery in an infant. Feel for the pulse for up to 10 seconds. The adult patient who has a pulse does not require chest compressions. If an infant or child has a pulse slower than 60 beats per minute, begin CPR (ventilations and chest compressions).

If the patient has a pulse but is not breathing, provide rescue breathing (artificial ventilations). If the adult patient is pulseless, begin CPR (ventilations and chest compressions).

Assessments of the ABCs are included in the steps described above. Keep the ABCs in mind throughout every patient encounter, whether or not resuscitation is underway. If the answer to any of the ABC questions is no, take the appropriate steps to correct the situation:

- Is the patient's *airway* open?
- Is the patient *breathing*?
- Does the patient have *circulation* of blood (a pulse)?

Activating EMS

If you have assistance, the other person should call 9-1-1 or otherwise activate the EMS system as soon as a patient collapses or is discovered in collapse. The quicker a defibrillator can reach the patient, the greater the patient's chances of survival.

If you are alone, and the patient is an adult, first determine unresponsiveness (as described earlier) and then activate EMS before returning to the patient to initiate the next steps. If the patient is a child or an infant, perform 2 minutes of resuscitation before activating EMS.

The reason for the difference in timing of EMS activation is that cardiac arrest in an adult is likely to be the result of a disturbance of the heart's electrical activity that will require defibrillation; so getting defibrillation equipment to the patient takes precedence over starting CPR. When cardiac arrest is probably not the result of a disturbance of the heart's electrical activity, however, (e.g., submersion, injury or drug overdose), it is more important to perform rescue breathing, so you perform CPR briefly before activating EMS.

Children and infants generally have healthy hearts, and cardiac arrest is likely to have resulted from respiratory arrest. In this situation, rescue breathing is more likely to be helpful in a child than in an adult, and defibrillation will not help. When the child has heart disease, however, cardiac arrest is more like that of an adult, so calling for a defibrillator is more important than performing rescue breathing.

So, in most cases, 2 minutes of resuscitation before activating EMS is recommended for children and infants, but immediate activation of EMS is recommended for adults.

Positioning the Patient

As soon as you have determined unresponsiveness and activated EMS, make sure that the patient is lying supine (on his back) before attempting to open the airway and assess breathing and circulation. If you find the patient in some other position, help him to the floor or stretcher. If the patient is already lying on the floor, move him onto his back. If you suspect that the patient may have been injured, you or a helper must support the patient's neck and hold the head still and in line with his spine while you are moving, assessing, and caring for him.

Opening the Airway

Most airway problems are caused by the tongue. As the head tips forward, especially when the patient is lying on his back, the tongue may slide into the airway. When the patient is unconscious, the risk of airway problems is worsened because unconsciousness causes the tongue to lose muscle tone and the muscles of the lower jaw (to which the tongue is attached) to relax.

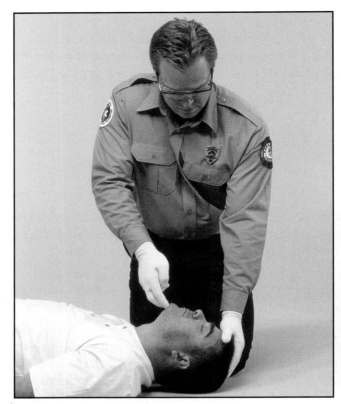
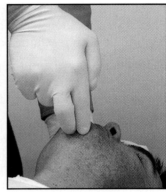

Figure 21.14 The head-tilt, chin-lift maneuver, side view. Insert shows EMT's fingertips under bony area at center of patient's lower jaw.

Two procedures can help to correct the position of the tongue and thus open the airway. These procedures are the head-tilt, chin-lift maneuver and the jaw-thrust maneuver.

Head-Tilt, Chin-Lift Maneuver

The head-tilt, chin-lift maneuver (**Figure 21.14**) provides for maximum opening of the airway. It is useful on all patients who need assistance in maintaining an airway or breathing. It is one of the best methods for correcting obstructions caused by the tongue. However, because it involves changing the position of the head, the head-tilt, chin-lift maneuver should be used only on a patient who you can be quite sure has not suffered a spinal injury.

Follow these steps to perform the head-tilt, chin-lift maneuver:

1. Once the patient is supine, place one hand on the forehead and the fingertips of the other hand under the bony area at the center of the patient's lower jaw.

2. Tilt the head by applying gentle pressure to the patient's forehead.

3. Use your fingertips to lift the chin and support the lower jaw. Move the jaw forward to a point where the lower teeth are almost touching the upper teeth. Do not compress the soft tissues under the lower jaw, which can press and close off the airway.

4. Do not allow the patient's mouth to close. To provide an adequate opening at the mouth, you may need to use the thumb of the hand supporting the chin to pull back the patient's lower lip. For your own safety (to prevent being bitten), do not insert your thumb into the patient's mouth.

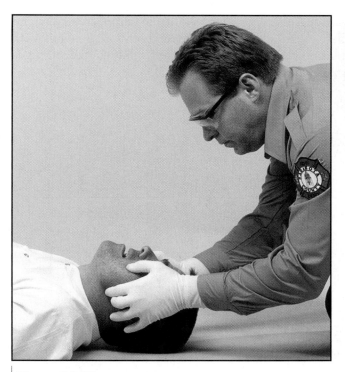

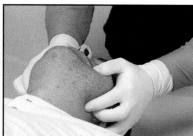

Figure 21.15 The jaw-thrust maneuver, side view. Insert shows EMT's finger position at angle of jaw just below the ears.

Jaw-Thrust Maneuver

The jaw-thrust maneuver is most commonly used to open the airway of an unconscious patient or one with suspected head, neck, or spinal injuries (**Figure 21.15**).

Follow these steps to perform the jaw-thrust maneuver.

1. Carefully keep the patient's head, neck, and spine aligned, moving him as a unit, as you place him in the supine position.

2. Kneel at the top of the patient's head. For greater comfort, you might rest your elbows on the same surface the patient is lying on.

3. Reach forward and gently place one hand on each side of the patient's lower jaw, at the angles of the jaw below the ears.

4. You can help to stabilize the patient's head by using your wrists or forearms.

5. Using your index fingers, push the angles of the patient's lower jaw forward.

6. You may need to retract the patient's lower lip with your thumb to keep the mouth open.

7. Do not tilt or rotate the patient's head. REMEMBER, THE PURPOSE OF THE JAW-THRUST MANEUVER IS TO OPEN THE AIRWAY WITHOUT MOVING THE HEAD OR NECK.

Initial Ventilations and Pulse Check

Deliver 2 breaths, each delivered over 1 second and of sufficient volume to make the chest rise. If the first breath is unsuccessful, reposition the patient's head before attempting the second breath. If the second ventilation is unsuccessful, assume that there is a foreign-body airway obstruction and perform airway clearance techniques (as described later).

WARNING!
The jaw-thrust maneuver is the only widely recommended procedure for use on patients with possible head, neck, or spinal injuries.

If the initial ventilations are successful, you have confirmed an open airway and should feel for a pulse. If the patient has no pulse, begin chest compressions with ventilations (as described later under CPR). If the patient has a pulse but breathing is absent or inadequate, perform rescue breathing (**Table 21.3**).

Table 21.3
Rescue Breathing

	Adult	**Child**	**Infant**
Age	Puberty and older	1 yr–puberty	Birth–1 yr
Ventilation duration	1/sec	1/sec	1/sec
Ventilation rate	10–12 breaths/min	12–20 breaths/min	12–20 breaths/min

Rescue Breathing
Mouth-to-Mask Ventilation

Mouth-to-mask ventilation is performed using a pocket face mask with a one-way valve. The pocket face mask is made of soft, collapsible material and can be carried in your pocket, jacket, or purse. The ventilation rates for mouth-to-mask ventilation are summarized in **Table 21.3**.

Gastric Distention

Rescue breathing can force some air into the patient's stomach, causing the stomach to become distended. This may indicate that the airway is blocked, that there is improper head position, or that the ventilations being provided are too large or too quick to be accommodated by the lungs or the trachea. This problem is seen more frequently in infants and children but can occur with any patient.

A slight bulge is of little worry, but major distention can cause two serious problems. First, the air-filled stomach reduces lung volume by forcing the diaphragm upward. Second, regurgitation (the passive expulsion of fluids and partially digested foods from the stomach into the throat) or vomiting (the forceful expulsion of the stomach's contents) is a strong possibility. This could lead to additional airway obstruction or aspiration of vomitus into the patient's lungs. When this happens, lung damage can occur and a lethal form of pneumonia may develop.

The best way to avoid gastric distention, or to avoid making it worse once it develops, is to position the patient's head properly, avoid too forceful and too quickly delivered ventilations, and limit the volume of ventilations delivered. The volume delivered should be limited to the size breath that causes the chest to rise. This is why it is so important to watch the patient's chest rise as each ventilation is delivered and to feel for resistance to your breaths.

When gastric distention is present, be prepared for vomiting. If the patient does vomit, roll the entire patient onto his side. (Turning just the head may allow for aspiration of vomitus as well as aggravation of any possible neck injury.) Manually stabilize the head and neck of the patient as you roll him. Be prepared to clear the patient's mouth and throat of vomitus with gauze and gloved fingers. Apply suction if you are trained and equipped to do so.

Recovery Position

Patients who resume adequate breathing and pulse after rescue breathing or CPR, and who do not require immobilization for possible spinal injury, are placed in the recovery position. The recovery position allows for drainage from the mouth and prevents the tongue from falling backward and causing an airway obstruction.

The patient should be rolled onto his side. This should be done moving the patient as a unit, not twisting the head, shoulders, or torso. The patient may be rolled onto either side; however, it is preferable to have the patient facing you so that monitoring and suctioning may be more easily performed.

If the patient does not have respirations that are sufficient to support life, the recovery position must not be used. The patient should be placed supine and his ventilations assisted.

CPR

Checking for Circulation

Before beginning CPR, you must confirm that the patient is pulseless. In an adult or child (not infant), check the carotid pulse **(Figure 21.16)**. While stabilizing the patient's head and maintaining the proper head tilt, use your hand that is closest to the patient's neck to locate his "Adam's apple" (the prominent bulge in the front of the neck). Place the tips of your index and middle fingers directly over the midline of this structure. Slide your fingertips to the side of the patient's neck closest to you. Keep the palm side of your fingertips against the patient's neck. Feel for a groove between the Adam's apple and the muscles located along the side of the neck. Very little pressure needs to be applied to the neck to feel the carotid pulse. Keep in mind that laypeople are taught not to check for a pulse, but to look for signs of circulation: normal breathing, coughing, or movement. In an infant, check for a brachial pulse **(Figure 21.17)**. If the adult patient is pulseless, begin CPR. If an infant or child has a pulse slower than 60 beats per minute, begin CPR (ventilations and chest compressions).

To provide chest compressions, place the patient supine on a hard surface and compress the chest by applying downward pressure with your hands. This action

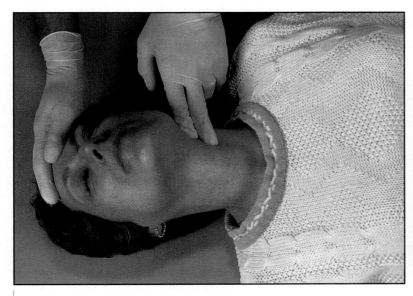

Figure 21.16 Check the carotid pulse to confirm circulation.

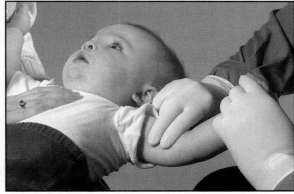

Figure 21.17 For infants, determine circulation by feeling for a brachial pulse.

causes an increase of pressure inside the chest and possible actual compression of the heart itself, one or both of which force the blood out of the heart and into circulation. When pressure is released, the heart refills with blood. The next compression sends this fresh blood into circulation and the cycle continues.

How to Perform CPR

CPR is a method of artificial breathing and circulation. When natural heart action and breathing have stopped, we must provide an artificial means to oxygenate the blood and keep it in circulation. This is accomplished by providing chest compressions and ventilations.

CPR can be done by one or by two rescuers. All of the information under "Providing Chest Compressions," and "Providing Ventilations" applies to both one-rescuer and two-rescuer CPR. Specific information about each type of CPR follows under "One-Rescuer CPR" and "Two-Rescuer CPR." These procedures are for an adult patient. Procedures for infants and children will be described later.

Providing Chest Compressions

After you have placed the patient supine on a hard surface and your hands are properly positioned on the CPR compression site (**Figure 21.18**):

1. Straighten your arms and lock your elbows. You must not bend the elbows when delivering or releasing compressions.

2. Make certain that your shoulders are directly over your hands (directly over the patient's sternum). This will allow you to deliver compressions straight down onto the site. Keep both of your knees on the ground or floor. Be sure that your hands are sufficiently above the xiphoid process at the bottom tip of the sternum. Pressing down on this piece of cartilage may break it, puncturing the liver below.

3. Deliver compressions STRAIGHT DOWN, with enough force to depress the sternum of a typical adult 1½ to 2 inches.

4. Fully release pressure on the patient's sternum, but *do not* bend your elbows and *do not* lift your hands from the sternum, which can cause you to lose correct positioning of your hands. Your movement should be from your hips. Compressions should be delivered in a rhythmic, not a "jabbing," fashion. THE AMOUNT OF TIME YOU SPEND COMPRESSING SHOULD BE THE SAME AS THE TIME FOR THE RELEASE. This is known as the 50:50 rule: 50% compression, 50% release.

Providing Ventilations

Ventilations are given between sets of compressions. The mouth-to-mask techniques described earlier for rescue breathing are used.

One-Rescuer and Two-Rescuer CPR

Figure 21.19 shows the techniques of one-rescuer CPR and two-rescuer CPR for the adult patient and describes compression rates and ratios for CPR on adults, children, and infants.

How to Join CPR in Progress

If CPR has been started by someone who is trained to perform CPR but is not part of the EMS system, and you join this person to perform CPR:

1. Identify yourself as someone who knows CPR and ask to help.

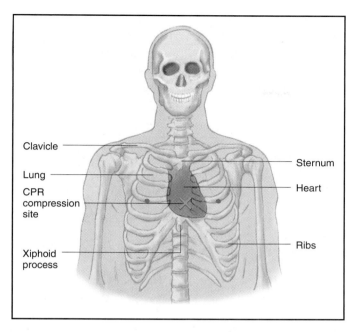

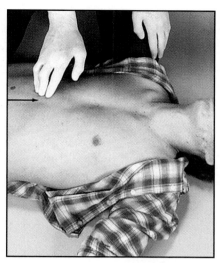

Locate the compression site for CPR using the steps outlined in steps 1 to 3. This is the preferred method for professional rescuers. NOTE: The American Heart Association has identified an alternative method that involves placing the heel of the hand on the sternum between the nipples. This correlates with the lower half of the sternum.

▲ **STEP 1:** Use the index and middle fingers of the hand that is closer to the patient's feet to locate the lower border of the rib cage.

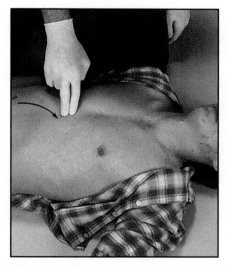

▲ **STEP 2:** Move your fingers along the rib cage to the point where the ribs meet the sternum, the substernal notch. Keep your middle finger at the notch and your index finger resting on the lower tip of the sternum.

Figure 21.18 Locating the CPR Compression Site.

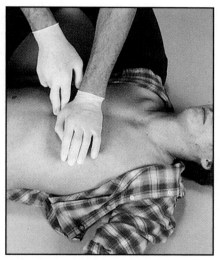

▲ **STEP 3:** Move your hand to the midline. Place its thumb side against the index finger of the lower hand.

2. Ensure that EMS has been activated.

3. Allow the first rescuer to complete a cycle of 30 compressions and 2 ventilations.

4. Check for a pulse. If there is no pulse, start one-rescuer CPR. If you tire before EMS arrives, let the other rescuer perform CPR. Alternate until EMS arrives.

ONE RESCUER	FUNCTIONS		TWO RESCUERS
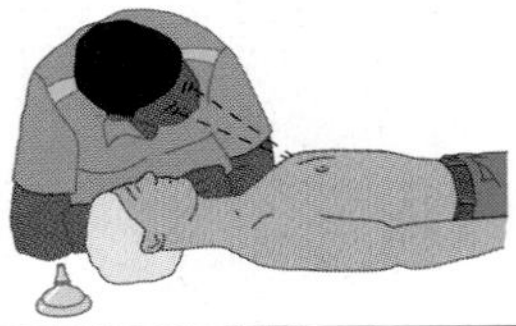	• Establish unresponsiveness • If there's no response, call 9-1-1 • Position patient • Open airway • Look, listen, and feel (no more than 10 seconds)		
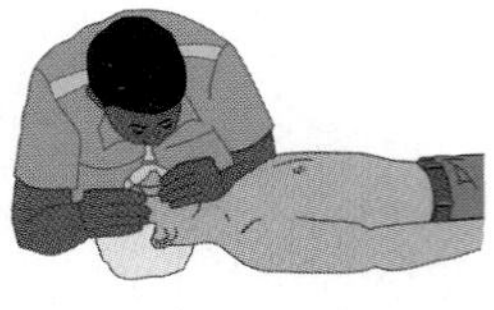	• Deliver 2 breaths (1 sec/ ventilation). If unsuccessful, reposition head and try again. Clear airway if necessary.		
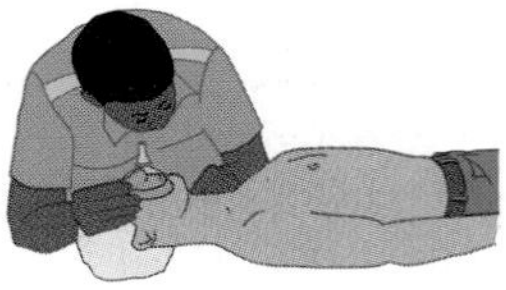	• Check carotid pulse. . . (5–10 seconds) If no pulse. . . • Begin chest compressions		
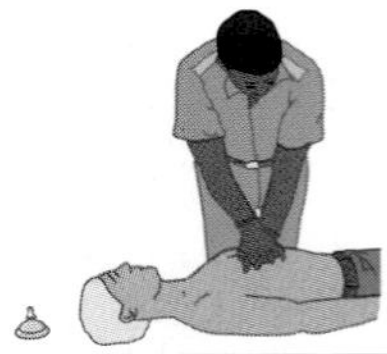	**DELIVER COMPRESSIONS** 1½–2 inches 100/min		
	DELIVER VENTILATIONS **10–12 breaths/min**		
	Compression: Ventilation ratio		
	30:2		
	• Do 4 cycles • Check pulse	• Ventilator checks effective- ness	
	CONTINUE PERIODIC ASSESSMENT		

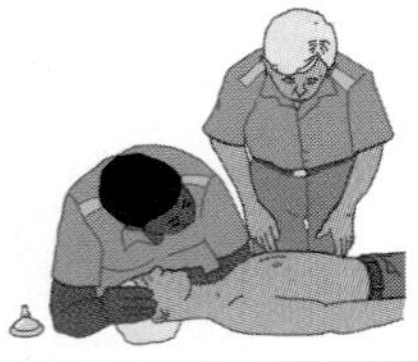

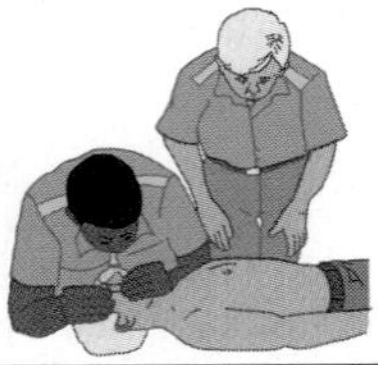

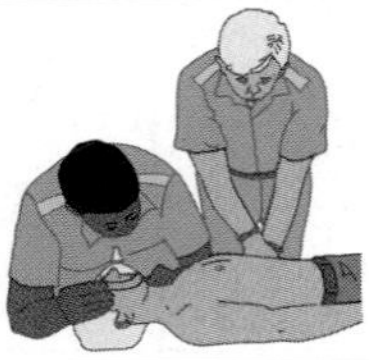

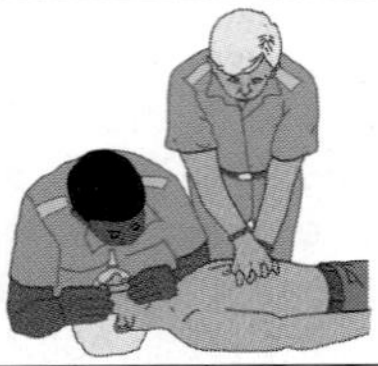

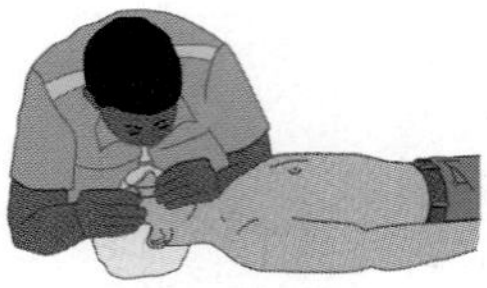

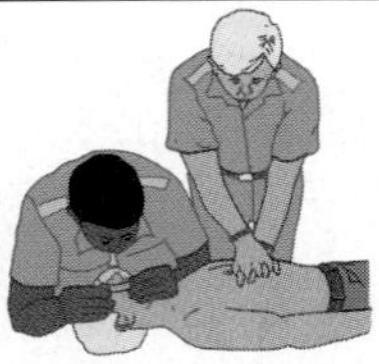

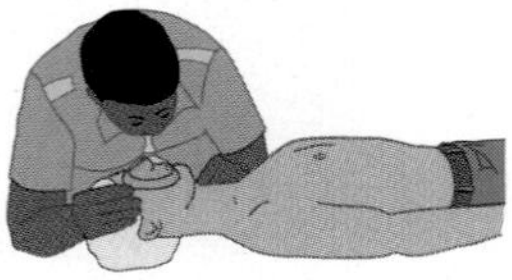

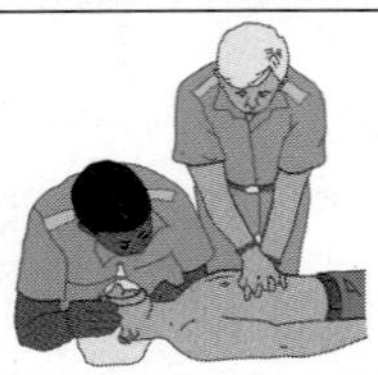

NOTE: Wear gloves and use either a pocket mask with one-way valve or bag-valve mask.

Figure 21.19 CPR Summary—Adult Patient.

If you wish to join another member of the EMS system who has initiated CPR, you should:

1. Identify yourself and your training, and state that you are ready to perform two-rescuer CPR.

2. While the first rescuer is providing compressions, spend 5 seconds checking for a carotid pulse produced by each compression. This is to determine if the compressions being delivered are effective. Inform the first rescuer if there is or is not a pulse being produced. (If the first rescuer cannot deliver effective compressions, you will have to take over for him when CPR is resumed.)

3. You should say, "Stop compressions," and check for spontaneous pulse and breathing. This should take only a few seconds.

4. If there is no pulse, you should state, "No pulse. Continue CPR."

5. The first rescuer resumes compressions, and the second rescuer provides two ventilations during a pause after every thirtieth compression (fifteenth in a child or infant). If desired, the second rescuer can start compressions and allow the first rescuer to provide the ventilations.

CPR Techniques for Children and Infants

The techniques of CPR for children and infants are essentially the same as those used for adults. However, some procedures and rates differ when the patient is a child or an infant. (If younger than 1 year of age, the patient is considered an infant. Between 1 year and puberty (about 12 to 14 years old), the patient is considered a child. Over the age of puberty, adult procedures apply to the patient. Keep in mind that the size of the patient can also be an important factor. A very small 13-year-old may have to be treated as a child.)

The techniques of CPR for an infant are shown in **Figure 21.20**. For a child, CPR is conducted as for an adult, the chief difference in procedure being the hand position—using the heel of one or two hands—for chest compressions, see **Figure 21.21**. To compare adult, child, and infant CPR, see **Table 21.4**.

When CPR must be performed, adults, children, and infants are placed on their back on a hard surface. For an infant, the hard surface can be the rescuer's hand or forearm. For an infant or a child, use the head-tilt, chin-lift or the jaw-thrust maneuver, but apply only a slight tilt for an infant. Too great a tilt may close off the infant's airway; however, make certain that the opening is adequate (note chest rise during ventilation). Always be sure to support an infant's head. Take these steps to establish a pulse in an infant or a child:

- *For an infant,* you should use the brachial pulse.

- *For a child,* determine circulation in the same manner as for an adult.

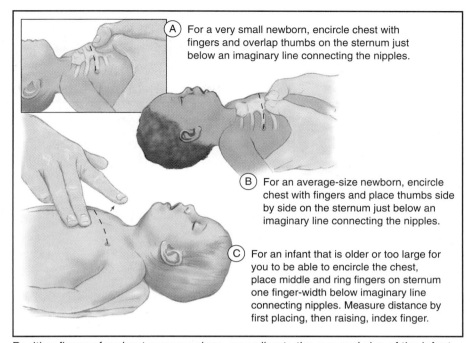

(A) For a very small newborn, encircle chest with fingers and overlap thumbs on the sternum just below an imaginary line connecting the nipples.

(B) For an average-size newborn, encircle chest with fingers and place thumbs side by side on the sternum just below an imaginary line connecting the nipples.

(C) For an infant that is older or too large for you to be able to encircle the chest, place middle and ring fingers on sternum one finger-width below imaginary line connecting nipples. Measure distance by first placing, then raising, index finger.

Position fingers for chest compressions according to the age and size of the infant. The two thumb-encircling hands method is preferred when two rescuers are present.

Figure 21.20 Infant CPR.

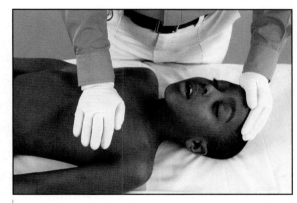

Figure 21.21 Performing chest compressions on a child.

Special Considerations in CPR

How to Know If CPR Is Effective

To determine if CPR is effective, *if possible have someone else feel for a carotid pulse* during compressions and watch to see the patient's chest rise during ventilations. *Listen for exhalation of air*, either naturally or during compressions, as additional verification that air has entered the lungs.

Table 21.4
CPR for Adults, Children, and Infants

	Adult	Child	Infant
Age	Puberty and older	1 yr–puberty	Birth–1 yr
Compression depth	1½ to 2 inches	⅓ to ½ depth of chest	⅓ to ½ depth of chest
Compression rate	100/min	100/min	100/min (newborn 120/min)
Each ventilation	second	1 second	1 second
Pulse check location	carotid artery (throat)	carotid artery (throat)	brachial artery (upper arm)
One-rescuer CPR compressions-to-ventilations ratio	30:2	30:2	30:2 (alone) 15:2 (2 rescuers) 3:1 (newborn)
When working alone: Call 9-1-1 or emergency dispatcher	After establishing unresponsiveness—before beginning resuscitation unless submersion, injury, or overdose	After establishing unresponsiveness and 2 minutes of resuscitation unless heart disease present	After establishing unresponsiveness and 2 minutes of resuscitation unless heart disease present

In addition, any of the following indications of effective CPR may be noticed:

- Pupils constrict
- Skin color improves
- Heartbeat returns spontaneously
- Spontaneous, gasping respirations are made
- Arms and legs move
- Swallowing is attempted
- Consciousness returns

Interrupting CPR

Once you begin CPR, you may interrupt the process for no more than a few seconds to check for pulse and breathing or to reposition yourself and the patient. The first recommended pulse and breathing check is after the first 2 minutes of CPR. You should continue to check for these vital signs every few minutes.

In addition to these built-in interruptions, you may interrupt CPR to:

- Move a patient onto a stretcher.
- Move a patient down a flight of stairs or through a narrow doorway or hallway.
- Move a patient on or off the ambulance.
- Suction to clear vomitus or airway obstructions.
- Allow for defibrillation or advanced cardiac life support measures to be initiated.

When CPR is resumed, begin with chest compressions rather than with ventilations.

When Not to Begin or to Terminate CPR

As discussed earlier in this chapter, CPR should not be initiated when you find that the patient—even though unresponsive and perhaps not breathing—does have a pulse. Usually, of course, you will perform CPR when the patient has no pulse. *However, there are special circumstances in which CPR should not be initiated even though the patient has no pulse:*

- *Obvious mortal wounds.* These include decapitation, incineration, a severed body, and injuries that are so extensive that CPR cannot be effectively performed (e.g., severe crush injuries to the head, neck, and chest).

- *Rigor mortis.* This is the stiffening of the body and its limbs that occurs after death, usually within 4 to 10 hours.

- *Obvious decomposition.*

- *A line of lividity.* Lividity is a red or purple skin discoloration that occurs when gravity causes the blood to sink to the lowest parts of the body and collect there. Lividity usually indicates that the patient has been dead for more than 15 minutes unless the patient has been exposed to cold temperatures. Using lividity as a sign requires special training.

- *Stillbirth.* CPR should not be initiated for a stillborn infant who has died hours prior to birth. This infant may be recognized by blisters on the skin, a very soft head, and a strong disagreeable odor.

In all cases, if you are in doubt, seek a physician's advice. Once you have started CPR, you must continue to provide CPR until:

- Spontaneous circulation occurs . . . then provide rescue breathing as needed.

- Spontaneous circulation and breathing occur.

- Another trained rescuer can take over for you.

- You turn care of the patient over to a person with a higher level of training.

- You are too exhausted to continue.

- You receive a "no CPR" order from a physician or other authority per local protocols.

If you turn the patient over to another rescuer, this person must be trained in basic cardiac life support.

Clearing Airway Obstructions

Not every airway problem is caused by the tongue (the situation in which you would use the head-tilt, chin-lift maneuver or the jaw-thrust maneuver, described earlier, to open the airway). The airway can also be blocked by foreign objects or materials. These can include pieces of food, ice, toys, or vomitus. This problem is often seen with children and with patients who have abused alcohol or other drugs. It also happens when an injured person's airway becomes blocked by blood or broken teeth or dentures or when a person chokes on food.

Airway obstructions are either partial or complete. Partial and complete obstructions have different characteristics that may be noted during assessment, and each type has a different procedure of care. It is important to understand the differences between partial and complete obstruction and the correct care for each.

Partial (Mild) Airway Obstruction

A conscious patient trying to indicate an airway problem will usually point to his mouth or hold his neck. Many do this even when a partial obstruction does not prevent speech. Ask the patient if he is choking, or ask if he can speak or cough. If he can, then the obstruction is mild.

For the conscious patient with an apparent mild airway obstruction, have him cough. A strong and forceful cough indicates he is exchanging enough air. Continue to encourage the patient to cough in the hope that such action will dislodge and expel the foreign object. *Do not* interfere with the patient's efforts to clear the obstruction by means of forceful coughing.

In cases where the patient has an apparent mild airway obstruction but he cannot cough or has a very weak cough, or the patient is blue or gray or shows other signs of poor air exchange, treat the patient as if there is a severe airway obstruction, as described below.

Complete (Severe) Airway Obstruction

Be alert for signs of a severe airway obstruction in the conscious or unconscious patient:

- *The conscious patient* with a severe airway obstruction will try to speak but will not be able to. He will also not be able to breathe or cough. Usually, he will display the distress signal for choking by clutching the neck between thumb and fingers.

- *The unconscious patient* with a severe airway obstruction will be in respiratory arrest. When ventilation attempts are unsuccessful, it becomes apparent that there is an obstruction.

Abdominal Thrusts

The use of abdominal thrusts (the Heimlich maneuver) to clear a foreign body from the airway of an adult or child (not infant) patient is performed as follows:

Use the following procedures for the conscious adult or child (not infant) who is standing or sitting:

1. Make a fist, and place the thumb side of this fist against the midline of the patient's abdomen between waist and rib cage. Avoid touching the chest, especially the area immediately below the sternum.

2. Grasp your properly positioned fist with your other hand and apply pressure inward and up toward the patient's head in a smooth, quick movement. Continue to deliver thrusts until the foreign body is dislodged.

Use the following procedures for the unconscious adult or child (not infant) or for a conscious patient who cannot sit or stand, or if you are too short to reach around the patient to deliver thrusts.

Place the patient in a supine position and begin CPR. Every time you open the airway, look in the mouth for an object. If, and only if, you see an object, remove it by sweeping your fingers in the patient's mouth from one side to the other.

Chest Thrusts

Chest thrusts are used in place of abdominal thrusts when the patient is in the late stages of pregnancy, or when the patient is too obese for abdominal thrusts to be effective. The use of chest thrusts to relieve an airway obstruction is described below.

Table 21.5
Airway Clearance Sequences

	Adult	Child	Infant
Age	Puberty and older	1 yr–puberty	Birth–1 yr
Conscious	Ask, "Are you choking?" Heimlich maneuver until obstruction is relieved or patient loses consciousness	Ask, "Are you choking?" Heimlich maneuver until obstruction is relieved or patient loses consciousness	Observe signs of choking (small objects or food, wheezing, agitation, blue color, not breathing). Series of: 5 back blows and 5 chest thrusts
Unconscious	Establish unresponsiveness. If alone, call for help. Then open airway. Attempt to ventilate. If unsuccessful, reposition head and attempt to ventilate again. If unsuccessful, perform CPR. Remove objects from the mouth if they become visible.	Establish unresponsiveness. Open airway. Attempt to ventilate. If unsuccessful, reposition head and attempt to ventilate again. If unsuccessful, perform CPR. Remove visible objects (NO blind sweeps). After 2 minutes, call for help if alone.	Establish unresponsiveness. Open airway. Attempt to ventilate. If unsuccessful, reposition head and attempt to ventilate again. If unsuccessful, perform CPR. Remove visible objects (NO blind sweeps). After 2 minutes, call for help if alone.

For the conscious adult who is standing or sitting

1. Position yourself behind the patient and slide your arms under his armpits, so that you encircle his chest.
2. Form a fist with one hand, and place the thumb side of this fist on the midline of the sternum about 2 to 3 finger widths above the xiphoid process. This places your fist on the lower half of the sternum but not in contact with the edge of the rib cage.
3. Grasp the fist with your other hand and deliver 5 chest thrusts directly backward toward the spine.

For the unconscious adult

1. Place the patient in a supine position.
2. Perform CPR. Every time you open the airway, look in the mouth for an object. If, and only if, you see an object, remove it by sweeping your fingers in the patient's mouth from one side to the other.

Airway Clearance Sequences

Table 21.5 lists sequences of procedures to use in the event of a severe airway obstruction or a mild airway obstruction. Note that, as discussed earlier, you should activate the EMS system as soon as unresponsiveness is determined, before carrying out the remainder of the airway clearance procedures.

Airway clearance procedures are considered to have been effective if any of the following happens:

- Patient re-establishes good air exchange or spontaneous breathing.
- Foreign object is expelled from the mouth.
- Foreign object is expelled into the mouth where it can be removed by the rescuer.

- Unconscious patient regains consciousness.
- Patient's skin color improves.

If a person has only a mild airway obstruction and is still able to speak and cough forcefully, do not interfere with his attempts to expel the foreign body. Carefully watch him, however, so that you can immediately provide help if this partial obstruction becomes a complete one.

Procedures for a Child or an Infant

The procedure for clearing a foreign body from the airway of a child is very similar to that used for an adult. The airway clearance procedure for an infant uses a combination of back blows and chest compressions as shown in **Figure 21.22**.

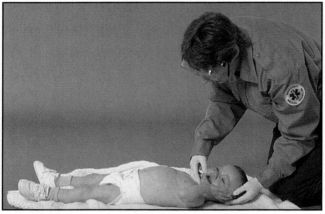

▲ **STEP 1:** Recognize and assess for choking. Look for breathing difficulty, ineffective cough, and lack of strong cry.

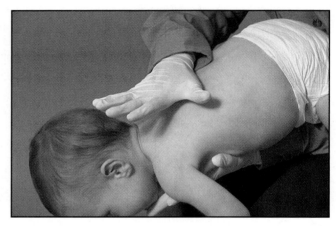

▲ **STEP 2:** Give up to 5 back blows and . . .

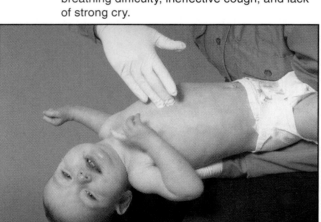

▲ **STEP 3:** 5 chest thrusts.

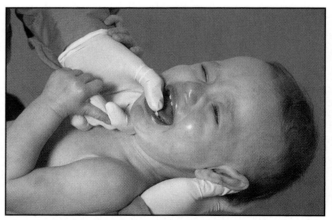

▲ **STEP 4:** If the infant becomes unresponsive, perform a tongue-jaw lift and look for a foreign body. If you see one, use a finger sweep to remove it. (Never do blind finger sweeps in an infant or child.) Attempt to ventilate. If this fails, reposition the head and try again. If you are not successful, repeat the sequence shown above. If you are working alone, after 1 minute activate the EMS system and continue airway clearance and ventilation efforts. Transport as quickly as possible.

Figure 21.22 Clearing the Airway—Infant.

To perform back blows, first place the infant facedown on your arm. Be sure to support the infant's head with your hand, as well as your own arm with your thigh. Deliver five back slaps in between the shoulder blades with the heel of your hand. While taking care to support the head, move the infant over to your other hand in a way that results in his being face up. Using the same technique as chest compressions, deliver five chest thrusts. Repeat this procedure as required, until either the object is dislodged or the infant becomes unconscious. While performing these procedures, allow gravity to assist by keeping the infant's head below the chest. If the infant loses consciousness, begin CPR.

Bleeding Control

Main Components of the Circulatory System

The circulatory (or cardiovascular) system is responsible for the distribution of blood to all parts of the body. This system has three main components: the heart, blood vessels, and the blood that flows through them. All components must function properly for the system to remain intact.

The heart is a muscular organ that lies within the chest, behind the sternum. Its job is to pump blood, which supplies oxygen and nutrients to the cells of the body. To provide a sufficient supply of oxygen and nutrients to all parts of the body, the heart must pump at an adequate rate and rhythm. The blood is circulated throughout the body through three major types of blood vessels (**Figure 21.23**):

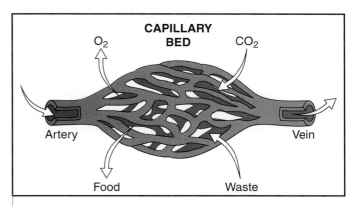

Figure 21.23 Blood vessels.

- *Arteries.* The arteries carry oxygen-rich blood away from the heart. They are under a great deal of pressure during the heart's contractions. (Taking the patient's blood pressure is a means of measuring arterial pressure.) An artery has a thick, muscular wall that enables it to dilate or constrict, depending on the need for oxygen and nutrients of the cells or organs it feeds.

- *Capillaries.* Oxygen-rich blood is emptied from the arteries into microscopically small capillaries, which supply every cell of the body. Where capillaries and body cells are in contact, a vital "exchange" takes place. Oxygen and nutrients are given up by the blood and pass through the extremely thin capillary walls into the cells. At the same time, carbon dioxide and other waste products given up by the cells pass through the capillary walls and are taken up by the blood.

- *Veins.* Blood that has been depleted of oxygen and loaded with carbon dioxide and other wastes in the capillaries empties into the veins, which carry it back to the heart. Veins have one-way valves that prevent the blood from flowing in the wrong direction. Blood in a vein is under much less pressure than blood in an artery.

The adequate circulation of blood throughout the body, which fills the capillaries and supplies the cells and tissues with oxygen and nutrients, is called perfusion. If, for some reason, blood is not adequately circulated, some of the cells and organs of the body do not receive adequate supplies of oxygen, and dangerous waste products build up. Inadequate perfusion of the body's tissues and organs is called hypoperfusion, which is also known as **shock.** (*Hypo-* means "low," so *hypoperfusion* means "low perfusion.") Recall that the heart, blood vessels, and blood are the three main components of the circulatory system. These components may be likened to a pump, pipes, and fluid in the pipes. For the circulatory system to

Perfusion — The supply of oxygen to and removal of wastes from the cells and tissues of the body as a result of the flow of blood through the capillaries.

function properly, all three components must function properly. If any component fails, or "leaks," the body will try in various ways to compensate and maintain adequate perfusion. However, if the problem is not corrected and the condition quickly reversed, adequate perfusion cannot be maintained and shock (hypoperfusion) will result.

Bleeding

Severe bleeding, or hemorrhage, is the major cause of shock. The body contains a certain amount of blood to circulate through the blood vessels. If enough blood volume is lost, perfusion of all cells will not occur. Inadequate perfusion of the body's cells will eventually lead to the death of tissues and organs. The cells and tissues that are the most sensitive to inadequate perfusion are those of the brain, the spinal cord, and the kidneys.

Bleeding, or hemorrhage, is classified as either external or internal, as explained below.

External Bleeding

Whenever bleeding is anticipated or discovered, the use of body substance isolation (BSI) precautions is essential to avoid exposure of the skin and mucous membranes. Blood and open wounds pose a high risk of infection to the firefighter. Protective gloves must be worn when caring for any bleeding patient. A mask and protective eyewear should also be worn if there is a chance of splattered blood. With profuse or spurting (arterial) bleeding, or if the patient is spitting or coughing blood, masks should be worn. Gowns should be considered if clothing may become contaminated.

While BSI precautions decrease the possibility of exposure to blood and body fluids, you should ALWAYS wash your hands with soap and water immediately after each call. Gloves may develop tears or small holes without your knowing it. Always remove the gloves carefully, turning them inside out as you take them off. This reduces the possibility of blood or fluid on the gloves coming in contact with your hands.

External bleeding may be classified according to which of the three types of blood vessel is injured and losing blood (**Figure 21.24**):

- Arterial bleeding
- Venous bleeding
- Capillary bleeding

Arterial bleeding is usually bright red in color because it is still rich in oxygen. It is often rapid and profuse, spurting with each heartbeat, though as the patient's systolic blood pressure drops, the strength of the spurting may decrease. As explained earlier, blood in an artery is under high pressure, and arteries have thick, muscular walls that maintain the pressure. For this reason, arterial bleeding is the most difficult bleeding to control.

Venous bleeding is usually dark red or maroon in color, because it has already passed its oxygen to the cells and picked up carbon dioxide and wastes. Bleeding from the veins has a steady flow. It is usually easy to control (although it can be profuse), because veins return blood to the heart under low pressure. Because venous pressure may be lower than atmospheric pressure, large veins may actually suck in debris or air bubbles. Bleeding from the large veins in the neck is an especially serious

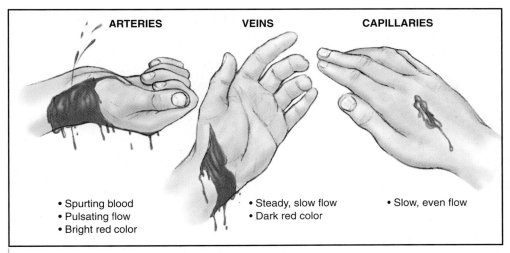

ARTERIES	VEINS	CAPILLARIES
• Spurting blood • Pulsating flow • Bright red color	• Steady, slow flow • Dark red color	• Slow, even flow

Figure 21.24 Three types of external bleeding.

problem. An air bubble, or embolism, may be carried directly to the heart, interfering with the regular heart rhythm, or even stopping it completely.

Capillary bleeding is usually slow and "oozing" due to their small size and low pressure. Most capillary bleeding is considered minor and is easily controlled. It often clots spontaneously, or with minimal treatment. As you would expect, the color of capillary blood is usually somewhere between the bright red of arterial blood and the darker red of venous blood. Capillary bleeding usually results from a minor injury such as a scrape. However, these wounds can easily become contaminated, leading to infection. You may also see capillary bleeding from the edges of more serious wounds.

Capillary Bleeding – Bleeding from capillaries that is characterized by a slow, oozing flow of blood.

Severity of External Bleeding

The severity of the bleeding is somewhat dependent on the amount of blood lost in relation to the physical size of the patient. While a sudden loss of 1 liter (1,000 cc) of blood is considered serious in the average adult, half that amount, or 500 cc, is considered serious in a child. In a one-year-old infant, with a total blood volume of only about 800 cc, a loss of even 150 cc is considered serious.

Severity is dependent on the patient's condition as well as on the relative amount of blood lost. If at any time the patient begins to exhibit signs and symptoms of shock, the bleeding is immediately considered serious.

The body's natural response to bleeding is constriction of the injured blood vessel and clotting. However, a serious injury may prevent effective clotting, allowing continued bleeding. Wounds that are large or deep may not clot well. Uncontrolled bleeding or significant blood loss can eventually lead to death.

Controlling External Bleeding

The control of external bleeding is one of the most important elements in the prevention and management of shock. If bleeding is not controlled, shock will continue to develop and worsen, leading to the death of the patient.

Patient assessment and care always begin with the ABCs and taking BSI precautions. The major methods of controlling external bleeding are (**Figure 21.25**):

- Direct pressure
- Elevation
- Pressure points

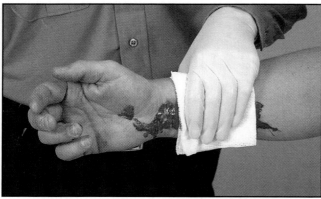

▲ **1.** Apply direct pressure to a bleeding wound with a gauze pad.

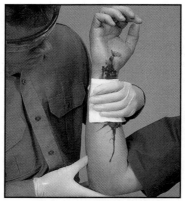

▲ **2.** Elevate a bleeding extremity above the level of the heart.

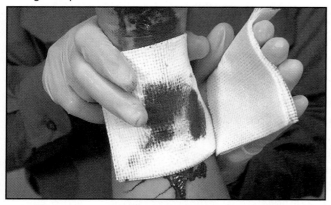

▲ **3.** If the wound continues to bleed, apply additional dressings over the first one.

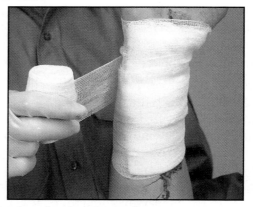

▲ **4.** Bandage the dressing in place.

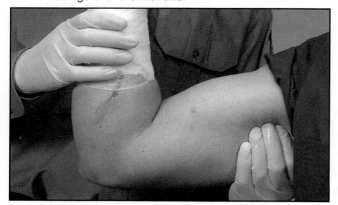

▲ **5.** If a wound to the arm continues to bleed, apply pressure to the brachial artery, or . . .

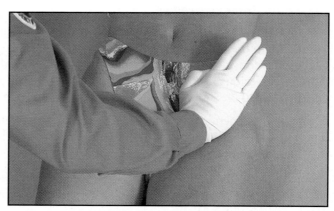

▲ **6.** If a wound to the leg continues to bleed, apply pressure to the femoral artery.

Figure 21.25 Controlling Bleeding.

BSI precautions and infection control are mandatory when attempting to control external bleeding. Always wear disposable gloves when caring for the patient and cleaning up after the call. Follow your local infection exposure control plan regarding cleaning and disposal of contaminated bandages, sheets, and other materials and supplies.

Direct pressure. The most common and effective way to control external bleeding is by applying direct pressure to the wound. This can be done with your gloved hand, a dressing and your gloved hand, or by a pressure dressing and bandage.

1. Apply pressure to the wound until bleeding is controlled. If the bleeding is mild, use a sterile dressing. If the bleeding is severe or spurting, immediately place your gloved hand directly on the wound. Do not waste time trying to find a dressing (**Figure 21.26**).

2. Hold the pressure firmly until the bleeding is controlled. Remember your goal of limiting additional blood loss.

3. Once the bleeding has been controlled, bandage a dressing firmly in place to form a pressure dressing.

4. Never remove a dressing once it has been placed on the wound. Removal of a dressing may destroy clots or cause further injury to the site. If a dressing becomes blood soaked, apply additional dressings on top of it and hold them firmly in place. You may need to continue holding direct pressure on the wound until arrival at the hospital.

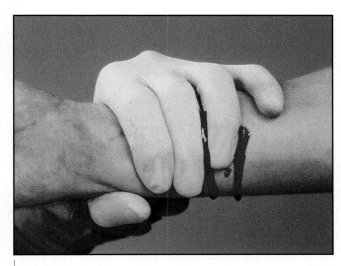

Figure 21.26 In cases of profuse bleeding, do not waste time finding a dressing. Instead, use your gloved hand to apply direct pressure.

Application of a pressure dressing will control most external bleeding. Place several gauze pads on the wound. Hold the dressings in place with a self-adhering roller bandage wrapped tightly over the dressings, and above and below the wound site. Enough pressure must be created to control the bleeding.

As noted above, dressings should not be removed from the wound. If bleeding continues, place additional dressings over the existing one and apply pressure by hand. A tighter bandage (which does not hinder circulation) may be applied. Several types of dressings may be used to control external bleeding (**Figure 21.27**).

You may be unable to apply an effective pressure dressing to some areas of the body. For example, bleeding from the armpit may require you to hold continuous direct pressure with your gloved hand and a dressing. Remember that direct pressure is usually the quickest and most effective method of controlling external bleeding.

Elevation. Elevation of an injured extremity may be used at the same time as direct pressure. When you elevate an injury above the level of the heart, gravity helps reduce the blood pressure in the extremity, slowing bleeding. However, do not use this method if you suspect possible musculoskeletal injuries, impaled objects in the extremity, or spine injury.

> **Pressure Dressing** — A bulky dressing held in position with a tightly wrapped bandage to apply pressure to help control bleeding.

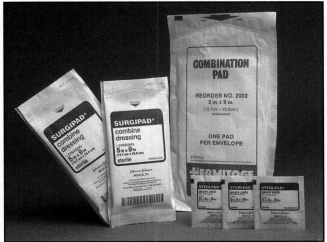

Figure 21.27 Various types of dressings.

To use elevation in controlling external bleeding, apply direct pressure to the injury site. Elevate the injured extremity, keeping the injury site above the level of the heart.

Pressure points. If direct pressure, or direct pressure with elevation, fails to control the external bleeding, your next method is the use of a pressure point. A pressure point is a site where a large artery lies close to the surface of the body and directly over a bone (**Figure 21.28**). Simply put, any site where a pulse can be felt is a pressure point. In emergency medical care, there are four sites (two on each side) used as pressure points to control profuse bleeding in extremities: the brachial arteries for

> **Pressure Point** — Site where a main artery lies near the surface of the body and directly over a bone. Pressure on such a point can stop distal bleeding.

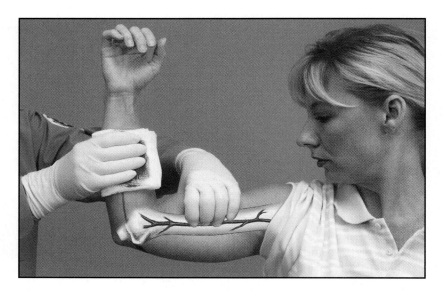

Figure 21.28 The use of pressure points can stop profuse bleeding from an arm or leg if direct pressure and elevation fail. The artery is constricted by pressing it firmly against a bone.

bleeding from the upper extremities, and the femoral arteries for bleeding from the lower extremities (**Figure 21.29**).

The pressure-point method of controlling external bleeding should be used only after direct pressure and elevation have failed. To properly use pressure points, you need to know exactly where the points are located and how much pressure to apply. Note that use of the pressure point may not be effective if the wound is at the distal end of the limb. Blood is being sent to this area of the arm from many other smaller arteries.

Brachial Artery — The major artery of the upper arm.

For bleeding from an upper extremity, apply pressure to a point over the brachial artery. To find the artery:

1. Hold the patient's arm out at a right angle to his body with the palm of the hand facing up. Do not use force to raise the arm, if the movement causes pain or may aggravate an injury. If it is not possible to raise the arm this far, do the best you can, leaving the arm in the position in which it was found.

2. Locate the groove between the biceps muscle and the upper arm bone (humerus) about midway between the elbow and armpit. Cradle the upper arm in the palm of your hand and position your fingers in this medial groove.

3. Compress the artery against the underlying bone by pressing your finger into the groove. If pressure is properly applied, no radial pulse should be felt.

Femoral Artery — The major artery supplying the thigh.

For bleeding from a lower extremity, apply pressure to a point over the femoral artery. Locate the femoral artery on the medial side of the anterior thigh where it joins the lower trunk of the body. You should be able to feel a pulse at a point just below the groin. Place the heel of your hand over the site and apply pressure toward the bone until the bleeding is controlled.

Due to the amount of muscle and tissue in the thigh, you will need to apply more pressure than you apply to the brachial artery pressure point. Even more force will be needed if the patient is very muscular or obese. As with the brachial artery pressure point, if pressure is properly applied, a distal pulse will not be felt. As noted for upper extremity injuries, this method may not control the bleeding of distal wounds.

Special Situations Involving Bleeding

Bleeding most often occurs from a wound caused by direct trauma (striking or being struck or cut by something, such as in a collision, a fall, a stabbing, or a shooting).

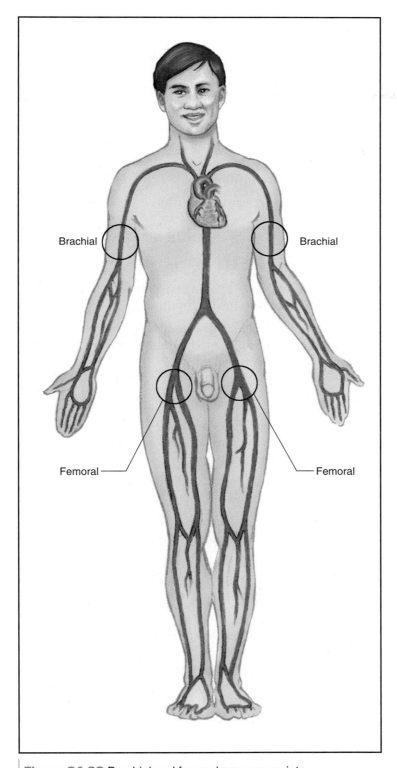

Figure 21.29 Brachial and femoral pressure points.

However, you may also find external bleeding from other causes, such as bleeding from the ears caused indirectly by a head injury or a nosebleed caused by high blood pressure.

Head injury. Traumatic injuries resulting in a fractured skull may cause bleeding or loss of cerebrospinal fluid (CSF) from the ears or nose. This fluid loss is not due to direct trauma to the ears or nose. Instead, the head injury results in increased pressure within the skull, which forces fluid out of the cranial cavity. You should not attempt to stop this bleeding or fluid loss. Doing so may increase the pressure in the skull. Do not apply pressure to the ears or nose. Allow the drainage to flow freely, using a gauze pad to collect it.

Nosebleed. Nosebleeds, also called *epistaxis*, may be caused by direct trauma to the nose. Bleeding from the nose may also be caused by medical problems such as high blood pressure (hypertension). Tiny capillaries in the nose may burst because of increased blood pressure, sinus infection, or digital trauma (nose picking). Controlling bleeding from the nose is sometimes more difficult if the patient is taking certain medications, such as an anticoagulant like Coumadin. To stop a nosebleed, follow these steps:

1. Have the patient sit down and lean forward.

2. Apply or instruct the patient to apply direct pressure to the fleshy portion around the nostrils.

3. Keep the patient calm and quiet.

4. Do not let the patient lean back. This can allow blood to flow down the esophagus to the stomach, resulting in nausea and vomiting.

5. If the patient becomes unconscious or is unable to control his own airway, place the patient in the recovery position (on his side) and be prepared to provide suction and aggressive airway management.

Internal Bleeding

Internal bleeding is bleeding that occurs inside the body. The bleeding itself is not visible, but many of the signs and symptoms are very apparent. There are several reasons why internal bleeding can be very serious:

- Damage to the internal organs and large blood vessels can result in loss of a large quantity of blood in a short period of time.

- Blood loss cannot be seen. While external bleeding is easy to identify, internal bleeding is hidden. Patients may die of blood loss without having any external bleeding.

- Severe internal blood loss may even occur from injuries to the extremities. Sharp bone ends of a fractured femur can cause enough tissue and blood vessel damage to cause shock (hypoperfusion).

Mechanism of Injury — A force or forces that have caused an injury.

Because internal bleeding is not visible, and may not be obvious, you must identify patients who may have internal bleeding by performing a thorough history of the event that caused the injury. Suspicion of internal bleeding and estimates of its severity should be based on the mechanism of injury as well as other signs. If a patient has a mechanism of injury that suggests the possibility of internal bleeding, treat as though the patient has internal bleeding.

Blunt trauma is the leading cause of internal injuries and bleeding. Mechanisms of blunt trauma that may cause internal bleeding include:

- Falls

- Motor-vehicle or motorcycle crashes

- Auto-pedestrian collisions

- Blast injuries

Penetrating trauma is also a common cause of internal injuries and bleeding. It is often difficult to judge the severity of the wound even when the size and length of the penetrating object are known. Always assess your patient for exit wounds. Mechanisms of penetrating trauma include:

- Gunshot wounds
- Stab wounds from a knife, ice pick, screwdriver, or similar object
- Impaled objects

Many of the signs of internal bleeding that you will see are also signs of shock. These signs have developed as a result of uncontrolled internal bleeding. They are late signs, indicating that a life-threatening condition has already developed. If you wait for signs of internal bleeding or shock to develop before beginning treatment, you have waited too long. Signs of internal bleeding are:

- Injuries to the surface of the body, which could indicate underlying injuries
- Bruising (**Figure 21.30**), swelling, or pain over vital organs (especially in the chest and abdomen)
- Painful, swollen, or deformed extremities
- Bleeding from the mouth, rectum, vagina, or other body orifice
- A tender, rigid, or distended abdomen
- Vomiting a coffee-ground-like substance or bright red vomitus, indicating the presence of blood. (Red blood is usually "new." Dark blood is usually "old.")
- Dark, tarry stools or bright red blood in the stool
- Signs and symptoms of shock. Remember that the signs listed in **Table 21.6** are late signs. They will appear only after internal bleeding has already resulted in significant blood loss.

Remember, your best clue to the possibility of internal bleeding may be the presence of a mechanism of injury that could have caused internal bleeding.

Care for the patient with internal bleeding centers on the prevention and treatment of shock. Definitive treatment for internal bleeding can only take place in the hospital. Patients with suspected internal bleeding must be considered serious and warrant immediate transport to the hospital.

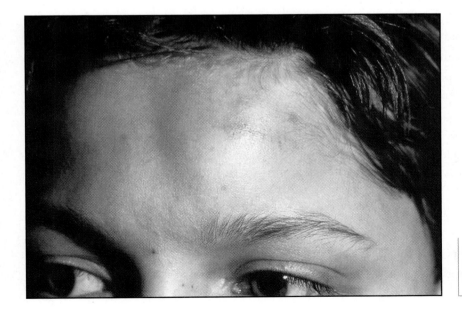

Figure 21.30 Bruising is one sign of internal bleeding. *Courtesy of Dr. P. Marazzi/Photo Researchers, Inc.*

Table 21.6
Signs of Shock

Signs (in order of appearance)	Description
Altered mental status	Altered mental status occurs because the brain is not receiving enough oxygen. The brain is very sensitive to oxygen deficiencies. When it is deprived of oxygen, even slightly, behavioral changes may be noted. These changes may begin as anxiety and progress to restlessness and sometimes combativeness.
Pale, cool, clammy skin	When the body senses low blood volume, natural mechanisms take over in an attempt to correct the problem. One of these mechanisms is to divert blood from non-vital areas to vital organs. Blood is quickly directed away from the skin to such organs as the brain and heart. This results in the loss of color and temperature in the skin. Infants and children may exhibit capillary refill times of greater than 2 seconds. Note: In neurogenic shock (rare) the skin is typically warm, flushed, and dry because the circulatory system has lost the ability to constrict blood vessels in the skin.
Nausea and vomiting	In the body's continuing effort to keep blood perfusing vital organs, blood is diverted from the digestive system. This causes feelings of nausea and occasionally vomiting.
Vital sign changes	The first vital signs to change are the pulse and respirations: • The pulse will increase in an attempt to pump more blood. As the pulse gradually increases, it becomes weak and thready. • Respirations also increase in an attempt to increase the amount of oxygen in the blood. The respirations will become more shallow and labored as shock progresses. • Blood pressure is one of the last signs to change. When blood pressure drops, the patient is clearly in a state of serious, life-threatening shock. Other signs of shock that you may encounter include thirst, dilated pupils, and in some cases cyanosis around the lips and nail beds.

As with all patients, your first priority is the standard ABCs; that is, ensure an open airway, adequate breathing, and circulation. Patients with internal bleeding may deteriorate quickly. Monitor the ABCs often. Be prepared to maintain the patient's airway, to provide or assist ventilations, or to administer CPR as needed.

1. Maintain the ABCs and provide support as needed.
2. If you are trained to do so and it is permitted by your organization, administer high-concentration oxygen by nonrebreather mask if oxygen administration has not already begun.
3. Control any external bleeding.
4. Provide prompt transport to an appropriate medical facility. Internal bleeding must often be controlled in the operating room.

Shock

Shock — The inability of the body to adequately circulate blood to the body's cells to supply them with oxygen and nutrients. A life-threatening condition.

Shock, also known as hypoperfusion, is inadequate tissue perfusion. (Perfusion and hypoperfusion were defined at the beginning of this chapter.) It is the inability of the circulatory system to supply cells with oxygen and nutrients. Hypoperfusion also causes the inadequate removal of waste products from the cells. The result of untreated shock is death.

Causes of Shock

As discussed previously, the circulatory system consists of three components: the heart, blood vessels, and blood. Failure of any of these components—the pumping of the heart, the supply of blood, the integrity of the blood vessels, or the ability of the vessels to dilate and constrict—means that perfusion of the brain, lungs, and other body organs will not be adequate.

Blood vessels can play an important role in the development of shock (if they are *not* functioning properly) or in the body's ability to compensate for shock (if they are functioning properly). Blood vessels can change their diameter, either dilating or constricting. These changes in size are governed by the need for blood in various areas of the body. In an area of the body that is doing more work, blood vessels dilate to allow more blood to flow to that area. At the same time, in another area of the body that is not working as hard, vessels will constrict. For example, if you are running, blood flow to the muscles in the legs increases through dilated arteries. At the same time, blood flow to the digestive system lessens because vessels supplying these organs have been constricted to compensate for the dilation of the arteries in the legs.

Because the amount of blood in the body does not change, this balance between dilation of some vessels and constriction of others is necessary to keep the system full. The same amount of blood is in the body, but it is distributed differently. If all the blood vessels in the body dilated at one time, there would not be enough blood to fill the entire circulatory system. Circulation would fail, and tissues would not be adequately perfused. The result would be the development of shock.

As shock develops, the failure of one component of the system may cause adverse effects on another. If blood is being lost through external bleeding, the heart rate will increase in an effort to circulate the remaining blood to all the tissues. However, the increased heart rate causes increased bleeding. As more blood is lost, the heart tries to compensate by increasing its rate even more. Left untreated, the process continues until the patient dies.

In early shock, the body attempts to compensate for blood loss. Most of the signs and symptoms of shock are caused by the body's compensating mechanisms attempting to adequately perfuse the tissues. Shock may develop if (1) the heart fails as a pump, or (2) blood volume is lost, or (3) blood vessels dilate, creating a vascular container capacity that is too great to be filled by the available blood.

There are three major types of shock:

- *Hypovolemic shock.* This is the type of shock most commonly seen by EMT-Bs. When it is caused by uncontrolled bleeding, or hemorrhage, it can be called hemorrhagic shock. The bleeding can be internal, external, or a combination of both. Hypovolemic shock may also be caused by burns or crush injuries, where plasma is lost.

- *Cardiogenic shock.* Patients suffering a myocardial infarction, or heart attack, may develop shock from the inadequate pumping of blood by the heart. The strength of the heart's contractions may be decreased because of the damage to the heart muscle. Or the heart's electrical system may be malfunctioning, causing a heartbeat that is too slow, too fast, or irregular. Other cardiac problems, such as congestive heart failure, may also cause shock.

- *Neurogenic/Vasodilatory shock.* Shock may result from the uncontrolled dilation (-dilatory) of blood vessels (vaso-) due to nerve paralysis caused by spinal-cord injuries. While there is no actual blood loss, the dilation of the blood vessels increases the capacity of the circulatory system to the point where the available blood can no longer adequately fill it. Sepsis (massive infection) or an anaphylactic (severe allergic) reaction may also cause neurogenic shock. Neurogenic shock is seen only occasionally in the field.

Keep in mind that, as a firefighter, you do not need to diagnose the type of shock. Instead, you must recognize and treat for shock whenever there is a mechanism of injury or signs that indicate the possibility of shock.

Severity of Shock

Shock is the body's reaction to decreased blood circulation to the organ systems. It is the result of the inadequate perfusion of tissues with oxygen and nutrients and the inadequate removal of metabolic waste products. If untreated, shock will cause cell and organ malfunction and, finally, it will result in death. Prompt recognition and aggressive treatment are vital to patient survival.

Signs of Shock

Many of the signs of shock are the same no matter what the cause (hypovolemic, cardiogenic, or neurogenic). The signs follow a logical progression as shock develops and worsens. The signs, in the order they appear, are as follows (**Table 21.6**):

- *Altered mental status.* The brain is very sensitive to any decrease in oxygen supply. When the brain is deprived of oxygen, even slightly, mental and behavioral changes may be seen. These may include anxiety, restlessness, and combativeness.

- *Pale, cool, clammy skin.* When the body senses inadequate tissue perfusion, it attempts to correct the problem by diverting, or shunting, blood from non-vital areas to the vital organs. Blood is quickly directed away from the skin and sent to organs such as the heart and brain. This results in loss of skin color and temperature.

- *Nausea and vomiting.* In the body's efforts to direct blood to the vital organs, blood is diverted from the digestive system, resulting in nausea and sometimes vomiting.

- *Vital sign changes.* The first vital signs to change are the respiratory and pulse rates:

  - The pulse will increase in an attempt to pump more blood. As it continues to increase and blood loss worsens, the pulse becomes weak and thready.

  - Respirations increase in an attempt to raise the oxygen saturation of the blood left in the system. As shock progresses, respirations become more rapid, labored, shallow, and sometimes irregular.

  - Blood pressure drops because the body's compensating mechanisms can no longer keep up with the decrease in perfusion or blood loss. Decreased blood pressure is a LATE sign of shock. By the time the blood pressure drops, the patient is in a truly life-threatening condition.

- *Other signs.* Other signs of shock include thirst, dilated pupils, and sometimes cyanosis (a blue or gray color resulting from lack of oxygen in the body).

Emergency Care for the Patient in Shock

Emergency care for the patient in shock includes airway maintenance and the administration of high-concentration oxygen. Increasing the oxygen saturation of the blood will improve oxygen supply to the tissues. You must also attempt to stop what is causing the shock, such as external bleeding, and attempt to maintain perfusion.

Remember that TRANSPORTATION IS AN INTERVENTION. Your most significant treatment for the shock patient may be early recognition of the problem and prompt transportation to a hospital where the patient will receive definitive care. A term frequently used in relation to trauma care is the golden hour. This term refers to the optimum limit of 1 hour from the time the patient's injury occurs until surgery takes place at a hospital.

Note that the clock begins running at the time of injury, not the time of your arrival on scene or your determination that the patient may be in shock. If the patient is not quickly found, or you have a long response time to the scene, much of the golden hour may have already ticked away. Many experts believe that survival rates from traumatic injuries are best if needed surgery takes place within this first hour. Survival rates rapidly decrease thereafter. Your goal in caring for trauma and shock patients is to limit on-scene time and provide immediate transportation to the hospital. Be sure to alert the receiving hospital as soon as possible because they may need to call in a surgeon. The clock does not stop with the patient's arrival at the hospital. It stops in the operating room.

The goal for on-scene time when caring for a trauma or shock patient is a maximum of 10 minutes (unless lengthy extrication is required). This on-scene time limit is referred to as the platinum ten minutes. In order to meet the golden hour, procedures done at the scene must be kept to a minimum. In patients showing signs of shock, or a mechanism of injury that suggests the possibility, some elements of patient assessment, such as detailed exams and treatments, are best done in an ambulance en route to the hospital.

Care for shock is similar to the care for bleeding described earlier. Remember BSI precautions when caring for any patient who is bleeding externally. The emergency care steps for shock are as follows:

1. Maintain an open airway and assess the respiratory rate. Assist ventilations or perform CPR if necessary. If the patient is breathing adequately, apply high-concentration oxygen by nonrebreather mask if you are trained to do so and allowed by your organization.

2. Control any external bleeding.

3. If there is no possibility of spine injury, elevate the legs 8 to 12 inches (200 mm to 300 mm). This position will assist the body in maintaining perfusion to the vital organs.

4. Prevent loss of body heat by covering the patient with a blanket.

5. Transport the patient immediately. Detailed exams and care procedures should be done en route to the hospital. Give the ambulance personnel information on the patient's injuries and condition.

6. If the patient is conscious, speak calmly and reassuringly throughout assessment, care, and transport. Fear increases the body's work and worsens developing shock.

Summary

Fire fighting is recognized as a dangerous profession. Each year thousands of firefighters are injured during fireground operations. In addition, many firefighters die from sudden cardiac events while on duty, both on the fireground and at the station. The availability of immediate emergency medical care greatly improves the chances of survival when serious injuries occur. The most effective strategy for ensuring that prompt, well-trained emergency medical care is available is to train all firefighters in basic prehospital emergency medical care. When that training occurs, every firefighter can be confident that help is only a shout away.

Fire Fighter I

1. What body substance isolation (BSI) precautions should firefighters take to protect against infection?

2. What are three communicable diseases of concern to firefighters?

3. What is the Ryan White CARE Act?

4. What are some causes of stress for emergency responders?

5. What are the links in the chain of survival for patients in respiratory and cardiac arrest?

6. Describe what actions are taken when assessing the patient during CPR.

7. What are the basic steps in performing CPR?

8. What are the major methods of controlling external bleeding?

9. What are the emergency care steps for shock?

Chapter Contents

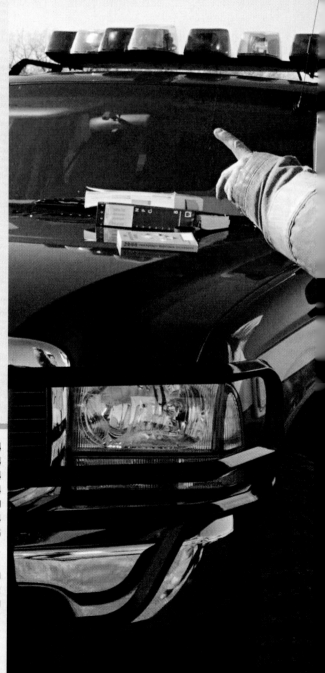

Chapter 22

Introduction to Hazardous Materials

Standards:

Combined, Chapters 22 and 23 meet the requirements of:

- **OSHA 1910.120 (q)(6)(i) and (ii),** Awareness and Operations levels

- **NFPA® 472,** *Standard for Competence of Responders to Hazardous Materials/ Weapons of Mass Destruction Incidents,* 2008 Edition, Chapters 4 and 5, Awareness level and Operations level core competencies

- **Office of Domestic Preparedness** *Emergency Responder Guidelines* for Awareness Level and Performance Level A for firefighters

Chapter Objectives

1. Summarize Awareness-Level and Operations-Level responsibilities at hazardous materials incidents.

2. Describe types of respiratory protection.

3. Summarize respiratory equipment limitations.

4. Describe types of protective clothing.

5. Discuss U.S. EPA levels of protective equipment.

6. Describe NFPA® 1994 PPE ensemble classifications.

7. Describe the U.S. military mission-oriented protective posture (MOPP) ensembles.

8. Discuss PPE selection factors.

9. Discuss health and safety issues when wearing PPE.

10. Explain proper procedures for inspection, testing, and maintenance of protective clothing and equipment.

11. Describe health and physical hazards that may be present at haz mat incidents.

12. Describe physical properties of hazardous materials.

13. Explain how the General Hazardous Materials Behavior Model (GEBMO) can help firefighters understand the likely course of an incident.

14. Explain locations or occupancies clues to the presence of hazardous materials.

15. Explain container shapes clues to the presence of hazardous materials.

16. Explain transportation placards, labels, and markings clues to the presence of hazardous materials.

17. Explain other markings and colors (non-transportation) clues to the presence of hazardous materials.

18. Explain how written resources can be used to assist firefighters in identifying hazardous materials.

19. Explain how the senses can provide clues to the presence of hazardous materials.

20. Explain how monitoring and detection devices can provide clues to the presence of hazardous materials.

21. Summarize indicators of terrorist attacks.

22. Discuss identifying illicit laboratories.

23. Discuss secondary attacks.

24. Obtain information about a hazardous material using the *Emergency Response Guide (ERG)*. (Skill Sheet 22-I-1)

Introduction to Hazardous Materials

Case History

Kansas City (MO) Fire Department Engine Companies 41 and 30 responded to a report of a pickup truck fire at a highway construction site in the early morning hours of November 29, 1988. Security guards who phoned in the call reported that explosives were stored on site; that information was communicated to the responding units. Arriving units found multiple fires on the site including two smoldering truck trailers that were loaded with explosives. The trailers' contents (which were not required to be marked or labeled) contained a total of nearly 50,000 pounds (22,680 kg) of ammonium nitrate, fuel-oil-mixture-based explosives. Other explosives stored on site were labeled. Shortly after firefighters arrived the smoldering trailers detonated. The first blast killed the six firefighters on scene instantly and destroyed the two pumping apparatus. The explosion left an 80-foot (24 m) diameter and 8-foot (2.4 m) deep crater. Sadly, the fire had been intentionally set. Nearly 10 years later, U.S. Department of Transportation regulations regarding when placards can be removed from vehicles were changed as a direct result of the incident.

To meet the needs of business, industry, and consumers, millions of tons of chemical substances, materials, and products are stored, manufactured, used, and transported throughout North America every year. In addition to their necessary and beneficial uses, however, many of these materials post considerable risks to the public and to the environment if they are uncontrolled and uncontained. Substances that possess harmful characteristics are called hazardous materials (or *haz mat*) in the United States and dangerous goods in Canada and other countries. When particularly dangerous hazardous materials such as certain chemical, biological, radiological, nuclear, or explosive (CBRNE) materials are used as weapons, they are sometimes referred to as *weapons of mass destruction (WMD)*.

Haz mat incidents are often more complex than other types of emergency incidents, in part because hazardous materials can be dangerous in many different ways, and sometimes in very small quantities. Their hazards may be extremely difficult to contain and/or control, requiring specialized equipment, procedures, and personal protective equipment (PPE) **(Figure 22.1)**. They may also be difficult to detect, needing sophisticated monitoring and detection equipment to identify and predict their severity. Incidents involving these materials can be caused by human error, mechanical breakdown or malfunction, container failure, transportation accidents,

Hazardous Material — Any material that possesses an unreasonable risk to the health and safety of persons and/or the environment if it is not properly controlled during handling, storage, manufacture, processing, packaging, use, disposal, or transportation.

Dangerous Goods — Any product, substance, or organism included by its nature or by the regulation in any of the nine United Nations classifications of hazardous materials. Used to describe hazardous materials in Canada and used in the United States and Canada for hazardous materials aboard aircraft.

Figure 22.1 Incidents involving hazardous materials may be more complex and difficult to mitigate than other types of emergencies. *Courtesy of Rich Mahaney.*

Awareness Level — Lowest level of training established by the National Fire Protection Association for first responders at hazardous materials incidents.

Operations Level — Level of training established by the National Fire Protection Association allowing first responders to take defensive actions at hazardous materials incidents.

or deliberate acts such as vandalism or terrorism. Firefighters must be aware of the potential for hazardous materials to be involved in fires, explosions, and criminal or terrorist activities. Because of the harmful potential of these substances, firefighters must possess the skills necessary to address incidents involving hazardous materials in a safe and effective manner.

In the United States and Canada, regulatory and standards organizations such as the Occupational Safety and Health Administration (OSHA), the National Fire Protection Association (NFPA®), and the Office of Domestic Preparedness (ODP) recognize two levels of training required for first responders in regard to hazardous materials and WMD incidents: Awareness Level and Operations Level. Persons trained to the Awareness Level are expected to be able to recognize a hazardous materials incident or terrorist attack, protect themselves from the hazards at the incident, call for additional help, and secure the incident scene. Operations Level personnel are expected to do all of these things, plus initiate defensive actions to protect the public, the environment, and property from the effects of the hazardous material(s) involved in the incident. Some Operations Level personnel may be trained to perform additional functions at a haz mat incident depending on their assigned missions or functions at such incidents. All responders must be trained to meet the legal requirements of the health and safety regulations of the local authority having jurisdiction (AHJ).

This chapter discusses the properties of hazardous materials including the ways in which they cause harm. Also covered are the clues that firefighters can use to recognize the presence of hazardous materials during an incident. Finally, it provides information to help firefighters differentiate between terrorist attacks and other types of emergency incidents, and to recognize illicit laboratories.

Personal Protective Equipment

Personnel responding to a haz mat incident must protect themselves with PPE appropriate to their mission at the incident. PPE may include anything from standard fire fighting protective clothing to chemical protective clothing (CPC) or body armor. An ensemble of appropriate PPE protects the skin, eyes, face, hearing, hands, feet, body, head, and respiratory system against a variety of hazards.

PPE Guards the Routes of Entry

Respiratory Protection Protects Against Inhalation and Ingestion

Protective Clothing Protects Against Skin Contact

Figure 22.2 Personal protective equipment is designed to prevent exposure to hazardous materials by guarding the routes of entry into the body.

Unfortunately, no one set of PPE will protect against all hazards. It is important for firefighters to understand that standard uniforms and traditional structural fire fighting clothing offer very limited protection against chemical hazards, whereas only body armor and bomb suits can provide limited protection against projectiles and explosives that might be a concern at a terrorist attack. A combination (ensemble) of CPC and respiratory protections may offer protection against hazardous materials by protecting the routes of exposure **(Figure 22.2)**, but it will not provide adequate protection against thermal hazards (fire) or explosives. The *Emergency Response Guidebook (ERG,* see *Emergency Response Guidebook* section) provides guidance on appropriate PPE for specific hazardous materials in the Protective Clothing section of the Orange-Bordered guide pages.

The correct use of PPE requires special training and instruction. While next generation turnout gear will provide improved protection against chemical, biological, and radiological weapons in the future, responders must be aware of how their current equipment is likely to perform in most situations. This chapter will discuss respiratory protection, types of protective clothing, PPE ensembles, classification, selection, inspection, testing, and maintenance of various types of PPE.

Emergency Response Guidebook (ERG) — A manual that aids emergency response and inspection personnel in identifying hazardous materials placards. It also gives guidelines for initial actions to be taken at hazardous materials incidents. Formerly the *North American Emergency Response Guidebook (NAERG).*

Respiratory Protection

Respiratory protection is a primary concern for first responders because inhalation is one of the major routes of exposure to hazardous substances (see Routes of Entry section). When correctly worn and used, protective breathing equipment protects the body from inhaling chemical, biological, and radiological materials. Each type of respiratory protection equipment is limited by capability; for example, self-contained breathing apparatus (SCBA) offers a limited working duration for supply of air. The basic types of protective breathing equipment used by responders at terrorist incidents are as follows:

- Self-Contained Breathing Apparatus (SCBA)
  — Closed-circuit SCBA
  — Open-circuit SCBA
- Supplied-air respirators (SARs)
- Air-purifying respirators (APRs)
  — Particulate-removing
  — Vapor- and gas-removing
  — Combination particulate-removing and vapor- and-gas-removing
- Powered air-purifying respirators (PAPRs)

Responders may also need to be familiar with powered-air hoods, escape respirators, and combined respirators. The following sections discuss these groups of respiratory equipment (including their basic limitations).

SCBAs

In the United States, the National Institute for Occupational Safety and Health (NIOSH) and Mine Safety and Health Administration (MSHA) must certify all SCBA for immediately dangerous to life or health (IDLH) atmospheres. SCBA that is not NIOSH/MSHA certified must not be used at haz mat incidents. The apparatus must also meet the design and testing criteria of NFPA® 1981, *Standard on Open-Circuit Self-Contained Breathing Apparatus (SCBA) for Emergency Services*, in jurisdictions that have adopted that standard by law or ordinance. In addition, American National Standards Institute (ANSI) standards for eye protection apply to the facepiece lens design and testing.

NIOSH classifies SCBA as either *closed-circuit* or *open-circuit*. Three types of SCBA are currently being manufactured in closed- or open-circuit designs: demand, pressure-demand, or positive-pressure. SCBA may also be either high- or low-pressure types. Only positive-pressure open-circuit or closed-circuit SCBA is allowed in incidents where personnel are exposed to hazardous materials. The advantages of using SCBA-type respiratory protection are independence and maneuverability; however, several disadvantages are as follows:

- Weight of the units
- Limited air-supply duration
- Change in profile that may hinder mobility because of the configuration of the harness assembly and the location of the air cylinder
- Limited vision caused by facepiece fogging
- Limited communications if the facepiece is not equipped with a microphone or speaking diaphragm

NIOSH has entered into a *Memorandum of Understanding* with the National Institute of Standards and Technology (NIST), OSHA, and NFPA® to jointly develop a certification program for SCBA used in emergency response to terrorist attacks. Working with the U.S. Army Soldier and Biological Chemical Command (SBCCOM), the group developed a new set of respiratory protection standards and test procedures for SCBA used in situations involving WMD. Under this voluntary program, NIOSH issues a special approval and label identifying the SCBA as appropriate for use against chemical, biological, radiological, and nuclear agents. The SCBA certified under this program must meet the following minimum requirements:

- Approval under NIOSH 42 *CFR* 84, Subpart H

- Compliance with NFPA® 1981

- Special tests under NIOSH 42 *CFR* 84.63(c):
    - Chemical Agent Permeation and Penetration Resistance Against Distilled Sulfur Mustard (HD [military designation]) and Sarin (GB [military designation])
    - Laboratory Respirator Protection Level (LRPL)

NIOSH maintains and disseminates an approval list for the SCBAs approved under this program. This list is entitled "CBRN (Chemical, Biological, Radiological, and Nuclear agents) SCBA (Self-Contained Breathing Apparatus)" and contains the name of the approval holder, model, component parts, accessories, and rated duration. This list is maintained as a separate category within the NIOSH Certified Equipment List.

NIOSH authorizes the use of an additional approval label on apparatus that demonstrate compliance to the CBRN criteria. This label is placed in a visible location on the SCBA backplate (for example, on the upper corner or in the area of the cylinder neck). The addition of this label provides visible and easy identification of equipment for its appropriate use (**Figure 22.3**).

Supplied-Air Respirators

The SAR or airline respirator is an atmosphere-supplying respirator in which the user does not carry the breathing air source. The apparatus usually consists of a facepiece, a belt- or facepiece-mounted regulator, a voice communications system, up to 300 feet (100 m) of air supply hose, an emergency escape pack or emergency breathing support system (EBSS), and a breathing air source (either cylinders mounted on a cart or a portable breathing-air compressor) (**Figure 22.4**). Because of the potential for damage to the

Figure 22.4 A complete supplied-air respirator (SAR) consists of the air-supply cart, manifold, air-supply hose, regulator, facepiece, and emergency breathing support system (EBSS).

CBRN Agent Approved

See Instructions for Required Component Part Numbers, Accessories, and Additional Cautions and Limitations of Use

Figure 22.3 SCBAs certified by NIOSH for use at CBRN incidents will bear a label.

air-supply hose, the EBSS provides enough air, usually 5, 10, or 15 minutes' worth, for the user to escape a hazardous atmosphere. SAR apparatus are not certified for fire fighting operations because of the potential damage to the airline from heat, fire, or debris.

NIOSH classifies SARs as Type C respirators. Type C respirators are further divided into two approved types: one type consists of a regulator and facepiece only, while the second consists of a regulator, facepiece, and EBSS. This second type may also be referred to as a SAR with escape (egress) capabilities. It is used in confined-space environments, IDLH environments, or potential IDLH environments. SARs used at haz mat incidents or terrorist events must provide positive pressure to the facepiece.

SAR apparatus have the advantage of reducing physical stress to the wearer by removing the weight of the SCBA. However, the air supply line is a limitation because of the potential for mechanical or heat damage. In addition, the length of the airline (no more than 300 feet [100 m] from the air source) restricts mobility. Other limitations are the same as those for SCBA: restricted vision and communications.

Air-Purifying Respirators

APRs contain an air-purifying filter, canister, or cartridge that removes specific contaminants found in ambient air as it passes through the air-purifying element. Based on which cartridge, canister, or filter is being used, these purifying elements are generally divided into the three following types:

- Particulate-removing APRs

- Vapor- and gas-removing APRs

- Combination particulate-removing and vapor- and gas-removing APRs

APRs may be powered (PAPRs) or non-powered. APRs do not supply oxygen or air from a separate source, and they protect only against specific contaminants at or below certain concentrations. Combination filters combine particulate-removing elements with vapor- and gas-removing elements in the same cartridge or canister.

Respirators with air-purifying filters may have either full facepieces that provide a complete seal to the face and protect the eyes, nose, and mouth or half facepieces that provide a complete seal to the face and protect the nose and mouth (**Figures 22.5 a and b**). It should be noted that half-face respirators will NOT protect against chemical, biological, and radiological (CBR) materials that can be absorbed through the skin or eyes and therefore are not recommended for use at terrorist incidents except in very specific situations (such as explosive attacks where the primary hazard is dust).

Disposable filters, canisters, or cartridges are mounted on one or both sides of the facepiece. Canister or cartridge respirators pass the air through a filter, sorbent, catalyst, or combination of these items to remove specific contaminants from the air. The air can enter the system either from the external atmosphere through the filter or sorbent or when the user's exhalation combines with a catalyst to provide breathable air.

No single canister, filter, or cartridge protects against all chemical hazards. Therefore, firefighters must know the hazards present in the atmosphere in order to select the appropriate canister, filter, or cartridge. Firefighters should be able to answer the following questions before deciding to use APRs for protection at an incident:

- What is the hazard?

- Is the hazard a vapor or a gas?

- Is the hazard a particle or dust?

Figure 22.5a Half-facepiece air-purifying respirators (APRs) are designed to protect against specific inhalation hazards, but they do not protect against chemicals that can be absorbed through the skin or eyes. *Courtesy of Federal Emergency Management Agency (FEMA) News Photo.*

Figure 22.5b Full-facepiece APRs provide greater protection for the face and eyes.

- Is there some combination of dust and vapors present?

- What concentrations are present?

Furthermore, responders should know that APRs do *not* protect against oxygen-deficient or oxygen-enriched atmospheres, and they must not be used in situations where the atmosphere is immediately dangerous to life and health (see Poisons/Toxic Chemicals section). The three primary limitations of an APR are as follows:

- Limited life of its filters and canisters

- Need for constant monitoring of the contaminated atmosphere

- Need for a normal oxygen content of the atmosphere before use

  The following precautions should be taken before using APRs:

- Know what chemicals/air contaminants are in the air.

- Know how much of the chemicals/air contaminants are in the air.

- Ensure that the oxygen level is between 19.5 and 23.5 percent.

- Ensure that atmospheric hazards are below IDLH conditions.

> **WARNING!**
> Do *not* wear APRs during emergency operations where unknown atmospheric conditions exist. Wear APRs only in controlled atmospheres where the hazards present are completely understood and at least 19.5 percent oxygen is present. <u>SCBA must be worn during emergency operations.</u>

At haz mat incidents, APRs may be used after emergency operations are over and the hazards at the scene have been properly identified. In some circumstances, APRs may also be used in other situations (law enforcement working perimeters of the scene or EMS/medical personnel) and escape situations. APRs used for these situations should utilize a combination organic vapor/high-efficiency particulate air (OV/HEPA) cartridge (see following sections).

Particulate-Removing Filters. Particulate filters protect the user from particulates (including biological hazards [see Etiological/Biological Hazards section]) in the air. These filters may be used with half facepiece masks or full facepiece masks. Eye protection must be provided when the full facepiece mask is not worn. Particulate-removing filters are divided into nine classes, three levels of filtration (95, 99, and 99.97 percents), and three categories of filter degradation. Particulate-removing filters may be used to protect against toxic dusts, mists, metal fumes, asbestos, and some biological hazards. HEPA filters used for medical emergencies must be 99.97 percent efficient, while 95 and 99 percent effective filters may be used depending on the health risk hazard.

Particle masks (also known as *dust masks*) are also classified as particulate-removing air-purifying filters **(Figure 22.6)**. These disposable masks protect the respiratory system from large-sized particulates. Dust masks provide very limited protection and should not be used to protect against chemical hazards or small particles such as asbestos fibers.

Vapor- and Gas-Removing Filters. As the name implies, vapor- and gas-removing cartridges and canisters are designed to protect against specific vapors and gases. They typically use some kind of *sorbent* material to remove the targeted vapor or gas from the air (see sidebar). Individual cartridges and canisters are usually designed to protect against related groups of chemicals such as *organic vapors* or *acid gases*. Many manufacturers color-code their canisters and cartridges so it is easy to see what contaminant(s) the canister or cartridge is designed to protect against **(Figure 22.7)**. Manufacturers also provide information about contaminant concentration limitations.

Figure 22.6 Particle masks are classified as particulate-removing air-purifying filters. They will protect against large particles such as concrete dust, but they will not protect against very small particulates such as asbestos.

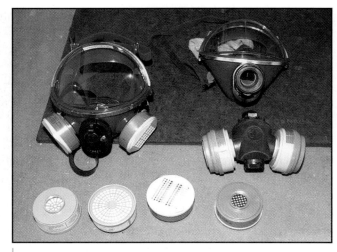

Figure 22.7 APR canisters are usually color coded to indicate what hazards they are designed to protect against.

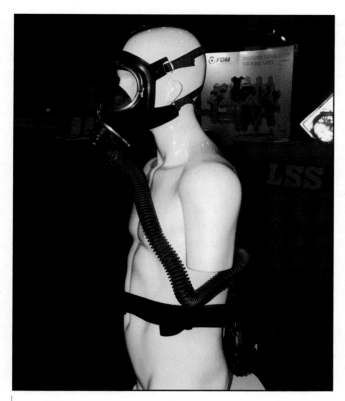

Figure 22.8 A blower forces ambient air through the filter canister and into the facepiece of a powered air-purifying respirator (PAPR).

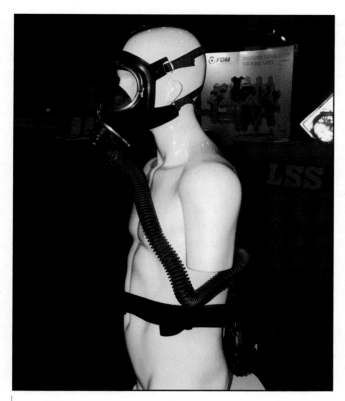

Supplied-Air Hood

Figure 22.9 Supplied-air hoods are simple to use and require no fit testing.

Powered Air-Purifying Respirators

The PAPR uses a blower to pass contaminated air through a canister or filter to remove the contaminants and supply the purified air to the full facepiece **(Figure 22.8)**. Because the facepiece is supplied with positive pressure, PAPRs offer a greater degree of safety than standard APRs in case of leaks or poor facial seals, and therefore may be of use at CBR incidents for personnel conducting decontamination operations and long-term operations. Airflow also makes PAPRs more comfortable to wear for many people.

Several types of PAPR are available. Some units are supplied with a small blower and are battery operated. The small size allows them to be worn on a belt. Other units have a stationary blower (usually mounted on a vehicle) that is connected by a long, flexible tube to the respirator facepiece.

As with all APRs, PAPRs should only be used in situations where the atmospheric hazards are understood and at least 19.5 percent oxygen is present. PAPRs are not safe to wear in atmospheres where potential respiratory hazards are unidentified, nor should they be used during initial emergency operations. **Do *not* use PAPRs in explosive or potentially explosive atmospheres.** Continuous atmospheric monitoring is needed to ensure the safety of the responder.

Supplied-Air Hoods

Powered- and supplied-air hoods provide loose fitting, lightweight respiratory protection that can be worn with glasses, facial hair, and beards **(Figure 22.9)**. Hospitals, emergency rooms, and other fire and emergency services organizations use these hoods as an alternative to other respirators, in part, because they require no fit testing and are simple to use.

Sorbent — Granular, porous filtering material used in vapor- or gas-removing respirators.

Figure 22.10 Escape respirators, like supplied-air hoods, require no fit testing, but they are designed to be used for only one, short-term use.

Escape Respirators

Escape respirators (sometimes called *personal escape canisters* or *escape hoods*) are designed for escaping the hot zone **(Figure 22.10)**. Escape respirators can be self-contained (usually utilizing rebreathing technology) or air-purifying (usually equipped with a combination of a HEPA filter and chemical filter such as activated carbon). Generally designed for a short duration of protection (such as 15 minutes), they are commonly designed in a hood style. The hood style is fast and easy to put on, and has a flexible seal around the neck. They can also accommodate glasses, facial hair, and beards. The filter canisters of APR-style escape respirators are usually not designed to be replaced (single use only). Some manufacturers provide escape respirator cases that can be strapped onto the body and worn as part of an emergency PPE ensemble.

Respiratory Equipment Limitations

Firefighters must consider the following limitations of equipment and air supply:

- *Limited visibility* — Facepieces reduce peripheral vision, and facepiece fogging can reduce overall vision.

- *Decreased ability to communicate* — Facepieces hinder voice communication.

- *Increased weight* — Depending on the model, the protective breathing equipment can add 25 to 35 pounds (11 kg to 16 kg) of weight to the emergency responder.

- *Decreased mobility* — The increase in weight and splinting effect of the harness straps reduce the wearer's mobility.

- *Inadequate oxygen levels* — APRs cannot be worn in IDLH or oxygen-deficient atmospheres.

- *Chemical specific* — APRs can only be used to protect against certain chemicals. The specific type of cartridge depends on the chemical the wearer is exposed to.

Additionally, open- and closed-circuit SCBA have maximum air-supply durations that limit the amount of time a first responder has to perform the tasks at hand. Non-NIOSH certified SCBAs may offer only limited protection in environments containing chemical warfare agents.

The following physical, medical, and mental limitations affect first responders' ability to use respiration protection equipment effectively:

- *Physical condition* — The wearer must be in good physical condition in order to maximize the work that can be performed and to stretch the air supply as far as possible.

- *Agility* — Wearing a protective breathing apparatus with an air cylinder or backpack restricts wearers' movements and affects their balance. Good agility can overcome these obstacles.

- *Facial features* — The shape and contour of the face affect the wearer's ability to get a good facepiece-to-face seal. Fit testing must be performed to ensure that facepieces fit properly.

- *Neurological functioning* — Good motor coordination is necessary for operating effectively in protective breathing equipment.

- *Mental soundness* — First responders must be of sound mind to handle emergency situations that may arise.

- *Muscular/skeletal condition* — First responders must have the physical strength and size required to perform necessary tasks while wearing protective breathing equipment.

- *Cardiovascular conditioning* — Poor cardiovascular conditioning can result in heart attacks, strokes, or other related problems during strenuous activity.

- *Respiratory functioning* — Proper respiratory functioning maximizes the wearer's operation time while wearing respiratory protection.

Do not assign a task requiring use of respirators and other PPE unless the individual is physically able to do the work.

Types of Protective Clothing

Protective clothing must be worn whenever a wearer faces potential hazards arising from exposure to hazardous materials. Skin contact with hazardous materials can cause a variety of problems, including chemical burns, allergic reactions and rashes, diseases, and absorption of toxic materials into the body (see Health and Physical Hazards sections). Protective clothing is designed to prevent these.

No single combination of protective equipment can protect against all hazards, so firefighters must be concerned with safety when choosing and using protective clothing. For example, it is extremely important to realize that fumes and vapors of many chemical weapons (see Chemical Attack Indicators section) can easily penetrate fire fighting turnout coats and pants. Similarly, CPC offers limited flash protection from fires.

Several types of personal protective clothing are available:

- Structural fire fighting protective clothing

- High-temperature protective clothing

- Chemical-protective clothing (CPC)

  — Liquid-splash protective clothing

  — Vapor-protective clothing

Structural Fire Fighting Protective Clothing

Structural fire fighting clothing is not a substitute for CPC, but it may provide limited protection against some hazardous materials. The multiple layers of the coat and pants may provide short-term exposure protection; however, to avoid harmful exposures, responders must recognize the limitations of this level of protection. For example, structural fire fighting clothing is neither corrosive-resistant (see Corrosives section) nor vapor-tight. Any liquids can soak through, acids and bases can dissolve or deteriorate the outer layers, and gases and vapors can penetrate the garment. Gaps in structural fire fighting clothing occur at the neck, wrists, waist, and the point where the pants and boots overlap (**Figure 22.11**).

Besides knowing what can deteriorate or destroy protective clothing, responders should be aware that some hazardous materials can permeate (pass through at the

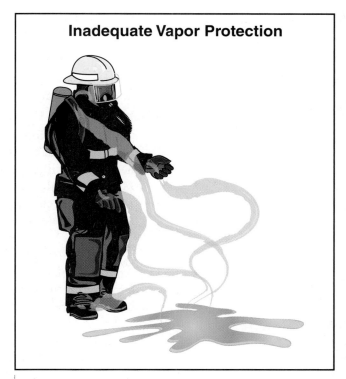

Inadequate Vapor Protection

Figure 22.11 While structural fire fighting protective clothing may provide limited protection against many hazardous chemicals, liquids can still soak through to come in contact with skin; acids and bases can dissolve or deteriorate the outer layers; and vapors, fumes, and gases can penetrate through gaps in the material and ensemble.

molecular level) and remain in the protective equipment. Chemicals absorbed into the equipment can subject the wearer to repeated exposure or to a later reaction with another chemical. In addition, the rubber or neoprene in boots, gloves, kneepads, and SCBA facepieces can become permeated by chemicals and rendered unsafe for use. It may be necessary to discard any equipment exposed to permeating types of chemicals.

Structural fire fighting protective clothing may be appropriate for use at haz mat incidents when the following conditions are met:

- Contact with splashes of extremely hazardous materials is unlikely
- The material's hazards have been identified, and they will not rapidly damage or permeate structural fire fighting protective clothing
- Total atmospheric concentrations do not contain high levels of chemicals that are toxic to the skin, and there are no adverse effects from chemical exposure to small areas of unprotected skin (see Chemical Hazards sections)
- There is a chance of fire or there is a fire (for example, a flammable liquid fire), and this type of protection is appropriate
- When structural fire fighting protective clothing is the only PPE available and CPC is not immediately available

At terrorism events, structural fire fighting protective clothing will provide protection against thermal damage in an explosive attack, but limited or no protection against projectiles, shrapnel, and other mechanical effects from a blast. It will provide adequate protection against some types of radiological materials, but not others. In cases where biological agents are strictly respiratory hazards, structural fire fighting protective clothing with SCBA may provide adequate protection. However, in any case where skin contact is potentially hazardous, it is not sufficient. Obviously, materials must be properly identified in order to make this determination, and any time a terrorist attack is suspected but not positively identified, it should be assumed that firefighters wearing structural fire fighting protective clothing with SCBA are still at some degree of risk (see Identification of Terrorist Attacks section).

Next generation turnout gear is being designed with greater chemical, biological, and radiological protection in mind. Better closures and interfaces (such as magnetic seals) will prevent exposure to hazardous materials penetrating through gaps. More chemical resistant materials will improve problems with permeation and degradation, thereby making it safer for firefighters to respond to incidents involving hazardous materials wearing structural fire fighting protective clothing alone.

High-Temperature Protective Clothing

Another type of special protective clothing that firefighters may encounter is high-temperature clothing. This clothing is designed to protect the wearer from short-term high-temperature exposures in situations where heat levels exceed the capabilities of standard fire fighting protective clothing. This type of clothing is usually of limited use in dealing with chemical hazards. Two basic types of high-temperature clothing that are available are as follows:

- *Proximity suits* — Permit close approach to fires for rescue, fire suppression, and property conservation activities such as in aircraft rescue and fire fighting or other fire fighting operations involving flammable liquids. Such suits provide greater heat protection than standard structural fire fighting protective clothing.
- *Fire-entry suits* — Allow a person to work in total flame environments for short periods of time; provide short-duration and close-proximity protection at radiant

heat temperatures as high as 2,000°F (1,093°C). Each suit has a specific use and is not interchangeable. Fire-entry suits are not designed to protect the wearer against chemical hazards.

Several limitations to high-temperature protective clothing are as follows:

- Contributes to heat stress by not allowing the body to release excess heat
- Is bulky
- Limits wearer's vision
- Limits wearer's mobility
- Limits communication
- Requires frequent and extensive training for efficient and safe use
- Is expensive to purchase

Chemical-Protective Clothing

The purpose of CPC and equipment is to shield or isolate individuals from the chemical, physical, and biological hazards that may be encountered during hazardous materials operations. Design and testing standards generally recognize two types of CPC: liquid-splash protective clothing and vapor-protective clothing.

CPC is made from a variety of different materials, none of which protects against all types of chemicals **(Figure 22.12)**. Each material provides protection against certain chemicals or products, but only limited or no protection against others. The

Chemical-Protective Clothing (CPC) — Clothing designed to shield or isolate individuals from the chemical, physical, and biological hazards that may be encountered during operations involving hazardous materials.

Figure 22.12 No one type of chemical protective clothing will protect against all hazardous materials.

manufacturer of a particular suit must provide a list of chemicals for which the suit is effective. Selection of appropriate CPC depends on the specific chemical and on the specific tasks to be performed by the wearer.

CPC is designed to afford the wearer a known degree of protection from a known type, concentration, and length of exposure to a hazardous material, but only if fitted properly and worn correctly (see Chemical Hazards sections). Improperly worn equipment can expose and endanger the wearer. One factor firefighters should remember during the selection process is that most protective clothing is designed to be impermeable to moisture, thus limiting the transfer of heat from the body through natural evaporation. This concern is particularly important in hot environments and when strenuous tasks cause such garments to increase the likelihood of heat injury. Regardless of the type of CPC worn at an incident, it must be decontaminated before storage or disposal.

Figure 22.13 Liquid-splash protective clothing may be nonencapsulated (left) or encapsulated (right). Most commonly it is nonencapsulated.

WARNING!
Responders must have sufficient training to operate in conditions that require use of vapor-protective ensembles.

Liquid-Splash Protective Clothing. *Liquid-splash protective clothing* is primarily designed to protect users from chemical liquid splashes, but not against chemical vapors or gases. Liquid-splash protective clothing can be encapsulating or nonencapsulating **(Figure 22.13)**.

One limitation common to both encapsulating and nonencapsulating liquid-splash protective clothing is that such clothing is not resistant to heat or flame exposure, nor does it protect against projectiles or shrapnel. The material of liquid-splash protective clothing is made from the same types of material used for vapor-protective suits (see following section).

When used as part of a protective ensemble, liquid-splash protective ensembles may use an SCBA, an airline (SAR), or a full-face, air-purifying, canister-equipped respirator. This type of protective clothing is a component of EPA Level B chemical protection ensembles (see PPE Ensembles: U.S. EPA Levels of Protection section), and Class 3 ensembles to be used at chemical and biological terrorist incidents as specified in NFPA® 1994, *Standard on Protective Ensembles for First Responders to CBRN Terrorism Incidents.*

Vapor-Protective Clothing. *Vapor-protective clothing* is designed to protect the wearer against chemical vapors or gases and offers a greater level of protection than liquid-splash protective clothing **(Figure 22.14)**. Vapor-protective ensembles must be worn with positive-pressure SCBA or combination SCBA/SAR. Vapor-protective ensembles are components of Class 1 and 2 ensembles to be used at chemical and biological terrorist incidents as specified in NFPA® 1994. These suits are also primarily used as part of an EPA Level A protective ensemble, providing the greatest degree of protection against respiratory, eye, or skin damage from hazardous vapors, gases, particulates, sudden splash, immersion, or contact with hazardous materials **(Figure 22.15)**. Several limitations to vapor-protective suits are as follows:

- Do not protect the user against all chemical hazards
- Impair mobility, vision, and communication
- Do not allow body heat to escape; can contribute to heat stress, which may require the use of a cooling vest

Figure 22.14 Vapor-protective clothing provides the highest degree of protection against hazardous materials. *Courtesy of the Illinois Fire Service Institute.*

Figure 22.15 Vapor-protective clothing is part of an EPA Level A ensemble.

PPE Ensembles: U.S. EPA Levels of Protection

The approach in selecting PPE must encompass an *ensemble* of clothing and equipment items that are easily integrated to provide both an appropriate level of protection and still allow one to perform activities involving hazardous materials **(Figure 22.16)**. For example, simple protective clothing such as gloves and a work uniform in combination with a faceshield (or safety goggles) may be sufficient to prevent exposure to certain etiological agents (such as bloodborne pathogens). At the other end of the spectrum, the use of vapor-protective, totally-encapsulating suits combined with positive-pressure SCBA is considered the minimum level of protection necessary when dealing with vapors, gases, or particulates of material that are harmful to skin or capable of being absorbed through the skin.

The U.S. EPA has established the following levels of protective equipment to be used at incidents involving CBR materials: Level A, Level B, Level C, and Level D. NIOSH, OSHA, and the U.S. Coast Guard (USCG) also recognize these levels. They can be used as the starting point for ensemble creation; however, each ensemble must be tailored to the specific situation in order to provide the most appropriate level of protection.

Figure 22.16 A PPE ensemble must be appropriate for the activities to be performed as well as the hazards present. PAPRs, as shown in this photo, should only be used on controlled atmospheres where the hazards present are completely understood and at least 19.5 percent oxygen is present.

Figure 22.17a The EPA Level A ensemble consists of a fully encapsulating vapor-protective suit worn over an SCBA. *Courtesy of Kenneth Baum.*

Level A

Level A Protection — Highest level of skin, respiratory, and eye protection that can be afforded by personal protective equipment. Consists of positive-pressure self-contained breathing apparatus, totally encapsulating chemical-protective suit, inner and outer gloves, and chemical-resistant boots.

The Level A ensemble provides the highest level of protection against vapors, gases, mists, and particles for the respiratory tract, eyes, and skin (**Figure 22.17 a**). Operations-Level responders are generally not allowed to operate in situations requiring Level A protection. However, Operations-Level personnel must be appropriately trained to wear Level A PPE if they are required to wear it. The elements of Level A ensembles are as follows:

- *Components* — Ensemble requirements are as follows:

  — Positive-pressure, full-facepiece, SCBA, or positive-pressure airline respirator with escape SCBA, approved by NIOSH

  — Vapor-protective suits: Totally encapsulated suits constructed of protective-clothing materials that meet the following criteria:

    • Cover the wearer's torso, head, arms, and legs

- Include boots and gloves that may either be an integral part of the suit or separate and tightly attached

- Enclose the wearer completely or in combination with the wearer's respiratory equipment, gloves, and boots

- Provide equivalent chemical-resistance protection for all components of a totally encapsulated suit (such as relief valves, seams, and closure assemblies)

- Meet the requirements in NFPA® 1991, *Standard on Vapor-Protective Ensembles for Hazardous Materials Emergencies.*

— Coveralls (optional)

— Long underwear (optional)

— Chemical-resistant outer gloves

— Chemical-resistant inner gloves

— Chemical-resistant boots with steel toe and shank

— Hardhat (under suit) (optional)

— Disposable protective suit, gloves, and boots (can be worn over totally encapsulating suit, depending on suit construction)

— Two-way radios (worn inside encapsulating suit)

- **Protection Provided** — Highest available level of respiratory, skin, and eye protection from solid, liquid, and gaseous chemicals

- **Use** — Ensembles are used in the following situations:

— Chemical hazards are unknown or unidentified.

— Chemical(s) have been identified and have high level of hazards to respiratory system, skin, and eyes

— Site operations and work functions involve a high potential for splash, immersion, or exposure to unexpected vapors, gases, or particulates of material that are harmful to skin or capable of being absorbed through the intact skin

— Substances are present with known or suspected skin toxicity or carcinogenicity

— Operations that are conducted in confined or poorly ventilated areas

Level B

Level B protection requires a garment that includes an SCBA or a SAR and provides protection against splashes from a hazardous chemical (**Figure 22.17 b**). This ensemble is worn when the highest level of respiratory protection is necessary but a lesser level of skin protection is needed. A Level B ensemble provides liquid-splash protection, but little or no protection against chemical vapors or gases to the skin. Level B CPC may be encapsulating or nonencapsulating. The elements of Level B ensembles are as follows:

Level B Protection — Personal protective equipment that affords the highest level of respiratory protection, but a lesser level of skin protection. Consists of positive-pressure self-contained breathing apparatus, hooded chemical-resistant suit, inner and outer gloves, and chemical-resistant boots.

Figure 22.17b The EPA Level B ensemble consists of a liquid-splash protective suit and SCBA. *Courtesy of Kenneth Baum.*

- *Components* — Ensemble requirements are as follows:
  — Positive-pressure, full-facepiece, SCBA, or positive-pressure airline respirator with escape SCBA approved by NIOSH
  — Hooded chemical-resistant clothing that meets the requirements of NFPA® 1992, *Standard on Liquid Splash-Protective Ensembles and Clothing for Hazardous Materials Emergencies* (overalls and long-sleeved jacket, coveralls, one- or two-piece chemical-splash suit, and disposable chemical-resistant overalls)
  — Coveralls (optional)
  — Chemical-resistant outer gloves
  — Chemical-resistant inner gloves
  — Chemical-resistant boots with steel toe and shank
  — Disposable chemical-resistant outer boot covers (optional)
  — Hardhat (outside or on top of nonencapsulating suits or under encapsulating suits)
  — Two-way radios (worn inside encapsulating suit or outside nonencapsulating suit)
  — Faceshield (optional)
- *Protection Provided* — Ensembles provide the same level of respiratory protection as Level A but have less skin protection. Ensembles provide liquid-splash protection, but no protection against chemical vapors or gases.
- *Use* — Ensembles are used in the following situations:
  — Type and atmospheric concentration of substances have been identified and require a high level of respiratory protection but less skin protection
  — Atmosphere contains less than 19.5 percent oxygen or more than 23.5 percent oxygen
  — Presence of incompletely identified vapors or gases is indicated by a direct-reading organic vapor detection instrument, but the vapors and gases are known not to contain high levels of chemicals harmful to skin or capable of being absorbed through intact skin
  — Presence of liquids or particulates is indicated, but they are known not to contain high levels of chemicals harmful to skin or capable of being absorbed through intact skin

Level C

Level C Protection — Personal protective equipment that affords a lesser level of respiratory and skin protection than levels A or B. Consists of full-face or half-mask APR, hooded chemical-resistant suit, inner and outer gloves, and chemical-resistant boots.

Level C protection differs from Level B in the area of equipment needed for respiratory protection (**Figure 22.17 c**). Level C is composed of a splash-protecting garment and an air-purifying device (APR or PAPR). Level C protection includes any of the various types of APRs. Emergency response personnel would not use this level of protection unless the specific material is known, it has been measured, and this protection level is approved by the incident commander (IC) after all qualifying conditions for APRs and PAPRs have been met (that is, the product is known, an appropriate filter is available, the atmospheric oxygen concentration is between 19 to 23.5 percent, and the atmosphere is not IDLH). Periodic air monitoring is required when using this level of PPE. The elements of Level C ensembles are as follows:

- *Components* — Ensemble requirements are as follows:

— Full-face or half-mask APRs, NIOSH approved

— Hooded chemical-resistant clothing (overalls, two-piece chemical-splash suit, and disposable chemical-resistant overalls)

— Coveralls (optional)

— Chemical-resistant outer gloves

— Chemical-resistant inner gloves

— Chemical-resistant boots with steel toe and shank

— Disposable, chemical-resistant outer boot covers (optional)

— Hardhat

— Escape mask (optional)

— Two-way radios (worn under outside protective clothing)

— Faceshield (optional)

- *Protection Provided* — Ensembles provide the same level of skin protection as Level B but have a lower level of respiratory protection. Ensembles provide liquid-splash protection but no protection from chemical vapors or gases on the skin.

- *Use* — Ensembles are used in the following situations:

— Atmospheric contaminants, liquid splashes, or other direct contact will not adversely affect exposed skin or be absorbed through any exposed skin

— Types of air contaminants have been identified, concentrations have been measured, and an APR is available that can remove the contaminants

— All criteria for the use of APRs are met

— Atmospheric concentration of chemicals does not exceed IDLH levels. The atmosphere must contain between 19.5 and 23.5 percent oxygen

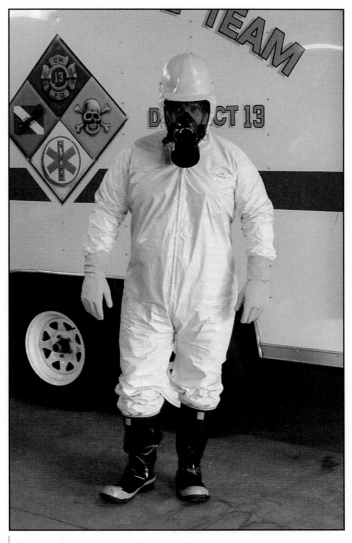

Figure 22.17c The EPA Level C ensemble consists of a liquid-splash protective suit and air-purifying respirator. *Courtesy of Kenneth Baum.*

Level D

Level D ensembles consist of typical work uniforms, street clothing, or coveralls **(Figure 22.17 d)**. This PPE level is used for nuisance contamination only. Level D protection can be worn only when no atmospheric hazards exist. The elements of Level D ensembles are as follows:

- *Components* — Ensemble requirements are as follows:

— Coveralls

— Gloves (optional)

— Chemical-resistant boots/shoes with steel toe and shank

— Disposable, chemical-resistant outer boot covers (optional)

— Safety glasses or chemical-splash goggles

Level D Protection — Personal protective equipment that affords the lowest level of respiratory and skin protection. Consists of coveralls, gloves, and chemical-resistant boots or shoes.

— Hardhat

— Escape device in case of accidental release and the need to immediately escape the area (optional)

— Faceshield (optional)

• **Protection Provided** — Ensembles provide no respiratory protection and minimal skin protection.

• *Use* — Ensembles may not be worn in the hot zone and are not acceptable for haz mat emergency response above the Awareness Level. Level D ensembles are used when both of the following conditions exist:

— Atmosphere contains no hazard.

— Work functions preclude splashes, immersion, or the potential for unexpected inhalation of or contact with hazardous levels of any chemicals

While the EPA has established this set of chemical-protective PPE ensembles that are commonly used by fire and emergency service organizations, other organizations such as law enforcement, industrial responders, and the military may have their own standard operating procedures or equivalent procedures guiding the choice and use of appropriate combinations of PPE. A special weapons attack team (SWAT) may be equipped with a far different PPE ensemble than a firefighter, haz mat technician, or environmental cleanup personnel responding to the same terrorist incident.

NFPA® 1994: PPE Ensemble Classifications

Figure 22.17d The EPA Level D ensemble consists of work uniforms, street clothing, or coveralls.

NFPA® 1994 establishes performance requirements for three classes of PPE ensembles that are used in situations involving chemical or biological terrorism agents. Unlike EPA levels that describe widely different combinations of PPE (from work uniforms to vapor-protective clothing), the NFPA® 1994 classes apply very specifically to the different performance standards of chemical (and biological) protective clothing.

The ensembles for all three classes must be designed to protect the wearer's upper and lower torso, head, hands, and feet; ensemble elements must include protective garments, protective gloves, and protective footwear. Ensembles can be either encapsulating or nonencapsulating and must accommodate appropriate respiratory protection.

Each ensemble (or component) bears a label describing the class to which it belongs and the circumstance for which it is designed to be used **(Figure 22.18)**. According to NFPA® 1994, manufacturers must provide a variety of information with liquid-splash protective clothing items or ensembles including safety considerations, PPE limitations, storage practices, recommended inspection procedures, directions for use, storage life, and care and maintenance instructions.

Class 1

Class 1 ensembles provide the highest degree of protection. They are designed to protect responders at chemical/biological terrorism incidents in the following situations:

- Whenever the identity or concentration of the vapor or liquid agent is undetermined or in question

- When vapor protection is needed

- Anytime liquid contact is expected and no direct skin contact can be permitted because exposure may present a serious health threat such as death or incapacitation (as with chemical nerve agents)

Class 2

Class 2 ensembles are designed to protect responders at chemical/biological terrorism incidents in the following situations:

- To provide necessary sufficient vapor protection for the intended operation

- When direct contact with liquid droplets is likely

- When victims are not ambulatory but are showing signs or symptoms of exposure

Class 3

Class 3 ensembles are designed to protect responders at chemical/biological terrorism incidents in the following situations:

- To provide necessary sufficient liquid protection for the intended operation is necessary

- When direct contact with liquid droplets is likely

- When victims are impaired but ambulatory

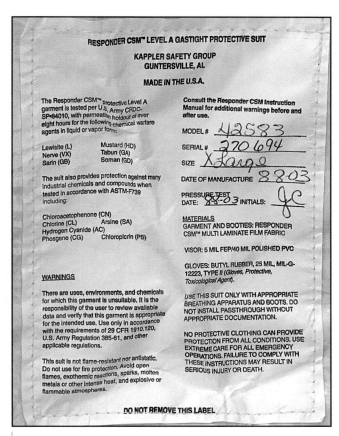

Figure 22.18 Vapor-protective ensembles that are designed to be used at chemical and biological terrorist incidents will be labeled accordingly. *Courtesy of Rich Mahaney.*

Mission-Oriented Protective Posture (MOPP)

The U.S. military uses MOPP ensembles to protect against chemical, biological, and radiological hazards. MOPP ensembles consist of an overgarment, mask, hood, overboots, and protective gloves (**Figure 22.19**). MOPPs provide six flexible levels of protection (0-4, plus Alpha) based on threat level, work rate for the mission, temperature and humidity (**Table 22.1**). The higher the MOPP level, the greater the protection (and the lower the rate of work productivity and efficiency). MOPP Level Alpha is designed for use in situations upwind from any threat with little danger of exposure to hazardous vapors.

Permeable garments such as the joint service lightweight integrated suit technology (JSLIST) protective overgarment provide protection against liquid, solid, and/or vapor CB

Figure 22.19 MOPP ensembles consist of an overgarment, mask, hood, overboots, and protective gloves.

Table 22.1
MOPP Levels

Non-firefighter MOPP Levels

	MOPP 0	MOPP 1	MOPP 2	MOPP 3	MOPP 4
JSLIST	Carried	Worn	Worn	Worn	Worn
Protective Mask	Carried	Carried	Carried	Worn	Worn
Cotton Insert Gloves	Carried	Carried	Carried	Carried	Worn
Butyl Rubber Gloves	Carried	Carried	Carried	Carried	Worn
Protective Overgarment Boots	Carried	Carried	Worn	Worn	Worn

Firefighter MOPP Levels*

	MOPP 0	MOPP 1	MOPP 4	MOPP 4 Firefighting Mode
JSLIST	Carried	Worn	Worn	Worn
Nomex Hood	Carried	Worn	Worn	Worn
Firefighting Protective Trousers	Carried	Carried	Worn	Worn
Firefighting Protective Jacket	Carried	Carried	Carried	Worn
Firefighting Protective Footwear	Carried	Carried	Worn	Worn
CW Mask	Carried	Carried	Worn	Worn
Fire & Chemical Protective Gloves	Carried	Carried	Worn	Worn
Cotton Insert Gloves *	Carried	Carried	Worn	Worn
Butyl Rubber Gloves *	Carried	Carried	Worn	Worn
Firefighting Gloves *	Carried	Carried	Carried	Worn
Structural ARFF Helmet	Carried	Carried	Carried	Worn
SCBA	Carried	Carried	Carried	Worn

*Per current military Technical Orders

agents and radioactive alpha and beta particles. JSLIST overgarments are manufactured of lightweight 50% Nylon and 50% cotton ripstop water repellant permeable materials. They are equipped with a charcoal/carbon lining designed to absorb harmful materials, much like the carbon filter/canister on an APR. The JSLIST protective garment will be degraded when in contact with certain solvents such as sweat and petroleum products. JSLIST can be laundered up to six times for personal hygiene. The JSLIST ensemble may be worn 45 consecutive days with a total out-of-the-bag available usage of 120 days. The protective mask is a hooded APR, and the boots and gloves are made of Butyl rubber.

Firefighters who wear the JSLIST overgarment will find that the MOPP levels are modified to suit their capabilities. With the additional wear and tear of fire fighting protective equipment that is commonly found by firefighters wearing the JSLIST, firefighters will use only three MOPP levels (0, 1, 4). When the typical wearer is in MOPP 2, firefighters will remain in MOPP 1 until directed to proceed to a higher level. When MOPP 3 is directed, firefighters will immediately proceed from MOPP 1 and don the appropriate MOPP 4 attire. When a need to conduct fire fighting operations arises, firefighters will proceed to MOPP 4 firefighter mode where they will don their fire fighting protective jacket and SCBA.

PPE Selection Factors

Many available sources can be consulted for determining which type and what level of PPE to use at terrorist attacks. First-arriving responders often rely upon information in the *ERG* (see *Emergency Response Guidebook* section) to determine the minimum type of protection required for defensive operations.

There are some things that first responders need to know and remember in the selection process. For example, the use of PPE itself can create significant wearer hazards such as heat stress and physical and psychological stress, in addition to impaired vision, mobility, and communication. In general, the higher the level of PPE is, the greater the associated risks. For any given situation, select equipment and clothing that provide an adequate level of protection. Overprotection as well as underprotection can be hazardous and should be avoided.

Climate Concerns and Health Issues

Most types of PPE inhibit the body's ability to disperse heat, which is magnified because the firefighter is usually performing strenuous work while wearing the equipment. Thus, wearing PPE usually increases firefighters' risks of developing heat-related disorders. However, when working in cold climates, considerations must be taken to protect responders from cold-related disorders, as well. For example, CPC is not designed to provide insulation against the cold. It is important to take preventive measures to reduce the effects of any temperature extreme. Medical monitoring of responders is required when they are at risk because of environmental hazards.

Heat Disorders

Wearing PPE or other special full-body protective clothing puts the wearer at considerable risk of developing heat stress. It is important for first responders to understand the effects of heat stress. This stress can result in health effects ranging from transient heat fatigue to serious illness (heat stroke) or death.

First responders need to be aware of several heat disorders, including heat stroke (the most serious), heat exhaustion, heat cramps, heat rashes, and heat fatigue. In addition, they should know how to prevent the effects of heat exposure.

Heat-Exposure Prevention

Firefighters wearing protective clothing need to be monitored for effects of heat exposure. Methods to prevent and/or reduce the effects of heat exposure include the following:

- *Fluid consumption* — Use water or commercial body-fluid-replenishment drink mixes to prevent dehydration. First responders should drink generous amounts of fluids both before and during operations. Drinking 7 ounces (200 ml) of fluid every 15 to 20 minutes is better than drinking large quantities once an hour. Balanced diets normally provide enough salts to avoid cramping problems. *Details:*

  — Before working, drinking chilled water is good.

  — After a work period in protective clothing and an increase in core temperature, drinking room-temperature water is better. It is not as severe a shock to the body.

Heat Stroke — Heat illness caused by heat exposure, resulting in failure of the body's heat regulating mechanism; symptoms include high fever of 105° to 106° F (40.5° C to 41.1° C); dry, red, hot skin; rapid, strong pulse; deep breaths; and convulsions. May result in coma or possibly death. Also called sunstroke.

Heat Exhaustion — Heat illness caused by exposure to excessive heat; symptoms include weakness, cold and clammy skin, heavy perspiration, rapid and shallow breathing, weak pulse, dizziness, and sometimes unconsciousness.

Heat Cramps — Heat illness resulting from prolonged exposure to high temperatures; characterized by excessive sweating, muscle cramps in the abdomen and legs, faintness, dizziness, and exhaustion.

Heat Rash — Condition that develops from continuous exposure to heat and humid air; aggravated by clothing that rubs the skin; reduces the individual's tolerance to heat.

Heat Stroke

Heat stroke occurs when the body's system of temperature regulation fails and body temperature rises to critical levels. This condition is caused by a combination of highly variable factors, and its occurrence is difficult to predict. **Heat stroke is a serious medical emergency and requires immediate medical treatment and transport to a medical care facility.** The primary signs and symptoms of heat stroke are as follows:

- Confusion
- Irrational behavior
- Loss of consciousness
- Convulsions
- Lack of sweating (usually)
- Hot, dry skin
- Abnormally high body temperature (for example, a rectal temperature of 105.8°F [41°C])

When the body's temperature becomes too high, it causes death. The elevated metabolic temperatures caused by a combination of workload and environmental heat load, both of which contribute to heat stroke, are also highly variable and difficult to predict. If a first responder shows signs of possible heat stroke, obtain professional medical treatment immediately.

Frostbite — Local freezing and tissue damage due to prolonged exposure to extreme cold.

Hypothermia — Abnormally low or decreased body temperature.

Figure 22.20 The use of cooling vests can help prevent heat illness when wearing chemical-protective clothing.

- ***Body ventilation*** — Wear long cotton undergarments or similar types of clothing to provide natural body ventilation.

- ***Body cooling*** — Provide mobile showers and misting facilities to reduce the body temperature and cool protective clothing. Wearing cooling vests beneath CPC can also help cooling **(Figure 22.20)**.

- ***Rest areas*** — Provide shaded and air-conditioned areas for resting.

- ***Work rotation*** — Rotate responders exposed to extreme temperatures or those performing difficult tasks frequently.

- ***Proper liquids*** — Avoid liquids such as alcohol, coffee, and caffeinated drinks (or minimize their intake) before working. These beverages can contribute to dehydration and heat stress.

- ***Physical fitness*** — Encourage responders to maintain good physical fitness.

Cold Disorders

Cold temperatures caused by weather and/or other conditions such as exposure to cryogenic liquids are also environmental factors that must be considered when selecting PPE. Prolonged exposure to freezing temperatures can result in health problems as serious as trench foot, frostbite, and hypothermia. Protection from the cold must be a priority when conditions warrant.

The four primary environmental conditions that cause cold-related stress are low temperatures, high/cool winds, dampness, and cold water. Wind chill, a combination of temperature and velocity, is a crucial factor to evaluate when working outside. For example, when the actual air temperature of the wind is 40°F (4.4°C) and its velocity is 35 mph (56 kph), the exposed skin experiences conditions equivalent to the still-air temperature of 11°F (–12°C). A dangerous situation of rapid heat loss may arise for any individual exposed to high winds and cold temperatures.

Medical Monitoring

It is very important to provide ongoing medical monitoring of responders who may be at risk because of environmental hazards (heat/cold stresses) as well as potential exposure to CBR materials. Medical monitoring should be conducted before responders wearing chemical liquid-splash or vapor-protective clothing enter the warm and hot zones (pre-entry monitoring) as well as after leaving these zones (post-entry monitoring) as directed by the AHJ **(Figure 22.21)**. The evaluation will check such things as vital signs, hydration, skin, mental status, and medical history. A postmedical monitoring follow-up is also recommended.

Inspection, Testing, and Maintenance

When wearing protective clothing and equipment, the end user must take all necessary steps to ensure that the protective ensemble performs as expected. During an emergency is not the right time to discover discrepancies in the protective clothing or respiratory protection. Following a standard program for inspection, proper storage, maintenance, and cleaning along with realizing PPE limitations is the best way to avoid exposure to dangerous materials during an emergency response. All inspections, testing, and maintenance of PPE must be conducted in accordance with manufacturer's recommendations.

Figure 22.21 Medical monitoring should be conducted before firefighters wearing chemical liquid-splash or vapor-protective clothing start to work as well as afterward.

Records must be kept of all inspection procedures. Periodic review of these records can provide an indication of protective clothing or equipment that requires excessive maintenance and can also serve to identify clothing or equipment that is susceptible to failure. At a minimum, record the following information at each inspection:

- Clothing or equipment item identification number
- Date of inspection
- Person making the inspection
- Results of the inspection
- Any unusual conditions noted

Respiratory equipment is initially inspected when it is purchased. Once the equipment is placed into service, the organization's personnel perform periodic inspections. Operational inspections of respiratory protection equipment occur after each use, daily or weekly, monthly, and annually. The organization must define the frequency and type of inspection in the respiratory protection policy, and should follow manufacturer's recommendations.

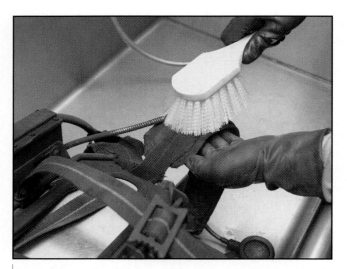

Figure 22.22 Following use, respiratory protection equipment such as SCBA should be cleaned to remove dirt, grime, and contaminants. Personnel should wear protective gloves and other garments to protect against exposure to unknown contaminants.

The care, cleaning, and maintenance schedules of respiratory protection equipment should be based on the manufacturer's recommendation, NFPA® standards, or OSHA requirements **(Figure 22.22)**. The minimum requirements are as follows:

- Clean respiratory protection equipment as needed during the daily/weekly inspections at the beginning of the work period.

- Inspect, clean, and disinfect respiratory protection equipment following use before placing it back into service.

- Clean respiratory protection equipment units that are used infrequently during the weekly inspection when required. Clean units before placing them in storage. Clean as needed when units are removed from storage.

Health and Physical Hazards

To safely mitigate hazardous materials incidents, firefighters must understand the variety of hazardous materials they may encounter, how these materials may potentially affect health, and the physical hazards associated with them. Knowing some of these basic concepts will help firefighters prevent or reduce injury, loss of life, and environmental/property losses caused by these materials.

Many hazardous materials have major effects on human health. Exposures to hazardous materials may be *acute* (single occurrence) or *chronic* (long-term, reoccurring). Health effects can also be acute or chronic. *Acute health effects* are short-term effects that appear within hours or days, such as vomiting or diarrhea. *Chronic health effects* are long-term effects that may take years to appear, such as cancer.

The following have the potential to cause harm (with acute and/or chronic effects) at a hazardous materials incident:

- Thermal hazards

- Radiological hazards

- Asphyxiation hazards

- Chemical hazards
  - Poisons/toxins
  - Corrosives
  - Irritants
  - Convulsants
  - Carcinogens
  - Sensitizers/allergens

- Etiological/biological hazards

- Mechanical hazards

Thermal Hazards

Cryogenics — Gases that are converted into liquids by being cooled below −150°F (−101°C).

Thermal hazards are related to temperature extremes. Hazardous materials themselves can cause temperature extremes such as with elevated-temperature materials, exothermic reactions, fires, explosions, or cryogenic (extremely cold) liquids

(Figure 22.23). Thermal hazards can also be caused by conditions on the scene such as extreme ambient air temperature, especially when using fully-encapsulating PPE.

Radiological Hazards

The potential for radiation exposure exists when firefighters respond to incidents at medical centers, certain industrial operations, nuclear power plants, and research facilities. And, of course, there is the potential for exposure during terrorist attacks. Different types of radiation exist, and some are more energetic than others **(Figure 22.24)**. The most energetic (and hazardous) form of radiation is ionizing radiation, and it is this type of radiation that is of greatest concern to firefighters. The types of ionizing radiation are as follows:

- *Alpha* — Energetic, positively charged alpha particles (helium nuclei) emitted from the nucleus during radioactive decay that rapidly lose energy when passing through matter. They are commonly emitted in the radioactive decay of the heaviest radioactive elements such as uranium and radium as well as by some manmade elements. *Details:*

    — Alpha particles lose energy rapidly in matter and do not penetrate very far; however, they can cause damage over their short path through human tissue. They are usually completely blocked by the outer dead layer of the human skin, so

Figure 22.23 Elevated temperature materials may present a thermal hazard. Transport trucks and railcars should be marked accordingly.

Ionizing Radiation — Radiation that has sufficient energy to remove electrons from atoms resulting in a chemical change in the atom.

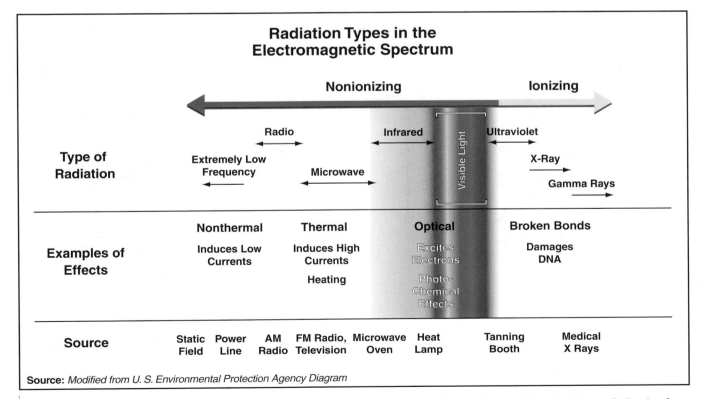

Figure 22.24 Ionizing radiation has more energy than nonionizing radiation. Generally speaking, ionizing radiation is of more concern to firefighters because of the potential health effects associated with exposure.

alpha-emitting radioisotopes are not a hazard outside the body. However, they can be very harmful if the material emitting the alpha particles is ingested or inhaled.

— Alpha particles can be stopped completely by a sheet of paper (**Figure 22.25**).

- **Beta** — Fast-moving, positively or negatively charged electrons (beta particles) emitted from the nucleus during radioactive decay. Humans are exposed to beta particles from manufactured and natural sources such as tritium, carbon-14, and strontium-90. *Details:*

    — Beta particles are more penetrating than alpha particles but less damaging over equally traveled distances. Some beta particles are capable of penetrating the skin and causing radiation damage; however, as with alpha emitters, beta emitters are generally more hazardous when they are inhaled or ingested.

    — Beta particles travel appreciable distances in air (up to 20 feet [6 m]) but can be reduced or stopped by a layer of clothing or by two or three millimeters (less than 0.08 inch) of a substance such as aluminum (Figure 22.25).

- **Gamma** — High-energy photons (weightless packets of energy like visible light and X-rays). Gamma rays often accompany the emission of alpha or beta particles from a nucleus. They have neither a charge nor a mass but are very penetrating. One

Electron — Minute component of an atom that possesses a negative charge.

Photon — Packet of electromagnetic energy.

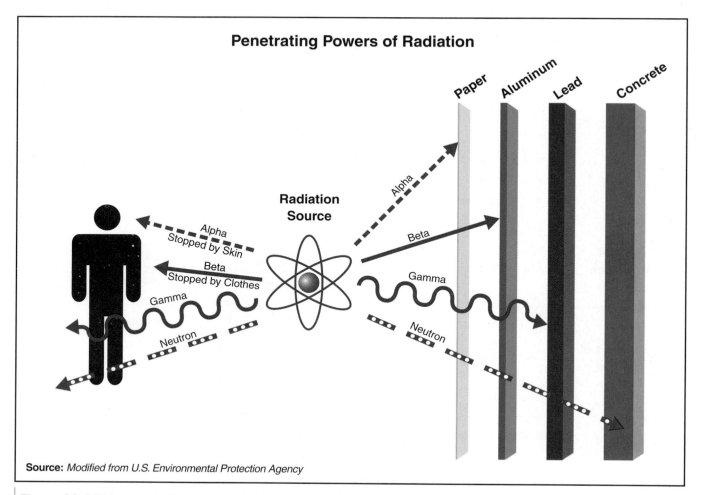

Penetrating Powers of Radiation

Paper Aluminum Lead Concrete

Alpha

Radiation Source

Beta

Gamma

Neutron

Alpha
Stopped by Skin

Beta
Stopped by Clothes

Gamma

Neutron

Source: *Modified from U.S. Environmental Protection Agency*

Figure 22.25 The penetrating powers of alpha and beta particles, gamma rays, and neutrons. Gamma rays and neutrons are the most difficult to protect against.

source of gamma rays in the environment is naturally occurring potassium-40. Common industrial gamma emitting sources include cobalt-60, iridium-192, and cesium-137. *Details:*

— Gamma rays can easily pass completely through the human body or be absorbed by tissue, thus constituting a radiation hazard for the entire body.

— Around two feet (about 0.6 m) of concrete or around two inches (about 50 mm) of lead may be required to stop the more energetic gamma rays. Standard fire fighting clothing provides *no* protection against gamma rays (Figure 22.25).

- *Neutron* — Ultrahigh energy particles that have a physical mass like alpha or beta radiation but have no electrical charge. Neutrons are highly penetrating. Fission reactions produce neutrons along with gamma radiation. Neutron radiation is difficult to measure in the field and is usually estimated based on gamma measurements. *Details:*

— Soil moisture density gauges, often used at construction sites, are a common source of neutron radiation. Neutrons may also be encountered in research

What's the Difference between Exposure to Radioactive Material and Radioactive Contamination?

Radioactive *contamination* occurs when a material that contains radioactive atoms is deposited on surfaces, skin, clothing, or any place where it is not desired. It is important to remember that radiation does *not* spread; rather, it is radioactive material/contamination that spreads. *Exposure* to radiation occurs when a person is near a radiation source and is exposed to the energy from that source. Exposure to radiation alone does *not* contaminate a person. Contamination can occur when a person comes in contact with radioactive material. A person can become contaminated externally, internally, or both. Radioactive material can enter the body via one or more routes of entry (see Routes of Entry section). An unprotected person contaminated with radioactive material receives radiation exposure until the source of radiation (radioactive material) is removed. Note the following situations:

- A person is *externally* contaminated (and receives external exposure) when radioactive material is on the skin or clothing.

- A person is *internally* contaminated (and receives internal exposure) when radioactive material is breathed, swallowed, or absorbed through wounds.

- The *environment* is contaminated when radioactive material is spread about or is unconfined. Environmental contamination is another potential source of external exposure.

Exposure to radioactive material is similar to exposure to sunlight; once you remove yourself from the source, you are no longer exposed. Radioactive contamination occurs when you get radioactive material on you or in you. Having the material on you makes it so you can't "remove yourself from the source" so you get exposure from the radioactive material that is on your body. Another way of thinking of it is to visualize a glow stick. The material inside the glow stick is the radioactive material. The light emitted from the glow stick is the radiation. If you are around the glow stick, you are exposed to the light. If you break open the glow stick and get the liquid (radioactive material) on you, you are contaminated. The material on your skin continues to give off the light—exposing you to the radiation (light).

laboratories or operating nuclear power plants.

— The health hazard that neutrons present arises from the fact that they cause the release of secondary radiation when they interact with the human body.

Radiation Health Hazards

The effects of ionizing radiation occur at the cellular level. The human body is composed of many organs, and each organ of the body is composed of specialized cells. Ionizing radiation can affect the normal operation of these cells.

Radiation may cause damage to any material by ionizing the atoms in that material—changing the material's atomic structure. When atoms are ionized, the chemical properties of those atoms are altered. Radiation can damage a cell by ionizing the atoms and changing the resulting chemical behavior of the atoms and/or molecules in the cell. If a person receives a sufficiently high dose of radiation and many cells are damaged, in rare cases there may be noticeable (observable) health effects including genetic mutations and cancer.

The biological effects of ionizing radiation depend on how much and how fast a radiation dose is received. There are two categories of radiation doses: acute and chronic radiation doses. **Table 22.2** shows the health effects associated with various doses of radiation.

Acute doses. Exposure to a large dose of radiation received in a short period of time is an acute dose. Some possible health effects from acute doses of radiation include reduced blood count, hair loss, nausea, vomiting, diarrhea, and fatigue. Extremely high levels of acute radiation exposure (such as those received by victims of a nuclear bomb) can result in death within a few hours, days, or weeks.

Chronic doses. Small amounts of radiation received over a long period of time are a chronic dose. The body is better equipped to handle a chronic dose of radiation than it is an acute radiation dose because the body has enough time to replace dead or nonfunctioning cells with healthy ones. Chronic doses do not result in the detectable health effects seen with acute doses. However, there is some concern that chronic exposure to radiation may cause cancer. Examples of chronic radiation doses include the everyday doses received from natural background radiation and those received by workers in nuclear and medical facilities.

NOTE: The exposures likely to be encountered by firefighters at most haz mat incidents are <u>very unlikely</u> to cause any health effects, especially if proper precautions are taken (Figure 22.26). Even at terrorist incidents, it is unlikely that firefighters will encounter dangerous or lethal doses of radiation. However, it is important to monitor for radiation at any incident involving explosions or suspected terrorist attacks.

Radiation Protection Strategies

Time, distance, and shielding are the three recognized ways to provide protection from external radiation during an emergency. These three protection principles are explained as follows:

- *Time* — The amount of radiation exposure increases or decreases according to the time spent near the source of radiation. The shorter the exposure time, the smaller the total radiation dose.

- *Distance* — The farther from the source, the less the exposure. Generally, doubling

Table 22.2
Health Effects of High Radiation Doses

Exposure (rem)*	Health Effect (cumulative)	Time to Onset
5 to 10**	• Radiation burns (more severe as exposure increases) • Changes in blood chemistry	
50	Nausea	Hours
55	Fatigue	
70	Vomiting	
75	Hair Loss	2 to 3 weeks
90	Diarrhea	
100	Hemorrhage	
400	Death from fatal doses	Within two months
1,000	• Destruction of intestinal lining • Internal bleeding • Death	1 to 2 weeks
2,000	• Damage to central nervous system • Loss of consciousness • Death	Minutes Hours to days

Source: U.S. Environmental Protection Agency, 2003.

*Roentgen equivalent man (rem) is a term used to express the absorbed dose equivalence as pertaining to a human body; applied to all types of radiation. This unit takes into account the energy absorbed (dose) and the biological effect on the body due to different types of radiation. Rem is used to set dose limits for emergency responders.

**Most people receive about 3 tenths of a rem (300 mrem) every year from natural background sources of radiation (mostly radon). The levels of radiation presented in this table are extremely high, and they are very unlikely to be encountered by emergency responders at radiological events. These levels of radiation might be present at nuclear events (such as a nuclear attack).

the distance from a gamma source reduces the exposure by a factor of four. Halving the distance increases the exposure by a factor of four. Since alpha and beta particles don't travel very far, distance is a prime concern when dealing with gamma rays.

- *Shielding* — Certain materials such as lead, earth, concrete, and water prevent penetration of some types of radiation. The thickness of the shielding required to stop the passage of the radiation depends on the type of shielding material

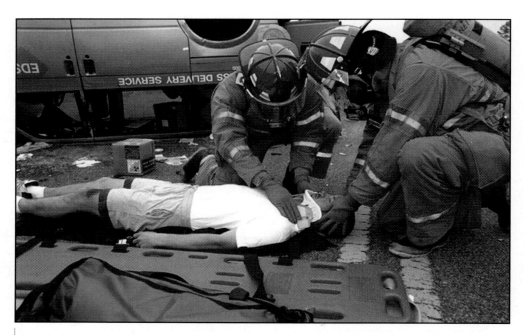

Figure 22.26 Radiation at emergency incidents is very unlikely to cause serious health effects as long as firefighters take proper precautions such as wearing their PPE. *Courtesy of Tom Clawson, Technical Resources Group, Inc.*

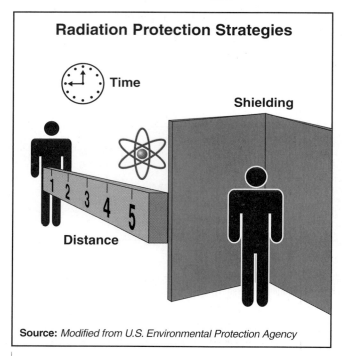

Radiation Protection Strategies

Time

Shielding

Distance

Source: *Modified from U.S. Environmental Protection Agency*

Figure 22.27 To reduce exposure to radiation, firefighters should use the protection strategies of time, distance, and shielding.

used, the type of radiation, and the distance from the source (**Figure 22.27**).

Asphyxiation Hazards

Asphyxiants are substances that affect the oxygenation of the body and generally lead to suffocation. Asphyxiants can be divided into two classes: simple and chemical. *Simple asphyxiants* are gases that displace the oxygen necessary for breathing. These gases dilute or displace the oxygen concentration below the level required by the human body. *Chemical asphyxiants* are substances that prohibit the body from using oxygen, and some of these chemicals may be used by terrorists for an attack. Even though oxygen is available, these substances starve the cells of the body for oxygen.

Chemical Hazards

Exposure to hazardous chemicals may produce a wide range of adverse health effects. The likelihood of an adverse health effect occurring and the severity of the effect depend on the following factors:

- Toxicity of the chemical

- Pathway or route of exposure

- Nature and extent of exposure

- Factors that affect the susceptibility of the exposed person such as age and the presence of certain chronic diseases

The following sections discuss the routes of entry that chemicals use to get into the body, toxicity and exposure limits, types of hazardous chemicals or substances (such as irritants, convulsants, corrosives, carcinogens, and sensitizers/allergens), and signs and symptoms of chemical exposure.

Routes of Entry

There are four main routes through which hazardous materials can enter the body and cause harm:

- *Inhalation* — Process of taking in materials by breathing through the nose or mouth. Hazardous vapors, smoke, gases, liquid aerosols, fumes, and suspended dusts may be inhaled into the body. When a hazardous material presents an inhalation threat, respiratory protection is required.

- *Ingestion* — Process of taking in materials through the mouth by means other than simple inhalation. Taking a pill is a simple example of how a chemical might be deliberately ingested.

- *Contact* — Occurrence when a chemical or hazardous material (in any state—solid, liquid, or gas) contacts the skin or exposed surface of the body. Contact with an acid, for example, can severely damage the skin.

- *Absorption* — Process of taking in hazardous materials through the skin or eyes. Some materials pass easily through the mucous membranes or areas of the body where the skin is the thinnest, allowing the least resistance to penetration. The eyes, nose, mouth, wrists, neck, ears, hands, groin, and underarms are areas of particular concern. Many poisons are easily absorbed into the body system in this manner. Absorption also includes taking in materials via punctures (such as with a contaminated needle) or tears in the skin.

Many hazardous materials have multiple routes of entry **(Figure 22.28)**. PPE is designed to shield the various routes of entry.

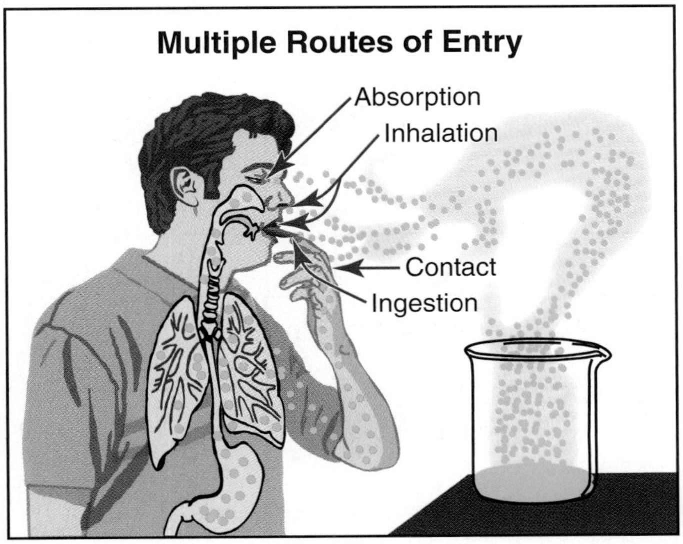

Multiple Routes of Entry

Absorption
Inhalation
Contact
Ingestion

Figure 22.28 Chemicals often have multiple routes of entry. Firefighters must ensure that all routes are protected when the potential for exposure exists.

Poisons/Toxic Chemicals

Toxic (poisonous) chemicals often produce injuries at the site where they come into contact with the body. A chemical injury at the site of contact (typically the skin and mucous membranes of the eyes, nose, mouth, or respiratory tract) is termed a *local toxic effect*. Irritant gases such as chlorine and ammonia can, for example, produce a localized toxic effect in the respiratory tract, while corrosive acids and bases can result in local damage to the skin.

In addition, a toxic chemical may be absorbed into the bloodstream and distributed to other parts of the body, producing systemic effects. Many pesticides, for example, absorb through the skin, distribute to other sites in the body, and produce adverse effects such as seizures or cardiac, pulmonary, or other problems.

Systemic Effect — Something that affects an entire system rather than a single location or entity.

Many toxins have fast-acting, acute toxic effects while others may have chronic effects that aren't manifested for many years. Exposure to chemical compounds can result not only in the development of a single systemic effect but also in the development of multiple systemic effects or a combination of systemic and local effects. Some of these effects may be delayed.

Exposures to poisons can cause damage to organs or other parts of the body and may even cause death. Types of toxins, their target organs, and chemical examples are given in **Table 22.3**.

Table 22.3
Types of Toxins and Their Target Organs

Toxin	Target Organ	Chemical Examples
Nephrotoxins	Kidney	Halogenated Hydrocarbons, Mercury, Carbon Tetrachloride
Hemotoxins	Blood	Carbon Monoxide, Cyanides, Benzene, Nitrates, Arsine, Naphthalene, Cocaine
Neurotoxins	Nervous System	Organophosphates, Mercury, Carbon Disulphide, Carbon Monoxide, Sarin
Hepatoxins	Liver	Alcohol, Carbon Tetrachloride, Trichloroethylene, Vinyl Chloride, Chlorinated HC
Immunotoxins	Immune System	Benzene, Polybrominated Biphenyls (PBBs), Polychlorinated Biphenyls (PCBs), Dioxins, Dieldrin
Endocrine Toxins	Endocrine System (including the pituitary, hypothalamus, thyroid adrenals, pancreas, thymus, ovaries, and testes)	Benzene, Cadmium, Chlordane, Chloroform, Ethanol, Kerosene, Iodine, Parathion
Musculoskeletal Toxins	Muscles/Bones	Fluorides, Sulfuric Acid, Phosphine
Respiratory Toxins	Lungs Acid, Chlorine	Hydrogen Sulfide, Xylene, Ammonia, Boric
Cutaneous Hazards	Skin	Gasoline, Xylene, Ketones, Chlorinated Compounds
Eye Hazards	Eyes	Organic Solvents, Corrosives, Acids
Mutagens	DNA	Aluminum Chloride, Beryllium, Dioxins
Teratogens	Embryo/Fetus	Lead, Lead Compounds, Benzene
Carcinogens	All	Tobacco Smoke, Benzene, Arsenic, Radon, Vinyl Chloride

Measuring Toxicity

Poisons and the measurements of their toxicity are often expressed in terms of *lethal dose* (LD) for amounts ingested and *lethal concentration* (LC) for amounts inhaled. These terms are often found on Material Safety Data Sheets (MSDS, see next section) and are used to express permissible exposure limits (PELs, average concentrations of materials to which workers can be exposed safely over a given period of time). As a general rule, the smaller the value presented, the more toxic the substance is. These values are normally established by testing the effects of exposure on animals (rats or rabbits) under laboratory conditions over a set period of time. Dose terms are defined as follows:

- **Lethal dose (LD)** — Minimum amount of solid or liquid that when ingested, absorbed, or injected through the skin will cause death. Sometimes the lethal dose is expressed in conjunction with a percentage such as LD_{50} (most common) or LD_{100}. The number refers to the percentage of an animal test group that was killed by the listed dose (usually administered orally).

- **Median lethal dose (LD_{50})** — Statistically derived single dose of a substance that can be expected to cause death in 50 percent of animals when administered by the oral route. The LD_{50} value is expressed in terms of weight of test substance per unit weight of test animal (mg/kg) (see information box). The term LD_{50} does not mean that the other half of the subjects are healthy. They may be very sick, but only half will actually die.

- **Lethal dose low (LD_{LO} or LDL)** — Lowest administered dose of a material capable of killing a specified test species.

Concentration terms are defined as follows:

- **Lethal concentration (LC)** — Minimum concentration of an inhaled substance in the gaseous state that will be fatal to the test group (usually within 1 to 4 hours). Like lethal dose, the lethal concentration is often expressed as LC_{50}, indicating that half of the test group was killed by concentrations at the listed value. LC may be expressed in the following ways:

  — Parts per million (ppm) **(Figure 22.29)**

  — Milligrams per cubic meter (mg/m³)

  — Micrograms of material per liter of air (μg/L)

  — Milligrams per liter (mg/L)

- **Lethal concentration low (LC_{LO} or LCL)** — Lowest concentration of a gas or vapor capable of killing a specified species over a specified time.

Lethal Dose — Concentration of an ingested or injected substance that results in the death of a certain percentage of the test population; the lower the dose the more toxic the substance; an oral or dermal exposure expressed in milligrams per kilogram (mg/kg).

Lethal Concentration — Concentration of an inhaled substance that results in the death of a certain percentage of the test population; the lower the value the more toxic the substance; an inhalation exposure expressed in parts per million (ppm), milligrams per liter (mg/liter), or milligrams per cubic meter (mg/m³).

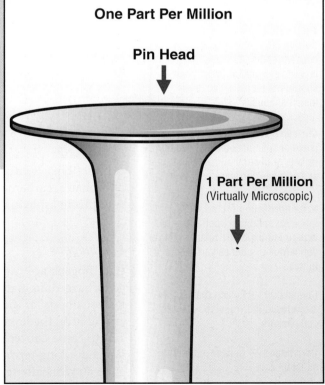

One Part Per Million

Pin Head

1 Part Per Million
(Virtually Microscopic)

Figure 22.29 Exposure limits are usually expressed in parts per million. Essentially, that translates to one molecule of the substance per million molecules of air. In terms of visibility to the human eye, a single part per million of most hazardous materials will be microscopic. Some exposure limits will even be a percentage of a single part per million.

Dose — Quantity of a chemical material ingested or absorbed through skin contact for purposes of measuring toxicity.

Concentration — (1) Quantity of a chemical material inhaled for purposes of measuring toxicity.

Threshold Limit Value (TLV®) — Concentration of a given material in parts per million (ppm) that may be tolerated for an 8-hour exposure during a regular workweek without ill effects.

Threshold Limit Value/Short-Term Exposure Limit (TLV®/STEL) — Fifteen-minute time-weighted average exposure that should not be exceeded at any time nor repeated more than four times daily with a 60-minute rest period required between each STEL exposure. These short-term exposures can be tolerated without suffering from irritation, chronic or irreversible tissue damage, or narcosis of a sufficient degree to increase the likelihood of accidental injury, impair self-rescue, or materially reduce worker efficiency. TLV/STELs are expressed in parts per million (ppm) and milligrams per cubic meter (mg/m³).

Threshold Limit Value/Ceiling (TLV®/C) — Maximum concentration of a given material in parts per million (ppm) that should not be exceeded, even instantaneously.

Corrosive Material — Gaseous, liquid or solid material that can burn, irritate, or destroy human skin tissue and severely corrode steel.

Irritant/Irritating Material — Liquid or solid that upon contact with fire or exposure to air emits dangerous or intensely irritating fumes.

Convulsant — Poison that causes an exposed individual to have convulsions.

Carcinogen — Cancer-producing substance.

The methods by which poisons attack the body vary depending on the type of poison. Irritants and chemical asphyxiants interfere with oxygen flow to the lungs and the blood. Nerve poisons act on the body's nervous system by disrupting nerve impulses. The effect produced by a toxic compound is primarily a function of the dose (amount of a substance ingested or administered through skin contact) and the concentration (amount of the substance inhaled in this context) of the compound. This principle, termed the *dose-response relationship*, is a key concept in toxicology. Many factors affect the normal dose-response relationship, but typically, as the dose increases, the severity of the toxic response increases. *Highly toxic materials* are deadly or very dangerous in very small doses or concentration.

Established measurements of toxicity may be used to set exposure limits to certain chemicals. Exposure limits may be expressed in terms such as threshold limit value (TLV®), short-term exposure limit (STEL), threshold limit value-ceiling (TLV®/C), and *permissible exposure limit (PEL)*. Concentrations that are high enough to kill or cause serious injury or illness are expressed in terms of *immediately dangerous to life or health* (IDLH). **Table 22.4** provides definitions and exposure periods for the various limits as well as the organizations responsible for establishing them.

Corrosives

[NFPA® 472: 5.2.3(1)(a)(iii), 5.2.3(8)(e)]

Chemicals that destroy or burn living tissues and have destructive effects by virtue of their *corrosivity* (ability to cause corrosion, particularly to metals) are often called corrosives. Corrosives are commonly divided into two broad categories: acids and bases (bases are sometimes called alkalis or caustics). However, it is important to note that some corrosives (such as hydrogen peroxide) are neither acids nor bases. The corrosivity of acids and bases is often measured or expressed in terms of pH (**Figure 22.30** on page 1100).

Irritants

Irritants are toxins that cause temporary but sometimes severe inflammation to the eyes, skin, or respiratory system. Irritants often attack the mucous membranes of the body such as the surfaces of the eyes, nose, mouth, throat, and lungs.

Convulsants

Convulsants are toxic materials that can cause convulsions (involuntary muscle contractions). Some chemicals considered to be convulsants are strychnine, organophosphates, carbamates, and infrequently used drugs such as picrotoxin. Death can result from asphyxiation or exhaustion.

Carcinogens

Carcinogens are cancer-causing agents. Examples of known or suspected carcinogenic hazardous materials are polyvinyl chloride, benzene, asbestos, some chlorinated hydrocarbons, arsenic, nickel, some pesticides, and many plastics. Exact data is not available on the level and duration or dose of exposure needed for individual chemicals to cause cancer. However, exposures to only small amounts of some substances may have long-term consequences. Disease and complications can occur as long as 10 to 40 years after exposure.

Table 22.4
Exposure Limits

Term	Definition	Exposure Period	Organization
IDLH Immediately Dangerous to Life or Health	An atmospheric concentration of any toxic, corrosive, or asphyxiating substance that poses an immediate threat to life. It can cause irreversible or delayed adverse health effects and interfere with the individual's ability to escape from a dangerous atmosphere.*	Immediate (This limit represents the maximum concentration from which an unprotected person can expect to escape in a 30-minute period of time without suffering irreversible health effects.)	**NIOSH** National Institute for Occupational Safety and Health
IDLH Immediately Dangerous to Life or Health	An atmosphere that poses an immediate threat to life, would cause irreversible adverse health effects, or would impair an individual's ability to escape from a dangerous atmosphere.	Immediate	**OSHA** Occupational Safety and Health Administration
LOC Levels of Concern	10% of the IDLH		
PEL Permissible Exposure Limit**	A regulatory limit on the amount or concentration of a substance in the air. PELs may also contain a skin designation.	8-hours Time-Weighted Average (TWA)*** (unless otherwise noted)	**OSHA** Occupational Safety and Health Administration
	The PEL is the maximum concentration to which the majority of healthy adults can be exposed over a 40-hour workweek without suffering adverse effects.		
PEL (C) PEL Ceiling Limit	The maximum concentration to which an employee may be exposed at any time, even instantaneously.	Instantaneous	**OSHA** Occupational Safety and Health Administration
STEL Short-Term Exposure Limit	The maximum concentration allowed for a 15-minute exposure period.	15 minutes (TWA)	**OSHA** Occupational Safety and Health Administration

Continued

Table 22.4 (continued)
Exposure Limits

Term	Definition	Exposure Period	Organization
TLV® Threshold Limit Value†	An occupational exposure value recommended by ACGIH® to which it is believed nearly all workers can be exposed day after day for a working lifetime without ill effect.	Lifetime	**ACGIH®** American Conference of Governmental Industrial Hygienists
TLV®-TWA Threshold Limit Value-Time-Weighted Average	The allowable time-weighted average concentration.	8-hour day or 40-hour workweek (TWA)	**ACGIH®** American Conference of Governmental Industrial Hygienists
TLV®-STEL Threshold Limit Value-Short-Term Exposure Limit	The maximum concentration for a continuous 15-minute exposure period (maximum of four such periods per day, with at least 60 minutes between exposure periods, provided the daily TLV®-TWA is not exceeded).	15 minutes (TWA)	**ACGIH®** American Conference of Governmental Industrial Hygienists
TLV®-C Threshold Limit Value-Ceiling	The concentration that should not be exceeded even instantaneously.	Instantaneous	**ACGIH®** American Conference of Governmental Industrial Hygienists
BEIs® Biological Exposure Indices	A guidance value recommended for assessing biological monitoring results.		**ACGIH®** American Conference of Governmental Industrial Hygienists
REL Recommended Exposure Limit	A recommended exposure limit made by NIOSH.	10-hours (TWA) ††	**NIOSH** National Institute for Occupational Safety and Health
AEGL-1 Acute Exposure Guideline Level-1	The airborne concentration of a substance at or above which it is predicted that the general population, including "susceptible" but excluding "hypersusceptible" individuals, could experience notable discomfort. †††	Multiple exposure periods: 10 minutes 30 minutes 1 hour 4 hours 8 hours	**EPA** Environmental Protection Agency

Continued

Table 22.4 (continued)
Exposure Limits

Term	Definition	Exposure Period	Organization
AEGL-2 Acute Exposure Guideline Level-2	The airborne concentration of a substance at or above which it is predicted that the general population, including "susceptible" but excluding "hypersusceptible" individuals, could experience irreversible or other serious, long-lasting effects or impaired ability to escape. Airborne concentrations below AEGL-2 but at or above AEGL-1 represent exposure levels that may cause notable discomfort.	Multiple exposure periods: 10 minutes 30 minutes 1 hour 4 hours 8 hours	**EPA** Environmental Protection Agency
AEGL-3 Acute Exposure Guideline Level-3	The airborne concentration of a substance at or above which it is predicted that the general population, including "susceptible" but excluding "hypersusceptible" individuals, could experience life-threatening effects or death. Airborne concentrations below AEGL-3 but at or above AEGL-2 represent exposure levels that may cause irreversible or other serious, long-lasting effects or impaired ability to escape.	Multiple exposure periods: 10 minutes 30 minutes 1 hour 4 hours 8 hours	**EPA** Environmental Protection Agency
ERPG-1 Emergency Response Planning Guideline Level 1	The maximum airborne concentration below which it is believed nearly all individuals could be exposed for up to one hour without experiencing other than mild transient adverse health effects or perceiving a clearly defined objectionable odor.	Up to 1 hour	**AIHA** American Industrial Hygiene Association

Continued

Table 22.4 (continued)
Exposure Limits

Term	Definition	Exposure Period	Organization
ERPG-2 Emergency Response Planning Guideline Level 2	The maximum airborne concentration below which it is believed nearly all individuals could be exposed for up to one hour without experiencing or developing irreversible or other serious health effects or symptoms that could impair an individual's ability to take protective action.	Up to 1 hour	**AIHA** American Industrial Hygiene Association
ERPG-3 Emergency Response Planning Guideline Level 3	The maximum airborne concentration below which it is believed nearly all individuals could be exposed without experiencing or developing life-threatening health effects.	Up to 1 hour	**AIHA** American Industrial Hygiene Association
TEEL-0 Temporary Emergency Exposure Limits Level 0	The threshold concentration below which most people will experience no appreciable risk of health effects.		**DOE** Department of Energy
TEEL-1 Temporary Emergency Exposure Limits Level 1	The maximum concentration in air below which it is believed nearly all individuals could be exposed without experiencing other than mild transient adverse health effects or perceiving a clearly defined objectionable odor.		**DOE** Department of Energy
TEEL-2 Temporary Emergency Exposure Limits Level 2	The maximum concentration in air below which it is believed nearly all individuals could be exposed without experiencing or developing irreversible or other serious health effects or symptoms that could impair their abilities to take protective action.		**DOE** Department of Energy

Continued

Table 22.4 (continued)
Exposure Limits

Term	Definition	Exposure Period	Organization
Teel-3 Temporary Emergency Exposure Limits Level 3	The maximum concentration in air below which it is believed nearly all individuals could be exposed without experiencing or developing life-threatening health effects.		**DOE** Department of Energy

*It should be noted that the NIOSH definition only addresses airborne concentrations. It does not include direct contact with liquids or other materials.

**PELs are issued in Title 29 *CFR* 1910.1000, particularly Tables Z-1, Z-2, and Z-3, and are enforceable as law.

***Time-weighted average means that changing concentration levels can be averaged over a given period of time to reach an average level of exposure.

†TLVs® and BEIs® are guidelines for use by industrial hygienists in making decisions regarding safe levels of exposure. They are not considered to be consensus standards by the ACGIH®, and they do not carry the force of law unless they are officially adopted as such by a particular jurisdiction.

††NIOSH may also list STELs (15-minute TWA) and ceiling limits.

†††Airborne concentrations below AEGL-1 represent exposure levels that could produce mild odor, taste, or other sensory irritation.

Sensitizers/Allergens

Allergens are substances that cause allergic reactions in people or animals. *Sensitizers* are chemicals that cause a substantial proportion of exposed people or animals to develop an allergic reaction after repeated exposure to the chemical. Common examples of sensitizers and allergens include latex, bleach, and urushiol (the chemical found in the sap of poison ivy, oak, and sumac). Some individuals exposed to a material may not be abnormally affected at first but may experience significant and dangerous effects when exposed to the material again.

Allergen — Material that can cause an allergic reaction of the skin or respiratory system.

Etiological/Biological Hazards

Etiological (or *biological*) hazards are microorganisms such as viruses or bacteria (or their toxins) that may cause severe, disabling disease or illness.

- *Viral agents* — Viruses are the simplest types of microorganisms that can only replicate themselves in the living cells of their hosts. Viruses do not respond to antibiotics.

- *Bacterial agents* — Bacteria are microscopic, single-celled organisms. Most bacteria do not cause disease in people, but when they do, two different mechanisms are possible: invading the tissues or producing poisons (toxins).

- *Rickettsias* — Rickettsias are specialized bacteria that live and multiply in the gastrointestinal tract of arthropod carriers (such as ticks and fleas). They are smaller than most bacteria, but larger than viruses. They have properties that are similar to

pH Scale

Concentration of Hydrogen Ions Compared to Distilled Water	pH	Examples of Solutions at this pH
10,000,000	0	Strong Hydrofluoric Acid
1,000,000	1	Hydrochloric Acid, Battery Acid
100,000	2	Lemon Juice, Vinegar
10,000	3	Apple Juice, Orange Juice
1,000	4	Acid rain, Wine
100	5	Black Coffee
10	6	Milk, Saliva
1	7	*Pure* Water, Human Blood
1/10	8	Seawater
1/100	9	Baking Soda
1/1,000	10	Milk of Magnesia
1/10,000	11	Ammonia
1/100,000	12	Lime
1/1,000,000	13	Lye
1/10,000,000	14	Sodium Hydroxide

Figure 22.30 The pH of a solution is a measurement of the hydrogen ions in a solution, indicating its strength. The higher the pH of bases, or the lower the pH of acids, the stronger the base or acid is, respectively. The pH scale ranges from 0 to 14. A pH of 7 is neutral. A pH less than 7 is acidic, and a pH greater than 7 is basic.

both. Like bacteria, they are single-celled organisms with their own metabolisms, and they are susceptible to broad-spectrum antibiotics. However, like viruses, they only grow in living cells. Most rickettsias are spread only through the bite of infected arthropods (such as ticks) and not through human contact.

- *Biological toxins* — Biological toxins are poisons produced by living organisms; however, the biological organism itself is usually not harmful to people. Some biological toxins have been manufactured synthetically and/or genetically altered in laboratories for purposes of biological warfare.

Infectious diseases are caused by the growth of microorganisms in the body and may or may not be contagious. (If a disease is contagious, it is readily communicable or easily transmitted from person to person.) Additionally, some biological hazards cause illness through their toxicity.

Examples of diseases associated with etiological hazards are malaria, acquired immunodeficiency syndrome (AIDS), tuberculosis, and typhoid. Exposure to biological hazards may occur in biological and medical laboratories or when dealing with people who are carriers of such diseases. Most of these diseases are carried in body fluids and are transmitted by contact with the fluids.

Firefighters may also be exposed to biological agents used as weapons in terrorist attacks and criminal activities. Biological weapons take the form of these disease-causing organisms and/or their toxins. Examples of potential biological weapons include:

- Smallpox (virus)
- Anthrax (bacteria)
- Botulism (toxin from the bacteria *Clostridium botulinum*)

Biological attacks could also be used to produce death and disease in animals and plants. The 2001 anthrax attacks in the United States were an example of biological attacks.

Mechanical Hazards

Mechanical hazards can cause trauma that occurs as a result of direct contact with an object. The two most common types are *striking* and *friction* exposures. Trauma can be mild, moderate, or severe and can occur in a single event.

In hazardous materials situations, a striking injury could be the result of an explosion caused by the failure of a pressurized container, a bomb (or improvised explosive device [IED]), or the reactivity of the hazardous material itself **(Figure 22.31)**. Friction injuries occur as a result of portions of the body rubbing against an abrasive surface, causing raw skin (abrasions), blisters, and burns. In hazardous material situations, contact between protective clothing and the skin sometimes causes friction injuries.

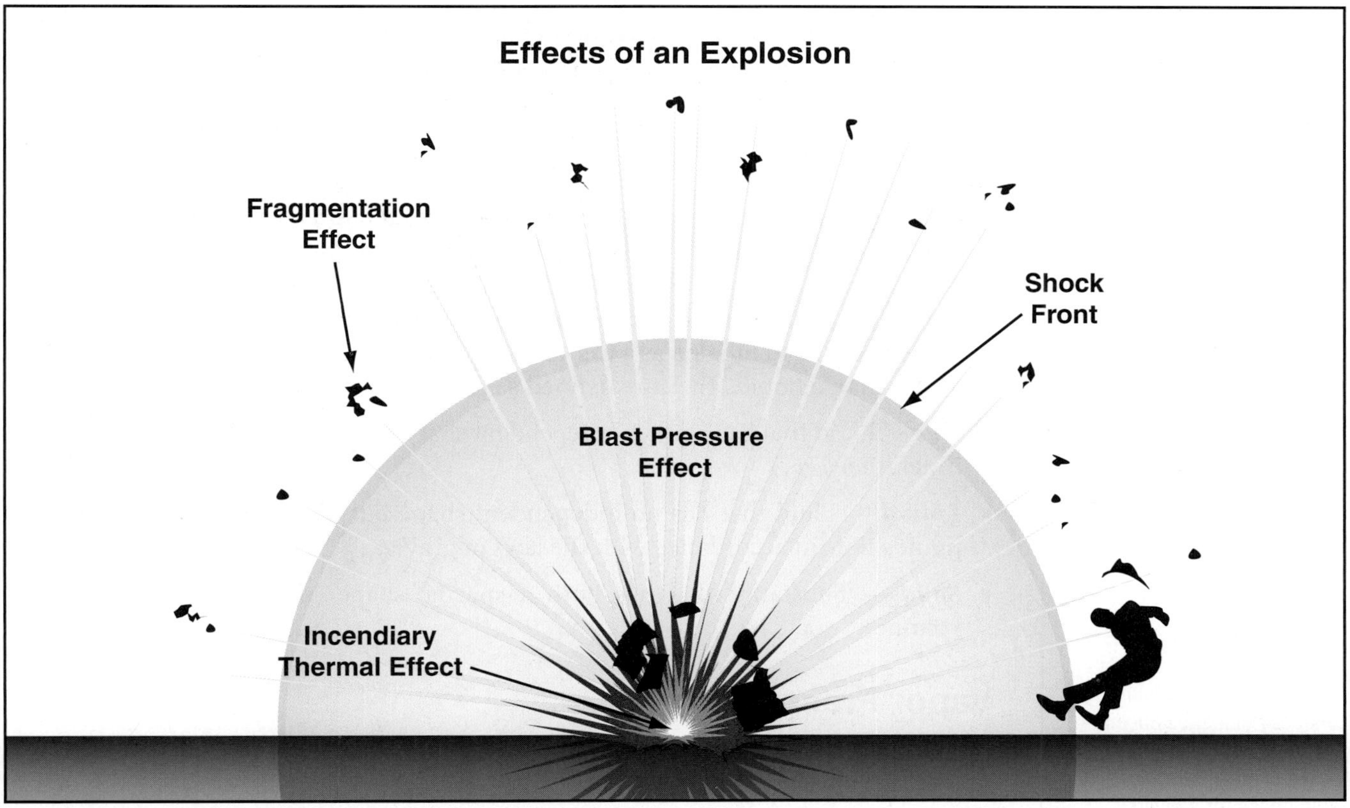

Improvised Explosive Device (IED) — Device that is categorized by its container and the way it is initiated; usually homemade, constructed for a specific target, and contained in almost anything.

Effects of an Explosion

Fragmentation Effect

Shock Front

Blast Pressure Effect

Incendiary Thermal Effect

Figure 22.31 Explosions can cause mechanical injuries.

An explosion can cause the following four hazards (three mechanical and one thermal):

- *Blast pressure wave (shock wave)* — Gases being released rapidly create a shock wave that travels outward from the center. As the wave increases in distance, the strength decreases. This blast pressure wave is the primary reason for injuries and damage.

- *Shrapnel fragmentation* — Small pieces of debris thrown from a container or structure that ruptures from containment or restricted blast pressure. Shrapnel may be thrown over a wide area and great distances (fragmentation), causing personal injury and other types of damage to surrounding structures or objects. Shrapnel can result in bruises, punctures, or even avulsions (part of the body being torn away) when the container or pieces of the container strike a person.

- *Seismic effect* — Earth vibration similar to an earthquake. When a blast occurs at or near ground level, the air blast creates a ground shock or crater. As shock waves move across or underground, a seismic disturbance is formed. The distance the shock wave travels depends on the type and size of the explosion and type of soil.

- *Incendiary thermal effect* — During an explosion, thermal heat energy in the form of a fireball is the result of burning combustible gases or flammable vapors and ambient air at very high temperatures. The thermal heat fireball is present for a limited time after the explosive event.

Properties and Behavior

The potential outcomes associated with haz mat incidents are often determined by the behavior of the materials involved. This behavior is a function of the material's physical properties. The material's physical state, vapor pressure, boiling point, chemical reactivity, and other properties affect how it behaves, determine the harm it can cause, and influence the effect it may have on its container (as well as the people, living organisms, and other chemicals it may contact).

Physical State

Everything in the world is made of substances called *matter*. Matter is found in three physical states: *gas, liquid,* and *solid*. Hazardous materials may come in any one of these states and each behaves differently **(Figure 22.32)**. This behavior in turn can influence the nature of the hazards presented by the material. For example, a liquid is not as likely to be inhaled as a gas, and a gas might travel much more quickly—and farther—than a liquid. They are defined as follows:

- *Gas* — Fluid that has neither independent shape nor volume; gases tend to expand indefinitely.

- *Liquid* — Fluid that has no independent shape but does have a specific volume; liquids flow in accordance with the laws of gravity.

- *Solid* — Substance that has both a specific shape (without a container) and volume.

Vapor Pressure

Vapor Pressure — Measure of the tendency of a substance to evaporate; pressure at which a vapor is in equilibrium with its liquid phase for a given temperature.

Vapor pressure is the pressure exerted by a saturated vapor above its own liquid in a closed container. More simply, it is the pressure produced or exerted by the vapors released by a liquid. Vapor pressure can be viewed as the measure of the tendency of a substance to evaporate.

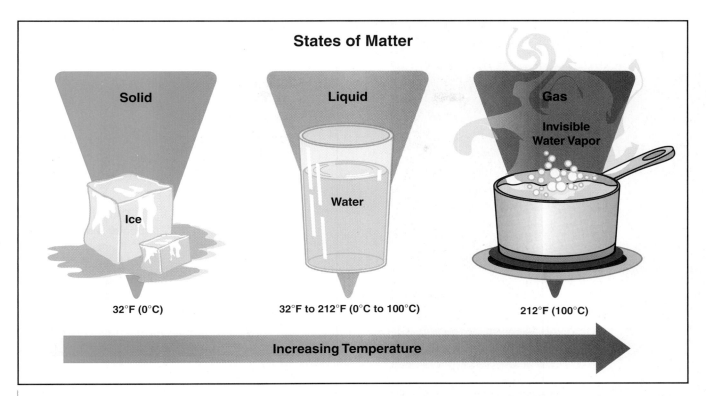

States of Matter

Solid	Liquid	Gas
Ice	Water	Invisible Water Vapor
32°F (0°C)	32°F to 212°F (0°C to 100°C)	212°F (100°C)

Increasing Temperature

Figure 22.32 The state of a material (solid, liquid, or gas) influences the way it behaves. Firefighters must be prepared to deal with hazardous materials in all three states.

Boiling Point

Boiling point is the temperature at which the vapor pressure of a liquid is equal to or greater than atmospheric pressure. In other words, it is the temperature at which a liquid changes to a gas at a given pressure. The boiling point is usually expressed in degrees Fahrenheit (Celsius) at sea level air pressure. For mixtures, the initial boiling point or boiling-point range may be given. Flammable materials with low boiling points generally present special fire hazards.

A *boiling liquid expanding vapor explosion (BLEVE)* (also called *violent rupture*) can occur when a liquid within a container is heated, causing the material inside to boil or vaporize (such as in the case of a liquefied petroleum gas tank exposed to a fire). If the resulting increase in internal vapor pressure exceeds the vessel's ability to relieve the excess pressure, it can cause the container to fail catastrophically **(Figure 22.33)**. As the vapor is released, it expands rapidly and ignites, sending flames and pieces of tank flying in a tremendous explosion. BLEVEs most commonly occur when flames contact a tank shell above the liquid level or when insufficient water is applied to keep a tank shell cool.

Vapor Density

Vapor density is the weight of a given volume of pure vapor or gas compared to the weight of an equal volume of dry air at the same temperature and pressure **(Figure 22.34)**. A vapor density less than 1 indicates a vapor lighter than air, while a vapor density greater than 1 indicates a vapor heavier than air. The majority of gases have a vapor density greater than 1. Examples of materials with a vapor density less than 1 include acetylene, methane (primary component of natural gas), and hydrogen.

Boiling Point — Temperature of a substance when the vapor pressure exceeds atmospheric pressure. At this temperature, the rate of evaporation exceeds the rate of condensation. At this point, more liquid is turning into gas than gas is turning back into a liquid.

Vapor Density — Weight of a given volume of pure vapor or gas compared to the weight of an equal volume of dry air at the same temperature and pressure. Vapor density less than 1 indicates a vapor lighter than air; vapor density greater than 1 indicates a vapor heavier than air.

Figure 22.33 When the internal vapor pressure of a container exceeds its ability to relieve the excess pressure, it can cause a boiling liquid expanding vapor explosion (BLEVE).

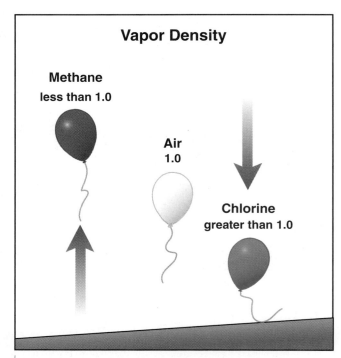

Vapor Density

Methane
less than 1.0

Air
1.0

Chlorine
greater than 1.0

Figure 22.34 Vapor densities compare the weight of a specific volume of air to an equal volume of another gas. Methane weighs less than air, so its specific vapor density is less than 1. Chlorine weighs more than air, so its vapor density is greater than 1.

All vapors and gases will mix with air, but the lighter materials tend to rise and dissipate (unless confined). Heavier vapors and gases are likely to concentrate in low places along or under floors; in sumps, sewers, and manholes; and in trenches and ditches where they may create fire or health hazards.

Solubility/Miscibility

Solubility in water is a term expressing the percentage of a material (by weight) that will dissolve in water at ambient temperature. Solubility information can be useful in determining spill cleanup methods and extinguishing agents. When a non-water-soluble liquid such as a *hydrocarbon* (gasoline, diesel fuel, pentane) is in the same container with water, the two liquids remain separate. When a water-soluble liquid such as a polar solvent (alcohol, methanol, methyl ethyl ketone [MEK]) is in the same container with water, the two liquids mix easily. A substance's solubility affects whether it mixes in water (**Figure 22.35**). *Miscibility* is the degree or readiness to which two or more gases or liquids are able to mix with or dissolve into each other.

Specific Gravity

Specific gravity is the ratio of the density (heaviness) of a material to the density of some standard material at standard conditions of pressure and temperature. The weight of a substance compared to the weight of an equal volume of water is an expression of the density of a material. For example, if a volume of a material weighs

Water Solubility — Ability of a liquid or solid to mix with or dissolve in water.

Polar Solvent Fuel — Flammable liquids that have an attraction for water, much like a positive magnetic pole attracts a negative pole.

Specific Gravity — Weight of a substance compared to the weight of an equal volume of water at a given temperature. Specific gravity less than 1 indicates a substance lighter than water; specific gravity greater than 1 indicates a substance heavier than water.

Figure 22.35 Water and fuel oil are immiscible, meaning they don't mix with each other. This fact can create a hazard because oil (which weighs less than water) rises to the surface and could ignite and burn on top of the water. In this example, the tank vessel *Haven*, wrecked and burned in 1991. *Courtesy of International Tanker Owners Pollution Federation Ltd., Houndsditch London.*

Persistence —Length of time a chemical agent remains effective without dispersing.

Reactivity — Ability of two or more chemicals to react and release energy and the ease with which this reaction takes place.

8 pounds (3.6 kg), and an equal volume of water weighs 10 pounds (4.5 kg), the material is said to have a specific gravity of 0.8. Materials with specific gravities less than 1 will float in (or on) water. Materials with specific gravities greater than 1 will sink in water.

Most (but not all) flammable liquids have specific gravities less than 1 and, if not soluble, will float on water (**Figure 22.36**). This fact is an important consideration for fire suppression activities.

Persistence

The persistence of a chemical is its ability to remain in the environment. Chemicals that remain in the environment for a long time are more persistent than chemicals that quickly dissipate or break down. For example, persistent nerve agents will remain effective at their point of dispersion for a much longer time than nonpersistent nerve agents.

Reactivity

The reactivity of a substance is its relative ability to undergo a chemical reaction with another material. Undesirable effects such as pressure buildup, temperature increase, and/or formation of noxious, toxic, or corrosive by-products may occur

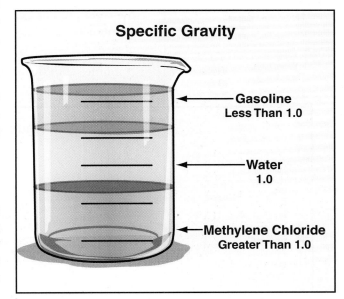

Specific Gravity

Gasoline
Less Than 1.0

Water
1.0

Methylene Chloride
Greater Than 1.0

Figure 22.36 In this example of specific gravity, equal volumes of water, gasoline, and methylene chloride do not weigh the same. Gasoline weighs less than water, so it has a specific gravity of less than 1. Methylene chloride weighs more than water, so it has a specific gravity greater than 1. In general, hydrocarbons like gasoline will float on water, and chlorinated solvents will sink.

as a result of a reaction. Substances referred to in the industrial world as *reactive* commonly react vigorously or violently with air, water, heat, light, each other, or other materials.

Many firefighters are familiar with the fire tetrahedron or the four elements necessary to produce combustion: oxygen, fuel, heat, and a chemical chain reaction. Fire is just one type of chemical reaction, and a *reactivity triangle* can be used to explain the basic components of many (though not all) chemical reactions: an oxidizing agent (oxygen), a reducing agent (fuel), and an activation energy source (often heat, but not always so) **(Figure 22.37)**.

All reactions require some energy to get them started, commonly referred to as activation energy. How much energy is needed depends on the particular reaction. The energy can be in the form of added heat from an external source. In some instances, radio waves, radiation, or another waveform of energy may provide the needed energy to the molecules. In other reactions, the energy could come from a shock or pressure change.

Reactions that have very low activation energies happen very easily or need very little help to begin the process. For example, materials that are generally classified as water-reactive typically react with water easily at room temperature simply because the heat being provided from the surroundings is sufficient to start the reaction. See **Table 22.5** for a summary of the different ways in which chemicals can be reactive. This table supplies the definition and chemical examples of nine reactive hazard classes.

The oxidizing agent in the reactivity triangle provides the oxygen necessary for the chemical reaction. Strong oxidizers are materials that encourage a strong reaction (by readily accepting electrons) from reducing agents (fuels). Many people are familiar with the concept that the greater the concentrations of oxygen present in the atmosphere, the hotter, faster, and brighter a fire will burn. The same principle applies to oxidation reactions – generally speaking, the stronger the oxidizer, the stronger the reaction. For example, many hydrocarbons (such as petroleum products) ignite spontaneously when they come into contact with a strong oxidizer.

The reducing agent in the fire tetrahedron acts as the fuel source for the reaction. It combines with the oxygen in such a way that energy is released. Oxidation-reduction

Activation Energy — Amount of energy that must be added to an atomic or molecular system to begin a reaction.

Strong Oxidizer — Material that encourages a strong reaction (by readily accepting electrons) from a reducing agent (fuel).

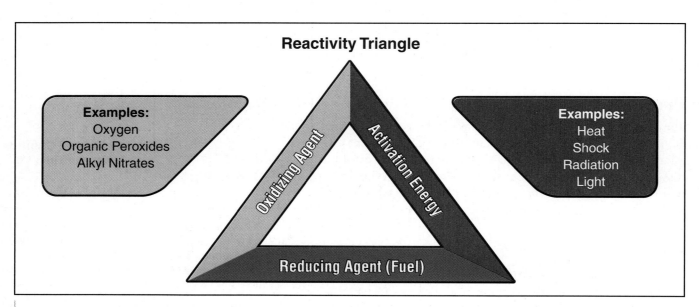

Figure 22.37 The reactivity triangle consists of an oxidizing agent, an activation energy source, and a reducing agent.

Table 22.5
Nine Reactive Hazard Classes

Reactive Hazard Class	Definition	Chemical Examples
Highly Flammable	Substances having flash points less than 100°F (38°C) and mixtures that include substances with flash points less than 100°F (38°C).	Gasoline, Acetone, Pentane, Ethyl Ether, Toluene, Methyl Ethyl Ketone (MEK), Turpentine
Explosive	A material synthesized or mixed deliberately to allow the very rapid release of chemical energy; also, a chemical substance that is intrinsically unstable and liable to detonate under conditions that might reasonably be encountered.	Dynamite, Nitroglycerin, Perchloric Acid, Picric Acid, Fulminates, Azide
Polymerizable	Capable of undergoing self-reactions that release energy; some polymerization reactions generate a great deal of heat. (The products of polymerization reactions are generally less reactive than the starting materials.)	Acrylic Acid, Butadiene, Ethylene, Styrene, Vinyl Chloride, Epoxies
Strong Oxidizing Agent	Oxidizing agents gain electrons from other substances and are themselves thereby chemically reduced, but strong oxidizing agents accept electrons particularly well from a large range of other substances. The ensuing oxidation-reduction reactions may be vigorous or violent and may release new substances that may take part in further additional reactions. Keep strong oxidizing agents well separated from strong reducing agents. In some cases, the presence of a strong oxidizing agent can greatly enhance the progress of a fire.	Hydrogen Peroxide, Fluorine, Bromine, Calcium Chlorate, Chromic Acid, Ammonium Perchlorate

Continued

reactions can be extremely violent and dangerous because they are releasing a tremendous amount of energy. Obviously, some reducing agents (fuels) are more volatile than others. Wood, for example, is not as prone to undergo rapid oxidation (that is, it won't burn as easily) as a highly flammable liquid.

Table 22.5 (continued)
Nine Reactive Hazard Classes

Reactive Hazard Class	Definition	Chemical Examples
Strong Reducing Agent	Reducing agents give up electrons to other substances and are thereby oxidized, but strong reducing agents donate electrons particularly well to a large range of other substances. The ensuing oxidation-reduction reactions may be vigorous or violent and may generate new substances that take part in further additional reactions.	Alkali metals (Sodium, Magnesium, Lithium, Potassium), Beryllium, Calcium, Barium, Phosphorus, Radium, Lithium Aluminum Hydride
Water-Reactive	Substances that may react rapidly or violently with liquid water and steam, producing heat (or fire) and often toxic reaction products.	Alkali metals (Sodium, Magnesium, Lithium, Potassium), Sodium Peroxide, Anhydrides, Carbides
Air-Reactive	Likely to react rapidly or violently with dry air or moist air; may generate toxic and corrosive fumes upon exposure to air or catch fire.	Finely divided metal dusts (Nickel, Zinc, Titanium), Alkali metals (Sodium, Magnesium, Lithium, Potassium), Hydrides (Diborane, Barium Hydrides, Diisobutyl Aluminum Hydride)
Peroxidizable Compound	Apt to undergo spontaneous reaction with oxygen at room temperature, to form peroxides and other products. Most such auto-oxidations are accelerated by light or trace impurities. Many peroxides are explosive, which makes peroxidizable compounds a particular hazard. Ethers and aldehydes are particularly subject to peroxide formation (the peroxides generally form slowly after evaporation of the solvent in which a peroxidizable material had been stored).	Isopropyl Ether, Furan, Acrylic Acid, Styrene, Vinyl Chloride, Methyl Isobutyl Ketone, Ethers, Aldehydes
Radioactive Material	Spontaneously and continuously emitting ions or ionizing radiation. Radioactivity is not a chemical property, but an additional hazard that exists in addition to the chemical properties of a material.	Radon, Uranium

Source: U.S. Environmental Protection Agency's CEPPO (Chemical Emergency Preparedness and Prevention Office) Computer-Aided Management of Emergency Operations (CAMEO) software was used to identify this information.

General Hazardous Materials Behavior Model (GEBMO)

An important element of hazard and risk assessment at haz mat incidents is predicting how a hazardous material is likely to behave in any given situation. The General Hazardous Materials Behavior Model (GEBMO) can help firefighters understand the likely course of an incident, thereby enabling them to successfully mitigate it.

The events in a hazardous materials incident follow a general pattern or model. In hazardous materials incidents the sequence shown in the chart generally occurs. This sequence is as follows:

- **Stress** — If a container is stressed beyond its design strength, it fails or breaches. The three types (or sources) of container stress are thermal, chemical, and mechanical.

- **Breach** — The way in which a container breaches is based on the material of which it is constructed, the type of stress it is exposed to, and pressure inside the container at the time it fails. A breach or failure of the container may be partial (as in a puncture) or total (as in disintegration). A breached container releases its contents. There are five types of breaches:

  — **Disintegration** — Container suffers a general loss of integrity such as a glass bottle shattering or a grenade exploding

  — **Runaway cracking** — A crack develops in a container as a result of some type of damage, which continues to grow rapidly, breaking the container into two or more relatively large pieces

  — **Attachments (closures) open or break** — Attachments (such as pressure-relief devices, discharge valves, or other related equipment) fail, open, or break off when subjected to stress, leading to a total failure of a container

  — **Puncture** — Mechanical stress coming into contact with a container causes a puncture; for example, a forklift puncturing a drum

  — **Split or tear** — Welded seam on a tank or drum fails or a seam on a bag of fertilizer rips. Mechanical or thermal stressors may cause splits or tears **(Figure 22.38)**.

- **Release** — When a container is breached or fails, its contents, stored energy, and pieces of the container may release. The released product disperses. The four ways in which containers release their contents are as follows:

  — **Detonation** — Instantaneous and explosive release of stored chemical energy of a hazardous material

  — **Violent rupture** — Immediate release of chemical or mechanical energy caused by runaway cracks. The results are ballistic behavior of the container and its contents and/or localized projection of container pieces/parts and hazardous material such as with a BLEVE.

  — **Rapid relief** — Fast release of a pressurized hazardous material through properly operating safety devices caused by damaged valves, piping, or attachments or holes in the container

  — **Spill/leak** — Slow release of a hazardous material under atmospheric or head pressure through holes, rips, tears, or usual openings/attachments

Figure 22.38 This cargo tank has been torn open.

Hemispheric Release

Figure 22.39 A hemispheric release consists of a semicircular or dome-shaped pattern of an airborne hazardous material that is still partially in contact with the ground or water.

- *Dispersion/Engulf* — When released, the product inside the container, any stored energy, and the container disperse. The patterns of dispersion are based on the laws of chemistry, physics, and the characteristics of the product. Seven common dispersion patterns are:

  — **Hemispheric** — Semicircular or dome-shaped pattern of airborne hazardous material that is still partially in contact with the ground or water (**Figure 22.39**).

  — **Cloud** — Ball-shaped pattern of the airborne hazardous material where the material has collectively risen above the ground or water (**Figure 22.40**).

  — **Plume** — Irregularly shaped pattern of an airborne hazardous material where wind and/or topography influence the downrange course from the point of release (**Figure 22.41**). Dispersion of a plume (generally composed of gases and vapors) is affected by vapor density and terrain (particularly if vapor density is greater than 1) as well as wind speed and direction.

  — **Cone** — Triangular-shaped pattern of a hazardous material with a point source at the breach and a wide base downrange (**Figure 22.42**).

  — **Stream** — Surface-following pattern of liquid hazardous material that is affected by gravity and topographical contours (**Figure 22.43**). Liquid releases flow downhill whenever there is a gradient away from the point of release.

  — **Pool** — Three-dimensional (including depth), slow-flowing liquid dispersion. Liquids assume the shape of their container and pool in low areas (**Figure 22.44**).

Vapor Cloud

Figure 22.40 A vapor cloud is a ball-shaped pattern of an airborne hazardous material where the material has collectively risen above the ground or water.

Plume

Figure 22.41 A plume is an irregularly shaped pattern of an airborne material where wind and/or topography influence the downrange course from the point of release.

Figure 22.42 A cone is a triangular-shaped pattern of a hazardous material with a point source and the breach and a wide base downrange.

Figure 22.43 A stream is a surface-following pattern of a liquid hazardous material affected by gravity and topographical contours.

Pool

Figure 22.44 A pool of a liquid hazardous material has assumed the shape of its container and/or accumulated in low areas.

— **Irregular** — Irregular or indiscriminate deposit of a hazardous material (such as that carried by contaminated responders) **(Figure 22.45)**.

- *Exposure/Contact* — Anything (such as persons, the environment, or property) that is in the area of the release is exposed. Exposure and contact with hazardous materials is often discussed in the context of the following timeframes:

  — **Short-term** – Seconds, minutes, and hours

  — **Medium-term** – Days, weeks, and months

  — **Long-term** – Years

- *Harm* — Depending on the container, product, energy involved, and duration of exposure/contact, exposures may be harmed. Estimations of potential harm should always begin with a worst-case scenario based on the hazards (or potential hazards) present at the incident.

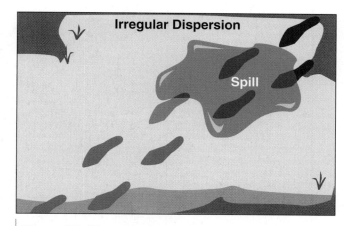

Irregular Dispersion

Spill

Figure 22.45 Indiscriminate deposits of hazardous materials (such as those carried around by contaminated individuals) are sometimes called irregular dispersion.

Various plume and dispersion modeling programs are available to emergency responders to help determine the behavior of hazardous materials once they have been released. These programs may also assist in estimating concentrations of the material and consequently the extent of hazards present within the affected area of the incident.

Hazardous Materials Identification

Firefighters should always be looking for clues or indicators that hazardous materials might be involved in an incident. These clues can be grouped into seven categories, and they approximate the order in which emergency responders might identify hazardous materials. Clues range from those where materials are most easily identified at a distance to ones where responders need to be much closer. The order also represents, in general, an increasing level of risk to firefighters **(Figure 22.46)**. The closer firefighters need to be in order to identify the material, the greater their chances of being in an area where they could be exposed to its harmful effects.

The seven clues to the presence of hazardous materials are as follows:

1. Locations and occupancies
2. Container shapes
3. Transportation placards, labels, and markings

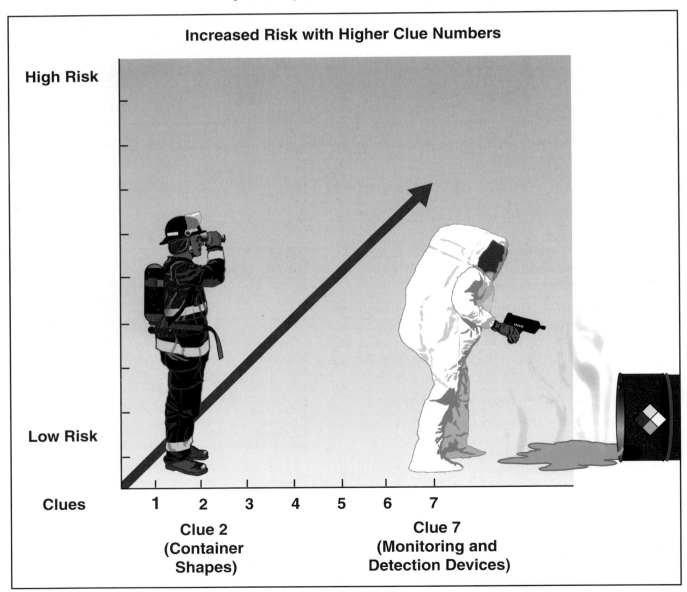

Increased Risk with Higher Clue Numbers

High Risk

Low Risk

Clues   1   2   3   4   5   6   7

Clue 2
(Container Shapes)

Clue 7
(Monitoring and Detection Devices)

Figure 22.46 The risk to firefighters increases as they move closer to the hazardous material. It is much safer to identify a material from a distance based on a container shape than it is to physically sample the substance with a monitoring device.

4. Other markings and colors (nontransportation)

5. Written resources

6. Senses

7. Monitoring and detection devices

While these clues provide many ways to identify hazardous materials involved in incidents, they do have limitations. Firefighters may encounter situations where they are unable to clearly see placards, markings, labels, and signs from a safe distance. Identifying markings can be destroyed in the incident. Inventories may change from those identified during preincident surveys, or containers may be improperly labeled. Mixed loads in transportation incidents may not be marked at all (**Figure 22.47**). Shipping papers may be inaccessible. Firefighters must always be prepared for the unexpected, including recognizing terrorist attacks.

Locations and Occupancies

Hazardous materials are found everywhere—from residential garages to the custodial closet in the local library (**Figure 22.48**). Not all locations or occupancies are as obvious as a local manufacturing plant, and firefighters may have little or no warning of hazardous materials being transported through their jurisdictions by road, rail, or waterway. However, preincident surveys (sometimes called *preplans*) and the occupancy type for a particular structure may provide the first clue to the emergency responder that hazardous materials may be involved in an incident.

Community emergency response plans can help firefighters identify specific sites where hazardous materials are located, used, and stored. However, certain occupancies are always highly probable locations for finding significant quantities of hazardous materials, including:

- Fuel storage facilities
- Gas/service stations (and convenience stores)
- Paint supply stores
- Plant nurseries, garden centers, and agricultural facilities
- Pest control and lawn care companies
- Medical facilities
- Photo processing laboratories
- Dry cleaners
- Plastics and high-technology factories
- Metal-plating businesses
- Mercantile concerns (hardware stores, grocery stores, certain department stores)
- Chemistry (and other) laboratories in educational facilities (including high schools)
- Lumberyards
- Feed/farm stores

Figure 22.47 Mixed loads containing hazardous materials in trucks may not be marked. For example, the presence of the one-ton cylinder in this truck was not indicated on a placard. Furthermore, truck drivers may not have very much information about the materials they transport. *Courtesy of Rich Mahaney.*

Figure 22.48 Many different commercial stores have hazardous materials.

- Veterinary clinics
- Print shops
- Warehouses
- Industrial and utility plants
- Port shipping facilities (with changing cargo hazards)
- Treatment storage disposal (TSD) facilities

Private property is not exempt from danger because hazardous chemicals in the form of drain cleaners, pesticides, fertilizers, paint products, and flammable liquids (such as gasoline) are common household products. In rural areas, propane tanks often provide fuel for heating, and farms may have large quantities of dangerous products such as pesticides.

Certain occupancies are also more likely to be targeted for terrorist attacks. Firefighters must be able to identify those locations where a targeted attack has the potential to do the greatest harm and predict the consequences of such an attack. *Harm* should be defined in terms of the following concerns (and others that are similar):

- Killing or injuring persons
- Causing panic and/or disruption
- Damaging the economy
- Destroying property
- Demoralizing the community

Obviously, when the goal is to kill as many people as possible, any place that has large public gatherings (such as football stadiums, sports arenas, theaters, and shopping malls) might become a potential target (**Figure 22.49**). Terrorists might also target places with historical, economic, or symbolic significance such as local monuments, high-profile buildings, or high-traffic bridges. Any of the following are considered potential terrorist targets:

- Airports
- Bridges
- Chemical manufacturing facilities
- Colleges and universities
- Convention centers
- Dams
- Electricity generation facilities and power plants
- Electricity substations
- Ferry terminals—buildings
- Financial facilities
- High-rise buildings
- Hospitals
- Hotel casinos
- Levees
- LNG terminals
- Maritime port facilities
- Mass transit stations

Figure 22.49 Places where people gather, including stadiums, sports arenas, theaters, and shopping malls, are potential terrorist targets.

- National health stockpile sites
- National monuments and icons
- NG compressor stations
- Non-power nuclear reactors
- Nuclear power plants
- Nuclear research labs
- Petroleum pumping stations
- Petroleum refineries
- Petroleum storage tank farms
- Primary and secondary schools
- Railroad bridges
- Railroad passenger stations
- Railroad tunnels
- Road tunnels
- Shipping facilities
- Shopping malls
- Stadiums
- Telcom-Telephone hotels
- Theme parks
- Trans-oceanic cable landings
- Water treatment facilities

Incidents reported at these occupancies should be scrutinized closely for potential terrorist involvement.

Container Shapes

While firefighters may recognize the location of an incident or type of occupancy as one that handles hazardous materials, the presence of certain storage vessels (as in pressure storage vessels or hollow utensils used for holding liquids), tanks, containers, packages, or vehicles are definitive indicators. These containers can also provide useful information about the materials inside (for example, cylinder-shaped containers often indicate the presence of pressurized gases or liquids). It is important for firefighters to be familiar with the shapes of the different types of packaging and containers in which hazardous materials are stored and transported.

This section discusses (1) bulk-capacity containment systems at fixed facilities, (2) bulk transportation packaging, (3) intermediate bulk containers, (4) ton containers, and (5) nonbulk containers in general. Pressure and nonpressure containers are discussed as appropriate under these categories.

Bulk and Nonbulk Packaging

Bulk packaging refers to a packaging, other than that on a vessel or barge, in which materials are loaded with no intermediate form of containment. This packaging type includes a transport vehicle or freight container such as a cargo tank, railcar, or portable tank. Intermediate bulk containers (IBCs) and intermodal (IM) containers are also examples. To be considered bulk packaging, one of the following criteria must be met:

- Maximum capacity is greater than 119 gallons (450 L) as a receptacle for a liquid

- Maximum net mass is greater than 882 pounds (400 kg) or maximum capacity is greater than 119 gallons (450 L) as a receptacle for a solid

- Water capacity is 1,001 pounds (454 kg) or greater as a receptacle for a gas

Nonbulk packaging is packaging that is smaller than the minimum criteria established for bulk packaging. Drums, boxes, carboys, and bags are examples. Composite packages (packages with an outer packaging and an inner receptacle) and combination packages (multiple packages grouped together in a single outer container such as bottles of acid packed inside a cardboard box) may also be classified as nonbulk packaging.

Bulk-Capacity Fixed-Facility Containers

Containers at fixed facilities include buildings, aboveground storage tanks, machinery, underground storage tanks, pipelines, reactors, open piles or bins, vats, storage cabinets, and other fixed, on-site containers. This section focuses on storage tanks holding bulk quantities of hazardous materials. Nonbulk packages that may be found at fixed facilities are discussed in the Nonbulk Packaging section. Pipeline identification and other labeling and marking systems used to identify the materials in storage cabinets, bins, vats, and the like are discussed under Other Markings and Colors.

In general, aboveground storage tanks are divided into two major categories:

- ***Nonpressure tanks (also called atmospheric tanks)*** — If these tanks are storing any product, they will normally have a small amount of pressure (up to 0.5 psi [3.45 kPa] {0.03 bar}) inside.

- *Pressure tanks* — These tanks are divided into the following two:
  — Low-pressure storage tanks that have pressures between 0.5 psi to 15 psi (3.45 kPa to 103 kPa) {0.03 bar to 1.03 bar}.
  — Pressure vessels that have pressures above 15 psi (103 kPa) {1.03 bar}.

Underground storage tanks may be atmospheric or pressurized. Cryogenic liquid storage tanks have varying pressures, but some can be very high. They are usually heavily insulated with a vacuum in the space between the outer and inner shells. **Table 22.6** provides pictures and examples of various atmospheric/nonpressure storage tanks and also describes underground storage caverns. **Table 22.7** provides pictures and examples of various pressure tanks.

Bulk Transportation Containers

Bulk transportation containers can be divided into three main categories determined by the mode of transportation:

- **Tank and other rail cars** (railroad) **(Tables 22.8, 22.9, 22.10, and 22.11)**
  — Nonpressure tank cars (also known as general service or low-pressure tank cars) with vapor pressures below 25 psi (172 kPa) {1.7 bar} at 105° to 115°F (41°C to 46°C)
  — Pressure tank cars with pressures greater than 25 psi (172 kPa) {1.7 bar} at 68°F (20°C)
  — Cryogenic liquid tank cars
  — Hopper cars
  — Box cars
  — Special service cars
- **Cargo tanks** (highway [MC and Department of Transportation (DOT)/Transport Canada (TC) designations indicate the construction specifications to which these cargo tanks are manufactured]) **(Table 22.12)**
  — Nonpressure liquid tanks (DOT/TC 406, MC 306)
  — Low-pressure chemical/liquid tanks (DOT/TC 407, MC 307)
  — Corrosive liquid tanks (DOT/TC 412, MC 312)
  — High-pressure tanks (MC/TC 331)
  — Cryogenic liquid tanks (MC/TC 338)
  — Compressed-gas/tube trailers
  — Dry bulk cargo tanks
  — Vacuum Loaded Tank (DOT/TC 407, DOT/TC 412)
- **Intermodal containers** (highway, railroad, or marine vessel) **(Table 22.13)**
  — Nonpressure intermodal tanks
  — Pressure intermodal tanks
  — Specialized intermodal tanks
    + Cryogenic intermodal tanks
    + Tube modules
  — Freight containers **(Figure 22.50)**

Table 22.6
Atmospheric/Nonpressure Storage Tanks

Tank Type

Horizontal Tank

Horizontal tanks: Cylindrical tanks sitting on legs, blocks, cement pads, or something similar; typically constructed of steel with flat ends. Horizontal tanks are commonly used for bulk storage in conjunction with fuel-dispensing operations. Old tanks (pre-1950s) have bolted seams, whereas new tanks are generally welded. A horizontal tank supported by unprotected steel supports or stilts (prohibited by most current fire codes) may fail quickly during fire conditions.

Contents: Flammable and combustible liquids, corrosives, poisons, etc.

Cone Roof Tank

Cone roof tanks: Have cone-shaped, pointed roofs with weak roof-to-shell seams that break when or if the container becomes overpressurized. When it is partially full, the remaining portion of the tank contains a potentially dangerous vapor space.

Contents: Flammable, combustible, and corrosive liquids

Open Top Floating Roof Tank

Floating Deck

Open top floating roof tanks (sometimes just called *floating roof tanks*): Large-capacity, aboveground holding tanks. They are usually much wider than they are tall. As with all floating roof tanks, the roof actually floats on the surface of the liquid and moves up and down depending on the liquid's level. This roof eliminates the potentially dangerous vapor space found in cone roof tanks. A fabric or rubber seal around the circumference of the roof provides a weather-tight seal.

Contents: Flammable and combustible liquids

Covered Top Floating Roof Tank

Vents around rim provide differentation from Cone Roof Tanks

Internal floating roof tanks (sometimes called *covered [or covered top] floating roof tanks*): Have fixed cone roofs with either a pan or deck-type float inside that rides directly on the product surface. This tank is a combination of the open top floating roof tank and the ordinary cone roof tank.

Contents: Flammable and combustible liquids

Continued

Table 22.6 (continued)
Atmospheric/Nonpressure Storage Tanks

Tank Type	Descriptions
Covered Top Floating Roof Tank with Geodesic Dome 	**Floating roof tanks:** Floating roof tanks covered by geodesic domes are used to store flammable liquids.

Contents: Flammable Liquids

Tank Type	Descriptions
	**Lifter roof tanks:** Have roofs that float within a series of vertical guides that allow only a few feet (meters) of travel. The roof is designed so that when the vapor pressure exceeds a designated limit, the roof lifts slightly and relieves the excess pressure. **Contents:** Flammable and combustible liquids
Vapordome Roof Tank 	**Vapordome roof tanks:** Vertical storage tanks that have lightweight aluminum geodesic domes on their tops. Attached to the underside of the dome is a flexible diaphragm that moves in conjunction with changes in vapor pressure. **Contents:** Combustible liquids of medium volatility and other nonhazardous materials
Atmospheric Underground Storage Tank 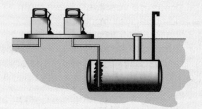 **Fill Connections Cover** 	**Atmospheric underground storage tanks:** Constructed of steel, fiberglass, or steel with a fiberglass coating. Underground tanks will have more than 10 percent of their surface areas underground. They can be buried under a building or driveway or adjacent to the occupancy. This tank has fill and vent connections located near the tank. Vents, fill points, and occupancy type (gas/service stations, private garages, and fleet maintenance stations) provide visual clues. Many commercial and private tanks have been abandoned, some with product still in them. These tanks are presenting major problems to many communities. **Contents:** Petroleum products
	Underground storage caverns: Rare and technically are not "tanks." First responders should be aware that some natural and manmade caverns are used to store natural gas. The locations of such caverns should be noted in local emergency response plans.

Table 22.7
Low-Pressure Storage Tanks and Pressure Vessels

Tank/Vessel Type

Dome Roof Tank

Spheroid Tank

Noded Spheroid Tank

Horizontal Pressure Vessel*

Spherical Pressure Vessel

Descriptions

Dome roof tanks: Generally classified as low-pressure tanks with operating pressures as high as 15 psi (103 kPa). They have domes on their tops.

Contents: Flammable liquids, combustible liquids, fertilizers, solvents, etc.

Spheroid tanks: Low-pressure storage tanks. They can store 3,000,000 gallons (11 356 200 L) or more of liquid.

Contents: Liquefied petroleum gas (LPG), methane, and some flammable liquids such as gasoline and crude oil

Noded spheroid tanks: Low-pressure storage tanks. They are similar in use to spheroid tanks, but they can be substantially larger and flatter in shape. These tanks are held together by a series of internal ties and supports that reduce stresses on the external shells.

Contents: LPG, methane, and some flammable liquids such as gasoline and crude oil

Horizontal pressure vessels:* Have high pressures and capacities from 500 to over 40,000 gallons (1 893 L to over 151 416 L). They have rounded ends and are not usually insulated. They usually are painted white or some other highly reflective color.

Contents: LPG, anhydrous ammonia, vinyl chloride, butane, ethane, compressed natural gas (CNG), chlorine, hydrogen chloride, and other similar products

Spherical pressure vessels: Have high pressures and capacities up to 600,000 gallons (2 271 240 L). They are often supported off the ground by a series of concrete or steel legs. They usually are painted white or some other highly reflective color.

Contents: Liquefied petroleum gases and vinyl chloride

Continued

Table 22.7 (continued)
Low-Pressure Storage Tanks and Pressure Vessels

Tank/Vessel Type	Descriptions
Cryogenic-Liquid Storage Tank 	**Cryogenic-liquid storage tanks:** Insulated, vacuum-jacketed tanks with safety-relief valves and rupture disks. Capacities can range from 300 to 400,000 gallons (1 136 L to 1 514 160 L). Pressures vary according to the materials stored and their uses. **Contents:** Cryogenic carbon dioxide, liquid oxygen, liquid nitrogen, etc.

*It is becoming more common for horizontal propane tanks to be buried underground. Underground residential tanks usually have capacities of 500 or 1,000 gallons (1 893 L or 3 785 L). Once buried, the tank may be noticeable only because of a small access dome protruding a few inches (millimeters) above the ground.

Table 22.8
Nonpressure Tank Cars

Photograph	Descriptions
Nonpressure Tank Car 	**Nonpressure tank car without an expansion dome:** Fittings are visible. **Carries:** Flammable liquids, flammable solids, reactive liquids, reactive solids, oxidizers, organic peroxides, poisons, irritants, corrosive materials, and similar products
Nonpressure Tank Car With an Expansion Dome 	**Nonpressure tank car with an expansion dome:** Old models have the expansion dome. **Carries:** Flammable liquids, flammable solids, reactive liquids, reactive solids, oxidizers, organic peroxides, poisons, irritants, corrosive materials, and similar products

Source: Information courtesy of Union Pacific Railroad.

Table 22.9
Pressure Tank Cars

Photograph	Descriptions
Typical Pressure Tank Car  	**Typical pressure tank car:** Fittings are inside the protective housing. **Carries:** Flammable, nonflammable, and poison gases as well as flammable liquids
Hydrogen Cyanide (Hydrocyanic Acid), HCN *Candy Stripe* Car 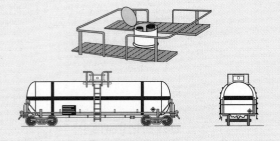 	**Hydrogen cyanide (hydrocyanic acid), HCN *candy stripe* car:** In the past, some HCN cars were painted white with red bands (called *candy stripes*). This practice is becoming obsolete due to concerns about terrorism.

Table 22.10
Cryogenic Liquid Tank Car

Photograph	Descriptions
Typical Cryogenic Liquid Tank Car	**Typical cryogenic liquid tank car:** Fittings are in ground-level cabinets at diagonal corners of the car or in the center of one end of the car. **Carries:** Argon, hydrogen, nitrogen, oxygen, LNG, and ethylene **NOTE:** Cryogenic liquid tanks can also be enclosed inside a boxcar.

Table 22.11
Other Railroad Cars

Illustration	Photograph
Covered Hopper Car 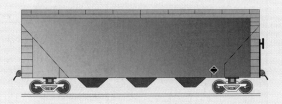	 **Carries:** Calcium carbide, cement, and grain
Open Top Hopper 	 **Carries:** Coal, rock, gravel, and sand
Miscellaneous: Boxcar 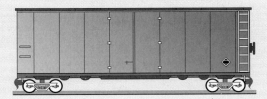	 **Carries:** All types of materials and finished goods
Miscellaneous: Gondola 	 **Carries:** Sand, rolled steel, and other materials that do not require protection from the weather

Continued

Table 22.11 (continued)
Other Railroad Cars

Illustration	Photograph

Miscellaneous: Flat Bed Car with Intermodal Containers

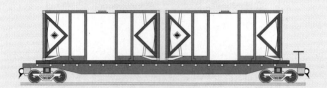

Carries: 1 ton containers, intermodal containers (shown), large vehicles, and other commodities that do not require protection from the weather

Pneumatically Unloaded Hopper Car

Carries: Dry caustic soda, ammonium nitrate fertilizer, other fine-powdered materials, plastic pellets, and flour

Table 22.12
Cargo Tank Trucks

Nonpressure Liquid Tank (DOT/TC 406/MC 306)

Low-Pressure Chemical Tank (DOT/TC 407/MC 307)

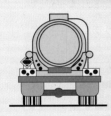

Corrosive Liquid Tank (MC 312)

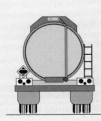

Nonpressure liquid tank (DOT/TC 406/MC 306):

- Pressure less than 3 psi (21 kPa)
- Typical maximum capacity: 9,000 gallons (34 069 L)
- New tanks made of aluminum
- Old tanks made of steel
- Oval shape
- Multiple compartments
- Recessed manways
- Rollover protection
- Bottom valves
- Vapor recovery likely

Carries: Gasoline, fuel oil, alcohol, other flammable/combustible liquids, other liquids, and liquid fuel products

Low-pressure chemical tank (DOT/TC 407/MC 307):

- Pressure under 40 psi (172 kPa to 276 kPa)
- Typical maximum capacity: 7,000 gallons (26 498 L)
- Rubber lined or steel
- Single- or double-top manway
- Typically double shell
- Stiffening rings
- Rollover protection
- Single or multiple compartments
- Horseshoe or round shaped

Carries: Flammable liquids, combustible liquids, acids, caustics, and poisons

Corrosive liquid tank (MC 312):

- Pressure less than 75 psi (517 kPa)
- Typical maximum capacity: 7,000 gallons (26 498 L)
- Rubber lined or steel
- Stiffening rings
- Rollover protection
- Rollover protection splash guard
- Top loading at rear or center
- Typically single compartment

Carries: Corrosive liquids (usually acids)

Continued

Table 22.12 (continued)
Cargo Tank Trucks

High-Pressure Tank
(MC 331)

High-pressure tank (MC 331):

- Pressure above 100 psi (689 kPa)
- Typical maximum capacity: 11,500 gallons (43 532 L)
- Single steel compartment
- Noninsulated
- Bolted manway at front or rear
- Internal and rear outlet valves
- Typically painted white or other reflective color
- May be marked FLAMMABLE GAS and COMPRESSED GAS
- Round/dome-shaped ends

Carries: Pressurized gases and liquids, anhydrous ammonia, propane, butane, and other gases that have been liquefied under pressure

High-Pressure Bobtail Tank

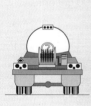

High-pressure bobtail tank: Used for local delivery of liquefied petroleum gas and anhydrous ammonia

Cryogenic Liquid Tank
(MC 338)

Cryogenic liquid tank (MC 338):

- Pressure less than 22 psi (152 kPa)
- Well-insulated steel tank
- Possibly discharging vapor from relief valves
- Loading/unloading valves enclosed at rear
- Possibly marked REFRIGERATED LIQUID
- Round tank with flat ends and some type of cabinet at rear

Carries: Liquid oxygen, liquid nitrogen, liquid carbon dioxide, liquid hydrogen, and other gases that have been liquefied by lowering their temperatures

Continued

Table 22.12 (continued)
Cargo Tank Trucks

Photograph	Descriptions

Photograph

Compressed-Gas/Tube Trailer

Dry Bulk Cargo Tanker

 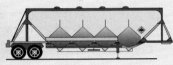

Descriptions

Compressed-gas/tube trailer:

- Pressure at 3,000 to 5,000 psi (20 684 kPa to 34 474 kPa) (gas only)
- Individual steel cylinders stacked and banded together
- Typically has over-pressure device for each cylinder
- Bolted manway at front or rear
- Valves at rear (protected)
- Manufacturer's name possibly marked on cylinders

Carries: Helium, hydrogen, methane, oxygen, and other gases

Dry bulk cargo tanker:

- Pressure less than 22 psi (152 kPa)
- Typically not under pressure
- Top side manway
- Bottom valves
- Air-assisted loading and unloading
- Shapes vary, but has hoppers

Carries: Calcium carbide, oxidizers, corrosive solids, cement, plastic pellets, and fertilizers

Table 22.13
Examples of Intermodal Tanks

Photograph	Descriptions
Nonpressure Intermodal Tank 	**Nonpressure intermodal tank:** • IM-101: 25.4 to 100 psi (175 kPa to 689 kPa) • IM-102: 14.5 to 25.4 psi (100 kPa to 175 kPa) **Contents:** Liquids or solids (both hazardous and nonhazardous)
Pressure Intermodal Tank 	**Pressure intermodal tank:** 100 to 500 psi (689 kPa to 3 447 kPa) **Contents:** Liquefied gases, liquefied petroleum gas, anhydrous ammonia, and other liquids
Cryogenic Intermodal Tank 	**Cryogenic intermodal tank:** **Contents:** Refrigerated liquid gases, argon, oxygen, helium
Tube Module Intermodal Container 	**Tube module intermodal container:** **Contents:** Gases in high-pressure cylinders (3,000 or 5,000 psi [20 684 kPa or 34 474 kPa]) mounted in the frame

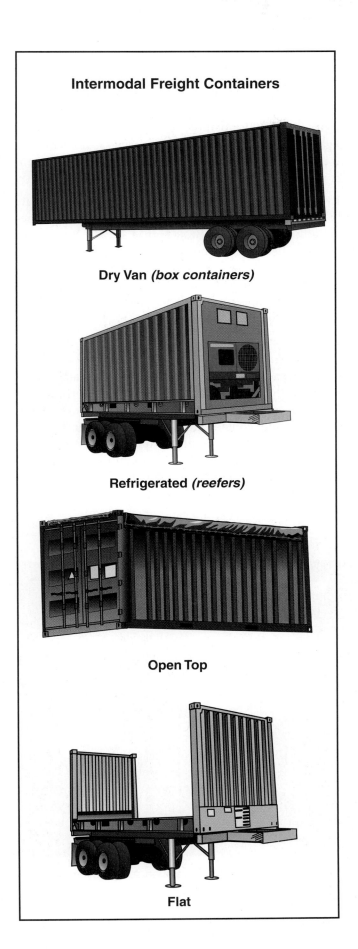

Intermodal Freight Containers

Dry Van *(box containers)*

Refrigerated *(reefers)*

Open Top

Flat

Figure 22.50 Intermodal freight containers come in a variety of different types including dry van containers (sometimes call *box containers*), refrigerated containers (sometimes called *reefers*), open top containers, and flat containers.

Figure 22.51 Flexible intermediate bulk containers (FIBCs) come in a variety of styles and carry both solid materials and fluids. *Courtesy of Rich Mahaney.*

Intermediate Bulk Containers

An *intermediate bulk container (IBC)* is either a rigid or flexible portable packaging (other than a cylinder or portable tank) designed for mechanical handling. The maximum capacity of an IBC is not more than 3 cubic meters (3,000 L, 793 gal, or 106 ft³). The minimum capacity is not less than 0.45 cubic meters (450 L, 119 gal, or 15.9 ft³) or a maximum net mass of not less than 400 kilograms (882 lbs).

IBCs are divided into two types: flexible intermediate bulk containers (FIBCs)(**Figure 22.51**) and rigid intermediate bulk containers (RIBCs)(**Figure 22.52**). Both types are often called *totes,* although correctly, only FIBCs are truly totes. Rigid portable tanks may be used to carry liquids, fertilizers, solvents, and other chemicals, and they may have capacities up to 400 gallons (1,514 L) and pressures up to 100 psi (689 kPa) {6.9 bar} (**Figure 22.53**).

Ton Containers

Ton containers are tanks that have capacities of 1 short ton or approximately 2,000 pounds (907 kg or 0.9 tonne). Typically stored on their sides, the ends (heads) of the containers are convex or concave and have two valves in the center of one end (**Figure 22.54**). Ton containers commonly contain chlorine and are often found at water treatment plants, commercial swimming pools, and so on. Ton containers may also contain other products such as sulphur dioxide, anhydrous ammonia, or Freon® refrigerant. Structural fire fighting gear does not provide adequate protection against the hazardous materials commonly stored in ton containers.

Figure 22.52 Rigid intermediate bulk containers (RIBCs) are often designed to be stacked, and they may contain a variety of products including both solids and liquids. *Courtesy of Rich Mahaney.*

Figure 22.53 Some RIBCs may have pressures up to 100 psi (689 kPa) {6.9 bar}. *Courtesy of Rich Mahaney.*

Nonbulk Packaging

Containers that are used to transport smaller quantities of hazardous materials than bulk or IBCs are called *nonbulk packaging*. The majority of these incidents occur during highway transport. **Table 22.14** shows that the most common types of nonbulk packaging include the following types of containers:

- Bags
- Carboys and jerry cans
- Cylinders
- Drums
- Dewar flasks (cryogenic liquids)

Containers for Radioactive Materials

In the United States, all shipments of radioactive materials (sometimes called *RAM*) must be packaged and transported according to strict regulations. The type of packaging used to transport radioactive materials is determined by the activity, type, and form of the material to be shipped. Depending on these factors, most radioactive material is shipped in one of four basic types of containers/packaging: Type A, Type B, industrial, or excepted (**Table 22.15**).

Type C packages are very rare. Type C packages are used for high-activity materials (including plutonium) transported by aircraft. They are designed to withstand

Figure 22.54 Ton containers often contain chlorine, but they may also contain other products such as anhydrous ammonia and other refrigerant gases. Their ends may be convex or concave, and they typically have two valves in the center of one end, one above the other. *Courtesy of Rich Mahaney.*

Table 22.14
Nonbulk Packaging

Bags

Bags:

- Made of paper, plastic, film, textiles, woven material, or others
- Sizes vary

Contents: Explosives, flammable solids, oxidizers, organic peroxides, fertilizers, pesticides, and other regulated materials

Carboys and Jerrycans

Carboys and Jerrycans:

- Made of glass or plastic
- Often encased in a basket or box
- Sizes vary

Contents: Flammable and combustible liquids, corrosives

Cylinders

Cylinders:

- Pressures higher than 40 psi (276 kPa) {2.76 bar} but vary
- Sizes range from lecture bottle size to very large

Contents: Compressed gases

Continued

Table 22.14 (continued)
Nonbulk Packaging

Drums

Dewar Flasks

Drums:

- Made of metal, fiberboard, plastic, plywood, or other materials
- May have open heads (removable tops) or tight (closed) heads with small openings
- Sizes vary from 55 gallons (208 L) to 100 gallons (379 L)

Contents: Hazardous and nonhazardous liquids and solids

Dewar Flasks:

- Vacuum insulated
- Made of glass, metal, or plastic with hollow walls from which the air has been removed
- Sizes vary

Contents: Cryogenic liquids; thermoses may contain nonhazardous liquids

Table 22.15
Radioactive Container Descriptions

Excepted Packaging	Industrial Packaging (IP)	Type A Packaging	Type B Packaging
Designed to survive normal conditions of transport	Designed to survive normal conditions of transport (IP-1) and at least the DROP test and stacking test for Type A packaging (IP-2 and IP-3)	• Designed to survive normal transportation, handling, and minor accidents • Certified as Type A on the basis of performance requirements, which means it must survive certain tests	Must be able to survive severe accidents
Used for transportation of materials that are either Low Specific Activity (LSA) or Surface Contaminated Objects (SCO) and that are limited quantity shipments, instruments or articles, articles manufactured from natural or depleted uranium, or natural thorium; empty packagings are also excepted (49 *CFR* 173.421–428)	Used for transportation of materials with very small amounts of radioactivity (Low Specific Activity [LSA] or Surface Contaminated Objects [SCO])	Used for the transportation of limited quantities of radioactive material (RAM) that would not result in significant health effects if they were released	Used for the transportation of large quantities of radioactive material
Can be almost any packaging that meets the basic requirements, with any of the above contents; they are excepted from several labeling and documentation requirements	Usually metal boxes or drums	• May be cardboard boxes, wooden crates, or drums • Shipper and carrier must have documentation of the certification of the packages being transported	• May be a metal drum or a huge, massive shielded transport container • Must meet severe accident performance standards that are considerably more rigorous than those required for Type A packages • Either has a Certificate of Compliance (COC) by the Nuclear Regulatory Commission (NRC) or Certificate of Competent Authority (COCA) by the Department of Transportation (DOT)

Source: Courtesy of U.S. Department of Energy (DOE)/Sandia National Laboratories.

severe accident conditions associated with air transport without loss of containment or significant increase in external radiation levels. The Type C package performance requirements are significantly more stringent than those for Type B packages. Type C packages are not authorized for domestic use but are authorized for international shipments of radioactive material.

Transportation Placards, Labels, and Markings

In the United States, transportation of hazardous materials is regulated by the U.S. Department of Transportation (DOT). In Canada they are regulated by Transport Canada (TC). The Ministry of Communications and Transport is responsible in Mexico. All countries that base their transportation placarding, labeling, and marking requirements on United Nations (UN) recommendations have very similar identification systems. Therefore, with a few country-specific variations, the majority of the placards, labels, and markings used to identify hazardous materials during transport are very similar in the United States, Canada, and Mexico.

Under the UN system, nine hazard classes are used to categorize hazardous materials:

Class 1:   Explosives

Class 2:   Gases

Class 3:   Flammable liquids

Class 4:   Flammable solids, substances liable to spontaneous combustion, substances that emit flammable gases on contact with water

Class 5:   Oxidizing substances and organic peroxides

Class 6:   Toxic and infectious substances

Class 7:   Radioactive materials

Class 8:   Corrosive substances

Class 9:   Miscellaneous dangerous substances and articles

Shreveport, Louisiana, Anhydrous Ammonia Incident

Placarded materials may have many hazards not reflected by the placard classification. A specific example is anhydrous ammonia; it is placarded in the United States as a nonflammable gas. However, under certain conditions (particularly inside where fumes can become concentrated), it will burn.

In 1984, in Shreveport, Louisiana, two hazardous materials response team members entered a cold-storage facility to stop a leak of anhydrous ammonia. With a lower explosive limit (LEL) of 16 percent and an explosive range of 16 to 25 percent, the fumes inside the facility reached a flammable concentration. Unfortunately, a spark ignited the vapors, and one team member was killed. The other was seriously burned. Anhydrous ammonia can catch fire even though it is not classified as a flammable gas by the DOT.

In other countries, anhydrous ammonia is classified as a corrosive (caustic) liquid and a poison gas because of its chemical effects. Anhydrous ammonia has a threshold limit value (TLV®) of 25 ppm in air. Inhaling concentrated fumes can kill a person even though it is not classified as an inhalation hazard by the DOT!

Table 22.16 provides the U.S. Department of Transportation (DOT) placard hazard classes and divisions. It also provides the hazards associated with each hazard class and provides examples of products for each division. Placards are required on specific bulk quantities of hazardous materials that are transported by tank cars, cargo tanks, and other bulk packages and portable tanks. Certain non-bulk containers may also require placarding. **Table 22.17** provides Canadian transportation placards, labels, and markings.

Table 22.18 provides examples of the unique U.S. DOT labels. Other labels for the nine hazard classes and subdivisions are essentially the same as the placards shown earlier in Table 22.16. Labels are 3.9-inch (100 mm), square-on-point diamonds, which may or may not have written text that identifies the hazardous material within the packaging (Class 7 Radioactive labels must always contain text).

Table 22.19 provides examples of U.S. DOT markings. A marking is a descriptive name, an identification number, a weight, or a specification and includes instructions, cautions, or UN marks (or combinations thereof) required on outer packagings of hazardous materials. Additional markings found on tank cars, intermodal containers, and other packaging will be discussed in later sections.

Four-Digit UN Identification Numbers

The UN has also developed a system of four-digit identification numbers that is used in conjunction with illustrated placards in North America. Each individual hazardous material is assigned a unique four-digit number. This number will often be displayed on placards, labels, orange panels, and/or white diamonds in association with materials being transported in cargo tanks, portable tanks, tank cars, or other containers and packages (**Figure 22.55** on page 1153). In some cases, on orange panels, the number may be preceded by the letters *NA* for North America (which means it is not recognized for international transportation except in North America) or *UN* for United Nations (which means it is recognized for international transportation) (**Figure 22.56** on page 1155).

Careful!

Don't be confused by an orange placard with two sets of numbers on intermodal tanks and containers. The four-digit UN ID number is on the bottom. The top number is a hazard identification number (or code) required under European and some South American regulations. These numbers* indicate the following hazards:

2 — Emission of a gas due to pressure or chemical reaction

3 — Flammability of liquids (vapors) and gases or self-heating liquid

4 — Flammability of solids or self-heating solid

5 — Oxidizing (fire intensifying) effect

6 — Toxicity or risk of infection

7 — Radioactivity

8 — Corrosivity

9 — Risk of spontaneous violent reaction

---

*Doubling a number (such as *33, 44,* or *88*) indicates an intensification of that particular hazard. When the hazard associated with a material is adequately indicated by a single number, it is followed by a *zero* (such as *30, 40,* or *60*). A hazard identification code prefixed by the letter *X* (such as *X88*) indicates that the material will react dangerously with water.

Table 22.16
U.S. DOT Placard Hazard Classes and Divisions

Class 1: Explosives (49 *CFR* 173.50)

EXPLOSIVES 1.1A

Compatibility Group Letter will vary

An **explosive** is any substance or article (including a device) that is designed to function by explosion (that is, an extremely rapid release of gas and heat) or (by chemical reaction within itself) is able to function in a similar manner even if not designed to function by explosion.

Explosive placards will have a *compatibility group* letter on them, which is a designated alphabetical letter used to categorize different types of explosive substances and articles for purposes of stowage and segregation. However, it is the division number that is of primary concern to first responders.

The primary hazards of explosives are thermal (heat) and mechanical, but may include the following:

- Blast pressure wave
- Shrapnel fragmentation
- Incendiary thermal effect
- Seismic effect
- Chemical hazards from the production of toxic gases and vapors
- Ability to self-contaminate with age, which increases their sensitivity and instability
- Sensitivity to shock and friction

Division 1.1 — Explosives that have a mass explosion hazard. A mass explosion is one that affects almost the entire load instantaneously.

Examples: dynamite, mines, wetted mercury fulminate

Division 1.2 — Explosives that have a projection hazard but not a mass explosion hazard.

Examples: detonation cord, rockets (with bursting charge), flares, fireworks

Division 1.3 — Explosives that have a fire hazard and either a minor blast hazard or a minor projection hazard or both, but not a mass explosion hazard.

Examples: liquid-fueled rocket motors, smokeless powder, practice grenades, aerial flares

Division 1.4 — Explosives that present a minor explosion hazard. The explosive effects are largely confined to the package and no projection of fragments of appreciable size or range is expected. An external fire must not cause virtually instantaneous explosion of almost the entire contents of the package.

Examples: signal cartridges, cap type primers, igniter fuses, fireworks

Division 1.5 — Substances that have a mass explosion hazard but are so insensitive that there is very little probability of initiation or of transition from burning to detonation under normal conditions of transport.

Examples: Prilled ammonium nitrate fertilizer/fuel oil (ANFO) mixtures and blasting agents

Continued

Table 22.16 (continued)
U.S. DOT Placard Hazard Classes and Divisions

Class 1: Explosives (continued)

Division 1.6 — Extremely insensitive articles that do not have a mass explosive hazard. This division is comprised of articles that contain only extremely insensitive detonating substances and that demonstrate a negligible probability of accidental initiation or propagation.

Class 2: Gases (49 *CFR* 173.115)

DOT defines *gas* as a material that has a vapor pressure greater than 43.5 psi (300 kPa) at 122°F (50°C) or is completely gaseous at 68°F (20°C) at a standard pressure of 14.7 psi (101.3 kPa).

NOTE: The DOT definition for gas is much more specific than the definition provided in Chapter 2, Hazardous Materials Properties and Hazards.

The potential hazards of gases may include thermal, asphyxiation, chemical, and mechanical hazards:

- Thermal hazards (heat) from fires, particularly associated with Division 2.1 and oxygen

- Thermal hazards (cold) associated with exposure to cryogens in Division 2.2

- Asphyxiation caused by leaking/released gases displacing oxygen in a confined space

- Chemical hazards from toxic and/or corrosive gases and vapors, particularly associated with Division 2.3

- Mechanical hazards from a boiling liquid expanding vapor explosion (BLEVE) for containers exposed to heat or flame

- Mechanical hazards from a ruptured cylinder rocketing after exposure to heat or flame

Division 2.1: Flammable Gas — Consists of any material that is a gas at 68°F (20°C) or less at normal atmospheric pressure or a material that has a boiling point of 68°F (20°C) or less at normal atmospheric pressure and that

(1) Is ignitable at normal atmospheric pressure when in a mixture of 13 percent or less by volume with air, or

(2) Has a flammable range at normal atmospheric pressure with air of at least 12 percent, regardless of the lower limit.

Examples: compressed hydrogen, isobutene, methane, and propane

Division 2.2: Nonflammable, Nonpoisonous Gas — Nonflammable, nonpoisonous compressed gas, including compressed gas, liquefied gas, pressurized cryogenic gas, and compressed gas in solution, asphyxiant gas and oxidizing gas; means any material (or mixture) which exerts in the packaging an absolute pressure of 40.6 psi (280 kPa) or greater at 68ºF (20ºC) and does not meet the definition of Divisions 2.1 or 2.3.

Examples: carbon dioxide, helium, compressed neon, refrigerated liquid nitrogen, cryogenic argon, anhydrous ammonia, pepper spray

Continued

Table 22.16 (continued)
U.S. DOT Placard Hazard Classes and Divisions

Class 2: Gases (continued)

	Division 2.3: Gas Poisonous by Inhalation — Material that is a gas at 68°F (20°C) or less and a pressure of 14.7 psi (101.3 kPa) (a material that has a boiling point of 68°F [20°C] or less at 14.7 psi [101.3 kPa]), and that is known to be so toxic to humans as to pose a hazard to health during transportation; or (in the absence of adequate data on human toxicity) is presumed to be toxic to humans because of specific test criteria on laboratory animals.

Division 2.3 has *ERG*-designated hazard zones associated with it, determined by the concentration of gas in the air:

- Hazard Zone A — LC_{50} less than or equal to 200 ppm
- Hazard Zone B — LC_{50} greater than 200 ppm and less than or equal to 1,000 ppm
- Hazard Zone C — LC_{50} greater than 1,000 ppm and less than or equal to 3,000 ppm
- Hazard Zone D — LC_{50} greater than 3,000 ppm and less than or equal to 5,000 ppm

Examples: cyanide, diphosgene, germane, phosphine, selenium hexafluoride, and hydrocyanic acid, arsine, cyanogen chloride, phosgene, chlorine

	Oxygen Placard — Oxygen is not a separate division under Class 2, but first responders may see this oxygen placard on containers with 1,001 lbs (454 kg) or more gross weight of either compressed gas or refrigerated liquid.

Class 3: Flammable and Combustible Liquids (49 *CFR* 173.120)

A *flammable liquid* is generally a liquid having a flash point of not more than 141°F (60.5°C), or any material in a liquid state with a flash point at or above 100°F (37.8°C) that is intentionally heated and offered for transportation or transported at or above its flash point in a bulk packaging.

A *combustible liquid* is any liquid that does not meet the definition of any other hazard class and has a flash point above 141°F (60.5°C) and below 200 °F (93 °C). A flammable liquid with a flash point at or above 100°F (38°C) that does not meet the definition of any other hazard class may be reclassified as a combustible liquid. This provision does not apply to transportation by vessel or aircraft, except where other means of transportation is impracticable. An elevated temperature material that meets the definition of a Class 3 material because it is intentionally heated and offered for transportation or transported at or above its flash point may not be reclassified as a combustible liquid.

The primary hazards of flammable and combustible liquids are thermal, asphyxiation, chemical, and mechanical, and may include the following:

- Thermal hazards (heat) from fires and vapor explosions
- Asphyxiation from heavier than air vapors displacing oxygen in low-lying, and/or confined spaces
- Chemical hazards from toxic and/or corrosive gases and vapors
- Chemical hazards from the production of toxic and/or corrosive gases and vapors during fires
- Mechanical hazards from a BLEVE, for containers exposed to heat or flame
- Mechanical hazards caused by a vapor explosion
- Vapors that can mix with air and travel great distances to an ignition source
- Environmental hazards (pollution) caused by runoff from fire control

When responding to a transportation incident, first responders must keep in mind that a flammable liquid placard can indicate a product with a flash point as high as 140°F (60°C).

Continued

Table 22.16 (continued)
U.S. DOT Placard Hazard Classes and Divisions

Class 3: Flammable and Combustible Liquids (continued)

FLAMMABLE 3	**Flammable Placard** *Examples:* gasoline, methyl ethyl ketone
GASOLINE 3	**Gasoline Placard** — May be used in the place of a flammable placard on a cargo tank or a portable tank being used to transport gasoline by highway
COMBUSTIBLE 3	**Combustible Placard** *Examples:* diesel, fuel oils, pine oil
FUEL OIL 3	**Fuel Oil Placard** — May be used in place of a combustible placard on a cargo tank or portable tank being used to transport fuel oil by highway.

Class 4: Flammable Solids, Spontaneously Combustible Materials, and Dangerous-When-Wet Materials (49 *CFR* 173.124)

This class is divided into three divisions: 4.1 Flammable Solids, 4.2 Spontaneously Combustible Materials, and 4.3 Dangerous When Wet (see definitions below).

First responders must be aware that fires involving Class 4 materials may be extremely difficult to extinguish. The primary hazards of Class 4 materials are thermal, chemical, and mechanical and may also include the following hazards:

- Thermal hazards (heat) from fires that may start spontaneously or upon contact with air or water
- Thermal hazards (heat) from fires and vapor explosions
- Thermal hazards (heat) from molten substances
- Chemical hazards from irritating, corrosive, and/or highly toxic gases and vapors produced by fire or decomposition
- Severe chemical burns
- Mechanical effects from unexpected, violent chemical reactions and explosions
- Mechanical hazards from a BLEVE, for containers exposed to heat or flame (or if contaminated with water, particularly for Division 4.3)
- Production of hydrogen gas from contact with metal
- Production of corrosive solutions on contact with water, for Division 4.3
- May spontaneously reignite after fire is extinguished
- Environmental hazards (pollution) caused by runoff from fire control

Continued

Table 22.16 (continued)
U.S. DOT Placard Hazard Classes and Divisions

Class 4 (continued)

Division 4.1: Flammable Solid Material — Includes (1) wetted explosives, (2) self-reactive materials that can undergo a strongly exothermal decomposition, and (3) readily combustible solids that may cause a fire through friction, certain metal powders that can be ignited and react over the whole length of a sample in 10 minutes or less, or readily combustible solids that burn faster than 2.2 mm/second:

- **Wetted explosives:** Explosives with their explosive properties suppressed by wetting with sufficient alcohol, plasticizers, or water
- **Self-reactive materials:** Materials liable to undergo a strong exothermic decomposition at normal or elevated temperatures due to excessively high transport temperatures or to contamination
- **Readily combustible solids:** Solids that may ignite through friction or any metal powders that can be ignited

Examples: phosphorus heptasulfide, paraformaldehyde, magnesium

Division 4.2: Spontaneous Combustible Material — Includes (1) a pyrophoric material (liquid or solid) that, without an external ignition source, can ignite within 5 minutes after coming in contact with air and (2) a self-heating material that, when in contact with air and without an energy supply, is liable to self-heat

Examples: sodium sulfide, potassium sulfide, phosphorus (white or yellow, dry), aluminum and magnesium alkyls, charcoal briquettes

Division 4.3: Dangerous-When-Wet Material — Material that, by contact with water, is liable to become spontaneously flammable or to release flammable or toxic gas at a rate greater than 1 liter per kilogram of the material per hour

Examples: magnesium powder, lithium, ethyldichlorosilane, calcium carbide, potassium

Class 5: Oxidizers and Organic Peroxides (49 *CFR* 173.127 and 128)

This class is divided into two divisions: 5.1 Oxidizers and 5.2 Organic Peroxides (see definitions below). Oxygen supports combustion, so the primary hazards of Class 5 materials are fires and explosions with their associated thermal and mechanical hazards:

- Thermal hazards (heat) from fires that may explode or burn extremely hot and fast
- Explosive reactions to contact with hydrocarbons (fuels)
- Chemical hazards from toxic gases, vapors, and dust
- Chemicals hazards from toxic products of combustion
- Chemical burns
- Ignition of combustibles (including paper, cloth, wood, etc.)
- Mechanical hazards from violent reactions and explosions
- Accumulation of toxic fumes and dusts in confined spaces
- Sensitivity to heat, friction, shock, and/or contamination with other materials

Continued

Table 22.16 (continued)
U.S. DOT Placard Hazard Classes and Divisions

Class 5 (continued)

OXIDIZER 5.1	**Division 5.1: Oxidizer** — Material that may, generally by yielding oxygen, cause or enhance the combustion of other material *Examples:* chromium nitrate, copper chlorate, calcium permanganate, ammonium nitrate fertilizer
ORGANIC PEROXIDE 5.2	**Division 5.2: Organic Peroxide** — Any organic compound containing oxygen (O) in the bivalent -O-O- structure and which may be considered a derivative of hydrogen peroxide, where one or more of the hydrogen atoms has been replaced by organic radicals *Examples*: liquid organic peroxide type B

Class 6: Poison (Toxic) and Poison Inhalation Hazard (49 *CFR* 173.132 and 134)

A poisonous material is a material, other than a gas, that is known to be toxic to humans. The primary hazards of Class 6 materials are chemical and thermal and may include the following:

- Toxic effects due to exposure via all routes of entry
- Chemicals hazards from toxic and/or corrosive products of combustion
- Thermal effects (heat) from substances transported in molten form
- Flammability and its associated thermal hazards (heat) from fires

TOXIC 6 **POISON** 6	**Division 6.1: Poisonous Material** — Material, other than a gas, that is known to be so toxic to humans as to afford a hazard to health during transportation or that is presumed to be toxic to humans based on toxicity tests on laboratory animals *Examples:* aniline, arsenic, mustard agents, nerve agents, hydrogen cyanide, most riot control agents
No Placard for Division 6.2, see labels	**Division 6.2: Infectious Substance** — Material known to contain or suspected of containing a pathogen. A pathogen is a virus or microorganism (including its viruses, plasmids, or other genetic elements, if any) or a proteinaceous infectious particle (prion) that has the potential to cause disease in humans or animals.
INHALATION HAZARD 6	**Inhalation Hazard Placard** — Used for any quantity of Division 6.1, Zones A or B inhalation hazard only (see Division 2.3 for hazard zones)

Continued

Table 22.16 (continued)
U.S. DOT Placard Hazard Classes and Divisions

Class 7: Radioactive Materials (49 *CFR* 173.403)

A radioactive material means any material having a specific activity greater than 70 becquerels per gram (0.002 microcurie per gram). The primary hazard of Class 7 materials is radiological, including burns and biological effects.

Radioactive Placard — Is required on certain shipments of radioactive materials; vehicles with this placard are carrying "highway route controlled quantities" of radioactive materials and must follow prescribed, predetermined transportation routes

Examples: solid thorium nitrate, uranium hexafluoride

Class 8: Corrosive Materials (49 *CFR* 173.136)

A corrosive material means a liquid or solid that causes full thickness destruction of human skin at the site of contact within a specific period of time or a liquid that has a severe corrosion rate on steel or aluminum. The primary hazards of Class 8 materials are chemical and thermal, and may include the following hazards:

- Chemical burns

- Toxic effects due to exposure via all routes of entry

- Thermal effects, including fire, caused by chemical reactions generating heat

- Reactivity to water

- Mechanical effects caused by BLEVEs and violent chemical reactions

Corrosive Placard

Examples: battery fluid, chromic acid solution, soda lime, sulfuric acid, hydrochloric acid (muriatic acid), sodium hydroxide, potassium hydroxide

Continued

Table 22.16 (continued)
U.S. DOT Placard Hazard Classes and Divisions

Class 9: Miscellaneous Dangerous Goods (49 *CFR* 173.140)

A miscellaneous dangerous good is a material that (1) has an anesthetic, noxious, or other similar property that could cause extreme annoyance or discomfort to flight crew members and would prevent their correct performance of assigned duties; (2) is a hazardous substance or a hazardous waste; or (3) is an elevated temperature material; or (4) is a marine pollutant.

Miscellaneous dangerous goods will primarily have thermal and chemical hazards. For example, polychlorinated biphenyls (PCBs) are carcinogenic, while elevated temperature materials may present some thermal hazards. However, hazardous wastes may present any of the hazards associated with the materials in normal use.

	Miscellaneous Placard *Examples:* blue asbestos, polychlorinated biphenyls (PCBs), solid carbon dioxide (dry ice)
	Dangerous Placard — A freight container, unit load device, transport vehicle, or railcar that contains nonbulk packaging with two or more DOT Chart 12, Table 2 categories of hazardous materials may be placarded *DANGEROUS*. However, when 2,205 lbs (1,000 kg) or more of one category of material is loaded at one loading facility, the placard specified in DOT Chart 12, Table 2 must be applied.

Table 22.17
Canadian Transportation Placards, Labels, and Markings

Class 1: Explosives

Placard and Label (1.1)	**Class 1.1** — Mass explosion hazard
Placard and Label (1.2)	**Class 1.2** — Projection hazard but not a mass explosion hazard
Placard and Label (1.3)	**Class 1.3** — Fire hazard and either a minor blast hazard or a minor projection hazard or both but not a mass explosion hazard
Placard and Label (1.4)	**Class 1.4** — No significant hazard beyond the package in the event of ignition or initiation during transport * = Compatibility group letter
Placard and Label (1.5)	**Class 1.5** — Very insensitive substances with a mass explosion hazard
Placard and Label (1.6)	**Class 1.6** — Extremely insensitive articles with no mass explosion hazard

Class 2: Gases

Placard and Label (2.1)	**Class 2.1** — Flammable Gases
Placard and Label (2.2)	**Class 2.2** — Nonflammable and nontoxic Gases

Continued

Table 22.17 (continued)
Canadian Transportation Placards, Labels, and Markings

Class 2: Gases (continued)

Placard and Label	Class 2.3 — Toxic Gases
Placard and Label	Oxidizing Gases

Class 3: Flammable Liquids

Placard and Label	Class 3 — Flammable Liquids

Class 4: Flammable Solids, Substances Liable to Spontaneous Combustion, and Substances that on Contact with Water Emit Flammable Gases (Water-Reactive Substances)

Placard and Label	Class 4.1 — Flammable Solids
Placard and Label	Class 4.2 — Substances Liable to Spontaneous Combustion
Placard and Label	Class 4.3 — Water-Reactive Substances

Class 5: Oxidizing Substances and Organic Peroxides

Placard and Label	Class 5.1 — Oxidizing Substances
Placard and Label	Class 5.2 — Organic Peroxides

Continued

Class 6: Toxic and Infectious Substances

Placard and Label	**Class 6.1 — Toxic Substances**
Label Only	**Class 6.2 — Infectious Substances** Text: INFECTIOUS In case of damage or leakage, Immediately notify local authorities AND INFECTIEUX En cas de Dommage ou de fuite communiquer Immédiatement avec les autorités locales ET CANUTEC 613-996-6666
Placard Only	**Class 6.2 — Infectious Substances**

Class 7: Radioactive Materials

Label and Optional Placard	**Class 7 — Radioactive Materials** **Category I** — White RADIOACTIVE CONTENTS.....................CONTENU ACTIVITY..........................ACTIVITÉ
Label and Optional Placard	**Class 7 — Radioactive Materials** **Category II** — Yellow RADIOACTIVE CONTENTS.....................CONTENU ACTIVITY..........................ACTIVITÉ INDICE DE TRANSPORT INDEX
Label and Optional Placard	**Class 7 — Radioactive Materials** **Category III** — Yellow RADIOACTIVE CONTENTS.....................CONTENU ACTIVITY..........................ACTIVITÉ INDICE DE TRANSPORT INDEX

Continued

Class 7: Radioactive Materials (continued)

Placard

Class 7 — Radioactive Materials

The word RADIOACTIVE is optional.

Class 8: Corrosives

Placard and Label

Class 8 — Corrosives

Class 9: Miscellaneous Products, Substances, or Organisms

Placard and Label

Class 9 — Miscellaneous Products, Substances, or Organisms

Other Placards, Labels, and Markings

Danger Placard

Elevated Temperature Sign

Fumigation Sign

Text is in both English and French

Marine Pollutant Mark

The text is MARINE POLLUTANT or POLLUANT MARIN

Table 22.18
Unique U.S. DOT Labels

Subsidiary Risk Labels

Subsidiary risk labels may be used for the following classes: Explosives, Flammable Gases, Flammable Liquids, Flammable Solids, Corrosives, Oxidizers, Poisons, Spontaneously Combustible Materials, and Dangerous-When-Wet Materials.

Class 1: Explosives

Explosive Subsidiary Risk Label

Class 3: Flammable Liquid

Flammable Liquid Label — Marks packages containing flammable liquids.

Examples: gasoline, methyl ethyl ketone

Class 6: Poison (Toxic), Poison Inhalation Hazard, Infectious Substance

Infectious Substances Label — Marks packages with infectious substances (viable microorganism, or its toxin, which causes or may cause disease in humans or animals).

This label may be used to mark packages of Class 6.2 materials as defined in 49 *CFR* 172.432.

Examples: anthrax, hepatitis B virus, *escherichia* coli (*E.* coli)

Biohazard Label — Marks bulk packaging containing a regulated medical waste as defined in 49 *CFR* 173.134(a)(5).

Examples: used needles/syringes, human blood or blood products, human tissue or anatomical waste, carcasses of animals intentionally infected with human pathogens for medical research

Etiological Agents Label — Marks packages containing etiologic agents transported in interstate traffic per 42 *CFR* 72.3 and 72.6

Examples: rabies virus, rickettsia, Ebola virus, salmonella bacteria

Continued

Table 22.18 (continued)
Unique U.S. DOT Labels

Class 7: Radioactive Materials

Packages of radioactive materials must be labeled on two opposite sides, with a distinctive warning label. Each of the three label categories — RADIOACTIVE WHITE-I, RADIOACTIVE YELLOW-II, or RADIOACTIVE YELLOW-III — bears the unique trefoil symbol for radiation.

Class 7 Radioactive I, II, and III labels must always contain the following additional information:

- Isotope name
- Radioactive activity

Radioactive II and III labels will also provide the transport Index (TI) indicating the degree of control to be excercised by the carrier during transportation. The number in the transport index box indicates the maximum radiation level measured (in mrem/hr) at one meter from the surface of the package. Packages with the Radioactive I label have a Transport Index of 0.

RADIOACTIVE I	**Radioactive I Label** — Label with an all-white background color that indicates that the external radiation level is low and no special stowage controls or handling are required.
RADIOACTIVE II	**Radioactive II Label** — Upper half of the label is yellow, which indicates that the package has an external radiation level or fissile (nuclear safety criticality) characteristic that requires consideration during stowage in transportation.
RADIOACTIVE III	**Radioactive III Label** — Yellow label with three red stripes indicates the transport vehicle must be placarded RADIOACTIVE.
FISSILE	**Fissile Label** — Used on containers of fissile materials (materials capable of undergoing fission such as uranium-233, uranium-235, and plutonium-239). The Criticality Safety Index (CSI) must be listed on this label. The CSI is used to provide control over the accumulation of packages, overpacks, or freight containers containing fissile material.
EMPTY	**Empty Label** — Used on containers that have been emptied of their radioactive materials, but still contain residual radioactivity.

Continued

**Table 22.18 (continued)
Unique U.S. DOT Labels**

Aircraft Labels

MAGNETIZED MATERIAL	**Magnetized Material Label**—Marks magnetized materials that could cause navigation deviations on aircraft.
DANGER	**Danger - Cargo Aircraft Only** — Used to indicate materials that cannot be transported on passenger aircraft.

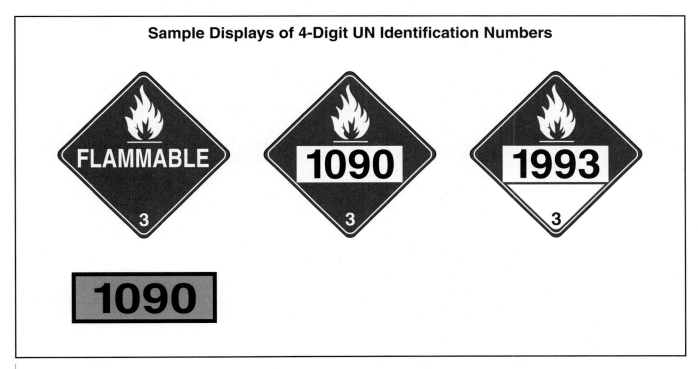

Sample Displays of 4-Digit UN Identification Numbers

Figure 22.55 The four-digit UN identification number can be used to identify hazardous materials in the *Emergency Response Guidebook* (*ERG*).

Table 22.19
Unique U.S. DOT Markings

Marking	Description
HOT	**Hot Marking** — Has the same dimensions as a placard and is used on elevated temperature materials. *Note:* Bulk containers of molten aluminum or molten sulfur must be marked MOLTEN ALUMINUM or MOLTEN SULFUR, respectively.
MARINE POLLUTANT	**Marine Pollutant Marking** — Must be displayed on packages of substances designated as marine pollutants. *Examples:* cadmium compounds, copper cyanide, mercury based pesticides
INHALATION HAZARD	**Inhalation Hazard Marking** — Used to mark materials that are poisonous by inhalation. *Examples:* anhydrous ammonia, methyl bromide, hydrogen cyanide, hydrogen sulfide
DANGER — THIS UNIT IS UNDER FUMIGATION — DO NOT ENTER	**Fumigant Marking** — Warning affixed on or near each door of a transport vehicle, freight container, or railcar in which the lading has been fumigated or is undergoing fumigation with any material. The vehicle, container, or railcar is considered a package containing a hazardous material unless it has been sufficiently aerated so that it does not pose a risk to health and safety.
↑↑ ↑↑	**Orientation Markings** — Markings used to designate the orientation of the package. Sometimes these markings will be accompanied by words such as "this side up."
CONSUMER COMMODITY ORM-D	**ORM-D** — Used on packages of ORM-D materials. *Examples:* consumer commodities, small arms cartridges
CONSUMER COMMODITY ORM-D-AIR	**ORM-D-AIR** — Used on packages of ORM-D materials shipped via air.
INNER PACKAGES COMPLY WITH PRESCRIBED SPECIFICATIONS	**Inner Packaging** — Used on authorized packages containing hazardous materials being transported in an overpack as defined in 49 *CFR* 171.8.

The *Emergency Response Guidebook (ERG)* provides a key to the four-digit identification numbers in the yellow section (see *Emergency Response Guidebook* section). Therefore, if the four-digit UN identification number of a hazardous material is identified, firefighters can use the *ERG* to determine appropriate response information based on the material involved. The four-digit UN identification number will also appear on shipping papers, and it should match the numbers displayed on the exteriors of tanks or shipping containers.

Other North American Highway Vehicle Identification Markings

[NFPA® 472: 5.2.1.2.1(1)]

In addition to UN commodity identification numbers, highway transportation vehicles may have other identification markings. These markings may include company names, logos, specific tank colors for certain tanks, stenciled commodity names (such as *Liquid Propane*), and manufacturers' specification plates (**Figure 22.57**). Specification plates provide information about the standards to which the container/tank was built.

North American Railroad Tank Car Markings

There may be a variety of markings on railroad tank cars that responders can use to gain valuable information about the tank and its contents such as the following:

* Initials (reporting marks) and number
* Capacity stencil
* Specification marking

The *ERG* provides a key to these markings in the railcar identification chart, and more information is provided in sections that follow. Additionally, manufacturers' names on cars may provide some contact information. Dedicated railcars transporting a single material must have the name of that material painted on the car. Likewise, a number of hazardous materials transported by rail are required to have their names stenciled on the sides of the car in 4-inch (102 mm) letters.

Initials and numbers (reporting marks). Initials and numbers may be used to get information about the car's contents from the railroad's computer or the shipper. Tank cars (like all other freight cars) are marked with their own unique sets of reporting marks. These reporting marks should match the initials and numbers provided on the shipping papers for the car. They are stenciled on both sides (to the left when facing the side of the car) and both ends (upper center) of the tank car tank (**Figure 22.58**). Some shippers and car owners also stencil the top of the car with the car's initials and numbers to help identify the car in case an accident turns it on its side.

Capacity stencil. The capacity stencil shows the volume of the tank car tank. The volume in gallons (and sometimes liters) is stenciled on both ends of the car under the car's initials and number. The volume in pounds (and sometimes kilograms) is stenciled on the sides of the car under the car's initials and number (**Figure 22.59**). The term *load limit* may be used to mean the same thing as *capacity*. For certain tank

Figure 22.56 Sometimes four-digit ID numbers are preceded by the letters *NA* or *UN* (abbreviations for North America and United Nations). *Courtesy of Rich Mahaney.*

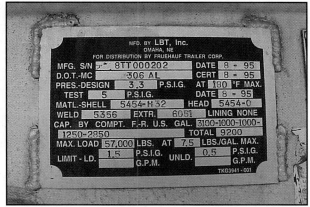

Figure 22.57 Manufacturer's specification plates provide a wealth of information about the standards to which the container/tank was built.

Initials and Numbers — Combination of letters and numbers stenciled on rail tank cars that may be used to get information about the car's contents from the railroad's computer or the shipper.

Capacity Stencil — Number stenciled on the exterior of tank cars to indicate the volume of the tank.

cars, the water capacity (water weight) of the tank, in pounds (and typically kilograms) is stenciled on the sides of the tank near the center of the car.

Specification Marking. The specification marking indicates the standards to which a tank car was built. The marking is stenciled on both sides of the tank. When facing the

Figure 22.58 *CELX 6430* are the reporting marks and car number on this fictional tank car. The other stencils depicted here indicate such things as the tank capacity (*CAPY* on the end view), various weights (load limit [*LD LMT*] and light weight [*LT WT*, the weight of the car unloaded]), and the *NEW* date (month and year). On a real car, these numbers will not all be zeros.

side of the car, the marking will be to the right (opposite from the initials and number) **(Figure 22.60 a)**. The specification marking is also stamped into the tank heads where it is not readily visible. Firefighters can also get specification information from the railroad,

Figure 22.59 An example of water capacity markings found on tank cars. On a real tank, the zeros will be replaced with actual weights.

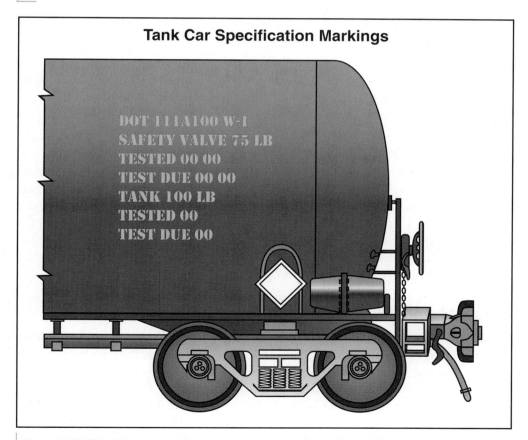

Figure 22.60a The top line of stencils (beginning with *DOT*) provides the specification markings indicating the standards to which the tank car was built.

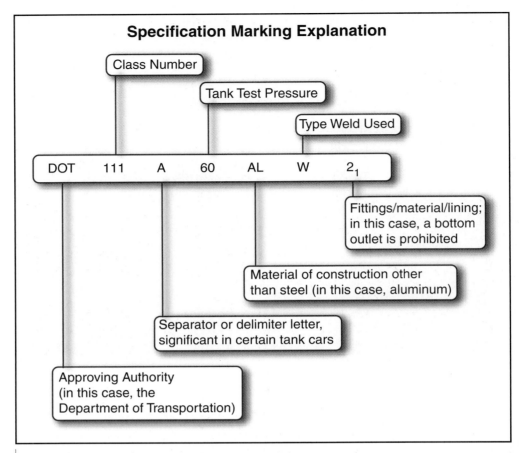

Specification Marking Explanation

Class Number

Tank Test Pressure

Type Weld Used

DOT 111 A 60 AL W 2_1

Fittings/material/lining; in this case, a bottom outlet is prohibited

Material of construction other than steel (in this case, aluminum)

Separator or delimiter letter, significant in certain tank cars

Approving Authority (in this case, the Department of Transportation)

Figure 22.60b Key to tank car specification markings.

shipper, car owner, or the Association of American Railroads (from the car's Certificate of Construction) by using the car's initials and number. **Figure 22.60 b** provides a brief explanation of tank car specification markings.

International Intermodal Container/Tank Markings

In addition to DOT-required placards, intermodal tanks and containers are marked with initials (reporting marks) and tank numbers **(Figure 22.61)**. These markings are generally found on the right-hand side of the tank or container as the emergency responder faces it from either the sides or the ends. The markings are either on the tank/container or the frame. As with tank car reporting marks, emergency responders can use this information in conjunction with the shipping papers or computer data to identify and verify the contents of the tank or container.

Other Markings and Colors

In addition to DOT transportation placards, labels, and markings, a number of other marking systems, labeling systems, color codes, and signs may indicate the presence of hazardous materials at fixed facilities, on pipelines, on piping systems, and on other containers. These other markings may be as simple as the word *chlorine* stenciled on the outside of a fixed-facility tank or as complicated as a site-specific hazard communication system using a unique combination of labels, placards, emergency contact information, and color codes. Some fixed-facility containers may have identification numbers that correspond to site or emergency plans that provide details on the product, quantity, and other pertinent information.

Figure 22.61 Intermodal tanks are also marked with initials (reporting marks) and tank numbers (top line of stencils: DCIU 843037 2). *Courtesy of Rich Mahaney.*

Firefighters need to be familiar with some of the more widely used specialized marking systems for hazardous materials. The sections that follow highlight the most common specialized systems in North America, including NFPA® 704, common hazardous communication labels, military markings, pipeline identifications, and pesticide labels.

NFPA® 704 System

The information in NFPA® 704, *Standard System for the Identification of the Hazards of Materials for Emergency Response*, gives a widely recognized method for indicating the presence of hazardous materials at commercial, manufacturing, institutional, and other fixed-storage facilities. Use of this system is commonly required by local ordinances for all occupancies that contain hazardous materials. It is designed to alert emergency responders to health, flammability, instability, and related hazards (specifically, oxidizers and water-reactive materials) that may present as short-term, acute exposures resulting from a fire, spill, or similar emergency. The NFPA® 704 system is *not* designed for the following situations or hazards:

- Transportation
- General public use
- Nonemergency occupational exposures
- Explosive and blasting agents, including commercial explosive materials
- Chronic health hazards
- Etiologic agents, and other similar hazards

Specifically, the NFPA® 704 system uses a rating system of numbers from *0* to *4*. The number *0* indicates a minimal hazard, whereas the number *4* indicates a severe hazard. The rating is assigned to three categories: health, flammability, and instability. The rating numbers are arranged on a diamond-shaped marker or sign. The

NFPA® 704 Limitations

NFPA® 704 markings provide very useful information but the system does have its limitations. For example, an NFPA® diamond doesn't tell exactly what chemical or chemicals may be present in specific quantities. Nor does it tell exactly where hazardous materials may be located when the sign is used for a building, structure, or area (such as a storage yard) rather than an individual container. Positive identification of the materials needs to be made through other means such as container markings, employee information, company records, and pre-incident surveys.

health rating is located on the blue background, the flammability hazard rating is positioned on the red background, and the instability hazard rating appears on a yellow background (**Figure 22.62**). As an alternative, the backgrounds for each of these rating positions may be any contrasting color, and the numbers (*0 to 4*) may be represented by the appropriate color (blue, red, and yellow).

Special hazards are located in the six o'clock position and have no specified background color; however, white is most commonly used. Only two special hazard symbols are presently authorized for use in this position by the NFPA®: *W* and *OX* (respectively, indicating unusual reactivity with water or that the material is an oxidizer). However, firefighters may see other symbols in the white quadrant on old placards, including the trefoil radiation symbol. If more than one special hazard is present, multiple symbols may be seen.

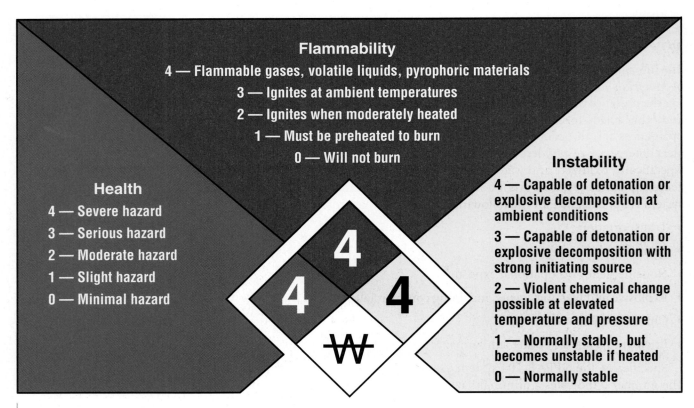

Figure 22.62 Key to the NFPA® 704 numerical ratings.

Hazard Communications Labels and Markings

The U.S. Occupational Safety and Health Administration's Hazard Communication Standard (HCS) requires employers to identify hazards in the workplace and train employees how to recognize those hazards. It also requires the employer to ensure that all hazardous material containers are labeled, tagged, or marked with the identity of the substances contained in them along with appropriate hazard warnings. The standard does not specify what system (or systems) of identification must be used, leaving that to be determined by individual employers. Firefighters, then, may encounter a variety of different (and sometimes unique) labeling and marking systems in their jurisdictions **(Figure 22.63)**. Conducting preincident surveys should assist firefighters in identifying and understanding these systems.

Canadian Workplace Hazardous Materials Information System

Like the U.S. HCS, the Canadian Workplace Hazardous Materials Information System (WHMIS) requires that hazardous products be appropriately labeled and marked. It also spells out requirements for MSDSs. As with the HCS, there are different ways for Canadian employers to meet the requirements of WHMIS; however, two types of labels will most commonly be used: the supplier label **(Figure 22.64)** and the workplace label. These labels will include information such as the product name, a statement that an MSDS is available, and other information that will vary depending on the type of label (supplier labels will include information about the supplier). **Table 22.20** shows the WHMIS Symbols and Hazard Classes.

Manufacturers' Labels and Signal Words

Under the HCS, chemical manufacturers and importers are required to provide appropriate labels on their product containers. Manufacturers' labels provide a variety of information to firefighters, including the name of the product, manufacturer's contact information, and precautionary hazard warnings. These labels may also provide directions for use and handling, names of active ingredients, first aid instructions, and other pertinent information.

Under the U.S. Federal Hazardous Substances Act (FHSA), labels on products destined for consumer households must incorporate one of the following four *signal words* to indicate the degree of hazard associated with the product **(Figure 22.65)**:

- *CAUTION* — Indicates the product may have minor health effects (such as eye or skin irritation)

- *WARNING* — Indicates the product has moderate hazards such as significant health effects or flammability

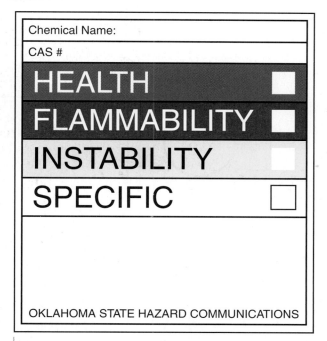

Figure 22.63 During preincident surveys and inspections, firefighters should familiarize themselves with the different labeling and marking systems used in their jurisdictions.

Figure 22.64 An example of a blank Canadian supplier label. Completed supplier labels must be provided on all controlled products received at Canadian workplaces.

Table 22.20
WHMIS Symbols and Hazard Classes

Symbol	Hazard Class	Description
	Class A: Compressed Gas	Contents under high pressure; cylinder may explode or burst when heated, dropped, or damaged
	Class B: Flammable and Combustible Material	May catch fire when exposed to heat, spark, or flame; may burst into flames
	Class C: Oxidizing Material	May cause fire or explosion when in contact with wood, fuels, or other combustible material
	Class D, Division 1: Poisonous and Infectious Material: Immediate and serious toxic effects	Poisonous substance; a single exposure may be fatal or cause serious or permanent damage to health
	Class D, Division 2: Poisonous and Infectious Material: Other toxic effects	Poisonous substance; may cause irritation; repeated exposure may cause cancer, birth defects, or other permanent damage
	Class D, Division 3: Poisonous and Infectious Material: Biohazardous infectious materials	May cause disease or serious illness; drastic exposures may result in death
	Class E: Corrosive Material	Can cause burns to eyes, skin, or respiratory system
	Class F: Dangerously Reactive Material	May react violently, causing explosion, fire, or release of toxic gases when exposed to light, heat, vibration, or extreme temperatures

Source: WHMIS = Canadian Workplace Hazardous Materials Information System. Table adapted from Canadian Centre for Occupational Health and Safety (CCOHS) with pictograms from Health Canada.

- *DANGER* — Indicates the highest degree of hazard (used on products that have potentially severe or deadly effects); also used on products that explode when exposed to heat

- *POISON* — Required in addition to DANGER on the labels of highly toxic materials

Military Markings

The U.S. and Canadian military services have their own marking systems for hazardous materials and chemicals in addition to DOT and TC transportation markings. These markings are used on fixed facilities, and they may be seen on military vehicles, although they are not required. Caution must be exercised, however, because the military placard system is not necessarily uniform. Some buildings and areas that store hazardous materials may not be marked due to security reasons. **Table 22.21** provides the U.S. and Canadian military markings for explosive ordnance and fire hazards, chemical hazards, and PPE requirements.

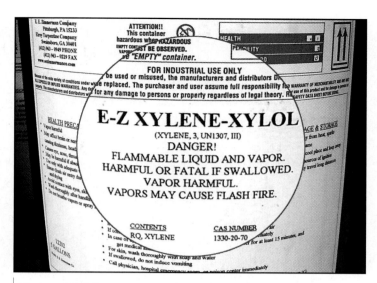

Figure 22.65 Signal words such as *CAUTION, WARNING,* and *DANGER* indicate the degree of hazard associated with the product. *DANGER* indicates the highest degree of hazard. An additional word, *POISON,* is required on highly toxic materials such as pesticides.

Pipeline Identification

Many types of materials, particularly petroleum varieties, are transported across both the United States and Canada in an extensive network of pipelines, most of which are buried in the ground. Where pipelines cross under (or over) roads, railroads, and waterways, pipeline companies must provide markers. They must be in sufficient numbers along the rest of the pipeline to identify the pipe's location. However, pipeline markers do not always mark the exact location of the pipeline. Pipeline markers in the United States and Canada include the signal words CAUTION, WARNING, or DANGER (representing an increasing level of hazard) and contain information describing the transported commodity and the name and emergency telephone number of the carrier (**Figure 22.66**).

CAUTION

The military ships some hazardous materials and chemicals by common carrier. When this is done they are not required to be marked with DOT and TC transportation markings!

Pesticide Labels

The EPA regulates the manufacture and labeling of pesticides. Each EPA label must contain the manufacturer's name for the pesticide and one of the following signal words: *DANGER/ POISON, WARNING,* or *CAUTION*. The words *DANGER/ POISON* are used for highly toxic materials, *WARNING* means moderate toxicity, and *CAUTION* is used for chemicals with relatively low toxicity (**Figure 22.67**). The words *EXTREMELY FLAMMABLE* are also displayed on the label if the contents have a flash point below 80°F (27°C).

The label also lists an EPA registration number. This number normally is used to obtain information about the product from the manufacturer's 24-hour emergency contact. Another requirement is an establishment number that identifies the manufacturing facility. Other information that may be found on these labels includes routes of entry into the body, precautionary statements (such as *Keep out of the reach of children*),

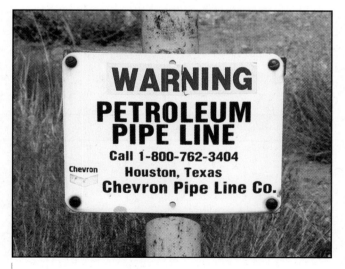

Figure 22.66 Pipeline markers in the United States and Canada include signal words, information describing the transported commodity, and the name and emergency telephone number of the carrier. *Courtesy of Rich Mahaney.*

Table 22.21
U.S. and Canadian Military Symbols

Symbol	Fire (Ordnance) Divisions
	Division 1: Mass Explosion Fire Division 1 indicates the greatest hazard. This division is equivalent to DOT/UN Class 1.1 Explosives Division **Also, this exact symbol may be used for:** **Division 5: Mass Explosion — very insensitive explosives (blasting agents)** This division is equivalent to DOT/UN Class 1.5 Explosives Division
	Division 2: Explosion with Fragment Hazard This division is equivalent to DOT/UN Class 1.2 Explosives Division **Also, this exact symbol may be used for:** **Division 6: Nonmass Explosion — extremely insensitive ammunition** This division is equivalent to DOT/UN Class 1.6 Explosives Division
	Division 3: Mass Fire This division is equivalent to DOT/UN Class 1.3 Explosives Division
	Division 4: Moderate Fire — no blast This division is equivalent to DOT/UN Class 1.4 Explosives Division

Symbol	Chemical Hazards
 "Red You're Dead"	**Wear Full Protective Clothing (Set One)** Indicates the presence of highly toxic chemical agents that may cause death or serious damage to body functions.
 "Yellow You're Mellow"	**Wear Full Protective Clothing (Set Two)** Indicates the presence of harassing agents (riot control agents and smokes).

Continued

Table 22.21 (continued)
U.S. and Canadian Military Symbols

Symbol	Chemical Hazards (continued)
"White is Bright"	**Wear Full Protective Clothing (Set Three)** Indicates the presence of white phosphorus and other spontaneously combustible material.
	Wear Breathing Apparatus Indicates the presence of incendiary and readily flammable chemical agents that present an intense heat hazard. This hazard and sign may be present with any of the other fire or chemical hazards/symbols.
	Apply No Water Indicates a dangerous reaction will occur if water is used in an attempt to extinguish the fire. This symbol may be posted together with any of the other hazard symbols.

Symbol	Supplemental Chemical Hazards
G	**G-Type Nerve Agents** — persistent and nonpersistent nerve agents *Examples: sarin (GB), tabun (GA), soman (GD)*
VX	**VX Nerve Agents** — persistent and nonpersistent V-nerve agents *Example: V-agents (VE, VG, VS)*
BZ	**Incapacitating Nerve Agent** *Examples: lacrymatory agent (BBC), vomiting agent (DM)*
H	**H-Type Mustard Agent/Blister Agent** *Example: persistent mustard/lewisite mixture (HL)*
L	**Lewisite Blister Agent** *Examples: nonpersistent choking agent (PFIB), nonpersistent blood agent (SA)*

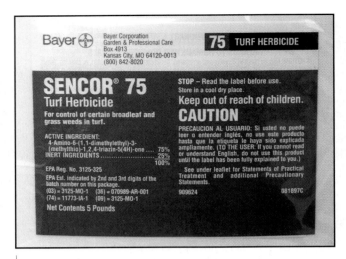

Figure 22.67 According to EPA requirements, pesticide labels must provide the name of the pesticide, the appropriate signal word, a precautionary statement, a hazard statement, and a list of active ingredients. Canadian pesticide labels will also have a PCP (Pest Control Products) Act number.

active ingredients, requirements for storage and disposal, first aid information, antidotes for poisoning (if known), and hazard statements indicating that the product poses an environmental hazard.

Materials originating in Canada carry a Pest Control Products (PCP) Act number. The Canadian Transport Emergency Centre (CANUTEC) operated by TC provides information about these materials when given the PCP registration number. Canadian products also have the same signal words and required information as the United States.

Written Resources

A variety of written resources are available to assist firefighters in identifying hazardous materials at both fixed facilities and transportation incidents. Fixed facilities should have MSDSs, inventory records, and other facility documents in addition to signs, markings, container shapes, and other labels. At transportation incidents, firefighters should be able to use the current *ERG* as well as shipping papers. These resources can be used to both identify and verify the materials involved in the incident.

Operations-Level responders needing response information directly from the manufacturer or shipper can gather contact information from shipping papers and MSDSs or by contacting an emergency response center such as Chemical Transportation Emergency Center (CHEMTREC®) of the American Chemistry Council, Canadian Transport Emergency Centre (CANUTEC), or Emergency Transportation System for the Chemical Industry (SETIQ) (see sidebar). These agencies can provide information about specific products as well as guidance on how to manage the incident.

Emergency Response Centers

Canada and Mexico have government-operated emergency response centers. The U.S. has several emergency response centers that are not government operated. These centers are staffed with experts who can provide 24-hour assistance to emergency responders dealing with haz mat emergencies.

The Canadian Transport Emergency Centre (CANUTEC) is operated by Transport Canada. This national bilingual (English and French) advisory center is part of the Transportation of Dangerous Goods Directorate. CANUTEC has a scientific data bank on chemicals manufactured, stored, and transported in Canada and is staffed by professional scientists who specialize in emergency response and are experienced in interpreting technical information and providing advice. Mexico has two emergency response centers: (1) National Center for Communications of the Civil Protection Agency (CENACOM) and (2) Emergency Transportation System for the Chemical Industry (SETIQ), which is operated by the National Association of Chemical Industries.

In the United States, several emergency response centers, such as the Chemical Transportation Emergency Center (CHEMTREC®), are not government operated. CHEMTREC®, for example, was established by the chemical industry as a public service hotline for firefighters, law enforcement responders, and other

Continued

Emergency Response Centers (continued)

emergency service responders to obtain information and assistance for emergency incidents involving chemicals and hazardous materials.

A full list of emergency response centers and their telephone numbers is provided in the *ERG*. Collect and provide to the center as much of the following information as safely possible:

- Caller's name, callback telephone number, and FAX number
- Location and nature of problem (spill, fire, etc.)
- Name and identification number of material(s) involved
- Shipper/consignee/point of origin
- Carrier name, railcar, or truck number
- Container type and size
- Quantity of material transported/released
- Local conditions (weather, terrain, proximity to schools, hospitals, waterways, etc.)
- Injuries and exposures
- Local emergency services that have been notified

The emergency response center will do the following:
- Confirm that a chemical emergency exists.
- Record details in writing and on tape.
- Provide immediate technical assistance to the caller.
- Contact the shipper of the material or other experts.
- Provide the shipper/manufacturer with the caller's name and callback number so that the shipper/manufacturer can deal directly with the party involved.

United States Centers
- CHEMTREC®: 800-424-9300 (Business CHEMTREC: 202-887-1255, 9–4 p.m. EST)
- CHEM-TEL, INC.: 800-255-3924
- INFOTRAC: 800-535-5053
- 3ECOMPANY: 800-451-8346
- National Response Center: 800-424-8802
- U.S. Army Operations Center (DOD shipments involving explosives and ammunition): 703-697-0218 (call collect)
- Defense Logistics Agency (DOD shipments involving other dangerous goods): 800-851-8061

Canada Centers
- CANUTEC: 613-996-6666 collect (Emergency only)
- Alberta: 800-272-9600
- British Columbia: 800-663-3456
- Manitoba: 204-945-4888
- New Brunswick: 800-565-1633
- Newfoundland: 709-772-2083
- Northwest Territories: 867-920-8130
- Nova Scotia: 800-565-1633
- Nunavut: 867-979-6262
- Ontario, Quebec: Local police
- Prince Edward Island: 800-565-1633

Continued

Shipping Papers

Shipments of hazardous materials must be accompanied by shipping papers that describe them. The information can be provided on a bill of lading, waybill, or similar document (**Figure 22.68**). The general location and type of paperwork change according to the mode of transport (**Table 22.22**). In addition, the exact location of the documents varies. The exceptions are hazardous waste shipments, which must be accompanied by a Uniform Hazardous Waste Manifest document. Instructions for describing hazardous materials are provided in the DOT/TC regulations as follows:

- Proper shipping name of the material

- Hazard class represented by the material

- Packing group assigned to the material

- Quantity of material

Bill of Lading — Shipping paper used by the trucking industry (and others) indicating origin, destination, route, and product; placed in the cab of every truck tractor. This document establishes the terms of a contract between shippers and transportation companies; serves as a document of title, contract of carriage, and receipt for goods.

Waybill — Shipping paper used by a railroad to indicate origin, destination, route, and product. Each car has a waybill that the conductor carries.

Table 22.22
Shipping Paper Identification

Transportation Mode	Shipping Paper Name	Location of Papers	Party Responsible
Air	Air Bill	Cockpit	Pilot
Highway	Bill of Lading	Vehicle Cab	Driver
Rail	Waybill/ Consist	Engine or Caboose	Conductor
Water	Dangerous Cargo Manifest	Bridge or Pilot House	Captain or Master

In addition, special description requirements apply to certain types of materials (for example, those that cause poison by inhalation, radioactive materials, and hazardous substances) and modes of transportation. Because of efforts to harmonize U.S. shipping regulations with international regulations, emergency responders may now see alternate basic description sequences, subsidiary hazard class or division numbers, types and number of packages, marking of limited quantity packagings, and other additional information on shipping papers.

When firefighters know that a close approach to an incident is safe, they can then examine the cargo shipping papers. They may need to check with the responsible party in order to locate these documents. If the responsible party is not carrying them, firefighters will need to check the appropriate locations. In trucks and airplanes,

THIS MEMORANDUM is an acknowledgement that a Bill of Lading has been issued and is not the Original Bill of Lading nor a copy or duplicate, covering the property named herein, and is intended solely for filing or record.
RECEIVED, subject to the classifications and lawfully filed tariffs in effect on the date of the receipt by the carrier of the property described in the Original Bill of Lading.

SHELL CHEMICAL COMPANY

BILLING OFFICE MEMORANDUM 1

SC-6084.4 (REV. 2-91)

AT	PAGE	OF	SHIPPER'S I.D. NO.	B/L SEQ. NO.	CARRIER NAME OR RAIL CAR INITIALS AND NO.	SCAC	SHIPPING DATE
DEER PARK	TX		141 A04602	978	MATLACK 6189	MTLK	11-05-94

SHIP VIA ▷

FOR CHEMICAL EMERGENCY
Spill, Leak, Fire, Exposure or Accident
CALL CHEMTREC Day or night
800-424-9300

ROUTE CODE | ROUTE

SEAL NO.'S 188288-90

Subject to Section 7 of Conditions of applicable bill of lading; if this shipment is to be delivered to the consignee without recourse on the consignor, the consignor shall sign the following statement:
The carrier shall not make delivery of this shipment without payment of freight and all other lawful charges.
SHELL CHEMICAL COMPANY.
(Signature of Consignor)

CUSTOMER NO. 39300001 CITY 100 STATE 09 DATE ORDERED CUSTOMER ORDER NO. 90576 R-4

CONSIGNED TO: JOHN OGORMAN
1123 DATELOG WAY
HOUSTON, TX 77090

BILL TO: JPO INDUSTRIES
P. O. BOX 90674
HOUSTON, TX 77090

FREIGHT TERMS
CODE 2

SR DIST.	CR DIST.	ORDERED BY AND DATE DIST.	MO	DAY	YR.	SUGGESTED SHIPPING DATE	REQUESTED DELIVERY DATE
82	82	47	11	01	93	11 05 93	11 05 AM

ORDERER'S INITIALS TELEPHONE & TWX: AMH 713-444-2430
SPLC CODES ORIGIN DEST.

DESCRIPTION OF ARTICLES, SPECIAL MARKS AND EXCEPTIONS

HM	NO. & KIND PACKAGES	TRUCK/ TRAILER NUMBER	TRUCK/ TRAILER OR T/C COMPT	CAPACITY	GALLONS OUTAGE AT LOADING TEMP.	LOADING TEMP. °F	LB./GAL AT LOAD TEMP.	GALLON CORR. FACTOR	NET WEIGHT OR GALLONS AT 60°F	GROSS WEIGHT (Subject to correction)

X 1 T/T EPICHLOROHYDRIN, 6.1, UN 2023, II, RQ, GUIDE 30

LINE 01 4,500 GAL EPICHLOROHYDRIN
BULK
CA 32410

GROSS	TARE	NET
70,240	25,960	44,280

SPECIAL INSTRUCTIONS: ANY UNLOADING DETENTION CHARGES BILL TO CONSIGNEE EQUIP. T/T WITH 2" CAMLOCK FITTING FOR UNLOADING AND 2" MALE CAMLOCK FITTING FOR VENTING VAPORS. DELIVER 10 AM - 3 PM 11/5.

IF SHIPMENT IS PREPAID, MAIL FREIGHT BILL IN DUPLICATE WITH NO. 4 COPY OF B/L TO: Shell Chemical Company P.O. Box 1876 Houston, Texas 77251 Attention: Chemical Products Accounting, Freight Accounting

SHIPMENTS VIA MOTOR CARRIER ▶ D.O.T. HAZARDOUS MATERIALS PLACARDS FURNISHED BY: ☐ SHIPPER ☐ CARRIER

1. The agreed or declared value of the property is hereby specifically stated by the shipper to be not exceeding 50 cents per pound where the rate is dependent on value. 2. Description and gross weight thereof as shown herein are correct, per Agreement filed with Weighing and Inspection Bureau, if applicable. 3. This is to certify that the above-named materials are properly classified, described, packaged, marked and labeled and are in proper condition for transportation according to the appl cable regulations of the Department of Transportation.
SHELL CHEMICAL COMPANY, Shipper — is a division of, and means herein, SHELL OIL COMPANY, a Delaware Corporation

Per ______

Carrier certifies that the container supplied by Carrier for this shipment is a proper container for transportation of the Materials as above described.
Carrier ______
Per Agent

DELIVERY RECEIPT — Received in good order Customer/Customer's Carrier certifies that the container supplied by it for this shipment is a proper container for transportation
For ______
By

Figure 22.68 An example of a hazardous material shipping paper. Note the *X* placed in the column captioned *HM* for hazardous material. *Courtesy of Shell Chemical Company.*

these papers are placed near the driver or pilot. On ships and barges, the papers are placed on the bridge or in the pilothouse of a controlling tugboat. On trains, the way-bills (each car's cargo lists), wheel reports, and/or consists (also called *wheels*) (entire train's cargo lists) may be placed in the engine, caboose, or both. During preincident surveys, the location of the papers for a specific rail line can be determined.

Material Safety Data Sheets

Material Safety Data Sheet (MSDS) — Form provided by the manufacturer and blender of chemicals that contains information about chemical composition, physical and chemical properties, health and safety hazards, emergency response procedures, and waste disposal procedures of a specified material.

A material safety data sheet (MSDS) or *safety data sheet (SDS)* is a detailed information bulletin prepared by the manufacturer or importer of a chemical that describes or gives the following information:

- Hazardous ingredients
- Physical and chemical properties (including fire and explosion hazard data)
- Physical and health hazards
- Routes of exposure
- Precautions for safe handling and use
- Emergency and first-aid procedures
- Control measures for the product

MSDSs/SDSs are often the best sources of detailed information about a particular material to which firefighters have access. Firefighters can acquire them from the manufacturer of the material, the supplier, the shipper, an emergency response center such as CHEMTREC®, or the facility hazard communication plan. They are sometimes attached to shipping papers and containers as well. Both OSHA and Canadian regulations have requirements on MSDS contents; however, both countries are switching to a Globally Harmonized System (GHS) SDS format. Until the switch is complete, responders may see MSDSs developed to American National Standards Institute (ANSI) standards, OSHA MSDS standards, Canadian MSDS standards, or the new GHS standard.

The GHS for Hazard Classification and Communication sets forth recommendations for minimum information to be provided on SDSs. These sheets are being used worldwide. SDSs must include the following sections:

1. Identification
2. Hazard(s) identification
3. Composition/information on ingredients
4. First aid measures
5. Fire fighting measures
6. Accidental release measures
7. Handling and storage
8. Exposure controls/personal protection
9. Physical and chemical properties
10. Stability and reactivity
11. Toxicological information
12. Ecological information
13. Disposal considerations
14. Transport information

15. Regulatory information

16. Other information

The old OSHA MSDS requirements do not prescribe the precise format for an MSDS. However, they must be written in English for use in the United States and are required to include the following information:

- **Top — Chemical Identity**
  - Chemical and common name(s) must be provided for single chemical substances
  - Identity on the MSDS must be cross-referenced to the identity found on the label

- **Section I — Manufacturer's Information and Chemical Identity**
  - Manufacturer's name and address
  - Emergency telephone number
  - Date prepared

- **Section II — Hazardous Ingredients**
  - Chemical and common names of hazardous components
  - Permissible exposure limits (PELs) and other recommended exposure limits

- **Section III — Physical and Chemical Characteristics**
  - Boiling point
  - Vapor pressure
  - Vapor density
  - Specific gravity
  - Melting point
  - Evaporation rate
  - Solubility in water
  - Physical appearance and odor

- **Section IV — Fire and Explosion Hazard Data**
  - Flashpoint
  - Flammability limits (lower explosive limit [LEL], upper explosive limit [UEL])
  - Extinguishing media
  - Special fire fighting procedures
  - Unusual fire and explosion hazards

- **Section V — Reactivity (Instability) Data**
  - Stability (stable/unstable conditions to avoid)
  - Incompatibility (materials to avoid)
  - Hazardous decomposition or by-products
  - Hazardous polymerization (may or may not occur, conditions to avoid)

- **Section VI — Health Hazard Data**
  - Routes of entry
  - Health hazards (acute and chronic)

Lower Explosive Limit (LEL) — Lower limit at which a flammable gas or vapor will ignite; below this limit the gas or vapor is too *lean* or *thin* to burn (too much oxygen and not enough gas).

Upper Explosive Limit (UEL) — Upper limit at which a flammable gas or vapor will ignite. Above this limit, the gas or vapor is too rich to burn (lacks the proper quantity of oxygen).

— Carcinogenicity

— Signs and symptoms of exposure

— Medical conditions generally aggravated by exposure

— Emergency and first aid procedures

- **Section VII — Precautions for Safe Handling and Use**

— Steps to be taken in case of a release or spill

— Waste disposal method

— Precautions to be taken in handling or storage, and other precautions

- **Section VIII — Control Measures**

— Engineering controls such as ventilation, safe handling procedures, and PPE

— Work/hygienic practices

Table 22.23 covers the information required on a pre-GHS format Canadian MSDS.

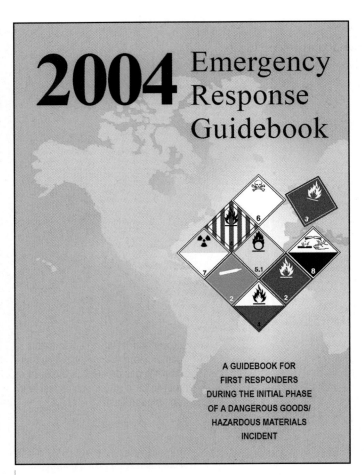

Figure 22.69 The current *Emergency Response Guidebook (ERG)* was developed by the U.S. DOT, Transport Canada (TC), and Mexico's Secretariat of Transport and Communications (SCT).

Emergency Response Guidebook

The current *ERG* was developed jointly by TC, DOT, and the Secretaría de Comunicaciones y Transportes (SCT) of Mexico for use by firefighters, law enforcement, and other emergency services personnel who may be the first to arrive at the scene of a transportation incident involving dangerous goods/hazardous materials (**Figure 22.69**). The *ERG* is primarily a guide to aid emergency responders in quickly identifying the specific or generic hazards of materials involved in an emergency incident and protecting themselves and the general public during the initial response phase of the incident.

Firefighters can locate the appropriate initial action guide page in several different ways. They can identify the four-digit U.N. identification number on a placard or shipping papers and then look up the appropriate guide in the yellow-bordered pages of the guidebook (see next section). They can also use the name of the material involved (if known) in the blue-bordered section of the guidebook (see *ERG* Material Name Index section). Firefighters must exercise extreme care when using this method because many chemical names differ only by a few letters, so exact spelling is important. A third method involves identifying the transportation placard of the material and then using the three-digit guide code associated with the placard in Table of Placards and Initial Response Guide to Use On-Scene located in the front of the *ERG*. **Skill Sheet 22-I-1** provides examples of how to use the *ERG*.

The *ERG* does not address all possible circumstances that may be associated with a dangerous goods/hazardous materials incident. It is primarily designed for use at a dangerous goods/hazardous materials incident occurring on a highway or railroad. There may be limited value in its application at fixed-facility locations.

Table 22.23
Information Disclosed on a Canadian MSDS

Item	Category	Information
1	Hazardous Ingredients	• Information required by Subparagraphs 13(a)(i) to (iv) of the Act • CAS registry number and product identification number • LC_{50} (species and route) • LD_{50} (species and route)
2	Preparation Information	• Name and phone of the group, department or party responsible for the preparation of the material safety data sheet • Date of preparation of the material safety data sheet
3	Product Information	• Manufacturer's information, name, street address, city, province, postal code, and emergency telephone number • Supplier identifier, the supplier's street address, city, province, postal code, and emergency telephone number • Product identifier • Product use
4	Physical Data	• Physical state (that is, gas, liquid, or solid) • Odour and appearance • Odour threshold • Specific gravity • Vapour pressure • Vapour density • Evaporation rate • Boiling point • Freezing point • pH • Coefficient of water/oil distribution
5	Fire or Explosion Hazard	• Conditions of flammability • Means of extinction • Flash point and method of determination • Upper flammable limit • Lower flammable limit • Autoignition temperature • Hazardous combustion products • Explosion data—sensitivity to mechanical impact • Explosion data—sensitivity to static discharge
6	Reactivity Data	• Conditions under which the product is chemically unstable • Name of any substance or class of substance with which the product is incompatible • Conditions of reactivity • Hazardous decomposition products

Continued

Table 22.23 (continued)
Information Disclosed on a Canadian MSDS

Item	Category	Information
7	Toxicological Properties	• Route of entry, including skin contact, skin absorption, eye contact, inhalation, and ingestion • Effects of acute exposure to product • Effects of chronic exposure to product • Exposure limits • Irritancy of product • Sensitization to product • Carcinogenicity • Reproductive toxicity • Teratogenicity • Mutagenicity • Name of toxicologically synergistic products
8	Preventative Measures	• Personal protective equipment to be used • Specific engineering controls to be used • Procedures to be followed in case of leak or spill • Waste disposal • Handling procedure and equipment • Storage requirements • Special shipping information
9	First Aid Measures	• Specific first aid measures

Firefighters at the scene of a haz mat incident should seek additional specific information about any material in question as soon as possible. The information received by contacting the appropriate emergency response agency, calling the emergency response number on the shipping document, or consulting the information on or accompanying the shipping document may be more specific and accurate than the guidebook in providing guidance for the materials involved.

Using Placards and Container Profiles in the *ERG*

Another method of using the *ERG* involves using the placards and container profiles provided in the white pages in the front of the book. Firefighters can identify placards (without the four-digit ID numbers) or container shapes, and then reference the guide number to the orange-bordered page provided in the nearest circle.

ERG *ID Number Index (Yellow Pages)*

The yellow-bordered pages of the *ERG* provide an index list of hazardous materials in numerical order of ID number. This index quickly identifies the Guide to consult for the ID Number/material involved. This list displays the four-digit UN/NA ID number of the material followed by its assigned emergency response Guide and the material's name (**Figure 22.70**).

The purpose of the yellow section in the *ERG* is to enable firefighters to quickly identify the Guide (orange-bordered pages) to consult for the ID number of the substance involved. If a material in the yellow or blue index is highlighted, it means that it releases gases that are toxic inhalation hazard (TIH) materials. These materials require the application of additional emergency response distances.

ID No.	Guide No.	Name of Material	ID No.	Guide No.	Name of Material
1087	116P	Vinyl methyl ether, inhibited	1127	130	Butyl chloride
1087	116P	Vinyl methyl ether, stabilized	1127	130	Chlorobutanes
1088	127	Acetal	1128	129	n-Butyl formate
1089	129	Acetaldehyde	1129	129	Butyraldehyde
1090	127	Acetone	1130	128	Camphor oil
1091	127	Acetone oils	1131	131	Carbon bisulfide
1092	131P	Acrolein, inhibited	1131	131	Carbon bisulphide
1092	131P	Acrolein, stabilized	1131	131	Carbon disulfide
1093	131P	Acrylonitrile, inhibited	1131	131	Carbon disulphide
1093	131P	Acrylonitrile, stabilized	1133	128	Adhesives (flammable)
1098	131	Allyl alcohol	1134	130	Chlorobenzene
1099	131	Allyl bromide	1135	131	Ethylene chlorohydrin
1100	131	Allyl chloride	1136	128	Coal tar distillates, flammable
1104	129	Amyl acetates	1139	127	Coating solution
1105	129	Amyl alcohols	1143	131P	Crotonaldehyde, inhibited
1105	129	Pentanols	1143	131P	Crotonaldehyde, stabilized
1106	132	Amylamines	1144	128	Crotonylene
1107	129	Amyl chloride	1145	128	Cyclohexane
1108	128	n-Amylene	1146	128	Cyclopentane
1108	128	1-Pentene	1147	130	Decahydronaphthalene
1109	129	Amyl formates	1148	129	Diacetone alcohol
1110	127	n-Amyl methyl ketone	1149	128	Butyl ethers
1110	127	Amyl methyl ketone	1149	128	Dibutyl ethers
1110	127	Methyl amyl ketone	1150	130P	1,2-Dichloroethylene
1111	130	Amyl mercaptan	1150	130P	Dichloroethylene
1112	140	Amyl nitrate	1152	130	Dichloropentanes
1113	129	Amyl nitrite	1153	127	Ethylene glycol diethyl ether
1114	130	Benzene	1154	132	Diethylamine
1120	129	Butanols	1155	127	Diethyl ether
1123	129	Butyl acetates	1155	127	Ethyl ether
1125	132	n-Butylamine	1156	127	Diethyl ketone
1126	130	1-Bromobutane	1157	128	Diisobutyl ketone
1126	130	n-Butyl bromide	1158	132	Diisopropylamine

Page 28

Figure 22.70 When provided with the four-digit UN identification number, firefighters can quickly identify a material using the yellow-bordered pages of the *ERG*. For example, the number *1090* on a placard would identify the material as acetone, and refer responders to Guide No. 127 for response information.

Name of Material	Guide No.	ID No.	Name of Material	Guide No.	ID No.
AC	117	1051	Acrolein dimer, stabilized	129P	2607
Accumulators, pressurized, pneumatic or hydraulic	126	1956	Acrylamide	153P	2074
			Acrylamide, solid	153P	2074
Acetal	127	1088	Acrylamide, solution	153P	3426
Acetaldehyde	129	1089	Acrylic acid, inhibited	132P	2218
Acetaldehyde ammonia	171	1841	Acrylic acid, stabilized	132P	2218
Acetaldehyde oxime	129	2332	Acrylonitrile, inhibited	131P	1093
Acetic acid, glacial	132	2789	Acrylonitrile, stabilized	131P	1093
Acetic acid, solution, more than 10% but not more than 80% acid	153	2790	Adamsite	154	1698
			Adhesives (flammable)	128	1133
Acetic acid, solution, more than 80% acid	132	2789	Adiponitrile	153	2205
			Aerosol dispensers	126	1950
Acetic anhydride	137	1715	Aerosols	126	1950
Acetone	127	1090	Air, compressed	122	1002
Acetone cyanohydrin, stabilized	155	1541	Air, refrigerated liquid (cryogenic liquid)	122	1003
Acetone oils	127	1091			
Acetonitrile	127	1648	Air, refrigerated liquid (cryogenic liquid), non-pressurized	122	1003
Acetyl bromide	156	1716			
Acetyl chloride	155	1717			
Acetylene	116	1001	Air bag inflators	171	3268
Acetylene, dissolved	116	1001	Air bag inflators, compressed gas	126	3353
Acetylene, solvent free	116	3374	Air bag inflators, pyrotechnic	171	3268
Acetylene, Ethylene and Propylene in mixture, refrigerated liquid containing at least 71.5% Ethylene with not more than 22.5% Acetylene and not more than 6% Propylene	115	3138	Air bag modules	171	3268
			Air bag modules, compressed gas	126	3353
			Air bag modules, pyrotechnic	171	3268
			Aircraft hydraulic power unit fuel tank	131	3165
			Alcoholates solution, n.o.s., in alcohol	132	3274
Acetylene tetrabromide	159	2504	Alcoholic beverages	127	3065
Acetyl iodide	156	1898	Alcohols, flammable, poisonous, n.o.s.	131	1986
Acetyl methyl carbinol	127	2621			
Acid, sludge	153	1906	Alcohols, flammable, toxic, n.o.s.	131	1986
Acid butyl phosphate	153	1718			
Acridine	153	2713	Alcohols, n.o.s.	127	1987
Acrolein, inhibited	131P	1092	Alcohols, poisonous, n.o.s.	131	1986
Acrolein, stabilized	131P	1092			
					Page 97

Figure 22.71 When provided with the name of a hazardous material, first responders can quickly locate the appropriate response information by using the blue-bordered pages of the *ERG*. For example, the name *acetal* refers responders to Guide No. 127.

ERG *Material Name Index (Blue Pages)*

The blue-bordered pages of the *ERG* provide an index of dangerous goods in alphabetical order of material name so that the firefighter can quickly identify the Guide to consult for the name of the material involved. This list displays the name of the material followed by its assigned emergency response Guide and four-digit ID number **(Figure 22.71)**.

ERG *Initial Action Guides (Orange Pages)*

[NFPA® 472: 4.2.3(2), 4.4.1(4)(a), 4.4.1(4)(b)]

The orange-bordered section of the book is the most important because it provides firefighters with safety recommendations and general hazards information. It comprises a total of 62 individual guides presented in a two-page format. The left-hand page provides safety related information, whereas the right-hand page provides emergency response guidance and activities for fire situations, spill or leak incidents, and first aid. Each Guide is designed to cover a group of materials that possesses similar chemical and toxicological characteristics. The Guide title identifies the general hazards of the dangerous goods covered.

Each Guide is divided into three main sections: The first section (*Potential Hazards*) describes potential hazards that the material may display in terms of fire/explosion and health effects upon exposure. The highest potential is listed first. Emergency responders consult this section first, which allows them to make decisions regarding the protection of the emergency response team as well as the surrounding population.

The second section (*Public Safety*) outlines suggested public safety measures based on the situation at hand. It provides general information regarding immediate isolation of the incident site and recommended type of protective clothing and respiratory protection **(Figure 22.72)**. This section also lists suggested evacuation distances for small and large spills and for fire situations (fragmentation hazard). When the material is one highlighted in the yellow-bordered and blue-bordered pages, this section also directs the reader to consult the tables on the green-bordered pages listing TIH materials and water-reactive materials.

The third section (*Emergency Response*) covers emergency response areas, including precautions for incidents involving fire, spills or leaks, and first aid. Several recommendations are listed under each of these areas to further assist the responder in the decision-making process. The information on first aid is general guidance before seeking medical care.

ERG *Table of Initial Isolation and Protective Action Distances (Green Pages)*

[NFPA® 472: 4.4.1(4)(c), 4.4.1(7), 4.4.1(8), 4.4.1(9), 4.4.1(9)(a), 4.4.1(9)(b), 4.4.1(10)]

This section of the *ERG* contains a table that lists (by ID number) TIH materials—including certain chemical warfare agents and water-reactive materials that produce toxic gases upon contact with water. The table provides two different types of

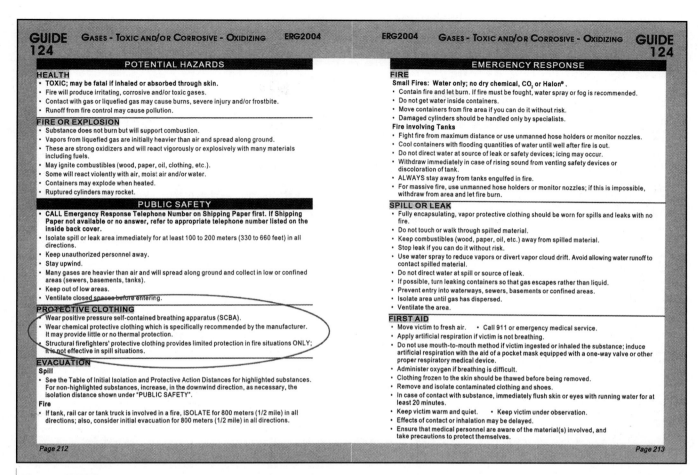

GUIDE 124   GASES - TOXIC AND/OR CORROSIVE - OXIDIZING   ERG2004

POTENTIAL HAZARDS

HEALTH
- TOXIC; may be fatal if inhaled or absorbed through skin.
- Fire will produce irritating, corrosive and/or toxic gases.
- Contact with gas or liquefied gas may cause burns, severe injury and/or frostbite.
- Runoff from fire control may cause pollution.

FIRE OR EXPLOSION
- Substance does not burn but will support combustion.
- Vapors from liquefied gas are initially heavier than air and spread along ground.
- These are strong oxidizers and will react vigorously or explosively with many materials including fuels.
- May ignite combustibles (wood, paper, oil, clothing, etc.).
- Some will react violently with air, moist air and/or water.
- Containers may explode when heated.
- Ruptured cylinders may rocket.

PUBLIC SAFETY
- CALL Emergency Response Telephone Number on Shipping Paper first. If Shipping Paper not available or no answer, refer to appropriate telephone number listed on the inside back cover.
- Isolate spill or leak area immediately for at least 100 to 200 meters (330 to 660 feet) in all directions.
- Keep unauthorized personnel away.
- Stay upwind.
- Many gases are heavier than air and will spread along ground and collect in low or confined areas (sewers, basements, tanks).
- Keep out of low areas.
- Ventilate closed spaces before entering.

PROTECTIVE CLOTHING
- Wear positive pressure self-contained breathing apparatus (SCBA).
- Wear chemical protective clothing which is specifically recommended by the manufacturer. It may provide little or no thermal protection.
- Structural firefighters' protective clothing provides limited protection in fire situations ONLY; it is not effective in spill situations.

EVACUATION
Spill
- See the Table of Initial Isolation and Protective Action Distances for highlighted substances. For non-highlighted substances, increase, in the downwind direction, as necessary, the isolation distance shown under "PUBLIC SAFETY".
Fire
- If tank, rail car or tank truck is involved in a fire, ISOLATE for 800 meters (1/2 mile) in all directions; also, consider initial evacuation for 800 meters (1/2 mile) in all directions.

Page 212

EMERGENCY RESPONSE

FIRE
Small Fires: Water only; no dry chemical, CO₂ or Halon®.
- Contain fire and let burn. If fire must be fought, water spray or fog is recommended.
- Do not get water inside containers.
- Move containers from fire area if you can do it without risk.
- Damaged cylinders should be handled only by specialists.
Fire involving Tanks
- Fight fire from maximum distance or use unmanned hose holders or monitor nozzles.
- Cool containers with flooding quantities of water until well after fire is out.
- Do not direct water at source of leak or safety devices; icing may occur.
- Withdraw immediately in case of rising sound from venting safety devices or discoloration of tank.
- ALWAYS stay away from tanks engulfed in fire.
- For massive fire, use unmanned hose holders or monitor nozzles; if this is impossible, withdraw from area and let fire burn.

SPILL OR LEAK
- Fully encapsulating, vapor protective clothing should be worn for spills and leaks with no fire.
- Do not touch or walk through spilled material.
- Keep combustibles (wood, paper, oil, etc.) away from spilled material.
- Stop leak if you can do it without risk.
- Use water spray to reduce vapors or divert vapor cloud drift. Avoid allowing water runoff to contact spilled material.
- Do not direct water at spill or source of leak.
- If possible, turn leaking containers so that gas escapes rather than liquid.
- Prevent entry into waterways, sewers, basements or confined areas.
- Isolate area until gas has dispersed.
- Ventilate the area.

FIRST AID
- Move victim to fresh air.   • Call 911 or emergency medical service.
- Apply artificial respiration if victim is not breathing.
- Do not use mouth-to-mouth method if victim ingested or inhaled the substance; induce artificial respiration with the aid of a pocket mask equipped with a one-way valve or other proper respiratory medical device.
- Administer oxygen if breathing is difficult.
- Clothing frozen to the skin should be thawed before being removed.
- Remove and isolate contaminated clothing and shoes.
- In case of contact with substance, immediately flush skin or eyes with running water for at least 20 minutes.
- Keep victim warm and quiet.   • Keep victim under observation.
- Effects of contact or inhalation may be delayed.
- Ensure that medical personnel are aware of the material(s) involved, and take precautions to protect themselves.

Page 213

Figure 22.72 The orange-bordered pages of the *ERG* provide a wealth of information about potential health hazards, safety recommendations, and general response guidelines for the material involved. The *Public Safety* section of a material's orange-bordered guide provides information about personal protective equipment.

recommended safe distances: initial isolation distances and protective action distances (**Figure 22.73**). These materials are highlighted for easy identification in both numeric (yellow-bordered) and alphabetic (blue-bordered) *ERG* indexes.

The table provides isolation and protective action distances for both small (approximately 53 gallons [200 L] or less) and large spills (more than 53 gallons [200 L]). Basically, a *small spill* is one that involves a single, small package (up to a 55-gallon [208 L] drum), small cylinder, or small leak from a large package. A *large spill* is one that involves a spill from a large package or multiple spills from many small packages.

The list is further subdivided into daytime and nighttime situations. This division is necessary because atmospheric conditions significantly affect the size of a chemically hazardous area, and differences can be generally associated with typical daytime and nighttime conditions. The warmer, more active atmosphere normal during the day disperses chemical contaminants more readily than the cooler, calmer conditions common at night. Therefore, during the day, lower toxic concentrations may be spread over a larger area than at night, when higher concentrations may exist in a smaller area. The quantity of material spilled or released and the area affected are both important, but the single most critical factor is the concentration of the contaminant in the air.

Initial Isolation Distance — Distance within which all persons are considered for evacuation in all directions from a hazardous materials incident.

Protective Action Distance — Downwind distance from a hazardous materials incident within which protective actions should be implemented.

		SMALL SPILLS (From a small package or small leak from a large package)				LARGE SPILLS (From a large package or from many small packages)			
		First ISOLATE in all Directions		Then PROTECT persons Downwind during-		First ISOLATE in all Directions		Then PROTECT persons Downwind during-	
ID No.	NAME OF MATERIAL	Meters (Feet)		DAY Kilometers (Miles)	NIGHT Kilometers (Miles)	Meters (Feet)		DAY Kilometers (Miles)	NIGHT Kilometers (Miles)
1953 / 1953	Liquefied gas, flammable, poisonous, n.o.s. / Liquefied gas, flammable, poisonous, n.o.s. (Inhalation Hazard Zone A)	120 m	(400 ft)	1.2 km (0.8 mi)	5.1 km (3.2 mi)	1000 m	(3000 ft)	8.7 km (5.4 mi)	11.0+ km (7.0+ mi)
1953	Liquefied gas, flammable, poisonous, n.o.s. (Inhalation Hazard Zone B)	30 m	(100 ft)	0.2 km (0.2 mi)	1.2 km (0.8 mi)	420 m	(1400 ft)	4.0 km (2.5 mi)	10.8 km (6.7 mi)
1953	Liquefied gas, flammable, poisonous, n.o.s. (Inhalation Hazard Zone C)	30 m	(100 ft)	0.2 km (0.1 mi)	0.8 km (0.5 mi)	240 m	(800 ft)	2.4 km (1.5 mi)	6.4 km (4.0 mi)
1953	Liquefied gas, flammable, poisonous, n.o.s. (Inhalation Hazard Zone D)	30 m	(100 ft)	0.1 km (0.1 mi)	0.2 km (0.1 mi)	90 m	(300 ft)	0.8 km (0.5 mi)	2.4 km (1.5 mi)
1953 / 1953	Liquefied gas, flammable, toxic, n.o.s. / Liquefied gas, flammable, toxic, n.o.s. (Inhalation Hazard Zone A)	120 m	(400 ft)	1.2 km (0.8 mi)	5.1 km (3.2 mi)	1000 m	(3000 ft)	8.7 km (5.4 mi)	11.0+ km (7.0+ mi)
1953	Liquefied gas, flammable, toxic, n.o.s. (Inhalation Hazard Zone B)	30 m	(100 ft)	0.2 km (0.2 mi)	1.2 km (0.8 mi)	420 m	(1400 ft)	4.0 km (2.5 mi)	10.8 km (6.7 mi)
1953	Liquefied gas, flammable, toxic, n.o.s. (Inhalation Hazard Zone C)	30 m	(100 ft)	0.2 km (0.1 mi)	0.8 km (0.5 mi)	240 m	(800 ft)	2.4 km (1.5 mi)	6.4 km (4.0 mi)
1953	Liquefied gas, flammable, toxic, n.o.s. (Inhalation Hazard Zone D)	30 m	(100 ft)	0.1 km (0.1 mi)	0.2 km (0.1 mi)	90 m	(300 ft)	0.8 km (0.5 mi)	2.4 km (1.5 mi)
1955 / 1955	Compressed gas, poisonous, n.o.s. / Compressed gas, poisonous, n.o.s. (Inhalation Hazard Zone A)	600 m	(2000 ft)	5.9 km (3.7 mi)	11.0+ km (7.0+ mi)	1000 m	(3000 ft)	11.0+ km (7.0+ mi)	11.0+ km (7.0+ mi)

Figure 22.73 The green-bordered pages of the Table of Initial Isolation and Protective Action Distances provide the downwind distances within which protective actions (such as evacuation) should be implemented. These distances vary depending on the size of the spill and whether it is day or night.

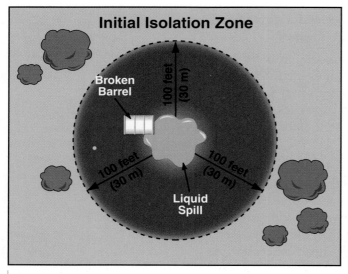

Figure 22.74 The smallest initial isolation distance for any chemical is 100 feet (30 m).

The *initial isolation distance* is a distance within which all persons should be considered for evacuation in all directions from an actual hazardous materials spill/leak source, according to the *ERG*. The *initial isolation zone* is a circular zone (with a radius equivalent to the initial isolation distance) within which persons may be exposed to dangerous concentrations upwind of the source and life-threatening concentrations downwind of the source (**Figure 22.74**).

Protective action distances represent a downwind distance from a spill/leak source within which protective actions should be implemented (**Figure 22.75**). Protective actions are those steps taken to preserve the health and safety of emergency responders and the public. People in this area could be evacuated and/or sheltered in-place (directed to stay inside until danger passes). For more information, consult the current *ERG*.

NOTE: If hazardous materials are on fire or have been leaking for longer than 30 minutes, this *ERG* table does not apply. Seek more detailed information on the involved

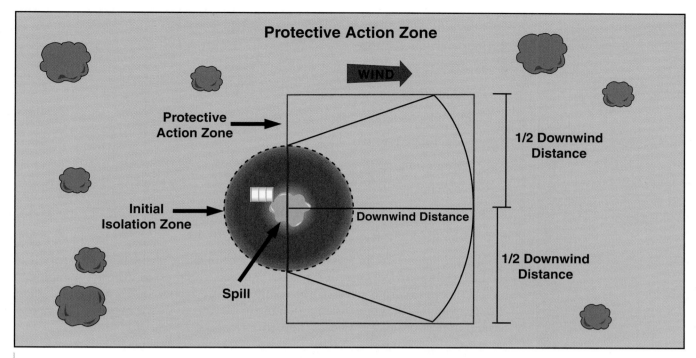

Protective Action Zone

WIND

Protective
Action Zone

1/2 Downwind
Distance

Initial
Isolation Zone

Downwind Distance

1/2 Downwind
Distance

Spill

Figure 22.75 The protective action zone is the area in imminent danger of being contaminated by airborne vapors within 30 minutes of the release.

material on the appropriate orange-bordered page in the *ERG*. Also, the orange-bordered pages in the *ERG* provide recommended isolation and evacuation distances for nonhighlighted chemicals with poisonous vapors and situations where the containers are exposed to fire.

Senses

Vision is definitely the safest of the five senses to use in the detection of a hazardous material. While it may be perfectly safe to observe an overturned cargo tank from a distance through binoculars, emergency responders have to come into close or actual physical contact with a hazardous material (or its mists, vapors, dusts, or fumes, and the like) in order to hear, smell, taste, or feel a release. While many products release odors well below dangerous levels, there is a good chance that if firefighters are this close to a hazardous material, they are *too* close for safety's sake. It should also be noted that many hazardous materials are invisible, have no odor, and cannot readily be detected by the senses. Hydrogen sulfide may cause olfactory fatigue (in other words, you may cease to smell it even though it is still present). However, any smells, tastes, or symptoms reported by victims and witnesses may prove to be helpful. Warning properties of chemicals include visible gas clouds, pungent odors, and irritating fumes.

Firefighters should be aware of visual/physical chemical indicators that provide tangible evidence of the presence of hazardous materials. Unusual noises (such as the hiss of a gas escaping a valve at high pressure) may also alert firefighters to the presence of hazards. Some hazardous materials have odorants added to them in order to aid in detection; for example, the distinct odor normally associated with natural gas (an odorless gas) is actually caused by mercaptan, an additive.

Direct visible evidence that physical and/or chemical actions and reactions are taking place includes such items as the following:

Initial Isolation Zone — Circular zone (with a radius equivalent to the initial isolation distance) within which persons may be exposed to dangerous concentrations upwind of the source and may be exposed to life-threatening concentrations downwind of the source.

Shelter in Place — Having occupants remain in a structure or vehicle in order to provide protection from a rapidly approaching hazard.

WARNING!
Deliberately using the human senses to detect the presence of hazardous materials is both unreliable and unsafe. <u>It could kill you!</u>

- Spreading vapor cloud or smoke
- Unusual colored smoke
- Flames
- Gloves melting
- Changes in vegetation
- Container deterioration
- Containers bulging
- Sick humans
- Dead or dying birds, animals, insects, or fish
- Discoloration of valves or piping

Physical actions are processes that do not change the elemental composition of the materials involved. One example is a liquefied, compressed material changing to a gas as it escapes from a vessel. The resulting white vapor cloud is another physical change, which is the condensation of moisture in the air by the expanding material. Several indications of a physical action are as follows:

- Rainbow sheen on water surfaces
- Wavy vapors over a volatile liquid
- Frost or ice buildup near a leak
- Containers deformed by the force of an accident
- Activated pressure-relief devices
- Pinging or popping of heat-exposed vessels

Chemical reactions convert one substance to another. There is much visual and sensory evidence of chemical reactions:

Exothermic — Chemical reaction between two or more materials that changes the materials and produces heat, flames, and toxic smoke.

- Exothermic heat
- Unusual or unexpected temperature drop (cold)
- Extraordinary fire conditions
- Peeling or discoloration of a container's finish
- Spattering or boiling of unheated materials
- Distinctively colored vapor clouds
- Smoking or self-igniting materials
- Unexpected deterioration of equipment
- Peculiar smells
- Unexplained changes in ordinary materials
- Symptoms of chemical exposure

Physical signs and symptoms of chemical exposure may also indicate the presence of hazardous materials. Symptoms can occur separately or in clusters, depending on the chemical. Be alert for these symptoms of chemical exposure:

- ***Changes in respiration*** — Difficult breathing, increase or decrease in respiration rate, tightness of the chest, irritation of the nose and throat, and/or respiratory arrest

- *Changes in level of consciousness* — Dizziness, lightheadedness, drowsiness, confusion, fainting, and/or unconsciousness

- *Abdominal distress* — Nausea, vomiting, and/or cramping

- *Change in activity level* — Fatigue, weakness, stupor, hyperactivity, restlessness, anxiety, giddiness, and/or faulty judgment

- *Visual disturbances* — Double vision, blurred vision, cloudy vision, burning of the eyes, and/or dilated or constricted pupils

- *Skin changes* — Burning sensations, reddening, paleness, fever, and/or chills

- *Changes in excretion or thirst* — Uncontrolled tears, profuse sweating, mucus flowing from the nose, diarrhea, frequent urination, bloody stool, and/or intense thirst

- *Pain* — Headache, muscle ache, stomachache, chest pain, and/or localized pain at sites of substance contact

Monitoring and Detection Devices

Monitoring and detection devices can be useful in determining the presence of hazardous materials as well as the concentration(s) present. As with the senses, effectively using the monitoring and detection devices requires actual contact with the hazardous material (or its mists, dusts, vapors, or fumes) in order to measure it and is therefore outside the scope of action for most Operations-Level responders unless additional training is provided and appropriate PPE used.

However, the use of monitoring equipment can help determine the scope of the incident, including the following:

- What materials are involved in the incident

- How far the materials have traveled (areas that have been contaminated)

- What concentrations are present (and how that might affect victims and exposures)

- Where the materials are absent (areas that are free of contamination, areas that are safe; this can also be used to determine the effectiveness of decontamination operations)

Monitoring and detection devices are commercially available that can detect most of the materials likely to be involved in haz mat incidents, including terrorist attacks involving biological agents, chemical agents, radiological agents, and explosives (**Figure 22.76**). However, **no single device will detect all materials.** It is important that response agencies be equipped with devices that will detect a variety of materials. At a minimum, chemical and radiological agents should be tested for at all explosive incidents (Figure 22.76).

There are many different types of monitoring and detection devices. All of them have advantages and disadvantages. For example, some can be easily used in the field, whereas others are less portable. Some instruments may not be very sensitive (able to identify low concentration/dose of materials), or specific (able to distinguish between different types of material) whereas others may be extremely so. Some may be user friendly; others may have a long response time; and

Figure 22.76 Chemical and radiological agents should be tested for at all explosive incidents because of the possibility of terrorism.

yet others may be extremely expensive. All of these factors must be considered when selecting detection equipment. The following is a partial list of monitoring and detection devices based on the substance being detected and/or the technology being used in the device:

- **Combustible gas indicators (CGIs)** — CGIs are used to detect the concentration of combustible gases and vapors in the air. Typically they measure the percentage of the LEL, percent of gas by volume, or parts per million of the material in air. They are designed to sound an alarm if hazardous or potentially hazardous concentrations are found (often set to sound at the equivalent of 10 percent LEL or higher).

Olfactory Fatigue — Gradual inability of a person to detect odors after initial exposure; may be extremely rapid in the case of some toxins such as hydrogen sulfide.

- **Two-, three-, and four-gas monitors** — CGI sensors are often combined with other common gas sensors (such as those that detect oxygen, carbon monoxide, or hydrogen sulfide) to form monitors that can detect two, three, or four gases, depending on the combination of sensors. Hydrogen sulfide is often included not only because of its toxicity, but because exposure to this substance causes "olfactory fatigue," inhibiting the individual's ability to smell it. A four-gas monitor would simultaneously check for combustible gases, oxygen concentrations (both high and low), carbon monoxide levels, and hydrogen sulfide levels. Single-sensor meters (also called oxygen meters) may also be found for oxygen, carbon monoxide, and hydrogen sulfide. Typically, an alarm is sounded when hazardous or potentially hazardous levels are detected. Four-gas monitors may sometimes be combined with photo-ionization detectors (PIDs).

- **Biological immunoassay indicators** — These detectors indicate the presence of biological agents and toxins by detecting the presence of specific antibodies.

- **Chemical agent monitors (CAMs)** — Chemical agent monitors utilize various technologies to specifically detect chemical warfare agents.

- **Flame ionization detectors (FIDs)** — Flame ionization detectors utilize a hydrogen flame to which gaseous materials are exposed. These devices will detect organic gases and vapors. FIDs may not be safe to operate in hazardous (explosive) atmospheres.

- **Photo-ionization detectors (PIDs)** — PIDs use an ultraviolet lamp to ionize samples of gaseous materials. They are used to detect the concentrations of many organic and some inorganic gases and vapors at the same time, and they make good general survey instruments. They are particularly useful when chemical hazards are unidentified or undetermined.

- **Infrared spectroscopy devices** — Devices using infrared spectroscopy technology compare the infrared spectra of chemical samples against a library of known spectral signatures. When the signature of the sample matches the signature of a known chemical in the library, identification is made. These devices cannot identify biological materials, but they can determine if a material is chemical or biological in nature.

- **Raman spectroscopy** — Similar to infrared spectroscopy, raman spectroscopy uses light (typically a laser) to illuminate a sample thereby creating a spectral signature that is unique to each material. By comparing the signatures to a library of known signatures, identification of the material can be made. This technology appears to have some potential for air-sampling.

- **Ion-mobility spectrometers** — These devices use a radioactive source to ionize samples in order to determine their spectra. Currently, these devices are designed to detect chemical warfare agents and explosives.

- **Mass spectrometers** — Mass spectrometers ionize samples in order to determine their composition. Like other spectroscopy devices, they compare test results to a library of known measurements in order to make a positive identification.

- **Specific chemical monitors** — Usually fixed devices used to sound an alarm when the presence of a specific chemical is detected. Carbon monoxide monitors are especially common, but one can also find monitors that detect chlorine, hydrogen sulfide, ammonia, ethylene oxide, and hydrogen cyanide.

- **Indicator papers and pH meters** (sometimes called color-change chemistry indicators) — Indicator papers change colors to indicate the presence of specific hazards such as oxidizers, hydrogen sulfide, and peroxides. Similar papers are now being used to detect the presence of biohazards such as anthrax. Indicator papers change color to indicate the pH of acids or alkalines. Meters may also measure pH.

- **Detector tubes (colorimetric tubes)** — These devices detect a variety of gases and vapors. They can be chemical specific or colorimetric and are most useful if a particular chemical is suspected as opposed to trying to identify a complete unknown.

- **Radiation monitors** — Devices such as Geiger Mueller tubes (commonly called Geiger counters) detect levels of alpha, beta, or gamma radiation by collecting and counting the number of ions present; readings are provided in cpm (counts per minute), μR/hr (microrem per hour), mR/hr (milliroentgens per hour), or rem (roentgens per hour).

- **Field chemistry testing kits** — These kits contain portable chemistry sets designed to enable logical and progressive testing of a sample in order to identify it. (Many response organizations use field test kits for white powder incidents.)

- **DNA fluoroscopy devices** — Detectors using DNA fluoroscopy technology have the ability to identify specific DNA sequences, thereby detecting and identifying types of biological agents.

- **Polymerase chain reaction devices (PCRs)** — PCR technology is used to identify DNA. Devices utilizing this technology are used to detect and identify biological agents and toxins.

- **Surface acoustical wave devices** — Devices utilizing surface acoustical wave technology are used to detect nerve agents, blister agents, blood agents, choking agents, and some toxic industrial materials/toxic industrial chemicals (TIMs/TICs).

- **Personal dosimeter** — Personal dosimeters are generally worn to measure (and sometimes identify) an individual's exposure to a particular radiation.

- **Other personal detection devices** — Include organic vapor badges (or film strips, wrist bands, stick pins, etc.), mercury badges, and formaldehyde badges or strips used to measure individual exposure to certain chemicals

Responders assigned monitoring, detection, and sampling duties must be trained to use the instruments available to them correctly. They must understand the capabilities and limitations of each device. They must also be able to interpret the data provided to them correctly. Additionally, they must be able to properly maintain, field test, and calibrate the devices per manufacturers' instructions. Finally, they must be able to use the devices in accordance with predetermined procedures based on the availability, capabilities, and limitations of personnel, appropriate PPE, and other resources available at the incident in accordance with the incident action plan.

Personnel who conduct sampling activities at terrorist attacks or crime scenes must follow appropriate protocols in regards to chain of custody, packaging, labeling, and

WARNING!

All personnel assigned to conduct monitoring and sampling must have proper training to do so, and they must wear appropriate personal protective equipment when operating in potentially hazardous areas.

transportation to the testing authority. Ensuring evidence preservation and chain of custody is a very important aspect of crime scene investigation.

Identification of Terrorist Attacks

Response to a terrorist incident is essentially the same as that for response to other haz mat incidents; however, there are critical differences that must be understood by firefighters. For example, terrorist incident sites are crime scenes; therefore, law enforcement organizations must be notified and included in the initial response. Terrorist incidents may also differ because of their size and complexity; the number of casualties; the presence of extremely hazardous materials; the potential for armed resistance, booby traps, secondary devices, and the use of weapons; and overall levels of increased risk due to contaminated victims, structural collapse hazards, and other dangers **(Figure 22.77)**.

Because of these important differences between terrorist incidents and other haz mat emergencies, firefighters must try to identify incidents involving terrorism as quickly as possible. The following are a few examples of situations that can cue the responder to consider the possibility of a terrorist attack:

- Report of two or more medical emergencies in public locations such as a shopping mall, transportation hub, mass transit system, office building, assembly occupancy, or other public buildings

Figure 22.77 Terrorist attacks may differ from other haz mat incidents because of the number of casualties, complexity, the presence of secondary devices, and other factors. *Courtesy of Federal Emergency Management Agency (FEMA) News Photo.*

- Unusually large number of people with similar signs and symptoms coming or being transported to physicians' offices or medical emergency rooms
- Reported explosion at a movie theater, department store, office building, government building, or a location with historical or symbolic significance

Additional information can provide clues as to the type of attack. Chemical, biological, radiological, nuclear, and explosive attacks (CBRNE attacks) each have their own unique indicators with which responders should be familiar. **Table 22.24** provides a very brief summary of these differences. Monitoring and detection devices also play an important role in determining which of these materials may be present at the incident scene.

Table 22.24 Terrorist Attacks at a Glance	
Chemical Attack	**Biological Attack**
• Victims in a concentrated area • Symptoms immediate (seconds to hours after exposure) • Symptoms very similar (SLUDGEM) • May have observable features such as chemical residue, dead foliage, dead animals/insects, and pungent odor	• Victims dispersed over a wide area • Symptons delayed (days — weeks after exposure) • Symptoms most likely vague and flu-like • No observable features
Explosive Attack	**Radiological Attack**
• Explosion self-evident (debris field, fire, etc.) • Victims in a concentrated area • Mechanical and thermal injuries • Potential radiation and chemical agent risk — monitoring for both is necessary	• Explosion self-evident (debris field, fire, etc.) • Victims in a concentrated area • Mechanical and thermal injuries initially, radiological symptoms (if any) will likely be delayed • Radiation detected through monitoring

Chemical Warfare Agent — Chemical substance that is intended for use in warfare or terrorist activities to kill, seriously injure, or seriously incapacitate people through its physiological effects.

Toxic Industrial Material (TIM)/ Toxic Industrial Chemical (TIC) — Industrial chemical that is toxic at certain concentration and is produced in quantities exceeding 30 tons per year at any one production facility; readily available and could be used by terrorists to deliberately kill, injure, or incapacitate people.

If criminal or terrorist activity is suspected at the scene of an incident, firefighters must forward that information to the incident commander (IC) as quickly as possible. It is also critical that this information be passed on to law enforcement representatives.

Chemical Attack Indicators

Chemical attacks may utilize chemical warfare agents (nerve agents, blister agents, blood agents, choking agents, and riot control agents) as well as toxic industrial materials/toxic industrial chemicals (TIMs/TICs) (**Table 22.25**). Chemical attacks usually result in readily observable features including signs and symptoms that develop very rapidly. Chemical attack indicators include:

Table 22.25
Chemical Attack Characteristics

Category	Agent	DOT Hazard	ERG Guide	Appearance and Characteristics	Signs and Symptoms*
Nerve Agents	Sarin (GB)	6.1	153	Clear, colorless, tasteless, and odorless liquid	Salivation, lacrimation (drooling), urination, defecation, gastrointestinal upset/aggravation (cramping), emesis (vomiting), miosis (pinpointed pupils), muscular twitching/spasms
	Soman (GD)	6.1	153	Clear, colorless, tasteless, liquid that discolors with age to dark brown; may have slight fruity or camphor odor	Salivation, lacrimation (drooling), urination, defecation, gastrointestinal upset/aggravation (cramping), emesis (vomiting), miosis (pinpointed pupils), muscular twitching/spasms
	Tabun (GA)	6.1	153	Clear, colorless, tasteless liquid; may have slight fruity odor when impure	Salivation, lacrimation (drooling), urination, defecation, gastrointestinal upset/aggravation (cramping), emesis (vomiting), miosis (pinpointed pupils), muscular twitching/spasms
	V Agent (VX)	6.1	153	Clear, amber-colored, odorless, oily liquid; similar to motor oil in appearance and volatility	Salivation, lacrimation (drooling), urination, defecation, gastrointestinal upset/aggravation (cramping), emesis (vomiting), miosis (pinpointed pupils), muscular twitching/spasms
Blister Agents	Mustard (H) and Distilled Mustard (HD)	6.1	153	Clear to yellow or brown liquid or solid; oily textured; potentially heavier than air vapor; odor sometimes like garlic or onions, sometimes odorless	Redness and irritation of skin changing to yellow blisters; eye irritation and damage including severe pain and temporary blindness; respiratory tract irritation including cough, runny nose, sneezing, and shortness of breath
	Nitrogen Mustard (HN)	6.1	153	Different forms can smell fishy, musty, soapy, or fruity; oily-textured liquid, vapor, or a solid (liquid at normal room temperature); clear, pale amber, or yellow colored when in liquid or solid form	Red skin, blistering, eye irritation, eye damage (including burns and blindness), respiratory tract irritation, shortness of breath, tremors, lack of coordination, seizures
	Lewisite (L)	6.1	153	Colorless liquid with odor like geraniums	Immediate burning and irritation to skin, red skin, blisters and lesions, eye irritation, tearing, coughing, airway irritation, shortness of breath
Blood Agents	Hydrogen Cyanide (AC)	6.1	117	Slightly lighter than air, colorless gas or liquid with bitter almond odor	Headache, dizziness, confusion, nausea, shortness of breath, convulsions, vomiting, weakness, anxiety, irregular heartbeat, tightness in chest, unconsciousness, death
	Cyanogen Chloride (CK)	2.3	125	Heavier than air, colorless gas with pungent odor	Runny nose, sore throat, nausea, vomiting, confusion, red skin, unconsciousness, convulsions, pulmonary edema, coma, death

Continued

Table 22.25 (continued)
Chemical Attack Characteristics

Category	Agent	DOT Hazard	ERG Guide	Appearance and Characteristics	Signs and Symptoms*
	Arsine (SA)	2.3	119	Colorless gas with mild garlic odor	Weakness, fatigue, headache, drowsiness, shortness of breath, rapid breathing, nausea, vomiting, muscle cramps, loss of consciousness, convulsions, paralysis, respiratory failure
Choking Agents	Chlorine (CL)	2.3	124	Yellow-green gas that smells like bleach and evaporates quickly	Coughing; chest tightness; burning nose, eyes, and throat; difficulty breathing; pulmonary edema
	Phosgene	2.3	125	Heavier than air, usually colorless gas (may be pale yellow to white); hay-like odor at low concentrations; pungent odor at high concentrations	Coughing, burning eyes and throat, tearing, difficulty breathing, nausea, vomiting, low blood pressure, bluish skin, lung damage, death
Riot Control Agents	Tear Gas (CS)	6.1	159	Dispersed as smoke, liquid, or powder; white crystalline solid with pungent pepper-like odor	Burning eyes, tearing, burning mouth and throat, choking
	Pepper Spray (OC)	2.2	159	Oily liquid dispersed as an aerosol with pepper-like odor	Irritated eyes, nose, mouth, throat, and lungs

*Signs and symptoms will vary depending on exposure factors; generally, the higher the concentration and the longer the duration, the more severe the symptoms.

- Warning or threat of an attack or received intelligence
- Presence of hazardous materials or laboratory equipment that is not relevant to the occupancy
- Intentional release of hazardous materials
- Unexplained patterns of sudden onset of similar, nontraumatic illnesses or deaths (the pattern could be geographic, by employer, or associated with agent dissemination methods)
- Unexplained odors or tastes that are out of character with the surroundings
- Multiple individuals exhibiting unexplained signs of skin, eye, or airway irritation
- Unexplained bomb or munition-like material, especially if it contains a liquid
- Unexplained vapor clouds, mists, and plumes, particularly if they are not consistent with their surroundings
- Multiple individuals exhibiting unexplained health problems such as nausea, vomiting, twitching, tightness in chest, sweating, pinpoint pupils (miosis), runny nose (rhinorrhea), disorientation, difficulty breathing, or convulsions. Some people teach the symptoms of exposure to chemical warfare agents with the acronym *SLUDGEM*:

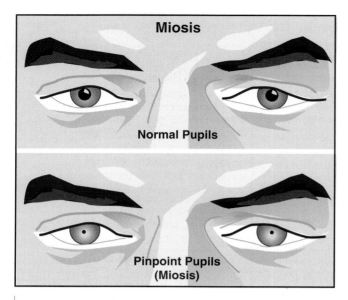

Figure 22.78 Miosis (pinpoint pupils) is a symptom of exposure to nerve agents.

— **S**alivation (drooling)

— **L**acrimation (tearing)

— **U**rination

— **D**efecation

— **G**astrointestinal upset/aggravation (cramping)

— **E**mesis (vomiting)

— **M**iosis (pinpointed pupils) or **M**uscular twitching/spasms (**Figure 22.78**)

• Unexplained deaths and/or mass casualties

• Casualties distributed downwind (outdoors) or near ventilation systems (indoors)

• Multiple individuals experiencing blisters and/or rashes

• Trees, shrubs, bushes, food crops, and/or lawns that are dead (not just a patch of dead weeds), discolored, abnormal in appearance, or withered (not under drought conditions)

• Surfaces exhibiting oily droplets or films and unexplained oily film on water surfaces

• Abnormal number of sick or dead birds, animals, and/or fish

• Unusual security, locks, bars on windows, covered windows, and barbed wire enclosures

TIMs/TICs used as chemical weapons may be identified through traditional methods such as identification of occupancy types and locations; container shapes; hazardous materials placards, labels, and markings; written resources; sensory indicators; and use of monitoring and detection devices.

Biological Attack Indicators

Biological Toxin — Poison produced by living organisms.

Biological attacks utilize viruses, bacteria, and/or biological toxins (**Table 22.26**). The effects of biological attacks may not be readily noticeable. Signs and symptoms may take many days to develop (**Figure 22.79**). Biological attack indicators include:

• Warning or threat of an attack or received intelligence

• Presentation of specific unusual diseases (such as smallpox)

• Unusual number of sick or dying people or animals (often of different species)

• Multiple casualties with similar signs or symptoms

• Dissemination of unscheduled or unusual spray

• Abandoned spray devices (devices may have no distinct odors)

• Non-endemic illness for the geographic area (for example, Venezuelan equine encephalitis in Europe)

• Casualty distribution aligned with wind direction

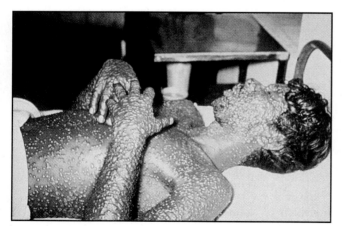

Figure 22.79 It may take days to develop the signs and symptoms associated with some diseases. This man has smallpox. *Courtesy of CDC Public Health Images Library.*

Table 22.26
Biological Agents

Name	Biological Agent	Common Signs and Symptoms*
Smallpox	Virus (*Variola major*)	• Acute rash, fever, fatigue, head and back aches • Rash lesions start red and flat, but fill with pus after first week
Anthrax	Bacteria (*Bacillus anthracis*)	• ***Inhalation anthrax:*** Acute respiratory distress with fever. Initial symptoms may resemble a common cold. • ***Intestinal anthrax:*** Nausea, loss of appetite, vomiting, and fever followed by abdominal pain, vomiting of blood, and severe diarrhea. • ***Cutaneous anthrax:*** Skin infection begins as a raised itchy bump that resembles an insect bite but within 1 to 2 days develops into a vesicle and then a painless ulcer, usually 0.4 to 1 inch (10 mm to 30 mm) in diameter, with a characteristic black necrotic (dying) area in the center.
Plague	Bacteria (*Yersinia pestis*)	Acute respiratory distress, fever, weakness and headache
Botulism	Toxin from the bacteria *Clostridium botulinum*	***Neurological syndromes:*** Double vision, blurred vision, drooping eyelids, slurred speech, difficulty swallowing, dry mouth, and muscle weakness
Tularemia	Bacteria (*Francisella tularensis*)	***Influenzalike illness:*** Sudden fever, chills, headaches, muscle aches, joint pain, dry cough, progressive weakness, and sometimes pneumonia
Hemorrhagic Fever	Virus (for example, Ebola, Marburg, and Lassa)	• Specific signs and symptoms vary by the type of viral hemorrhagic fever (VHF), but initial signs and symptoms often include marked fever, fatigue, dizziness, muscle aches, loss of strength, and exhaustion. • Patients with severe cases of VHF often show signs of bleeding under the skin, in internal organs, or from body orifices such as the mouth, eyes, or ears. • Rash and red eyes may be seen in some patients infected with the Ebola virus.
Ricin Toxin	From *Ricinus Communis* (castor beans)	• ***Inhalation:*** Acute onset of fever, chest pain, and cough, progressing to severe respiratory distress. • ***Ingestion:*** Nausea, vomiting, abdominal pain and cramping, diarrhea, gastrointestinal bleeding, low or no urinary output, dilation of the pupils, fever, thirst, sore throat, headache, vascular collapse, and shock.

*List of symptoms is *not* all-inclusive.

- Electronic tracking of signs and symptoms (syndromic surveillance) reported to hospitals, pharmacies, and other health-care organizations

- Illnesses associated with a common source of food, water, or location

- Large numbers of people exhibiting flu-like symptoms during non-flu months

Depending on the agent used and the scope of an incident, emergency medical services (EMS) responders and health-care personnel may be first to realize that there has been a biological attack. In some cases there may be reliable evidence to implicate terrorist activity such as a witness to an attack or the discovery of an appropriate delivery system (such as finding a contaminated dissemination device from which an infectious agent is subsequently isolated and identified).

Radiological Attack Indicators

Radiological attacks utilize weapons that release radiological materials, most likely in the form of dust or powder. Dispersal may be accomplished by including the material in a bomb or explosive device, i.e., a radiological dispersal device (RDD), sometimes called a *dirty bomb*. Radiological attack indicators include:

- Warning or threat of an attack or received intelligence

- Individuals exhibiting signs and symptoms of radiation exposure

- Radiological materials packaging left unattended or abandoned in public locations

- Suspicious packages that appear to weigh more than they should (such packages may contain lead to shield a radiation source)

- Activation of radiation detection devices, with or without an explosion

- Material that is hot or seems to emit heat without any sign of an external heat source

- Glowing material—strongly radioactive material may emit or cause radioluminescence

Radiological Dispersal Device (RDD) — Device that spreads radioactive contamination over the widest possible area by detonating conventional high explosives wrapped with radioactive material.

Nuclear Attack Indicators

Nuclear attacks are the intentional detonation of a nuclear weapon. Nuclear attack indicators include:

- Warning or threat of an attack or received intelligence

- Mushroom cloud

- Exceptionally large/powerful explosion

- Electromagnetic pulse (EMP)

Explosive/Incendiary Attack Indicators

The majority of terrorist attacks involve the use of explosive materials and incendiary devices, and normally these are considered conventional attacks (**Figure 22.80**). However, when used to inflict high casualties and large-scale damage (such as with a car or truck bomb destroying an occupied building), explosives may be classified as weapons of mass destruction. Explosives may also be used to disseminate chemical, biological, and radiological materials.

Explosive/incendiary attack indicators include:

- Warning or threat of an attack or received intelligence

- Reports of an explosion

Figure 22.80 The majority of terrorist attacks use explosives and incendiary materials.

- Explosion
- Accelerant odors (gasoline smells and other similar odors)
- Multiple fires or explosions
- Incendiary device or bomb components (such as broken glass from a Molotov cocktail or wreckage of a car bomb)
- Unexpectedly heavy burning or high temperatures
- Unusually fast burning fires
- Unusually colored smoke or flames
- Presence of propane or other flammable gas cylinders in unusual locations
- Unattended packages/backpacks/objects left in high traffic/public areas
- Fragmentation damage/injury
- Damage that exceeds that usually seen during gas explosions, including shattered reinforced concrete or bent structural steel
- Crater(s)
- Scattering of small metal objects such as nuts, bolts, and/or nails used as shrapnel

Table 22.27 provides examples of evacuation distances for various types of car and vehicle bombs.

Identifying Illicit Laboratories

Illicit laboratories are established to produce or manufacture illegal or controlled substances such as drugs, chemical warfare agents, explosives, or biological agents. Illegal clandestine labs can be found virtually anywhere. They may be located in abandoned buildings, hotel rooms, rural farms, urban apartments, rental storage units, or upscale residential neighborhoods. Illegal methamphetamine (meth) labs can be so

Table 22.27
Vehicle Bomb Explosion Hazard and Evacuation Distances

Vehicle Profile	Vehicle Description	Maximum Explosives Capacity	Lethal Air Blast Range	Minimum Evacuation Distance	Falling Glass Hazard
	Compact Sedan	500 pounds (227 kg) [In trunk]	100 feet (30 m)	1,500 feet (457 m)	1,250 feet (381 m)
	Full Size Sedan	1,000 pounds (454 kg) [In trunk]	125 feet (38 m)	1,750 feet (533 m)	1,750 feet (533 m)
	Passenger Van or Cargo Van	4,000 pounds (1 814 kg)	200 feet (61 m)	2,750 feet (838 m)	2,750 feet (838 m)
	Small Box Van (14 ft [4 m] Box)	10,000 pounds (4 536 kg)	300 feet (91 m)	3,750 feet (1 143 m)	3,750 feet (1 143 m)
	Box Van or Water/Fuel Truck	30,000 pounds (13 608 kg)	450 feet (137 m)	6,500 feet (1 981 m)	6,500 feet (1 981 m)
	Semitrailer	60,000 pounds (27 216 kg)	600 feet (183 m)	7,000 feet (2 134 m)	7,000 feet (2 134 m)

Source: *U.S. Bureau of Alcohol, Tobacco, Firearms and Explosives (ATF).*

portable that they have even been found in campgrounds, highway rest stops, and vehicles (**Figure 22.81**).

These labs can present numerous threats to firefighters. In addition to the chemical, explosive, or etiological hazards present, responders may face armed resistance, booby traps, and/or other weapons when responding to emergencies at illegal labs. Many of the products used in clandestine labs are toxic, explosive, and/or highly flammable.

Drug Labs

It is estimated that a significant majority (80 to 90 percent) of all illegal clandestine drug labs are set up to produce meth. These labs may also make ecstasy, cocaine, phenyl-2-propanone (P2P), phencyclidine (PCP), heroine, LSD, amphetamines, and other illegal drugs. However, because meth labs represent the most common type of clandestine drug lab, this section primarily focuses on them.

Meth is easy to make and uses a variety of ingredients commercially available in local stores. Because of the increasing hazard of meth labs, some U.S. states have placed restrictions on the purchase of items used in making meth. The process of making meth is called *cooking*, and many different recipes or methods exist. Two of the most common are known as the *Red P* method and *Nazi/Birch* method. The various recipes differ slightly in the process and the chemicals used, but all of them

Figure 22.81 Meth labs can be very portable. They may be found in hotel rooms, campgrounds, and even vehicles. *Courtesy of Joan Hepler.*

are potentially very dangerous because the chemicals are often highly flammable, corrosive, and toxic **(Figures 22.82 a and b)**. Meth labs present a danger to the meth cook, the community surrounding the lab, and emergency response personnel who discover the lab.

Flammability is perhaps the most serious hazard associated with meth labs, and many labs are discovered only after a fire or explosion has occurred. Other products used in making meth are highly corrosive acids or bases, while others are extremely toxic. One of the by-products, phosphine gas, is sometimes classified as a chemical warfare choking agent. Some products are oxidizers. Meth production processes generate hydrogen chloride gas and hydrogen iodide gas. Meth lab locations may remain a serious health and environmental hazard for years after the lab is removed unless they are properly decontaminated (often an extremely expensive process). A summary of the products commonly used in making meth and the hazards associated with them is provided in **Table 22.28**.

In addition to recognizing the types of chemicals typically found in meth labs, firefighters should also be familiar with the types of equipment used in the process. They may expect to find the following items:

Choking Agent — Chemical warfare agent that attacks the lungs causing tissue damage.

- *Condenser tubes* — Used to cool vapors produced during cooking
- *Filters* — Coffee filters, cloth, and cheesecloth
- *Funnels/turkey basters* — Used to separate layers of liquids
- *Gas containers* — Propane bottles, fire extinguishers, self-contained underwater breathing apparatus (SCUBA) tanks, plastic drink bottles (often attached to some sort of tubing) **(Figure 22.83** on page 1200)
- *Glassware* — Particularly Pyrex® or Visions® cookware, mason jars, and other laboratory glassware that can tolerate heating and violent chemical reactions
- *Heat sources* — Burners, hot plates, microwave ovens, and camp stoves

Figure 22.82a and b Meth lab products and equipment.

- *Grinders* — Used to grind up ephedrine or pseudoephedrine tablets
- *pH papers* — Used to test the pH levels of the reactions
- *Tubing* — Glass, plastic, copper, or rubber

Table 22.28
Methamphetamine Sources and Production Hazards

Chemical Name	Common Sources/Uses	Hazards	Production Role
Acetone	• Paint solvent • Nail polish remover	• Highly flammable • Vapor is irritating to eyes and mucous membranes • Inhalation may cause dizziness, narcosis, and coma • Liquid may do damage upon contact • Ingestion may cause gastric irritation, narcosis, and coma	• Pill extraction • Cleaning glassware • Cleaning finished methamphetamine (meth)
Anhydrous Ammonia	• Sold as fertilizer; also used as a refrigerant gas • Stolen from farms and other locations for illegal meth production • Often stored in propane tanks or fire extinguishers at illegal meth labs, which causes the fittings of the tank or extinguisher to turn blue (Figure 9.31, p. 491)	• Toxic • Corrosive • Flammable • Severe irritant; may cause severe eye damage, skin burns and blisters, chest pain, cessation of breathing, and death	Meth production process
Ethyl Alcohol/ Denatured Alcohol/ Ethanol/Grain Alcohol	• Sold as solvents • Is the alcohol found in beverages at greatly reduced concentrations	• Highly flammable • Toxic; may cause blindness or death if swallowed • Inhalation may affect central nervous system causing impaired thinking and coordination • Skin and respiratory tract irritant (may be absorbed through the skin) • May affect the liver, blood, kidneys, gastrointestinal tract, and reproductive system	• Used with sulfuric acid to produce ethyl ether (see Ethyl Ether/Ether entry) • Cleaning glassware

Continued

Table 22.28 (continued)
Methamphetamine Sources and Production Hazards

Chemical Name	Common Sources/Uses	Hazards	Production Role
Ephedrine	Over-the-counter cold and allergy medications	Harmful if swallowed in large quantities	Primary precursor for meth
Ethyl Ether/Ether	Starting fluids	• Highly flammable • Oxidizes readily in air to form unstable peroxides that may explode spontaneously • Vapors may cause drowsiness, dizziness, mental confusion, fainting, and unconsciousness at high concentrations	Separation of the meth base before the *salting-out* process begins, primarily in the *Nazi/Birch* method
Hydrochloric Acid/ Muriatic Acid (Other acids can be used as well, including sulfuric acid and phosphoric acid)	Commerical or industrial strength cleaners for driveways, pools, sinks, toilets, etc.	• Toxic; ingestion may cause death • Corrosive; contact with liquid or vapors may cause severe burns • Inhalation may cause coughing, choking, lung damage, pulmonary edema, and possible death • Reacts with metal to form explosive hydrogen gas	Production of water-soluble salts
Hydrogen Peroxide	• Common first aid supply • Used for chemical manufacturing, textile bleaching, food processing, and water purification	• Strong oxidizer • Eye irritant	Extrication of iodine crystals from Tincture of Iodine
Hypophosphorous Acid	Laboratory Chemical	• Corrosive • Toxic • Generates deadly phosgene during initial reaction	Source of phosphorous in *Red P* method

Continued

Table 22.28 (continued)
Methamphetamine Sources and Production Hazards

Chemical Name	Common Sources/Uses	Hazards	Production Role
Iodine	Tincture of iodine	• Toxic • Vapors irritating to respiratory tract and eyes • May irritate eyes and burn skin	• Meth production process • Can be mixed with hydrogen sulfide to make hydriodic acid (strong reducing agent) • Can be mixed with red phosphorus and water to form hydriodic acid
Isopropyl Alcohol	Rubbing Alcohol	• Flammable • Vapors in high concentrations may cause headache and dizziness • Liquid may cause severe eye damage	• Pill extraction • Cleaning finished meth
Lithium Metal	Lithium batteries	• Flammable solid • Water-reactive (reacts with water to form lithium hydroxide, which can burn the skin and eyes)	Reacts with anhydrous ammonia and ephedrine/pseudoephedrine in the *Nazi/Birch* method
Methyl Alcohol	HEET® Gas-Line Antifreeze and Water Remover	• Highly flammable • Vapors may cause headache, nausea, vomiting, and eye irritation • Vapors in high concentrations may cause dizziness, stupor, cramps, and digestive disturbances • Highly toxic when ingested	Pill extraction
Mineral Spirits/ Petroleum Distillate	• Lighter fluid • Paint thinner	• Flammable • Toxic when ingested • Vapors may cause dizziness • May affect central nervous system and kidneys	Separation of meth base before *salting-out* process begins

Continued

Table 22.28 (continued)
Methamphetamine Sources and Production Hazards

Chemical Name	Common Sources/Uses	Hazards	Production
Naphtha	Camping fuel for stoves and lanterns	• Highly flammable • Toxic when ingested • May affect the central nervous system • May cause irritation to the skin, eyes, and respiratory tract	• Separation of meth base before *salting-out* process begins • Cleaning preparation
Pseudoephedrine	Over-the-counter cold and allergy medications	Harmful if swallowed in large quantities	Production of meth (same as Ephedrine)
Red Phosphorous	Matches	• Flammable solid • Reacts with oxidizing agents, reducing agents, peroxides, and strong alkalis • When ignited, vapors are irritating to eyes and respiratory tract • Heating in a reaction or cooking process generates deadly phosphine gas • Can convert to white phosphorous (air reactive) when overheated	Mixed with iodine in the *Red P* method; serves as a catalyst by combining with elemental iodine to produce hydriodic acid (HI), which is used to reduce ephedrine or pseudoephedrine to meth
Sodium Hydroxide (Other alkaline materials may also be used such as sodium, calcium oxide, calcium carbonate, and potassium carbonate)	Drain openers	• Very corrosive; burns human skin and eyes • Generates heat when mixed with an acid or dissolved in water	After cooking, an alkaline product such as sodium hydroxide turns the very acidic mixture into a base
Sulfuric Acid	Drain openers	• Extremely corrosive • Inhalation of vapors may cause serious lung damage • Contact with eyes may cause blindness • Both ingestion and inhalation may be fatal	Creates the reaction in the salting phase; combines with salt to create hydrogen chloride gas, which is necessary for the *salting-out* phase

Continued

Table 22.28 (continued)
Methamphetamine Sources and Production Hazards

Chemical Name	Common Sources/Uses	Hazards	Production Role
Toluene	Solvent often used in automotive fuels	• Flammable • Vapors may cause burns or irritation of the respiratory tract, eyes, and mucous membranes • Inhalation may cause dizziness; severe exposure may cause pulmonary edema • May react with strong oxidizers	Separation of meth base before the *salting-out* process begins
Hydrogen Chloride		• Toxic • Corrosive • Eye irritant • Vapor or aerosol may produce inflammation and may cause ulceration of the nose, throat, and larynx	• Created by adding sulfuric acid to rock salt • Used to *salt out* meth from base solution
Phosphine Gas		• Very toxic by inhalation • Highly flammable; ignites spontaneously on contact with air and moisture, oxidizers, halogens, chlorine, and acids • May be fatal if inhaled, swallowed, or absorbed through skin • Contact causes burns to skin and eyes	• By-product • Produced when red phosphorous and iodine are combined during the cooking process
Hydrogen Iodide/ Hydriodic Acid Gas		• Highly toxic • Attacks mucous membranes and eyes	• By-product • Produced when red phosphorous and iodine are combined during the cooking process • Causes the reddish/orange staining commonly found on the walls, ceilings, and other surfaces of meth labs
Hydriodic Acid		• Corrosive • Causes burns if swallowed or comes in contact with skin	• By-product • Produced when red phosphorous and iodine are combined during the cooking process

Figure 22.83 Anhydrous ammonia, used in the meth production process, will turn the brass fittings on propane cylinders and other containers blue.

Other clues to the presence of meth labs in structures include the following:

- Windows covered with plastic or tinfoil
- Knowledge of renters who pay landlords in cash
- Unusual security systems or other devices
- Excessive trash
- Increased activity, especially at night
- Unusual structures
- Discoloration of structures, pavement, and soil
- Strong odor of solvents
- Smell of ammonia, starting fluid, or ether
- Iodine- or chemical-stained bathroom or kitchen fixtures

It is estimated that for every pound (0.5 kg) of meth produced, 6 pounds (2.7 kg) of hazardous waste is generated. This waste may then be dumped in the regular residential trash, down the drain to the septic system, beside roadways, on vacant properties, and in streams or ponds and lakes. It may also be buried. Disposal of this waste is very expensive, and the cleanup process is potentially very dangerous. Many law enforcement departments have contracts with private hazardous materials waste disposal contractors to handle the cleanup and decontamination of seized illegal meth labs and dumps.

Chemical Agents Labs

Some chemical warfare agents are surprisingly easy to make. While the recipes may be easy to find, the actual materials necessary to make them may not be. Some ingredients may be common, but access to others is restricted. Chemical agent labs might be identified by the presence of military manuals, underground cookbooks, chemicals (such as organophosphate pesticides) that would not normally be used to make meth or other illegal drugs, and more sophisticated lab equipment that is necessary to conduct some of the chemical reactions needed to make the agents. Responders should be suspicious if a laboratory seems to have unusual equipment or chemicals.

Explosives Labs

Explosives labs may lack the glassware, tubing, Bunsen burners, chemical bottles, and other trappings traditionally associated with the term *laboratory*. For example, a work area in a garage used to assemble custom fireworks or pyrotechnics would be considered an explosives lab for purposes of this manual. Generally, most explosives labs do not need to heat or cook any of their materials; however, a lab established to make explosive chemical mixtures might look more like a traditional industrial or university chemistry lab.

Obvious clues to the presence of an explosives lab might include literature on how to make bombs, significant quantities of fireworks, ammunition like shotgun shells, black powder, smokeless powder, blasting caps, commercial explosives, incendiary materials, or other explosive chemicals. Finding these items in conjunction with components that can be used to make improvised explosive devices (IEDs) (pipes, activation devices, empty fire extinguishers, propane containers, and the like) would give even more evidence of an explosives lab.

Biological Labs

Biological labs may look very similar to chemical agent or drug labs, but a trained and experienced person would see some significant differences. Biological labs are in the business of working with etiological agents; therefore, these labs are more likely to be equipped with petri dishes (used to grow biological organisms and materials), microscopes and glass slides (etiological agents are microscopic), autoclaves, and centrifuges. Generally, biological labs may also be cleaner than any other type of lab because keeping a sterile environment may be important in avoiding contamination of samples. The presence of castor beans might also be a clue because castor beans are needed to make ricin **(Figure 22.84)**.

Figure 22.84 Ricin is made from castor beans.

Secondary Attacks

The use of secondary devices at terrorist attacks or illicit laboratories is always a possibility. Usually, secondary devices are explosives of some kind, most likely an IED. However, booby traps utilizing other weapons are also possible, and some may use chemical, biological, or radiological materials. Secondary devices are often designed to impact an ongoing emergency response in order to create more chaos and injure responders and bystanders. Usually they will explode after a primary explosion, but they may also be used at other major emergency responses that do not involve explosives. Secondary devices may also be deployed as a diversionary tactic to route emergency responders away from the primary attack area.

Like many IEDs, secondary devices will be hidden or camouflaged. The devices may be detonated by a time delay, but radio-controlled devices, cell phone-activated devices, and other activation devices are also used. The majority of IEDs used in Iraq are command detonated (in other words, someone nearby, usually hidden but within line of sight, physically sends the signal to detonate the device). In some cases, an obvious IED may be used to lure personnel to a specific area where a less obvious IED is hidden.

NOTE: *[taken from US OSHA info sheet]*

The Transportation Security Administration has issued warnings related to a possible terrorist event in which a small explosive or nuisance device might be activated on a subway platform during the height of rush hour. This initial explosion would result in an evacuation to a nearby street where a secondary explosive device may be detonated to target responders and the public.

Use of Secondary Devices

Bombings on January 27, 1997 outside the Atlanta Northside Family Planning Service and on February 21, 1997 at the Otherside Lounge in Atlanta both had secondary explosive devices placed in proximity to the primary explosion. In another incident, a Puerto Rican peace officer was reported to have been killed by a flashlight bomb rigged with an improvised motion switch at a crime scene.

In other parts of the world, usage of secondary devices varies. In Israel, secondary devices have only been used rarely; however, they are commonly used in Iraq. Regardless of historical patterns, secondary devices are part of the terrorist's toolbox. **A good rule of thumb is: if there has already been one explosion (or one device has been found), always expect another.**

Guidelines for protecting against possible secondary devices include the following:

- Anticipate the presence of a secondary device at any suspicious incident.

- Visually search for a secondary device (or anything suspicious) before moving into the incident area.

- Avoid touching or moving anything that may conceal an explosive device (including items such as backpacks and purses).

- Effectively manage the scene with cordons, boundaries, and scene control zones.

- Evacuate victims and nonessential personnel as quickly as possible.

- Preserve the scene as much as possible for evidence collection and crime investigation.

While IEDs can be disguised as almost anything, responders should look for things that may seem out of place **(Figure 22.85)**. Items or circumstances that "send up a red flag" should be noted and treated with appropriate caution. Responders should be very cautious of any item(s) that arouse curiosity, including the following:

- Containers with unknown liquids or materials

Figure 22.85 IEDs can be disguised as anything. This is a replica of an actual IED.

- Unusual devices or containers with electronic components such as wires, circuit boards, cellular phones, antennas, and other items attached or exposed

- Devices containing quantities of fuses, fireworks, match heads, black powder, smokeless powder, incendiary materials, or other unusual materials

- Materials attached to or surrounding an item such as nails, bolts, drill bits, marbles, and so on that could be used for shrapnel

- Ordnance such as blasting caps, detcord, military explosives, commercial explosives, grenades, and so on.

- Any combination of the above described items

WARNING!
Firefighters should NEVER approach or move suspicious objects. Notify appropriate personnel (law enforcement/bomb techs) and evacuate the area immediately.

Summary

Because hazardous materials could be involved in virtually any emergency to which you and your fellow firefighters are called, and because these materials may be highly toxic – even in tiny amounts – it is critical that you have at least a basic understanding of the potential threats and the possible solutions. You should be aware of the vast quantities of these materials that are shipped, stored, and used every day in North America. You should also be aware of the various placards, labels, and signs that are required by law to make it easier for you to identify these materials from a safe distance. You should be familiar with the various references that are available to assist you in identifying specific materials, establishing safe control zones, conducting evacuations, and taking other appropriate precautions to stabilize the situation. Finally, you must know what specialized resources will be needed to mitigate a hazardous materials release and be prepared to assist in that mitigation effort.

End-of-Chapter Review Questions

Fire Fighter I

1. What are persons trained to the Awareness Level expected to do?

2. What are persons trained to the Operations Level expected to do?

3. What is a supplied-air respirator?

4. What U.S. EPA level of protective equipment provides the highest level of protection?

5. List three methods to prevent and/or reduce the effects of heat exposure while wearing protective clothing.

6. Describe the four main routes through which hazardous materials can enter the body.

7. What are the seven clues to the presence of hazardous materials?

8. What are the three ways to use the *ERG* to locate the appropriate orange-bordered guide page?

9. List four chemical attack indicators.

10. What are clues to the presence of meth labs?

Using the U.N. Identification Number

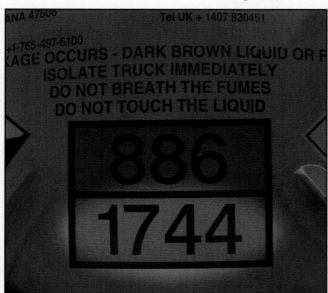

Step 1: Identify the four-digit U.N. identification number.

Step 2: Refer to the appropriate yellow-bordered pages reference guide number.

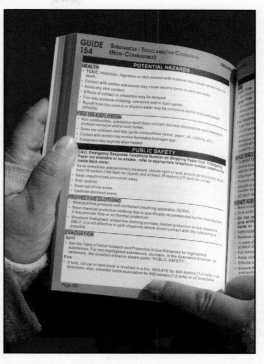

Step 3: Refer to the orange-bordered page with the appropriate guide number for information on managing the incident.

Step 4: For highlighted chemicals refer to the green-bordered pages for initial isolation by looking up the identification number.

NOTE: The green-bordered pages provide initial isolation and protective action distances. The action taken will vary according to whether or not the spill is large/small and whether it is night/day.

Using the Material Name

Step 1: Identify the name of the material.

Step 2: Refer to the name of the material in the blue-bordered pages to locate the correct guide number.

Step 3: Refer to the orange-bordered page with the appropriate guide number for information on managing the incident.

Step 4: For highlighted chemicals refer to the green-bordered pages for initial isolation by looking up the identification number.

Using the Container Profile

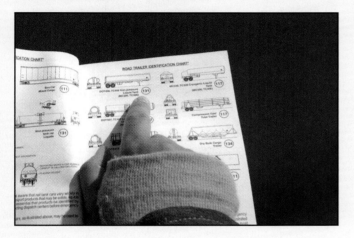

Step 1: Identify the profile of the container and locate the profile in the white pages of the ERG.

Step 2: Refer to the appropriate guide number in the circle and go to the appropriate orange-bordered page.

NOTE: Using this method does not identify whether or not the chemical is a highlighted chemical.

Skill SHEETS

22-I-1
Obtain information about a hazardous material using the *ERG*.

Using the Placard

Step 2: Refer to the appropriate guide number in the circle and go to the appropriate orange-bordered page.

Note: Using this method does not identify whether or not the chemical is a highlighted chemical.

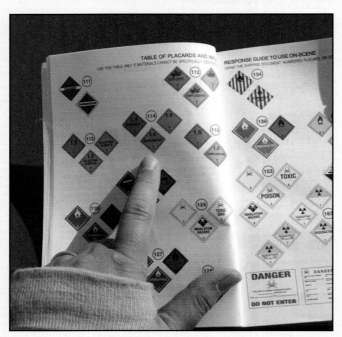

Step 1: Identify the placard and locate it in the white pages of the *ERG*.

Chapter Contents

Operations at Haz Mat Incidents

Chapters 22 and 23 of this manual meet the requirements of the following:

Combined, Chapters 22 and 23 meet the requirements of:

- **OSHA 1910.120 (q)(6)(i) and (ii),** Awareness and Operations levels

- **NFPA® 472,** *Standard for Competence of Responders to Hazardous Materials/ Weapons of Mass Destruction Incidents,* 2008 Edition, Chapters 4 and 5, Awareness level and Operations level core competencies

- **Office of Domestic Preparedness** *Emergency Responder Guidelines* for Awareness Level and Performance Level A for firefighters

Chapter Objectives

1. Summarize incident priorities for all haz mat and terrorist incidents.

2. Discuss the management structure at haz mat or terrorist incidents.

3. Describe the problem-solving stages at haz mat and terrorist incidents.

4. Explain how the strategic goal of isolation and scene control is achieved.

5. Explain how the strategic goal of notification is achieved.

6. Explain how the strategic goal of ensuring the safety of responders and the public is achieved.

7. Summarize general guidelines for decontamination operations.

8. Describe the three types of decontamination.

9. Discuss implementing decontamination.

10. Discuss rescue at haz mat incidents.

11. Explain how the strategic goal of spill control and confinement is achieved.

12. Discuss crime scene management and evidence preservation.

13. Explain actions taken during the recovery and termination phase of a haz mat or terrorist incident.

14. Perform emergency decontamination. (Skill Sheet 23-I-1)

15. Perform defensive control functions — absorption. (Skill Sheet 23-I-2)

16. Perform defensive control functions — diking. (Skill Sheet 23-I-3)

17. Perform defensive control functions — damming. (Skill Sheet 23-I-4)

18. Perform defensive control functions — diversion. (Skill Sheet 23-I-5)

19. Perform defensive control functions — retention. (Skill Sheet 23-I-6)

20. Perform defensive control functions — dilution. (Skill Sheet 23-I-7)

21. Perform defensive control functions — vapor dispersion. (Skill Sheet 23-I-8)

Operations at Haz Mat Incidents

Case History

On January 6, 2005 at approximately 02:45 a.m., two Norfolk Southern freight trains collided in Graniteville, South Carolina. The collision resulted in a catastrophic release of chlorine gas to the atmosphere from a tank car damaged in the derailment. This release rapidly vaporized to form a dense and highly toxic airborne cloud affecting Graniteville residents and employees of nearby Avondale Mill. Other hazardous materials cars involved in the derailment included 2 additional chlorine cars, 1 sodium hydroxide car, and 1 creosol car. More than 500 people sought medical evaluation, approximately 70 people were admitted to hospitals and 9 people were killed due to chlorine exposure.

Multiple agencies were involved in the response to this incident, requiring a unified command structure. The initial response was provided by agencies from the local community and county, including the local volunteer fire department. Eventually, approximately 600 federal, state, and local personnel participated in the response. Cargos were identified; gas concentrations were monitored; plume models were generated to assist in evacuation operations; and public protection operations were initiated. Over 5,000 residents were eventually displaced from their homes. Three separate decon locations were established with EMS support and mitigation efforts continued for days.

In many ways, haz mat incidents are not unlike other emergencies to which firefighters respond. Successful incident mitigation is achieved through incident management elements such as setting priorities, implementing an incident management structure, and following a problem-solving process that includes emphasis on firefighter safety from beginning to end **(Figure 23.1)**. However, if these incident management elements are not in place or are used improperly, a haz mat incident can rapidly spiral out of control. *Mistakes made in the initial response to the incident can mean the difference between solving the problem and becoming part of it.*

Incident management elements enable first responders to mitigate the incident effectively and safely. In general, these elements are composed of the following:

- *Priorities*
  - Life safety
  - Incident stabilization
  - Protection of property

Figure 23.1 Implementing an incident command system from the very beginning is very important at haz mat incidents. *Courtesy of Rich Mahaney.*

- *Management Structure*
  - Incident command system
  - Predetermined procedures and guidelines such as emergency response plans, standard operating procedures (SOPs) or operating instructions (OIs [military]) that outline operational procedures regarding communications, equipment, personnel, resources, mutual aid, and other needs

- *Incident Mitigation Process*
  - Analyzing the incident through scene analysis, information gathering, and/ or size-up to try to understand and identify the problem(s) and assess hazards and risks
  - Planning the response by determining specific strategic goals
  - Implementing the response using tactical objectives and assignment of specific tasks
  - Evaluating progress (feedback loop)

Priorities

There are three incident priorities for all haz mat and terrorist incidents:

- Life safety
- Incident stabilization
- Protection of property

All decisions must be made with these priorities in mind.

The first priority is the safety of emergency responders because if responders do not protect themselves first, they cannot protect the public. A dead, injured, or unexpectedly contaminated firefighter becomes part of the problem, not the solution. For this reason, with the exception of certain situations involving flammable and combustible liquids, firefighters are limited to defensive actions (or nonintervention) at haz mat and terrorist incidents unless they receive additional training. They should not come in contact with the hazardous materials involved. Hazardous materials response teams and personnel with additional training are needed to conduct offensive operations (see Modes of Operation section).

If there is no immediate threat to either responders or civilians, the next consideration is stabilizing the incident. Stabilizing actions can include such things as establishing scene perimeters to prevent the spread of contamination and constructing dams and dikes to prevent hazardous liquids from entering streams or water supplies. Stabilizing the incident can also minimize environmental and property damage.

At haz mat incidents, protection of both property and the environment must be addressed. Hazardous materials can seriously damage the environment, and cleanup from an incident can be extremely expensive. The hazards presented by these materials can continue to cause problems for a long time once released, so it is important for environmental protection to be a consideration from the very beginning of the mitigation process.

Management Structure

Firefighters will initiate and operate within their standard incident command system at haz mat incidents. There may, however, be some differences to note:

- Haz mat incidents may require inclusion of a haz mat division or group led by a Haz Mat Supervisor.

- Because of their complexity and scope, haz mat and terrorist incidents will often utilize a unified command structure.

- At all incidents in the United States that involve terrorism or a federal response (such as at a large oil spill), National Incident Management Center-Incident Command System (NIMS-ICS) will be the incident command system used at the incident **(Figure 23.2)**.

In addition to the incident command system, firefighters must also operate in accordance with predetermined procedures such as SOPs and their local emergency response plan (LERP). According to 29 CFR 1910.120, *Hazardous Waste Operations and Emergency Response* (HAZWOPER), all fire and emergency services organizations that respond to haz mat incidents are required to have an emergency response plan that includes predetermined guidelines or procedures for managing incidents involving terrorist attacks and hazardous materials releases.

Haz Mat Incident Mitigation

Incident priorities, NIMS-ICS, and predetermined procedures provide a management structure for first responders. But the incident also needs to be mitigated. The problem must be solved through a process of problem-solving and decision-making. While you and your fellow firefighters are not likely be in a command position responsible for planning an appropriate response to a terrorist attack, it is important to understand this process.

While there are many problem-solving and decision-making models, many contain four common elements: (1) an information gathering, input, and analysis stage (analyzing the situation); (2) a processing and/or planning stage (planning an appropriate response); (3) an implementation or output stage (implementing the appropriate response); and (4) a review or evaluation stage.

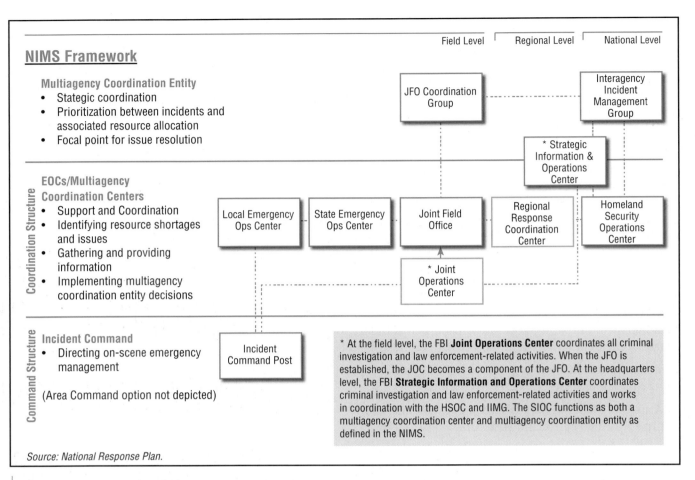

NIMS Framework

Multiagency Coordination Entity
- Stategic coordination
- Prioritization between incidents and associated resource allocation
- Focal point for issue resolution

Field Level | Regional Level | National Level

JFO Coordination Group

Interagency Incident Management Group

* Strategic Information & Operations Center

Coordination Structure

EOCs/Multiagency Coordination Centers
- Support and Coordination
- Identifying resource shortages and issues
- Gathering and providing information
- Implementing multiagency coordination entity decisions

Local Emergency Ops Center

State Emergency Ops Center

Joint Field Office

Regional Response Coordination Center

Homeland Security Operations Center

* Joint Operations Center

Command Structure

Incident Command
- Directing on-scene emergency management

(Area Command option not depicted)

Incident Command Post

* At the field level, the FBI **Joint Operations Center** coordinates all criminal investigation and law enforcement-related activities. When the JFO is established, the JOC becomes a component of the JFO. At the headquarters level, the FBI **Strategic Information and Operations Center** coordinates criminal investigation and law enforcement-related activities and works in coordination with the HSOC and IIMG. The SIOC functions as both a multiagency coordination center and multiagency coordination entity as defined in the NIMS.

Source: National Response Plan.

Figure 23.2 The NIMS framework for terrorist incidents.

Incident Problem-Solving Process Models

Many incident management process models are in use to provide first responders a series of steps or actions to take at haz mat incidents and terrorist attacks. Some of the better known systems are as follows:

- **GEDAPER** by David Lesak:

    G — Gather information

    E — Estimate potential course and harm

    D — Determine strategic goals

    A — Assess tactical options and resources

    P — Plan of action implementation

    E — Evaluate operations

    R — Review the process

Continued

Incident Problem-Solving Process Models (continued)

- *DECIDE* by Ludwig Benner
    - *D* — Detect the presence of hazardous materials
    - *E* — Estimate likely harm without intervention
    - *C* — Choose response objectives
    - *I* — Identify action options
    - *D* — Do best option
    - *E* — Evaluate progress
- *APIE* by the International Association of Fire Fighters (IAFF)
    - *A* — Analyze
    - *P* — Plan
    - *I* — Implement
    - *E* — Evaluate (and repeat)
- *IFSTA* by International Fire Service Training Association (IFSTA)
    - *I* — Identify the nature of the problem
    - *F* — Formulate objectives based on available information
    - *S* — Select the desired alternatives from the available options
    - *T* — Take appropriate action
    - *A* — Analyze outcomes continually
- *SISIACMRD* – Source: New South Wales Fire Brigades (NSWFB) Australia. This acronym is easily remembered by the using the following mnemonic: **S**ick In **S**ide **I A**lways **C**all **M**y **R**egular **D**octor
    - *S* — Safe approach
    - *I* — Incident command/control
    - *S* — Secure the scene
    - *I* — Identify risk/haz mat
    - *A* — Assess potential harm
    - *C* — Call in resources
    - *M* — Monitor information
    - *R* — Render safe
    - *D* — Decontaminate
- *OODA* – U.S. Military
    - *O* — Observe
    - *O* — Orient
    - *D* — Decide
    - *A* — Act

Gregory G. Noll, Michael S. Hildebrand, and James G. Yvorra developed the Eight Step Incident Management Process©, which is a tactical decision-making model that focuses on haz mat incident safe operating practices.* The eight steps are as follows:

1. Site management and control

2. Identify the problem

Continued

In emergency incidents, problem solving and decision making are fluid processes (**Table 23.1**), and an incident commander's (IC's) understanding of a problem (and subsequent plans to address it) may change as more information becomes available and/or conditions change. For example, the incident command system should be initiated from the very beginning of the incident, and information gathering may continue long after initial response options have been planned and implemented, resulting in changes to strategies and tactics.

Analyzing the Situation

The first thing emergency responders must do upon arrival at an incident is determine the type of incident (i.e., fire, emergency medical incident, structural collapse, haz mat incident, terrorist attack, etc.) and the scope (size) of the incident (**Figure 23.3**). Responders will then need to follow internal notification protocols depending on the type of incident to provide critical information and/or request additional resources. The

Figure 23.3 Determining the type and scope of the incident are top priorities in the initial response. *Courtesy of Tom Clawson, Technical Resources Group, Inc.*

Table 23.1
Four-Step Problem-Solving Process

Problem-Solving Stages at Haz Mat and Terrorist Incidents

Analysis Stage and Information Gathering.
- Recognize the type of incident (Is it a haz mat incident or terrorist attack? If so, what materials are involved?)
- Identify all the hazards presented by the incident (chemical, mechanical, electrical, etc.)
- Predict the likely behavior of chemical, explosive, biological, and radiological materials
- Estimate potential harm

Planning Stage.
- Determine if additional help is needed
- Identify actions to protect emergency responders and others from hazards
- If appropriate, consult the *ERG* for instructions and guidelines regarding hazardous materials
- Determine strategies and tactics to stabilize the incident
- Determine appropriate personal protective equipment
- Determine appropriate decontamination methods
- Devise the Incident Action Plan

Implementation Stage.
- Implement the incident management system
- Transmit information to an appropriate authority and call for appropriate assistance
- Establish and enforce scene control perimeters
- Implement the Incident Action Plan
- Implement strategies and tactics appropriate to training such as isolating the area and denying entry, conducting rescues, conducting mass decontamination, etc.
- Identify and preserve evidence

Evaluation and Review Stage.
- Evaluate effectiveness of approach (is the incident stabilizing?)
- Process and provide feedback to IC

following sections discuss the initial size-up of the incident including **hazard** and risk assessment and incident levels.

Initial Size-up

Size-up is the initial and ongoing assessment of incident conditions. It is the mental process of considering all available factors that will affect an incident during the course of the operation. Hazard and risk assessment is part of the size-up process, focusing particularly on the dangers, hazards, and risks presented by the materials involved in an incident.

Hazard Assessment — Formal review of the hazards that may be encountered while performing the functions of a firefighter or emergency responder; used to determine the appropriate level and type of personal and respiratory protection that must be worn.

After approaching the incident from upwind, uphill, and upstream, if possible, size-up up of a haz mat or terrorist incident initially focuses on identifying the material(s) involved. Once the size of the endangered area has been estimated utilizing available resources (including the *Emergency Response Guidebook* [*ERG*] and plume modeling software), it is possible to predict the potential exposures, including the number of people in the area, the number of threatened/affected buildings/property, and any environmental concerns.

Understanding of the potential area of effect and number of exposures, a number of strategic goals and tactical objectives (response options or objectives) can be considered for achieving incident stabilization. Decisions about which options to use are based on the availability of resources and a risk/benefit analysis. For example, rescue may be an initial priority, but evacuation and sheltering in place may be the best way to save additional lives (see following sections).

Figure 23.4 Some incidents may be so large that it is impossible to see the entire incident scene from one location, complicating hazard assessment and management of the incident. Additional resources may need to be utilized, such as aerial reconnaissance.

Evacuation — Process of leaving or being removed from a potentially hazardous location.

Size-up is frequently complicated by a lack of information. The IC's view of the incident may be limited by the size of the hazard area or location of the event (that is, inside a vehicle or structure). Sometimes, the scope of the incident is so great that hazard assessment becomes difficult because it is impossible to see the entire incident scene **(Figure 23.4)**. In such cases, more than one agency may be involved in the assessment of hazards, and additional resources such as aerial reconnaissance may be needed. In addition, limited or conflicting information regarding the nature and scope of the incident is possible. Initial assessment may be based on anticipated conditions and updated as additional information becomes available.

Hazard and risk assessment is a continuous and ongoing process. It starts with preincident planning, extends through the receipt of an alarm, and continues throughout the course of an incident. Once on the scene, the IC conducts a hazard assessment and repeats this process as the incident continues. In some cases, witnesses and other on-site personnel can provide specific, detailed information. Final pieces of the hazard and risk assessment are added to the information made available before and after arrival.

Initial assessment by first responders is extremely important. Information to be gathered includes:

- Type of incident (and identification of the product involved)
- Exact location
- Type, size, and occupancy of structure
- Building construction
- Atmospheric conditions and risks
- Utilities
- Potential for secondary attacks at criminal or terrorist incidents
- Best means of exterior and interior access to the incident scene
- Vertical/horizontal openings, shafts, voids, tunnels, etc.
- Existing hazards (for example, fire, hazardous materials, explosion, floodwaters)

- Structural damage and collapse

- Special hazards

- Fire spread and direction of travel

- Distribution, numbers, and types of dead, injured, or uninjured

- Common symptoms and injuries

- Number of trapped victims

- Estimated resources required for search and rescue

- Estimated resources required for fire suppression

- Estimated resources for hazardous materials operations

- Specialized equipment required for recovery operations

Because size-up is a continual and on-going process, responders can gather additional information relating to size-up by taking the following actions:

- Liaison with other on-scene agencies

- Rapid formation of a unified command structure (or multi-agency command team)

- Review of preincident surveys, emergency response plans, and sketches

- Review of geographical information systems (GIS), topography maps and utility plans (for drains/sewers, rivers, streams, etc.)

- Note exposure types, hazards, and distances

- Review existing water supply systems

- Consider multiple access and egress routes

- Consider multiple resource staging areas

- Consider current and forecast weather conditions and their effect upon the incident

- Ensure correct apparatus placement on scene

- Secure any additional information from the dispatcher

- Decide if and what additional resources are needed for response and recovery operations

- Initiate crime scene preservation procedures

With this combined information, a formal plan of action (Incident Action Plan [IAP]) is decided upon and implemented.

Topography — Physical configuration of the land or terrain.

Incident Levels

After the initial size-up has determined the scope of an incident, the level of the incident can be determined in accordance with the definitions in the LERP. Most incident level models identify three levels of response graduating from Level I (least serious) to Level III (most serious). By predefining the levels of response, an increasing level of involvement and necessary resources can be identified quickly based on the severity of the incident.

Level I. This type of incident is within the capabilities of the fire or emergency services organization or other first responders. A Level I incident is the least serious and the easiest to handle. It may pose a serious threat to life or property, although this is

Ignition Sources at an Incident Scene

[NFPA® 472: 4.4.1(3)(b)]

Many potential ignition sources may exist at the scene of a hazardous materials incident. Where flammable liquids or gases are involved, it is extremely important for hazard and risk assessments to consider these items (and measures taken to control them):

- Open flames
- Static electricity
- Existing pilot lights
- Electrical sources, including non-explosion-proof electrical equipment
- Internal combustion engines in vehicles and generators
- Heated surfaces
- Cutting and welding operations
- Radiant heat
- Heat caused by friction or chemical reactions
- Cigarettes and other smoking materials
- Cameras
- Road flares

Explosive atmospheres can be ignited by several simple actions:

- Opening or closing a switch or electrical circuit (for example, a light switch)
- Turning on a flashlight
- Operating a radio
- Activating a cell phone

Plug — Patch to seal a small leak in a container.

Figure 23.5 Level II incidents require the services of a hazardous materials response team to perform activities such as plugging and patching.

not usually the case. Evacuation (if required) is limited to the immediate area of the incident. The following are examples of Level I incidents:

- Small amount of gasoline or diesel fuel spilled from a vehicle
- Leak from domestic natural gas line on the consumer side of the meter
- Broken containers of consumer commodities such as paint, thinners, bleach, swimming pool chemicals, and fertilizers (owner or proprietor is responsible for cleanup and disposal)

Level II. This type of incident is beyond the capabilities of the first responders on the scene and may be beyond the capabilities of the first response agency/organization having jurisdiction. Level II incidents require the services of a formal haz mat response team. A properly trained and equipped response team could be expected to perform the following tasks:

- Use chemical protective clothing (CPC).
- Dike and confine within the contaminated areas.
- Perform plugging and patching activities **(Figure 23.5)**.

Figure 23.6 Sampling and testing should be conducted by haz mat technicians. *Courtesy of Sherry Arasim.*

Figure 23.7 A leaking tanker is an example of a Level II incident. *Courtesy of District Chief Chris E. Mickal, NOFD Photo Unit.*

- Sample and test unknown substances (**Figure 23.6**).
- Perform various levels of decontamination.

  The following are examples of Level II incidents:

- Spill or leak requiring large-scale evacuation
- Any major accident, spillage, or overflow of flammable liquids (**Figure 23.7**)

Decontaminate — To remove a foreign substance that could cause harm; frequently used to describe removal of a hazardous material from the person, clothing, or area.

- Spill or leak of unfamiliar or unknown chemicals
- Accident involving extremely hazardous substances
- Rupture of an underground pipeline
- Fire that is posing a boiling liquid expanding vapor explosion (BLEVE) threat in a storage tank

Unified Command — In the Incident Management System, a shared command role that allows all agencies with responsibility for the incident, either geographical or functional, to manage the incident by establishing a common set of incident objectives and strategies. In unified command there is a single incident command post and a single operations chief at any given time.

Level III. This type of incident requires resources from state/provincial agencies, federal agencies, and/or private industry in addition to unified command. A Level III incident is the most serious of all hazardous material incidents. A large-scale evacuation may be required. Most likely, the incident will not be concluded by any one agency. Successful handling of the incident requires a collective effort from several of the following resources/procedures.

- Specialists from industry and governmental agencies
- Sophisticated sampling and monitoring equipment
- Specialized leak and spill control techniques
- Decontamination on a large scale

The following are examples of Level III incidents:

- Incidents that require an evacuation extending across jurisdictional boundaries
- Incidents beyond the capabilities of the local hazardous material response team
- Incidents that activate (in part or in whole) the federal response plan

Planning the Appropriate Response

As part of the problem-solving process, ICs must select the strategic goals and the tactical objectives used to mitigate a terrorist incident. Strategic goals are broad statements of what must be done to resolve the incident, while tactical objectives are the specific operations that must be done in order to accomplish those goals. Response objectives should focus on controlling current events, preventing future problems, and conducting other actions that will assist in stabilizing the incident.

Strategic goals are prioritized depending on available resources and the particulars of the incident. Some of these goals may not be needed if the hazard is not present at the incident. For example, if the incident involves nonflammable chemicals, fire control may not be an issue. Some goals may require the use of specialized resources (such as CPC or a bomb squad) that are not yet available and therefore must be postponed. Others may require the use of so many available resources that the ability to complete other goals in an expedient timeframe might be compromised. The strategies and tactics commonly used at haz mat incidents are discussed in greater detail in following sections.

Terrorism Preparedness

Planning for terrorist attacks can reduce the on-scene decision making at an actual incident. Operations-Level responders participate in hazard and risk analysis of potential targets in many communities, and are informed of the plans for managing the consequences of such attacks. Preincident plans can include possible locations for command posts, conducting decontamination, evacuation centers, and many other aspects of a response to such an event. Joint training exercises are also beneficial.

Modes of Operation

Strategies (and their resulting tactical operations) are divided into three options that relate to modes of operation: defensive, offensive, and nonintervention. A *defensive strategy* provides confinement of the hazard to a given area by performing defensive actions such as containing the hazard, for example, by diking or damming a storm drain if a toxic industrial material is involved **(Figure 23.8)**. An *offensive strategy* includes actions to actively control the hazard, for example, haz mat technicians plugging or patching a leaking container. A *nonintervention strategy* isolates the area to protect the public and emergency responders, but allows the incident to run its course on its own. Nonintervention is the only safe strategy in many types of incidents and the best strategy in certain types of incidents when mitigation is failing or otherwise impossible.

Incident Action Plans

IAPs are critical to the rapid, effective control of emergency operations. An IAP is a well-thought-out, organized course of events developed to address all phases of incident control within a specified time. The timeframe specified is one that allows the least negative action to continue. Written IAPs may not be necessary for short-term, routine operations; however, large-scale or complex incidents require the creation and maintenance of a written plan for each operational period. Regardless of the form of the IAP, its contents must be disseminated throughout the incident organization.

Action planning starts with identifying the strategy to achieve a solution to the confronted problems. Strategy is broad in nature and defines what has to be done. Once the strategy has been defined, the Command Staff needs to select the tactics (how, where, and when) to achieve the strategy. Tactics are measurable in both time and performance. An IAP also provides for necessary support resources such as water supply, utility control, self-contained breathing apparatus (SCBA) cylinder filling, and the like.

The IAP essentially ties the entire problem-solving process together by stating what the analysis has found, what the plan is, and how it shall be implemented. Once the plan is established and resources are committed, it is necessary to assess its effectiveness. Information must be gathered and analyzed so that necessary modifications may be made to improve the plan if necessary. This step is part of a continuous size-up process. Elements of an IAP include the following:

- Strategies/incident objectives
- Current situation summary
- Resource assignment and needs
- Accomplishments
- Hazard statement
- Risk assessment
- Safety plan and message
- Protective measures

Figure 23.8 Damming and diking are examples of defensive strategies. *Courtesy of Rich Mahaney.*

- Current and projected weather conditions
- Status of injuries
- Communications plan
- Medical plan

All incident personnel must function according to the IAP. Company officers or group/division supervisors should follow predetermined procedures, and every action should be directed toward achieving the goals and objectives specified in the plan.

Implementing the Incident Action Plan

After strategic goals have been selected and the IAP formulated, the IC can begin to implement the plan. Strategic goals are met by achieving tactical objectives. Tactical objectives are accomplished or conducted by performing specific tasks.

Tactics follow their respective strategies in that they can be nonintervening, offensive, or defensive in nature. Tactics related to controlling chemical and radiological releases basically fall into two categories: confinement and containment, with the majority of defensive control options being related to confinement. Other tactics may include such things as establishing scene control zones, monitoring for radiation and other hazards, calling for additional resources, wearing appropriate personal protective equipment (PPE), conducting decontamination, and extinguishing fires.

Evaluating Progress

The final aspect of the problem-solving process is reviewing or evaluating progress. If an IAP is effective, the IC should receive favorable progress reports from tactical and/or task supervisors and the incident should begin to stabilize. If, on the other hand, mitigation efforts are failing or the situation is getting worse (or more intense), the plan must be reevaluated and very possibly revised. The plan must also be reevaluated as new information becomes available and circumstances change. If the initial plan isn't working, it must be changed either by selecting new strategies or by changing the tactics used to achieve them. In accordance with predetermined communication procedures (such as designated radio channels), it is important for firefighters to communicate the status of the planned response and the progress of their actions to the IC.

Strategic Goals and Tactical Objectives

In addition to determining the incident type and scope, the following are strategic goals commonly used to mitigate haz mat and terrorist incidents:

- Isolation and scene control
- Notification
- Identification
- Protection of responders and the public
- Rescue
- Spill control and leak containment
- Fire control
- Crime scene management and evidence preservation
- Recovery and termination

Confinement — The process of controlling the flow of a spill and capturing it at some specified location.

Containment — Act of stopping the further release of a material from its container.

Radiation — Energy from a radioactive source emitted in the form of waves or particles.

It should be understood that this list is not all-inclusive, nor must these strategies be undertaken in the order presented. ICs set the goals they deem appropriate by using strategies that will achieve successful mitigation of the incident. For example, rescue might be considered an important strategic goal at one incident but not at another. If conditions at an incident change suddenly for the worse, evacuation and accountability might become strategic goals that spring to the top of the priority list.

Previous sections in this manual have discussed several of these response objectives including identification of materials/hazards and fire control. The following sections will discuss the remaining objectives in the context of haz mat and terrorist incidents. PPE and decontamination, although falling under the umbrella of *Protection*, are discussed in detail in their own sections.

Isolation and Scene Control

Isolation and scene control is one of the primary strategic goals at haz mat incidents and one of the most important means by which responders can ensure the safety of themselves and others. Separating people from the potential source of harm is necessary to protect the life safety of all involved. It is also necessary to prevent the spread of hazardous materials through cross contamination (sometimes called *secondary contamination*, see Decontamination section). Isolation involves physically securing and maintaining the emergency scene by establishing isolation perimeters (or cordons) and denying entry to unauthorized persons. It also includes preventing contaminated or potentially contaminated individuals from leaving the scene in order to stop the spread of hazardous materials. The process may continue with either evacuation, defending in place, or shelter in place (sometimes called *protection in place*) of people located within protective-action zones (see Protection section). Controlling the scene of a haz mat incident may also include the establishment of hazard-control zones.

CAUTION
Whenever possible, personnel should operate within their respective disciplines. Firefighters should conduct fire and spill control activities, EMS personnel should provide medical aid, and law enforcement personnel should provide scene control as appropriate.

Isolation Perimeter

The *isolation perimeter* (sometimes called the *outer perimeter* or *outer cordon*) is the boundary established to prevent access by the public and unauthorized persons. It may be established even before the type of incident or attack is positively identified. If an incident is inside a building, the isolation perimeter might be set at the outside entrances, accomplished by posting personnel to deny entry. If the incident is outside, the perimeter might be set at the surrounding intersections with response vehicles or law enforcement officers diverting traffic and pedestrians **(Figure 23.9)**. In some cases, a traffic cordon may be set up beyond the outer cordon to prevent unauthorized vehicle access only, whereas pedestrian traffic is still allowed.

The isolation perimeter can be expanded or reduced as needed, and it is used to control both access and egress from the scene. For example, in some cases, the initial isolation perimeter established by first responders is expanded outward as additional help arrives. Law enforcement officers are often used to establish and maintain isolation perimeters. Once hazard-control zones are established, the isolation perimeter is generally considered to be the boundary between the public and the cold (safe) zone (see Hazard-Control Zones section).

Figure 23.9 Perimeter control is a vital aspect of scene control and overall incident management.

Hazard-Control Zones

Hazard-control zones provide for the scene control required at haz mat and terrorist incidents to protect responders from interference by unauthorized persons, help regulate movement of first responders within the zones, and minimize contamination (including secondary contamination from exposed or potentially exposed victims). At large, multi-agency response incidents, establishing hard perimeters for hazard-control zones can assist in the challenge of ensuring accountability for all personnel involved in the response.

These control zones are not necessarily static and can be adjusted as the incident changes. It should be noted that different agencies may have different needs in terms of establishing control zones. For example, law enforcement may designate a zone to incorporate an entire crime scene at a terrorist attack that would not correspond to traditional fire service activities. These law enforcement zones might change as evidence is processed and the crime scene is released. These dynamics create an extreme need for flexibility on the part of all agencies in the unified command when establishing these zones.

Zones divide the levels of hazard of an incident, and what a zone is called generally depicts this level. These zones are often referred to as hot, warm, and cold (**Figure 23.10**). It should be noted that there may be multiple control zones within the perimeter of the incident. For example, at a massive explosion, there may

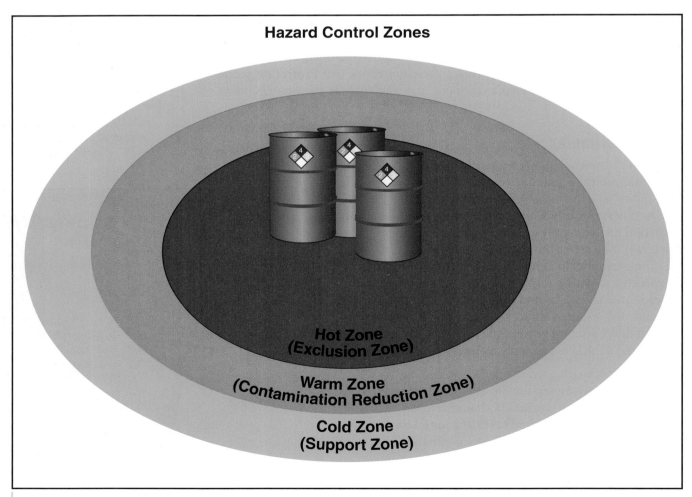

Figure 23.10 Hazard-control zones divide the levels of hazard at an incident into hot, warm, and cold zones, with the hot zone indicating the highest degree of danger.

be multiple building collapses that might require their own set of control zones within the overall incident area.

U.S. Occupational Safety and Health Administration (OSHA) and the U.S. Environmental Protection Agency (EPA) refer to these zones collectively as *site work zones*. They are sometimes called *scene-control zones* as well. Other countries may use different terminology for these zones with which responders must be familiar.

Hot Zone

The *hot zone* (also called *exclusion zone*) is an area surrounding an incident that is potentially very dangerous either because it presents a threat in the form of a hazardous material or the effects thereof, or there are armed and dangerous individuals present (such as in a hostage situation). The area may be contaminated by hazardous materials, or it may have the potential to become contaminated by a release (the area has been or could be exposed to the gases, vapors, mists, dusts, or runoff of the material).

The hot zone extends far enough to prevent people outside the zone from suffering ill effects from the released material, explosion, or other threat. Work performed inside the hot zone is often limited to highly trained personnel such as hazardous materials technicians, special weapons and tactics (SWAT) teams, urban search and rescue (USAR) teams, joint hazard assessment teams (JHAT), and bomb technicians.

Warm Zone

The *warm zone* (also called *contamination reduction zone* or *corridor*) is an area adjoining the hot zone and extending to the cold zone (see following section). The warm zone is used as a buffer between the hot and cold zones and is the place to decontaminate personnel and equipment exiting the hot zone **(Figure 23.11)**. Decontamination usually takes place within a corridor (decon corridor) located in the warm zone. Parts of the warm zone may also be considered to be part of the crime scene with emphasis on minimal disturbance. PPE will normally be required in this zone, although in some circumstances it may be at a reduced level from the hot zone. Command will determine the level of PPE to be required for work within this zone.

Cold Zone

The *cold zone* (also called *support zone*) encompasses the warm zone and is used to carry out all logistical support functions of the incident. Workers in the cold zone are not required to wear personal protective clothing because the zone is considered safe, although some personnel may still be wearing PPE to ensure safe evacuation in the case of rapid expansion of the hot zone (i.e., in body armor in case of secondary devices and/or attacks). The multi-agency command post (CP), staging area, donning/doffing area, backup teams, research teams, logistical support, criminal investigation teams, triage/treatment/rehabilitation (rehab), and transportation areas are located within the cold zone.

Additional Zones

Additional areas or zones may be required within the isolation perimeter. These are as follows:

> **WARNING!**
> Responders must have proper training and appropriate PPE to work in the hot zone or to support work being done inside the hot zone. There will be established access and egress points to ensure both accountability and designated PPE prior to entry.

Figure 23.11 Decontamination takes place in the warm zone. *Courtesy of District Chief Chris E. Mickal, NOFD Photo Unit.*

Staging Area — Prearranged, temporary strategic location, away from the emergency scene, where units assemble and wait until they are assigned a position on the emergency scene and from which these resources (personnel, apparatus, tools, and equipment) must be able to respond within three minutes of being assigned. Staging area managers report to the IC or operations section chief if established.

Transportation Area — Location where accident casualties are held after receiving medical care or triage before being transported to medical facilities.

- *Decontamination zone* — Area located in the warm zone where contaminated clothing, people, and equipment can be cleaned or secured; see Decontamination section.

- *Area of safe refuge* — Traditionally in the warm zone (with greater flexibility at explosive incidents), this is primarily an area serving as a safe place to wait for evacuation and/or decon; used in the haz mat world as a safe location (or locations) where evacuated persons are directed to gather while potential emergencies are assessed, decisions are made, and mitigating activities are begun; in bombings, hostage situations, or sniper attacks, law enforcement will create areas of refuge that are rendered safe.

- *Staging area* — Area(s) where personnel and equipment awaiting assignment to the incident are held (which keeps responders and equipment out of the way and safe until tasked), minimizing confusion and freelancing at the scene; located at an isolated spot in the cold zone where occupants cannot interfere with ongoing operations.

- *Rehabilitation area (rehab area)* — Safe location where emergency personnel can rest, sit or lie down, have food and drink, and have medical conditions evaluated; located in the cold zone.

- *Triage/treatment area* — Area where persons are brought for medical assessment (triage) and stabilization (treatment); located in the cold zone.

NOTE: Triage and treatment is delegated to personnel who are trained and equipped to provide medical aid.

The size, nature, and scope of an incident determine which of these zones and areas are needed. Predetermined procedures help identify which zones and areas to establish under what circumstances.

Notification

Emergency response plans must ensure that responders understand their role in notification processes and predetermined procedures such as SOPs. Notification may be as simple as dialing 9-1-1 (in North America) to report an incident and get additional help dispatched (**Figure 23.12**). The strategic goal of notification may also include such items as incident-level identification and public emergency information/notification. It is better to dispatch more resources than necessary in an initial response to ensure appropriate weight of attack to combat incident conditions.

Figure 23.12 Contacting the telecommunicator to request additional help may be all that is necessary to meet the strategic goal of notification.

Emergency Operations Center (EOC) — Disaster management center for government agencies. Can be located within a communications center or in separate facilities.

Notification also involves contacting law enforcement whenever a terrorist or criminal incident is suspected (or to request their assistance in handling suspicious objects and/or suspected explosives), as well as notifying other agencies that an incident has occurred (public works, the local emergency operations center [EOC], etc.). Procedures will differ between military and civilian agencies as well as from country to country. Always follow SOPs/OIs and emergency response plans for notification procedures. Methods to notify the public are listed in the Evacuation section of this chapter.

Because some haz mat incidents and terrorist attacks have the potential to overwhelm local responders, it is important to know how to request additional resources. This process should be spelled out through local, district, regional, state and national emergency response plans as well as through mutual/automatic aid agreements.

In the United States the notification process is spelled out in the National Response Plan (NRP), and all local, state, and federal emergency response plans must comply with these provisions. Per the NRP, all incidents are handled at the lowest geographic, organizational, and jurisdictional level. In other words, emphasis is on local response. However, when local agencies need additional assistance, they must contact their state authorities to request help. State governors have the authority to mobilize state resources and may include activation of mutual aid agreements with other states. If state resources are overwhelmed, state governors have the authority to request federal aid.

The LERP should be the first resource a responder in the United States should turn to if they need to request outside assistance for an incident of significance. Per the NRP, the local response agency should be closely tied with the community's EOC. If local assets are insufficient to manage the emergency, requests for additional assistance (such as activation of National Guard units) will be made to the state EOC. States may then request federal assistance through the Department of Homeland Security (DHS). Even if additional assistance is not required for an incident, the proper authorities (local, state, and federal) must be informed that an incident has occurred.

Protection

Protection is the overall (*catch-all*) goal of ensuring safety of responders and the public. Protection goals also include measures taken to protect property and the environment. Protection goals are accomplished through such tactics as the following:

- Identifying and controlling materials and hazards

- Using and wearing appropriate PPE (see Personal Protective Equipment section, Chapter 22)

- Conducting rescues (see Rescue section)

- Implementing shoring and stabilization at incidents involving structural collapses

- Providing decontamination (see Decontamination section)

- Providing emergency medical care and first aid

- Taking any other measures to protect responders and the public, including conducting evacuations and sheltering in place

Protection of Responders

The protection and safety of emergency responders is the first priority at any incident. Measures to protect responders include wearing the appropriate level of PPE and decontaminating responders when necessary **(Figure 23.13)**. Additionally, ensuring accountability of all personnel at a haz mat or terrorist incident is critical. First responders must be familiar with systems for tracking and identifying all personnel working at an incident. All personnel working in hazardous areas at terrorist incidents must also be working as part of a team or buddy system. Evacuation and escape procedures must be in place, and safety officers must be assigned.

Mutual Aid — Reciprocal assistance from one fire and emergency services agency to another during an emergency based upon a prearrangement between agencies involved and generally made upon the request of the receiving agency.

Automatic Aid — Written agreement between two or more agencies to automatically dispatch predetermined resources to any fire or other emergency reported in the geographic area covered by the agreement. These areas are generally where the boundaries between jurisdictions meet or where jurisdictional "islands" exist.

Stabilization — Stage of an incident when the immediate problem or emergency has been controlled, contained, or extinguished.

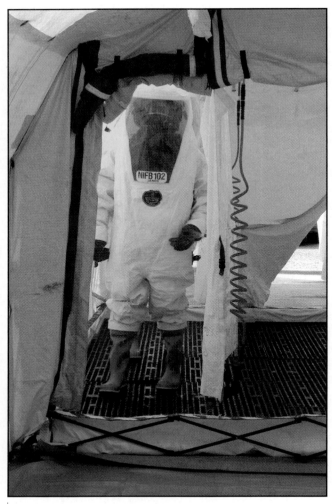

Figure 23.13 Wearing PPE and conducting decontamination are examples of strategies used to protect emergency responders.

Figure 23.14 This accountability system also doubles as an SCBA air management system.

Accountability Systems

One of the most important functions of NIMS-ICS is to provide a means of tracking all personnel and equipment assigned to the incident (**Figure 23.14**). Most units responding to an incident arrive fully staffed and ready to be assigned an operational objective; other personnel may have to be formed into units at the scene. To handle these and other differences in the resources available, the IAP must contain a tracking and accountability system that has the following elements:

- Procedure for checking in at the scene

- Way of identifying and tracking the location of each unit and all personnel on scene, including those entering and exiting hazard control zones such as the hot zone

- Procedure for releasing people, equipment, and apparatus no longer needed

Accountability systems are especially important for incidents where multiple agencies/organizations may be responding, all of which may have different levels of PPE and training. The agency/organization in command is responsible for tracking all responders, not just their own. Methods for tracking accountability should be determined in preplans and implemented as soon as possible at the incident scene.

Preplan — Document, developed during preincident planning, that contains the operational plan or set procedures for the safe and efficient handling of emergency situations at a given location (such as a specific building or occupancy).

Types of accountability systems include traditional systems such as the fire service passport system, T-card systems for wildland incidents as well as systems that utilize newer technologies such as GIS systems and global positioning systems (GPSs). NFPA® 1500 and 1561 address the requirements of accountability systems.

Buddy Systems

In addition to accountability systems, use of buddy systems and backup personnel are mandated by NFPA® and OSHA at incidents involving hazardous materials. A *buddy system* is a system of organizing personnel into workgroups in such a manner that each member has a *buddy* or partner, so that nobody is working alone **(Figure 23.15)**. The minimum number of personnel necessary for performing work in the hot zone or hazard area is four—two working in the hot zone itself and two standing by as backup in case of emergency.

Evacuation/Escape Procedures

Because haz mat and terrorist incidents have a high potential for rapidly changing conditions, secondary devices, secondary collapse, or discovery of other hazardous situations, it is important to adopt and use a signaling system that will advise personnel inside the danger area when to evacuate and/or withdraw. If mitigation efforts fail, the situation deteriorates rapidly, or conditions change in such a fashion as to endanger personnel, the evacuation system must be activated and personnel withdrawn. The Federal Emergency Management Agency (FEMA) Urban Search and Rescue (US&R) Task Force program has developed a system for evacuating rescuers from dangerous areas. Notification can be made using devices such as handheld carbon dioxide (CO_2) air horns, air horns on fire apparatus, or vehicle horns. The US&R designated signals and their meanings are as follows:

- Cease Operations/All Quiet: one long blast (three seconds)
- Evacuate the Area: three short blasts (one second each)
- Resume Operations: one long and one short blast

Other communication methods in the event of an emergency can include portable radios, voice, or the use of predetermined signals.

Safety Officers

While IC's may take responsibility for overseeing all aspects of safety at small incidents, at larger, more complex incidents a Safety Officer may be appointed to ensure the safety of operations **(Figure 23.16)**. Safety Officers are responsible for monitoring and identifying hazardous and unsafe situations and developing measures for ensuring personnel safety. The

Figure 23.15 Personnel operating in or near the hazard area should always be working with a buddy, never alone. *Courtesy of Phil Linder.*

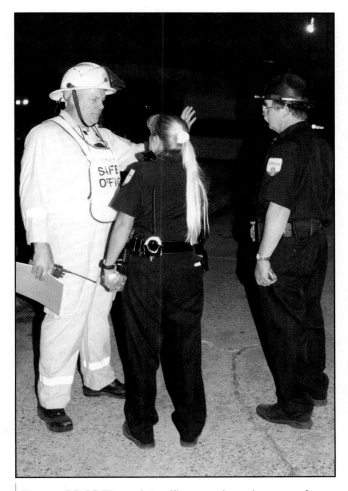

Figure 23.16 The safety officer monitors the scene for unsafe conditions.

Safety Officer must be trained to the level and scope of operations conducted at the incident and is required to perform the following duties:

- Obtain a briefing from the IC.
- Review IAPs for safety issues.
- Identify and assess hazards at the incident scene and report them to the IC.
- Provide the IC with risk assessments.
- Participate in the preparation and monitoring of incident safety considerations, including medical monitoring of entry team personnel before and after entry, and rehabilitation.
- Communicate information about injuries, illnesses, or exposure to the IC.
- Maintain communications with the IC, and advise the IC of deviations from the incident safety considerations and of any dangerous situations.
- Monitor incident communications for events that might pose safety concerns.
- Alter, suspend, or terminate any activity that is judged to be unsafe.

The Safety Officer also needs to ensure that safety briefings are conducted for personnel entering hazard zones. Safety briefings include information about the status of the incident (based on the preliminary evaluation and subsequent updates), the hazards identified, a description of the site, the tasks to be performed, the expected duration of the tasks, the PPE requirements, monitoring requirements, notification of identified risks, and any other pertinent information. At incidents involving potential criminal or terrorist activities, the safety briefing should also cover the following items:

- Being alert for secondary devices
- Not touching or moving any suspicious-looking articles (bags, boxes, briefcases, soda cans, etc.)
- Not touching or entering any damp, wet, or oily areas
- Wearing full protective clothing, including SCBA
- Limiting the number of personnel entering the crime scene
- Documenting all actions
- Not picking up or taking *any* souvenirs
- Photographing or videotaping anything suspicious
- Not destroying any possible evidence
- Seeking professional crime-scene assistance

Protection of the Public

Measures to protect the public include such things as isolating the area and denying entry, conducting rescues, performing mass decontamination, and providing emergency medical care and first aid. Additional measures include evacuation, sheltering in place, and protecting/defending in place.

Evacuation

Evacuate means to move all people from a threatened area to a safer place. To perform an evacuation, there must be enough time to warn people, for them to get ready, and for them to leave the area. Generally, evacuation is the best

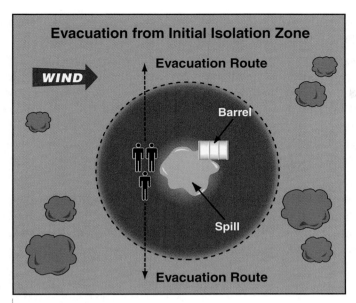

Figure 23.17 Whenever possible, evacuate in a crosswind direction, which is especially important when a person is downwind of the spill or leak.

protective action if there is enough time to carry it out. Firefighters should begin evacuating people who are most threatened by the incident in accordance with distances recommended by the *ERG*, preincident surveys, or other sources **(Figure 23.17)**. Even after people move the recommended distances, they are not necessarily completely safe from harm. Do not permit evacuees to congregate at these safe distances. Send them by a specific route to a designated place (or area of safe refuge).

The number of responders needed to perform an evacuation varies with the size of the area and number of people to evacuate. Evacuation can be an expensive, labor-intensive operation, so it is important to assign enough personnel resources to an incident to conduct it.

Evacuation and traffic-control activities on the downwind side could cause responders and evacuees to become contaminated and, consequently, need decontamination. Responders may also need to wear PPE to safely conduct the evacuation. Evacuation plans (including casualties) for likely terrorist targets such as stadiums and other public gathering places should be made in advance as part of the LERP.

Large-scale evacuations present many factors that must be addressed by the IC including:

- **Notification** — The public must be alerted of the need to evacuate, plus they must understand where to go. Notification methods include: knocking on doors, public address systems, radio, TV, sirens, building alarms, short message service (SMS) through cell phones, reverse 911, the Emergency Alerting System (EAS, in the United States), loudspeakers mounted on helicopters or emergency vehicles, electronic billboards, and others. Methods to be used should be spelled out in the LERP. When notifications are made, clear and concise information must be relayed to avoid confusion or additional panic.

- **Transportation** — As evidenced in the evacuation of New Orleans prior to hurricane Katrina, not everybody owns a car. Alternate means of transportation must be planned in advance using school buses, public transit systems, or other means (planes, trains, boats, barges, ferries, etc.).

- **Relocation Facilities and Temporary Shelters** — Evacuees must have some place to go. These places must be able to provide and care for the people relocated to them, including providing food, water, medicine, bathroom and shower facilities, and places to sleep (for evacuations of long duration). Many evacuees may bring pets. Appropriate evacuation shelters must be designated in the LERP, and staffing arrangements should be determined in advance. An information/registration system should also be established to track the whereabouts of evacuees so their friends and relatives can find them.

- **Prevention of Looting** — Protection of property, while not as high a priority as life safety, must be taken into consideration. When home and business owners are no longer able to protect their property, the IC must ensure that steps are taken to prevent looting by unscrupulous individuals who are willing to risk staying in the evacuation area in order to take advantage of the situation.

Evacuation of Contaminated Victims

Individuals who have been exposed or potentially exposed to chemical, biological, or radiological agents must be decontaminated (sometimes called *decon,* see Decontamination section). While it may be impossible to keep these individuals at the scene, efforts to keep them in place should be made in order to prevent the spread of harmful or potentially deadly materials to other locations. Keeping people from leaving the incident scene is just as important as keeping unauthorized people out. Evacuate contaminated or potentially contaminated individuals to an area of safe refuge (or a triage and treatment area as appropriate) within the isolation perimeter to await decontamination. The safe refuge area should be located upwind and upgrade of the hazardous area.

Because victims may leave the scene before emergency responders arrive (or ignore requests to stay in order to undergo decon), shelters, hospitals, and other public health care facilities must be prepared to conduct decon of *self-presenters* (individuals who have not gone through victim registration processes at the incident scene) at their facilities. See the Rescue section for more information on rapid extraction and evacuation of contaminated persons.

Sheltering in Place

Sheltering in place means to direct people to go quickly inside a building and remain inside until danger passes. It may be determined that shelter in place is the preferred option over evacuation. The decision to shelter in place may be guided by the following factors:

- The population is unable to initiate evacuation because of health care, detention, or educational occupancies.

- The material is spreading too rapidly to allow time for evacuation.

- The material is too toxic to risk any exposure.

- Vapors are heavier than air, and people are in a high-rise structure.

When protecting people inside a structure, directions should be given to close all doors, windows, and heating, ventilation and air-conditioning systems. Shelter in place may not be the best option if vapors or gases are explosive as it may take a long time for vapor or gases to dissipate. Additionally, vapor or gases may permeate into any building that cannot be sealed from the outside atmosphere. Vehicles are not as effective as buildings for shelter in place, but they can offer temporary protection if windows are closed and the ventilation system is turned off.

Whether using evacuation or shelter in place, the public needs to be informed as early as possible and receive additional instructions and information throughout the course of an emergency. Shelter in place may be more effective if public education has been done ahead of time through emergency planning.

Whenever an indoor environment is suspected to be insufficiently sealed against possible outdoor concentration/dose, the public may be instructed to enhance sealing by use of polyethylene sheets and adhesive tapes to improve sealing of openings. Such instruction may be provided in advance and the public instructed to keep the necessary materials on hand along with a basic supply of drinking water, medication, etc.

Firefighters should also pay attention to the condition of surrounding buildings before ordering sheltering in place. For example, some areas may have old and dilapidated structures without air-conditioning or with openings between floorboards. Sheltering in place might not provide sufficient protection in such cases, making evacuation the better option.

Protecting/Defending in Place

Protecting/defending in place is an active (offensive) role or aggressive posture to physically protect those in harm's way. For example, using hose streams to diffuse a plume, or sending in law enforcement to secure a neighborhood or area. When appropriate and safe to do so, defending in place eliminates the need for unnecessary evacuations which, if initiated, will require additional logistical support to ensure the health and safety of evacuees.

Decontamination

Decontamination (decon) is performed at haz mat incidents to remove hazardous materials from victims, PPE, tools, equipment, and anything else that has been contaminated. Realistically, since removing *all* contaminants may be impossible in many cases, decontamination is done simply to reduce contamination to a level that is no longer harmful **(Figure 23.18)**. Decon operations prevent harmful exposures and reduce or eliminate the spread of contaminants outside the hot zone.

Decon also provides victims with psychological reassurance. Some individuals who have been potentially exposed to hazardous materials may develop psychologically-based symptoms (i.e., shortness of breath) even if they have not actually been exposed to harmful levels of contamination. Conducting decon can reduce or prevent these types of problems.

The type of decon operations conducted at an incident will be determined by the size of the incident, the type of hazardous materials involved, weather (washing off contaminants

Figure 23.18 Even animals that have entered the hot zone must be decontaminated.

with a hose stream may not be a viable option in subzero temperatures), personnel available, and a variety of other factors. However, regardless of the many variables that may be encountered at the incident, the basic principles of any decontamination operation are:

(1) Get it off

(2) Keep it off

(3) Contain it (prevent cross-contamination)

Other decon rules of thumb include the following:

- Start decon operations as quickly as possible — the longer the exposure to hazardous materials, generally the worse the victim outcome.

- **Always** wear appropriate PPE for rescue and decon operations.

- Avoid contacting hazardous materials, including contaminated victims.

- Decontaminate all persons (public or emergency responders) who move from the hot zone to the cold zone regardless of symptoms or exposure.

- Establish clearly designated decon entry points so that victims and responders both know where to go.

- Decon emergency responders separately from victims (establish separate decon lines when possible and practical).

- When conducting decon of victims, the more clothing removed the better (disrobing is effective decon by itself).

- Communicate with victims by using hand signals, signs with pictures, apparatus public address systems, megaphones or other methods to direct them to decon gathering areas as well as through the decon process itself. It is very important to provide clear and easily understood directions since people may be traumatized and/or suffering from exposures.

- Provide privacy whenever possible (including from overhead vantage points, for example, from circling news helicopters or upper stories of nearby buildings).

- If possible, provide warm water for washing. If water is cold, allow victims to gradually get wet in order to acclimate to the temperature and avoid cold shock.

Because decon procedures, terminology, and other details may differ greatly from organization to organization, responders must be familiar with their organization's decon policies and procedures.

There are three types of decontamination with which firefighters should be familiar:

- *Emergency decon* — Removing contamination on individuals in potentially life-threatening situations with or without the formal establishment of a decontamination corridor. Emergency decon can consist of anything from simply removing contaminated clothing to flushing a person with water from a safety shower or hoseline.

- *Mass decon* — Conducting rapid decontamination of multiple people at one time. Mass decon may be conducted with or without a formal decon corridor or line and usually involves removing clothing and flushing contaminated individuals with large quantities of water from low-pressure hoselines or other water sources.

- *Technical decon* — Using chemical or physical methods to thoroughly remove contaminants from responders (primarily entry team personnel) and their equipment. It may also be used on incident victims in non-life-threatening situations.

Decontamination Terminology

Decon terminology varies greatly from country to country and region to region. However, for purposes of this book, the following terms will be used:

- *Contamination* — The direct transfer of a hazardous material to persons, equipment, and the environment in greater than acceptable quantities.

- *Decontamination (or contamination reduction)* — The process of removing hazardous materials to prevent the spread of contaminants beyond a specific area and reduce the level of contamination to levels that are no longer harmful.

- *Cross (or secondary) contamination* — Indirect contamination of people, equipment, or the environment outside the hot zone. In secondary contamination, the contaminant is transferred to previously uncontaminated surfaces, persons, or equipment, and so on. For example, contaminated victims who drive themselves to a hospital prior to decontamination could cause cross (or secondary) contamination of medical personnel at the hospital.

- *Exposure* — The process by which people, property, animals, or the environment are put at risk of harm from contact (or potential contact) with hazardous materials.

It should be noted that a person who has been exposed to a hazardous material has not necessarily been contaminated by it. It's a fine distinction, but, essentially, in the case of firefighters (or other living organisms), exposure implies that individuals have been in a situation where the hazardous material had the potential to enter their bodies (or has actually entered their bodies) by one or more routes of entry.

CPC is designed to protect a person from exposure to chemicals, but in the process of doing so, it (and the person, in the sense that the person is wearing the clothing) may become contaminated. Just because the CPC has become contaminated does not mean that the person has been exposed to the material. When a person is exposed to a chemical or other product, it is the *hazard* of the material (or the harm it can do) based on the nature of the exposure (amount inhaled, duration of exposure, and the like) that determines how it may ultimately affect a person's health.

Technical decon is normally conducted within a formal decon line or corridor. The type and scope of technical decon is determined by the contaminants involved at the incident.

These will be discussed in greater detail in following sections. Information about decon implementation will also be discussed.

Emergency Decon

The goal of *emergency decontamination* is to remove the threatening contaminant from the victim as quickly as possible — there is no regard for the environment or property **(Figure 23.19)**. Emergency decon may be necessary for both victims and rescuers. If either is contaminated, individuals are stripped of their clothing and washed quickly. Victims may need immediate medical treatment, so they cannot wait for a formal decontamination corridor to be established. The following situations are examples of instances where emergency decontamination is needed:

- Failure of protective clothing

- Accidental contamination of first responders

Figure 23.19 Emergency decon is conducted whenever and wherever necessary. The object is to get the contamination off as quickly as possible. *Courtesy of Judy Halmich.*

- Heat illness or other injury suffered by emergency workers in the hot zone
- Immediate medical attention is required

Emergency decontamination could be considered a *quick fix,* which is a definite limitation. Removal of all contaminants may not occur, and a more thorough decontamination must follow. Emergency decontamination can definitely harm the environment. However, the advantage of eradicating a life-threatening situation far outweighs any negative effects that may result.

There are times when what appears to be a normal incident really involves hazardous materials. First responders may become contaminated before they realize what the situation really is. When this situation occurs, first responders need to withdraw immediately. They need to remove their turnout clothing and get emergency decon even if there is no apparent contamination evidence. These responders should remain isolated until someone with the proper expertise can ensure that they have been adequately decontaminated.

Emergency decontamination procedures may differ depending on the circumstances and hazards present at the scene. However, a basic set of emergency decontamination procedures is provided in **Skill Sheet 23-I-1**.

Mass Decon

Mass decontamination is the physical process of rapidly reducing or removing contaminants from multiple persons (victims and responders) in potentially life-threatening situations, with or without the formal establishment of a decon corridor **(Figure 23.20)**. Mass decon is initiated when the number of victims and time constraints do not allow the establishment of an in-depth decontamination process. Ideally, a soap-and-water

Emergency Decontamination: Advantages and Limitations

Advantages:

- Requires minimal equipment (usually just a water source such as a hoseline)
- Reduces contamination quickly
- Does not require a formal contamination reduction corridor or decon process

Limitations:

- Does not always totally decontaminate the victim
- Creates contaminated runoff that can harm the environment and other exposures

Figure 23.20 Many different ways are available to conduct mass decon, but flushing with water from handheld hose lines or apparatus master streams can usually be accomplished quickly. *Courtesy of Rich Mahaney.*

solution or universal decontamination solution would be more effective at many incidents; however, availability of such solutions in sufficient quantities cannot always be ensured. Therefore, mass decon can be most readily and effectively accomplished with a simple water shower system. It is a decon process that uses large volumes of low-pressure water to quickly reduce the level of contamination.

Goal of Mass Decon and Chemical, Biological, and Radiological (CBR) Decon Prioritization:

Do the greatest good for the greatest number!

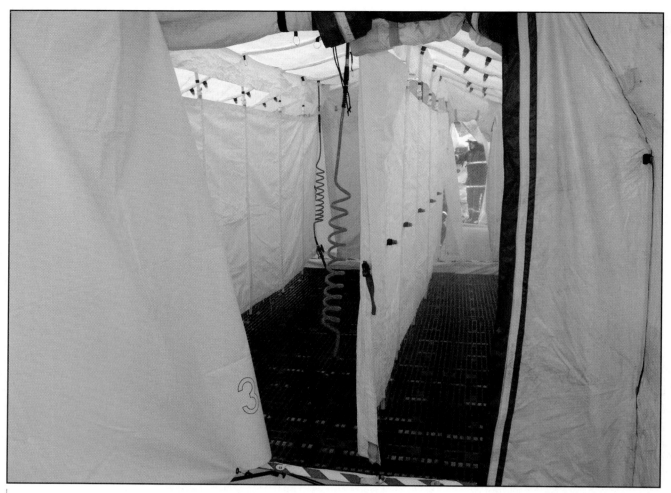

Figure 23.21 It is important to use a high volume of water delivered at a minimum of water pressure to ensure the showering process physically removes viscous agents. *Courtesy of Steven Baker.*

Mass decon showers should use a high volume of water delivered at a minimum of 60 psi (420 kPa) {4.14 bar} water pressure to ensure the showering process physically removes viscous agent **(Figure 23.21)**. Standard household shower water pressures usually average between 60 to 90 psi (420 kPa to 630 kPa) {4.14 bar to 6.2 bar}. The actual showering time is an incident-specific decision but may be as long as 2 to 3 minutes per individual under ideal situations. When large numbers of potential casualties are involved and queued for decon, showering time may be significantly shortened. This time may also depend upon the volume of water available in the showering facilities.

Emergency responders should not overlook existing facilities when identifying means for rapid decontamination methods. For example, although water damage to a facility might result, the necessity of saving victims' lives would justify the activation of overhead fire sprinklers for use as showers. Similarly, having victims wade and wash in water sources such as public fountains, chlorinated swimming pools, swimming areas, and the like provides an effective, high-volume decontamination technique, although consideration must be given to the persistence of chemical agents in contained and contaminated water.

It is recommended that all victims undergoing mass decon remove clothing at least down to their undergarments before showering **(Figure 23.22)**. Victims should

Mass Decon

Disrobing

Showering

Figure 23.22 Disrobing and showering are two of the most important elements of chemical agent decon.

be encouraged to remove as much clothing as possible, proceeding from head to toe. Victims unwilling to disrobe should shower clothed before leaving the decon area.

Many new innovations and products have been developed to assist in mass decon operations, from decon trailers and portable tents (that help alleviate privacy concerns) to portable water heaters and tagging and bagging systems (**Figure 23.23 a–c**). Emergency responders should be familiar with the equipment and resources available as well as the mass decon procedures used by their agency. **Figure 23.24** provides an example of apparatus placement for a generic mass decon schematic; however, many agencies have tents or trailers available for use.

To determine victim priority during decon, responders must consider factors related both to medical triage and decontamination. For maximum effectiveness, it is recommended that patients be divided into two groups: ambulatory and nonambulatory.

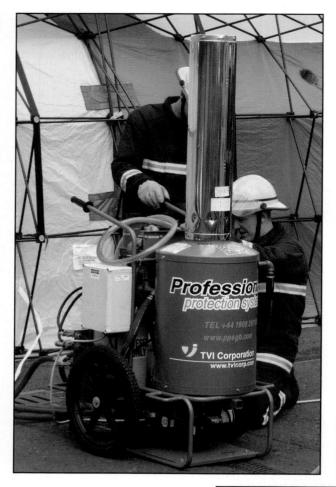

Figure 23.23b This forced air heater can warm the interior of the decon tent.

Figure 23.23a Portable water heaters can make the decon process much more comfortable, particularly in cool weather.

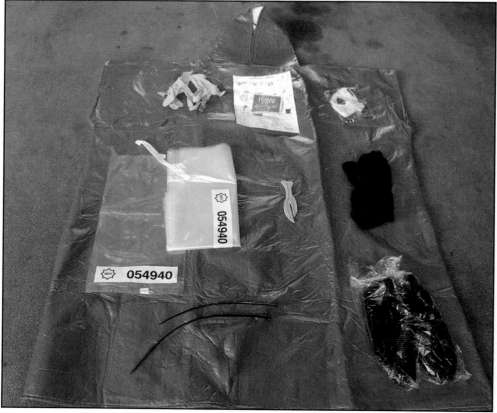

Figure 23.23c Pre-assembled decon kits may include everything from individual bags and ID tags to disposable garments, towels, and shoes.

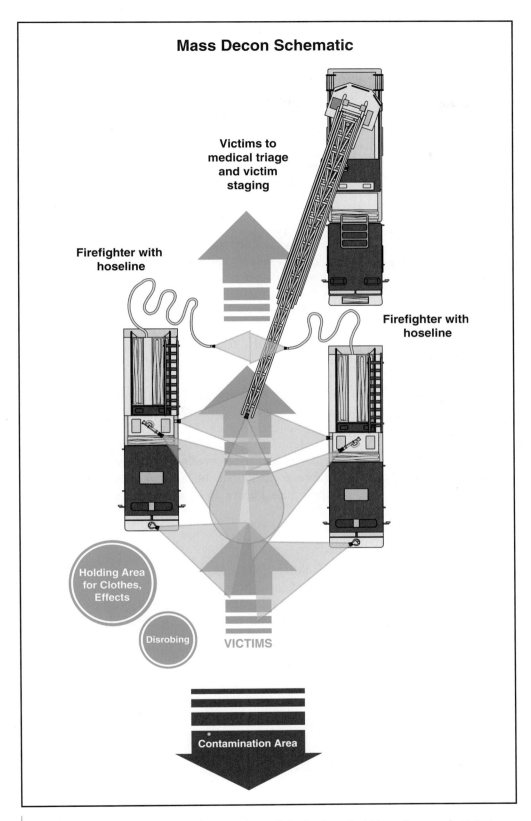

Mass Decon Schematic

Victims to medical triage and victim staging

Firefighter with hoseline

Firefighter with hoseline

Holding Area for Clothes, Effects

Disrobing

VICTIMS

Contamination Area

Figure 23.24 A sample mass decon schematic using handheld hoselines and master streams. *Original source courtesy of the U.S. Army Soldier and Biological Chemical Command (SBCCOM).*

Figure 23.25 A system of rollers can be used to push nonambulatory patients through the decon corridor.

Incidents involving a large number of incapacitated victims may require additional resources for decon corridors because nonambulatory victims will not be able to walk through the decon line (**Figure 23.25**). A separate decon line for emergency response personnel should also be provided.

It may be beneficial to separate victims by gender for privacy reasons; however, families should be allowed to stay together. Children, the elderly, and/or the disabled should not be separated from their parents or caretakers.

Triage

When a mass casualty incident occurs it is likely to overwhelm local resources. The availability of personnel and supplies will govern triage and treatment options. Many patients will take themselves to hospitals or be taken by convergent volunteers, and hospitals will provide most of the initial care. However, responders will need to conduct triage of patients at the scene under the philosophy of doing the greatest good for the greatest number. **Figure 23.26** provides four Simple Triage and Rapid Treatment/Transport (START) categories as provided in the Soldier and Biological Chemical Command (SBCCOM) "Guidelines for Mass Casualty Decontamination During a Terrorist Chemical Agent Incident." These categories can be adopted for use at any haz mat incident. However, many different triage systems exist, and responders must be familiar with the system(s) in use in their area. **Figure 23.27** provides a sample layout for patient decon.

Ambulatory Victims

Victims who are able to understand directions, talk, and walk unassisted are considered ambulatory, and they should be directed to an area of safe refuge (or a triage and treatment area as appropriate) within the isolation perimeter to await prioritization for decontamination. Several of the following factors determine the highest priority for ambulatory patients:

- Casualties closest to the point of release
- Casualties reporting exposure to the hazardous material
- Casualties with evidence of contamination on their clothing or skin
- Casualties with serious medical symptoms (such as shortness of breath or chest tightness)
- Casualties with conventional injuries (broken bones, open wounds, and the like)

Nonambulatory

Nonambulatory patients are victims who are unconscious, unresponsive, or unable to move unassisted. These patients may be more seriously injured than ambulatory patients. They will remain in place while further prioritization for decontamination occurs. Prioritization of nonambulatory patients for decontamination can be done using medical triage systems such as START and others (**Figure 23.28**).

START Medical Triage System

START Category	Decon Priority	Classic Observations	Chemical Agent Observations
IMMEDIATE Red Tag	1	Respiration is present only after repositioning the airway. Applies to victims with respiratory rate >30. Capillary refill delayed more than 2 seconds. Significantly altered level of consciousness.	• Serious signs/symptoms • Known liquid agent contamination
DELAYED Yellow Tag	2	Victim displaying injuries that can be controlled/treated for a limited time in the field.	• Moderate to minimal signs/symptoms • Known or suspected liquid agent contamination • Known aerosol contamination • Close to point of release
MINOR Green Tag	3	Ambulatory, with or without minor traumatic injuries that do not require immediate or significant treatment.	• Minimal signs/symptoms • No known or suspected exposure to liquid, aerosol, or vapor
DECEASED/ EXPECTANT Black Tag	4	No spontaneous effective respiration present after an attempt to reposition the airway.	• Very serious signs/symptoms • Grossly contaminated with liquid nerve agent • Unresponsive to autoinjections

Figure 23.26 While the example here was designed for chemical warfare agent incidents, responders can adapt these four START categories to triage patients at many haz mat incidents. *Courtesy of the U.S. Army Soldier and Biological Chemical Command (SBCCOM).*

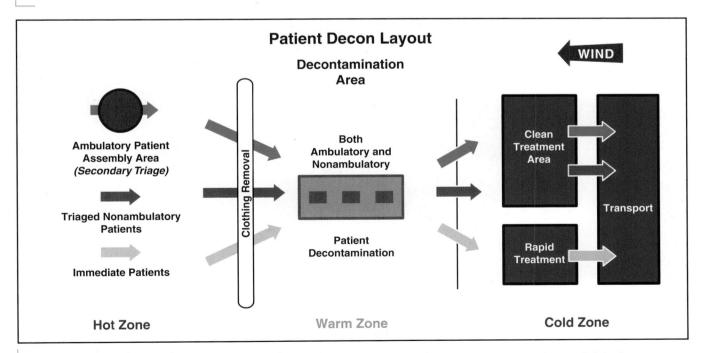

Figure 23.27 A sample schematic for moving ambulatory and nonambulatory patients through decon. *Original source courtesy of the U.S. Army Soldier and Biological Chemical Command (SBCCOM).*

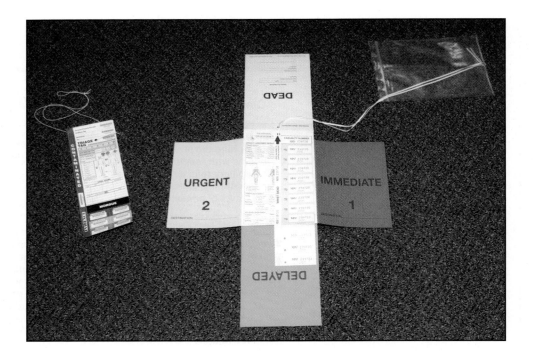

Figure 23.28 Many different commercial triage systems exist.

Figure 23.29 Technical decon is used in non-life-threatening situations. *Courtesy of Rich Mahaney.*

Technical Decon

Technical decontamination uses chemical or physical methods to thoroughly remove contaminants from responders (primarily entry team personnel) and their equipment. It may also be used on incident victims in non-life-threatening situations. Technical decon is usually conducted within a formal decon line or corridor **(Figure 23.29)**. The type and scope of technical decon is determined by the contaminants involved at the incident. Technical decon may use the following techniques:

- **Absorption** — Process of picking up liquid contaminants with absorbents. Some examples of absorbents used in decon are soil, diatomaceous earth, baking powder, ashes, activated carbon, vermiculite, or other commercially available materials. Most absorbents are inexpensive and readily available. They work extremely well on flat surfaces. Some of their disadvantages are as follows:

  — They do not alter the hazardous material.

  — They have limited use on protective clothing and vertical surfaces.

  — Their disposal may be a problem.

- **Adsorption** — Process in which a hazardous liquid interacts with (or is bound to) the surface of a sorbent material (such as activated charcoal).

- **Brushing and scraping** — Process of removing large particles of contaminant or contaminated materials such as mud from boots or other PPE. Generally, brushing and scraping alone is not sufficient decontamination. This technique is used before other types of decon.

- **Chemical degradation** — Process of using another material to change the chemical structure of a hazardous material. For example, household liquid bleach is commonly used to neutralize spills of etiological agents. The interaction of the bleach with the agent almost instantaneously kills the dangerous germs and makes the material safer to handle. Several of the following materials are commonly used to chemically degrade a hazardous material:

  — Household bleach (sodium hypochlorite)

  — Isopropyl alcohol

  — Hydrated lime (calcium oxide)

  — Household drain cleaner (sodium hydroxide)

  — Baking soda (sodium bicarbonate)

  — Liquid detergents

Chemical degradation can reduce cleanup costs and the risk posed to the first responder. Some of the disadvantages of the process are as follows:

  — Takes time to determine the right chemical to use (which should be approved by a chemist) and set up the process.

  — Can be harmful to first responders if the process creates heat and toxic vapors. For this reason, chemical degradation is rarely used on people who have been contaminated.

- **Dilution** — Process of using water to flush contaminants from contaminated victims or objects and diluting water-soluble hazardous materials to safe levels. Dilution is advantageous because of the accessibility, speed, and economy of using water. However, there are disadvantages as well. Depending on the material, water may cause a reaction and create even more serious problems. Additionally, runoff water from the process is still contaminated and must be confined and then disposed of properly.

- **Neutralization** — Process of changing the pH of a corrosive; raising or lowering it towards 7 (neutral) on the pH scale.

- **Sanitization, disinfection, or sterilization** — Processes that render etiological contaminates harmless:

  — *Sanitization:* Reduces the number of microorganisms to a safe level (such as by washing hands with soap and water).

Diatomaceous Earth — A light siliceous material consisting chiefly of the skeletons (minute unicellular algae) and used especially as an absorbent or filter.

Vermiculite — Expanded mica used for loose fill insulation and as aggregate in concrete.

— *Disinfection:* Kills most of the microorganisms present. In a decon setting, disinfection may be accomplished by a variety of chemical or antiseptic products. Most first responders are familiar with the disinfection procedures used to kill bloodborne pathogens such as wiping contaminated surfaces with a bleach solution.

— *Sterilization:* Kills all microorganisms present. Sterilization is normally accomplished with chemicals, steam, heat, or radiation. While sterilization of tools and equipment may be necessary before they are returned to service, this process is usually impossible or impractical to do in most onsite decon situations. Such equipment will normally be disinfected on the scene and then sterilized later.

- *Solidification* — Process that takes a hazardous liquid and treats it chemically so that it turns into a solid.

- *Vacuuming* — Process using high efficiency particulate air (HEPA) filter vacuum cleaners to vacuum solid materials such as fibers, dusts, powders, and particulates from surfaces. Regular vacuums are not used for this purpose because their filters are not fine enough to catch all of the material.

- *Washing* — Process similar to dilution in that they are both wet methods of decontamination. However, washing also involves using prepared solutions such as solvents, soap, and/or detergents mixed with water in order to make the contaminant more water-soluble before rinsing with plain water (**Figure 23.30**). The difference is similar to simply rinsing a dirty dinner dish in the sink versus washing it with a dishwashing liquid. In some cases the former may be sufficient; in others the latter may be necessary. Washing is an advantageous method of decontamination because of the accessibility, speed, and economy of using water and soap. As with the dilution process, runoff water from washing must be contained and disposed of properly.

While the following techniques are not always considered technical decon techniques, first responders should also be familiar with them:

- *Evaporation* — Some hazardous materials evaporate quickly and completely. In some instances, effective decontamination can be accomplished by simply waiting long enough for the materials to evaporate. Evaporation is generally not a technique used during emergency operations. However, it can be used on tools and equipment when extending exposure time is not a safety issue.

- *Isolation and disposal* — This process isolates the contaminated items (such as clothing, tools, or equipment) by collecting them in some fashion and then disposing of them in accordance with applicable regulations and laws. All equipment that cannot be decontaminated properly must be disposed of correctly. All spent solutions and wash water must be collected and disposed of properly. Disposing of equipment may be easier than decontaminating it; however, disposal can be very costly in circumstances where large quantities of equipment have been exposed to a material.

Emergency responders must know what to do when assigned to a decontamination corridor or line. They should be briefed before being assigned. Technical decon corridors vary in the number of stations, depending on the needs of the situation. Corridors may be set up for wet methods or dry methods. **Figure 23.31** provides a sample technical

High Efficiency Particulate Air (HEPA) Filter — Respiratory protection filter designed and certified to protect the user from particulates in the air. The HEPA filter must be at least 99.97 percent efficient in removing monodisperse particles of 0.3 micrometers in diameter.

Figure 23.30 Washing involves using water mixed with some other prepared solution or detergent (such as soap).

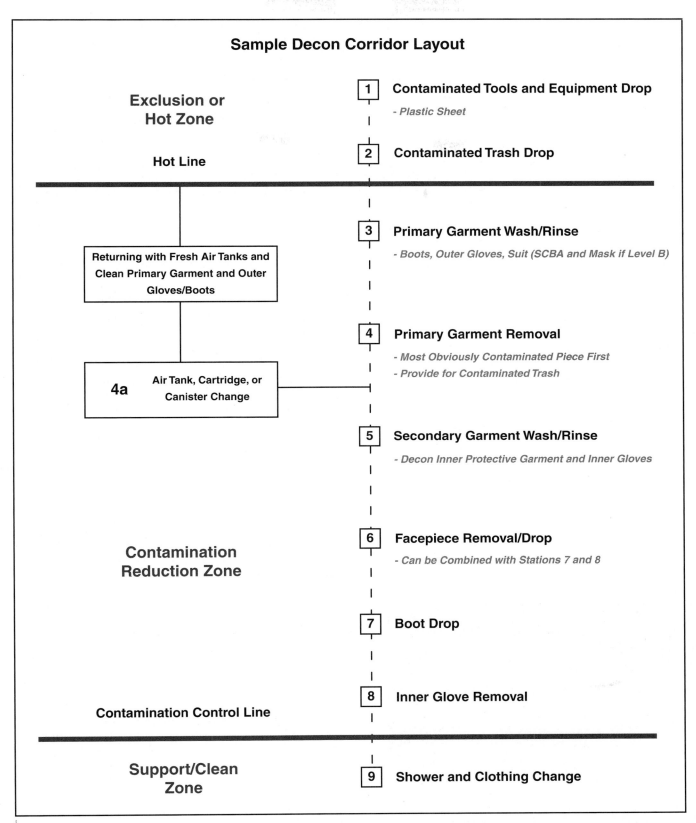

Sample Decon Corridor Layout

Exclusion or Hot Zone

1 **Contaminated Tools and Equipment Drop**
- *Plastic Sheet*

2 **Contaminated Trash Drop**

Hot Line

3 **Primary Garment Wash/Rinse**
- *Boots, Outer Gloves, Suit (SCBA and Mask if Level B)*

Returning with Fresh Air Tanks and Clean Primary Garment and Outer Gloves/Boots

4 **Primary Garment Removal**
- *Most Obviously Contaminated Piece First*
- *Provide for Contaminated Trash*

4a Air Tank, Cartridge, or Canister Change

5 **Secondary Garment Wash/Rinse**
- *Decon Inner Protective Garment and Inner Gloves*

Contamination Reduction Zone

6 **Facepiece Removal/Drop**
- *Can be Combined with Stations 7 and 8*

7 **Boot Drop**

8 **Inner Glove Removal**

Contamination Control Line

Support/Clean Zone

9 **Shower and Clothing Change**

Figure 23.31 A sample nine-step technical decon corridor layout. The number of stations varies depending on the needs of the incident. *Original source courtesy of the U.S. Agency for Toxic Substances and Disease Registry (ATSDR).*

decon corridor layout with decon steps as advocated by the U.S. Agency for Toxic Substances and Disease Registry (ATSDR).

As with mass decon, monitoring should be conducted to determine if decon operations are effective. In some cases, equipment may need to be disposed of due to permeation of contaminants or other factors. Such equipment must be removed from service and properly contained before disposal according to established policies and procedures.

Decon Implementation

Many things must be considered when implementing decontamination. An appropriate site must be selected, the number of stations and setup of the decon corridor or line during technical decon must be decided, methods for collecting evidence must be determined, and termination procedures must be followed.

Site Selection

The following factors are considered when choosing a decontamination site:

- **Accessibility** — The site must be away from the hazards, but adjacent to the hot zone so that persons exiting the hot zone can step directly into the decontamination corridor. An adjacent site eliminates the chance of contaminating clean areas. It also puts the decontamination site as close as possible to the actual incident. Time is a major consideration in the selection of a site. The less time it takes personnel to get to and from the hot zone, the longer personnel can work. Four crucial time periods are as follows:

  — Travel time in the hot zone

  — Time allotted to work in the hot zone

  — Travel time back to the decontamination site

  — Decontamination time

- **Terrain and surface material** — The decontamination site ideally slopes toward the hot zone; thus, anything that may accidentally get released in the decontamination corridor would drain toward or into the contaminated hot zone. If the site slopes away from the hot zone, contaminants could flow into a clean area and spread contamination. Finding the perfect topography is not always possible, and first responders may have to place some type of barrier to ensure confinement of an unintentional release. *Details:*

  — Diking around the site prevents accidental contamination escaping.

  — It is best if the site has a hard, nonporous surface to prevent ground contamination.

  — When a hard-surfaced driveway, parking lot, or street is not accessible, some type of impervious covering may be used to cover the ground. Salvage covers or plastic sheeting will prevent contaminated water from soaking into the earth.

  — Use covers or sheeting to form the decontamination corridor regardless of whether the surface is porous **(Figure 23.32)**.

Figure 23.32 Tarps, plasic sheeting, or salvage covers should be used for flooring even when the decon corridor is set up on a hard surface such as concrete or asphalt.

- *Lighting (and electrical supply)* — The decontamination corridor should have adequate lighting to help reduce the potential for injury to personnel in the area. Selecting a decontamination site illuminated by streetlights, floodlights, or other type of permanent lighting reduces the need for portable lighting (**Figure 23.33**). If permanent lighting is unavailable or inadequate, portable lighting will be required. Ideally, the decontamination site will have a ready source of electricity for portable lighting (as well as heaters, water heaters, and other needs). However, if such a source is not available, portable generators will be needed.

- *Drains and waterways* — Locating a decontamination site near storm and sewer drains, creeks, ponds, ditches, and other waterways should be avoided. If this situation is not possible, a dike can be constructed to protect the sewer opening, or a dike may be constructed between the site and a nearby waterway. Protect all environmentally sensitive areas if possible.

- *Water supply* — Water must be available at the decontamination site. This situation is probably not a problem for most fire and emergency services organizations. Fire apparatus carry water on board, or an apparatus can easily hook to fire hydrants. Water is more of a concern for private emergency response groups and cleanup companies that are working without the assistance of a fire and emergency services organization. They may not have the ability to provide adequate amounts of water to the scene. *Basic materials:*

  — Water and detergent are most commonly used for decontamination. They dilute and remove most hazardous materials.

  — Some jurisdictions use other solutions that degrade the contaminant. The use of such solutions, however, normally takes place in a formal decontamination setting and requires a higher level of training than an Operations-Level first responder.

- *Weather* — The decontamination site needs to be upwind of the hot zone to help prevent the spread of airborne contaminants into clean areas. If the decontamination site is improperly located downwind, wind currents will blow mists, vapors, powders, and dusts toward responders. Ideally, during cold weather, the site should be protected from blowing winds, especially near the end of the corridor. Certainly victims should be shielded from cold winds when they are removing protective clothing. *Solutions:*

  — Set up a portable decontamination shelter or tent (**Figure 23.34**).

  — Set up decontamination trailers at the scene.

  — Use a remote building such as a fire station away from the site. This practice extends travel time and may not be practical.

Figure 23.33 Portable lighting is needed if other light sources are unavailable.

Figure 23.34 Decon shelters, tents, and trailers provide protection from the elements during cool or cold weather.

Decon Corridor Layout

Establish the decontamination corridor before performing any work in the hot zone. First responders are often involved with setting up and working in the decontamination corridor. The types of decontamination corridors vary as to the numbers of sections or steps used in the decontamination process. Corridors can be straightforward and require only a few steps, or they can be more complex and require a handful of sections and a dozen or more steps. Emergency responders must understand the process and be trained in setting up the type of decontamination required by different materials. Some factors to consider are as follows:

- *Ensure privacy* — Decon tents or decon trailers allow more privacy for individuals going through the decon corridor. Decon officers and ICs need to be particularly sensitive to the needs of women being asked to remove their clothing in front of men (regardless of whether they are other victims or emergency responders) **(Figure 23.35)**. Lawsuits have resulted from situations in which women have felt uncomfortable or even humiliated while going through decon. Providing a private, restricted area such as a tent or trailer in which to conduct decon may prevent similar litigation. Use female responders to assist whenever possible when decontaminating women.

- *Bag and tag contaminated clothing/effects* — Various methods can be used. Place clothing and/or personal effects in bags into trash barrels. Label the bags whenever possible. Separate personal effects (wallets, rings, watches, identification cards, and the like) into clear plastic bags clearly marked with the person's name or a unique identifying number (triage tag, ticket, and the like). These items may need to be decontaminated before being returned. Have some sort of system in place to label or mark all personal effects so that they can be returned to their proper owners after the incident without confusion.

Figure 23.35 It is important to allow men and women to shower separately—and in private—if at all possible. However, family units and others with personal ties (such as an elderly person and their caregiver or children with their babysitter) should not be separated if they wish to remain together. *Courtesy of Steven Baker.*

Commercial *tagging systems* may be used for this purpose or multiple-part plastic hospital identification bracelets for example. Be aware that many individuals will find it very stressful to be separated from their personal belongings. For this reason, responders assisting with disrobing should be sympathetic and empathic about the process to help alleviate anxiety.

To the degree possible, keep track of the status of people and personal belongings. This is important for public information (victim relatives will want to know who went where and why) as well as for crime scene investigation.

The decontamination corridor may be identified with barrier tape, safety cones, or other items that are visually recognizable. Coverings such as a salvage cover or plastic sheeting may also be used to form the corridor. Aside from delineating the corridor and providing privacy, protective covering ensures against environmental harm if contaminated rinse water splashes from a containment ba-

Figure 23.36 Garden hoses can provide low-volume, low-pressure spray for decon operations in many situations. *Courtesy of Joan Hepler.*

sin. Containment basins can be constructed of salvage covers and fire hose or ladders. Some organizations use wading pools or portable drafting tanks as containment basins. Also needed at the site are recovery drums or other types of containers and plastic bags for stowing contaminated tools and PPE. A low-volume, low-pressure hoseline such as a garden hose or booster hoseline can be used for decontamination in many situations **(Figure 23.36)**. Pump pressure at idling speed is adequate and minimizes splashing.

How firefighters are protected when working in the decontamination area depends on the hazards of the material. In some cases, standard firefighter personal protective clothing and SCBA may be adequate. In other cases, CPC may be necessary. Often, those conducting decon are dressed in an ensemble classified one level below that of the entry team. Thus, if the entry team is dressed in an EPA Level A ensemble, the decon team is dressed in Level B. In either case, chemical gloves are necessary; fire fighting gloves should not be used in decontamination procedures. Because there is a possibility that responders in decontamination operations will become contaminated themselves, they need to pass through decontamination before leaving the corridor.

Cold Weather Decon

For obvious reasons, conducting wet decon operations in freezing weather can be difficult to execute safely. Even if showers utilize warmed water, run-off water can quickly turn to ice, creating a serious slipping hazard for both victims and responders. If pre-warmed water is not available, susceptible individuals (elderly, very young, individuals with chemical injuries or pre-existing health conditions such as diabetes) can suffer cold shock or hypothermia.

If temperatures are 64 degrees or lower, consideration should be given to protecting victims from the cold. Factors to consider include:

- Are wet methods necessary, or can disrobing and dry methods accomplish effective decon?

- Is wind chill a factor?

- Is shelter available for victims during and after decon?

- Is it possible to conduct decon indoors (sprinkler systems, indoor swimming pools, locker room showers, etc.)?

- If decon will be conducted indoors (at preplanned facilities, for example), how will victims be transported?

- If decon must be conducted outside in freezing temperatures, how will icy conditions be managed (i.e., sand, sawdust, kitty litter, salt)?

Rescue

Due to the potential of extreme hazards at a haz mat incident and the defensive nature of actions at the Operations Level, rescue can be a difficult strategy to implement for firefighters, particularly in the initial stages of a response. Search and rescue attempts must be made within the framework of the IAP with appropriate PPE, backup personnel, and other safety considerations in place.

In many emergency incidents, rescue of victims is the IC's first priority, but it is important to balance the vulnerability of firefighters against the lives of victims. Hesitation to rush into a situation to assist casualties may directly conflict with the fire fighting strategic priority of rescue first as well as with many firefighters' natural desire to help victims as quickly as possible. However, because of the dangers presented by hazardous materials, responders who rush to the rescue may quickly require the need to be rescued themselves. Firefighters who contact contaminated victims may become contaminated too, thereby complicating and spreading the problem.

WARNING!

Individuals who have been exposed to deadly levels of chemical warfare agents should undergo emergency decon immediately, regardless of ambient temperatures. They should disrobe and thoroughly shower. Dry clothing and warm shelter should be provided as soon as possible after showering.

Sample Rapid Extraction Policy

One of the primary missions that first responders may need to accomplish at a haz mat or weapons of mass destruction (WMD) incident is rescue of exposure victims. The objective of rescue at hazardous material incidents is to rapidly extract people from unsafe areas and deliver them for decontamination and treatment, while controlling the associated risks to firefighters.

To differentiate victim rescue from rescue of first responders, the term used to describe actions directed at retrieving civilian exposure victims is "rapid extraction." ICs may assemble units to perform extraction missions in extraction crews or groups.

Under ideal conditions, everyone in the area of a hazardous materials release would self-rescue, gather in a location upwind and uphill, and initiate self-decontamination. In the case of industrial accidents, responders may find that the initial reaction by trained people on the scene of the release may closely match the ideal. This will not be the case, however, where the release is committed intentionally or accidentally on an unsuspecting and unprepared population. In the event of such a criminal act, the need to perform extraction of victims is more likely.

Types of Extraction

Responders have a choice among different extraction methods, and should focus on using a method that minimizes risk but still allows the mission to be completed.

Extractions can be categorized by who performs the rescue: self-rescue or bystander rescue, which represents the least risk to the responder; responder rescue, where the level of risk and potential benefit may be initially unknown; and technical rescue, such as that performed by USAR or the Hazardous Materials Task Force (HMTF).

Continued

Sample Rapid Extraction Policy   (continued)

Assessment of Risk Vs. Benefit

Type	Risk	Benefit
Self-Rescue	Low	High
Bystander Rescue	Low	High
Responder Rescue	Low-High	Unknown
Technical Rescue	Med-High	Low

Using remote methods to contact the people in the unsafe area, such as the apparatus PA, a bullhorn, or in-place public address systems is the least risk method to get people to leave an unsafe area. Where this method is employed, those people less impacted by the release should be encouraged to assist those viable victims who have been impacted to a greater degree by the release (bystander rescue).

In some cases, responders may find that victims, though ambulatory, may need assistance to get to a safe environment due to the symptoms associated with the exposure, such as dimness of vision or confusion. This type of assisted rescue may require responders to be in proximity or even direct contact with the victims.

Extraction of unresponsive victims may result in exposure to levels of contamination that exceed the capability of the protective clothing available to first responders. As a result, this type of physical rescue or extrication may require the specialized equipment and PPE available to the HMTF.

Risk Assessment

First responders must quickly assess whether there are victims needing extraction, the number of victims and what resources will be needed. In order to accomplish this assessment, ICs should assign a company to this task as soon as possible, in spite of the demands of the incident.

Since the primary protective equipment assigned to firefighters consists of structural fire fighting PPE and SCBA, risk assessment should determine whether this level of protection is appropriate for the environment.

The primary factor in determining whether the environment is appropriate for structural PPE and SCBA is the viability of the victims. Where the hazardous material is unknown, exposure victims are considered viable when they are able to control their movement (standing, sitting, or kneeling), or are able to respond verbally or physically.

Those victims who are unable to respond, or are unable to voluntarily control movement, may still be viable under certain conditions. If the number of victims is low, they are not located in an enclosed space or the enclosed space can be adequately ventilated, and access to decontamination and treatment is immediately available, viability is more likely.

The assessment to determine the presence of viable victims may include direct observation, interview of witnesses or employees exiting the venue, and reports from other first responders. The existence of unprotected viable victims is a strong indication that the toxicity of the environment is below that which would impact a firefighter protected by full structural PPE and an SCBA for short exposures.

Even though an IC may determine that there are viable victims who need extraction, the scale of the incident will directly impact the decision process.

Continued

Sample Rapid Extraction Policy (continued)

Similar to multi-casualty incidents, the priority must be on providing the most good for the most people. Directing initial arriving units to provide emergency decontamination to self and bystander-rescued exposure victims may produce a better outcome than devoting all units to extraction efforts.

Rapid Extraction Guidelines

A successful extraction will minimize the exposure for both the victims and the rescuers by limiting exposure time and providing immediate decontamination. To accomplish this, decontamination must be available prior to entry, and extraction teams shall use no more than one 45-minute SCBA bottle prior to evaluation for exposure by Haz Mat personnel.

Prior to initiating rapid extraction, first responders shall take the following actions:

1. Establish that there are viable victims and a need for extraction operations.
2. Establish decontamination. At minimum, a charged hoseline must be in the designated location.
3. Establish rapid intervention. A single engine company, following the 2-in 2-out rule, can fulfill the rapid intervention requirement. At larger incidents, ICs may designate entire companies as Rapid Intervention (may be assigned to HMTF).
4. Announce the intent to initiate rapid extraction to IC or incoming companies:
   a. Identify the access point.
   b. Identify the number of personnel on team.
5. Maintain continuous communications during extraction efforts.

Upon encountering any of the following exit conditions, first responders must immediately leave the rescue site, make appropriate notifications, and proceed to decontamination:

1. Any signs or symptoms of exposure for any members of an extraction team
2. Encountering an area with ONLY nonviable victims
3. The first SCBA audible alarm (approximately 25% reserve for exiting and decontamination)

Until the exit conditions are encountered, and while viable victims remain, first responders may continue extraction operations.

Upon completion of the extraction missions, first responders shall:

1. Proceed through decontamination. Be guided by directions from the HMTF.
2. Consider PPE used during extraction as contaminated, and keep it segregated from other equipment until it is determined to be safe.
3. Be examined by advanced life support (ALS) personnel for post-entry symptoms and vitals.
4. Proceed to Rehab as needed.

Courtesy of Los Angeles Fire Department

The IC makes decisions about rescue based on a variety of factors at the incident, including a risk–benefit analysis that weighs the possible benefits of taking certain actions versus the potential negative outcomes that might result. The following factors affect the ability of personnel to perform a rescue:

- Nature of the incident, hazardous materials present (including the possibility of secondary devices), and incident severity
- Availability of appropriate PPE and training of response personnel
- Number of victims and their conditions
- Time needed (including a safety margin) to complete a rescue
- Tools, equipment, and other devices needed to effect the rescue
- Availability of monitoring equipment

Decisions about rescue may differ depending on country and jurisdiction, based on a variety of factors. For example, SOPs on levels of PPE necessary to conduct rescue at various types of terrorist incidents may vary from place to place. Responders must understand what actions are appropriate for them given their SOPs, training, and available equipment and resources.

Without coming in contact with the hazardous material, Operations Level firefighters may instruct people on what to do and where to go at the incident. They may assist with searches in the warm zone and decontamination procedures. Because they may be allowed to do reconnaissance and take other actions within the warm zone, Operations Level firefighters may also perform the following actions with appropriate PPE for the hazardous material:

- Direct people to an area of safe refuge or evacuation point located in a safe place that is upwind and uphill of the hazard area.
- Instruct victims to move to an area that is less dangerous before moving them to an area that offers complete safety.
- Direct contaminated or potentially contaminated victims to an isolation point, safe refuge area, safety shower, eyewash facility, or decontamination area **(Figure 23.37)**.
- Give directions to a large number of people for mass decontamination.

> **WARNING!**
> Never rush to conduct a rescue without proper training, PPE, planning, and coordination under the direction of the IC!

Figure 23.37 Firefighters can direct victims to decon locations. *Courtesy of Rich Mahaney.*

- Conduct searches during reconnaissance or defensive activities.
- Conduct searches on the edge of the hot zone.
- Assist with decontamination while not coming into contact with the hazardous material itself.
- Assist with the identification of victims.

Exceptions to the Rule

In most circumstances, Awareness- and Operations-Level responders should not enter the hot zone or come into physical contact with hazardous materials (except in those cases where they are trained to do so in regards to certain petroleum products). When it comes to rescue operations, however, a risk-benefit analysis may determine that it is acceptable to conduct a rescue despite these usual restrictions.

For example, conducting a rescue in the hot zone might be appropriate *if* the following conditions are met:

- The product is positively identified.
- Responders have appropriate PPE to protect from any possible hazards presented.
- The risk to responders is minimal.
- A life can be saved or serious injury prevented.

In these cases, the IC may decide that it is appropriate to take action and conduct a rescue before haz mat technicians arrive. However, this is a decision that only the IC should make and only after a serious risk–benefit assessment has been conducted. If ever in doubt, play it safe.

Responders can begin to assess the following *risks** of conducting a rescue at a hazardous materials incident based on the DOT hazard classes of the material:

- ***Class 1 Risk:*** Explosives that may involve thermal injury, due to the heat generated by the detonation, mechanical injury from the shock, blast overpressure, fragmentation, shrapnel, or structural damage, and chemical injuries from associated contamination. During a rescue, contact with the blood or other bodily fluids from the victim may result in etiological harm. Because burning depletes oxygen, asphyxiation should always be a consideration.

- ***Class 2 Risk:*** Gases that are stored in containers that may be under pressure can rupture violently and, depending upon their contents, could create a thermal, asphyxiant, chemical, or mechanical hazard. Cryogenic materials (Class 2.2), such as liquid oxygen, may cause thermal harm because of its extremely cold temperature. Asphyxiation is always a concern with chemical vapors in a confined space because chemical reactions may deplete oxygen or create gases that displace oxygen such as carbon dioxide (CO_2).

- ***Class 3 Risk:*** Flammable liquids that could also disseminate shrapnel caused by a forceful explosion, resulting in thermal hazards from heat and fire as well as associated chemical and mechanical injuries.

- ***Class 4 Risk:*** Flammable solids, spontaneously combustible materials, and dangerous when wet materials that could cause thermal harm from heat and flammability. Mechanical harm could occur as some materials react spontaneously, creating slip, trip, and fall hazards, while other water reactive, toxic, and/or corrosive materials can cause chemical harm.

Continued

- **Class 5 Risk:** Oxidizers and organic peroxides that can create thermal, chemical, and mechanical harm, because they supply oxygen to support combustion and are sensitive to heat, shock, friction and contamination.

- **Class 6 Risk:** Toxic/poisonous materials and infectious substances that cause chemical harm due to toxicity by inhalation, ingestion, and skin and eye contact. Etiological harm can come from either disease-causing organisms or toxins derived from living organisms. Because these products may be flammable, thermal injuries are a potential hazard.

- **Class 7 Risk:** Radioactive material that may cause thermal harm as well as radiological harm from alpha, beta, and neutron particles, and gamma rays, although depending upon the proximity of the responder to the source and the cumulative dose rate, delayed effects would most likely occur. Radiological substances should not be overlooked as chemical hazards. For Class 7 (radioactive) material, a great deal of information regarding risk to response personnel can be obtained by determining the type of package present. Excepted, industrial, and Type A packages contain non-life-endangering amounts of radioactive material. Type B packages may contain life-endangering amounts of radioactive material; however, life-endangering conditions would only exist if the contents of a Type B package are released or if package shielding fails. Type B packages are designed to withstand severe accident conditions.

- **Class 8 Risk:** Corrosive materials that have chemical and thermal hazards associated with the disintegration of contacted tissues and the recognition that chemical reactions create heat, especially if the material is fuming and/or water reactive. Corrosive chemicals, like strong acids, can weaken structural elements, causing the potential for mechanical harm.

- **Class 9 Risk:** Miscellaneous dangerous goods incidents that would be agent and situation specific but could encompass a multitude of potential hazards.

*Reprinted with permission from *Hazardous Materials Response Handbook,* 4th Edition © 2002, National Fire Protection Association, Quincy, MA 02269.

Spill Control and Confinement

In order to understand the principles behind spill control/confinement, it is useful to return to the information in the General Hazardous Materials Behavior Model section in the previous chapter. Container stress can result in a breach (rupture). When the container is breached, it may release its contents. The released material then disperses according to its chemical and physical properties, topography, prevailing weather conditions, and the amount and duration of the release. Depending on its hazardous properties, the material can then harm whatever it contacts.

The strategic goal of spill control involves controlling the product that has already been released from its container. In other words, *spill control* minimizes the amount of contact the product makes with people, property, and the environment by limiting or confining the dispersion and/or reducing the amount of harm caused by contact with the material **(Figure 23.38)**. Tactics and tasks relating to spill control are determined by the material involved and type of dispersion and are generally defensive in nature.

The main priority of spill control is confinement and the prevention of further contamination or contact with the hazardous material. Firefighters trained to the Operations Level may perform spill-control activities as long as they do not come in

Figure 23.38 Spill control is aimed at confining the material that has been (or is being) released from a container. Damming, diking, and vapor suppression are examples of spill control measures that minimize the amount of contact the product makes with people, property, and the environment.

contact with the product or have appropriate training and PPE for the specific substance such as flammable liquids. In the latter case, they still limit their contact with the material as much as possible. Spill control is a defensive operation with the most important issue being the safety of the firefighters performing these actions.

Spills may involve gases, liquids, or solids, and the product involved may be released into the air (as a vapor or gas), into water, and/or onto a surface such as the ground or a bench top. The type of dispersion determines the defensive measure(s) needed to control it. For example, in the event of a flammable liquid spill, first responders must address not only the liquid spreading on the ground, but also the vapors being released into the air. If the spill pours into a sanitary sewer system or stream, responders need to address the contamination of the water, too.

Spill Control and Confinement Tactics

Hazardous materials may be confined by building dams or dikes near the source, catching the material in another container, or directing (diverting) the flow to a remote location for collection. Generally, the fire apparatus carries the following necessary tools:

- Shovels for building earthen dams
- Salvage covers for making catch basins
- Charged hoselines for creating diversion channels

Before using the equipment to confine spilled materials, ICs need to seek advice from technical sources to determine if the spilled materials will adversely affect the equipment. Large or rapidly spreading spills may require the use of heavy construction-type equipment, floating confinement booms, or special sewer and storm drain plugs.

Confinement is not restricted to controlling liquids. Dusts, vapors, and gases can also be confined. A protective covering consisting of a fine spray of water, a layer of earth, plastic sheets, or a salvage cover can keep dusts from blowing about at incidents. Foam blankets can be used on liquids to reduce the release of vapors. Strategically placed fire streams can direct gases, or allow the water to absorb them. Check reference sources for the proper procedures for confining gases. The material type, rate of release, speed of spread, number of personnel available, tools and equipment needed, weather, and topography dictate confinement efforts.

The following defensive confinement and spill control actions are discussed in the following sections:

- Absorption
- Blanketing/covering
- Dam, dike, diversion, and retention
- Dilution
- Dissolution
- Vapor dispersion
- Vapor suppression
- Ventilation

CAUTION

Confinement actions are undertaken only if firefighters are reasonably certain they will not come in contact with or be exposed to the hazardous material.

Absorption

Absorption is a physical and/or chemical event occurring during contact between materials that have an attraction for each other. This event results in one material being retained in the other. The bulk of the material being absorbed enters the cell structure of the absorbing medium. An example of absorption is soaking an axe head in water to make the handle swell. Some of the materials typically used as absorbents are sawdust, clays, charcoal, and polyolefin-type fibers **(Figure 23.39)**. The absorbent is spread directly onto the hazardous material or in a location where the material is expected to flow. After use, absorbents must be treated and disposed of as hazardous materials themselves because they retain the properties of the materials they absorb. The steps described in **Skill Sheet 23-I-2** are an example of how to perform a basic form of absorption.

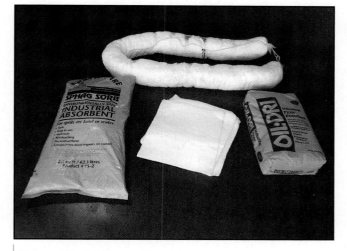

Figure 23.39 Many different materials are used as absorbents.

Blanketing/Covering

As the title implies, this spill-control measure involves blanketing or covering the surface of the spill to prevent dispersion of materials such as powders or dusts. Blanketing or covering of solids can be done with tarps, plastic sheeting, salvage covers, or other materials (including foam), but consideration must be given to compatibility between the material being covered and the material covering it. Covering may also be done as a form of temporary mitigation for radioactive and biological substances.

Blanketing of liquids is essentially the same as vapor suppression (see Vapor Suppression section), because it primarily uses an appropriate aqueous (water) foam agent to cover the surface of the spill. Operations-Level responders may or may not be allowed to perform blanketing/covering actions, depending on the hazards of the material, the nature of the incident, and the distance from which they must operate to ensure their safety.

Dike, Dam, Diversion, and Retention

Diking, damming, diverting, and retaining are ways to confine a hazardous material. These actions are taken to control the flow of liquid hazardous materials away from the point of discharge. Firefighters can use available earthen materials or materials carried on their response vehicles to construct curbs that direct or divert the flow away from gutters, drains, storm sewers, flood-control channels, and outfalls **(Figure 23.40)**. In some cases, it may be desirable to direct the flow into certain locations in order to capture and retain the material for later pickup and disposal. Dams may be built that permit surface water or runoff to pass over the dam while holding back the hazardous material. Any construction materials that contact the spilled material must be properly disposed of. **Skill Sheets 23-I-3 through 23-I-6** describe procedures for performing basic diking, damming, diversion, and retention.

Dilution

Dilution is the application of water to a water-soluble material to reduce the hazard. Dilution of liquid materials rarely has practical applications at haz mat incidents in terms of spill control; dilution is often used during decontamination operations. The amount of water needed to reach an effective dilution increases overall volume and creates a runoff problem. This situation is especially true of slightly water-soluble liquids. Dilution may be useful when very small amounts of corrosive material are involved such as in cases of irregular dispersion or a minor accident in a laboratory, but even then, it is generally considered for use only after other methods have been rejected. **Skill Sheet 23-I-7** describes a procedure to perform dilution.

Dissolution

The process of dissolving a gas in water is called *dissolution*. This tactic can only be used on such water-soluble gases as anhydrous ammonia or chlorine and is generally conducted by applying a fog stream to a breach in a container or directly onto the spill. Ideally, the escaping gas then passes through the water and dissolves. When considering this option, firefighters must remember that it may create additional problems with contaminated runoff water and other issues. For example, using water spray at a chlorine incident to bring

Figure 23.40 These responders are building an overflow dam to trap a hazardous material that has a specific gravity greater than 1. The water will flow out the pipes at the top of the dam while the heavier hazardous material will be trapped behind it. *Courtesy of Rich Mahaney.*

the vapors to the ground may have the beneficial effect of reducing or eliminating a toxic plume, but it may also create hydrochloric acid on the ground with all the complications associated with that chemical.

Vapor Dispersion

Vapor dispersion is the action taken to direct or influence the course of airborne hazardous materials. Pressurized streams of water from hoselines or unattended master streams may be used to help disperse vapors **(Figure 23.41)**. These streams create **turbulence,** which increases the rate of mixing with air and reduces the concentration of the hazardous material. After using hoselines for vapor dispersion, it is necessary for

Turbulence — Irregular motion of the atmosphere usually produced when air flows over a comparatively uneven surface such as the surface of the earth; when two currents of air flow pass or over each other in different directions or at different speeds.

Vapor Dispersion

Figure 23.41 Vapor dispersion actions using pressurized water streams can prevent toxic gases or vapors from moving downwind, but contaminated runoff water from this procedure can create its own problems.

first responders to confine and analyze runoff water for possible contamination. The steps listed in **Skill Sheet 23-I-8** may be followed to perform vapor dispersion.

Vapor Suppression

Vapor suppression is the action taken to reduce the emission of vapors at a haz mat spill. Fire fighting foams are effective on spills of flammable and combustible liquids if the foam concentrate is compatible with the material. Water-miscible (capable of being mixed) materials such as alcohols, esters, and ketones destroy regular fire fighting foams and require an alcohol-resistant foam agent. In general, the required application rate for applying foam to control an unignited liquid spill is substantially less than that required to extinguish a spill fire.

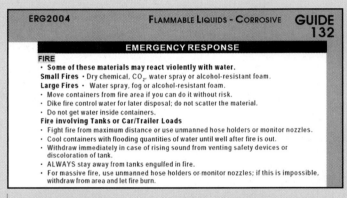

What Type of Foam Concentrate Should Be Used?

The orange section of the current *ERG* provides information on what type of foam to use. For polar/water-miscible liquids, it recommends alcohol-resistant foam. For nonpolar/water-immiscible liquids, it recommends regular foam **(Figure 23.42)**.

Figure 23.42 The orange-bordered pages of the *ERG* provide information on what type of foam to use. Alcohol-resistant foams should be used on polar/water-miscible liquids. Regular foam may be used on spills of nonpolar/water-miscible liquids.

Drainage Time — The amount of time it takes foam to break down or dissolve; also called drainage rate, drainage dropout rate, or drainage.

Expansion Ratio — Ratio of the finished foam volume to the volume of the original foam solution.

Foam concentrates vary in their finished-foam quality and, therefore, their effectiveness in suppressing vapors. Foam quality is measured in terms of its 25-percent-drainage time and its expansion ratio. *Drainage time* is the time required for one-fourth (25 percent or quarter) of the total liquid solution to drain from the foam. *Expansion ratio* is the volume of finished foam that results from a unit volume of foam solution.

Long drainage times result in long-lasting foam blankets. The greater the expansion ratio is, the thicker the foam blanket that can be developed in a given period of time. All Class B foam concentrates, except the special foams made for acid and alkaline spills, may be used for both fire fighting and vapor suppression. Air-aspirating nozzles produce a larger expansion ratio than do water fog nozzles.

Firefighters must exercise care when applying any of the foams onto a spill. All foams (except fluoroprotein types) should not be plunged directly into the spill; instead, they should be applied onto the ground at the edge of the spill and rolled

Figure 23.43 Never plunge alcohol-resistant AFFF into the surface of a polar solvent fuel. Instead, spray it over the top of the fuel in a rain-down method.

gently onto the material. If the spill surrounds some type of obstacle, the foam can be banked off the obstacle. Another foam application method is the rainfall method **(Figure 23.43)**. Foam is sprayed into the air over the target area in a fog pattern. The tiny foam droplets rain down over the spill. As the foam bubbles burst, the foam bleeds together to form a film over the fuel. When possible, first responders should use air-aspirating nozzles for vapor suppression rather than water fog nozzles.

Selection of the proper foam concentrate for vapor suppression is important. Because finished foam is composed principally of water, it should not be used to cover water-reactive materials. Some fuels destroy foam bubbles, so a foam concentrate must be selected that is compatible with the liquid as mentioned earlier. Other points to consider when using foam for vapor suppression are as follows:

- Water destroys and washes away foam blankets; do not use water streams in conjunction with the application of foam.

- A material must be below its boiling point; foam cannot seal vapors of boiling liquids.

- If the film that precedes the foam blanket is not visible (such as with AFFF blankets), the foam blanket may be unreliable. Reapply aerated foam periodically until the spill is completely covered.

First responders must be trained in the techniques of vapor suppression. Training for extinguishment of flammable liquid fires does not necessarily qualify a first responder to mitigate vapors produced by haz mat spills. More information on foam application is contained in the Fire Control section.

Ventilation

Ventilation involves controlling the movement of air by natural or mechanical means. Ventilation is used to remove and/or disperse harmful airborne particles, vapors, or gases when spills occur inside structures. The same ventilation techniques used for smoke removal can be used for haz mat incidents. When conducting negative-pressure ventilation, fans and other ventilators must be compatible with the atmosphere where they are being operated. Equipment must be explosion-proof in a flammable atmosphere. When choosing the type of ventilation to use, remember that positive-pressure ventilation is usually more effective than negative-pressure ventilation when it comes to removing atmospheric contaminants. Some experts consider ventilation to be a type of vapor dispersion.

Leak Control and Containment

A leak involves the physical breach in a container through which product is escaping. The goal of leak control is to stop or limit the escape or to contain the release either in its original container or by transferring it to a new one (**Figure 23.44**). The type of container involved, the type of breach, and properties of the material determine tactics and tasks relating to leak control. Leak control and containment are generally considered offensive actions. Offensive actions are not attempted by personnel trained below the Technician Level with two exceptions: The first is turning off a remote valve (**Figure 23.45**). The second involves situations dealing with gasoline, diesel, liquefied petroleum gas (LPG), and natural gas fuels. With these fuels, Operations-Level responders can take offensive actions provided they have appropriate training, procedures, equipment, and PPE.

Leak control dictates that personnel enter the hot zone, which puts them at great

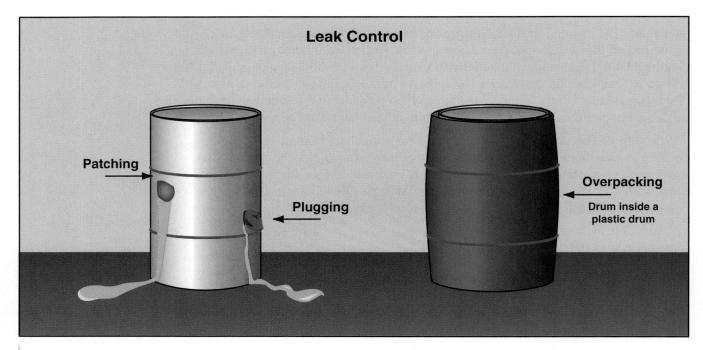

Figure 23.44 Leak control is aimed at containing the material to prevent further release by such tactics as plugging, patching, and/or overpacking.

risk. The IC must remember that the level of training and equipment provided to personnel are limiting factors in performing leak control. A list of leak control/containment tactics follows; however, the scope of this book does not include discussing each of them (except for remote valve shutoff activation):

- Remote valve shutoff activation (control of release)
- Patching/plugging
- Overpacking
- Product transfer
- Crimping
- Valve actuation
- Vacuuming
- Pressure isolation and reduction
- Solidification

In some situations it may be safe and acceptable for Operations-Level responders to operate emergency remote shutoff valves on cargo tank trucks (specifically, low-pressure liquid tanks and high-pressure tanks). The locations and types of remote shutoff valves vary depending on the truck (**Figures 23.46 a and b**).

Piping systems and pipelines carrying hazardous materials may have remote shutoff or control valves that can be operated to stop the flow of product to the incident area without entering the hot zone. For example, by closing remote valves feeding a broken natural gas line, the flow of gas to the break can be stopped. A significant amount of product may release for some time before the flow stops. In most cases, onsite maintenance personnel or local utility workers know where these valves are located and can be given the authority and

Figure 23.45 Operations-Level firefighters may stop the flow of hazardous materials by operating remote shut-off valves.

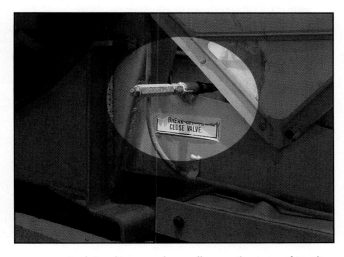

Figure 23.46a and b Locations and types of remote shutoff valves on cargo tank trucks vary, depending on the type of truck. (a) Shutoff valve on a nonpressure liquid tank. (b) Shutoff valve on a corrosive liquid tank.

responsibility for closing them under the direction of the IC. However, Operations-Level first responders who are trained and authorized to operate shutoff valves at their facilities in the event of emergency may do so in accordance with their SOPs.

Crime Scene Management and Evidence Preservation

[NFPA® 472: 4.4.1(12), 5.4.2, 5.1.2.2(3)(b)] [ODP: Awareness IV.a., IV.b., IV.c., IV.d., V.c., V.d., V.e.; Operations I.d., IV.a.]

The framework for a response to a terrorist or criminal incident is essentially the same as that used for a response to any other hazardous materials incident. However, because a crime is involved, law enforcement organizations must be notified and included in the response. Notifying law enforcement ensures that the proper state/province and federal/national agencies respond to the incident (such as the Federal Bureau of Investigation (FBI), who has jurisdiction over terrorist incidents in the U.S.) **(Figure 23.47)**. See Chapter 18 for more information on crime scene management and evidence preservation.

Recovery and Termination

The recovery and termination phase of a haz mat or terrorist incident (also called the *demobilization phase*) occurs when the IC determines that all victims have been

Figure 23.47 Law enforcement must be notified whenever criminal activity is suspected at an incident.

accounted for and all hazards have been controlled. This phase of an operation includes the following:

- Securing the scene

- Accounting for all emergency response personnel and equipment

- Monitoring exposures to hazardous substances

- Meeting the media

- Conducting post-incident analysis

- Returning personnel and equipment to readiness status

- Criminal investigation of the incident

- Releasing the scene to the owner or other authority

- Performing the post-incident medical analysis for all emergency response personnel

- Conducting interagency post-incident analysis

To facilitate the termination (demobilization) phase, an Incident Termination (Demobilization) Checklist should be developed and included as a standard part of the local NIMS-ICS operating procedures. An Incident Termination (Demobilization) Checklist for major incidents is likely to cover some of the following types of information:

- Ensure that personnel rehabilitation (rehab) is completed.

- Assess need for Critical Incident Stress Defusing for responders, if available.

- Ensure scene stabilization.

 — Determine how agency equipment will be retrieved.

 — Determine any needs of the investigating agency.

 — Prevent further injury by ensuring that the scene is secure (barricades, security fencing, temporary coverings, etc.).

- Compile appropriate documentation.

 — IAP

 — Scene notes/diagrams

 — Photos/videos

 — Incident management organization chart

 — Dispatch voice recordings

 — Incident report/patient care report

 — Transcripts of voice recordings

 — Casualty reports

 — Incident management organization chart

 — Building inspection documents

 — Hazardous materials documents (if applicable)

 — Lost/damaged equipment lists

 — Personnel interviews

Figure 23.48 Equipment must be decontaminated before being returned to service. *Courtesy of FEMA News Photo, Photographer Marvin Nauman.*

- Transition the incident to investigating agency.

- Determine the final disposition of incident scene (after investigation is complete).

- Establish a time for incident analysis (post-incident critique).

Decontamination/Disposal Issues

Equipment that cannot be decontaminated at the incident site must be isolated in sealed containers and taken to a decontamination site for proper cleaning (**Figure 23.48**). Equipment that is too contaminated to be cleaned must be disposed of according to agency policy. Hazardous waste must be disposed of in accordance with SOPs and applicable legislation. Medical waste must be collected and disposed of in biohazard bags according to agency policy.

On-Scene Debriefing

On-scene debriefing, conducted in the form of a group discussion, gathers information from all operating personnel, including law enforcement, public works, and EMS responders. During the debriefing stage, obtain the following information from responders:

- Important observations

- Actions taken

- Timeline of those actions

In addition, one very important step in this process is to provide information to personnel concerning the signs and symptoms of overexposure to the hazardous materials, which is referred to as the hazardous communication briefing (required by OSHA in the United States). It is extremely important that this debriefing process be thoroughly documented. Each person attending must receive and understand the instructions and sign a document stating those facts. The information provided to responders before they leave the scene includes the following:

- Identity of material involved

- Potential adverse effects of exposure to the material

- Actions to be taken for further decontamination

- Signs and symptoms of an exposure

- Mechanism by which a responder can obtain medical evaluation and treatment

- Exposure documentation procedures

Post-Incident Analysis/Critique

The post-incident analysis or critique (sometimes called the *after-action analysis*) serves many purposes. It provides the responding agencies with an opportunity to evaluate, review, and refine:

- Operational tactics

- Training methods

- SOPs/SOGs/OIs and other policies

- Resource management/allocation

It can be used to evaluate the effectiveness of the response, identify problem areas, and correct deficiencies. This information can be used to modify and improve operations at future similar incidents.

The IC is responsible for assigning someone to write a post-incident report (or after-action report) and provide it to the health and safety officer and the chief of the agency. This individual will also participate in the post-incident analysis critique. In addition to standard operational issues, the report provides information on all aspects of the incident that involved safety issues. If the incident resulted in the death or injury of a rescue agency member, this individual is responsible for investigating the circumstances and reporting the findings.

Once all the necessary interviews have been made and documents gathered, a post-incident analysis should be scheduled for all agencies involved in the incident. This should take place within two weeks of the incident if possible. The agency training facility will usually provide the necessary space and materials needed for the analysis. Map tables, chalkboards, flip charts, and audio/visual equipment all play a part in a critical analysis. If the agency does not have a suitable location to conduct such an analysis, other local area occupancies may be used.

The formal analysis of the incident will be based on the information gathered for the post-incident analysis. In order for an incident critique to be successful, the top management of the emergency response organization must support the process. If the incident involved a response, representatives of all the responding agencies, as well as dispatchers, should be present.

Summary

For firefighters, hazardous materials incidents are similar in many ways to other emergencies to which they respond. The same universal priorities apply: life safety, incident stabilization, and property conservation; and an incident management system is needed. However, there are also some major differences compared to structure fires, for example. In a structure fire, as long as firefighters are located outside of the collapse zone, they are relatively safe. In a hazardous materials incident, personnel can be at risk a considerable distance from the point of release, especially if topography or prevailing winds spread the material beyond its original location. One major difference between the property conservation priority at hazardous materials incidents is the increased need for environmental protection.

There are also differences in the size-up process compared to structure fires. Since many hazardous materials are highly toxic and can be spread over a wide area, the initial identification of the materials involved may have to be done from a considerable distance away. Binoculars may have to be used to read labels and placards on buildings, tanks, or vehicles.

Finally, because of the highly toxic nature of some hazardous materials, fire officers and their crews may be untrained and unequipped to mitigate a hazardous materials release. In these cases, they must establish and maintain a safe perimeter around the incident scene and call for hazardous materials specialists who are trained and equipped to handle such incidents. In support of these specialists, firefighters must provide fire protection, and be capable of assisting with containment efforts such as damming and diking, and setting up and operating decontamination stations.

Fire Fighter I

1. What are the three incident priorities at all haz mat and terrorist incidents?

2. What information should be gathered by first responders during the initial assessment of an incident?

3. Describe incident levels.

4. What are hazard-control zones?

5. What factors must be addressed in large-scale evacuations?

6. What are the three types of decontamination?

7. What actions can Operations-Level firefighters perform during rescue operations?

8. List defensive confinement and spill control actions.

9. Who must be notified and included in the response to a terrorist or criminal incident?

10. What information should be given to responders at an on-scene debriefing?

Step 1: Confirm order to perform decontamination with officer.

Step 2: Ensure that all responders involved in decontamination operations are wearing appropriate PPE for performing emergency decontamination operations.

Step 3: Remove the victim from the contaminated area.

Step 4: Wash immediately any contaminated clothing or exposed body parts with flooding quantities of water.

Step 5: Remove victims clothing and/or PPE rapidly—if necessary, cutting from the top down in a manner that minimizes the spread of contaminants.

Step 6: Perform a quick cycle of head-to-toe rinse, wash, and rinse.

Step 7: Transfer the victim to treatment personnel for assessment, first aid, and medical treatment.

Step 8: Ensure that ambulance and hospital personnel are told about the contaminant involved.

Step 9: Decontaminate tools.

Step 10: Proceed to decontamination line for decontamination.

Step 3: Select a location to efficiently and safely perform the absorption operation.

NOTE: The deployment of sorbent materials is usually conducted in locations where the hazardous liquid has pooled or for small spills where the hazardous liquid threatens to enter areas where it will be difficult to control (sewers or a water course).

Step 4: Select the most appropriate sorbent.

Step 5: Deploy the sorbent in a manner that most efficiently controls the spill.

NOTE: The hazardous chemical is contained and absorbed within the immediate location of the spill without additional environmental damage.

WARNING !

Hazardous materials incidents can be extremely dangerous. Hazardous materials can cause serious injury or fatality. Appropriate personal protective equipment (PPE) must be worn and safety precautions must be followed. This skill sheet demonstrates general steps; specific haz mat incidents may differ in procedure. Always follow departmental procedures for specific incidents.

NOTE: Prior to performing the defensive control function, the incident commander or other qualified responder must identify the material and determine the appropriate level of PPE required at the incident based on the hazardous material, training of responders, terrain, weather, and other size-up factors.

Step 1: Confirm order to perform function with officer.

Step 2: Verify that all responders involved in the control function are wearing appropriate PPE for performing absorption operations and that appropriate hand tools have been selected.

Step 6: Upon mitigation of the incident, place any contaminated material, such as clothing, in an approved container for transportation to a disposal location. Seal and label the container and document appropriate information for department records.

Step 7: Decontaminate tools.

Step 8: Advance to decontamination line for decontamination.

Step 3: Select a location to efficiently and safely perform the diking operation.

NOTE: The construction of a dike is usually done when the hazardous liquid spill threatens to contaminate a large area if it is not contained, or if the spill can be more easily managed at another location further downstream. It is used to direct the flow to another location. Most often the location of the dike is determined by the natural terrain where the spill has occurred relative to potential exposure and damage if the spill threatens a high value location.

WARNING !

Hazardous materials incidents can be extremely dangerous. Hazardous materials can cause serious injury or fatality. Appropriate personal protective equipment (PPE) must be worn and safety precautions must be followed. This skill sheet demonstrates general steps; specific haz mat incidents may differ in procedure. Always follow departmental procedures for specific incidents.

NOTE: Prior to performing the defensive control function, the incident commander or other qualified responder must identify the material and determine the appropriate level of PPE required at the incident based on the hazardous material, training of responders, terrain, weather, and other size-up factors.

Step 1: Confirm order to perform function with officer.

Step 2: Verify that all responders involved in the control function are wearing appropriate PPE for performing diking operations and that appropriate hand tools have been selected.

Step 4: Construct the dike in a location and manner that most efficiently controls and directs the spill to a desired location.

Step 6: Decontaminate tools.

Step 7: Advance to decontamination line for decontamination.

Step 5: Upon mitigation of the incident, place any contaminated material, such as clothing, in an approved container for transportation to a disposal location. Seal and label the container and document appropriate information for department records.

NOTE: Prior to performing the defensive control function, the incident commander or other qualified responder must identify the material and determine the appropriate level of PPE required at the incident based on the hazardous material, training of responders, terrain, weather, and other size-up factors.

Step 1: Confirm order to perform function with officer.

Step 2: Verify that all responders involved in the control function are wearing appropriate PPE for performing damming operations and that appropriate hand tools have been selected.

Step 3: Select a location to efficiently and safely perform the damming operation.

NOTE: The construction of a dam is usually done when the hazardous liquid spill threatens to contaminate a large area if it is not contained. It can also be used to buy time. For smaller spills, the dam can hold the material for a period of time while sorbent is deployed or another method of recovery can be utilized. Many times the location of the dam is determined by the natural terrain where the spill has occurred.

NOTE: The location that is selected should be far enough downstream from the spill so that the dam is not "overrun" during its construction.

Step 4: Construct the dam in a location and manner that most efficiently controls the spill.

Step 5: Upon mitigation of the incident, place any contaminated material, such as clothing, in an approved container for transportation to a disposal location. Seal and label the container and document appropriate information for department records.

Step 6: Decontaminate tools.

Step 7: Advance to decontamination line for decontamination.

NOTE: Prior to performing the defensive control function, the incident commander or other qualified responder must identify the material and determine the appropriate level of PPE required at the incident based on the hazardous material, training of responders, terrain, weather, and other size-up factors.

Step 1: Confirm order to perform function with officer.

Step 2: Verify that all responders involved in the control function are wearing appropriate PPE for performing diversion operations and that appropriate hand tools have been selected.

Step 3: Select a location to efficiently and safely perform the diversion operation.

NOTE: Hazardous liquid spills are usually diverted when they threaten to contaminate a stream or other high value area. Diversion is used to divert the flow to another location, to buy time. Most often the diversion is used to divert the flow to a containment area.

WARNING!

Hazardous materials incidents can be extremely dangerous. Hazardous materials can cause serious injury or fatality. Appropriate personal protective equipment (PPE) must be worn and safety precautions must be followed. This skill sheet demonstrates general steps; specific haz mat incidents may differ in procedure. Always follow departmental procedures for specific incidents.

Step 4: Construct the diversion in a location and manner that most efficiently controls and directs the spill to a desired location. Working as a team, use hand tools to break the soil, remove the soil, pile the soil, and pack the soil tightly.

Step 5: Upon mitigation of the incident, place any contaminated material, such as clothing, in an approved container for transportation to a disposal location. Seal and label the container and document appropriate information for department records.

Step 6: Decontaminate tools.

Step 7: Advance to decontamination line for decontamination.

NOTE: Prior to performing the defensive control function, the incident commander or other qualified responder must identify the material and determine the appropriate level of PPE required at the incident based on the hazardous material, training of responders, terrain, weather, and other size-up factors.

Step 1: Confirm order to perform function with officer.

Step 2: Verify that all responders involved in the control function are wearing appropriate PPE for performing retention operations and that appropriate hand tools have been selected.

Step 3: Select a location to efficiently and safely perform the retention operation.

Step 4: Evaluate the rate of flow of the leak to determine the required capacity of the retention vessel.

Step 5: Working as a team, retain the hazardous liquid so that it can no longer flow.

Step 6: Upon mitigation of the incident, place any contaminated material, such as clothing, in an approved container for transport to a disposal location. Seal and label the container and document appropriate information for department records.

Step 7: Decontaminate tools.

Step 8: Advance to decontamination line for decontamination.

NOTE: Prior to performing the defensive control function, the incident commander or other qualified responder must identify the material and determine the appropriate level of PPE required at the incident based on the hazardous material, training of responders, terrain, weather, and other size-up factors.

Step 1: Confirm order to perform function with officer.

Step 2: Verify that all responders involved in the control function are wearing appropriate PPE for performing dilution operations.

Step 3: Select a location to efficiently and safely perform dilution operations.

Step 4: Evaluate the rate of flow of the leak to determine the required capacity of the retention area and the quantity of water required to dilute the material.

Step 5: Working as a team, monitor and assess the leak, and advance hose lines and tools to retention area.

Step 6: Flow water to dilute spilled material.

Step 7: Monitor any diking or dams to ensure integrity of retention area.

Step 8: Upon mitigation of the incident, place any contaminated material, such as clothing, in an approved container for transportation to a disposal location. Seal and label the container and document appropriate information for department records.

Step 9: Decontaminate tools.

Step 10: Advance to decontamination line for decontamination.

WARNING!
Hazardous materials incidents can be extremely dangerous. Hazardous materials can cause serious injury or fatality. Appropriate personal protective equipment (PPE) must be worn and safety precautions must be followed. This skill sheet demonstrates general steps; specific haz mat incidents may differ in procedure. Always follow departmental procedures for specific incidents.

Step 1: Confirm order to perform function with officer.

Step 2: Verify that all responders involved in the control function are wearing appropriate PPE for performing vapor dispersion operations.

Step 3: Select a location to efficiently and safely perform the vapor dispersion operation.

Step 4: Apply agent through vapor cloud to disperse vapors.

Step 5: Constantly monitor the leak concentration, wind direction, exposed personnel, environmental impact, and water stream effectiveness.

Step 6: Upon mitigation of the incident, place any contaminated material, such as clothing, in an approved container for transportation to a disposal location. Seal and label the container and document appropriate information for department records.

Step 7: Decontaminate tools.

Step 8: Advance to decontamination line for decontamination.

WARNING!

Hazardous materials incidents can be extremely dangerous. Hazardous materials can cause serious injury or fatality. Appropriate personal protective equipment (PPE) must be worn and safety precautions must be followed. This skill sheet demonstrates general steps; specific haz mat incidents may differ in procedure. Always follow departmental procedures for specific incidents.

NOTE: Prior to performing the defensive control function, the incident commander or other qualified responder must identify the material and determine the appropriate level of PPE required at the incident based on the hazardous material, training of responders, terrain, weather, and other size-up factors.

Appendices

Contents

Photo courtesy of FEMA.

NFPA® Job Performance Requirements (JPRs) with Chapter and Page References

NFPA® 1001 JPR Numbers	Chapter References	Page References
5.1.1	1, 2, 5, 7, 18	14-34, 40-43, 59-78, 167-180, 263-283, 910
5.1.2	5, 7	167-180, 200-205, 263-283
5.2.1	19	926-941
5.2.2	19	936
5.2.3	19	943-949
5.3.1	5	180-205, 207-212
5.3.2	2	53-54,
5.3.3	2	53-54, 69-78
5.3.4	4, 9	141-150, 397-450
5.3.5	2, 5, 8	69-78, 180-200, 207-212, 319-327
5.3.6	10	471-507
5.3.7	15	796-799
5.3.8	3, 15, 18	130-132, 769-772, 799, 912-914, 918
5.3.9	5, 8, 10	207-212, 306-330, 490-494, 505-506
5.3.10	3, 4, 13, 14, 15, 17	86-112, 130-132, 152-158,633-634, 671-680, 718-
733,		762-769, 799, 883-886
5.3.11	3, 10, 11	86-130, 481-487, 541-579
5.3.12	3, 4, 10, 11	112-132, 151-158, 481-494, 498-500, 501-505, 541-579
5.3.13	17, 18	882-886, 912-914, 918
5.3.14	16, 17,	842-857, 867-882,
5.3.15	12,	594-616,
5.3.16	3, 6, 15	110-112, 234-254, 780-787
5.3.17	2, 8	65-67, 530-532
5.3.18	15	778-787
5.3.19	15	800-806
5.5.1	5, 7, 9, 10, 17	205-207, 272-272, 411-413, 479-481, 873
5.5.2	13	634-642, 653-663, 680-682
6.1.1	1, 2, 15	24-25, 35-40, 63-78, 792
6.1.2	15	792
6.2.1	19	949
6.2.2	19	926-949
6.3.1	14, 15	734-749, 773-780

NFPA® 472 Competencies	Chapter References	Page References
4.2.1	22	1114-1203
4.2.1(1)	22	1059
4.2.1(2)	22	1137-1154
4.2.1(3)	22	1137-1154
4.2.1(4)	22	1059-1060
4.2.1(5)	22	1115-1117
4.2.1(6)	22	1118-1137
4.2.1(7)	22	1137-1166
4.2.1(7)(a)	22	1154
4.2.1(7)(b)	22	1159-1160
4.2.1(7)(c)	22	1163, 1164-1165
4.2.1(7)(d)	22	1161-1162
4.2.1(7)(e)	22	1163
4.2.1(7)(f)	22	1155-1166
4.2.1(8)	22	1159-1160
4.2.1(9)	22	1137-1153
4.2.1(10)	22	1170-1172
4.2.1(10)(a)	22	1170
4.2.1(10)(b)	22	1170-1171
4.2.1(10)(c)	22	1168
4.2.1(10)(d)	22	1168
4.2.1(10)(e)	22	1168
4.2.1(10)(f)	22	1168
4.2.1(10)(g)	22	1168
4.2.1(11)	22	1179-1181
4.2.1(12)	22	1179
4.2.1(13)	22	1116-1117
4.2.1(14)	22	1185-1190
4.2.1(15)	22	1185-1188
4.2.1(16)	22	1188-1190
4.2.1(17)	22	1190
4.2.1(18)	22	1191-1201
4.2.1(19)	22	1190-1191
4.2.1(20)	22	1201-1203
4.2.2	22	1114-1203
4.2.2(1)	22	1115
4.2.2(2)	22	1138-1155
4.2.2(3)	22	1166
4.2.3	22	1172, 1174-1179

NFPA® 472 Competencies	Chapter References	Page References
4.2.3(1)	22	1172
4.2.3(2)	22	1176
4.4.1(1)	23	1215
4.4.1(2)	22	1060
4.4.1(3)(a)	22, 23	1060-1061, 1237-1239, 1256-1261
4.4.1(3)(b)	23	1222
4.4.1(3)(c)	22	1084-1102
4.4.1(3)(d)	22	1091
4.4.1(4)	22	1176-1179
4.4.1(4)(a)	22	1176
4.4.1(4)(b)	22	1176-1177
4.4.1(4)(c)	22	1176-1179
4.4.1(5)(a)	22	1077-1078, 1176-1177
4.4.1(5)(b)	22	1069-1070, 1176-1177
4.4.1(5)(c)	22	1062-1063, 1176-1177
4.4.1(5)(d)	22	1071-1072, 1176-1177
4.4.1(6)	23	1227-1229, 1234-1237
4.4.1(6)(a)	23	1227-1229
4.4.1(6)(b)	23	1234-1236
4.4.1(6)(c)	23	1236-1237
4.4.1(7)	22	1176-1179
4.4.1(8)	22	1177
4.4.1(9)	22	1176-1179
4.4.1(9)(a)	22	1176-1179
4.4.1(9)(b)	22	1176,1179
4.4.1(10)	22	1176-1179
4.4.1(11)	23	1227
4.4.1(12)	18, 23	909-918, 1270
4.4.2	23	1230-1231
5.2.1	22, 23	1109-1113, 1118-1137, 1218-1221
5.2.1.1	22	1102, 1118-1137
5.2.1.1.1	22	1119, 1123-1124
5.2.1.1.1(1)	22	1119, 1124
5.2.1.1.1(2)	22	1119, 1123
5.2.1.1.1(3)	22	1119, 1124
5.2.1.1.2	22	1119, 1130-1131
5.2.1.1.2(1)	22	1119, 1130
5.2.1.1.2(2)	22	1119, 1130
5.2.1.1.2(3)	22	1119, 1130

NFPA® 472 Competencies	Chapter References	Page References
5.2.1.1.2(3)(a)	22	1119, 1130
5.2.1.1.2(3)(b)	22	1119, 1130
5.2.1.1.3	22	1119, 1127-1129
5.2.1.1.3(1)	22	1119, 1129
5.2.1.1.3(2)	22	1119, 1127
5.2.1.1.3(3)	22	1119, 1128
5.2.1.1.3(4)	22	1119, 1129
5.2.1.1.3(5)	22	1119, 1128
5.2.1.1.3(6)	22	1119, 1127
5.2.1.1.3(7)	22	1119, 1127
5.2.1.1.4	22	1118-1123
5.2.1.1.4(1)	22	1119, 1123
5.2.1.1.4(2)	22	1118-1121
5.2.1.1.4(3)	22	1119, 1122
5.2.1.1.5	22	1133-1135
5.2.1.1.5(1)	22	1133-1134
5.2.1.1.5(2)	22	1133-1134
5.2.1.1.5(3)	22	1133-1134
5.2.1.1.5(4)	22	1133, 1135
5.2.1.1.5(5)	22	1133, 1135
5.2.1.1.6	22	1133, 1136-1137
5.2.1.1.6(1)	22	1133, 1136
5.2.1.1.6(2)	22	1133, 1136
5.2.1.1.6(3)	22	1133, 1136
5.2.1.1.6(4)	22	1133, 1136
5.2.1.1.6(5)	22	1133, 1137
5.2.1.2	22	1137-1166
5.2.1.2.1	22	1137-1158
5.2.1.2.1(1)	22	1137-1155
5.2.1.2.1(2)	22	1137-1158
5.2.1.2.1(3)	22	1137-1158
5.2.1.2.2	22	1158
5.2.1.3	22	1152, 1163, 1166
5.2.1.3.1	22	1163
5.2.1.3.1(1)	22	1163
5.2.1.3.1(2)	22	1163
5.2.1.3.1(3)	22	1163
5.2.1.3.2	22	1163, 1166
5.2.1.3.2(1)	22	1163, 1166

NFPA® 472 Competencies	Chapter References	Page References
5.2.1.3.2(2)	22	1163, 1166
5.2.1.3.2(3)	22	1163, 1166
5.2.1.3.2(4)	22	1163, 1166
5.2.1.3.2(5)	22	1163, 1166
5.2.1.3.2(6)	22	1163, 1166
5.2.1.3.3	22	1152
5.2.1.4	23	1218-1221
5.2.1.5	22	1166-1184
5.2.1.6	22	1184
5.2.2(1)	22	1137-1146
5.2.2(2)	22	1170
5.2.2(3)	22	1170-1172
5.2.2(3)(a)	22	1170-1172
5.2.2(3)(b)	22	1170-1172
5.2.2(3)(c)	22	1170-1172
5.2.2(3)(d)	22	1170-1172
5.2.2(3)(e)	22	1170-1172
5.2.2(3)(f)	22	1170-1172
5.2.2(3)(g)	22	1170-1172
5.2.2(3)(h)	22	1170-1172
5.2.2(3)(i)	22	1170-1172
5.2.2(3)(j)	22	1170-1172
5.2.2(4)(a)	22	1166-1167
5.2.2(4)(b)	22	1166-1167
5.2.2(4)(c)	22	1166-1167
5.2.2(5)	22	1166
5.2.2(6)	23	1231
5.2.2(7)	23	1230-1231
5.2.2(8)	22	1085-1088
5.2.2(8)(a)	22	1085
5.2.2(8)(b)	22	1086
5.2.2(8)(c)	22	1086-1087
5.2.2(8)(d)	22	1087-1088
5.2.3(1)(a)(i)	22	1103
5.2.3(1)(a)(ii)	22	1105-1108
5.2.3(1)(a)(iii)	22	1094, 1100
5.2.3(a)(iv)	3	105
5.2.3(1)(a)(v)	3	98
5.2.3(1)(a)(vi)	3	90-91

NFPA® 472 Competencies	Chapter References	Page References
5.2.3(1)(a)(vii)	22	1085-1087
5.2.3(1)(a)(viii)	22	1105
5.2.3(1)(a)(ix)	22	1102, 1103
5.2.3(1)(a)(x)	22	1085
5.2.3(1)(a)(xi)	22	1104-1105
5.2.3(1)(a)(xii)	3	109
5.2.3(1)(a)(xiii)	22	1103-1104
5.2.3(1)(a)(xiv)	22	1102
5.2.3(1)(a)(xv)	22	1104
5.2.3(1)(b)(i)	23	1239
5.2.3(1)(b)(ii)	23	1239
5.2.3(1)(b)(iii)	23	1239
5.2.3(1)(b)(iv)	22	1100
5.2.3(1)(b)(v)	22	1084
5.2.3(1)(b)(vi)	22	1084
5.2.3(2)	22	1109
5.2.3(3)	22	1109
5.2.3(4)	22	1109
5.2.3(5)	22	1110-1113
5.2.3(6)	22	1113
5.2.3(7)	22	1084-1102
5.2.3(8)(a)	22	1085-1188
5.2.3(8)(b)	22	1090
5.2.3(8)(c)	22	1094
5.2.3(8)(d)	22	1094
5.2.3(8)(e)	22	1094
5.2.3(8)(f)	22	1094
5.2.3(8)(g)	22	1094
5.2.3(8)(h)	22	1099
5.2.3(8)(i)	22	1091-1092
5.2.3(8)(j)	22	1091
5.2.3(9)(a)	22	1186
5.2.3(9)(b)	22	1151
5.2.3(9)(c)	22	1187
5.2.3(9)(d)	22	1187
5.2.3(9)(e)	22	1186
5.2.3(9)(f)	22	1186
5.2.3(9)(g)	22	1152
5.2.4(1)	22, 23	1166-1179, 1220

NFPA® 472 Competencies	Chapter References	Page References
5.2.4(2)	23	1220
5.2.4(3)	22	1181-1184
5.2.4(4)	22, 23	1109-1113, 1219-1221
5.2.4(5)	22	1088-1090
5.3.1(1)	23	1220
5.3.1(2)	23	1224
5.3.1(3)	23	1260-1261
5.3.1(4)	22	1201-1203
5.3.2(1)	23	1226-1227
5.3.2(2)	23	1246-1247
5.3.3(1)	22	1061-1069
5.3.3(1)(a)	22	1061-1068
5.3.3(1)(a)(i)	22	1062-1063
5.3.3(1)(a)(ii)	22	1063-1064
5.3.3(1)(a)(iii)	22	1062
5.3.3(1)(a)(iv)	22	1067
5.3.3(1)(a)(v)	22	1064-1066
5.3.3(1)(a)(vi)	22	1066
5.3.3(1)(b)	22	1068-1069
5.3.3(2)(a)	22	1069
5.3.3(2)(b)	22	1069-1072
5.3.3(2)(b)(i)	22	1071-1072
5.3.3(2)(b)(ii)	22	1070-1071
5.3.3(2)(b)(iii)	22	1069-1070
5.3.4(1)	23	1237-1240
5.3.4(2)	23	1239
5.3.4(3)	23	1237-1241
5.3.4(4)	23	1239-1240, 1241
5.3.4(5)	23	1237
5.3.4(6)	23	1241
5.4.1(1)	23	1227-1229
5.4.1(2)	23	1227-1230
5.4.1(3)	23	1234-1237
5.4.1(3)(a)	23	1234-1236
5.4.1(3)(b)	23	1236-1237
5.4.1(4)	23	1239-1240, 1276
5.4.1(5)	23	1234
5.4.1(5)(a)	23	1234
5.4.1(5)(b)	23	1234

NFPA® 472 Competencies	Chapter References	Page References
5.4.2	18	909-918
5.4.3(1)	22	1060
5.4.3(2)	23	1221-1224
5.4.3(3)	1, 23	35-40, 1215
5.4.3(4)(a)	1, 23	36, 1233
5.4.3(4)(b)	23	1215
5.4.3(5)	23	1229
5.4.3(6)	23	1230-1231
5.4.3(7)	1	40-43
5.4.4(1)	23	1233
5.4.4(2)	23	1233
5.4.4(3)	23	1231
5.5.4(4)	22	1081-1083
5.4.4(5)	22	1068-1069
5.4.4(6)	22	1083-1084
5.4.4(7)	22	1083-1084
5.5.1(1)	23	1226
5.5.1(2)	23	1233
5.5.2(1)	23	1226, 1233
5.5.2(2)	23	1233
6.1.2	15	792, 821
6.2.1	19	949, 955
6.2.2	15	792, 821
6.3.2	15	792, 821
6.3.3	15	779, 820
6.3.4	18	918, 920
6.4.1	8	349 -361, 387-390
6.4.2	8	361-374, 391
6.5.1	20	976-982, 996
6.5.2	20	982-992, 997-998
6.5.3	12, 20	609-610, 626, 970-975, 995
6.5.4	8	330-334, 386
6.5.5	13	680-682, 712

Essentials V Chapter Titles

1 Firefighter Orientation
2 Firefighter Safety and Health
3 Fire Behavior
4 Building Construction
5 Firefighter Personal Protective Equipment
6 Portable Fire Extinguishers
7 Ropes and Knots
8 Rescue and Extrication
9 Forcible Entry
10 Ground Ladders
11 Ventilation
12 Water Supply

13 Fire Hose
14 Fire Streams
15 Fire Control
16 Fire Detection, Alarm, and Suppression Systems
17 Loss Control
18 Protecting Fire Scene Evidence
19 Fire Department Communications
20 Fire Prevention and Public Education
21 Basic Prehospital Emergency Care for Firefighters
22 Introduction to Hazardous Materials
23 Operations at Haz Mat Incidents

PREHOSPITAL 9-1-1 EMERGENCY MEDICAL RESPONSE:

The Role of the United States Fire Service
in Delivery and Coordination

Authors and Contributors

Franklin D. Pratt, M.D., FACEP

Medical Director
Los Angeles County Fire Department

Medical Director
Emergency Department
Torrance Memorial Medical Center
Torrance, CA 90505

Assistant Clinical Professor
Geffen School of Medicine
UCLA

Steven Katz, M.D., FACEP, EMT-P

Associate Medical Director
Palm Beach County Fire Rescue
West Palm Beach, FL

President
National Paramedic Institute
Boynton Beach, FL

Chairman
Department of Emergency Medicine
Memorial Hospital West
Pembroke Pines, FL

Paul E. Pepe, MD, MPH

Professor of Medicine, Surgery, Public
Health and Chair, Emergency Medicine
University of Texas Southwestern
Medical Center and the Parkland Health
and Hospital System

Director, City of Dallas Medical
Emergency Services
for Public Health, Public Safety and
Homeland Security

David Persse, MD, EMT-P, FACEP

Physician Director
Houston Fire Department
Emergency Medical Services

Public Health Authority
Houston Department of Health and
Human Services

Associate Professor of Surgery
Baylor College of Medicine

Associate Professor of Emergency
Medicine
University of Texas Medical School
Houston

2

PREHOSPITAL 9-1-1 EMERGENCY MEDICAL RESPONSE: THE ROLE OF THE UNITED STATES FIRE SERVICE IN DELIVERY AND COORDINATION

ABSTRACT

Prehospital 9-1-1 emergency response is one of the essential public safety functions provided by the United States fire service in support of community health, security and prosperity. Fire service-based emergency medical services (EMS) systems are strategically positioned to deliver time critical response and effective patient care. Fire service-based EMS provides this pivotal public safety service while also emphasizing responder safety, competent and compassionate workers, and cost-effective operations. As the federal, state, and local governments consider their strategic plans for an 'all hazards' emergency response system, EMS should be included in those considerations and decision makers should recognize that the U.S. fire service is the most ideal prehospital 9-1-1 emergency response agency.

INTRODUCTION TO FIRE SERVICE-BASED EMS

EMS is an essential component of the public services provided in the United States. The Federal EMS Act of 1973 defined an EMS system as "an entity that provides for the arrangement of personnel, facilities, and equipment for the effective and coordinated delivery of health care services under emergency conditions in an appropriate geographic area" (EMS Act 1973, (P.L. 93-154)). Much of the dialogue in the public arena today concerning prehospital 9-1-1 emergency medical care often focuses on ambulance services and, accordingly, may ignore the important distinction between prehospital 9-1-1 emergency medical response and the other key uses of the ambulance-based, out-of-hospital providers for non-emergency medical and transportation services.

3

The primary purpose of this discussion is to underscore the reality today that the fire service has become the first-line medical responder for critical illness and injury in virtually every community in America. Regardless of whatever agency provides medical transportation services, the fire service is the agency that first delivers on-scene health care services under most true emergency conditions. Therefore, prehospital 9-1-1 emergency response, in support of community health, security and prosperity, is not only a key function of each community; it has become, almost universally, a principal duty of the fire service as well. In addition, fire service-based EMS systems are strategically positioned to deliver time critical response and effective patient care rapidly. Furthermore, the fire service-based EMS accomplishes this rapid first response while emphasizing responder safety, sending competent and compassionate workers, and delivering cost-effective operations.

Although the role of the fire service is central in 9-1-1 emergency medical response, financial, political, cultural and organizational factors often can make the conversation about prehospital care providers confusing and complex for many decision makers in local communities. The goal of this discussion is to resolve and demonstrate that the use of fire service equipment and personnel to provide 9-1-1 emergency response is the best approach for a community regardless of size. This basic premise is consistent with recent Institute of Medicine publications that have placed EMS at the intersection of public safety, public health, and medical care. The U.S. Fire Service is uniquely qualified to be at that intersection and in the following pages, the history, evolution, and current medical capabilities of the fire service will be reviewed.

The Maltese Cross and Its Legacy for Fire-Service Based EMS

During the Middle Ages, the Knights of Malta, the forerunners of the fire service, took care of travelers and specifically burn victims from the Crusades and associated battles. Eventually, the Knights of Malta adopted the Maltese Cross as their emblem and it has created a revered legacy for fire departments.

4

The Knights originally began their work as the creators, administrators and care givers in a hospital in Jerusalem. As such, they were known as the Hospitallers of Jerusalem, starting their work before the year 1000 AD. For the next two hundred years, they helped the sick and poor and they set up hospitals and hospices across Europe.

Eventually, the Hospitallers became firefighters out of necessity. The conflict of the Crusades often threatened the hospitals that they had founded. So, they adapted and even engaged in battle to protect their hospitals. As a result, they also became firefighters because one of the weapons of war at that time was the glass fire bomb. The fire bomb, thrown by the enemy, created a horrendous inferno. After rescuing a fellow knight from the inferno and extinguishing the fire, a Hospitaller was awarded a medal, shaped like a Maltese Cross to honor those actions.

As conflict continued, the Hospitallers needed an identifying mark for their armor. This was necessary because without identifying markings, it was difficult to tell who was who because everyone was wearing similar armor in battle. They adopted the Maltese Cross as their identifying mark. (Maltese Cross, 2007, Foster, 2007)

In essence, more than 1200 years ago, some of the earliest ancestors of the fire service were "all-hazards responders." They initially started as caregivers for the sick and then became firefighters to protect their own. These are two of the concepts firefighters still believe in today and hold as their most sacred responsibilities—caring for the sick and caring for their own.

Longstanding History of Fire Service-Based Medical Care in the U.S.

The fire service has formally been part of the 9-1-1 emergency care delivery system since EMS began in the late 1960's. Many of the original prehospital EMS providers were firefighters, who had "special" additional training in providing medical services during emergencies that occurred outside the hospital. Today, essentially every firefighter receives emergency medical training and the fire service provides the majority of medical services during emergencies that occur out of the hospital, just as it has done for the past

5

four decades. Of the 200 largest cities in the United States, 97% have fire service-based prehospital 9-1-1 emergency medical response (*JEMS* 200-City Survey, 2006) and the fire service provides advanced life support (ALS) response and care in 90% of the 30 most populated U.S. jurisdictions (cities and counties) (IAFF/IAFC Fire Operations Survey, 2005).

Although the origin of the modern relationship between emergency medicine and fire departments is cited as the 1960's, the involvement of the fire service in patient care began much earlier. For example, in 1937 a fire department ambulance in New York transported famous song writer Cole Porter to the hospital after a horseback riding accident.

While the fire service was involved in many famous anecdotal events, other accounts demonstrate its profound effect on public safety and patient care procedures. In 1921 Claude Beck, M.D., a surgeon at Western Reserve University in Cleveland, called the fire department so he could apply a "pulmotor," an artificial breathing apparatus, to attempt resuscitation in a patient who died unexpectedly during surgery (Beck, 1941). Dr. Beck continued to be involved in resuscitation and today is recognized as one of the founders of the science of resuscitation.

The following quote from the *Journal of the American Medical Association* in 1928 summarized the evolving relationship between fire department-based out-of-hospital emergency care and subsequent resuscitation in the hospital.

> "...inhalators are introduced: Cases of gas asphyxiation occur; the rescue crew of the fire department is called and resuscitates the patient. A physician sees the resuscitation and is impressed by the effectiveness of the treatment. Some time thereafter he finds himself confronted with a child which he has delivered, and which has come through a prolonged labor. It refuses to breath effectively, in spite of the application of all the ancient practices. The respiratory center has been depressed by the diminished blood supply to the brain resulting from compression of the head, and needs more than the

6

normal amount of carbon dioxide to stimulate breathing. So the physician calls in the fire department. If, as is often the case, the fire department succeeds where his medical skill and knowledge have failed, he calls for it again the next time. Now the hospitals in some cities are adopting the practice of calling for the inhalator whenever they have a baby who breathes poorly. In effect, *they add the rescue crew of the fire department to their board of consultants, and these new consultants thus contribute another service to the community over and above that for which the fire department is primarily organized.* Obviously, it is the hospitals that should be equipped to treat asphyxia -- asphyxia of every form -- and thus to help firemen overcome by smoke and gas, instead of relying on the fire department to help the hospital in such a matter as asphyxia of the newborn." (Henderson, *JAMA* 1928. note: italics added).

According to a historical account on the City's website, in 1947 in the city of Virginia, Minnesota, "the Fire Department took full possession of the ambulance along with a Pulmotor Resuscitator … This would be the first time that ambulance personnel would be properly trained in first aid, and resuscitation procedures of that time" (City of Virginia, MN, 2007).

Such widespread anecdotes not only indicate longstanding involvement of the fire service in medical care, but it demonstrates the often-quoted mission of the fire service established in the 19th Century, "To Protect and Save Lives and Property." Clearly, protecting and saving lives is the first and foremost mission for these dedicated first responders.

Growth and Specialization of Fire Service-Based EMS

As illustrated by its history, the fire service has continuously adapted and changed to meet the current needs of a community. As EMS developed, the fire service was integrally involved. In the early stages, firefighters were chosen by expert physicians to take on the role of paramedic. This era of EMS in the fire service is represented well by

7

looking at the City of Miami Fire Rescue Department nearly half a century ago. The age-old firefighter mantra to "protect and save lives and property" is well-illustrated within the history books of the City of Miami Fire Rescue and serves as an important example of Fire Rescue today in the United States. Miami was the first city to call itself a "Fire Rescue" department. Miami Fire Rescue was also revolutionary in using the advancements of technology in 2-way radios to bridge physicians in the hospital with firefighter-paramedics in the prehospital setting.

In fact, the Rescue Division of Miami Fire Rescue was established in 1939 in order to give first aid to firefighters. Rescue One, the department's first rescue truck used to treat citizens, came on-line in 1941. In these early days, "Rescue" services were limited to basic first aid with transportation usually performed by funeral homes.

In 1964, Dr. Eugene Nagel, started to teach first aid and basic cardiopulmonary resuscitation (CPR) to the firefighters of Miami Fire Rescue. Dr. Nagel's goal was to improve out-of-hospital cardiac arrest survival in the community by using lessons learned from the "quick response" system in the hospitals and apply it to the prehospital setting. Dr. Nagel still reflects, "We chose firefighters because they were there, they were available, they were willing, and they were motivated. It was really quite simple" (Nagel interview, February 2007). According to Dr. Nagel, "The fire service is dispersed throughout America and is everywhere in our country. It is an efficient method for offering emergency care rather than creating a completely separate service with separate communications, vehicles, housing, and personnel. It worked well in Miami in the 1960's and continues to work well when integrated into the fire service. It is a natural fit" (Nagel interview, February 2007). Firefighters in Miami clearly demonstrated in the pioneer days of EMS that firefighters are ideal candidates and willing dispensers of high-quality EMS. In Miami, this started with basic first aid, and progressed to CPR, intravenous therapy, electrocardiographs, telemetry, and advanced airway intervention. During this same time period, similar efforts were underway using firefighters in the cities such as Baltimore, Columbus, Seattle, and Los Angeles. Providing firefighters training in lifesaving techniques and procedures has allowed them to deliver advances in medicine to

8

the prehospital 9-1-1 emergency care patient in a cost-effective and time-sensitive manner. Just as fire departments have evolved since the 1960's to provide prehospital emergency medical care, government oversight must evolve to cohesively organize, coordinate, and supervise the integrated delivery of emergency medical care from the scene to the hospital and even the rehabilitation and recovery phase. A critical link in that chain of survival and recovery is the rapid on-scene response of the Fire service, a service that cannot be underestimated and truly emphasized in planning, funding, support, research, and quality assurance.

The protection of life and property has been the mission of the fire service for over 200 years, but the fire department of the 21st century is evolving into a multidisciplinary public safety department. It not only handles most aspects of public safety (beyond law enforcement security issues), but it also will continue to provide advances in emergency medical care and many developing public health needs such as preparations for pandemics, disasters, and weapons of mass effect.

Today, the community-based fire station, with its ready availability of personnel 24 hours a day, coupled with the unique nature of medicine outside of the hospital, creates a symbiotic blend of the traditional public concepts and duties of the fire service with the potential for the most rapid delivery of advanced prehospital 9-1-1 emergency response and care. Traditionally, fire stations are strategically placed across geographic regions, typically commensurate with population densities and workload needs. This creates an all-hazard response infrastructure meeting the routine and catastrophic emergency needs of all communities regardless of the nature of the emergency. Accordingly, the fire service helps ensure the prosperity and security of all communities and providing prehospital 9-1-1 emergency medical care is consistent with its legacy going back 1200 years.

Types of Fire Service-Based EMS Systems

The fire service can be configured many ways to deliver prehospital 9-1-1 emergency medical care such as the following general configurations:

9

- Fire service-based system using cross-trained/multi-role firefighters. Firefighters are all-hazards responders, prepared to handle any situation that may arise at a scene including patient care and transport.

- Fire service-based system using employees who are not cross-trained as fire suppression personnel. Single role EMS-trained responders accompanying firefighter first-responders on 9-1-1 emergency medical calls.

- Combined system using the fire department for emergency response and a private or "third service" (police, fire, EMS) provider for transportation support. Single role emergency medical technicians and paramedics accompany firefighter first responders to emergency scenes to provide patient transport in a private or third service ambulance.

While there are pros and cons to the various system approaches, the emergency medicine (EM) literature indicates that the most likely time to create error in medical care is when care is transferred from one provider to another in a relatively short encounter time. Such circumstances require that the fire service regularly exercise the leadership needed to ensure that integration of the parts of the prehospital emergency care system are coordinated well, with maximum benefit to the patient and minimum risk to the community. For example, in the fire service-based EMS model in which the fire department provides extrication, triage and treatment services, and a separate private provider transports the patients, appropriate quality assurance measures must be in place. This quality assurance is most effective when the fire department, as the public agency, administers and monitors the performance requirements on-scene and within the transportation agreement.

10

National Incident Management System

The U.S. Fire Service-based emergency response and medical care system is the most effective, coordinated system worldwide. The National Incident Management System (NIMS) and other nationally-defined coordination plans ensure that fire service-based 9-1-1 emergency response and medical care always provides skilled medical services to the patient regardless of the circumstances surrounding the location and condition of the patient. In addition, the fire service has the day-to-day experience and ability to work smoothly with other participants in the prehospital 9-1-1 emergency medical care arena: private ambulance companies, law enforcement agencies, health departments, public works departments, the American Red Cross and other government and non-government agencies involved in medical care, disaster response and patient services. This type of universal coordination takes leadership, work, and the willingness to subordinate fire service prerogatives to those of the greater public need. The fire service is the creator of the unified command concept that brings everyone to the table, at the same time. Using the National Incident Management System, the fire service has superior ability to coordinate incidents of any size. As a result, it provides the best return on investment of public dollars to provide the delivery of prehospital 9-1-1 emergency medical service.

Emergency 9-1-1 Response is Different from Non-emergency and In-hospital Care

For government decision makers who do not work in the public safety environment on a day-to-day basis, it may be difficult to appreciate the differences between emergency response and ambulance transport. Unless one actually has used the EMS system in a medical emergency, he or she might be likely to define a call to 9-1-1 in a medical emergency as 'needing an ambulance.' However, with the recent advances in resuscitative medical care, particularly in cardiac emergencies, we now know that what occurs in the first few minutes after onset of the medical emergency will change the long term outcome. In many of these critical circumstances, what happens on-scene determines whether the patient lives or dies. Therefore, rapid, efficient and effective delivery of emergency response and care is dependent on immediately sending nearby

11

trained personnel to the scene of an emergency regardless of the vehicle or mode of transportation.

Ambulances, of course, are necessary to transport patients to a hospital where more definitive care may be needed. However, because ambulances are often busy evacuating, transporting and turning over patients at the hospital, the most reliable vehicle to ensure a rapid response generally is the neighborhood fire truck. It should be realized that the first emergency care provider who is responsible for competent care may arrive on a fire truck separate from an ambulance. This is the case in most communities in America.

There are sub-specialties of ambulance service in the out-of-hospital arena that must not be confused with 9-1-1 emergency response. For example, ambulance services are often employed for interfacility transfers for specialty care or the need to transfer patients from one hospital to another can provide a higher level of required care. These transfers may include critical care transfers between hospitals, but more often they may also be non-emergent interfacility transports or day transport for persons with home-delivered chronic care services. Such services typically are not performed by fire departments as a fundamental public policy device to better ensure dedicated 9-1-1 emergency services and thus provide security and prosperity for the community served.

Multi-Role Firefighters: Patient Safety from Multiple Perspectives

To further emphasize that the prehospital 9-1-1 emergency care patient should be considered a separate and distinct type of patient in the continuum of health care, consider the setting and the circumstances of emergency medical care delivery. These patients not only have medical needs, but they also need simultaneous physical rescue, protection from the elements and the creation of a safe physical environment as well as management of non-medical surrounding sociologic concerns. The fire service is uniquely equipped to simultaneously address all of these needs.

The mission of the fire service is to protect and save lives and property. There are no other conflicting agendas. The fire service-based prehospital, 9-1-1 emergency response

12

medical care system is designed to be part of society's safety net. Fire and prehospital 9-1-1 emergency response medical care are intimately intertwined. Separating them from the EMS focus only serves to polarize our country's already fragmented emergency response system.

All out-of-hospital emergency care and ambulance transport professionals are taught that scene safety is the primary objective at every emergency scene. However, many of today's non-fire service-based EMS professionals do not have the additional resources and often do not have the training to effectively secure a scene. When there is a strict medical orientation in their professional training and practice, adequate preparation to appropriately and safely provide emergency medical care to an emergency patient may be compromised. Scene safety issues are often not apparent until a crew is on-scene to assess the incident.

Decision makers should consider, 'What does a non-fire based EMS crew do on the scene of a motor vehicle accident when the car is engulfed in flames and occupants are trapped inside, and fire crews were not dispatched?' In many cases, a non-fire service-based EMS provider would need to request dispatch of a fire company after the initial scene size-up, further delaying care, and further increasing risk to rescuers and victims. Streamlining this approach into the fire service-based prehospital 9-1-1 emergency medical care system is quite arguably more effective from the perspective of scene safety, short response time, integrated rescue and treatment, and then transport to a medical facility. Regardless, the firefighter response is a key element of patient safety, both medically and environmentally.

In the era of homeland security threats and the spiraling growth of the commercial transport industry, the threat of hazardous materials (Haz-Mat) is center-stage. Again, fire service Haz-Mat teams are the front-line of protection and rapid delivery of medical care can be pre-empted by such chem-bio threats, but where rapid care can be given, it can be expedited directly by cross-trained fire-service Haz-Mat care providers.

13

Fire Service-Based EMS as the Health Care System Safety Net

Prehospital 9-1-1 emergency patient medical care is a major part of the safety net for the American healthcare system. They may be the provider of last resort for the needy, yet they can be one more mechanism for overloading the health care system. Nevertheless, to its credit, the fire service-based, prehospital 9-1-1 emergency patient medical care provides unconditional service to all members of our population. Therefore, the fire service must now become an integral part of the public health system and work closely with medical and public health experts to help alleviate unnecessary burdens on already overburdened hospital, medical and public health systems. Already part of local government, the fire service may be best positioned to sit at the table and help provide important data to facilitate creating solutions to pressing health care public policy issues.

Above all, rapid response times are a pivotal advantage of fire service-based, prehospital 9-1-1 emergency EMS systems. Now equipped with automated defibrillators to reverse sudden cardiac arrest, the fire truck, coupled with bystander CPR, has become one of the greatest life-saving tools in medical history. With stroke centers to treat stroke within the golden 3 hour window, cardiac catheterization centers to treat heart attack in the 90 minute door-to-balloon time, and trauma centers to treat hemorrhaging patients, time efficiency is a key component of the best designed EMS systems. The service most capable of rapid multi-faceted response, rapid identification and triage to the appropriate facility is a fire service-based EMS system.

EMS is Not an Ambulance Ride

One of the central themes of this discussion is concern over the common misconception that EMS begins with the transport of a patient in an ambulance to a hospital. This misunderstanding resulted essentially in funding of transport service providers but not providers of emergency medical care rendered at the scene. This funding aberrancy occurred in the 1960s as Medicare provided reimbursement for transportation of trauma patients to the hospital, long before the contemporary EMS system developed. About the same time, fire service delivery of 9-1-1 emergency medical care was becoming part of the fabric of the fire service. It was managed and funded as an integral component of

14

public safety service provided by a fire department. Thus, it was funded solely as part of the fire department budget.

Payment for transportation does not fairly portray the full picture of 9-1-1 emergency response and medical care. As the need to pay for EMS was realized, federal dollars for "emergency medical services" went to the perceived greatest area of need at that time, the need for transportation. These federal dollars even provided payment of non-emergency ambulance transport for the care of chronic medical problems. Even though much of the life-saving effect of EMS in today's circumstances will play out routinely on the scene long before ambulance arrival, the focus on transport and not medical care delivery remains. This distinction has been lost and, to this date, never totally reconciled. Especially considering the resource impact, educating the public and government officials about this distinction within the EMS system in the U.S. is a critical and timely issue in the era of homeland security and Haz-Mat threats.

Funding for Prehospital EMS

The fire service supports the recent Institute of Medicine recommendations for ensuring federal payment for emergency medical care not associated with transport. Although not labeled specifically for EMS activities, grant funds are received by fire departments and emergency management agencies to enhance EMS response capabilities throughout the United States. It is deceptive to imply that only funds awarded to single function EMS delivery agencies are the only dollars benefiting those receiving prehospital 9-1-1 emergency medical care services.

For example, Assistance to Firefighter Grants (AFG) are essential to ensuring that fire departments have the baseline response capability that prepares them to respond not only to local incidents but also to effectively participate in broader, national responses. Fire department 'response' is considered 'all-hazards', inclusive of emergency, prehospital 9-1-1 medical care services. The program is extraordinarily cost-effective, with low administrative overhead and direct payments to local fire departments. As almost all fire departments provide EMS at some level, AFG dollars support equipment purchases,

15

training efforts as well as public safety education and injury prevention efforts. In fiscal year 2006 (FY 2006), 4,726 grants were awarded to fire departments throughout the United States totaling $461,092,358.

Another example of federal funding of local emergency response systems is the Staffing for Adequate Fire and Emergency Response (SAFER) Grants. The single most important obligation the federal government should fulfill to enhance local preparedness and protect Americans against all-hazards—natural and man-made—is to assure that every fire department in the nation has sufficient numbers of adequately trained and equipped fire fighter/ EMS responders. In FY 2006, there were 242 SAFER awards totaling $96,151,433 provided to fire departments throughout the United States.

Both AFG and SAFER grants present the federal government with its best opportunity to assure a strong, emergency response component in every community in America.

Federal Oversight and Administration of EMS

EMS has many voices at the federal level including the Department of Health and Human Services, Department of Transportation, Department of Justice, and Department of Homeland Security. Each voice advocates for specific entities that provide EMS as part of its services. Congress appropriately has empowered all EMS-related agencies under the Federal Interagency Committee on Emergency Medical Services (FICEMS). Recently, the FICEMS has been strengthened and provides the mechanism to accomplish this "coordination of the voices." The leadership challenge is to bring all of the voices together. The FICEMS can do this, if given a chance and a mandate.

Conclusion

In terms of the rapid delivery of emergency medical care in the out-of-hospital environment, fire departments have the advantage of having a free-standing army ready to respond anytime and anywhere. Prehospital, 9-1-1 emergency response in support of community prosperity and security is one of the essential public safety functions provided by the United States fire service. Fire service-based EMS systems are strategically

16

positioned to deliver time critical response and effective patient care and scene safety. Fire service-based EMS accomplishes this while emphasizing responder and patient safety, providing competent and compassionate workers, and delivering cost-effective operations.

References

Beck CS. Resuscitation for cardiac standstill and ventricular fibrillation occurring during operation. Am J Surg 53 (4):273-279, 1941.

City of Virginia, Minnesota, VFD History, Emergency Medical Services History, http://www.virginiamn.us/VFD%20History.htm , April 2007

Emergency Medical Services Systems Act of 1973. (P.L. 93-154). 93rd Congress S 2410.

Eugene Nagel, Personal Interview, February 2007

Foster, M. *History of the Maltese Cross, as used by the Order of St John of Jerusalem* http://www2.prestel.co.uk/church/oosj/cross.htm   April 2007.

Henderson Y. *The Prevention and Treatment of Asphyxia in the Newborn.* JAMA 90(8):383-386, 1928.

Maltese Cross, http://en.wikipedia.org/wiki/Maltese_Cross_(symbol) April 2007.

Moore-Merrell, L., IAFF/IAFC Fire Department Operations Survey, March 2007

Pepe PE, Roppolo LP, Cobb LA. Successful systems for out-of-hospital resuscitation. In: Cardiopulmonary Arrest. Ornato JP and Peberdy MA, (eds); Humana Press, Totowa, NJ 2004; pp 649-681.

Williams, D.M., *2006 JEMS 200-City Survey: EMS From All Angles.* 2007, 38-53

17

Firefighter Life Safety Initiatives

The Firefighter Life Safety Summit held in Tampa, Florida, in March 2004, produced 16 major initiatives that will give the fire service a blueprint for making changes.

1. Define and advocate the need for a cultural change within the fire service relating to safety, incorporating leadership, management, supervision, accountability and personal responsibility.

2. Enhance the personal and organizational accountability for health and safety throughout the fire service.

3. Focus greater attention on the integration of risk management with incident management at all levels, including strategic, tactical, and planning responsibilities.

4. Empower all firefighters to stop unsafe practices.

5. Develop and implement national standards for training, qualifications, and certification (including regular recertification) that are equally applicable to all firefighters, based on the duties they are expected to perform.

6. Develop and implement national medical and physical fitness standards that are equally applicable to all firefighters, based on the duties they are expected to perform.

7. Create a national research agenda and data collection system that relate to the initiatives.

8. Utilize available technology wherever it can produce higher levels of health and safety.

9. Thoroughly investigate all firefighter fatalities, injuries, and near misses.

10. Ensure grant programs support the implementation of safe practices and/or mandate safe practices as an eligibility requirement.

11. Develop and champion national standards for emergency response policies and procedures.

12. Develop and champion national protocols for response to violent incidents.

13. Provide firefighters and their families access to counseling and psychological support.

14. Provide public education more resources and champion it as a critical fire and life safety program.

15. Strengthen advocacy for the enforcement of codes and the installation of home fire sprinklers.

16. Make safety a primary consideration in the design of apparatus and equipment.

National Fallen Firefighters Foundation

www.firehero.org

Foam Concentrate Characteristics/Application Techniques

Appendix D **Foam Concentrate Characteristics/Application Techniques**					
Type	**Characteristics**	**Storage Range**	**Application Rate**	**Application Techniques**	**Primary Uses**
Protein Foam (3% and 6%)	• Protein based • Low expansion • Good reignition (burnback) resistance • Excellent water retention • High heat resistance and stability • Performance can be affected by freezing and thawing • Can freeze protect with antifreeze • Not as mobile or fluid on fuel surface as other low-expansion foams	35–120°F (2°C to 49°C)	0.16 gpm/ft^2 (6.5 L/min/m^2)	• Indirect foam stream; do not mix fuel with foam • Avoid agitating fuel during application; static spark ignition of volatile hydrocarbons can result from plunging and turbulence • Use alcohol-resistant type within seconds of proportioning • Not compatible with dry chemical extinguishing agents	• Class B fires involving hydrocarbons • Protecting flammable and combustible liquids where they are stored, transported, and processed
Fluoroprotein Foam (3% and 6%)	• Protein and synthetic based; derived from protein foam • Fuel shedding • Long-term vapor suppression • Good water retention • Excellent, long-lasting heat resistance • Performance not affected by freezing and thawing • Maintains low viscosity at low temperatures • Can freeze protect with antifreeze • Use either freshwater or saltwater • Nontoxic and biodegradable after dilution • Good mobility and fluidity on fuel surface • Premixable for short periods of time	35–120°F (2°C to 49°C)	0.16 gpm/ft^2 (6.5 L/min/m^2)	• Direct plunge technique • Subsurface injection • Compatible with simultaneous application of dry chemical extinguishing agents • Deliver through air-aspirating equipment	• Hydrocarbon vapor suppression • Subsurface application to hydrocarbon fuel storage tanks • Extinguishing in-depth crude petroleum or other hydrocarbon fuel fires

Continued on next page.

Type	Characteristics	Storage Range	Application Rate	Application Techniques	Primary Uses
Film Forming Fluoroprotein Foam (FFFP) (3% and 6%)	• Protein based; fortified with additional surfactants that reduce the burnback characteristics of other protein-based foams • Fuel shedding • Develops a fast-healing, continuous-floating film on hydrocarbon fuel surfaces • Excellent, long-lasting heat resistance • Good low-temperature viscosity • Fast fire knockdown • Affected by freezing and thawing • Use either freshwater or saltwater • Can store premixed • Can freeze protect with antifreeze • Use alcohol-resistant type on polar solvents at 6% solution and on hydrocarbon fuels at 3% solution • Nontoxic and biodegradable after dilution	35–120°F (2°C to 49°C)	**Ignited Hydrocarbon Fuel:** 0.10 gpm/ft² (4.1 L/min/m²) **Polar Solvent Fuel:** 0.24 gpm/ft² (9.8 L/min/m²)	• Cover entire fuel surface • May apply with dry chemical agents • May apply with spray nozzles • Subsurface injection • Can plunge into fuel during application	• Suppressing vapors in unignited spills of hazardous liquids • Extinguishing fires in hydrocarbon fuels
Aqueous Film Forming Foam (AFFF) (1%, 3%, and 6%)	• Synthetic based • Good penetrating capabilities • Spreads vapor-sealing film over and floats on hydrocarbon fuels • Can use nonaerating nozzles • Performance may be adversely affected by freezing and storing • Has good low-temperature viscosity • Can freeze protect with antifreeze • Use either freshwater or saltwater • Can premix	25–120°F (-4°F to 49°C)	0.10 gpm/ft² (4.1 L/min/m²)	• May apply directly onto fuel surface • May apply indirectly by bouncing it off a wall and allowing it to float onto fuel surface • Subsurface injection • May apply with dry chemical agents	• Controlling and extinguishing Class B fires • Handling land or sea crash rescues involving spills • Extinguishing most transportation-related fires • Wetting and penetrating Class A fuels • Securing unignited hydrocarbon spills

Continued on next page.

Type	Characteristics	Storage Range	Application Rate	Application Techniques	Primary Uses
Alcohol-Resistant AFFF (3% and 6%)	• Polymer has been added to AFFF concentrate • Multipurpose: Use on both polar solvents and hydrocarbon fuels (use on polar solvents at 6% solution and on hydrocarbon fuels at 3% solution) • Forms a membrane on polar solvent fuels that prevents destruction of the foam blanket • Forms same aqueous film on hydrocarbon fuels as AFFF • Fast flame knockdown • Good burnback resistance on both fuels • Not easily premixed	25–120°F (-4°C to 49°C) (May become viscous at temperatures under 50°F [10°C])	**Ignited Hydrocarbon Fuel:** 0.10 gpm/ft² (4.1 L/min/m²) **Polar Solvent Fuel:** 0.24 gpm/ft² (9.8 L/min/m²)	• Apply directly but gently onto fuel surface • May apply indirectly by bouncing it off a wall and allowing it to float onto fuel surface • Subsurface injection	Fires or spills of both hydrocarbon and polar solvent fuels
High-Expansion Foam	• Synthetic detergent based • Special-purpose, low water content • High air-to-solution ratios: 200:1 to 1,000:1 • Performance not affected by freezing and thawing • Poor heat resistance • Prolonged contact with galvanized or raw steel may attack these surfaces	27–110°F (-3°C to 43°C)	Sufficient to quickly cover the fuel or fill the space	• Gentle application; do not mix foam with fuel • Cover entire fuel surface • Usually fills entire space in confined space incidents	• Extinguishing Class A and some Class B fires • Flooding confined spaces • Volumetrically displacing vapor, heat, and smoke • Reducing vaporization from liquefied natural gas spills • Extinguishing pesticide fires • Suppressing fuming acid vapors • Suppressing vapors in coal mines and other subterranean spaces and concealed spaces in basements • Extinguishing agent in fixed extinguishing systems • Not recommended for outdoor use

Continued on next page.

Type	Characteristics	Storage Range	Application Rate	Application Techniques	Primary Uses
Class A Foam	• Synthetic • Wetting agent that reduces surface tension of water and allows it to soak into combustible materials • Rapid extinguishment with less water use than other foams • Use regular water stream equipment • Can premix with water • Mildly corrosive • Requires lower percentage of concentration (0.2 to 1.0) than other foams • Outstanding insulating qualities • Good penetrating capabilities	25–120°F (-4°C to 49°C) (Concentrate is subject to freezing but can be thawed and used if freezing occurs)	Same as the minimum critical flow rate for plain water on similar Class A Fuels; flow rates are not reduced when using Class A foam	• Can propel with compressed-air systems • Can apply with conventional nozzles	Extinguishing Class A combustibles only

Glossary

A

Absorption — (1) Penetration of one substance into the structure of another such as the process of picking up a liquid contaminant with an absorbent. Also see Contaminant and Absorbent. (2) Passage of materials (such as toxins) through some body surface into body fluids and tissue. *Also see* Routes of Entry and Toxin.

Acceptance Testing (Proof Test) — Preservice tests on fire apparatus or equipment performed at the factory or after delivery to assure the purchaser that the apparatus or equipment meets bid specifications.

Accordion Load — Arrangement of fire hose in a hose bed or compartment in which the hose lies on edge with the folds adjacent to each other.

Acetylene [c₂h₂] — Colorless gas that has an explosive range from 2.5 percent to 81 percent in air; used as a fuel gas for cutting and welding operations.

Acquired Building (Structure) — Structure acquired by the authority having jurisdiction from a property owner for the purpose of conducting live fire training or rescue training evolutions.

Activation Energy — Amount of energy that must be added to an atomic or molecular system to begin a reaction.

Acute — (1) Characterized by sharpness or severity; having rapid onset and a relatively short duration. (2) Single exposure (dose) or several repeated exposures to a substance within a short time period. *Also see* Chronic.

Adapter — Fitting for connecting hose couplings with dissimilar threads but with the same inside diameter. *Also see* Fitting, Increaser, and Reducer.

Aeration — Introduction of air into a foam solution to create bubbles that result in finished foam.

Aerial Device — General term used to describe the hydraulic aerial ladder, elevating platform, or telescoping or articulating booms attached to an automotive fire apparatus that have the means to reach at least 65 feet (20 m) above grade, with many being capable of reaching more than 100 feet (30 m).

Aerial Fuels — Standing and supported live and dead combustibles not in direct contact with the ground and consisting mainly of foliage, twigs, branches, stems, cones, bark, and vines.

Aerial Ladder — A rotating, power-operated (usually hydraulically) ladder mounted on a self-propelled automotive fire apparatus.

Aerial Ladder Platform — A power-operated (usually hydraulically) aerial device which combines an aerial ladder with a personnel carrying platform supported at the end of the ladder.

A-Frame Collapse — Pattern of collapse that occurs when the floor and/or roof assemblies on both sides of a center wall collapse into what might be seen as opposing lean-to collapses. This pattern offers a good chance of habitable void spaces on both sides of the center wall.

Agency for Toxic Substances and Disease Registry (ATSDR) — Lead U.S. public health agency responsible for implementing the health-related provisions of the Comprehensive Environmental Response, Compensation and Liability Act (CERCLA), and charged with assessing health hazards at specific hazardous waste sites, helping to prevent or reduce exposure and the illnesses that result, and increasing knowledge and understanding of the health effects that may result from exposure to hazardous substances.

Air-Aspirating Foam Nozzle — Foam nozzle especially designed to provide the aeration required to make the highest quality foam possible; most effective appliance for the generation of low-expansion foam.

Air Lifting Bag — Inflatable, envelope-type device that can be placed between the ground and an object and then inflated to lift the object. Can also be used to separate objects. Depending on size, may have lifting capabilities in excess of 75 tons (68 040 kg). Also called pneumatic lifting bag.

Alarm Assignment — Predetermined number of fire units assigned to respond to an emergency.

All Clear — Signal given to the Incident Commander that a specific area has been checked for victims and none have been found or all found victims have been extricated from an entrapment.

Allergen — Material that can cause an allergic reaction of the skin or respiratory system.

Alloy — Substance or mixture composed of two or more metals (or a metal and nonmetallic elements) fused and dissolved into each other in order to enhance the properties or usefulness of the base metal.

Alpha Radiation — Consists of particles having a large mass and positive electrical charge; least penetrating of the three common forms of radiation. It is normally not considered dangerous to plants, animals, or people unless it gets into the body. *Also see* Beta Radiation, Gamma Radiation and Radiation (2).

Ambient Temperature — Temperature of the surrounding environment.

Americans with Disabilities Act (ADA) of 1990 - Public Law 101-336 — A U.S. federal statute intended to remove barriers — physical and otherwise — that limit access by individuals with disabilities.

Aqueous Film Forming Foam (AFFF) — Synthetic foam concentrate that, when combined with water, can form a complete vapor barrier over fuel spills and fires and is a highly effective extinguishing and blanketing agent on hydrocarbon fuels. Also called light water.

Arc — A luminous discharge of electricity across a gap. Arcs produce very high temperature.

Arched Roof — Any of several different types of roofs of which all are curved or arch shaped, resembling the top half of a horizontal cylinder. Typical applications are on supermarkets, auditoriums, bowling centers, sports arenas, and aircraft hangars.

Around-the-Pump Proportioner — Apparatus-mounted foam proportioner; small quantity of water is diverted from the apparatus pump through an inline proportioner where it picks up the foam concentrate and carries it to the intake side of the pump; most common apparatus-mounted foam proportioner in service.

Arson — Crime of willfully, maliciously, and intentionally starting a fire (firesetting) or causing an explosion to destroy property. Precise legal definitions vary among jurisdictions, wherein it is defined by statutes and judicial decisions.

Arterial Bleeding – Bleeding from an artery that is characterized by bright red blood and as rapid, profuse, and difficult to control.

Ascender — Mechanical contrivance used when climbing rope that allows upward movement but not downward movement.

Aspect — (1) Position facing a particular direction; exposure. (2) Compass direction toward which a slope faces.

Asphyxiation — Condition that causes death because of a deficient amount of oxygen and an excessive amount of carbon monoxide and/or other gases in the blood.

A-Tool — A prying tool with a sharp notch with cutting edges machined into it. The notch resembles the letter A. This tool is designed to cut behind the protective collar of a lock cylinder and maintain a hold so that the lock cylinder can be pried out.

ATSDR — Abbreviation for Agency for Toxic Substances and Disease Registry.

Autoignition — Ignition that occurs when a substance in air, whether solid, liquid, or gaseous, is heated sufficiently to initiate or cause self-sustained combustion without an external ignition source.

Autoignition Temperature — Same as ignition temperature except that no external ignition source is required for ignition because the material itself has been heated to ignition temperature. The temperature at which autoignition occurs through the spontaneous ignition of the gases or vapor given off by a heated material.

Automatic Aid — Written agreement between two or more agencies to automatically dispatch predetermined resources to any fire or other emergency reported in the geographic area covered by the agreement. These areas are generally where the boundaries between jurisdictions meet or where jurisdictional "islands" exist.

Automatic Alarm — (1) Alarm actuated by heat, gas, smoke, flame-sensing devices, or waterflow in a sprinkler system conveyed to local alarm bells or the fire station. (2) Alarm boxes that automatically transmit a coded signal to the fire station to give the location of the alarm box.

Automatic-Closing Door — Self-closing door normally held in the open position by an automatic releasing device such as a magnetic hold-open device. When the door is released by the hold-open device, it closes.

Automatic Location Identification (ALI) — Enhanced 9-1-1 feature that displays the address of the party calling 9-1-1 on a screen for use by the public safety telecommunicator. Usually used in tandem with automatic number identification (ANI) services. This feature is also used to route calls to the appropriate public safety answering point (PSAP) and can even store information in its database regarding the appropriate emergency services (police, fire, and medical) that respond to that address.

Automatic Nozzle — Fog stream nozzle that automatically corrects itself to provide a good stream at the proper nozzle pressure.

Automatic Number Identification (ANI) — Enhanced 9-1-1 feature that displays the phone number of the party calling 9-1-1 on a screen for use by the public safety telecommunicator. Usually used in tandem with automatic location identification (ALI) services.

Automatic Sprinkler System — System of water pipes, discharge nozzles, and control valves designed to activate during fires by automatically discharging enough water to control or extinguish a fire. Also called Sprinkler System.

Automatic Vehicle Locator (AVL) — System that uses global positioning satellites to determine the exact location of units in the field. This information is relayed to public safety communications centers to help determine the closest unit to send during an emergency.

Auxiliary Alarm System — System that connects the protected property with the fire department alarm communications center by a municipal master fire alarm box or over a dedicated telephone line.

Awareness Level — Lowest level of training established by the National Fire Protection Association for first responders at hazardous materials incidents.

Awning Window — Window consisting of large sections of glass, usually double-strength, set in a metal or wood frame and are hinged along the top rail while the bottom rail swings out.

Axe — Forcible entry tool having a pick or flat head and a blade attached to a wood or fiberglass handle. Also called firefighter's axe.

B

Backdraft —Very rapid, often explosive burning of hot gases that occurs when oxygen is introduced into an oxygen-depleted confined space. It may occur because of inadequate or improper ventilation procedures.

Balanced Pressure Proportioner — A foam concentrate proportioner that operates in tandem with a fire water pump to ensure a proper foam concentrate-to-water mixture.

Bale – Valve handle.

Bale Hook/Baling Hook — Tool used for moving bales or boxed goods and for moving and overhauling stuffed furniture or mattresses. Also called Hay Hook.

Ball Valve — Valve having a ball-shaped internal component with a hole through its center that permits water to flow through when aligned with the waterway.

Bam-Bam Tool — Tool used to disassemble padlocks and trip its lock mechanism.

Bank-Down Application Method (Deflect) — Method of foam application that may be employed on an un-ignited or ignited Class B fuel spill. The foam stream is directed at a vertical surface or object that is next to or within the spill area. The foam deflects off the surface or object and flows down onto the surface of the spill to form a foam blanket.

Bar-Screw Jack — Jack used to hold loads under compression. It is commonly used in shoring work or other similar evolutions.

Batch-Mixing — Production of foam solution by adding an appropriate amount of foam concentrate to a water tank before application. The resulting solution must be used or discarded following the incident.

Battering Ram — Large metal pipe with handles and a blunt end used to break down doors or create holes in walls.

Beam — (1) Structural member subjected to loads, usually vertical, perpendicular to its length. (2) Main structural member of a ladder supporting the rungs or rung blocks. Also called Side Rail or Rail.

Becket Bend — Knot used for joining two ropes. It is particularly well suited for joining ropes of unequal diameters or joining a rope and a chain. Also called Sheet Bend.

Bed Section — Bottom section of an extension ladder. Also called Base Section.

Belay — Climber's term for a safety line.

Beta Radiation — Type of radiation that can cause skin burns. *Also see* Alpha Radiation, Gamma Radiation, and Radiation (2).

Bight — Element of a knot formed by simply bending the rope back on itself (creating a loop) while keeping the sides parallel.

Bill of Lading — Shipping paper used by the trucking industry (and others) indicating origin, destination, route, and product; placed in the cab of every truck tractor. This document establishes the terms of a contract between shippers and transportation companies; serves as a document of title, contract of carriage, and receipt for goods.

Bimetallic — Strip or disk composed of two different metals that are bonded together; used in heat detection equipment.

Biological Toxin — Poison produced by living organisms.

Black — Area already burned by a wildland fire. Also called Burn.

Blast Pressure Wave — Shock wave created by rapidly expanding gases in an explosion.

BLEVE — Acronym for Boiling Liquid Expanding Vapor Explosion.

Blitz Attack — To aggressively attack a fire from the exterior with a large diameter (2½-inch [65 mm] or larger) fire stream.

Block and Tackle — Series of pulleys (sheaves) contained within a wood or metal frame. They are used with rope to provide a mechanical advantage for pulling operations.

Bloodborne Pathogens — Pathogenic microorganisms that are present in the human blood and can cause disease in humans. These pathogens include (but are not limited to) hepatitis B virus (HBV) and human immunodeficiency virus (HIV).

Body Substance Isolation (BSI) — A form of infection control based on the presumption that all body fluids are infectious.

Boiling Liquid Expanding Vapor Explosion (BLEVE) — Rapid vaporization of a liquid stored under pressure upon release to the atmosphere following major failure of its containing vessel. The failure of the containing vessel is the result of over-pressurization caused by an external heat source causing the vessel to explode into two or more pieces when the temperature of the liquid is well above its boiling point at normal atmospheric pressure.

Boiling Point — Temperature of a substance when the vapor pressure exceeds atmospheric pressure. At this temperature, the rate of evaporation exceeds the rate of condensation. At this point, more liquid is turning into gas than gas is turning back into a liquid.

Bolt Cutters — Cutting tool designed to make a precise, controlled cut; used for cutting wire, fencing, bolts, and small steel bars.

Booster Hose or Booster Line — Noncollapsible rubber-covered, rubber-lined hose usually wound on a reel and used for the initial attack and extinguishment of incipient and smoldering fires. This hose is most commonly found in 1-inch (13 mm, 19 mm, and 25 mm) diameters.

Booster Reel — Mounted reel on which booster hose is carried.

Bored Lock — Lock installed within right angle holes bored in a door. Also called Cylindrical Lock.

Bowline Knot — Knot used to form a loop in rope.

Bowstring Truss — Lightweight truss design noted by the bow shape, or curve, of the top chord.

Brachial Artery – The major artery of the upper arm.

Braided Rope — Rope constructed by uniformly intertwining strands of rope (similar to braiding a person's hair).

Braid-on-Braid Rope — Rope constructed with both a braided core and a braided sheath. The appearance of the sheath is that of a herringbone pattern.

British Thermal Unit (Btu) — Amount of heat energy required to raise the temperature of one pound of water one degree Fahrenheit. One Btu = 1.055 kilo joules (kJ).

Broken Stream — Stream of water that has been broken into coarsely divided drops.

Burglar Block — A metal rod or similar device that sometimes is used to bar or block patio sliding doors. This feature can easily be seen from the outside, and it virtually eliminates any possibility of forcing the door without breaking the glass.

Burn Building — Training structure specially designed to contain live fires for the purpose of fire suppression training.

Burning Bars — Ultra-high temperature cutting devices capable of cutting through virtually any metallic, nonmetallic, or composite material. Also called exothermic cutting rods.

Butterfly Roof — V-shaped roof style resembling two opposing shed roofs joined along their lower edges.

Butterfly Valve — Type of control valve that uses a flat circular plate in the pipe which rotates ninety degrees across the cross section of the pipe to control flow.

Butt Spurs — Metal safety plates or spikes attached to the butt end of ground ladder beams.

C

Cantilever Collapse — Pattern of collapse that occurs when one or more walls of a multistory building collapse, leaving the floors attached to and supported by the remaining walls. This pattern also offers a good chance of habitable voids being formed above and below the supported ends of the floors. This collapse pattern is perhaps the least stable of all the patterns and is the most vulnerable to secondary or subsequent collapse.

Cantilever Walls – Walls that extend beyond the structure that supports them.

Capacity Stencil — Number stenciled on the exterior of tank cars to indicate the volume of the tank.

Capillary Bleeding — Bleeding from capillaries, which is characterized by a slow, oozing flow of blood.

Carabiner — A steel or aluminum D-shaped snap-link device for attaching components of rope rescue systems. In rescue work, carabiners should be of a positive locking type, with a 5,000-pound minimum breaking strength. They are also called biners, crabs, or snap links.

Carbon Dioxide (CO$_2$) — Colorless, odorless, heavier-than-air gas that neither supports combustion nor burns. CO$_2$ is used in portable fire extinguishers as an extinguishing agent to extinguish Class B or C fires by smothering or displacing the oxygen. CO$_2$ is a waste product of aerobic metabolism.

Carbon Monoxide (CO) — Colorless, odorless, dangerous gas (both toxic and flammable) formed by the incomplete combustion of carbon. It combines more than 200 times as quickly with hemoglobin as oxygen, thus decreases the blood's ability to carry oxygen.

Carbon Monoxide Poisoning — Sometimes lethal condition in which carbon monoxide molecules attach to hemoglobin, decreasing the blood's ability to carry oxygen.

Carboxyhemoglobin (COHB) — Hemoglobin saturated with carbon monoxide and therefore unable to absorb needed oxygen.

Carcinogen — Cancer-producing substance.

Cardiac Arrest – The sudden, abrupt loss of heart function.

Cardiopulmonary Resuscitation (CPR) — Application of rescue breathing and external cardiac compression used on patients in cardiac arrest to provide an adequate circulation and oxygen to support life.

Carryall — Waterproof carrier or bag used to carry and catch debris or used as a water sump basin for immersing small burning objects.

Casement Window — Window hinged along one side, usually designed to swing outward, with the screen on the inside. Also called Crank-Out Window.

Catchall — Retaining basin, usually made from salvage covers, to impound water dripping from above.

CBRNE — Abbreviation for Chemical, Biological, Radiological, Nuclear, and Explosive.

Celsius Scale — Temperature scale on which the freezing point is 0 degrees and the boiling point at sea level is 100 degrees. Also known as Centigrade scale.

Central Station Alarm System — System that functions through a constantly attended location (central station) operated by an alarm company. Alarm signals from the protected property are received in the central station and are then retransmitted by trained personnel to the fire department alarm communications center.

Chafing Block — Block placed under hoselines to protect the hose covering from damage due to rubbing against the ground or concrete.

Chain of Command — (1) Order of rank and authority in the fire service. (2) The proper sequence of information and command flow as described in the Incident Management System.

Chain of Custody — Continuous changes of possession of physical "evidence" that must be established in court to admit such material into evidence.

Chain of Survival – The four critical elements that impact the survival of cardiac arrest patients: early access, early CPR, early defibrillation, and early advanced care.

Chain Saw — Cutting tool designed for cutting wood; most commonly used as a ventilation tool; useful during natural disasters.

Charged Line — Hose loaded with water under pressure and prepared for use.

Cheater — A piece of pipe slipped over the handle of a prying tool to lengthen the handle, thus providing additional leverage.

Checkrail Window — Type of window usually consisting of two sashes, known as the upper and lower sashes, that meet in the center of the window.

Chemical Degradation — Process that occurs when the characteristics of a material are altered through contact with chemical substances.

Chemical Flame Inhibition — The extinguishment of a fire by interruption of the chemical chain reaction.

Chemical Heat Energy — Heat produced from a chemical reaction including combustion, spontaneous heating, heat of decomposition, and heat of solution; sometimes occurs as a result of a material being improperly used or stored. Some materials may simply come in contact with each other and react, or they may decompose and generate heat.

Chemical Protective Clothing (CPC) — Clothing designed to shield or isolate individuals from the chemical, physical, and biological hazards that may be encountered during operations involving hazardous materials. *Also see* Personal Protective Equipment (PPE).

Chemical Warfare Agent — Chemical substance that is intended for use in warfare or terrorist activities to kill, seriously injure, or seriously incapacitate people through its physiological effects.

Chimney Effect — Created when a ventilation opening is made in the upper portion of a building and air currents throughout the building are drawn in the direction of the opening; also occurs in wildland fires when the fire advances up a V-shaped drainage swale.

Chock — Wooden, plastic, or metal block constructed to fit the curvature of a tire; placed against the tire to prevent apparatus rolling.

Choking Agent — Chemical warfare agent that attacks the lungs causing tissue damage.

Chronic — Marked by long duration; recurring over a period of time. *Also see* Acute.

Churning — (1) Movement of smoke being blown out of a ventilation opening only to be drawn back inside by the negative pressure created by the ejector because the open area around the ejector has not been sealed. Also called Recirculation.

Chute — Salvage cover arrangement that channels excess water from a building. A modified version can be made with larger sizes of fire hose.

Circular Saw — Originally designed for use in construction; especially useful in situations where electrical power is readily available and heavier and bulkier power saws are too difficult to handle.

Circulating Feed — Fire hydrant that receives water from two or more directions.

Citizens Band (CB) Radio — Low-power radio transceiver that operates on frequencies authorized by the Federal Communications Commission (FCC) for public use with no license requirement.

Clapper Valve — Hinged valve that permits the flow of water in one direction only.

Class A Fire — Fire involving ordinary combustibles such as wood, paper, cloth, and similar materials.

Class A Foam — Foam specially designed for use on Class A combustibles; increasingly popular for use in wildland and structural fire fighting. Designed to reduce the surface tension of water and allow it to soak into combustible materials more easily than plain water. *Also see* Foam Concentrate.

Class B Fire — Fire of flammable and combustible liquids and gases such as gasoline, kerosene, and propane.

Class B Foam Concentrate — Foam fire-suppression agent designed for use on unignited or ignited Class B flammable or combustible liquids.

Class C Fire — Fire involving energized electrical equipment.

Class D Fire — Fire involving combustible metals such as magnesium, sodium, and titanium.

Class K Fire — Fire in cooking appliances that involves combustible cooking media (vegetable or animal oils and fats). Commonly occurring in commercial cooking facilities such as restaurants and institutional kitchens.

Class I Harness — Harness that fastens around the waist and around the thighs or under the buttocks and is intended to be used for emergency escape with a load of up to 300 pounds (1.33 k/N). Also known as a seat harness.

Class II Harness — Harness that fastens around the waist and around the thighs or under the buttocks and is intended to be used for emergency escape with a load of up to 600 pounds (2.67 k/N). Class II harness looks exactly like Class I harness so the attached label must be used to verify its rating.

Class III Harness — Harness that fastens around the waist, around the thighs or under the buttocks, and over the shoulders. Class III harness is rated for loads of up to 600 pounds (2.67 k/N). Also known as full body harness.

Clear Text — Use of plain English, including certain standard words and phrases, in radio communications transmissions.

Closed-Circuit Breathing Apparatus — Respiratory protection system in which the exhalations of the wearer is rebreathed after carbon dioxide has been effectively removed and a suitable oxygen concentration restored from resources composed of compressed oxygen, chemical oxygen, or liquid oxygen; usually long-duration device systems; not approved for fire fighting operations.

Clove Hitch — Knot that consists essentially of two half hitches. Its principal use is to attach a rope to an object such as a pole, post, or hose.

CNG — Abbreviation for Compressed Natural Gas.

Codes — Body of laws arranged systematically that usually pertain to one subject area such as a mechanical code, a building code, an electrical code, or a fire code.

Collapse Zone – The area extending horizontally from the base of the wall to one and one-half times the height of the wall.

Combination Attack — Method of battling a fire by using both a direct and an indirect attack. This method combines the steam-generating technique of a ceiling level attack with an attack on the burning materials near floor level.

Combination Detector — Alarm-initiating device capable of detecting an abnormal condition by more than one means. The most common combination detector is the fixed-temperature/rate-of-rise heat detector.

Combination Ladder — Ladder that can be used as either a single, extension, or A-frame ladder.

Combination Lay — Hose lay in which two or more hoselines are laid in either direction—water source to fire or fire to water source.

Combination Smoke Detector — Smoke-sensing device consisting of both photoelectric and ionization smoke detectors.

Combination Spreader/Shears — Powered hydraulic tool consisting of two arms equipped with spreader tips that can be used for pulling or pushing. The insides of the arms contain cutting shears.

Combination System — Water supply system that is a combination of both gravity and direct pumping systems. It is the most common type of municipal water supply system.

Combustible Liquid — Liquid having a flash point at or above 100°F (37.8°C) and below 200°F (93.3°C). *Also see* Flammable Liquid.

Combustion — An exothermic chemical reaction that is a self-sustaining process of rapid oxidation of a fuel, that produces heat and light.

Come-Along — Manually operated pulling tool that uses a ratchet/pulley arrangement to provide a mechanical advantage.

Common Fire Hazard — Condition prevalent in almost all occupancies that increases the likelihood of a fire starting.

Communicable Disease — Disease that is transmissible from one person to another.

Communications/Dispatch Center — Telecommunications facility (a building or portion of a building) specifically configured for the primary purpose of providing emergency communication services or public safety answering point (PSAP) services to one or more public safety agencies under the authority or authorities having jurisdiction.

Compressed Air Foam System (CAFS) — Generic term used to describe a high-energy foam-generation system consisting of an air compressor (or other air source), a water pump, and foam solution that injects air into the foam solution before it enters a hoseline.

Compressed Natural Gas (CNG) — Natural gas that is stored in a vessel at pressures of 2,400 to 3,600 psi (16 800 kPa to 25 200 kPa).

Compression — Vertical and/or horizontal force that tends to push the mass of a material together. For example, the force exerted on the top chord of a truss.

Concentration — Quantity of a chemical material inhaled for purposes of measuring toxicity.

Conduction — Physical flow or transfer of heat energy from one body to another through direct contact or an intervening medium from the point where the heat is produced to another location or from a region of high temperature to a region of low temperature. *Also see* Convection, Radiation (1), and Law of Heat Flow.

Conductivity — The ability of a substance to conduct an electrical current.

Confinement — (1) Process of controlling the flow of a spill and capturing it at some specified location. (2) Operations required to prevent fire from extending from the area of origin to uninvolved areas or structures. *Also see* Containment.

Consensus Standard — Rules, principles, or measures that are established through agreement of members of the standards-setting organization.

Containment — Act of stopping the further release of a material from its container. *Also see* Confinement.

Contamination — The introduction of dangerous chemicals, diseases, or infectious materials.

Convection — Transfer of heat by the movement of heated fluids or gases, usually in an upward direction.

Convulsant — Poison that causes an exposed individual to have convulsions.

Corrosive Materials — Gaseous, liquid, or solid materials that can burn, irritate, or destroy human skin tissue and can severely corrode steel. Also called Corrosives.

CPC — Abbreviation for Chemical Protective Clothing.

CPR — Abbreviation for Cardiopulmonary Resuscitation.

Cribbing — (1) Varying lengths of hardwood, usually 4 × 4 inches (100 mm by 100 mm) or larger, used to stabilize vehicles and collapsed buildings during extrication incidents. (2) Process of arranging planks into a crate-like construction. Also called Shoring.

Critical Incident Stress Debriefing (CISD) — A process in which teams of professional and peer counselors provide emotional and psychological support to EMS personnel who are or have been involved in a critical (highly stressful) event.

Cross Contamination — *See* Secondary Contamination.

Crowd Control — Limiting access to an emergency scene by curious spectators and other nonemergency personnel.

Cryogen — Gas that is cooled to a very low temperature, usually below –150°F (–101°C), to change to a liquid. Also called refrigerated liquid and cryogenic liquid.

Cryogenics — Gases that are converted into liquids by being cooled below –150°F (–101°C).

Customer Service — Quality of an organization's relationship with individuals who have contact with the organization. There are internal customers such as the various levels of personnel and trainees and external customers such as other organizations and the public.

D

Dangerous Good — (1) Any product, substance, or organism included by its nature or by the regulation in any of the nine United Nations classifications of hazardous materials. (2) Term used to describe hazardous materials in Canada. (3) Term used in the U.S. and Canada for hazardous materials on board aircraft. *Also see* Hazardous Material (1).

Dead-End Hydrant — Fire hydrant located on a dead-end main that receives water from only one direction.

Decay — State of fire development when fuel is consumed and energy release diminishes, and temperatures decrease. During this stage the fire goes from ventilation-controlled to fuel controlled.

Decon — Abbreviation for Decontamination.

Decontaminate — To remove a foreign substance that could cause harm; frequently used to describe removal of a hazardous material from the person, clothing, or area.

Decontamination (Decon) — Process of removing a hazardous, foreign substance from a person, clothing, or area.

Defensive Strategy — Overall plan for incident control established by the Incident Commander (IC) that involves protection of exposures as opposed to aggressive, offensive intervention.

Defusing — Informal discussion with incident responders conducted after the incident has been terminated, either at the scene or after the units have returned to quarters. Here commanders address possible chemical and medical exposure information, identify damaged equipment and apparatus that requires immediate attention, identify unsafe operating procedures, assign information gathering responsibilities to prepare for the Post-Incident Analysis and reinforce the positive aspects of the incident.

Deluge Sprinkler System — Fire-suppression system consisting of piping and open sprinklers. A fire detection system is used to activate the water or foam control valve. When the system activates, the extinguishing agent expels from all sprinkler heads in the designated area.

Department of Transportation (DOT) — Administrative body of the executive branch of the state/provincial or federal government responsible for transportation policy, regulation, and enforcement.

Dewatering — Process of removing water from a vessel.

Diatomaceous Earth — A light siliceous material consisting chiefly of the skeletons (minute unicellular algae) and used especially as an absorbent or filter.

Dike — Temporary or permanent barriers that contain or direct the flow of liquids.

Dilution — Application of water to a water-soluble material to reduce the hazard.

Direct Attack (Structural) — Attack method that involves the discharge of water or a foam stream directly onto the burning fuel.

Direct Attack (Wildland) — Operation where action is taken directly on burning fuels by applying an extinguishing agent to the edge of the fire or close to it.

Direct Pumping System — Water supply system supplied directly by a system of pumps rather than elevated storage tanks.

Discipline — Setting the limits or boundaries for expected performance and enforcing them.

Dispatcher — Person who works in the communications center and processes information from the public and emergency responders. Also known as Telecommunicator.

Dissolution — Act or process of dissolving one thing into another; process of dissolving a gas in water.

Distributors — Grid arrangement of smaller mains serving individual fire hydrants and blocks of consumers.

Donut Roll — Length of hose rolled up for storage and transport.

Dose — Quantity of a chemical material ingested or absorbed through skin contact for purposes of measuring toxicity.

Double-Hung Window — Window having two vertically moving sashes.

Draft — (1) Process of acquiring water from a static source and transferring it into a pump that is above the source's level; atmospheric pressure on the water surface forces the water into the pump where a partial vacuum was created. (2) Vertical distance between the water surface and the lowest point of a vessel; depth of water a vessel needs in order to float. Draft varies with the amount of cargo, fuel, and other loads on board.

Drainage Time — The amount of time it takes foam to break down or dissolve; also called drainage rate, drainage dropout rate, or drainage.

Drop Bar — Metal or wooden bar that serves as a locking device when placed or dropped into brackets across a swinging door.

Dry-Barrel Hydrant — Fire hydrant that has its operating valve at the water main rather than in the barrel of the hydrant. When operating properly, there is no water in the barrel of the hydrant when it is not in use. These hydrants are used in areas where freezing could occur.

Dry Chemical — (1) Any one of a number of powdery extinguishing agents used to extinguish fires. The most common include sodium or potassium bicarbonate, monoammonium phosphate, or potassium chloride. (2) Extinguishing system that uses dry chemical powder as the primary extinguishing agent; often used to protect areas containing volatile flammable liquids.

Dry-Pipe Sprinkler System — Fire-suppression system that consists of closed sprinklers attached to a piping system that contains air under pressure. When a sprinkler activates, air is released that activates the water or foam control valve and fills the piping with extinguishing agent. Dry systems are often installed in areas subject to freezing.

Dutchman — Extra fold placed along the length of a section of hose as it is loaded so that its coupling rests in proper position.

Dynamic — The amount of stretch built into a rope. Dynamic ropes have a large amount of stretch, in order to reduce the shock on the climber and anchor systems. They are used for recreational climbing, where long falls may occur.

E

EDITH — Abbreviation for Exit Drills In The Home, a life-safety home education program.

Eductor (Inductor) — (1) Portable proportioning device that injects a liquid, such as foam concentrate, into the water flowing through a hoseline or pipe. (2) Siphon used to remove water from flooded basements. (3) Venturi device that uses water pressure to draw foam concentrate into a water stream for mixing; also enables a pump to draw water from an auxiliary source.

Egress — Place or means of exiting a structure.

Electrical Heat Energy — Heat energy that is electrical in origin including resistance heating, dielectric heating, heat from arcing, and heat from static electricity. Poorly maintained electrical appliances, exposed wiring, and lightning are sources of electrical heat energy.

Electron — Minute component of an atom that possesses a negative charge.

Emergency Decontamination — Removing contamination on individuals in potentially life-threatening situations with or without the formal establishment of a decontamination corridor.

Emergency Medical Services (EMS) — Initial medical evaluation/treatment provided to employees and others who become ill or are injured in the workplace. These services may be provided by an in-house emergency response team or by an outside provider.

Emergency Operations — Activities involved in responding to the scene of an incident and performing assigned duties in order to mitigate the emergency.

Emergency Operations Center (EOC) — Disaster management center for government agencies. Can be located within a communications center or in separate facilities.

Emergency Response Guidebook (ERG) — Guide developed jointly by Transport Canada (TC), U.S. Department of Transportation (DOT), and the Secretariat of Transport and Communications of Mexico (SCT) for use by firefighters, police, and other emergency services personnel who may be the first to arrive at the scene of a transportation incident involving dangerous goods. It is primarily a guide to aid first responders in quickly identifying the specific or generic hazards of the material involved and protecting themselves and the general public during the initial response phase.

Emergency Traffic — Urgent radio traffic; a request for other units to clear the radio waves for an urgent message. Also called Priority Traffic.

Employee Assistance Program (EAP) — Any one of many programs that may be provided by an employer to employees and their families to aid in solving work or personal problems.

Endothermic Heat Reaction — Chemical reaction in which a substance absorbs heat energy.

Energy — the capacity to perform work.

Entry Point (Entry Opening) — Ventilation opening through which replacement air enters the structure. This is usually the same opening that rescue or attack crews use to enter the structure.

Environmental Protection Agency (EPA) — U.S. government agency that creates and enforces laws designed to protect the air, water, and soil from contamination; responsible for researching and setting national standards for a variety of environmental programs.

Escape Route — Pathway to safety. It can lead to an already burned area, a previously constructed safety area, a meadow that will not burn, or a natural rocky area that is large enough to take refuge without being burned. When escape routes deviate from a defined physical path, they must be clearly marked (flagged).

Escape Rope — Rope (not considered life-safety or utility rope) that is 19/64-inch (7.5 mm) in diameter or greater, but less than 3/8-inch (9.5 mm) and is constructed in the same manner as life-safety rope. It is intended to be used one time only and then destroyed.

Etiologic Agents — Living microorganisms, such as germs, that can cause human disease; a biologically hazardous material.

Evacuation — Process of leaving or being removed from a potentially hazardous location.

Evaporation — Process of a solid or a liquid turning into gas.

Evidence — (1) One of three requirements of evaluation. The information, data, or observation that allows the instructor to compare what was expected to what actually occurred. (2) In law, something legally presented in court that bears on the point in question. (3) Information collected and analyzed by an investigator.

Evolution — Operation of fire service training or suppression covering one or several aspects of fire fighting. Also called Practical Training Evolution.

Exothermic — Chemical reaction between two or more materials that changes the materials and produces heat, flames, and toxic smoke.

Expansion Ratio — Ratio of the finished foam volume to the volume of the original foam solution.

Expansion Ring — Malleable metal band that binds fire hose to a threaded coupling by compressing the hose tightly against the inner surface of the coupling.

Explosive — (1) Any material or mixture that will undergo an extremely fast, self-propagation reaction when subjected to some form of energy. *Also see* High Explosive and Detonation. (2) Material capable of burning or bursting suddenly and violently. *Also see* Detonation and Low Explosive.

Exposure — (1) Structure or separate part of the fireground to which a fire could spread. (2) People, properties, systems, or natural features that are or may be exposed to the harmful effects of a hazardous materials emergency.

Extension Ladder — Sectional ladder of two or more parts that can be extended to various heights.

Extension Ram — Powered hydraulic tool designed especially for straight pushing operations that may extend as far as 63 inches (1 600 mm). Also called Ram.

Extinguishing Agent — Any substance used for the purpose of controlling or extinguishing a fire.

Extrication — Actions involving the removal and treatment of victims who are trapped by some type of machinery or equipment.

F

Fahrenheit Scale — Temperature scale on which the freezing point is 32°F (0°C) and the boiling point at sea level is 212°F (100°C) at normal atmospheric pressure.

Family of Eight Knots — A series of rescue knots based on a figure-eight knot.

Fascia — Flat horizontal or vertical board located at the outer face of a cornice or a broad flat surface over a storefront or below a cornice.

Fax — (1) Term for facsimile. (2) To send a hard copy (facsimile) of print or illustrative material via telephone.

Federal Communications Commission (FCC) — U.S. government agency charged with the control of all radio and television communications. Acts as the main regulator of radio frequencies in the United States in both the public and private sector.

Feed Main — Pipe connecting the sprinkler system riser to the cross mains. The cross mains directly service a number of branch lines on which the sprinklers are installed.

Female Coupling — Threaded swivel device on a hose or appliance with internal threads designed to receive a male coupling having external threads of the same thread and diameter.

Femoral Artery — The major artery supplying the thigh.

Figure-Eight — Forged metal device in the shape of an eight; used to help control the speed of a person descending a rope.

Figure-Eight Knot — Knot used to form a loop in the end of a rope; should be used in place of the bowline knot when working with synthetic fiber rope.

Film Forming Fluoroprotein Foam (FFFP) — Foam concentrate that combines the qualities of fluoroprotein foam with those of aqueous film forming foam.

Finger(s) — Long, narrow extension(s) of a fire projecting from the main body of a natural cover fire.

Finish — (1) Arrangement of hose usually placed on top of a hose load and connected to the end of the load. Also called Hose Load Finish. (2) Fine or decorative work required for a building or one of its parts. (3) Finishing material used in painting.

Fire — Rapid oxidation of combustible materials accompanied by a release of energy in the form of heat and light.

Fire Alarm — (1) Call announcing a fire. (2) Bell or other device summoning a fire company to respond to a fire or other emergency.

Fire Department Connection (FDC) — Point at which the fire department can connect into a sprinkler or standpipe system to boost the water flow in the system. This connection consists of a clappered siamese with two or more 2½-inch (65 mm) intakes or one large-diameter (4-inch [100 mm] or larger) intake. Also called Fire Department Sprinkler Connection.

Fire Door — A specially constructed, tested, and approved fire-rated door assembly designed and installed to prevent fire spread by automatically closing and covering a doorway in a fire wall during a fire to block the spread of fire through the door opening.

Fire Edge — Boundary of a fire at a given moment.

Fire Extinguisher — Portable fire fighting device designed to combat incipient fires.

Fire-Gas Detector — Device used to detect gases produced by a fire within a confined space.

Fire Growth — The formation of a fire plume above the burning fuel after ignition.

Fire Hazard — Any material, condition, or act that contributes to the start of a fire or that increases the extent or severity of fire.

Fire Hose —Type of flexible tube used by firefighters to carry water or other extinguishing agents under pressure from a source of supply to a point of application.

Fire Inspector — Fire personnel assigned to inspect property with the purpose of enforcing fire regulations.

Fire Mark — Distinctive metal marker once produced by insurance companies for identifying their policyholders' buildings.

Fire Marshal — Highest fire prevention officer of a state, province, county, or municipality. In Canada, this officer is sometimes called the fire commissioner.

Fire Point — Temperature at which a liquid fuel produces sufficient vapors to support combustion once the fuel is ignited. The fire point is usually a few degrees above the flash point.

Fire Prevention Bureau — Division of the fire department responsible for conducting fire prevention programs of inspection, code enforcement, education, and investigation.

Fire Resistant — Capacity of structural components to resist higher heat temperatures for certain periods of time. Lesser degree of resistance to fire than "fireproof."

Fire-Resistive Construction — Another term for Type I construction; construction that maintains its structural integrity during a fire.

Fire Stream — Stream of water or other water-based extinguishing agent after it leaves the fire hose and nozzle until it reaches the desired point.

Fire Tetrahedron — Model of the four elements/conditions required to have a fire. The four sides of the tetrahedron represent fuel, heat, oxygen, and chemical chain reaction.

Fire Triangle — Plane geometric figure of an equilateral triangle that is used to explain the conditions necessary for fire. The sides of the triangle represent heat, oxygen, and fuel. The fire triangle was used prior to the general adaptation of the fire tetrahedron that includes a chemical chain reaction.

Fire Wall — Fire-rated wall with a specified degree of fire resistance, built of fire-resistive materials and usually extending from the foundation up to and through the roof of a building, that is designed to limit the spread of a fire within a structure or between adjacent structures.

First-Due — Apparatus that should reach the scene of an emergency first based on the pre-fire attack plan. Also referred to as First-In.

First Responder (EMS) — (1) First person arriving at the scene of an accident or medical emergency who is trained to administer first aid and basic life support. (2) Level of emergency medical training, between first aider and emergency medical technician levels, that is recognized by the authority having jurisdiction.

Fitting — Device that facilitates the connection of hoselines of different sizes to provide an uninterrupted flow of extinguishing agent.

Fixed-Temperature Heat Detector — Temperature-sensitive device that senses temperature changes and sounds an alarm at a specific point, usually 135°F (57°C) or higher.

Flame Detectors — Detection and alarm devices used in some fire detection systems (generally in high-hazard areas) that detect light/flames in the ultraviolet wave spectrum (UV detectors) or detect light in the infrared wave spectrum (IR detectors). Also called Light Detectors.

Flameover — Condition that occurs when a portion of the fire gases trapped at the upper level of a room ignite, spreading flame across the ceiling of the room.

Flammable Liquid — Any liquid having a flash point below 100°F (37.8°C) and having a vapor pressure not exceeding 40 psi absolute (276 kPa).

Flammable Range — The range between the upper flammable limit and lower flammable limit in which a substance can be ignited.

Flank — Sides of a natural cover fire.

Flash Point — Minimum temperature at which a liquid gives off enough vapors to form an ignitable mixture with air near the liquid's surface.

Flashback — Spontaneous reignition of fuel when the blanket of extinguishing agent breaks down or is compromised through physical disturbance.

Flashover — Stage of a fire at which all surfaces and objects within a space have been heated to their ignition temperature and flame breaks out almost at once over the surface of all objects in the space.

Flat-Head Axe — Axe with a cutting edge on one side of the head and a blunt or flat head on the opposite side.

Flat Load — Arrangement of fire hose in a hose bed or compartment in which the hose lies flat with successive layers one upon the other.

Flow Pressure — Pressure created by the rate of flow or velocity of water coming from a discharge opening. Also called Plug Pressure.

Fly Section — Extendable section of ground extension or aerial ladder. Also referred to as Fly.

Foam — Extinguishing agent formed by mixing a foam concentrate with water and aerating the solution for expansion; for use on Class A and Class B fires. Foam may be protein, synthetic, aqueous film forming, high expansion, or alcohol type. Also known as Finished Foam.

Foam Concentrate — (1) Raw chemical compound solution that is mixed with water and air to produce foam. (2) The raw foam liquid as it rests in its storage container prior to the introduction of water and air.

Foam Proportioner — Device that injects the correct amount of foam concentrate into the water stream to make the foam solution.

Foam Solution — Result of mixing the appropriate amount of foam concentrate with water. Foam solution exists between the proportioner and the nozzle or aerating device that adds air to create finished foam.

Fog Nozzle — Nozzle that can provide either a fixed or variable spray pattern. The nozzle breaks the foam solution into small droplets that mix with air to form finished foam.

Fog Stream — Water stream of finely divided particles used for fire control.

Folding Ladder — Short, collapsible ladder easy to maneuver in tight places such as reaching through openings in attics and lofts.

Foot Pads — Feet mounted on the butt of the ladder by a swivel to facilitate the placement of ladders on hard surfaces.

Forced Ventilation — Ventilation caused by any means other than natural ventilation. May involve the use of fans, blowers, smoke ejectors, and fire streams. Also called Mechanical Ventilation.

Forcible Entry — (1) Techniques used by fire personnel to gain entry into buildings, vehicles, aircraft, or other areas of confinement when normal means of entry are locked or blocked. (2) Entering a structure or vehicle by means of physical force, characterized by prying open doors and breaking windows, leaving visible indicators of illegal entry if pry marks and certain window breakage are present upon the first firefighters' arrival.

Forward Lay — Method of laying hose from the water supply to the fire scene.

Four-Way Hydrant Valve — Device that permits a pumper to boost the pressure in a supply line connected to a hydrant without interrupting the water flow.

Frangible Bulb — Small glass vial fitted into the discharge orifice of a fire sprinkler. The glass vial is partly filled with a liquid that expands as heat builds up. At a predetermined temperature, vapor pressure causes the glass bulb to break, which permits water to flow.

Friction Loss — That part of the total pressure lost as water moves through a hose or piping system, caused by water turbulence and the roughness of interior surfaces of hose or pipe.

Frostbite — Local freezing and tissue damage due to prolonged exposure to extreme cold.

Fuel — Flammable and combustible substances available for a fire to consume.

Fulcrum — Support or point of support on which a lever turns in raising or moving something.

Fully Developed Stage — Stage of burning process where energy release is at maximum rate and is limited only by availability of fuel and oxygen.

Fusible Link — (1) Connecting link device that fuses or melts when exposed to heat. Used in sprinklers, fire doors, dampers, and ventilators. (2) Two-piece link held together with a metal that melts or fuses at a specific temperature.

G

Gabled Roof — Style of pitched roof with square ends in which the end walls of the building form triangular areas beneath the roof.

Gamma Radiation — Very high-energy ionizing radiation composed of gamma rays. *Also see* Alpha Radiation, Beta Radiation, Ionizing Radiation, and Radiation (2).

Gang Nail — Form of gusset plate. These thin steel plates are punched with acutely V-shaped holes that form sharp prongs on one side that penetrate wooden members to fasten them together.

Gas — Compressible substance, with no specific volume, that tends to assume the shape of a container. Molecules move about most rapidly in this state. *Also see* Compressed Gas.

Gate Valve — Control valve with a solid plate operated by a handle and screw mechanism. Rotating the handle moves the plate into or out of the waterway.

Gated Wye — Hose appliance with one female inlet and two or more male outlets with a gate valve on each outlet.

General-Use Rope — Life-safety rope that is 7/16-inch (11 mm) in diameter or greater, but less than or equal to 5/8-inch (16 mm) and is intended to support the weight of two persons.

Generator — Auxiliary electrical power generating device. Portable generators are powered by small gasoline or diesel engines and generally have 110- and/ or 220-volt capacities.

Global Positioning System (GPS) — System for determining position on the earth's surface by calculating the difference in time for the signal from a number of satellites to reach a GPS receiver on the ground.

Golden Hour – Refers to the optimum limit of one hour between the time of injury and surgery at a hospital.

GPS — Abbreviation for Global Positioning System.

Gravity System — Water supply system that relies entirely on the force of gravity to create pressure and cause water to flow through the system. The water supply, which is often an elevated tank, is at a higher level than the system.

Green — Area of unburned fuels, not necessarily green in color, adjacent to but not involved in a wildland fire.

Green Wood —Wood with high moisture content.

Grid System — Water supply system that utilizes lateral feeders for improved distribution.

Ground Gradient — When an energized electrical wire comes in contact with the ground, current flows outward in all directions from the point of contact. As the current flows away from the point of contact, the voltage drops progressively.

Growth Stage — The early stage of a fire during which fuel and oxygen are virtually unlimited. This phase is characterized by a rapidly increasing release of heat.

Guides — Devices to hold sections of an extension ladder together while allowing free movement.

Gusset Plates — Metal or wooden plates used to connect and strengthen the joints or two or more separate components (such as metal or wooden truss components or roof or floor components) into a load-bearing unit.

H

Half Hitch — Knot that is always used in conjunction with another knot. The half hitch is particularly useful in stabilizing tall objects that are being hoisted.

Halligan Tool — Prying tool with a claw at one end and a spike or point at a right angle to a wedge at the other end.

Halogenated Agent System — Extinguishing system that uses a halogenated gas as the primary extinguishing agent; usually installed to protect highly sensitive electronic equipment.

Halogenated Agents — Chemical compounds (halogenated hydrocarbons) that contain carbon plus one or more elements from the halogen series. Halon 1301 and Halon 1211 are most commonly used as extinguishing agents for Class B and Class C fires. Also called Halogenated Hydrocarbons.

Halon — Halogenated agent; extinguishes fire by inhibiting the chemical reaction between fuel and oxygen. *See* Halogenated Agents.

Halyard — Rope used on extension ladders to extend the fly sections. Also called Fly Rope.

Handline Nozzle — Any nozzle that can be safely handled by one to three firefighters and flows less than 350 gpm (1 400 L/min).

Handsaw — A saw that is operated by hand rather than a power source. Especially useful for cutting objects that require a controlled cut but are too big to fit in the jaws of a scissors-type cutter or unsuitable for cutting with a power saw.

Hard-Suction Hose — A flexible rubber hose reinforced with a steel core to prevent collapse from atmospheric pressure when drafting; connected between the intake of a fire pump and a water supply and must be used when drafting. Also called hard hose, hard sleeve, or hard intake/suction hose.

Hazard Area — Established area from which bystanders and unneeded rescue workers are prohibited.

Hazard Assessment — Formal review of the hazards that may be encountered while performing the functions of a firefighter or emergency responder; used to determine the appropriate level and type of personal and respiratory protection that must be worn.

Hazard-Control Zones — System of barriers surrounding designated areas at emergency scenes intended to limit the number of persons exposed to a hazard and to facilitate its mitigation; major incident has three zones: restricted (hot), limited access (warm), and support (cold). U.S. EPA/OSHA term: *site work zones.* Also called scene-control zones and control zones. *Also see* Hazard Area.

Hazardous Material — (1) Any substance or material that possesses an unreasonable risk to health and safety of persons and/or the environment if it is not properly controlled during handling, storage, manufacture, processing, packaging, use, disposal, or transportation. (2) Substance or material in quantities or forms that may pose an unreasonable risk to health, safety, or property when stored, transported, or used in commerce (U.S. Department of Transportation definition). *Also see* Dangerous Good (1), Corrosive Material, Hazardous Substance, Hazardous Waste, Hazardous Chemical, Material, and Product.

Hazardous Waste Operations and Emergency Response (HAZWOPER) — U.S. regulations in Title 29 (Labor) *CFR* 1910.120 for cleanup operations involving hazardous substances and emergency response operations for releases of hazardous substances. *Also see Code of Federal Regulations (CFR).*

Head — (1) Front and rear closure of a tank shell. (2) Alternate term for pressure, especially pressure due to elevation. For every 1-foot increase in elevation, 0.434 psi is gained (for every 1-meter increase in elevation, 9.82 kPa is gained). Also called Head Pressure. (3) Top of a window or door frame. (4) Most active part of a ground cover fire; the forward advancing part.

Health and Safety Officer (HSO) — Member of the fire and emergency services organization who is assigned and authorized by the administration as the manager of the health and safety program and performs the duties, functions, and responsibilities in NFPA 1521, *Standard for Fire Department Safety Officer.*

Hearsay — Evidence presented by a witness who did not see or hear the incident in question but heard about it from someone else.

Heat Cramps — Heat illness resulting from prolonged exposure to high temperatures; characterized by excessive sweating, muscle cramps in the abdomen and legs, faintness, dizziness, and exhaustion. *Also see* Heat Exhaustion, Heat Stroke, Heat Rash, and Heat Stress.

Heat Detector — Alarm-initiating device designed to be responsive to a predetermined rate of temperature increase or to a predetermined temperature level.

Heat Exhaustion — Heat illness caused by exposure to excessive heat; symptoms include weakness, cold and clammy skin, heavy perspiration, rapid and shallow breathing, weak pulse, dizziness, and sometimes unconsciousness. *Also see* Heat Cramps, Heat Stroke, Heat Rash, and Heat Stress.

Heat of Combustion — Total amount of thermal energy (heat) that could be generated by the combustion (oxidation) reaction if a fuel were completely burned. The heat of combustion is measured in British Thermal Units (Btu) per pound or calories per gram.

Heat of Compression — Heat generated when a gas is compressed.

Heat of Friction — Heat created by the movement of two surfaces against each other.

Heat Rash — Condition that develops from continuous exposure to heat and humid air; aggravated by clothing that rubs the skin; reduces an individual's tolerance to heat. *Also see* Heat Exhaustion, Heat Cramps, Heat Stroke, and Heat Stress.

Heat Release Rate (HRR) — Total amount of heat produced or released to the atmosphere from the convective-lift fire phase of a fire per unit mass of fuel consumed per unit time.

Heat Sensor Label — Label affixed to the ladder beam near the tip to provide a warning that the ladder has been subjected to excessive heat.

Heat Stroke — Heat illness caused by heat exposure, resulting in failure of the body's heat regulating mechanism; symptoms include (a) high fever of 105 to 106°F (40.5°C to 41.1°C), (b) dry, red, and hot skin, (c) rapid, strong pulse, and (d) deep breaths or convulsions; may result in coma or possibly death. Also called sunstroke. *Also see* Heat Exhaustion, Heat Cramps, Heat Rash, and Heat Stress.

Heavy Fire Loading — Presence of large amounts of combustible materials in an area or a building

Heel — (1) Base or butt end of a ground ladder. (2) To steady a ladder while it is being raised. (3) Rear portion of a wildland fire. Also called Rear. (4) Angle a vessel leans to one side due to wind, waves, or turning of the vessel; measured in degrees.

Heeler — The firefighter at the butt end of the ladder responsible for placing it at the desired distance from the building and determining whether the ladder will be raised parallel with or perpendicular to the building.

Hemorrhage – Bleeding, especially severe bleeding.

HEPA — Acronym for High Efficiency Particulate Air.

Higbee Cut — Special cut at the beginning of the thread on a hose coupling that provides positive identification of the first thread to eliminate cross threading. Also called Blunt Start.

Higbee Indicators — Notches or grooves cut into coupling lugs to identify by touch or sight the exact location of the Higbee Cut.

High Efficiency Particulate Air (HEPA) Filter — Respiratory protection filter designed and certified to protect the user from particulates in the air. The HEPA filter must be at least 99.97 percent efficient in removing monodisperse particles of 0.3 micrometers in diameter.

High-Expansion Foam — Foam concentrate that is mixed with air in the range of 200 parts air to 1 part foam solution (200:1) to 1,000 parts air to 1 part foam solution (1,000:1).

High-Rise/High-Rise Building — Any building that requires fire fighting on levels above the reach of the department's equipment. Various building and fire codes have written definitions of what is to be considered a high rise. For example, the Uniform Building Code (UBC) defines a high-rise building as any building of more than 75 feet (23 m) in height.

Hip Roof — Pitched roof that has no gables. All facets of the roof slope down from the peak to an outside wall.

Hooks — Curved metal devices installed on the tip end of roof ladders to secure the ladder to the highest point on the roof of a building.

Horizontal Ventilation — Any technique by which heat, smoke, and other products of combustion are channeled horizontally out of a structure by way of existing or created horizontal openings such as windows, doors, or other holes in walls.

Horseshoe Load — Arrangement of fire hose in a hose bed or compartment in which the hose lies on edge in the form of a horseshoe.

Hose Bed — Main hose-carrying area of a pumper or other piece of apparatus designed for carrying hose. Also called Hose Body.

Hose Bridge — Device placed alongside or astride hose that is laid across a street to permit traffic to drive over the hose without damaging it. Also called Hose Ramp.

Hose Cap — Threaded female fitting used to cap a hoseline or a pump outlet.

Hose Clamp — Mechanical or hydraulic device used to compress fire hose to stop the flow of water.

Hose Hoist — Metal device having a roller that can be placed over a windowsill or roof's edge to protect a hose and make it easier to hoist. Also called Hose Roller.

Hose Jacket — (1) Outer covering of a hose. (2) Device clamped over a hose to contain water at a rupture point or to join hose with damaged or dissimilar couplings.

Hose Plug — Threaded male fitting used to cap off a pump intake.

Hose Strap — Strap or chain with a handle suitable for placing over a ladder rung; used to carry and secure a hoseline.

HVAC System — Abbreviation for Heating, Ventilating, and Air-Conditioning System. (1) The mechanical system used to provide environmental control within a structure. (2) Air-handling system within a building consisting of fans, ducts, dampers, and other equipment necessary to make it function. Also called Air-Handling System.

Hydrant Wrench — Specially designed tool used to open or close a hydrant and to remove hydrant caps.

Hydraulic Jack — Lifting jack that uses hydraulic fluid power supplied from a manually operated hand lever.

Hydraulic Prying Tools — Can be either powered hydraulic or manual hydraulic, both types receive their power from hydraulic fluid pumped through special high-pressure hoses.

Hydraulic Ventilation — Method of ventilating a fire building by directing a fog stream of water out a window to increase air and smoke movement.

Hydraulics — Branch of fluid mechanics dealing with the mechanical properties of liquids and the application of these properties in engineering.

Hydrocarbon Fuel — Petroleum-based organic compound that contain only hydrogen and carbon.

Hydrogen Cyanide (HCN) — Colorless, toxic gas with a faint odor similar to bitter almonds; produced by the combustion of nitrogen-bearing substances.

Hydrostatic Test — A testing method that uses water under pressure to check the integrity of pressure vessels.

Hypothermia — Abnormally low or decreased body temperature.

Hypoxia — Condition caused by a deficiency in the amount of oxygen reaching body tissues.

I

IAP — Abbreviation for Incident Action Plan.

IC — Abbreviation for Incident Commander.

IDLH — Abbreviation for Immediately Dangerous to Life or Health.

IED — Abbreviation for Improvised Explosive Device.

Ignition — Beginning of flame propagation or burning; the start of a fire.

Immediately Dangerous to Life and Health (IDLH) — Any atmosphere that poses an immediate hazard to life or produces immediate irreversible, debilitating effects on health.

Improvised Explosive Device (IED) — Device that is categorized by its container and the way it is initiated; usually homemade, constructed for a specific target, and contained in almost anything.

In Service — Operational and available for an assignment.

Incendiary — (1) An incendiary agent such as a bomb. (2) A fire deliberately set under circumstances in which the responsible party knows it should not be ignited. (3) Relating to or involving a deliberate burning of property.

Incendiary Device — Material or chemicals designed and used to start a fire.

Incident Action Plan (IAP) — Written or unwritten plan for the disposition of an incident. The IAP contains the overall strategic goals, tactical objectives, and support requirements for a given operational period during an incident. All incidents require an action plan. On relatively small incidents, the IAP is usually not in writing. On larger, more complex incidents, a written IAP is created for each operational period and is disseminated to units assigned to the incident.

Incident Command System (ICS) — (1) System by which facilities, equipment, personnel, procedures, and communications are organized to operate within a common organizational structure designed to aid in the management of resources at emergency incidents. (2) Management system of procedures for controlling personnel, facilities, equipment, and communications so that different agencies can work together toward a common goal in an effective and efficient manner. (3) Recommended method of establishing and maintaining command and control of an incident. It is an organized approach to incident management, adaptable to any size or type of incident.

Incident Commander (IC) — Person in charge of the incident management system and responsible for the management of all incident operations during an emergency.

Incident Investigation — Act of investigating or gathering data to determine the factors that contributed to a fatality, injury, or property loss or to determine fire cause and origin.

Incident Safety Officer (ISO) — Member of the Command Staff responsible for monitoring and assessing safety hazards and unsafe conditions during an incident, and developing measures for ensuring personnel safety. The ISO is responsible for the enforcement of all mandated safety laws and regulations and departmental safety-related SOPs. On very small incidents, the IC may act as the ISO.

Incipient Stage — First stage of the burning process in a confined space in which the substance being oxidized is producing some heat, but the heat has not spread to other substances nearby. During this phase, the oxygen content of the air has not been significantly reduced.

Indicating Valve — Water main valve that visually shows the open or closed status of the valve.

Indirect Attack (Structural) — Directing fire streams toward the ceiling of a room or building in order to generate a large amount of steam. Converting the water to steam absorbs the heat of the fire and cools the area sufficiently for firefighters to safely enter and make a direct attack on the fire.

Indirect Attack (Wildland) — Method of controlling a wildland fire where a control line is constructed or located some distance from the edge of the main fire and the fuel between the two points is burned.

Industrial Fire Brigade — Team of employees organized within a private company, industrial facility, or plant who are assigned to respond to fires and emergencies on that property.

Industrial Occupancy — Industrial, commercial, mercantile, warehouse, utility power station, and institutional or similar facilities.

Ingestion — Taking in food or other substances through the mouth.

Inhalation — Taking in materials by breathing through the nose or mouth.

Initial Isolation Distance — Distance within which all persons are considered for evacuation in all directions from a hazardous materials incident.

Initial Isolation Zone — Circular zone (with a radius equivalent to the initial isolation distance) within which persons may be exposed to dangerous concentrations upwind of the source and may be exposed to life-threatening concentrations downwind of the source. *Also see* Initial Isolation Distance, Isolation Perimeter, and Hazard-Control Zones.

Initials and Numbers — Combination of letters and numbers stenciled on rail tank cars that may be used to get information about the car's contents from the railroad's computer or the shipper.

Injection — Method of proportioning foam that uses an external pump or head pressure to force foam concentrate into the fire stream at the correct ratio for the flow desired.

In-Line Eductor — Eductor that is placed along the length of a hoseline.

Inspection — Formal examination of an occupancy and its associated uses or processes to determine its compliance with the fire and life safety codes and standards.

Intake Hose — Hose used to connect a fire department pumper or a portable pump to a nearby water source. It may be soft sleeve or hard suction hose.

Ionization Detector — Type of smoke detector that uses a small amount of radioactive material to make the air within a sensing chamber conduct electricity.

Ionizing Radiation — Radiation that has sufficient energy to remove electrons from atoms resulting in a chemical change in the atom.

Inverter — Auxiliary electrical power generating device. The inverter is a step-up transformer that converts the vehicle's 12- or 24-volt DC current into 110- or 220-volt AC current.

Irritant/Irritating Material — Liquid or solid that upon contact with fire or exposure to air emits dangerous or intensely irritating fumes.

Island — Unburned area within a fire perimeter.

Isolation Perimeter — Outer boundary of an incident that is controlled to prevent entrance by the public or unauthorized persons. *Also see* Initial Isolation Distance and Initial Isolation Zone.

J

Jalousie Window — Window consisting of narrow, frameless, glass panes set in metal brackets at each end that allow the panes a limited amount of axial rotation for ventilation.

J-Tool — A device made of rigid, heavy gauge wire and designed to fit through the space between double-swinging doors equipped with panic hardware.

Jet-Assist Device — A siphon used to transfer water from one portable tank to another.

Joule (J) — Unit of work or energy in the International System of Units; the energy (or work) when unit force (1 newton) moves a body through a unit distance (1 meter); takes the place of calorie for heat measurement (1 calorie = 4.19 J).

Jurisdiction — (1) Legal authority to operate or function. (2) Boundaries of a legally constituted entity.

K

Kerf Cut — A single cut the width of the saw blade made in a roof to check for fire extension.

Kernmantle Rope — Rope that consists of a protective shield (mantle) over the load-bearing core strands (kern).

Kinetic Energy — The energy possessed by a moving object.

Knot — (1) Term used for tying a rope around itself. (2) International nautical unit of speed; 1 knot = 6,076 feet or 1 nautical mile per hour (1.15 miles or 1.85 kilometers per hour).

K-Tool — V-blade tool that is designed to pull lock cylinders from a door with only minimal damage to the door itself.

L

Ladder Belt — Belt with a hook that secures the firefighter to the ladder.

Laid Rope — Rope constructed by twisting several groups of individual strands together.

Lamella Arch — A special type of arch constructed of short pieces of wood called lamellas.

Large-Diameter Hose (LDH) — Relay-supply hose of 3½- to 6 inches (90 mm to 150 mm) in diameter; used to move large volumes of water quickly with a minimum number of pumpers and personnel.

Large Handline — Fire hose/nozzle assembly capable of flowing up to 300 gpm (1 140 L/min).

Latent Heat of Vaporization — Quantity of heat absorbed by a substance at the point at which it changes from a liquid to a vapor.

LC$_{50}$ — Abbreviation for Lethal Concentration.

LD$_{50}$ — Abbreviation for Lethal Dose.

Lean-to Collapse — Type of structural collapse where one end of a floor or roof section support fails while the other end remains secured to a wall. The floors and roof drop in large sections and form voids.

Ledge Door — Door constructed of individual boards joined within a frame. Also called batten door.

Leeward — Protected side; the direction opposite from which the wind is blowing.

LERP — Abbreviation for Local Emergency Response Plan.

Lethal Concentration (LC$_{50}$) — Concentration of an inhaled substance that results in the death of 50 percent of the test population; the lower the value the more toxic the substance; an inhalation exposure expressed in parts per million (ppm), milligrams per liter (mg/liter), or milligrams per cubic meter (mg/m^3). *Also see* Concentration (1).

Lethal Dose (LD$_{50}$) — Concentration of an ingested or injected substance that results in the death of 50 percent of the test population; the lower the dose the more toxic the substance; an oral or dermal exposure expressed in milligrams per kilogram (mg/kg). *Also see* Dose.

Level A Protection — Highest level of skin, respiratory, and eye protection that can be afforded by personal protective equipment. Consists of positive-pressure self-contained breathing apparatus, totally encapsulating chemical-protective suit, inner and outer gloves, and chemical-resistant boots.

Level B Protection — Personal protective equipment that affords the highest level of respiratory protection, but a lesser level of skin protection. Consists of positive-pressure self-contained breathing apparatus, hooded chemical-resistant suit, inner and outer gloves, and chemical-resistant boots.

Level C Protection — Personal protective equipment that affords a lesser level of respiratory and skin protection than levels A or B. Consists of full-face or half-mask APR, hooded chemical-resistant suit, inner and outer gloves, and chemical-resistant boots.

Level D Protection — Personal protective equipment that affords the lowest level of respiratory and skin protection. Consists of coveralls, gloves, and chemical-resistant boots or shoes.

Lever — Device consisting of a bar turning about a fixed point (fulcrum), using power or force applied at a second point to lift or sustain an object at a third point.

Lexan® — Polycarbonate plastic used for windows. It has one-half the weight of an equivalent-sized piece of glass, yet is 30 times stronger than safety glass and 250 times stronger than ordinary glass. It cannot be broken using standard forcible entry techniques.

Life Safety — Refers to the joint consideration of the life and physical well-being of individuals, both civilians and firefighters.

Life Safety Rope — Rope that meets the requirements of NFPA 1983, *Standard on Fire Service Life Safety Rope, Harness, and Hardware,* and is dedicated solely for the purpose of constructing lines to be used for the raising, lowering, or supporting people during rescue, fire fighting, or other emergency operations, or during training. Also called Lifeline.

Light-Use Rope — Life-safety rope that is $^3/_8$-inch (9.5 mm) in diameter or greater, but less than ½-inch (12.5 mm) and is intended to support the weight of one person.

Line-of-Duty Death (LODD) — Firefighter or emergency responder death resulting from the performance of fire department duties.

Liquefied Petroleum Gas (LPG) — Any of several petroleum products, such as propane or butane, stored under pressure as a liquid.

Live Fire or Burn Exercises — Training exercises that involve the use of an unconfined open flame or fire in a structure or other combustibles to provide a controlled burning environment.

Load-Bearing Wall — Wall that is used for structural support.

Local Alarm Systems — (1) Alarm systems that alert and notify only occupants on the premises of the existence of a fire so that they can safely exit the building and call the fire department. If a response by a public safety agency (police or fire department) is required, an occupant hearing the alarm must notify the agency. (2) Combination of alarm components designed to detect a fire and transmit an alarm on the immediate premises.

Local Emergency Response Plan (LERP) — Plan required by U.S. Environmental Protection Agency (EPA) that is prepared by the Local Emergency Planning Committee (LEPC) detailing how local emergency response agencies will respond to community emergencies.

Loop System — Water main arranged in a complete circuit so that water will be supplied to a given point from more than one direction. Also called Circle System, Circulating System, or Belt System.

Loss Control — The practice of minimizing damage and providing customer service through effective mitigation and recovery efforts before, during, and after an incident.

Loss Control Risk Analysis — The process in which specific potential risks (potential because the incident has not occurred yet) are identified and evaluated. The goal of this process is to develop strategies to minimize the impact of these risks.

Louver Cut or Vent — Rectangular exit opening cut in a roof, allowing a section of roof deck (still nailed to a center rafter) to be tilted, thus creating an opening similar to a louver. Also called center rafter cut.

Low-Pressure Alarm — Bell, whistle, or other audible alarm that warns the wearer when the SCBA air supply is low and needs replacement, usually 25 percent of full container pressure.

Lower Explosive Limit (LEL) — Lower limit at which a flammable gas or vapor will ignite; below this limit the gas or vapor is too *lean* or *thin* to burn (too much oxygen and not enough gas).

Lower Flammable Limit (LFL) — Lower limit at which a flammable gas or vapor will ignite; below this limit the gas or vapor is too lean or thin to burn (too much oxygen and not enough gas). *Also see* Upper Flammable Limit.

LPG — Abbreviation for Liquefied Petroleum Gas.

M

Main Section — Bottom section of an extension ladder.

Mains — A network of underground pipes.

Maintenance — Keeping equipment or apparatus in a state of usefulness or readiness.

Male Coupling — Hose coupling with external threads that fit into the threads of a female coupling of the same pitch and appropriate diameter and thread count.

Malicious — With intent to cause harm.

Manifold — (1) Hose appliance that divides one larger hoseline into three or more small hoselines. Also called portable hydrant. (2) Top portion of the pump casing. (3) Used to join a number of discharge pipelines to a common outlet.

Manual Prying Tools — Prying tools that use the principle of lever and fulcrum to provide mechanical advantage

Masonry — Bricks, blocks, stones, and unreinforced and reinforced concrete products.

Master Stream — Large-caliber water stream usually supplied by siamesing two or more hoselines into a manifold device or by fixed piping that delivers 350 gpm (1 400 l/min) or more. Also called Heavy Stream.

Material Safety Data Sheet (MSDS) — Form provided by the manufacturer and blender of chemicals that contains information about chemical composition, physical and chemical properties, health and safety hazards, emergency response procedures, and waste disposal procedures of the specified material.

Matter — Anything that occupies space and has mass.

Mayday — International distress signal broadcast by voice.

Mechanical Heat Energy — Heat generated by friction or compression. Moving parts on machines, such as belts and bearings, are a source of mechanical heating.

Mechanism of Injury — A force or forces that have caused an injury.

Medium-Expansion Foam — Foam concentrate that is mixed with air in the range of 20 parts air to 1 part foam solution (20:1) to 200 parts air to 1 part foam solution (200:1).

Miscible — Materials that are capable of being mixed.

Mitigate — To cause to become less harsh or hostile; to make less severe, intense or painful; to alleviate.

Mortar — Cement-like liquid material that hardens and bonds individual masonry units into a solid mass.

Mortise — (1) Notch, hole, or space cut into a door to receive a lock case, which contains the lock mechanism. (2) Hole, groove, or slot cut into a wooden ladder beam to receive a rung tenon.

Motivation — Internal process in which energy is produced by needs or expended in the direction of goals. Motivation usually occurs in someone who is interested in achieving some goal.

Multiple-Casualty Incident (MCI) — An emergency involving multiple patients.

Mushrooming — Tendency of heat, smoke, and other products of combustion to rise until they encounter a horizontal obstruction. At this point they will spread laterally until they encounter vertical obstructions and begin to bank downward.

Mutual Aid — Reciprocal assistance from one fire and emergency services agency to another during an emergency based upon a prearrangement between agencies involved and generally made upon the request of the receiving agency.

N

National Fire Incident Reporting System (NFIRS) — One of the main sources of information (data, statistics) about fires in the United States; under NFIRS, local fire departments collect fire incident data and send these to a state coordinator; the state coordinator develops statewide fire incident data and also forwards information to the USFA; begun by FEMA.

National Institute of Standards and Technology (NIST) — A federal technology agency that develops and promotes measurement, standards, and technology.

Natural Fiber Rope — Rope made of hemp or cotton used for utility purposes. Natural fiber rope is not accepted for use in life-safety applications.

Natural Ventilation — Techniques that use the wind, convection currents, and other natural phenomena to ventilate a structure without the use of fans, blowers, or other mechanical devices.

Negative-Pressure Ventilation — Technique using smoke ejectors to develop artificial circulation and to pull smoke out of a structure. Smoke ejectors are placed in windows, doors, or roof vent holes to pull the smoke, heat, and gases from inside the building and eject them to the exterior.

Noncombustible Construction — Another term for Type II construction; construction made of the same materials as fire-resistive construction except that the structural components lack the insulation or other protection of Type I construction.

Nonintervention Strategy — Strategy for handling fires involving hazardous materials where the fire is allowed to burn until all of the fuel is consumed. Also called nonintervention mode.

Nonload-Bearing Wall — Wall, usually interior, that supports only its own weight.

Nonpiloted — Ignition caused when a material reaches its autoignition temperature as the result of self-heating.

Nonthreaded Coupling — Coupling with no distinct male or female components. Also called Storz Coupling or sexless coupling.

Nozzle Pressure — Velocity pressure at which water is discharged from the nozzle.

Nuclear Heat Energy — Creation of heat through the splitting apart or combining of atoms.

O

Occupational Safety and Health Administration (OSHA) — U.S. federal agency that develops and enforces standards and regulations for occupational safety in the workplace.

ODP — Abbreviation for Office for Domestic Preparedness.

Offensive Strategy — Overall plan for incident control established by the Incident Commander (IC) in which responders take aggressive, direct action on the material, container, or process equipment involved in an incident. *Also see* Defensive Strategy, Strategy, and Nonintervention Strategy.

OI — Abbreviation for Operating Instruction.

Olfactory Fatigue — Gradual inability of a person to detect odors after initial exposure; may be extremely rapid in the case of some toxins such as hydrogen sulfide.

Open-Circuit Airline Equipment — Airline breathing equipment that allows exhaled air to be discharged to the open atmosphere.

Open-Circuit Self-Contained Breathing Apparatus — An SCBA that allows the wearer's exhaled air to be discharged or vented to the atmosphere.

Operational Level — Level of training established by the National Fire Protection Administration allowing first responders to take defensive actions at hazardous materials incidents. *Also see* Awareness Level and Operations Plus.

Operations Level — Level of training established by the National Fire Protection Association allowing first responders to take defensive actions at hazardous materials incidents.

Ordinary Construction — Another term for Type III construction; construction that requires that exterior walls and structural members be made of noncombustible or limited combustible materials.

OS&Y Valve — Outside stem and yoke valve; a type of control valve for a sprinkler system in which the position of the center screw indicates whether the valve is open or closed. Also known as Outside Screw and Yoke Valve.

Overcurrent — Any current that is in excess of the rated current of equipment or the ampacity of a conductor, and may be caused by an overload, short circuit, or ground fault. Also called Overload.

Overhand Safety Knot — Knot used in conjunction with other knots to eliminate the danger of the running end of the rope slipping back through a knot, causing the knot to fail.

Overhaul — Those operations conducted once the main body of fire has been extinguished that consist of searching for and extinguishing hidden or remaining fire, placing the building and its contents in a safe condition, determining the cause of the fire, and recognizing and preserving evidence of arson.

Oxidizer — Any substance or material that yields oxygen readily and may stimulate the combustion of organic and inorganic matter.

Oxyacetylene Cutting Torch — A commonly used torch that burns oxygen and acetylene to produce a very hot flame. Used as a forcible entry cutting tool for penetrating metal enclosures that are resistant to more conventional forcible entry equipment.

Oxygasoline Cutting Torch — A relatively new cutting system, fueled by gasoline and oxygen.

Oxygen (O_2) — A chemical element. Colorless, odorless, tasteless gas constituting 21 percent of the atmosphere.

Oxygen-Deficient Atmosphere — An atmosphere containing less than 19.5 percent oxygen.

Oxyhemoglobin — Combination of oxygen and hemoglobin.

P

Packaging — (1) Broad term the U.S. Department of Transportation uses to describe shipping containers and their markings, labels, and/or placards. (2) Readying a patient for transport.

Padlock — Detachable, portable lock with a hinged or sliding shackle.

Pancake Collapse — Situation where the weakening or destruction of bearing walls causes the floors or the roof to collapse, which allows the debris to fall as far as the lower floor or basement. Typically, there is little void space with these types of collapse.

Panel Door — Door inset with panels, which are usually of wood, metal, glass, or plastic.

Parapet — Portion of the exterior walls of a building that extends above the roof. A low wall at the edge of a roof.

Partition Wall — Interior non-load bearing wall that separates a space into rooms.

Pathogens — The organisms that cause infection such as viruses and bacteria.

Pawls — Devices attached to the inside of the beams on fly sections used to hold the fly section in place after it has been extended. Also called Dogs or Ladder Locks.

Pendant Sprinkler — Automatic sprinkler designed for placement and operation with the head pointing downward from the piping.

PEL — Acronym for Permissible Exposure Limit.

Perfusion —The supply of oxygen to and removal of wastes from the cells and tissues of the body as a result of the flow of blood through the capillaries.

Perimeter — The perimeter of a wildland fire is the boundary of the fire. It is the total length of the outer edge of the burning or burned area.

Permissible Exposure Limit (PEL) — Maximum time-weighted concentrations at which 95 percent of exposed healthy adults suffer no adverse effects over a 40-hour workweek; an 8-hour time-weighted average unless otherwise noted; expressed in either parts per million (ppm) or milligrams per cubic meter (mg/m³).

Persistence —Length of time a chemical agent remains effective without dispersing.

Personnel Accountability Report (PAR) — A roll call of all units (crews, teams, groups, companies, sectors) assigned to an incident. Usually by radio, the supervisor of each unit reports the status of the personnel within the unit at that time. A PAR may be required by SOP at specific intervals during an incident, or may be requested at any time by the IC or the ISO.

Personal Alert Safety System (PASS) — Electronic lack-of-motion sensor that sounds a loud tone when a firefighter becomes motionless. It can also be manually triggered to operate.

Personal Protective Equipment (PPE) — General term for the equipment worn by firefighters and rescuers; includes helmets, coats, pants, boots, eye protection, gloves, protective hoods, self-contained breathing apparatus, and personal alert safety systems (PASS devices). Also called Bunker Clothes, Protective Clothing, Turnout Clothing, or Turnout Gear.

Photoelectric Smoke Detector — Type of smoke detector that uses a small light source, either an incandescent bulb or a light-emitting diode (LED), to detect smoke by shining light through the detector's chamber: smoke particles reflect the light into a light-sensitive device called a photocell.

Photon — Packet of electromagnetic energy.

Pick-Head Axe — Forcible entry tool that has a chopping blade on one side of the head and a sharp pick on the other side.

Pike Pole — Sharp prong and hook of steel, on a wood, metal, fiberglass, or plastic handle of varying length, used for pulling, dragging, and probing.

Piloted Ignition — Occurs when a mixture of fuel and oxygen encounter an external heat (ignition) source with sufficient heat energy to start the combustion reaction.

Pin Lug Couplings — Hose couplings with round lugs in the shape of a pin.

Pitched Roof — Roof, other than a flat or arched roof, that has one or more pitched or sloping surfaces.

Pitot Tube — Instrument that is inserted into a flowing fluid (such as a stream of water) to measure the velocity pressure of the stream; commonly used to measure flow. A pitot tube functions by converting the velocity energy to pressure energy that can then be measured by a pressure gauge. The gauge reads in units of pounds per square inch (psi) or kilopascals (kPa).

Plasma Cutters — Ultra-high temperature metal-cutting devices that produce cutting temperatures as high as 25,000°F (13 871°C).

Plaster Hook — Barbed collapsible hook on a pole used to puncture and pull down materials.

Plug — Patch to seal a small leak in a container.

Pneumatic Chisel/Hammer — Pneumatic chisel useful for extrication work; designed to operate at air pressures between 100 and 150 psi (700 kPa and 1 050 kPa). During periods of normal consumption, it will use about 4 to 5 cubic feet (113 L to 142 L) of air per minute. Also called Air Chisels, Pneumatic Hammers, or Impact Hammers.

Pneumatic Tools — Tools that receive their operating energy from compressed air.

Pocket Doors — Door that moves laterally into a "pocket" framed into the wall, usually suspended from an overhead track. When this type of door is in the pocket, only the leading edge of the door is visible.

Point of Origin — Exact physical location where the heat source and fuel come in contact with each other and a fire begins.

Point of No Return — That time at which the remaining operation time of the SCBA is equal to the time necessary to return safely to a nonhazardous atmosphere.

Poison — Any material that when taken into the body is injurious to health. *Also see* Convulsant and Toxin.

Polar Solvent Fuel — Flammable liquids that have an attraction for water, much like a positive magnetic pole attracts a negative pole.

Polar Solvents — (1) Flammable liquids that have an attraction for water, much like a positive magnetic pole attracts a negative pole; examples include alcohols, ketones, and lacquers. (2) A liquid having a molecule in which the positive and negative charges are permanently separated resulting in their ability to ionize in solution and create electrical conductivity. Water, alcohol, and sulfuric acid are examples of polar solvents.

Pole Ladder — Large extension ladder that requires tormentor poles to steady the ladder as it is raised and lowered. Also called Bangor Ladder.

Policy — Guide to decision making in an organization.

Polychlorinated Biphenyl (PCB) — Toxic compound found in some older oil-filled electric transformers.

Pompier Ladder — Scaling ladder with a single beam and a large curved metal hook that can be put over windowsills for climbing.

Porta-Power — Manually operated hydraulic tool that has been adapted from the auto body business to the rescue service. This device has a variety of tool accessories that allows it to be used in numerous applications.

Portable Tank — Storage tank used during a relay or shuttle operation to hold water from water tanks or hydrants. This water can then be used to supply attack apparatus. Also called Catch Basin, Fold-a-Tank, Portable Basin, or Porta-Tank.

Positive-Pressure Ventilation (PPV) — Method of ventilating a confined space by mechanically blowing fresh air into the space in sufficient volume to create a slight positive pressure within and thereby forcing the contaminated atmosphere out the exit opening.

Post-Incident Analysis — General overview and critique of the incident by members of all responding agencies (including dispatchers) that should take place within two weeks of the actual incident.

Post-Incident Stress — Psychological stress that affects emergency responders after returning from a stressful emergency incident.

Post-Indicator Valve (PIV) — A type of valve used to control underground water mains that provides a visual means for indicating "open" or "shut" position; found on the supply main of installed fire protection systems. The operating stem of the valve extends above ground through a "post" and a visual means is provided at the top of the post for indicating "Open" or "Shut."

Potential Energy — Stored energy possessed by an object that can be released in the future to perform work once released.

Power Saw — A saw that is operated using a power source. When equipped with the proper blade, these devices make fast and efficient cuts in a variety of materials.

Powered Hydraulic Shears — Large rescue tool whose two blades open and close by the use of hydraulic power supplied through hose from a power unit.

Powered Hydraulic Spreaders — Large rescue tool whose two arms open and close by the use of hydraulic power supplied through hose from a power unit. This device is capable of exerting in excess of 20,000 pounds (900 kg) of force at its tips. Also called Jaws.

Preaction Sprinkler System — Fire-suppression system that consists of closed sprinklers attached to a piping system that contains air under pressure and a secondary detection system; both must operate before the extinguishing agent is released into the system; similar to the dry-pipe sprinkler system.

Precipitation — Any or all forms of water particles, liquid or solid, that fall from the atmosphere.

Preconnect — (1) Attach hose connected to a discharge when the hose is loaded; this shortens the time it takes to deploy the hose for fire fighting. (2) Soft-sleeve intake hose that is carried connected to the pump intake. (3) Hard suction hose or discharge hose carried connected to a pump, eliminating delay when hose and nozzles must be connected and attached at a fire.

Predischarge Alarm — Alarm that sounds before a total flooding fire extinguishing system is about to discharge. This gives occupants the opportunity to leave the area.

Preincident Inspection — Thorough and systematic inspection of a building for the purpose of identifying significant structural and/or occupancy characteristics to assist in the development of a preincident plan for that building.

Preincident Plan — Document, developed during preincident planning, that contains the operational plan or set procedures for the safe and efficient handling of emergency situations at a given location (such as a specific building or occupancy). Also called Preplan.

Premixing — Mixing premeasured portions of water and foam concentrate in a container. Typically, the premix method is used with portable extinguishers, wheeled extinguishers, skid-mounted twin-agent units, and vehicle-mounted tank systems.

Preplan — Document, developed during pre-incident planning, that contains the operational plan or set procedures for the safe and efficient handling of emergency situations at a given location (such as a specific building or occupancy).

Pressure — Force per unit area exerted by a liquid or gas measured in pounds per square inch (psi) or kilopascals (kPa).

Pressure Dressing — A bulky dressing held in position with a tightly wrapped bandage to apply pressure to help control bleeding.

Pressure Point — Site where a main artery lies near the surface of the body and directly over a bone. Pressure on such a point can stop distal bleeding.

Primary Feeder — Large pipes (mains), with relatively widespread spacing, that convey large quantities of water to various points of the system for local distribution to the smaller mains.

Private Branch Exchange (PBX) — Telephone switching system found in large businesses or organizations that allows many users to be reached by dialing a seven-digit number. Internal users of a PBX system have special features available such as hold, conference calling, transfer, etc.

Procedure – A written communication closely related to a policy.

Projected Window — Type of swinging window that is hinged at the top and swings either outward or inward.

Proportioning — Mixing of water with an appropriate amount of foam concentrate to form a foam solution.

Proprietary Alarm System — Fire protection system owned and operated by the property owner.

Protected Premises Alarm System — Alarm system that alerts and notifies only occupants on the premises of the existence of a fire so that they can safely exit the building and call the fire department. If a response by a public safety agency (police or fire department) is required, an occupant hearing the alarm must notify the agency.

Protection Plates — Plates fastened to a ladder to prevent wear at points where it comes in contact with mounting brackets.

Protective Action Distance — Downwind distance from a hazardous materials incident within which protective actions should be implemented. *Also see* Initial Isolation Distance, Protective Action Zone, and Protective Actions.

Protective Action Zone — Area immediately adjacent to and downwind from the initial isolation zone, which is in imminent danger of being contaminated by airborne vapors within 30 minutes of material release. *Also see* Initial Isolation Zone, Protective Action Distance, and Protective Actions.

Protective Actions — Steps taken to preserve health and safety of emergency responders and the public. *Also see* Protective Action Distance and Protective Action Zone.

Protective Clothing — Includes the helmet, protective coat, protective trousers, protective hood, boots, gloves, self-contained breathing apparatus, and eye protection where applicable. *See* Personal Protective Equipment.

Protective Hood — Hood designed to protect the firefighter's ears, neck, and face from exposure to extreme heat. Hoods are typically made of Nomex®, Kevlar®, or PBI® and are available in long or short styles.

Public Safety Answering Point (PSAP) — Any location or facility at which 9-1-1 calls are answered either by direct calling, rerouting, or diversion.

Pulley — (1) Small, grooved wheel through which the halyard is drawn on an extension ladder. (2) Wheel used to transmit power by means of a band, belt, cord, rope, or chain passing over its rim. (3) Steel or aluminum rollers used to change direction and to reduce friction in rope rescue systems.

Pulmonary Edema — Accumulation of fluids in the lungs.

Pulse – The rhythmic beats caused as waves of blood move through and expand the arteries.

Purlins — Horizontal members that connect the steel, concrete, or laminated wood arches of trussless arched roofs.

Pyrolysis (Pyrolysis Process or Sublimation) — Thermal or chemical decomposition of fuel (matter) because of heat that generally results in the lowered ignition temperature of the material. The pre-ignition combustion phase of burning during which heat energy is absorbed by the fuel, which in turn gives off flammable tars, pitches, and gases. Pyrolysis of wood releases combustible gases and leaves a charred surface.

Q

Quint — Fire apparatus equipped with a fire pump, water tank, ground ladders, and hose bed in addition to the aerial device.

R

Rabbeted Jamb — Jamb into which a shoulder has been milled to permit the door to close against the provided shoulder.

Rabbit Tool — Hydraulic spreading tool that is specially designed to open doors that swing inward.

Radiation — (1) Transmission or transfer of heat energy from one body to another at a lower temperature through intervening space by electromagnetic waves such as infrared thermal waves, radio waves, or X-rays. Also called *radiated heat*. (2) Energy from a radioactive source emitted in the form of waves or particles; emission of radiation as a result of the decay of an atomic nucleus; process know as *radioactivity*. Also called *nuclear radiation*.

Rails — The two lengthwise members of a trussed ladder beam that are separated by truss or separation blocks.

Rain-Down Method — This method of foam application directs the stream into the air above the unignited or ignited spill or fire and allows the foam to float gently down onto the surface of the fuel.

Rain Roof – A second roof constructed over an older roof.

Radiation —The transmission or transfer of heat energy from one body to another body at a lower temperature through intervening space by electromagnetic waves such as infrared thermal waves, radio waves, or X-rays.

Radiological Dispersal Device (RDD) — Device that spreads radioactive contamination over the widest possible area by detonating conventional high explosives wrapped with radioactive material.

Rapid Intervention Crew (RIC) — Two or more fully equipped and immediately available firefighters designated to stand by outside the hazard zone to enter and effect rescue of firefighters inside, if necessary. Also known as Rapid Intervention Team.

Ratchet-Lever Jack — Type of lifting jack that uses the principles of leverage to operate. It is capable of lifting moderately heavy loads but tends to be unstable. It is generally recognized as the most dangerous of all jacks. Also called High-Lift Jack.

Rate-Compensated Heat Detector — Temperature-sensitive device that sounds an alarm at a preset temperature, regardless of how fast temperatures change.

Rate-of-Rise Heat Detector — Temperature-sensitive device that sounds alarm when the temperature changes at a preset value, such as 12°F to –15°F per minute.

Reactivity — Ability of two or more chemicals to react and release energy and the ease with which this reaction takes place.

Rebar Cutter — Cutting tool used for cutting rebar and security bars on windows or doors.

Reciprocating Saw — Electric saw that uses a short, straight blade that moves back and forth.

Recovery — Situation where the victim is most probably dead and the goal of the operation is to recover the body.

Reducer — Adapter used to attach a smaller hose to a larger hose. The female end has the larger threads, while the male end has the smaller threads.

Reducing Agent — The fuel that is being oxidized or burned during combustion.

Rehab — Term for a rehabilitation station at a fire or other incident where personnel can rest, rehydrate, and recover from the stresses of the incident.

Rekindle — Reignition of a fire because of latent heat, sparks, or smoldering embers; can be prevented by proper overhaul.

Relative Humidity — Measure of the moisture content (water quantity expressed in a percentage) in both the air and solid fuels.

Relay Operation — Using two or more pumpers to move water over a long distance by operating them in series. Water discharged from one pumper flows through hoses to the inlet of the next pumper, and so on. Also called Relay Pumping.

Remote Station Alarm System — System in which alarm signals from the protected premises are transmitted over a leased telephone line to a remote receiving station with a 24-hour staff; usually the municipal fire department's alarm communications center.

Repair — To restore or put together that which has become inoperable or out of place.

Rescue — Saving a life from fire or accident; removing a victim from an untenable or unhealthy atmosphere.

Residual Pressure — Pressure at the test hydrant while water is flowing. It represents the pressure remaining in the water supply system while the test water is flowing and is that part of the total pressure that is not used to overcome friction or gravity while forcing water through fire hose, pipe, fittings, and adapters.

Resistance Heating — Heat generated by passing an electrical current through a conductor such as a wire or an appliance.

Respiratory Arrest – When breathing stops completely.

Respiratory Hazards — Any exposure to products of combustion, superheated atmospheres, toxic gases, vapors, or dust, or potentially explosive or oxygen-deficient atmospheres or any condition that creates a hazard to the respiratory system.

Reverse Lay — Method of laying hose from the fire scene to the water supply.

Revolving Door — Door made of three or four sections, or wings, arranged on a central pivot that operates by rotating within a cylindrical housing.

Rim Lock — Type of auxiliary lock mounted on the surface of a door.

Riser — (1) Vertical part of a stair step. (2) Vertical water pipe used to carry water for fire protection systems above ground such as a standpipe riser or sprinkler riser. (3) Pipe leading from the fire main to the fire station (hydrants) on upper deck levels of a vessel.

Risk Management Plan — Written plan that identifies and analyzes the exposure to hazards, selection of appropriate risk management techniques to handle exposures, implementation of chosen techniques, and monitoring of the results of those risk management techniques.

Roll-On Application Method (Bounce) — Method of foam application in which the foam stream is directed at the ground at the front edge of the unignited or ignited liquid fuel spill; foam then spreads across the surface of the liquid.

Rollover — Condition in which the unburned combustible gases released in a confined space (such as a room or aircraft cabin) during the incipient or early steady-state phase and accumulate at the ceiling level. These superheated gases are pushed, under pressure, away from the fire area and into uninvolved areas where they mix with oxygen. When their flammable range is reached and additional oxygen is supplied by opening doors and/or applying fog streams, they ignite and a fire front develops, expanding very rapidly in a rolling action across the ceiling.

Roof Covering — Final outside cover that is placed on top of a roof deck assembly. Common roof coverings include composition or wood shake shingles, tile, slate, tin, or asphaltic tar paper.

Roof Ladder — Straight ladder with folding hooks at the top end. The hooks anchor the ladder over the roof ridge.

Rope Hose Tool — Piece of rope spliced to form a loop through the eye of a metal hook; used to secure hose to ladders or other objects.

Rope Log — A record kept by a department throughout a rope's working life. The date of each use and the inspection/maintenance records for the rope should be entered into the log, which should be kept in a waterproof envelope and placed in a pocket that is usually sewn on the side of the rope's storage bag.

Rope Rescue — The use of rope and related equipment to perform rescue.

Rotary Saw — Fire service version of this device is usually gasoline powered and has changeable blades for cutting wood, metal, and masonry.

Round Turn — Element of a knot that consists of further bending one side of a loop.

Rung — Step portion of a ladder running from beam to beam.

Running End — Part of the rope that is to be used for work such as hoisting, pulling, or belaying.

S

Safety Glass (Laminated Glass) — Special glass composed of two sheets of glass that are laminated to a sheet of plastic sandwiched between them under high temperature and pressure. Primarily used for automobile windshields and some rear windows.

Safety Policy — Written policy that is designed to promote safety to departmental members.

Safety Program — Program that sets standards, policies, procedures, and precautions to safely purchase, operate, and maintain the department's equipment and to educate employees on how to protect themselves from personal injury.

Safety Shoes — Protective footwear meeting OSHA requirements.

Salvage — Methods and operating procedures associated with fire fighting by which firefighters attempt to save property and reduce further damage from water, smoke, heat, and exposure during or immediately after a fire by removing property from a fire area, by covering it, or other means.

Salvage Cover — Waterproof cover made of cotton duck, plastic, or other material used by fire departments to protect unaffected furniture and building areas from heat, smoke, and water damage; a tarpaulin. Also called Tarp.

Saponification — A phenomenon that occurs when mixtures of alkaline based chemicals and certain cooking oils come into contact resulting in the formation of a soapy film.

SCBA — Abbreviation for Self-Contained Breathing Apparatus.

Search — Techniques that allow the rescuer to identify the location of victims and to determine access to those victims in order to remove them to a safe area.

Search Warrant — Written order, in the name of the People, State, Province, Territory, or Commonwealth, signed by a magistrate, commanding a peace officer to search for personal property or other evidence and return it to the magistrate.

Secondary Collapse — A collapse that occurs after the initial collapse of a structure. There are many possible causes, but an aftershock (earthquake), the movement of structural members, or weather conditions are common causes.

Secondary Contamination — Contamination of people, equipment, or the environment outside the hot zone without contacting the primary source of contamination; sometimes called cross contamination. *Also see* Contamination, Decontamination, and Hazard-Control Zones.

Secondary Feeder — Network of intermediate-sized pipes that reinforce the grid within the loops of the primary feeder system and aid in providing the required fire flow at any given point in a sprinkler system.

Seismic Effect — Earth vibration created by an explosion that is similar to an earthquake.

Self-Closing Door — (1) Door equipped with a door closer. (2) On a ship, an installation in which watertight doors are remotely operated by a hydraulic pressure system, allowing them to be closed simultaneously from the bridge or separately at the doors from either side of the bulkhead.

Self-Contained Breathing Apparatus (SCBA) — Respirator worn by the user that supplies a breathable atmosphere that is either carried in or generated by the apparatus and is independent of the ambient atmosphere. Respiratory protection is worn in all atmospheres that are considered to be Immediately Dangerous to Life and Health (IDLH). Also called Air Mask or Air Pack.

Separating — Act of creating a barrier between the fuel and the fire.

Service Test — Series of tests performed on apparatus and equipment in order to ensure operational readiness of the unit. These tests should be performed at least yearly or whenever a piece of apparatus or equipment has undergone extensive repair.

Setback — Distance from the street line to the front of a building.

Shank — Portion of a coupling that serves as a point of attachment to the hose. Also called tailpiece, bowl, or shell.

Shed Roof — Pitched roof with a single sloping aspect, resembling half of a gabled roof.

Shelter in Place — Having occupants remain in a structure or vehicle in order to provide protection from a rapidly approaching hazard. *Also see* Evacuation.

Shock — The inability of the body to adequately circulate blood to the body's cells to supply them with oxygen and nutrients. A life-threatening condition.

Shoe — Metal plate used at the bottom of heavy timber columns.

Shoring — General term used for lengths of timber, screw jacks, hydraulic and pneumatic jacks, and other devices that can be used as temporary support for formwork or structural components or used to hold sheeting against trench walls. Individual supports are called shores, cross braces, and struts. *Also see* Cribbing.

Short-Term Exposure Limit (STEL) — Fifteen-minute time-weighted average that should not be exceeded at any time during a workday; exposures should not last longer than 15 minutes nor be repeated more than four times per day with at least 60 minutes between exposures.

Shove Knife — Tool for opening a latch on a lock.

Siamese — Hose appliance used to combine two or more hoselines into one. The siamese generally has female inlets and a male outlet and is commonly used to supply the hose leading to a ladder pipe.

Sidewall Sprinkler — A sprinkler designed to be positioned at the wall of a room rather than in the center of a room. It has a special deflector that creates a fan-shaped pattern of water that is projected into the room away from the wall. Also called Wall Sprinkler.

Single Ladder — *See* Straight Ladder.

Situational Awareness — Awareness of immediate surroundings.

Size-Up — Ongoing mental evaluation process performed by the operational officer in charge of an incident to evaluate all influencing factors and to develop objectives, strategy, and tactics for fire suppression before committing personnel and equipment to a course of action. Size-up results in a plan of action that may be adjusted as the situation changes. It includes such factors as time, location, nature of occupancy, life hazard, exposures, property involved, nature and extent of fire, weather, and fire fighting facilities.

Skid Load — System of loading fire hose such that the top layer can be pulled off at the fire.

Slab Door — Door that has the appearance of being made of a single piece, or slab, of wood; may be of two types — either hollow core or solid core.

Sliding Door — Door that opens and closes by sliding across its opening; usually on rollers.

Slopover — (1) Fire edge that crosses a control line at a wildland fire. Also called Breakover. (2) Situation that occurs when burning oil that is stored in a tank is forced over the edge of the tank by water that is heated to the boiling point accumulates under the surface of the oil.

Small-Diameter Hose (SDH) — Hose of ¾-inches to 2 inches (20 mm to 50 mm) in diameter; used for fire fighting purposes. Also called Small Line.

Smoke Alarm — a device designed to sound an alarm when the products of combustion are present in the room where the device is installed. The alarm is built into the device rather than being a separate system.

Smoke Detector — Alarm-initiating device designed to actuate when visible or invisible products of combustion (other than fire gases) are present in the room or space where the unit is installed.

Smoke Ejector — Gasoline, electrically, or hydraulically driven blower (ducted fan) device used primarily to expel (eject) smoke from burning buildings although it is sometimes used to blow fresh air into a building to assist in purging smoke or other contaminants. They may be used in conjunction with a flexible duct.

Smoke explosion — a form of fire gas ignition; the ignition of accumulated flammable products of combustion.

Smothering — Act of excluding oxygen from a fuel.

Soft-Sleeve Hose — Large diameter, collapsible piece of hose used to connect a fire pump to a pressurized water supply source; sometimes incorrectly referred to as "soft suction hose."

Solid Stream — Hose stream that stays together as a solid mass as opposed to a fog or spray stream. A solid stream is produced by a smooth bore nozzle and should not be confused with a straight stream.

Solubility — Degree to which a solid, liquid, or gas dissolves in a solvent (usually water).

SOP — Abbreviation for Standard Operating Procedure.

Sorbent — Granular, porous filtering material used in vapor- or gas-removing respirators.

Spanner Wrench — Small tool primarily used to tighten or loosen hose couplings; can also be used as a prying tool or a gas key.

Special Fire Hazard — Fire hazard arising from the processes or operations that are peculiar to the individual occupancy.

Specific Gravity — Weight of a substance compared to the weight of an equal volume of water at a given temperature. A specific gravity less than 1 indicates a substance lighter than water; a specific gravity greater than 1 indicates a substance heavier than water.

Specification Marking — Stencil on the exterior of tank cars indicating the standards to which the tank car was built.

Split Lay — Hose lay deployed by two pumpers, one making a forward lay and one making a reverse lay from the same point.

Spontaneous Heating — Heating resulting from chemical or bacterial action in combustible materials that may lead to spontaneous ignition. Also known as Self-heating.

Spot Fire — Natural cover fires started outside the perimeter of a main fire; typically caused by flying sparks or embers landing outside the main fire area.

Sprinkler — Waterflow device in a sprinkler system. The sprinkler consists of a threaded nipple that connects to the water pipe, a discharge orifice, a heat-actuated plug that drops out when a certain temperature is reached, and a deflector that creates a stream pattern that is suitable for fire control. Formerly referred to as a sprinkler head.

Sprinkler System — A fixed piping system that is designed to discharge water in the event of a fire.

Stabilization — Process of providing additional support to key places between an object of entrapment and the ground or other solid anchor points to prevent unwanted movement.

Stack Effect — Tendency of any vertical shaft within a tall building to act as a chimney or "smokestack" by channeling heat, smoke, and other products of combustion upward due to convection. Also called Chimney Effect.

Staging Area — Prearranged, temporary strategic location, away from the emergency scene, where units assemble and wait until they are assigned a position on the emergency scene and from which these resources (personnel, apparatus, tools, and equipment) must be able to respond within three minutes of being assigned. Staging area managers report to the IC or operations section chief if established.

Standard Operating Procedures (SOPs) — Standard methods or rules in which an organization or a fire department operates to carry out a routine function. Usually these procedures are written in a policies and procedures handbook and all firefighters should be well versed in their content. An SOP may specify the functional limitations of fire brigade members in performing emergency operations.

Standing Part — That part of a rope between the working end and the running part.

Static — (1) Stationary; without movement. (2) Refers to the amount of stretch built into a rope. Static ropes have relatively little stretch. They are ideal for rescue work where stretch becomes a hazard and a nuisance. They should not be used where long falls could be anticipated before being stopped by the rope.

Static Pressure — (1) Potential energy that is available to force water through pipes and fittings, fire hose, and adapters. (2) Pressure at a given point in a water system when no water is flowing.

Stops — Wood or metal pieces that prevent the fly section of a ladder from being extended too far.

Storz Coupling — Nonthreaded (sexless) coupling commonly found on large-diameter hose.

Straight Ladder — One-section ladder. Also called Single Ladder.

Straight Lay — Hose laid from the hydrant or water source to the fire.

Straight Stream — The most compact discharge pattern that a fog nozzle can produce; similar to but not as compact as a solid stream.

Strategic Goals — Broad statements of desired achievement to control an incident; achieved by the completion of tactical objectives. *Also see* Strategy, Tactical Objectives, and Tactics.

Strategy — Overall plan for incident attack and control established by the Incident Commander (IC). *Also see* Tactics, Nonintervention Strategy, Defensive Strategy, and Offensive Strategy.

Stratification — Formation of smoke into layers as a result of differences in density with respect to height with low density layers on the top and high density layers on the bottom.

Stress – A state of physical and/or psychological arousal to a stimulus.

Strong Oxidizer — Material that encourages a strong reaction (by readily accepting electrons) from a reducing agent (fuel).

Subsurface Fuel — All combustible materials below the surface litter, such as tree or shrub roots, peat, and sawdust, that normally support smoldering combustion without flame.

Supine — Lying horizontal in a face upward position.

Supplied Air Respirator — An atmosphere-supplying respirator for which the source of breathing air is not designed to be carried by the user; not certified for fire fighting operations.

Suppressing — Sometimes referred to as smothering. The act of preventing the release of flammable vapors and therefore reducing the possibility of ignition or reignition.

Surface Fuel — Fuel that contacts the surface of the ground; consists of duff, leaf and needle litter, dead branch material, downed logs, bark, tree cones, and low-stature living plants. These are the materials normally scraped away to construct a fireline. Sometimes called Ground Fuel.

Surface-To-Mass Ratio — The ratio of the surface area of the fuel to the mass of the fuel.

Survey — Evaluation instrument used to identify the behavior and/or attitude of an individual or audience both before and after a presentation.

Swivel — Free-turning ring on all Storz fire hose couplings and on the female coupling of threaded couplings.

Synthetic Fiber Rope — Rope featuring continuous fibers running the entire length of the rope; has excellent resistance to mildew and rotting, has excellent strength, and is easy to maintain.

Systemic Effect — Something that affects an entire system rather than a single location or entity.

T

Tactical Objectives — Specific operations that must be accomplished to achieve strategic goals. *Also see* Strategic Goals and Tactics.

Tactics — Methods of employing equipment and personnel on an incident to accomplish specific tactical objectives in order to achieve established strategic goals. *Also see* Tactical Objectives, Strategy, and Strategic Goals.

Tag Line — Nonload-bearing rope attached to an object to help steer it in a desired direction or act as a safety line.

Target Fire Hazard — Facility in which there is a great potential likelihood of life or property loss.

Telecommunications Device For The Deaf (TDD) — Special type of phone used by hearing-impaired individuals that sends text messages. Letters are transmitted in a specific code to be decoded by another similar device at the communications center. The Americans with Disabilities Act of 1990 requires that all public safety communications centers have one available for use.

Teletype (TTY) — Similar to a TDD device in which messages are sent over phone lines. However, usually teletype machines use dedicated phone lines and have other features such as the ability to send messages to many users at once, query certain databases, and provide a secure network.

Temperature — Measurement of the intensity of the heating of a material.

Tempered Glass — Type of glass specially treated to become harder and more break-resistant than plate glass or a single sheet of laminated glass. Tempered glass is most commonly used in side windows and some rear windows.

Tension — Those vertical or horizontal forces that tend to pull things apart; for example, the force exerted on the bottom chord of a truss.

Thermal Column — Updraft of heated air, fire gases, and smoke directly above the involved fire area.

Thermal Layering (of Gases) — Outcome of combustion in a confined space in which gases tend to form into layers, according to temperature, with the hottest gases are found at the ceiling and the coolest gases at floor level. Also called Thermal Balance or Heat Stratification.

Threaded Coupling — Male or female coupling with a spiral thread.

Threshold Limit Value (TLV®) — Concentration of a given material in parts per million (ppm) that may be tolerated for an 8-hour exposure during a regular workweek without ill effects.

Threshold Limit Value/Ceiling (TLV®/C) — Maximum concentration of a given material in parts per million (ppm) that should not be exceeded, even instantaneously.

Threshold Limit Value/Short-Term Exposure Limit (TLV®/STEL) — Fifteen-minute time-weighted average exposure that should not be exceeded at any time nor repeated more than four times daily with a 60-minute rest period required between each STEL exposure. These short-term exposures can be tolerated without suffering from irritation, chronic or irreversible tissue damage, or narcosis of a sufficient degree to increase the likelihood of accidental injury, impair self-rescue, or materially reduce worker efficiency. TLV/STELs are expressed in parts per million (ppm) and milligrams per cubic meter (mg/m^3).

TIC — Abbreviation for Toxic Industrial Chemical.

Tie Rods — Metal rods running from one beam to the other.

TIM — Abbreviation for Toxic Industrial Material.

Tip — (1) Extreme top of a ladder. Also called Top.

Topography — Physical configuration of the land or terrain.

Total Flooding System — Fire-suppression system designed to protect hazards within enclosed structures. Foam is released into a compartment or area and fills it completely to extinguish the fire.

Tourniquet — Any wide, flat material wrapped tightly around a limb to stop bleeding; used only for severe, life-threatening hemorrhage that cannot be controlled by other means.

Toxic Industrial Chemical (TIC) — *See* Toxic Industrial Material (TIM).

Toxic Industrial Material (TIM) — Industrial chemical that is toxic at certain concentration and is produced in quantities exceeding 30 tons per year at any one production facility; readily available and could be used by terrorists to deliberately kill, injure or incapacitate people. Also called toxic industrial chemical (TIC).

Toxic Industrial Material (TIM)/Toxic Industrial Chemical (TIC) — Industrial chemical that is toxic at certain concentration and is produced in quantities exceeding 30 tons per year at any one production facility; readily available and could be used by terrorists to deliberately kill, injure, or incapacitate people.

Toxic Material — Substance classified as a poison, asphyxiant, irritant, or anesthetic that can be poisonous if inhaled, swallowed, absorbed, or introduced into the body through cuts or breaks in the skin.

Toxin — Substance that has the property of being poisonous.

Trailer — (1) Combustible material, such as rolled rags, blankets, newspapers, or flammable liquid, often used in intentionally set fires to connect remote fuel packages (combustible materials, pools of ignitable liquid, etc.) in order to spread fire from one point or area to other points or areas. Frequently used in conjunction with an incendiary device. (2) Highway or industrial-plant vehicle designed to be hauled/pulled by a tractor.

Transportation Area — Location where accident casualties are held after receiving medical care or triage before being transported to medical facilities.

Trench Ventilation — Defensive tactic that involves cutting an exit opening in the roof of a burning building, extending from one outside wall to the other, to create an opening at which a spreading fire may be cut off. Also called Strip Vent.

Triage — System used for sorting and classifying accident casualties to determine the priority for medical treatment and transportation.

Truss Block — Used to separate the beams of a truss beam ladder. Also called Beam Block and Run Block.

Turbulence — Irregular motion of the atmosphere usually produced when air flows over a comparatively uneven surface such as the surface of the earth; when two currents of air flow pass or over each other in different directions or at different speeds.

Turnout Gear — Term used to describe personal protective clothing made of fire resistant materials that includes coats, pants, and boots. Also referred to as Turnout Clothing, Turnout Boots, Turnout Coat, Turnout Pants, Turnouts, or Bunker Gear. *See also* Personal Protective Equipment.

Two-Way Radio — Voice network that provides an always-on connection that enables the user to just "push the button and talk." Also called "dispatch radio or transceiver." The radio design allows either transmitting or receiving — not both at once unless it is a full-duplex system.

Tying In — (1) Securing oneself to a ladder; accomplished by using a rope hose tool or belt or by inserting one leg between the rungs. (2) Securing a ladder to a building or object.

U

Underwriters Laboratories Inc. (UL) — Independent fire research and testing laboratory with headquarters in Northbrook, Illinois that certifies equipment and materials. Equipment and materials are approved only for the specific use for which they are tested.

Unibody Construction — Method of automobile construction used for most modern cars in which the frame and body of a vehicle is all one integral unit.

Unified Command — In the Incident Management System, a shared command role that allows all agencies with responsibility for the incident, either geographical or functional, to manage the incident by establishing a common set of incident objectives and strategies. In unified command there is a single incident command post and a single operations chief at any given time.

Upper Explosive Limit (UEL) — Upper limit at which a flammable gas or vapor will ignite. Above this limit, the gas or vapor is too rich to burn (lacks the proper quantity of oxygen).

Upper Flammable Limit (UFL) — Upper limit at which a flammable gas or vapor will ignite. Above this limit the gas of vapor is too rich to burn (lacks the proper quantity of oxygen).

Upright Sprinkler — Sprinkler that sits on top of the piping and sprays water against a solid deflector that breaks up the spray into a hemispherical pattern that is redirected toward the floor.

USFA — Abbreviation for the United States Fire Administration, which operates within the Federal Emergency Management Agency (FEMA). Formerly the National Fire Prevention and Control Administration (NFPCA).

Utility Rope — Rope to be used in any situation that requires a rope — except life safety applications. Utility ropes can be used for hoisting equipment, securing unstable objects, and cordoning off an area. *See* Non-lifeline Rope.

V

Vapor Density — Weight of a given volume of pure vapor or gas compared to the weight of an equal volume of dry air at the same temperature and pressure. A vapor density less than 1 indicates a vapor lighter than air; a vapor density greater than 1 indicates a vapor density heavier than air.

Vapor Pressure — Measure of the tendency of a substance to evaporate; pressure at which a vapor is in equilibrium with its liquid phase for a given temperature.

Vaporization — Process of evolution that changes a liquid into a gaseous state. The rate of vaporization depends on the substance involved, heat, and pressure.

Veneer Walls — Walls with a surface layer of attractive material laid over a base of common material.

Venous Bleeding — Bleeding from a vein that is characterized by dark red or maroon blood and has a steady flow; easy to control.

Ventilation — (1) Systematic removal of heated air, smoke, gases or other airborne contaminants from a structure and replacing them with cooler and/or fresher air to reduce damage and to facilitate fire fighting operations. (2) Process of replacing foul air in any of a vessel's compartments with pure air.

Venturi Principle — Physical law stating that when a fluid, such as water or air, is forced under pressure through a restricted orifice, there is an increase in the velocity of the fluid passing through the orifice and a corresponding decrease in the pressure exerted against the sides of the constriction. Because the surrounding fluid is under greater pressure (atmospheric), it is forced into the area of lower pressure. Also called Venturi Effect.

Vermiculite — Expanded mica used for loose fill insulation and as aggregate in concrete.

Vertical Ventilation — Ventilating at the highest point of a building through existing or created openings and channeling the contaminated atmosphere vertically within the structure and out the top. Done with holes in the roof, skylights, roof vents, or roof doors. Also called Top Ventilation.

Violation — Infringement of existing rules, codes, or laws.

Volatility — Ability of a substance to vaporize easily at a relatively low temperature. *Also see* Vapor, Volatile, and Vaporization.

Volunteer — Anyone, inside or outside the fire department, who helps with fire and life safety education programs.

V-Type Collapse — (1) Situation where the excess weight of heavy loads, such as furniture and equipment, concentrated near the center of the floor causes the floor to give way. A V-type collapse will result in void spaces near the walls. (2) A type of collapse void in which a floor section fails at the center and falls to the floor below. This results in two void spaces along the supporting outer wall.

W

Water Hammer — Force created by the rapid deceleration of water causing a violent increase in pressure that can be powerful enough to rupture piping or damage fixtures. It generally results from closing a valve or nozzle too quickly.

Water Mist — In the fire service, water mist is associated with a fire extinguisher capable of atomizing water through a special applicator. Water-mist fire extinguishers use distilled water, while back-pump type water-mist extinguishers use ordinary water.

Water Shuttle Operation — Method of water supply by which tenders/tankers continuously transport water between a fill site and the dump site located near the emergency scene.

Water Solubility — Ability of a liquid or solid to mix with or dissolve in water.

Water Thief — Any of a variety of hose appliances with one female inlet for 2½ -inch (65 mm) or larger hose and with three gated outlets, usually two 1½-inch (38 mm) outlets and one 2½-inch (65 mm) outlet.

Water Tower — Aerial device primarily intended for deploying an elevated master stream. Not generally intended for climbing operations. Also known as an Elevating Master Stream Device.

Water Vacuum — Appliance similar to a household vacuum cleaner designed to pick up water.

Waybill — Shipping paper used by a railroad to indicate origin, destination, route, and product. Each car has a waybill that the conductor carries.

Webbing — Synthetic nylon, spiral weave, tubular material used for creating anchors, lashings, and for packaging patients and rescuers.

Wet-Barrel Hydrant — Fire hydrant that has water all the way up to the discharge outlets. The hydrant may have separate valves for each discharge or one valve for all the discharges. This type of hydrant is only used in areas where there is no danger of freezing weather conditions.

Wet-Chemical System — Extinguishing system that uses a wet-chemical solution as the primary extinguishing agent; usually installed in range hoods and associated ducting where grease may accumulate.

Wet-Pipe Sprinkler System — Fire-suppression system is built into a structure or site; piping contains either water or foam solution continuously; activation of a sprinkler causes the extinguishing agent to flow from the open sprinkler.

Winch — Pulling tool that consists of a length of steel chain or cable wrapped around a motor-driven drum.

These are most commonly attached to the front or rear of a vehicle.

Wind — Horizontal movement of air relative to the surface of the earth.

Windward Side — (1) The unprotected side of the building the wind is striking. (2) The side or direction from which the wind is blowing.

Wood-Frame Construction — Another term for Type V construction; construction that has exterior walls, bearing walls, floors, roofs, and supports made completely or partially of wood or other approved materials of smaller dimensions than those used for heavy-timber construction.

Working End — Part of the rope that is to be used in forming the knot. Also called Bitter End or Loose End.

Wye — Hose appliance with one female inlet and two or more male outlets, usually smaller than the inlet. Outlets are also usually gated.

Index

A

Abdominal thrusts, 1038

Absorption of hazardous materials, 1091, 1249, 1263, 1277–1278

Abuse, as cause of stress, 1017

Acceptance testing fire hose, 680

Accidental fire, defined, 907

Accordion load, 657, 689

Accountability of members
  in confined space rescues, 368, 369
  emergency operations, 74–76
  at haz mat incidents, 1232–1233
  NFPA standards, 54
  passport system, 75, 76
  personnel accountability report, 71, 210, 948–949

Acetylene, 399–400

Acid gases, 1066

Acquired building, 69

Activation energy, 1106

ADA (Americans with Disabilities Act of 1990), 836

Adapter, 647

Adhesive-sprayed glass, 358–359

Adsorption of hazardous materials, 1249

Advancing hoselines, 672–678
  burst sections, replacing, 678, 710
  control of loose hoseline, 677, 710
  down a stairway, 673–674, 703–706
  extending hoselines, 677, 709
  from a standpipe, 674–675
  into a structure, 673, 702
  up a ladder, 675–677, 707, 708, 711
  up a stairway, 673

Aerial apparatus
  aerial ladder platform, 772, 773
  aerial ladders, 771–772
  ladder requirements, 482, 486

AFFF. See Aqueous Film Forming Foam (AFFF)

A-frame collapse, 363

After-analysis analysis, 1273

AIDS, 1004, 1008

Air bags, 354–355

Air chisel, 340–341

Air hammer, 340

Air knife, 341

Air monitoring, 368

Air vacuum, 342

Air-aspirating foam nozzle, 240, 745

Airborne pathogens
  HEPA respirators, 1013
  notification of exposure, 1012
  spread of, 1004

Aircraft labels, 1153

Airport firefighter, 28, 29

Air-pressurized water extinguishers, 237

Air-purifying respirators, 1064–1068
  escape respirators, 1068
  particulate-removing filters, 1066
  powered, 1067
  supplied air-hoods, 1067
  vapor- and gas-removing filters, 1066

Airway, opening of, 1027–1029
  abdominal thrusts, 1038
  chest thrusts, 1038–1039
  clearance sequences, 1039–1040
  clearing obstructions, 1037–1041
  complete (severe) obstructions, 1038–1041
  head-tilt, chin-lift maneuver, 1028
  jaw-thrust maneuver, 1029
  partial (mild) obstructions, 1038

Alarm assignment, 947

Alarm reaction, 1015

Alarm systems
  alarm-receiving equipment, 931–934
  automatic systems, 837–840, 841
  auxiliary systems, 837–838
  bimetallic detector, 828–829
  central station systems, 839–840, 841
  combination detectors, 835
  continuous line detector, 828, 829
  fire-gas detectors, 835
  fixed-temperature heat detectors, 826–829
  flame detectors, 834
  frangible bulbs, 827–828
  functions, 825
  fusible devices, 827–828
  heat detectors, 826–831
  indicating devices, 836–837
  ionization smoke detectors, 833
  local warning systems, 826
  photoelectric smoke detectors, 832
  pneumatic rate-of-rise line detector, 830, 831
  pneumatic rate-of-rise spot detector, 829–830
  predischarge alarm, 841
  proprietary systems, 839, 840
  protected premises alarm system, 826
  rate-compensated detector, 830, 831
  remote station systems, 838–839
  reporting procedures, 940–941
  supervision, 840–841
  thermoelectric detector, 830–831
  waterflow alarms, 837, 838, 850

Alcohol-based hand cleaners, 1005–1006

Alerting of emergencies
  staffed stations, 941–942
  unstaffed stations, 942–943

"All Clear", 71, 308

Allergens, 1099

Alloy, 237

Alpha radiation, 1085–1086

Ambient conditions of compartment fires, 129–130

Ambient temperature, 827

American Conference of Governmental Industrial Hygienists (ACGIH), toxin exposure limits, 1096

American Industrial Hygiene Association (AIHA), toxin exposure limits, 1097–1098

Americans with Disabilities Act (ADA) of 1990, 836

Ammonia
  anhydrous ammonia for meth production, 1200
  anhydrous ammonia incident, 1137
  toxicity, 185